Geology

AN INTRODUCTION TO PHYSICAL GEOLOGY

Fourth Edition

Stanley Chernicoff
University of Washington, Seattle

Donna Whitney
University of Minnesota, Twin Cities

PEARSON

Prentice
Hall

Upper Saddle River, New Jersey 07458

Library of Congress Cataloging-in-Publication Data
Chernicoff, Stanley.
    Geology : an introduction to physical geology.--4th ed. /
        Stanley Chernicoff, Donna Whitney.
        p.   cm.
    Includes index
    ISBN 0-13-147464-2
    1. Physical geology--Textbooks. I. Whitney, Donna. II. Title.
QE28.2.C48 2007
550--dc22

2005053493

Executive Editor: Patrick Lynch
Editor-in-Chief, Development: Carol Trueheart
Executive Managing Editor: Kathleen Schiaparelli
Assistant Managing Editor: Beth Sweeten
Production Editor: Edward Thomas
Managing Editor, Science and Math Media: Nicole M. Jackson
Media Editor: Chris Rapp
Project Manager: Dorothy Marrero
Manufacturing Manager: Alexis Heydt-Long
Manufacturing Buyer: Alan Fischer
Director of Creative Services: Paul Belfanti
Creative Director: Juan R. López
Art Director: Maureen Eide
Interior Design: Susan Anderson
Cover Design: Maureen Eide
Development Editor: Rebecca Strehlow
Copy Editor: Barbara Booth
Editorial Assistant: Sean Hale

Director of Image Resource Center: Melinda Patelli-Reo
Research Manager: Beth Brenzel
Image Permission Coordinator: Debbie Hewitson
Photo Researcher: Diane Austin
Proofreader: Alison Lorber
Color Scanning Supervisor: Joseph Conti
Production Assistant: Nancy Bauer
Marketing Assistant: Renee Hanenberg
Composition: Pine Tree Composition
Senior Managing Editor, Art Production and Management: Patricia Burns
Production Manager, Artworks: Ronda Whitson
Manager, Production Technologies: Matthew Haas
Managing Editor, Art Management: Abigail Bass
Art Production Editor: Jess Einsig
Illustrations: J.B. Woolsey
Art Development: Charlotte Miller
Cover and Title Page Photo: Chris Johns/National
    Geographic Image Collection

© 2007 by Pearson Education, Inc.
Pearson Prentice Hall
Pearson Education, Inc.
Upper Saddle River, New Jersey 07458

Earlier editions © 2002 and © 1999 by Houghton Mifflin Company, and © 1995 by Worth Publishers, Inc.

Printed in the United States of America

10 9 8 7 6 5

ISBN: 0-13-147464-2 (student edition)
ISBN: 0-13-173719-8 (instructor edition)

Pearson Education Ltd., *London*
Pearson Education Australia Pty., Limited, *Sydney*
Pearson Education *Singapore,* Pte. Ltd
Pearson Education North Asia Ltd., *Hong Kong*
Pearson Education Canada, Ltd., *Toronto*
Pearson Educación de Mexico, S.A. de C.V.
Pearson Education—Japan, *Tokyo*
Pearson Education Malaysia, Pte. Ltd

About the Authors

Stanley Chernicoff

Born in Brooklyn, New York, Stan Chernicoff began his academic career as a political science major at Brooklyn College of the City University of New York. Upon graduation, he intended to enter law school and pursue a career in constitutional law. He had, however, the good fortune to take geology as his last requirement for graduation in the spring of his senior year, and he was so thoroughly captivated by it that his plans were forever changed.

After an intensive post-baccalaureate program of physics, calculus, chemistry, and geology, Stan entered the University of Minnesota–Twin Cities, where he received his doctorate in Glacial and Quaternary Geology under the guidance of one of North America's preeminent glacial geologists, Dr. H. E. Wright. Stan launched his teaching career as a senior graduate student teaching physical geology to hundreds of bright Minnesotans.

Stan has been a member of the faculty of the Department of Geological Sciences at the University of Washington in Seattle since 1981. He was awarded the University's Distinguished Teaching Award in 2000, an honor that has been bestowed on fewer than 150 professors in the University's history. At Washington, he has taught Physical Geology, the Great Ice Ages, and the Geology of the Pacific Northwest to more than 40,000 students, and he has trained hundreds of graduate teaching assistants in the art of bringing geology alive for non-science majors. Stan is currently a Dean in the UDub Office of Undergraduate Education. He lives in Seattle with his wife, Dr. Julie Stein, a professor of archaeology and the director of The Burke Museum of Natural History and Culture. The couple has two wonderful sons, Matthew (a UW graduate, sports journalist, and documentary filmmaker) and David (UW senior studying Biology and Biological Anthropology, planning on a career in Public Health).

Donna Whitney

Donna Whitney grew up in Maine, scrambling around on the rocky coast, collecting rocks and shells. She did not intend to study geology—or any science—when she went to college, but then she took an introductory geology course during her freshman year at Smith College, and she knew she had discovered exactly what she wanted to do. After graduation, Donna went directly to graduate school at the University of Washington, where she studied how and why continental crust partially melts during mountain-building, and obtained her Ph.D. in 1991.

During her doctoral studies in Seattle, Donna met Stan Chernicoff and worked with him as a teaching assistant for many classes. Following postdoctoral research at the University of Calgary, a year as a Visiting Assistant Professor at Vassar College, and four years as an Assistant Professor at the University of North Carolina–Chapel Hill, Donna joined the faculty at the University of Minnesota–Twin Cities. She is now a Professor at the University of Minnesota, and lives in Minneapolis with her husband, Marc Hirschmann (also a geology professor at the U of M), and their daughter Naomi.

Her main research interests are metamorphic geology and tectonics; in particular, how mountain belts are constructed and destroyed, and how the microstructures of metamorphic minerals can be used to understand continental tectonics. At Minnesota, she works with a dynamic group of colleagues and students on a wide range of projects, including field research in the Cordilleran Mountains of western North America (British Columbia, Washington); the Appalachians in New England and North Carolina; and the Alpine-Himalayan belt in Turkey and Greece. Donna teaches undergraduate courses in Introductory Geology, Mineralogy, and Petrology, and a freshman seminar on Geology and Civilization.

Brief Contents

Contents

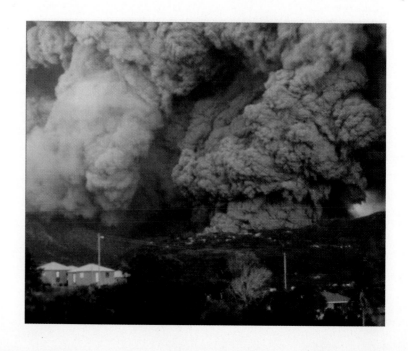

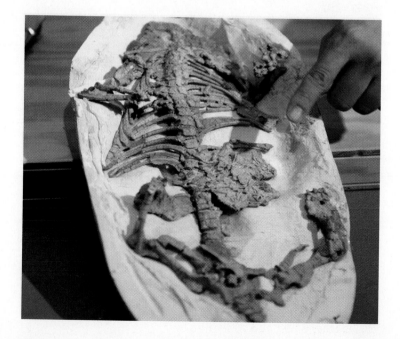

PART 3

Sculpting the Earth's Surface 435

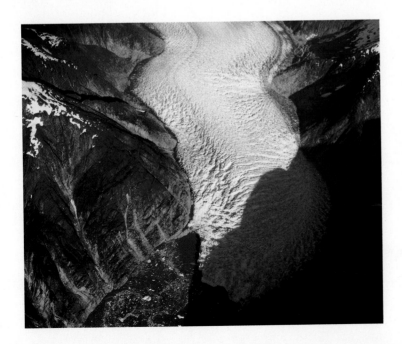

Preface

The introductory course in physical geology, taken predominantly by non-science majors, may be the only science course some students take during their college years. What a wonderful opportunity to introduce students to the field we love and show them how fascinating and useful it is. Indeed, much of what students will learn in their physical geology course will be recalled throughout their lives, as they travel across this and other continents, dig in their backyard gardens, walk along a beach, or sit by a mountain stream. And as has become abundantly evident in late 2005 in the wake of the Hurricane Katrina tragedy and in the face of soaring fuel prices, the concepts and principles our students discover in their physical geology class may affect their lives profoundly—throughout their lives—in numerous and substantial ways. For this reason, our book team—authors, illustrators, photo researchers, and editors—have expended the best of our abilities to craft an exciting, stimulating, informative, and enduring introduction to our field.

This book has been written expressly for your students, with their interests and learning styles in mind. The authors have spent a combined four decades teaching physical geology at a wide range of colleges and universities, public and private, large and small. Both of the authors have been acknowledged on their campuses as award-winning instructors.

The Book's Goal

The book's goal is simple and basic: to help teach what *everyone* should know about geology in a way that will engage and stimulate our students to discover and explore this field on their own beyond our classrooms. The book embodies the view that this is perhaps the most useful college-level science class a non-science major can take—one that we believe all students should take. Physical geology can show students the essence of how science and scientists work, while at the same time nurturing their interest in understanding, appreciating, and protecting their surroundings. In this course they can learn to prepare for any number of geologic and environmental threats, and see how our Earth can continue providing all of our needs for food, shelter, and material well-being as long as we don't squander these resources.

Content and Organization

The unifying themes of plate tectonics, environmental geology, natural resources, natural disasters, and the interconnectedness of the Earth's various systems are introduced in Chapter 1 and discussed in their proper context within virtually every chapter. Chapter 1 also presents the important groups of rocks, the rock cycle, and geological time—building a foundation for the succeeding chapters. Chapters 1 through 8 introduce the basics—minerals, rocks, and time—to prepare the reader for the in-depth discussions of structural geology, geophysics, and tectonics that follow in Chapters 9 through 13. After the Earth's first-order features—ocean basins, continents, and mountain systems—have been discussed, the processes that sculpt these large-scale features are addressed. Chapters 14 through 19 present the principal geomorphic processes of mass movement, streams, groundwater, cave formation, glacial flow, arid region and eolian processes, and coastal evolution. The final chapter, "Human Use of the Earth's Resources", ties together earlier discussions from throughout the book. It reinforces principles that relate to the origins of resources, especially energy-producing ones, and stresses our responsibility to manage them wisely.

New to the Fourth Edition of *Geology*

A fourth edition of a textbook is a prime opportunity to build on the strengths of the first three editions, eliminate their flaws, and strike out in new conceptual directions. In this editing, the authors:

- Weave in the latest discoveries in the geosciences.

- Offer up-to-the-minute examples of exciting geological processes, such as the most recent volcanic eruptions, earthquakes, tsunami, floods, and discoveries on Mars and on the Earth's own seafloors.

- Rethink how concepts have been presented in the earlier editions and clarify and illustrate them more effectively. Expand the use of learning and teaching devices—such as the *Geology at a Glance* feature—to help students engage in the text's content and retain it.

The fourth edition of *Geology* attempts to accomplish all of these goals, all to ensure that our students have the very best introductory experience with our science. Toward these ends, the following is just a sampling of the new and exciting areas that this edition of *Geology* explores:

- The causes and tragic effects of the Indonesian earthquake and tsunami of 2004 and the Pakistan earthquake of 2005.

- An in-depth discussion of the search for water in the rocks and soils of Mars and the implications of the discovery of hematite "blueberries" on the Martian surface.

- The human tragedy and the underlying geological causes of the flooding that followed Hurricane Katrina.

- Increased emphasis on Earth systems interactions.

- A new chapter on continental tectonics and mountain-building explains how the great mountain systems of the world are created and how they collapse, and gives a brief geologic tour of the evolution of the North American continent through time.

- A focus on the renewed eruptive activity of Mount St. Helens, and a discussion about our increasing concern for the potential of a Yellowstone eruption.

- A new highlighted discussion on the dangers of life in the shadow of a looming eruption—"Living in the shadow of Vesuvius."

- A vastly expanded treatment of the geologic timescale and the evolution of life, including the richly illustrated history of the "greatest moments on Earth," prepared by the Smithsonian Institution.

- An updated highlighted discussion on the effect of rock weathering and the carbon cycle on global climatic change.

- An up-to-date discussion of the Earth's most recent earthquakes and the proposals for new early-warning systems in the Pacific Ocean and Caribbean Sea.

- A new highlighted discussion of the SAFOD project—the effort to drill through the San Andreas fault zone in order to capture earthquakes in progress.

- A focus on our recent efforts to pinpoint earthquakes and provide ample warning to quake-prone communities, including an introduction to the Trinet seismic network.

- An expanded discussion of the use of geochronology to date past tectonic events.

- A new highlighted discussion of the hypothetical scheme to study the Earth's core . . . directly.

- An introduction to D ″, a remote layer of the Earth's deep interior.

- A highlighted essay on the remarkably constant position of the Earth's oceans through our planet's long history.

- An introduction to Project Neptune, geoscience's bold initiative to study *directly* the Earth's final frontier—the tectonics of the deepsea structure—with seafloor observatories and monitoring systems.

- Expanded discussions of the plate tectonics of the seafloors and continents, including a speculative essay about the resumption of volcanism along North America's East Coast.

- An expanded discussion of the tectonic history of North America.

- An update on the tragic mass-movement events in Santa Barbara's La Conchita region and in the Laguna Hills area of southern California.

- A highlighted discussion of the alarming prospects for a massive landslide and the potential for a monumental sea wave generated on the eastern slopes of the Cumbre Vieja volcano on the Canary Island, La Palma.

- A highlighted discussion of the prospects for an ice-free Northwest Passage in the wake of protracted global warming.

- A modern treatment of the potential for rapid climate deterioration in the wake of the disruption of the Northern Atlantic Deep Water current.

- A highlight that focuses on the plight of low-lying Pacific Islands—such as Tuvalu—that are threatened with elimination from the world's maps with the warming-induced rise in sea level.

- A highlight on hydrogen fuel cells and their potential for meeting our burgeoning twenty-first-century energy needs.

These topics and many more constitute a substantial effort to ensure that a new edition of *Geology* brings a vast array of new ideas to its readers.

The Artwork

The drawings in this book are unique. Ramesh Venkatakrishnan, the text's inaugural artist, is both an experienced and respected geology professor and consultant as well as a highly gifted artist. His drawings evolved along with the earliest drafts of the first draft of the manuscript, sometimes leading the way for the text discussions. Chernicoff and Venkatakrishnan worked together as graduate teaching assistants at the University of Minnesota—Twin Cities. The desire to illuminate what we wanted introductory students to know about the Earth has been the central theme of this book's art program, even in its fourth edition.

As you will see when you leaf through this book, the art explains, describes, stimulates, and teaches. It is not schematic; it shows how the Earth and its geological features actually look. It is also not static; it shows geological processes in action, allowing students to discover how geological features evolve through time. Every effort has been made to illustrate accurately a wide range of geological and geomorphic settings, including vegetation and wildlife, weathering patterns, and even the shadows cast by the Sun at various latitudes. The artistic style is consistent throughout so that students may become familiar with the appearance of some features even before reading about them in subsequent chapters. For example, the stream drainage patterns appearing on volcanoes in Chapter 4, "Volcanoes and Volcanism", set the stage for the discussion of drainage patterns in Chapter 15, "Streams and Floods". The colors used and the map symbols keyed to various rock types follow international conventions and are consistent throughout.

The fourth edition builds on the strengths of the art program of the first three. The images in this edition have been enhanced by renowned scientific illustrator John Woolsey to sharpen their forces, deepen their colors, and lend additional clarity and simplicity to their subjects.

The art program in the fourth edition has undergone an exhaustive revision to ensure that core concepts are more effectively communicated. The result is clearer, less complex illustrations that explain geologic principles and processes in a straightforward yet stimulating manner.

Pedagogy

The fourth edition of *Geology* includes four key learning tools and several other pedagogical enhancements. This edition enhances the use of the *Geology at a Glance* feature, a series of "concept map" figures that enable the readers to take a breath and review major bodies of information as they read through lengthy stretches of new knowledge. These figures summarize the text's major discussions through a series of diagrams, complete with photographs and line art. In a sense, the *Geology at a Glance* feature provides a visual means of reviewing text material to enhance your student's prospects for comprehension and retention of the book's major principles. This feature should help students substantially as they review for course exams.

A second learning tool can be found at the end of each chapter—a section called *Learning Actively*. In recent days, educators have confirmed that most students learn more from "doing and discovering independently" than simply by sitting passively and absorbing their professors' lectures. *Learning Actively* provides several ways for a student to experience the content of a chapter in a more active, personal way. These exercises are designed to be simple, fun, and informative. The goal is greater understanding and retention of the chapter's concepts.

The third and fourth new features work in conjunction with the text to focus student efforts on two primary indicators of success in the course—lecture attendance and assessment.

The *Student Lecture Notebook* reproduces all of the line art from the text and transparency set in full color with space for taking notes. Students can now focus fully on the lecture instead of attempting to replicate projected figures. Included with the *Student Lecture Notebook* are *Concept Map Exercises*—using the *Geology at a Glance* feature of the text as a springboard, the students can construct their own concept maps of the chapter. Educational research shows that concept mapping can significantly increase retention and mastery of new topics.

The *Online Study Guide,* referenced at the end of each chapter, refers the student to an open-access Website containing multiple quizzes for each chapter. The quizzes are *organized by concept* so the students can self-assess their mastery of the material before taking the exam. The quizzes are auto-graded, and the results can be e-mailed to the instructor.

Nearly every chapter contains one or more *Highlights*—in–depth discussions of topics of popular interest that provide a broader view of the relevance of geology. In many cases, the *Highlights* comprise a late-breaking story that also shows your students that the Earth's geology and its effects on us are changing daily.

To help readers learn and retain the important principles, every chapter ends with a *Summary*—a narrative discussion replete with bulleted lists of key concepts that recall all of the important chapter concepts. Also at the end of every chapter are sets of *Questions for Review* to help students gauge their retention of the facts and principles presented.

One hallmark of the first three editions of *Geology* has been its thoroughly integrated environmental theme. For the fourth edition, the authors have worked diligently to enhance this rich environmental emphasis. The authors have also tried to help the reader develop a deep appreciation for the interconnectedness of the Earth's various systems. Throughout the text, students are shown how the Earth's lithosphere, hydrosphere, atmosphere, and biosphere interact virtually everywhere to drive the geologic processes that shape the Earth's surface and produce the geologic features we see around us.

The authors and illustrators have tried to introduce readers to world geology. This book emphasizes, however, the geology of North America (including the offshore state, Hawai'i), while acknowledging that geological processes do not stop at national boundaries or at the continent's coasts. Wherever data are available—from the distribution of coal to the survey of seismic hazards—we have tried to show our readers as much of this continent, and beyond, as feasible. Photos and examples have been selected from throughout the United States and Canada and from many other regions of the world.

The metric system is used for all numerical units, with their English equivalents in parentheses, so that U.S. students can become more familiar with the units of measurement used by virtually every other country in the world.

Instructor Resources

Prentice Hall is pleased to offer a suite of resources to accompany *Geology, 4e* that will help you both save time in preparing for lectures and maximize new media capabilities during your lectures.

Prentice Hall Environmental Geology Video DVD

Focusing on first-person experiences with geologic processes, this DVD contains **more than 35 30-second to 3-minute video segments** featuring:

- Hurricane storm surge
- Surveillance camera earthquake video
- Volcanic destruction
- Urban flooding
- Dust storms
- Landslide devastation
- Tornado formation and destruction
- Giant sinkholes . . . and much more

This DVD is designed to function in computer-based DVD players, in addition to traditional component DVD players. The video is offered at the highest quality to allow for full-screen viewing on your computer and projection in large lecture classrooms.

Instructor Resource Center (IRC) on DVD:

The IRC puts all your lecture resources in one easy-to-access place:

- Three **PowerPoint™ presentations** for each chapter (see below)
- **Prentice Hall Geoscience Animation Library**—more than 100 animations of geologic processes (see next page)
- **All of the drawn illustrations, tables and photos** from the text as .jpg files (Are illustrations central to your lecture? Check out the *Student Lecture Notebook.*)
- **"Images of Earth"** photo gallery (see xxi)
- **Instructor's Manual with Tests** in Microsoft Word™
- **TestGenEQ** test generation and management software

PowerPoints™

The *IRC on DVD* contains three PowerPoint™ files for each chapter. Cut down on your preparation time, no matter what your lecture needs:

1. **Art and Animations**—all of the line art, tables, and photos from the text, along with the Prentice Hall Geoscience Animation Library, *pre-loaded into PowerPoint™ slides* for easy integration into your presentation

2. **Lecture Outline**—this set includes customizable lecture outlines with supporting art

3. **Clicker Questions**—allows you to easily incorporate assessment into your lectures

Prentice Hall Geoscience Animation Library

The *IRC on DVD* contains more than 100 animations illuminating many difficult-to-visualize geologic topics. Created through a unique collaboration among six of Prentice Hall's leading geoscience authors, these animations represent a truly significant leap forward in lecture presentation aids. The animations are grouped by topic for easy reference. They are provided as both Flash™ files and, for your convenience, pre-loaded into PowerPoint™ slides for both Mac and PC users. The animations include:

Angular Unconformities and Nonconformities
Atmospheric Energy Balance
Atmospheric Stability
Beach Drift and Longshore Currents
Beach Drifting and Longshore Currents
Breakup of Pangaea
Changing Shoreline and Sedimentation
Coastal Stabilization Structures
Cold Fronts and Warm Fronts
Convection and Tectonics
Convection in a Lava Lamp
Convergent Margins: India-Asia Collision
Coriolis Force
Cyclones and Anticyclones
Debris Avalanche and Eruption of Mount St. Helens, Washington
Dimensions of the Mantle and Core
Dry Compaction and Liquefaction
Earth's Water and the Hydrologic Cycle
Earthquake Waves
Earth-Sun Relations
Effects of Groins and Jetties
El Niño and La Niña
Elastic Rebound
Erosion of Deformed Sedimentary Rock
Exposing Metamorphic Rock
Fault Motion
Faults
Floodplain and Natural Levee Development
Flowing of Ice Within a Glacier
Folding Rock
Folding
Foliation
Formation of a Cone of Depression
Formation of a Water Table
Formation of Crater Lake
Formation of Cross-Bedding
Forming a Divergent Boundary
Forming Igneous Features and Landforms
Forming Stream Terraces
Fractional Crystallization
Glacial Advance and Retreat
Glacial Isostasy
Glacial Processes and Budget
Glacial Processes
Global Atmospheric Circulation Model

Global Warming
Global Wind Patterns
Graphing Stress and Strain
Hot Spot Volcano Tracks
How Calderas Form
How Magma Rises
How Streams Move Sediment
How Tides Work
Hurricane Wind Patterns
Hydrologic Cycle
Intrusive Igneous Features
Kelvin Calculation of Earth Age
Mass Movements
Metamorphic Rock Foliation
Midlatitude Cyclones
Midlatitude Productivity
Monthly Tidal Cycle
Motion at Plate Boundaries
Motion at Transform Boundaries
Nebular Hypothesis of Solar System Formation
Ocean Circulation
Oxbow Lakes
Ozone Depletion
Physical Weathering
Plate Boundary Features
Plate Motions Through Time
Plate Tectonics and Magma Generation
Properties of Waves
Radioactive Decay
Relative and Absolute Motion
Relative Dating Principles
Relative Geologic Dating
Seafloor Spreading and Plate Boundaries
Seafloor Spreading and Seafloor Magnetization
Seasonal Pressure and Precipitation Patterns
Sediment Transport by Wind
Seismic Wave Motion
Seismographs
Stream Processes and Floodplain Development
Tectonic Settings and Volcanic Activity
Terrain Formation
The Jet Stream and Rossby Waves
Tornado Wind Patterns
Transform Faults
Tsunami
Unconformity Types
Understanding Inclination and Declination
Understanding Tuttle and Bowen's Data
Uplift and Mass Movement
Using Graphs to Understand Mantle Melting
Volcano Types
Water Phase Changes
Wave Motion and Refraction
Wave Reflection and Refraction
Wind Pattern Development

Images of Earth Photo Gallery

Supplement your personal and text-specific slides with this amazing collection of more than 300 geologic photos contributed by Marli Miller (University of Oregon) and other professionals in the field. The photos are grouped by geologic concept and available on the IRC on DVD.

Transparency Set

Every drawn illustration in *Geology, 4e* is available on full-color, projection-enhanced transparency—more than 400 in all. (Are illustrations central to your lecture? Check out the *Student Lecture Notebook*.)

Instructor's Manual with Tests

The *Instructor's Manual* contains learning objectives, chapter outlines, answers to end-of-chapter questions, and pre-tests designed to assess student understanding of each chapter. The *Test Item File* incorporates art and contains multiple-choice, true/false, short-answer, and critical-thinking questions for each chapter.

TestGenEQ

Use this electronic version of the *Test Item File* to customize and manage your tests. Create multiple versions, add or edit questions, and add illustrations—your customization needs are easily addressed by this powerful software.

Student Resources

The student resources to accompany *Geology, 4e* have been created with the goal of focusing the students' efforts and improving their understanding of the concepts of geology.

Online Study Guide

www.prenhall.com/chernicoff

The open-access *Online Study Guide* contains numerous chapter review quizzes from which students can get immediate feedback. Organized by chapter concept, the quizzes employ varied question types to help students self-check both their retention and understanding.

Student Lecture Notebook

All of the line art from the text, transparency set, and Power-Points™ are reproduced in this full-color notebook, with space for taking notes. Students can now focus more fully on the lecture instead of attempting to replicate figures. Each page is three-hole punched for easy integration with other course materials. New to the *Student Lecture Notebook* are *Concept Map Exercises*—using the *Geology at a Glance* feature of the text as a springboard, the students can construct their own concept maps of the chapter concepts. Educational research shows that concept mapping can significantly increase retention and mastery of new topics. **The *Student Lecture Notebook* can be packaged with the text for no additional cost.**

Acknowledgments

Some remarkably talented, dedicated people have helped us accomplish far more than we could have done alone. A "committee" of top-flight geologists have been assembled, who have dramatically clarified definitions and explanations, eliminated ambiguities, corrected factual errors and fuzzy logic, and, in general, helped the authors hone the manuscript in countless ways and helped the illustrator select what to show and how best to do it. Special thanks must go to Kurt Hollocher of Union College and L. B. Gillett of SUNY-Plattsburgh for their extremely insightful critiques of the first edition. In addition, for their constructive criticism at various stages along the way, we wish to thank these excellent reviewers:

From the first edition of *Geology:*

Gail M. Ashley, *Rutgers University, Piscataway*
David M. Best, *Northern Arizona University*
David P. Bucke, Jr., *University of Vermont*
Michael E. Campana, *University of New Mexico*
Joseph V. Chernosky, Jr., *University of Maine, Orono*
Michael Clark, *University of Tennessee, Knoxville*
R. Danner, *University of British Columbia*
Paul Frederick Edinger, *Coker College (South Carolina)*
Robert L. Eves, *Southern Utah University*
Stanley C. Finney, *California State University, Long Beach*
Roberto Garza, *San Antonio College*
Charles W. Hickox, *Emory University*
Kenneth M. Hinkel, *University of Cincinnati*
Darrel Hoff, *Luther College (Iowa)*
David T. King, Jr., *Auburn University*
Peter T. Kolesar, *Utah State University*
Albert M. Kudo, *University of New Mexico*
Martin B. Lagoe, *University of Texas, Austin*
Lauretta A. Miller, *Fairleigh Dickinson University*
Robert E. Nelson, *Colby College (Maine)*
David M. Patrick, *University of Southern Mississippi*
Terry L. Pavlis, *University of New Orleans*
John J. Renton, *West Virginia University*
Vernon P. Scott, *Oklahoma State University*
Dorothy Stout, *Cypress College (California)*
Daniel A. Sundeen, *University of Southern Mississippi*
Allan M. Thompson, *University of Delaware*
Charles P. Thornton, *Pennsylvania State University*

From the second edition of *Geology:*

William W. Atkinson, *University of Colorado, Boulder*
Joseph Chernosky, *University of Maine, Orono*
Cassandra Coombs, *College of Charleston*

Peter Copeland, *University of Houston*

Katherine Giles, *New Mexico State University*

B. Gillett, *SUNY-Plattsburgh*

Kurt Hollocher, *Union College*

Kathleen Johnson, *University of New Orleans*

Judith Kusnick, *Cal-State Sacramento*

Bart Martin, *Ohio Wesleyan University*

Ronald Nusbaum, *College of Charleston*

Meg Riesenberg

Roger Stewart, *University of Idaho*

Donna L. Whitney, *University of Minnesota*

From the third edition of *Geology:*

Scott Babcock, *Western Washington University*

David M. Best, *Northern Arizona University*

Paul Bierman, *University of Vermont*

Timothy J. Bralower, *University of North Carolina*

Lynn A. Brant, *University of Northern Iowa*

Robert E. Boyer, *University of Texas, Austin*

Lloyd Burckle, *Rutgers University*

Edward E. Chatelain, *Valdosta State University*

Habte Giorgis Churnet, *University of Tennessee, Chattanooga*

Dean A. Dunn, *University of Southern Mississippi*

John Ebel, *Boston College*

Robert L. Eves, *Southern Utah University*

Robert H. Filson, *Green River Community College*

Richard A. Flory, *California State University, Chico*

Katherine A. Giles, *New Mexico State University*

Ernest H. Gilmour, *Eastern Washington University*

Pamela Gore, *Georgia Perimeter College*

Solomon A. Isiorho, *Indiana University-Purdue University Fort Wayne*

Kathleen E. Johnson, *University of New Orleans*

Elizabeth A. McClellan, *Western Kentucky University*

Christine Massey, *University of Vermont*

Jeffrey B. Noblett, *Colorado College*

Mark R. Noll, *State University of New York, Brockport*

Robert Nusbaum, *College of Charleston*

Kirsten Peters, *Washington State University*

David R. Schwimmer, *Columbus State University*

Susan C. Slaymaker, *California State University, Sacramento*

Patrick K. Spencer, *Whitman College*

Dion C. Stewart, *Adams State College*

John J. Thomas, *Skidmore College*

Mari Vice, *University of Wisconsin, Platteville*

Thomas H. Wolosz, *State University of New York, Plattsburgh*

Nick Zentner, *Central Washington University*

From the fourth edition of *Geology:*

James R. Albanese, *State College of Oneonta*

Douglas E. Allen, *Salem State College*

Pranoti M. Asher, *Georgia Southern University*

Jerry Bartholomew, *University of Memphis*

Mark M. Baskaran, *Wayne State University*

David Black, *University of Akron*

Anya Butt, *Central College*

Richard C. Capps, *Augusta State University*

Beth A. Christensen, *Georgia State University*

Kevin Cole, *Grand Valley State University*

Bernard Coakley, *University of Alaska, Fairbanks*

Linda Lee Davis, *Washington and Lee University*

Diane Doser, *University of Texas at El Paso*

Martha Cary Eppes, *University of North Carolina, Charlotte*

Mark Feigenson, *Rutgers University*

Danny Glenn, *Wharton County Junior College*

Alan Goldin, *Westminster College*

Francisco Gomez, *University of Missouri*

David Gonzales, *Fort Lewis College*

Scott Hippensteel, *University of North Carolina-Charlotte*

Jeffrey Howard, *Wayne State University*

Melinda Hutson, *Portland Community College*

Solomon A. Isiorho, *Indiana, Purdue Fort Wayne*

Gary L. Kinsland, *University of Louisiana at Lafayette*

Giuseppina Kysar Mattietti, *George Mason University*

Veronique Lankar, *Pace University*

Ronald Martino, *Marshall University*

Clive Neal, *University of Notre Dame*

Mark R. Noll, *SUNY College at Brockport*

Philip Novack-Gottshall, *University of West Georgia*

Cassi Paslick, *Rock Valley College*

Hans-Olaf Pfannkuch, *University of Minnesota, Twin Cities*

John C. Ridge, *Tufts University*

Jennifer Rivers, *Northeastern University*

Jennifer A. Roberts, *University of Kansas*

Richard B. Schultz, *Elmhurst College*

John Sharp, *University of Texas, Austin*

Gary S. Solar, *Buffalo State College*

George C. Stephens, *George Washington University*

Jennifer Thomson, *Eastern Washington University*

Benjamin F. Turner, *University of Notre Dame*

Stephen Thorne, *Buffalo State College*

Paul Umhoefer, *Northern Arizona University*

David Vinson, *University of North Carolina-Charlotte*

Christopher Woltemade, *Shippensburg University*

Nick Zentner, *Central Washington University*

We would also like to thank William A. Smith of Charleston Southern State University for his sharp eye in reviewing the art for accuracy for the second edition.

At Prentice Hall, we would like to thank the following people for their extraordinary effort in overseeing this revision: Rebecca Strehlow, Development Editor; Edward Thomas, Production Editor; Charlotte Miller, Art Developer; Jessica Einsig, Art Production Editor; Sean Hale, Editorial Assistant; Chris Rapp, Senior Media Editor; Maureen Eide, Art Director/Designer; Diane Austin, Photo Researcher; Barbara Booth, Copyeditor. We would also like to thank Dorothy Marrero, Project Manager, who coordinated the extensive supplements package.

Donna Whitney would like to thank all those who gave her advice and information that helped with the revision process. In particular, Allen Glazner (UNC-Chapel Hill), Laurel Goodwin (Univ. Wisconsin-Madison), Paul Umhoefer (Northern Arizona Univ.), Jay Ague (Yale), and her friends and colleagues at the Univ. of Minnesota: Marc Hirschmann, Annia Fayon, Christian Teyssier, Peter Hudleston, and Justin Revenaugh. She did much of the revision work while a Guest Professor at the Geological Institute of the Eidgenössische Technische Hochschule (ETH) in Zürich, and she thanks her host, Professor Jean-Pierre Burg, for sponsoring her sabbatical visit there.

Finally, Stan Chernicoff also wishes to acknowledge with deep appreciation the role of Ron Pullins (formerly of Little, Brown and now of Focus Publishing) and Worth Publisher's Kerry Baruth, who championed the cause of *Geology* with their respective companies during its early gestation. Our new editor at Prentice Hall, Patrick Lynch, has helped to shepherd this latest edition of *Geology* through the editorial maze of his company and has sensed and believed that geology students across this continent might enjoy learning about the Earth from these pages. We are deeply indebted to his support and guidance.

After they leave our classrooms, students may well forget specific facts and terminology of geology, but they will still retain the general impressions and attitudes they formed during our courses. We hope that our words and illustrations will help advance the goals of those teaching this course and contribute to their students' learning experience. We have used our own teaching experiences to craft a textbook we think our own students will learn from effectively and enjoy. We hope your students will, too. We invite your comments; please send them to the authors, whose email addresses are sechern@u.washington.edu and dwhitney@umn.edu

STAN CHERNICOFF

DONNA WHITNEY

To the Student

Geology is the scientific study of the structure and origin of the Earth, and of the processes that have formed it over time. This book was created to bring you some of the excitement of that study through words, images, and illustrations. Over the years, we have derived much pleasure from introducing geology to thousands of students at the Universities of Washington and Minnesota. We have also been students ourselves and understand that some topics will be more interesting to you than others. We have worked diligently to make every aspect of this book as fascinating and useful to you as possible.

Unlike some subjects you might study, your physical geology course is not over after the final exam. Wherever you live or travel, geology is far more than just the scenery—although your appreciation for the landscape will be much enhanced by a basic knowledge of geology. When you drink from a kitchen tap, dig in garden soil, see a forest—or see it being cut down for a construction project—geology has a role to play. If you gaze at a waterfall, swim in coastal currents, endure an earthquake—or read about those who have—you will understand more about the experience after taking this course. As you will learn, geology is everywhere—in the products you buy, the food you eat, the quality of your environment. The materials and processes encountered in the study of geology supply all of our needs for shelter, food, and warmth. This course will help you understand why earthquakes occur in some places and not in others, where we should build our homes and businesses to avoid floods and landslides, where we can find safe drinking water, and why building the city of New Orleans below sea level was perhaps not the wisest urban-planning decision. Knowledge of geology can also help make us better citizens as we learn how to prevent further damage to our environment and clean up some of our past mistakes.

You will also learn about more distant matters: the age of the Earth, how the planet has changed over its long lifetime, and how some of its creatures have changed along with it. You will explore some geological ideas about our Moon and the planets with which we share our solar system. *Highlight* boxes in each chapter provide additional information about particularly interesting topics. These include some of the geo-news events you've heard or read about in recent months.

The language that geologists use helps us describe natural phenomena with precision and accuracy. We have minimized the new terminology you will need to learn, but to do well in the course you will still need to master some technical terms. We have also included some of their etymology (mostly Latin) so you can learn from where these unusual terms are derived. The key terms are in bold type when introduced. They are also listed at the end of each chapter and defined in the glossary at the end of the book.

The drawings in this book are unique, showing you how the Earth really looks both on and below the surface. As you read and examine each illustration, you will find that the words, the photographs, and the drawings are all important—they are expressly designed to give you a full picture of geology. The text and illustrations, the key terms, and even the chapter summaries work together to help you learn the concepts and terms. When you read the summary, ask yourself if you know these key points; can you cite examples beyond those given in the brief summary?

Each chapter also includes sets of questions to test your retention of the facts. Test yourself, and then go back and reread any material you are not sure you understand.

The fourth edition of *Geology* includes three new learning tools to help you master the course material. This edition introduces *Geology at a Glance,* a series of "concept maps" that will enable you to review a major body of information as you read through a chapter of new knowledge. In that way, you won't have to wait until the chapter-end summary to see if you understood all the key points. These figures summarize the text's major discussions through a series of diagrams, complete with photographs and line art. In a sense, *Geology at a Glance* provides a visual means of reviewing text material to enhance your prospects for understanding and retaining the book's major principles. This feature should really help as you review for course exams.

A second new learning tool can be found at the end of each chapter—a section called *Learning Actively.* In recent days, educators have confirmed that most students learn more from "doing" than from just sitting passively and absorbing their professors' lectures. *Learning Actively* provides several ways for you to experience the content of a chapter in a more active, personal way. These exercises are designed to be simple, fun, and informative. The goal is greater comprehension and retention of the chapter's concepts.

The third and fourth new features work in conjunction with the text to help you focus your efforts on the activities that matter most—lecture attendance and test-taking.

The *Student Lecture Notebook* reproduces all of the line art from the text and transparency set in full color with space for note-taking. You can now focus fully on the lecture instead of attempting to replicate projected figures. Included with the Student Lecture Notebook are *Concept Map Exercises*—using the *Geology at a Glance* feature of the text as a springboard, you can construct your own concept maps of the chapter concepts. Educational research shows that concept mapping can significantly increase retention and mastery of new topics.

The *On-line Study Guide,* referenced at the end of each chapter, refers to an open-access Website containing multiple quizzes for each chapter. The quizzes are *organized by concept* so you can check mastery of the material before taking the exam. The quizzes are auto-graded and the results can be e-mailed to the instructor.

We hope you will enjoy *Geology, Fourth Edition* and gain an appreciation for geology that will enrich your life. This book has been written, illustrated, and designed to convince you to keep it on your bookshelf (although a few of you may actually choose to sell it to "Used Books"). It is our hope that 5, 10, or 20 years from now, when you see something that sparks your geological curiosity, you'll use your old geology book as a reference. If you have any comments, complaints, or compliments, please send them to us or to Prentice Hall. Our email addresses are sechern@u.washington.edu and dwhitney@ umn.edu

A First Look at Planet Earth—Its Origins, Systems, and Tectonic Plates

Welcome to Geology! If this is your first day of classes, your geology professor is probably sitting in an office on campus trying to think of ways to make this scientifically exciting course relevant and timely for you, the new geology student. The relevance of geology could not be better demonstrated today than by recalling the horrific events of December 26, 2004, the day the entire world became painfully aware of the power of the Earth's geological forces (Fig. 1-1). At 7:59 A.M. local time, a 9.0-magnitude earthquake struck the seafloor of the Indian Ocean about 200 km (120 mi) off the northern coast of the Indonesian island of Sumatra with the power of a billion tons of TNT. The quake tore open the seafloor for 1200 km (about 750 mi), which is comparable to the length of California, pushed up the Earth's crust as much as 20 m (66 ft), and sped the rotation of the Earth three millionths of a second, shortening our day by roughly 1/10,000 of a second. The entire Earth vibrated, the tilt of the planet's axis shifted about 2 cm (nearly an inch), and the ground surface rose and fell as much as 10 cm (about 4 inches) as far away as 1600 km (1000 mi) from the quake. As told on the nightly news and in the local papers, the quake also spawned a killer tsunami (a quake-generated sea wave) that raced across the Indian Ocean at a speed of 800 km per hour (500 mph), rose to a height of 25 m (more than 80 ft) in the shallow coastal waters, and struck the coastlines of eleven countries, rushing on land at about 50 km per hour (30 mph). The lives of 300,000 people in the coastal towns and cities that surround the Indian Ocean (Fig. 1-2) were lost that day, and the greatest tragedy . . . this unimaginable loss of life, the worst from a natural disaster in more than five centuries, was completely avoidable, if only the geological warning signs had been monitored and heeded and the coastal zone evacuated. Thousands of lives were lost to the tsunami more than seven hours *after* the actual earthquake occurred, more than enough time for the people to have headed for higher ground.

FIGURE 1-1 Residents of the town of Galle in southern Sri Lanka make their way through the debris left in the aftermath of the tsunami of December 27, 2004. More than 12,000 died in Sri Lanka that day. ▶

3

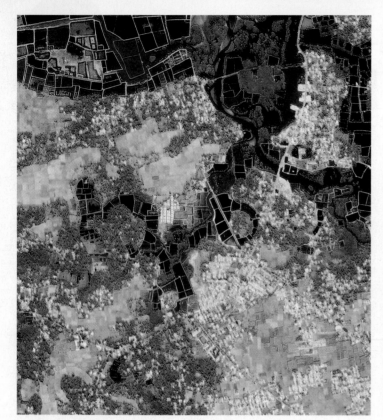 

▲ FIGURE 1-2 Satellite images—before and after—the tragic tsunami that struck Indonesia on December 27, 2004. The images show the northern coast of Aceh province on the island of Sumatra. "Before" image was taken on January 10, 2003 and shows the lush green vegetation of the island. The "after" image was taken on December 29, 2004 and shows how the tsunami stripped bare the region's landscape. More than 100,000 Sumatrans perished in the catastrophe.

This event dramatically illustrates the grim possibility that virtually all corners of the Earth are vulnerable to such catastrophes (although some far more than others)—be they powerful hurricanes, monumental volcanic eruptions, enormously powerful earthquakes, great floods and landslides, or cataclysmic meteorite strikes. So is this geology course going to be relevant to you? Your professor and your authors certainly believe so, and together we will be showing you the excitement, beauty, wonder, and yes, *danger* of the Earth's geological forces throughout this course of study.

Clearly, the events of late 2004 and early 2005 in the Indian Ocean should have grabbed your attention, but if you need something a bit closer to home, consider the following scenario: You're a student at the University of Miami in Coral Gables, Florida, or the University of North Carolina at Wilmington, or The University of Houston in Houston, Texas or one of the dozens of universities and colleges located along North America's coastlines. You've signed up to study physical geology this semester. You've bought or borrowed this textbook and you're sitting down for the first time to do some geology reading. In Chapter 1 you learn of an impending crisis that might lead to the complete disappearance of your campus, many other coastal cities and towns, and much if not all of the state of Florida at some time during the next 200 years or so. What powerful geological force could wipe away the beautiful Sunshine State from our maps? A monumental earthquake? Not likely—Florida hardly ever has them. A catastrophic volcanic explosion? Again, no—Florida's geology contains no volcanoes. The answer lies 8000 km (about 5000 mi) to the south and north of Florida in the icy wastelands of Antarctica (Fig. 1-3) and Greenland—a clear indication that some geological events affect the *entire* Earth, sometimes with dire consequences.

▲ FIGURE 1-3 A massive iceberg, 68 km (41 mi) long and 35 km (21 mi) wide, as viewed from the space shuttle *Discovery.* This ice slab is seen drifting northward in the South Atlantic after breaking off from an Antarctic ice shelf in 1991.

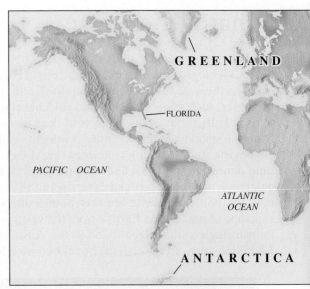

◄ **FIGURE** 1-4 Massive ice floes from the breakup of the Larsen Ice Shelf are shown here drifting in the southern Pacific Ocean.

In January 1995 an iceberg 37 km wide (about 23 mi) and 77 km long (about 48 mi) broke off from Antarctica's Larsen-B ice shelf (Fig. 1-4). For months afterward, numerous other icebergs, like so many giant slabs of Styrofoam, were shed from the edge of the ice shelf as it began to break apart. And in March of 2002, after existing for perhaps 12,000 years, much of the remaining 200-m-thick (650 ft) Larsen-B ice shelf collapsed and distintegrated, a result of a 50-year warming trend in the region. The volume of the resulting floating icebergs would be enough to fill 29,000,000,000,000 (that's 29 trillion) 5-lb bags of ice, like the ones you pick up at the local convenience store.

The breakup of the Larsen-B ice shelf and the massive icebergs now floating in the southern Pacific Ocean may foretell the complete collapse of the West Antarctic Ice Sheet. Recent studies indicate that West Antarctica's entire ice sheet is in real danger of disintegrating and melting. If and when that happens, worldwide sea level would rise roughly 5 m (about 17 ft) as the ice melts in the adjacent oceans. And because virtually the entire state of Florida sits at elevations of less than 10 m (33 ft) above sea level, the shape of Florida would most definitely change, as much of the state—especially its coasts—submerges beneath the rising seas.

Today, geologists are also beginning to look with concern at the ice sheet that covers the island of Greenland in the North Atlantic (Fig. 1-5). Some scientists predict that as the Earth's climate warms—in part, because humans are adding certain heat-trapping gases to the atmosphere (the so-called greenhouse effect)— Greenland's vast ice cover could melt as well, raising global sea level by an additional 6 m (20 ft). With such a rise in global sea level, major coastal areas, including much of Florida, Alabama, Louisiana, and Texas, would find themselves underwater, and populous cities such as Miami, Mobile, Alabama, and Corpus Cristi, Texas, would resemble scenes from the movie *Waterworld*.

What could be causing this potential human catastrophe? Some scientists point to the natural cycles of climate that have been cooling and warming the Earth for several million years. Others point the finger of blame at the mirror—humans' burning of fossil fuels has undoubtedly warmed the Earth's climate dur-

ing the last 100 years or so and accelerated ice-sheet melting. If humanity finds ways to curtail its effects on the Earth's atmospheric composition, we may yet reduce the likelihood and magnitude of such a catastrophic rise in sea level. Yet, regardless of the culprit in this particular case, such global changes are the rule on Earth rather than the exception. The study of geology helps us to understand the possibilities of such natural disasters so that we may find ways to deal with or perhaps even prevent such potential crises. Keeping the practical aspects of geology in mind, let's look at what geology is and what geologists do.

▲ **FIGURE** 1-5 Icebergs being shed by the Greenland Ice Sheet in June 2000.

What Is Geology?

Geology is the scientific study of the Earth's processes and materials. Geologists speculate about the origin of the Earth. They examine the wide variety of materials the Earth contains and the processes, such as earthquakes and volcanic eruptions, that act at or near its surface. The subjects of their investigations range from the smallest atoms to entire continents and ocean basins. Geologists study erupting volcanoes (Fig. 1-6a) and windswept sand dunes; raging floods and creeping glaciers; the slow, silent enlargement of underground caverns; and the remains left behind by titanic meteorite impacts. Some collect and interpret evidence of past life on Earth—from the vestiges of the earliest known single-celled algae found in the Australian outback in 3.5-billion-year-old rocks to the bones of our earliest human an-cestors excavated from 4-million-year-old volcanic-ash layers in East African valleys. Geologists today are not even Earthbound: They study the dust from the tails of comets collected by far-flung space probes; they search for water on Mars and study the Red Planet's rocks and soils from afar (Fig. 1-6b); and they analyze data from the other planets and moons in our solar system.

Everything we use in our daily lives comes from the Earth, so geology has an enormous practical impact on us. The natural resources that shape modern life have come to us, in part, through geological knowledge. Geologists help locate the ingredients for cement, concrete, and asphalt, with which we build our cities and highways. They find the oil, gas, and coal that fuel our cars and light and heat our homes. In recent years, many geologists have searched for a rapidly dwindling resource—fresh, uncontaminated underground water for industrial, agricultural, and domestic use. (Think of what your life would be like without a cool refreshing bottle of water.) Geological study also helps us to predict and avoid some of nature's life-threatening hazards. Geologists identify slopes that may produce the landslides that can destroy our homes and sites that are too unstable to build dams or nuclear-power plants safely. They detect potential earthquake locations and recommend ways to avoid damage from flooding rivers. They warn us about eroding shorelines and devise ways to clean up the mess we've made through decades of environmental ignorance or neglect.

Other geologists probe the most fundamental mysteries of our planet: How old is the Earth? How did it form? When did life first appear here? Why do some areas suffer from devastating earthquakes, while others are spared? Why are some regions endowed with breathtaking mountains and others with flat but fertile plains? Some of the same questions have engaged human curiosity for thousands of years. Perhaps you asked yourself some of these very same questions as you walked into the lecture hall for your first geology class. We will begin to answer these questions by describing how the various elements of the Earth (its rocks, its waters, its atmosphere, its life) interact to create all that we see around us. We will then explore how the science of geology operates and introduce some of geology's basic concepts and standards. We will discuss how our planet may have formed and speculate about some changes it has gone through since. Finally, we will examine some of the Earth's large-scale geological processes—those that create mountains, continents, and oceans—and determine how scientists deduced the nature of these processes.

(a)

(b)

▲ **FIGURE 1-6** Geologists at work. **(a)** Sampling 1200°C lava from Hawaiian volcanoes; **(b)** A robotic "geologist" called Sojourner studying a large boulder (called Yogi) on the surface of Mars, July, 1997. This robot weighed about 9 kg (nearly 20 lbs) and was powered by a solar panel on its back.

The Earth as a System

Ten years ago, students taking a physical geology course would have learned a great deal about the various aspects of the Earth in a sequence of topics—one week a discussion about earthquakes, another week volcanoes, a third week landslides or floods. Each topic within the vast subject of geology tended to be self-contained; you learned about its specifics and then moved on to the next topic. During the past decade or so, geologists have begun to appreciate fully the *interrelatedness* of different aspects of our physical environment—looking more closely at how one geological phenomenon may affect several others. They are now sharing these insights with their students by introducing the concept of **Earth systems.**

In a generic way, a *system* is a group of interdependent parts that together form a more complex whole. You might want to think of something complex—like a Boeing 747 (Fig. 1-7)—as a system (the whole plane) with various subsystems. Within this plane you would find a navigational system, the jet's engines, the hydraulics that control its wing flaps, the plane's electrical system, and its landing gear. Clearly, if one of these systems fails (such as the electrical system), the flaps won't flap, the landing gear won't deploy, and the plane won't survive. Similarly, we can consider the Earth as a highly complex system, containing four principal interrelated subsystems that interact with one another constantly and with substantial consequences (Fig. 1-8). The

▲ **FIGURE 1-7** A Boeing 747, a complex system of interrelated subsystems.

Earth's subsystems include (from the ground up) its *lithosphere* (the solid Earth beneath our feet), its *hydrosphere* (its waters . . . in their various states and locations), and its *atmosphere.* The fourth subsystem, which is intimately linked to the workings of the

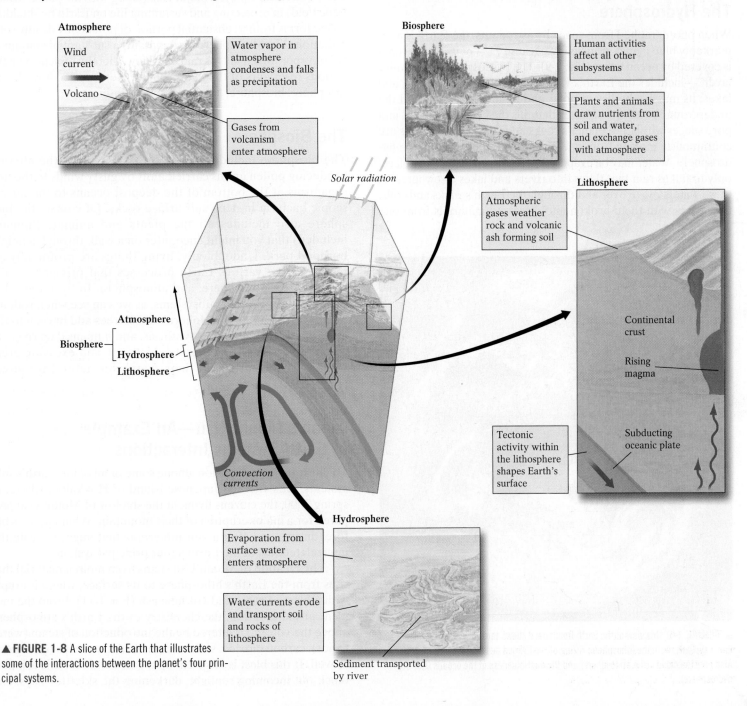

▲ **FIGURE 1-8** A slice of the Earth that illustrates some of the interactions between the planet's four principal systems.

other three, is the Earth's *biosphere*—its inventory of living animal and plant organisms, including us.

The Lithosphere

The Earth's lithosphere is its relatively strong, rocky external layer—the outermost 100 km (60 mi). Here is where much of the planet's dynamic geological activity, such as mountain building, volcanism, ocean-basin formation, and earthquakes, occurs. Virtually all of the Earth's atmospheric gases and thus the air we breathe have come from the Earth's lithosphere, expelled principally by volcanism. The relationship between the lithosphere and the atmosphere offers a clear example of how the Earth's subsystems interact.

The Hydrosphere

When photographed from space by astronauts, the Earth is a remarkably blue planet, principally because 71% of its surface area is covered by ocean water (Fig. 1-9). The hydrosphere—or "water layer"—includes the Earth's vast oceans; its rivers, streams, and lakes; its massive bodies of polar and alpine glacial ice; and the underground reservoir of water that fills the cracks, crevices, and pore spaces within the planet's rocks and soils. These waters are continuously moving from one place to another through the hydrosphere. Water may be evaporating from an ocean in one place, only to fall as rain that flows into rivers and lakes elsewhere. As these waters move over and through the Earth's rocks and soils, they carry with them broken and dissolved materials from one

▲ **FIGURE 1-9** This view of the Earth from space shows three of the four principal Earth systems. Note the lithospheric rifting of East Africa near the top of the image. Also note the clouds (the atmosphere) and the vast blueness of the oceans (the hydrosphere).

place that is then deposited in another, thus changing the shape of the landscape. **Erosion**—the loosening and transporting of materials at the Earth's surface—illustrates another distinct connection between two of the Earth's subsystems—the hydrosphere and the lithosphere.

The Atmosphere

Take a deep breath and you'll appreciate the life-sustaining composition of the Earth's atmosphere. The atmosphere is an envelope of gases that surrounds the planet to an altitude as high as 10,000 km (6000 mi). Roughly 99% of these gases are concentrated within about 30 km (18 mi) of the Earth's surface. Composed almost entirely of nitrogen (78%) and oxygen (21%), this gaseous envelope plays a major role (along with the planet's magnetic field) in protecting and sustaining life on Earth by shielding its life forms from a substantial portion of potentially deadly solar radiation. The atmosphere also contains a substantial amount of water vapor, which falls as rain and snow, thereby replenishing the various reservoirs of the hydrosphere. Here we see how the atmosphere, biosphere, and hydrosphere interact.

The Biosphere

The biosphere includes all life on Earth—from the allergy-producing pollen grains carried aloft by gusty winds to the tiny organisms at the bottom of the deepest oceans to the microscopic bacteria in deep subsurface rocks. Of course, the biosphere also includes all the plants and animals (humans included) that you might encounter on a walk through a neighborhood park. Undoubtedly, living things are profoundly affected by the various Earth processes that originate in the lithosphere, hydrosphere, and atmosphere. In turn, these living things affect other subsystems, as we can see when rodents burrow through the lithosphere's soils, trees add oxygen to the atmosphere through photosynthesis, and humans dam rivers to create reservoirs and generate electricity and excavate great cavities in the surface in their search for natural resources (Fig. 1-10).

Japan's Mount Usu—An Example of Earth Systems Interactions

To illustrate the interaction among some or all of the Earth's subsystems, let's visit the Japanese island of Hokkaido, where, in spring 2000, the citizens living in the shadow of Mount Usu prepared for a major eruption of their mountain. What happens before, during, and after a volcano erupts that might indicate the interrelatedness of the Earth's four principal systems?

The creation of Mount Usu stems from molten material that rises from the Earth's lithosphere to its surface, where it erupts with a blast of gases and volcanic ash (Fig. 1-11). From the moment of the eruption, the chemistry of the Earth's atmosphere above the volcano is altered by the introduction of steam (water vapor), carbon dioxide, sulfurous gases, and perhaps other gases as well. If this blast is large enough, gases and volcanic ash may block out incoming sunlight, darkening the skies for days, and

The Earth Systems' Power Supply

The Earth's various subsystems derive their energy from two primary sources: one *external:* the Sun; and one *internal:* the Earth's internal heat engine. When solar radiation heats the atmosphere, it produces the Earth's winds, evaporates water from the hydrosphere, and creates a comfortable environment for life on Earth to thrive. Essentially, this *external* energy source generates the Earth's weather and climate and drives the forces of erosion, powered principally by flowing water. The Earth's *internal* heat—derived primarily from residual heat from the planet's initial formation and from radioactive decay of elements such as uranium—drives the motions of the lithosphere that cause volcanoes to erupt, the Earth to quake, and mountains to rise. Both of these energy sources are enhanced in large part by the force of gravity, which compels water and glaciers to flow downhill and which, in certain places, drags slabs of the Earth's lithosphere into the planet's interior. We will explore the Earth's power supply extensively throughout the upcoming chapters.

▲ FIGURE 1-10 The Carajás iron mine in the Amazon Region of Brazil.

perhaps even cooling both the local and global climates for several years. Rainfall during the ensuing weeks and months may show signs of increased acidity as the sulfurous gases in the atmosphere combine with water vapor to form sulfuric acid. This acidified rain may eventually reach the region's rivers, lakes, and underground water.

Meanwhile, volcanic ash, pulled back to Earth by gravity, rains down on the surrounding landscape, burying the surface rocks with a new layer of geological material added to the Earth's lithosphere. The ash layer, in turn, may replenish the nutrients in the local soils with fresh minerals, increasing the soil's fertility. The initial effects on the local human population include an initial period of darkness and naturally "polluted" air, perhaps a longer period of cooler, rainier weather, and the possibility of acid rain. The long-term effect will be the gift of rejuvenated soils that may lead to improved local agriculture. So what began within the lithospheric subsystem with a volcanic eruption ultimately affects the composition of the atmosphere and hydrosphere, with several negative and positive effects on the biosphere.

▲ FIGURE 1-11 Mount Usu, Hokkaido, Japan. This volcano, like others around the Pacific Rim, will continue to erupt in the foreseeable future.

A Final Comment About Earth Systems

In the opening pages of this chapter in our discussion about the possibility of a catastrophic collapse of the polar ice sheets, we discussed the possibility that the burning of fossil fuels such as coal (a rock from the *lithosphere*) may be changing the composition of the *atmosphere*. This change appears to be warming the planet, melting polar ice sheets (part of the *hydrosphere*), and raising global sea level (another part of the *hydrosphere*). The end result—coastal cities may someday be swamped, and humans and other living things (the *biosphere*) will be forced to migrate elsewhere or perish. This sequence of events is a classic illustration of how the Earth systems relate to one another. Throughout this book, we will discuss numerous other examples of these systems' interactions. Be on the lookout for these relationships—they will most definitely enhance your understanding of the complexity of our planet. But first, let's discuss how geoscientists unravel the Earth's geological mysteries and learn how these various systems operate.

Geologists—How They Do Their Work

The principal objective of science is to discover the fundamental patterns of the natural world and understand the processes that shaped it. Scientists make one basic assumption: The world works in an orderly fashion in which every effect has a cause. Whenever scientists observe effects, such as earthquakes or landslides, they're eager to know the causes.

Geologists, like other scientists, begin their investigations with a question or set of questions about how some part of the natural world works. Why do earthquakes occur in some places and not in others? Why are some places rich with great mineral wealth—diamonds, gold, emeralds, platinum—and others not? Using **scientific methods,** scientists gather all available data about their subject. They measure things and describe what they observe, compile the results of laboratory experiments, and create computer models of natural phenomena. They then propose an **hypothesis,** a logical but *tentative* explanation that fits all the data collected and *is expected to account for future observations as well.* Often a number of hypotheses are proposed by different scientists to explain the same set of data. For example, within the past 100 years, more than 50 hypotheses have been proposed to explain why the Earth's climate periodically plunges into ice ages.

Once an hypothesis is introduced to the scientific community, interested parties begin to test it—often vigorously. Scientists conduct further experiments or make further observations to see if the hypothesis can stand up to intense scrutiny. Some hypotheses change substantially over many years until they fully account for all relevant findings. Others simply fail the test and are cast aside. Good scientists must be receptive to *whatever* they discover, even when, in light of their findings, their own preferred hypotheses must be rejected. New technologies capable of more accurate observations often reveal fatal flaws in once reasonable hypotheses. For example, the invention of high-powered tele-

scopes in the early 1700s shattered the hypothesis that the Earth is the center of our solar system, a popular notion for 2000 years. The history of science is littered with discarded hypotheses that were once quite popular but after closer analysis proved to be false. On a positive note, a flawed hypothesis usually gives way to one supported by more of the available data; that's how science progresses.

An hypothesis that is repeatedly confirmed by observation and experimentation eventually becomes widely accepted within the scientific community. Ultimately, an hypothesis that consistently explains all available data may be elevated to the status of a **theory**—a generally accepted explanation for a given set of data or observations. An example of how multiple hypotheses are developed and tested, possibly to become theories, is given in Highlight 1-1. Spend a few minutes with this essay and learn what many scientists believed caused the demise of the Earth's dinosaurs.

Even when an hypothesis survives testing and becomes an established theory, new technology and new data may still require it to be updated—perhaps even replaced. A theory that meets rigorous testing over a long period of time may ultimately be declared a **scientific law,** but there are very few of these. Let go of that pen with which you're highlighting the textbook and watch it fall to the floor. Do you need to do this 100 times before you accept that every free-falling object within the Earth's gravitational field *always* falls toward the Earth's surface? Here you have demonstrated the *law* of gravitational attraction. Obviously, it's unlikely that this law will ever be disproved and replaced.

In many ways, geology is not like the other sciences on your campus. Geologists must sometimes apply scientific methods in ways that are unique to their science. Unlike some chemistry or physics experiments that may involve reactions that last but one millionth of a second and that occur on an atomic scale, geologists must often deal with the effects of processes acting over long periods of time, across vast areas, and sometimes out of sight deep within the Earth. Furthermore, geologists can't always test their hypotheses through direct observation or experimentation. To supplement their field and laboratory work, they sometimes use scaled-down models to study large-scale geological phenomena, or they rely on computers to calculate the results of mathematical models.

How the Science of Geology First Developed

Almost two centuries of observation and hypothesis-testing have contributed to our current understanding of how our planet formed. For a long time people believed that the Earth had evolved through episodes of dramatic and rapid change ("catastrophes") separated by long periods of relative stability and little change. Since natural scientists understood best those processes that they could directly observe, such as titanic volcanic eruptions, monumental earthquakes, and raging floods, they believed that these violent events alone explained the origin of such seemingly inactive Earth features as mountains and valleys. **Catastrophism,** the hypothesis that the Earth evolved through a series of immense worldwide upheavals, was the accepted view until the mid-eighteenth century.

During the latter part of the eighteenth century, the Scottish naturalist James Hutton (1726–1797) recognized that slow, steady processes, such as the action of rivers cutting through valley floors, *acting over a vast amount of time,* may affect the Earth more than do occasional catastrophic events. This idea met with great resistance because it implied that the Earth was much older than most Christians at the time believed. Reckoning based on the Bible put the Earth's age at only a few thousand years. Hutton proposed that the physical, chemical, and biological processes that anyone could see changing the Earth in small subtle ways during one's own lifetime must have worked in a similar manner throughout a very long history. His hypothesis, called **uniformitarianism,** proposed that our observations of current geological processes could be used to interpret the rock record of long-past geological events. By the 1830s, after much heated debate, uniformitarianism prevailed over catastrophism. Its acceptance—and its belief in a very old Earth—has been hailed as the birth of modern geology.

Hutton maintained that "the present is the key to the past." Geologists do acknowledge that the physical processes we see shaping the Earth's appearance today (such as erosion, glaciation, volcanism, mountain-building) have probably acted throughout the Earth's history, although probably at significantly variable rates. They also know that some geological events are indeed catastrophic and that much geological change does occur during these brief, spectacular events. In 1989, for example, Hurricane Hugo eroded more of the Carolina coast *in one day* than had the preceding century of slow, steady wave action. Thus, slow but consistent processes *as well as* catastrophic events are continuously reshaping our planet together.

The Earth in Space

In some respects, the Earth seems like nothing special. The planet is a slightly flattened sphere with an average diameter of 12,742 km (7914 mi), orbiting approximately 150 million km (93 million mi) from the medium-sized star we call the Sun. Our Sun is only one of about 100 billion stars in the Milky Way galaxy, a pancake-shaped cluster of stars that itself is only one of about 100 billion such galaxies in the observable universe. Despite Earth's relative smallness in the universe, it is perfectly positioned to receive just the right amount of the Sun's radiant energy to support life. Because of Earth's composition and geologic past, it has developed a watery envelope and a protective atmosphere on which countless living species have relied for billions of years. But how did the Earth come to be what *may* be the only life-sustaining planet in the solar system?

The Probable Origin of the Sun and its Planets

Cosmologists (scientists who study the origin of the universe) have proposed that the present universe began as a very small, very hot volume of space containing an enormous amount of energy. Many scientists believe the birth of all the matter in the universe occurred when this space expanded rapidly with a "big bang" roughly 13 billion years ago. The timing of the hypothetical Big Bang Theory has been estimated recently with the aid of the Hubble Space Telescope. The Hubble observes the current positions and speeds of the visible galaxies as they move away from one another. By tracing their paths backward, cosmologists estimate their point and time of origin.

Immediately after the "big bang," the universe began to expand and cool, as it continues to do today. Within a few minutes after "the bang," the universe had cooled to a temperature of about 1 billion degrees Celsius. At this time, the universe consisted only of three kinds of subatomic particles—protons, neutrons, and electrons—remnants of its original composition. These particles eventually combined to form atoms, the building blocks of all the matter in the universe today. At this early stage, however, any atomic particles that may have formed immediately broke apart in violent collisions with other particles.

After about a million years had passed, the universe cooled to approximately 3000°C (about 6000°F). At that temperature, atoms of hydrogen and helium—the simplest, lightest elements—begin to exist without being torn apart. At that time the universe consisted of about 75% hydrogen gas and 25% helium gas by weight, a composition that has not changed much to this day. As the universe continued to expand, its matter became more concentrated in space. Pockets of relatively high gas concentrations began to attract more gas by the force of gravity. Where enough gas gathered, the resulting gas clouds collapsed inward because of gravity. These accumulations became galaxies (large disk-shaped structures) and clusters of galaxies. Although cosmologists have amassed a great deal of evidence to support these details of the Big Bang hypothesis, galaxy formation is still not well understood.

Within each galaxy, such as our own Milky Way, some gas clouds collapsed further to form stars. Stars continue to be born in this way in all galaxies, including our own. The heat released by the compression resulting from the collision of infalling gas particles causes the particles in the core of each star to move faster and faster. Eventually they move fast enough to collide and engage in nuclear reactions. These reactions release the energy that keeps the star hot and glowing brightly. In these nuclear reactions, the particles fuse, forming larger particles that will become the nuclei of helium and other, heavier elements.

Stars are not only born, they die as well—some slowly and some rapidly. A star that is dying very rapidly is called a *nova* (from the Latin for "new") because it appears as a very bright new star in the heavens. Dying stars are important because they heat up so much that new nuclear reactions occur, producing the nuclei of heavier elements. In this way, dying stars act as manufacturing plants for heavy elements such as iron. They are also distribution centers, dispersing these nuclei over the space surrounding each dying star. It is believed that most of the matter of the Earth—*including the atoms that make up our bodies*—comes from such dying stars.

Our own Sun is a star that came into being well after our galaxy was created, even after earlier stars had already died. Their remnants contributed to the gas cloud, or *nebula,* that eventually

Highlight 1-1

What Killed the Dinosaurs?

For years, paleontologists (geologists who study ancient life forms) have wondered what might have caused more than 75% of all the forms of life on Earth to vanish about 65 million years ago. The most dramatic loss was the extinction of the dinosaurs, a group of animals that had roamed the planet for 150 million years. Numerous other life forms vanished as well—large and small, water- and land-dwelling, plant and animal.

Some early hypotheses proposed that epidemic diseases eliminated dinosaur populations or that egg-stealing mammals, on the rise 65 million years ago, ravaged unprotected dinosaur nests. But, of course, neither of these hypotheses accounted for the loss of two-thirds of all the microscopic forms of life in the Earth's oceans. These marine extinctions then led some scientists to propose that the oceans became lethally salty at this time, but the "salty sea" hypothesis failed to explain why some marine creatures survived, and of course, the salty sea couldn't explain what killed off the dinosaurs. To explain the simultaneous extinction of gigantic land-based reptiles, tiny marine organisms, and many life forms in between, a sound hypothesis needed to invoke some form of *global* environmental catastrophe. Did the Earth suffer from a period of drastic cooling 65 million years ago? Did a shift in the planet's protective magnetic field allow harmful solar radiation to reach land and sea, eliminating a wide variety of life forms? Did a nearby star explode, bathing the Earth in cosmic radiation? Surely, each of these events would have affected *all* life on Earth simultaneously. But then, why were 25% of the planet's species spared?

Several hypotheses suggest that wholesale extinction followed a catastrophic disruption of the global food chain. The food chain refers to nature's predator–prey relationships, basically, "who eats whom." In this chain, meat-eating dinosaurs such as *Tyrannosaurus rex* would die off if their prey, typically another plant-eating dinosaur, became scarce. The plant-eaters would die off if their food supply somehow diminished. Some scientists looking for indications of a global event that could have caused such a disruption in the food chain have noted evidence in the geological record of widespread fires followed by very rapid cooling 65 million years ago. Yet they disagree on what might have caused these events. One group has cited evidence that the massive volcanic eruptions of India's Deccan Plateau sent a cloud of volcanic ash and gas around the Earth at this time, blocking out sunlight, cooling the planet, and leading to a worldwide decline in vegetation, including microscopic marine plants. These scientists reason that without plants to eat, plant-eating animals would have died out and their demise would have wiped out the meat-eaters who ate them.

Another group of scientists, led by geologist Walter Alvarez and his father, Nobel prize-winning physicist Luis Alvarez, have proposed another scenario: A meteorite at least 10 km (6 mi) in diameter crashed into the Earth with the force of 100 million hydrogen bombs, releasing a shower of pulverized rock into the atmosphere. The resulting dust veil and the accompanying smoke from widespread fires set by the red-hot fragments would have blocked sunlight (in much the same way as volcanic ash would have), sent global temperatures plummeting, and led to an "impact winter" that may have lasted for decades—long enough to devastate the global food chain.

Strong evidence supports this impact hypothesis. An unusual 2.5-cm (1 in.) thick layer of a special type of clay has been documented at many localities and on all continents in rocks that date from about 65 million years ago (Photo a). The clay contains iridium, an element that is extremely rare in Earth-bound rocks but much more abundant in meteorites. The Alvarezes and their associates contend that the iridium-rich layer resulted from the global fallout of an enormous pulverized meteorite. Fossils of

▼ **(a)** An iridium-containing layer of clay (marked by coin) found in Gubbio, Italy. **(b)** Tektites from Atlantic City, New Jersey. Such aerodynamic spheres, found around the world in sediments of about 65 million years of age, are believed to have been formed when rock melted by meteor impact was hurled into the air and cooled and solidified before falling back to Earth.

(a)

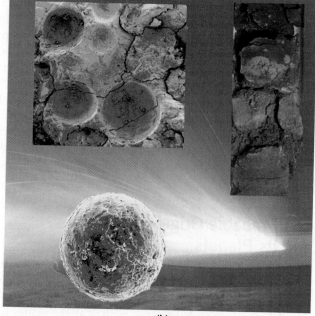

(b)

numerous species, including many now extinct, have been found in the rocks that formed *just before* the iridium-rich layer was deposited, but only about one-fourth as many species appear in the rocks that formed *just after* this layer was deposited. This finding suggests that many extinctions occurred during the time of iridium deposition.

Further evidence of a meteorite impact includes the presence of small glassy spheres called *tektites* in sediment layers around the world that date from this period (Photo b). These may have been formed when superheated rock at the impact site was hurled into the air in a molten state, dispersed in the atmosphere, and cooled rapidly as they fell back to Earth. A 5 cm (2 in)-thick layer of such glassy spheres was recently discovered in a drilling project near Atlantic City, New Jersey. Mineral grains shattered by very high pressures—as would occur if they had been struck by a meteorite—have also been found at proposed impact sites and throughout the world in rock layers dated at about 65 million years (Photo c). Furthermore, the high concentration of carbon soot (a product of burned vegetation) found within the iridium layer could be evidence of global wildfires, which may have been touched off when countless flaming bits of falling debris ignited the Earth's vegetation.

Although this evidence of a 65-million-year-old meteorite impact is certainly compelling, most of it is scattered in various locations around the world. In 1996, a fragment that may be the final piece to the puzzle was extracted during the coring of the Atlantic Ocean floor, 320 km (200 miles) east of Jacksonville, Florida. The 40-cm (16-in.) core of mud contains a whitish fossil-rich layer at the bottom of the core that is overlain by a thin gray-green layer of impact debris topped by an iron-rich band that may constitute remains from the meteorite itself. This layer, in turn, is overlain by a fossil-poor layer that may have been deposited over a period of about 5000 years. The interpretation? Life was abundant before the meteorite struck and severely diminished afterward.

As yet, no extinction hypothesis has achieved theory status. Analysis of the Earth's 65-million-year-old deposits continues today, to document further the percentage of organisms that became extinct at that time and to search for additional evidence of a meteorite strike or a catastrophic volcanic eruption. For more than a decade, the proponents of the impact hypothesis awaited confirmation of a key piece of evidence—definite location of an impact site. A meteorite capable of such vast environmental disruption would have left behind a

crater at least 160 km (100 mi) across. The search for the impact site has led geologists to a huge, partially submerged crater along the coast of Mexico's Yucatán Peninsula. Because of the age of its rocks and the presence of tektites in nearby sediments, the Yucatán's Chicxulub crater, 300 km (180 mi) in diameter, is widely considered to be the proverbial "smoking gun" (Photo d and inset map). Within the past several years, other smaller craters, each about 65 million years old, have been discovered—one in the Russian Ukraine, about 25 km (15 mi) in diameter, another off the coast of Great Britain at the bottom of the North Sea, about 19 km (12 mi) in diameter and of the same age. These craters suggest that the dinosaurs' extinction may have resulted from a swarm of large impacts over the course of several thousand years, yet another twist in the plot of this mystery.

And, in 1998, a core of Pacific Ocean seafloor mud that dates from 65 million years ago yielded an actual sliver of what may be a remnant of the main dinosaur-killing culprit. This speck of carbon–iron meteorite (Photo e) may have been part of the killer projectile that struck the Yucatán Peninsula. Still, the debate over the cause of this mass extinction is expected to rage for years to come. It is important to remember that all geologists do not agree with the meteorite hypothesis, and more importantly, it is acceptable . . . *no, make that essential* . . . that they do *not* agree. Such lively debate is the driving force of scientific progress.

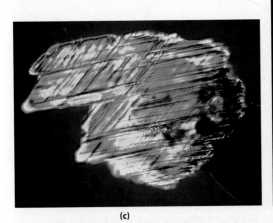

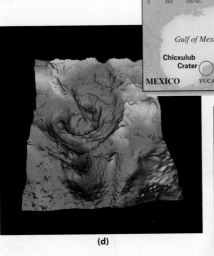

▲ (c) A magnified photograph of a mineral grain that has been shattered by the shock of a meteorite impact. (d) This radar image shows the circular outline of an enormous impact crater in Mexico's Yucatán Peninsula. The Chicxulub crater is large enough to account for the effects of a massive meteorite strike. (e) A 2.5-millimeter speck of a 65-million-year-old meteorite, retrieved from the Pacific Ocean floor.

became our entire solar system (Fig. 1–12). The interstellar ("between stars") nebula that would become our Sun and planets was originally dispersed across a vast area of space. It probably extended well beyond what would become the outermost planets, Pluto and our solar system's newly *proposed* tenth planet, Sedna, which sits about 13 billion km (8 billion mi) from the Sun. (*Note:* There is much debate raging about whether Sedna is a planet or not; it may in fact be one of many large bodies of interstellar debris that make up a "cloud" of material out beyond Pluto.) About 5 billion years ago, this nebula began to collapse inwardly, perhaps due to a shock wave from a nearby exploding star. As its component materials were compressed and heated by the shock wave, they were drawn by gravity toward a hot, denser center. There they collided in nuclear reactions and generated enormous heat, forming the infant Sun.

As heat became more and more concentrated in the center of this new star, some material in the nebula surrounding it began to cool and condense into small grains of matter. Lighter, un-

condensed substances were swept outward by strong solar "winds"—the streams of high-speed particles that flowed from the infant Sun. In this way, the first solid materials to form in our solar system became separated into a hot inner zone of denser substances, such as iron and nickel, and a cold outer zone of low-density gases, such as hydrogen and helium. Ultimately, this compositional partitioning would evolve into the four rocky inner planets (Mercury, Venus, Earth, and Mars) and the four gaseous outer planets (Jupiter, Saturn, Uranus, and Neptune). (Pluto, believed to be an escaped moon of Neptune, is an icy mixture of gas and rock. The composition of the recently discovered Sedna is believed to be similar to Pluto's.)

As the first bits of matter condensed, they continued to collide and coalesce, forming aggregates that grew to a few kilometers in diameter. As they grew, the increasing gravity of these planetary seeds, or *planetesimals,* attracted other bodies of solid matter. Cosmologists originally believed that this process of planetary growth, or *accretion,* was slow and gradual, much like the

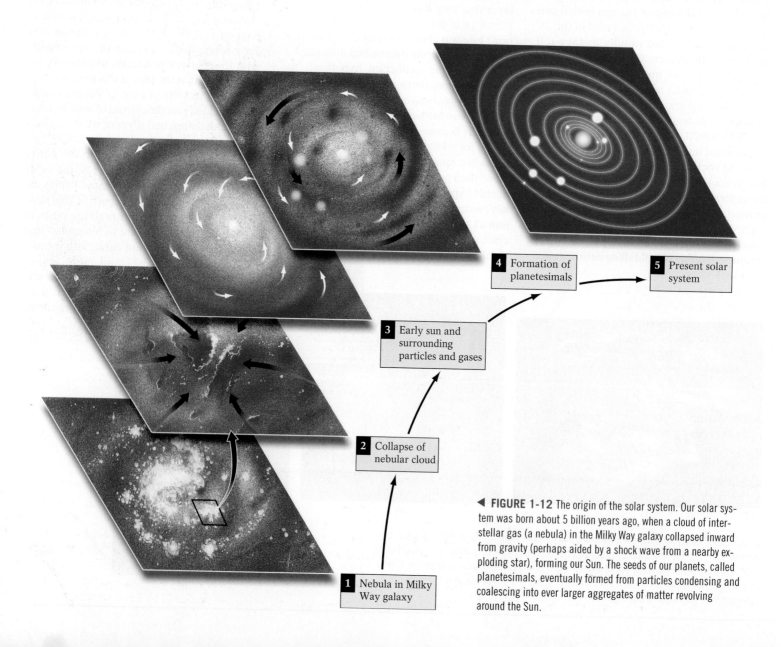

4 Formation of planetesimals

5 Present solar system

3 Early sun and surrounding particles and gases

2 Collapse of nebular cloud

1 Nebula in Milky Way galaxy

◀ **FIGURE 1-12** The origin of the solar system. Our solar system was born about 5 billion years ago, when a cloud of interstellar gas (a nebula) in the Milky Way galaxy collapsed inward from gravity (perhaps aided by a shock wave from a nearby exploding star), forming our Sun. The seeds of our planets, called planetesimals, eventually formed from particles condensing and coalescing into ever larger aggregates of matter revolving around the Sun.

way one might create a large aluminum-foil ball by the steady addition of small lumps. Recently, our view of planetary accretion has changed drastically. We now believe that as they grew, huge planetesimals, easily the size of Mercury or even Mars, collided violently. Such violent collisions ejected great masses of molten material into space, perhaps forming some of the moons that today orbit the planets of our solar system. (More on this topic shortly.)

As intense solar radiation warmed the four protoplanets closest to the Sun, their surfaces were heated to high temperatures. Nearly all of their light gases—hydrogen, helium, ammonia, and methane—vaporized and were carried away by solar winds. The matter remaining in these four small, dense inner planets consisted primarily of iron, nickel, and silicate minerals, those that contain a large amount of silicon and oxygen. Together these heavier substances make up most of Mercury, Venus, Earth, and Mars—the inner, or terrestrial, planets.

The frigid outer planets developed so far from the warmth of the Sun that much less of their matter vaporized. These planets—Jupiter, Saturn, Uranus, and Neptune—consist predominantly of low-density masses of icy condensed gases, principally hydrogen, helium, methane, and ammonia.

The Early Days of Planet Earth

No Hollywood disaster movie with the greatest special effects could begin to depict the extreme violence and chaos of the first 20 million years or so of Earth history. During this time our planet became heated to the melting point of iron, developed its layered internal structure, collided with one or more other massive planetary bodies, and spewed out a mass of its own substance to form our nearest neighbor, the Moon. Quite a story.

During its first few million years, the gravitational pull of the proto-Earth attracted other speeding planetesimals. The resulting collisions converted the enormous energy of motion to thermal energy upon impact. Some of this energy radiated back into space, but much was retained by the new planet as succeeding planetesimals struck its surface and "buried" the heat from earlier strikes. As the proto-Earth grew, its deep rocks became compressed under the weight of the growing mass of overlying rock. Such "compressional heating" is akin to the heating produced in a bicycle pump as the air in the pump is compressed. Think about how warm the barrel of a bicycle pump becomes as you inflate your tires.

Another source of heat were the atoms of radioactive substances, such as uranium, that released heat as their nuclei split apart, a process known as *fission*. As heat from these two sources accumulated and became trapped within the Earth's interior, the planet's internal temperature began to rise. This heat set in motion the process of *differentiation* that created our layered Earth (Fig. 1-13).

Differentiation   What were the initial substances that made up the early Earth? For a clue, geologists look to some of the meteorites that have struck the Earth . . . and survived. Many meteorites contain **chondrules** (KON-drools), small nuggets of rocky material believed to be droplets of matter that condensed directly from the original solar nebula. The composition of the Earth itself, coalescing from the same nebula, should be similar to that of chondritic meteorites. Yet when we analyze such meteorites, we find a puzzling disparity: They are roughly 35% iron in composi-

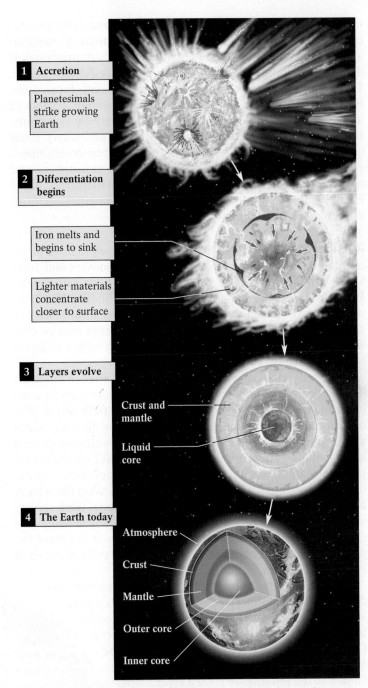

1   **Accretion**

Planetesimals strike growing Earth

2   **Differentiation begins**

Iron melts and begins to sink

Lighter materials concentrate closer to surface

3   **Layers evolve**

Crust and mantle

Liquid core

4   **The Earth today**

Atmosphere

Crust

Mantle

Outer core

Inner core

▲ **FIGURE 1-13** Heating and differentiation of the early Earth. Increased temperatures in the planet's interior, caused largely by planetesimal impacts and decay of radioactive substances, are believed to be responsible for the layering of the Earth's interior.

tion, whereas the rocks at the Earth's surface average only about 6% iron. If our assumptions regarding chondrules are correct, where is all the Earth's missing iron?

We now believe that during the Earth's first 10 to 20 million years of existence, as the planet was still accreting, its internal temperature rose to the melting point of iron. As a result, much of the iron liquefied. Because the iron was denser than the surrounding materials, it sank to the proto-Earth's center by the pull of gravity. (*Density* expresses the quantity of matter in a given volume of a substance. Because its chemical structure is more compact, 1 cm^3 of iron is far denser than 1 cm^3 of cotton.) While the iron sank, lighter materials rose and became concentrated closer to the surface. Thus, the matter that had originally made up a *homogeneous* Earth became separated into concentric zones of differing densities. This separation process is called *differentiation.* The densest materials, probably iron and nickel, formed the Earth's core at the planet's center. Lighter materials, composed largely of silicon and oxygen and other relatively light elements, formed the Earth's outer layers (the mantle and crust). Even lighter materials—gases that had been trapped in the interior— escaped, combining to form the Earth's first atmosphere and oceans. Let's look more closely for a moment at the differentiated Earth before we consider the Earth's next great event—the collision that spawned the Moon.

A Glimpse of the Earth's Interior

Contrary to those memorable but misleading scenes in Hollywood movies and Jules Verne novels, the Earth is not a hollow ball filled with jungles and dinosaurs or forests of giant crystals and floating Hummer-size diamonds (as depicted in 2003's geologically absurd flick *The Core*). Scientists have determined that the Earth's interior consists of three principal concentric layers, each with a different basic composition and density (Fig. 1-14).

The outermost layer of the Earth is a thin **crust** of relatively low-density, silicon-and-oxygen-based rocks. Underlying the crust is the **mantle,** a thicker layer of denser rocks (still silicon-and-oxygen-based but containing some heavier elements, such as iron and magnesium). At the Earth's center is its **core,** the densest layer of all, consisting primarily of metals such as iron and nickel. It is important to note that these three layers have three very different compositions. The Earth's interior is somewhat like a hard-boiled egg—with its thin brittle shell (the crust), thick spongy white (the mantle), and small central solid yolk (the core). The egg model, however, does not show a number of the other important *sublayers* that are fundamental to our understanding of our dynamic Earth.

The near-surface segment of the Earth's interior is subdivided into two distinct layers—*based not on their composition but rather on how strong they are.* The outer 100 km (60 mi) of the Earth, encompassing both the Earth's crust and the uppermost portion of the mantle, is a solid, relatively strong, rocky layer known as the **lithosphere** ("rock layer," from the Greek *lithos,* "rock"). Underlying the lithosphere is the **asthenosphere** or "weak layer" (from the Greek *aesthenos,* "weak"). The asthenosphere is a layer of heat-softened, slow-flowing yet still-solid rock located in the upper mantle. It extends from about 100 to 350 km (60–220 mi) beneath the Earth's surface. The processes active within the lithosphere and asthenosphere build the Earth's mountains, fuel its volcanoes, trigger its earthquakes, and create its ocean basins.

Overlying the lithosphere is the Earth's atmosphere, composed largely of the gases that have escaped over time from the Earth's interior, primarily during volcanic eruptions. The ongoing *degassing* of the Earth's interior (yes, it's still happening today) has largely created the oceans that cover 71% of the Earth's surface (one prominent gas emitted during volcanic eruptions is water vapor) and the thick atmospheric envelope of gases that support our planet's living organisms. Fortunately, the Earth's interior probably did not melt completely during the period of planetary differentiation. If it had, the Earth might have lost *all* of its gases during its first 20 million years, and its life-supporting atmosphere might never have accumulated. The Earth did, however, lose its earliest atmosphere, which is believed to have been composed mainly of hydrogen, helium, ammonia, and methane. This was no great loss to us, because these gases are lethal to most life forms.

With this knowledge of the composition of the differentiated Earth, we can now consider the dramatic manner in which we may have lost our first atmosphere—and gained our nighttime lantern, the Moon.

The Origin of the Moon

The birth of our Moon has sparked lively debate for centuries. Did it form as a companion planet coalescing independently from the solar nebula at the same time as

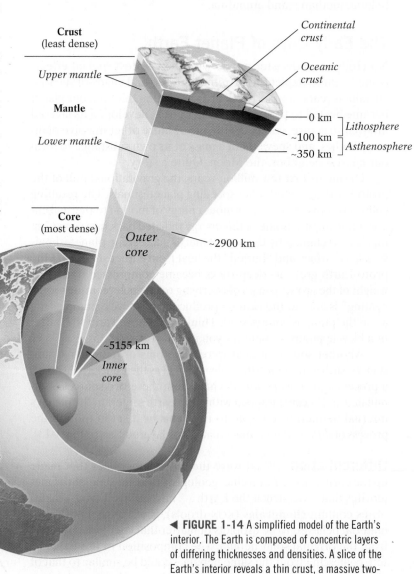

◀ **FIGURE 1-14** A simplified model of the Earth's interior. The Earth is composed of concentric layers of differing thicknesses and densities. A slice of the Earth's interior reveals a thin crust, a massive two-part mantle, and a two-part core.

Earth? Did it form elsewhere, only to be drawn into Earth's orbit and held hostage there by our planet's relatively strong gravitational pull? Or did it form, as some scientists now believe, in a massive head-on collision between the Earth and another nearby planetesimal?

The answer may lie in the Moon's composition. The Moon is 36% *less dense* than the Earth, suggesting that it contains much less iron. This would apparently rule out independent accretion from the solar nebula. If the Moon did form in the same way as the Earth, its composition would be similar. The Moon's composition, confirmed in part by the rock-collecting efforts of U.S. *Apollo* astronauts, is actually quite similar to that of the Earth's mantle . . . perhaps for the following reason.

By roughly 4.55 billion years ago, the Earth had probably attained much of its size and had become layered, with most of its iron dispatched to its core. Its mantle and crust were composed largely of a variety of lighter, silicon-and-oxygen-based minerals. The solar system at this time was still far more crowded than it is today. Numerous high-speed objects—many larger than Mercury, some as large as Mars—still hurtled through Earth's neighborhood. Thus the stage was set for a monumental collision between Earth and a Mars-sized impactor that jolted and tilted the Earth's axis, reheated the planet's outer layers to their melting points, and most probably propelled massive molten chunks of the Earth into space. Many planetary scientists believe that these chunks ultimately coalesced as our familiar orbiting companion, the Moon.

Picture this scene. The gravitational pull of the growing Earth, the largest body in this area of the solar system, attracts a Mars-size planetesimal. The speeding impactor is traveling at supersonic speeds, perhaps as fast as 14 km per second (31,500 mi per hour). At the moment of impact, the Earth's early atmosphere is blown away, replaced by a sky filled with a rain of molten iron blobs, remnants of the impactor's iron core. (It, too, must have undergone some differentiation.) The collision vaporizes much of the crust of both the Earth and the impactor and a good portion of their mantles as well (Fig. 1-15), but does not penetrate all the way to the Earth's iron core. As the remnants of the impactor's dense iron core are pulled back to Earth, jets of vaporized crust and mantle are shot into space. The Earth's gravitational pull, however, captures this material before it escapes the area altogether, holding it in orbit at a distance of roughly 400,000 km (240,000 mi). Over a few tens of thousands of years, this debris coalesces to form the Moon.

This hypothesis (and it is *only* an hypothesis) explains why the Moon is relatively iron-poor. The impactor failed to reach the iron core of the already differentiated Earth, and the impactor's dense iron core failed to reach sufficient velocity to escape the Earth's gravitational pull. Thus, the Moon may be composed largely of some fraction of the Earth's vaporized mantle and crust that was cast into space during what certainly would have been the most spectacular moment of Earth's early history.

So where are the "scars" of this great collision? Unfortunately, the Earth's surface is perhaps the worst place to look for evidence of the planet's violent past. The Earth's dynamic internal processes—such as volcanism, earthquake activity, and mountain-building—would have wiped away much of the evi-

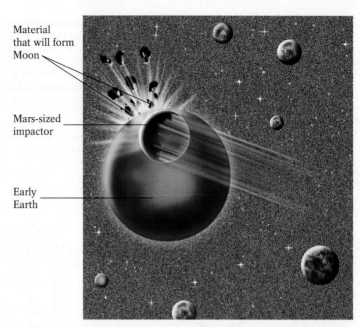

▲ **FIGURE 1-15** A catastrophic impact between the proto-Earth and a Mars-size impactor is now believed by many scientists to have spawned the Earth's Moon.

dence. And the Earth's atmosphere, which causes the planet's surface rocks to weather away and erode, would have removed the rest. Evidence of much more recent meteorite impacts, such as the Yucatán crater discussed in Highlight 1-1, still remains, not yet erased by the Earth's active geological processes. But a search for the evidence of the greatest collision in the Earth's history is unlikely to yield a clue. We are forced to look to the composition of Moon rocks and chondritic meteorites to speculate about this fascinating story.

Thermal Energy in the Earth

We have seen how the buildup of thermal energy in the Earth's interior caused some of its densest substances to melt and sink to the Earth's center during the planet's earliest days—before the colossal collision that produced the Moon. The phenomenon of heat, which is a form of energy transfer, is closely tied to many of the processes that continue to shape our planet today. Thermal energy is transferred from place to place within the Earth, *always moving from warmer to cooler areas*. There are three primary methods by which thermal energy is transmitted through the Earth: conduction, convection, and radiation (Fig. 1-16).

In *conduction*, minute particles such as atoms become "excited" by thermal energy from an outside source. This energy causes the particles to vibrate rapidly, colliding with neighboring particles and setting them in motion. This activity generates a chain reaction of vibration that transfers thermal energy. (This is the process that burns the fingers of anyone foolish enough to grab a metal spoon that has been heating up by conduction in a hot pot of soup.) Rocks, however, are generally poor conductors of thermal energy. In fact, if the thermal energy produced in the Earth's deep interior during the planet's creation had flowed by conduction alone, it would not yet have come to the surface today, about 4.6 billion years later.

Conduction of heat involves the passage of thermal energy from atom to neighboring atom. Thus, the atoms in the hot part of the nail, which is directly in the flame, vibrate very rapidly. Farther from the heat source, the atoms vibrate less rapidly. And at the end of the nail, which is still cool, the atoms vibrate very slightly. Eventually the heat will be conducted throughout the length of the nail.

Convection involves the movement of heat from place to place by a flowing medium. Because the soup at the bottom of the pot is closer to the flame, it is the first part of the soup to become hot. As it heats up, it expands (becoming less dense) and rises, and the cooler (more dense) soup above it sinks to the bottom, displacing it and forcing it upward. When the warm soup arrives at the top, it encounters the relative coolness of the air and contracts in volume (becomes more dense) as it begins to cool. The cooled soup then sinks back toward the bottom, to be reheated and then to rise again.

Radiation involves the transfer of heat from a hot object to its cooler surroundings. The hot radiator heats up the cool air that surrounds it.

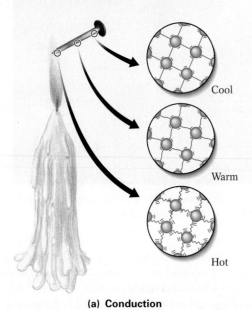

Cool

Warm

Hot

(a) Conduction

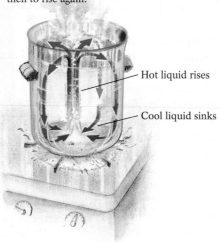

Hot liquid rises

Cool liquid sinks

(b) Convection

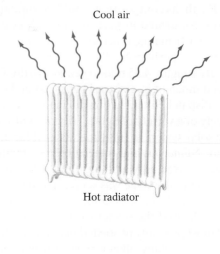

Cool air

Hot radiator

(c) Radiation

▲ **FIGURE 1-16** Three ways in which heat is transmitted. These phenomena, occurring on a much larger scale within the Earth, are responsible for many geological processes.

When thermal energy is transferred by *convection,* material actually moves from one place to another, carrying thermal energy with it. When the temperature in the Earth's interior became high enough to melt some of its components, or soften them sufficiently so that they were able to flow, thermal energy began to be transported by moving warm fluids. Eventually this energy heated surrounding substances as well, causing hot, low-density materials to rise toward the surface, carrying thermal energy with them. Convection is a much faster, more efficient way of transferring energy than conduction. The rising hot material carried by convection probably caused the planet's first volcanic eruptions.

All heated objects also *radiate* energy in one form or another. Energy transmitted by radiation moves in the form of one or more different types of electromagnetic waves, such as radio waves, microwaves, infrared waves, visible light waves, ultraviolet light waves, and X-rays. These forms of energy are then converted into thermal energy when they strike and are absorbed by an object. In this way, a microwave oven's radiant energy is converted to thermal energy to thaw a frozen pizza. Similarly, light radiated by the Sun as ultraviolet rays is transformed into the thermal energy that heats the Earth's atmosphere when it is absorbed by the rocks, soils, and waters at the Earth's surface.

As the Earth's internal heat caused it to differentiate into its major concentric layers, some upwelling molten material reached the surface. There it cooled and solidified, forming the Earth's earliest crust. Among the low-density substances that rose toward the surface were oxygen and silicon. These substances combined to form the silicate minerals that abound in the Earth's crust and upper mantle. Some heat-producing radioactive substances, such as uranium and thorium, also moved toward the surface, incorporated within the silicates and other light minerals. The heat radiating from these elements repeatedly melts the rocks of the Earth's lithosphere and reforms them into a wide variety of rocks. This "cycling" of crustal materials is summarized briefly in the next section and discussed more thoroughly in upcoming chapters.

Cycling Rocks

A **rock** is a naturally formed grouping of one or more **minerals,** which are naturally occurring *inorganic* solids that originated within the Earth. Three types of rocks exist in the Earth's crust and at its surface, each reflecting a different fundamental origin. **Igneous rocks** have cooled and solidified from molten material either at or beneath the Earth's surface. **Sedimentary rocks** form when preexisting rocks are exposed to the rigors of the rain, wind, water, and the gases of the Earth's atmosphere and broken down into fragments (known as sediment) that later accumulate and become compacted or cemented together. They may also form from

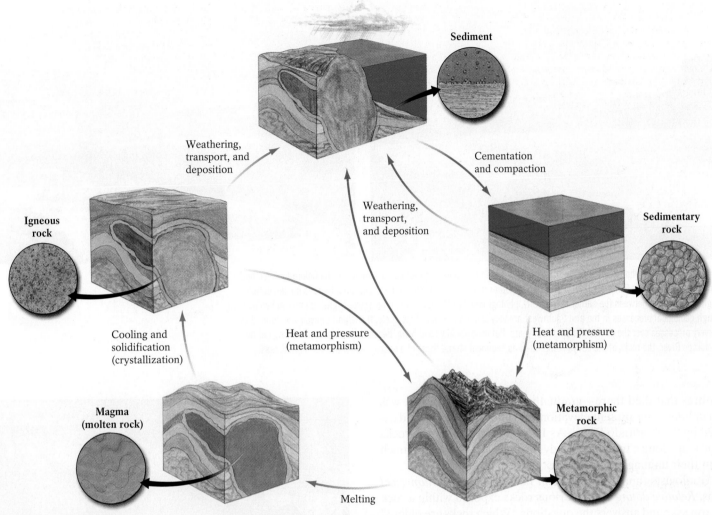

Sediment

Weathering,
transport, and
deposition

Cementation
and compaction

**Igneous
rock**

Weathering,
transport,
and deposition

**Sedimentary
rock**

Cooling and
solidification
(crystallization)

Heat and pressure
(metamorphism)

Heat and pressure
(metamorphism)

**Magma
(molten rock)**

**Metamorphic
rock**

Melting

▲ **FIGURE 1-17** The rock cycle—a simplified scheme illustrating the variety of ways that the Earth's rocks may change into other types of rocks. For example, an igneous rock may weather away, and its particles eventually consolidate, to become a sedimentary rock. The same igneous rock may remain buried deep beneath the Earth's surface, where heat and pressure might convert it into a metamorphic rock. The same igneous rock, if it is buried even deeper, may actually melt and form a new igneous rock once it has recooled and solidified. There is no prescribed sequence to the rock cycle. A given rock's transformation may be altered at any time by a change in the geological conditions around it.

the accumulated and compressed remains of certain plants and animals, or from dissolved chemicals that precipitate from water. **Metamorphic rocks** form when heat, pressure, or chemical reactions in the Earth's interior change the mineralogy, chemical composition, and structure of *any* type of preexisting rock.

Over the great stretch of geologic time, rocks of any one of these basic types may eventually be transformed into either of the other two types or into a different form of their own type. Any rock exposed at the Earth's surface can be worn away by rain, wind, crashing waves, flowing glaciers, or other means. The resulting fragments may then be carried elsewhere to be deposited as new sediment. This sediment might eventually become new sedimentary rock. Sedimentary rocks, in turn, may become buried so deeply in the Earth's hot interior that they are changed into metamorphic rocks, or they may melt and eventually form new igneous rocks. Under heat and pressure, igneous rocks can also

become metamorphic rocks. The processes by which the various rock types change over time are illustrated in the **rock cycle** (Fig. 1–17).

Time and Geology

To change sedimentary rocks to metamorphic rocks may require a vast stretch of time. Clearly, the Earth, which is believed to be about 4.6 billion years old, has had plenty of time to work on this. Over so vast a period of time, even processes that operate at imperceptibly slow rates (Fig. 1–18) can change the Earth's appearance dramatically. Fossils of ancient sea creatures in the rocks of the Grand Canyon show that those rocks once lay at the bottom of an ocean. Today they sit near the top of a hot, dry plateau, 2300 m (7500 ft) above sea level. Over millions of years, the mud at the bottom of this ocean—still containing the remains of sea

▲ **FIGURE 1-18** Two photos of the Grand Canyon, taken from the same perspective (looking down the Colorado River, one half kilometer below Lees Ferry) about 100 years apart. Other than signs of encroaching vegetation, there is no perceptible geological difference between the earlier scene (left), photographed in 1873, and the later scene (right), photographed in 1972. Although geological processes in the arid Southwest are very slow on a human time scale, the Grand Canyon was formed by these very processes over the course of millions of years. Before gradually solidifying, becoming uplifted, and being cut by the Colorado River, the rocks of the canyon originated as sediment on the floors of a succession of ancient inland seas.

creatures that died there—gradually solidified into rock. It was then uplifted to its present elevation, and later cut through and exposed by the Colorado River. As James Hutton knew, the rocks we pick up along a trail, if we study them closely, can tell us much about their unimaginably long history.

Geologists think about time in both *relative* and *numerical* terms. *Relative dating* of the various rocks exposed within a rock outcrop asks and answers the questions "Which rocks are older?" and "Which rocks are younger?" The answers are typically found by observing the spatial relationships of rock bodies to one another. For example, the *principle of superposition* states that where layers of flat-lying sedimentary rocks or a series of solidified lava flows have not been disturbed since their formation, *younger rocks overlie older rocks.* Quite logically, the older rocks had to be there first for the younger rocks to be laid on top of them (Fig. 1–19).

Numerical dating of rocks (also known as **absolute dating**) answers the question: "How old?" We are not satisfied by simply knowing which is older or younger; we want a specific age—*in years.* Rock-dating techniques developed during the twentieth century, based on the constant decay of radioactive elements, now let geologists specify the numerical ages of some rocks. The oldest rocks found on Earth, near Yellowknife Lake in Canada's Northwest Territories, have been dated by the decay rate of uranium at 4.03 billion years. We discuss the methods for dating rocks in Chapter 8.

If Earth's history were viewed as a great textbook, its rocks would be the pages on which most of the events are written. Almost everything we know about the Earth's past we know because it was preserved in rock. Geologists initially divided Earth's history into four major *eras* of varying length and a dozen smaller *periods,* based largely on fossil evidence of various key organisms in sedimentary rocks. With the advent of modern rock-dating technology, however, the geologic time scale now includes numerical ages for its eras and periods (Fig. 1–20).

▲ **FIGURE 1-19** These sedimentary rock layers from Grand Junction, Colorado, illustrate the principle of superposition. The child's feet rest on the outcrop's oldest rocks; the thick layer at the top is the outcrop's youngest.

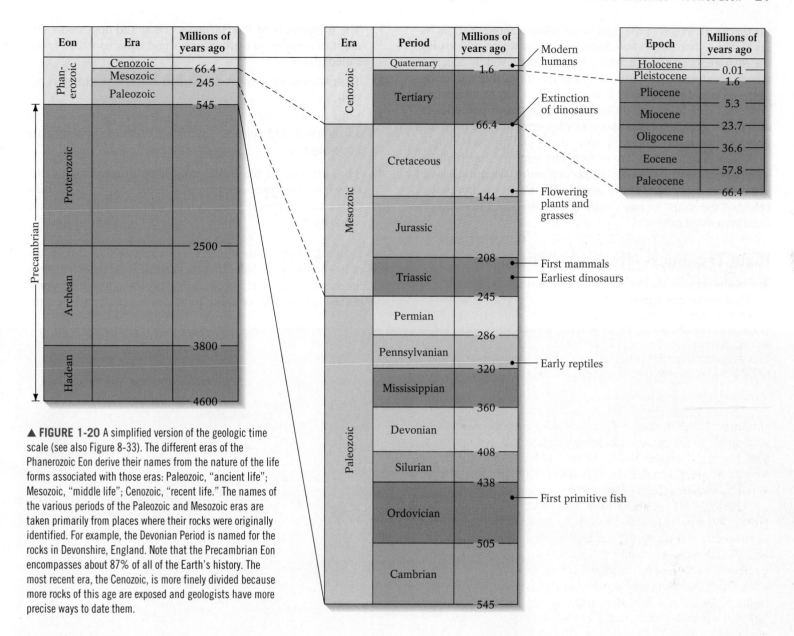

Eon	Era	Millions of years ago
Phanerozoic	Cenozoic	66.4
	Mesozoic	245
	Paleozoic	545
Proterozoic		2500
Archean		3800
Hadean		4600

Precambrian

Era	Period	Millions of years ago	
Cenozoic	Quaternary	1.6	Modern humans
	Tertiary	66.4	Extinction of dinosaurs
Mesozoic	Cretaceous	144	Flowering plants and grasses
	Jurassic	208	First mammals
	Triassic	245	Earliest dinosaurs
Paleozoic	Permian	286	
	Pennsylvanian	320	Early reptiles
	Mississippian	360	
	Devonian	408	
	Silurian	438	
	Ordovician	505	First primitive fish
	Cambrian	545	

Epoch	Millions of years ago
Holocene	0.01
Pleistocene	1.6
Pliocene	5.3
Miocene	23.7
Oligocene	36.6
Eocene	57.8
Paleocene	66.4

▲ FIGURE 1-20 A simplified version of the geologic time scale (see also Figure 8-33). The different eras of the Phanerozoic Eon derive their names from the nature of the life forms associated with those eras: Paleozoic, "ancient life"; Mesozoic, "middle life"; Cenozoic, "recent life." The names of the various periods of the Paleozoic and Mesozoic eras are taken primarily from places where their rocks were originally identified. For example, the Devonian Period is named for the rocks in Devonshire, England. Note that the Precambrian Eon encompasses about 87% of all of the Earth's history. The most recent era, the Cenozoic, is more finely divided because more rocks of this age are exposed and geologists have more precise ways to date them.

Plate Tectonics—A First Look

When we tell the life story of the Earth, we discover that certain regions of the world have been periodically devastated by earthquakes, while others have remained unscathed. We puzzle over why a chain of volcanoes stretches along the west coast of North America from northern California to Alaska but no such chain of volcanic peaks can be found mid-continent in Minnesota, Iowa, or Missouri. And we all surely know that the "ski slopes" of Ohio and Indiana stir far less excitement among snowboarders, who flock instead to Sun Valley, Idaho; Vail, Colorado; and Killington, Vermont. Each of these examples highlights a definite pattern in the Earth's geological makeup and a fundamental truth about the geology of our planet—*it is not random.*

For centuries, geologists sought to explain such large-scale patterns with a host of *independent* hypotheses, one to explain earthquakes, another to explain volcanoes . . . and so forth. Each hypothesis was tailored to a specific location. That's why geologists in the Swiss Alps proposed that mountain building occurs when rocks are pushed together, whereas those studying the Grand Tetons of Wyoming proposed that mountains form when rocks are pulled apart. Such hypotheses might have given reasonable pictures of regional processes, but they failed to explain some common underlying cause of *all* mountain ranges. And no *unifying* hypothesis adequately explained the observed combination of geological phenomena—why, for example, earthquakes typically occur where mountains rise and volcanoes erupt.

In the 1960s, an exciting new hypothesis called **plate tectonics** (from the Greek adjective *tektonikos,* or "built," as in archi*tec*ture) revolutionized our understanding of how the outer portion of the Earth functions. This hypothesis changed the way geologists viewed the world as dramatically as, a century earlier, the theory of evolution changed how biologists thought about living things. After only a few decades of observation and testing, the *hypothesis* of plate tectonics has become a widely accepted *theory* because it provides answers to questions that earlier hypotheses could not resolve. It has given us a way to understand processes such as mountain building, predict such potential catastrophes as earthquakes and volcanic eruptions, and even find underground reservoirs of oil, natural gas, and precious metals.

According to the theory of plate tectonics, the Earth's lithosphere consists of large rigid plates *that move.* Over millions of years, plate movements have created the Earth's ocean basins, changed the shape of our continents, and crafted the planet's great mountain ranges.

Plate Tectonics—Its Basic Concepts

The wide-ranging theory of plate tectonics can be summed up with four basic concepts:

1. The lithosphere of the Earth—its crust and uppermost segment of mantle—is composed of rigid massive slabs of rock called plates.

2. The plates move slowly on the order of a few centimeters per year.

3. Most of the world's large-scale geological activity, such as earthquakes and volcanic eruptions, occurs at or near plate boundaries.

4. The interiors of plates are relatively quiet geologically; they have far fewer and typically milder earthquakes than those that occur at plate boundaries and little volcanic activity.

Figure 1–21 shows the Earth's seven major plates and a number of its smaller ones. Note that the continents generally are not independent plates, but instead are typically parts of composite plates—ones that contain both continental and oceanic

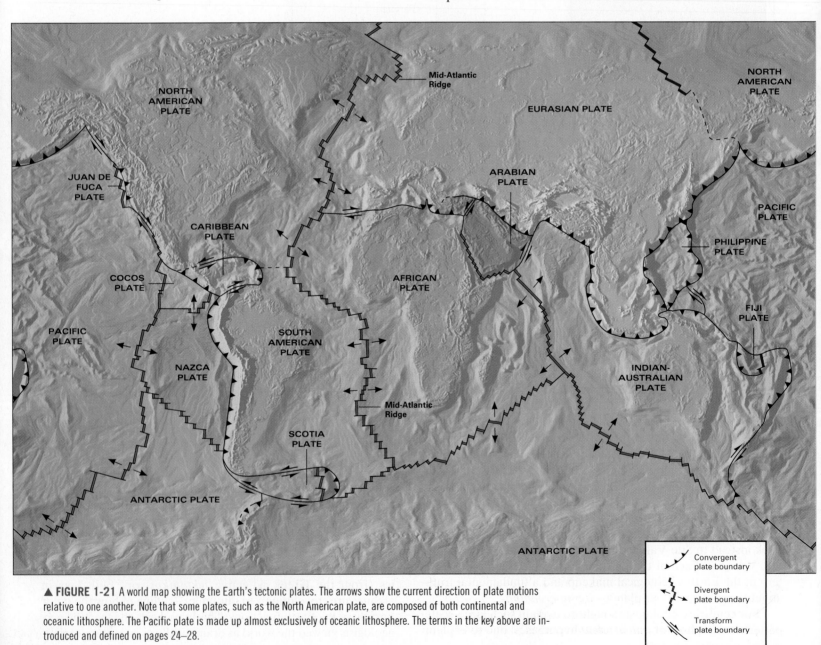

▲ **FIGURE 1-21** A world map showing the Earth's tectonic plates. The arrows show the current direction of plate motions relative to one another. Note that some plates, such as the North American plate, are composed of both continental and oceanic lithosphere. The Pacific plate is made up almost exclusively of oceanic lithosphere. The terms in the key above are introduced and defined on pages 24–28.

lithosphere. For example, the North American plate includes the North American continent and the adjacent western half of the Atlantic Ocean. The continental segments of plates are composed of thicker, lower-density lithosphere, whereas oceanic segments are thinner and of higher density. Plates are typically more than 120 km (75 mi) thick in continental regions but only about 80 km (50 mi) thick within ocean basins.

As the Earth's plates move, everything on them, even features as large as continents and oceans, moves with them. As you read these words, most of you in North America are moving westward at about 4 cm (1.5 in.) per year. Those of you living in Los Angeles and the surrounding communities of westernmost California live on a different plate, the Pacific plate, which is headed northward at a rate of about 6 cm (2.5 in.) per year. The North American plate is moving westward away from the Eurasian plate, its neighbor to the east, which in turn is moving eastward. The increasing distance between cities on these two plates, however, will not cause airfares between New York and London to increase—the length of a trans-Atlantic flight is extended by only about 5 to 10 cm (2–4 in.) per year as a result of plate movements. (By the way, for a human perspective, this is about two times the rate at which your toenails grow, a fact that should impress your friends and relatives alike.) If Columbus were crossing the Atlantic today, 500 years after his famous voyage, he would have to sail only an extra 25 to 50 m (80–160 ft) to reach shore.

Although this rate of plate motion may seem insignificant in human terms, over the vast course of geologic time it can have huge consequences. In just a few tens of millions of years, the Atlantic Ocean has grown 1000 km (about 600 mi) wider. But does this mean that the Earth's overall size is increasing? No, our planet's size has remained essentially constant since its early formative years, 4.6 billion years ago. Thus, if plates diverge and grow at certain boundaries, they *must* converge and be consumed at others.

When plates move together or past each other, heat and pressure build at their boundaries. Depending on the direction of their movement, plate edges may break and generate earthquakes, buckle to form mountains, melt and erupt volcanically, or do all three. In fact, the locations of earthquakes are used to map the outlines of plates in a kind of geological connect-the-dots game (Fig. 1–22).

Plate interiors are generally quiet geologically, typically unaffected by distant plate-edge activity. San Francisco has frequent earthquakes because it is located at a moving plate boundary; Chicago has very few because it is located within a relatively inactive plate interior.

Moving Plates and Their Dynamic Boundaries

There are three major types of plate motions and corresponding boundaries:

- divergent plate boundaries—where plates move apart;
- convergent plate boundaries—where plates move together; and
- transform plate boundaries—where plates move past one another in opposite directions.

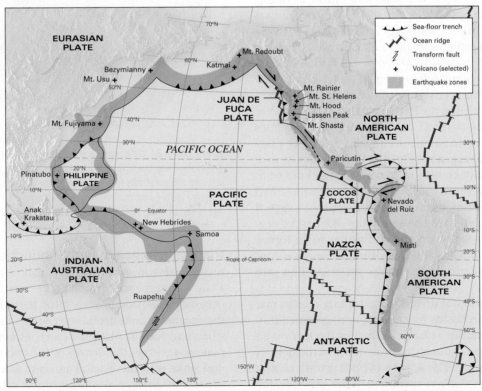

▲ **FIGURE 1-22** A map of the Pacific Ocean and surrounding continents, indicating some of the sites of volcanic eruptions and zones of earthquake activity recorded over the last 100 years or so. Note the proportion of volcanoes and earthquakes that occur at plate boundaries.

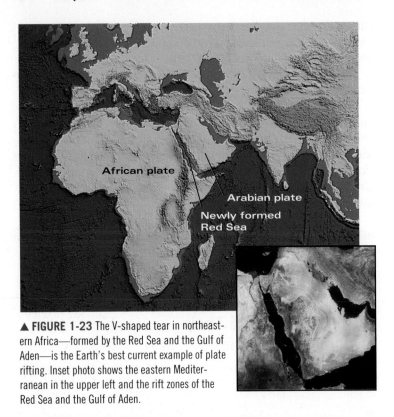

▲ **FIGURE 1-23** The V-shaped tear in northeastern Africa—formed by the Red Sea and the Gulf of Aden—is the Earth's best current example of plate rifting. Inset photo shows the eastern Mediterranean in the upper left and the rift zones of the Red Sea and the Gulf of Aden.

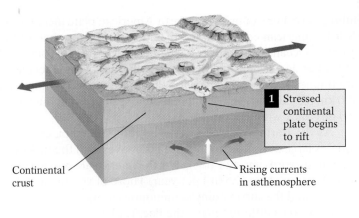

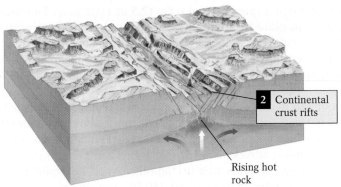

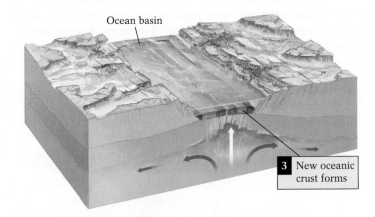

▲ **FIGURE 1-24** Plate rifting and divergence. When currents in the underlying asthenosphere pull one of the Earth's plates in opposite directions (top), the plate is stressed and eventually rifts. (middle) As the plate fragments continue to move (diverge) farther from one another, molten material from the mantle rises into the gap and solidifies along the edges of the plates (bottom), forming new oceanic crust that is eventually covered with water to form a new ocean.

Divergent Plate Boundaries Typically quiet plate interiors may become geologically exciting if slow-flowing, rising currents in the Earth's asthenosphere pull apart and tear a preexisting plate into two or more smaller plates. This process is known as **rifting.** The Great Rift Valley of East Africa, where the African plate has been coming apart, is a prime example of an early stage in rifting (Fig. 1–23). If this rifting continues, there may be two Africas on the world map in the not-too-distant *geological* future. Farther north, such rifting has completely separated the Arabian plate from the African plate over the last 20 million years, forming the crustal depressions occupied by the Red Sea and the Gulf of Aden.

Once a plate has been rifted, the resulting smaller plates may continue separating from one another by a type of plate motion described as **divergence** (Fig. 1–24). Divergence proceeds typically at a rate of about 1 to 10 cm (0.5–4 in.) per year. As molten material rises into the fractures between the rifted plates, it cools and solidifies, becoming attached to the edges of the rifted plates. As divergence continues, the older rifted segments separate further and eventually an ocean basin forms. The ocean basin fills with seawater as rifting opens new connections to adjacent oceans, enabling their waters to flow into the newly rifted basin.

Throughout the period of divergence, erupting molten material expands the ocean basins by creating new oceanic lithosphere. The center of all this volcanic activity is the *mid-ocean ridge,* a continuous chain of submarine mountains on the floor of a growing ocean (Fig. 1–25). The process of plate growth at mid-ocean ridges is known as **seafloor spreading.** If, for example, the young Red Sea between the African and Arabian plates continues to grow at its present rate, it may someday become a full-blown ocean like the Atlantic or Pacific.

Plate Convergence and Subduction Boundaries Some plates move *toward* each other—a motion known as **convergence.** Plate convergence may involve two continental plates, two oceanic plates, or one of each (or, in the case of composite plates, the oceanic or continental segments of the two plates). Most often when one oceanic plate and one continental plate or two oceanic plates converge, the *denser* of the two dives beneath the other and sinks into the Earth's interior where it is consumed. This process of diving plates is known as **subduction.** Because plates of oceanic lithosphere are denser than those of continental lithosphere, the oceanic plate always subducts when it converges with a continental plate (Fig. 1–26). Some oceanic plates, however, are more dense than others; thus, when two oceanic plates converge, the denser one—usually the older, colder one—subducts beneath the less dense one.

Continental plates are generally too buoyant to subduct into the denser, underlying mantle. Thus, when two continental plates collide, neither plate can subduct completely. Their edges may be *temporarily* dragged down to depths perhaps as deep as 200 km (120 mi) before being thrown back toward the Earth's surface. Instead of subducting, colliding continental plates become welded together, forming a much larger *single* plate. Such convergence of continental plates is known as a **continental collision** (Fig. 1–27), the process that has created many of the Earth's highest mountain ranges. Northern Africa's Atlas Mountains and southern Europe's Alps were formed by past collisions of the African and Eurasian plates. The Himalaya owe their great height to the collision of the Indian and Eurasian plates. The Appalachian Mountains of eastern North America formed from a series of past collisions of the African, Eurasian, and North American plates and several other smaller plates. Those collisions took place between about 400 and 250 million years ago.

When one oceanic plate subducts beneath another, the molten material produced as the subducting plate sinks deep into the Earth's hot mantle creates a chain of volcanoes. Initially such a chain forms under water. Eventually, after millions of years, the volcanoes may grow high enough above the seafloor to emerge as a chain of volcanic islands, known as an *island arc*. This has happened in the northern Pacific, where the Pacific plate descends beneath the northwestern oceanic edge of the North American plate to form the earthquake-wracked, explosively volcanic Aleutian Islands of Alaska. The 1990 eruption of Mount Redoubt and the 1996 eruption of Mount Pavlov, both in Alaska, indicate ongoing subduction in this region (Fig. 1–28). In August of 2003, marine geologists discovered what

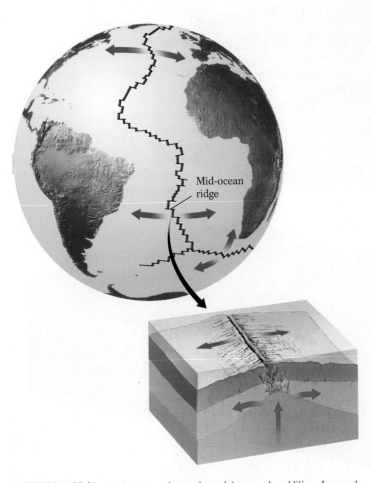

Mid-ocean ridge

▲ **FIGURE 1-25** Divergent zones—places where plates grow by addition of new volcanic rock—stretch for 65,000 connected kilometers (40,000 mi) principally along the Earth's seafloors. Here, divergence between South America and Africa has created the southern Atlantic Ocean basin.

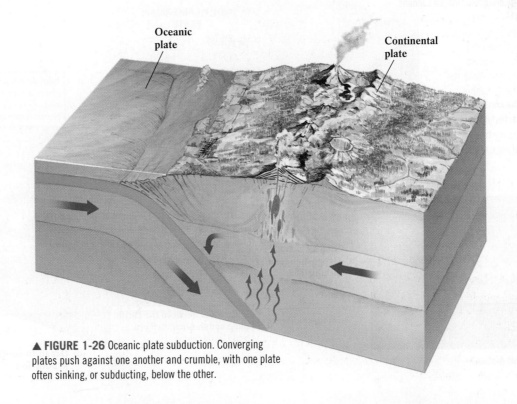

Oceanic plate

Continental plate

▲ **FIGURE 1-26** Oceanic plate subduction. Converging plates push against one another and crumble, with one plate often sinking, or subducting, below the other.

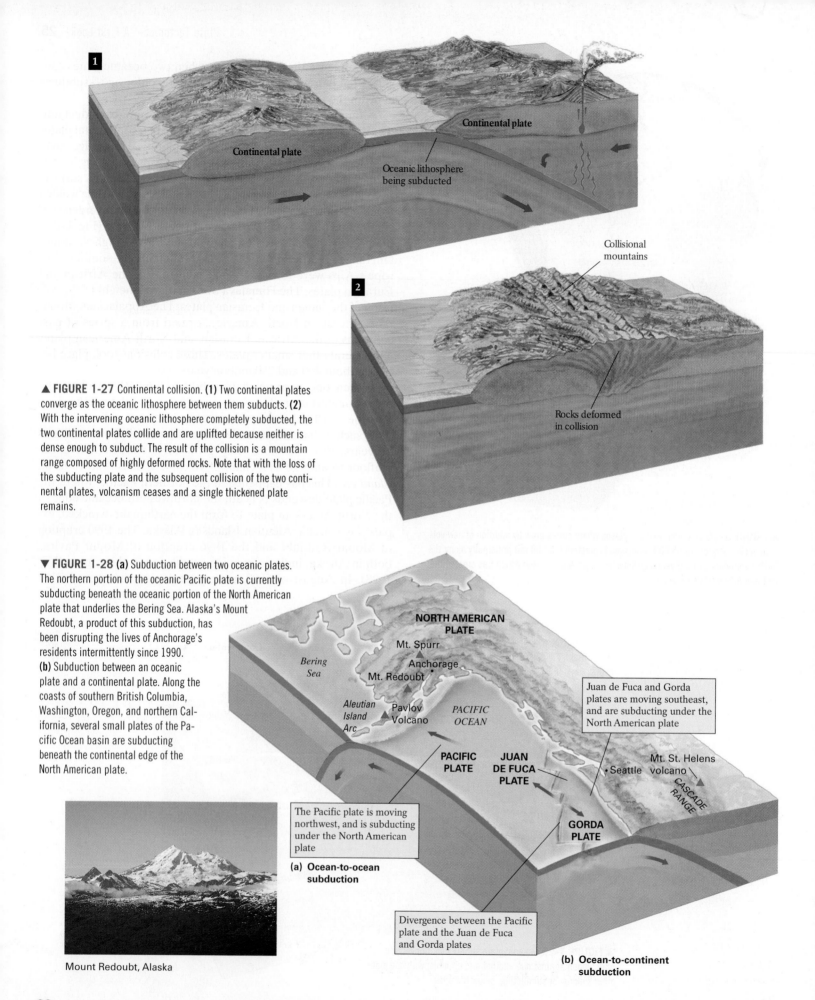

▲ **FIGURE 1-27** Continental collision. **(1)** Two continental plates converge as the oceanic lithosphere between them subducts. **(2)** With the intervening oceanic lithosphere completely subducted, the two continental plates collide and are uplifted because neither is dense enough to subduct. The result of the collision is a mountain range composed of highly deformed rocks. Note that with the loss of the subducting plate and the subsequent collision of the two continental plates, volcanism ceases and a single thickened plate remains.

▼ **FIGURE 1-28 (a)** Subduction between two oceanic plates. The northern portion of the oceanic Pacific plate is currently subducting beneath the oceanic portion of the North American plate that underlies the Bering Sea. Alaska's Mount Redoubt, a product of this subduction, has been disrupting the lives of Anchorage's residents intermittently since 1990. **(b)** Subduction between an oceanic plate and a continental plate. Along the coasts of southern British Columbia, Washington, Oregon, and northern California, several small plates of the Pacific Ocean basin are subducting beneath the continental edge of the North American plate.

Mount Redoubt, Alaska

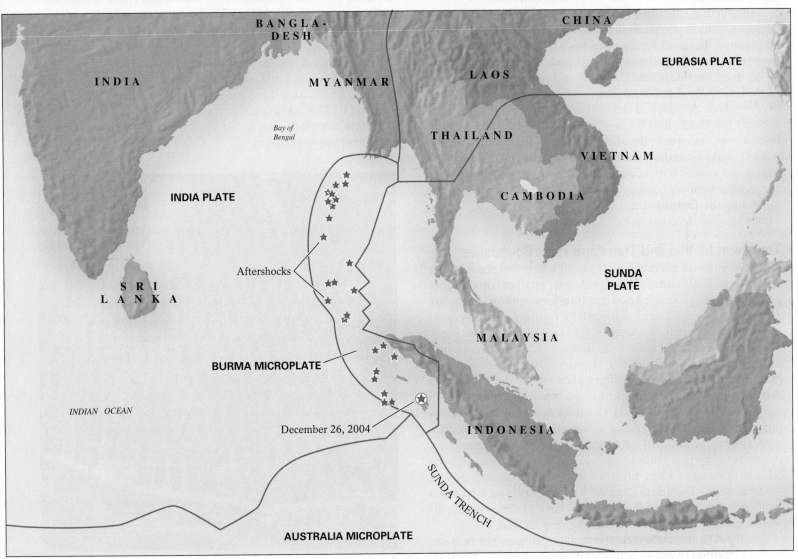

▲ **FIGURE 1-29** Subduction of the Indian, Australian, and Burma plates beneath the adjacent Sunda Plate has produced the vast number of volcanoes in Indonesia and is also the cause of the devastating earthquake and tsunami that killed 300,000 in late December, 2004. The photo above illustrates the aftermath of the deadly tsunami.

may become the "next" Aleutian Island, an undersea volcano, rising more than 570 m (1900 ft) above the seafloor of Amchitka Pass at the westernmost edge of the Aleutians—another clear indication of ongoing subduction there. And in the summer of 2005, volcanologists are monitoring Mount Spurr very closely, watching and waiting for what they believe is an impending eruption. The most dramatic recent geological activity associated with island-arc formation and tectonics has been the tragic events in 2005 in Indonesia—a classic island arc (Fig. 1-29).

When an oceanic plate subducts beneath a continental plate, a chain of explosive volcanoes also grows—but on land. At the western shores of North America, the edges of the oceanic Juan de Fuca and Gorda plates are subducting beneath the advancing continental edge of the North American plate (see Fig. 1-28). As these subducting plates descend to warmer depths, melting occurs that fuels the chain of volcanoes of the Cascade Range, which stretches from northern California to southern British Columbia and which includes some of North America's volcanic "all-stars"—Mount Rainier, Mount St. Helens, and Mount Shasta.

Subduction rates can be as high as 15 to 25 cm (6–10 in.) per year. This is considerably faster than the average rate of divergence. These differing rates might lead us to expect that over time, oceanic plates would eventually disappear and the Earth itself would shrink. This does not happen, however, because *less of the Earth's surface is devoted to subduction than to divergence.* As maps of the seafloor show, subduction zones account for about 40,000 km (25,000 mi) of the Earth's plate boundaries, whereas about 65,000 km (40,000 mi) of the world's plate boundaries are diverging. It is this balance between the creation of oceanic plates at divergent zones and the destruction of oceanic plates at subduction zones that maintains the Earth's size.

Transform Motion and Transform Plate Boundaries
The third major type of plate boundary occurs where two plates, either oceanic or continental, slide past one another in opposite directions. This process is known as **transform motion** (Fig. 1-30). At transform boundaries, plates neither grow (as at divergent boundaries) nor are they consumed (as at convergent boundaries). Because no mantle material wells upward and no plates are subducting into the Earth's interior, transform boundaries generally produce little volcanism. Transform plate motion does break and displace any feature, natural or human-made, that cuts across a transform boundary (Fig. 1-31). Great friction results as the moving plates grind past each other, so these boundaries do produce a great deal of powerful earthquake activity. Any community caught between two plates at a transform boundary may be periodically devastated. Certainly, the folks in San Francisco and Los Angeles, who live within the San Andreas transform zone between the North American and Pacific plates, have to deal with the always looming prospect of a catastrophic earthquake.

So those are the basic facts about plate tectonics. The next important question to ask is, Why should we believe such a fantastic story? What evidence supports the theory that the Earth's outer layer moves and, as it does, creates and consumes oceans and shuffles the position of the Earth's continents? Before mov-

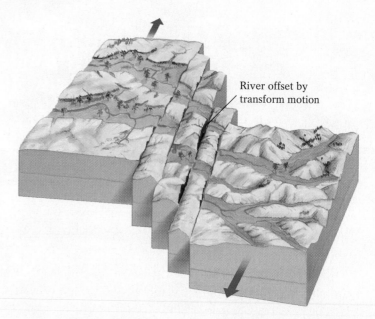

River offset by transform motion

▲ **FIGURE 1-30** Transform motion. Where or when the Earth's plates move past one another in opposite directions, friction builds up at their edges but the plates are neither uplifted nor subducted.

ing on to learn about the evidence that has led geologists to believe in all of these Earth-shaping plate-tectonic processes, spend a moment reviewing these processes in the Geology at a Glance on page 30. This is the first of many such illustrations in this book that have been designed to help you master and retain text material and ace your exams.

▲ **FIGURE 1-31** An orange grove along the San Andreas transform plate boundary, with its tree rows offset by plate motion there. Since any feature that cuts across an active transform plate boundary will be broken and displaced, this type of plate motion confounds precise planting in orchards.

In Support of Plate Tectonics— A Theory Develops

In 1782 the American scientist–philosopher–statesman Benjamin Franklin hypothesized: "The crust of the Earth must be a shell floating on a fluid interior. Thus the surface of the globe would be capable of being broken and disordered by the violent movements of the fluids on which it rested." Almost 200 years later that remarkable insight by one of history's finest scientific thinkers embodies some of the key concepts of the theory of plate tectonics. But how did modern plate-tectonic theory develop? And what causes all this geological action to happen?

Alfred Wegener and Continental Drift

In the early part of the twentieth century the German geophysicist–meteorologist Alfred Wegener (1880–1930) proposed a controversial hypothesis that laid the foundation for the revolutionary theory of plate tectonics. Despite blistering ridicule from the leading geologists of his day, Wegener proposed that the continents float on the denser underlying interior of the Earth and periodically break up and drift apart. He asserted that all of the Earth's continents had been joined together about 200 million years ago as a supercontinent he called *Pangaea* ("all lands") (Fig. 1-32). Wegener hypothesized that Pangaea had covered about 40% of the Earth's surface, most of it in the Southern Hemisphere. While Pangaea existed, what is now the area of New York City sweltered in the lush tropics near the equator, and most of eastern Africa, Australia, and India shivered under a dome of glacial ice

near the South Pole. Pangaea was surrounded by an immense ocean, which Wegener called *Panthalassa* (named for the Greek goddess of the sea). Over a vast period of time beginning about 180 million years ago, Pangaea broke up, forming a number of continents that migrated to all regions of the globe. Wegener called this dispersal **continental drift.**

Wegener supported his hypothesis with the observations he made on the shapes of continental margins, patterns of present-day animal life, similarities among far-distant fossils and rocks, and evidence of past climates at odds with the climates at present locations.

Continental Fit   The English philosopher Sir Francis Bacon was among the first to note that the outlines of the continents of the world could be pieced together jigsaw-puzzle style. Bacon wrote of this in 1620, soon after seeing the new maps that came out of the global explorations of the sixteenth century. This concept reappeared periodically for the next three centuries (Fig. 1-33). Almost 300 years later, Wegener tinkered with the puzzle

▲ **FIGURE 1-33** An early observation of the continental fit of South America and Africa, which was first proposed by Sir Francis Bacon in 1620. French naturalist Antonio Snider-Pelligrini sketched this diagram in 1858 for his work *La Création et ses Mystères Dévoiles,* which suggested that the biblical account of Noah's Deluge was responsible for the movement of the continents.

▲ **FIGURE 1-32** A reconstruction of the proposed supercontinent Pangaea, with present-day coastlines and continent names shown for reference. This configuration of plates dates from about 225 million years ago.

1.1   Geology at a Glance

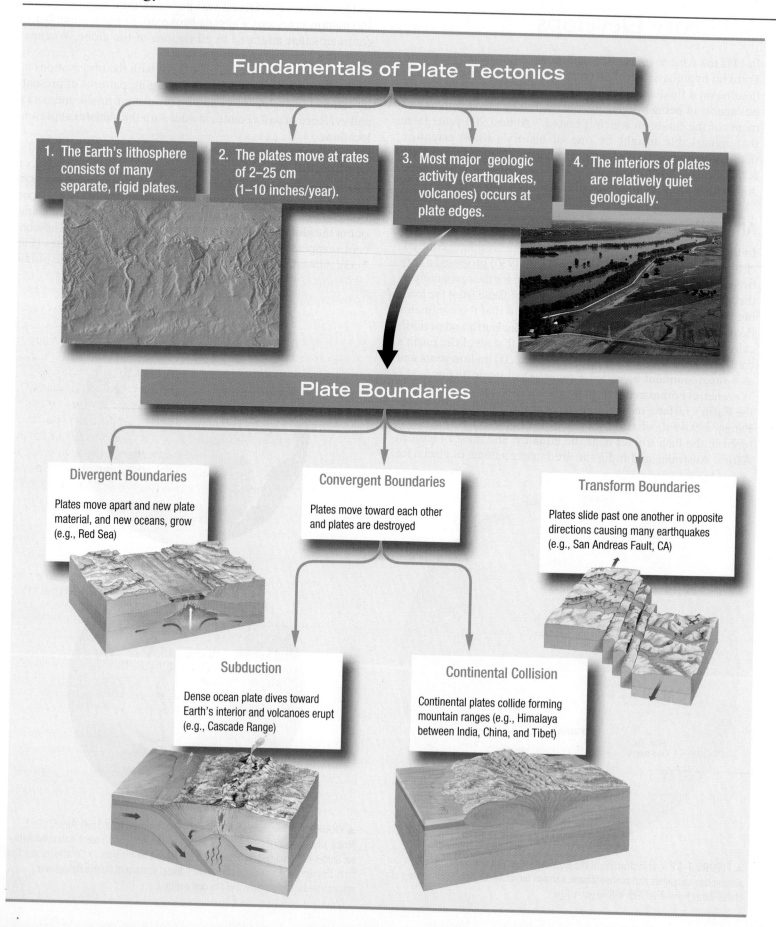

Fundamentals of Plate Tectonics

1. The Earth's lithosphere consists of many separate, rigid plates.

2. The plates move at rates of 2–25 cm (1–10 inches/year).

3. Most major geologic activity (earthquakes, volcanoes) occurs at plate edges.

4. The interiors of plates are relatively quiet geologically.

Plate Boundaries

Divergent Boundaries

Plates move apart and new plate material, and new oceans, grow (e.g., Red Sea)

Convergent Boundaries

Plates move toward each other and plates are destroyed

Transform Boundaries

Plates slide past one another in opposite directions causing many earthquakes (e.g., San Andreas Fault, CA)

Subduction

Dense ocean plate dives toward Earth's interior and volcanoes erupt (e.g., Cascade Range)

Continental Collision

Continental plates collide forming mountain ranges (e.g., Himalaya between India, China, and Tibet)

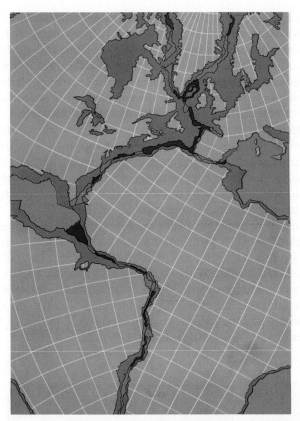

▲ **FIGURE 1-34** Precise matching of the continental shelves of circum-Atlantic continents by computer analysis. (From *The Origin of the Oceans,* by Sir Edward Bullard, published 1969. Copyright © 1969 by *Scientific American, Inc.* All rights reserved.) Because the seaward (below sea level) edges of the continental shelves, shaded dark brown, represent the true edges of the continents, their fitting is more precise than simple shoreline fitting. The only places at which the plates appear to overlap, shaded red, are marked by geologic materials that have been deposited *since* the breakup of Pangaea. The precision of this fitting is one of the compelling lines of evidence that led to the wide acceptance of the theory of plate tectonics.

of continental shapes until he fit them together, forming a hypothetical landmass he called Pangaea. Today, precise computer fitting has confirmed Bacon's and Wegener's hypotheses, and we can see that the reunited continents would indeed fit together remarkably well. For example, when the borders of South America and Africa are reunited, South America tucks almost perfectly into the niche in the western coast of Africa (Fig. 1-34).

Habitats of Living Animals
Studies of the distribution patterns of certain modern animals helped to convince Wegener that now separated landmasses were once united as a supercontinent. He noted, for example, the unusual wildlife native to Australia, the only continent with kangaroos, wallabies, and koalas. Wegener hypothesized that this continent was once part of a large Southern Hemisphere landmass that rifted about 40 million years ago. Drifting alone on their island continent, Australia's fauna began to evolve along their own distinctive path, gradually becoming the unique animals we see there today.

Habitats of Ancient Animals
Wegener also examined the fossil record for rare occurrences of past life forms (Fig. 1-35). Fossils of *Mesosaurus,* a small reptile that lived 240 million years ago, have been found *only* in Brazil and South Africa, which are separated today by 5000 km (3000 mi) of the southern Atlantic Ocean. The skeletal structure of *Mesosaurus* and the composition of the deposits in which its fossils have been found show that it paddled around in shallow lakes and estuaries (the part of a river that empties into an ocean). It probably did not—or could not—swim long distances in open oceans. Paleontologists conclude from this that *Mesosaurus* lived on what was then a single landmass. It simply traveled overland between the sites of what are now Africa and South America before the landmass rifted. A similar explanation accounts for the occurrence in Antarctica, Africa, Madagascar, and India of the fossil remains of another non–ocean-swimming prehistoric creature: *Lystrosaurus.*

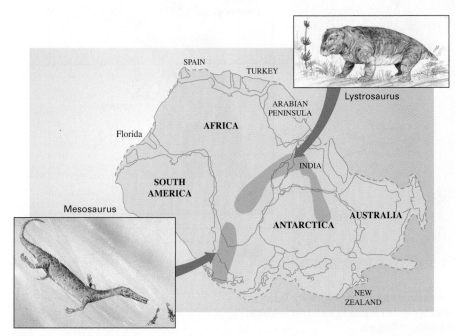

◄ **FIGURE 1-35** The distribution of *Mesosaurus* and *Lystrosaurus* when the Southern Hemisphere continents were joined together as part of Pangaea. Fossils of these reptiles date from the Permian (*Mesosaurus*) and Triassic (*Lystrosaurus*) Periods (between about 260 and 230 million years ago). The less-than-sleek *Lystrosaurus,* a sheep-size reptile, lived in what are now Antarctica, Africa, Madagascar, and India. Both this creature and *Mesosaurus* (discussed in the text) were not long-distance swimmers, and thus simply strolled to the spots where we find their remains today. That was, of course, long before these once contiguous lands rifted and diverged, spreading the seafloors that were to become the Indian and the southern Atlantic Oceans.

Related Rocks   In the Northern Hemisphere, the 390-million-year-old rocks of the mountains of eastern North America are remarkably similar in mineral composition, structure, and fossil content to rocks of the same age in eastern Greenland, western Europe, and western Africa (Fig. 1-36a). Wegener recognized that if North America, Africa, and Europe had been joined in the past, there would have been a continuous chain of mountains from Alabama all the way to Scandinavia (Fig. 1-36b). Iceland would have been a puzzle for Wegener, for it shows no evidence of this mountain chain. We now know, however, that Iceland formed atop the diverging mid-Atlantic ridge only about 60 million years ago, long after Pangaea broke up to form the separate continents of Europe and North America.

Ancient Climates   In his attempt to prove that the continents drifted, Wegener also pointed out puzzling geological evidence of past climates that were very different from the current climates of certain locations. For example, as glaciers slowly flow across a landscape, they pick up rocks, boulders, and sand grains and, as they melt, deposit a jumble of debris of all sizes (Fig. 1-37a). The boundary material shown in Figure 1-37b is glacial debris found on the hot, dry western coast of South Africa. Its presence there tells us that South Africa was glaciated about 250 million years ago (the estimated age of that deposit). An icy past for this region would be plausible only if South Africa once sat close to one of the Earth's icy poles. In the case of South Africa, it apparently once occupied the area now claimed by the frigid continent of Antarctica.

As further evidence of continental drift, the ancient rocks of such warm places as India, Australia, Africa, and South America are scored by a distinctive pattern of aligned scratches and grooves (Fig. 1-38). Such scratches form as glaciers drag debris along their beds across the underlying surface rock. These warm regions, then, must have once been located in colder glaciated climes.

Similar reasoning explains the discovery of coal in presently cold climates. Coal forms after a great accumulation of swamp vegetation in a warm, wet environment has been buried, compressed, and heated. When we find a cliff in arctic Spitsbergen, Norway, containing a seam of coal surrounded by kilometers of ice, we infer that the coal must have formed elsewhere in a tropical swamp and then journeyed a great distance to the icy North as part of a drifting continent (Fig. 1-39).

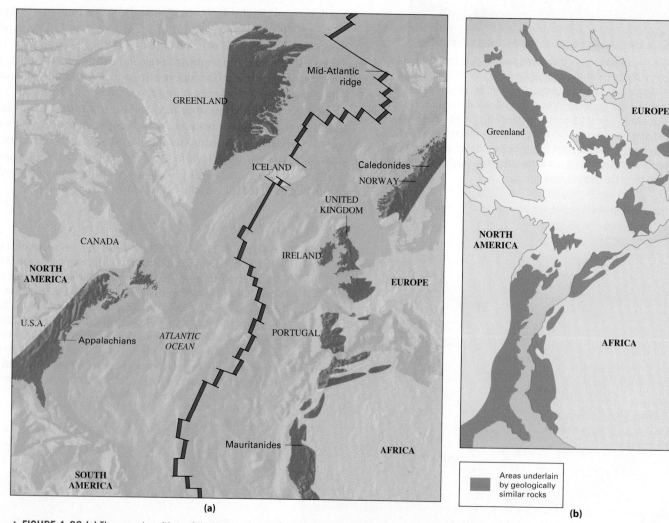

(a)

(b)

▲ **FIGURE 1-36 (a)** The current positions of the Northern Hemisphere continents surrounding the Atlantic Ocean, and **(b)** their pre-rifting positions as part of Pangaea. Note the absence of Iceland in the Pangaea reconstruction—Iceland originated from volcanic eruptions at the mid-Atlantic ridge *after* the breakup of Pangaea.

▲ **FIGURE 1-37** (a) Glacial debris at the margin of the Reid Glacier, Glacier Bay National Park, Alaska. The debris is in the foreground and on the slope left of the ice flow. (b) Glacial debris from the Dwyka area, Cape Province, South Africa.

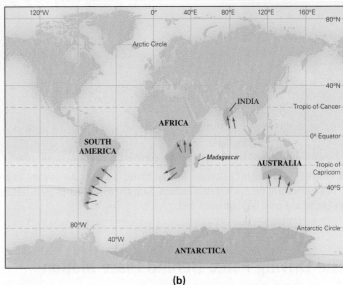

▲ **FIGURE 1-38** (a) Glacial striations in rock surfaces. (b) Striation patterns as they appear today on separated Southern Hemisphere continents: The striations are oriented in what appears to be a random pattern. (c) Striation patterns as they would appear on Pangaea, with "reunited" Southern Hemisphere continents occupying the South Pole: The striations form a systematic pattern resembling the spokes of a bicycle wheel, a pattern that typifies the outward flow of a large glacier.

▲ **FIGURE 1-39** Coal in Spitsbergen, Norway, formed in a warm, wet swamp, not exactly the climate at the coal's present location. Such climatically out-of-place features are one form of proof that continents drift.

The Driving Force Behind Plate Motion

With such strong evidence to support it, why was Wegener's hypothesis of continental drift initially so thoroughly rejected? The best answer is that Wegener's proposal simply did not seem physically plausible. He proposed that the continents, driven by the Earth's rotation, plowed through denser oceanic rocks. At the time, no one knew of the existence of lithospheric plates and the underlying heat-softened, convecting asthenosphere. Consequently, Wegener was unable to propose a scientifically acceptable driving mechanism for continental drift. The scientific community of Wegener's day could not accept so radical an idea that contested the long-held notion of immovable continents and ancient, featureless seafloors—without a reasonable driving force. In November 1928, only two years before he died, Wegener endured his final rejection when a meeting of esteemed American geologists concluded, "If we are to believe Wegener's hypothesis, we must forget everything which has been learned in the last 70 years and start all over again."

That's where Wegener's hypothesis of continental drift stood for the next three decades. With the advent of deep-sea drilling and the development of other ways to study the Earth's interior, scientists discovered the existence of the lithospheric plates and identified several ways in which the plates might move. During the 1960s and 1970s, a great deal of evidence appeared in support of Wegener's lines of reasoning. Ultimately, this evidence enhanced the acceptability of the Pangaea hypothesis, and the notion of continental drift was replaced with the theory of plate tectonics. Since the 1980s, new technologies have further increased our knowledge of the physics of the Earth's interior, enabling geologists to propose plausible hypotheses to explain what drives plate motion.

Many geologists now believe that heat-driven currents in the Earth's mantle are principally responsible for plate movements. These currents, known as **convection cells,** develop when portions of a heated substance become less dense and rise toward the surface, displacing cooler, more dense portions. These cooler areas are in turn pulled down by gravity (Fig. 1-40). As convection cells move heated mantle rocks toward the surface in this way, the convecting mantle material encounters the overlying solid lithospheric plates. Unable to reach the surface, the rising material moves laterally beneath the plates, dragging them along as it flows.

Sometimes the drag produced by neighboring convection cells pulls the lithosphere in opposite directions. If the pull is great enough to cause the lithosphere to rift, the rifted plates are then carried along in opposite directions by the slowly convecting currents. A relatively small amount of mantle-derived material does escape to the surface as the flowing lava at divergent plate boundaries. This lava gradually cools to form the new outer boundaries of the plates and new oceanic crust. The plates continue to diverge from the site of the rising hot mantle over millions of years, becoming cooler and more dense. Eventually, now far removed from the spreading mid-ocean ridge, the old, cold oceanic lithosphere has become dense enough to sink back into the Earth's interior at a subduction zone. The subducted lithosphere may eventually be reheated and reenter the cycle.

Some geologists believe that convection alone drives plate-tectonic processes. Others believe that gravity assists convection by causing plates that are uplifted at mid-ocean ridges to slide down and away from the ridge, and eventually helping the dense plates to be pulled down further in subduction zones into the Earth's interior. These processes are known respectively as *ridge push* and *slab pull*. Convection and gravity, then, may be working together to cycle the Earth's plates. We will discuss the driving forces of plate motion in greater detail in Chapter 12 after you've learned more about rocks, the Earth's interior, and a host of other processes involved in plate-tectonic processes.

The Earth's Plate-Tectonic Future

Our knowledge of the current rates and directions of plate motion enables us to speculate about the Earth's continental configuration millions of years into the future, *assuming that those rates and directions remain relatively constant*. In Figure 1-41, we see the Earth as it *may* look 100 million years from now. This scenario, of course,

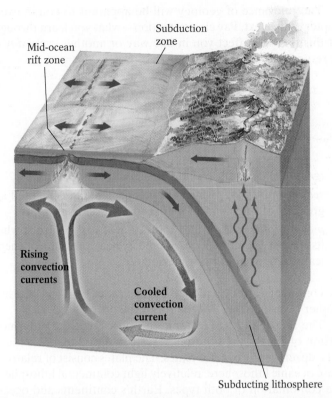

▲ FIGURE 1-40 Convection cells and plate motion. Heat within the Earth's mantle results in rising ("convecting") currents of warm mantle material that drag the lighter lithospheric plates along with them as they flow beneath the Earth's surface. As rising mantle material spreads beneath the lithospheric plates, it cools, becomes more dense, and begins to sink back to the deeper interior, where it is reheated to rise again. Such a cycle, known as a convection cell, is believed by many scientists to be a principal driving mechanism of plate tectonics.

assumes that the rates and directions of all plate movement will remain constant. Plate motions, however, have changed in the past and may do so again. Nevertheless, the reunion of the continents to form another supercontinent in the future is a very strong likelihood.

A Preview of Things to Come

The brief introduction to the principles of plate tectonics here in Chapter 1 should prepare you well for the next 12 chapters—the first two major parts of this text. As Part 1 continues, Chapters 2 through 7 focus on the origins of various Earth materials—from minerals to igneous, sedimentary, and metamorphic rocks—and explain some important processes that produce those rocks, such as volcanism and weathering, the processes by which solid rock can be turned to dust or dissolved away by the Earth's climate. Chapter 8 explains how geologists interpret the rock record to unravel the Earth's long history. In Part 2, Chapters 9 through 11 are devoted to the large- and small-scale geologic structures at the Earth's surface and the dynamics of the Earth's interior. Chapter 12 provides a more detailed look at the global effects of plate tectonics at the Earth's ocean floors. Chapter 13 focuses on the plate tectonics of the Earth's continental lands.

Finally, Part 3 describes how surface processes sculpt the Earth's large-scale features created principally by tectonic forces. In Chapters 14 through 19, we will explore why some slopes are more stable than others and where to buy—and not buy—a house in a landslide-prone region; how streams modify the landscape and why they occasionally flood our homes; where to find clean, drinkable, underground water; how caves form; why glaciers once covered much of North America and when they will be

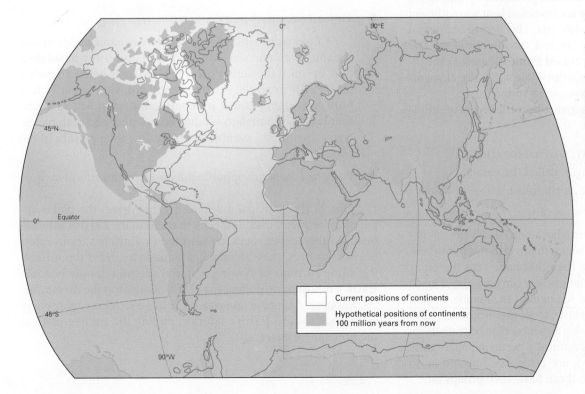

◄ FIGURE 1-41 The projected position of the Earth's continents 100 million years from now. These positions are based on the assumption that plate velocities and directions will remain as they are today. Note the possible collisions of Africa and Europe (with the loss of the Mediterranean Sea) and Australia and Indonesia, and the movement of western California toward Alaska along a continent-long transform plate boundary.

back; where sand dunes form and why the world's deserts grow larger every day; and why crashing waves are eroding some of our coasts and beaches at an alarming rate. The concluding chapter focuses on the ways humans use and manage the Earth's natural resources—from the oil that heats our homes and fuels our cars to the fertilizers that enrich our crops and guarantee our next meal.

The relevance of geology will be apparent to you in every chapter of this text. Pay close attention—what you learn throughout this text will affect you in one way or another *every day of the rest of your life.*

Chapter 1 Summary

Geology is the scientific study of the Earth. Geologists analyze and interpret their findings and develop hypotheses to explain how the forces of nature work. When hypotheses are consistently supported by further study, they may be elevated to the status of a widely accepted explanation, or theory. A theory that withstands rigorous testing over a long period of time may be declared a scientific law. The hypotheses that have undergone such scrutiny include two that attempted to explain the evolution of Earth's geologic features. Catastrophism held that Earth had evolved through a series of immense worldwide upheavals; uniformitarianism proposed that Earth has evolved slowly and gradually by small-scale processes that can still be seen operating. Today scientists recognize the effects of both slow processes and catastrophic events in the evolution of Earth.

The present universe began with the "Big Bang" roughly 13 billion years ago. The Sun formed about 5 billion years ago from the collapse of a gas cloud, the center of which heated up as particles drawn inward by gravity collided and produced nuclear reactions. As the outer region of the gas cloud cooled, Earth and other planets developed (Earth about 4.6 billion years ago) by accretion of colliding masses of matter, some perhaps as large as the planet Mars. Earth's collision with one such mass may have spawned the Moon.

During Earth's first few tens of millions of years, the impact of these accreted masses along with the heat produced by radioactive decay warmed its interior until the accumulated heat was sufficient to melt much of the planet's constituents. This period of internal heating caused Earth to become layered, or differentiated. The early period of heating must have caused Earth's densest elements, primarily iron, to sink toward its interior while its lightest elements rose and became concentrated closer to its surface.

Today, Earth has three principal concentric layers of different densities:

- the thin, least dense outer layer, called the *crust;*
- a thick, denser underlying layer, called the *mantle;*
- a much smaller *core,* which is the most dense of Earth's layers.

Over the last 4 billion years, Earth has cooled slowly from its initial higher temperatures. Enough heat remains in its interior to generate currents of flowing mantle rock that have kept the outer portion of Earth mobile. Earth's lithosphere, a composite layer made up of the crust and the outermost segment of the mantle, is solid and brittle and forms large rocky plates; these plates move along at Earth's surface atop the warm, flowing asthenosphere beneath them.

Rocks, which are defined as naturally occurring aggregates of inorganic materials (minerals), are categorized according to the way in which they form. The three basic rock groups are:

- *igneous rocks,* which solidify from molten material;
- *sedimentary rocks,* which are compacted and cemented aggregates of fragments of preexisting rocks of any type;
- *metamorphic rocks,* which form from any type of rock when its chemical composition is altered by heat, pressure, or chemical reactions in Earth's interior.

The continual transformation of Earth's rocks from one type into another over time is called the rock cycle.

The theory of plate tectonics proposes that the outermost portion of Earth (the lithosphere) is composed of seven major and a dozen or more minor plates. The plates consist of relatively dense oceanic lithosphere, relatively light continental lithosphere, or a combination of both types. Earth's continents and oceans drift from place to place atop the moving plates. The plates move in three ways:

- Away from one another, by *divergence;*

  Divergence is preceded by rifting, the process in which a preexisting plate is torn into a number of smaller plates. New oceanic lithosphere forms between rifted plates as molten material from Earth's mantle wells upward, cools, and solidifies.

- Toward one another, by *convergence;*

  Convergence involving either two oceanic plates or one oceanic plate and one continental plate results in subduction, in which the denser oceanic plate sinks below the other into Earth's mantle. A continental collision occurs when both converging plates are continental, in which case neither is dense enough to subduct completely and both plates' edges are instead eventually uplifted by the pressure of the collision.

- Past one another in opposite directions, by *transform motion.*

One of the principal mechanisms that drives the plates appears to be the convection cells that circulate Earth's internal heat. As the deeper rocks are warmed by Earth's various internal heat sources, they become lighter and rise. Near the surface, the rising currents of flowing material spread laterally beneath Earth's lithosphere, pulling the plates along atop the flowing asthenosphere. At subduction zones, Earth's cool, dense lithosphere sinks back toward Earth's interior, pulling the trailing plate as the lithosphere descends.

KEY TERMS

absolute dating *(p. 20)*
asthenosphere *(p. 16)*
catastrophism *(p. 10)*
chondrules *(p. 15)*
continental collisions *(p. 25)*
continental drift *(p. 29)*
convection cells *(p. 34)*
convergence *(p. 24)*

core *(p. 16)*
crust *(p. 16)*
divergence *(p. 24)*
Earth systems *(p. 7)*
erosion *(p. 8)*
geology *(p. 6)*
hypothesis *(p. 10)*
igneous rocks *(p. 18)*

lithosphere *(p. 16)*
mantle *(p. 16)*
metamorphic rocks *(p. 19)*
minerals *(p. 18)*
plate tectonics *(p. 21)*
rifting *(p. 24)*
rock *(p. 18)*
rock cycle *(p. 19)*

scientific law *(p. 10)*
scientific methods *(p. 10)*
seafloor spreading *(p. 24)*
sedimentary rocks *(p. 18)*
subduction *(p. 24)*
theory *(p. 10)*
transform motion *(p. 28)*
uniformitarianism *(p. 11)*

QUESTIONS FOR REVIEW

1. Briefly explain the differences between a scientific hypothesis, a scientific theory, and a scientific law.

2. Contrast the principles of catastrophism and uniformitarianism.

3. Draw a simple sketch of the major layers that make up the Earth's interior. What parts of the Earth's internal structure form the Earth's plates?

4. Describe the three major types of rocks in the Earth's rock cycle. Give an example of how each of these rocks can evolve into another type of rock.

5. Draw simple sketches of divergent plate boundaries, two kinds of convergent plate boundaries, and transform plate boundaries.

6. At which type of plate boundary do oceanic plates grow? At which type of plate boundary are oceanic plates consumed?

7. How could one use the distribution of modern and ancient animals to support the theory of continental drift?

8. Briefly describe how oceanic plates vary in age as they proceed outward from a diverging mid-ocean ridge.

9. Draw a simple sketch of a convection cell and explain how it works. Explain the difference between convection and conduction of heat.

10. When geologists find ancient glacial deposits in equatorial Africa, they usually interpret them as polar deposits that have drifted as part of the lithospheric plate from a cold place to a warm place. Formulate another hypothesis to explain this phenomenon.

LEARNING ACTIVELY

1. Get a copy of today's newspaper and go through it page by page. List at least five articles that are related to geology, Earth science, natural resources, and the environment. (*Hint:* Include articles on the price of gasoline, earthquakes, floods, volcanic eruptions, food prices, the rise or fall in gold prices, and global warming.)

2. Get a world atlas and locate the major mountain ranges of the Earth's surface. Speculate about the plate-tectonic processes (such as subduction, collisions, and divergence) that produced those mountains. How do you think the Andes of South America and the Cascades of the Pacific Northwest formed? What about the Urals between Europe and Asia? The Himalayas between

Asia and India? The Alps of southern Europe and the Atlas Mountains of northern Africa?

3. Take a field trip somewhere in your city or town. Find a natural outcropping of rock and try to identify it as belonging to one of the major rock categories (igneous, sedimentary, metamorphic). You should be able to do this activity regardless of where you live—New York City, Los Angeles, or Philadelphia—virtually every place has some geology exposed. If you can't find exposed rock, look for a geologic process in action: the waves at a beach, a stream flowing through your town, a slope that has soil moving downward from the pull of gravity.

ONLINE STUDY GUIDE

Practice makes you better. Make the time to take the *Online Study Guide* quiz for this chapter. It should take you about 20 minutes. Automatic grading and instant feedback will help you quickly and accurately identify the concepts you have mastered and the areas in which you need more study.
www.prenhall.com/chernicoff

Minerals

2

Many cultures have long valued minerals for their sheer visual appeal—for their stunning colors or luster, or their fascinating shapes and symmetry (Fig. 2-1). But the importance of minerals is hardly just aesthetic, and it is not restricted to the "pretty" specimens found in museums or jewelers' showcases.

Every day we use a vast array of minerals in a remarkable number of ways. A common absorbent mineral called talc (used in talcum powder) dries and soothes our skin. Minerals provide the sulfur used to manufacture fertilizers, paints, dyes, detergents, explosives, rubber, synthetic fibers, or a simple book of matches. Certain minerals yield the valuable metal aluminum, an ideal component for airplanes, garden furniture, and beer and soft-drink cans because of its lightness, strength, and resistance to corrosion. When you're laid low by a common intestinal problem, you may run for a spoonful of Kaopectate, a remedy whose active ingredient is the mineral kaolinite. (For another everyday example, see Figure 2-2.)

Even our good health is tied to the minerals we encounter in our physical environment. Certain minerals are to be avoided at all costs—they have been linked conclusively to the onset of certain diseases. For example, erionite, a relatively rare mineral found in several regions of Turkey, has been linked directly to the onset of mesothelioma, a deadly form of lung cancer. This mineral forms typically in warm climates where dried-out lake beds develop atop volcanic soils—a highly specific set of conditions. Here, as we discussed in Chapter 1, is a place where a variety of the Earth's systems interact: The local composition of the lithosphere (the volcanic soils) and the conditions of the hydrosphere (the evaporating lake) profoundly affect the biosphere (the townspeople exposed to this cancer-causing mineral).

As you will see, minerals are vastly important to geologists because they make up the rocks we study to interpret the Earth's past. **Rocks** are naturally occurring combinations of one or more minerals, with each mineral retaining its own discrete characteristics within the rock (Fig. 2-3).

In this chapter, we will examine what minerals are and how they are formed, and discuss methods of identifying different kinds of minerals. We will also look at the distinctive properties of some important minerals and see how their characteristics determine how we use them in our daily lives.

FIGURE 2-1 Aquamarine crystals with muscovite mica. Aquamarine is a blue-green variety of the mineral beryl. Emerald is the green variety of beryl. The beautiful color is caused by trace amounts of iron within the beryl crystals. ▶

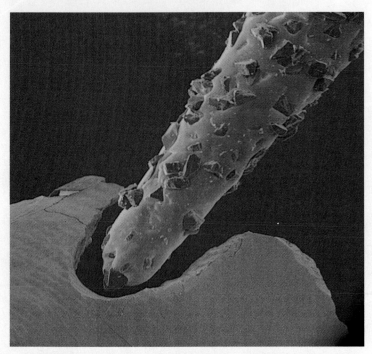

▲ FIGURE 2-2 Today, because of technological advances and the existence of diamonds, a visit to the dentist is not nearly as long or as painful as it once was. Here we see a drill bit studded with diamonds (the hardest-known natural substance) cutting swiftly through the relatively soft enamel of a human molar composed largely of the calcium–phosphorus mineral, apatite (magnified 25×).

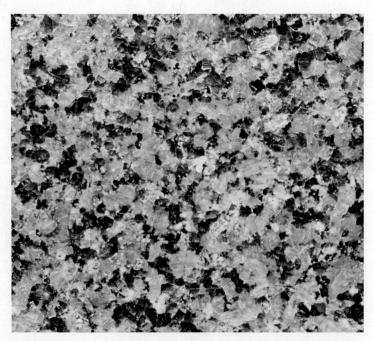

▲ FIGURE 2-3 One can clearly see the individual mineral grains in this piece of granite, an igneous rock. Can you see at least four different minerals in this specimen?

What Is a Mineral?

Consider the following: Ice is a mineral, but water is not. The diamonds that form deep in the Earth are minerals, but the diamonds created in laboratories are (technically) not. Quartz (SiO_2) is a mineral, but silica glass (SiO_2) is not. So what is a mineral? A **mineral** is a solid, naturally occurring substance that has a systematic internal organization. Each mineral is made up of chemical elements in specific proportions for that mineral. So, for example, quartz is a mineral because it is a solid substance that occurs in nature, and it always has the formula SiO_2 because silicon and oxygen are arranged in a very particular way in *every* quartz crystal.

Minerals contain one or more **elements**—the forms of matter that cannot be broken down into a simpler form by heat, cold, or reaction with other chemical elements. Aluminum and oxygen are two common elements. **Atoms** (from the Greek *atomos*, or "indivisible") are the smallest particles of an element that retain all of the element's chemical **properties.** A property is a characteristic of a substance that enables us to distinguish it from other substances, such as its color or its hardness (physical properties) or its ability to bond with other atoms (a chemical property). All atoms of a given element are identical, and the atoms of one element differ in fundamental ways from the atoms of every other element. There are 115 known elements, of which 90 occur naturally and 25 are laboratory creations. Chemists have arranged the 115 elements into the Periodic Table of Elements (Fig. 2-4), which contains vital information about these building blocks of the Earth's matter.

We can describe a mineral using its name (quartz, diamond, halite, calcite) or the elements that make up the mineral (SiO_2, C, NaCl, $CaCO_3$). The numbers indicate the number of atoms of each element relative to others in the mineral's formula. So, quartz contains two atoms of oxygen for every one atom of silicon; diamond consists entirely of carbon; halite (common table salt) has equal numbers of sodium (Na) and chlorine (Cl) atoms; and calcite has one calcium and one carbon atom for every three oxygen atoms.

Intro geology students sometimes wonder why so much discussion about minerals involves chemistry. The principal reason—virtually all minerals are chemical **compounds,** so we need to talk a bit about chemistry to understand the composition and organization of the atoms in minerals. The composition and internal arrangement of their atoms determine the different characteristics of minerals—how they look, where they occur, and how people use them. To know why diamonds are hard, why gold can be pounded into wafer-thin leaves, and why we build skyscrapers with skeletons of titanium steel, we must first understand the chemical makeup of minerals.

The Structure of Atoms

An atom is incredibly small, approximately 0.00000001 cm (one hundred-millionth of a centimeter) in diameter. This line of type would contain about 800,000,000 atoms laid side by side. As small as an atom is, it consists of even smaller particles—**protons** and

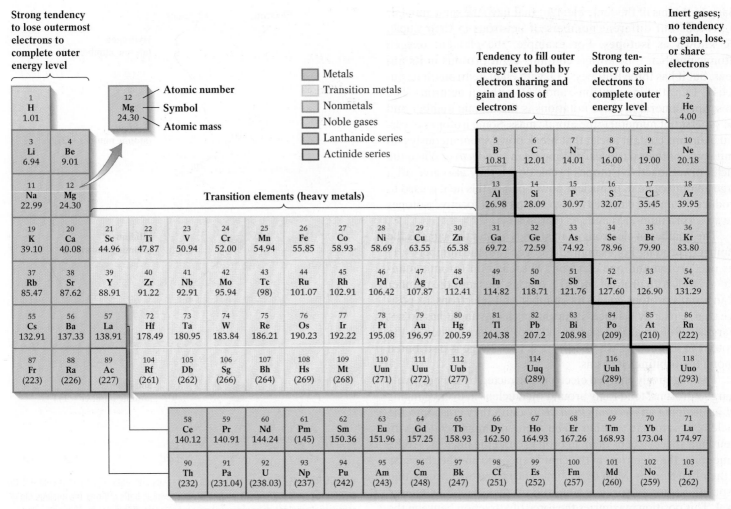

Strong tendency to lose outermost electrons to complete outer energy level

Tendency to fill outer energy level both by electron sharing and gain and loss of electrons

Strong tendency to gain electrons to complete outer energy level

Inert gases; no tendency to gain, lose, or share electrons

Metals						
Transition metals						
Nonmetals						
Noble gases						
Lanthanide series						
Actinide series						

Atomic number
Symbol
Atomic mass

Transition elements (heavy metals)

▲ **FIGURE 2-4** The Periodic Table of Elements. The periodic table groups all the elements by similarities in their atomic structures, which result, in turn, in similarities in their chemical properties.

neutrons—which are located in the atom's **nucleus,** or center, and **electrons,** which orbit around the nucleus. Most of an atom's volume is, in fact, just empty space. If, for example, the volume of an atom's nucleus were expanded to the size of a basketball, the orbiting electrons would be located up to 3 km (2 mi) away from the basketball! A simplified model of an atom is shown in Figure 2-5.

Each proton in an atom's nucleus carries a single positive charge, expressed as +1, and has a mass of 1.67×10^{-24} g, which for convenience is referred to as an *atomic mass unit*

(AMU) of 1. Neutrons are nearly identical to protons in size and mass, but as their name suggests, they have no charge—they are *neutral.* They do, however, contribute to the **atomic mass** of the atom, defined as the sum of protons and neutrons within an atom's nucleus. Thus, an atom containing one proton and one neutron has an atomic mass of 2 AMU. The third type of atomic particle, the electron, travels around the nucleus at a speed so great that it could orbit the Earth in less than a second. Each electron carries a single negative charge, expressed as −1, and has a mass about 1/1836 that of a proton or neutron. Thus an electron's contribution to an atom's mass is quite minor.

The number of protons in an atom's nucleus is its **atomic number.** *Every* atom of an element has the same number of protons in its nucleus. This number differs from the number of protons in the nuclei of all other elements. For example, every magnesium atom has 12 protons in its nucleus, or an atomic number of 12. An atom with 13 protons in its nucleus is an entirely different element, aluminum. Thus the number of protons determines an atom's identity.

The number of neutrons in an atom's nucleus can vary without causing the atom's identity to change; only the atomic mass

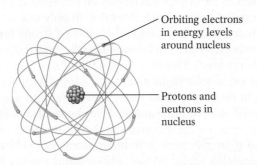

Orbiting electrons in energy levels around nucleus

Protons and neutrons in nucleus

▲ **FIGURE 2-5** A simplified model of an atom. Protons and neutrons compose the nucleus; electrons move around the nucleus at high speed.

changes. Atoms of the same element that have the same number of protons but different numbers of neutrons in their nuclei are known as **isotopes.** For example, the element oxygen (atomic number = 8) always contains eight protons in its nucleus, but it has three isotopes: $^{16}_{8}O$, with eight neutrons in its nucleus; $^{17}_{8}O$, with nine neutrons; and $^{18}_{8}O$, with ten neutrons. The subscript numeral in these notations is the atomic number and the superscript numeral is the atomic mass. Some isotopes of certain elements contain nuclei that break down spontaneously and emit some of their particles, making them *radioactive.* When the nuclei of radioactive isotopes break down, they also give off a large amount of heat. In nuclear power plants, this heat is used to produce steam and electricity. Two common radioactive isotopes are $^{235}_{92}U$ (uranium-235) and $^{14}_{6}C$ (carbon-14).

The number of electrons in an atom is *always* the same as the number of protons. For example, hydrogen (atomic number = 1) has one proton and one electron, whereas iron (atomic number = 26) has 26 protons and 26 electrons. Because the number of an atom's protons equals the number of its electrons, its positive charge exactly balances its negative charge and it has no net charge. All of an atom's positive charge is concentrated in the protons in its nucleus, whereas its negative charge is distributed among the orbiting electrons.

The negatively charged electrons, attracted by the positively charged protons, tend to fly around the nucleus in ever changing orbits, forming an *electron cloud.* Why don't they crash into the nucleus or into one another? The momentum associated with their high speeds keeps them in orbit around the nucleus. At the same time, the electrons in the cloud repel one another because of their like charges. Most of the time, each electron moves within a specific region of space around the nucleus, called an **energy level.** This position maximizes the force of attraction between the electron and the nucleus while minimizing the force of repulsion between the electron and all the other electrons.

Electrons fill an atom's lowest, or first, energy level before any enter the higher energy levels more distant from the nucleus. The lowest energy level in any atom always has a maximum capacity of two electrons. The second energy level can hold a maximum of eight electrons, and succeeding energy levels can each hold eight or more electrons. The number and energy-level positions of some atoms' electrons are shown schematically in Figure 2-6.

Making Minerals—How Atoms Bond

To form chemical compounds, atoms combine, or **bond.** They do this by losing, gaining, or sharing one or more electrons during a chemical reaction. The transfer or sharing of electrons between bonded atoms changes the electron configuration of each. Two key factors determine which atoms will unite with which others to form compounds: How will each atom achieve chemical stability? Will the resulting compound be electrically neutral?

An atom is chemically stable when its outermost energy level is filled with electrons. Thus atoms bond by transferring or sharing electrons to fill their outermost energy levels. For hydrogen and helium atoms, which have only the lowest energy level, this requires two electrons in the single energy level. For all other atoms, chemical stability is achieved when their outermost en-

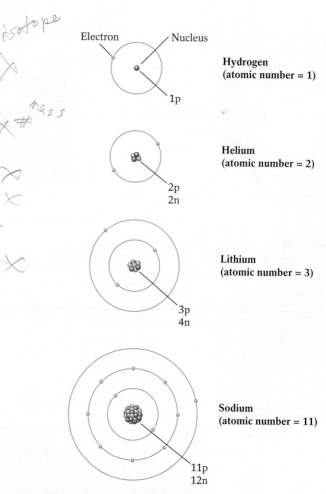

▲ **FIGURE 2-6** Energy-level diagrams of various elements. The nucleus contains the protons (p) and neutrons (n); electrons are shown as balls orbiting the nucleus along concentric circular tracks representing energy levels. (Electron size is exaggerated for visibility; electrons are actually much smaller than protons and neutrons.) Hydrogen and helium, because they have 2 or fewer electrons, have only 1 energy level; lithium, with 3 electrons, has 2 energy levels; sodium, because it has 11 electrons, has 3 energy levels (2 electrons in the first level, 8 electrons in the second level, and 1 electron in the third level). Note that the number of electrons in an atom always equals the number of protons.

ergy level contains eight electrons. Because chemical stability almost always requires eight electrons in the outermost energy level, this requirement is known as the **octet rule.**

Whether an atom of an element loses outer electrons, gains them, or shares them depends largely on the number of electrons in its outermost energy level. Atoms with only one or two electrons in their outermost energy levels have a strong tendency to give up those electrons, thus eliminating their partially filled outermost energy level. This then leaves their "new" outermost energy level filled with eight electrons. Atoms with six or seven electrons in their outermost energy levels tend to acquire electrons to fulfill the octet rule and achieve chemical stability. Atoms with three, four, or five electrons in their outermost energy levels tend to *share* electrons with other atoms instead of transferring or receiving them. Atoms whose outer energy levels are already filled with eight electrons (or, in the case of helium, with two electrons) are very stable chemically—we describe these atoms (and their elements) as *inert.* Because these atoms generally do not need to lose, gain, or share electrons, they are far less likely to bond with other atoms.

Elements vary a great deal in the electron configurations of their atoms, so various types of bonding are possible. The atoms that make up the vast majority of the Earth's minerals are most often linked by ionic bonding or covalent bonding, and less commonly, metallic bonding.

Ionic Bonding An atom does not change its identity when it loses or gains an electron (it still has the same number of protons in its nucleus), but it does lose its electrical neutrality and becomes positively or negatively charged. When a potassium atom loses one electron, it becomes a positively charged particle because it then has one more proton than it has electrons. When a chlorine atom gains a single electron, it becomes a negatively charged particle because it then has one more electron than it has protons. An atom that has lost or gained one or more electrons to become a charged particle is called an **ion.** A positively charged ion is indicated by the symbol for the element followed by a superscript "+" (for example, K^+). A negatively charged ion is indicated by the symbol for the element followed by a superscript "−" (for example, Cl^-). The number of electrons lost or gained appears as a superscript before the charge (for example, Ca^{2+}).

When an atom with a strong tendency to lose electrons comes into contact with an atom with a strong tendency to gain electrons, one or more electrons are transferred so that each atom achieves the chemical stability of a full outer energy level. The donor atom loses one or more electrons and becomes a positively charged ion, and the receiving atom gains one or more electrons and becomes a negatively charged ion. These oppositely charged ions then attract each other to form an **ionic bond.** The result is an electrically neutral, chemically stable compound.

Figure 2-7 shows the ionic bonding of sodium and chlorine atoms to form an electrically neutral compound. Sodium has a strong tendency to lose the lone electron in its third, or outer, energy level, thereby completely eliminating that energy level. Loss of an electron turns the sodium atom into a positively charged ion with only two energy levels. The second energy level, now the outer one, is full, with eight electrons. Chlorine has a strong tendency to gain a single electron to add to the seven in its outer energy level, which then becomes filled with eight electrons. Addition of an electron turns the chlorine atom into a negatively charged ion. The electrically neutral compound that results from ionic bonding of sodium (Na) and chlorine (Cl) is sodium chloride (NaCl).

The physical and chemical properties of ionic compounds often differ from those of their component elements. Pure sodium is a soft, silvery metal that reacts vigorously when mixed with water and may even burst into flame. Chlorine usually occurs as Cl_2, a green poisonous gas that was used as a chemical weapon during World War I. Their bonded compound, sodium chloride, neither ignites nor poisons. It is, however, the white crystalline mineral *halite,* the common table salt that we sprinkle on french fries—a substance that regulates some of the biochemical processes essential to all animal life.

Covalent Bonding Atoms whose outer energy levels are about half full, containing three, four, or five electrons, tend to achieve chemical stability by *sharing* electrons with other similarly equipped atoms. In such a case, both atoms fill their outer energy levels with the shared electrons rather than transferring electrons from one to the other. Sharing electrons produces a **covalent bond,** in which the outer energy levels of the atoms overlap. Covalent bonds are generally stronger than any other type of bond. In some cases, two or more atoms of a *single* element may bond covalently with each other. For example, overlapping carbon atoms bond covalently in the *all-carbon* mineral, diamond (Fig. 2-8).

Metallic Bonding The atoms of some electron-donating elements tend to pack together closely, with each typically surrounded by either eight or twelve others. This produces a cloud of electrons that roam independently among the positively charged nuclei, unattached to any specific nucleus. The attraction of a negatively charged electron cloud to a cluster of positively charged nuclei is called **metallic bonding,** and is responsible for the properties that define metals. For example, metals are efficient conductors of electricity, which requires freely moving electrons. The mobile electrons of metallically bonded substances explain (among other phenomena) why copper wiring can be used to transmit an electrical current to the kitchen toaster.

▲ **FIGURE 2-7** Ionic bonding of sodium (Na) and chlorine (Cl). When a sodium atom (with one electron in its outer energy level) donates its outermost electron to a chlorine atom (with seven electrons in its outer energy level), the outer energy level in the new configuration of each atom has eight electrons. The two resulting ions (Na^+ and Cl^-) unite to form sodium chloride (NaCl), a neutral ionic compound. This simple diagram is designed to show the process of ionic bonding. In reality, the internal structure of halite is far more complex.

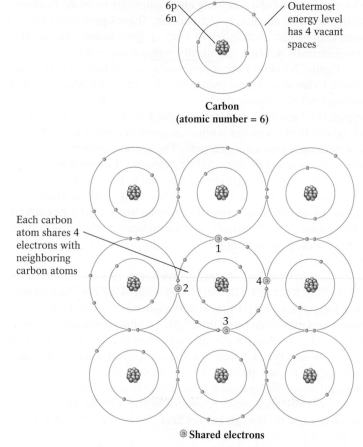

▲ **FIGURE 2-8** Covalent bonding in diamond, a mineral that consists entirely of the element carbon. Each carbon atom in diamond is bonded covalently to four neighboring carbon atoms; the great strength of these bonds accounts for the fact that diamond is the hardest known substance on Earth. This simple diagram is designed to show the process of covalent bonding. In reality, the internal structure of diamond is far more complex. See Figure 2-18 for a look at how each carbon atom bonds with four neighboring carbon atoms.

Intermolecular Bonding

Some atoms bond covalently to form *molecules*, stable groups of bonded atoms that exhibit distinct identifiable properties as a group. A prime example is water. One molecule of water, H_2O, consisting of exactly two atoms of hydrogen bonded to one atom of oxygen, has all the properties of water, whereas its component elements alone do not.

The water molecules, which are held together internally by strong covalent bonds, often weakly attach themselves to other molecules by **intermolecular bonds.** This bonding is a result of an uneven distribution of electrons. The positive charge of the oxygen atom's nucleus is greater than that of the hydrogen atoms' nuclei, which means the shared electrons are more attracted to, and thus spend more time near, the oxygen nucleus. The oxygen side of a water molecule therefore develops a weak negative charge from the presence of these negatively charged electrons. Because the electrons spend less time near the hydrogen nuclei, a weak positive charge develops on that side of the molecule. These weakly charged regions attract *oppositely* charged regions of nearby molecules, forming weak **hydrogen bonds** with these molecules.

The formation of hydrogen bonds explains why so many substances, even those that are strongly bonded, dissolve in water. Consider what happens to salt: When halite (NaCl) is mixed with water, the positively charged sides of the water molecules attract the chlorine ions and the negatively charged sides attract the sodium ions. Although hydrogen bonds are weak compared to the ionic bonds that hold the Na and Cl together in halite, the combined effect of many water molecules is strong enough to overcome the ionic bonds and split halite into its component Na^+ and Cl^- ions, thus dissolving the salt (Fig. 2-9).

Like molecules, some minerals have an uneven distribution of their charges. These minerals may be subject to a type of intermolecular bond in which a number of electrons temporarily group on the same side of an atom's nucleus, giving that side of the atom a slight negative charge and the electron-poor side a slight positive charge. The positive side may briefly attract electrons of neighboring atoms, and the negatively charged side may attract the nuclei of neighboring atoms. This type of weak intermolecular attraction is called a **van der Waals bond,** after Johannes van der Waals, the Dutch physicist who discovered it in the late nineteenth century. Although weak, van der Waals bonds can be sufficient to bond atoms or layers of atoms together in certain minerals, such as the weakly held carbon atoms in graphite (the "lead" in your pencil).

Hydrogen and van der Waals bonds are much weaker than ionic, covalent, and metallic bonds, and can be broken by the addition of small amounts of heat. For example, the intermolecular bonds between water molecules in ice break when the temperature is raised to 0°C (32°F)—the melting point of ice—changing the water from a solid to a liquid. The strong covalent bonds that form the individual water molecules, however, are unaffected.

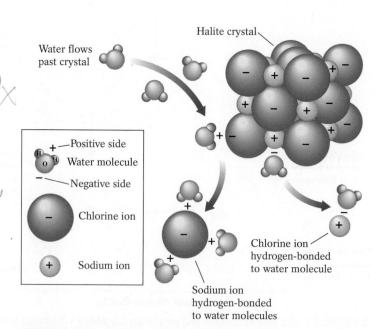

▲ **FIGURE 2-9** Water molecules each have a positively charged side (the hydrogen side) and a negatively charged side (the oxygen side). These regions attract oppositely charged ions within other compounds. The attraction of sodium ions to the negative side of water molecules combined with the attraction of chlorine ions to the positive side causes halite (table salt) to dissolve in water.

The thousands of minerals that exist in nature owe their physical and chemical properties to the ways in which their atoms bond. In the following section, we'll learn more about these distinguishing properties.

A Mineral's Structure

When atoms and ions bond to form a mineral, the mineral typically develops a regular geometric shape known as a **crystal** (from the Greek *kryos,* or "ice"). This crystal form is most often a combination of common geometric shapes such as cubes, pyramids, and prisms (see Fig. 2-1). The geometric shape is the external expression of the mineral's microscopic internal **crystal structure,** the orderly arrangement of its ions or atoms into a latticework of repeated three-dimensional units (Fig. 2-10). Every crystal of a given mineral has the same crystal structure, as can be seen externally by the orientation of the planar surfaces, or *faces,* in well-formed crystals.

More often than not in nature, minerals grow in restricted regions, often competing with other growing minerals for the available space. In a rock crowded with many crystals and little or no open space, minerals grow together in shapes that may be different from well-formed crystals. The internal arrangement of atoms in a mineral—even one with an irregular shape on the outside—will be ordered in a particular way, just the same as it is in the most beautifully shaped crystal.

Sometimes molten rock cools too rapidly for its atoms to form even a semblance of an orderly arrangement. The resulting solid, called *glass,* does not qualify as a true mineral. It may have a relatively constant composition, but it lacks a specific crystal structure. Geologists call glass and other similar mineral-like substances, **mineraloids.**

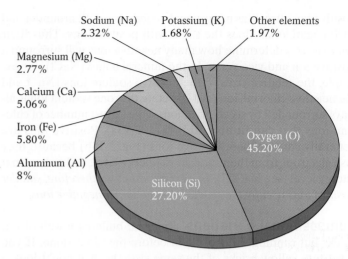

▲ **FIGURE 2-11** The most abundant elements in the Earth's continental crust.

What Controls a Mineral's Composition and Structure?

The kinds of minerals that form in a particular time and place depend on:

- The relative abundances of available elements.
- The relative sizes and charges of these elements' atoms and ions.
- The temperature and pressure at the time of formation.

Only eight of the 90 naturally occurring elements in the Earth's continental crust are relatively abundant, with oxygen and silicon dominating (Fig. 2-11). In the entire Earth, iron and oxygen dominate, silicon is fairly abundant, and magnesium is more significant deeper in the Earth than it is in the crust (Fig. 2-12).

Given two elements of equal abundance, the element that will contribute more readily to mineral formation is the one whose atoms "fit" better with the atoms of the other elements present. Atoms and ions in minerals tend to become packed together as closely as their sizes permit. In an ionically bonded mineral, each

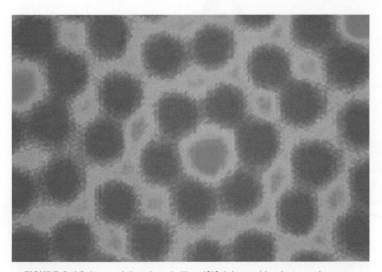

▲ **FIGURE 2-10** Atoms of the mineral silica (SiO_2) imaged by electron microscope (magnified 1,000,000×). The structure of this mineral—and its properties—are related to the way in which the atoms are arranged and the strength of the bonds between the atoms.

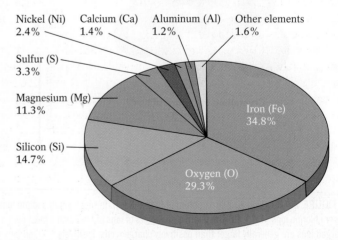

▲ **FIGURE 2-12** The most abundant elements in the entire Earth.

positive ion attracts as many negative ions as can fit around it and each negative ion does the same with positive ions. Thus their relative sizes determine how many negative ions will surround a positive ion and vice versa. In the mineral halite (NaCl), for example, the relative sizes of the positive sodium ions (Na^+) and the negative chlorine ions (Cl^-) dictate that one sodium ion is always surrounded by six chlorine ions—the largest number of chlorine ions that can fit around it (Fig. 2-13a). Positive ions are generally smaller than negative ions (Fig. 2-13b) because they lose electrons from their outermost energy level, and so lose that energy level (see Fig. 2-7). *In a crystal structure, therefore, smaller positive ions occupy the spaces between larger negative ions.*

Compositional Variations   Imagine building a wall of red bricks but running short of them before the job is done. If you substitute yellow bricks of the same size, the wall might look a little weird . . . but it's still a wall. The same is true of ionic substitution—when certain ions of similar size and charge replace one another within a crystal structure. As a result of **ionic substitution,** some minerals that have the same internal arrangement of ions may vary in composition.

Iron (Fe^{2+}) and magnesium (Mg^{2+}), which are nearly identical in size and charge, substitute freely for one another in the mineral olivine $(Fe,Mg)_2SiO_4$. (*Note:* In a mineral's chemical formula, the elements that can substitute for one another in a mineral's crystal structure appear in parentheses and are separated by commas.) The color, melting point, and other physical characteristics of olivine differ depending on whether Fe^{2+} or Mg^{2+} predominates, but its crystal structure is unaffected.

Structural Variations   Two minerals may have the same chemical composition but different crystal structures because they formed under different temperature and pressure conditions. Such minerals

are known as **polymorphs** (Greek for "many forms"). Graphite and diamond are polymorphs that consist entirely of carbon. Graphite's structure forms under relatively low pressure at shallow depths, whereas diamond's structure results from intense pressure at depths greater than 150 kilometers (about 90 miles). Theoretically, from the moment it reaches the surface, diamond should begin to change into graphite, the carbon polymorph that is more stable in the Earth's low-temperature/low-pressure surface environment. Fortunately for owners of diamond jewelry, a great number of *very strong* carbon–carbon bonds must be broken before a diamond becomes graphite. At room temperature, the energy needed to break these bonds—called the activation energy—is simply too high, and thus diamonds are effectively stable. (Geologists describe minerals that are effectively stable as *metastable*.) So whether you give or receive a diamond gift, you can rest assured that in human terms, "diamonds *are* forever."

Identifying Minerals

A mineral's chemical composition and crystal structure give it a unique combination of chemical and physical properties so we can distinguish it from all other minerals. Geologists can observe

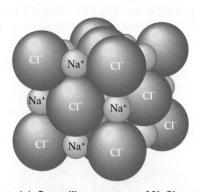

(a) **Crystalline structure of NaCl**

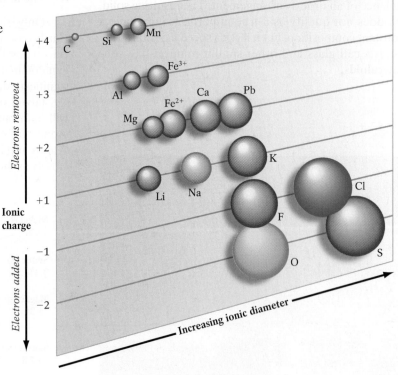

(b) **Ionic diameters**

▲ **FIGURE 2-13 (a)** The crystal structure of halite (NaCl), in which small sodium ions (Na^+) are tucked in between six larger chlorine ions (Cl^-). **(b)** The relative diameters of ions of the Earth's most common elements. Note that negatively charged ions are generally larger than positively charged ions. Positively charged ions have given up their outermost electrons, have lost an energy level, and thus are smaller.

many of these properties readily in the field, but other properties require high-powered microscopes or other machines that analyze mineral compositions and structures in detail. We can seldom identify a mineral accurately on the basis of just one property, so geologists must understand a variety of chemical and physical properties of minerals to identify them correctly.

In the Field

Most geologists and dedicated rockhounds can identify a great many minerals in the field by examining them with the naked eye and performing some very simple tests.

Color   Color may be the first thing you notice about a mineral, but it is perhaps the least reliable identifying characteristic. Different-colored minerals may contain the same elements (as is the case with the polymorphs diamond and graphite), yet minerals that are similar in color may be completely different in composition. As a result, we cannot determine the identity of an unknown mineral solely on the basis of its color.

A mineral's color depends on how much light is absorbed by the chemical makeup and its crystal structure. When white light, which contains all the colors of the spectrum, strikes a mineral, part of the spectrum is absorbed and the remainder is reflected back from the mineral and seen by our eyes. For example, when we see an emerald, only the green part of the spectrum is transmitted; the mineral absorbs the rest of the colors of the spectrum. Certain elements, such as iron, manganese, chromium, and nickel, absorb a great deal of light and transmit little; minerals

containing these elements are commonly dark in color or black. Elements such as sodium, potassium, calcium, and silicon absorb little light; minerals containing them are characteristically light in color.

Another reason why we can't simply trust color as the key to mineral identification—a few stray atoms in a mineral's structure may completely change its color. For example, pure corundum (Al_2O_3) is generally white or light gray, but even a trace of the element chromium produces a brilliant red variety of corundum called *ruby*. The addition of a little titanium and iron produces another variety of corundum, a deep blue *sapphire*. Depending on the types of impurities they contain, different samples of a given mineral may exhibit an array of different colors (Fig. 2-14).

Heating or exposure to radiation displaces some electrons, atoms, or ions from their designated sites in a crystal structure and can cause significant color variations. Unprincipled jewel merchants have been known to "improve" the colors of poor-quality gems with heat or X-rays. Those colors may fade, however, as the displaced ions and atoms gradually return to their original sites within the crystal structure.

Because a mineral's color is rarely unique to that mineral and because a mineral's true color can be "disguised" in so many ways, color by itself is not a reliable way to identify a mineral.

Luster   Luster describes how a mineral's surface reflects light. Minerals can exhibit metallic or nonmetallic lusters. When light shines on a metal, its energy stimulates the metal's loosely held electrons and causes them to vibrate. The vibrating electrons emit a diffuse light, giving metallic surfaces their characteristic shiny

(a)

(b)

▲ **FIGURE 2-14** Variations in the color of minerals. **(a)** These two mineral samples have one thing in common, although you'd never guess it from their colors—they are both quartz, composed of silicon dioxide (SiO_2). Their colors differ because they each contain minute traces of different impurities. **(b)** Variations in color *within* a single crystal of a variety of tourmaline called elbaite (or "watermelon" tourmaline). The impurities incorporated into their structures during formation create the color of the crystals of this mineral, typically including pink and green.

▲ FIGURE 2-15 Silver exhibits the shiny luster characteristic of metals. This specimen was collected on the Keweenaw Peninsula of northern Michigan.

luster (Fig. 2-15). Thus metals, with their great number of "free" electrons, often display a shiny, metallic luster. Nonmetallic lusters are more varied; they can be pearly, earthy, silky, vitreous (glassy), or adamantine ("like a diamond") (Fig. 2-16).

Streak Streak is the color of a mineral in its powdered form. We can observe streak when we rub a mineral across an unglazed porcelain slab known as a streak plate, which pulverizes the mineral's surface. The color of a mineral's streak often differs from that of the intact mineral sample. For instance, the mineral

hematite (Fe_2O_3), commonly steel-gray in color, has a distinctive reddish-brown streak (Fig. 2-17). Streak is usually a better tool for identifying a mineral than color. Powdering the mineral reduces the nonchemical effects on color—such as minor defects in crystal structure that affect absorption and transmission of light.

A simple streak test can sometimes save us a batch of money and severe embarrassment by helping us distinguish between similar-appearing minerals that may have considerably different monetary values. Pyrite (FeS_2), often called "fool's gold" because of its brassy yellow color, has a black streak, in contrast with the golden-yellow streak of *true* gold. When similar-appearing minerals have similar streaks, additional tests are necessary to identify them.

Hardness Geologists test the **hardness** of a mineral by determining how difficult (or easy) it is to scratch its crystal surface with a series of other minerals or substances with known hardnesses. (Hardness is *not* a function of how easily a mineral breaks—a solid rap with a hammer will easily shatter a diamond, the world's "hardest" natural substance.) Because every scratch removes atoms from the surface of the mineral, and thus breaks the bonds holding these atoms, a mineral's hardness indicates the relative strength of its bonds.

The Mohs scale, named for its developer, German mineralogist Friedrich Mohs (1773–1839), assigns relative hardnesses to several common and a few rare and precious minerals. Here's how it's used: An unknown mineral that can be scratched by topaz (hardness = 8) but not quartz (hardness = 7) has a hardness between 7 and 8 on the Mohs scale. Table 2.1, which lists the minerals and common testing standards used in the Mohs scale, explains why geologists are often found with a few copper pennies, a pocketknife, and well-worn fingernails.

The hardness of a mineral depends on the strength of its weakest bonds. Minerals with many covalent bonds are generally harder than those with ionic or other bonds. Because every atom of carbon in a diamond forms a strong covalent bond with four neighboring carbon atoms, diamond is exceptionally hard (that is, resistant to scratching) and durable (Mohs hardness = 10). (On the 2004 mission to Mars, the Mars rover *Spirit* was outfitted with a diamond-tipped abrasion tool that spun 3000 times per minute to drill into the Martian rocks.) Graphite, diamond's polymorph, is one of the softest minerals (Mohs hardness = 1). Its sturdy layers of covalently bonded carbon atoms are only weakly bonded to one another by van der Waals forces (Fig. 2-18). When you write with a pencil, pressure on the point breaks these weak graphite bonds, leaving a trail of carbon layers as words on the paper.

▲ FIGURE 2-16 Nonmetallic lusters can be: (a) pearly, as in gypsum; (b) earthy, as in realgar; (c) silky, as in chrysotile; (d) vitreous (glassy), as in rose quartz; or (e) adamantine, as in anglesite.

▲ **FIGURE 2-17** Although many samples of hematite (Fe_2O_3) are steel-gray, the streaks of all samples are reddish-brown.

Table 2.1 The Mohs Hardness Scale

MINERAL	HARDNESS	HARDNESS OF SOME COMMON OBJECTS
Talc	1	
Gypsum	2	
		Human fingernail (2.5)
Calcite	3	
		Copper penny (3.5)
Fluorite	4	
Apatite	5	
		Glass (5–6)
		Pocketknife blade (5–6)
Orthoclase (potassium feldspar)	6	
		Steel file (6.5)
Quartz	7	
Topaz	8	
Corundum	9	
Diamond	10	

▼ **FIGURE 2-18** The structures of graphite and diamond. Unlike its polymorph, diamond (below), in which all bonds are covalent, the graphite structure (right) contains weak van der Waals bonds between layers of covalently bonded carbon atoms. Graphite is easily broken at the sites of these weak bonds, making it a very soft mineral.

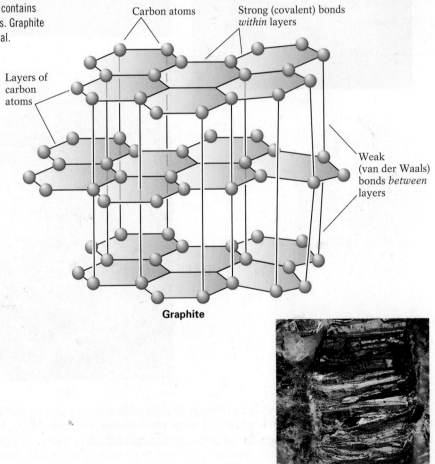

Carbon atoms

Strong (covalent) bonds *within* layers

Layers of carbon atoms

Weak (van der Waals) bonds *between* layers

Graphite

Strong (covalent) bonds

Carbon atoms

Diamond

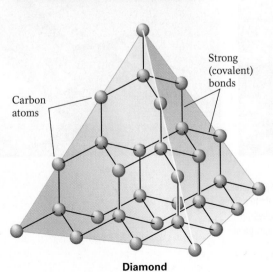

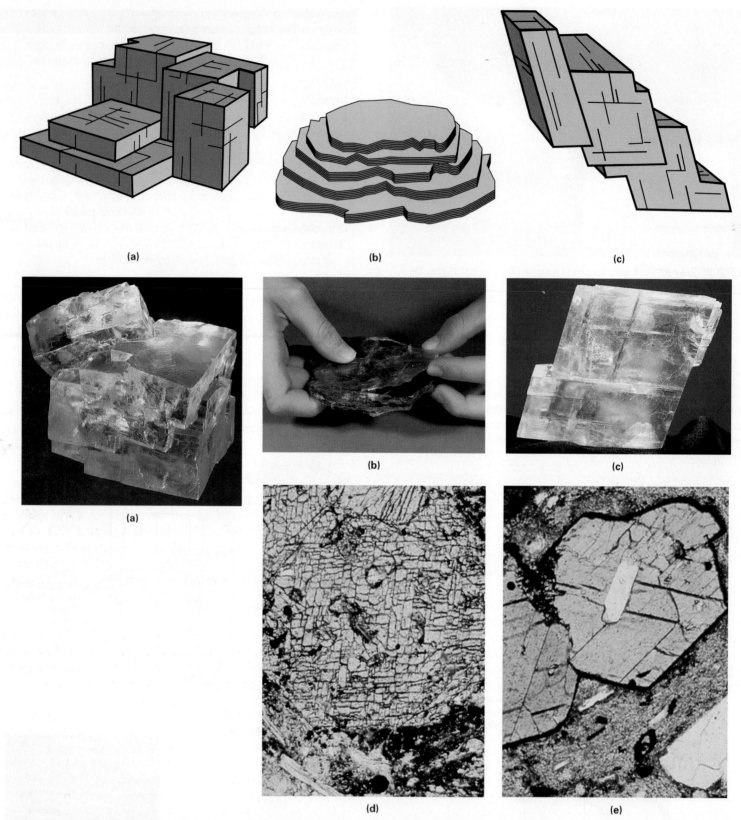

▲ **FIGURE 2-19** Distinguishing minerals by their cleavage. **(a)** Halite has three mutually perpendicular cleavage planes, forming cubes, rectangles, and stepped shapes. **(b)** Mica has one perfect cleavage plane, forming sheets. **(c)** Calcite's three cleavage planes are not mutually perpendicular; calcite cleavage produces a geometrical shape called a rhombohedron. **(d)** A photomicrograph of cleavage in augite shows its planes intersecting at nearly 90° angles. **(e)** A photomicrograph of cleavage in hornblende shows its planes intersecting at angles of about 60° and 120°.

▲ FIGURE 2-20 A conchoidal fracture surface on a quartz crystal. Quartz, with equally strong covalent bonds in all directions, has no planes of weakness. It therefore fractures irregularly instead of cleaving.

Cleavage   **Cleavage** is the tendency of some minerals to break consistently along distinct planes in their crystal structures. These planes typically occur where the bonds are weakest, or fewer in number. Such breaks form smooth, flat surfaces, called *cleavage planes,* on the mineral. When a mineral that tends to cleave is struck along a plane of cleavage, every fragment that breaks off will have the same general shape.

In characterizing cleavage, geologists consider the number of cleavage surfaces produced and the angles between adjacent surfaces. Halite, for example, cleaves in three mutually perpendicular directions, forming cubes, rectangles, and stepped shapes with 90° angles between the fragments' sides (Fig. 2-19a). Mica cleaves in only one direction, forming sheets (Fig. 2-19b). Calcite's ($CaCO_3$) three cleavage planes are not perpendicular to one another (Fig. 2-19c).

Minerals that appear similar by other diagnostic criteria can have different cleavage-plane angles. For example, augite and hornblende are two common black minerals that are similar in external form, hardness, and other characteristics. For reasons we'll discuss later in this chapter, when a crystal of each is broken, however, the two prominent cleavage planes of augite intersect at an angle of about 90°, whereas the two cleavage planes of hornblende intersect at about 120° and 60° angles. The different cleavage angles indicate the different arrangements of atoms and strengths of bonds within the crystal structures of these minerals (Figs. 2-19d and 2-19e).

The cleavage planes in diamonds and certain other minerals enable gem cutters to fashion beautiful jewels from raw, uncut stones. Highlight 2-1 explains how valuable specimens of the

Earth's hardest substance can be cut without destroying them, and goes on to highlight some key facts about our favorite gemstone—the diamond.

Fracture   Some minerals do not cleave—because their bonds are equally strong in all directions and are distributed uniformly throughout the crystal. Instead, these minerals break randomly, or **fracture.** Unlike a straight, smooth-faced cleavage plane, a fracture appears as a jagged irregular surface or as a curved, shell-shaped (*conchoidal*) surface. In the mineral quartz, composed exclusively of silicon and oxygen, all of the atoms bond covalently in a three-dimensional framework with equal bond strengths in all directions. Strike a crystal of quartz with a hammer and it fractures (Fig. 2-20). Because quartz fractures instead of cleaving, we can distinguish it from similar-looking minerals that cleave, such as calcite and fluorite.

Smell and Taste   Experienced geologists occasionally sniff and lick rocks to identify minerals with a distinctive smell or taste—not always a pleasant task. Some sulfur-containing minerals, when treated with dilute hydrochloric acid (HCl), give off the familiar rotten-egg stench associated with hydrogen sulfide gas (H_2S). Halite's salty taste distinguishes it from similar-looking minerals such as calcite. Sylvite (KCl) is distinctively bitter. Kaolinite absorbs liquid rapidly—when licked, it absorbs saliva and sticks to the tongue. But taste unknown minerals cautiously: The minerals realgar and orpiment, which both smell like garlic especially when heated, are rich in the *poisonous* metal arsenic.

Effervescence   Certain minerals, particularly those that contain carbonate ions (CO_3^{2-}), *effervesce,* or fizz, when sprayed with an acid. A few drops of dilute hydrochloric acid (HCl) on calcite ($CaCO_3$) produce a rapid chemical reaction that releases bubbles of carbon dioxide gas. This fizzing helps to distinguish calcite from similar-looking minerals such as halite, which does not effervesce.

Crystal Form   The three-dimensional geometric form of a crystal is an *external* expression of a mineral's *internal* structure. Thus the shape of a well-formed crystal may be distinctive enough to identify the mineral. For example, the angles between adjacent faces of a given mineral crystal are the same in every well-formed, unbroken sample of that mineral. The adjacent faces on a perfect quartz crystal from Herkimer, New York (a good place to find them), are at angles of 120°, the same as on perfect quartz crystals from Hot Springs, Arkansas (another good place to find them).

The crystals of some minerals, however, are not always geometrically shaped. Instead, some crystals may grow together into forms resembling distinctive non-mineral objects, making the minerals relatively easy to identify (Fig. 2-21). A flowerlike rosette shape characterizes barite ($BaSO_4$); botryoidal malachite [$Cu_2(OH)_2CO_3$] looks like "a bunch of grapes," from which it gets its name (derived from the Greek *botruoeides,* for "grapes"); stibnite (Sb_2S_3) typically forms a cactuslike splay of delicate needles.

Highlight 2-1

Diamonds, Diamonds, Diamonds

Diamonds consist entirely of carbon, which is not a particularly attractive element in its other forms (such as graphite and charcoal, for example). *Why, then, are diamonds so beautiful? Why are they the most valuable of gemstones? Why are they by far the hardest natural substance?*

The answers to these questions lie not in the composition of diamond, but in the structure of the diamond crystal. In a diamond, the carbon atoms are arranged in a tight network in which each carbon atom is tightly bonded to four other nearby carbon atoms, forming a "tetrahedral" structure (a pyramid with four sides). Here are some facts and points of interest—related to the diamond's structure and its extraordinary properties—that students are always asking their geology professors.

What is a carat? A carat corresponds to 200 mg (1/5 of a gram), but this standard weight was first adopted only about 100 years ago. Before that, a carat was based on the weight of one seed from a locust-pod tree (called *kerat*ion in Greek). These seeds vary naturally weighing slightly more or slightly less than 200 mg.

▶ Diamond before and after cutting. **(a)** A raw, uncut diamond. **(b)** A cut and faceted diamond. Some very rare diamonds are found in nature in perfect "octahedral" shapes (two pyramids combined together base-to-base).

What is the largest diamond ever found? Although difficult to imagine, the largest diamond ever found was 3105 carats . . . about the size of a bowling ball. Found in South Africa in 1905, this massive diamond has been cut many times; its main gem, the Cullinan diamond (530 carats, or about the size of a tennis ball), sits today atop the royal scepter of the British crown jewels. During the summer of 2004, another spectacular diamond was found in Africa, this time in the west African forests of Guinea. The stone—182 carats (about the size of a computer mouse)—is worth millions and was immediately confiscated by the Guinean government from the miner who found it. It now resides in the vault of the country's central bank.

If diamond is so hard, how can it be cut into beautifully faceted gemstones? Although hard, diamonds are brittle ("stiff") and can be made to break along smooth planes that correspond to planes of atoms in the crystal structure. Gem cutters may not actually "cut" diamonds, but by using knowledge of the crystal structure of diamond, they can force diamonds to cleave along planes that contain relatively fewer bonds. Owing to the tetrahedral arrangement of the carbon atoms, diamonds can be cut along four main planes. The combination of these planes gives the gem cutter a vast number of options for shaping the cut stone. Diamonds may also be cut by ultrafine saw blades that have diamond dust embedded into their cutting edges

(a)

(b)

(a)

(b)

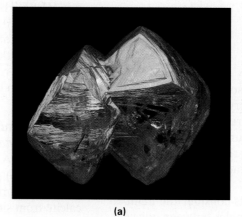

(c)

▲ **FIGURE 2-21** Unusual crystal aggregates. **(a)** Barite rosettes; **(b)** botryoidal malachite (green); **(c)** stibnite needles.

(what other mineral could cut diamond . . . other than diamond itself?). Although a diamond will become smaller during the cutting and polishing process, it will gain in value as it becomes an exquisitely faceted gem than can disperse light into brilliant flashes of color. (The polishing process again involves diamond dust—the only substance hard enough to abrade a diamond—mixed into a paste of pulverized industrial-quality diamonds with various types of oil, including olive oil.)

Have people always valued diamonds? The name "diamond" may have been derived from the Greek word *adamas,* which means "invincible," suggesting that the remarkable hardness of diamond has been known for thousands of years. Diamonds are mentioned in the Bible, in a fourth-century B.C. Indian text (that refers to rules for taxing diamonds), and in the works of various Roman authors. Alexander the Great worked hard to acquire large quantities of diamonds, and diamonds figure prominently in many myths and curses (see Highlight 2-4 about the Hope Diamond). For example, the Koh-i-Noor diamond, found in India and acquired by Queen Victoria in 1849, could only be worn by gods and women (never by men!). The Koh-i-Noor now resides safely in the Tower of London in England.

Where are diamonds found? Can they be found just about anywhere? Sorry to disappoint, but diamonds that are large enough to see (and wear) form in the Earth's mantle under continents and are then brought to the Earth's surface *only* by a rare type of very deep, superexplosive volcanic eruption. Thus the conditions for their appearance at the Earth's surface are rare . . . and thus so are the prospects for finding them. When diamond-bearing magmas are exposed at the surface, they can then be eroded, and the diamonds can be carried away from their source rocks by rivers and glaciers. Because diamonds are so hard and thus they resist erosion, they can travel enormous distances from their original exposures. Several large diamonds have been found in the farmlands of southern Canada and the Great Lakes region of the United States within the long-traveled glacial deposits of the region (see figure 17-17 for a map of these diamond finds). Geologists have searched for many years to locate the sources of these diamonds. In recent years, they have successfully identified diamond-bearing kimberlites in the far north of Canada (in the Northwest Territories, Nunavut, and northern Ontario).

If diamonds form deep in the Earth's interior, why don't they transform back to graphite at the Earth's surface? That is, how can diamonds be "forever" if they are not stable at the Earth's surface? As we mentioned earlier, diamonds are described as *meta*stable at the surface. This means that although they are not stable at low pressures and would "prefer" to transform into graphite (the "lead" in pencils), they require an activitation energy to begin to change to graphite and even then, they transform so slowly that on a human timescale, they are effectively stable. Diamonds are therefore not "forever," but the billions of years it would take diamonds to turn to graphite at the Earth's surface might as well be forever to us humans. So go ahead and give those diamond earrings as a gift—they won't be turning into graphite any time soon.

Have wars ever been fought over diamonds? Certainly battles have been fought over ownership of diamonds and diamond mines, and control over diamonds has caused considerable misery in various places around the world throughout history. In recent times, diamonds from parts of west Africa have been labeled "conflict diamonds" and many lives have been lost as rival factions seek to control these treasures. International laws have been established to make smuggling and illegal mining of diamonds more difficult, but the conflicts and misery continue.

Density A mineral's **density** is a measure of its weight for a given volume of the mineral, typically measured in grams per cubic centimeter. This property is a function of the elements in a mineral and how tightly packed the atoms in a mineral's structure. It can be very useful for distinguishing between otherwise similar-looking minerals. Gold and pyrite (fool's gold) are a good example of minerals that can be distinguished from each other by their densities. Even though they look similar (except for their streaks), gold has a density of 19.3 gm/cm^3 and pyrite, a density of 5 gm/cm^3. Gold's high density is the reason prospectors can "pan" for it. When they slosh gold-bearing river sand about in a pan, the heavier gold particles sink to the bottom, while lighter particles, usually of such common low-density minerals as quartz and feldspar, readily wash away.

How tightly packed the atoms are in a mineral also influences its density, even for minerals with the same composition. Quartz (SiO_2) has a density of 2.65 gm/cm^3, whereas its polymorph, stishovite, has a density of 4.28 gm/cm^3. The reason? Stishovite forms at very high pressures and has a more tightly packed structure. Its density is extremely high for a mineral composed exclusively of the lightweight elements, silicon and oxygen.

In the Laboratory

So many minerals have similar physical characteristics that even experienced geologists can only make an educated guess as to their identity. Often they must bring mineral samples back from the field to test them in the laboratory. There, they can use special equipment to analyze a variety of physical and chemical properties with greater precision than is ever possible in the field.

Laboratory Tests Although geologists commonly examine minerals and rocks in the field with a handheld lens, there are many important features that can only be seen with a *petrographic* microscope—a special microscope that has a light source at the bottom. The best way to study minerals under a microscope is to make a very thin slice of a rock, cutting it with a diamond saw to

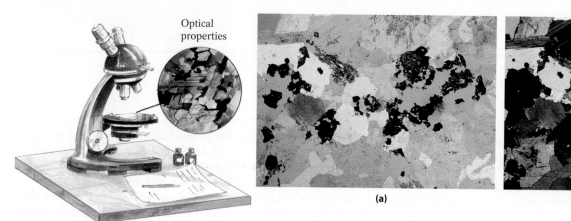

Optical properties

▲ **FIGURE 2-22** The colors that appear in a mineral sample while it is magnified under polarized light can be used to identify the mineral. **(a)** A thin section of granite as seen in polarized light (magnified 70×). **(b)** The same section as seen under cross-polarized light (with polarizers at 90° from each other). The rock is identified as granite by the types of minerals that appear and their abundances within the rock.

(a)

(b)

▲ **FIGURE 2-23** Minerals that glow in distinctive colors when exposed to ultraviolet light are said to be *fluorescent*. **(a)** The minerals willemite and calcite in this rock specimen **(b)** glow bright green and red, respectively, when exposed to UV light.

a thickness of 0.03 mm. In order to make and study such a "thin section," the rock is glued to a glass slide and then ground to the correct thickness. Except in the case of minerals that reflect light (such as metals), light will pass through these thin slices of minerals, and you can then look down the microscope and see textures and fine-grained minerals that couldn't be detected without high magnification and this type of light transmission. For example, you can only see the surface of a dark-colored mineral like a garnet when you look at a specimen in the lab or in an outcrop, but if you use a petrographic microscope, you can actually see inside the crystals (Fig. 2-22). Some of the photographs throughout this book will show you what rocks and minerals look like in thin section. In some cases, you will see a range of colors that help highlight the minerals; these colors result from passing polarized light (light that has passed through a filter and thus travels in only one plane) through minerals with different crystal structures, chemistries, and orientations in thin section. (The chemistry and internal structures of the mineral split the incoming light in ways that produce various diagnostic colors.) Geologists can identify most minerals by their display of unique colors and optical properties under polarized light conditions.

When exposed to ultraviolet light, certain minerals glow in distinctive colors. This property, called **fluorescence,** is common in the minerals fluorite (CaF_2), calcite, scheelite ($CaWO_4$), and willemite (Zn_2SiO_4) (Fig. 2-23). A mineral that continues to glow *after* the ultraviolet light has been removed is said to exhibit **phosphorescence.**

In **X-ray diffraction,** the X-rays passed through a mineral sample become scattered, or *diffracted,* into distinctive patterns. The arrangement of the atoms and ions in a mineral's crystal structure determines these patterns, which are unique to each mineral (Fig. 2-24). X-ray diffraction is a powerful tool for identifying minerals in the laboratory.

The Earth's Rock-Forming Minerals

Rock-forming minerals are those that make up most of the common rocks of the Earth's crust and mantle. Five groups of minerals predominate. Most are *silicates,* which contain silicon, oxygen, and usually one or a few other common elements. Im-

X-ray diffraction

Principles of X-ray diffraction
(interaction of X-rays with a mineral's crystal structure)

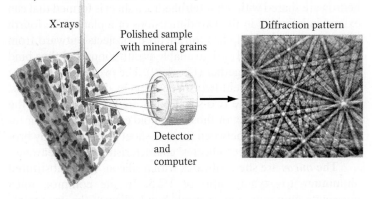

X-ray diffraction in an electron microscope

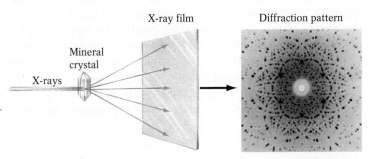

X-ray diffraction using photographic film

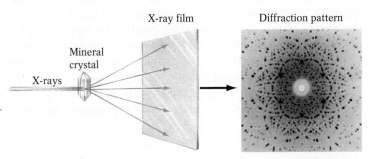

▲ **FIGURE 2-24** Mineral identification by X-ray diffraction. X-rays passed through a mineral scatter in a manner that is unique to each mineral. A *diffractogram* is the photographic record of the pattern of X-rays produced by a mineral.

portant nonsilicates include *carbonates,* containing carbon, oxygen, and other elements; *oxides,* containing oxygen and various metallic ions; *sulfates,* containing sulfur, oxygen, and various other metallic ions; and **sulfides,** containing sulfur and other metallic ions but no oxygen.

The Silicates and Their Structures

Pick up any rock—in a park, at the beach, from a riverbank—and the most common minerals in that rock will be the silicate minerals. Because silicon and oxygen are so abundant and unite so readily, **silicates** make up more than 90% of the weight of the Earth's crust (see Fig. 2-11). Silicates are the dominant component of all types of crustal rocks—igneous, sedimentary, or metamorphic.

But why oxygen and silicon? Oxygen alone makes up almost 50% of the Earth's crust by weight and roughly 90% by volume. It is also the only common crustal element whose atoms readily accept electrons to form negative ions. Silicon, the second most abundant element in the Earth's crust, is very compatible with oxygen: Its small, positively charged ion fits snugly in the niches among large, closely packed oxygen ions. The crystal structure of all silicates contains repeating groupings of four negative oxygen ions (that's how many can fit around a silicon ion) grouped around a single positive silicon ion to form a four-faced, pyramid-like structure called a *tetrahedron* (Fig. 2-25).

The four oxygen ions, each with a −2 charge, have a combined negative charge of −8; the one silicon ion has a +4 charge. The result is a **silicon–oxygen tetrahedron** (SiO_4^{4-}) with a −4 charge. Still, to form an electrically neutral mineral (remember, all minerals are electrically neutral), a silicon–oxygen tetrahedron must either acquire four positive charges or find another way to eliminate its negative charge. Positive ions of one or more other elements may bond with the tetrahedron to balance its charge, or adjacent tetrahedra may share oxygen ions, claiming half of their negative charge.

The number of oxygen ions shared with other silicate tetrahedra, expressed as a silicon-to-oxygen ratio, determines the type of silicate mineral and its crystal structure. A tetrahedron that does not share its oxygen ions at all has a silicon-to-oxygen ratio of 1:4, whereas a tetrahedron that shares all four of its oxygen ions has a silicon-to-oxygen ratio of 1:2.

The Earth's crust contains more than a thousand different silicate minerals, testimony to the ease with which SiO_4 tetrahedra combine in nature with various positive ions and with each other. There are five principal silicate crystal structures: *independent tetrahedra, single chains, double chains, sheets,* and *three-dimensional frameworks.* (Illustrations of these structures, which stretch across the next three pages, are summarized on pages 64–65). Each structure represents a different means of sharing oxygen ions, and each has its own silicon-to-oxygen ratio. Consequently, each displays distinctive physical characteristics and properties.

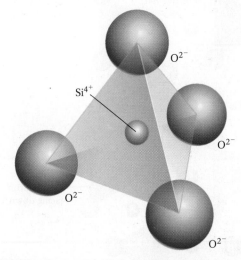

▲ **FIGURE 2-25** The silicon–oxygen tetrahedron. Four oxygen ions occupy the corners of this structure, with a lone silicon ion embedded in the open space at the center. (If you have access to four basketballs and a racquetball, you can easily recreate the SiO_4 tetrahedron.) The silicon–oxygen tetrahedron has an overall charge of −4.

NESO SILICATES

Independent Tetrahedra   Independent tetrahedra can bond with positive ions of other elements yet share *no* oxygen ions (Fig. 2-26). Olivine, for example, contains two positive iron and/or magnesium ions whose +2 charges balance the −4 charge of the tetrahedron. Silicates with independent tetrahedra are noted for their hardness (6.5–8 on the Mohs scale), a consequence of the strong ionic bonds between the tetrahedra and the interspersed positive ions. Independent tetrahedra always maintain a silicon-to-oxygen ratio of 1:4.

INOSILICATES (chain silicates)

Single Chains   A tetrahedron that shares two corner oxygen ions and still has an excess −2 charge forms a linear chain of tetrahedra (Fig. 2-27). The remaining negative charges cause the chain itself to act as a negative ion complex, attracting positive ions that neutralize its negative charge and bind adjacent chains together. Because the bonds *within* chains are strong and those *between* chains are relatively weak, single-chain silicates tend to cleave parallel to the chains—showing again how a mineral's internal crystal structure is reflected in its external physical properties. A prominent group of single-chain silicates are the *pyroxenes*. These silicates typically contain positive calcium, iron, and magnesium ions that bind the negatively charged chains together. This pattern of oxygen-sharing produces a silicon-to-oxygen ratio of 1:3.

Double Chains   A double chain forms when adjacent tetrahedra share two corner oxygens in a linear chain, and, in addition, some share a third oxygen with a tetrahedron in a neighboring chain (Fig. 2-28). This shared oxygen binds the two chains together. Because they share more oxygen than do single chains, double chains have a silicon-to-oxygen ratio of 1:2.75. Any of several positive ions (commonly Na, Ca, Mg, Fe, and Al) may link double chains to adjacent double chains, producing the complex silicates known as the *amphiboles*. The most common amphibole is *hornblende*, in which aluminum substitutes for some

of the silicon within the tetrahedron because their ions are similar in size and charge.

Hornblende's relatively high iron and/or magnesium content gives it a dark green-to-black color, similar to that of the common pyroxene augite. These two similar-looking types of silicates, however, have different crystal structures, distinctive internal planes of weakness, and readily distinguishable cleavage angles (pyroxenes have clevage angles of about 90°; amphiboles have cleavage angles of 60° and 120°). The difference lies in the internal planes of weakness created by the single chains versus double chains (Fig. 2-29).

PHYLLO SILICATES

Sheet Silicates   When all three oxygens at the base of a tetrahedron are shared with other tetrahedra, a sheet is formed that can extend indefinitely in the two dimensions of a plane. The fourth oxygen, at the peak of each tetrahedron, projects outward from the sheet, free to bond with available positive ions and thus bind two adjacent sheets together (Fig. 2-30). The two-sheet layers are in turn bound to adjacent two-sheet layers by other positive ions. The bonds *between* the two-sheet pairs are far weaker than the bonds *within* each sheet or those that hold each two-sheet pair together. These bonds (between the two-sheet pairs) are easily broken, producing the sheet silicates' characteristic planar cleavage.

The *micas* are sheet silicates with a silicon- (and substituted aluminum-)to-oxygen ratio of 1:2.5. In the common mica *muscovite,* aluminum ions securely bond adjacent sheets of tetrahedra to form a two-sheet pair, but the two-sheet pairs are weakly bonded by intervening potassium ions. These weak bonds readily break when we attempt to peel the layers of muscovite, causing sheets of this mica to separate easily. The layers of other micas, such as black biotite mica, also peel in this same fashion.

To visualize the structure of a sheet silicate like that of muscovite, consider a stack of peanut-butter sandwiches separated by pieces of wax paper. Each sandwich is analogous to a two-sheet layer of mica; the peanut butter represents the strong

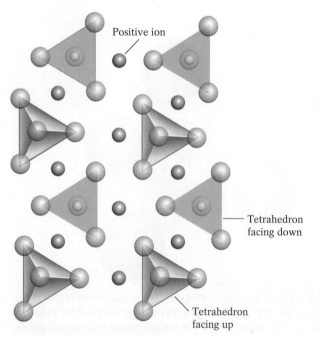

Positive ion

Tetrahedron facing down

Tetrahedron facing up

Independent tetrahedra

◀ **FIGURE 2-26** Silicate structures—independent tetrahedra. Positive ions are positioned between tetrahedra such that each tetrahedron, with a −4 charge bonds to two positive ions, each with a +2 charge, thereby neutralizing their combined charge. No oxygen ions are shared between the tetrahedra; therefore the silicon-to-oxygen ratio is 1:4. Photo: Olivine. As is typical of silicates with independent tetrahedral structures, this mineral fractures rather than cleaving.

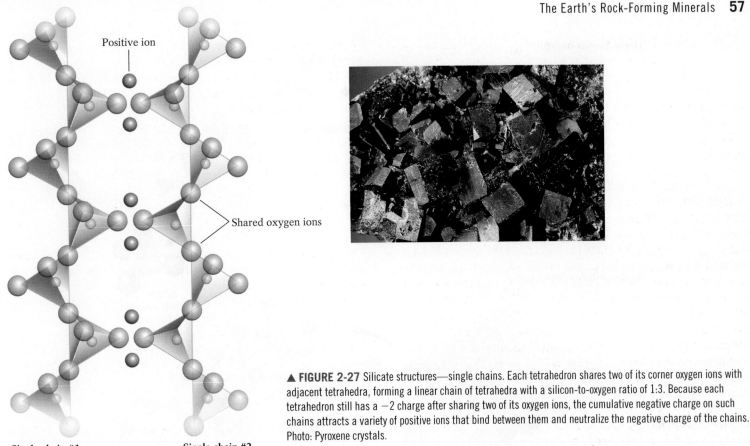

Positive ion

Shared oxygen ions

Single chain #1   Single chain #2

Single chains

▲ **FIGURE 2-27** Silicate structures—single chains. Each tetrahedron shares two of its corner oxygen ions with adjacent tetrahedra, forming a linear chain of tetrahedra with a silicon-to-oxygen ratio of 1:3. Because each tetrahedron still has a −2 charge after sharing two of its oxygen ions, the cumulative negative charge on such chains attracts a variety of positive ions that bind between them and neutralize the negative charge of the chains. Photo: Pyroxene crystals.

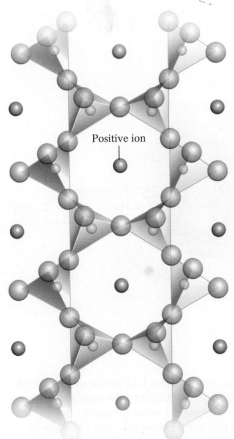

Positive ion

Double chains

▲ **FIGURE 2-28** Silicate structures—double chains. Each tetrahedron shares two of its corner oxygens with adjacent tetrahedra, forming a linear chain, and in addition some of the tetrahedra share a third oxygen with tetrahedra in an adjacent chain, thus joining the chains together. Positive ions are interspersed between and within the double chains. The silicon-to-oxygen ratio of double chains is 1:2.75. Photo: Hornblende (an amphibole).

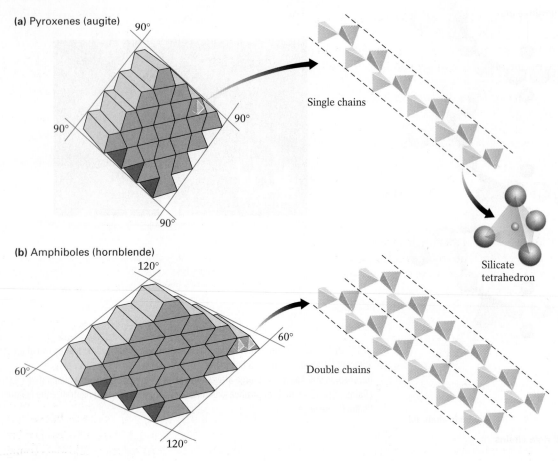

(a) Pyroxenes (augite)

90° 90° 90° 90°

Single chains

Silicate
tetrahedron

(b) Amphiboles (hornblende)

120° 60° 60° 120°

Double chains

▲ **FIGURE 2-29** Differing cleavage angles in pyroxenes (single chains) and amphiboles (double chains).

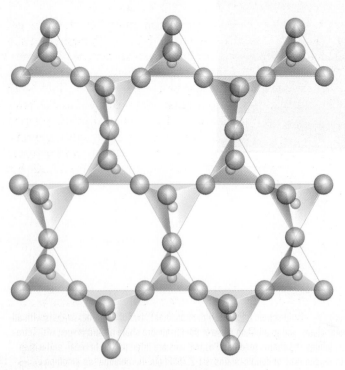

Sheet silicates

▲ **FIGURE 2-30** Silicate structures—sheet silicates. Each tetrahedron shares all three of its corner oxygens, forming a sheet of adjoined tetrahedra. The fourth oxygen in each tetrahedron extends upward to bond with positive ions and, subsequently, another sheet. (Other positive ions bind two-sheet pairs to adjacent two-sheet pairs.) The silicon-to-oxygen ratio of sheet silicates is 1:2.5. Photo: Muscovite mica, showing the planar cleavage of the sheet silicates.

aluminum bonds between the sheets, and the wax paper represents the weak potassium bonds between the two-sheet layers. If you pick up the top sandwich by grasping its upper slice of bread, the whole sandwich, bound strongly by the peanut butter, remains intact. However, each sandwich separates readily from the one below. The same is true for each two-sheet pair in a sheet-silicate mineral such as muscovite or biotite mica.

Another variety of sheet silicates is the clay minerals. These practical minerals generally form from the chemical breakdown of the feldspars in the Earth's surface environment. Highlight 2-2 illustrates how clay minerals may affect us daily, in both positive and negative ways. You will learn much more about these important sheet silicates in Chapter 5.

TECTOSILICATES

Framework Silicates   When a tetrahedron shares all four of its oxygen ions with adjacent tetrahedra, a structure with a three-dimensional framework results. This highest degree of oxygen-sharing results in the lowest silicon-to-oxygen ratio, 1:2. This is the structure of the two most abundant minerals in the Earth's continental crust—quartz and feldspar.

Quartz (SiO_2) and its various polymorphs are the second most abundant group of minerals in continental crust. They are also the only silicates composed entirely of silicon and oxygen. In quartz, every oxygen ion in a tetrahedron is shared with an adjacent tetrahedron. Thus quartz achieves electrical neutrality without the need for other ions (Fig. 2-31). Its strong bonds (which include both covalent and ionic bonds) make quartz one of the hardest of the common rock-forming minerals (7 on the Mohs scale). Because all of its bonds are equally strong, quartz has no weak internal planes and therefore does not cleave. Instead, it breaks by fracturing.

Pure quartz is transparent and colorless, due to the absence of other ions that would absorb light. Because of its open structure, however, quartz often traps a few stray ions when it crystallizes. These impurities give quartz a wonderful range of colors, as is seen in the gemstones amethyst, rose quartz, and tigereye. When quartz—which typically crystallizes from molten rock—has room to grow undisturbed, it assumes its characteristic crystal shape—long prisms with pyramid-like faces on top and bottom (Fig. 2-31).

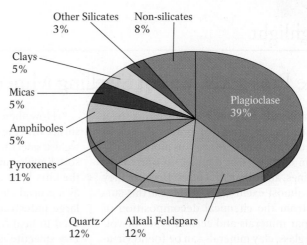

▲ **FIGURE 2-32** The relative percentages of minerals in the Earth's crust. Note that the silicates comprise roughly 92% of all crustal minerals, with the framework minerals (feldspars and quartz) dominating (>60%).

The *feldspars,* which account for about 50% by volume of continental crust, also crystallize from molten rock, across a wide range of temperature and pressure. In feldspars, as in quartz, every oxygen ion in a tetrahedron is shared with an adjacent tetrahedron. Unlike in quartz, however, atoms of aluminum regularly replace some of the silicon atoms in the feldspars, and potassium, sodium, and calcium ions occupy open spaces between the tetrahedra.

There are two types of feldspars, classified based on their chemical composition. The *plagioclase* feldspars contain varying proportions of sodium and calcium ions, which substitute for one another. The all-sodium plagioclase feldspar *albite* is at one end of the range, and the all-calcium plagioclase feldspar *anorthite* is at the other. The *alkali* feldspars are rich in potassium; the large size of the potassium ion produces a slightly different internal structure. All feldspars have two prominent cleavage planes that intersect at nearly 90° angles. They are relatively hard, about 6 on the Mohs scale.

Framework silicates dominate continental crust; chain and sheet silicates are also common (Fig. 2-32). Although other silicates and nonsilicates make up but a small proportion of the crust, this does not mean that they are unimportant. Despite their low abundances, these minerals may be economically precious (such as diamonds, platinum, gold) or may contain a lot of useful information about the geological processes that formed them (such as carbonates, garnets, olivine). The following section describes some of the common and most valuable *non*silicate minerals.

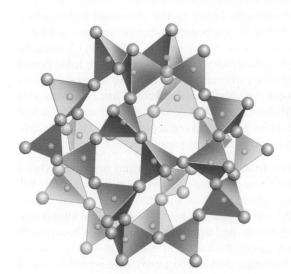

Framework silicates

◀ **FIGURE 2-31** Silicate structures—framework silicates. Each tetrahedron shares all four of its oxygens, forming a three-dimensional framework structure. Because the charge on each tetrahedron is neutralized by the sharing of all its negative charges, no positive ions bond with the structure. The silicon-to-oxygen ratio of framework silicates is 1:2. Photo: Quartz crystals.

Highlight 2-2

The Pros and Cons of Swelling Minerals

Clay minerals, a common variety of the sheet silicates, have extensive practical uses in everyday life. They also cost us more than $3 billion each year in damage to our roads, buildings, and other structures. Because they form almost exclusively at the Earth's surface from the chemical decomposition of feldspar minerals and certain types of volcanic glass, clay minerals can be found virtually everywhere at the surface. They blanket the seafloor, make up much of our soils, are a major component of most sedimentary rocks, and show up in a vast array of everyday human-made products. The crystal structure of the clay minerals is relatively simple—they are essentially stacks of millions of microscopic, negatively charged sheets of silicon–oxygen and aluminum–oxygen tetrahedra held together by interspersed positive ions, such as sodium and calcium.

One variety of the clay minerals forms typically from chemical attack on volcanic ash by water. This mineral, called *smectite*, stands out from the other clays—or rather it *swells up* compared to other clays—because its crystal structure can absorb a large volume of water. It does so because the negative charges on the surfaces of the tetrahedral sheets attract the positive sides (the two-hydrogen side) of the water molecules. The result? Smectite can expand by as much as eight times its dry volume when wet.

How, then, does this ability to swell (and shrink with the loss of water) determine how we use smectite clay? Given its ability to absorb liquids, smectite is a prime ingredient in the kitty litter in your cat's litter box as well as a crucial cleansing agent in the event of large industrial spills. It is also commonly used to heal fractures in rock or concrete. Dry smectite is mixed with water and then injected into the fracture, where it subsequently swells and seals the crack.

Smectite is also used extensively as a lubricant, especially in the oil-well industry. Vast quantities of *bentonite*, a clayey mixture of mostly smectite and small traces of residual feldspar and quartz, are mined throughout the Black Hills of northeastern Wyoming and southwestern South Dakota. These clays record past explosive volcanism in western Montana that blanketed the region with volcanic ash between 70 and 20 million years ago. The dry bentonite, when mixed with water, becomes extraordinarily slimy as tiny water-soaked silicate sheets slide past neighboring sheets (see photo at right). In the past, westward-traveling pioneers used Wyoming bentonite to quiet their squeaky wagon wheels. More recently, oil-well drillers have lubricated their drill bits with a constant slurry of bentonite mud. The swelling smectite in the bentonite lines the drill hole, expands, and seals the hole, preventing cave-ins and blowouts.

Strangely enough, swelling smectite is even used in foods—primarily for farm animals but also occasionally for the rest of us. Cattle ranchers spike their cattle feed with this swelling clay because it's cheap and filling. It also makes a reasonable filler for human food. Those thick, creamy "shakes" at certain popular fast-food chains (they can't call them "milkshakes" because they contain no milk) owe their unusual creaminess to smectite. (Ever try to suck that stuff up through a straw?)

On the negative side, structures built directly atop swelling soils may be damaged or destroyed by the upward push of expanding smectite. As much as 50,000 kg/m^2 of upward force may be generated as smectite swells, enough to lift and destroy a substantial building. Although not as frightening as a tornado, hurricane, or earthquake, ground movements from swelling soils are one of our most costly natural hazards. Every year roughly 250,000 new homes are built on swelling soils. Sixty percent will suffer minor structural damage (such as the cracked wall on the facing page), while 10% will suffer a total loss. Soils such as these should be mapped by geologists and avoided by builders. Subsurface drainage structures and well-constructed storm-water gutters help reduce the threat to buildings located on swelling soils by limiting the amount of water that enters them beneath a building.

Nonsilicates

Because the common rock-forming silicates are so abundant, no one gets very excited about the discovery of a fresh supply of plagioclase feldspar—we've got plenty already. But nonsilicate minerals constitute only about 8% of the Earth's continental crust. They include native elements—those that are not combined in nature with other elements—such as the precious metals gold, silver, and platinum. Nonsilicates also include such useful metals as iron, aluminum, copper, nickel, zinc, lead, and tin, in various compounds. They also include a few ever-popular gems, such as diamonds, rubies, and sapphires.

Nonsilicates are constructed from a variety of negative-ion groups, most containing oxygen. The most common nonsilicates are the carbonates, oxides, and sulfates, as well as a few groups without oxygen, such as the sulfides (Table 2.2).

Carbonates   The **carbonate**-ion complex (CO_3^{2}) contains one central carbon atom with strong covalent bonds to three neighboring oxygen atoms. Because it has an overall negative charge, it typically combines with one or more positive ions to form the carbonate minerals. When it combines with calcium, it forms *calcite* ($CaCO_3$), commonly found in the abundant sedimentary rock, *limestone*. When the carbonate complex combines with calcium *and* magnesium, it forms *dolomite* [$CaMg(CO_3)_2$], found in another common sedimentary rock, *dolostone*. The ionic bonds between the negative carbonate-ion group and the positive ions are relatively weak, so calcite and dolomite are relatively soft (3 to 4 on the Mohs scale). They also dissolve readily in acidic water, with important geological consequences, such as the creation of most of the world's caves (discussed in Chapter 16).

Oxides   Mineral **oxides** are produced when negative oxygen ions combine with one or more positive metallic ions. The resulting minerals include some of our major sources of iron—hematite (Fe_2O_3) and magnetite (Fe_3O_4)—as well as aluminum, tin, titanium, chromium, and uranium. Hematite has been much in the news lately (along with several other non-silicate minerals) because it has appeared in the newly discovered rocks of Mars. Spend a moment with the following Highlight on page 64 to learn how these minerals *may* indicate the presence of surface

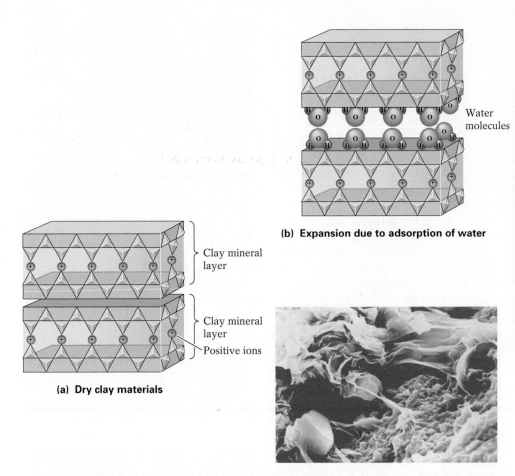

(b) **Expansion due to adsorption of water**

Water molecules

Clay mineral layer

Clay mineral layer

Positive ions

(a) **Dry clay materials**

▲ When water molecules enter the spaces between the layers within clay minerals, the positively charged (hydrogen) sides of the molecules bond with the negatively charged surfaces of the clay layers. The addition of water to the clay-mineral structure causes it to expand in volume. The crack in the building's wall is produced by swelling clays.

water in the Martian past and hint at the possibility of past life there as well.

Sulfides and Sulfates With six electrons in its outer energy level, sulfur can either accept or donate electrons. When sulfur accepts electrons, it becomes a negative ion that bonds with various positive ions to form the sulfides. This valuable group of minerals, including copper sulfides such as chalcocite (Cu_2S) and lead sulfides such as galena (PbS), is the source of a number of important metals.

When sulfur donates electrons, it becomes a positive ion that bonds with oxygen to form the negative **sulfate**-ion complex (SO_4^{2-}). *Gypsum* ($CaSO_4 \cdot 2\,H_2O$), a common and useful sulfate mineral, is used to manufacture sheetrock and plaster of Paris, staples in the building-construction trade.

Finding and Naming Minerals

Between 30 and 60 new minerals are discovered every year. Most are so small they can hardly be seen without the aid of a microscope. Most are very rare and contain exotic elements such as niobium, yttrium, and germanium, so it is highly unlikely that you'll encounter these minerals in your geology class or on a hike. Most of the major rock-forming minerals were identified and named hundreds or even thousands of years ago. (The Roman historian Pliny named the shiny silvery mineral galena (PbS_2) in 77 B.C.) Once characterized mineralogically and chemically, each new mineral joins the encyclopedia of more than 4000 known minerals. (Don't panic—even the toughest geology professor out there won't make you memorize all 4000 mineral names . . . probably no more than 20 or 30 or so.)

In order for a new mineral to be accepted by the International Mineralogical Association (the official organization responsible for classifying and keeping track of mineral names and mineral properties), the newly discovered mineral must fit the strict definition of a mineral (a solid, inorganic substance with a definite crystal structure and composition) and must have formed by a natural process on Earth (without any help from humans) *or in space*.

To propose a new mineral, the discoverer (perhaps you) must fill out a very extensive application that includes the following information:

Table 2.2 Common Nonsilicate Minerals

MINERAL TYPE	COMPOSITION	EXAMPLES	USES
Carbonates	Metallic ion(s) plus carbonate ion complex (CO_3^{2-})	Calcite ($CaCO_3$)	Cement
		Dolomite ($CaMg(CO_3)_2$)	Cement
Oxides	Metallic ion(s) plus oxygen ion (O^{2-})	Hematite (Fe_2O_3)	Iron ore
		Magnetite (Fe_3O_4)	Iron ore
		Corundum (Al_2O_3)	Gems, abrasives
		Cassiterite (SnO_2)	Tin ore
		Rutile (TiO_2)	Titanium ore
		Ilmenite ($FeTiO_3$)	Titanium ore
		Uraninite (UO_2)	Uranium ore
Hydroxides	Metallic ion(s) plus hydroxide ion (OH^-)	Goethite ($FeO(OH)$)	Iron ore, Aluminum ore
		Bauxite ($(Al, Fe)O(OH)$)	
Sulfides	Metallic ion(s) plus sulfur (S^{2-})	Galena (PbS)	Lead ore
		Pyrite (FeS_2)	Sulfur ore
		Cinnabar (HgS)	Mercury ore
		Sphalerite (ZnS)	Zinc ore
		Molybdenite (MoS_2)	Molybdenum ore
		Chalcopyrite ($CuFeS_2$)	Copper ore
Sulfates	Metallic ion(s) plus sulfate ion (SO_4^{2-})	Gypsum ($CaSO_4 \cdot 2\,H_2O$)	Plaster
		Anhydrite ($CaSO_4$)	Plaster
		Barite ($BaSO_4$)	Drilling mud
Native elements	Minerals consisting of a single element	Gold (Au)	Jewelry, coins, electronics
		Silver (Ag)	Jewelry, coins, photography
		Platinum (Pt)	Jewelry, catalyst for gasoline production
		Diamond (C)	Jewelry, drill bits, cutting tools

- Proposed name of the mineral
- Geographic and geologic occurrence (exactly where did you find it?)
- Chemical composition (what's the chemical formula?)
- Crystal structure (exact atomic arrangement, including location of the elements within the crystal)
- Physical properties (including hardness, cleavage, density, color, streak, etc.)
- Other data (optical properties, magnetism, radioactivity, etc.)

Obtaining all those data involves a lot of careful work, but choosing a new name also requires much thought. A new mineral name must be distinct from existing mineral names, and you cannot name a mineral after yourself! (Sorry.) If you want to honor a favorite friend, relative, pet, or professor—but they have a common name (e.g., Smith, Johnson)—you can include both their first and last names in the new mineral name. For example, in 1977, Dr. James Thompson, a professor of Geology at Harvard University, was honored by the naming of the mineral jimthompsonite [$(Mg,Fe)_5Si_6O_{16}(OH)_2$], a soft, silky, colorless mineral found in the mountains of Vermont and produced by weathering of certain metamorphic rocks. You may also be interested to know that a mineral cannot be named after you without your permission, and the association specifies in its mineral-naming rules that minerals should not be named after a political figure, a star athlete, or an artist (including movie stars, famous painters, poets, musicians, etc.) So you won't find any bradpittite, halleberryite, lebronjamesite, or jay-z-

ite. This rule, however, has not been strictly adhered to in the past; there is a mineral—nyerereite [$Na_2Ca(CO_3)_2$], a sodium carbonate associated with the volcanoes of the East African rift zone—named in honor of the former president of Tanzania, Julius Nyerere.

The International Mineralogical Association prefers mineral names that are informative and descriptive with regard to a physical property, chemical composition, or geographic location of the mineral. Kyanite (Al_2SiO_5) is blue, and the name is derived from the ancient Greek word *kyanos* for "blue"); tunisite was discovered in the northern African country, Tunisia; and the lunar mineral *armalcolite* honors the Moon's first explorers, Neil *Arm*strong, Edwin *Al*drin, and Michael *Col*lins.

The association also strongly discourages mineral names that are too difficult to pronounce ... but that hasn't stopped the naming of such minerals as yugawaralite, ammoniojarosite, and anabohitsite. Why do some minerals get such tongue-twisting names? Why not simply refer to them in a systematic way based on the elements that form them, such as "sodium chloride" for halite (NaCl)? The main reason—some minerals' chemical compositions are so complex that a name based on their formulas would not be possible. Consider pumpellyite, a mineral found in rocks that have been slightly heated and compressed during certain types of mountain-building events—its chemical formula is $Ca_4(Mg,Fe^{2+})(Al,Fe^{3+})_5O\ (OH)_3(Si_2O_7)_2(SiO_4)_2 \cdot 2\,H_2O$. Try referring to that one by its chemical formula. Finally, many minerals were known by common names for centuries before their chemical compositions were identified, and today few of us would choose to say, "Please pass the sodium chloride" (that is, "salt") at the dinner table.

Highlight 2-3

Martian Minerals, Water, and Life

The recent images and data obtained in 2004 by the Mars Exploration rovers (*Spirit* and *Opportunity*–check out the mission website—*marsrovers.jpl.nasa.gov/home*) and other missions to Mars (*Global Surveyor, Odyssey, Viking*) have ignited debates about Mars: Did Mars once have oceans and rivers and lakes? If so, were there living creatures in the water and sediments? How much of the Martian landscape was sculpted by flowing water? Some features on the surface of Mars (canyons, channels) suggest that the landscape was shaped in part by flowing water, but what about the rocks and minerals at the planet's surface: Can we learn anything about the possibility of life on Mars from its minerals?

"Gray hematite is a mineral indicator of past water." (Joy Crisp, project scientist for the Rover mission)

The Rover mission landed purposefully in regions of Mars believed to be rich in iron oxides, in particular a form called specular (or gray) hematite (Fe_2O_3). Is gray hematite really present, and if so, does it prove that Mars had lakes and rivers at its surface?

To examine this hypothesis, we can consider the different places where hematite forms on Earth. Hematite can take several forms. Specular hematite is a shiny gray mineral, but hematite can also be rusty and deep red (the name *hematite* comes from the Greek word, *haemo*, for "blood"). Very fine-grained hematite particles or similar minerals (such as iron hydroxides) give Mars its reddish appearance.

■ *Lava or Water?* But, if there was no water in the Martian past, the hematite could have crystallized from lava or volcanic ash, or formed when iron oxides were heated in the Martian crust (a metamorphic process).

Although the landforms and minerals could have been formed by water or volcanic eruptions, many scientists agree that the first one or two billion years of Martian history were marked by a warmer, wetter environment than Mars' current cold, dry conditions. The mystery concerns how much water there was back then, when it disappeared, and whether a more moderate climate in the distant past created conditions for life to exist on Mars.

If the presence of the gray crystalline hematite in Martian rocks does indicate the past occurrence of water, there are at least three possibilities for the form taken by the water:

■ It existed as a body of water (such as a lake) containing so much iron that the iron precipitated and was left behind when the water dried up.

■ Iron-rich water trickled through the ground, depositing iron between the sediment grains.

■ The hematite formed during the weathering of rocks at or near the surface of Mars, perhaps in the presence of a small amount of liquid water.

To test ideas about the origin of the Martian hematite, scientists are examining the textures of the minerals—that is, their arrangements, shapes, and sizes. Is the hematite in layers (suggesting precipitation from water) or scattered about randomly in the rocks (suggesting trickling water and deposition between sediment grains)? Does the hematite occur as actual individual grains (suggesting crystallization from magma) or does it coat other minerals (suggesting the action of circulating water)? Are there other features or minerals that might help us distinguish hematite associated with volcanic activity or with the action of free water?

The presence of hematite alone is not enough to prove the former presence of water, but other recent findings—the presence of salt, high levels of sulfur (including *hydrous* iron sulfate that forms in the presence of water), and small cavities and spherical concretions (rounded, concentrically layered accumulations of mineral matter)—are other indicators of past Martian waters. When these factors are detected together in some combination, they strengthen the case for the past presence of water on Mars.

We should keep in mind that the rovers exploring Mars are only examining a tiny portion of the planet's surface. Imagine exploring a small corner park in your neighborhood and applying what you learn about its geology to the rest of the Earth! What if the Martian rovers' landing sites, selected because Earth-bound scientists suspected they contained gray hematite, are atypical of the Martian surface? Imagine trying to understand the geology of the entire Earth by studying only the rock outcrops in Central Park in New York City.

Eventually, we will have explored enough of the surface of Mars to understand better its geologic history. As you read reports of new findings, keep an open mind and think critically about the scientific interpretations. If someone claims that "mineral X proves that Mars had enormous lakes," ask yourself, "Why does it prove that? Are there other environments in which mineral X forms? Now that you are studying geology, you can answer some of these questions on your own from what you are learning this semester in your geology class.

▲ Photo of Martian landscape taken by the rover, Spirit, on Jan. 10, 2004. The view is from the landing site at Gusev Crater.

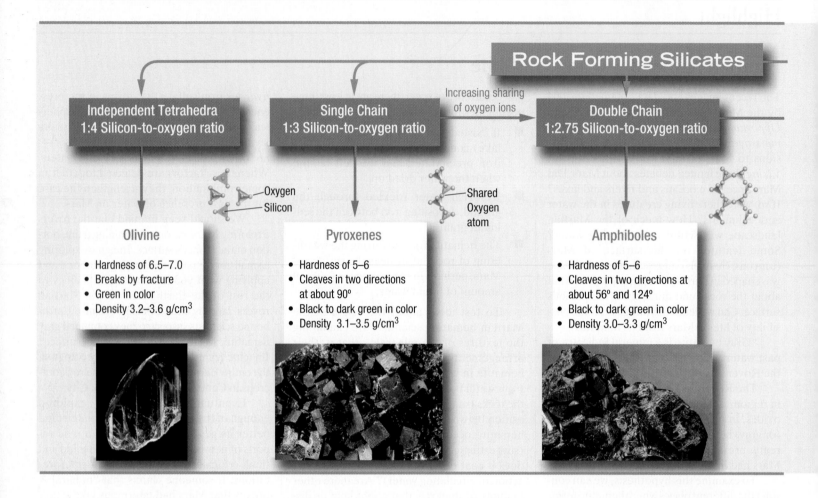

Rock Forming Silicates

Independent Tetrahedra
1:4 Silicon-to-oxygen ratio

Single Chain
1:3 Silicon-to-oxygen ratio

Increasing sharing
of oxygen ions

Double Chain
1:2.75 Silicon-to-oxygen ratio

— Oxygen
— Silicon

Shared
Oxygen
atom

Olivine

- Hardness of 6.5–7.0
- Breaks by fracture
- Green in color
- Density 3.2–3.6 g/cm^3

Pyroxenes

- Hardness of 5–6
- Cleaves in two directions
  at about 90°
- Black to dark green in color
- Density 3.1–3.5 g/cm^3

Amphiboles

- Hardness of 5–6
- Cleaves in two directions at
  about 56° and 124°
- Black to dark green in color
- Density 3.0–3.3 g/cm^3

Gemstones—The Earth's Most Valuable Minerals

Several minerals lead a glamorous life as *gemstones*—precious or semiprecious minerals that display particularly appealing color, luster, or crystal form and can be cut or polished into jewelry or other ornaments. Some gems, such as diamonds and emeralds, are quite rare. Others may be unusually perfect crystals of common rock-forming silicates. *Amethyst* is a variety of quartz whose appealing purple tones are produced by a small number of iron atoms scattered throughout the crystal structure. *Opal* is a mineraloid made of simple silica and water. A modern-day "gold rush" of opal prospectors is under way in south-central Wyoming where a new find of this precious gem has triggered hundreds of land claims in Fremont County, about 150 km (100 mi) west of Casper. (Multicolored, gem-quality opal may be more valuable than diamond, sometimes fetching up to $10,000 per carat.) The valued gemstone *amazonite* (KAlSi$_3$O$_8$) is a variety of the abundant alkali feldspar *microcline* that owes its vivid blue-green color to the substitution of a few lead ions for potassium ions in its crystal structure. Some com-

mon nonsilicates also can be gems. The aluminum oxide *corundum*, with its hardness of 9, is a popular abrasive used in emery cloth and sandpaper. When its crystals are perfectly formed, and derive color from a few other ions trapped in their structures, ordinary corundum becomes not-so-ordinary *sapphires* and *rubies* (Fig. 2-33).

How Gemstones Form

Minerals of gemstone quality form under conditions that promote the development of perfect, large crystals. This happens most often in two ways: when molten rock cools and crystallizes underground, or when preexisting rock is subjected to very high pressure and heat.

(a)

(b)

▲ **FIGURE 2-33** The minerals corundum, sapphire, and ruby each have the chemical formula Al$_2$O$_3$. Their different colors are produced by trace impurities in their crystal structure. **(a)** Sapphire **(b)** Ruby

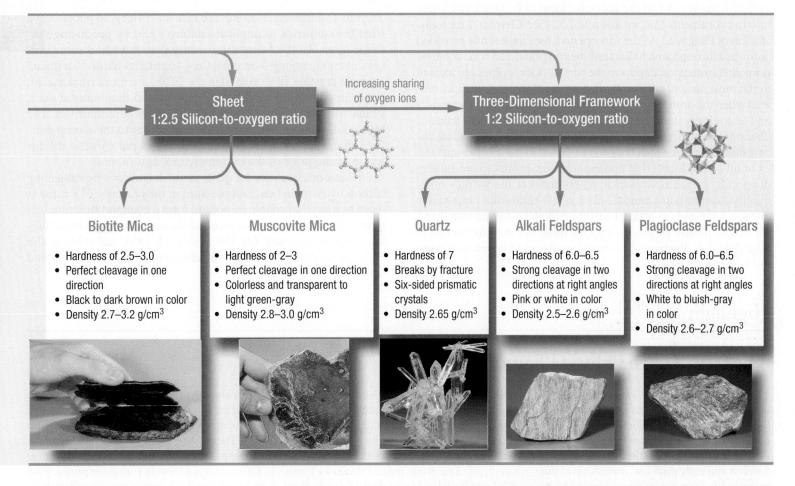

| Sheet 1:2.5 Silicon-to-oxygen ratio | | Increasing sharing of oxygen ions → | Three-Dimensional Framework 1:2 Silicon-to-oxygen ratio | |

Biotite Mica	Muscovite Mica	Quartz	Alkali Feldspars	Plagioclase Feldspars
• Hardness of 2.5–3.0 • Perfect cleavage in one direction • Black to dark brown in color • Density 2.7–3.2 g/cm³	• Hardness of 2–3 • Perfect cleavage in one direction • Colorless and transparent to light green-gray • Density 2.8–3.0 g/cm³	• Hardness of 7 • Breaks by fracture • Six-sided prismatic crystals • Density 2.65 g/cm³	• Hardness of 6.0–6.5 • Strong cleavage in two directions at right angles • Pink or white in color • Density 2.5–2.6 g/cm³	• Hardness of 6.0–6.5 • Strong cleavage in two directions at right angles • White to bluish-gray in color • Density 2.6–2.7 g/cm³

Molten rock often migrates into fractures in surrounding cooler rocks. There its ions and atoms crystallize in reasonably large spaces, and are thus free to produce perfect crystals and, if the space is big enough, enormous crystals (Fig. 2-34). A single pyroxene crystal excavated in South Dakota was more than 12 meters (40 feet) long and 2 meters (6.5 feet) wide; it weighed more than 8000 kilograms (8 tons). Along with these massive crystals that form from molten rock, other gems may crystallize from hot fluids containing large ions that don't fit easily into common mineral structures. The fluids may crystallize silicate gemstones such as topaz, tourmaline, and beryl.

Gemstones may also form when the heat, pressure, and circulating hot fluids at the edges of colliding plates cause the ions and atoms in their rocks to migrate and recombine. The resulting new minerals are more stable under the new conditions. For example, heating and compression of carbonate rocks that contain aluminum ions can cause the aluminum to combine with oxygen liberated from calcite. We know some of these newly formed aluminum oxides as rubies and sapphires.

And then there is that rare stone, the diamond, which is transformed from unspectacular carbon into a brilliant crystal by extremely high pressures at great depths (greater than 150 km, or 90 mi). Diamonds most often occur where hot gas has propelled fluidized rock rapidly and explosively from great depth to the surface, carrying deep-forming diamonds along with it. The resulting diamond-rich structures, known as *kimberlite pipes* (named for Kimberley, South Africa), are typically a few hundred meters to a kilometer across and many kilometers deep. Diamond-bearing kimberlite pipes are found in Siberia, India, Australia, Brazil, the Northwest Territories of Canada, southern and central Africa, and scattered throughout the Rockies of Colorado and Wyoming.

▲ **FIGURE 2-34** Recently discovered gigantic quartz crystals in a mine in southern Chihuahua, Mexico.

Many kimberlite pipes apparently formed between 70 and 150 million years ago, when the supercontinent of Pangaea was rifting into the continents that we know today. (See Chapter 1 for a discussion of Pangaea.) As the rifts opened, they apparently provided pathways through which diamond-bearing fluidized rock could be propelled from great depths to the surface. Other pipes are located in the most ancient rocks of each continent's central nucleus, created when the hotter early Earth was far more dynamic volcanically and tectonically than it is today. The scarcity and thus the value of diamonds is in large part related to the rare set of geological circumstances under which they were blasted to the Earth's surface.

Only a few kimberlite bodies, however, produce gem-quality diamonds. In southern Africa, where most of the world's high-quality diamonds are mined, only 1 in 200 kimberlite pipes yields enough diamonds to make it worth the high cost of mining them. The story of one of South Africa's most famous diamonds, in Highlight 2-4, illustrates the mystique this most coveted of gems holds for many of us.

Synthetic Gems—Can We Imitate Nature?

The most valuable gemstones are rare and pricey, so people have tried for centuries to duplicate nature's feat by producing synthetic gems. In the twentieth century they have had some success, even managing—on occasion—to surpass nature. Artificial emerald crystals, first created in the 1930s, are more transparent, richer in color, and more exquisitely shaped than natural ones, which are often marred by gas bubbles and other impurities. The high quality of synthetic emeralds contributes to their great market value, which, at several hundred dollars per carat, is still far less than the price of the extremely rare natural ones.

Diamonds can now be made in the laboratory by subjecting carbon to extreme heat and pressure in the presence of a catalyst (such as iron) that enable carbons to form a diamond structure. Almost any carbon-rich substance will do as a starting point, including sugar or peanuts. On December 12, 1954, scientists in the General Electric research lab in Schenectady, New York, created

Highlight 2-4

The Curse of the Hope Diamond

The exquisite 44.5-carat Hope Diamond is the world's largest blue diamond and one of the most popular attractions in the National Museum of Natural History in Washington, D.C. (see photo). Yet, although it is virtually flawless and truly priceless, ownership of this remarkable gem has at times proved to be somewhat less than a stroke of great fortune.

The Hope Diamond is believed to be a remnant of a 112-carat Indian diamond stolen from a statue of the Hindu goddess Sita by a Brahman priest. Legend has it that the angry goddess cast an eternal spell of misfortune on any and all who acquired the gem. The spell apparently took effect shortly after the stone was smuggled to France, where it was sold to the French "Sun King," Louis XIV, who named it the "French Blue." Louis

had the gem cut into a 67-carat teardrop-shaped jewel. Unfortunately for Louis, merely changing the gem's name didn't protect him from the curse. After wearing it but once, he was stricken by a deadly strain of smallpox. The stone passed down to Louis XVI and his wife, Marie Antoinette, both of whom, while possessing the gem, literally lost their heads in the French Revolution.

The gem was stolen again during the chaos at the end of the French Revolution. It reappeared in London 38 years later, recut and bought by a British banker and gem collector, Henry Thomas Hope. We know little about any notable misfortune Hope may have suffered, but in 1890 the Hope Diamond was inherited by Lord Francis Hope, whose wife, an American actress, soon ran off with another man.

Lord Hope was later forced to sell the gem to avoid bankruptcy. His unfaithful wife, who had frequently worn the stone, died in poverty.

The ongoing woeful saga of the diamond's subsequent owners continued for decades. One owner, an Eastern European prince, gave it to an exotic dancer—but later began to doubt her affections and shot her to death. Another, a Greek gem merchant, drove off a cliff and perished with his wife and children. Finally, in 1911, the Hope Diamond was purchased for $154,000 by Evalyn Walsh McLean, a wealthy and eccentric American socialite who sometimes had her Great Dane wear the diamond to greet her party guests. (Happily, there are no reports of misfortune befalling the pooch.) Although McLean scoffed at the diamond's curse, she too suffered personal tragedy during her years of ownership: A son died in a car accident; a daughter died from an overdose of sleeping pills; her husband went insane and was institutionalized.

The diamond's final private owner, diamond merchant Harry Winston, acquired the stone from the late Dame McLean's estate in 1947. For unknown reasons, he donated it to the Smithsonian Institution 11 years later, inexplicably sending it in a plain brown wrapper by registered mail. (The actual package is a popular item in the National Postal Museum.) The Hope Diamond now holds a place of honor in the Smithsonian's remodeled Hall of Gems, *perhaps* bringing an end to its colorful yet sometimes tragic history.

▲ The exquisite blue Hope Diamond. The gem owes its unusual blue color to a few stray atoms of chromium within its crystal lattice.

what they believed were the first tiny synthetic diamonds. (They were dismayed to learn that synthetic diamonds had been made in Sweden in 1953 . . . but had not been reported.) Today, more than 20 tons of industrial-grade synthetic diamonds are produced each year, destined for such practical uses as drill bits in oil-well drilling and modern high-speed dentistry (a diamond-studded drill bit can remove tooth enamel so quickly that a visit to the dentist is not nearly as painful as it once was).

In 1970, the first *gem-quality* diamonds were synthesized. Today, some synthetic diamonds even outshine the original. Strontium titanate, a synthetic mineral sold under the trade names of Fabulite and Wellington Diamond, glitters four times more vividly than a real diamond. Unfortunately, being only moderately hard—between 5 and 6 on the Mohs scale—it is not very durable. And because it can be manufactured by the ton, this synthetic gem is only as rare as its creators choose to make it. Another synthetic gem, cubic zirconia, can be manufactured in batches of 50 kg (110 lb) and sold wholesale for a few cents per carat. Its optical qualities are virtually indistinguishable from those of nature's diamonds, and cubic zirconia is quite durable. How, then, do you know if your diamond is real or fake? One of the principal ways to tell is to check the minerals' densities, which is higher in zirconia than in a natural diamond.

Minerals as Clues to the Past

After they identify their minerals, geologists then use their knowledge of how different minerals form to reconstruct the geological events and environmental conditions that may have produced them. For example, large deposits of halite, such as those in Michigan, Kansas, and Louisiana, suggest the evaporation of an ancient saltwater sea. Glaucophane, a blue variety of amphibole, is known to form only in high-pressure, low-temperature conditions that occur only in subduction zones where oceanic plates descend beneath adjacent plates at convergent plate boundaries. Thus a rock containing glaucophane is believed to have formed in an ancient subduction zone.

Polymorphs, with identical chemical compositions but different crystal structures, form under different heat and pressure conditions. Ordinary quartz, with a density of 2.65 gm/cm^3, crystallizes at relatively low temperature and pressure, about 600–700°C (1100–1300°F). One of its polymorphs, *stishovite,* with a density of 4.28 gm/cm^3 and a densely compressed structure, crystallizes at temperatures higher than 1200°C (2200°F) and pressures more than 130,000 times that at sea level. Geologists believe that these extreme conditions were probably produced by geological events of enormous force, such as meteorite impacts. This hypothesis gains support from the presence of stishovite in the fractured rocks of Meteor Crater near Winslow, Arizona, and at Manicouagan Lake, Quebec, both of which are ancient meteorite-impact sites (Fig. 2-35).

Now that we have introduced the basic structure of minerals and explained how they form and are identified, we can begin to examine more closely the common types of rocks that make up the Earth's crust. In the next six chapters, we will dis-

cuss how igneous, sedimentary, and metamorphic rocks form and how we use the minerals in these rocks to interpret past geological events.

One final thought about minerals: As a geology student, you may now be tempted to browse from time to time through the large bins of generic minerals at rock shops and museums. Not long ago, a man in Tucson, Arizona, was doing just that. From a barrel of unspectacular stones, he pulled a potato-size lavender specimen, paid the proprietor $10, and went home with what may have been the largest sapphire (1905 carats) ever found. Although the value and quality of the stone have been hotly debated, a quote from the keen-eyed rockhound still rings true: "The only reason more gems aren't found in this country is that no one is looking for them." Now that you know something about minerals, you too can start looking.

(a)

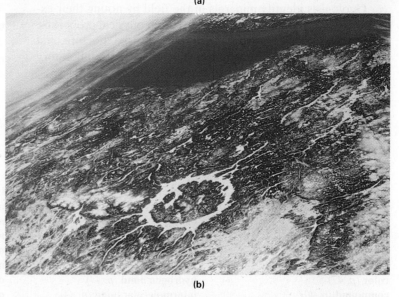

(b)

▲ **FIGURE 2-35** **(a)** Meteor Crater, near Winslow, Arizona, is believed to have been produced by the impact of a relatively small meteorite about 50,000 years ago. The crater is about 1.6 km (1 mi) across and 170 m (570 ft) deep. **(b)** The circular basin that forms Manicouagan Lake (75 km, or 40 mi, in diameter), in Quebec, Canada, is also thought to be the product of an ancient meteorite strike, dating from about 214 million years ago. Studies of rocks near both sites have revealed the presence of the extremely high-pressure variety of SiO_2, stishovite. Our knowledge of the conditions that form stishovite reinforces hypotheses that propose impact origins for Meteor Crater and Manicouagan Lake.

Chapter 2 Summary

Minerals are naturally occurring solids with specific chemical compositions and definite internal structures. Rocks are naturally occurring groupings of minerals. Minerals are composed of one or more chemical elements, the form of matter that cannot be broken down to a simpler form by heat, cold, or reaction with other elements. Each element, in turn, is made up of atoms, submicroscopic particles that retain all of an element's properties, the characteristics of a substance that enable us to distinguish it from other substances. When atoms of two or more elements combine in specific proportions, they form chemical compounds that have properties different from any of their constituent elements.

Atoms combine to form chemical compounds in a variety of ways known as bonding. Many minerals form when an atom donates or acquires electrons, and it becomes an electrically charged particle called an ion. An atom that donates electrons becomes positively charged; an atom that acquires electrons becomes negatively charged. Positive and negative ions are attracted to one another, forming electrically neutral chemical compounds; the result is an *ionic bond*. Other minerals form when electrons are shared by several atoms; the result is a *covalent bond*.

As a mineral forms through chemical bonding, all of its ions or atoms occupy specific positions within a mineral's crystal structure, a three-dimensional pattern that repeats throughout the mineral. When mineral growth is not limited by space, a crystal may form with a regular geometric shape that reflects the mineral's internal crystal structure. Naturally occurring inorganic solids that lack systematic crystal structures are called mineraloids. Opal, with its specific chemical composition of silica and water but with a disorderly internal arrangement of its atoms, is an example of a mineraloid.

Geologists identify minerals in the field by noting their external characteristics and measuring certain physical properties, including color, luster, streak, hardness, cleavage, fracture, and crystal form. Methods used to identify minerals in the laboratory include:

- determining their density;
- examining them under a petrographic microscope;
- assessing whether they fluoresce or phosphoresce under ultraviolet light;
- analyzing them by X-ray diffraction.

Although many substances may have orderly internal patterns, the number of naturally occurring minerals is limited by the availability of Earth's elements, their electrical charges, and the relative sizes of their ions. Earth's crust is composed primarily of only eight common elements that combine to produce the rock-forming minerals. The two most prominent elements, oxygen and silicon, readily combine to form the silicon–oxygen tetrahedron. This structure is the basic building block of the most abundant group of minerals in the Earth's crust—the *silicates*. Silicon–oxygen tetrahedra may be linked in a variety of crystal structures:

- independent tetrahedra (olivine is an example);
- single chains of tetrahedra (such as the pyroxenes);
- double chains (the amphiboles);
- sheet structures (the micas);
- framework structures (feldspars and quartz).

A number of *nonsilicates* are also common rock-forming minerals. Among them:

- the *carbonates* (such as calcite and dolomite);
- the *oxides* (such as iron-rich magnetite and hematite);
- the *sulfides* (such as lead-based galena and iron-based pyrite);
- the *sulfates* (such as calcium-based gypsum).

Minerals that form under specific geological circumstances, when identified in ancient rocks, can provide clues to the past. The quartz polymorph stishovite, for instance, is known to form under high-temperature/high-pressure conditions; thus it is considered to be evidence of a possible past meteoric impact.

KEY TERMS

atomic mass *(p. 41)*
atomic number *(p. 41)*
atoms *(p. 40)*
bond *(p. 42)*
carbonates *(p. 60)*
cleavage *(p. 51)*
color *(p. 47)*
compound *(p. 40)*
covalent bond *(p. 43)*
crystal *(p. 45)*
crystal structure *(p. 45)*
density *(p. 53)*

electron *(p. 41)*
elements *(p. 40)*
energy level *(p. 42)*
fluorescence *(p. 54)*
fracture *(p. 51)*
hardness *(p. 48)*
hydrogen bond *(p. 44)*
intermolecular bond *(p. 44)*
ion *(p. 43)*
ionic bond *(p. 43)*
ionic substitution *(p. 46)*
isotope *(p. 42)*

luster *(p. 47)*
metallic bonding *(p. 43)*
minerals *(p. 40)*
mineraloid *(p. 45)*
neutron *(p. 41)*
nucleus *(p. 41)*
octet rule *(p. 42)*
oxides *(p. 60)*
phosphorescence *(p. 54)*
polymorph *(p. 46)*
property *(p. 40)*
protron *(p. 40)*

rocks *(p. 38)*
rock-forming minerals *(p. 54)*
silicates *(p. 55)*
silicon–oxygen tetrahedron *(p. 55)*
streak *(p. 48)*
sulfates *(p. 61)*
sulfides *(p. 55)*
van der Waals bond *(p. 44)*
X-ray diffraction *(p. 54)*

QUESTIONS FOR REVIEW

1. What is a mineral? How does a mineral differ from a rock?

2. Briefly describe the structure of an atom. What is an isotope? What is an ion?

3. How does an atom achieve chemical stability? How does a chemical compound achieve electrical neutrality?

4. Describe four properties of minerals that you could use to identify an unknown mineral.

5. Briefly discuss why silicon and oxygen are so compatible in nature.

6. List four different silicate structures and give a specific mineral example of each.

7. List two types of common nonsilicates and give a specific mineral example of each.

8. If some calcium (Ca^{2+}) substitutes for sodium (Na^+) in the plagioclase feldspars, why must some aluminum (Al^{3+}) replace some silicon (Si^{4+}) in the mineral's structure?

9. Sulfur forms a small ion with a high positive charge. Why, then, doesn't sulfur unite universally with oxygen to form the basic building blocks of most crustal minerals?

10. Why doesn't the mineral quartz exhibit the diagnostic property of cleavage? Considering its physical beauty, why isn't a quartz crystal a more valuable gemstone?

11. In the photos at right you can see crystals of real gold and "fool's gold" (pyrite). How would you distinguish between these two similar-looking minerals?

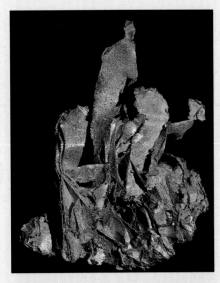

LEARNING ACTIVELY

1. On a nice day, take a stroll around your campus and look—really look—at the rocks that make up the buildings. Within them, especially if you're equipped with a magnifying glass or hand lens, you should be able to see a host of minerals, many of which were discussed in this chapter. Try to locate several of the common rock-forming minerals. A flight of granite steps in front of a campus library should help. (Try not to feel too self-conscious when your schoolmates ask you what you're doing.)

2. Visit a local rock shop (nearly every town and city has at least one; some have dozens) and buy something. Poke around in the barrels of everyday specimens. Hidden in these grungy rocks may be a rock filled with beautiful, even valuable, crystals. Look for "thunder eggs"—fist-size, wave-rounded beach rocks that reveal their crystals after you open them with a hammer. You might also find some beautiful, multicolored agates.

3. Visit the mineral and gem exhibit at a nearby museum. Make a pilgrimage to the amazing collections at the Smithsonian Institution in Washington, D.C., the New York Museum of Natural History, Chicago's Field Museum, or Los Angeles's Museum of Natural History. These collections are some of the very best in the world and contain the likes of the Hope Diamond, the Star of India sapphire, and other invaluable gems.

4. Go to a quarry and collect some minerals "in the field." Most rock shops have guidebooks to local mineral-collecting sites.

5. Look around your dwelling for minerals used in daily life. Read a vitamin label or a toothpaste label. Check the ingredients in a bag of kitty litter. You'll notice that minerals are practical . . . and practically everywhere.

ONLINE STUDY GUIDE

Practice makes you better. Make the time to take the *Online Study Guide* quiz for this chapter. It should take you about 20 minutes. Automatic grading and instant feedback will help you quickly and accurately identify the concepts you have mastered and the areas in which you need more study.

www.prenhall.com/chernicoff

Igneous Rocks and Processes

3

On the big island of Hawai'i, you can watch a volcano erupt and later touch the warm rock that passed through the volcano in a molten state just hours or days before (Fig. 3-1). On the beautiful rocky coast of Maine, you can walk on rocks that cooled 300 million years ago from molten rock 10 kilometers (6 miles) below the Earth's surface. Nearly every state and province in North America contains rocks that formed during volcanic eruptions or that crystallized from molten rock material underground. As we saw in Chapter 1, rocks that crystallized from molten material are called **igneous rocks** (from the Latin *ignis,* or "fire").

In certain places inside the Earth, rocks melt and because the melt is less dense than the surrounding rock, it rises. Finally it cools and crystallizes, forming the primary crustal rocks of the continents and ocean basins. The melting process releases water and various gases from the Earth's interior. Early in the Earth's history, this process formed the first oceans and atmosphere, and the composition of the seas and the air today is still connected to rock melting and crystallization. The history (and perhaps the origin) of life is also intimately tied to rock-melting processes, as is the motion of the Earth's plates, the location of our volcanoes, and the formation of many precious metals and other valuable minerals.

Human evolution and history have been closely related to igneous processes for millions of years. Without igneous processes, there would be no continents, no air, no water . . . and thus, no humans. Early human evolution occurred in close association with volcanic activity in east Africa and has been preserved for us to study in volcanic deposits that contain the bones, footprints, and relics of our early human ancestors. Throughout history, humans have used volcanic products (such as the volcanic glass obsidian) to make their cutting tools and have grown their crops in or foraged for their next meals from the fertile soils common to volcanic deposits. Of course, humans have also had to cope with the ever-present dangers of life in the shadow of an active volcano for thousands of years (Fig. 3-2).

We give different names to molten rock inside the Earth and molten rock at the Earth's surface. **Magma** is molten rock *within* the Earth. It is a fluid mixture of liquid, solid crystals, and dissolved gases. When magma reaches the Earth's surface, we call it **lava,** molten rock that flows *above* ground. (Lava and volcanoes are discussed in the next chapter.)

FIGURE 3-1 This red-hot lava flowing from K'ilauea volcano at Hawai'i Volcanoes National Park solidified as volcanic rock within hours of this photograph. ▶

▲ **FIGURE 3-2** Wall painting in Catal Hoyuk, Turkey, one of humankind's earliest cities (about 8000 years ago). Note that the artist has depicted the local twin-peaked volcano hurling massive volcanic projectiles onto the homes in the surrounding village.

Most igneous processes occur underground, hidden from our view by the loose soils, sediments, or sedimentary rocks that cover most of the Earth's surface. Although we can see the lava spewing from a volcano, we cannot see magma moving underground. Thus, to investigate igneous processes, geologists often rely on geophysical studies of the Earth's interior (discussed in Chapter 11), computer models that simulate the movement of magma, and laboratory experiments that reproduce pressures, temperatures, compositions and other factors that affect the melting and crystallization of rocks. Geologists also look for regions where surface rock layers have been eroded, physically worn away by rivers, glaciers, winds, or ocean waves. Erosion of these upper layers of the Earth lets us see rocks that originally formed underground.

We can see igneous rocks that solidified underground in some of North America's most scenic places, from Acadia National Park in Maine to the Yosemite Valley of eastern California. Some of our continent's most ancient rocks are the igneous roots of one-time mountains that have long since been eroded by downcutting rivers and glaciers. Such rocks can be seen in Minnesota, Ontario, and Quebec, and northern Michigan and Wisconsin. Igneous rocks are essentially the framework for nearly all of the Earth's crust:

All the Earth's oceans are underlain by igneous rocks, and all of its continental crust originated as igneous rocks as well.

A Brief Introduction to Igneous Rocks and Processes

Our discussion of the Earth's igneous processes and rocks describes first how, why, and where rocks melt, some characteristics of molten rock, and then the types of rocks that form when molten rock cools and solidifies. We will pay special attention to how plate tectonics affects the origin and distribution of igneous rocks on Earth. Finally, we will compare igneous rocks and processes on Earth with those of some of the other rocky planets and moons of our solar system. We will also see that valuable materials—such as gold, silver, and copper—are often associated with igneous processes.

Before we talk about melting and crystallization processes, it helps to know a few things about common igneous-rock compositions and names. In the discussion below, we will refer to silica-poor compositions as **basalt** (ba-SALT), if it crystallizes at the Earth's surface from lava, and *gabbro*, if it crystallizes underground. These types of rock make up most of the oceanic crust and oceanic islands such as Hawai'i. We will refer to the rocks at the other end of the compositional spectrum—those that are silica-*rich*—as **granite** (silica-rich rocks that crystallize underground) and **rhyolite** (silica-rich rocks that crystallize at the Earth's surface from lava), rocks typically found on our continents.

Melting Rocks and Crystallizing Magma

A crystalline solid melts when the bonds between its ions break, allowing the charged particles to move freely (Fig. 3-3). When a rock melts, bonds in minerals break. When the ions no longer form crystalline solids, they become part of a molten material called a *magma*. Magma is liquid rock that may still contain some crystals or interconnected ions. Rocks do not entirely melt at one temperature and pressure because each mineral in the rock has a particular temperature and pressure at which it melts. Therefore, rocks *partially melt*, with the amount of melt relative to remaining minerals increasing as melting proceeds. At the same time, the composition of the magma changes as minerals of different composition begin to melt and contribute different elements to the magma.

When magma cools, it crystallizes, new bonds form between ions, and tiny crystals begin to appear. Additional ions and atoms bond to the surfaces of these crystals, and the crystals grow until they touch the edges of other crystals. As cooling progresses, different minerals crystallize out of the magma at different temperatures. The magma's composition again changes as each crystallized mineral is removed from the magma. Ultimately, if cooling continues, the entire body of magma becomes solidified.

Now that you know how the melting process works, we need to consider *why* rocks melt (what drives melting inside the Earth) and *where* melting is likely to occur.

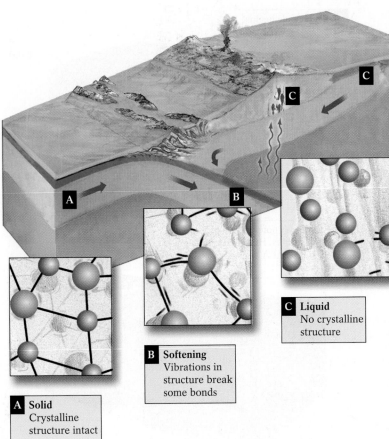

▲ FIGURE 3-3 Heat—a form of energy—vibrates a solid's atoms until some of its chemical bonds weaken and break, causing melting. Here, a subducting oceanic plate moves from a region of relative coolness (**A**), then begins to descend into the Earth's warmer interior. The minerals begin to heat up, converting them from a crystalline solid (at point **A**) to a liquid magma (at point **C**).

The Creation of Magma

When rocks melt to produce new magma, the process is not quite as simple as that of the warm midday sun melting last night's snowfall or a chocolate bar melting in your backpack. Snow consists of a single solid mineral, ice (H_2O), and therefore it all melts at one temperature, 0°C (32°F). Most rocks are composed of several minerals, each with a different melting point that reflects the strength of its own chemical bonds. For example, the bonds between sodium and oxygen in the plagioclase mineral albite are weaker than those between calcium and oxygen in the plagioclase anorthite. Consequently, at Earth-surface pressures, albite's bonds break (and albite melts) at 1118°C (2050°F), whereas anorthite's bonds can withstand a temperature of 1553°C (2800°F) before they break (and anorthite melts).

Because each mineral has its own melting point, a rock consisting of several different minerals typically undergoes **partial melting** when it reaches the melting points of some but not all of its minerals. When rocks melt *partially,* some minerals remain solid (the minerals with higher melting points) while the melted portion (the minerals with lower melting points) may flow away as magma. To visualize partial melting, consider what might

happen to a Mr. Goodbar if it were left out on the sidewalk on a warm sunny day. The chocolate (with a lower melting point) would melt and ooze away; the peanuts (with a high melting point), however, would remain solid.

Numerous factors control the melting of rocks and the creation of magmas. Heat, pressure, and the amount of water in the rocks all combine to influence whether and how much a rock will melt. These factors therefore also affect what rocks eventually form from magmas and where on Earth they occur.

Heat Temperatures in the Earth increase with depth. This increase in temperature as a function of depth is called the *geothermal gradient.* As you can see in Figure 3-4, interior temperatures rise from about 50°C (125°F) to more than 1200°C (2200°F) within the Earth's outer 250 km (150 mi). Because the high-temperature rocks at great depth are also under very high pressure, the rocks may not melt unless other factors help to drive melting. We typically think of materials melting solely because they are heated (for example, snow and chocolate), but it is important to remember that in the Earth, most partial melting of rocks

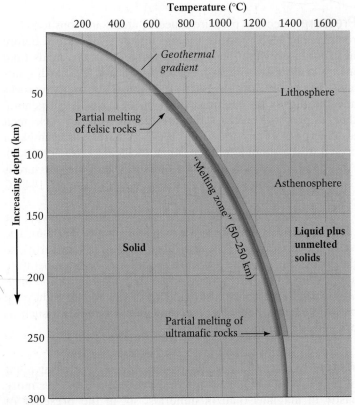

▲ FIGURE 3-4 A graph showing the geothermal gradient—the rate at which the Earth's internal temperature increases with depth within continental lithosphere and the underlying asthenosphere. The temperatures between 50 and 250 km (30–150 mi) in depth exceed the 700°C (1300°F) melting point of felsic rock and the 1300°C (2400°F) melting point of ultramafic rock. Thus rocks tend to melt at these temperatures in the Earth's lower crust and upper mantle.

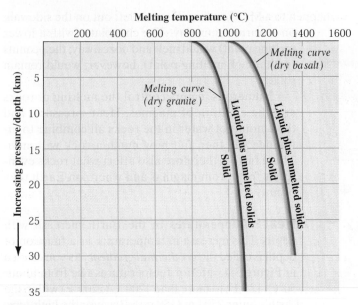

▲ FIGURE 3-5 Melting-temperature curves for basalt and granite. For both basalt and granite, melting temperatures increase with increasing depth, because the pressure at greater depths raises a rock's melting point.

happens because of changes in other variables—pressure and water content—and not simply by heating.

Pressure Because pressure squeezes atoms and ions together, rocks under high pressure require higher temperature before heat-produced vibrations weaken and break their bonds (and melt a rock). The deeper a rock lies beneath the Earth's surface, the greater the pressure from the weight of overlying rocks and the higher the melting points of its minerals. In general then, as pressure increases, the temperature at which rocks melt also increases (Fig. 3-5). At the Earth's surface, for example, a crystal of the sodium feldspar albite melts at 1118°C (2050°F). At a depth of 100 km (60 mi), however, where the pressure is 35,000 times that of the surface, a temperature of 1440°C (2650°F) is required to melt albite.

If the pressure on rock is somehow reduced or removed, the rock may reach conditions at which melting occurs (Fig. 3-6). The rocks are already hot (because they are deep in the Earth), so the decrease in pressure leads to a condition of high temperature and low pressure. This process—known as *decompression melting*—partially melts the rocks of the Earth's mantle to produce the basaltic magma beneath divergent plate boundaries such as oceanic and continental rift zones.

Water Water, even a small amount, *lowers* the melting point of rocks. Water enhances melting because it dissolves more easily in magmas than in minerals, so, in the presence of water, magma may be more stable than solid rock. The undissolved water within solid rocks tends to destabilize the crystal structures within a rock's minerals, promoting the breaking of their bonds and lowering their minerals' melting points. Under high pressure, water has an even greater effect on the melting

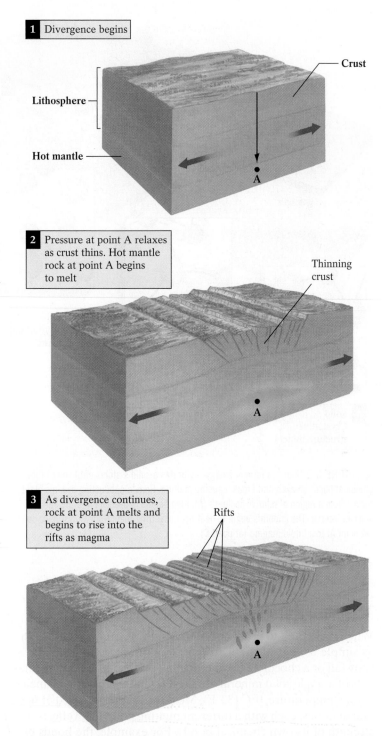

▲ FIGURE 3-6 The effect of lithospheric thinning on rock in the Earth's upper mantle. Hot mantle rocks remain solid primarily because of the pressure applied by the weight of overlying rock. This pressure may be reduced through the plate-tectonic process of rifting. Directly below the rift zone, where pressure from overlying rocks is reduced, the hot mantle rocks begin to rise and melt to produce new mantle-derived magmas.

point of a mineral: Whereas dry rocks become more resistant to melting with greater depth, wet rocks become less resistant (Fig. 3-7). As we will see, this combination of high pressure and high water content is an important factor in the production of magmas at plate boundaries where oceanic plates subduct (Fig. 3-8).

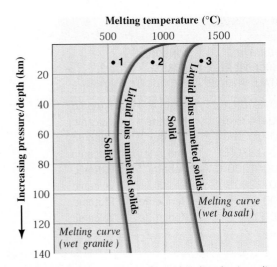

FIGURE 3-7 Melting-temperature curves for wet basalt and wet granite. For both, melting temperatures decrease with increasing depth, because highly pressured water destabilizes the crystal structures of the minerals in these rocks, lowering their melting points. At point 1, both granite and basalt would be solid; at point 3, both would be partially liquid; at point 2, granite would be partially liquid but basalt would still be solid.

Motion of Magma

Once partial melting produces a magma, the magma then tends to rise toward the Earth's surface because it is less dense than the solid rock that surrounds it. Just as hot air rises relative to cold air, hot rocks (and magma) rise relative to cold rocks. Magma can rise through solid rock because it is less dense than the rock. The rising magma migrates upward through fractures in overlying rocks (some created by the magma itself) or by melting and incorporating some of the rock it encounters as it rises.

A magma's ability to rise is largely controlled by its *fluidity*—its ability to flow—which is governed by its temperature and composition. Increased heat invariably increases the fluidity of any substance—be it maple syrup, molasses, or magma. When the temperature of a substance increases, its ions and atoms move about more rapidly, thus breaking the temporary bonds that inhibit flow. Think of what happens when you pour cold maple syrup—thick and gooey—direct from the refrigerator into a pot, warm it up on the stovetop, and then pour hot, runny syrup on your

pancakes. **Viscosity**—the term generally applied to describe magma flow—is a fluid's *resistance* to flow (the opposite of fluidity): Viscous fluids flow quite sluggishly; less viscous fluids flow more easily. Viscosity *increases* with *decreasing* temperature, and *decreases* with *increasing* temperatures. Thus relatively cool magmas flow sluggishly because they have high viscosity; relatively hot magmas flow more easily because they have low viscosity.

The viscosity of magma also generally increases with silica content (see Table 3.1), because the oxygen ions at the corners of the unbonded silicon-oxygen tetrahedra form temporary bonds with other ions in the magma. Silica-rich magma is also very viscous because it tends to be relatively cool (it crystallizes at low temperatures). Conversely, because basaltic magma is relatively hot and has a low silica content, it is much less viscous and flows easily. Therefore, low-silica, basaltic magmas are more likely to rise to the surface and erupt than are silica-rich granitic magmas, which tend to cool underground. There are other factors that account for the scarcity of silica-rich lavas at the surface, but this viscosity difference may explain in part why there are more low-silica lavas (basalt) at the Earth's surface than high-silica lavas (rhyolite). The low-viscosity basaltic lava flows to the surface with relative ease which is why we find so much basalt at the Earth's surface. High-viscosity, silica-rich magma is less likely to reach the surface, instead cooling underground to form granite.

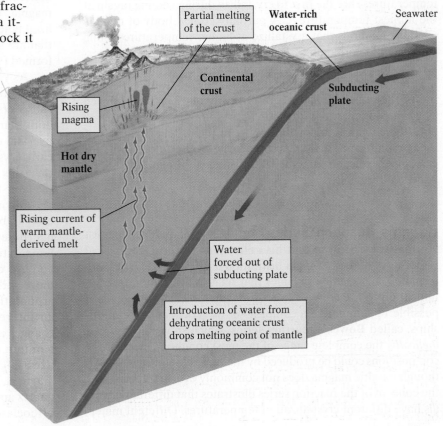

FIGURE 3-8 As a water-rich oceanic plate subducts, it descends into the Earth's warmer, higher-pressure interior. Increasing pressure and temperature during subduction drive water from the plate's sediments and basalts into the dry, hot mantle rocks above it. Before this water enters the mantle rocks, high pressure prevents them from melting; when water is driven into these rocks, their melting point drops and they begin to melt partially, producing the new magmas that fuel subduction-zone volcanism.

Table 3.1 Common Igneous Compositions

COMPOSITION TYPE	PERCENTAGE OF SILICA	OTHER MAJOR ELEMENTS	RELATIVE VISCOSITY OF MAGMA	TEMPERATURE AT WHICH FIRST CRYSTALS SOLIDIFY	IGNEOUS ROCKS PRODUCED	TYPES
Felsic	>65%	Al, K, Na	High	~600–800°C (1100–1475°F)	Granite Rhyolite	Plutonic Volcanic
Intermediate	55–65%	Al, Ca, Na, Fe, Mg	Medium	~800–1000°C (1475–1830°F)	Diorite Andesite	Plutonic Volcanic
Mafic	45–55%	Al, Ca, Fe, Mg	Low	~1000–1200°C (1830–2200°F)	Gabbro Basalt	Plutonic Volcanic
Ultramafic	<40%	Mg, Fe, Al, Ca	Very low	>1200°C (2200°F)	Peridotite Komatiite	Plutonic Volcanic

The Crystallization of Magma

Eventually, every magma cools and solidifies. As it does so, its composition may change significantly. The temperature at which a mineral melts is roughly the same as the temperature at which it crystallizes. Minerals that melt first (at the lowest temperatures) are thus the last to crystallize during cooling (again at the lowest temperatures). Minerals that melt last (at the highest temperatures) are the first to crystallize during cooling (again at the highest temperatures). A partially cooled body of magma contains minerals that crystallize at higher temperatures along with a liquid containing the ions that will crystallize into later-formed minerals when the temperature is lowered further. As the magma continues to cool, additional ions and atoms crystallize out of the melt, leaving progressively less liquid. Thus at each stage of cooling, the proportion of crystal to liquid changes, as does the chemical interaction between them. The chemistry of the remaining melt also changes as early-forming atoms and ions crystallize, concentrating the late-forming atoms and ions in the uncrystallized melt.

Bowen's Reaction Series In 1922, Canadian geochemist Norman Levi Bowen and his colleagues at the Geophysical Laboratory of the Carnegie Institution in Washington, D.C., determined some basic principles that relate silicate-mineral crystallization to the composition of magmas. Their work made it possible to summarize a complex set of geochemical relationships, called **Bowen's reaction series,** which shows how, *hypothetically,* the complete range of the most common igneous-rock compositions could be produced by the same basaltic magma. Although basaltic magma does not commonly evolve into rhyolite, the concept of the reaction series illustrates that different minerals have different crystallization temperatures. Different minerals will therefore crystallize from a magma in a sequence as magma cools and its composition changes (Fig. 3-9).

Consider an example for how a magma and the crystals that form from it evolve during cooling. The crystallization of olivine removes some iron and magnesium from the original magma, changing its composition by increasing the proportion of the other as-yet-uncrystallized major ions (such as Ca) in the magma. New minerals—such as pyroxene—can then grow from the resulting magma. We can see some evidence for this evolution in crystallized igneous rocks that contain early-crystallized olivine grains that are surrounded by the later-formed pyroxene crystals that formed from the evolving magma (Fig. 3-10).

As you learned in Chapter 2, some minerals can have ionic substitution between elements with atoms of similar sizes and charges (such as iron and magnesium). Olivine and pyroxene are both examples of minerals that have ionic substitution involving iron and magnesium. During crystallization of a basaltic magma, early-crystallized olivine and pyroxene will be more magnesium-rich (this composition is stable at the highest temperature) than the later-formed iron-rich olivine and pyroxene.

Olivine and pyroxene aren't the only minerals to crystallize early in basaltic magma. Calcium plagioclase (a feldspar called anorthite) also crystallizes at the highest temperatures. As the temperature decreases and the magma becomes more silica-rich and less calcic, more sodium-rich plagioclase crystallizes. The more sodic plagioclase can form new crystals, or it can grow around the earlier-formed calcium plagioclase. Plagioclase is an interesting mineral because a single crystal can record changes in composition through time, reflecting changing temperature and composition of the magma from which it crystallizes (Fig. 3-11). In the lowest-temperature, most silica-rich magmas, sodium plagioclase (albite) crystallizes along with quartz and typically with potassium feldspar (a feldspar called orthoclase) as well.

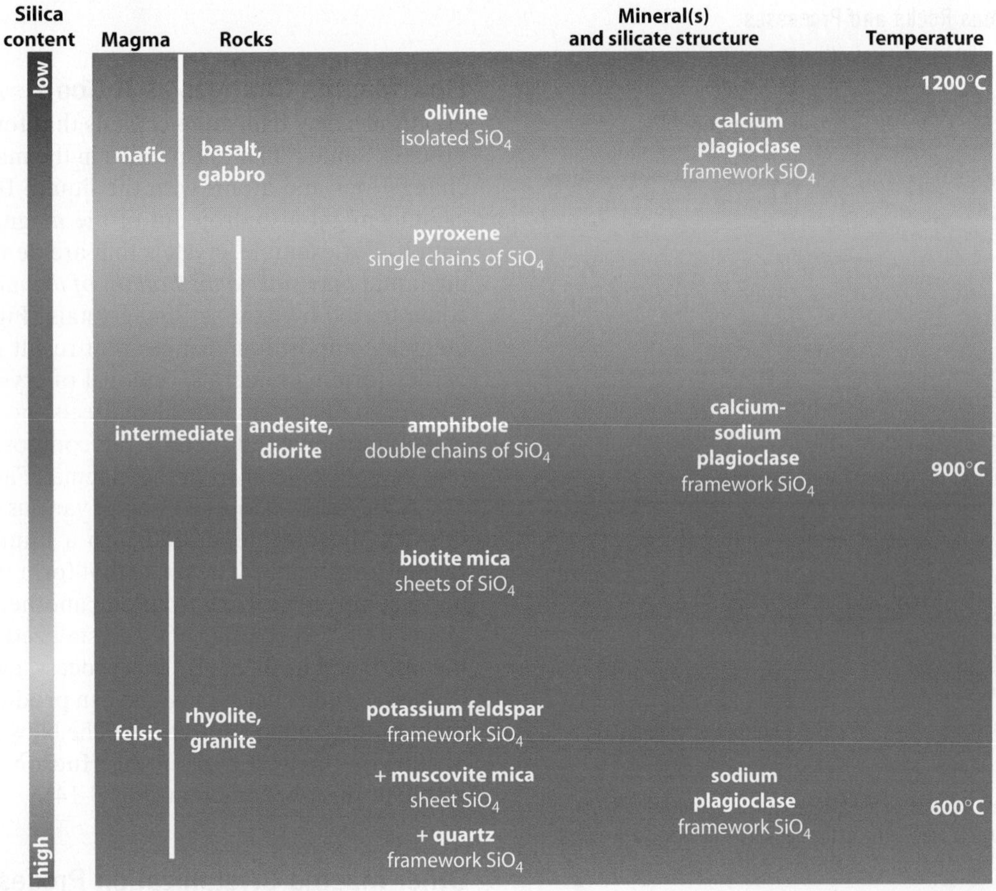

Silica content | **Magma** | **Rocks** | **Mineral(s) and silicate structure** | | **Temperature**

low ... high (vertical scale on left)

1200°C

mafic | basalt, gabbro | olivine — isolated SiO₄ | calcium plagioclase — framework SiO₄

pyroxene — single chains of SiO₄

intermediate | andesite, diorite | amphibole — double chains of SiO₄ | calcium-sodium plagioclase — framework SiO₄ | 900°C

biotite mica — sheets of SiO₄

felsic | rhyolite, granite | potassium feldspar — framework SiO₄

+ muscovite mica — sheet SiO₄ | sodium plagioclase — framework SiO₄ | 600°C

+ quartz — framework SiO₄

▲ **FIGURE 3-9** An illustration of the concept that different minerals melt (and therefore crystallize) at different conditions, accounting for the partial melting of rocks and the sequence of crystallization of minerals in a cooling magma.

▲ **FIGURE 3-10** Early-crystallized olivine crystals (brightly colored, fractured crystals) surrounded by later-crystallized pyroxene crystals (gray). The minerals are in a rock thin section, photographed on a microscope with cross-polarized light. The field of view is 3mm.

▶ **FIGURE 3-11** A zoned plagioclase crystal formed under rapid cooling conditions. The crystal's outer layer is dominated by sodium ions, which replaced earlier-bonding calcium ions as the crystal reacted with ions in the cooling magma. The crystal cooled too rapidly, however, to allow for complete replacement of calcium ions by sodium ions in the interior. The different bands inside the crystal may have formed as magma composition or other conditions changed through time as the crystal grew. (Crystal is 2.2 mm [0.15 in.] long.)

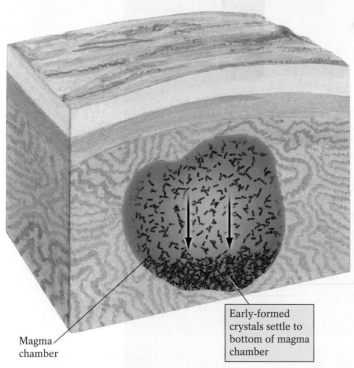

Magma chamber

Early-formed crystals settle to bottom of magma chamber

▲ **FIGURE 3-12** Early-forming crystals do not always remain in contact with the liquid magma. Instead, the crystals may settle to the bottom of the magma chamber.

How Magma Changes as It Cools As a magma cools, several things may happen to crystals that form early in the cooling process. Some remain suspended in the magma, continuing to exchange ions and atoms with the liquid. But some early-formed crystals might also drop out of the magma and no longer react with it. For example, crystals that are denser than the surrounding liquid *may sink to the bottom of the magma chamber* and become buried by later-settling crystals (Fig. 3-12). In addition to magma-composition changes that result from crystallization of earlier-forming minerals, removal of crystals from contact with later-crystallizing magma also affects the composition of the remaining magma and thereby the composition of any rocks that may crystallize later from the magma (Fig. 3-13).

A magma that loses crystals at various stages of its cooling has in effect become separated into a number of independently crystallizing bodies. The rocks that form from such a magma differ in composition both from one another and from the original magma. Each successive body crystallized is more silica-rich than the last (see Fig. 3-13). By this process, called **fractional crystallization,** a single parent magma can produce a variety of igneous rocks of different compositions. The New Jersey Palisades, a line of cliffs on the west bank of the Hudson River, are a classic example of this phenomenon (Fig. 3-14).

Other Magma Crystallization Processes Bowen believed that all igneous rocks form by fractional crystallization of basaltic magmas. Geologists building on Bowen's work, however, have realized that these processes alone do not account for all igneous rocks. For example, although silica-rich igneous rocks can crystallize from magmas that started with a basaltic composition, too little magma remains after the minerals such as olivine, pyroxene, and calcium plagioclase crystallize to produce large bodies of silica-rich igneous rocks. How, then, does one explain the 1000-km (600-mi) stretches of silica-rich and intermediate igneous rock (such as California's Sierra Nevada range) on our continents? To account for this much granitic rock by fractional crystallization would require that there be ten times more basaltic rock as there is granite . . . and this volume of silica-poor magma is simply not observed. Although fractional crystallization may partly explain the origin of these rocks, other processes must have operated as well.

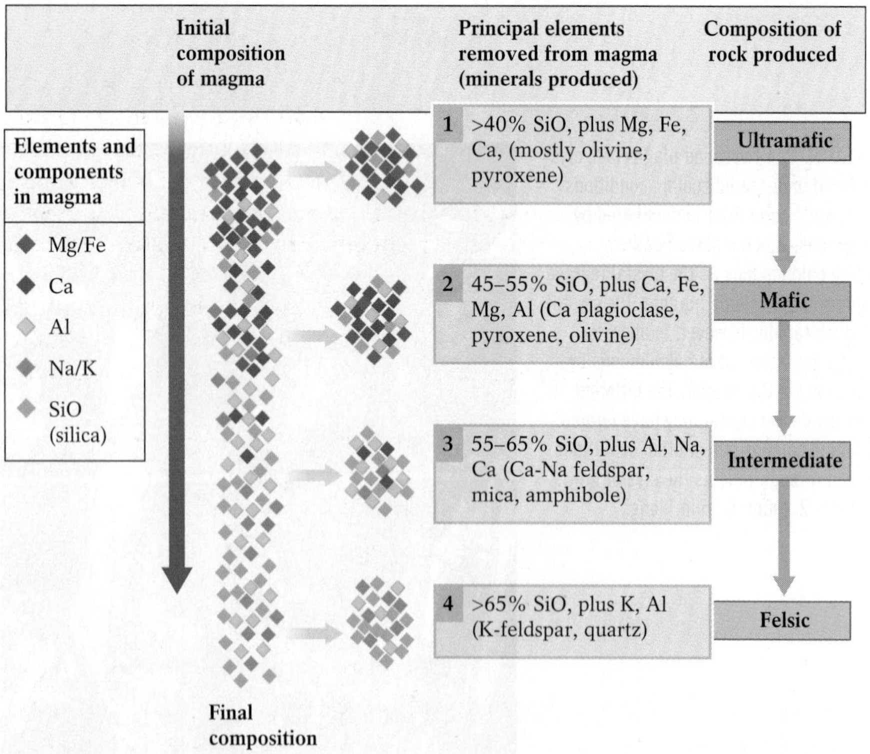

Magmas that are more silica rich than basalt may be produced by a combination of processes: some fractionation of basaltic magma as described above; some further melting of already-formed oceanic or continental crust; and some mixing of magmas of different compositions. This latter process, in which two or more different bodies of magma flow together to form a magma of hybrid composition is called *magma mixing* (Fig. 3-15). The 1912 volcanic eruption in Alaska's Aleutian Islands produced rocks containing minerals from very different magma sources, suggesting that two distinct bodies of magma had combined to fuel the eruption. A similar style of eruption occurred at Mount Pinatubo in the Philippines in 1991.

Initial composition of magma	Principal elements removed from magma (minerals produced)	Composition of rock produced
	1 >40% SiO, plus Mg, Fe, Ca, (mostly olivine, pyroxene)	Ultramafic
	2 45–55% SiO, plus Ca, Fe, Mg, Al (Ca plagioclase, pyroxene, olivine)	Mafic
	3 55–65% SiO, plus Al, Na, Ca (Ca-Na feldspar, mica, amphibole)	Intermediate
	4 >65% SiO, plus K, Al (K-feldspar, quartz)	Felsic

Elements and components in magma

◆ Mg/Fe
◆ Ca
◇ Al
◆ Na/K
◇ SiO (silica)

Final composition

▲ **FIGURE 3-13** When earlier-forming minerals are removed from a magma, they deplete the remaining magma of their elements. In turn, the magma contains a higher proportion of later-crystallizing elements. Thus igneous rocks that crystallize later from such an evolving magma have different compositions (progressively more felsic) than rocks that crystallize earlier (which are more mafic). Note that the parent magma can have any composition; the magma will become more felsic as this process of fractionation proceeds.

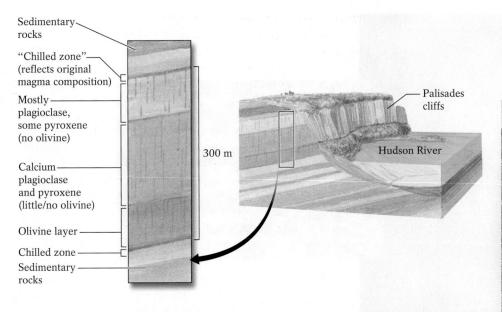

Sedimentary rocks

"Chilled zone" (reflects original magma composition)

Mostly plagioclase, some pyroxene (no olivine)

Calcium plagioclase and pyroxene (little/no olivine)

Olivine layer

Chilled zone

Sedimentary rocks

300 m

Palisades cliffs

Hudson River

▲ **FIGURE 3-14** The New Jersey Palisades, a line of cliffs in the northeastern part of the state, demonstrate the result of fractional crystallization and crystal settling. The rocks of the Palisades crystallized from a 300-m (1000-ft)-thick body of magma that intruded preexisting rocks at temperatures of at least 1200°C (2200°F). The top and bottom margins ("chilled zone") of the Palisades solidified very rapidly without undergoing fractional crystallization because the magma came into contact with cold surrounding rocks. These rocks provide us with a glimpse of the magma's original composition (that haven't been fractionated). The bottom third of the Palisades has a high concentration of olivine crystals, the central third is a mixture of calcium plagioclase and pyroxene with no appreciable olivine, and the upper third consists largely of plagioclase with no olivine and little pyroxene. It appears that early-forming olivine crystallized and then settled to the bottom of the magma body. Pyroxene and plagioclase crystallized next, with the denser pyroxenes settling and concentrating in the center, and the lighter plagioclases occupying the uppermost section. Since all the magma of the Palisades cooled and solidified fairly quickly, there was no residual magma from which lower-temperature minerals could crystallize.

Igneous Textures

Geologists classify igneous rocks by the two most obvious properties of the rocks; the *size* of the mineral crystals in the rock and the rock's *composition*. A rock's *texture* refers to the size and shape of its minerals. The most important factor controlling these features in igneous rocks is the rate at which magma or lava cools. When a magma's minerals crystallize slowly over hundreds or thousands of years, there is ample time for crystals to grow large enough to be seen by the unaided eye. When we can see the crystals clearly, the rock texture is called *phaneritic* (from the Greek *phaneros,* or "visible") (Fig. 3-16). Slow cooling typically occurs when magmas enter, or *intrude,* preexisting solid rocks underground; thus rocks with phaneritic textures are called **intrusive rocks.** They are also known as **plutonic rocks** (for Pluto, the Greek god of the underworld).

Some igneous rocks develop at relatively low temperatures from magmas with a high proportion of water. Under these watery conditions, ions are free to move quite readily to bond

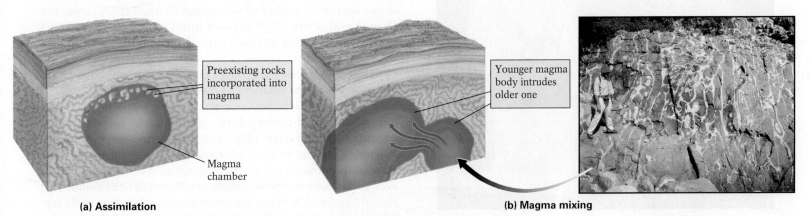

Preexisting rocks incorporated into magma

Magma chamber

Younger magma body intrudes older one

(a) Assimilation

(b) Magma mixing

▲ **FIGURE 3-15** Magmas of varying composition can develop in several ways: **(a)** as hot magma rises, it may incorporate masses of the preexisting rock through which it moves. When these masses melt, they contribute their chemical elements to the magma, changing its composition. **(b)** Several bodies of magma—with differing compositions—may mix to form a magma with its own composition. Photo (right) is from the coast of Maine.

▲ **FIGURE 3-16** Rocks that cool slowly underground like these in Yosemite National Park, California, have *phaneritic* (coarse-grained) textures. (Sample of granitic rock is 6 cm [2.2 in.] long.)

▲ **FIGURE 3-17** Extremely coarse-grained *pegmatites,* such as the one shown here, form from ion-rich magmas having a high water content. (Rock sample is 7.5 cm [3 in.] high and 11 cm [4.5 in.] long.) The greyish glassy mineral is quartz; the light-colored mineral is plagioclase feldspar.

to growing crystals, enabling crystals to grow unusually large (sometimes several meters long). Rocks with such exceptionally large crystals are called **pegmatites** (from the Greek *pegma,* or "fastened together") (Fig. 3-17). Most pegmatites consist primarily of such common minerals as quartz, feldspar, and mica. Some rare elements, such as beryllium, boron, and lithium, also occur in pegmatites and crystallize in minerals such as beryl, emerald (a green variety of beryl), and aquamarine. In western Maine, near the towns of Bethel and Rumford, rocks with pegmatitic textures contain 5-m (17-ft)-long crystals of the mineral beryl. Pegmatites are an excellent place to look for interesting and beautiful crystals, including gemstones.

Some igneous rocks solidify so quickly that their crystals do not have time to grow large. In these *aphanitic* textures (from the Greek *a phaneros,* or "not visible"), crystals are so small they can barely be seen by the naked eye (Fig. 3-18). Rocks with aphanitic textures are typically **extrusive rocks** because they commonly form from lava that has flowed out, or been *extruded,* onto the Earth's surface, where they cool too quickly to form visible crystals. These rocks are also known as **volcanic rocks,** because lava is a product of volcanoes (named for Vulcan, the Roman god of fire). Aphanitic textures may also form when a thin body of magma intrudes cool near-surface rocks and thus cools rapidly.

In some igneous rocks, large crystals are surrounded by regions with much smaller or even invisible grains. These *porphyritic* textures form as a result of slow cooling followed abruptly by rapid cooling. First, gradual underground cooling produces large crystals (Fig. 3-19) that grow slowly within a magma. Then the mixture of remaining liquid magma and the large early-formed crystals rises nearer to the surface, where it encounters cooler rocks or erupts into the cooler environment at the surface. In either case, the remaining liquid magma cools rapidly to produce the matrix of smaller grains.

Any magma can produce igneous rocks having the full range of igneous textures—the amount of time available for cooling and crystal growth determines whether the rock textures will be

◀ **FIGURE 3-18** Volcanic rocks such as this basalt typically have *aphanitic* textures; the mineral grains within the rock are not visible to the naked eye because they cool rapidly at the Earth's surface. (Basalt sample is 10 cm [4 in.] by 12 cm [4.75 in.].)

▲ **FIGURE 3-19** Some rocks have a *porphyritic* texture, marked by large crystals surrounded by an aphanitic matrix. This texture develops as early-forming crystals grow slowly to large, visible sizes. This slow growth is interrupted when the uncrystallized magma rises rapidly, causing the remaining magma to crystallize rapidly into a matrix of small, invisible crystals. (Sample size is 10 cm [4 in.] by 8 cm [3.2 in.].)

aphanitic, phaneritic, or porphyritic. Aphanitic igneous rocks that cooled rapidly at the Earth's surface, for example, are commonly underlain by phaneritic rock, derived from the same magma but cooled slowly underground.

Volcanic Glass   When lava from a volcano erupts into the air or flows into a body of water, it cools so quickly that its ions don't have enough time to form any crystals at all. The ions are essentially frozen in place randomly, bonded to any available ions nearby. The texture of the resulting rock is described as *glassy* (Fig. 3-20).

There are two common types of volcanic glass. *Pumice* (from the Latin *spuma,* or "foam") forms when dissolved gases (mostly H_2O and CO_2) escape rapidly from silica-rich magma. The gases cannot escape easily because the magma is so viscous, and so are

▲ **FIGURE 3-20** The volcanic glass obsidian has a *glassy* texture (containing no crystals) because it solidifies instantaneously from hot lava. Inset: Photo of obsidian. (Sample is 7.5 cm [3 in.] across.)

trapped as numerous gas bubbles. Expansion of the frothy magma driven by the growth of these gas bubbles produces tremendous pressures, leading to explosive volcanic eruptions. The frothy magma is blown out of exploding volcanoes and, on cooling, becomes fragments of pumice. Some pumice has so many tiny cavities that it can float. Large rafts of pumice blown from coastal and island volcanoes have been known to float out to sea for hundreds of kilometers. Pumice can float for a long time without becoming waterlogged because many of its cavities are not connected with each other, so they remain filled with air rather than water. Floating pumice may be a major means (second only to transport on or in migrating birds) by which small organisms are transported great distances across the water among islands or even continents.

A second type of volcanic glass is *obsidian*. Obsidian forms when very silica-rich lavas, containing less gas than those that produce pumice, cool very quickly. The disordered arrangement of its ions means that obsidian lacks an organized crystal structure as well as the systematic internal planes of weakness that characterize most crystals. For this reason, obsidian breaks by fracture rather than by cleavage. Early humans worked obsidian to fashion projectile points and sharp-edged cutting tools. To demonstrate the quality of such ancient tools, in 1986 David Pokotylo, the curator of archaeology at the University of British Columbia's Museum of Anthropology, underwent hand surgery with an obsidian microblade scalpel prepared in the ancient style of obsidian tool-making. To the surprise of the surgical team, Dr. Pokotylo's incisions healed more rapidly and more cleanly than those made with conventional steel blades.

Igneous Compositions

The Earth's magmas consist largely of the most common elements: oxygen, silicon, aluminum, iron, calcium, magnesium, sodium, and potassium. The relative proportions of these components within

▼ **FIGURE 3-21** An igneous-rock classification chart, showing the range of compositional types among the igneous rocks, from felsic to ultramafic. The colored areas in the body of the chart indicate the mineral components of the rocks. The sample segment shows how to interpret the chart, using as an example a rock with a composition of granite. (The samples are all roughly 12.5 cm [5 in.] by 10 cm [4 in.].) All the rock photos show plutonic rocks with phaneritic (coarse-grained) textures.

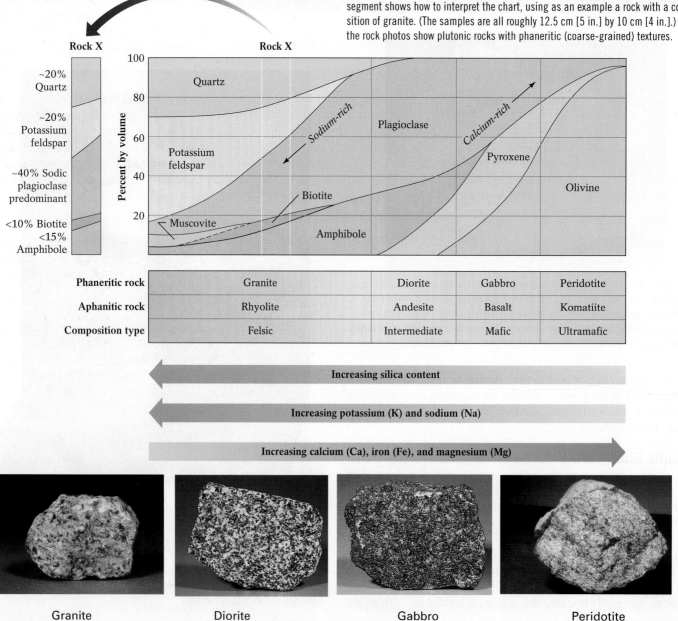

Phaneritic rock	Granite	Diorite	Gabbro	Peridotite
Aphanitic rock	Rhyolite	Andesite	Basalt	Komatiite
Composition type	Felsic	Intermediate	Mafic	Ultramafic

Increasing silica content

Increasing potassium (K) and sodium (Na)

Increasing calcium (Ca), iron (Fe), and magnesium (Mg)

Granite   Diorite   Gabbro   Peridotite

a body of magma, along with the prevailing temperature, pressure, and its water content, distinctly characterize a magma. Ultimately these factors determine the mineral content of the rocks it will form. Water vapor (H_2O), carbon dioxide (CO_2), and sulfur dioxide (SO_2) are the major dissolved gases in molten rock, accounting for a small percentage of a magma's total volume.

We saw in Chapter 2 that the Earth's crust consists primarily of silicon- and oxygen-based minerals. These silicates are also the major constituents of igneous rocks. Based on their silica content (the amount of silicon and oxygen), igneous rocks and magmas are classified as ultramafic, mafic, intermediate, or felsic (see Table 3.1). Ultramafic rocks have the smallest proportion of silica to other ions, whereas felsic rocks have the largest. Figure 3-21 illustrates how the mineralogical composition of the common igneous rocks varies in each of these categories.

Ultramafic Igneous Rocks

The term "mafic" is derived from *magnesium* and *ferrum* (Latin for "iron"). Ultramafic rocks are dominated by the iron–magnesium silicate minerals olivine and pyroxene and contain relatively little silica (less than 40%) and no free quartz. The most common ultramafic rock, **peridotite** (pe-RID-o-tite), contains between nearly 60% and almost 100% olivine. Ultramafic rocks such as peridotite aren't *technically* igneous rocks because they did not crystallize from a magma. (The relatively rare *extrusive* equivalent of peridotite is called *komatiite*, and most komatiite erupted early in the Earth's history when the Earth was much hotter than it is today.) Peridotites are essential to a discussion of igneous rocks and processes, however, because they are the dominant portion of the Earth's mantle, and they are the source rocks that melt to form basalt. Because they are rich in iron and magnesium, poor in silica (silicon and oxygen),

and occur at great depth in the Earth (that is, under great pressures), ultramafic rocks are very dense. Many ultramafic rocks are green because they contain so much olivine. Although it is characteristic of the mantle, peridotite can also be found on continents where converging tectonic plates have collided, and deep rocks have been thrust or otherwise brought up to the surface.

Mafic Igneous Rocks

Mafic igneous rocks have a silica content between 45% and 55%. These are the most abundant rocks of the Earth's crust, and the mafic rock basalt is the single most abundant of them all. Basalt, whose principal minerals include pyroxene, calcium feldspar, and olivine, is dark in color and relatively dense. Because it forms from molten rock that has cooled fairly rapidly at or near the surface, its texture is typically aphanitic. As noted above, basalt is the dominant rock of the world's oceanic crust, making up most of the ocean floor and many islands, including the entire Hawaiian chain (Fig. 3-22).

When a magma containing the same mix of minerals as basalt cools more slowly underground, its plutonic equivalent, the coarse-grained phaneritic rock **gabbro,** is produced. Gabbro outcrops are generally seen only where extensive erosion has removed surface rocks. Geologists, drilling deep along mid-ocean ridges, have found that gabbro lies beneath the basalts of the ocean floor, and therefore makes up the deep parts of oceanic crust.

Intermediate Igneous Rocks

Intermediate igneous rocks contain more silica than mafic rocks—between about 55% and 65%. They typically consist of iron and magnesium silicates, such as pyroxene and amphibole, along with sodium- and aluminum-rich minerals, such as calcium-sodium plagioclase and mica, and a small amount of quartz. They are generally lighter in color than

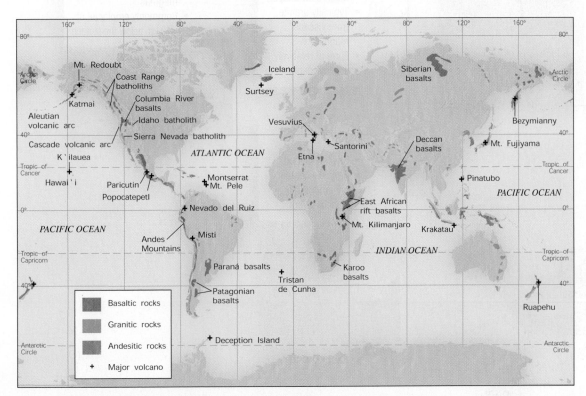

◀ **FIGURE 3-22** Distribution of the Earth's major continental igneous-rock provinces. As well as being the most abundant igneous rock of the ocean floors, the mafic igneous rock basalt is widespread on the continents. Basalt covers several large areas of our continents, most notably in Brazil, India, South Africa, Siberia, and the Pacific Northwest of North America.

3.1 Geology at a Glance

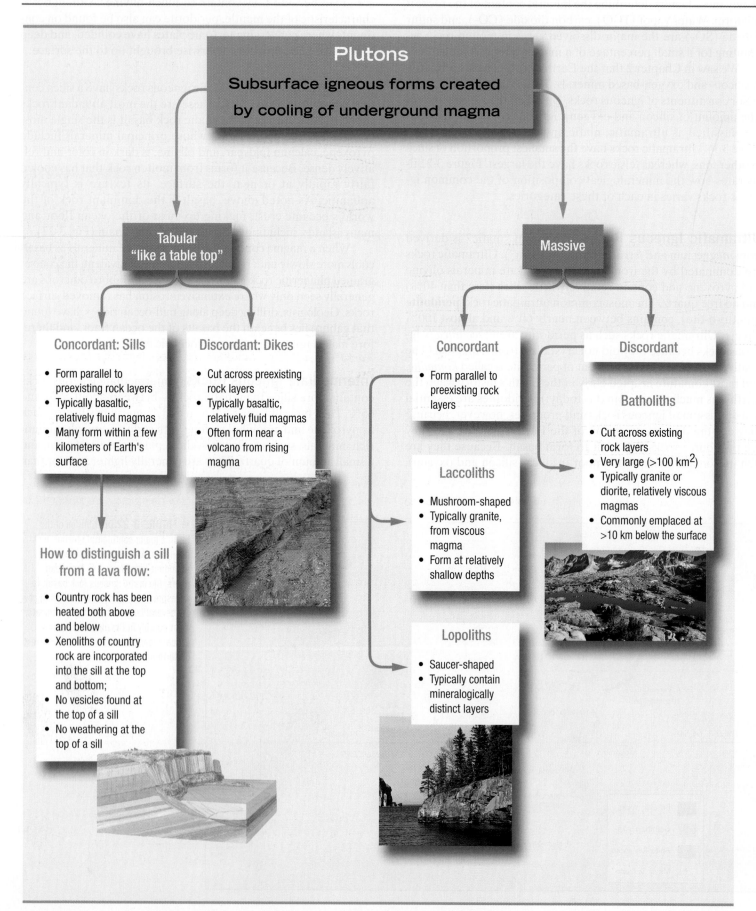

Plutons

Subsurface igneous forms created by cooling of underground magma

Tabular
"like a table top"

Massive

Concordant: Sills

- Form parallel to preexisting rock layers
- Typically basaltic, relatively fluid magmas
- Many form within a few kilometers of Earth's surface

Discordant: Dikes

- Cut across preexisting rock layers
- Typically basaltic, relatively fluid magmas
- Often form near a volcano from rising magma

How to distinguish a sill from a lava flow:

- Country rock has been heated both above and below
- Xenoliths of country rock are incorporated into the sill at the top and bottom;
- No vesicles found at the top of a sill
- No weathering at the top of a sill

Concordant

- Form parallel to preexisting rock layers

Laccoliths

- Mushroom-shaped
- Typically granite, from viscous magma
- Form at relatively shallow depths

Lopoliths

- Saucer-shaped
- Typically contain mineralogically distinct layers

Discordant

Batholiths

- Cut across existing rock layers
- Very large (>100 km^2)
- Typically granite or diorite, relatively viscous magmas
- Commonly emplaced at >10 km below the surface

mafic rocks. The intermediate igneous rock **andesite,** which is commonly porphyritic, is the world's second most abundant type of volcanic rock, and commonly occurs in the volcanoes that form on continental margins. Andesite is named for the Andes mountains of South America, where it dominates the local geology (see Fig. 3-22). Andesite is also found in the Cascade Mountains of North America—the volcanic chain that contains Mount Rainier and Mount St. Helens. Andesites contain pyroxene, hornblende (an amphibole), a minor amount of quartz, and an abundance of andesine, a plagioclase mineral with about 60% sodium and 40% calcium ions. Andesite's plutonic equivalent is **diorite,** which can be recognized by its coarse-grained, salt-and-pepper appearance (because both dark- and light-colored grains are present).

Felsic Igneous Rocks

The term "felsic" is derived from *fel*dspar and *si*lica. Felsic igneous rocks contain more silica—65% or more—than mafic or intermediate igneous rocks. They are generally poor in iron, magnesium, and calcium silicates, but rich in potassium feldspar, aluminum-rich micas, and quartz. The most common felsic igneous rock is the plutonic rock granite, the major rock of continental crust. Potassium feldspars typically impart granite's usual pinkish cast. In many specimens, sodium feldspars contribute a porcelain-like look to these rocks. Quartz grains and flakes of biotite mica are also scattered throughout granite. The magma that produces granite flows and crystallizes very slowly underground, seldom reaching the surface. Thus outcrops of rhyolite, granite's volcanic equivalent, are far less common than those of granite. Granitic rocks can generally be seen at the surface only after erosion removes overlying rocks. As you can see in Figure 3-22, granitic igneous rocks are exposed sporadically on the Earth's continents. Rocks of felsic composition have a greater variety of textures than any other igneous rock. They range from aphanitic rhyolites and several glassy rocks to ultra-coarse pegmatites.

One note about the felsic glass obsidian—don't be fooled by its dark color. Typically, felsic rocks are very light in color, but the presence of tiny amounts of iron give obsidian a dark color even though it is one of the most silica-rich igneous rocks. The reason why obsidian is black—its smattering of iron scatters and traps the light that enters the glass rather than transmitting it back to our eyes.

Intrusive Rock Formations

No one has ever actually seen magma move underground, so we can only speculate about how it does so from the igneous formations that we can see at the surface. In some areas, erosion has exposed thousands of square kilometers of solidified intrusive magma. Did these vast regions result from magma flowing into preexisting subterranean cavities? That is unlikely: The weight of overlying rocks at depths below about 10 km (6 mi) would have collapsed and destroyed any large underground spaces. The most likely scenario is that magma moved into fractured or faulted zones, in some cases incorporating parts of the surrounding rock

into the magma. Rising magma may force overlying rocks to bulge upward. The resulting igneous rock forms a domed intrusion within other rocks, an igneous structure known as a *diapir* (DI-a-pir).

When moving magma incorporates preexisting rock as it rises, some of the incorporated rocks melt (those with relatively low melting points) and become assimilated into the magma, while other unmelted rocks (those with relatively high melting points) are carried within the magma. When this magma eventually solidifies, we can see such "foreign" incorporated rocks within plutonic igneous rock as distinctly different rock masses called **xenoliths** (ZEEN-o-liths; from the Greek *xenos,* or "stranger," and *lithos,* or "stone") (Fig. 3-23).

The subsurface intrusive igneous forms created by solidified underground magma are known as **plutons.** Plutons may be classified by their position relative to the preexisting rock, or *country rock,* that surrounds it. *Concordant* plutons lie parallel to layers of country rock; *discordant* plutons cut across layers of country rock. Plutons of both varieties come in a range of shapes and sizes, as shown in Figure 3-24. Erosion of overlying rocks may eventually expose these formations.

Tabular Plutons

Tabular plutons are slablike intrusions of igneous rock that are broader than they are thick, like a table top. If magma flows into a relatively thin fracture in country rock, or pushes into the space between sedimentary rock layers, a tabular pluton will result when the magma cools. Tabular plutons may be as small as a few centimeters thick or as large as several hundred meters thick.

A **dike** is a *discordant* tabular pluton that *cuts across* preexisting rocks. Dikes are generally steeply inclined or nearly vertical, suggesting that they formed from rising magma, which tends to follow the most direct route upward. Dikes often occur in swarms where magma has infiltrated and solidified in a network of fractures.

Many dikes result from magma that rises into volcanoes and then solidifies. We get to see these dikes when the less-resistant volcanic material at the surface above them wears away, exposing them. Some dikes diverge like the spokes of a bicycle wheel from a *volcanic neck,* the vertical pluton of hard, erosion-resistant

▲ **FIGURE 3-23** A mafic xenolith within diorite. The dioritic magma incorporated this preexisting mafic rock (seen here as a dark gray mass within the lighter gray diorite) but was not hot enough to melt it. As a result, the mafic rock was preserved as a discrete mass when the magma eventually solidified. (Xenolith is 40 cm [12 in.] long.)

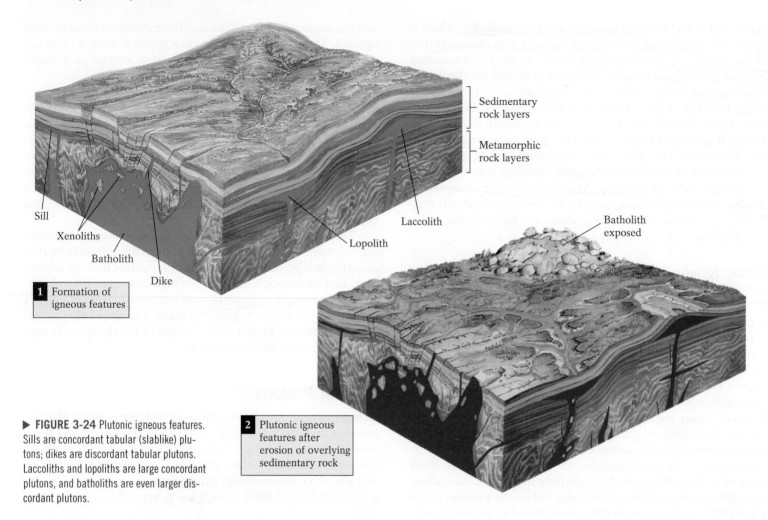

Sill

Xenoliths

Batholith

Dike

Sedimentary
rock layers

Metamorphic
rock layers

Laccolith

Lopolith

Batholith
exposed

1 Formation of
igneous features

2 Plutonic igneous
features after
erosion of overlying
sedimentary rock

▶ **FIGURE 3-24** Plutonic igneous features.
Sills are concordant tabular (slablike) plutons; dikes are discordant tabular plutons.
Laccoliths and lopoliths are large concordant
plutons, and batholiths are even larger discordant plutons.

volcanic rock that remains in what was once a volcano's central
magma pathway (Fig. 3-25). If you travel along Route 64, which
runs through the Navajo and Hopi lands of America's Four
Corners area (the intersection of Colorado, New Mexico, Arizona, and Utah), you can see more than a hundred such volcanic
necks, remnants of the region's ancient volcanic "plumbing."
Some of these volcanic necks, called **diatremes,** are the products
of deep-seated, extremely gaseous, highly explosive eruptions.
Diatremes contain highly fragmented chunks of lava mixed with
rock ripped from the vent wall, and masses of rock from the
Earth's deep crust and mantle. To picture such an explosive blast,
imagine the power of a space-shuttle booster rocket sticking vertically from the Earth's surface—upside down—blasting rocks
and gas skyward.

A **sill**—a *concordant* tabular pluton—forms parallel to layers
of preexisting rocks. Sills are produced when intruding magma
enters a space between layers of rock, sometimes melting and incorporating adjacent sedimentary material. Sills generally form
within a few kilometers of the Earth's surface; at greater depths,
the great weight of the overlying rocks compresses and closes off
any spaces into which magma might flow.

A sill and a dike in the south-central Pennsylvania town of
Gettysburg provided the setting for an event that affected the
course of American history. This event is recounted in Highlight
3-1, "Tabular Plutons Save the Union."

Sills pose an interesting challenge to geological detectives. How can we distinguish a sill intruded between layers
of preexisting rock from a lava flow that is buried under
subsequent flows or sedimentary rocks? There are several critical clues (Fig. 3-26):

1. Look at the adjacent surfaces of the surrounding rocks. When
    lava is extruded (that is, it flows out at the surface), there are
    no overlying rocks and only the rocks *beneath* it will be
    heated. The subsequent sedimentary layers that bury the lava
    flow are laid down *after* the lava has cooled and thus, they're
    not "baked" (that is, they show no evidence of heating).
    When hot magma intrudes between two layers of rock to
    form a sill, it heats the adjacent surfaces of *both* layers—
    above and below—before cooling.

2. Look at the top and bottom surfaces of the igneous layer.
    Both surfaces of a sill will contain fragments of the sur-
    rounding rock that were pried loose as magma intruded,
    whereas only the bottom of a lava flow incorporates chunks
    of the preexisting rock.

3. Look at the top surface of the igneous layer. Because the top
    of a lava flow is exposed to the air for some time before being
    overlain by other lava or sediment, its gases are free to es-
    cape. Consequently, the surface of a lava flow is often pock-

▲ **FIGURE 3-25** Shiprock Peak, in New Mexico, is believed to be a diatreme, a form of volcanic neck. The rock that makes up Shiprock Peak is highly fragmented, a consequence of an explosive gas-charged eruption about 27 million years ago that shot chunks of the Earth's interior toward the surface. The peak stands 515 m (1700 ft) above the surrounding plain. Erosion of the sedimentary rocks of the surrounding plain has exposed this diatreme and its radial dikes as can be seen in the adjacent sequence of events that produced today's view at Shiprock.

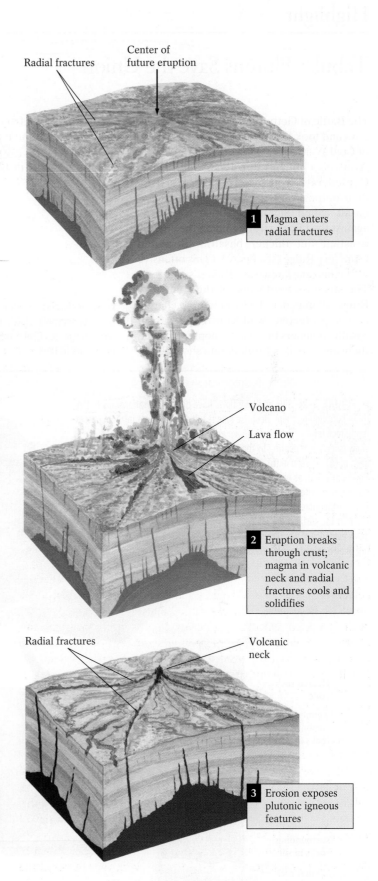

marked by cavities called *vesicles* that were formerly occupied by escaping gas bubbles. The tops of most sills—not exposed to the air while they're cooling—display no vesicles.

4. Look for signs of weathering (discussed in Chapter 5). The upper surface of a lava flow would appear somewhat weathered from its exposure to the atmosphere before being buried by subsequent layers. A sill would show no signs of weathering because it was not exposed to the weathering environment at the Earth's surface.

Other Plutons and Batholiths

Large plutons are commonly several kilometers thick and tens or even hundreds of kilometers across. When thick, viscous magma intrudes between two parallel layers of rock and lifts the overlying one, it cools to form a mushroom-shaped, or domed, concordant pluton, or **laccolith** (from the Greek *lakkos*, or "reservoir") (see Fig. 3-24). Many laccoliths are granitic, formed from felsic magma that flows so slowly that it tends to bulge upward instead of spreading laterally (Fig. 3-27).

Unlike the upward-bulging laccoliths, **lopoliths** (from the Greek *lopas*, or "saucer") are saucer-shaped concordant plutons that sag downward (see Fig. 3-24). These are probably produced when mafic magma rises from a deep magma source. The weight of the dense magma then depresses the underlying rock, pushing it downward into the space from which the magma has risen. One such structure is exposed along the western shore of Lake Superior, where the overlying country rock has been eroded away (Fig. 3-28). Lopoliths often contain layers of concentrated minerals—including some with great economic value. The layers near the base of the lopolith often hold the densest

Highlight 3-1

Tabular Plutons Save the Union

The Battle of Gettysburg, which lasted three days and took the lives of tens of thousands of Civil War soldiers, was won by the Union Army on a hot July 3 in 1863. On this day, Confederate troops ventured forth from their outpost on a narrow dike of resistant basalt called Seminary Ridge to charge against the Union stronghold on the equally resistant but thicker basaltic sill called Cemetery Ridge (see figure). (This offensive would become known as "Pickett's charge.") The steep forward slope of the Cemetery Ridge sill impeded the Confederate charge, and a protective wall constructed from basaltic boulders by Union troops concealed them and repelled Confederate shots. Thus,

with an assist from a well-placed basaltic sill, Union forces defeated the Confederate offensive at Gettysburg, a turning point in the American Civil War.

▶ The mafic sills and dikes of Gettysburg, Pennsylvania, played an important role in the pivotal Civil War Battle of Gettysburg in 1863.

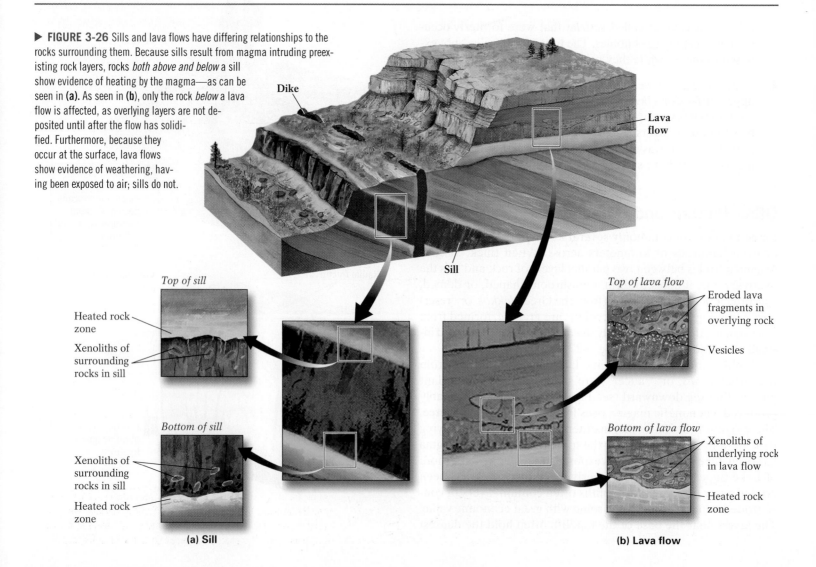

▶ **FIGURE 3-26** Sills and lava flows have differing relationships to the rocks surrounding them. Because sills result from magma intruding preexisting rock layers, rocks *both above and below* a sill show evidence of heating by the magma—as can be seen in (a). As seen in (b), only the rock *below* a lava flow is affected, as overlying layers are not deposited until after the flow has solidified. Furthermore, because they occur at the surface, lava flows show evidence of weathering, having been exposed to air; sills do not.

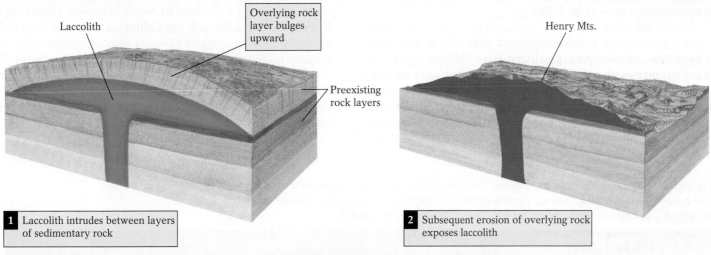

Laccolith

Overlying rock layer bulges upward

Preexisting rock layers

Henry Mts.

1 Laccolith intrudes between layers of sedimentary rock

2 Subsequent erosion of overlying rock exposes laccolith

▲ **FIGURE 3-27** The best laccolith-viewing in North America is along Route 95, through the Henry Mountains of southeastern Utah. The laccolith is the massive gray-blue structure in the background. The diagram shows the general mushroom shape of a laccolith.

◄ **FIGURE 3-28** The rocks of the gabbroic Duluth lopolith are exposed along the shores of the western end of Lake Superior in Minnesota. The lopolith is more than 250 km (150 mi) in diameter and about 15 kilometers (10 mi) thick. This photo shows only a small portion of the lopolith.

early-forming crystals. Several layers near the base of the Bushveld lopolith of South Africa contain the Earth's richest concentration of the dense metal platinum, which is more valuable per ounce than gold.

Batholiths (from the Greek *bathos,* or "deep") are massive plutons with surface areas (when exposed) of 100 km² (40 mi²) or more (see Fig. 3-24) that typically form from intermediate- to high-viscosity magmas (diorites and granites). These magmas flow so slowly, they are more likely to remain underground and crystallize than to rise to the surface because the viscous magma can't flow readily into small cracks and fissures in near-surface rocks. Most batholiths are a complex of many large plutons that have intruded one another in stages, often over tens of millions of years. The Sierra Nevada batholith in east-central California is an excellent example of this, as it consists of numerous plutons packed so closely as to appear to be a continuous mass (Fig. 3-29).

Batholiths also reside in the ancient centers of continents. These deep, erosion-resistant rocks form the cores of long-gone mountains crystallized 2 to 3 billion years ago. These mountains may have risen sharply at what were then the ancient convergent boundaries of the Earth's earliest plates. In North America, they have since been relegated to their mid-continental positions in what are now Minnesota, Wisconsin, Michigan, and Ontario by vast collisions of ancient lithospheric plates that added new crust to the margins of the most-ancient continents.

Plate Tectonics and Igneous Rock

The worldwide distribution of igneous structures and rocks *is not random.* Certain structures and rocks consistently appear in some geological settings but not in others. In general, most igneous rocks form at or near plate boundaries. For example, intermediate and granitic batholiths form where oceanic plates have subducted, marking many modern and ancient plate boundaries. The chain of western batholiths in North America, stretching from British Columbia through the California Sierras to Baja California in Mexico, developed through tens of millions of years of oceanic-plate subduction beneath the western edge of North America's continental crust. If you find igneous rocks that are not near a modern plate boundary, they may have formed near an ancient plate boundary. There are prominent exceptions to this principle—such as intraplate volcanoes of Hawai'i and Yel-

▼ FIGURE 3-29 Virtually all the rock exposed in Kings Canyon National Park, California, is composed of only one rock type—granodiorite (an intrusive igneous rock similar to granite but which contains more plagioclase feldspar than potassium feldspar). The panorama of granodioritic rock here is part of the Sierra Nevada batholith.

lowstone National Park—but most igneous rocks do form near plate boundaries. To explain this, we will integrate what you have learned about how rocks melt with what you have learned about igneous-rock compositions.

The Origin of Basalts and Gabbros

Basalt and gabbro are virtually the only igneous rocks in oceanic crust. Because the thin oceanic crust lies directly above the mantle, the only source for the mafic ocean-floor magma must be the ultramafic mantle. So if the mantle is the source of these mafic rocks, why are the basalts and gabbros that form at one plate-tectonic setting different compositionally from the basalts and gabbros at another? These differences lie in the variations in the mantle composition that are related to the mantle's history and to the depth in the mantle from which the magmas are rising.

When the uppermost parts of the Earth's mantle partially melt, the buoyant magma rises to become the Earth's oceanic

and continental crust. Geologists debate whether most continental crust formed early in Earth's history during large-scale mantle melting or whether continental formation has continued throughout Earth history. Whichever the case, continental-crustal formation depletes the mantle in elements such as potassium and uranium, which tend to be concentrated in melts. A basalt or gabbro that is relatively rich in these elements compared to other basalts and gabbros is probably derived from deep, relatively undepleted regions of the mantle—the regions that have not lost potassium and uranium through past continent formation. Basalts and gabbros that are depleted in those elements must have formed at shallower levels—that is, the parts of the mantle that have been depleted of these elements by continent formation (Fig. 3-30).

Basalts are grouped into two main categories—oceanic and continental—based on the general setting in which they formed. Within each of the categories, basalts can be further distinguished by the related factors of composition, magma source, and plate-tectonic setting.

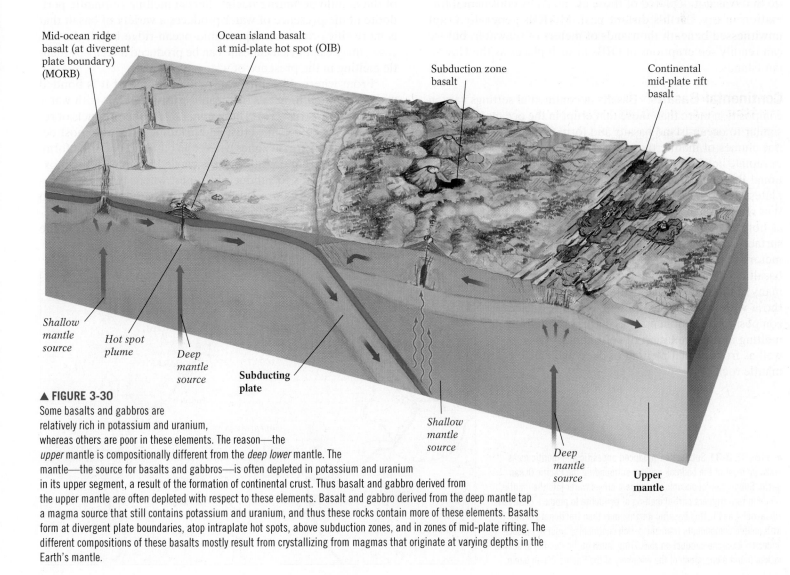

▲ FIGURE 3-30
Some basalts and gabbros are relatively rich in potassium and uranium, whereas others are poor in these elements. The reason—the *upper* mantle is compositionally different from the *deep lower* mantle. The mantle—the source for basalts and gabbros—is often depleted in potassium and uranium in its upper segment, a result of the formation of continental crust. Thus basalt and gabbro derived from the upper mantle are often depleted with respect to these elements. Basalt and gabbro derived from the deep mantle tap a magma source that still contains potassium and uranium, and thus these rocks contain more of these elements. Basalts form at divergent plate boundaries, atop intraplate hot spots, above subduction zones, and in zones of mid-plate rifting. The different compositions of these basalts mostly result from crystallizing from magmas that originate at varying depths in the Earth's mantle.

Oceanic Basalts *M*id-*o*cean *r*idge *b*asalts, or MORBs, the Earth's most abundant volcanic rock, account for about 65% of the Earth's surface area. Eruptions at oceanic divergent boundaries produce these rocks. Because they have low concentrations of elements easily depleted by melting (such as potassium and uranium), MORBs probably formed from partial melting of the *upper* mantle, which has already been depleted by melting events in the mantle earlier in Earth's history. Recall from Figure 3-6 that melting may occur without heating if already-hot deep rocks rise to shallower levels into temperature and pressure conditions suitable for melting. This happens beneath oceanic ridges where solid mantle peridotite rises up and partially melts at depths of 15–80 km (10–50 mi), creating the basaltic magma that then rises and solidifies as oceanic crust. About 20% of this basalt erupts on the seafloor as MORBs; the rest crystallizes in sub-seafloor magma chambers, forming oceanic gabbros.

*O*cean *i*sland *b*asalts, or OIBs, are found not at divergent plate boundaries but atop **hot spots**—volcanic zones, generally *intra*plate (that is, located *within* a plate and not at its margin), that lie over *deep*-mantle heat sources. Because they contain larger amounts of potassium and uranium than MORBs, OIBs probably originated from a deep part of the mantle that was not depleted of those elements by continental formation in the Earth's distant past. MORBs generally erupt unwitnessed beneath thousands of meters of seawater, but we can readily see eruptions of OIBs in such places as the Hawaiian Islands.

Continental Basalts Basalts in continental settings vary in composition more than those that erupt in the oceans. Some are similar to ocean-island basalts and form from partial melting of hot plumes of mantle that rise from great depths—typically at intraplate hotspots within continents, such as at Yellowstone National Park. Others form where new rifts tear at old continental plates. Still others form where oceanic plates subduct, and are thus related to andesites (described in the next section). All form as hot basaltic magma in the mantle, and their eruption at the surface is preceded by ascent through tens of kilometers of continental crust. As a result, these basaltic magmas (typically quite hot) incorporate many of the materials along their path (especially those with low melting points). Thus the varied composition of continental basalts may result from melting and assimilation of continental rocks as well as from partial melting of deep and shallow mantle rocks.

The Origin of Andesites and Diorites

Whereas basalts and gabbros are the most common igneous rocks of the divergent zones and hot spots within ocean basins, andesites and diorites are the primary igneous products of subducting oceanic plates. Vast regions of andesitic rock are found on virtually all the lands that border the Pacific Ocean, tracing the nearly continuous pattern of subduction zones surrounding the Pacific Ocean basin (Fig. 3-31). Some of these regions are old edges of continents that have had younger magmatic rocks added to them (for example, the western coasts of North and South America). Some of these regions are islands built from the seafloor almost entirely by subduction-related magmatism (for example, Japan, the Philippines, the Aleutian Islands of Alaska, and many of the other islands in the northern and western Pacific).

Geologists believe that a number of processes combine to produce rocks of intermediate composition in a subduction zone. One likely factor is water. The mantle is hot, but it is solid because of the high pressures applied by overlying crustal rocks. Recall that water *lowers* the melting point of rocks under pressure. When water driven from the subducting plate enters the hot, solid mantle, its melting point is lowered and partial melting of the mantle peridotite begins. Partial melting of mantle peridotite in the presence of water produces a variety of basalt that is more silica-rich than typical mid-ocean-ridge basalt. In rare cases, magnesium-rich andesite can be produced *directly* by mantle melting in the presence of water.

From where does this water come? Some of it is bonded within hydrated minerals of oceanic crust (minerals with water bound to their structure). These minerals are the products of reactions between oceanic crust and invading seawater, most occurring near divergent zones where the basaltic rocks first form. The water-bearing oceanic crust is carried along with the moving oceanic plates to subduction zones, where the water is released as the rocks are subducted and heated. Water is also trapped in spaces within the sediments that blanket the descending plate, within the crystal structures of minerals in the

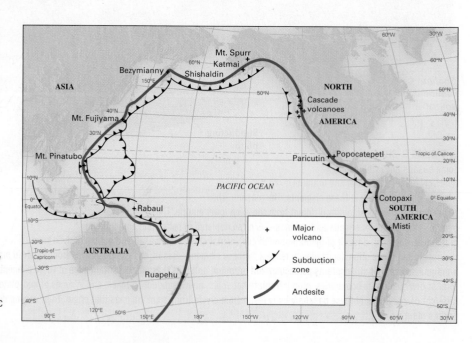

▶ **FIGURE 3-31** Subduction-produced andesitic and dioritic rocks make up most of the bedrock geology surrounding the Pacific Ocean basin. Subduction of oceanic lithosphere drives water into the mantle, which in turn triggers partial melting of peridotite to produce relatively silica-rich basalt. This basaltic magma may then fractionate or mix with melted components from silica-rich continental crust to produce andesite. Because subduction zones ring much of the Pacific, andesitic rock is found along much of the periphery of the Pacific Ocean basin.

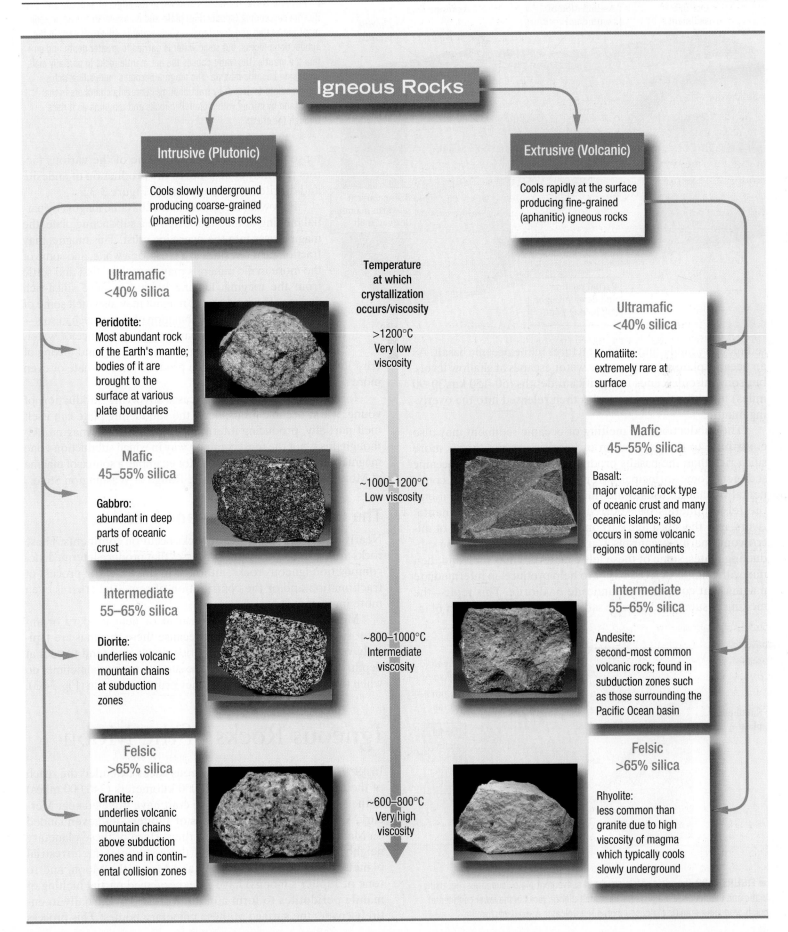

Igneous Rocks

Intrusive (Plutonic)

Cools slowly underground producing coarse-grained (phaneritic) igneous rocks

Ultramafic
<40% silica

Peridotite:
Most abundant rock of the Earth's mantle; bodies of it are brought to the surface at various plate boundaries

Mafic
45–55% silica

Gabbro:
abundant in deep parts of oceanic crust

Intermediate
55–65% silica

Diorite:
underlies volcanic mountain chains at subduction zones

Felsic
>65% silica

Granite:
underlies volcanic mountain chains above subduction zones and in continental collision zones

Extrusive (Volcanic)

Cools rapidly at the surface producing fine-grained (aphanitic) igneous rocks

Ultramafic
<40% silica

Komatiite:
extremely rare at surface

Mafic
45–55% silica

Basalt:
major volcanic rock type of oceanic crust and many oceanic islands; also occurs in some volcanic regions on continents

Intermediate
55–65% silica

Andesite:
second-most common volcanic rock; found in subduction zones such as those surrounding the Pacific Ocean basin

Felsic
>65% silica

Rhyolite:
less common than granite due to high viscosity of magma which typically cools slowly underground

Temperature at which crystallization occurs/viscosity

>1200°C
Very low viscosity

~1000–1200°C
Low viscosity

~800–1000°C
Intermediate viscosity

~600–800°C
Very high viscosity

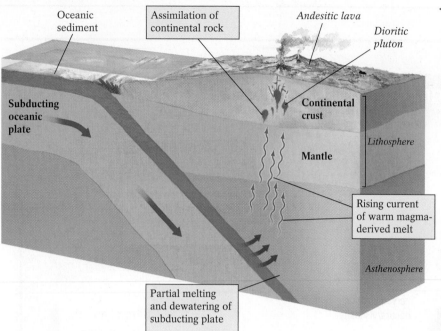

◀ **FIGURE 3-32** The factors involved in the origin of andesite and diorite. The descending (subducting) plate and its associated oceanic sediment are heated and dewatered—some water is driven off early in the subduction process, but some water is carried to greater depth and enters the mantle. The water causes the hot mantle rocks to partially melt, producing basaltic magma. The magma becomes more silica-rich (intermediate-to-felsic) by fractionating in magma chambers in the crust, and by mixing with more felsic rocks and magmas as it rises through the crust.

tense debate and research. Some of the various factors that may contribute to the production of andesite and diorite are summarized in Figure 3-32.

Following the production of basaltic magma by partial melting of the mantle above a subducting plate, the magma rises into the overlying crust. This magma may fractionate if it resides in a crust for a while, and some of the more mafic minerals may crystallize first and settle from the magma, leaving behind a more silica-rich magma. The hot basaltic magma may also melt some of the existing crust. A combination of these mechanisms—partial melting of the mantle in the presence of water, fractionation of the magma, and melting and mixing of crust into the original magma—can produce intermediate or even more silica-rich magma.

In rare, somewhat isolated cases involving subduction of young, warm oceanic lithosphere, the subducting plate can itself melt partially, producing intermediate-composition magma. Although this is not the most common way in which subduction-zone magmas form, it does occur, and is, for example, a source of magma at Mount St. Helens in the Cascade Range of Washington State.

The Origin of Rhyolites and Granites

Nearly all rhyolitic and granitic rocks occur on continents. These rocks are created by the same processes that produce intermediate-composition igneous rocks, although in this case, the process of fractionation and/or the contribution of remelted crust is even more important.

Most granitic intrusions appear at or near modern or ancient subduction plate margins. Because these magmas are typically very viscous, they generally rise slowly and tend to cool at depth. When such magmas reach the surface, as they sometimes do when their water content is high, they erupt as rhyolite (Fig. 3-33).

Igneous Rocks of the Moon

In 1969, geologists who study igneous rocks extended the reach of their rock hammers some 400,000 kilometers (240,000 miles) to the Moon. More recently, space craft have traveled near Mercury, Venus, Mars, and the moons of Jupiter—and even landed on Mars—to analyze the rocks at the surface of these planetary neighbors. A major conclusion of these studies—the terrestrial planets (Mercury, Venus, Earth, and Mars), the Moon, and Io (one of Jupiter's moons) have all experienced partial melting of mantle peridotites to form a basaltic crust that partially to entirely covers the surface of these planetary bodies. This process

sedimentary muds, and within fractures in the oceanic basalt. As an oceanic plate subducts, some water escapes at shallow levels, but some also descends to significant depths (50–150 km/30–90 miles) in subduction zones and is then released into the overlying mantle.

The subduction and melting of oceanic sediment may also contribute to the production of basaltic magma that is more silica-rich than the basalts produced at oceanic ridges. Oceanic sediment comes principally from the airborne debris of continental volcanism (usually intermediate-to-felsic in composition), the felsic minerals transported to the oceans by continental rivers, and the silica and carbonate shells and skeletons of microscopic marine organisms. Carried into the hot mantle on subducting plates, some of these felsic materials may melt and then mix with mantle-derived basalt to help produce an intermediate magma that cools to form andesite or diorite. This topic—the melting of subducted oceanic sediment—is still the subject of intense

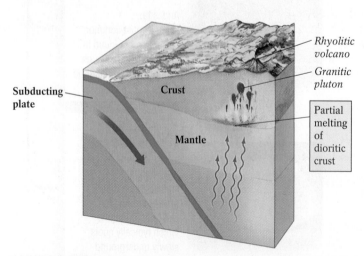

▲ **FIGURE 3-33** The origin of felsic rocks at convergent plate boundaries. Hot rising mafic and intermediate magmas partially melt dioritic rocks in the lower continental crust, producing granitic plutons and granite's volcanic equivalent, rhyolite.

of mantle melting to form basaltic crust is therefore a general process of the rocky bodies in the solar system.

Moon rocks collected by *Apollo* astronauts in the 1970s indicate that the Moon's surface contains at least two distinct types of geological/geographical provinces—the highlands and the *maria* (MAR-ee-a) (plural of Latin *mare,* or "sea") (Fig. 3-34). The rocks of the lunar highlands date from 4.0 to 4.5 billion years ago, a time when the early Moon's interior was apparently hot enough to develop a multilayered internal structure. The rocks of the lunar highlands, which probably crystallized as the Moon's earliest crust, consist principally of anorthosite, a coarse-grained plutonic igneous rock composed almost exclusively of the calcium plagioclase anorthite.

The plagioclase-rich rocks of the lunar highlands are fascinating because they are the prime reason the Moon shines so brightly in the night sky—the sunlight reflects very well off the bright gray-white anorthosite. The anorthosites formed within a magma ocean on the Moon as plagioclase crystals—less dense than the surrounding basaltic magma—floated to the ocean's surface, just as icebergs float on denser water.

Unlike the Earth's *watery* oceans, the lunar maria, or "seas," are actually vast solidified basalt flows. They were named when Galileo and other early astronomers, using the primitive telescopes of the time, believed they were looking at true seas. From about 4 to 3.3 billion years ago, intense meteorite activity gouged numerous craters in the Moon's predominantly anorthositic surface. The impacts removed great volumes of overlying rock, instantaneously reducing the pressure on the Moon's ultramafic upper mantle, and triggering partial melting. This process produced basaltic magma that rose to the surface through the fractures in the crust created by the meteorite impacts. Basaltic lava flowed into and filled the craters, forming the lunar maria (Fig. 3-35). Recent studies of the lunar maria suggest that small basaltic eruptions continued on the Moon until about 1.2 billion years ago.

▲ **FIGURE 3-34** A telescopic view of the near side of the Moon, showing its highlands and maria. (Dark areas are the maria; light areas are the highlands.) The light areas reflect sunlight from the surfaces of anorthite crystals.

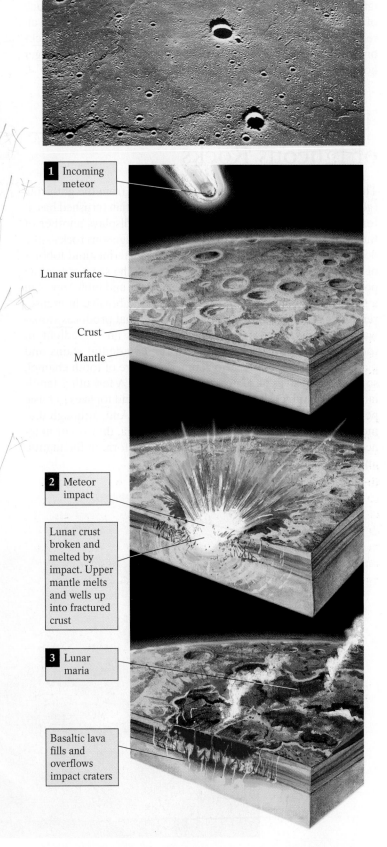

1 Incoming meteor

Lunar surface

Crust

Mantle

2 Meteor impact

Lunar crust broken and melted by impact. Upper mantle melts and wells up into fractured crust

3 Lunar maria

Basaltic lava fills and overflows impact craters

▲ **FIGURE 3-35** Lunar maria—the vast solidified basalt flows of the Moon—form from meteor impacts. Above: A photo of lunar craters, mountains, and mare taken during the *Apollo 10* mission of May 1969.

Although minor cratering continues today, the eruptions of the mare basalts apparently marked the final episodes of major igneous activity on the Moon. By the close of this period, much of the debris remaining from the origin of the solar system had already been swept up. The relative lack of younger igneous rocks on the Moon's surface today suggests that its interior is no longer hot enough to produce new magmas.

The Economic Value of Igneous Rocks

The practical uses of igneous materials range from the glittering (gemstones and precious metals) to the utilitarian (crushed basalt for road construction). Every urban center displays another of the principal uses we have found for plutonic igneous rocks—the decorative building stone that adorns the exteriors and lobbies of many banks and office buildings. We see the same appealing polished igneous rocks in kitchens as counter and table tops. On a smaller scale, glassy pumice is a common abrasive in grease-removing cleansers, and is also the "stone" that produces stone-washed jeans. Until recently, pumice was an ingredient in toothpaste because of its ability to remove dental stains and plaque. Unfortunately, it also claimed its share of tooth enamel, so milder abrasives have now taken its place. A few other familiar desirable minerals, such as the emeralds and topazes in felsic pegmatitic rocks, are also of igneous origin. And, although diamonds don't crystallize directly from a magma, they occur in igneous rocks (typically kimberlite dikes) that form in the mantle and intrude ancient continental crust.

Gold and silver often appear in or around granitic rocks, as do less shiny but highly valuable ores of copper, lead, and zinc. These late-crystallizing metallic ions typically become concentrated in the hot, watery fluids that circulate after most other minerals have crystallized. When these fluids enter fractures in adjoining rocks, the elements crystallize as metal-rich veins. Recently, geologists studying the eruptive potential of Colombia's Galeras volcano discovered what prospectors have always counted on: Some volcanoes manufacture gold deposits. Highlight 3-2 illustrates this useful but *dangerous* phenomenon.

Groundwater that percolates down through an igneous region may redeposit and thus concentrate valuable minerals. When the water comes in contact with a magma chamber or a body of still-warm plutonic rock, it heats up and some or all of it may be converted to steam. This steam may then invade surrounding rocks and dissolve soluble minerals in them. When the steam eventually cools and condenses, its dissolved load recrystallizes to form mineral-rich deposits. The gold of the Homestake Mine in the Black Hills of South Dakota, the lead, silver, and zinc of northern Idaho and the copper of northern Michigan and Bingham Canyon in northern Utah (Fig. 3-36) all accumulated from the action of such heat-driven subterranean fluids, known as *hydrothermal fluids.*

We can now build on your general knowledge of the Earth's igneous processes and rocks and expand our discussion of igneous activity to what we can actually see *above ground.* The following chapter focuses on igneous phenomena we can observe—volcanic eruptions and the rocks produced when magma reaches the Earth's surface and escapes as lava.

▲ **FIGURE 3-36** The Bingham Canyon copper mine near Salt Lake City, Utah, where plutonic igneous rocks are exposed over a 5- by 8-km (3- by 5-mi) area. Subterranean magmas and groundwater deposited the abundant copper found throughout these rocks.

Highlight 3-2

Making Gold in an Active Volcano

Galeras volcano looms ominously over the peaceful city of Pasto (population 300,000) in the Andes of western Colombia. One of the world's most dangerous volcanoes, Galeras erupted unexpectedly in January 1993, killing six geologists and three tourists and inspiring an urgent vigil that continues today. Teams of geologists regularly probe the volcano's crater (see photo), collecting gases, water from hot springs, and samples of lava from past eruptions, searching for signs of an impending blast to avert a volcanic disaster. A team from New Mexico's Los Alamos National Laboratory recently visited Galeras for just that purpose but came away with a valuable discovery—the rocks of Galeras volcano are accumulating gold at a surprisingly rapid rate.

Laboratory tests of rock samples collected around gas vents in the valley revealed that veins in some old volcanic deposits were studded with tiny pencil-point-sized flecks of gold—as much as 7.8 ounces of gold per ton of rock. This concentration is quite impressive, especially when compared with concentrations of less than 1 ounce per ton within numerous active gold mines of the western United States. Geologists estimate that if the volcano remains active for roughly 10,000 years, as much as 200 tons of gold could accumulate within the volcano—a future gold mine in the making.

The volcano's gold, like the gold in Alaska, the Colorado Rockies, and California's Sierra Nevada range, was probably dissolved from crustal rocks by the heat and associated fluids from rising magma beneath the volcano. Water vapor and other gases, charged with dissolved gold, rise from the volcano's subterranean magma chamber and seep upward through the rocks of the volcano, disseminating the precious golden flakes. Compared with other volcanoes around the world, Galeras appears to be one of the richest. And more recently, geologists have discovered what may be one of the top-three gold deposits in the world—the young, and growing Ladolam gold deposit within the Luise caldera on Lihir Island, a tiny 15-km-long (10-mi), 10-km-wide (6-mi) section of Papua New Guinea, in the southwestern Pacific Ocean. But remember—with the lure of these riches comes the ever-present threat of a volcanic disaster without warning.

▶ Colombia's Galeras volcano—one of the world's most active and potentially explosive—is a veritable factory of gold production. Geologists' enthusiasm for searching for and mining this treasure, however, is tempered by Galeras' recent history of tragic eruptions.

Chapter 3 Summary

Igneous rocks are the most abundant type of rock in Earth's crust. They form when molten rock cools and crystallizes. Molten rock contained beneath Earth's surface is called magma; when it erupts onto the surface, it is called lava. The texture of an igneous rock reflects the rate at which its parent magma or lava cooled, and its mineral content reflects the composition and evolution of the molten rock from which it formed.

Intrusive, or plutonic, igneous rocks form from magma that cools slowly underground. These rocks are generally coarse-grained, or *phaneritic,* because ample cooling time allows crystals to grow to visible sizes. Extrusive, or volcanic, igneous rocks form when lava cools quickly at Earth's surface. These rocks are generally fine-grained, or *aphanitic,* because rapid cooling limits crystal growth.

The most common igneous rocks include:

- The *mafic* rocks—*basalt* (volcanic) and *gabbro* (plutonic). These rocks are rich in iron, magnesium, and calcium, and low in silica (45–55%); they crystallize and melt at high-temperature (1000–1200°C), form low-viscosity magmas, and are the major component of oceanic crust.

- The *intermediate* rocks—*andesite* (volcanic) and *diorite* (plutonic). These rocks are rich in iron, aluminum, calcium, and sodium and have intermediate silica content (about 60%); they crystallize at moderate temperatures (800–1000°C) from fairly high-viscosity magmas. Andesite and diorite are the major components of continental arcs (volcanoes on continental margins), associated with the subduction of oceanic plates.

■ The *felsic* rocks—*rhyolite* (volcanic) and *granite* (plutonic). These rocks are rich in potassium, sodium, and aluminum and are high in silica (>70%); they crystallize from very viscous magma at relatively low temperatures (600–800°C). Granite is a major component of continental crust.

It is also important to know about ultramafic mantle peridotite, an iron- and magnesium-rich rock containing less than 40% silica. Peridotite forms most of the mantle, and is the source rock of basalt and gabbro.

Most magmas are created when preexisting rocks partially melt—that is, the minerals with the lower melting points liquefy first and start to flow as a molten mass, leaving behind the unmelted portion of rock. The factors that affect a rock's melting point include:

■ local heat and pressue conditions;
■ the rock's water content;
■ the composition of the parent rock that melts.

Magma's viscosity—its resistance to flow—increases with decreasing temperature (and increasing silica content) and decreases with increasing temperature (and decreasing silica content). As magma cools, different minerals crystallize from it at different temperatures.

Bodies of magma that cool underground form plutons, igneous features that are distinct from surrounding rocks. Plutons are classified by their shapes, sizes, and orientation relative to the rocks they intrude. Common plutons include:

■ Dikes, which are thin, tabular plutons that cut across the host rock;
■ Sills, which are thin, tabular plutons that follow preexisting layers or other features of the host rock;
■ mushroom-shaped laccoliths and saucer-shaped lopoliths; large discordant plutons;
■ batholiths, which are large discordant plutons that can grow to a huge size by addition of different plutons over time.

Dikes are thin, tabular plutons that cut across the host rock. *Sills* are thin, tabular plutons that follow preexisting layers or other features of the host rock. Large concordant plutons include mushroom-shaped *laccoliths* and saucer-shaped *lopoliths;* large discordant plutons are called *batholiths,* which can grow to a huge size by addition of different plutons over time.

The principal igneous rock types are typically associated with specific plate-tectonic settings. Basalts and gabbros form at divergent plate boundaries (for example, mid-ocean ridge basalts, or MORBs), atop intraplate hot spots (the ocean island basalts, or OIBs), and where a continental plate is rifting. Andesites and diorites form where oceanic plates have subducted, creating chains of volcanic mountains. Volcanic belts may also contain rhyolite, and are commonly underlain by granite. Intermediate and silica-rich igneous rocks form from a combination of processes, including *partial melting* of the mantle in the presence of water, fractionation of mafic magma to produce more silica-rich compositions, and melting of preexisting continental crust.

KEY TERMS

andesite *(p. 85)*
basalt *(p. 72)*
batholith *(p. 90)*
Bowen's reaction series *(p. 76)*
diatremes *(p. 86)*
dike *(p. 85)*
diorite *(p. 85)*

extrusive rocks *(p. 80)*
fractional crystallization *(p. 78)*
gabbro *(p. 83)*
granite *(p. 72)*
hot spots *(p. 92)*
igneous rocks *(p. 70)*
intrusive rocks *(p. 79)*

laccolith *(p. 87)*
lava *(p. 70)*
lopolith *(p. 87)*
magma *(p. 70)*
partial melting *(p. 73)*
pegmatites *(p. 80)*
peridotite *(p. 83)*

plutons *(p. 85)*
plutonic rocks *(p. 79)*
rhyolite *(p. 72)*
sill *(p. 86)*
viscosity *(p. 75)*
volcanic rocks *(p. 80)*
xenolith *(p. 85)*

QUESTIONS FOR REVIEW

1. Why do granites and rhyolites have different textures? What are these textures called?

2. What elements would you expect to be most abundant in a mafic igneous rock? In a felsic igneous rock?

3. Name the common *extrusive* igneous rocks in which you would expect to find each of the following mineral types: calcium feldspar; potassium feldspar; muscovite mica; olivine; amphiboles; sodium feldspars. Which *plutonic* igneous rock contains abundant quartz and sodium plagioclase, but no olivine or pyroxene?

4. Describe two ways that rocks can melt without heating?

5. What is the difference between a sill and a dike?

6. Briefly discuss two specific types of plate-tectonic boundaries and the igneous rocks that are associated with them.

7. What type of igneous feature is shown in the photo to the right?

8. Felsic rocks such as rhyolite may occur together with basaltic rocks at locations where continents are undergoing rifting. Give one possible explanation for this.

9. How does an andesite form from a magma that is originally basaltic?

10. Describe and explain the relationship between a magma's silica content and its viscosity (compare and contrast mafic and felsic magmas).

11. Why is the Moon dominated by basalt and peridotite? That is, why aren't there granitic plutons and andesitic volcanoes on the Moon? Why does the Moon appear so bright in our night sky?

LEARNING ACTIVELY

1. Speculate about how the distribution of the Earth's igneous rocks will change when the Earth's internal heat is exhausted and plate-tectonic movement stops. Will the place where you live be affected?

2. Igneous rocks are everywhere. If you live in a town or city with a significant number of banks and other office buildings and libraries, walk down a city street and try to identify the various plutonic igneous rocks that have been cut and polished as decorative stone at the street level of most buildings. Try to identify the various rock-forming minerals (feldspars, quartz, micas, amphiboles, pyroxenes) in the stones with the coarsest textures.

3. If you live near a gravel pit, a pebbly beach, or a river that flows from some mountains, look for igneous rocks in these sediments. If you live in the northern U.S., southern Canada, or northern Europe, your region's gravels—probably glacial deposits—represent a sampling of many of the bedrock outcrops that the glaciers crossed as they flowed southward during the Earth's last major glacial advance about 15,000 years ago. If you live near a beach or stream that has received sediment from a mountainous area, you may also see abundant igneous rocks. A survey of the rock types provides a quick inventory of the various igneous rocks exposed in the region. Collect representative samples of each distinct igneous rock that you find, and sort them by texture or composition—either from most mafic to most felsic, or from coarsest-grained to finest-grained. Speculate about the possible plate-tectonic settings that are represented in your new igneous-rock collection.

4. Watch a lava lamp for a few minutes (or a few hours, depending on your state of mind). How does it work? Why do the blobs rise and sink? Imagine that the rising blobs are magma. What kinds of shapes do you see? Do you think this is a good approximation of magma rising in the Earth? Why . . . or why not?

ONLINE STUDY GUIDE

Practice makes you better. Make the time to take the *Online Study Guide* quiz for this chapter. It should take you about 20 minutes. Automatic grading and instant feedback will help you quickly and accurately identify the concepts you have mastered and the areas in which you need more study.
www.prenhall.com/chernicoff

Volcanoes and Volcanism

4

In 1979, pop-music icon Jimmy Buffett visited the Caribbean island paradise of Montserrat to record a reggae piece entitled "Volcano." The locals, who had lived peaceful, productive lives on the lush, fertile slopes of the long-resting Soufriere Hills volcano for centuries, laughed indifferently at the song's line—"I don't know where I'm gonna go when the volcano blow." After all, it had been more than 400 years since the hills had rumbled at all, and 10,000 years since the last major blow. But in the mid-1990s, Montserrat's Soufriere Hills (from the French *soufriere,* meaning sulfur— a reference to the region's deposits of volcanically produced sulfur) had become no laughing matter. Hot swirling ash had scorched half of the island's land and frightened away nearly all of the 11,000 permanent residents. Then, on June 25, 1997, dozens of the remaining brave souls lost their lives as a massive avalanche of red-hot volcanic particles swept across the island at a speed of more than 160 km/h (100 mph). That tragedy was overshadowed six weeks later by the volcano's "main event."

On August 20, 1997, during a series of blasts—several hours apart—the volcano, like a well-shaken bottle of champagne, "popped its cork," belching out a 10-km (6-mi)-high column of ash. The eruption shrouded the island in darkness and sent a red-hot plume of glowing volcanic ash racing across the island's emerald-green farmlands (Fig. 4-1) and through the abandoned capital city of Plymouth. During a brief lull in the eruption, helicopters flew in to survey the damage: Luxury tourist hotels lost; the island's new world-class golf course buried in ash; the region's brand-new $20 million hospital collapsed; and Plymouth, a quaint seaport town of beautiful pastel-colored homes, lay wasted beneath a steaming blanket of gray ash. Still standing was the just-completed $29 million cruise-ship terminal—Montserrat's ill-fated attempt to capture its share of the Caribbean's growing tourist trade.

In the aftermath of the eruption, nearly 7000 of the island's 11,000 residents have abandoned their lives on the island, opting to start over on nearby islands such as Antigua. The future of Montserrat is certainly grim: its tourism is dead, and its unemployment is greater than 60%. The island's economy is barely surviving on the enormous charity of its former colonizers in Great Britain. And the volcanic activity continues. Sulfur-dioxide gas continues to rise from the Soufriere Hills. July 2003 was marked by large pyroclastic flows— glowing swirling avalanches of hot volcanic ash. And a new ash column rose from the

FIGURE 4-1 The 1997 eruption of the Soufriere Hills volcano, on the Caribbean island of Montserrat. ▶

volcano to an altitude of 9 km (nearly 6 mi) on March 29, 2004. Montserrat's story offers clear evidence of the inevitable outcome when humanity chooses to live too close to a powerful volcano.

Yet the ongoing story of Montserrat's Soufriere Hills, which was, in truth, just a minor eruption, simply can't compare with the great volcanic blasts of recent history. Consider what happened on one summer day in 1883, when the Indonesian island of Krakatau all but vanished in a spectacular volcanic eruption. Krakatau, an uninhabited volcanic island in the Sunda Straits of the southwest Pacific Ocean, had for many years been a landmark for clipper ships carrying tea from China to England. The volcano, standing 792.5 m (2601 ft) high, had been inactive for more than 200 years. On the morning of August 27, Krakatau suddenly exploded in one of modern history's most violent volcanic eruptions, leaving a crater 300 m (1000 ft) *below* sea level. Krakatau's eruption equaled in force the detonation of 100 million tons of TNT. The sound of the explosion was heard as far away as central Australia, 4802 km (2983 mi) away. This is akin to the residents of San Diego, California, hearing an explosion in Boston, Massachusetts. Sadly, between 36,000 and 100,000 lives were lost when the resulting waves, up to 37 m (121 ft) high, pounded coastal villages on the nearby islands of Java and Sumatra.

The Krakatau eruption produced a black cloud of volcanic debris that rose to an altitude of 80 km (50 mi), blocked out all sunlight, and plunged the region into darkness for three days. The cloud's finest particles, swept aloft by wind currents, reduced incoming solar radiation by as much as 10% worldwide, causing a drop of more than 1°C (1.8°F) in global temperatures. The suspended particles also produced years of spectacular crimson sunsets. A few months after the eruption, on October 30, 1883, residents of Poughkeepsie, New York, and New Haven, Connecticut, summoned fire brigades to douse blazes that were, in fact, only the fiery glow of the brilliant evening sky.

As Montserrat and Krakatau taught us, powerful volcanic eruptions and their aftereffects are among the Earth's most destructive natural events. **Volcanoes** are the landforms created when magma escapes from the Earth's interior through openings, or **vents,** in the Earth's surface and flows out as *lava.* The lava then cools and solidifies around the vents, forming volcanic rock. A recent survey by the Smithsonian Institution found that about 600 volcanoes have erupted in the past 2000 years, some of them many times over. In a single year, approximately 50 volcanoes erupt around the world. Yet volcanoes, despite their dangers, also provide some of the world's most breathtaking scenery. Each year, millions are drawn to the potentially dangerous slopes of Mount Rainier in Washington state, Fujiyama in Japan, and Mount Vesuvius in Italy.

The geological processes that result in the expulsion of lava at the Earth's surface are collectively known as **volcanism.** Volcanoes are like windows into the Earth, providing us with information that would otherwise be inaccessible. Rising magma carries subterranean rock fragments to the surface, giving us a glimpse of actual rocks from the Earth's interior. Volcanoes can also show us a glimpse of the past, wherever volcanic deposits buried and preserved evidence of past organisms. Footprints of the earliest upright-walking hominids, for example, were pre-

▲ **FIGURE 4-2** A footprint in volcanic ash from East Africa. Paleoanthropologists have learned a great deal about human evolution from imprints such as this one, recording the passage of some of our early ancestors more than 3.6 million years ago.

served in fresh volcanic ash in East Africa 3.6 million years ago (Fig. 4-2), much like stepping in wet cement.

Volcanoes and Life

It is easy to focus on the death and devastation caused by volcanic eruptions; although it's definitely grim, it's still pretty exciting stuff. But volcanoes are also essential to life on Earth. We owe much of the air we breathe and the water we drink to volcanic eruptions, which throughout the Earth's history have released useful gases from the planet's interior. The water liberated by past volcanic eruptions makes up much of the Earth's *hydrosphere*—its oceans, lakes, rivers, underground waters, glaciers, and clouds. Nitrogen and oxygen from past eruptions have combined with other components to produce the Earth's gaseous *atmosphere.*

Volcanic activity constantly adds to the Earth's inventory of habitable real estate. Iceland, Hawai'i, Tahiti, many islands of the Pacific and Caribbean, and nearly all of Japan and Central America are products of volcanism. A new volcanic island is even growing where Krakatau once stood (Fig. 4-3). In the wake of Indonesia's tragic series of powerful earthquakes in 2004 and 2005, its volcanoes—including Krakatau—are being watched very closely. Several have already stirred back to life, perhaps in response to their recent seismic jostling. And on May 24, 2000, geologists witnessed the explosive birth of a brand-new island—the Kavachi volcano—growing from the seafloor in the southwestern Pacific.

Volcanic terrains often become prime agricultural lands as fresh volcanic ash replenishes the nutrients in nearby soils. The

▲ **FIGURE 4-3** Anak Krakatau ("Child of Krakatau"), the small island that emerged from the remains of the volcanic island Krakatau during eruptions in the 1920s. The original Krakatau volcano was demolished in a monumental eruption in 1883. In 2005, in the wake of the catastrophic seismic activity in the region, Anak Krakatau may be rumbling back to life and preparing for an eruption. Geologists are watching closely.

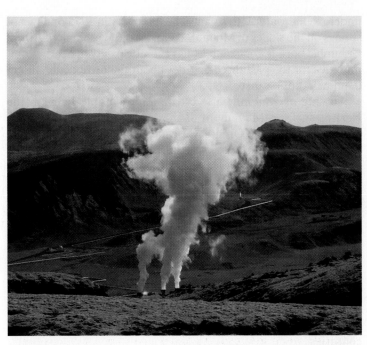

▲ **FIGURE 4-4** Steam emissions and pipelines on the outskirts of Reykjavik, Iceland

rich coffees grown in South and Central America sprout from fertile volcanic soils. On the Indonesian island of Java, fine volcanic ash retains water and the abundant nutrients (such as potassium, calcium, and sodium) that nourish plants. This allows Java's population density to be about 200 times that of neighboring Borneo, whose far-less-fertile soils derive from solid volcanic rock, such as andesite, produced at subduction zones.

Volcanic landscapes may also contain a source of inexpensive, clean energy. During the dark polar winter in Reykjavik, Iceland, the world's northernmost capital, people stroll in shirt-sleeves through warm shopping malls. Iceland has no oil, no coal, no natural gas, and very few trees for fuel, but it does have an abundant underground hot-water supply, thanks to the molten material that fuels Iceland's nearly constant volcanic activity. By tapping the scalding water just meters beneath their feet, Icelanders can heat more than 80% of their homes and businesses (Fig. 4-4). Similar *geothermal* ("Earth heat") resources are being tapped in other recently active volcanic settings, among them the Geysers area, 150 km (100 mi) north of San Francisco, California.

Volcanism may even be credited with "giving birth" to life on Earth. Some geoscientists believe that the Earth's earliest forms of life first appeared in the warm, nutrient-rich vents of ancient submarine volcanoes. Thus volcanoes and volcanism are at once both a great hazard and a great gift to humankind. In this chapter, we will explore the causes and characteristics of different types of volcanoes. We will also describe the threats they pose to us and how we have learned to cope with them. Finally, we will examine volcanism on some of our neighboring planets in the solar system.

The Nature and Origin of Volcanoes

The current status of a volcano is the key to its threat to human life and property. Is it active, dormant, or extinct? An *active* volcano is either currently erupting or has erupted within the last 200 years or so. Certain active volcanoes, such as K'ilauea in Hawai'i or Stromboli in the Mediterranean, erupt almost continuously. Others erupt periodically, such as Lassen Peak in northern California, which last erupted in 1917. Active volcanoes can be found on all the continents except Australia and on the floors of all the major ocean basins. Indonesia with 76 active volcanoes, Japan with 60, and the United States with 53 are the world's most volcanically active nations. In the early years of the 21st century, major eruptions are occurring or brewing in the northern Pacific (Kamchatka, Russia; the Aleutian Islands, Alaska); the western and southern Pacific (Japan, Papua New Guinea, Indonesia, Philippines, Vanuato); South and Central America (Ecuador, Guatemala, Mexico); Italy (Etna, Stromboli, Vesuvius); and K'ilauea and Mauna Loa in Hawai'i. In this list of some of the Earth's most active volcanoes, the style of eruption varies from the quiet oozing of molten basalt from K'ilauea to the cataclysmic explosions of Mexico City's fearsome "El Popo" (Popocatepetl) (Fig. 4-5).

A *dormant* volcano is one that has not erupted recently (within the past few thousand years) but is considered likely to do so in the future (Fig. 4-6). A number of signs may suggest that a dormant volcano is reawakening. When we encounter new hot-water springs, we suspect that a shallow heat source exists—such as rising magma or heated rocks—that might be converting underground water into steam. The discovery in 1975 of new hot springs on the slopes of Mount Baker near the Washington–British Columbia border, for example, has caused anxiety in nearby communities for the past three decades about the prospect of an impending eruption there.

Rising magma may also create a measurable bulge at the Earth's surface. Movement of the Earth's crust at the South Sister volcano in the Oregon Cascades, and a bulge beneath the Mammoth Lakes area of eastern California have heightened concern of impending volcanic eruptions among volcanologists, local residents, and thousands of vacationing skiers and hikers. The presence of relatively fresh volcanic rocks in a volcano's vicinity (erupted within the past 1000 years) also suggests that it is capable of a repeat performance, as are frequent small earthquakes that may signal that magma is moving underground toward the surface. Thousands of small earthquakes (2000 in November 1997 alone), emissions of carbon-monoxide gas, and a bulge at the Earth's surface are all reason to suspect that the Mammoth Lakes region may be building toward another volcanic eruption, although perhaps not for decades or even centuries.

A volcano is considered *extinct* if it has not erupted for a very long time (perhaps tens of thousands of years) and is considered unlikely to do so in the future. One indication that a volcano is probably extinct is extensive erosion since its last eruption

(a)

(b)

▲ FIGURE 4-6 Which of these two volcanoes seems more likely to erupt? **(a)** The slopes of Washington state's Mount Rainier remain deeply scored by repeated episodes of glacial erosion that occurred over hundreds of thousands of years and appear not to have received a fresh covering of lava in thousands of years. **(b)** The slopes of Mount St. Helens (shown here before its 1980 eruption), believed to be the youngest of Washington's Cascade Range volcanoes, are relatively uneroded. Although intermittently dormant for hundreds or thousands of years, its volcanic cone has continued to grow, and thus it's more likely to erupt in the near future.

▲ FIGURE 4-5 Popocatepetl ("El Popo")—Mexico City's neighboring volcano—is one of several volcanoes around the world that threatens highly populated surrounding communities.

(see Figure 3-25—Shiprock Peak, the erosional remains of an extinct volcano). A truly extinct volcano is one that is no longer fueled by a magma source, such as the Hawaiian volcanoes on the islands that have moved far from Hawai'i's island-forming hot spot.

The Causes of Volcanism

Volcanism begins when magma created by the melting of preexisting rock in the Earth's interior reaches the surface of the Earth. Magma will erupt if it flows upward rapidly enough to reach the surface before it can cool and solidify. Two characteristics of magma determine its potential to erupt: its gas content and its viscosity.

Gas in Magma

Magmatic gases make up from less than 1% to about 9% of most magmas. The principal gases are water vapor and carbon dioxide, with smaller quantities of nitrogen, sulfur dioxide, and chlorine. When the magma is tens of kilometers underground, the gases are dissolved in the magma, held in solution by the pressure of surrounding rocks. But this pressure decreases as magma rises toward the surface, allowing the gases to separate, or *unmix,* from the magma. Because the gases are less dense than their surrounding magma, they migrate upward through the magma and expand outward, pushing any overlying magma before them.

The rising, expanding, unmixed gases collect near the top of a magma body and press against the overlying rock. Where their passage to the surface is blocked, such as by a plug of old hardened lava, these gases accumulate and exert great pressure against the overlying rock until it ultimately shatters. As soon as the overlying rock is removed, the pent-up gases expand rapidly, much as the gases in a shaken soft-drink bottle fizz vigorously and bubble out when the bottle is opened. A volcano's initial blast typically removes any overlying obstructions, hurling masses of older rock skyward. Shreds of the liberated, gas-driven lava are then sprayed violently into the air as their gases expand. The eruption may continue violently for hours or even days as the gases escape. It may then settle down to a relatively quiet outpouring of degassed magma, or it may cease altogether.

Magma Viscosity

A magma's *viscosity* is its *resistance to flow.* High-viscosity magma is less likely to make it to the surface and erupt than a more fluid, low-viscosity magma. However, high-viscosity lavas that do erupt tend to do so far more explosively, for reasons we'll discuss shortly. The key factors that control a magma's viscosity are its temperature and composition.

Increased heat typically *reduces* the viscosity of any fluid— be it motor oil, maple syrup, or magma. Hot fluids flow more easily because their ions and atoms move about more rapidly, breaking the temporary bonds that inhibit flow. Conversely, decreasing temperature typically *increases* a fluid's viscosity, causing it to move more sluggishly, allowing its slower-moving ions to bond.

The viscosity of a magma also generally increases with silica content (see Table 3.1). In silica-rich (felsic) magma (which tends to be a lower-temperature magma than higher-temperature silica-poor, mafic magma), the oxygen ions at the corners of the unbonded silicon–oxygen tetrahedra (discussed in Chapter 2) form relatively strong temporary bonds that link clusters of chains of tetrahedra. Felsic magma, for example, tends to be high-viscosity, and thus slow-flowing because it has both a high silica content *and* it is relatively cool (Fig. 4-7a). On the other hand, mafic magma, with its low silica content and high temperature, is much lower in viscosity, flowing easily (Fig. 4-7b). That is why mafic magmas are more likely to rise to the surface and erupt than are felsic magmas, which tend to cool underground and form plutonic rocks.

How does viscosity affect explosiveness? Gases do not escape from all magmas with equal ease. In the more fluid, low-

(a)

(b)

▲ **FIGURE 4-7** The viscosity of a volcano's lava typically controls its eruptive style (quiet or explosive) and the shape of the volcano. Here **(a)** at the Valley of the Ten Thousand Smokes in Katmai National Park in Alaska, high in viscosity, virtually nonflowing, felsic lava produces this steaming dome. **(b)** at Hawai'i's K'ilauea Volcano, low-viscosity, basaltic lava flows rapidly as a red-hot lava stream.

viscosity mafic magmas, migrating gases meet with relatively little resistance and therefore escape easily when the magma reaches the surface. The gases don't accumulate to build up the high pressure that causes explosive eruptions. Mafic magmas, then, tend to erupt *quietly,* with a relatively gentle outpouring of lava. On the other hand, because the movement of gases in high-viscosity, felsic magma is impeded, gas pressure builds up within the magma and it tends to erupt explosively when it reaches the surface. The viscosity of intermediate magmas, such as those that produce andesites, falls somewhere between mafic and felsic. But these magmas are cool enough and sufficiently rich in silica to erupt explosively.

The Products of Volcanism

A volcanic eruption can produce a flowing stream of red-hot lava, a shower of ash particles as fine as talcum powder, a hail of volcanic blocks the size of minivans, or any number of intermediate-sized products. The quantity of lava produced by volcanoes ranges from small spurts to vast floods. An immense submarine lava flow, believed to have been extruded within the last 25 years, was recently discovered in the vicinity of the East Pacific rise, off the coast of South America. The flow contains approximately 15 km³ (9 mi³) of basalt, enough to pave over the entire U.S. interstate-highway system to a depth of 10 m (33 ft).

Both the type and the amount of material produced by a volcano depend largely on the composition of its lava. We will first consider the different types of lava and their properties and then describe the various forms in which volcanic materials accumulate at the surface.

Types of Lava

The composition of a lava reflects the composition of its parent magma. The lava, however, contains less dissolved gas, as most gases escape into the atmosphere during an eruption. The most common type of lava is basalt—a mafic lava. As we've learned, mafic magmas are the most likely to erupt because they tend to be hot and highly fluid, moving readily to the surface before solidifying. Magmas of felsic composition tend to be cooler and much higher in viscosity, only rarely reaching the surface—as rhyolite lava—before solidifying. For this reason, there is far less rhyolite than granite (its plutonic equivalent) in crustal rocks. Andesitic lavas are intermediate between basaltic and rhyolitic lavas in both composition and fluidity.

Basaltic Lava   For nearly a century, the Hawaiian Volcano Observatory on the Big Island of Hawai'i has been observing basaltic eruptions. These eruptions have taught us much of what we know about *subaerial* ("under air," as opposed to under water) basaltic lava flows and the resulting rocks. The observatory has found the temperature of Hawaiian flows to be as high as 1175°C (2150°F). Such hot, low-viscosity lava cools to produce two principal types of basalt—*pahoehoe* (pa-HOY-hoy) and *'a'a* (AH-ah). Pahoehoe, which means "ropy" in a Polynesian dialect, is aptly named. Highly fluid basaltic lava moves swiftly down a steep slope at speeds that may exceed 30 km/h (20 mph), spreading out rapidly into sheets about 1 m (3 ft) thick (Fig. 4-8). The surface of such a flow cools to form an elastic skin that is then dragged into ropelike folds by the continuing movement of the still-fluid lava beneath it. The ropy surface of pahoehoe basalt is generally quite smooth. Native islanders refer to it as "ground you can walk on barefoot," and most old Hawaiian foot trails follow ancient pahoehoe flows.

As basaltic lava flows farther from the vent, it cools, loses much of its dissolved gas, and flows more slowly. A thick, brittle crust develops at its surface and continues to move forward slowly, carried along by the warmer, more fluid lava below it. The cooler outer region breaks up to produce a rough, jagged surface sharp enough to cut animals' hoofs or a geologist's field shoes. Such flows are called 'a'a flows (Fig. 4-9). ('A'a is a local term of unknown origin that may recall the cries of a barefoot islander who strayed onto its surface. Ancient foot trails typically avoid 'a'a fields.) 'A'a is often found downstream from pahoehoe, the product of the same flow that has cooled and slowed.

Basalt flows may also contain *lava tubes,* subwaylike tunnels that form when lava solidifies into a crust at its surface but continues to flow underneath (Fig. 4-10). Eventually, as the eruption wanes, the still-molten lava drains from the cooled

▲ FIGURE 4-8 These ropy pahoehoe-type lavas from Hawai'i's K'ilauea volcano cooled only days or even hours before these tourists began trekking around on them.

▲ FIGURE 4-9 This relatively slow-moving basaltic lava cools to form a blocky, jagged, 'a'a-type surface texture.

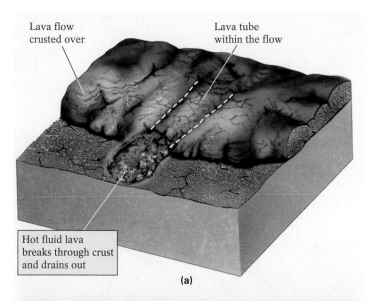

Lava flow crusted over

Lava tube within the flow

Hot fluid lava breaks through crust and drains out

(a)

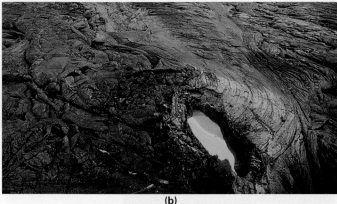

(b)

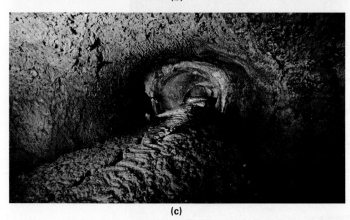

(c)

▲ **FIGURE 4-10 (a)** Lava tubes form when a lava flow's surface cools and solidifies but the lava continues to flow under the surface in tunnels. When the lava drains from the tunnel, it leaves behind the empty tube. **(b)** A lava tube forming at K'ilauea. The opening here is about 1.5 m (5 ft) long. **(c)** These lava tubes in Northern California were used by the Modoc Indians to escape detection by the U.S. Army.

tubes, leaving them hollow. These tubes enable lava to travel great distances from a volcano's vent. Lava Beds National Monument in northeastern California contains 300 or more tubes within the area's pahoehoe flows. This natural maze sheltered the Modoc Indians in 1872 as they battled several hundred troops from the United States Army in an unsuccessful attempt to reclaim their native lands. Fewer than 60 Modoc warriors, who knew every underground passage in the lava field, inflicted serious harm on their pursuers from their sanctuary within the lava tubes but eventually were overcome and forced to leave their territory.

Subaerial basalt flows may produce several other distinctive features as they cool. Escaping gas often leaves small, pea-size *vesicles* (gas holes), which are preserved at the top of the basalt when it cools. Vesicle-rich basalt is known as *scoria* (Fig. 4-11). As basaltic lava cools, it shrinks in volume, often producing a pattern of cracks known as *columnar jointing*. As cooling proceeds, these joints extend inward from the top and bottom surfaces of the flow into its interior, creating six-sided polygonal columns of basaltic rock (Fig. 4-12a). A side view of these columns suggests

▲ **FIGURE 4-11** Vesicular basalt, or scoria. These pores remain in the cooled rock after gases burst from bubbles on the surface of highly fluid lava.

Hot flowing
basaltic lava

Fissure volcano

Hexagonal
columns form
as basalt contracts
while cooling

Older lava flow
has columnar
jointed structure

Cooler

Hotter

(a)

(b)

(c)

▲ **FIGURE 4-12** Contraction of basaltic lava flows as they cool **(a)** sometimes produces geometrically patterned joints.
(b) Such structures can be found in North America in eastern California, eastern Oregon and Washington, southern Idaho,
and eastern Wyoming, such as those shown here at Devil's Tower National Monument. **(c)** Devil's Postpiles National Monument, California. Note the polygonal fracture pattern—similar to the floor tiles in a restroom.

▲ FIGURE 4-13 (a) Patterned ground in the Arctic develops polygons at the surface as the ground freezes and shrinks. (b) Mudcracks develop similar geometric shapes as the mud dries out, shrinks, and cracks. Both of these patterns resemble the joints in cooling basalt flows.

a large bunch of pencils (Fig. 4-12b); an aerial view suggests oversized ceramic bathroom floor tiles (Fig. 4-12c).

The six-sided nature of the basaltic columns with their 120° angles is an efficient way for shrinking rocks to break that minimizes the amount of energy expended to break them as they cool and shrink. Similar patterns can be seen in other places where Earth materials shrink, such as the cracks that develop when the ground freezes in the frigid Arctic (Fig. 4-13a) or when mud dries out and shrinks on an exposed lakeshore mudflat (Fig. 4-13b).

In North America, Devil's Tower in northeastern Wyoming (site of the climax of the film *Close Encounters of the Third Kind*) is a spectacular example of basaltic columns. Devil's Tower is a sacred place in the lives of the Cheyenne and the Lakota Sioux of the Dakotas and Wyoming. These tribes call the tower Mateo Tepee, meaning the "Grizzly Bear Lodge." According to one Native American legend, the tower rose when the Great Spirit created it to aid fleeing children from the sharp claws of an enormous grizzly bear. The tribes believe that the deep "scratches" in the tower's rocks (Fig. 4-14) record the bear's desperate attempt to reach its human prey.

Subaqueous ("underwater") basaltic eruptions are much more common than subaerial eruptions (after all, 71% of the Earth's surface is devoted to the world's oceans . . . and basalt is the principal rock of the ocean floor). They are, however, less well understood, because they usually form beneath thousands of meters of seawater. We do know that when basaltic lava erupts beneath the sea, it develops a distinctive *pillow structure* that, as its name suggests, resembles a stack of bedroom pillows. As cold water instantly chills the extruded lava, a thin deformable skin forms that stretches to resemble an elongated pillow as additional hot lava enters it. As the pillow shape expands, its surface cracks, allowing more lava to flow out from it and form another pillow, and so on (Fig. 4-15).

▲ FIGURE 4-14 This drawing depicts the Cheyenne and Sioux legends that explain the columnar jointing that appears on the Devil's Tower in Wyoming.

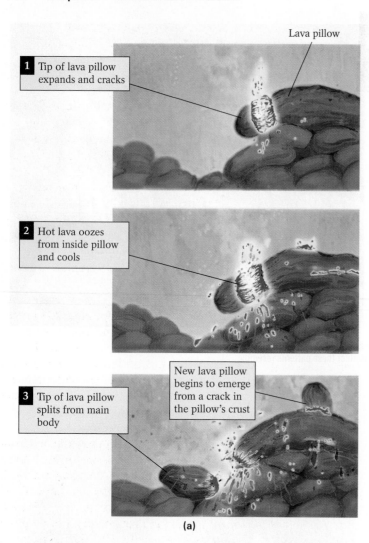

1. Tip of lava pillow expands and cracks

Lava pillow

2. Hot lava oozes from inside pillow and cools

3. Tip of lava pillow splits from main body

New lava pillow begins to emerge from a crack in the pillow's crust

(a)

(b)

▲ **FIGURE 4-15 (a)** The formation of basaltic pillow lavas. Our knowledge of these structures comes from studying preserved ancient pillows, **(b)** and from observing modern pillows, such as these in the Galapagos, from deep-sea submersibles.

Andesitic Lava Andesitic lava is intermediate in silica content, temperature, and viscosity between mafic basaltic lava and felsic rhyolitic lava. It typically flows more slowly than basaltic lava, solidifying relatively close to its vent. Like basaltic lavas, andesitic lavas may develop vesicles and 'a'a-type surface textures. We rarely see pahoehoe-type andesitic flows, however, because these lavas are simply too high in viscosity to stretch into a ropy structure. Andesitic lavas, particularly those closer to basalt in composition, can also produce columnar jointing. More felsic andesitic magmas can be sufficiently high in viscosity to impede the passage of rising gases and thus typically erupt in major volcanic explosions.

Rhyolitic Lava Silica-rich felsic magma is the most felsic and the highest in viscosity. It typically moves so slowly that it tends to cool and solidify underground as plutonic granite; it is less likely to erupt as lava at the Earth's surface. When rhyolitic lava does erupt, however, it usually explodes violently, producing an enormous volume of solid airborne fragments in place of a lava flow. High-viscosity rhyolitic lava rarely flows far from the vent and does not produce the volcanic structures that typify lower-viscosity lavas (such as pahoehoe structure, pillows, or columnar joints). Rhyolitic lava flows are typically short and thick, such as those that flowed as recently as 1000 years ago from the Newberry caldera in central Oregon (Fig. 4-16).

Felsic magmas with high water and gas content may bubble out of a vent as a froth of lava. This froth quickly solidifies into an extremely porous volcanic rock called *pumice*. Pumice—which looks like solidified beer foam—may contain so many vesicles from which its gases have escaped that it can actually be lighter than water. When Krakatau erupted in 1883, sailing ships were trapped for three days in the waters of the Sunda Straits by huge

▲ **FIGURE 4-16** This young rhyolite flow is located at Paulina Lake in eastern Oregon near the town of Bend.

rafts of floating pumice. When the pumice finally became waterlogged and sank, the ships could finally escape the volcano's destructive shower of hot ash.

Pyroclastic Materials

An explosive eruption propels lava forcefully into the atmosphere, shattering it into countless shreds of airborne lava. As these shreds fly through the air, they cool rapidly, solidifying into fragments of volcanic glass of various sizes and shapes. Such an eruption might also shatter some of a volcano's preexisting cone. *Volcanic blocks*, for instance, are chunks of igneous rock ripped from the throat of a volcano during an eruption. Such blocks tend to be angular and range from the size of a baseball to the size of a house. Blocks weighing as much as 100 tons have been found as far as 10 km (6 mi) from their source volcano. All such fragmental volcanic products—both the shattered old rock and the shredded new lava—are known collectively as **pyroclastics** (from the Greek *pyro,* meaning "fire," and *klastos,* meaning "fragments"). Pyroclastic materials may travel through the air as independent particles, or they may hug the ground as dense flows.

Tephra

Those pyroclastic materials that cool and solidify from lava as they fly through the air are called **tephra.** Tephra particles are classified by size, ranging from fine dust to massive chunks (Fig. 4-17). *Volcanic dust* particles are only about one-thousandth of a millimeter in diameter, the consistency of cake flour. The dust can travel great distances downwind from an erupting volcano and remain in the upper atmosphere for as long as a year or two. Somewhat grittier is *volcanic ash,* with particles less than 2 mm in diameter, ranging from the size of fine sand to that of a grain of rice. Ash generally stays in the air for only a few hours or days.

Tephra and modern technology clearly don't mix. In the past 15 years alone, the jet engines of more than 90 commercial aircraft have been seriously damaged by inadvertently flying through gritty tephra clouds that fouled their engines. Some of the craft even stalled in mid-flight. On December 15, 1989, KLM flight 867 from Amsterdam to Anchorage, Alaska, flew through the cloud of ash from the Mount Redoubt eruption, causing all four engines to fail. The plane plummeted 3000 m (nearly two mi) before the pilots were able to restart the engines and land the plane—and its 231 passengers—safely.

Coarser tephras fall sooner and closer to the volcanic vent, pulled to Earth by gravity. *Cinders,* or *lapilli* (Italian for "little stones"), range from the size of peas to that of walnuts (2 to 64 mm in diameter). *Volcanic bombs* (64 mm or more) form when sizable blobs of lava erupt and solidify in mid-air.

Tephras typically accumulate in distinct layers that record the frequency and magnitude of volcanic activity. The first step in "reading" a tephra layer is to link it to a specific volcano, usually by comparing the tephra's chemical composition to samples of known volcanic origin. We can also determine the source of a given tephra layer by mapping its thickness and grain size. In general, the coarser-grained and thicker a tephra layer, the closer it is to its source (Fig. 4-18). Once we can link a region's tephra layers to specific volcanoes, we can date the layers and reconstruct the individual eruptive histories of each volcano.

Pyroclastic Flows

When the amount of pyroclastic material expelled by a volcano is great and the particles are too large to be carried aloft, gravity almost immediately pulls the material back down onto the slopes of the volcano. There it rushes downslope as a **pyroclastic flow,** or **nuée ardente** (noo-AY AR-dant) (French for "glowing cloud"). Trapped air and magmatic gases keep the flow quite buoyant. The gases sharply reduce the friction between

◄ FIGURE 4-17 A thick layer of fine-grained felsic tephra along the Ukak River, Katmai, Alaska.

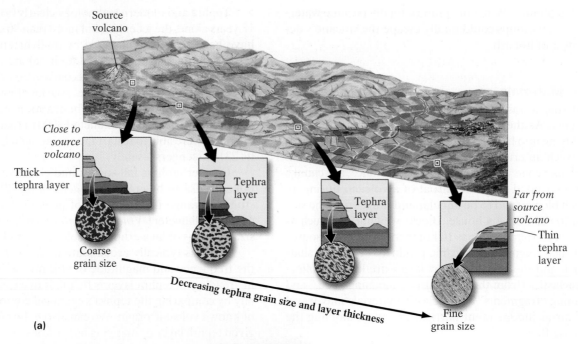

(a)

(b)

▲ **FIGURE 4-18 (a)** The thickness and texture (grain size) of tephra layers decrease with distance from their volcanic source. **(b)** Note the variations in color, texture, and thickness of these Cascade Mountain tephras from Washington state. Their thicknesses and textures vary because the sources of these tephras include at least five different volcanoes at varying distances from this site. For example, the lowest layer here is the coarsest (note the large chunks of pumice) and thickest; thus, it probably erupted from the nearest volcano. The variations in tephra color are a function of variations in the chemistry of the magmas from the various volcanoes—in large part a consequence of the different rocks encountered and incorporated into the magmas as they rose toward the surface.

the flow and the ground, enabling some pyroclastic flows to reach speeds exceeding 150 km/h (almost 100 mph), even on gentle slopes (Fig. 4-19).

Particles in a pyroclastic flow travel through the air so briefly that they do not cool significantly. Thus they may still be red-hot when they settle to the ground at temperatures capable of melt-ing glass or burning an apple into a smear of carbon. As the flow travels downslope, its gases escape. When the warm particles finally come to rest, they may still be soft enough to fuse with one another. This forms a **welded tuff,** a solid volcanic rock composed of solidified tephra. Tuffs also form when a thick pile of accu-mulated tephra solidifies under the weight of overlying material.

Volcanic Mudflows Pyroclastic material that accumulates on the slope of a volcano may mix with water, forming a volcanic mudflow, or **lahar.** This term was coined on the Indonesian island of Java, where explosive eruptions and abundant loose, moist soil cause frequent disastrous mudflows. A lahar may contain a range of particle sizes from the finest ash to enormous 100-ton boulders. Lahars often form when an explosive eruption occurs on a snowcapped volcano, and hot pyroclastic material melts a large volume of snow or glacial ice. In November 1985, a small eruption of Nevado del Ruiz in the Andes of Colombia caused the devastating lahar that buried the highland town of Armero and took the lives of 23,000 people (Fig. 4-20).

The fine dust and ash sprayed from explosive eruptions may actually produce rain clouds, as atmospheric moisture, plus the water vapor from the eruption, condenses around the cooling tephra particles. These clouds may then produce torrential rains, triggering lahars that carry loose ash and soil down a volcano's slope. A saturated mudflow may travel tens of kilometers per hour on a steep slope. Even the swiftest among us could not outrun such flows. Lahars may continue to develop long after a volcanic eruption has ended if normal heavy rains mobilize loose, fresh volcanic ash.

Secondary Volcanic Effects

As if suffocating ash falls, searing nuées ardentes, and enveloping hot lahars were not enough, volcanic eruptions produce a variety of other secondary effects that alter the environment and affect human, animal, and plant life. These effects may also

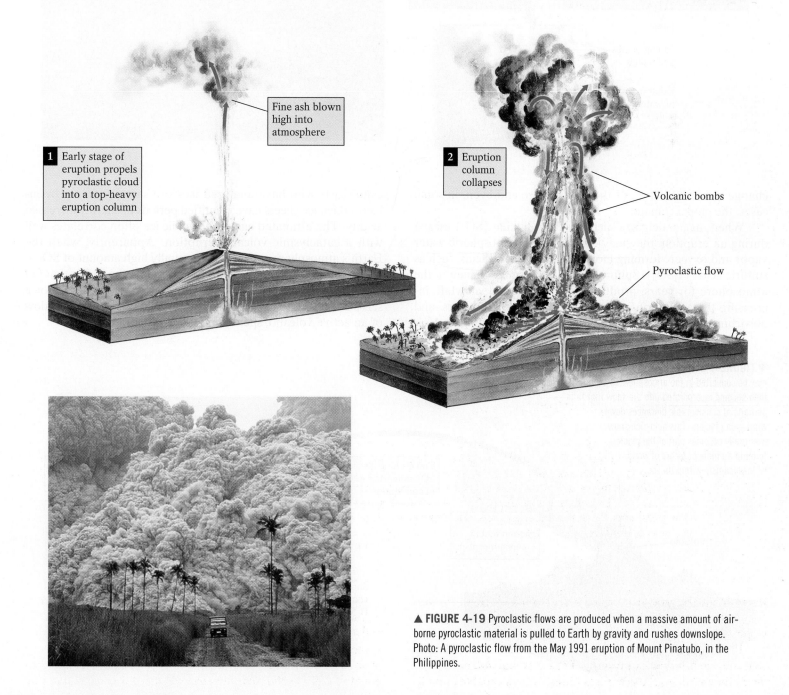

Fine ash blown high into atmosphere

1 Early stage of eruption propels pyroclastic cloud into a top-heavy eruption column

2 Eruption column collapses

Volcanic bombs

Pyroclastic flow

▲ **FIGURE 4-19** Pyroclastic flows are produced when a massive amount of airborne pyroclastic material is pulled to Earth by gravity and rushes downslope. Photo: A pyroclastic flow from the May 1991 eruption of Mount Pinatubo, in the Philippines.

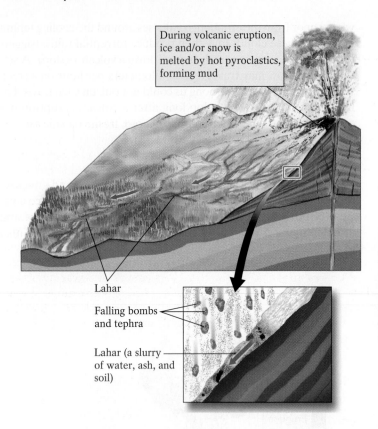

During volcanic eruption, ice and/or snow is melted by hot pyroclastics, forming mud

Lahar

Falling bombs and tephra

Lahar (a slurry of water, ash, and soil)

◄ **FIGURE 4-20** Some lahars occur when pyroclastic eruptions melt snow and ice on volcanic slopes, producing torrents of mud. Photo: This lahar resulted when Colombia's Nevado del Ruiz volcano erupted in 1985, melting about 10% of the snow on the slopes of the volcano and producing a 40-m (137-ft)-high wall of mud that buried the town of Armero and approximately 23,000 of its residents. Armero is located about 50 km (30 mi) from the summit of Nevado del Ruiz.

change the composition of the atmosphere and even, in some cases, the global climate.

When magmatic gases such as sulfur dioxide (SO_2) escape during an eruption, they may combine with atmospheric water vapor and oxygen, forming molecules of airborne acids, such as sulfuric acid (H_2SO_4). Sulfuric-acid droplets may remain in the atmosphere for years, producing acidic rain and snowfall, increasing the acidity of local, regional, and global waters, and absorbing incoming solar radiation, cooling the climate. Vol-

canologists who have analyzed ice-core samples from Greenland's thick ice sheets have identified periods of sharply increased acidity. The estimated age of the acidic ice often correlates well with a cataclysmic volcanic eruption. Apparently, when the Earth's atmosphere contains an unusually high amount of SO_2—possibly due to a massive volcanic eruption—the snows that fall around the world are markedly more acidic (Fig. 4-21). For a sense of the environmental challenges posed by life in the shadow of an active volcano, spend a moment with Highlight 4-1.

▼ **FIGURE 4-21** Gas emissions—such as sulfur dioxide—from volcanic eruptions may be converted in the atmosphere to acids. These acids may then become incorporated into the snow that falls—perhaps at considerable distances downwind—on glaciers. This acid-rich snow eventually becomes part of the glacier, forming an indirect record of worldwide volcanism within the ice.

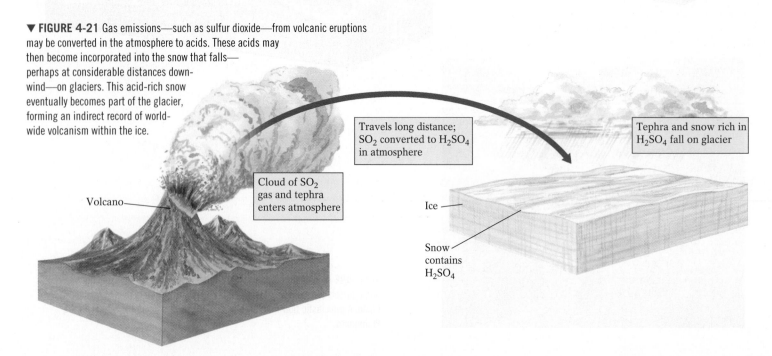

Volcano

Cloud of SO_2 gas and tephra enters atmosphere

Travels long distance; SO_2 converted to H_2SO_4 in atmosphere

Tephra and snow rich in H_2SO_4 fall on glacier

Ice

Snow contains H_2SO_4

Highlight 4-1

Hawai'i's Dreaded VOG—Coping with "Natural" Air Pollution

Sulfur-dioxide concentrations in the air over the Big Island of Hawai'i often rise to 1000 parts per billion, almost as high as the killer smog that choked London in December 1952, taking 4000 lives. This "pollution" has been blamed for a host of physical complaints among the local residents and is a potential threat to agriculture and the island's economy. Yet the Big Island sits thousands of kilometers from the nearest smokestack and has no coal-burning power plants or copper smelters. There are only a few small communities, several daily busloads of tourists . . . and a hyperactive volcano named K'ilauea. Here is the culprit responsible for the clouds of sulfur-dioxide gas that form Hawai'i's "VOG"—the locals' nickname for volcanic smog.

Finding the source of sulfur dioxide is not too hard—simply strap on a gas mask and stroll across the still-warm, freshly hardened black lavas at K'ilauea's summit crater. There, mounds of yellow sulfur crystals encrust the lava's fractures wherever sulfurous gas plumes have been expelled. K'ilauea, the world's most active volcano, has been erupting almost continuously since the mid-1980s, producing a steady stream of basaltic lava and a variety of noxious gases (see photo). The most abundant—CO_2—quickly dissipates, but the next most common—SO_2—does not.

Typically, K'ilauea's gaseous emissions last only a few minutes or hours. The gases oxidize quickly as they drift away, affecting only the immediate vicinity of the volcano. On calm, windless days these brief episodes may foul the air for the scores of scientists and other employees of Hawai'i Volcanoes National Park and the U.S. Geological Survey's volcanic observatory, but not at lethal levels. As the sulfur dioxide reacts with water in the atmosphere, tiny molecules of sulfuric acid form, scattering light and creating the haziness reminiscent of unpleasant smoggy days in the Los Angeles basin. During much of the year, prevailing winds from the east sweep K'ilauea's VOG to the island's western coast, where it often becomes trapped against the mountains that overlook the city of Kailua-Kona (population 10,000). From time to time, the VOG even reaches Honolulu, about 260 km (160 mi) to the west.

Hawaiian VOG has yet to be blamed directly for any serious health hazards, although episodes of high sulfur emissions appear to correlate with increased incidence of bronchitis and other respiratory problems. Short-term exposure need not worry tourists, but longer-term exposure, especially to senior citizens, may be of greater concern. As in air-polluted industrial centers, local residents are advised to refrain from outdoor exercise and avoid smoking on particularly bad VOG days. And unlike at industrial sources, where emissions can be cleansed by scrubbers and filtration systems, little can be done about Hawai'i's natural pollution source.

A final note: Here again is another clear example of the interconnectedness of the Earth's various systems. A volcanic emission from the *lithosphere* passes through the *atmosphere* and affects the local *biosphere* (the residents of the island).

▲ Gases emitted regularly from Hawai'i's K'ilauea volcano, especially sulfur dioxide, are responsible for the Big Island's air "pollution."

Volcanic gas and tephra emissions also affect worldwide climate. The dust and ash from a large tephra column can rise into the stratosphere and remain suspended there for as long as five years. These particles reflect incoming sunlight back into space, thus lowering the amount of radiation that reaches and warms the Earth's surface. Droplets of SO_2 in the atmosphere also reflect and absorb radiation. The combination of dust and gas from a single large eruption can drop the Earth's temperature by as much as 2° to 3° Celsius (4–6°F), an effect that may last for more than a decade.

The first recorded scientific demonstration of the effect of volcanism on climate took place in May 1784, when a blue haze and dry fog hovered over Europe and the weather was unusually cool. Scientist–philosopher–statesman Benjamin Franklin, living in Paris as America's ambassador to France, hypothesized that Europe's unusual climate was caused by an enormous eruption of gassy lava during the previous year in Iceland. He tested his hypothesis by using a magnifying glass to focus the Sun's rays on a sheet of paper. Normally this activity would burn a hole in the paper, but in the spring of 1784 it did not. Franklin concluded correctly that less sunlight was reaching Europe as a result of the Icelandic eruption.

The greater the eruption, the longer its effects will last. The spectacular eruption of Indonesia's Mount Tambora in 1815, which took an estimated 50,000 lives, was followed by what became known as the "Year Without a Summer." In all, 150 to 200 km³ (60–80 mi³) of tephra were ejected into the atmosphere. Airborne tephra blocked the Sun's rays and brought two days of total darkness to a 600-km (400-mi) area surrounding the volcano. Globally, summer air temperatures dropped 1.0° to 2.5°C (2–5°F). Snow fell in upstate New York the following June. People in Connecticut celebrated the Fourth of July in overcoats. August frosts destroyed crops from the Midwest to Maine. Tambora's dust and gas reduced sunlight by more than 10% and produced a long succession of dreary days. The depressing weather that summer at Lake Geneva in Switzerland may have caused the morbid mood that inspired Mary Shelley to write the gruesome classic, *Frankenstein.*

Like Krakatau, major volcanic eruptions of oceanic islands and volcanic seamounts (submerged volcanoes) may produce another potentially catastrophic secondary effect—giant seawaves. No place better illustrates this threat than the seamount Loihi, the next Hawaiian island, discussed in Highlight 4-2.

Eruptive Styles and Associated Landforms

When we think of volcanoes, highly photogenic, snow-capped peaks usually come to mind (Fig. 4-22a). We do not often picture scruffy little hills, such as those of Figure 4-22b. Despite their different appearances, nearly all volcanoes have the same two major components: a mountain or hill, known as a **volcanic cone,** constructed from the products of numerous eruptions over time, and a steep-walled, bowl-shaped depression, or **volcanic crater,** sur-

rounding the vent from which those volcanic products erupted. A volcano's crater forms when the area around the vent subsides to form a depression as the underlying magma chamber empties. Volcanic cones and craters have a variety of shapes and dimensions, depending on the eruptive style and the composition of their volcanic products.

Effusive Eruptions

Effusive eruptions are relatively quiet, nonexplosive events that generally involve basaltic lava. They are seldom accompanied by powerful volcanic blasts and rarely produce large volumes of pyroclastic materials. Basaltic lava, which is highly fluid, flows freely from central volcanic vents, as well as from elongated cracks on land and from fractures at submarine plate boundaries and intraplate hot spots.

(a)

(b)

▲ **FIGURE 4-22** Volcanoes come in various shapes and sizes. **(a)** Lofty, symmetrical Mount Augustine, in Alaska's Cook Inlet; **(b)** small, stubby volcanoes from the Flagstaff area of northern Arizona.

Highlight 4-2

Loihi—Growing Pains of the Next Hawaiian Island

About 30 km (20 mi) southeast of the Big Island of Hawai'i, geologists are watching nervously as a new oceanic island comes to life. There, a submerged shield volcano named Loihi (low-EE-hee, which means "the long one" in a Polynesian dialect) has grown to a height of roughly 4500 m (15,000 ft) by erupting a vast volume of basaltic lava over the last 100,000 years. Its summit still sits 1000 m (3300 ft) below the Pacific Ocean's surface, and local geologists estimate that it won't emerge from the sea for another 50,000 to 100,000 years. Still, Loihi is more than capable of making its presence felt, both to the local islanders and to distant points around the Pacific Ocean basin—as the summer of 1996 dramatically demonstrated.

The most intense swarm of earthquakes ever recorded in Hawai'i—more than 5000 quakes—occurred during July and August of 1996 beneath Loihi's summit. Following this activity, oceanographers and geologists from the University of Hawai'i visited Loihi to assess any changes. They began their underwater investigation along the southern rim of Loihi's summit at Pele's Dome, a peak named for the Hawaiian fire goddess. To their amazement, the dome, formerly standing 300 m (1000 ft) high atop the broad shield volcano, had vanished, only to be replaced with a crater more than a kilometer (about six-tenths of a mile) in diameter and more than 300 m (1000 ft) deep. Further inspection of the area showed that bus-sized boulders littered Loihi's slopes and the surrounding seafloor over a 10-km (6-mi) area.

Apparently, Loihi's summit had collapsed catastrophically, spawning huge rock avalanches and forming a deep crater—Pele's Pit (see figure below). Geologists hypothesize that the summit's collapse and the associated earthquake swarm occurred when lava erupting from deep fractures near Loihi's base partially drained the magma reservoir beneath the summit, allowing it to collapse inward. The mass of rock that fell into the abyss—some 300 million tons—would have filled 50 million dump trucks.

The violent rearrangement of Loihi's summit left countless steaming cracks, some as large as 6 m (20 ft) across, spewing a mixture of boiling water and dissolved minerals. Of greater concern, however, was the instability of the entire volcano. The immense cave-in of the summit could have triggered a large seawave, or *tsunami*, with the potential to destroy nearby populated coastal areas, such as Honolulu's Waikiki Beach. Fortunately, no tsunami developed, perhaps because the collapse occurred over several days rather than instantaneously. In the future, Hawaiians may not be so lucky. The seafloor around Loihi and along the southeastern coast of the Big Island of Hawai'i is littered with the rocky remains of numerous submarine slides, some of which caused giant waves that may have reached as high as 300 m (1000 ft). And unlike other tsunami that have struck Hawai'i after long journeys from distant Pacific Rim earthquakes, a giant wave spawned by another, perhaps swifter, collapse of a portion of Loihi could arrive with a warning of only minutes or seconds.

Thus Loihi grows higher in fits and starts—building itself up volcanically and then collapsing and shedding its substance in great rockslides. These processes broaden the volcano's base as a foundation for future upward growth. This pattern may be the blueprint for the creation of many of the thousands of volcanic islands—from Hawai'i to Tahiti—that adorn the Pacific Ocean floor.

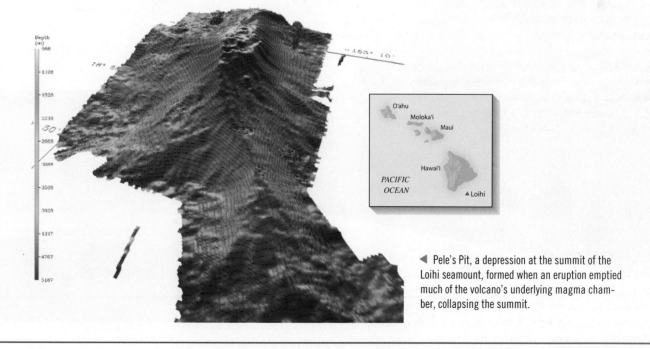

◀ Pele's Pit, a depression at the summit of the Loihi seamount, formed when an eruption emptied much of the volcano's underlying magma chamber, collapsing the summit.

Central-Vent Eruptions In central-vent eruptions, basaltic lava flows out—sometimes as a spectacular red-hot fountain—in all directions from one main vent. The lava then solidifies in more or less the same volume all around the vent. Over time, successive flows accumulate to form a low, broad, cone-shaped structure, known as a **shield volcano** because it resembles a warrior's shield lying on the ground (Fig. 4-23). Shield volcanoes owe their gentle slopes and broad summits to the low viscosity of basaltic lava. Such lava flows swiftly, sometimes within a network of lava tubes, and far enough from the vent to produce gentle slopes. An example of a shield volcano is Mauna Loa on the island of Hawai'i.

The classic location for the study of central-vent eruptions and shield-volcano development is on the Big Island of Hawai'i. The Big Island is located near the southeastern end of the 2590-km (1600-mi)-long chain of islands and undersea mountains in the north-central Pacific Ocean. This chain developed as the Pacific plate moved over a *relatively* stationary hot spot in the Earth's mantle (the heat-softened rock of the mantle is always moving . . . just *very* slowly) that has acted like a blowtorch, burning vast welts into the overlying plate.

Hawai'i's shield volcanoes are the largest single objects on Earth. The Big Island of Hawai'i consists of five major shields, the two largest being Mauna Kea and Mauna Loa, whose summits are 4205 m (13,792 ft) and 4170 m (13,678 ft) above sea level, respectively. Because the bases of these volcanoes are on the floor of the Pacific Ocean, thousands of meters below sea level, their total heights—more than 9000 m (30,000 ft)—are greater than that of Mount Everest, the Earth's highest continental peak.

At the start of an effusive eruption, lava begins to accumulate in the volcanic crater, sometimes forming a lava lake that may eventually overflow the rim of the crater. If enough lava has erupted to empty or partially empty the volcano's subterranean reservoir of magma, the unsupported summit of the volcanic cone may collapse inward, forming a larger, more-or-less circular summit depression known as a **caldera** (cal-DER-a). A typical summit crater may be 300 m (1000 ft) in diameter and several hundred meters deep; a caldera, however, may be a few kilometers to more than 15 km (10 mi) in diameter and more than 1000 m (3300 ft) deep.

The weight of a collapsed summit may close off a volcano's central vent and divert the remaining magma laterally, producing a *flank eruption* from the side of the volcano. Flank eruptions also occur when hardened lava plugs up the central vent or when the volcanic cone has grown so high that rising magma seeks a lower, more direct route to the surface. Pressurized lava spewing from a secondary vent on a volcano's flank often forms spectacular lava fountains, 100 to 200 m (330 to 660 ft) high. Hawai'i's K'ilauea is a volcanic cone built up by flank eruptions on Mauna Loa's southeastern slope. One of the world's most active volcanoes, K'ilauea's supply of fresh basaltic lava is continuously replenished because it is still located above the Hawaiian hot spot.

Fissure Eruptions on Land Rising low-viscosity basaltic magma will exploit any available route to the surface. That route may be through a series of linear fractures, or *fissures,* in the Earth's crust, which develop most often where plates are diverging. A large volume of highly fluid basaltic lava may gush out at speeds of 40 km/hr (25 mph) from kilometer-long cracks in the crust. These are called *flood basalts.* As lava flows away from a fissure, it may spread out over thousands of square kilo-

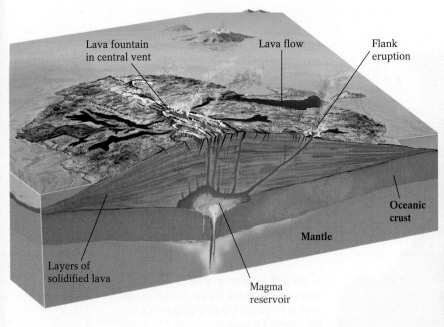

▲ **FIGURE 4-23** Low, broad shield volcanoes form their gentle slopes by the gradual accumulation of numerous basaltic lava flows. Photo: Summit of Hawai'i's Mauna Kea volcano, a shield volcano viewed here from the harbor at Hilo, Hawai'i.

meters, building thick *lava plateaus* (Fig. 4-24). A disastrous basaltic flood began in Iceland on June 8, 1783, as a series of lava flows spread out from one 32-km (20-mi)-long fissure. The lava blocked and diverted rivers, melted snow and ice into raging torrents, and destroyed much of Iceland's scarce agricultural land. Many livestock were poisoned after grazing on grass contaminated by the hydrogen-fluoride gas emitted during the eruption. As a result, 10,000 people—20% of Iceland's population—perished from starvation.

Shield volcanoes, such as Hawai'i's K'ilauea, can also produce fissure-type eruptions when a linear section of the volcano's flank rifts open. Along K'ilauea's southeastern flank, low-viscosity basaltic lava issues from the volcano's eastern rift zone.

Subaqueous (Underwater) Eruptions

Most subaqueous eruptions—especially oceanic, or *submarine,* eruptions—are quiet and effusive. This is especially true in deep water because below about 300 m (1000 ft) of water, water pressure prevents the water vapor and other gases in lava from expanding or escaping. Effusive submarine eruptions of basalt may form submarine shield volcanoes, when lava flows from a central vent, or flood basalts, when it erupts from a linear fissure or network of fissures. In both cases, the basalt exhibits the pillow structures that typify submarine eruptions of basalt. In late February and early March of 1996, the Gorda ridge, an oceanic di-

vergent zone off the Oregon coast, entered an eruptive phase that will probably culminate in another substantial submarine flood basalt, the third in the past three years along the Pacific Northwest coast (Fig. 4-25). And in 2005, a swarm of shallow earthquakes have alerted geologists to the possibility of yet another flood-basalt eruption in the region—this one off the coast of Canada's Vancouver Island.

Some subaqueous eruptions, however, are far from quiet and effusive. Occasionally, shallow submarine eruptions are driven by jets of steam that propel lava above the ocean's surface, where it shatters and cools rapidly to form a torrent of falling tephra. Even more explosive submarine eruptions occur when seawater enters a magma chamber through the ruptured walls of an island volcano. The cold water in contact with the red-hot magma produces a great deal of superheated steam that may then expand violently, shatter the volcanic cone, and propel magma and cone fragments skyward (Fig. 4-26).

Subglacial Eruptions

When polar regions coincide with zones of volcanic activity, two of geology's most pronounced extremes—fire *and* ice—collaborate in spectacular geological events. In areas where large ice sheets overlie volcanically active fissures, hot mafic lava periodically erupts beneath the ice, melting vast amounts of it and producing catastrophic outbursts of meltwater. One such drama captivated geologists for six weeks in

▲ **FIGURE 4-24** Lava-plateau formation. Photo: The 15-million-year-old basaltic Columbia River Plateau of eastern Washington. Geologists have identified more than 60 individual lava flows in these basalts, in some places totaling more than 2 kilometers (1 mile) in thickness. The lava *from just one of these flows* could pave Interstate 90 from Boston to Seattle to a depth of 175 m (575 ft), about the height of the Washington Monument in Washington, D.C. The photo here—Palouse Falls near the campus of Washington State University—shows five different basalt flows.

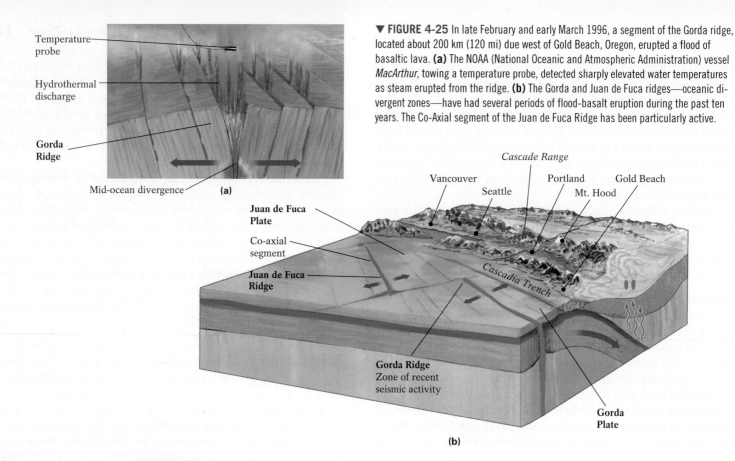

Temperature probe

Hydrothermal discharge

Gorda Ridge

Mid-ocean divergence

(a)

▼ **FIGURE 4-25** In late February and early March 1996, a segment of the Gorda ridge, located about 200 km (120 mi) due west of Gold Beach, Oregon, erupted a flood of basaltic lava. **(a)** The NOAA (National Oceanic and Atmospheric Administration) vessel *MacArthur,* towing a temperature probe, detected sharply elevated water temperatures as steam erupted from the ridge. **(b)** The Gorda and Juan de Fuca ridges—oceanic divergent zones—have had several periods of flood-basalt eruption during the past ten years. The Co-Axial segment of the Juan de Fuca Ridge has been particularly active.

Cascade Range

Vancouver

Seattle

Portland

Mt. Hood

Gold Beach

Juan de Fuca Plate

Co-axial segment

Juan de Fuca Ridge

Cascadia Trench

Gorda Ridge
Zone of recent seismic activity

Gorda Plate

(b)

◄ **FIGURE 4-26** When seawater enters fractures in a submarine volcano and makes contact with the underlying magma chamber, it is converted to steam. Pressure from the pent-up steam may build until there is an explosive eruption, such as this one of a submarine volcano near Japan in 1986.

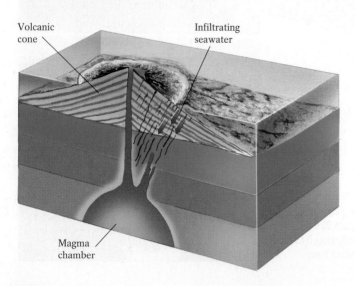

Volcanic cone

Infiltrating seawater

Magma chamber

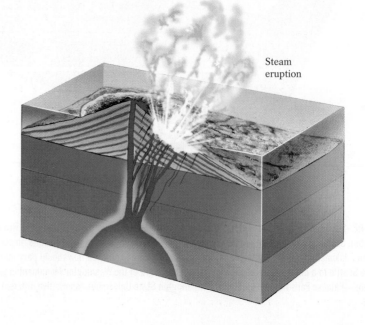

Steam eruption

the autumn of 1996 when a volcano erupted beneath a massive icecap in Iceland (Highlight 4-3).

Pyroclastic Eruptions

Pyroclastic eruptions usually involve high-viscosity, gas-rich felsic magmas. They typically vary from moderately to catastrophically explosive and tend to produce a great volume of solid volcanic fragments. Pyroclastic eruptions may periodically produce spectacular vertical columns of tephra rising tens of kilometers into the atmosphere. The surrounding countryside is simultaneously smothered by suffocating showers of hot ash, swirling pyroclastic flows, and hot lahars.

Residents of the prosperous Roman resort town of Pompeii, for example, lived in the shadow of the craggy peak of Mount Vesuvius, which looms 1220 m (4002 ft) above the Bay of Naples. Unfortunately, most did not realize that this vine-clad mountain was a volcano until it exploded on August 24, A.D. 79, sending a tephra column into the atmosphere, obscuring the midday sun, and shrouding the region in darkness. Most Pompeiians, inhaling the still-warm particles, perished by asphyxiation. The entire city was entombed beneath a layer of more than 6 m (20 ft) of volcanic ash. This layer preserved the most minute details of Roman life until it was stripped away by archaeologists in the mid-eighteenth century (Fig. 4-27).

If it is fed by magma containing less gas, a pyroclastic eruption may produce dense clouds of hot ash and pumice instead of towering tephra columns. These pyroclastic materials tend to fall back to the surface almost immediately, with little chance for air-cooling, and then race downslope as destructive, ground-hugging pyroclastic flows.

A felsic magma that contains relatively little gas or water may become a lava so viscous that it does not even flow out of the volcano's crater. Instead, it cools and hardens within the crater, forming a **volcanic dome** that caps the vent. Volcanic domes trap the volcano's gases and build pressure for another explosive erup-

▲ FIGURE 4-27 When Mount Vesuvius erupted in A.D. 79, the people in nearby Pompeii were trapped and suffocated beneath a layer of volcanic ash as much as 6 m (20 ft) thick. Archaeologists who excavated Pompeii found cavities lined with imprints of the decomposed bodies. By pouring plaster into the cavities, they were able to make casts that displayed the victims' musculature, agonized facial expressions, and in some cases even the folds in their clothing.

tion. This pressure might be relieved by the escape of gases laterally through the flank of the volcano, or it might build to such a point that it finally shatters the volcanic dome in a particularly explosive eruption (Fig. 4-28). In October 2004, through the time of this writing in September 2005, Washington's Mount St. Helens

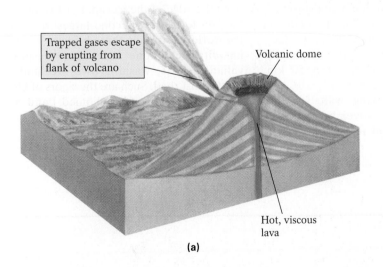

Trapped gases escape by erupting from flank of volcano

Volcanic dome

Hot, viscous lava

(a)

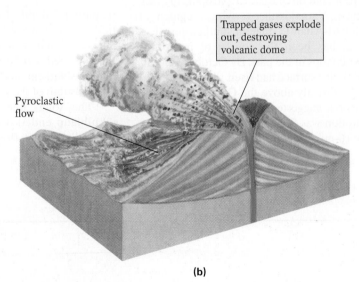

Trapped gases explode out, destroying volcanic dome

Pyroclastic flow

(b)

▲ FIGURE 4-28 Volcanic domes are caused by lavas with extremely high viscosity that solidify before they leave their crater, plugging the volcanic vent. Buildup of pressure beneath a dome **(a)** may result in gases escaping from the flank of the volcano, or **(b)** eventual explosive destruction of the dome.

Highlight 4-3

Iceland—Where Hot Volcanoes Meet Cold Glaciers

On September 29, 1996, an earthquake shook the ground beneath the 600-m (2000-ft)-thick Vatnajökull icecap of southeastern Iceland. Vatnajökull, covering 8300 km² (3200 mi²), is the largest glacier in Europe. The quake signaled the onset of a fissure eruption that emanated from Iceland's Bardarbunga and Grimvötn volcanic centers. On the morning of October 1, scientists discovered a deep depression in the glacier's surface, close to the location of a major subglacial eruption in 1938. Throughout the day, the depression grew and three additional depressions in the ice surface appeared as well. The development of these aligned depressions indicated that intensive melting was occurring at the base of the glacier along a 5- to 6-km (3- to 4-mi)-long fissure. By October 2, steam rising from the glacier's surface indicated that the eruption had broken through the ice. Over the next several days, a steam column rose to more than 10 km (6 mi), the surface fissure extended roughly the same distance, and black ash was being thrown intermittently into the air to heights of 300 to 500 m (1000 to 1600 ft) (see photo).

The greatest danger from a subglacial eruption is a *jökulhlaup*, or outburst flood. These floods occur when a great volume of subglacial meltwater, draining through subglacial tunnels, finds an outlet at the glacier's margin and discharges catastrophically (see figure). Vatnajökull, overlying a segment of the mid-Atlantic ridge in southeastern Iceland, has been the source of numerous jökulhlaups in the past. By October 12, Vatnajökull's surface had risen 15 to 20 m (50 to 65 ft) directly above the Grimvötn caldera. This rise suggested that meltwater from the eruption was flowing into and filling the 10 km (6-mi)-diameter caldera. Engineers, an-

▲ Steam and ash rising from the subsidence basin on the surface of the Vatnajökull icecap, directly above the subglacial Grimvötn volcano.

ticipating a flood of monumental proportions, worked around the clock to reinforce dikes and cut diversionary channels to steer the expected torrents away from Iceland's main, island-ringing road, about 50 km (30 mi) south of the icecap.

The long, unnerving wait for the jökulhlaup ended on November 5, when Iceland's largest flood in 60 years burst from the edge of the icecap. For two days, the flow formed the second-largest river in the world, destroying three bridges, severing vital telephone lines, and, despite valiant efforts, washing away Iceland's south-coast road. Such are the rigors of life in Iceland—the self-proclaimed Land of Fire *and* Ice.

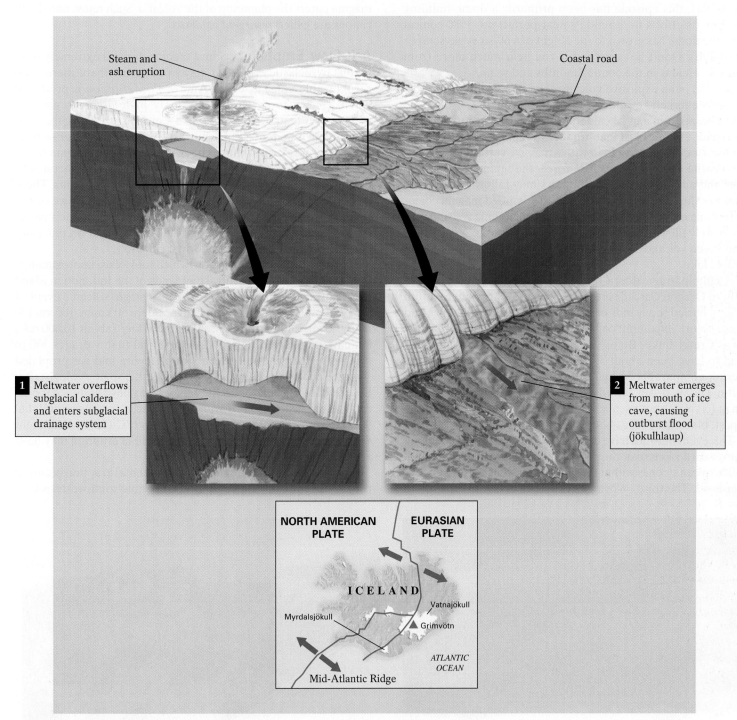

Steam and ash eruption

Coastal road

1 Meltwater overflows subglacial caldera and enters subglacial drainage system

2 Meltwater emerges from mouth of ice cave, causing outburst flood (jökulhlaup)

NORTH AMERICAN PLATE

EURASIAN PLATE

I C E L A N D

Vatnajökull

Myrdalsjökull

Grimvötn

ATLANTIC OCEAN

Mid-Atlantic Ridge

▲ Muddy meltwater bursts catastrophically from the margin of the Vatnajökull icecap, producing an outburst flood, or jökulhlaup.

has reawakened from its 19-year volcanic slumber and teased geologists with the prospects of a major eruption. While the volcano did send plumes of tephra skyward on several occasions (Fig. 4-29a), this episode has been primarily a dome-building period. In its efforts to rebuild its near-perfect pre-1980 conical shape, the dome has grown to the height of a 60-story skyscraper (Fig. 4-29b). Don't be surprised if you hear much more from Mount St. Helens in the coming months and years.

The life of an explosive volcano sometimes culminates in an extremely energetic eruption, followed by collapse of the cone and the creation of a huge caldera. This final explosion typically begins with a titanic blast that produces a rapidly moving blanket of pyroclastic debris combined with a massive vertical tephra column. As the escaping gases lose force, the tephra column collapses and additional pyroclastic flows may follow. Finally, the summit of the cone may subside into the emptied magma chamber, forming a large caldera. A caldera in an oceanic volcano may immediately fill with seawater, as it did at Krakatau. A caldera in a land-based volcano might eventually fill with freshwater to form a sizable lake.

A culminating caldera-forming eruption took place in the southern Oregon Cascades roughly 7700 years ago, when Mount Mazama buried thousands of square kilometers with tephra and pumice before its summit collapsed to form the caldera we know now as Crater Lake. Before this eruption, the mountain is believed to have been about 3700 m (12,000 ft) high and glacier-covered, comparable in majesty to Mount Rainier or Mount Shasta. Today, it is a sawed-off, goblet-shaped mountain only 1836 m (6058 ft) above sea level, but it contains North America's deepest, bluest lake (Fig. 4-30).

This event may not have ended Mount Mazama's excitement, however. Eruptions over the past 1000 years have produced three smaller volcanic cones within the lake, two of which are still below water level. The third, Wizard Island, rises above the surface at the western shore of the lake. Recent dives to the lake bottom reveal that hot water and steam are being vented continuously. Mount Mazama may be entering a new eruptive sequence as new magma enters the plumbing of the volcano. Such rejuvenated volcanoes are called *resurgent calderas.*

Ash-Flow Eruptions   Some spectacular and extremely dangerous pyroclastic eruptions may even occur where there is no recognizable volcanic cone. An *ash-flow eruption* is produced when a large reservoir of extremely high-viscosity, gas-rich magma rises to just below the Earth's surface, stretching the crust and forcing it up into a bulge marked by a series of ringlike fractures (Fig. 4-31). The thinned bedrock over such a magma reservoir may collapse, forcing magma upward into the fractures to produce a circular pattern of tremendous tephra columns. These soon collapse to form numerous ground-hugging flows of hot swirling ash and pumice that incinerate everything in their paths. As the magma reservoir empties, the surface crust becomes undermined until it collapses to form a caldera.

The Toba ash-flow eruption, which took place approximately 75,000 years ago in the central highlands of the Indonesian island of Sumatra, may have been the greatest single volcanic event of the past million years. It produced a caldera 50 km by 100 km (30 mi by 60 mi) and buried an area of 25,000 km^2 (about 10,000 mi^2) beneath a blanket of pyroclastic material that averages 300 m (1000 ft) in thickness. The amount of tephra and gas propelled skyward cooled the Earth so profoundly that some geologists believe that the ensuing "volcanic winter" may have hastened the arrival of the last period of worldwide glacial expansion. Global temperatures plummeted by as much as 6°C (12°F). Highlight 4-4 on page 130 describes how this incredible eruption may have affected the course of human affairs.

In North America, a similarly monumental eruption occurred roughly 28 million years ago in what is now southwestern Col-

(a)

(b)

▲ **FIGURE 4-29** The ongoing eruption of Mount St. Helens that began in 2004 produced both airborne tephra **(a)** and substantial dome-building eruptions **(b)**.

▼ **FIGURE 4-30** The culminating eruption of Mount Mazama, a volcano in the southern Oregon Cascades, occurred roughly 7700 years ago and lowered its height by more than about 1.5 km (about 1 mi). Because very little rock from the shattered cone of the volcano has been found in the areas surrounding Mount Mazama—including Oregon, Washington, and British Columbia—geologists have concluded that the summit of the volcano must have collapsed inward during its last eruption, rather than exploding outward. To this day, the ancient summit remains buried beneath the caldera produced by its collapse. This caldera (above) now contains Crater Lake, North America's deepest lake (more than 600 m [1900 ft] deep). Wizard Island, a new cone growing from the lake bottom, appears in this photo in the left-hand side of the lake.

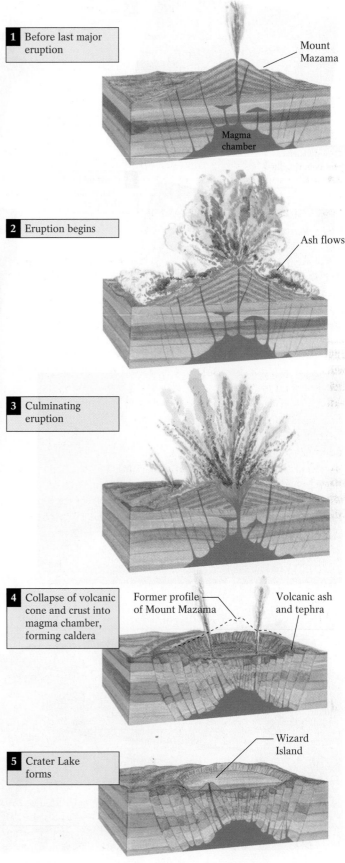

1 Before last major eruption — Mount Mazama — Magma chamber

2 Eruption begins — Ash flows

3 Culminating eruption

4 Collapse of volcanic cone and crust into magma chamber, forming caldera — Former profile of Mount Mazama — Volcanic ash and tephra

5 Crater Lake forms — Wizard Island

orado. About 50 km (30 mi) northeast of Durango sits a crumpled eroded caldera named La Garita, a 75-km (45-mi)-diameter depression that resulted from an eruption of mythic proportions. The La Garita ash flow was at least 20,000 times larger than the 1980 eruption of Mount St. Helens.

Ash-flow eruptions may recur when a collapsed magma chamber refills with new magma, forcing the caldera floor to arch upward again. At least three times in the last 2 million years, the Yellowstone plateau in Wyoming has erupted in devastating ash flows after being pushed up like a blister by an enormous mass of felsic magma. These events felled entire forests, changed the course of the region's rivers, and left much of the Great Plains blanketed with a mantle of fine tephra. The last of these events, about 650,000 years ago, created the huge caldera—measuring 50 km by 70 km (35 by 45 mi)—that now contains beautiful Yellowstone Lake. Today, only a few kilometers below Yellowstone's caldera, a new mass of felsic magma may be pooling. New technologies enable geologists to monitor a rising bulge beneath the Yellowstone Plateau. A robotic submarine has been dispatched to the bottom of Yellowstone Lake to study hydrothermal vents that suggest that magma may be relatively close to the surface beneath the lake. Yellowstone National Park's thermal features— its geysers (periodic steam blasts), hot springs (quieter flows of hot water), and gurgling mudpots (pools of boiling mud)—are all heated by this shallow subterranean magma reservoir (Fig. 4-32). Another ash-flow eruption may not be imminent here, but

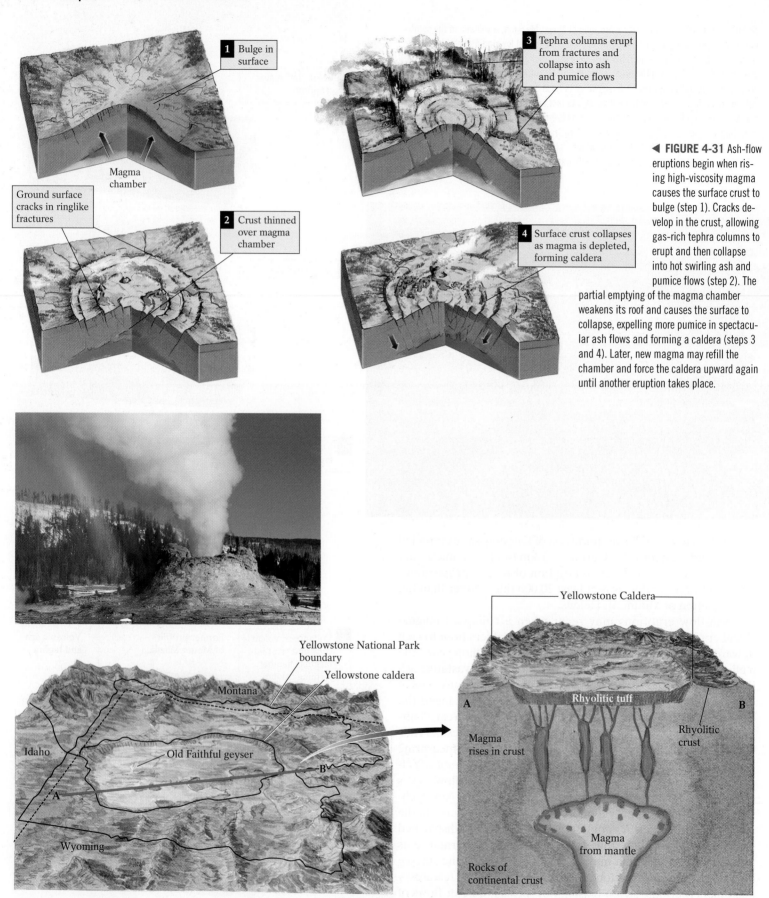

1 Bulge in surface

Magma chamber

Ground surface cracks in ringlike fractures

2 Crust thinned over magma chamber

3 Tephra columns erupt from fractures and collapse into ash and pumice flows

4 Surface crust collapses as magma is depleted, forming caldera

◀ **FIGURE 4-31** Ash-flow eruptions begin when rising high-viscosity magma causes the surface crust to bulge (step 1). Cracks develop in the crust, allowing gas-rich tephra columns to erupt and then collapse into hot swirling ash and pumice flows (step 2). The partial emptying of the magma chamber weakens its roof and causes the surface to collapse, expelling more pumice in spectacular ash flows and forming a caldera (steps 3 and 4). Later, new magma may refill the chamber and force the caldera upward again until another eruption takes place.

Yellowstone National Park boundary

Yellowstone caldera

Montana

Old Faithful geyser

Idaho

A

B

Wyoming

Yellowstone Caldera

Rhyolitic tuff

A

B

Rhyolitic crust

Magma rises in crust

Magma from mantle

Rocks of continental crust

▲ **FIGURE 4-32** The thermal features of Yellowstone National Park, which sits above an intraplate hot spot. These features indicate that Yellowstone is quite capable of erupting at any time.

1 Beginning of development of composite cone

Lava flow

2 Lava flow

3

Blast cloud

Pyroclastic flow

Eruption on flank of upbuilding composite cone

4 Composite volcanic cone

Summit crater

Volcanic neck

Layers of lava flows and pyroclastics

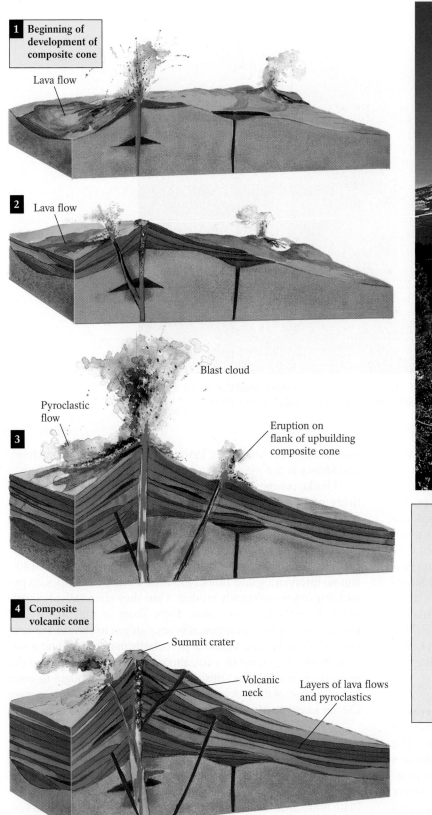

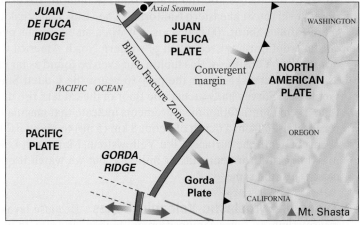

Axial Seamount

JUAN DE FUCA RIDGE

JUAN DE FUCA PLATE

WASHINGTON

Blanco Fracture Zone

Convergent margin

NORTH AMERICAN PLATE

PACIFIC OCEAN

PACIFIC PLATE

OREGON

GORDA RIDGE

Gorda Plate

CALIFORNIA

▲ Mt. Shasta

▲ **FIGURE 4-33** The origin of composite cones, steep-sided volcanoes consisting of layers of predominantly andesitic lava flows alternating with deposits of pyroclastic materials (and often associated with one or more flank eruptions). The composite cones in the Cascade Mountains of western North America, such as Mount Shasta shown here, have been built up over tens or even hundreds of thousands of years. Note: Mount Shasta has a satellite cone growing on its flank called Shastina (the second, lower peak on the left).

▲ **FIGURE 4-34** Sunset Crater, north of Flagstaff, Arizona. This cinder cone grew to a height of about 300 m (1000 ft) sometime around A.D. 1065, toward the end of the region's recent period of volcanic activity.

geologists monitor the area continually for signs of increasing volcanic activity.

The most immediate threat of an ash-flow eruption in North America has focused on the area around Mammoth Lakes, the popular ski resort in eastern California. The Long Valley caldera there formed about 700,000 years ago when an enormous pyroclastic eruption showered much of western North America with hot tephra. A dusting of this tephra can even be found as far east as central Nebraska! Over the past 20 years, the United States Geological Survey has watched the floor of the caldera rise more than 25 cm (9 in.). Other measurements indicate that magma has recently risen from a depth of about 8 km (5 mi) to about 3.2 km (2 mi) beneath the surface. Like Yellowstone, Mammoth Lakes may not erupt for centuries or millenia, but we watch it quite closely just the same.

Types of Pyroclastic Volcanic Cones   Because lavas of intermediate-to-felsic composition are high in viscosity, they resist flowing and thus they solidify close to the volcano's vent. The composition of such a magma, and consequently the volcano's eruptive style, may change over time so that the cone intermittently ejects a large quantity of tephra instead of lava. The larger tephra particles from these eruptions fall near the summit to form steep cinder piles, which are then covered by the next lava flow. This sequence of volcanic events produces the characteristic landform of pyroclastic eruptions—the **composite cone,** or **stratovolcano** (*strato* means "layered"), built up from alternating layers of lava and pyroclastics (Fig. 4–33). Because each pyroclastic deposit produces a steep slope that is then protected from erosion by a succeeding layer of lava, composite cones grow to be very large and have steep 10° to 25° slopes (a typical shield volcano has a slope of 5° or less). Stratovolcanoes include some of the Earth's

most picturesque volcanoes, among them Mounts Rainier, Hood, and Shasta in the western United States.

Unlike composite cones, **pyroclastic cones** consist almost entirely of loose pyroclastic material. When the dominant pyroclastics are cinders, they form **cinder cones** (Fig. 4–34). Pyroclastic cones are typically the smallest volcanoes (less than 450 m [1500 ft] high) and are generally steep-sided, because pyroclastic materials can pile up to form slopes between 30° and 40°. But no intervening lava flows bind their loose pyroclastics, so pyroclastic cones are readily eroded. Thus they are relatively short-lived. Pyroclastic cones may form from a magma of any composition as long as it contains enough gas to shower the lava into the air and create a cone of pyroclastic material around a volcanic vent. Once most of a volcano's gases have escaped and the pyroclastic activity has subsided, lava flows may begin to ooze from the base of the pyroclastic cone. Look at Figure 4–41 on page 137, an example of a late-stage lava flow from a pyroclastic cone.

Plate Tectonics and Volcanism

Plate boundaries and intraplate hot spots often coincide with volcanic activity. About 80% of the Earth's volcanoes surround the Pacific Ocean, where several oceanic plates are subducting beneath adjacent landmasses. Another 15% are similarly situated above subduction zones in the Mediterranean and Caribbean seas. The remainder are scattered along the ridges of divergent plate boundaries (as in the case of Iceland's 22 active volcanoes) and atop continental and oceanic intraplate hot spots (such as Yellowstone National Park and the Hawaiian volcanoes). Each type of plate-tectonic setting is associated generally with the

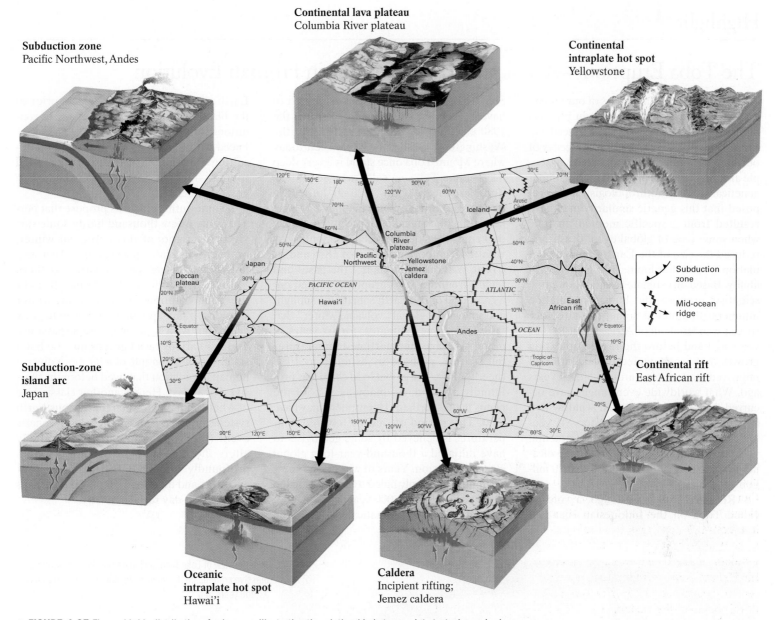

Continental lava plateau
Columbia River plateau

Subduction zone
Pacific Northwest, Andes

Continental
intraplate hot spot
Yellowstone

Subduction-zone
island arc
Japan

Oceanic
intraplate hot spot
Hawai'i

Caldera
Incipient rifting;
Jemez caldera

Continental rift
East African rift

▲ FIGURE 4-35 The worldwide distribution of volcanoes, illustrating the relationship between plate tectonics and volcanism. Note the coincidence of volcanic activity with plate boundaries, forming an alignment of the volcanoes surrounding the Pacific Ocean basin, the "ring of fire."

distinct eruptive style and physical appearance of its volcanoes (Fig. 4–35).

Subduction zones foster explosive pyroclastic volcanism because of the intermediate and felsic magmas produced there. As we saw in Chapter 3, water expelled from the descending plate enters the warm, overlying mantle, lowering its melting point and causing it to melt partially. When these materials mix with some of the descending plate's melted silica-rich sediments, an intermediate-to-felsic magma forms. Thus subduction-zone volcanism tends to produce steep-sided composite cones composed primarily of the intermediate volcanic rock andesite.

Explosive volcanism also occurs at intracontinental hot spots and where felsic continental plates are stretched thin and begin to rift. In both of these cases, felsic and intermediate rocks at the

base of the continental crust are melted by contact with hot mafic magma rising from the mantle. Eruptions of high-viscosity felsic magmas create the volcanic domes, calderas, and ash-flow deposits characteristic of intracontinental rift zones and hot spots. Rift zones and hot spots may also produce mafic lava, although not necessarily simultaneously with the felsic eruptions. This type of *bimodal* (two-type) volcanism can be found in the stretched-and-thinned Basin and Range region of southwestern North America. There, powerful blasts of felsic ash and pumice have alternated with quiet effusions of basaltic lava throughout the last 20 million years.

At divergent zones (such as along the Juan de Fuca and Gorda ridges), continental rifts (such as the East African rift valley), and oceanic intraplate hot spots (such as Hawai'i), effusive

Highlight 4-4

The Toba Eruption—A Volcano's Possible Role in Human Evolution

When we look at the members of our human family, we sometimes focus on our physical differences—body type, skin color, facial features. When biologists look at the *genes* of our human family, they conclude that people around the world are remarkably similar genetically. Some anthropologists have proposed that this genetic similarity may have resulted from a specific moment in time when some type of global event drastically reduced the size of the Earth's human population . . . and thus limited our genetic variability. Based on estimates of mutation rates, scientists propose that human-population numbers plummeted sometime after our species wandered out of Africa about 100,000 years ago and before the rapid population growth that accompanied the invention of improved stone tools—around 50,000 years ago. What worldwide event might have occurred during this 50,000-year window that killed off so many of our ancestors? A geological answer: perhaps a monumental volcanic eruption, believed by some volcanologists to be the largest in the past 20 million years.

Roughly 71,000 years ago, a volcano called Toba on the Indonesian island of Sumatra erupted with a blast estimated to have been at least 4000 times the size of the 1980 eruption of Mount St. Helens in the Washington Cascades. (All that's left today where Mount Toba once stood is a very deep lake—Lake Toba—a caldera) (see photo). The blast shot 800 km³ (320 mi³)of tephra into the atmosphere, burying much of Asia beneath a thick blanket of ash and probably darkening the skies for weeks or months. As with the Tambora eruption of 1815 and its "Year Without a Summer," the gases and tephra from the Toba eruption blocked out a great deal of the Earth's incoming sunlight and spawned a global "volcanic winter" that lasted, by some estimates, for six years or more. Average summertime temperatures fell sharply, perhaps by as much as 10°C (nearly 20°F). The global food chain was profoundly affected: As many as 75% of the plants in the Northern Hemisphere may have died. This period of sharply lowered temperatures is believed by some geologists to have initiated a thousand-year-long global glacial expansion. Years of wintertime snow accumulation simply failed to melt during the frigid summers of 71,000 years ago. The deteriorating climate dramatically affected the Earth's human population, which, prior to the blast, had enjoyed relatively warm conditions for the preceding 50,000 years or so. Faced with the frigid conditions and the resulting elimination of many of their food sources, our ancestors probably died off in great numbers.

Some anthropologists propose that perhaps only a few thousand hardy souls survived the rigors of this perpetual winter, probably in isolated pockets spread across a few warmer places, such as Africa, southern Europe, and southern Asia. As the effects of the Toba eruption and its volcanic winter subsided, human populations began to grow again, but with some major repercussions. Today's various racial groups may be but a small surviving sample of a far more diverse human population that lived before the massive Toba blast. Here is another classic case of what can happen when the various Earth systems interact: A monumental lithospheric event (the Toba eruption) fundamentally alters the atmosphere (and the climate), profoundly affecting the biosphere (our ancestors and their food supply) and perhaps our modern-day racial diversity.

◄ Aerial view of Lake Toba, a caldera that formed during a massive eruption on the Indonesian island of Sumatra about 71,000 years ago.

4.1 Geology at a Glance

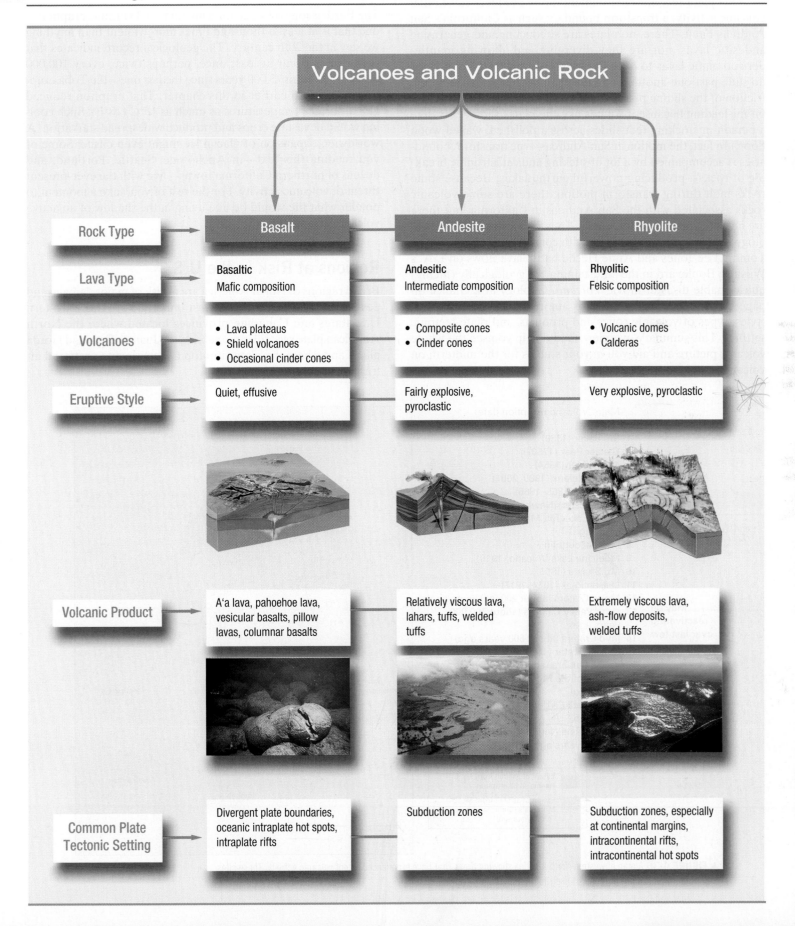

Volcanoes and Volcanic Rock

	Basalt	Andesite	Rhyolite
Rock Type	Basalt	Andesite	Rhyolite
Lava Type	**Basaltic** Mafic composition	**Andesitic** Intermediate composition	**Rhyolitic** Felsic composition
Volcanoes	• Lava plateaus • Shield volcanoes • Occasional cinder cones	• Composite cones • Cinder cones	• Volcanic domes • Calderas
Eruptive Style	Quiet, effusive	Fairly explosive, pyroclastic	Very explosive, pyroclastic
Volcanic Product	A'a lava, pahoehoe lava, vesicular basalts, pillow lavas, columnar basalts	Relatively viscous lava, lahars, tuffs, welded tuffs	Extremely viscous lava, ash-flow deposits, welded tuffs
Common Plate Tectonic Setting	Divergent plate boundaries, oceanic intraplate hot spots, intraplate rifts	Subduction zones	Subduction zones, especially at continental margins, intracontinental rifts, intracontinental hot spots

eruptions of low-viscosity basaltic lava generally produce near-horizontal lava plateaus and gently sloping shield volcanoes.

One plate-tectonic environment rarely produces substantial volcanic activity: a transform boundary such as California's San Andreas Fault. There, no plates are subducting and generating andesitic lavas, nor are they diverging and allowing mantle-derived mafic lavas to rise. The plates at transform boundaries do slide past one another, and magma may be produced by the friction of the sliding plates. (The word "slide" here is actually a bit misleading because it sounds like the plates slide easily, like a person in stocking feet slides across a polished, waxed wood floor.) In fact, the motion at San Andreas–type transform boundaries is accompanied by a lot of sticking and catastrophic breaking of rocks—producing powerful earthquakes. Because some rocks melt during transform motion, there are some volcanic rocks associated with the San Andreas in California, but there are no noteworthy volcanic cones. Thus, as a knowledgeable geology student, you would watch with considerable skepticism as Tommy Lee Jones and Anne Heche battle lava flows on L.A.'s Wilshire Boulevard in the entertaining but geologically (highly) questionable disaster movie *Volcano*. Geology at a Glance on page 131 shows the relationships among lava types, eruptive styles, types of volcanic cones and products, and plate-tectonic settings. This summary overview should help you see the whole volcanic picture and aid you in your studies for the midterm on volcanoes and volcanism.

Coping with Volcanic Hazards

The Earth is long overdue for a monumental volcanic eruption—one that is at least a thousand times more violent than anything we saw in the 20th century. The geological record indicates that such events occur at least once, perhaps twice, every 100,000 years. It has been 75,000 years since the last one—the Toba eruption—discussed earlier in this chapter. That eruption reduced global surface temperatures as much as 10°C (18°F). Such cooling would devastate crops and produce widespread starvation. A worldwide expansion of glacial ice might even ensue. Some of you reading this text—in Anchorage, Seattle, Portland, and dozens of northern California towns—live with the ever-present threat of volcanic activity. For the rest of you, take a moment to ponder what life would be like living in the shadow of an active volcano—Highlight 4–5.

Regions at Risk in the U.S.

Broad regions of North America are at risk of major damage and destruction from volcanism. Most imperiled are those western U.S. states and Canadian provinces located where the North American plate overrides the subducting Juan de Fuca and Gorda plates and where the plate sits atop the continent's scattered intraplate hot spots (Fig. 4–36).

Major Volcano (Eruption date)

1 Mt. Baker (1880)
2 Glacier Peak (1750?)
3 Mt. Rainier (1854)
4 Mt. St. Helens (1980, 2004)
5 Mt. Hood (1865–1866)
6 Three Sisters/Newberry Caldera (1853?)
7 Crater Lake (Mt. Mazama) (about 7700 years ago)
8 Mt. McLoughlin
9 Medicine Lake Volcano (1910)
10 Mt. Shasta (1855)
11 Lassen Peak (1914–1917)
12 Mono Craters (about 200,000 years ago)
13 Long Valley Caldera (about 700,000 years ago)
14 Inyo Craters (about 600 years ago)
15 Sunset Crater (1000 years ago)
16 Valles Caldera (about 60,000 years ago)
17 Craters of the Moon (about 2000 years ago)
18 Island Park Caldera (1.3 million years ago)
19 Yellowstone National Park (about 650,000 years ago)

May be reactivating over last few millennia {12, 13, 14}

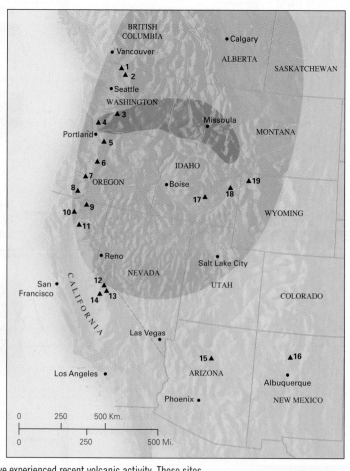

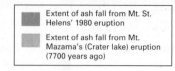

Extent of ash fall from Mt. St. Helens' 1980 eruption

Extent of ash fall from Mt. Mazama's (Crater lake) eruption (7700 years ago)

▲ **FIGURE 4-36** A map of western North America showing areas that have experienced recent volcanic activity. These sites may be considered some of the most likely for future volcanic events.

Highlight 4-5

Living in the Shadow of Vesuvius

If you had taken a stroll through a U.S. airport during the winter of 2004, you would have seen a considerable number of travelers toting around a thoroughly captivating novel by Robert Harris, entitled *Pompeii*. This novel chronicles—in vivid detail—the catastrophic eruption in 79 A.D. of Pompeii's looming volcano—Monte Vesuvius. Written as a first-person account by a Roman water engineer, *Pompeii* effectively captures the terror of an explosive volcanic eruption and its associated pyroclastic flows. Today, hundreds of thousands of Italian citizens continue to live in the shadow of Vesuvius in scores of scenic cities and towns along the Bay of Naples. (The modern city of Naples and its suburbs, located on the flanks of Vesuvius, is one of the largest cities in the world today that is built in the high-danger zone of an active volcano.) Here's a bit about the history of the Neapolitans' love-hate relationship with their mountain.

Monte Vesuvius is one of the most famous volcanoes in the world because of its role in destroying—and preserving—the Roman cities of Pompeii and Herculaneum in the year 79 A.D. (Figure 4-27). This eruption was the first ever to be described *in writing*. From a relatively safe distance of 20 km (12 mi) from the volcano, Pliny the Younger watched the eruption that killed his uncle, historian/aristocrat Pliny the Elder, and he later wrote down what he saw. (Much of Pliny's account is the source for author Harris's 21st-century novel.) He described the earthquakes before the eruption, the vertical plume of tephra that rose from the volcano, the falling tephra, the pyroclastic flows, and the pounding sea waves that struck the coast. In honor of this excellent account, geologists now use the term "plinian" to refer to the column of tephra that rises to high altitudes in explosive eruptions.

Monte Vesuvius *is not extinct*. It sits atop the subduction zone formed as the African plate moves northward at a rate of 2 to 3 cm (1 in) per year under southern Europe, and volcanologists consider Vesuvius to be extremely dangerous. The volcano has erupted at least 50 times since 79 A.D. Mudflows and tephra falls from an eruption in 1631 killed 3500 people, more than the death toll from the 79 A.D. eruption. There is some evidence for a 50-year cycle of activity in the last few hundred years . . . so Neapolitans view their mountain warily because the last eruption occurred 60-plus years ago in 1944.

Despite the dangers of the volcano, the population in this region has been growing steadily. Volcanic eruptions are widely spaced on a human timescale, so memories fade of the last disaster. Besides, the region has highly fertile soils (volcanically produced), an excellent harbor, and great natural beauty. As a result, approximately one million people live within a 7-km (4.5-mi)-high danger zone surrounding Vesuvius. (The high-danger zone is the region that would be destroyed within the first 15 minutes of a major eruption.) Another two million Italians live within reach of the pyroclastic flows, tephra clouds, and gas emissions that almost always accompany major eruptions of stratovolcanoes like Vesuvius.

Because we can't slow subduction, stop divergence, or chill intraplate hot spots, we will probably never be able to prevent such catastrophic volcanic eruptions. In certain cultures, people rely on faith and prayer to protect them from volcanoes. Hawaiian legend tells us of the fire goddess Pele, who flies into fits of rage when she is defeated in competition with the islands' young athletic chiefs. When angry Pele stamps her feet, she causes earthquakes and summons rivers of destructive lava in revenge. Some Hawaiians attempt to appease Pele by offering her bushels of sacred ohelo berries, flowered leis, and bottles of gin. For their part, geologists are developing ways to predict volcanic eruptions more accurately, so as to prevent casualties and minimize damage.

The Valles caldera, one of North America's largest volcanoes, looms over the Los Alamos National Laboratories (the birthplace of the atomic bomb) near Santa Fe, New Mexico. The massive crater—more than 25 km (15 mi) in diameter—developed during two major eruptions, 1.6 and 1.2 million years ago. Since that time, it has been the site of numerous explosive ash-flow eruptions, the last occurring about 60,000 years ago. This relatively recent event suggests that the caldera may be entering a new phase of activity. Thus geologists monitor the caldera very closely, looking for signs that the crust beneath the caldera is heating up and rising.

Many sites in the Cascade Range are more immediately threatening. Although most of the range's volcanoes have been relatively quiet in recent decades, Mount Baker and Mount St. Helens in Washington, Mount Hood in Oregon, and Mount Shasta and Lassen Peak in northern California were all actively erupting between 1832 and 1880. Today, Cascade volcanoes threaten major metropolitan areas in southern British Columbia, Washington, Oregon, and northern California. Mount McLoughlin, 50 km (30 mi) from Medford and Klamath Falls in Oregon, Mount Hood near Portland, which last erupted in 1865, and the South Sister volcano near Bend, Oregon, (which may have new magma rising beneath it) are closely watched for signs that they are awakening. Mount Baker, near the Washington–British Columbia border, resumed intermittent rumbling and steam emissions so actively that in 1975, the U.S. Geological Survey predicted that it would be the next Cascade volcano to erupt. It still hasn't, and in 1980, Mount St. Helens awoke instead (Fig. 4-37). Mount Rainier, the Cascades' grandest peak, last erupted in 1882 with a small whiff of brown ash, but its last major eruption occurred some 2000 years ago. In a future eruption, Rainier's greatest threat to nearby towns and cities—such as Tacoma, Washington—would come from large lahars spawned by steam and tephra emissions onto its glacier-clad slopes.

(a)

(c)

(b)

◀ **FIGURE 4-37 (a)** Mount St. Helens came to life with an audible boom on March 30, 1980, after 123 years of silence. **(b)** Six weeks of public anxiety and small ashy eruptions ended on May 18, when a massive eruption blasted about 400 m (1300 ft) of rock from the volcano's summit. What many had considered North America's most beautiful peak changed in just a matter of seconds into a squat gray crater. **(c)** A jet of hot (500°C [900°F]) tephra produced a vast pyroclastic flow that raced downslope with hurricane force. The flow, moving at greater than 300 km/h (200 mph), cut a swath of complete destruction 30 km (20 mi) wide. Entire stands of mature trees were blown down like matchsticks at a distance of 20 km (12 mi) from the blast zone. In all, 60 human lives were lost, along with 500 blacktail deer, 200 brown bear, and 1500 elk. The effects of the eruption were felt throughout the region. In Spokane, Washington, 400 km (250 mi) east of Mount St. Helens, a dense cloud of gray ash darkened the skies, dropping visibility to less than 3 m (10 ft) and switching on automatic street lights at noon.

Defense Plans

Today, as much as 10% of the Earth's population lives in the shadow of one of the planet's 600 active volcanoes. Any day now you may hear of a catastrophic eruption of such potentially dangerous volcanoes as Popocatepetl outside Mexico City, Soufriere Hills on the Caribbean island of Montserrat, or Mount Unzen in Japan. We should need no other reminder of the destructive power of volcanoes than the memory of the 23,000 who perished in 1985 in Armero, Colombia, victims of a huge lahar flowing from the slopes of Nevado del Ruiz.

Here in North America (and within clear view of one of your authors' office windows on one of Seattle's rare sunny days), tens of thousands live uneasily within the reach of a similar potential lahar from Mount Rainier. Much to the chagrin of its growing number of neighbors, this magnificent Cascade peak has recently been cited as the most volcanically hazardous terrain in the continental United States. With all this volcanic danger lurking, an effective plan to avert volcanic disasters is most definitely needed. Such a plan must start with keeping people out of the most hazardous regions. We must also build structures that protect life and property, and learn how to predict eruptions accurately so as to provide ample warning and allow for timely evacuations.

Volcanic Zoning

Many volcanic regions practice *volcanic zoning*. Under such a policy, areas with great potential for danger are set aside as national parks, monuments, and recreation areas, and are closed to residential and commercial development. As a result, there are no large metropolitan areas near Hawai'i Volcanoes National Park, nor are there any on the slopes of Mount Rainier, Mount St. Helens, Crater Lake, Yellowstone National Park, or other hazardous areas throughout the volcanic western states. Occasionally, however, our efforts to zone volcanic areas are less successful. Much of east and south Hilo, Hawai'i, for example, is built atop 30-year-old lava flows. Despite governmental efforts, developmental pressures have placed some residents "in harm's way." Local insurance companies have helped to limit dangerous development in Hilo by simply refusing to insure properties in high-risk zones.

Battling Lava and Lahars

Where preventive zoning is impractical, other means can avert or reduce potential damage, especially from lava flows and lahars. The first recorded attempt to fight a lava flow took place in Sicily during an eruption of Mount Etna on March 25, 1669. About 50 residents of the town of Catania, wearing wet cowskins for protection, used iron bars to poke holes through the hardening crust at the sides of the advancing flow. They did this to divert the lava, allowing it to drain along a different path away from their homes. The idea was basically sound, except for the fact that the lava's new course headed directly toward the watchful town of Paterno. Staunch defenders of Paterno rushed into a somewhat violent confrontation with the Catanians, "convincing" them (somewhat forcefully) to discontinue their innovative lava-control efforts.

The 20th century has provided more effective methods for diverting lava flows. Flows have been bombed from the air to disperse them over a wider area so that they thin out, cool, and solidify more rapidly. The hardening lava blocks the path of still-flowing lava behind it, forcing it to accumulate upstream and flow along a less damaging route. In 1935, a strategically placed bomb coaxed a flow to detour around Hilo.

In 1973, Icelanders on the coastal island of Vestmannaeyjar saved the local port city of Heimaey by cooling a vast lava flow issuing from a 1600-m (1 mi)-long fissure with seawater pumped by 47 large barge-mounted pumps anchored in a nearby harbor (Fig. 4-38). Even a simple lava wall—a barrier of boulders or

◀ **FIGURE 4-38** The residents of the Icelandic port of Heimaey fought the encroaching lava flows that threatened to fill the harbor with cold seawater.

▲ **FIGURE 4-39** A lava wall should be a minimum of 3 m (10 ft) high and have a wide, sturdy base. Because the wall is meant to divert lava away from inhabited areas, it should be built at an angle to the expected direction of the lava flow. The wall in this photo, hand-built from lava rocks, has successfully diverted a basaltic flow on the island of Hawai'i.

rocks (Fig. 4-39)—can sometimes protect an individual homestead effectively.

In some parts of Indonesia, a series of ropes are strung across valleys that have a history of catastrophic lahars. The first movement of mud triggers a siren, warning downstream communities to evacuate. In Japan, video cameras vigilantly monitor the slopes of some of that country's most potentially dangerous volcanoes.

Studying Volcanoes and Predicting Their Eruptions

In any given year, roughly 50 of the Earth's active volcanoes erupt—usually with some warning. Before they blow, they typically shake, swell, warm up, and belch a variety of gases. Because developing countries rarely have the necessary equipment to monitor an awakening volcano, they call the Volcano Disaster Assistance Program (VDAP)—a scientific SWAT team that rushes to volcanoes to assess their potential for violence and to predict when they might ignite. The VDAP, created a decade ago after the Armero disaster, is based at the U.S. Geological Survey's Cascade Volcanic Observatory in Vancouver, Washington. There, team members wait for a call that sends them jetting around the world armed with lasers, seismometers, and other devices used to monitor volcanoes. The VDAP was dramatically depicted in Hollywood's 1997 epic, *Dante's Peak* (compared to *Volcano*, a geological Oscar winner).

Geologists have had fair success predicting individual eruptive episodes when they concentrate on a specific volcano *after* an eruptive phase has begun. They monitor changes in a volcano's surface temperature, watch for the slightest expansion in its slope,

and keep track of regional earthquake activity (Fig. 4-40). A laboratory at the University of Washington in Seattle is staffed 24 hours a day to monitor the rumblings of Mount St. Helens. From time to time, the lab will issue a volcanic-hazard warning such as "We anticipate a quiet, dome-building event within the next 72 hours." If this sounds like we've mastered the art of volcano prediction, remember that the U.S. Geological Survey missed the call on Mount St. Helens' 1980 blast, despite the fact that a large team of scientists armed with the latest in prediction technology was closely watching the mountain. And several times during the winter and spring of 2005, Mount St. Helens "surprised" local

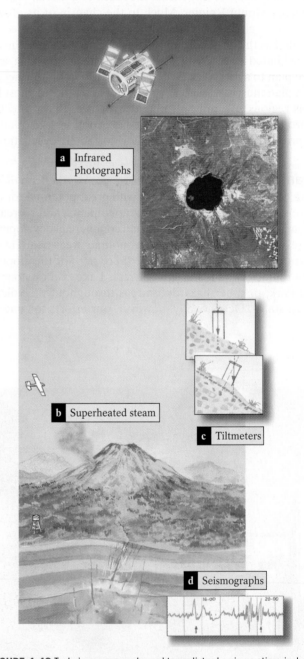

▲ **FIGURE 4-40** Techniques commonly used to predict volcanic eruptions include **(a)** satellite monitoring of volcanic-cone temperature (here, Crater Lake in southern Oregon), **(b)** watching for escape of superheated steam, **(c)** detecting volcanic-cone bulges via tiltmeters, **(d)** and locating increased tremor activity via seismographs.

volcanologists with several energetic eruptions that may foretell even more forceful activity in the near future. On a more encouraging note, volcanologists sharply limited loss of life when they successfully predicted the eruption of Mount Pinatubo in the Philippines. Two hundred and fifty thousand residents (including 16,000 on Clark Air Force Base) were evacuated from the area within 25 km (15 mi) of the mountain before the volcano's powerful blast of June 15, 1991. Tens of thousands were saved from the eruption's massive lahars and pyroclastic flows.

Before a volcano erupts, hot magma rises toward the surface, giving signs of increased heat. Ongoing surveys can identify new surface hot springs and measure the temperature of the water and steam in existing ones. If the escaping steam isn't much hotter than the boiling point of water, surface water is probably seeping into the mountain and being heated by contact with hot subsurface rocks. If the steam is "superheated," with temperatures as high as 500°C (900°F), it probably is rising from shallow water-rich magma, a sign that an eruption may be brewing. Impending eruptions may also be preceded by increased gas emissions from rising magmas. For this reason, volcanologists continuously monitor sulfur-dioxide and carbon-dioxide emissions on potentially active volcanoes.

As magma rises, the volcanic cone itself begins to heat up. Its overall temperature can be monitored from an orbiting satellite equipped with infrared-heat sensors to detect the slight changes in surface temperature. This high-altitude technology serves as a simultaneous early-warning system for many of the Earth's 600 or so active volcanoes.

Active volcanoes also expand in volume as they swell with a new supply of magma from below. An increase in the steepness or the bulging of a volcano's slope may signal an impending eruption. A *tiltmeter,* a highly sensitive device somewhat like a carpenter's level, detects the inflation of a volcanic cone and is sensitive enough to measure a change in slope angle equal to the thickness of a dime. Changes in the geographic position of the volcano itself, perhaps in response to movement along faults, are monitored via satellite using the Global Positioning System (the same system that guides the navigation systems in cars). To illustrate how the changing shape of a volcano heightens our concerns about the prospects for eruption, consider Mauna Loa on the Big Island of Hawai'i. Mauna Loa has been quiet for 20 years and in the interim, resorts and housing developments have sprung up around the island. In 2002, volcanologists began to see ominous signs of potential renewed activity on Mauna Loa—the summit of the volcano is rising and its caldera is widening, suggesting that new magma is filling the magma reservoir within the mountain, causing it to swell. Based on past eruptions, Mauna Loa's lava flows could travel in virtually any direction and reach the island's most populated areas.

Perhaps the most common way to predict volcanic eruptions involves monitoring the accelerated pulse of earthquake activity that often precedes an eruption. As magma rises, it pushes aside fractured rock, enlarging the fractures as it moves. Because this fracturing causes earthquakes, a distinctive pattern of earthquake activity occurs called *harmonic tremors,* a continuous rhythmic rumbling that can warn us of an impending eruption. Sensitive earthquake-detecting equipment locates these tremors and documents the changing position of the rising magma. Knowing the rate at which the magma rises enables volcanologists to estimate when an eruption will occur.

But what if we don't know a specific region's volcanic potential? Occasionally a *new* volcano may even appear suddenly—literally out of nowhere. In 1943, the volcano Parícutin developed *overnight* in the quiet cornfield of a farmer named Señor Pulido in the Mexican state of Michoacan, 334 km (207 mi) west of Mexico City (Fig. 4-41). This area was known to be volcanic, lying just northeast of a subducting segment of the Pacific Ocean, but one day Señor Pulido was farming corn and several months later, he owned his own volcano. Parícutin is the only volcano we've ever seen grow completely from scratch, so clearly such an event would be impossible to predict.

Make no mistake about it—studying volcanoes up close is a dangerous business. Within the past decade or so, several prominent volcanologists have lost their lives as they worked in the craters of active volcanoes. Two

▲ **FIGURE 4-41** The remains of the town of San Juan Parangaricutiro, which was engulfed during June and July of 1944 with lava from the eruption of Mexico's Parícutin volcano (visible in the background). The eruption, which lasted nine years, began in February 1943 with the sudden appearance of a small cinder cone in a cornfield about 300 km (200 mi) west of Mexico City. Because the flow that buried the town moved only a few feet per hour, the Mexican Army was able to evacuate the 4000 or so residents before any lives were lost.

geologists died in a pyroclastic flow spawned by the 1991 eruption of Japan's Mount Unzen, and six others perished during the 1993 blast from the crater of Galeras volcano in Colombia. Because the principal way to study a volcano is on the ground in the active zone, the U.S. Geological Survey and NASA have been working to create "robotic" volcanologists to reduce the human death toll. Designed to navigate the rubbly slopes of craters, map surfaces with lasers, collect samples, and sniff and analyze gases, these 850-kg (1700-lb), spider-legged mechanical geologists have thus far produced decidedly mixed results. The original robo-geologist broke down after "hiking" just a few tens of meters into the crater of Antarctica's Mount Erebus in 1993. In 1994, a new and improved model completed most of its mission into the crater of Alaska's Mount Spurr, even scrambling over a field littered with boulders and a 3-m (10-ft)-deep snowpack. Unfortunately, it lost its footing and keeled over just 150 meters shy of the crater rim. The plucky little android, lying like a turtle on its back with four of its eight legs broken, had to be retrieved by two human geologists.

Extraterrestrial Volcanism

Volcanism in our solar system occurs (or has occurred in the past) not only on Earth but also on the Moon, Mars, Venus, and the moons of Jupiter and Neptune. There are now no active volcanoes on the Moon, but at least one-fourth of its surface is covered by ancient flood basalts known as lunar *maria* ("seas"). As we saw in Chapter 3, early in its existence, the Moon was struck repeatedly by large meteorites that left deep craters and fractures in its crust. Basaltic lava flowed through those cracks, filling the craters and forming the maria. The smoother-appearing parts of the maria often contain winding trenches, called *rilles,* stretching for hundreds of kilometers. These are probably collapsed lava tubes

in the basalt. The Moon's surface also includes a few distinct shield volcanoes, such as those in the Marius Hills.

As much as 60% of the surface of Mars is covered by volcanic rock derived from approximately 20 volcanic centers. Virtually all Martian volcanoes are shields of incredible size. Olympus Mons, for example, is approximately 23 km (14 mi) high, more than twice the height of Mount Everest. Its diameter is about equal to the length of the state of California (Fig. 4-42). The size of this and other Martian shields suggests that they had remained stationary over their underlying hot spots for a very long time, allowing enormous mountains of volcanic rock to accumulate. Thus we can hypothesize that either the lithosphere of Mars does not move or it moves far more slowly than the Earth's plates. The volume of volcanic rock in Olympus Mons exceeds the total volume of the Hawaiian Islands, a *chain* of volcanoes produced because the Earth's Pacific plate moves over its underlying hotspot. Photos taken by the Mars *Global Surveyor* spacecraft in September 1997 suggest that the northern lowlands of Mars may have been volcanically active as recently as 10 million years ago. The eruptions may have been similar to vast fissure eruptions on Earth.

Venus also contains large shield volcanoes, some stretching in long chains along great faults in the surface. We know much about the volcanoes of Venus thanks to the remarkably sharp radar images from NASA's Magellan satellite, which began orbiting Venus in August 1990. The Venutian shield volcano of Rhea Mons, for example, would cover all of New Mexico and much of adjacent parts of Colorado, Texas, and Arizona. Radar also indicates that molten lava lakes may still exist on the Venutian surface. A cluster of seven lava domes discovered in 1991, each more than 15 km (10 mi) in diameter, may be similar in structure and composition to the one growing today in the crater of Mount St. Helens (Fig. 4-43). A series of lava flows, in what appear to be rift valleys similar to those of East Africa, suggest a long sequence of multiple volcanic events on Venus. We have not

▲ **FIGURE 4-42** An artist's rendition, based on images from NASA's *Viking* mission to Mars, of the Olympus Mons volcano. This enormous shield volcano is large enough to cover most of the northeastern section of the United States.

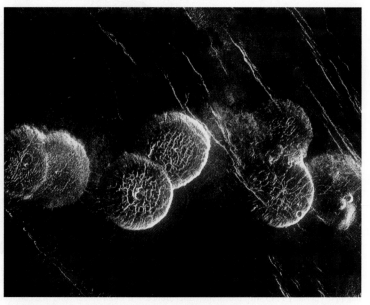

▲ **FIGURE 4-43** Radar images of volcanic domes on Venus taken by the *Magellan* spacecraft.

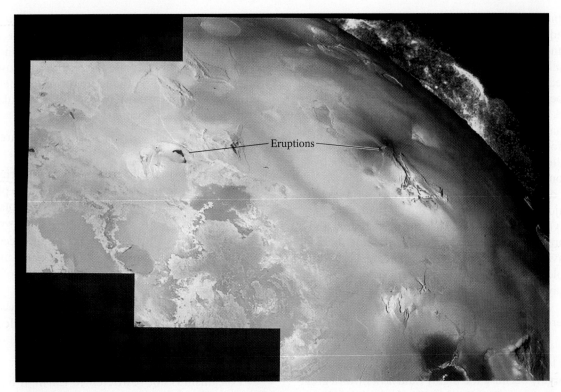

Eruptions

▲ **FIGURE 4-44** A computer-enhanced photo of a volcanic landscape on Jupiter's moon, Io, from data gathered by *Voyager 1*.

yet determined the ages of these flows or whether they record plate activity on Venus. In all, we have identified roughly 22,000 volcanic landforms on Venus.

One of Jupiter's moons, Io, and Neptune's largest moon, Triton, are believed to be the only other bodies in our solar system showing direct evidence of volcanic activity. NASA's *Voyager* and *Galileo* probes have detected as many as 120 hot spots on Io, some containing lava lakes, others spewing volcanic plumes (Fig. 4-44). Io erupts with molten sulfur and enormous clouds of sulfurous gas, which are propelled skyward at speeds approaching 3200 kph (2000 mph) to heights of 500 km (300 mi) above the surface. Consequently, Io is noted for its yellow-red snowfalls of sulfur, lakes of molten sulfur, and huge multicolored lava flows of black, yellow, orange, red, and brown. Geologists believe that volcanism on Io may result from the frictional heat generated as

this moon's surface rises and falls in response to the enormous gravitational pull of Jupiter. Triton has "ice" volcanoes. Our best guess of what they erupt is liquid nitrogen, dust, or methane compounds. As icy Triton's eruptions aren't driven by internal heat (like those on Earth and Venus), they are probably caused by seasonal heating by the Sun (which melts Triton's ice compounds).

We have now examined igneous activity at relatively shallow depths and the impact it has on the Earth's surface. In later chapters we will explore the more deep-seated processes that generate such near-surface activity and along with it, the movement of the Earth's plates. In the next few chapters, however, we will continue to look at the Earth's crust and the variety of rocks that exist there. In Chapter 5, we will see how these rocks are changed by environmental conditions at the planet's surface.

Chapter 4 Summary

Volcanism is the set of geological processes that occur as magma rises to the Earth's surface and erupts as lava. A volcano is the landform that results from the accumulation of lava and rock particles around an opening, or vent, in Earth's surface from which lava is extruded.

Lavas flow and erupt in distinctive ways, depending on their gas content and their viscosity.

- *Basaltic lava* has low viscosity (that is, it's highly fluid) because of its high temperature and relatively low silica content. It generally erupts relatively quietly, because its gases can readily escape from the magma and don't build up high pressure.

- *Andesitic lavas,* derived from magmas of intermediate temperature and composition, are higher in viscosity. They typically trap their gases, causing pressures to build up until an explosive eruption occurs.

- *Rhyolitic lavas,* with their high silica content and relatively low temperature, are highest in viscosity and generally erupt most explosively.

When basaltic lavas solidify, they create such distinctive features as pahoehoe- and 'a'a-type surface textures, columnar joints, lava tubes, and pillow structures (if erupted under water). The explosive volcanic eruptions characteristic of andesitic and rhyolitic lavas typically eject pyroclastic material—fragments of solidified lava and shattered preexisting rock—into the atmosphere. Lava fragments that cool and solidify as they fall back to the surface are called tephra.

Explosive ejection of pyroclastic material is usually accompanied by a number of life-threatening effects, such as pyroclastic flows (high-speed, ground-hugging avalanches of hot pyroclastic material), and lahars (hot volcanic mudflows). The secondary effects that often follow explosive eruptions include short-term water and air pollution and longer-term, sometimes global, climatic changes from wind-borne tephra and gas emissions.

Despite their different appearances, nearly all volcanoes have the same two major components: a mountain, or volcanic cone, composed of the volcano's solidified lava and pyroclastics, and a bowl-shaped depression, or volcanic crater, containing the vent from which the lava and pyroclastics erupt. If enough lava erupts and empties a volcano's subterranean reservoir of magma, the cone's summit may collapse, forming a much larger depression, or caldera.

Lavas of different composition form distinctly different landforms.

- Eruptions of low-viscosity basaltic lava form gently sloping landforms, largely because the highly fluid lava flows a great distance from the vent. Broad-based, gently sloping volcanic cones are called *shield volcanoes.* At divergent plate boundaries, such as in Iceland, basaltic magma generally reaches the surface through long linear cracks, or fissures, in the Earth's crust. These lavas then spread to produce nearly horizontal lava plateaus.

- The characteristic landform of pyroclastic eruptions is the *composite cone, or stratovolcano,* which is composed of alternating layers of pyroclastic deposits and solidified lava, typically of intermediate (andesitic) composition. Pyroclastic eruptions may also produce pyroclastic cones or cinder cones, composed almost entirely from the accumulation of loose pyroclastic material around a vent.

The various types of volcanic eruptions are associated with different plate-tectonic settings.

- Explosive pyroclastic eruptions of felsic (rhyolitic) lava generally occur within continental areas where plate rifting is taking place, or atop intracontinental hot spots such as the one beneath the Yellowstone plateau of northwestern Wyoming.

- Most intermediate (andesitic) eruptions occur where oceanic plates are subducting, for example, along most of the rim of the Pacific Ocean, such as in the Cascades of Washington and Oregon.

- Effusive eruptions of basalt occur generally at divergent plate margins, such as the mid-Atlantic ridge in Iceland, and above oceanic intraplate hot spots, such as the one beneath the Hawaiian Islands.

Volcanism is not restricted to Earth; it has occurred elsewhere in the solar system in the past and continues to do so today. Ancient (3-billion-year-old) volcanism is responsible for much of the rock and landform development on the surface of Earth's Moon. Relatively recent volcanic activity has been detected on Mars and Venus; Io, one of the moons of Jupiter, and Triton, one of Neptune's moons, are currently active.

KEY TERMS

caldera *(p. 118)*
cinder cone *(p. 128)*
composite cone *(p. 128)*
lahar *(p. 113)*
nuée ardente *(p. 111)*

pyroclastics *(p. 111)*
pyroclastic cone *(p. 128)*
pyroclastic eruption *(p. 121)*
pyroclastic flow *(p. 111)*
shield volcano *(p. 118)*

stratovolcano *(p. 128)*
tephra *(p. 111)*
vent *(p. 102)*
volcanic cone *(p. 116)*
volcanic crater *(p. 116)*

volcanic dome *(p. 121)*
volcanism *(p. 102)*
volcano *(p. 102)*
welded tuff *(p. 112)*

QUESTIONS FOR REVIEW

1. Briefly compare basaltic, andesitic, and rhyolitic lava in terms of their composition, viscosity, temperature, and eruptive behavior.

2. Within a single basaltic lava flow issuing from a Hawai'ian volcano, why is pahoehoe lava found closer to the vent than 'a'a lava?

3. Contrast the nature and origin of pyroclastic flows and lahars.

4. How does a composite cone form? What type of lava is associated with it? How could you distinguish a composite cone from a pyroclastic cone?

5. What types of volcanoes and volcanic landforms are associated with subduction zones? With divergent plate boundaries?

6. Identify three sites in North America that pose a volcanic threat to nearby residents and describe the nature of the volcanism at each one.

7. Describe three techniques that geologists use to predict volcanic eruptions.

8. Look at the photograph below and speculate about the plate-tectonic setting where these volcanoes are found, the composition of the rocks that make up the volcanoes, and whether eruptions of this volcanoes tend to be explosive or effusive.

9. Why do we find andesite throughout the islands of Japan but not throughout the islands of Hawai'i?

10. How would you explain the origin of a volcanic structure composed of 10,000 m of pillow lava covered by 3000 m of basalt containing vesicles, columnar jointing, and 'a'a and pahoehoe structure?

LEARNING ACTIVELY

1. If you have access to a VCR or DVD player, head down to your local video store and rent *Volcano* with Tommy Lee Jones and Anne Heche, and *Dante's Peak* with Pierce Brosnan and Linda Hamilton. Watch each of these Hollywood attempts to show volcanism in action and, while watching, assess the following:

   Are the eruptions depicted in these films geologically reasonable? Are the eruptions consistent with their plate-tectonic settings? Are the lava flows consistent with the nature of their source volcanoes? Would you expect to find volcanism in the San Andreas fault zone of southern California as portrayed in *Volcano*? Should Anne and Tommy Lee have been covered with snowflakelike tephra from this type of volcano? Is the lava flow that Pierce drives over in his truck consistent with the nature of the *Dante's*

   Peak eruption? Why did the lakes in *Dante's Peak* become so acidic that the stubborn Grandma disintegrated? Were the depictions of the lahar and pyroclastic flow in *Dante's Peak* reasonable geologically? Of the two films, which is the more geologically accurate?

2. Do a kitchen experiment to illustrate the relationship between temperature and the viscosity of fluids. Assemble a variety of "fluid" foods: pancake syrup, peanut butter, margarine, olive oil, and ice cream. Compare their viscosities when cold (straight from the refrigerator), warm (sitting out in the Sun at room temperature), and hot (heated in a saucepan on a stovetop). (Sorry for the mess.) Which of the fluids has the highest viscosity? Which of the fluids has the lowest viscosity?

ONLINE STUDY GUIDE

Practice makes you better. Make the time to take the *Online Study Guide* quiz for this chapter. It should take you about 20 minutes. Automatic grading and instant feedback will help you quickly and accurately identify the concepts you have mastered and the areas in which you need more study.
www.prenhall.com/chernicoff

Weathering— The Breakdown of Rocks

Where are the enormous volcanoes that used to sit above the granitic rocks of Yosemite National Park in California? Have you ever seen a marble statue or tombstone that was so worn and pitted that its words and features were all but gone? Will George Washington's substantial nose ever fall off South Dakota's Mount Rushmore? How could water alone carve the mile-deep Grand Canyon? The answers to all of these seemingly random questions involve the processes by which rocks and minerals break down at or near the Earth's surface, in response to environmental factors—processes known collectively as **weathering.** Weathering is a slow but powerful process that attacks even the hardest rocks largely *in place,* where the rocks originally formed. Rocks weakened by weathering are then more vulnerable to **erosion,** the host of processes by which moving water, wind, ice, and crashing surf (or, sometimes, the sheer force of gravity alone) carries away pieces of rock from the exposed weathered bedrock and deposits them elsewhere.

Like many geological processes, weathering and erosion are interrelated, often working together to produce vast landforms such as Arizona's Grand Canyon. They also produce *sediment,* the loose, fragmented surface material that is the raw material for many types of sedimentary rock. We see the effects of weathering and erosion all around us, on both geologic and human-made surfaces. In this chapter, we will focus on weathering; the various processes of erosion (rivers, landslides, glaciers, crashing waves, etc.) will be discussed in detail throughout the second half of this text.

Some scientists think of the thin layer of soil and sediment formed by weathering of rocks as the Earth's "final frontier." Digging in soil might not sound as exciting as sending a spacecraft to Mars or a submarine to the bottom of the deepest ocean, but in some ways the top few centimeters of the Earth's surface are currently as unknown and mysterious to us as those exotic faraway places. Our very existence, and that of all land creatures, depends largely on the Earth's soils and sediments, so we need to study how this material forms— that is, how solid rock breaks down into soil and sediment—and how the activities of humans and other organisms change these materials with time.

Weathering plays a vital role in our daily lives—with both positive and negative outcomes. It frees life-sustaining minerals and elements from solid rock so they can become part of our soils, later to pass into the foods we eat and the water we drink. Weathering is also responsible for some of the Earth's most spectacular scenery (Fig. 5-1). But weathering can also wreak havoc on the structures we build, including the roads and bridges we take to work and school. Countless monuments, from the pyramids of Egypt to ordinary tombstones

FIGURE 5-1 Weathered sandstone outcrop in the Vermillion Cliffs Wilderness near Page, Arizona. ▶

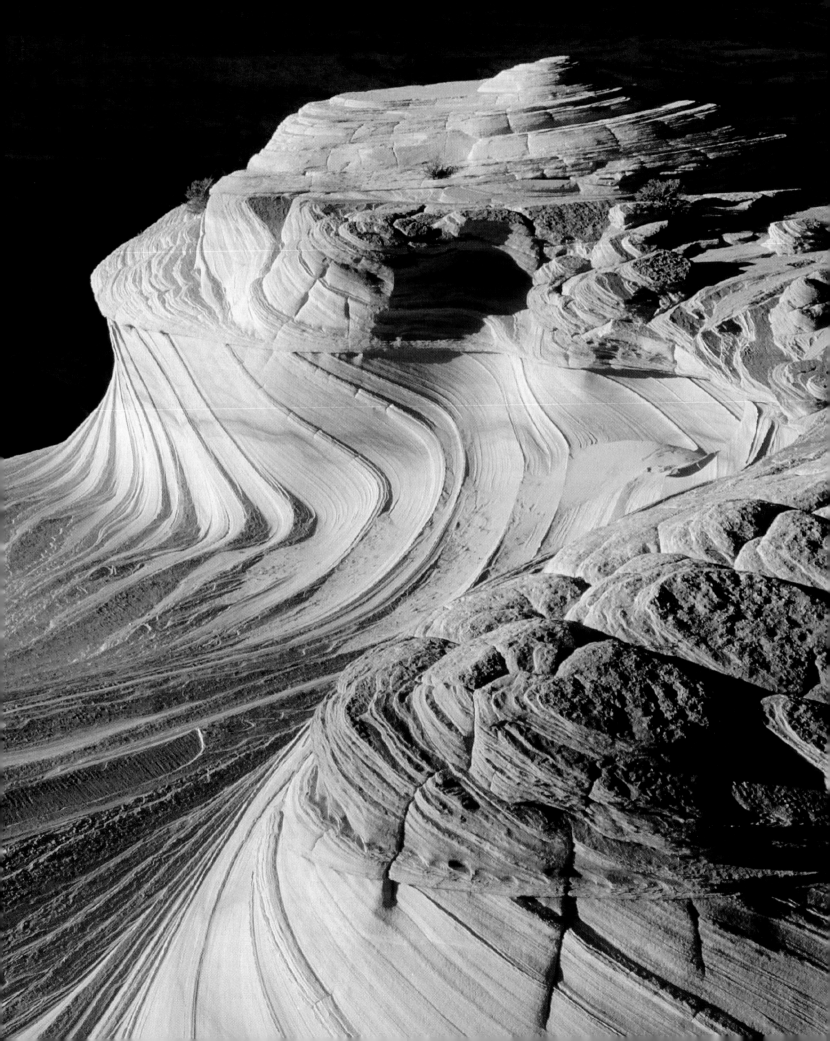

in the town cemetery, undergo drastic deterioration from freezing water, hot sunshine, and other climatic forces. Engineers and stonemasons are called upon regularly to secure loosened rocks on dams and downtown office buildings. They are also being called upon to sharpen the fading features of the faces on Mount Rushmore (Fig. 5-2). Weathering operates virtually everywhere at the Earth's land surface, both on human timescales (altering our statues, tombstones, and buildings) and on geologic timescales (weathering has claimed the lofty mountains that once towered over what are now Minnesota, Michigan, and Wisconsin).

In Chapter 3, you learned that there is a predictable sequence of crystallization and melting of minerals as a function of temperature. Minerals like olivine are stable at high temperatures and will therefore melt last in a heated rock and crystallize first (at the highest temperature) from a cooling mafic magma. Low-temperature minerals like quartz will melt at low temperatures and crystallize last (at the lowest temperature) relative to other minerals in a magma. This way of understanding the conditions under which minerals are stable in the Earth's crust can also be used to predict—in a general way—how unstable minerals will be at the Earth's surface and thus how vulnerable they will be to the weathering process. For example, since olivine is stable in the hot, dry conditions of the Earth's mantle, it follows that olivine will be very unstable in the much cooler, moister environment at the Earth's surface. It therefore weathers quite readily at the surface. Likewise, we can use these relationships to predict that quartz will be more stable at the surface of the Earth because of quartz's lower-temperature stability. This way of looking at how likely a mineral is to weather at the Earth's surface is meant only as a general illustration (and basically applies more to the silicate minerals than to others, such as the carbonate minerals). Many other factors play important roles in the weathering of minerals and rocks, such as the chemical composition of the mineral and the details of the local environment of the rock at or near the Earth's surface (such as the climate, landscape, and relative abundance of animal and plant life). In the following pages, we will explore the different ways in which solid rock weathers, the factors that control weathering, and the various practical products that result from it.

Weathering Processes

Rocks weather in two ways: **Mechanical weathering** breaks a mineral or rock into smaller pieces (*disintegrates* it) but does not change the pieces' chemical makeup; **chemical weathering** actually changes the chemical composition of minerals and rocks that are unstable at the Earth's surface (*decomposes* them), converting them into more stable compounds, or dissolving them away. Minerals and rocks that are chemically stable at the Earth's surface (such as quartz) resist chemical weathering.

Mechanical and chemical weathering go hand-in-hand in most environments. Disintegrating a rock by mechanical weathering makes it more susceptible to chemical weathering by creating more surface area for chemical attack (Fig. 5-3). Think about how crushing a sugar cube with a spoon causes it to dissolve more rapidly in hot water. By mechanically disintegrating the sugar cube, you vastly increase the surface area exposed to the hot water. Similarly, the area of a boulder exposed to weathering agents consists only of its outer surface until mechanical weathering breaks it down and increases its total surface area.

For a map of the many weathering concepts in upcoming section, see Geology at a Glance on page 154.

Mechanical Weathering—Disintegrating Rocks

Several natural processes reduce rocks to smaller sizes without causing any change in their chemical makeup. In any given location, one or more of the following processes may be working at the same time.

Frost Wedging   If you have ever put a full water bottle in the freezer, only to find later it has popped its top, you have learned that water expands in volume (by about 9%) when it freezes. This means that when water trapped in the pores or cracks in a rock turns to ice in freezing temperatures, it has to expand. The upper surface of the water freezes first because it is in direct contact with the cold atmosphere. As the cold penetrates downward, the rest of the water freezes as well and, because the surface has already frozen, it cannot expand upward. This ice expands outward, exerting a force greater than that needed to fracture even solid granite, enlarging the cracks and often loosening or dislodging fragments of rock (Fig. 5-4). If water freezes in a confined space, such as in a crack in a rock, the outward force against the walls of the crack can be enormous—up to 4.3 *million* pounds per square foot. This process, called **frost wedging,** is one of the most effective ways to weather a rock mechanically.

Frost wedging is most common in environments where surface water is abundant and temperatures often fluctuate around the freezing point of water (0°C, 32°F). When water frozen in a rock's cracks thaws each day in the warm afternoon sun and then refreezes each night, the cracks expand rapidly, causing blocks or slivers of rock to fall and collect as *scree* (a deposit of shattered rock) on a **talus slope** (the apron of shattered rock often found at the foot of a steep slope) (Fig. 5-5). Ice wedging is largely

▲ **FIGURE 5-2** More than half a century after they were sculpted, the seemingly eternal faces of Presidents Washington, Lincoln, Jefferson, and Theodore Roosevelt, carved out of the side of South Dakota's Mount Rushmore, are in real danger of losing their noses, lips, and mustaches. Here a worker repairs weathering damage to the granitic sculptures in Mount Rushmore National Monument, South Dakota.

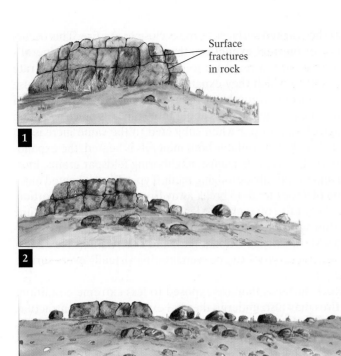

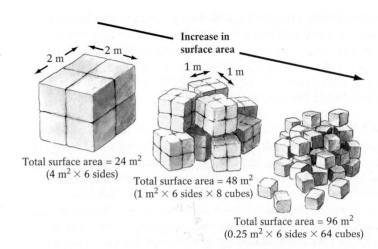

▲ **FIGURE 5-3** Mechanical weathering breaks rocks and minerals into fragments, increasing the total surface area exposed to the processes of chemical weathering.

▲ **FIGURE 5-5** Frost wedging is particularly evident on mountains in moist temperate regions, where, even in summer, nighttime temperatures often fall below freezing. Here at Wheeler Peak in Great Basin National Park, Nevada, a cycle of freezing and thawing occurs daily, creating the fresh rock rubble, or scree, that carpets the slopes of many mid- to high-latitude mountains.

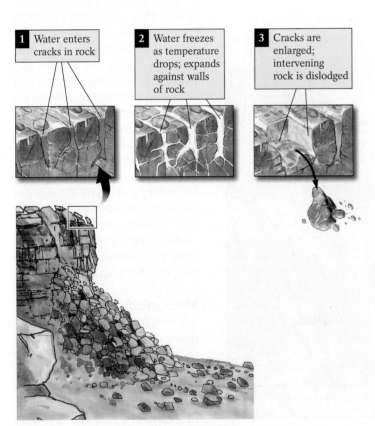

▲ **FIGURE 5-4** Frost wedging occurs when water freezes and expands within cracks in rock, enlarging them. The cracks get bigger with each freeze and may eventually dislodge intervening pieces of rock.

responsible for creating the notorious pothole "obstacle courses" found in most northern cities that form when water seeps into roadway cracks and then freezes. And on May 3, 2003, this same process caused the demise of one of New England's most beloved tourist attractions—*The Old Man of the Mountain,* in New Hampshire's White Mountains (Fig. 5-6).

Salt Wedging   When saltwater enters a crack in a rock and evaporates, the growing salt crystals apply great pressure to the walls of the crack, prying them farther apart. Coastal cliffs are particularly susceptible to this process. Salty sea spray soaks into rocks, evaporates, and leaves behind growing salt crystals. This process has also damaged some well-known stone structures of human design. Among them is Cleopatra's Needle, a granite obelisk that stood for more than 2000 years in Egypt without much loss of the finely engraved hieroglyphics on its sides. Unfortunately, before being relocated to Central Park in New York City in 1880, the obelisk was stored at a site where salty groundwater readily penetrated the column. Salt crystallized within small water-filled cracks in the granite as the water evaporated in the hot Middle Eastern air, and this salt-crystal growth caused the initial disintegration of the needle's surface. On arrival in New York's humid environment (as a gift from Egypt to commemorate the opening of the Suez Canal), the same salt crystals then absorbed water and expanded, further disrupting the obelisk's surface. Today, little remains of the ancient message etched into the obelisk's sides (Fig. 5-7).

Thermal Expansion and Contraction   If you have ever sat around a campfire, you may have noticed that thin layers of rock fall off the cracked surfaces of rocks close to the fire. This occurs because of **thermal expansion,** the enlargement of a mineral's crystal structure in response to heat. Minerals vary in the rate and extent to which they expand when heated, largely because of the different strengths of their chemical bonds. A grain of quartz, for example, expands about three times as much as a grain of plagioclase feldspar when subjected to the same increase in heat. If a rock that contains both minerals is heated, the expanding quartz grains push against neighboring feldspar grains, loosening and eventually dislodging them. Typically, the heated outer portion of a rock tends to break away from the cool inner portion. The effect of fire on rock was demonstrated dramatically in June 2000 during the forest fire that devastated a vast area of Colorado's Mesa Verde National Park and Los Alamos, New Mexico. Fire-cracked rock can be seen littering virtually every surface (Fig. 5-8).

Rock surfaces that are exposed to less extreme conditions than fires may also undergo thermal expansion and contraction, but at a much slower pace. Would the minerals at the surfaces of rocks exposed to high daytime and low nighttime temperatures expand with repeated heating and contract upon cooling, eventually causing the rock's outer layer to break apart? Geologists have studied this possibility in deserts where the rocks are exposed to daily temperature fluctuations of up to 56°C (100°F). To test the effects of thermal expansion and contraction in deserts, geologists heated and then cooled a block of highly polished granite through a range of 38°C (68°F) *89,500 times,* equal to 244 years of daily temperature fluctuations in a desert. The rock didn't change at all! Is 244 years enough time for a rock to weather by this mechanism? The geologists repeated the experiment, but this time they sprayed the rock with a fine mist of water during the cooling period as if the rocks were coated with a bit of morning dew during the cool desert nights. With water in the experiment, the granite quickly showed signs of weathering; it lost its polish, the surfaces of some grains became rough and irregular, and some cracks appeared near the rock's surface. These results suggest that extreme daily temperature variations *in combination with water* may weather rocks in deserts.

Mechanical Exfoliation   Consider what happens when you sit down on the soft cushions of a comfortable couch. While you're sitting there, your weight depresses the cushions, but when you rise and remove your weight, the cushions expand upward, restoring their former shape. When erosion of overlying rock

▲ **FIGURE 5-6** The Old Man in the Mountain—"Before and After." This popular tourist attraction in New Hampshire's White Mountains disintegrated from frost wedging in 2003.

(a)

(b)

▲ **FIGURE 5-7** Cleopatra's Needle **(a)** as it looked for thousands of years in Egypt and **(b)** as it looks today, after a century of weathering in New York City's Central Park. Acid rainfall and air pollution also contributed to the Needle's rapid weathering.

▲ **FIGURE 5-8** The intense heat of a forest fire in Colorado's Mesa Verde National Park in June of 2000 caused these chips of rock to fall from the rock surface as its outer layers expanded.

and soil removes a great weight from deeper rocks, a similar process occurs. This can be seen, for example, where erosion exposes an igneous intrusion, such as a batholith, that crystallized at some depth within the Earth's crust. The overlying pressure on the intrusion is reduced as the weight of the eroded rock is removed, and the rocks near the surface respond by expanding—upward. As rock expands, it fractures into sheets parallel to its exposed surface. These sheets may then break loose and slip from the sloping surface of the exposed rock, a weathering process known as **mechanical exfoliation** (also known as "unloading") (Fig. 5-9). Many granitic mountains, and other types of rocks as well, have a dramatic "stepped" appearance that is produced when large thin slabs of rock, several meters thick, "exfoliate" from underlying rocks (Fig. 5-10).

Other Mechanical Weathering Processes   Almost all rocks contain some cracks and crevices. In the case of surface rocks, plants and trees often take root in these cracks. Although a rock would seem strong enough to withstand it, the force applied to a crack by a growing tree root is surprisingly powerful and quite capable of enlarging the crack (Fig. 5-11). The buckled and broken sidewalks of some tree-lined streets testify to the weathering power of such root growth.

Mechanical weathering by **abrasion** occurs when rocks and minerals collide during transport or when loose transported material scrapes across exposed bedrock surfaces. For example, sediment fragments carried in swirling streams or by gusty winds collide or grind against one another and other rock surfaces, breaking loose fragments into smaller particles and shearing off new fragments as well. Similarly, rocks carried along at the base of a glacier "abrade" underlying rocks and are ground to even smaller sizes.

Chemical Weathering—Decomposing Rocks

Chemical weathering alters the composition of minerals and rocks, principally through chemical reactions involving water, oxygen, and microscopic organisms. The water may come from

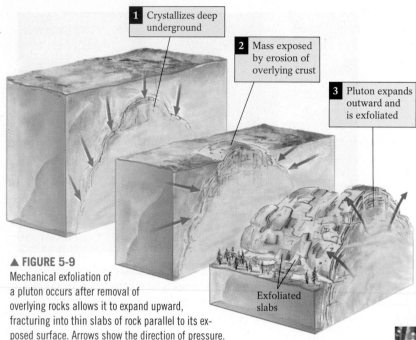

1 Crystallizes deep underground

2 Mass exposed by erosion of overlying crust

3 Pluton expands outward and is exfoliated

Exfoliated slabs

▲ FIGURE 5-9
Mechanical exfoliation of a pluton occurs after removal of overlying rocks allows it to expand upward, fracturing into thin slabs of rock parallel to its exposed surface. Arrows show the direction of pressure.

Chapter 2) and its tendency to break down into its component H^+ and OH^- ions. The three major processes by which rocks—with an assist from water—weather chemically are dissolution, oxidation, and hydrolysis.

Dissolution   A mineral experiences **dissolution** when ionic bonds between ions are broken and the now separated ions are carried away in water. (Such ions are said to be *in solution.*) When the water carrying them eventually evaporates, these compounds *precipitate* from the solution—that is, the ions combine again into solid form as new minerals.

Why do some minerals dissolve in water? The slightly charged sides of the water molecule are what attract and remove oppositely charged ions from the surfaces of minerals. The mineral halite (NaCl, or salt), for example, readily dissolves. Dissolution of halite occurs when the positive (H^+) side of a water molecule attracts and dislodges

oceans, lakes, rivers, glaciers, or the underground water system. It may also come directly from the atmosphere as rain, snow, or dew. All of these sources of water—as well as the oxygen from the air—highlight again the interaction between the hydrosphere, atmosphere, and the rocky lithosphere at or near the surface of the Earth.

Water is the single most important factor controlling the rate of chemical weathering, because it carries ions to the reaction sites, participates in the reactions, and then carries away the products of the reactions. This role is made possible by the oppositely charged nature of the two sides of a water molecule (discussed in

▲ FIGURE 5-11 Tree roots growing in rock fractures exert an outward force that expands the fractures, mechanically weathering the rocks. Pressure from the growing birch root has split this rock.

▲ FIGURE 5-10 The "steps" of this mountain (imagine walking upslope toward the trees on this terrain) in California's Yosemite National Park were produced by mechanical exfoliation of the mountain's granitic rock.

chloride ions (Cl^-) and the negative (OH^-) side of a water molecule attracts and dislodges sodium ions (Na^+). Different minerals have different solubilities depending on the chemistry and bonding characteristics of the minerals and the chemistry of the fluid.

The water that reacts with minerals is typically not pure H_2O; the water may be slightly to very acidic. Water in the air or in the ground may combine with carbon dioxide (CO_2) to produce carbonic acid (H_2CO_3), which is very effective at dissolving limestone, a common sedimentary rock.

Although relatively weak, this acid effectively decomposes all minerals, but it is especially effective at dissolving calcite, the principal mineral in limestone, to form calcium and bicarbonate ions:

$$H_2O \; + \; CO_2 \; \longrightarrow \; H_2CO_3$$
Water   Carbon dioxide   Carbonic acid

$$CaCO_3 \; + \; H_2CO_3 \; \longrightarrow \; Ca^{2+} \; + \; 2\,HCO_3^-$$
Calcite   Carbonic acid   Calcium ion   Bicarbonate ions

The calcium and bicarbonate ions are then carried off in solution by circulating groundwater, often leaving holes in the parent rock (Fig. 5-12).

These two simple reactions have created most of the world's caves (discussed in Chapter 16). Microscopic fractures in soluble limestone grow into large underground passages through the gradual dissolution of the surrounding rock. Also as a result of

▲ **FIGURE 5-12** Weathered and unweathered limestone boulders. In humid environments, such as those in most southeastern states, limestone dissolves extensively along cracks and crevices, leaving behind thousands of cavities that range from minor surface depressions to large underground cave systems. In this photo, the limestone boulder above the hammer is weathered; the one below is not.

these reactions, dissolved calcium and bicarbonate move from the continents to the oceans through the Earth's groundwater and surface-water systems. In the ocean these ions provide many types of marine creatures with the raw materials for their shells. When these sea creatures die, they ultimately discard their carbonate shells onto the seafloors, where they may become consolidated into new carbonate rock. These rocks may then rise above sea level, perhaps uplifted by plate-tectonic forces or exposed by receding seas, or they may be carried down a subduction zone as the underlying plate descends into the mantle. If exposed above sea level, the newly exposed layers of rock will weather and supply future sea creatures with their calcium and bicarbonate ions, continuing this carbonate cycle between land and sea. If subducted, the limestone may release the CO_2 trapped in calcite into the mantle, where it can eventually reach the atmosphere or oceans as CO_2 emissions from arc volcanoes or seafloor basalts.

Limestone buildings and statues, like natural limestone structures, are also subject to dissolution. The formation of carbonic acid in the atmosphere makes rainwater naturally slightly acidic (natural, unpolluted rainfall has an average pH of 5.5). The more acidic the rain in a given locality, the more rapidly its limestone features, natural or human-made, will dissolve. Downwind of some industrial centers, the rainfall is particularly acidic: Industrial processes such as the burning of sulfurous coal and the smelting of sulfur-rich copper, iron, and nickel ore release sulfur-dioxide gas (SO_2). This gas combines chemically with oxygen and water in the atmosphere to form sulfuric acid (H_2SO_4), which is a much stronger acid than naturally occurring carbonic acid. When this acid rain falls, it dissolves carbonate rock and building stone quite rapidly. Highlight 5-1 focuses on the ongoing problem with acid rain.

Hydrolysis is one of the most important chemical-weathering mechanisms because it involves alteration of silicate minerals, the main constituents of the Earth's crust. In **hydrolysis**, H^+ or OH^- ions from water molecules displace other ions from a mineral's structure, forming a different mineral such as a clay mineral or an hydroxide. For example, aluminum-rich silicates such as the feldspars (the most abundant minerals in the Earth's crust) are weathered primarily by hydrolysis to form clay minerals.

In a typical hydrolysis reaction, potassium feldspar reacts with water's hydrogen ions. The resulting products include a stable clay mineral, dissolved silica in the form of silicic acid, and potassium ions liberated by the following reaction:

$$2\,KAlSi_3O_8 + 2\,H^+ + 9\,H_2O$$
Potassium   Hydrogen ions
feldspar   (from water)

$$\rightarrow Al_2Si_2O_5(OH)_4 + 4\,H_4SiO_4 + \quad 2\,K^{2+}$$
Kaolinite clay   Silicic acid   Potassium ions
(in solution)   (in solution)

The released potassium ions also form some of the soluble salts that are ultimately the source of the salt in the world's seawater.

The clays formed during hydrolysis accumulate within the upper few meters of the Earth's surface. They are either incorporated into soils or washed out to sea to become oceanic mud. The silicic acid and the potassium ions are generally transported

Highlight 5-1

Coping with Acid Rain

The good people of Seattle, Washington (where one of your authors lives), have learned to deal with a life into which a little rain must fall—almost every day from October to May. Yet how would they feel if the gentle mists of Seattle stripped the paint from their cars, killed all of their beautiful forests, and rendered their lakes completely devoid of aquatic life? Fortunately for Seattleites, few polluting smokestacks lie upwind of the city. Thus, while we're often wet, the city does not suffer from the extreme effects of acid rain. But if you stroll around many of the world's cities, you may notice that the statues in some have suffered greatly from exposure to local environment conditions (see photo below). In other cities, the statues appear as fresh as when they first left the sculptor's studio. This assault on these public works—by the relentless attack of acid rain—is primarily an unfortunate by-product of the Industrial Revolution of the past 200 years, which has added sulfurous gases to the atmosphere when we burn certain fossil fuels.

Natural rainwater that falls far from areas of industrial development has a pH of about 5.5. The pH is the measure of the concentration of hydrogen ions in solution. (*Note:* Each change of one pH unit represents a 10 times increase or decrease in the amount of H^+ present.) As you can see in the figure below, the pH of distilled water is 7. Distilled water is described as "neutral." A solution with a pH higher than 7 is described as *alkaline* or *basic.* A solution with a pH lower than 7 is described as *acidic.* Solutions with very low pH measurements—in the 1 to 2 range—are very strong acids. These solutions have a very high number of H^+ ions available for chemical reactions. Natural rainfall is slightly acidic for the following reason: As the carbon dioxide in the Earth's atmosphere combines with water, a weak acid, carbonic acid (H_2CO_3), forms. The breakup ("dissociation") of the carbonic-acid molecule releases H^+ ions to the water. Acid rain has a pH lower than 5. What is the source of its H^+ ions?

When we burn sulfur-rich coals in the furnaces of our factories and smelters, the sulfur combines with oxygen in the atmosphere to form sulfur-dioxide gas. This gas, in turn, combines with atmospheric water to form sulfuric acid.

$$\text{S} + \text{O}_2 \longrightarrow \text{SO}_2$$
Sulfur   Oxygen   Sulfur dioxide

$$2\,\text{SO}_2 + 2\,\text{H}_2\text{O} + \text{O}_2 \longrightarrow 2\,\text{H}_2\text{SO}_4$$
Sulfur   Water   Oxygen   Sulfuric
dioxide   acid

When this acid combines with atmospheric moisture, the consequences of the re-

▲ Acid rainfall has damaged this and many other statues throughout Europe.

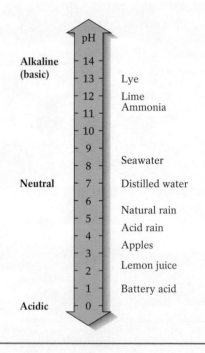

▶ This is the pH scale for some common substances. The lower the pH, the more acidic the substance. Note that "natural" water is slightly acidic (a pH of 7 is neutral—halfway between an acid and a base).

in solution by underground and surface water. Much of the dissolved silica either cements loose grains of sediment to form sedimentary rocks or serves as the raw material for the shells and skeletons of many marine organisms. Growing plants may extract the dissolved potassium ions from soil water, or the ions may be transported out to sea.

Oxidation   In **oxidation,** a mineral's positive ions combine with oxygen to form an *oxide.* For example, when iron ions bond with oxygen in the atmosphere, the reaction forms the reddish iron oxide Fe_2O_3 (hematite):

$$4\,\text{Fe}^{3+} + 3\,\text{O}_2 \longrightarrow 2\,\text{Fe}_2\text{O}_3$$

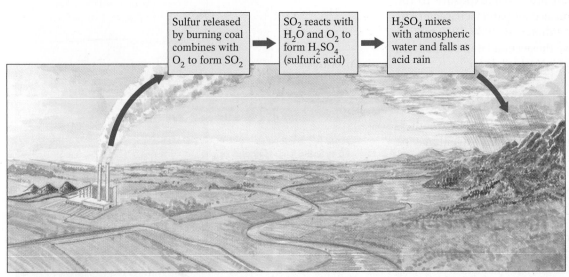

Sulfur released by burning coal combines with O_2 to form SO_2	→	SO_2 reacts with H_2O and O_2 to form H_2SO_4 (sulfuric acid)	→	H_2SO_4 mixes with atmospheric water and falls as acid rain

▲ The emissions from our factory smokestacks often contain sulfur dioxide (SO_2), especially when coal is being burned for fuel. This gas then combines with water in the atmosphere to form the sulfuric acid (H_2SO_4) that can turn raindrops highly acidic.

sulting acidic rainfall at the Earth's surface range from the unfortunate (the accelerated destruction of our monuments and statues) to the dire (the destruction of our crops, forests, and lakes) (see figure and photo on this page).

Acid rain corrodes the paint on our cars and the steel in our bridges. It also dissolves the concrete of our buildings and marble of our statues. Quite tragically, it also kills our forests and the fish in our lakes. Downwind of the world's great industrial centers, the highly acidic rain spawned by sulfurous emissions from belching smokestacks has extensively damaged trees, destroyed the root systems of vast forests, and dissolved away the nutrients in the soils that support these forests. At the same time, the water in many lakes in these regions has become so acidic that their aquatic life has been completely eliminated—thus, no insects, no fish, and no amphibians remain. These "ghost" lakes, despite their crystal-clear water, are devoid of all life.

And alas, the problem of acid rain knows no regional or international boundaries. The areas responsible for creating the acidic conditions often export the problem to their unsuspecting, nonpolluting neigh-

▲ Dead spruce trees killed in Krkonose National Park, Poland. These trees were killed by acid rain emanating from the highly industrialized areas of Poland that are located upwind of the park.

bors, sometimes with unpleasant international consequences. The smokestacks of industrial Michigan, Ohio, and western Pennsylvania, for example, have damaged or destroyed the lakes and forests of New England and Quebec. The factories of the United Kingdom have done the same to the lakes and forests of Sweden and Norway. What can we do about this environmental disaster? More affluent countries have substantially reduced their industrial emissions through both legislative and technological measures. In the United States, a concerned Congress passed the Clean Air Act of 1990 to address the problem of acid rainfall. The act requires that smokestacks be fitted with devices that "scrub" sulfur from their emissions before releasing them to the atmosphere. These efforts have substantially reduced acid rainfall throughout northeastern North America. The natural pH of targeted lakes and forest tracts can even be restored by introducing substances—such as lime ($CaCO_3$)—that neutralize the acids in lakes and streams in much the same way that certain over-the-counter antacids do when our digestive juices become too acidic.

Nevertheless, the problem of acid rainfall remains far from solved. Only a concerted international effort that involves *both* the wealthy industrialized nations that can afford these remedies *and* the poorer emerging nations will save our forests and lakes from this environmental threat.

Oxidation and hydrolysis may work together to weather a rock chemically. In the presence of water, iron hydroxides form: Two of the most common are yellow-brown goethite, $FeO(OH)$, and yellowish limonite, $FeO(OH) \cdot nH_2O$ (where n represents any integer). These iron hydroxides, which form wherever iron-rich minerals come into contact with water, are known commonly as rust. Rust often stains the surfaces of minerals with a high iron content, such as olivine, pyroxene, and amphibole. Rust also coats rocks made up of these minerals, such as basalt, gabbro, and peridotite. Thus, if you go out to look for an outcrop of green peridotite or black basalt, unless the rock is young and fresh, newly exposed, or located in a dry climate (where little chemical weathering takes place), you should probably keep an eye out for a rusty orange or reddish outcrop. Like other products of chemical weathering, rust is very stable at the Earth's surface. Thus (much to the chagrin of the owners of aging cars) after rust forms, it does not dissolve away.

Other metals are also subject to oxidation. Copper is more resistant than iron to oxidation, but after long exposure to the atmosphere, it develops a green surface, or patina, characteristic of copper carbonates and copper sulfates. In 1986, $230 million were spent to renovate and remove the accumulated patina from the copper Statue of Liberty in New York's harbor.

Factors That Influence Chemical Weathering

We have seen that water is a key player in all chemical-weathering reactions. But water's function as a weathering agent may be further enhanced by other factors: local temperature and moisture patterns, the activity of living organisms, the amount of time that the mineral or rock has been exposed to weathering, and the chemical stability of a rock's minerals.

Climate   Water drives all of the three chemical-weathering processes just discussed, so clearly a place with a wet climate is more likely to experience more substantial chemical weathering than a dry climate. Another key climatic factor that accelerates virtually all chemical reactions is heat. Thermal energy from heat excites bonded atoms and ions, causing the bonds to vibrate rapidly until they break. The greater the heat applied, the more bonds that are broken. Thus, more atoms and ions are freed to participate in other chemical reactions. *In general, the rate of chemical reaction doubles with every 10°C (18°F) increase in temperature.*

Because of the combined effects of water and heat, chemical weathering occurs more readily in warm, moist climates than in cold, arid ones. Warm, moist climates promote the growth of lush vegetation and support abundant animal life. How do these conditions promote greater chemical weathering? Photosynthesis by plants produces O_2, promoting oxidation, and after they die, the decayed plants produce the organic acids that strongly enhance chemical weathering. Animal respiration produces CO_2, which combines with water in soil to form chemically reactive carbonic acid. Thus, the hot, steamy climate of low-latitude tropical areas such as Puerto Rico is optimal for chemical weathering. Cold, dry places such as Antarctica experience very little chemical weathering (Fig. 5-13).

Living Organisms   Organisms can affect the chemical-weathering rate of rocks and minerals by exposing them to weathering agents. Burrowing animals, such as groundhogs, prairie dogs, and even ant colonies, commonly transport unweathered materials from below ground to the surface, where they can then be weathered. The common earthworm, which churns up underground minerals in the course of consuming the organic matter in soil, is particularly effective (Fig. 5-14). In humid, temperate climates, an average earthworm population brings 7 to 18 tons of soil per acre to the surface each year.

An extremely significant way in which organisms influence chemical weathering is the effect of bacteria and other microorganisms on the dissolution (and precipitation) of minerals. Bacteria thrive on the elements supplied by crustal minerals—they

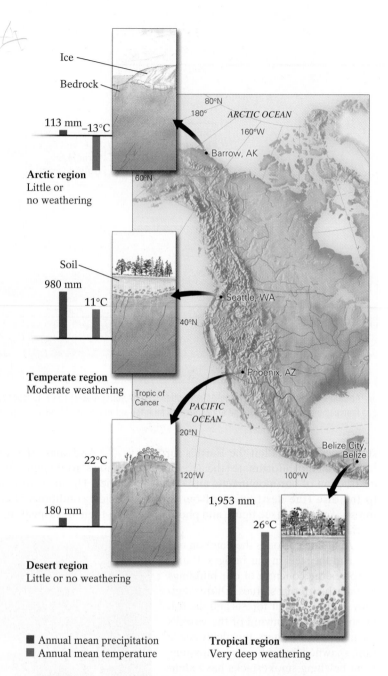

■ Annual mean precipitation
■ Annual mean temperature

▲ **FIGURE 5-13** The relationship between climate and weathering, shown as a function of the depth to which a region's bedrock (the solid rock immediately underlying the surface soil) is typically weathered. The deepest weathering occurs in the warm, moist tropical zone near the equator where temperature and precipitation are greatest. Weathering is minimal in the arctic and in the deserts, where water is in short supply as an agent of chemical weathering.

actually "eat" the mineral as we eat food. Bacteria can also use elements from the atmosphere (such as nitrogen) or from a mineral (such as sulfur from pyrite), combine these elements with oxygen and hydrogen, and from these produce highly corrosive acids (such as sulfuric acid) that can then dissolve minerals. Some types of bacteria can oxidize metals (such as iron and manganese) in minerals, while still others can attack silicate minerals with the organic acids they produce.

Bacteria occur almost everywhere that rocks and water interact, including in fresh and salt water on land and in oceans, in

▲ **FIGURE 5-14** Earthworms churn up soil as they move through it, bringing underground minerals to the surface to be exposed to weathering.

the underground water circulating within the soils and sediments of continents, and in hydrothermal systems at oceanic-divergent zones. These organisms thrive even in some the Earth's most extreme conditions, such as in the hot springs at Yellowstone National Park, the deep-sea hydrothermal systems where new oceanic crust is formed, and in the highly acidic waters near mineral-ore deposits (Fig. 5-15).

Time   A direct relationship exists between time and weathering: The longer a rock or sediment is exposed to a weathering environment, the more it will disintegrate and decompose. Within some of the hills of sediment left behind by the glaciers that once covered much of North America's Great Lakes region, pebbles containing minerals susceptible to weathering have been so thor-

oughly weathered that you can rub them between your fingers into balls of soft clay. Pebbles of the same composition in other glacially deposited hills in the region, however, are still fresh and solid. Assuming the other factors that control weathering affected these hills equally, the hills with the crumbling pebbles must have been exposed to the weathering environment for a much longer time (that is, they are much older) than those with the fresh, unweathered pebbles.

Mineral Structure and Composition   The key factor governing the chemical stability of minerals—and thus, their resistance to weathering—is the strength and number of a mineral's chemical bonds. At the beginning of this chapter, we noted that the stability of a mineral with respect to melting and crystallization could be used as a general predictor of how easily a mineral weathered at the Earth's surface. Recall that the low-temperature mineral quartz is far more stable (and thus far more resistant to weathering) than the high-temperature mineral olivine (which is thus more susceptible to weathering). This relationship is a product of a mineral's internal structure and composition. Quartz consists of a strongly bonded framework of silicon-oxygen tetrahedra, and the bonds do not break as easily in the weathering environment as those of olivine, which is made up of isolated silicon-oxygen tetrahedra bonded to iron and magnesium ions. Both quartz and feldspar, however, are both framework silicate minerals, yet feldspars weather much more easily than quartz, so we must also consider the effect of the chemical composition on a mineral's susceptibility to weathering (Fig. 5-16).

▲ **FIGURE 5-16** Two gravestone inscriptions from the same cemetery in Williamstown, Massachusetts, showing differential weathering of rock types. The marble gravestone (above), though exposed to the same climate as the granite gravestone (below) and erected 50 years *later*, has suffered noticeably more weathering damage. Marble, composed predominantly of the chemically reactive mineral calcite, is much more susceptible to chemical weathering than is granite, a rock composed of relatively stable minerals such as quartz and orthoclase feldspar.

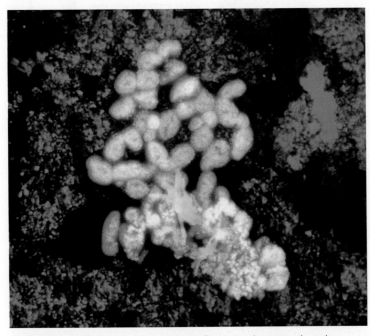

▲ **FIGURE 5-15** Bacteria "eating" a mineral. These organisms occur in such astounding numbers in some environments, they may appear as coatings on mineral grains. This microcolony of unidentified slime bacteria can be found deep in rocks. *(Photo provided by Kevin Cole, Grand Valley State)*

5.1 Geology at a Glance

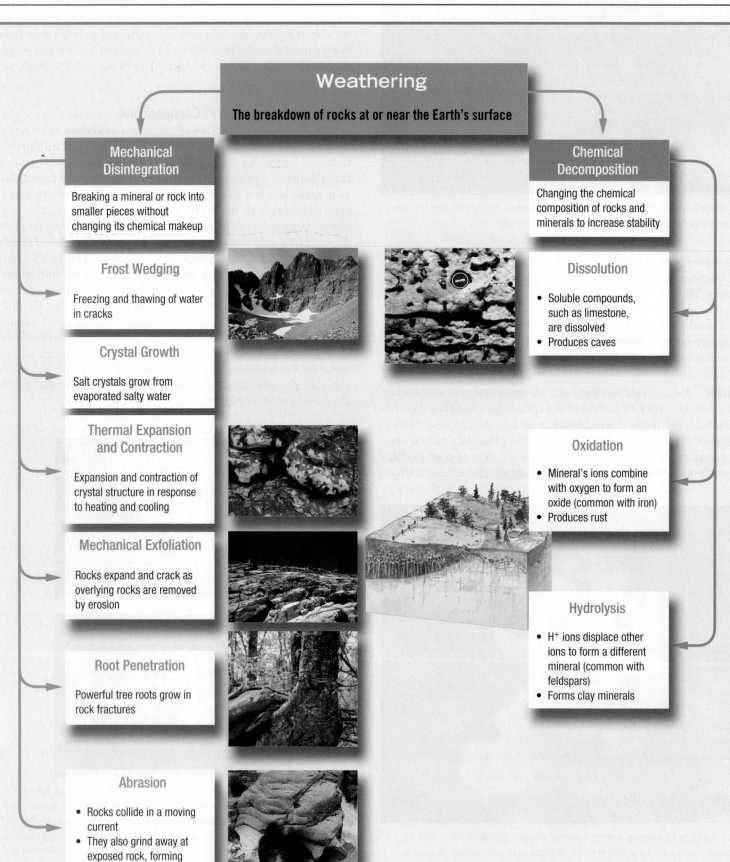

Weathering

The breakdown of rocks at or near the Earth's surface

Mechanical Disintegration

Breaking a mineral or rock into smaller pieces without changing its chemical makeup

Frost Wedging

Freezing and thawing of water in cracks

Crystal Growth

Salt crystals grow from evaporated salty water

Thermal Expansion and Contraction

Expansion and contraction of crystal structure in response to heating and cooling

Mechanical Exfoliation

Rocks expand and crack as overlying rocks are removed by erosion

Root Penetration

Powerful tree roots grow in rock fractures

Abrasion

- Rocks collide in a moving current
- They also grind away at exposed rock, forming potholes in bedrock

Chemical Decomposition

Changing the chemical composition of rocks and minerals to increase stability

Dissolution

- Soluble compounds, such as limestone, are dissolved
- Produces caves

Oxidation

- Mineral's ions combine with oxygen to form an oxide (common with iron)
- Produces rust

Hydrolysis

- H^+ ions displace other ions to form a different mineral (common with feldspars)
- Forms clay minerals

When chemically unstable minerals and rocks are exposed at the surface, they quickly begin to react with the water, gases, and organisms in the atmosphere and soils. The end products of these chemical reactions—such as the clay minerals and certain metallic oxides—are stable at the Earth's surface. Thus, through the process of chemical weathering, unstable minerals are replaced by the chemically stable weathering products that we see all around us.

Some Products of Chemical Weathering

After dissolution, oxidation, and hydrolysis remove the soluble ("dissolvable") components in rocks, a combination of relatively insoluble, unreactive, and stable new products of chemical weathering remains, and in some settings, unusual landforms are created. The processes of chemical weathering operate much like a household coffee maker (Fig. 5-17). As hot water passes through the ground coffee (which acts like sediment grains with lots of surface area), the coffee itself is dissolved ("leached") from the coffee grinds and both the grinds and the water are changed chemically. The coffee-rich water flows away (into your cup), leaving behind a residue that is now depleted grounds. Note that this process would not work if you tried to pass the hot water directly through unground coffee beans. Our soils are constantly being depleted of their soluble ions in the same way.

Clay Minerals Clay minerals are produced primarily by the hydrolysis of feldspars and mafic minerals. There are several varieties of clay minerals, each with a distinctive chemical composition and set of physical properties that determine its practical uses. The clay type produced from a given rock depends on the climatic conditions at the time of hydrolysis. For example, kaolinite, $Al_2Si_2O_5(OH)_4$, is the common product of the hydrolysis of feldspars in a warm, humid climate. To produce kaolinite, H^+ ions from water must completely displace the large positive ions (K^+, Na^+, Ca^{2+}) in the parent feldspars during hydrolysis.

In drier, cooler climates, some large positive ions remain after hydrolysis, producing other types of clay minerals. Smectites, a highly absorbent group of clay minerals, form when micas and amphiboles are only partially hydrolyzed, leaving some of their calcium, sodium, and magnesium ions in place. Smectites are often used to filter impurities from beer and wine. The negatively charged surfaces of smectite particles attract and bond with the positively charged impurities in those beverages. The particles then settle from the liquids, leaving them clear and pure. The water-absorbing properties of smectites were discussed in Chapter 2, in Highlight 2-2.

Clay minerals are also used widely in industry as agricultural fertilizers, as lubricants in the boreholes of oil-drilling rigs, in the manufacture of bricks and cement, and in the production of paper. (The slick finish of this book's paper is a clay product.) Because of its absorbency and deodorizing properties, the clay mineral smectite is a key ingredient in kitty litter. Kaolinite is also the active ingredient in Kaopectate, absorbing the toxins and bacteria that cause intestinal irritation.

Metal Ores *Ores* are aggregates of minerals that have economic value and can be mined profitably from their surrounding rocks. Most economically valuable minerals—usually metals—rarely occur in pure form. Aluminum, although quite abundant in the Earth's crust, is generally widely dispersed in clays, feldspars, and micas. Pure aluminum metal was once so highly prized that Napoleon had his banquet cutlery fashioned from it instead of from silver or gold.

Intense chemical weathering of feldspar-rich rocks in hot, moist climates produces the aluminum ore bauxite. Bauxite forms because aluminum, being insoluble in water, remains concentrated in soils after most common elements, such as calcium, sodium, and silicon, have dissolved out of them. This important ore is found most often in tropical areas, such as the Caribbean islands and parts of Australia, Africa, and South America. Substantial *ancient* bauxite deposits can also be found in Georgia, Alabama, North Carolina, and Arkansas, suggesting that the climate in those places was once even warmer and moister than today.

Rounded Boulders Mechanically weathered rocks (and those that are also fractured by other geologic stresses) are initially quite angular. But in an active chemical-weathering environment, they don't remain that way for long. Because the corners and edges of angular rock masses offer more surface area than a smooth, rounded surface, the corners and edges are more likely to weather chemically and at a faster rate than the rocks' other surfaces. (Remember from Figure 5-3 that when more surface area is available for chemical attack, a rock will weather more rapidly by chemical-weathering processes.) As their corners and edges decompose by chemical reactions, angular rocks eventually become smoothly rounded boulders. This process is known as **spheroidal weathering** (Fig. 5-18).

Once a spheroidal shape is attained, it remains essentially unchanged, because weathering agents (particularly water) then act uniformly across the boulder's entire rounded surface. Weathering of the feldspars on the surface of the boulder produces clay minerals. As these minerals absorb water and increase in

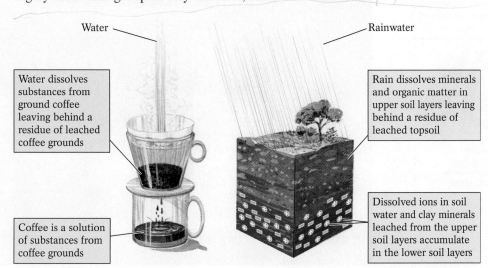

▲ **FIGURE 5-17** The processes of chemical weathering operate much like a household coffee maker. As hot water passes through the ground coffee, the coffee itself is dissolved ("leached") from the coffee grounds, leaving behind a depleted residue.

Weathering attacks corners on three sides.

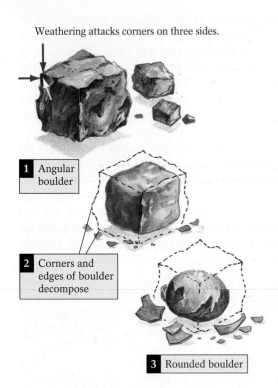

1 Angular boulder

2 Corners and edges of boulder decompose

3 Rounded boulder

◄ **FIGURE 5-18** Spheroidal weathering. Rounded boulders, such as these in Australia, evolve from angular ones as chemical and mechanical weathering gradually eliminate the rocks' sharp corners and edges. The corners and edges are more susceptible to weathering because they offer more surface area for attack by chemical weathering.

volume, they expand outward and separate from the boulder's unweathered interior. As they do so, they peel off in concentric layers much like the layers of an onion, a process called *chemical exfoliation* (Fig. 5-19).

Rounded boulders, new clays from old feldspars, valuable metal deposits—all are products of weathering. Another weathering product—and one that is more subtle but much more influential in humans' lives—is widespread climate change. Highlight 5-2 explains how the processes of mechanical and chemical weathering can alter global climates.

Soils and Soil Formation

Mechanical and chemical weathering of the Earth's crust produces a loose "rock blanket" or **regolith** (from the Greek *rhegos*, or "cover"), the fragmented material covering much of the Earth's land surface (Fig. 5-20). Geologists refer to the upper few meters of regolith *that supports plant life* as **soil.** Soil—another product of weathering—contains both mineral and organic material (typically decaying vegetation) along with water and air.

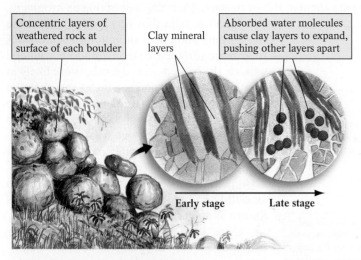

Concentric layers of weathered rock at surface of each boulder

Clay mineral layers

Absorbed water molecules cause clay layers to expand, pushing other layers apart

Early stage   Late stage

▲ **FIGURE 5-19** A rounded boulder maintains its shape because chemical weathering occurs uniformly over its surface. The surface of the boulder falls off, in concentric layers, because the clay minerals produced by chemical weathering of feldspars at the boulder's surface expand with water and occupy a greater volume than the unweathered feldspars in underlying layers.

▲ **FIGURE 5-20** A roadcut in warm, wet Sarawak, Borneo, showing layers of regolith (weathered surface sediment and bedrock). Extreme chemical weathering can produce regolith that is more than 60 m (200 ft) deep.

Soils may be the most valuable of all natural resources (besides clean water) simply because they feed us. We all depend on fertile soils to support plant growth and provide the life-sustaining nutrients in foods. Because they play such an important role in our lives, we need to answer the following fundamental questions: How does soil form? Why do different regions have markedly different types of soils, and thus different levels of soil fertility and productivity?

Influences on Soil Formation

Five factors are particularly important in determining the nature of an area's soil: parent material, climate, topography, biological activity, and time.

Parent Material A soil's **parent material** is the rock or sediment from which soil develops. The parent material's mineral content determines both the nutrient richness of the resulting soil and the amount of soil produced (Fig. 5-21).

We can clearly see the relationship between parent material and soil development in Java and Borneo, two neighboring Indonesian islands *with the same climate*. The parent materials for Java's soils are largely fresh, nutrient-rich volcanic-ash deposits, whereas the parent materials for Borneo's soils consist of numerous granitic batholiths, gabbroic intrusions, and andesitic lava flows. Ongoing volcanism on Java replenishes the island's posi-

tive-ion nutrients—such as potassium, magnesium, and calcium—in its soils. Borneo, lacking the replenishing effect of fresh ash deposits, has soils depleted of nutrients. The islands' population densities reflect their dramatically different soil fertility and agricultural productivity: Java, with its great soils, has approximately 460 inhabitants per square kilometer; Borneo, with its depleted soils, has about two inhabitants per square kilometer.

Climate An area's climate—the amount of precipitation it receives and its prevailing temperatures—controls the rate of chemical weathering and consequently the rate of soil formation. Climate also regulates the growth of vegetation and the abundance of microorganisms that contribute CO_2 and O_2 to the processes of dissolution, hydrolysis, and oxidation. Like chemical weathering, soil formation is most rapid in warm, moist climates, and slowest in cold, dry climates.

Topography **Topography** refers to the physical features of a landscape, such as mountains and valleys, the steepness of its slopes, and the shapes of its landforms. Topography influences the availability of water as well as the rate of soil accumulation. For example, steep slopes cause rainfall and snowmelt to flow away swiftly, leaving little water to penetrate the surface. Without water, little chemical weathering occurs and thus little or no soil develops. Any soil that does form on steep slopes is usually transported downslope by the pull of gravity before it can

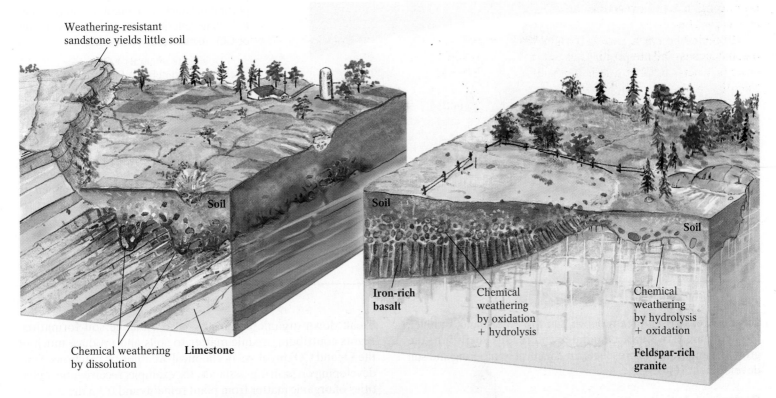

Weathering-resistant sandstone yields little soil

Soil

Chemical weathering by dissolution **Limestone**

(a) Weathering of sedimentary rocks

Soil

Iron-rich basalt Chemical weathering by oxidation + hydrolysis

Soil

Chemical weathering by hydrolysis + oxidation

Feldspar-rich granite

(b) Weathering of igneous rocks

▲ **FIGURE 5-21** The effect of bedrock composition on soil development in a moist, temperate environment. Sandstone, however, tends to remain relatively unweathered because it consists primarily of weathering-resistant quartz grains; as a result, any soil produced on sandstone will be relatively sparse. Advanced chemical weathering of fractured limestone (dissolution), iron-rich basalts (oxidation and hydrolysis), and feldspar-rich granite (hydrolysis and oxidation) produces thick soils on these types of rock.

Highlight 5-2

Rock Weathering and Global Climates

The weathering of rock is not only important for shaping the landscape, creating fertile soils, and feeding us, it may also change dramatically the Earth's climates—both regionally and globally. One way that weathering affects climate is by reducing the height of mountain ranges. Regions of high elevation, such as the Himalaya, change the pattern of atmospheric circulation in their vicinity. By doing so, these highlands may affect such climatic factors as temperature, rain and snowfall, and wind patterns over a wide area. The Tibetan plateau, for example (see photo), occupies an area half the size of the continental United States and has an average elevation of nearly 5 km (3 mi). The plateau is so high and wide that it controls the monsoons in Asia and affects atmospheric circulation in a large part of the Northern Hemisphere. Weathering—especially the mechanical-weathering process of frost wedging—is constantly attacking the rocks in this and other elevated regions, breaking down the solid rock. Over time, the mountains and plateaus will be worn down, and once the high elevations are gone, atmospheric circulation will no longer be disrupted and related climatic effects will end.

Another weathering-related process that affects global and regional climates involves the chemical weathering of silicate minerals, which in turn strongly influences the amount of carbon dioxide (CO_2) in the atmosphere. When the atmosphere is particularly rich in CO_2, the climate warms by the so-called greenhouse effect. As shown in the figure on the facing page, the greenhouse effect develops when atmospheric CO_2 allows *incoming* short-wave (ultraviolet) radiation from the Sun to pass to the Earth's surface. At the same time, the very same CO_2 traps *outgoing* heat (long-wave) radiation near the surface, warming the Earth's atmosphere. This effect is similar to the warming of a greenhouse, which occurs even on a cold winter day. The glass of the greenhouse roof allows ultraviolet sunlight to enter but traps the reradiated long-wave (infrared) radiation energy from the plants, pots, and other greenhouse surfaces. When the amount of heat-trapping atmospheric CO_2 is relatively low, the global climate typically cools. That's a lot like opening the windows of the greenhouse and allowing the heat energy to escape.

CO_2 from the Earth's interior is continuously pumped into the Earth's atmosphere by volcanic eruptions and extrusion of mantle-derived magma at oceanic divergent zones. This CO_2 is involved in several processes at the Earth's surface, the most important of which is chemical weathering. The following general reaction illustrates the role of CO_2 in chemical weathering:

$$\text{silicate minerals} + CO_2 \longrightarrow$$
$$\text{carbonate minerals} + SiO_2 \text{ (in solution)}$$

What might cause the rate of chemical weathering to increase, thereby increasing the amount of CO_2 *removed* from the atmosphere and stored for millions of years in carbonate minerals in rocks such as limestone? Rising temperatures may be one factor. Consider also what may happen to chemical-weathering rates when a vast region of elevated continental crust is created through the collision of tectonic plates. In mountainous regions, mechanical-weathering processes, such as frost wedging, breaks

▲ A view along the Himalaya as seen from a space shuttle. These snowcapped peaks are in Nepal. The great height of the range strongly affects the regional climate.

accumulate to a significant depth. On the other hand, in level, low-lying areas, water accumulates and readily seeps into the ground, enhancing the prospects for chemical weathering and soil development (Fig. 5-22).

Biological Activity   Plants, animals (especially burrowing animals from gophers to earthworms), and a host of microorganisms affect soil formation. As we discussed earlier in this chapter in the section on chemical weathering, microorganisms such as bacteria play a central role in producing the organic acids that promote chemical weathering. These acids—and their ability to break down organic materials—also enhance soil formation. Plants contribute organic matter to soils and produce much of the O_2 and CO_2 involved in chemical-weathering reactions. Soils developing on prairie grasslands, for example, receive large quantities of organic matter from plant remains and from the decay of extensive subsurface root systems.

Plants also contribute H^+ ions that help weather soils. The H^+ ions, weakly attached to plant roots, replace the large positive ions—such as calcium, potassium, and sodium—in feldspars and other minerals, hydrolyzing them into clay minerals. This exchange helps develop the soil while providing the plants with ions that are

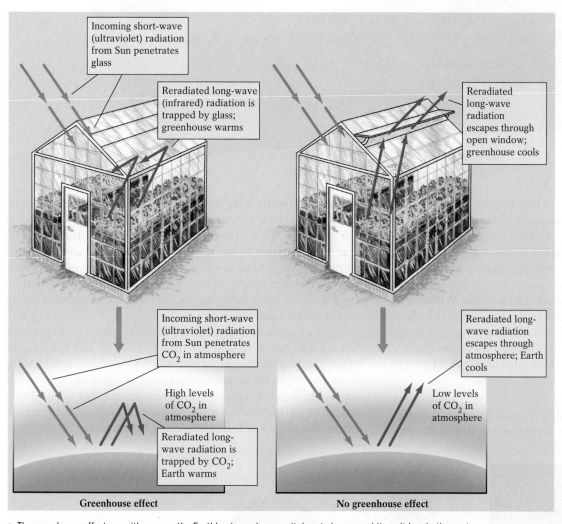

▲ The greenhouse effect can either warm the Earth's atmosphere, as it does today, or cool it, as it has in the past.

down rocks, exposing more surface area to chemical-weathering attack. Landslides and swift-flowing streams on steep mountain slopes constantly remove weathered material, exposing even more fresh rock to attack. The rate and extent of chemical weathering in mountain regions are typically extremely high, as can be seen by analyzing the composition of water in rivers that flow out of the mountains. Mountain waters generally contain much more dissolved mineral matter than rivers that flow through low-lying regions.

Thus, the buildup of high mountainous areas may alter global climate, both by disrupting atmospheric-circulation patterns and by changing the amount of CO_2 in the atmosphere. Increased chemical weathering removes CO_2 from the atmosphere, perhaps driving global cooling. Quite possibly, periods of plate convergence, continental collision, and mountain building may cause global climatic cooling in large part through this CO_2 effect. Whether these processes are powerful enough to plunge the Earth into a full-fledged ice age, however, remains a topic of lively scientific debate.

nutritionally beneficial to the animals that eat them, including humans (Fig. 5-23). These ions return to the soil when the plants die and decompose, becoming available for the next generation of plants. If the plants are harvested before they die, however, these ions are *permanently* removed from the soil. As a result, continuous farming eventually depletes a soil's supply of calcium, sodium, potassium, and other important ions. Such depleted soils must then be supplemented with natural or synthetic fertilizers.

Time If other factors in soil formation were equal, a landscape weathering over a long period of time would contain a thicker,

more well-developed soil than a younger landscape. Other factors, however, almost always vary considerably. Climate is a major factor: Soils may begin to develop within a few hundred years in warm, wet environments but may require thousands or hundreds of thousands of years in arid or polar regions. A thicker soil, then, is not necessarily an older one.

How long does it take for fertile soils and abundant vegetation to develop on a fresh lava flow on the Big Island of Hawai'i? Geologists estimate that only a few hundred years are necessary on the iron-rich basalts exposed in the island's rainy eastern side.

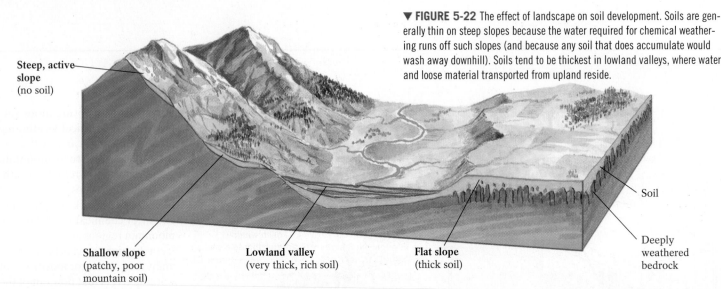

▼ **FIGURE 5-22** The effect of landscape on soil development. Soils are generally thin on steep slopes because the water required for chemical weathering runs off such slopes (and because any soil that does accumulate would wash away downhill). Soils tend to be thickest in lowland valleys, where water and loose material transported from upland reside.

Steep, active slope (no soil)

Shallow slope (patchy, poor mountain soil)

Lowland valley (very thick, rich soil)

Flat slope (thick soil)

Soil

Deeply weathered bedrock

One final word about the factors that control soil formation and how all this affects tonight's dinner: Before we go on to discuss the specifics of soil structures, think for a moment about the various Earth systems that must join forces to produce the cheese pizza you may eat tonight. Water from the Earth's *hydrosphere* must combine with gases in the *atmosphere* to weather the rocks of the Earth's *lithosphere* to produce fertile soils. Wheat and tomato plants and grazing dairy cows from the Earth's *biosphere* provide the "raw materials" for the crust, the sauce, and the cheese of the pizza. So with every bite, consider how all of the Earth's systems contribute to every single thing you eat (Fig. 5-24).

Typical Soil Profile

The effects of weathering are greatest at a soil's surface, which is directly exposed to the weathering environment. Below the sur-

face, distinct weathering zones develop as infiltrating water dissolves, or **leaches,** substances from the upper layers of the soil and transports them to lower levels—either in solution or as suspended fragments. These transported substances are then precipitated or deposited in the lower soil layers. Thus, the upper part of a developing soil loses some of its original materials, while the lower part gains new components. Each distinct weathering zone is called a **soil horizon.**

Soil scientists have identified dozens of distinct horizon types. Among the most common are the O, A, E, B, and C horizons, which can be found in any soil but which develop most extensively in temperate zones. Subhorizons of these main horizons develop in different specialized environments. The vertical succession of soil horizons in a given location is called the location's **soil profile** (Fig. 5-25). Because different localities vary in their

▲ **FIGURE 5-23** Vegetation contributes to soil development through the exchange of H⁺ ions (from the surfaces of plant roots) with positive ions (from soil minerals such as feldspars). This process increases the soil's clay content while providing the plant with nutrients that return to the soil when the plant dies.

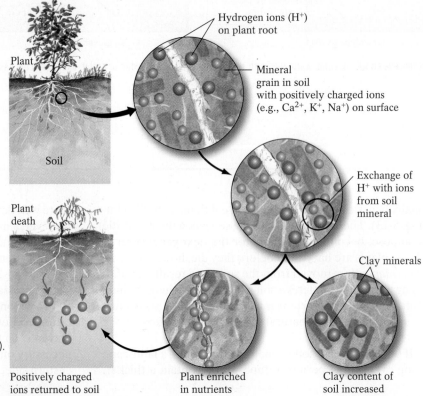

Hydrogen ions (H⁺) on plant root

Plant

Soil

Mineral grain in soil with positively charged ions (e.g., Ca²⁺, K⁺, Na⁺) on surface

Exchange of H⁺ with ions from soil mineral

Plant death

Clay minerals

Positively charged ions returned to soil

Plant enriched in nutrients

Clay content of soil increased

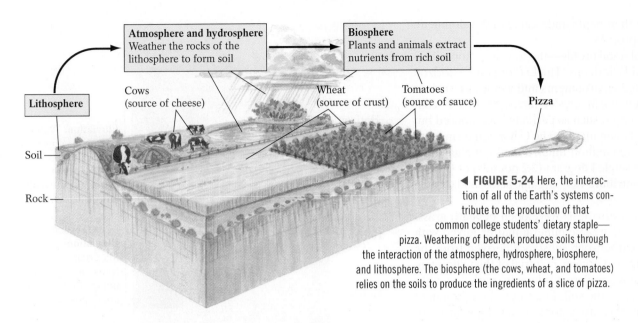

Lithosphere

Soil

Rock

Cows
(source of cheese)

Atmosphere and hydrosphere
Weather the rocks of the
lithosphere to form soil

Biosphere
Plants and animals extract
nutrients from rich soil

Wheat
(source of crust)

Tomatoes
(source of sauce)

Pizza

◀ **FIGURE 5-24** Here, the interaction of all of the Earth's systems contribute to the production of that common college students' dietary staple— pizza. Weathering of bedrock produces soils through the interaction of the atmosphere, hydrosphere, biosphere, and lithosphere. The biosphere (the cows, wheat, and tomatoes) relies on the soils to produce the ingredients of a slice of pizza.

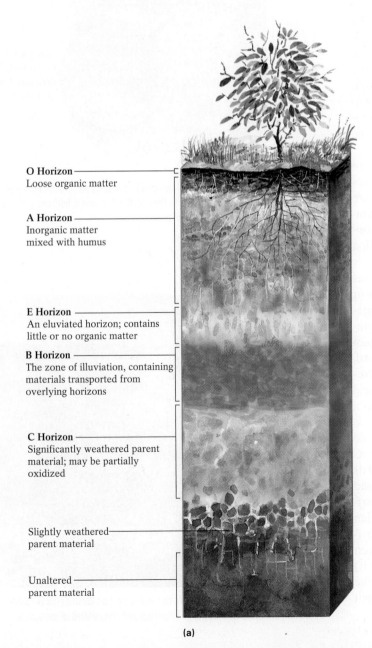

O Horizon
Loose organic matter

A Horizon
Inorganic matter
mixed with humus

E Horizon
An eluviated horizon; contains
little or no organic matter

B Horizon
The zone of illuviation, containing
materials transported from
overlying horizons

C Horizon
Significantly weathered parent
material; may be partially
oxidized

Slightly weathered
parent material

Unaltered
parent material

(a)

(b)

▲ **FIGURE 5-25** Profile of a typical mature soil. **(a)** The features of a typical cool, forested temperate-zone soil profile, such as in the Great Lakes states, New England, and southeast Canada, including the O, A, E, B, and C horizons. **(b)** A vertical succession of soil horizons is clearly visible even in this small section of soil in Australia.

types of horizons and in their depth and degree of development, they have different soil profiles.

The upper portion of a soil profile—the so-called "topsoil"— consists of the O, A, and E horizons. The O ("organic") horizon typically forms in forested environments and wet areas, such as bogs and swamps. This horizon in forested temperate regions consists mainly of organic matter, such as partially decomposed but still recognizable fibers of plant matter. The O horizon teems with life—2 trillion bacteria, 400 million fungi, 50 million algae, and thousands of insects in a single kilogram (2.2 pounds) of O horizon. These organisms contribute CO_2, organic acids, and oxygen to the developing soil.

The A horizon—most obviously developed beneath grasslands—consists mainly of inorganic mineral matter mixed with *humus,* a dark-colored, carbon-rich substance derived from the more completely decomposed organic material of the O horizon. The thickness of the A horizon depends largely on the quantity of decomposed vegetation in the soil. In a tropical environment with lush vegetation, an obvious A horizon develops far more rapidly than in an area with sparse vegetation.

The E horizon is a light-colored zone below the A horizon with little or no organic material. Its light color results from the dissolution and removal ("leaching") of soluble compounds from the upper few meters of regolith. The E stands for *eluviation,* the process by which water removes material from a soil horizon. Freshwater containing organically produced CO_2 seeps downward from the surface, dissolving soluble inorganic soil components and transporting them to lower horizons along with fine soil particles.

The B horizon, under the O, A, and E horizons, is a zone of *illuviation,* where the materials dissolved or transported mechanically from the upper horizons end up (Fig. 5-26). There are many types of B-horizons, classified according to their predominant components: A Bh horizon, for example, has a high concentration of added humus; a Bo horizon has a high concentration of oxides.

In arid and semi-arid areas, where surface water evaporates quickly, a distinct carbonate-rich horizon develops within or below the B horizon at the depth to which annual rainfall penetrates. This so-called **caliche** layer (ka-LEE-chee, from the Spanish *cal* for "lime") forms when brief, heavy rains dissolve calcium carbonate in the upper layers of soil and transport it downward. As the water rapidly evaporates, the carbonate precipitates, creating a distinct white layer in the soil (Fig. 5-27). Particularly arid regions may require hundreds of thousands of years to form a well-developed layer of caliche.

The O, A, E, and B horizons bear little resemblance to the original parent rock and mineral material. In contrast, the C horizon—the lowest zone of significant weathering—consists of the parent material that has only partially weathered and thus still retains most of its original appearance. The C horizon may show signs of oxidation from penetrating oxygen-rich groundwater, or it may be completely unoxidized. The C horizon is very thin where little chemical weathering has taken place, such as in a desert. It can be as much as 100 m (330 ft) thick where chemical weathering is extensive, such as in the warm, wet tropics. Below the C horizon lies unaltered parent material.

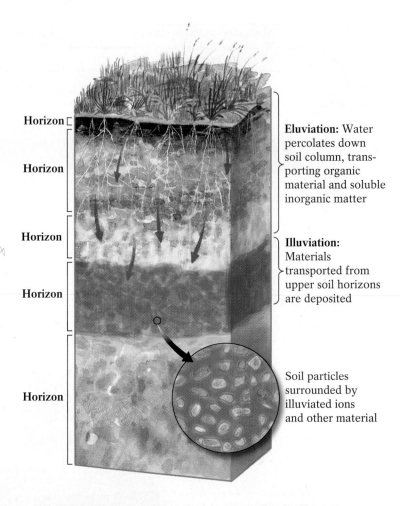

Horizon

Horizon

Horizon

Horizon

Horizon

Eluviation: Water percolates down soil column, transporting organic material and soluble inorganic matter

Illuviation: Materials transported from upper soil horizons are deposited

Soil particles surrounded by illuviated ions and other material

▲ **FIGURE 5-26** Eluviation and illuviation in a mature soil profile. Material is removed, or eluviated, as water passes through the O, A, and E horizons and is deposited, or illuviated, as water infiltrates the B horizon.

▲ **FIGURE 5-27** The white layer in this soil in eastern Washington is caliche, calcium carbonate dissolved from upper soil horizons and precipitated in lower ones.

Classifying Soils

As we have seen, the complex interaction of parent material, climate, topography, biological activity, and time causes local soil development to vary in significant ways. Distinct soil types have specific physical and chemical characteristics that affect the ways we use them. Where we put landfills, how we design buildings, and the ways we cultivate soils for food all depend on accurate soil evaluation. Our current classification scheme used in the United States (Table 5.1) names soils according to obvious physical characteristics and, among other things, describes a soil's clay content and indicates the degree of nutrient depletion. Classification terminology also provides information about moisture content, mean annual air temperature, horizon development, soil chemistry, organic matter content, and even the origin and relative age of the soil. Other countries have developed their own classification schemes; the U.S. scheme emphasizes the origins of the soils rather than the visual description of the soil.

If you were to eavesdrop on a couple of soil scientists talking soils, you would learn that they speak a fascinating language, routinely exchanging exotic references such as "That soil must be a *typic fragiaquept* (*typic* = common, or typical; *fragi* = fragile, or brittle and crumbly; *aqu* = wet, from aqua; *ept* = inceptisol—thus, it's a common wet, immature, relatively undeveloped soil that crumbles). Quite a mouthful. Every syllable describes a characteristic of the soil, and the last syllable is derived from the soil type and tells us much about the soil's origins. For example, an *entisol* (the root "ent" is derived from *recent*) is a soil that has not yet experienced significant horizon development. It may be a recent flood deposit, fresh volcanic ash, or any other recently exposed surface.

Extreme weathering in tropical areas produces *oxisols* (named for their high concentration of insoluble iron *ox*ides) or *ultisols* (for their *ult*imate, or most advanced, degree of soil development). In oxisols, even ordinarily insoluble quartz has been

Table 5.1 Classification of Soils

SOIL ORDER	GENERAL PROPERTIES	TYPICAL GEOLOGIC OR GEOGRAPHIC SETTING
Andisols	Soils develop principally on fresh volcanic ash; high fertility resulting from high weathering rates of glass shards	Young surfaces, in volcanic terranes
Entisols	Minimal development of soil horizons; first appearance of O and A horizons; some dissolved salt in subsurface	Young, newly exposed surfaces, such as new flood or landslide deposits, fresh volcanic ash, or recent lava flow; also found in very cold and very dry climates, or wherever bedrock strongly resists weathering
Gelisols	Young soils with little horizon development that occur in regions of permafrost—perennially frozen ground.	Occur in polar landscapes and at the tops of high mountains where low temperatures and freezing conditions slow chemical-weathering processes.
Inceptisols	Well-developed A horizon; weak development of B horizon, which still lacks clay enrichment; some evidence of oxidation in B horizon; little evidence of eluviation or illuviation	Relatively young surfaces; cold climates where chemical weathering is minimal, or on very young volcanic ash in tropics, in resistant bedrock, and on very steep slopes
Mollisols	Thick, dark, highly organic A horizon; B horizon may be enriched with clays; first appearance of E horizon	Semi-arid regions; generally grass-covered areas having adequate moisture to support grasses but not to cause significant dissolution of soluble materials in upper horizons
Alfisols	Relatively thin A horizon overlying clay-rich B horizon; strongly developed E horizon	Most common in forested, moist environments
Spodosols	Much eluviation of A and E horizons, leaving a light-colored, grayish topsoil; aluminum/iron-enriched B horizon stained by dissolved organic material	Cool, moist climates, usually on sandy parent materials (which allow water to infiltrate readily); grasses or trees may provide the organic matter
Aridisols	Thin A horizon with little organic matter overlying thin B horizon with some clay enrichment; caliche layer present in B or C horizons	Arid lands with sparse plant growth
Histosols	Wet, organic-rich soil dominated by thick O horizons	Found where production of organic matter exceeds addition of mineral matter, generally where surface is continuously water-saturated; often found in coastal environments
Vertisols	Very high clay content; soil shrinks (upon drying) and swells (upon wetting) with moisture variations	Equatorial and tropical areas with pronounced wet and dry seasons
Oxisols	Shows extensive weathering; highly oxidized B horizon is deep red from layer of oxidized iron	Generally older landscapes in moist climates having tropical rainforests
Ultisols	Shows extensive weathering; highly weathered clay-rich B horizons with high concentrations of aluminum	Very moist, lushly forested climates, often subtropical and tropical

◀ **FIGURE 5-28** The Angkor Wat temple in Cambodia, showing the varying degrees of durability exhibited by different soil types. The bricks that form the temple's foundation have been fashioned from oxisols, soils that have undergone advanced chemical weathering. These bricks remain remarkably fresh because they are composed of the stable products of this weathering that resist additional chemical weathering. The intensity of chemical weathering in this warm, moist climate is obvious from the eroded condition of the temple's columns and statuary, carved from normally resistant sandstone.

▼ **FIGURE 5-29** Soils of the western hemisphere. Look at each color on the map and check the characteristics of the corresponding soil type in Table 5.1. For each, answer the question: "Why does this specific soil type develop at this location?"

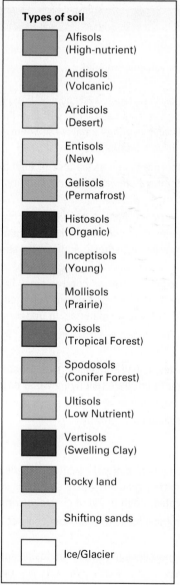

Types of soil

Alfisols
(High-nutrient)

Andisols
(Volcanic)

Aridisols
(Desert)

Entisols
(New)

Gelisols
(Permafrost)

Histosols
(Organic)

Inceptisols
(Young)

Mollisols
(Prairie)

Oxisols
(Tropical Forest)

Spodosols
(Conifer Forest)

Ultisols
(Low Nutrient)

Vertisols
(Swelling Clay)

Rocky land

Shifting sands

Ice/Glacier

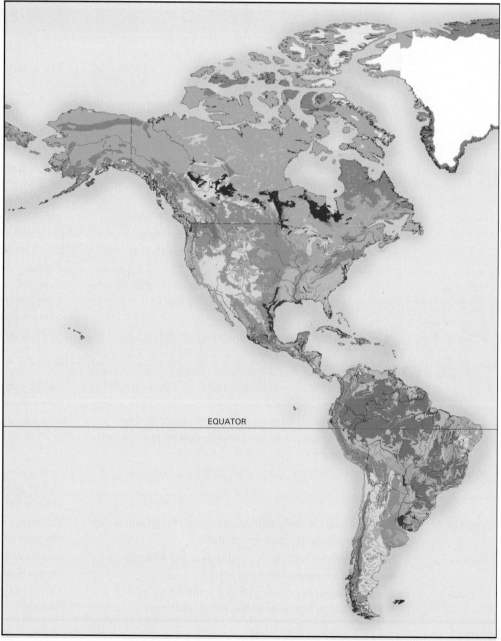

EQUATOR

dissolved away, leaving only the most insoluble elements, such as iron and aluminum. (Oxidized iron makes these soils dark red.) When these soils dry out, they are strong enough to serve as bricks (Fig. 5-28). Oxisols and ultisols are depleted of potassium, calcium, sodium, magnesium, and other nutrients and so are poor prospects for agriculture. The crops of the tropics that grow in oxisols and ultisols therefore tend to be the so-called "cash crops" from which we can make plenty of money but that are typically poor in nutritional value—such as coffee, tobacco, sugarcane, marijuana, and cacao (the prime ingredient in chocolate). While some college students may live on such a diet (except the marijuana, of course), they are not going to get the nutrients that less-weathered and more agriculturally beneficial soils would provide. Spend a moment with Figure 5-29, a map of the soils that occur throughout the Western Hemisphere (and be sure to check the soils in and around your home and campus). A quick check of the soil types will tell you much about the climates of the various regions of this part of the world.

Paleosols   Sometimes previously buried soils are uncovered that differ from other soils in their regions. These soils may have formed under different, perhaps ancient conditions. Buried soils that are substantially older than modern soil formations are called **paleosols** ("old soils"). Examples include aluminum-rich bauxite deposits (oxisols and ultisols rich in residual aluminum) in Georgia, caliche horizons in Connecticut (typically found in arid, desertlike environments), and deep, highly weathered regolith in southwestern Minnesota. All these soils presumably formed under climates with very different temperatures and moisture conditions from those of today. Similarly, oxisols and ultisols, which indicate humid semitropical or tropical climates, have been found beneath the soils forming today in the warm, dry climate of Australia (Fig. 5-30). Thus,

paleosols enable us to identify climate changes that have occurred in the geologic past.

Some *truly ancient* paleosols have recently been discovered in the *2-billion-year-old* rocks of South Africa. These "fossilized" soils contain a red rust-stained horizon, a sure sign that the atmosphere at that time was oxygen-rich. (Iron doesn't rust without oxygen.) Overlying this iron-enriched zone is a light-colored, iron-poor zone from which virtually all the iron was leached (Fig. 5-31). Some scientists believe that such leaching strongly suggests the presence of early organic matter. They reason that when organic matter decomposes, it uses up all of the available oxygen. This, in turn, causes iron to dissolve and seep in solution downward to lower horizons. Such a discovery tells us much about the early Earth atmosphere and the possible timing of the earliest organic life on land.

Aside from satisfying our curiosity about the ancient Earth and its early atmosphere, buried paleosols and other sediments have a very modern, very practical application. Highlight 5-3 shows how geologists use buried soils to help predict earthquakes.

▲ **FIGURE 5-31** This red, rust-stained soil horizon from South Africa and the light-colored, iron-leached layer above it date from about 2 billion years ago. Leaching and precipitation of iron here suggests that the Earth's atmosphere at that time was already oxygen-rich. Decomposition of some of the planet's earliest organic matter may have produced the organic acids and chemical conditions that leached the iron from the upper horizon and precipitated it in the lower horizon.

▲ **FIGURE 5-30** These ultisols of Australia indicate warm, wet, *tropical* conditions. We find them today in central Australia, a dry region. This "out-of-place" soil is actually a paleosol—an artifact of an earlier time and a different, wetter, climate.

Highlight 5-3

Using Soils to Predict the Next Large Earthquake

Earthquakes are basically unpredictable events. Although, as we discussed in Chapter 1, most quakes occur along the edges of the Earth's tectonic plates (such as in California, Japan, Chile, Turkey, and Iran), they can happen virtually anywhere without warning. Despite the difficulties, scientists try to predict them so that thousands of lives may be saved and widespread destruction of property may be minimized.

There are many different ways to try to predict earthquakes, most of which we'll be talking about in Chapter 10. One way that relates to soils involves reconstructing a region's earthquake history by looking at how the ground surface has been affected by ancient quakes. If we can determine the age of a disrupted layer of Earth material (soil, sediment, or rock) or a linear feature such as a stream channel, then perhaps we can figure out when an earthquake occurred in the past. Combining this ancient information with modern historical data for earthquakes in the same region may help us predict the likelihood and timing of future quakes.

An excellent way to study ancient earthquake history is to find old, buried soil layers or other deposits that contain organic material. Soils and other sediments formed from the weathering of rock commonly contain organic material that can be dated by radiocarbon-dating (carbon-14) methods. Formerly continuous soil layers may preserve a record of past earthquakes when they are offset by faulting during an earthquake, or disrupted and offset by earth-shaking (see photo).

This method has been applied to several areas around the world, including the San Andreas Fault in California and the Mississippi Valley area of Missouri–Arkansas–Tennessee. In California, geologists have cut trenches through the upper layers of soil and sediment, and then examined how the layers are disrupted by past motion along segments of the San Andreas Fault. Layers containing organic material are analyzed to determine their age. In one excavated section of the fault, geologists used this method to identify 12 ancient earthquakes, the oldest occurring more than 1500 years ago. A particularly interesting result of this research is that past earthquakes may have occurred in clusters: Several powerful earthquakes occurred within 100 years of each other, followed by a "quiet" time with no earthquakes for 200 to 300 years, followed by another, century-long cluster. The San Andreas Fault is a plate boundary between the Pacific and North American plates, so we know it is seismically active. Although we can expect frequent large earthquakes to occur along the San Andreas, it helps to know how often these earthquakes occur so we might predict the next "Big One."

The Mississippi Valley is not located anywhere near a modern plate boundary. Yet during the winter of 1811–1812, the New Madrid area of southeastern Missouri was the site of the largest historical earthquakes to strike the continental United States. Have other such large earthquakes occurred in the past in the otherwise stable midsection of the North American plate, or were the New Madrid earthquakes unique? If such earthquakes occurred in the past, can future ones be predicted? One way to answer these questions is to dig into the ground and look at old soil and sediment layers. Disturbed soil layers and other evidence from past earthquakes have been exposed in trenches dug into the sediments near New Madrid. Scientists have dated organic material (including wood) from these disrupted layers and determined that prior to the 1811–1812 earthquakes, the region was also shaken by major seismic events twice in relatively recent times—approximately 1000 and 600 years ago. Using this information,

▲ Note the break in the sediment layers at mid-slope. If we can date the sediment layers, we know that they were around at the time of the earthquake that broke and moved the layers along the fault.

some geologists have suggested that the New Madrid region, which includes metropolitan St. Louis, may experience large earthquakes every 400 to 500 years. That's important information to have when we design and build structures such as St. Louis's Gateway Arch.

Weathering in Extraterrestrial Environments

Weathering as we know it on Earth does not take place on our celestial neighbors—the terrestrial planets and moons. Long suspected, this fact has been confirmed by recent discoveries about the surface conditions of the Moon and Mars. The reasons for the absence of weathering, however, differ in each case.

The Moon has no atmosphere. Without atmospheric water, oxygen, or biological activity, virtually no chemical weathering occurs, although so-called "space weathering" may include cosmic-ray bombardment and radiation damage from sunlight. A mechanical-weathering process—the impact of meteorites and micrometeorites—produced the Moon's regolith, which consists primarily of shattered bedrock and glassy fragments expelled from impact craters. The sharp edges of lunar craters—even the oldest ones—suggest the absence of Earth-like chemical-weathering and water-related mechanical-weathering processes as well. Thus the footprints left in the lunar dust by the *Apollo 11* astronauts—the first to walk on the Moon—are likely to retain their freshness for millions of years (Fig. 5-32).

Lunar impacts also generate sufficient heat to melt the Moon's surface rocks, forming a vapor that sprays from a lunar-impact site, cools and solidifies as glassy tephra, and is then deposited nearby. That's how such materials have been added to the Moon's regolith. These impact-produced materials are rich in iron-silicon minerals (they are both felsic and mafic), an un-

▲ **FIGURE 5-33** An unenhanced color photo of the surface of Mars, taken by the *Sojourner* lander of the United States *Pathfinder* probe in July 1997. The redness of the Martian regolith is most likely due to oxidation of iron-rich rocks and sediments.

usual combination as you might recall from our discussions of igneous-rock compositions in Chapter 3. These materials also contain a new mineral recently discovered from a chunk of the Moon's surface that arrived here on Earth as a meteorite (dislodged from the lunar surface by a meteorite strike there). This new mineral is called *hapkeite* (Fe_2Si), named for the geologist—Bruce Hapke of the University of Pittsburgh—who first proposed this unusual lunar-weathering process. (Recall our discussion in Chapter 2 about the naming of new minerals.) Although this process is rare here on Earth because our atmosphere burns up most incoming meteorites, materials with similar iron-silicon compositions are apparently produced when lightning strikes the Earth's surface rocks, sediments, and soils.

Of all the planets in our solar system, the surface conditions on Mars most closely resemble those on Earth, although its non-Earth-like surface temperatures range from −108°C (−225°F) to 18°C (64°F). There is, however, clear evidence of chemical weathering in the planet's past. The oxidation of iron-rich bedrock apparently produced the reddish-brown color of the Martian landscape (Fig. 5-33), which we can see easily with a high-quality telescope from a distance of 78,000,000 km (47,000,000 mi).

As we have seen, various mechanical- and chemical-weathering processes can convert solid bedrock to loose, transportable fragments and dissolved ions. Once liberated from their parent bedrock, these fragments and ions are free to move under the influence of such surface-shaping forces as gravity, stream and glacial flow, gusting winds, and pounding coastal waves. The next chapter discusses some of the processes that transport weathered material, and then focuses on their subsequent deposition and conversion to sedimentary rocks, the most common rocks exposed at the Earth's surface.

▲ **FIGURE 5-32** A footprint left on the lunar surface by one of the Apollo 11 astronauts in July 1969. Because no chemical weathering occurs on the Moon, this footprint will remain for millennia. Micrometeorite bombardment and the debris it ejects, however, may obscure the print over millions of years.

Chapter 5 Summary

Weathering is the slow but steady process whereby environmental factors gradually break down bedrock at the Earth's surface. There are two basic types of weathering:

- *Mechanical weathering* physically disintegrates a rock into smaller pieces without changing its chemical composition;
- *Chemical weathering* decomposes the rock by changing its chemical composition.

Rocks and minerals with structures that are chemically unstable at the Earth's surface are most susceptible to chemical weathering, which changes them into more stable compounds.

Mechanical weathering may involve a variety of processes:

- *frost wedging*—the expansion of cracks in rock as water in the cracks freezes and expands;
- *salt wedging* within cracks, which forces the crack's walls farther apart as salt crystals grow;
- *thermal expansion and contraction,* the alternating enlargement and shrinking of rock as it is repeatedly heated and cooled;
- *mechanical exfoliation,* the fracturing and removal of successive rock layers as deep rocks expand upward after overlying rocks have eroded away;
- *penetration of growing plant roots,* which expands existing cracks in rock;
- *abrasion* of transported particles as they collide with one another or with stationary rock surfaces.

Chemical weathering is largely controlled by climatic factors, such as temperature and the availability of water. Chemical-weathering processes include:

- *Dissolution, or carbonation,* most effective on soluble rocks such as limestone, occurs when acidic water decomposes minerals or rocks;
- *Oxidation* is the reaction of certain chemical compounds with oxygen. It works most effectively on iron-rich rocks, such as basalt and ultramafic peridotites;
- *Hydrolysis* is the replacement of major positive ions in minerals (particularly the feldspars) with H^+ ions from water. This process produces clay minerals, the most common products of chemical weathering.

The rate at which a given rock or mineral weathers chemically depends on a number of factors:

- climate (hot, moist regions exhibit more chemical weathering than cold, dry regions);
- the activity of living organisms;
- the length of time it has been exposed to weathering;
- the chemical stability of its components at the Earth's surface.

Mechanical and chemical weathering together produce the Earth's regolith, the loose, fragmented material that covers much of the Earth's surface. Soils are the uppermost, organic-rich portion of the regolith that supports plant growth. Soil development is governed by five factors:

- parent material (the bedrock or sediment from which a soil develops);
- climate;
- topography (the physical features of a landscape);
- vegetation cover;
- time.

A developing soil consists of distinct layers having different compositions; these layers are called *soil horizons.* The vertical succession of soil horizons in a given location is the location's *soil profile.* Soil classification helps guide land-use decisions. Soil scientists now distinguish 12 different orders of soil within North America.

Weathering takes place on other planets in our solar system and on the Moon, although it operates differently than it does on Earth. No chemical weathering occurs on the Moon, because it has neither atmosphere (and thus no O_2 or CO_2) nor surface water. Instead, frequent meteorite impacts weather lunar-surface rocks mechanically. Although most of the surface water of Mars is now trapped in the ground as ice, the characteristic redness of the planet's surface suggests that in the past climatic conditions did enable water to oxidize its iron-rich bedrock.

KEY TERMS

abrasion *(p. 147)*
caliche *(p. 162)*
chemical weathering *(p. 144)*
dissolution *(p. 148)*
erosion *(p. 142)*
frost wedging *(p. 144)*

hydrolysis *(p. 149)*
leach *(p. 160)*
mechanical exfoliation *(p. 147)*
mechanical weathering *(p. 144)*
oxidation *(p. 151)*
paleosols *(p. 165)*

parent material *(p. 157)*
regolith *(p. 156)*
soil *(p. 156)*
soil horizon *(p. 160)*
soil profile *(p. 160)*
spheroidal weathering *(p. 155)*

talus slope *(p. 144)*
thermal expansion *(p. 146)*
topography *(p. 157)*
weathering *(p. 142)*

QUESTIONS FOR REVIEW

1. Describe the fundamental difference between mechanical and chemical weathering.

2. Discuss three ways in which rocks can be weathered mechanically.

3. Of granite, limestone, and basalt, which would be most susceptible to the chemical-weathering process of oxidation? To the process of dissolution? Which would weather to produce the most clay?

4. What role does climate play in chemical weathering?

5. Discuss how soils vary in a region of irregular topography.

6. Describe the principal characteristics of the major soil horizons, and explain how those characteristics develop.

7. Since the Industrial Revolution, we have been burning coal, heating oil, and gasoline at an ever increasing rate. Combustion of these fuels produces carbon dioxide. What would you expect to be the effect of burning these fuels on weathering rates? Explain.

8. Imagine that the Earth someday becomes devoid of water. How would the nature of chemical weathering change in polar regions? In the arid subtropical deserts? In the equatorial tropics?

9. Above is a photo of a soil developed on a lava flow in eastern Washington. Judging from the appearance of the soil, what rock type constitutes the lava flow? What weathering processes and products are responsible for the color of the soil? Describe the climate that was most likely responsible for this type of weathering.

LEARNING ACTIVELY

1. Look around your community or campus, identify the major building stones, and compare their relative states of weathering. Did the local builders choose wisely when selecting their building materials considering your local climate? Which rocks would work best for construction in your area? Which would be poor choices?

2. Pay a visit to a local cemetery—especially an old one. Examining the gravestones and the condition of their inscribed names and dates, which rock types have fared the best over the course of time? Look for stones made of limestone and granite. Which of these has retained its sharpness? Which processes are weathering these stones?

3. Where it is safe to stop, look at a roadcut along a road or highway, and study the soil developed near the top of the exposure. Check for the O, A, E, and B horizons. Guess the type of soil and the degree of soil-horizon development. Pick out some pebbles from the soils and check their relative degree of freshness. Are the pebbles of some rock types more decomposed than others?

4. Look back through this chapter and try to identify examples of how the Earth's various systems (lithosphere, hydrosphere, atmosphere, biosphere) interact to produce the various weathering products.

 ## ONLINE STUDY GUIDE

Practice makes you better. Make the time to take the *Online Study Guide* quiz for this chapter. It should take you about 20 minutes. Automatic grading and instant feedback will help you quickly and accurately identify the concepts you have mastered and the areas in which you need more study.
www.prenhall.com/chernicoff

Sedimentary Processes, Environments, and Rocks

- **The Origins of Sedimentary Rocks**

- **Classifying Sedimentary Rocks**

- *Highlight 6-1*
  Finding "Blueberries" on Mars

- *Highlight 6-2*
  When the Mediterranean Sea Was a Desert

- **"Reading" Sedimentary Rocks**

- *Highlight 6-3*
  Telling the Earth's History from Its Seafloor Mud

Most of us don't live where we can readily view a volcanic eruption or the movement of an active fault, but we probably can walk on a sandy beach, fish in a quiet lake, or sit by a muddy stream a short distance from home. Even in the heart of New York City, we can watch rivulets of rain collect leaves and dirt from downtown streets and deposit them into a corner storm drain. Much of that rainwater will eventually find its way to the Atlantic Ocean, and some of that dirt will come to rest on the Atlantic seafloor. No matter where you live—in a major city, on a farm, on a remote island, or high in the mountains—you don't have to travel far to find **sediment** (from the Latin *sedimentum,* meaning "settling"). Sediment forms at the Earth's surface and consists of the products of mechanical and chemical weathering—loose rock and mineral fragments that are deposited, typically in layers, after being transported some distance by water, wind, ice, or gravity. Sediment also includes the dissolved minerals that precipitate out of solution in water.

Sediment accumulates virtually anywhere on the Earth's surface—from the glaciated summits of the Himalaya, 10 km (6 mi) *above* sea level, to the deep trenches on the floor of the Pacific Ocean, 10 km *below* sea level, and everywhere in between. Sediments are continually deposited in lakes, streams, deserts, swamps, beaches, lagoons, and caves, on continental shelves (the underwater extensions of continents), and at the bases of glaciers throughout the world.

Most sediment is ultimately converted to solid **sedimentary rock** (Fig. 6-1), which makes up only the thin top layer of the Earth's crust. Yet, although sedimentary rocks account for barely 5% by volume of the Earth's outer 15 km (10 mi), they blanket much of the Earth's surface; 75% of all the rocks exposed at the Earth's land surface are sedimentary, so they are the most common rocks you'll encounter as you travel our continental landscapes. Sedimentary rocks are also quite valuable to our human interests—they are our principal source of coal, oil, and natural gas, they contain much of our iron and aluminum ores, all of our cement and much of our other natural building materials. Sedimentary rocks also store nearly all of our underground freshwater—a vital resource for the health and welfare of our communities.

FIGURE 6-1 Sedimentary-rock formation in the Paria Canyon, Utah. The rocks represent ancient sand dunes frozen in time. ▶

From a geologist's perspective, sedimentary rocks are fascinating because they contain the essential clues about the condition of the Earth's surface as it existed in the far-distant past. They record the presence of great mountains in areas now monotonously flat, and tell tales of vast seas that once covered the now-dry interior of North America. Some sedimentary rocks hold the fossil remains of past life, thus telling us much about the evolution of life on Earth through its history. It's important to realize that it generally takes a *long* time—millions of years—for sedimentary rocks to form and then be reshaped into lofty mountains or spectacular canyons. The layers of rock must first be laid down as loose sediment, then be turned into solid rock, perhaps to be uplifted into mountains and then eroded, only to be buried again by later periods of sedimentation. The scenic grandeur of the Grand Canyon simply could not have formed in a matter of days; it takes time—lots of it.

This chapter examines the origins of sedimentary rocks, their classification, and the ways in which geologists use them to reconstruct past surface environments. The chapter concludes by showing the relationship of various sedimentary rocks to common plate-tectonic settings.

The Origins of Sedimentary Rocks

As we saw in Chapter 5, mechanical weathering breaks down rocks into smaller fragments without changing their chemical composition. Chemical weathering converts unstable minerals to new, more stable compounds through various chemical reactions. Our concern in this chapter is what happens to these weathering products. How are they moved from one place to another? How are they deposited as sediment at a new location, and buried by subsequent deposition? How do they eventually become new rocks? What clues can we look for in these rocks that might tell us about past life and past climates?

Geologists classify sediments according to the source of their constituent materials (Fig. 6-2). **Detrital sediment** is composed of transported solid fragments, or *detritus,* weathered from preexisting igneous, sedimentary, or metamorphic rocks. **Chemical sediment** forms when previously dissolved minerals either precipitate from solution or are extracted from water by living organisms and converted to shells, skeletons, or other organic substances. They are deposited as sediment when the organisms die or discard their shells. The different types of sediments and the rocks they form are discussed in more detail later in this chapter.

Sediment Transport and Texture

The vast majority of sediments are detrital. They are composed primarily of the solid fragments, or *clasts* (from the Greek *klastos,* meaning "broken"), produced by mechanical weathering or released by erosion from preexisting rocks. During transport, detrital sediments move from high places to low places, largely drawn by the pull of gravity. Gravity is usually assisted by a transporting medium, such as running water, wind, or glacial ice. Detrital particles are deposited when the transporting medium loses its capacity to carry the sediment farther. For example, when a river ceases to flow upon entering relatively still marine water at a coast, it drops its sediment load on the seafloor. Each year, an estimated 10 billion tons of detrital sediment, most of it carried by rivers, arrive at the world's coastlines.

If you've ever held a handful of beach sand or lake-bottom mud, you can appreciate the variety of *textures* of detrital sediments—the size, shape, and arrangement of the particles. Sediment texture depends on the source rocks of the sediment particles, the energy of the medium that transported them, and their environment of deposition.

The raw materials for chemical sediments—the *dissolved* products of chemical-weathering processes—are transported by the water in which they are dissolved. These ions remain in solution until a change in the water's temperature, pressure, or chemical composition causes them to precipitate. These ions may also be extracted from solution by a living organism that manufactures some type of biological structure, such as a protective outer shell or an internal skeleton. These extracted ions and the compounds they form are deposited (perhaps after further transport by the organism) only after the organism dies or discards its shell.

Grain Size Sediment clasts are commonly broken and worn away while being carried by turbulent streams, crushed under glaciers, and tossed about by crashing waves at the shore. The amount that they are worn down in part relates to the composition and texture of the parent rock. The key questions: What minerals are present in the parent rock? How hard are those minerals? What is the parent rock's grain size? When a *coarse*-grained granite weathers, it generally produces larger sediment grains than does a *fine*-grained volcanic tuff. (Sediment derived from granite typically contains relatively large, hard quartz grains; the sediment derived from a fine-grained tuff contains small, soft particles of volcanic glass.) The grain size of a sediment also depends on the nature and energy level of the transport medium. A pebble that would be pulverized at the base of a creeping glacier might remain unchanged when transported within an oozing mudflow. The same pebble would be worn down by tumbling in a raging whitewater river, but might be unaffected if carried along gently by a trickling stream.

The texture of sediment relates not just to the transport medium's ability to weather down its rock fragments, but also to its ability to carry them at all. A flooding river can transport large boulders along with tiny particles and all sized particles in between. By contrast, a gentle wind can carry aloft only tiny silt-sized grains (the size of cooking flour). **Sorting** is the process by which a transport medium "selects" particles of different sizes, shapes, or densities. Wind is the most selective of the transport processes; a deposit of windblown silt, which contains particles that are nearly all within a narrow size range, is considered *well sorted* (Fig. 6-3a). At the other extreme, glacial ice and flooding rivers are relatively unselective, transporting particles with a wide range of sizes. Glacial deposits, which may contain the finest particles (as fine as talcum powder) along with boulders the size of a two-story building are said to be *poorly sorted* (Fig. 6-3b). Geologists can often determine what medium transported a sediment by a quick visual estimate of its sorting.

When a moving current (wind or water) loses its energy and can no longer carry its suspended sediment load, the sediment

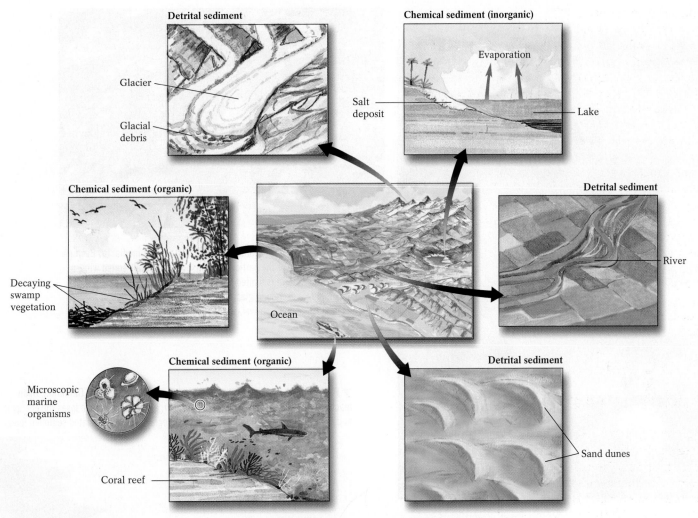

▲ FIGURE 6-2 Various types of sediments and their origins. Detrital sediments consist of preexisting rock fragments, such as glacial debris, river-channel sand, and sand dunes. Chemical sediments often consist of minerals precipitated directly from water, such as salt deposits produced by the evaporation of small, temporary lakes. They may also be composed of organic debris, such as partially decayed swamp vegetation or the shells of small marine organisms.

particles settle out and drop to the Earth's surface. Because their transport requires more energy, the larger, heavier particles are deposited first; the smallest, lightest particles are typically carried farthest and deposited last. When a rushing stream emerges from a steep narrow mountain gorge onto a broad flat valley plain, it fans out and loses much of its energy, and thus it can no longer carry its sediment. The stream drops its sediment where the angle of slope changes sharply, producing a wedge-shaped body of poorly sorted sediment called an *alluvial fan* (Fig. 6-4). Within an alluvial fan, the coarsest grains are deposited first (near the mountain) and the smaller grains are deposited slightly farther downstream.

Grain Shape   Particles released from rock by mechanical weathering may be jagged and angular, particularly if they originate as irregular grains in a plutonic igneous rock. Abrasion during transport, however, wears off a grain's jagged edges. Some transport media are particularly effective in rounding particles. Swiftly flowing rivers, for instance, bounce pebbles and sand grains around vigorously, so that they collide with other particles and with the river bottom. Thus they become ever smoother and smaller as they move downstream. Sediment particles *at the base*

of a glacier may be ground to a fine rocky powder, called "glacial flour." Other glacial sediments, embedded in hundreds of meters of ice, are cushioned from such collisions and retain their initial size and shape. *In general, the more vigorous collisions a particle experiences, and the farther it moves away from its parent rock, the more rounded it becomes* (Fig. 6-5).

Grains of softer minerals, such as gypsum and calcite, become rounded more readily than harder minerals, such as quartz. In a revealing field study, fragments of soft sedimentary rock became well rounded after only 11 km (6.6 mi) of stream transport. More durable fragments of granite, carried in the same stream, required 85 to 335 km (53–208 mi) of transport to become comparably rounded.

Sedimentary Structures

Detrital sediments commonly contain **sedimentary structures,** physical features that tell us how the sediments were deposited and what the environment was like during deposition. For example, a seemingly featureless layer of sandstone might contain structures that let a geologist determine whether the sand was

(a)

(b)

▲ **FIGURE 6-3** Differential sediment sorting by transport media. **(a)** Because wind is highly selective of the particles it transports, this well-sorted wind-blown silt in western Iowa is limited to very fine sediment particles. **(b)** Stream eroding a section of glacial till in central Colorado. Note the numerous rounded cobbles, gravel, sand, and silt that make up the till.

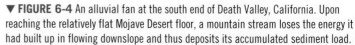

▼ **FIGURE 6-4** An alluvial fan at the south end of Death Valley, California. Upon reaching the relatively flat Mojave Desert floor, a mountain stream loses the energy it had built up in flowing downslope and thus deposits its accumulated sediment load.

deposited in a desert or a river. Some features even tell us which way the desert wind was blowing or the river was flowing. Let's examine some common sedimentary structures and see how they can be used to interpret past environmental conditions.

Bedding (Stratification) **Bedding,** or **stratification,** is the arrangement of sediment particles into distinct layers *(beds,* or *strata)* of different sediment compositions or grain sizes. A clearly visual break, or *bedding plane,* generally separates adjacent beds, marking the end of one depositional event and the beginning of the next. Bedding planes most often signal a change in the nature of the sediment itself, a shift in the energy with which it was transported, or both. Consider, for example, a typical river bed. Its layers consist principally of the river's usual sediment load of medium-sized particles. Interspersed between these beds, however, you might find occasional layers of heavier, coarser-grained particles. How did these particles arrive there? They were probably deposited during high-energy flooding episodes, the only times when the river would be capable of carrying the larger, heavier sediment. Similarly, a flooding river typically deposits a particle load of heavy, coarse-grained sediment on top of finer preexisting sediments in its surrounding *floodplain* (the area adjacent to a river where its excess waters go during a flood). Such a difference in grain size would appear as a bedding plane in a cross section of the resulting sediment layers (Fig. 6-6).

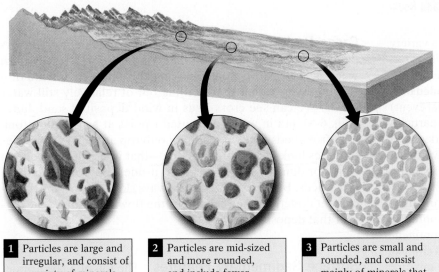

1 Particles are large and irregular, and consist of a variety of minerals, including some that weather easily

2 Particles are mid-sized and more rounded, and include fewer minerals that weather easily

3 Particles are small and rounded, and consist mainly of minerals that resist weathering, such as quartz

◀ FIGURE 6-5 As you can see here, a stream's sediment changes along its course. Its particles become more rounded as they collide with other particles and the stream's bed. Sediments' compositions change as well, with resistant minerals surviving the journey while the softer ones disintegrate.

1 Pre-flood

Floodplain sediment
Old bedding plane
Older sediment

2 Flood stage

Flood water
Erosion of uppermost sediment

3 Post-flood

Coarse-grained flood deposit
New bedding plane

(a)

(b)

▲ FIGURE 6-6 (a) Development of a bedding plane due to river flooding. Any depositional event that leaves sediment that differs (either in grain size or composition) from the preexisting sediment leaves a bedding plane between the resulting sediment layers. (b) These sediments in Wayne County, Pennsylvania—very coarse and boulderlike on top, much finer below—are separated by a bedding plane that signals a change in the physical nature of the sediments and the variations in energy of their transport media.

175

Graded Bedding   When a sediment load containing a variety of sediment sizes is suddenly dumped into relatively still water, its particles will settle at different rates, depending on their sizes, densities, and shapes. This settling process produces a **graded bed,** a single sediment layer (formed during a single depositional event) in which particle size varies gradually—with the coarsest particles on the bottom and the finest at the top (Fig. 6-7). To understand how graded beds form, drop a handful of unsorted backyard dirt into a tall glass of water and note how the largest particles settle quickly to the bottom, while the finest particles settle last.

Graded sediments commonly occur on ocean floors, near shores where muddy streams shed their sediment loads onto the continental shelf. Offshore sediment accumulates on a continental shelf as an unstable mass that can easily be dislodged (for example, by an earthquake). When this happens, a dense mixture of sediment and seawater called a *turbidity current* (from the Latin *turbidus,* meaning "disturbed") flows rapidly downslope toward the deepsea floor at 60 km/hr (40 mph) or more (Fig. 6-7b). Upon reaching the horizontal ocean floor, the turbidity current slows to a virtual standstill, losing its transport energy and dropping its sediment load. The particles from this mixture settle out to form a graded bed. (The subsequent sedimentary rock that forms from a turbidity current is called *a turbidite.*)

Cross-Bedding   **Cross-beds** consist of sedimentary layers deposited *at an angle to the underlying set of beds.* They form when particles drop from a moving current, such as wind or a flowing river, rather than settling out of relatively still water or air. We can see cross-beds in wind-deposited sand dunes (Fig. 6-8) and in water-deposited ripples at a river's bottom. The cross-beds that we see in an outcrop are what the ripples or dunes look like in cross-section—that is, if you could slice a ripple or dune open and look at it inside, you would see these cross-beds. Because cross-beds always slope toward the down-current direction, they can tell us the flow direction of the current that deposited them.

Surface Sedimentary Features   The top of a layer of detrital sediment often provides clues to the environmental conditions to which it was exposed during or after deposition. For instance, **ripple marks** preserved on top of a sediment bed indicate that wind or water currents shaped the particles into a series of shallow curving ridges during deposition. The configuration of these ridges, which are often visible on sandy surfaces, reflects the nature of the current that produced them (Fig. 6-9). When geologists look closely at ripple marks, they are often trying to

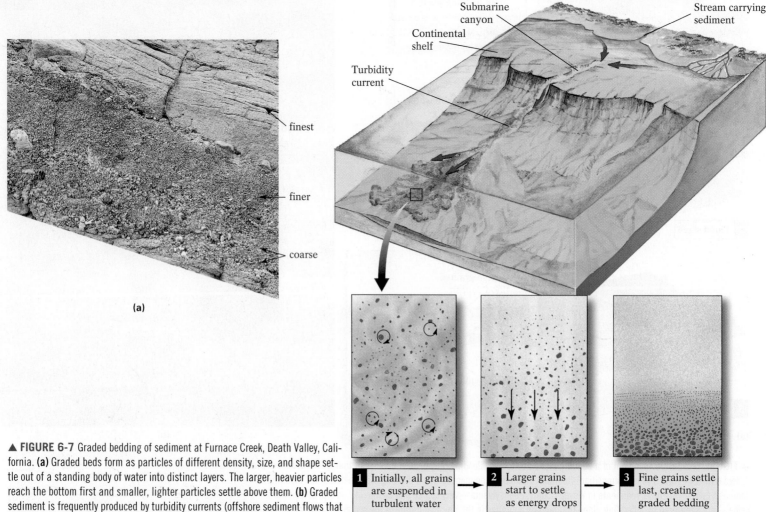

▲ **FIGURE 6-7** Graded bedding of sediment at Furnace Creek, Death Valley, California. **(a)** Graded beds form as particles of different density, size, and shape settle out of a standing body of water into distinct layers. The larger, heavier particles reach the bottom first and smaller, lighter particles settle above them. **(b)** Graded sediment is frequently produced by turbidity currents (offshore sediment flows that abruptly lose their energy and drop their particle loads onto the ocean floor).

finest

finer

coarse

(a)

Submarine canyon

Continental shelf

Stream carrying sediment

Turbidity current

1 Initially, all grains are suspended in turbulent water

2 Larger grains start to settle as energy drops

3 Fine grains settle last, creating graded bedding

(b)

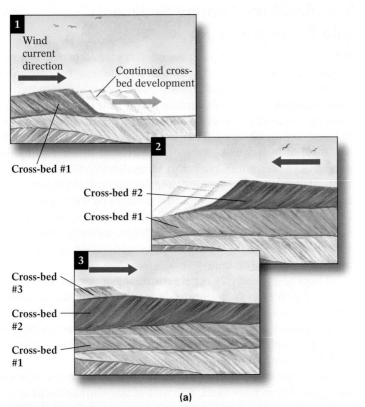

▲ **FIGURE 6-8 (a)** The development of cross-bedding in sand dunes. The sets of cross-beds with different orientations that we see here form as wind directions shift through time. **(b)** Ancient cross-bedded sand dunes from Zion National Park in southern Utah.

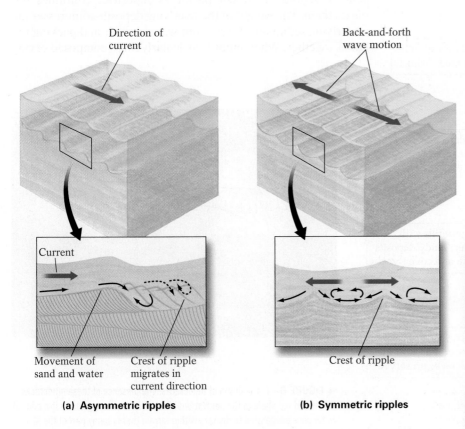

◄ **FIGURE 6-9** Different types of currents produce different ripple patterns. **(a)** A current that generally flows in one direction, such as a stream, produces asymmetric ripples. Sand grains roll up the gently sloping upstream side of each ridge and then cascade down the steeper downstream side. **(b)** Symmetric ripples form from the back-and-forth motion of waves in shallow surf zones at the coast or at the water's edge in a lake. **(c)** Exposed rocks show asymmetric ripple marks, evidence of past current flow (water or wind).

(a) Asymmetric ripples

(b) Symmetric ripples

(c)

determine whether the sediment was deposited by waves—such as in a lake or along an ocean coast—or by a single-direction current—such as a stream, the prevailing winds, or certain ocean currents (like the Atlantic's Gulf Stream). When we locate an ancient stream deposit, for example, we can figure out which way the stream was flowing. Armed with this knowledge, we can look for the highlands or mountains that were the stream's source and the ocean into which the stream eventually flowed. Thus, from the sedimentary rocks, we can sometimes reconstruct much of a now vanished landscape.

Mudcracks are fractures that develop when the surface of fine-grained sediment becomes exposed to the air, dries out, and shrinks (Fig. 6-10). They indicate that the watery environment in which the sediment was deposited dried up at some point, as happens, for example, when shallow lakes evaporate. The mudcracks typically form polygonal shapes, much like the shapes we talked about in Chapter 4 in our discussion of columnar jointing of cooling basalts—another example of how Earth materials crack as they shrink.

Lithification—Turning Sediment into Sedimentary Rock

When a sediment layer is deposited, it buries all previous layers at that location. In time, a sedimentary pile may become thousands of meters thick. Sediments buried several kilometers or more beneath the Earth's surface retain heat (produced largely from the decay of radioactive mineral grains and conducted from the deep interior below). They are also compressed by the accumulation of overlying sediment and invaded by circulating underground water, which typically carries dissolved ions. Together, the heat, pressure, and the ions alter the physical and chemical nature of both detrital and chemical sediments by a set of processes known collectively as **diagenesis.** Sometimes these processes result in **lithification** (from the Greek *lithos,* meaning "rock," and Latin *facere,* meaning "to make"), the conversion of loose sediment into solid sedimentary rock.

Diagenesis generally occurs within the upper few kilometers of the Earth's crust at temperatures less than about 200°C (400°F). Thus, it differs from the intense heat- and pressure-related processes that change rocks deep in the Earth's interior by metamorphism or melting. During lithification, sediment grains are compacted and commonly cemented.

To illustrate how a diagenetic process might work, squeeze some wet clay in your hand. The wad of clay shrinks because you are forcing water out of it and packing the particles of clay more closely together. This example illustrates **compaction,** the process by which pressure reduces the volume of a sediment during diagenesis. Sediments are compacted as deposition continues and buries them. The weight of the overlying deposits compresses the underlying sediments, forcing out water and air and packing the grains together. When muds (particularly those composed of clay

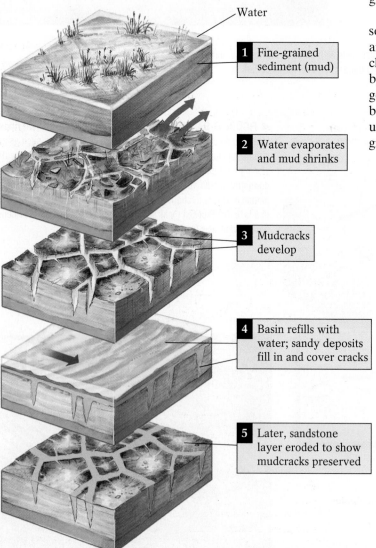

Water

1 Fine-grained sediment (mud)

2 Water evaporates and mud shrinks

3 Mudcracks develop

4 Basin refills with water; sandy deposits fill in and cover cracks

5 Later, sandstone layer eroded to show mudcracks preserved

◀ **FIGURE 6-10** The origin of mudcracks. The presence of these mudcracks in oxidized red shale in Glacier National Park, Montana, suggests that a body of water once evaporated to dryness within terrain that is today part of the Montana Rockies. The photo corresponds to part 5 of the illustration to the left. Note the pocketknife at upper right for scale.

minerals) are compacted in this way, weak attractive forces between the particles cause them to adhere to one another, converting loose sediment into more cohesive sedimentary rock such as, in this case, a mudstone.

Sediment grains can also be cemented together. **Cementation** takes place as water circulates through a sediment, and the elements dissolved in the water during chemical weathering precipitate and bind the sediment grains together (Fig. 6-11). Precipitation of the cementing agents occurs when the circulating groundwater encounters a change in its physical or chemical environment. Common cementing agents include calcium carbonate, silica, and several iron compounds. Calcium carbonate forms when calcium ions released by chemical weathering from calcium-rich minerals combine with carbon dioxide and water in soil. Silica cements are produced primarily by chemical weathering of the feldspars in igneous rocks. Iron carbonates (principally siderite: $FeCO_3$), iron sulfides (such as pyrite: FeS_2), and iron oxides (such as hematite: Fe_2O_3) also act as cements.

In 2004, in yet another remarkable discovery by NASA's *Opportunity* rover, hematite nodules—presumably water-deposited as cementing agents in Martian soils—were photographed and analyzed chemically. Spend a moment with Highlight 6-1, "Finding 'Blueberries' on Mars."

Compaction and cementation can also lithify shells, shell fragments, and other remains of dead organisms. Any rock that consists of preexisting solid particles compacted and cemented

together—whether they're preexisting rock fragments or a hash of organic debris—has a **clastic** texture (the term clastic comes from the Greek *klastos* for "broken").

Classifying Sedimentary Rocks

We generally classify sedimentary rocks as either detrital or chemical, depending on their source material. Each of these broad categories, however, includes a wide variety of rock types, each rock reflecting the diverse transport, deposition, and lithification processes that formed it.

Detrital Sedimentary Rocks

Classification of detrital sedimentary rocks depends on their particle sizes rather than the composition of their particles (Table 6.1). Shales and mudstones are the finest-grained; sandstones have grains of intermediate size; and conglomerates and breccias contain the largest grains. Note that all detrital rocks—whether composed of mineral grains or shell fragments—have a *clastic* texture because they consist of solid particles cemented or compacted together.

Mudstones More than half of all sedimentary rocks are **mudstones,** the detrital sedimentary rocks containing the smallest particles (less than 0.002 mm in diameter). Such fine particles only settle out of relatively still waters; more energetic waters keep them suspended and carry them away. Thus most mudstones originate in lakes and lagoons, in deep-ocean basins, and on flooded river floodplains. The extremely fine particles in mudstones consist largely of clay minerals and micas; they are so small that their mineral composition is best analyzed by X-ray diffraction (discussed in Chapter 2). When these flat or tabular particles become buried beneath hundreds of meters of sediment, compaction

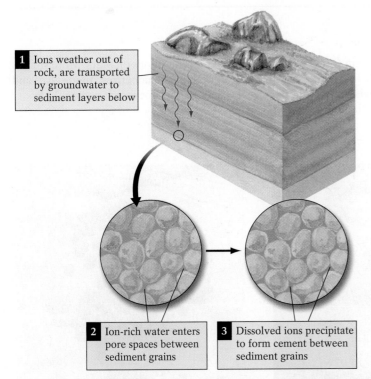

▲ FIGURE 6-11 Lithification of sediment by cementation. Weathering of source rocks releases ions in solution that then circulate with the groundwater through coarse-grained sediments. When these dissolved elements precipitate as solid compounds in the spaces between sediment grains (called *pore spaces*), they form a cement that binds the grains together.

Table 6.1 Detrital Sediments and Rocks

PARTICLE SIZE (MM)	PARTICLE NAME	NAME OF ROCK FORMED
<0.002	Clay* } Mud	Shale } Mudstone
0.002–0.05	Silt	Siltstone
0.05–2	Sand	Sandstone
2–4	Granule } Gravel	Breccia (if particles are angular)
4–64	Pebble	
64–256	Cobble	Conglomerate (if particles are rounded)
>256	Boulder	

1 mm = 0.039 inch

*Note that the term "clay," when used in the context of sediment size, denotes very fine particles of any rock or mineral (as opposed to the term "clay mineral," which refers to a compositionally specific group of minerals); all clay minerals have clay-size particles, but not all clay-size particles are composed of clay minerals.

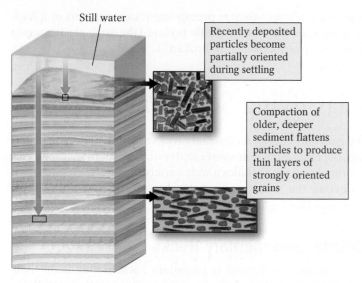

Still water

Recently deposited particles become partially oriented during settling

Compaction of older, deeper sediment flattens particles to produce thin layers of strongly oriented grains

▲ **FIGURE 6-12** Deposits of flat or tabular clay and mica grains may be initially oriented randomly. The weight of subsequent overlying deposits causes these grains to "collapse" into a parallel orientation, producing the typical layered appearance of shales.

arranges them into parallel layers resembling a deck of cards (Fig. 6-12). **Shale** is a type of mudstone noted for its ability to split easily along very thin parallel surfaces.

A geologist will often nibble on bits of fine-grained sedimentary rock to distinguish clay-rich mudstones from slightly coarser, quartz-rich siltstones (with grain sizes between 0.002 mm and 0.05 mm). The particles of the clay minerals feel smooth against the teeth, whereas the abrasive quartz grains of siltstones feel noticeably gritty.

Shales vary considerably in color, depending on their mineral composition. Red shales contain iron oxides that precipitated from water containing both dissolved iron and abundant oxygen. In contrast, green shales contain iron minerals that precipitated in an oxygen-poor environment. Black shales form in water with insufficient oxygen to decompose all of the organic matter in its sediment, leaving a black, carbon-rich residue. Such conditions might occur in the still waters of a swamp, lagoon, or deep-marine environment, where little circulation of oxygen-rich surface water takes place.

Shales have numerous practical uses. They are a source of the clays we use to make bricks and ceramics, such as pottery, fine china, and tile. Mixing shale at various stages with sand, gypsum (hydrous calcium sulfate), and calcium carbonate produces Portland cement, a staple of the construction industry. Oil shale (fine-grained rocks that contain abundant tar) may someday provide a key source of energy.

Highlight 6-1

Finding "Blueberries" on Mars

First . . . no . . . not those kinds of blueberries, the ones we buy at the local farmers' market. Early in 2004, NASA scientists discovered another type of "blueberry"—a geological type in the images sent back from Mars by the Mars Exploration Rover mission. These blueberries (so-named by the scientists because they resembled the scattered blueberries in their morning blueberry muffins) are BB-size spherical grains of hematite (see *photo on the right*)—some embedded in Martian sedimentary rocks, others lying freely at the surface, eroded from those sedimentary rocks. These spherules are yet another line of evidence that the Martian surface and its underlying regolith harbored substantial water—probably in the form of underground water—at some time in the Red Planet's past.

Similar iron-oxide features have been found on Earth in many locations—many associated with precipitation from circulating groundwater rich in dissolved iron. The iron precipitated when chemical conditions changed—such as the groundwater's flow rate, flow path, its salinity, pH, and in some cases, the presence of certain forms of bacteria. Iron-oxide nodules, like those shown here, may provide geologists with valuable information about past climates and environmental conditions on Mars. More studies await both the "blueberries" themselves and the sedimentary rocks that encase them.

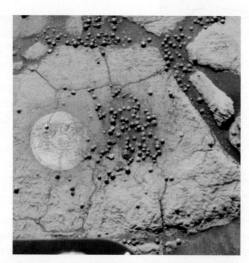

▲ Marjorie A. Chan, a geologist at the University of Utah in Salt Lake City, has identified similar iron-oxide "blueberries" in Utah's sedimentary rocks. Here we see the Utah "blueberries" on the right (The largest is about 5 cm–2 in.—across. The Martian "blueberries" are on the left (the largest is about 5 mm–0.2 in—across.)

Sandstones **Sandstones** are detrital sedimentary rocks whose grains range from 0.05 mm to 2 mm in diameter. Sandstones account for approximately 25% of all sedimentary rocks. Silica or carbonate cement generally surrounds the mineral grains in sandstones.

There are three major types of sandstones, classified on the basis of the mineralogy of their sand grains. A sandstone composed predominantly of quartz grains (90%), with very little surrounding *matrix* (the finer material between the larger grains), is a *quartz arenite* (from the Latin *arena*, or "sand"). Quartz arenites generally have a light color, varying from white to red depending on the cementing agent (Fig. 6-13, top left). Their grains are rounded and well sorted, suggesting that they were transported a long distance from their source rocks.

Arkoses (named for a Greek word that denotes a rock created by consolidation of debris) are distinctive pink-to-red sandstones containing more than 25% feldspar (Fig. 6-13, rightmost photo). (For this reason, arkoses are also known as *feldspathic sandstones*.) The grains, which are typically derived from feldspar-rich granitic source rocks, are generally poorly sorted and angular. This pattern suggests that the grains underwent short-distance transport, minimal chemical weathering, and rapid deposition and burial.

Graywackes (derived from the German *wacken*, or "waste," and also known as *lithic* sandstones) are dark gray-to-green sandstones that contain a mixture of quartz and feldspar grains, abundant dark rock fragments (commonly of volcanic origin), and a fine-grained clay-and-mica matrix (Fig. 6-13, bottom left). The poor sorting, angular grains, and presence of such easily weathered minerals as feldspar suggest that the sediments were not transported far from their source.

Sandstone's durability has made it a popular building stone. You can see it in the Gothic-style buildings found on many college

Rounded quartz grain

Angular feldspar grain

▲ **FIGURE 6-13** The three major types of sandstone. (top left) Quartz arenite, composed predominantly of highly rounded quartz grains. One of North America's most celebrated quartz arenites is the St. Peter Sandstone of Minnesota, Iowa, and Wisconsin. It is most prominently exposed in the Minneapolis–St. Paul area, where the rock is so pure it has been mined, melted, and used in manufacturing glass. Inset: Quartz arenite in thin section. Scale bar = 1 mm. (above) Arkose, containing an abundance of angular feldspar grains. Some of North America's classic arkoses can be found in the Red Rocks area along Highway I-70, west of Denver, Colorado. Inset: Arkose in thin section. Scale bar = 1 mm. (bottom left) Graywacke, distinguished by an abundance of dark volcanic fragments and relatively poor sorting of its particles. We can find good examples of graywacke in the Ouachita Mountains of Oklahoma and Arkansas, and in the coastal mountains of California, Oregon, and Washington. Inset: Graywacke in thin section. Scale bar = 1 mm.

▲ **FIGURE 6-14** Beautiful Pillsbury Hall houses the geology department at the University of Minnesota. The building's exterior is made from Hinckley Sandstone, an arkose.

campuses (Fig. 6-14). Sandstones also hold much of the world's crude oil, natural gas, and drinkable groundwater, thanks to the pore space between their sand grains, which become easily saturated by migrating fluids.

Conglomerates and Breccias **Conglomerates** and **breccias** (BRETCH-ahs), the coarsest of detrital sedimentary rocks, contain grains larger than 2 mm in diameter. Conglomerates are characterized by rounded grains; breccias have angular grains (Fig. 6-15). Both contain fine matrix material, which is typically cemented by silica, calcium carbonate, or iron oxides. The size of conglomerate and breccia grains makes it relatively easy to identify their parent rocks. Likewise, the shape of the grains provides clues to their transport path: The rounded particles in conglomerates suggest lengthy transport by vigorous currents. The angular grains in breccias suggest short-distance transport, such as when shattered rock debris accumulates at the base of a cliff.

Chemical Sedimentary Rocks

Detrital sedimentary rocks are made of fragments of preexisting grains that have been bound together by compaction or cementation. In contrast, chemical sedimentary rocks typically consist of interlocking mineral crystals that form in a rock as it is deposited. There are two kinds of chemical sediments: **inorganic,** which precipitate *directly* from water; and **biogenic,** which form from the remains of animals and plants.

▼ **FIGURE 6-15** Conglomerates and breccias. The grains in these coarse sedimentary rocks reveal much about their history. **(a)** The roundness of the grains in conglomerates suggests long-distance transport by vigorously moving water. The various rocks represented in this conglomerate indicate a plutonic-igneous source. **(b)** The angularity of the grains in breccias suggests short-distance transport.

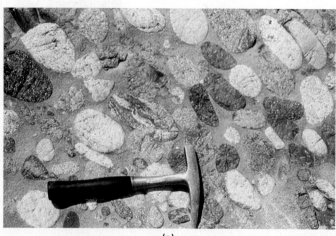

(a)

(b)

Inorganic Chemical Sedimentary Rocks

Inorganic chemical sedimentary rocks form when the dissolved products of chemical weathering precipitate from solution. Typically precipitation occurs when the water in which they are dissolved evaporates or undergoes a significant temperature change. Four common types of inorganic chemical-sedimentary rocks form by varying processes: *inorganic limestones* and *cherts* precipitate directly from both seawater and freshwater; *evaporites* precipitate when ion-rich water (freshwater or seawater) evaporates; *dolostone* is a rock whose origin remains the subject of much debate.

Limestones, which are composed largely of calcium carbonate ($CaCO_3$), account for 10% to 15% of all sedimentary rocks. Most limestones forming today are biogenic. Under certain conditions, however, limestone-forming sediments also precipitate inorganically, directly from calcium-rich water. Here's how: Soluble materials usually dissolve more rapidly as water temperature increases—hot coffee, for example, dissolves sugar more quickly than cold. The solubility of calcium carbonate, on the other hand, is directly proportional to the amount of carbon dioxide (CO_2) in the water, and warm water typically holds less CO_2 in solution than does cold water. In general, as water warms and the proportion of CO_2 decreases, calcium carbonate in the water becomes *less* soluble. Thus, *more* calcium carbonate precipitates, taking the form of *inorganic limestone.* On the other hand, as water cools and the proportion of CO_2 in it increases, *more* calcium carbonate dissolves and *less* inorganic limestone precipitates.

The amount of $CaCO_3$ that remains in solution is also affected by factors other than temperature (Fig. 6-16). Agitation of the water, the presence of photosynthesizing plants, and water depth and pressure also determine if $CaCO_3$ will precipitate or dissolve. When water is stirred up or agitated, as by wave action, CO_2 escapes to the atmosphere and calcium carbonate therefore tends to precipitate. Aquatic plants remove CO_2 from water during photosynthesis, also promoting precipitation of calcium carbonate. A decrease in water pressure allows CO_2 to escape from the water into the atmosphere; because water pressure increases with depth, more calcium carbonate precipitation tends to occur in shallow water than in deep water.

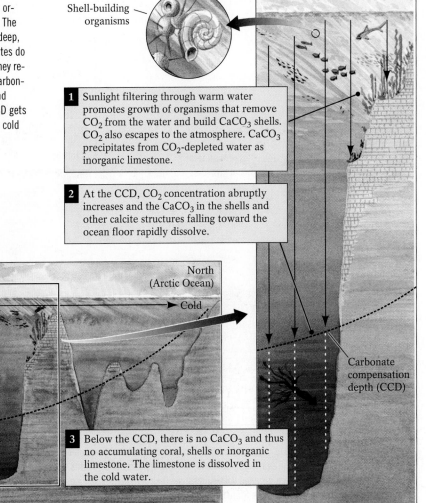

▼ **FIGURE 6-16** The role of CO_2 in $CaCO_3$ dissolution and precipitation. The optimal conditions for precipitating $CaCO_3$ include warm, shallow, agitated water rich in organic nutrients and vegetation, all of which reduces the volume of CO_2 in water. The optimal conditions for dissolving $CaCO_3$ and holding it in solution include cold, deep, still water, devoid of organic nutrients and vegetation. Here we see that carbonates do not accumulate in deep, cold water below about 4 km (2.5 mi) of water depth. They remain dissolved in the cold carbon dioxide–rich water. This depth is called the Carbonate Compensation Depth (CCD). Above this depth, carbonates can precipitate and accumulate as they are doing so atop the undersea mountains. Note that the CCD gets shallow toward the icy polar regions because at that point even shallow water is cold enough to dissolve carbonates.

Shell-building organisms

1 Sunlight filtering through warm water promotes growth of organisms that remove CO_2 from the water and build $CaCO_3$ shells. CO_2 also escapes to the atmosphere. $CaCO_3$ precipitates from CO_2-depleted water as inorganic limestone.

2 At the CCD, CO_2 concentration abruptly increases and the $CaCO_3$ in the shells and other calcite structures falling toward the ocean floor rapidly dissolve.

3 Below the CCD, there is no $CaCO_3$ and thus no accumulating coral, shells or inorganic limestone. The limestone is dissolved in the cold water.

South (Antarctica) Equator North (Arctic Ocean)

Sea level

Cold Warm Cold

Carbonate compensation depth (CCD) Undersea mountains

Depth (km)

Carbonate compensation depth (CCD)

Inorganic limestone precipitates most readily when several of these factors act together. The Grand Bahamas Banks, for example, is a *shallow* submarine platform separated from Florida by the Straits of Florida. Here relatively pure carbonate muds accumulate as carbonate-rich marine waters wash across the broad continental shelves within 30° of the equator. The conditions—a combination of warm water temperatures, active waves and currents, shallow water, and abundant marine plant life—act to remove CO_2 from solution. The resulting calcium-carbonate precipitation coats sand grains or other particles on the seafloor. Successive coats of calcite form concentric layers around the growing grains as tidal currents roll them back and forth along the ocean floor (Fig. 6-17). The result is a bed of spheres called *ooliths* (from the Greek *oo,* or "egg," a reference to the sediment's resemblance to fish eggs) about 2 mm in diameter. The chemical sedimentary rock formed from this sediment is called *oolitic limestone.* We find great outcrops of oolitic limestone in the states of Indiana, Kentucky, and Ohio—there's even a town called Oolitic, Indiana, 110 km (70 mi) south of Indianapolis. These rocks remain from a warm, shallow inland sea that covered Indiana and much of the Midwest about 350 million years ago.

Inorganic limestone can also form on land. *Tufa* (too-fah) is a soft, spongy inorganic limestone that accumulates where underground water emerges at the surface as a natural spring. At the surface, water encounters a low-pressure environment, warms in the sunshine, and bubbles out in an agitated fashion. These factors promote the loss of CO_2, hastening carbonate precipitation. Inorganic limestone, in the form of travertine (limestone deposited by groundwater) forms both in natural hotsprings, such as those at Yellowstone National Park (Fig. 6-18), and by precipitation from groundwater in caves. The travertine in caves (Fig. 6-19) typically accumulates as droplets of carbonate-rich water on the ceilings, walls, and floors lose CO_2 to the cave atmosphere and precipitate carbonate rock.

Evaporites are inorganic chemical sedimentary deposits that accumulate when salty water evaporates (Fig. 6-20). The world's seawater contains, by volume, almost 3.5% dissolved salts. Where marine water is shallow and the climate is warm, evaporation increases the concentration of these salts. When evaporation exceeds the inflow of water into the marine basin, solid crystals precipitate and accumulate at the sea bottom.

Evaporites precipitate from seawater in an orderly sequence: The least soluble crystallize first and the most soluble stay in

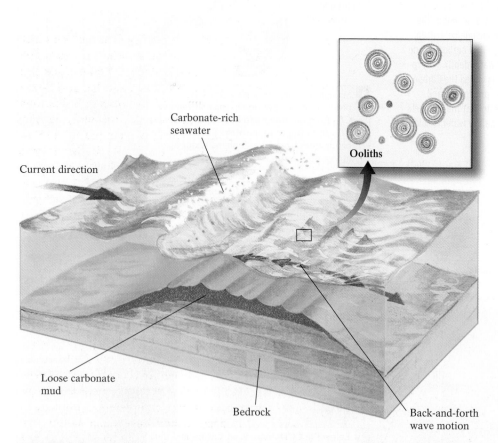

▲ **FIGURE 6-17** The formation of ooliths on a tropical carbonate platform. Calcium carbonate precipitates and coats sand grains as they are rolled along the seafloor by currents, forming spheres called ooliths (shown above right). The resulting inorganic chemical sedimentary rock is called oolitic limestone (below right). These limestones have been quarried and shaped to form the building blocks of the Empire State Building in New York and the Holocaust Museum and National Cathedral in Washington, D.C.

Labels in figure:
Carbonate-rich seawater
Current direction
Ooliths
Loose carbonate mud
Bedrock
Back-and-forth wave motion

▲ FIGURE 6-18 Travertine cliffs in Yellowstone National Park. These calcite deposits precipitate inorganically from hot water.

solution until little water remains. Relatively insoluble gypsum ($CaSO_4 \cdot 2\,H_2O$), for example, begins to precipitate when roughly two-thirds of a volume of saltwater has evaporated. More soluble salts, such as halite (NaCl), require about 90% evaporation. We rarely see the most soluble salts, such as sylvite (KCl) and magnesium chloride ($MgCl_2$), in evaporites; they precipitate only after more than 99% of the water has dried up.

Because evaporites dissolve so readily in water, we rarely find them at the surface in moist climates, such as in the United States' rainy Northwest or humid Southeast. You can see gypsum at the surface in Colorado, Wyoming, and the arid Southwest, where it forms the dunes of White Sands National Monument, New Mexico. Subsurface salt deposits underlie many areas of the continental United States (for example, in the East, Midwest, Great Plains, Rocky Mountains, and Southwest), wherever shallow inland seas and warmer climates existed in the past (Fig. 6-21). The salt deposits in central Michigan, for example, are more than 750 m (2475 ft) thick.

If today's seawaters average only 3.5% salt by volume, how did such great bodies of salt accumulate from Michigan's long-departed seas? Were those seas much saltier than the Earth's oceans today? If, for example, 300 m (1000 ft) of average seawater evaporated today, only 10.5 m (35 ft) of salt would remain. To produce the volume of salt now found in Michigan, an ocean about 22 km (13 mi) deep with today's salt content would have had to evaporate completely. Although ancient oceans may have been saltier than oceans today, no *single* ocean is likely to have contained enough salt at one time to produce this thick deposit. Michigan's salt probably resulted from *long-term* precipitation of gypsum and rock salt (a rock composed largely

▲ FIGURE 6-19 Subterranean calcite deposits from the Lehman Caves in Great Basin National Park, Nevada.

▶ **FIGURE 6-20** Evaporite deposits at the Bonneville Salt Flats, west of Salt Lake City, Utah. The modern-day Great Salt Lake is a small remnant of Lake Bonneville, a vast lake that existed in Utah about 15,000 years ago when the local climate was cooler, cloudier, and more humid. A warm, clear, dry climate evaporated most of Lake Bonneville, leaving these salt deposits.

▶ **FIGURE 6-21** The locations of major subsurface evaporite deposits in North America. These deposits were produced in the ancient past when salty marine water invaded topographic low spots on the North American continent during times of high sea level. These shallow seas later evaporated during periods of climatic warming, leaving behind thick evaporite deposits. Four hundred million years ago, evaporating seas occupying what is now Michigan, Ohio, West Virginia, Pennsylvania, New York, and Ontario produced great deposits of gypsum and rock salt. Younger sedimentary rocks have covered most of these deposits.

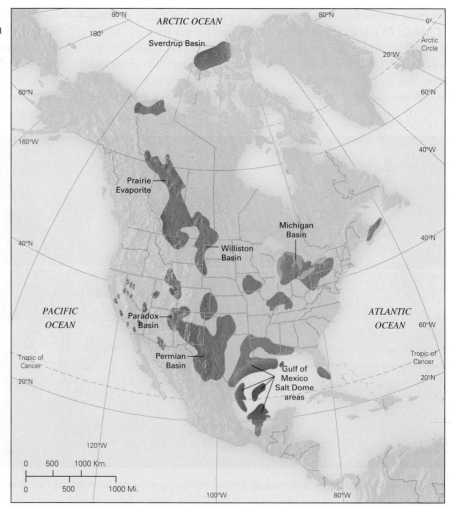

Highlight 6-2

When the Mediterranean Sea Was a Desert

The climate in the lands bordering the Mediterranean Sea, such as Morocco, Algeria, Libya, and Greece, can be oppressively hot. Every year this heat evaporates more than 4000 km^3 (960 mi^3) of the Mediterranean's water, with only 400 km^3 (96 mi^3) being replaced by rainfall and inflowing rivers. If these two sources were the only means of maintaining its water level, the Mediterranean would be a dried-up lowland between northern Africa and southern Europe in just 1000 years. Today, a massive inflow of Atlantic Ocean water through the Strait of Gibraltar balances the Mediterranean's water deficit. This balance has not always been present, however.

Deep-sea exploration in the early 1970s revealed a massive evaporite layer more than 2000 m (6600 ft) thick beneath the floor of the Mediterranean Sea. This layer is more than 25 times the thickness that would accumulate if the current Mediterranean evaporated completely to dryness. What past conditions could have produced such a great salt accumulation?

One hypothesis proposes that these deposits formed when a barrier restricted water circulation between the Atlantic Ocean and the Mediterranean Sea (see figure). According to this hypothesis, about 8 million years ago, ongoing convergence between the African and Eurasian plates, which meet beneath the Mediterranean, gradually raised the seafloor in the area of what is now the Strait of Gibraltar. The natural limestone dam created by this movement increasingly blocked the inflow of Atlantic waters. As the replenishing water supply diminished, the Mediterranean rapidly evaporated. With less water to moderate the region's semitropical heat, daytime temperatures may have risen to 65°C (150°F), fur-

ther hastening evaporation and producing desertlike conditions. Occasional pulses of Atlantic water over or through the Gibraltar barrier, like a leaky faucet, may have provided the then-drying basin with a periodic supply of salty water that evaporated and added to the thick Mediterranean salt deposits.

Analysis of sediment samples taken from the Mediterranean's floor indicates that deposition of the evaporites ceased 5.5 million years ago, perhaps when a break in the limestone dam expanded dramatically, boosting the trickle of Atlantic water into a spectacular waterfall (20-km [13 mi] wide) spanning what is now the Strait of Gibraltar. Such a force would have removed the barrier completely by 4 million years ago. A flow 100 times the volume of Niagara Falls would have been required just to balance regional water loss by evaporation and begin to refill the empty Mediterranean basin. To provide enough water to support the sea life of the time (whose extent and makeup are revealed in the fossil record), a flow in excess of 1000 Niagara Falls would have been required! Even at that flow rate, it would have taken more than a century to refill the basin to its present level. In contributing the water that refilled the Mediterranean, the other oceans of the world probably fell about 10 m (35 ft) over that time.

The Rock of Gibraltar is a remnant of the ancient limestone dam, a reminder of the time when an intermittently dry Mediterranean basin separated Europe and Africa. Today, although the heat of the region causes much evaporation, the Mediterranean fails to reach the point of gypsum deposition, which would require about two-thirds of the sea to evaporate. If the African and Eurasian plates continue to converge, however, as they do today, a *new* rocky barrier to Atlantic-water inflow may once again rise from the seafloor and the Mediterranean basin may again dry out, fill with a great thickness of salt, and return the region to uninhabitable climatic conditions of extreme heat and aridity.

▲ The thick evaporite deposits that underlie much of the Mediterranean basin today accumulated during a time (8–5.5 million years ago) when a topographic barrier stretched across what is now the Strait of Gibraltar, cutting off the water influx from the Atlantic Ocean. With no water supply to replenish it, the Mediterranean Sea experienced prolonged periods of evaporation lasting several million years, precipitating vast amounts of salt.

of halite), either from a succession of ancient shallow inland seas that periodically evaporated and then were refilled, or from continuous *partial* evaporation and refilling of a single long-lasting inland sea (evaporating to about 90% dryness, enough to precipitate gypsum and halite but not highly soluble salts such as sylvite). In at least one case, an entire ocean apparently did evaporate completely as recounted in Highlight 6-2.

Apparently, the accumulation of salty evaporites is not restricted to the Earth's ancient evaporating seas. In 2004, both Martian rovers—*Spirit* and *Opportunity*—roving independently on opposite sides of the planet, discovered large amounts of crystallized salt in Martian rocks. These findings are yet another signal that the Martian landscape was watery at some time in its past.

Dolostone is rock composed of the mineral *dolomite*, a calcium-and-magnesium carbonate [$CaMg(CO_3)_2$] similar to limestone in appearance and chemical structure. The origin of dolostone has inspired lively debate for decades. As yet, no strong consensus favors any single mechanism for its origin.

For many years, geologists believed that most dolostone formed simply when magnesium ions replaced calcium ions in a preexisting body of limestone. This substitution could happen when ocean water evaporated enough to produce a magnesium-rich solution that then circulated through a limestone bed (Fig. 6-22). This hypothesis predicted dolostone production to occur wherever a tropical or semitropical climate promoted evaporation of salty marine water near a porous body of limestone. But could this process really explain the great thicknesses of dolostone found today in places such as Platteville, Wisconsin, Chicago, Illinois, and near Rolla, Missouri? Were these places once located along the shores of tropical inland seas? And why

is so little dolostone forming today, even where these ideal environmental conditions currently exist? Certainly, the rock's ingredients—calcium, magnesium, and carbonate ions—are common enough in seawater. Perhaps other factors, more prevalent in the past, promoted far greater dolostone accumulation than occurs in today's oceans.

Chert is the general name for a group of sedimentary rocks that consist largely of silica (SiO_2) and whose crystals are so small, they can be seen only through a microscope. Although many cherts have biogenic origins, the most common are inorganic, forming as a chemical precipitate from silica-rich water. *Chert nodules,* fist-size masses of silica that commonly occur in bodies of limestone and dolostone, are believed to form when portions of these rocks are dissolved by circulating groundwater and replaced by precipitated silica. Because chert is easy to chip and forms sharp edges, early humans often shaped it into weapons, cutting blades, and other tools (Fig. 6-23). Some archaeologists are forever searching for the sources of chert tools that they find in their excavations. By identifying chert sources, they can learn much about possible trade activity between groups of people from different regions (some with chert ..., some without) and about the ways in which ancient people exploited the natural resources in their regions.

Biogenic Chemical-Sedimentary Rocks
Biogenic chemical-sedimentary rocks are derived from living organisms. The principal rocks of this type are *biogenic limestones* and *cherts,* which are composed largely of the skeletal remains of marine animals and plants, and *coal,* which formed from the compressed remains of terrestrial plants. In each case, biogenic chemical

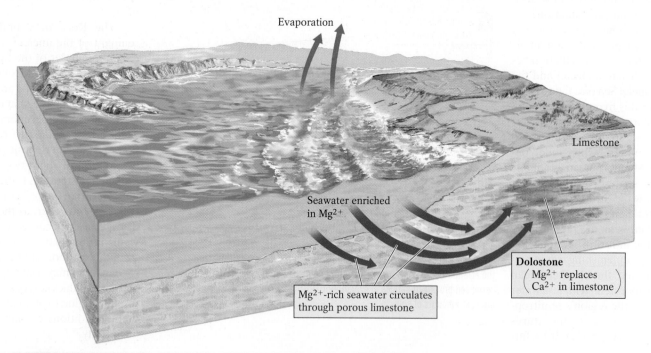

▲ **FIGURE 6-22** One hypothesis for the formation of dolostone suggests that seawater enriched in magnesium ions (which increase in concentration as the water evaporates) circulates through porous limestone; dolostone then forms as the magnesium ions replace calcium ions in the limestone.

▲ **FIGURE 6-23** Stone axes made from chert 150,000 years ago. Because of the sharp edge that forms when chert is chipped, our human ancestors constructed many of their tools from this hard, silica-rich rock. These axes are about 5.5 in. (133 mm) long, and 2.5 in. (63 mm) wide.

(a)

(b)

▲ **FIGURE 6-24** Chalk is a shallow-marine limestone formed from the carbonate secretions of countless microscopic organisms. Accumulation of chalk can eventually produce deposits of impressive size, such as the famed White Cliffs of Dover in England [see photo **(b)**]. The cliffs are composed mainly of the shells of microscopic marine plants and animals [photo **(a)**] that accumulated 100 million years ago, when the global sea level was apparently higher and coastal England was underwater.

sediments undergo the same diagenetic processes of compaction and cementation that produce detrital sedimentary rocks.

Nearly all biogenic limestones consist of calcite ($CaCO_3$) that formed in a warm marine environment. Warm seawater is nearly saturated with calcium ions that have been chemically weathered from calcium-rich rocks on land and then transported to the sea by rivers and streams. Numerous ocean-dwelling organisms extract the ions to form their calcium-carbonate shells and internal skeletons. When these organisms die, their hard parts settle to the seafloor, where they accumulate in great thicknesses and then lithify as clastic biogenic limestone.

Most biogenic limestones form in shallow water along the continental shelves of equatorial landmasses where the presence of warm water, plentiful sunlight, and abundant nutrients enable marine life to flourish. For example, an abundance of microscopic algae that secrete needlelike calcium-carbonate crystals inhabit the waters around most Caribbean islands, the Florida Keys, and the east coast of Australia. The deaths of these organisms produce a rain of delicate calcite stalks and branches that falls to the shelf floor and accumulates to produce a calcite-rich mud. These particles are so small that 600,000 laid end to end would span this line of type. Other algae and tiny carbonate-secreting animals living within the mud add to the growing volume of carbonate sediment, which comprises a soft white biogenic deposit called *chalk* (Fig. 6-24).

Biogenic limestone can also form in deep-marine environments. Although few carbonate-secreting organisms live at the cold, lightless, deep-sea floor, microscopic animals called *foraminifera,* or "forams," swim and float in an ocean's upper 50 m (160 ft), which is warmer and sunlit (Fig. 6-25). When the forams die, their protective shells accumulate on the seafloor, where they mix with marine mud to form a deep-sea *ooze.* Burial of an ooze by subsequent sedimentation causes its clays and carbonates to become cemented as biogenic limestone. When an

ocean is very deep—deeper than about 4 km (2.5 mi)—the water down there is so cold that the carbonate-based forams dissolve rather than accumulate because of the higher CO_2 content in cold water. The noticeable absence of limestones from thick sequences of deep-marine sedimentary rocks suggests that the rocks were deposited in a very deep (very cold) oceanic environment.

Biogenic chert is a chemical sedimentary rock composed mainly of silica derived from the remains of a variety of marine

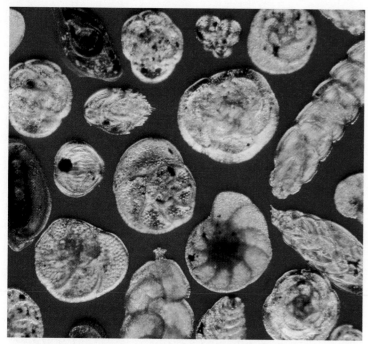

▲ **FIGURE 6-25** Foraminifera, microscopic marine animals whose calcium-carbonate shells are an important component of deep-marine limestone (magnified 115×).

▲ **FIGURE 6-26** Biogenic chert, found in the form of layered beds, develops from accumulation and lithification of the silica-based remains of marine organisms.

organisms. These siliceous remains may also form an ooze. Chert that takes the form of layered beds (rather than the nodules characteristic of inorganic cherts) is believed to be of biogenic origin (Fig. 6-26). Microscopic examination of such biogenic cherts typically reveals silica-based organic debris, such as the shells of single-celled animals called *radiolaria,* the internal structures of single-celled plants called *diatoms,* and the skeletons of larger, more complex animals such as marine sponges. One variety of chert, known as *flint,* gets its dark-gray-to-black color from the presence of carbon-rich matter.

Coal is a biogenic sedimentary rock composed largely of plant remains. You can often see original plant structures, such as fragments of leaves, bark, wood, and pollen, in a lump of coal under magnification, and sometimes even with the naked eye. Although vegetation generally decomposes quickly at the Earth's surface, in an environment with little oxygen (and thus little oxidation) and rapid sedimentation (and thus rapid burial), it may be preserved until it's converted to coal. The stagnant water of a warm, lushly vegetated swamp is ideal for the production of coal. The windless conditions often found in swamps produce calm water surfaces, and relatively little atmospheric oxygen circulates down to reach the organic debris accumulating at the swamp's bottom. Plant decay by oxidation quickly depletes any oxygen dissolved in the swamp water, leaving the remaining vegetation to accumulate undecayed.

As time passes, increasing pressure from the weight of overlying sediments expels water, CO_2, and other gases from the accumulating mass of vegetation. This process increases the percentage of carbon in the plant residue. Early in the process, when much of the original plant structure remains intact, a soft brown material called *peat* is produced. Over time, increased heat and pressure create ever harder and more compact forms of coal, ranging from soft brown *lignite* to moderately hard *bituminous* coal. Ultimately, heat and pressure may convert coal, a sedimentary rock, into dense, lustrous *anthracite*—a metamorphic rock (Fig. 6-27).

Rising seas generally create and sustain swamps along coasts. When sea levels rise, the swamps may be submerged. Plant debris then ceases to accumulate and is buried by sediments from the marine environment. Falling sea levels restore the swamp, allowing accumulation of plant debris to resume. The cyclical nature of deposition associated with rising and falling sea levels typically produces alternating coal beds and detrital sedimentary rock layers.

Much of North America's coal developed during two periods in its geologic past: approximately 280 million years ago, forming the coal in the Appalachians of Pennsylvania and West Virginia, and nearly 75 million years ago, producing the coal found today in the Rockies and throughout the plains of Montana, Wyoming, North Dakota, and Saskatchewan. Where are North America's *future* coal deposits forming today? The warm, richly vegetated candidates include the Dismal Swamp of coastal North Carolina, the barrens of the Florida Everglades, and the colorful bayous of Louisiana. All of the details about the wonderful array of sedi-

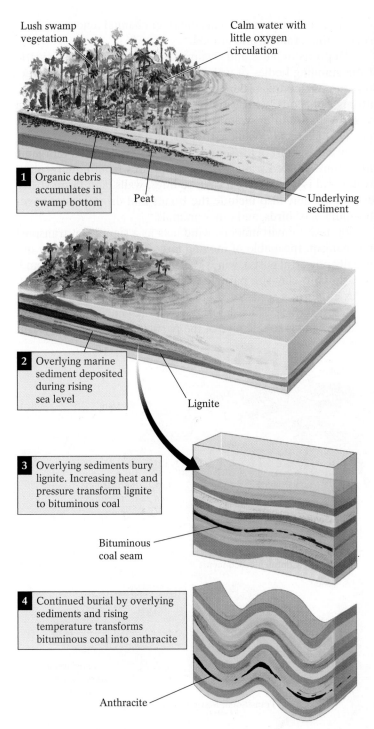

1 Organic debris accumulates in swamp bottom

Lush swamp vegetation

Calm water with little oxygen circulation

Peat

Underlying sediment

2 Overlying marine sediment deposited during rising sea level

Lignite

3 Overlying sediments bury lignite. Increasing heat and pressure transform lignite to bituminous coal

Bituminous coal seam

4 Continued burial by overlying sediments and rising temperature transforms bituminous coal into anthracite

Anthracite

▲ **FIGURE 6-27** The formation of coal from swamp deposits. Abundant organic debris accumulates at the bottom of a swamp and becomes buried before it can decay in oxygen-poor swamp water. The weight of subsequent deposits and the increased temperatures at greater depths transform the debris to progressively harder forms of coal. Above: A body of coal (a "coal seam") in central Utah and a modern coal-producing environment (in the Florida Everglades east of Fort Myers, Florida).

mentary rocks, their minerals and their properties, are summarized in a Geology at a Glance feature on page 193.

"Reading" Sedimentary Rocks

Many sedimentologists spend much of their time at beaches, along stream banks and lakeshores, and hiking through deserts and atop glaciers. They do so to study modern depositional environments and the processes that create the various sedimentary structures (such as mudcracks, cross-bedding, ripplemarks) they see forming today, and to catalogue the life forms that inhabit these places. They

can then analyze ancient sedimentary-rock formations, sedimentary structures, study their fossils, and determine the depositional environment that produced each formation. Ultimately, it is possible to deduce much of a region's unique geologic history—including the sequence of its major geological events, such as the rise and fall of sea levels, and possibly even past plate-tectonic activity.

Depositional Environments

Where can we find sediment deposits? At virtually any spot on the Earth's surface—from atop the highest peak (pushed there by powerful plate collisions) to the depths of the deepest ocean (dragged down there by subducting oceanic plates). Figure 6-28 shows just a small sample of the diversity of depositional environments. **Sedimentary environments** may be *continental* (on a landmass), *marine* (at sea), or *transitional* (in the zone in-between). In this section, we will discuss the principal characteristics of these three categories; in later chapters, we will examine individual environments—such as streams, glaciers, deserts, and coasts—in greater detail.

Continental Environments
Sedimentation in continental environments is mostly detrital. Sediment layers contain numerous indicators of past water-, wind-, and glacier-flow directions, and plant and freshwater fossils abound. Rivers, lakes, caves, deserts, and glaciers are all continental depositional environments, and as you will see, they produce sedimentary rocks that are highly variable in texture, structure, and composition.

Swift river currents carry and then deposit coarse-grained sediments (sand and gravel), forming rippled and cross-bedded structures. Sediments that settle from "standing" floodwaters on floodplains after they've dropped their initial high-energy load of coarse sediment in and near the river channel tend to be well-sorted, fine-grained, and graded.

Deposits in lakes and swamps, where sediments also settle from standing bodies of water, share some of the same characteristics as floodplain deposits. Lake deposits often contain diatoms (siliceous algae) and other organic matter. The lush vegetation found in swamps may appear in their sediment deposits as peat and organic-rich mud.

In caves, most of the sedimentary rock is chemical. Inorganic limestone—called travertine—precipitates from underground water and is deposited on cave ceilings, walls, and floors. Cave sediments may also include the bones and droppings of cave-dwelling bats, birds, and other animals.

In desert environments, wind acts as a significant transport mechanism. Incapable of moving large particles (except by unusually powerful gusts), winds typically carry aloft only silt- and, occasionally, clay-size particles. They transport larger sand particles either in a series of short jumps or by rolling them along the desert surface. Thus, desert deposits are generally well-sorted because wind is so selective of the particle sizes it can carry. But

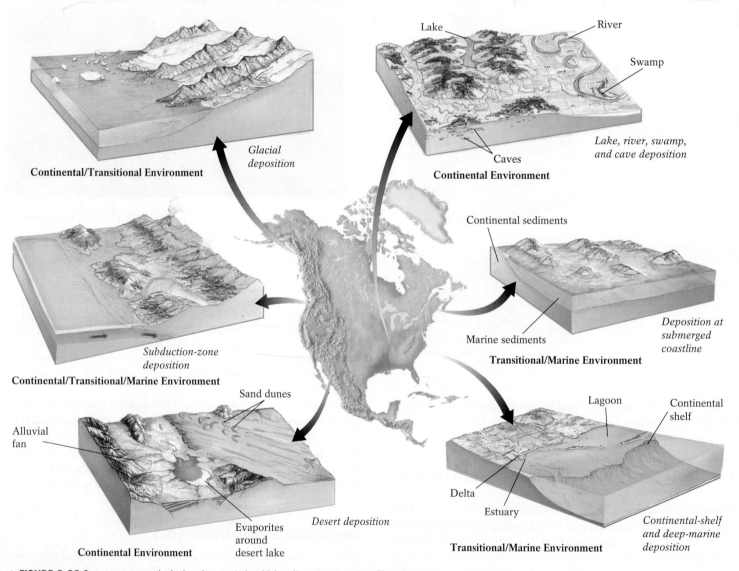

▲ **FIGURE 6-28** Some common geological environments in which sediments accumulate. These sediments may ultimately form sedimentary rocks.

6.1 Geology at a Glance

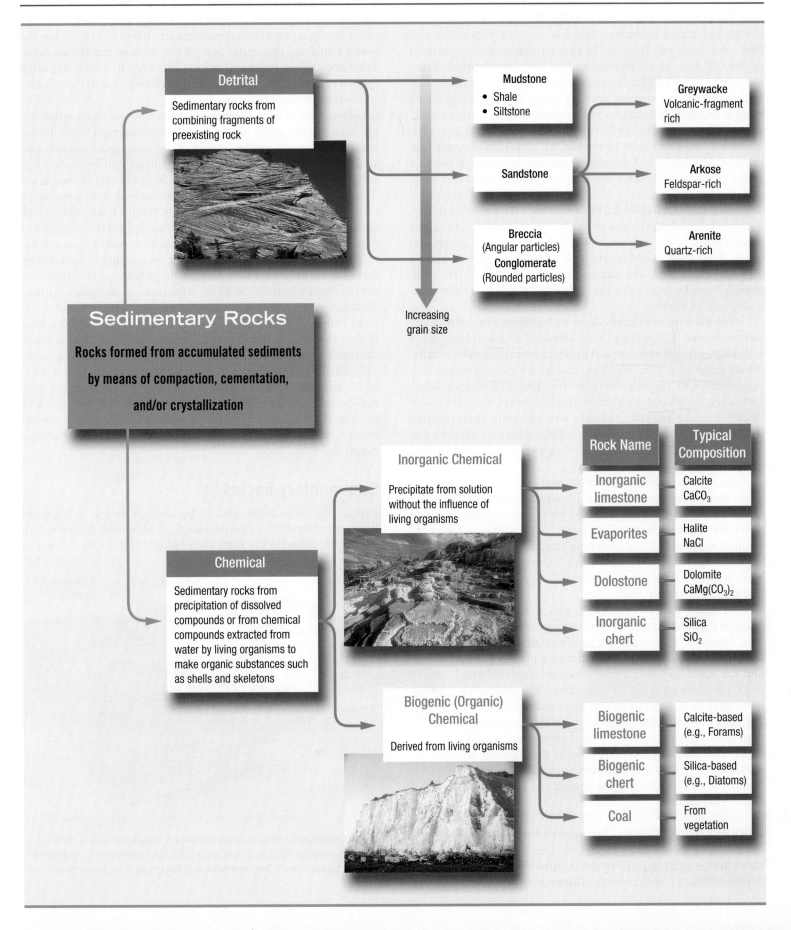

Detrital

Sedimentary rocks from combining fragments of preexisting rock

Mudstone
- Shale
- Siltstone

Sandstone

Breccia (Angular particles)
Conglomerate (Rounded particles)

Increasing grain size

Greywacke
Volcanic-fragment rich

Arkose
Feldspar-rich

Arenite
Quartz-rich

Sedimentary Rocks

Rocks formed from accumulated sediments by means of compaction, cementation, and/or crystallization

Inorganic Chemical

Precipitate from solution without the influence of living organisms

Chemical

Sedimentary rocks from precipitation of dissolved compounds or from chemical compounds extracted from water by living organisms to make organic substances such as shells and skeletons

Rock Name	Typical Composition
Inorganic limestone	Calcite $CaCO_3$
Evaporites	Halite $NaCl$
Dolostone	Dolomite $CaMg(CO_3)_2$
Inorganic chert	Silica SiO_2

Biogenic (Organic) Chemical

Derived from living organisms

Biogenic limestone	Calcite-based (e.g., Forams)
Biogenic chert	Silica-based (e.g., Diatoms)
Coal	From vegetation

where steep mountain streams abruptly drop their loads onto flat desert floors, coarse, poorly sorted, poorly stratified, alluvial-fan deposits form. Evaporites also occur in deserts where the extreme heat has evaporated temporary bodies of water.

In a glaciated landscape, deposits vary in composition, texture, and structure. Because flowing ice can transport particles of virtually any size, its deposits are typically poorly sorted. As the ice melts, it drops its load in an unstratified heap. Sediments carried by meltwater streams and deposited beyond the glacier's margin are generally coarse, well-rounded, and relatively well-sorted. These streams' swift currents can carry large particles, and their turbulent flow promotes the forceful collisions that round sediment grains. Glacial lakes are still, however, so their deposits are characteristically fine-grained, graded, and well-sorted.

Transitional (Coastal) Environments

Along ocean shores, continental- and marine-sedimentary processes merge. Breaking waves, tides, and ocean currents crush soft mineral grains and shells. They then sweep fine particles out to sea, leaving behind a beach, most often marked by deposits of well-sorted, well-rounded, sand-size, durable mineral grains of relatively hard minerals, such as quartz.

When a river enters an ocean, its freshwater mixes with the salty seawater. This mixing forms a body of *brackish* (somewhat salty) water called an *estuary*. If the estuary water is not too salty, it may support marine, brackish, and freshwater organisms that then contribute biogenic debris to its sediment. Estuaries sometimes contain *deltas,* fan-shaped accumulations of well-sorted sediment formed when river currents slow suddenly upon entering the sea. Coastal deltas grow seaward if sediment is deposited at a faster rate than it is removed by waves, tides, and coastal currents.

Sometimes sediment of continental origin is deposited offshore to form narrow islands that lie parallel to the coastline. The shallow body of water between the coast and an offshore island is another sedimentary environment, called a *lagoon*. Fine continental sediments from inflowing streams as well as peat and organic mud may accumulate there. Storm surges that wash seawater over the narrow island may also dump marine sediment into the lagoon.

All active subduction zones occur in coastal areas because the subducting plate is always an oceanic plate. Vast amounts of sediment are produced in subduction zones, as explosive volcanoes produce tephra deposits and the volcanic cones are attacked by chemical and mechanical weathering. At the same time, powerful earthquakes shake the mountains, dislodging great quantities of loose sediment. Some of this sediment is deposited at the shore, carried there by rivers and mudflows. Turbidity currents carry some of these sediments offshore where they may become caught up at the plate boundary and bulldozed back toward the land by the converging plates.

Marine Environments

Marine environments vary according to their depth (Fig. 6-29). The *shallow-marine* environment lies above the continental shelf, at depths of 200 m (700 ft) or less. Although a continental

shelf may extend several hundred kilometers into the sea, most are much narrower, with some being less than 1 km (0.6 mi) wide. This narrow zone bordering all of the world's continents receives land-derived sediment carried seaward by waves and tides. Because sunlight penetrates approximately 100 m (330 ft) below the water's surface, the upper part of the shallow-marine environment abounds in plant and animal life. Thus its sediments often consist of carbonate-rich sands and muds mixed with the remains of diverse marine life forms.

The *deep-marine* environment lies beyond the continental shelf, beginning at the foot of the steep continental slope. In this region, sediments consist largely of the remains of calcium carbonate- and silica-secreting microorganisms that have died and settled to the bottom from the upper 50 m (165 ft) of the ocean. They also contain red and brown clays derived from weathering of windblown silts from the continents, tephra from continental, ocean-island, and submarine volcanoes, land-derived deposits carried to the deep-sea floor by submarine landslides, and meteoritic fragments from outer space. These sediments contain few large fossils because too little sunlight penetrates this region to support many bottom-dwelling organisms. In the cold, lightless conditions below a depth of about 4 km (2.5 mi), carbonates dissolve rather than being deposited. Thus, as we discussed earlier in this chapter, in the absence of carbonates, silica-rich muds abound on the floors of very deep ocean basins. These sediments accumulating at the floors of the Earth's oceans are perhaps the best source of information about the Earth's most recent 200-million-year history. Highlight 6-3 on pages 196–197 explains how geologists probe ocean-floor mud for clues about global climate change, tectonic movements, mass extinctions, and great volcanic eruptions.

Sedimentary Facies

Just as sediments deposited in the same place but at different times are recorded as a *vertical* stack of distinct rock layers, sediments deposited at the same time but in different places appear

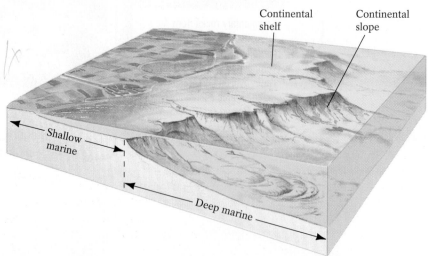

▲ **FIGURE 6-29** Marine sedimentary environments are described as being either shallow marine or deep marine. Shallow-marine environments are present above continental shelves and the shallow platforms that surround many oceanic islands. Deep-marine environments begin at the foot of the continental slope.

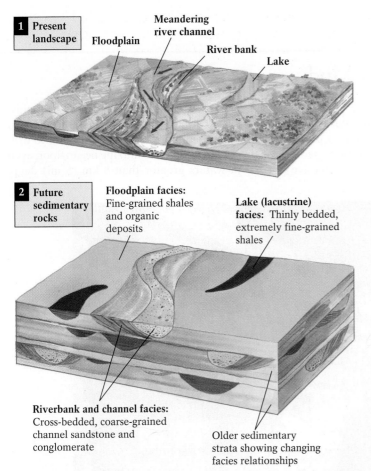

1 Present landscape

Floodplain
Meandering river channel
River bank
Lake

2 Future sedimentary rocks

Floodplain facies: Fine-grained shales and organic deposits

Lake (lacustrine) facies: Thinly bedded, extremely fine-grained shales

Riverbank and channel facies: Cross-bedded, coarse-grained channel sandstone and conglomerate

Older sedimentary strata showing changing facies relationships

▲ **FIGURE 6-30** An example of sedimentary-facies formation. The hypothetical modern river shown in **(1)** is surrounded by a lake-studded floodplain. If we could freeze this scene and convert today's sediments to their future horizontal sedimentary-rock layer **(2)**, different types of sedimentary rocks would represent the river channel, the riverbank, the floodplain, and the lakes. Although these rocks are closely spaced geographically and formed at the same time, they would differ in composition and appearance because they were deposited in four different sedimentary settings. The changing relationships of facies over time, such as when a meandering river alters its course, can be seen by examining the *vertical* succession of rock layers over a broad area.

as a *horizontal* continuum of distinct rock types. The set of characteristics—such as mineral content and particle size, shape, and sorting—that distinguish one sedimentary-rock formation from neighboring formations *deposited at the same time* is called a **sedimentary facies** (*facies* comes from the Latin for "form" or "aspect"). The nature of a sedimentary facies depends on the particular conditions under which it was deposited. Thus the pattern of facies in a given rock layer (which represents a single time period) reflects the different environments that existed across the landscape when the sediments were deposited.

This is a difficult concept, so let's explain it further. Think of, for example, the different sediments associated with a modern coastal river. If they were instantly lithified—turned to stone—each distinct depositional environment each with its distinctive sediments and sedimentary structures would appear as a separate facies (Fig. 6-30). Coarse, cross-bedded sandstones and conglomerates would record the flowing water in the river's channel. Fine shales and organic deposits would mark the river's floodplain. Thinly bedded, graded shales, perhaps containing fossils,

would be left behind by floodplain lakes. Remember—all of these deposits accumulated *at the same time,* but under different *local* sedimentary conditions. Thus at some time in the distant future, the geological record of this modern river environment would include channel *facies,* floodplain *facies,* and lacustrine ("lake") *facies*—all of the same age, but each located at a different position within this hypothetical sedimentary-rock layer associated with the early 21st century.

As we saw earlier, the vertical succession of rock layers at a specific location shows how depositional conditions at that spot changed over time. Similarly, the vertical succession of rock layers encompassing various sedimentary environments would reveal how these environments changed *in relation to one another* over time. For example, if the course of a river shifted over time, its channel facies might be succeeded by its floodplain facies. This shift would appear in the rock record as a noticeable change in the nature of the sediment—from coarse and cross-bedded to fine and thinly bedded (Fig. 6-30).

Shifting Sedimentary Facies Sedimentary environments move as bodies of water dry up, climate warms or cools, sea levels rise and fall, and so on. These changes alter the particle size and composition of the deposited sediments. After lithification, these sediments then appear as distinctive facies in the rock record.

In Figure 6-31, note the position of the river, beach, shallow-marine, and deep-marine environments at time A. If sea level were to rise, the shoreline would migrate inland and a marine environment would replace the river environment. Gradually, a coastal beach would replace the river's original floodplain, a shallow-marine environment would replace the old beach, and a deep-marine environment would replace the former shallow-marine environment. By time B, these environments would in turn be replaced by even deeper marine environments. The sediments deposited at any given location at time B would differ from those deposited at the same location at time A, reflecting the new environment of deposition. Such shifting sedimentary facies preserved in a vertical sequence of sedimentary rock can provide clues to changes in ancient environmental conditions.

Sedimentary Rocks and Plate Tectonics

Some of the Earth's most majestic mountains—the Rockies, the European Alps, and the Himalaya of China, Nepal, and Tibet—contain sedimentary rock strata that were clearly deposited under water. Within the sedimentary rocks of these mountains, we can easily see the ripple marks that record rising and falling tides and coastal waves, evaporites and mudcracks from past sea-level fluctuations, and thousands of meters of seafloor limestone. These rocks lithified from sediments that accumulated in shallow-marine environments. They ended up in mountains when plate-tectonic forces uplifted them during continental collisions.

Sedimentary rocks also provide evidence that lofty mountains once stood where no such range exists today. The squat Taconic Hills of western Connecticut and Massachusetts and eastern New York, for example, are thought to be the remnants of mountains that formed when the plates containing what are now Europe and North America collided about 400 million years ago.

Highlight 6-3

Telling the Earth's History from Its Seafloor Mud

Although the ocean floor might seem far removed from events that occur here on land, clues about the timing and nature of important events on the continents and in the atmosphere are widely preserved in the Earth's oceanic sediments. Seafloor geology is linked to the Earth's continents in many ways:

- The products of erosion of the continents (sediments and ions) are carried to the oceans and accumulate there as clastic- or chemical-sedimentary rocks.
- Changes in global climate influence the temperature and circulation of the oceans.
- These changes in temperature, in turn, affect the organisms that populate the oceans, the composition of the shells they grow, and the sediments that form from the accumulation of these shells.
- The motion of tectonic plates in and around ocean basins dramatically changes global geography and topography.
- Much of the tephra from the Earth's great volcanic eruptions lands on ocean surfaces and settles to the seafloor.

Whereas evidence for these and other major events is commonly erased from the continents by erosion, the very same evidence is typically preserved in the sediments of the seafloor. In the oceans, these thick deposits provide a nearly continuous sedimentary record about climate change and ice ages, the evolution of life, monumental meteorite impacts, great volcanic eruptions, and the history of the tectonic events that have occurred during the lifespan of the ocean basins. To understand the occurrence of these events through time, geologists study the composition, texture, sedimentary structures, and fossil contents of seafloor deposits.

Of course, it is much more difficult to retrieve samples of seafloor sediments than it is to walk along a trail and pick up rock samples on land. Early marine geologists could merely scratch the surface of the seafloor using dredges or hollow pipes that were dropped to the ocean floor to bring back small sediment samples. Modern geologists, however, have access to more extensive samples of ocean sediment recovered by sophisticated drilling ships. Such ships are equipped with technology that enables them to drill several kilometers into the ocean floor, even in waters greater than 8 km (5 mi) deep,

▲ Geologists studying the large collection of ocean-floor cores.

Rivers carried a great volume of coarse detrital sediment from those now departed mountains into shallow seas to the west. The cross-bedded sandstones derived from these sediments were themselves uplifted in a later tectonic event and then cut through by streams, forming the Catskill Mountains of east-central New York.

Specific plate-tectonic settings are often associated with specific sedimentary facies. Recently rifted plate margins, such as the East African rift zone, commonly contain large alluvial-fan deposits, graywacke sandstones, and extensive lake deposits and associated evaporites. Such a grouping of rocks is typical of

while remaining perfectly stationary over the hole. Global positioning systems even enable ships to return to the same hole some time later and redrill it, even though the hole is just a tiny speck on the vast ocean floor. The goal of most deep-sea drilling projects is to recover long cylinders of sediment and rock that represent tens of millions of years of *continuous* Earth history. These cylinders, called *cores*, are examined in great detail, both on board and in shore-based laboratories.

As mentioned earlier, oceanic clastic sediments tell us much about geologic processes taking place on land. For example, sediment cores in the Caribbean region contain tephra layers from numerous explosive volcanic eruptions related to the subduction zones that ring the Caribbean Sea. Some volcanic-ash layers are very thick (10 cm or more), even though they were collected from cores drilled hundreds of kilometers from the nearest volcano. The thickness of these layers indicates that in prehistoric time, this region experienced volcanic eruptions—far more explosive than anything we see there today. Predictably, these eruptions are not recorded by tephra layers on land, because they have either completely eroded or become incorporated within the region's thick soils in its tropical, weathering-intensive climate. The nearby marine sediments, however, have faithfully preserved these events.

Another major land event recorded for us in the oceans is the advance and retreat of ice sheets, which is indicated in cores by the occurrence of coarse, ice-rafted debris. Periods of worldwide glacial expansion are also charted by the amount of windblown dust carried to the oceans. In cold times, vast areas of ice-covered lands create great storm-producing instabilities in the Earth's atmosphere. These storms are accompanied by high winds. During these chilly times, large clouds of fine sediment (dust) blow over the oceans, creating deposits of fine clastic sediment that settle to the seafloor.

Changes in marine life that occur in response to a variety of climatic, tectonic, and environmental shifts are also recorded in ocean sediments. Certain marine fossils may be present in the lower, older part of a sediment core and then either gradually or suddenly disappear from the younger sediments above them. This change indicates that the organisms either migrated away or died off, perhaps catastrophically. For example, in an event that marks the end of the Paleocene Epoch, approximately 55 million years ago, marine life was devastated by a major global warming, while these organisms' continental counterparts continued to thrive. Thus, while marine organisms were perishing in great numbers in warm seas apparently depleted of oxygen, flamingoes and crocodiles were flourishing in what was then warm, swampy North Dakota.

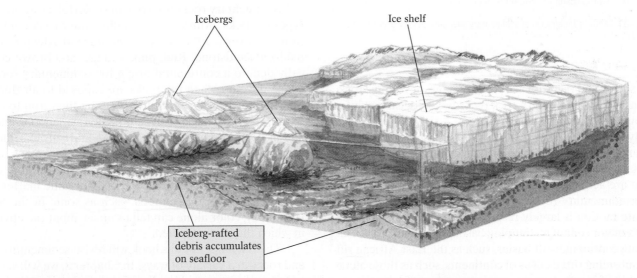

▲ Floating ice masses release their sediments onto the seafloor. There they are preserved along with the history of land-based glaciations.

volcanically active areas that have been faulted downward during rifting. Transform boundaries are marked by rapid sedimentation of angular, feldspar-rich arkosic sands, formed when igneous rock is crushed at the plate boundaries and transported a short distance before being dumped into basins within the fault zones.

Rapid sedimentation also takes place near sites of continental collisions, such as in the shadow of the still rising Himalaya, where a 15-km (10-mi)-thick tongue of sediment extends 2500 km (1500 mi) into the Indian Ocean south of Calcutta, India. Volcanoes that rise above a subducting plate supply sediment that tends to

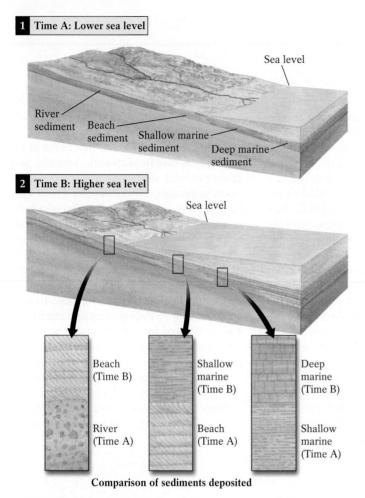

1 Time A: Lower sea level

Sea level

River sediment Beach sediment Shallow marine sediment Deep marine sediment

2 Time B: Higher sea level

Sea level

Beach (Time B)

Shallow marine (Time B)

Deep marine (Time B)

River (Time A)

Beach (Time A)

Shallow marine (Time A)

Comparison of sediments deposited

▲ **FIGURE 6-31** Landward migration of sedimentary environments associated with rising sea level.

be poorly sorted and rich with angular fragments of volcanic rock. By studying these sedimentary-rock records, geologists can infer much about past plate margins.

Finally, tectonic events create most of the *sedimentary basins* in which much of the Earth's sediment accumulates. Great sediment thicknesses can only accumulate in basins (surface depressions) that are actively subsiding (that is, they're sinking). How else could sedimentary-rock thicknesses reach 20 km (12 mi) or more? Plate motion is largely responsible for creating many of the Earth's major sediment-trapping basins. These include: relatively young continental-rift basins, such as the East African rift zone; old subsiding rifted edges of continents, such as those along the east coast of North America; major transform-fault systems, such as the San Andreas fault zone of California; and *intra-continental* ("within" the continent) basins that occupy old "failed" continental rifts (where, for some reason, rifting stopped before the plate was completely rifted and diverged to form a new ocean). Sedimentary basins of the southern Mississippi River valley of Louisiana, Mississippi, Tennessee, and Arkansas, may have formed in this way. Sedimentologists analyze the sediments, sedimentary facies, and sedimentary structures throughout an an-

cient basin and address questions such as: How did the basin fill with sediment? What types of sedimentary environments are represented within the basin's sedimentary rocks? Where were the sources for the basin's sediment and from which directions were the sediments transported? What were the tectonic conditions surrounding and within the basin? Such questions can only be addressed by a broad basin analysis.

Sedimentary Rocks from a Distance

Driving down a highway (of course, at or below the legal speed limit), we may catch glimpses of great outcrops of layered sedimentary rocks off in the distance. Even at Interstate-highway speeds and from many miles away, these glimpses can tell us much about a sedimentary rock's composition and history. For instance, in moist temperate regions such as the Eastern Seaboard of North America, accelerated chemical weathering and erosion remove much of the soft, erodible shale and soluble limestone. This process leaves prominent ridges of durable, well-cemented sandstone, such as those in the Appalachians of central Pennsylvania. The less-resistant rocks form the region's long gentle slopes and underlie its broad valley bottoms (Fig. 6-32a).

In the Southwest, durable sandstones form the cliffs, and soft shales make up the slopes. Thick limestone beds may form steep cliffs as well, because the arid climate prevents much weathering and erosion (Fig. 6-32b). From the scenic overlooks along the South-Rim road of the Grand Canyon, for example, we can readily distinguish slope-forming shales from cliff-forming sandstones and limestones, even from great distances.

Sedimentary rocks are typically colorful, more so than other types of rocks. Almost every sedimentary rock contains some iron, and even a 0.1% iron content can lend a deep-red hue to an oxidized sandstone. Red, pink, orange, and brown colors typically denote a continental origin for sedimentary rocks. These colors develop when the rocks are exposed to air during transport or shortly after deposition, causing their iron to oxidize. In oxygen-poor settings, such as organic-rich lagoons and swamps and in deep-marine settings, the iron minerals in sediment take on a characteristic green or purple hue. Low-oxygen, aqueous conditions promote the formation of black and dark-gray sediments containing undecomposed organic matter. In many sedimentary-rock formations, such as some in the Southwest (Fig. 6-33), color alone can tell us much about the environments in which the layers formed.

The remainder of this book will revisit sedimentary processes and rocks in a variety of ways. In Chapter 8, we will see how sedimentary rocks record local, regional, and even global geological events, as well as the evolution of life on Earth. In chapters 14 through 19, we will explore how gravity, rivers and streams, underground water, glaciers, desert winds, and crashing surf move sediment from place to place, reshaping the landscape. First, however, Chapter 7 will examine how all kinds of rocks—including sedimentary rocks—are heated and pressed deep within the Earth to form the remaining major rock group of the rock cycle—metamorphic rocks.

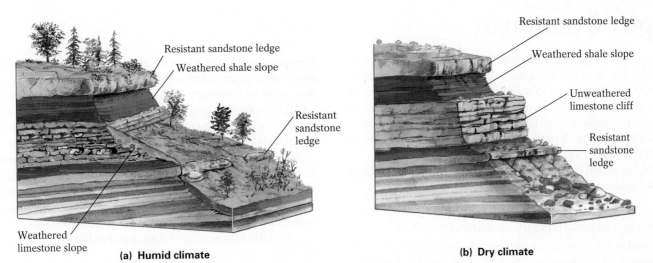

Resistant sandstone ledge
Weathered shale slope
Resistant sandstone ledge
Weathered limestone slope

(a) Humid climate

Resistant sandstone ledge
Weathered shale slope
Unweathered limestone cliff
Resistant sandstone ledge

(b) Dry climate

▲ **FIGURE 6-32** Different sedimentary rocks impart different types of surface expressions to the landscape, often depending on the climate to which they are exposed. In both humid and dry climates, well-cemented, weathering-resistant, quartz-rich sandstones form prominent ridges, and poorly cemented, less-resistant shales form slopes and valleys. Limestones form steep cliffs in dry climates but dissolve away to form slopes in humid climates. Thus, the same sequence of rocks that would form a gently sloping, soil-rich topography in a humid climate **(a),** such as those found in the northeastern United States, might form soil-poor ridges and cliffs in a dry climate **(b),** such as that found in the southwestern United States.

▲ **FIGURE 6-33** The American Southwest affords numerous vistas of cliffs and canyons in brilliant reds, oranges, pinks, greens, and purples. Here, at Little Painted Desert Park north of Winslow, Arizona, the sedimentary rocks owe their vivid colors to their mineral content, the compounds that cement their grains, the nature of their iron compounds, the presence or absence of organic matter, and the past and present climates in which these rocks originally formed and subsequently weathered.

Chapter 6 Summary

Sediment consists of fragments of solid material derived from pre-existing rock, the remains of organisms, or the direct precipitation of dissolved minerals from solution in water. A vast amount of sediment accumulates continuously at the Earth's surface, most of which is eventually converted to *sedimentary rocks.* The thin layer of sedimentary rock in the Earth's crust accounts for about 75% of all the rocks exposed on land.

Geologists classify sediments according to the source of their constituent materials.

- *Detrital sediment* is composed principally of fragments of pre-existing igneous, sedimentary, or metamorphic rock.
- *Chemical sediment* consists of minerals that dissolved during the process of chemical weathering and then either precipitated directly out of solution or were extracted from solution by organisms and deposited later in the form of shells, skeletons, and other organic materials.

Detrital-sediment texture—the size, shape, and arrangement of particles in a deposit—is determined largely by transport and depositional processes. During transport, sediment grains undergo sorting, a process by which they are carried or deposited selectively, based on the energy of their transport medium and the grain's size, density, and shape. A well-sorted deposit consists of particles of one size; a poorly sorted deposit contains particles of widely varying sizes.

Detrital sediments often display sedimentary structures, physical features that reflect the conditions of deposition. These include:

- *Bedding,* or *stratification,* is the arrangement of sediment particles into distinct layers that mark separate depositional events. *Graded bedding* forms when sediment settles through relatively still water. The coarsest grains settle to the bottom first, and grain size decreases gradually toward the finest grains at the top of a layer. *Cross-bedding* refers to sediment layers that are oriented at an angle to the underlying sets of beds, typical of sediments deposited by wind and moving currents of water.
- *Ripple marks* are small surface ridges produced when water or wind flows over sediment after it is deposited.
- *Mudcracks* occur at the top of a sediment layer when muddy sediment dries and contracts.

After a body of sediment has been buried by subsequent deposits, increased heat, pressure, and circulating groundwater change the composition and texture of the sediment by a set of processes collectively known as *diagenesis.* The end result of diagenesis is often *lithification,* the conversion of loose sediment into solid sedimentary rock. Detrital sediments are most often lithified by either compaction or cementation. *Compaction* is the diagenetic process by which the weight of overlying materials reduces the volume of a sedimentary body by forcing water from the sediment and compressing the particles together. *Cementation* of sediment grains occurs when dissolved ions in circulating groundwater are precipitated as "cement" in the pore spaces between sediment grains. A rock that is formed by compaction and cementation of sediment particles has a clastic texture.

Detrital sedimentary rocks are classified by grain size. They include:

- fine-grained *mudstones* (such as shale);
- intermediate-grained *sandstones;*
- coarse-grained *conglomerates* (which contain large *rounded* grains) and *breccias* (which contain large *angular* grains).

We base classification of *chemical-sedimentary rocks* not on grain size, but on the composition of the sediment. Chemical-sedimentary rocks are classified as inorganic or biogenic.

- *Inorganic chemical-sedimentary rocks* precipitate directly from water, usually when the water evaporates or undergoes a significant temperature change. They include inorganic limestone, evaporites, and dolostone.
- *Biogenic chemical-sedimentary rocks* form when organisms extract dissolved compounds from water, convert them into biological hard parts (such as shells and skeletons), and subsequently deposit them as sediment when they die. They include biogenic limestone, biogenic chert, and coal.

Sediment accumulates in numerous sedimentary environments, which may be continental, transitional (coastal), or marine. Because the properties of any sedimentary rock stem from the specific conditions under which it develops, geologists can distinguish rocks formed in one depositional setting from rocks formed in another. A sedimentary facies is the set of unique properties that distinguish a rock in a given layer from surrounding rocks formed in different depositional settings at the same time. By noting how sedimentary facies change over distance as well as over time, we can interpret not only individual sedimentary environments of the past but also their changing relationships to one another as environments shift with the passage of time.

KEY TERMS

bedding (stratification) *(p. 174)*
biogenic chemical sediments *(p. 182)*
breccias *(p. 182)*

cementation *(p. 179)*
chemical sediment *(p. 172)*
chert *(p. 188)*
clastic *(p. 179)*

coal *(p. 190)*
compaction *(p. 178)*
conglomerates *(p. 182)*
cross-beds *(p. 176)*

detrital sediment *(p. 172)*
diagenesis *(p. 178)*
dolostone *(p. 188)*
evaporites *(p. 184)*

graded bed (*p. 176*)
inorganic chemical sediments
 (*p. 182*)
limestones (*p. 183*)
lithification (*p. 178*)

mudcracks (*p. 178*)
mudstones (*p. 179*)
ripple marks (*p. 176*)
sandstones (*p. 181*)

sediment (*p. 170*)
sedimentary environments
 (*p. 192*)
sedimentary facies (*p. 195*)

sedimentary rock (*p. 170*)
sedimentary structures (*p. 173*)
shale (*p. 180*)
sorting (*p. 172*)

QUESTIONS FOR REVIEW

1. What are the major types of sedimentary rocks? On what bases are they classified?

2. Briefly describe how sorting and rounding of detrital sediment vary with different transport media, such as wind, rivers, and glaciers.

3. Describe the differences between graded beds and cross-beds. Which indicate the flow direction of ancient currents? Give an example of a setting in which each forms.

4. Explain the processes involved in diagenesis, and describe how each affects the physical properties of sediment.

5. Explain the relationship between carbon dioxide and precipitation of inorganic limestone.

6. Name and describe the origin of three different types of biogenic chemical sedimentary rocks.

7. Name some types of sedimentary facies associated with each of two different plate-tectonic boundaries.

8. Why are the evaporites halite and gypsum much more common than the evaporite sylvite? Why are dolostone sites often associated with evaporite deposits?

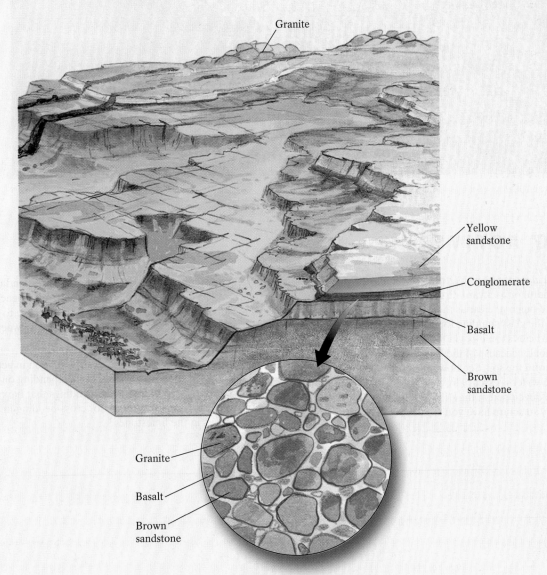

Section of conglomerate

9. Study the photo of a modern salt flat below and speculate about the environmental conditions that existed when these sediments were first deposited. How would the appearance of the present sediments change if the climate in this area became very moist?

10. In the figure on the previous page, what can you tell about the path taken by the stream that deposited the conglomerate?

LEARNING ACTIVELY

1. Pour yourself a tall glass of water in a transparent glass. Dump into the glass a handful of backyard dirt and observe. The initially murky water should begin to clear as the coarser particles sink to the bottom. Leave the glass standing for a few hours and check again. The finer particles should begin to settle out, but the water will still be quite cloudy because of the finest clay-size particles and clay minerals from the soil. See how much time passes (it may be days) before the water is essentially clear. Look at the layers of sediment that have developed in the still water of the glass. What type of sedimentary structure have you created? Leaving the backyard sediment at the bottom of the glass and the glass filled with water, dump in a handful of sand from a beach or riverbank. How does the settling rate of the sand compare to the soil? What type of sedimentary feature did you create between the dirt and the sand?

2. Look around your campus for sedimentation in action. Look for mudcracks as mud puddles dry out. Depending on the architecture of your campus, find some buildings that have been built from sedimentary rock. Look for cross-bedding in the sandstones and fossils in the limestones.

ONLINE STUDY GUIDE

Practice makes you better. Make the time to take the *Online Study Guide* quiz for this chapter. It should take you about 20 minutes. Automatic grading and instant feedback will help you quickly and accurately identify the concepts you have mastered and the areas in which you need more study.
www.prenhall.com/chernicoff

Metamorphism and Metamorphic Rocks

7

When basalt that crystallizes at the ocean floor is dragged down a subduction zone, at some point thereafter (perhaps millions of years later), it is no longer basalt. When shale deposited on the edge of a continent is caught in the vise between colliding continents, it is no longer shale. Rocks that experience conditions different from those at which they originally formed will change—that is, they will *metamorphose*. The minerals in these changing rocks will react with each other to produce new minerals that are stable under the new geologic conditions, and the texture (size and shape of the mineral grains) may change.

Thus far, you have learned how continental and oceanic crust are created by magmatism, how rocks exposed to the atmosphere break down, and how sediments accumulate at the top of the Earth's crust by mechanical and chemical processes. When we look at most rocks of the Earth's crust, however, we do not see igneous or sedimentary rocks; we see rocks that were *once* igneous or sedimentary, but that have been changed (metamorphosed) from their original form (Fig. 7-1). Recalling the rock-cycle diagram in Chapter 1, we are reminded that all rocks—even those that have existed in more or less the same form for millions of years—are changeable. They change in response to changes in the conditions around them. We saw in Chapter 3 that high temperatures deep in the Earth's interior create the magmas that then cool to form igneous rocks. We learned in Chapter 6 that sediments lithify to become sedimentary rocks in the relatively low-temperature environment near the Earth's surface. The third major type of rock, **metamorphic rocks,** generally forms at conditions *between* those that produce igneous and sedimentary rocks (Fig. 7-2). **Metamorphism** (from the Greek *meta*, meaning "change," and *morphe*, meaning "form") is the process by which heat, pressure, and chemical reactions deep within the Earth alter the mineral content and/or structure of preexisting rock *without melting it*. Any rock—whether igneous, sedimentary, or metamorphic—is a candidate for metamorphism.

Metamorphic rocks are well-exposed in many mountain belts, but elsewhere, vast regions of metamorphic rock are buried beneath thousands of meters of sedimentary rock. Near Topeka, Kansas, for example, the rocks extending for thousands of meters below the surface formed from shallow-marine sediments left behind by the inland seas that invaded North America more than 100 million years ago. The rocks under these sedimentary rocks, however, are largely metamorphic. In some regions deep erosion by grinding glaciers or energetic rivers has stripped away the surface rocks, exposing the metamorphic rocks

FIGURE 7-1 Highly contorted and metamorphosed rock on Sledge Island in Alaska. ▶

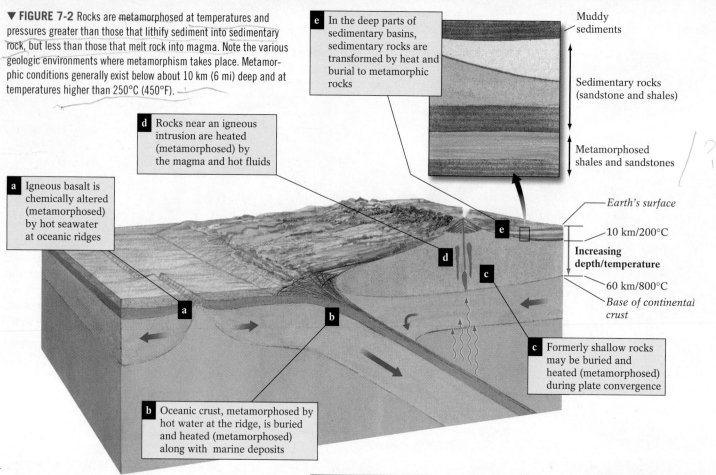

▼ **FIGURE 7-2** Rocks are metamorphosed at temperatures and pressures greater than those that lithify sediment into sedimentary rock, but less than those that melt rock into magma. Note the various geologic environments where metamorphism takes place. Metamorphic conditions generally exist below about 10 km (6 mi) deep and at temperatures higher than 250°C (450°F).

e In the deep parts of sedimentary basins, sedimentary rocks are transformed by heat and burial to metamorphic rocks

Muddy sediments

Sedimentary rocks (sandstone and shales)

Metamorphosed shales and sandstones

d Rocks near an igneous intrusion are heated (metamorphosed) by the magma and hot fluids

a Igneous basalt is chemically altered (metamorphosed) by hot seawater at oceanic ridges

Earth's surface

10 km/200°C

Increasing depth/temperature

60 km/800°C

Base of continental crust

c Formerly shallow rocks may be buried and heated (metamorphosed) during plate convergence

b Oceanic crust, metamorphosed by hot water at the ridge, is buried and heated (metamorphosed) along with marine deposits

beneath. A geologic map of North America reveals metamorphic rock exposed at the surface throughout much of glaciated Canada, the northern portions of the midwestern Great Lakes states, and many of the continent's mountainous regions (Fig. 7-3). Keep in mind, though, that even where metamorphic rocks are not exposed on continents, the rocks below the top layer of sediments are typically metamorphic.

Can we view firsthand the Earth's metamorphic processes? We can observe the destructiveness of lava flowing from Hawai'i's K'ilauea volcano and watch it solidify into another acre of Hawaiian basalt. We can observe the Mississippi River carrying a load of sand and then depositing it on the river's floodplain during a flood; that sand may someday form a layer of sandstone. No one has ever seen metamorphism in action, though, because metamorphic processes involve such high temperatures and pressures, they only take place *unseen* beneath the Earth's surface. We can study metamorphic rocks

▶ **FIGURE 7-3** Exposed and near-surface metamorphic rocks of North America. Generally buried beneath kilometers of sedimentary rocks, the continent's metamorphic rocks can be seen where erosion has removed the overlying rock, such as in the Grand Canyon. They can also be found in the cores of mountain ranges, such as the North Cascades of Washington state, the Appalachians of eastern North America, and in the glacially scoured Canadian Shield region of Ontario, Quebec, and the adjacent Great Lakes states of Minnesota, Wisconsin, and Michigan.

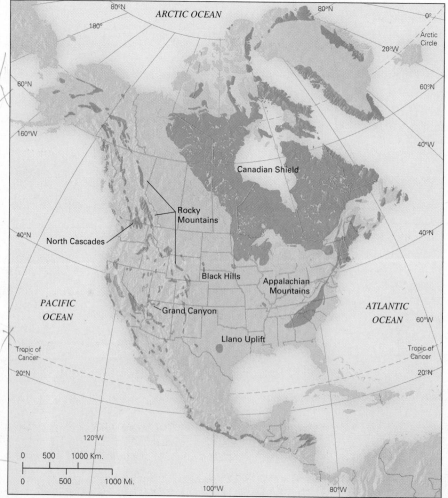

only after tectonic events, uplift, and erosion have stripped away the overlying rocks. Thus, we must be geological "detectives," using clues from the metamorphosed rocks to learn about the deep interior processes that produced them. Much of our knowledge about the causes of metamorphism comes from laboratory experiments, theoretical models that simulate conditions in the Earth's interior, and field studies of exposed metamorphic rocks in different tectonic settings.

We use metamorphic rocks in everyday life in many ways—from the raw materials for building and road construction to key elements in flame-resistant firefighting material to the marble that Michelangelo used to sculpt his famous statue of David. We'll touch on these uses later in this chapter. We must first consider the conditions that promote metamorphic processes, the various types of metamorphism, and the common rocks produced. We will then examine some of the most important features of metamorphic rocks—their minerals and their textures—as these allow us to interpret the tectonic history of the rocks. We'll use these metamorphic clues to determine how intensely the rock was metamorphosed (How hot did it get? How deeply was it buried?) and where it was metamorphosed (Did it change in a subduction zone? A continental-collision zone? A divergent zone?).

What Is Metamorphism?

Minerals are most stable in the places where they originally form—in the relatively cool environment at the Earth's surface for sedimentary rocks, and at high temperature (and perhaps pressure) for igneous rocks. If a rock moves to a different position in the Earth, such as by tectonic motion, it will experience a different environment than the one in which it is most stable. The minerals in this rock will respond to the new conditions by changing because they are no longer stable, and new minerals will form that are stable in the new environment. As you saw in our discussion of weathering in Chapter 5, minerals such as feldspar that may originally form deep within the Earth as part of a granite will break down to other minerals (such as clays) at the Earth's surface. Abundant water and atmospheric gases are found at the Earth's surface—very different conditions from the hot, dry environment where the feldspars originally crystallized in the Earth's interior. Change the environment, and new minerals will replace the old ones. Similarly, if surface-forming minerals (such as weathering-produced clays) experience higher temperatures than those at which they were originally stable, the minerals will react to form new ones (perhaps even feldspars) that are stable at higher temperatures. Thus, when clays are buried beneath a thick overlying pile of sediment, the higher temperatures at depth change the clays into new, more stable minerals.

Metamorphism involves the transformation of *solid rock* to form new minerals and textures. All of this action takes place in a solid state; the metamorphosed rock does not melt in order to produce new minerals. The preexisting minerals can change in a number of ways: A relatively stable mineral such as quartz will remain quartz, but the size and shape of the individual quartz grains will change to achieve a new, more stable configuration. Minerals that are very unstable when heated, such as clays, will break down and

their elements will recombine into new minerals. At the elevated temperatures associated with metamorphism, atoms and ions can migrate within a rock and rebond into more stable forms. The composition of the metamorphic rock as a whole may be the same as the original rock (minus some water or carbon dioxide), but the types of minerals and their textures (the size, shape, and spatial distribution of minerals) have changed. In this way, a clay-rich (and therefore aluminum-rich) sedimentary rock will metamorphose into a metamorphic rock containing new Al-rich minerals such as garnets and micas in place of the unstable clays.

Remember, the rocks undergoing metamorphism remain in a solid state. The unstable minerals in a metamorphosed rock either recrystallize into new, more stable forms or react with other unstable minerals to produce new minerals with stable atomic arrangements. Such rearrangement is possible because heat and pressure break the bonds between *some* atoms or ions in an unstable mineral. Once freed, these atoms and ions migrate to other sites within that mineral, or to another mineral, and then rebond. *Metamorphic* processes never break all the bonds in a rock's minerals—if all its bonds were broken, the rock would melt (an *igneous* process) and become a magma.

What Drives Metamorphism?

You already know that metamorphism occurs when a rock experiences different environmental conditions, such as different temperature and pressure, from the conditions at which it originally formed. But what causes these changes? And how can changes in variables such as temperature change a rock? We now consider the factors that drive metamorphism, and explore how they change a rock.

Heat

Mix the appropriate proportions of flour, milk, sugar, and water, and you are on your way to baking a cake. But unless you turn on the oven, nothing will happen; you'll just have a mixture of the ingredients. Similarly, little happens to the ingredients (atoms and ions) for making new metamorphic minerals until the rock is heated.

Heat, which accelerates the pace of nearly all chemical reactions, is the most important factor driving metamorphism. Beneath the Earth's surface, temperature increases with depth. In the Earth's continental crust, temperature increases at an average rate of 20° to 30°C per kilometer of depth (approximately 72°F per mile). To see how the geothermal gradient can be related to metamorphism, consider a rock that moves from the cool temperatures of the Earth's surface to the high temperatures of the Earth's interior. Such movement of rocks (in this case *burial*) occurs during tectonic activity, especially at convergent plate margins.

Rocks at depth in the Earth are *always* at high temperatures and therefore do not change (metamorphose) as long as they remain at the same temperature and depth and in the same chemical environment. Unless you live at an active plate boundary, the rocks 10 km (6 mi) below your feet are not being actively metamorphosed today. Therefore, although there might be metamorphic rocks beneath your feet—at any depth, from the surface to tens of kilometers of depth—that doesn't mean that the rocks beneath your house are metamorphosing today. What drives metamorphism is not simply "temperature," but rather it's the heating of rock from a lower temperature to a higher temperature. Some rocks are very stable at high temperatures and pressures, so unless conditions change, the minerals in the rocks will not change. Metamorphism involves *action* and a change in a rock's environment.

Heat sources within the Earth's crust include the energy emitted as radioactive isotopes decay and the upward conduction of heat from the underlying mantle. Changing a rock's position with respect to these heat sources will metamorphose a rock. For example, plate-tectonic activity, such as subduction, can move a crustal rock closer to the hot mantle. The rock will then heat up. Rocks in continental crust can also metamorphose if they are intruded by a nearby igneous body—the heat from the magma will drive metamorphism in rocks near the intrusion. (See Figure 7-2.)

Pressure

Although heat is the most important factor driving metamorphism, changes in pressure may also profoundly affect the minerals in a rock. Rocks deep in the Earth's crust are under more pressure than the rocks at shallow levels. Rocks in the interior of the Earth are under **lithostatic pressure** (*litho,* meaning "rock," and *static* from the Greek *statikos,* or "causing to stand in place"). Lithostatic, or *confining,* pressure pushes on rocks *equally* from all directions (Fig. 7-4). As a result of lithostatic pressure, a rock that has a particular volume at a shallow depth will become compressed into a smaller, denser form as it is buried. The overall shape of the rock body does not change as it experiences higher lithostatic pressure (Fig. 7-5a); it just becomes more compressed.

In some geologic settings, such as at plate boundaries, pressure may be greater in one direction than in others. Such **directed**

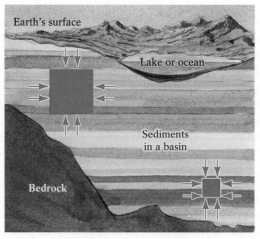

(a) Confining pressure

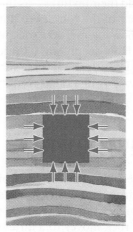

Before compression   After compression

(b) Directed Pressure

▲ **FIGURE 7-5 (a)** A rock subjected to confining pressure becomes compressed without changing shape. **(b)** A rock subjected to directed pressure has its shape distorted, becoming thinner in the direction of the greatest stress and elongated in the direction perpendicular to this stress.

pressure changes the shape of a rock (Fig. 7-5b). It typically flattens a rock in the direction at which the greatest pressure is applied, and it lengthens the rock in the direction perpendicular to the greatest pressure direction.

Directed pressure may change the appearance of a rock by causing existing minerals or other features (such as layering or fossils) to stretch or bend (Fig. 7-6), but these pressures are not great enough to affect the mineralogy of the rock. New minerals grow during metamorphism primarily in response to changes in lithostatic pressure, not directed pressure.

Directed pressure also affects the shape and arrangement of the *new* minerals that grow during metamorphism in response to lithostatic pressure. Minerals growing in a rock that is under greater pressure in one direction than in others will tend to be aligned perpendicular to the direction of the greatest pressure. This preferred alignment is called **foliation** (from the Latin *foliatus,* or "leaflike"). When the mineral grains in a metamorphic rock are platy (that is, thin and flaky, like micas), their orientation will create a layering in the rock (Fig. 7-7).

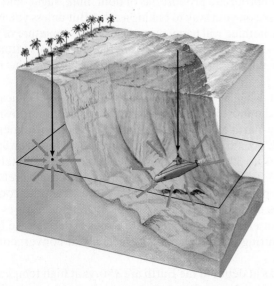

▲ **FIGURE 7-4** Deep burial of rocks generally subjects them to lithostatic, or confining, pressure, an inward-pressing force that acts equally from all directions. Any object immersed in water, such as this submarine, is also subjected to equal pressure on all sides.

(a)

(b)

▲ FIGURE 7-6 **(a)** Stretched fossil (left) associated with directed pressure; photo (right) shows similar fossil, undeformed. **(b)** Folded rocks in the Blue Ridge Mountains of North Carolina, showing the deformation typical of rocks subjected to directed pressure at great depth in the crust.

▲ FIGURE 7-7 Directed pressure creates foliation in rock. These foliated rocks are from the Tien Shan Mountains in Kyrgyzstan. The inset photo is a microscopic view of aligned mica minerals in a foliated metamorphic rock.

Geologists measure pressure with various units, including bars and pascals. A *bar* is equal to the pressure applied to the Earth's surface at sea level by the atmosphere (1 bar = 1.02 kg/cm^2 or 14.7 lbs/in^2; 1 bar also equals 10^5 pascals). Metamorphism typically occurs at more than 1 *kilo*bar (1000 bars, or 10^8 pascals) of pressure, which is approximately equal to the pressure prevailing about 3 km (2 mi) beneath the Earth's surface. As the *temperatures* needed for metamorphism do not normally occur within the upper 10 km (6 mi) of the Earth's interior, metamorphism rarely takes place in such shallow depths unless heat is carried up from greater depths by rising magma.

Circulating Fluids

Put two dry chemicals at room temperature into a test tube, and what happens? Nothing. Shake the chemicals in the tube. Still nothing. Heat the test tube with a Bunsen burner, and still no activity results. But add one more ingredient—a fluid, such as water—and a significant reaction will almost always occur. Because dissolved ions move about easily in fluids, the presence of a liquid or gas in or around hot rock under pressure helps unbonded atoms and ions migrate from place to place. Thus, the presence of a fluid dramatically increases the potential for metamorphic reactions. Water surrounding a mineral grain promotes the exchange of atoms and ions with adjacent grains. Without water, unbonded ions migrate *very slowly* along grain boundaries or through invisible pathways within a mineral's atomic structure.

Water may also drive the metamorphism of rocks that did not originally form in the presence of water (thus a change in their environment). Mid-ocean-ridge basalts, for example, formed by mantle melting, are thus stable at high temperatures; they contain minerals with no water in their crystal structures. Basalt is therefore unstable in the presence of water, and hot water in particular will react with basalt, transforming its anhydrous ("water-free") minerals (olivine, pyroxene, plagioclase) into new hydrous minerals (such as serpentine and chlorite) and creating a metamorphic rock. This is an example of metamorphism in which the process is driven not by heating a rock to a temperature above its stable temperatures (basalts are very "content" at high temperatures), but by changing something else in the rock's environment—in this case, by adding water.

When water drives metamorphism, the water may come from various sources. In the case of metamorphosing oceanic crust, the water is seawater. In continental crust, the water may have percolated down from the Earth's surface through fractures or faults, or become trapped in the spaces between sediment grains or in the cracks of a subducting plate. The cooling and crystallization of magmas may release water and other fluids (such as carbon dioxide or sulfur-dioxide gas) to surrounding rocks. As water-rich minerals (such as clays and amphiboles) decompose during metamorphism, they too may release water.

Water-rich fluid (and in the case of some rocks, carbon-dioxide-rich fluid) is therefore an important variable in metamorphism because it can drive metamorphism of formerly dry rocks and it can enhance the rate of metamorphic transformation of minerals. Eventually, as rocks heat up, the fluid is driven off and the highest-temperature metamorphic rocks can be as dry as the driest igneous rocks.

What Controls the Mineralogy of a Metamorphic Rock?

The factors that control the mineralogy of a metamorphic rock include some of the variables that you have just learned about—temperature, lithostatic pressure, and the amount and composition of fluids. There is one other critical factor: the composition of the parent rock. Calcium-rich parent rocks are transformed into calcium-rich metamorphic rocks. The presence of *specific* calcium-rich minerals is controlled by temperature, pressure, and fluids, but no matter what the temperature and pressure, a parent rock that contains little or no calcium cannot be transformed into a calcium-rich metamorphic rock. (Similarly, you can't make a batch of chocolate-chip cookies from the ingredients for a Caesar salad.)

A parent rock that consists of only one mineral will be transformed into a metamorphic rock consisting of that same mineral. The only change will be the texture of the rock, because no chemical reactions occur to change the rock's mineralogy. If you put a pan of flour in the oven and heat it up, you will end up with baked flour. If you add other ingredients, you might end up with a cake. Similarly, a limestone consisting *entirely* of the mineral calcite will recrystallize into a coarser-grained marble, but calcite will still be the only mineral in the rock. A pure quartz sandstone will metamorphose to produce quartzite, a metamorphic rock consisting of recrystallized quartz. In a parent rock that contains several minerals, the elements in those minerals will be recombined during metamorphism to produce new minerals that were not present in the parent rock, in addition to some minerals that were already present (such as quartz).

Different plate-tectonic and other settings are marked by different temperatures, pressures, and amounts of chemically active fluids. Each setting therefore can produce distinctive types of metamorphism and distinctive types of metamorphic rocks. Once you learn to recognize metamorphic rocks, you will be able to identify the tectonic and other settings in which they formed.

Types of Metamorphism

Heat, pressure, and chemically active fluids interact differently in different geological settings to produce metamorphic rocks. These settings yield distinctive types of metamorphism. Under some conditions, only a narrow area of rock is affected; in other settings, rocks may be changed over a vast region.

Contact Metamorphism

Preexisting rock touched by the intense heat of magma may undergo **contact metamorphism.** That is, solid rocks near magma will change in response to heating and the circulation of hot fluids without melting themselves. Pressure is not a significant factor, because the metamorphosing rocks are not changing depth and are not typically being subjected to directed pressure. Because rocks are poor conductors of heat (you would not want a woodstove in a house made of granite walls because the heat would never reach the other rooms), the effects of contact metamorphism decrease sharply a short distance away from the magma. Only rocks near the intruding magma will be highly metamorphosed. Those located farther away are largely unaffected by magmatic heat and will be changed only slightly if at all (Fig. 7-8).

The effects of contact metamorphism vary with the size of the igneous intrusion. Although contact metamorphism may extend only a few centimeters or meters around a small igneous dike or sill, it may affect an area of several kilometers around a large intrusion (such as a batholith).

Another important factor that determines the extent of contact metamorphism is the difference in temperature between the magma and the intruded rock at the time of the intrusion. If a magma intrudes into relatively cool rock in the upper parts of the Earth's continental crust, contact metamorphism will be very dramatic because the cool intruded rocks will be heated to a much higher temperature than their original condition. Deep in the crust, however, rocks are already fairly hot. They may, in fact, be at temperatures only slightly lower than the temperature of the magma. As a result, the minerals and texture of these rocks will not be much affected by the intrusion of magma because they are already stable at a high temperature.

Regional Metamorphism

Contact metamorphism has relatively local effects, whereas **regional metamorphism** alters rocks for thousands of square kilometers. Regional metamorphism produced the vast regions of

exposed metamorphic rock in central Canada and the tracts of metamorphic rock found in the Appalachians of the eastern United States. This process is also responsible for the metamorphic rocks found in the mountains of western North America—the North Cascades of northwestern Washington and southwestern British Columbia, many of the mountain ranges of Alaska, and a belt of metamorphic rocks extending from eastern British Columbia to Mexico (see Fig. 7-3). As already noted, metamorphic rocks also underlie large regions of the continents that are now covered by sedimentary rocks. *Regional metamorphic rocks make up most of the continental crust.*

Regional metamorphism, also called **dynamothermal metamorphism** to highlight the importance of dynamic pressures on the rocks and the key role of thermal energy (heat), occurs when converging plates squeeze rock caught between them (Fig. 7-9). The crust thickens (sometimes even doubles), and some rocks and sediments that were at or near the Earth's surface prior to convergence are tectonically buried to great depths. Because of the directed pressure on rocks at convergent boundaries, regional metamorphic rocks are typically foliated.

If you look back at Figure 7-3, the map of exposed and near-surface metamorphic rocks in North America, a few questions may come to mind: If metamorphic rocks form at high temperatures and pressures and are stable only at depth in the Earth, why do we find such vast areas of metamorphic rocks at the Earth's

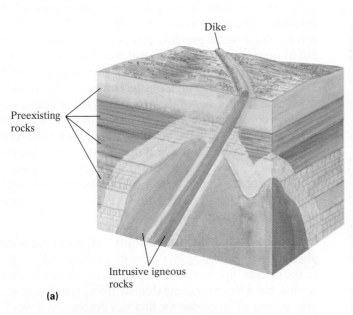

(a)

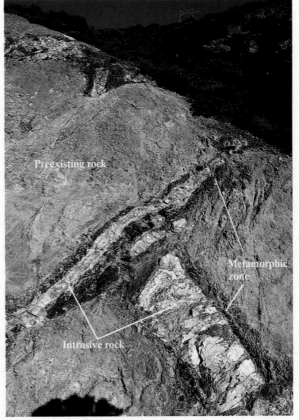

(b)

▲ **FIGURE 7-8 (a)** Contact metamorphism occurs when magma intrudes preexisting rocks, especially those that are significantly cooler than the magma. The rocks adjacent to the magma become strongly metamorphosed; those farther from the region of direct contact are less affected, although they may still receive metamorphic heat from circulating ion-rich fluids. *You can determine how close a rock was to a magma by looking at the following clues:* Regarding the minerals in the metamorphic rock, when we find metamorphic minerals that are stable at very high temperatures, we conclude that those rocks were very close to the intrusion. Lower-temperature minerals must have formed farther away from the magma. Texturally, the coarsest-grained metamorphic minerals will be closest to the intrusion. Finally, when we find water-bearing minerals in the rocks, we conclude that those rocks were relatively distant from the intrusion; the closer a rock was to the magma, the more water was "baked out" of the rock during contact heating. **(b)** An igneous dike in preexisting rock. Note the metamorphic effects that surround the intrusive rock.

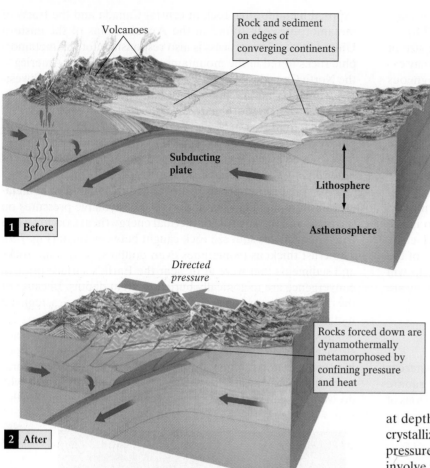

Volcanoes

Rock and sediment
on edges of
converging continents

Subducting
plate

Lithosphere

Asthenosphere

1 Before

*Directed
pressure*

Rocks forced down are
dynamothermally
metamorphosed by
confining pressure
and heat

2 After

▲ **FIGURE 7-9** Dynamothermal metamorphism is generally associated with the directed pressures and heat of convergent plate boundaries. Lithostatic pressure from deep tectonic burial also plays an important role in the metamorphism.

surface today? That is, given that a parent rock goes through a series of mineralogical and textural changes as it is heated and buried, why don't these changes reverse themselves so that the metamorphic rocks are transformed back to their parent rocks as they rise from the depths to the surface? Why do metamorphic rocks preserve a record of the high temperatures and pressures at which they formed?

One answer to these questions involves *time.* Rocks may take millions of years to be heated and buried, but a process such as faulting may bring them to the surface relatively rapidly—on the order of a few million years or less. Erosion, too, may occur rapidly in certain climates, such as in areas scoured by glaciers or incised deeply by rivers. The combination of high mountains, rushing rivers, and powerful glaciers creates the ideal set of conditions for eroding and removing a lot of crustal rock, thereby exposing deep metamorphic rocks at the surface. Such is the case in parts of the Himalaya. Faulting and rapid erosion can expose deep rocks "quickly" (in geologic terms) *before the effects of metamorphism can be undone.* Reactions can occur when rocks are hot, but once they cool, the reactions switch off and the new high-temperature minerals will not break down. Another reason that dynamothermal metamorphic rocks are preserved is that as the rocks are heated during metamorphism, they dehydrate, or in the case of carbon-

ate-rich rocks, they release CO_2. Once the water and CO_2 leave the rocks, the metamorphic reactions cannot be reversed unless enough water or CO_2 is added back to the rocks and the temperature remains high enough for reactions to occur. Because this set of conditions is rarely met, this reverse process rarely occurs. We are fortunate that metamorphism is so difficult to "undo," so we can use metamorphic rocks to decipher past plate-tectonic processes.

Other Types of Metamorphism

Regional increases in heat and pressure and contact with intruding hot magmas are not the only ways in which rock may be metamorphosed. Metamorphism can also occur in the deep parts of sedimentary basins, at sites where rocks are invaded by hot water, within active fault zones, and where meteorites and lightning strike the Earth's surface rocks.

Burial Metamorphism   As the sediments deposited early in the history of a sedimentary basin are buried beneath a thick pile of younger sediments, heat and pressure build deep within the growing pile and metamorphic reactions begin to occur. **Burial metamorphism** occurs deep in sedimentary basins—at depths greater than 10 km (6 mi)—where metamorphic recrystallization occurs in response to the combination of lithostatic pressure and geothermal heating. Because the process does not involve the *directed* pressure associated with plate convergence, burial-metamorphic rocks are not well-foliated, if at all.

Burial metamorphism is occurring today in the Gulf Coast of Louisiana, where clay minerals lie beneath 12 km (8 mi) of sediment near the bottom of the Mississippi River's deltaic deposits. Samples collected from deep drill holes in the delta indicate that these clay minerals have already begun to metamorphose into minerals that are more stable at that depth, such as mica and chlorite. Note that burial metamorphism does not involve a plate-tectonic process. In this case, metamorphism occurs simply from deposition of a thick pile of sediments; this type of metamorphism does not create mountain belts.

Hydrothermal Metamorphism   **Hydrothermal metamorphism** is the chemical alteration of preexisting rocks by hot water. In the oceans, this water is of course salty seawater. On the continents, hot water may come directly from crystallizing magma, or may consist of groundwater that has percolated down from the surface and been heated by contact with hot rock or magma.

Most hydrothermal metamorphism occurs beneath ocean floors, where seawater penetrates cracks near a divergent plate boundary. Descending until it encounters hot basaltic magma or warm young oceanic rocks, the seawater heats up to about 300°C (572°F). As it passes through basalt and gabbro fractured as the plates move apart, the hot water dissolves soluble elements and transforms "dry" minerals such as olivine, pyroxene, and plagioclase (that is, the minerals with no water in their crystal structures) into hydrous minerals such as serpentine, chlorite (a micalike iron–magnesium silicate), epidote (an aluminum–iron–calcium silicate), and talc (Fig. 7-10). These new hydrous, metamorphic min-

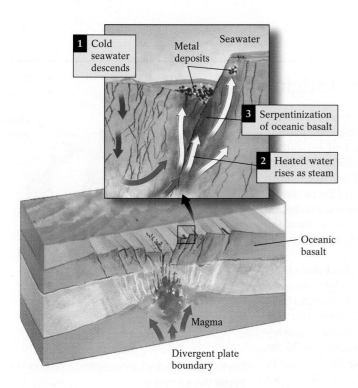

◀ **FIGURE 7-10** Hydrothermal metamorphism occurs most often at mid-ocean ridges, where tectonic plates diverge. Cold seawater comes into contact with hot basaltic magma, heats up, and rises as hot water and steam. Along the way, it dissolves metals and other ions from mantle peridotites, basalts, and oceanic sediments. This process converts dry olivine and pyroxene minerals to the hydrous magnesium silicate, serpentine and several other iron–magnesium silicates.

erals are more stable in the presence of hot water than are the dry igneous minerals. The heated water, now laden with soluble elements dissolved from the rock during metamorphism, eventually rises back up toward the seafloor. When the mineral-rich steam reaches the seafloor, it cools and precipitates its dissolved load, creating valuable concentrations of copper, nickel, iron, and lead. The hydrated oceanic crust produced by this type of metamorphism at the seafloor is the key to subduction-zone magmatism; it is the water given off by these rocks as they are buried and heated in a subduction zone that drives melting (and therefore volcanism) in subduction zones.

Fault-Zone Metamorphism   Rocks grinding past one another along a fault generate a large amount of directed pressure and considerable frictional heat. The pressure and heat in the immediate vicinity of an active fault can be intense enough to produce **fault-zone metamorphism,** although the effects of directed pressure and heating taper off a short distance from the fault.

Shock Metamorphism   A dramatic form of metamorphism, called **shock metamorphism,** occurs where a meteorite strikes rocks at the Earth's surface. The tremendous pressures and temperatures generated at meteorite-impact sites "shock" the minerals in the stricken rocks, causing them to shatter and recrystallize. The minerals produced commonly in meteorite impacts include stishovite and coesite, two types of SiO_2 that are found near and within impact craters. Some minerals created at meteorite-impact sites, such as stishovite, occur in no other geological setting.

The search for impact sites, both on land and undersea, has produced extensive information on shock metamorphism. Such study has led researchers to hypothesize that the Earth's dinosaurs were driven to extinction by an enormous meteorite impact on the Gulf Coast of Mexico, about 65 million years ago.

(See Highlight 1-1, "What Killed the Dinosaurs?") There is also speculation that meteorites "ferried" the Earth's first organic molecules to our planet, perhaps sparking the creation of life on Earth (refer to Highlight 7-1).

Pyrometamorphism   Pyrometamorphism is caused by ultra-high temperatures that occur occasionally in relatively low-pressure environments. We can see this type of metamorphism where bolts of lightning have recrystallized surface rocks and sediments and where burning subterranean coal seams recrystallized adjacent rocks. The various types of metamorphism are summarized and illustrated to help you study in the following "Geology at a Glance—Types of Metamorphism" on page 215.

Common Metamorphic Rocks

When you examine a metamorphic rock, one of the first things you may notice is whether it is foliated or nonfoliated. Foliation forms in response to directed pressure, but keep in mind that platy or elongated minerals must be present for foliation to develop. A rock that consists only of quartz or calcite may not be foliated even if it was subjected to directed pressure. You may have seen beautiful white marble in sculptures or building stones. Such marbles consist of interlocking crystals of calcite, but they may well be regionally metamorphosed rocks. Marble and quartz-rich rocks may have a foliation or other oriented texture if the grains were flattened or stretched out during metamorphism, but these textures might not be as obvious as in rocks containing platy minerals such as the micas. The presence or absence of foliation is one means of identifying and interpreting metamorphic rocks, so it is important to note whether this feature is present in a rock. Do so cautiously with metamorphic sandstones (quartzites) and metamorphic limestones (marbles).

Highlight 7-1

Shock Metamorphism, Buckyballs . . . and the Origin of Life on Earth?

For many years, a host of geoscientists and paleobiologists shared the belief that life on Earth originated during the planet's infancy when a random bolt of lightning struck a primordial stew of inorganic molecules at the Earth's surface. This energy was thought to have sparked a chemical reaction that converted simple *inorganic* molecules into the first amino acids—the building blocks of complex *organic* molecules. In more recent years, other scientists have proposed an alternative hypothesis: The organic molecules from which all living things evolved were deposited here by huge meteorites that collided with our planet during its earliest days.

How could those molecules survive the searing heat and unimaginable shock of a high-speed meteorite impact? Recent discoveries suggest that such molecules need not have perished upon impact. Perhaps they were encased in some type of protective structure. A group of researchers from the University of California, San Diego, walking the hills of a meteorite-impact site at Sudbury, Ontario, collected samples of shocked rocks that contained countless *fullerenes*—or buckyballs—minute carbon-based structures named for the great twentieth-century thinker

Buckminster Fuller. Some scientists believe that these microscopic yet exceedingly sturdy molecular cages, composed of 60 carbon atoms arranged in a manner similar to Fuller's geodesic domes (or a soccer ball; see figure), reached the Earth's surface via the meteorite that struck what is now Sudbury, two billion years ago. This conclusion stems from the unique chemical makeup of helium gas found trapped within the buckyballs.

The helium in Sudbury's buckyballs is believed to be of extraterrestrial origin, quite possibly manufactured, along with the buckyballs, in the core of a dying "Red Giant" star. According to this fascinating hypothesis, gases—including helium—from a bloated, carbon-rich star were sprayed, along with the buckyballs, into interstellar space, joining the gas and dust from which our solar system and its meteorites ultimately formed. If we accept that one such meteorite ferried stellar gases and fullerenes to Sudbury, we may also speculate that other kinds of complex organic molecules—such as those that formed the basic building blocks of life—may have taken the same route to our planet. Undoubtedly, their arrival here would have been smashing, as the metamorphically shocked rocks at various impact sites attest.

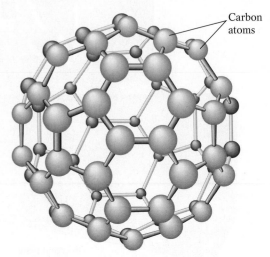

Carbon atoms

This buckyball molecule is composed of 60 covalently bonded atoms of carbon.

As a side note, the recent discovery of a global layer of these extraterrestrial buckyballs now suggests that the Earth's greatest extinction event—about 250 million years ago—may have been caused by a monumental meteorite impact. This event, reported in February 2001 is believed to have killed 90% of all life on earth.

Table 7.1 summarizes the classification of metamorphic rocks. Note that some metamorphic rocks can be produced from a number of different parent rocks. Such metamorphic rocks are distinguished *by their appearance and by the conditions under which they formed, rather than by their composition.* Other metamorphic rocks are produced by a specific type of parent rock and thus are classified according to their composition.

Some rock compositions are particularly useful in the study of metamorphic rocks because these compositions produce interesting metamorphic minerals that are particularly sensitive to changes in metamorphic variables (temperature, pressure, fluids). We can therefore use metamorphic rocks of these compositions to understand tectonic processes. Perhaps the most useful rock compositions of all for this purpose are shale (mudstone) and basalt. In the next section, we will describe the metamorphism of shale and how the rock changes with increasing temperature and pressure, so that you can see how metamorphic minerals and textures can be related to metamorphic conditions.

Metamorphism of Mud

Metamorphosed clay-rich rocks (that is, mudstones and shales) are among the most informative rocks for reconstructing metamorphic conditions because the minerals in their parent rocks

are so unstable at high temperatures that these rocks undergo spectacular changes in mineral assemblages and textures. As temperatures and pressures rise beyond the range of sedimentary diagenesis, clay minerals in shale recrystallize or react to form flat mica flakes that become aligned perpendicular to the direction of the applied pressure. This produces a strongly foliated rock known as **slate** (from the Old French *esclat,* or "splinter"). Slate tends to break parallel to the mica-rich planes into relatively thin, flat fragments, a pattern known as *slaty cleavage* (Fig. 7-11). (The difference between rock cleavage and mineral cleavage was discussed in Chapter 2. *Mineral cleavage* refers to minerals breaking between planes of atoms or ions in a crystal; *rock cleavage* refers to rocks breaking between planes of oriented minerals.)

Slate comes in a rainbow of colors. The color depends on the chemical composition of the parent shale. Red slates are rich in hematite, an iron oxide; green slates contain a significant amount of chlorite; purple slates are stained by manganese oxides; and black slates are rich in carbon-rich organic material. Whatever their color, slates exhibit slaty cleavage, producing smooth slabs of rock that are commonly used as blackboards, roofing materials, floor tiles, and pool-table tops. Slate still resembles its parent rock, shale, in that both are fine-grained, but you can tell that slate is a metamorphic rock because of its cleavage.

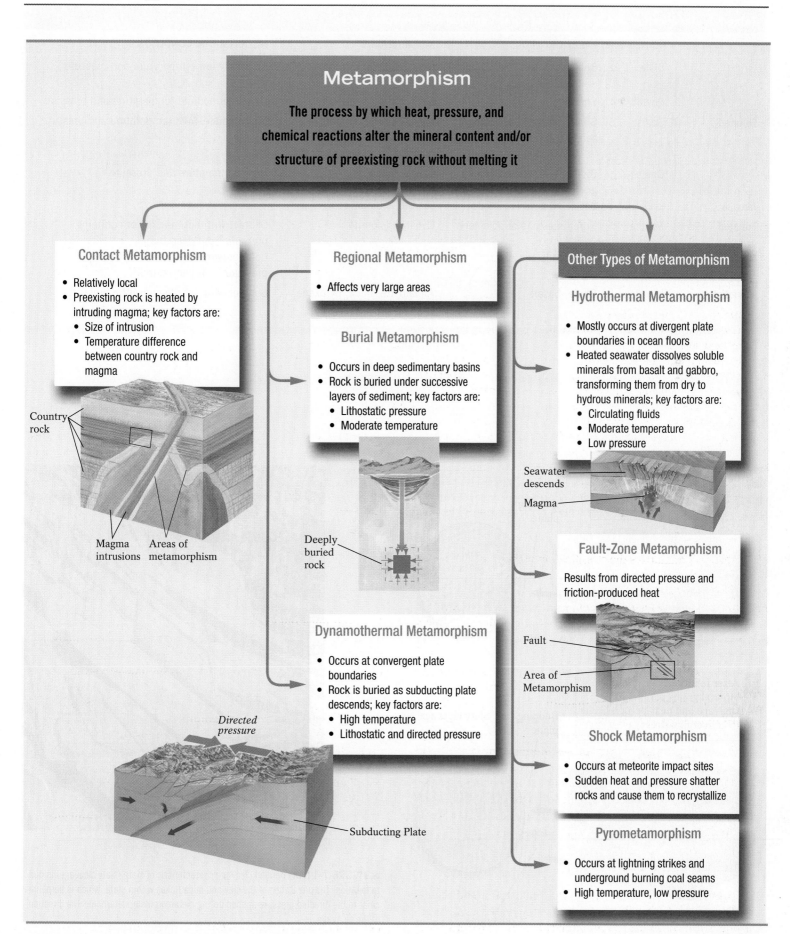

Metamorphism

The process by which heat, pressure, and chemical reactions alter the mineral content and/or structure of preexisting rock without melting it

Contact Metamorphism

- Relatively local
- Preexisting rock is heated by intruding magma; key factors are:
  - Size of intrusion
  - Temperature difference between country rock and magma

Country rock

Magma intrusions   Areas of metamorphism

Regional Metamorphism

- Affects very large areas

Burial Metamorphism

- Occurs in deep sedimentary basins
- Rock is buried under successive layers of sediment; key factors are:
  - Lithostatic pressure
  - Moderate temperature

Deeply buried rock

Dynamothermal Metamorphism

- Occurs at convergent plate boundaries
- Rock is buried as subducting plate descends; key factors are:
  - High temperature
  - Lithostatic and directed pressure

Directed pressure

Subducting Plate

Other Types of Metamorphism

Hydrothermal Metamorphism

- Mostly occurs at divergent plate boundaries in ocean floors
- Heated seawater dissolves soluble minerals from basalt and gabbro, transforming them from dry to hydrous minerals; key factors are:
  - Circulating fluids
  - Moderate temperature
  - Low pressure

Seawater descends

Magma

Fault-Zone Metamorphism

Results from directed pressure and friction-produced heat

Fault

Area of Metamorphism

Shock Metamorphism

- Occurs at meteorite impact sites
- Sudden heat and pressure shatter rocks and cause them to recrystallize

Pyrometamorphism

- Occurs at lightning strikes and underground burning coal seams
- High temperature, low pressure

Table 7.1 Classification and Derivation of Some Common Metamorphic Rocks

	ROCK	PARENT ROCK(S)	KEY MINERALS	METAMORPHIC CONDITIONS
Foliated	Slate	Shale	Clay minerals, micas, chlorite, graphite	Relatively low temperature and pressure
	Phyllite	Shale	Mica, chlorite, graphite	Low–intermediate temperature and pressure
	Schist	Shale, basalt, graywacke sandstone, impure limestone	Mica, chlorite, epidote, garnet, talc, hornblende, graphite, staurolite, kyanite	Intermediate–high temperature and pressure
	Gneiss	Shale, felsic igneous rocks, graywacke sandstone, basalt, impure limestone	Garnet, mica, augite, hornblende, staurolite, kyanite, sillimanite	High temperature and pressure
May be Foliated or Nonfoliated	Marble	Pure limestone or dolostone	Calcite, dolomite	Contact with hot magma, or confining pressure from deep burial
	Quartzite	Pure sandstone, chert	Quartz	Contact with hot magma, or confining pressure from deep burial
Nonfoliated	Hornfels	Shale, basalt	Andalusite, mica, pyroxene, sillimanite	Contact with hot magma; little pressure

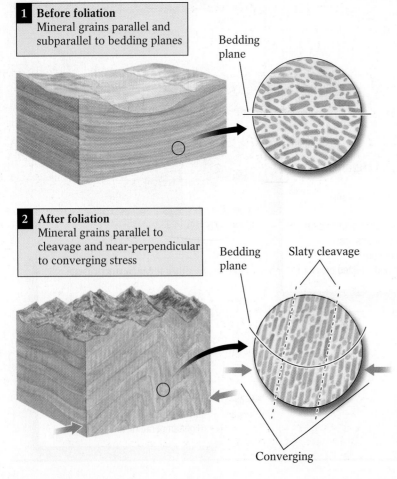

1 Before foliation
Mineral grains parallel and subparallel to bedding planes

Bedding plane

2 After foliation
Mineral grains parallel to cleavage and near-perpendicular to converging stress

Bedding plane

Slaty cleavage

Converging

▲ **FIGURE 7-11** The parallel cleavage characteristic of slate (slaty cleavage) is due to foliation. The orientation of the clay and mica flakes within slate, which is perpendicular to the directed pressure applied during metamorphism, determines the direction of the cleavage.

As temperatures increase to 300° to 400°C (575–750°F), the sizes of the microscopic mica, chlorite, and graphite flakes in slate increase as heat breaks their bonds, their ions migrate, and then rebond at new sites, producing the foliated metamorphic rock **phyllite** (from the Greek *phyllon,* or "leaflike," so-named for its pronounced thin, wavy foliation). This wavy foliation is evidence of metamorphism under directed pressure. Light reflected off the surfaces of these mica and graphite grains gives phyllite its characteristic sheen. The flakes in phyllites continue to grow as heat and pressure increase. Phyllites are still relatively fine-grained, like their parent rock, but you may be able to pick out a few minerals without a magnifying lens.

When some of the minerals—typically the mica—become clearly visible to the unaided eye, the metamorphic rock is called **schist** (from Greek *schistos,* or "split"), a medium- to coarse-grained, strongly foliated rock (Fig. 7-12). Because schists are coarse-grained and well-foliated, they do not much resemble their parent rock, shale. Keep in mind that although schists contain new minerals and a new texture, they (and the slates and phyllites) and their parent rock have the same overall composition (minus some water). The temperature range for schists is 450°C (840°F) to about 650°C (1200°F).

At temperatures above 650°C (1200°F), the minerals become even coarser. At these conditions, the chemical components of minerals such as quartz and feldspars become mobile and migrate from regions of higher stress to regions of lower stress. This process, called **metamorphic differentiation,** separates light-colored minerals, such as quartz and feldspar, from dark-colored minerals, such as biotite mica, garnet, amphibole, and pyroxene. This separation produces rocks with alternating dark and light layers, or dark rocks containing light-colored blobs. These coarse rocks, many of which contain alternating light and dark layers, are called **gneiss** (from the German *gneisto,* meaning "sparkle,"

and pronounced "nice"). Gneisses can form from many different types of parent rock, but in the case of those that form from shales and mudstones, the gneisses do not look at all like their parent rocks.

Where do gneisses form? The high temperatures and pressures required to form gneiss from shale tell us that these rocks were buried to great depths. The two settings in which sedimentary rocks can be metamorphosed to such a great extent are in the deep roots of mountains that form when continents collide, and in continental crust that is dragged down a subduction zone as it trails along behind subducting oceanic crust. Because we can collect metamorphosed shales in the form of gneiss at the Earth's surface today, these rocks must have traveled from the surface (as shales) to great depths—in some cases 30 to 60 km or even >100 km (20 to >60 mi)—and back again (returning to the surface as gneiss).

We find gneisses at the Earth's surface once they have been exposed by extensive erosion and faulting. We can see beautiful examples of gneiss in the Front Range of the Colorado Rockies, in the North Cascades of Washington, in the Adirondacks of New York, and in the Appalachians of Virginia. These rocks also occur in places that don't have high mountains today, but where geologists believe lofty mountains once stood. The 70-million-year-old Tonasket Gneiss (Fig. 7-13) exposed in northern Washington, the 600- to 700-million-year-old Fordham Gneiss exposed in New York's Central Park, and much of the rock in the Canadian Shield are remnants of these former mountain ranges. (See Fig. 7-3 to locate these regions of rock.)

Gneisses are subjected to such high temperatures, they may melt partially, marking a transition to igneous activity. The rocks

▲ FIGURE 7-12 Photograph of muscovite and biotite mica (brightly colored crystals) with quartz (gray crystals) in a schist as seen in thin section with a microscope under cross-polarized light. The field of view is 2 mm.

▲ FIGURE 7-13 The Tonasket Gneiss from the Okanogan region of Washington. The rock's layering formed under high directed pressure and high temperature during metamorphism of continental rocks in western North America about 70 million years ago. The light-colored layers consist of plagioclase feldspar and quartz. The dark layers consist mostly of biotite mica.

do not melt completely, however, so they are part igneous and part metamorphic. The magma may leave the rock, or it may form quartz- and feldspar-rich layers in a process similar to metamorphic differentiation. Gneisses with light-colored layers and dark-colored layers are called **migmatites** (Greek for "mixed rock"), and many migmatites form by partial melting. The light-colored layers, which represent crystallized melt, have mineralogies and textures similar to those of granitic intrusions. Because the rocks deform at high temperatures under directed pressure, they commonly have highly contorted, swirled patterns (Fig. 7-14). Migmatites are very important rocks for understanding the dynamics of mountain-building because they may make up a large part of the continental crust involved in convergent plate boundaries. These rocks may even contain enough melt to allow them to flow easily (similar to magma underground).

Metamorphism of Sandstones and Limestones

Sedimentary rocks composed of primarily one mineral—such as quartz or calcite—may recrystallize as coarse-grained rocks consisting of interlocking *equant* grains (that is, grains having sides of roughly equal length). Strong foliation may not develop in these rocks if they lack *platy* minerals (such as mica). Metamorphosed limestones and dolostones contain coarsely granular mosaics of calcite and/or dolomite grains—a rock called **marble** (Fig. 7-15).

When metamorphosed, quartz sandstone transforms to a rock called **quartzite,** a very durable rock. Quartzite can be foliated or unfoliated. If the rock contains some mica or if the quartz grains are stretched or flattened, the rock will be foliated. Often it's tough to distinguish a hard, well-cemented sandstone from a nonfoliated quartzite. To do so, strike both rocks with a rock hammer and then study their fragments under a binocular microscope. When the sandstone is shattered, it tends to break *around* its sand grains because quartz grains are generally stronger than their surrounding cement. During metamorphism, however, the cement recrystallizes. Thus, when a quartzite is shattered, the rock may break directly *through* the grains (Fig. 7-16).

Metamorphism of Igneous Rocks

Like sedimentary rocks that contain more than one mineral, igneous rocks may metamorphose to form schists and gneisses when exposed to a new set of temperature and pressure conditions. The type of metamorphic rock that forms depends on the composition of the parent igneous rock; basalt and gabbro produce different metamorphic rocks than do felsic volcanic rocks and granite.

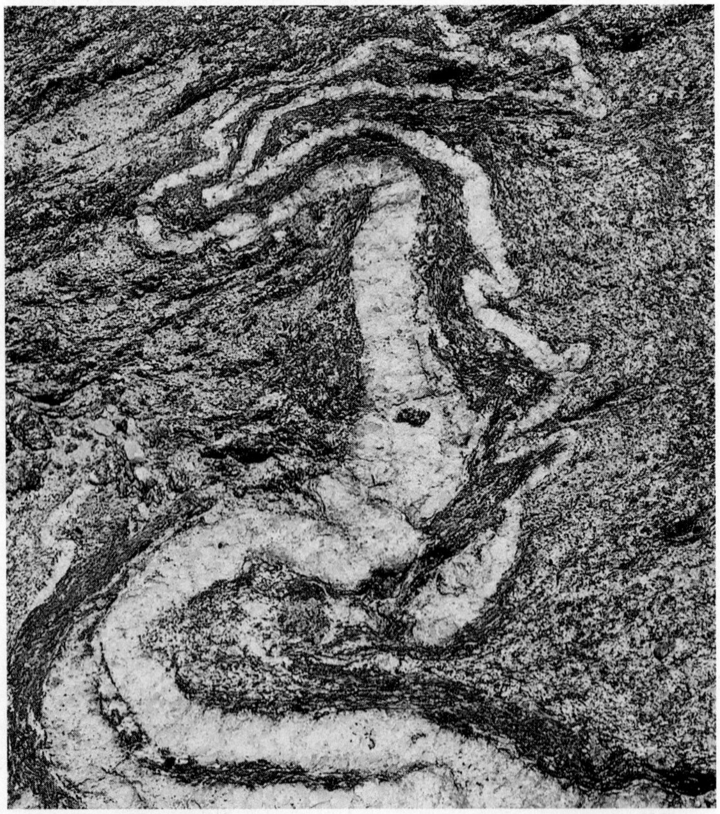

▲ **FIGURE 7-14** Migmatite from the Chuckwalla Mountains of southeastern California. The light-colored layers may have compositions and textures similar to those of intrusive igneous rocks such as granite. Note that the region adjacent to the light-colored layer in the photo is particularly rich in dark minerals (biotite). It indicates that the quartz and feldspar in the light-colored layer have been extracted from this zone, depleting it in felsic minerals.

▲ **FIGURE 7-15** Metamorphic limestone, or marble, is a common material for sculpture. (Abraham Lincoln in the Lincoln Memorial)

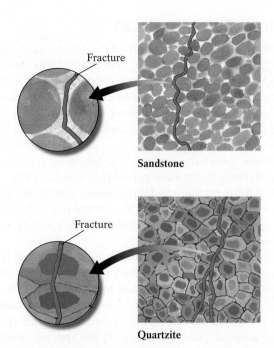

FIGURE 7-16 When a quartz sandstone is fractured, it tends to break within the cement that surrounds the sand grains. After undergoing metamorphism to quartzite, the rock may break right through both the grains and the recrystallized cement.

As you saw in Chapter 3, basalt crystallizes at high temperature, and thus the major minerals in basalt are *anhydrous* ("water-free"). When basalt, a major constituent of oceanic crust, is hydrothermally metamorphosed at divergent zones, the anhydrous minerals hydrate. When this hydrated oceanic crust reaches a subduction zone, it is buried there and thus metamorphoses again—under a new set of temperature conditions. In the subduction zone, the basalt that hasn't already been metamorphosed (perhaps because it did not come in contact with hydrothermal fluids at the divergent zone) will also start to change. The calcium plagioclase in the basalt will be converted to a lower-temperature variety of plagioclase—sodium plagioclase—and anhydrous olivine and pyroxene will be transformed into hydrous minerals such as chlorite, epidote, and amphibole. Because chlorite, epidote, and the amphibole are green, the resulting foliated metamorphic rock is called *greenschist.*

At even higher pressures (but still relatively low temperatures), the green minerals will react to form new minerals such as glaucophane, a blue amphibole that lends its color to *blueschist,* another foliated metamorphic rock. Blueschists are formed only in a subduction zone because it is the only place in the Earth with the right combination of high pressure and low temperature required to form this type of sodium-rich amphibole. Most blueschists are metamorphosed basaltic oceanic crust.

When granite and diorite, the most common plutonic rocks of the Earth's continental crust, are metamorphosed under directed pressure, they form gneisses. Because their parent rocks are originally coarse-grained, metamorphosed granites and diorites do not go through such dramatic textural changes as fine-grained sedimentary rocks do. Because they contain minerals that remain stable at high temperatures (such as quartz) and may not encounter conditions that are very different from the conditions at which the magmas originally crystallized, the mineralogy of these rocks does not change much either. The most obvious metamorphic effect is the development of foliation. Foliation in this type of gneiss is characterized by alignment of micas or amphiboles and formation of light- and dark-colored layers (Figs. 7-14 and 7-15).

A Non-foliated Metamorphic Rock—Hornfels

When magma invades cooler rock, the heat drives mineral reactions and textural changes in the surrounding solid rock. The metamorphic rocks that form next to intrusions typically contain anhydrous minerals that remain stable at high temperatures. These contact-metamorphosed rocks are nonfoliated because no directed pressure was involved in their metamorphism. They tend to be very fine-grained because the magmatic-heating episode doesn't last long enough for the mineral grains to grow very large. The resulting nonfoliated, fine-grained rock is called **hornfels,** and it can form from a variety of rock compositions.

Temperatures and Pressures of Metamorphism

Metamorphic rocks provide key information about how continental and oceanic crust change through time during subduction, plate collisions, and divergence. To extract this information, we need to be able to relate the minerals in the rock to specific temperature and pressure conditions. When geologists collect a regional-metamorphic rock or a contact-metamorphic rock today, we know that it appeared at the surface only after extensive faulting or erosion. But how do we know how hot the rock was when it was being actively metamorphosed? And at what depth in the crust did the metamorphism occur?

Metamorphic Grade

You already know about one way to estimate temperature and pressure conditions—by comparing the progressive stages of metamorphism of a known parent rock such as a shale. When you happen upon an outcrop of slate, you know just from its fine grain size that it couldn't have been heated or buried very much because it still looks very much like its parent rock, shale. Slate is an example of a *low-grade* metamorphic rock. When you see a gneiss or migmatite, you know that the temperatures and perhaps pressures under which it formed were very high. These are *high-grade* metamorphic rocks. **Metamorphic grade** describes the temperature conditions experienced by a metamorphic rock, and is expressed in very general terms—low, medium, and high. The highest-grade metamorphic rocks so far found on Earth were metamorphosed at temperatures in the neighborhood of 1000°C (1825°F).

Index Minerals and Metamorphic Zones

In some places, among them the Scottish Highlands, Dutchess County (north of New York City), southeast British Columbia in western Canada, and various spots around New England (Fig. 7-17), as you walk along looking at the metamorphic rocks exposed in outcrops, you can see a *systematic* change in metamorphic grade over a wide area. If you look at the rocks exposed in and around Poughkeepsie, New York (Dutchess County), for example, you will see layered sedimentary rocks that formed as turbidites along the east coast of North America more than 400 million years ago. You can still recognize the different sedimentary layers in the turbidite, including sandy layers and more shaly layers. If you walk east of Poughkeepsie, similar rocks appear to be slightly folded and have a slaty cleavage, although you can still see the sedimentary layering. When viewed with a microscope, metamorphic minerals such as micas can be identified in these rocks. Even further east, you will see garnet in phyllite and then schist, and so on, all the way up to coarse, high-grade gneiss containing several key high-temperature minerals. Each new metamorphic mineral that appears in sequence from low to high grade is called a **metamorphic index mineral.** It is evidence that a reaction produced a new mineral in response to increasing temperature and pressure. Index minerals also appear in a predictable sequence near intrusions in contact-metamorphosed rocks (Fig. 7-18).

Index minerals are useful for determining how metamorphic grade changes within a region, but the best way to estimate conditions of metamorphism is to look at *assemblages* (groups) of minerals in a rock. Minerals that coexist in a metamorphic rock may have formed together at the highest grade of metamorphism, when metamorphic reactions are most effective. Therefore, coexisting minerals can be used as an indicator of the lithostatic pressure and the temperature of metamorphism.

Experimenting with Rocks

Geologists want to know more about the temperature and pressure conditions of metamorphism than simply "low" and "high." If we find a metamorphic rock today exposed in the core of an eroded mountain range, we want to know whether it was metamorphosed by 6 kilobars of pressure (equivalent to a depth of about 18 km, or 12 mi, in the crust) or 40 kilobars (120 km or 80 mi depth). This information helps us to know how thick the crust was when the mountains formed, and this in turn may relate to how high the mountains were before they eroded away. We also want to know the pressure and temperature conditions of a metamorphic rock because it helps us determine the tectonic setting. Certain conditions indicate particular tectonic settings, such as subduction zones.

One way to determine pressures and temperatures is to simulate metamorphism in a laboratory experiment. The chemical compounds that make up common metamorphic rocks can be combined, and the resulting mixtures heated and pressurized. We can then observe which minerals formed during this episode of simulated metamorphism and identify the specific temperature and pressure conditions at which they formed.

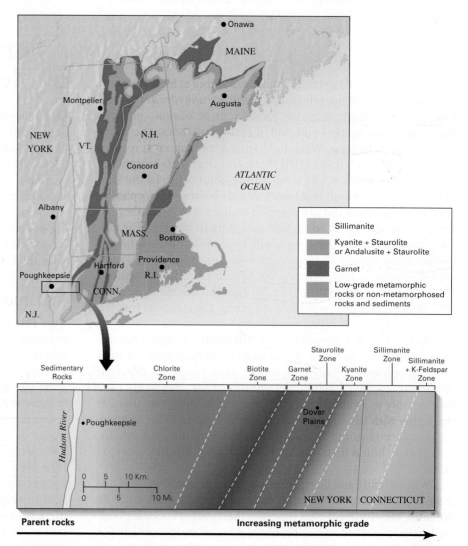

▶ **FIGURE 7-17** The metamorphic mineral zones of the northeastern United States. These zones resulted from regional metamorphism associated with one of the major Appalachian mountain-building episodes. In a few places, metamorphic zones from earlier mountain-building episodes are preserved. For example, metamorphic mineral zones between the Hudson River and the New York-Connecticut border range from the parent rocks (very low grade metamorphosed sediments) near Poughkeepsie to high-grade metamorphic rocks near the state line. These zones represent an increase in metamorphic temperature of at least 600°C (1100°F) from west to east. This sequence of rocks formed during metamorphism and mountain-building over 400 million years ago.

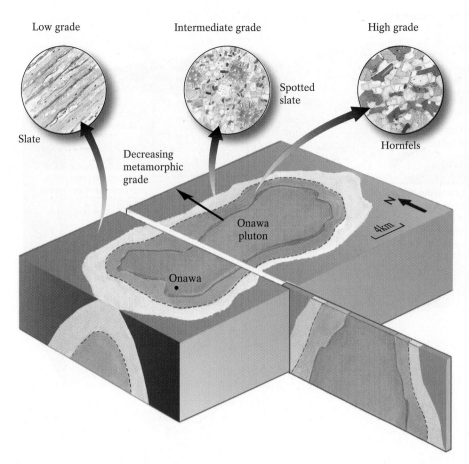

▲ **FIGURE 7-18** Heat from the intrusion of the Onawa pluton in central Maine created a series of metamorphic mineral zones typical of large-scale contact metamorphism. (See map in Figure 7-17 for location of Onawa, Maine.) The metamorphic rocks associated with this pluton range from high-grade hornfels to intermediate-grade "spotted slate" (slate containing a number of large, recrystallized grains of andalusite) to low-grade slate. The slate was formed by an earlier regional-metamorphic event.

periments with knowledge of the physics and chemistry of groups of coexisting minerals. This technique, called *thermobarometry,* involves using rocks as "thermometers" and "barometers" for determining past temperatures and pressures. It allows geologists to calculate metamorphic temperatures and pressures to within an accuracy of 50°C and less than 2 kilobars for some rocks.

A limitation of the use of index minerals and even thermobarometry for determining metamorphic conditions is that not all parent rocks have suitable compositions that contain minerals useful for pressure–temperature determinations. A typical limestone, for example, does not have enough aluminum to produce kyanite, andalusite, or sillimanite during metamorphism. It is very difficult to determine the conditions of metamorphism for rocks such as marble, quartzite, and metamorphosed granite, because they generally lack index minerals and groups of minerals that can be used as thermobarometers. Instead, geologists must survey all of the rock types in a region, find as many indicators of pressure–temperature conditions as possible, and look for trends in metamorphic grade, mineral assemblages (groups of minerals), and associated structural features in the rocks. Together, these findings can be combined to identify the plate-tectonic setting and metamorphic history of the rock.

Pressure–Temperature Paths and Plate Tectonics

As described above, the entire history of a metamorphic rock from start (the original location of the

Figure 7-19 shows the pressure–temperature stability of the minerals kyanite, sillimanite, and andalusite. All three minerals have the same composition (Al_2SiO_5) but different crystal structures. The stability field of each mineral shown on this diagram was determined by experimentation. Because these minerals are so useful for determining pressure and temperature conditions, geologists who study metamorphism are very excited when they find rocks that contain them. This diagram is a favorite tool of geologists trying to estimate the pressures and temperatures of metamorphism of rocks containing these minerals. An even more powerful technique for determining pressures and temperatures combines information from such ex-

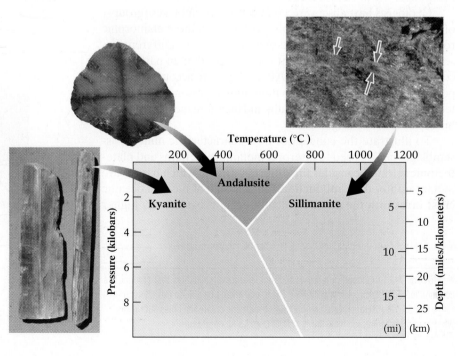

▶ **FIGURE 7-19** The depth, pressure, and temperature relationships that are responsible for the crystallization of the aluminum silicates kyanite, andalusite, and sillimanite. Andalusite is generally found only in rocks that form at relatively low pressures.

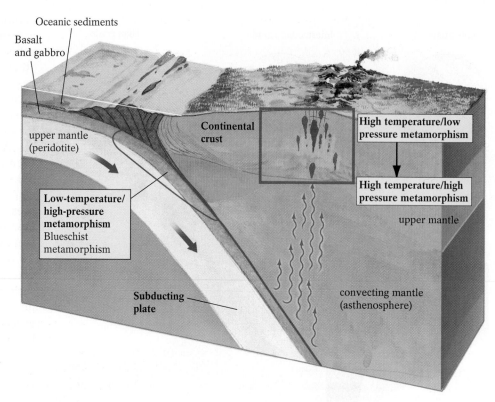

▲ **FIGURE 7-20** The three main metamorphic environments associated with subduction zones: low temperature/high pressure, high temperature/high pressure, and high temperature/low pressure.

parent rock) to finish (exposure of the rock at the Earth's surface) involves changes in pressure and temperature. The sum of these changes is called the *metamorphic path*. During dynamothermal metamorphism, the first part of the path involves heating and burial, followed by cooling and the upward motion of the rock toward the Earth's surface. During contact metamorphism, the path involves heating and cooling.

Geologists have discovered that certain assemblages (groups) of minerals are found predictably worldwide in the metamorphic rocks formed in specific plate-tectonic settings. Different assemblages are found in different rocks, depending in part on the composition of the parent rock. After taking into account the composition of the rock, we can use these mineral assemblages to determine pressure–temperature paths and therefore the metamorphic history of the rock (Fig. 7-20).

To illustrate the concept of how characteristic mineral assemblages can be used to interpret metamorphic paths and plate-tectonic settings, consider the different ways in which basalts are metamorphosed. In subduction zones, a slab of basaltic oceanic crust dives into the mantle, encountering progressively higher

temperatures and pressures as it descends. As mentioned earlier, temperature in subduction zones does not increase as much with depth as it does in other settings, such as within continental crust. Figure 7-21 shows the pressure–temperature path that might be experienced by subducting oceanic rocks. This steep path, which shows little change in temperature but a huge change in pressure (depth), is characteristic of subduction-zone metamorphism. It

▶ **FIGURE 7-21** The depth, pressure, and temperature relationships that produce the common metamorphic rocks. With such a chart, one should be able to extract such information as: What temperature and pressure conditions create the blueschist rocks? What kind of metamorphic rocks form at 500°C and 3 kb of pressure?

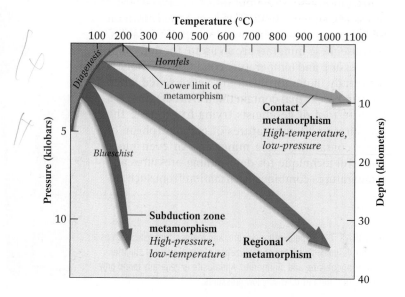

is called high-pressure/low-temperature metamorphism, or *blueschist metamorphism.*

Continental crust above a subduction zone or continents experiencing dynamothermal metamorphism during plate collisions are metamorphosed along a path characterized by changing temperature and pressure. Figure 7-21 shows the intermediate path followed by many continental rocks in these settings. These rocks make up most of the world's great mountain belts.

Contact metamorphism, hydrothermal metamorphism, and metamorphism at shallow levels of the crust in volcanic arcs are characterized by high temperatures at low pressures. Figure 7-21 shows a metamorphic path for these rocks that involves heating without much (if any) change in pressure.

Plate Tectonics and Metamorphic Rocks

Plate movements create the changes in heat and pressure (depth) that drive most metamorphism. Directed pressure at plate boundaries creates the common textures such as foliation in metamorphic rocks, with the orientation of the foliation giving clues to the direction of past plate motion. (Remember, the foliation lies perpendicular to the directed stress, and thus it lies perpendicular to the principal direction of plate convergence, the source of the directed stress.) The new magmas generated at subduction zones are responsible for most contact metamorphism.

Blueschist rocks represent a particularly valuable clue to the location of ancient subduction zones and convergent plate boundaries. These rocks form at the high-pressure, low-temperature conditions that occur *only* in subduction zones. The blueschist rocks in western California, for example, most likely mark the subduction zone that produced the intrusions of the Sierra Nevada, some of which are visible in Yosemite National Park. Blueschist rocks have also been discovered in the Appalachians of northern Vermont; this area was a zone of converging plates, volcanism, and great earthquake activity 300 to 400 million years ago as Pangaea was being assembled (discussed in Chapter 1, page 29). Extensive belts of blueschists also occur in the Greek Isles, Turkey, and Japan.

Another interesting association between metamorphism and tectonics is the link between serpentinites and plate boundaries. This rock is produced by hydrothermal metamorphism of mantle rocks (peridotite) at divergent plate boundaries (see Fig. 7-10 on page 213). Some geologists have recently proposed that the heat-induced seafloor reactions that produce serpentinite may have been important in the origins of the Earth's early life. Serpentinites are also important at subduction zones—the stresses related to the dehydration of the water-bearing serpentine minerals may produce some subduction-zone earthquakes. And finally, serpentinites may be important at transform boundaries (like the San Andreas Fault); many of the transform faults in the San Francisco Bay area occur within layers of serpentinite. Perhaps because it is such a soft, weak rock, it creates zones where the Earth's crust can break easily, creating faults and triggering earthquakes.

Metamorphic Rocks in Daily Life

High-grade metamorphic rock (such as gneiss)—a hard and durable rock—is a popular building material for the weather-resistant exteriors of office buildings and other large structures and as foundation stones for bridges and dams. Building and road construction also uses roughly 1.6 billion metric tons of slate and quartzite every year in the United States.

We value certain metamorphic rocks highly for their appearance as well as their durability. Serpentinite is among architecture's most prized decorative building stones. Called *verd antique* ("ancient green") by architects, serpentinite can be seen in the interiors of such stately buildings as the United Nations and the National Gallery of Art, as well as many office towers, banks, and libraries (Fig. 7-22). The white streaks that commonly vein the green serpentinite are carbonate minerals such as calcite.

Sculptors prize the metamorphic rock marble for its appearance, softness, and texture. Because pure marble is snow-white, it shows best the details of a sculptor's carving. Because it is soft, it responds extremely well to the sculptor's chisel. (Calcite, marble's main constituent, measures only 3 on the Mohs Hardness Scale.) The stone from which Michelangelo fashioned his renowned sculptures was produced by metamorphism when the Apennine mountains of Italy, which contained a layer of pure marine limestone, were squeezed between colliding plates (Fig. 7-23). In contrast, metamorphism of impure limestones yields marbles colored by their impurities—pink, red, or brown from

▲ **FIGURE 7-22** Verd antique, a popular decorative stone, is composed of the hydrothermally metamorphosed rock, serpentinite.

▲ **FIGURE 7-23** The photo shows the Carrara marble quarry in northern Italy. The purity of Carrara marble, which stems from the purity of its parent limestone, has long made it a favored source of sculpting material.

Some metamorphic rocks contain economically important mineral deposits (discussed in Chapter 20). Zinc (in the mineral sphalerite), lead (in galena), copper (in chalcopyrite and bornite), iron (in magnetite), and gold are just a few of the valuable minerals found in metamorphic rocks.

Potential Hazards from Metamorphic Rocks

While some characteristics of metamorphic rocks have great aesthetic or practical value, others can produce conditions that are hazardous to human safety. Foliation, in particular, can prove quite dangerous. Because it reduces the strength of metamorphic rocks, slaty cleavage and schistosity can weaken slopes and trigger landslides, particularly where the cleavage planes lie parallel to steep slopes (Fig. 7-24). Building in foliated metamorphic terrains requires careful planning and thorough geological investigation to determine the orientation of the foliation.

Weak planes of metamorphic foliation are especially susceptible to earthquake damage. Such was the case at midnight on August 17, 1959, near Yellowstone National Park in Montana, when a powerful earthquake shook the Madison Canyon area. A deeply weathered body of gneiss and schist, the foliations of which paralleled the canyon's south wall (Fig. 7-25), slid away from adjacent rock during the shaking. The landslide picked up debris as it raced downslope at about 100 km/hr (60 mph), burying a campground and its 28 visitors beneath 45 m (150 ft) of rock, soil, and vegetation.

Similar disasters have occurred when dams built into foliated metamorphic rock—with foliation planes parallel to the valley wall containing the reservoir behind the dam—have failed when the foliation plane gave way. Metamorphic foliation must be analyzed when designing large-scale engineering projects such as dams. Today, teams of geologists and engineers routinely study local and regional rock formations for evidence of inherent weakness before beginning any major construction. In the past, however, this step was seldom taken—sometimes with disastrous consequences. On the night of March 12, 1928, the St. Francis

iron oxides; gray or black from carbon-rich material; or green from the calcium–magnesium pyroxene mineral, diopside, or the common amphibole, hornblende.

Low-grade regional metamorphism of ultramafic rocks may produce talc, used in talcum powder, as well as flame-resistant asbestos, formerly used in automobile brake linings, the safety garb favored by firefighters, and toddler's pajamas (the latter use has ended because of asbestos toxicity). Highlight 7-2 explains some of the issues surrounding asbestos use in North America. Garnet, an industrial abrasive valued for its hardness (and also a semiprecious gem and January's birthstone), is produced by intermediate- to high-grade metamorphism. The high-grade metamorphic minerals kyanite and sillimanite are key components of porcelain casings for spark plugs because the high temperatures at which they crystallize enable them to withstand intense heat.

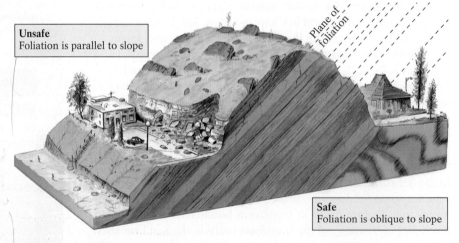

Unsafe
Foliation is parallel to slope

Plane of foliation

Safe
Foliation is oblique to slope

▲ **FIGURE 7-24** Building in areas with metamorphic surface rocks requires consideration of the direction of the rocks' foliation planes. The slope on the left side of this figure is unstable, because the rock's foliation lies parallel to the slope and can readily fail, causing a landslide of thick slabs of rock. The slope on the right is stable, because the foliation is not oriented parallel to the slope.

Highlight 7-2

Asbestos, Cancer, and Crayons—A Study in Environmental Health

For more than 30 years, the Crayola Corporation has been manufacturing its crayons with a recipe that includes the mineral talc from the R. T. Vanderbilt mine in upstate New York near the town of Waterville. In late May 2000, the talc industry and numerous physicians' groups were locked in a legal–scientific battle to determine whether the talc from this mine contains cancer-causing asbestos particles. With this dispute in mind, let's shed some light on asbestos, a serious environmental threat to human health.

Asbestos isn't actually a specific type of mineral. Several different minerals produce long, thin crystals that are called collectively by the general name *asbestos*. The flexibility of these fiberlike crystals and their useful insulating properties give them many industrial uses. Asbestos crystals are so fine (finer than human hair) that they may be woven into fabrics, such as in fireproof clothing worn by firefighters. Some forms of asbestos can withstand temperatures of nearly 2800°C (5000°F)! Asbestos is also used to strengthen cement, insulate heating pipes and boilers, soundproof walls and ceilings, and line brake and clutch pads in cars and trucks.

"White asbestos" is a mineral called serpentine that forms when magnesium-rich silicate minerals, such as olivine and pyroxene, react with hot water and are converted to hydrous minerals. Hard, blocky olivine crystals can be changed into soft, silky serpentine by the addition of water to the crystal structure. More than 90% of all asbestos in buildings in the United States, and 99% of all the asbestos used worldwide, is a type of serpentine called *chrysotile* (see figure).

Other types of asbestos belong to the amphibole group of silicates. Amphiboles are minerals whose crystal structures consist of long chains of silicon–oxygen tetrahedra, and a few types can crystallize as asbestos. Amphibole asbestos occurs as stiff, needlelike crystals that are more iron-rich than serpentine asbestos. The asbestos detected in Crayola crayons is *tremolite* asbestos, an amphibole.

Health hazards caused by exposure to asbestos were first identified as early as 1900. In the early 1970s, the U.S. federal government ordered major restrictions on the manufacture of asbestos. Nevertheless, widespread use of asbestos-containing materials continued into

The white fibers are chrysotile asbestos.

the 1980s. Many court battles have been fought over asbestos regulation and health effects.

What are the health hazards caused by breathing in asbestos fibers? Several diseases may be caused by inhaling asbestos, including permanent scarring of lung tissue, which makes breathing difficult, and some types of deadly lung cancer, including mesothelioma. These diseases do not occur immediately after exposure to asbestos—they may take 15 to 40 years to develop.

Why is asbestos so dangerous? Asbestos minerals are chemically inert, so they aren't poisonous like lead and other environmental hazards. But because asbestos crystals are so small and light, they can remain suspended in the air for long periods of time. People living or working in environments containing large quantities of airborne asbestos will breathe in the crystals, and the fibers can become lodged in lung tissue. There they inflame the cells and cause several types of infection, some of which are believed to release cancer-causing chemicals. The most serious health threat is apparently related to the length and diameter of the asbestos crystals—factors that determine whether the crystals will remain long-term in the lungs. Tremolite asbestos—with its long needlelike crystals—is particularly likely to become lodged in the human lung.

Asbestos fibers are everywhere in the air and water, and most of us breathe and eat thousands to millions of asbestos crystals every day without serious health effects. Asbestos is a major health problem, however, for people who inhale lots of fibers over a long period of time, such as miners and people living or working in buildings with old, deteriorated, or damaged asbestos-containing materials. Because the health effects of long-term exposure to low levels of asbestos are not yet known, the Environmental Protection Agency has determined that there is *no* safe level of asbestos exposure. Consequently, extensive measures have been taken to remove asbestos from the environment. Because of these concerns, asbestos is one of the most highly regulated mineral substances in the world.

Much research is currently being done on the health effects of asbestos. Recent work has shown that the different types of asbestos have different health effects, with the amphibole forms being the most dangerous. All asbestos must be handled with care, and steps should be taken to minimize the number of fibers in the air we breathe. Special caution must be taken while removing asbestos during destruction or renovations of buildings. In particular, filtration masks should be worn during this work.

There are many places to seek information and advice about asbestos, including a booklet published by the EPA called *Asbestos in Your Home*. This booklet can be read on the Internet at **www.epa.gov// asbestos/ashome.html.**

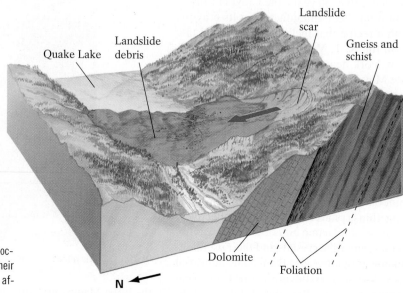

▲ **FIGURE 7-25** The disastrous rock slide of 1959 at Madison Canyon, Montana, occurred when an earthquake jolted weak, highly fractured gneiss and schist along their foliation planes parallel to the canyon's south wall. Above: A photo of the landslide aftermath, taken from the northern canyon wall.

▲ **FIGURE 7-26** (above) The geology of the St. Francis dam, which collapsed in 1928. The weak, foliated rocks that supported one side of the dam and the faulted bedrock below the dam led to the dam's failure. (below) The aftermath of the St. Francis dam collapse.

dam in Saugus, California, failed, claiming 420 lives in the ensuing flood. Its eastern side was attached to loose landslide debris from a weak schist whose foliation planes lay parallel to the canyon wall. This structure, combined with a fault that traversed the dam site, was largely responsible for the dam's failure (Fig. 7-26). The disaster led to subsequent improvements in the way we investigate sites for prospective dams, although community opinion and political pressure may still cause the geologists' protective warnings to go unheeded.

The Rock Cycle Revisited

In our preview of the Earth's rocks in Chapter 1, we saw the principal pathways of the rock cycle and the variety of processes that create them. We have now examined in detail the three basic types of rocks—igneous, sedimentary, and metamorphic—and the major processes (melting, crystallization, weathering, lithification, and metamorphism) that form and alter them. We can now fully appreciate the relationships among these processes. Any rock, including an existing igneous rock, can be melted and form a new igneous rock. Any rock, including an existing sedimentary rock, can be weathered, eroded, transported, deposited, and lithified to form a new sedimentary rock. Any rock, including an existing metamorphic rock, can be heated and buried to form a new metamorphic rock. The cycle has no end. Every body of rock on Earth is constantly changing—although, as we have seen, some processes change rocks more slowly than others.

We have also seen how plate-tectonic processes control the cycling of the Earth's rocks. The igneous processes associated with rifting and divergent zones produce new oceanic lithosphere. Subduction recycles mafic oceanic lithosphere and its overlying sediments, producing plutons of intermediate and felsic rock in the overlying crust. In this chapter, we have seen how the heat and directed pressure at convergent plate boundaries can convert a soft sedimentary rock (such as shale) to a hard metamorphic rock (such as gneiss). Clearly, plate tectonics plays a pivotal role in the evolution of the Earth's rocks.

In the next chapter, we will examine what the Earth's rocks tell us about the past. In succeeding chapters, we will look more closely at how some of the processes that create rocks also shape the Earth's mountains, continents, and ocean floors.

Chapter 7 Summary

Metamorphic rocks, the third major type of rock in the Earth's rock cycle, are formed at temperatures and pressures intermediate between those that form igneous and sedimentary rocks. They are created by metamorphism of preexisting rock, when the mineral content and texture of solid rock are altered by heat, pressure, and chemically active fluids, *but the rock does not melt.* During metamorphism the rock may lose many of its original features: its mineral grains will generally grow larger and develop an orderly parallel arrangement, and they may even recrystallize into new minerals that are more stable under the prevailing conditions.

Most metamorphism occurs at depths greater than 10 kilometers (6 miles) below the Earth's surface, where the Earth's internal heat and the weight of the overlying rock alter solid rock. Water-rich fluids also circulate at these depths, enhancing migration of unbonded ions in rocks undergoing metamorphism. Under any given set of conditions, the composition of the parent rock determines which metamorphic rocks can form.

Rocks at great depths are subjected to *lithostatic,* or confining, *pressure,* which pushes at the rock equally from all sides, making it smaller and denser but maintaining its general shape. Some rocks—particularly those at convergent plate boundaries—are also subjected to *directed pressure,* which distorts their shapes, flattening them in a plane perpendicular to the directed pressure. The minerals in a rock subjected to directed pressure align in parallel planes perpendicular to the direction of the pressure, giving the rock a preferred mineral orientation, or *foliation.*

Metamorphism varies according to the prevailing geological conditions. The various types of metamorphic processes include:

- *Contact metamorphism,* which occurs where rocks are heated by direct or close contact with magma;
- *Regional metamorphism,* which occurs over a broad area of rock, and involves a change in heat and pressure (both directed and confining) generally in an area of plate convergence;
- *Hydrothermal metamorphism,* which involves hot circulating fluids – either seawater (at oceanic divergent zones) or fluids from various sources in the continents;
- *Fault-zone metamorphism,* which occurs where rocks grind against one another at active faults;
- *Shock metamorphism* occurs where rocks become violently heated and compressed by the impact of a meteorite.

Increased heat and directed pressure are responsible for the sequence of foliated rocks that forms from a typical marine shale. This sequence begins with the fine-grained *slate* and *phyllite,* and progresses to coarse-grained *schist* and *gneiss.* The mineral bands in gneisses result, in part, from *metamorphic differentiation,* in which unstable minerals recrystallize as new, more stable minerals in alternating bands of light (felsic) and dark (mafic) minerals. When temperature and pressure increase beyond metamorphic conditions to the threshold of igneous conditions, a "mixed" rock—called *a migmatite*—develops with both metamorphic and igneous parts.

The degree to which a metamorphic rock differs from its parent material is described in terms of *metamorphic grade.* The more extreme the metamorphic conditions (primarily temperature), the higher the grade—and the more a metamorphic rock differs from its parent rock.

The minerals in metamorphic rocks can provide clues to the conditions under which the metamorphism occurred. Some minerals, which remain stable only within a narrow range of temperatures and pressures, indicate specific metamorphic conditions; they are called *metamorphic index minerals.*

Different types of metamorphic rocks form in different plate-tectonic settings. Some examples of this include:

- *Blueschists,* which are produced in the high-pressure, low-temperature environment in a subduction trench;
- *Serpentinites,* which are produced at divergent plate boundaries and are created by hydrothermal metamorphism. These relatively low-temperature reactions involve the interaction between warm ocean-ridge basalts and circulating seawater.

KEY TERMS

burial metamorphism *(p. 212)*
contact metamorphism *(p. 210)*
directed pressure *(p. 208)*
dynamothermal metamorphism
 (p. 211)
fault-zone metamorphism
 (p. 213)

foliation *(p. 208)*
gneiss *(p. 217)*
hornfels *(p. 219)*
hydrothermal metamorphism
 (p. 212)
lithostatic pressure *(p. 208)*
marble *(p. 218)*

metamorphic differentiation
 (p. 217)
metamorphic grade *(p. 219)*
metamorphic index mineral
 (p. 220)
metamorphic rocks *(p. 204)*
metamorphism *(p. 204)*

migmatite *(p. 218)*
phyllite *(p. 217)*
quartzite *(p. 218)*
regional metamorphism *(p. 211)*
schist *(p. 217)*
shock metamorphism *(p. 213)*
slate *(p. 214)*

QUESTIONS FOR REVIEW

1. What are the major factors that cause metamorphism?

2. Describe four ways that rocks may change during metamorphism.

3. Define metamorphic foliation, explain how it develops, and list three metamorphic rocks that are foliated.

4. List the sequence of metamorphic rocks that develops during the progression of shale to migmatite. Describe the successive changes in mineralogy and structure that the rocks undergo.

5. Under what conditions do nonfoliated metamorphic rocks generally form? Give examples.

6. List the six common types of metamorphism, and discuss the geological conditions under which they occur.

7. Kyanite, andalusite, and sillimanite are considered metamorphic index minerals. What is a metamorphic index mineral, and why do these three minerals qualify?

8. What is the plate-tectonic significance attached to the presence of the blueschist mineral glaucophane in a rock? What type of metamorphism generally occurs at mid-ocean ridges?

9. Why do minerals such as chlorite rarely form during high-grade contact metamorphism?

10. The Earth's internal temperature was apparently much higher during the first billion years of its history. Speculate about how metamorphism, metamorphic environments, and the distribution of metamorphic rocks would have differed during that period.

11. From the photo below, estimate the orientation of the directed pressures that metamorphosed these rocks.

12. Are the rocks 20 km beneath your campus metamorphosing today? Why? Or, why not?

LEARNING ACTIVELY

1. Metamorphic rocks are the most decorative of building stones. Look around your campus—especially in the libraries—and also check out the decorative stone on any nearby office buildings, banks, and train stations. They are most likely to be metamorphic. Try to identify them. Speculate about the metamorphic conditions that produced them (High or low grade? Foliated or nonfoliated? Directed or confining pressure?). Could they be from your state or region?

ONLINE STUDY GUIDE

Practice makes you better. Make the time to take the *Online Study Guide* quiz for this chapter. It should take you about 20 minutes. Automatic grading and instant feedback will help you quickly and accurately identify the concepts you have mastered and the areas in which you need more study.

www.prenhall.com/chernicoff

Telling Time Geologically

D o you remember the names of your classmates who sat next to you in kindergarten when you were first learning to tell time? Seems like a long time ago, no? From our human perspective, fifteen or twenty years seem like ages ago, but in a geological time sense, they are less than a blink of an eye. In this chapter, you, like all beginning geology students, will *relearn* how to "tell time," but in this case, you'll learn how to tell time *geologically*—not in minutes, hours, and days, but in thousands, millions, and billions of years.

Throughout the first seven chapters of this text, we've asked and answered some basic geological questions: How do plates move? How and where do rocks melt, in some cases to erupt at the surface from volcanoes? How do solid surface rocks weather to form loose grains of sand and then turn to stone as sedimentary rocks? How do heat and pressure convert rocks underground into metamorphic rocks with new minerals and textures? Missing thus far from these topics are the critical questions of time: How long do these Earth-changing processes take? How do we know the actual age of a rock, a mountain, or the Earth itself?

When thinking about geologic time, we are often overwhelmed by its vastness. Most of us have little difficulty remembering our high school days (although for some of us it's quite a while ago); we are used to thinking about events in our own lives and in human history on the scale of years or decades. But many geologic events take place over millions or billions of years, and these are lengths of time that are tough for us humans—who live for 90 or so years—to conceive. Earth processes that occur rapidly on a human timescale include volcanic eruptions, floods, landslides, earthquakes, and meteorite impacts. Most large-scale tectonic events, such as the creation of a mountain range, occur very slowly. The opening of ocean basins as continents break apart and drift away from each other and the collision of plates to produce mountains take tens or hundreds of millions of years. The Himalaya have been building up to their present great height for the past 55 million years. That might sound old, but these mountains are actually considered geologically "young."

How do we know that the Himalaya started forming 55 million years ago? How do we know that the dinosaurs became extinct 65 million years ago (Fig. 8-1)? And for that matter, why do we believe that the Earth is 4.6 billion years old? Determining the timing of geologic events and the ages of geologic materials is the subject of **geochronology,** the study of "Earth time."

FIGURE 8-1 This fossil of a million-year-old mammal called *Repenomamus robustus* contains the remains of a very young *Psittacosaurus* in its stomach. Some paleontologists suggest that the animal's last meal probably is the first proof that mammals hunted small dinosaurs some 130 million years ago. It contradicts conventional evolutionary theory that early mammals were timid, chipmunk-sized creatures that scurried in the looming shadow of the giant reptiles.

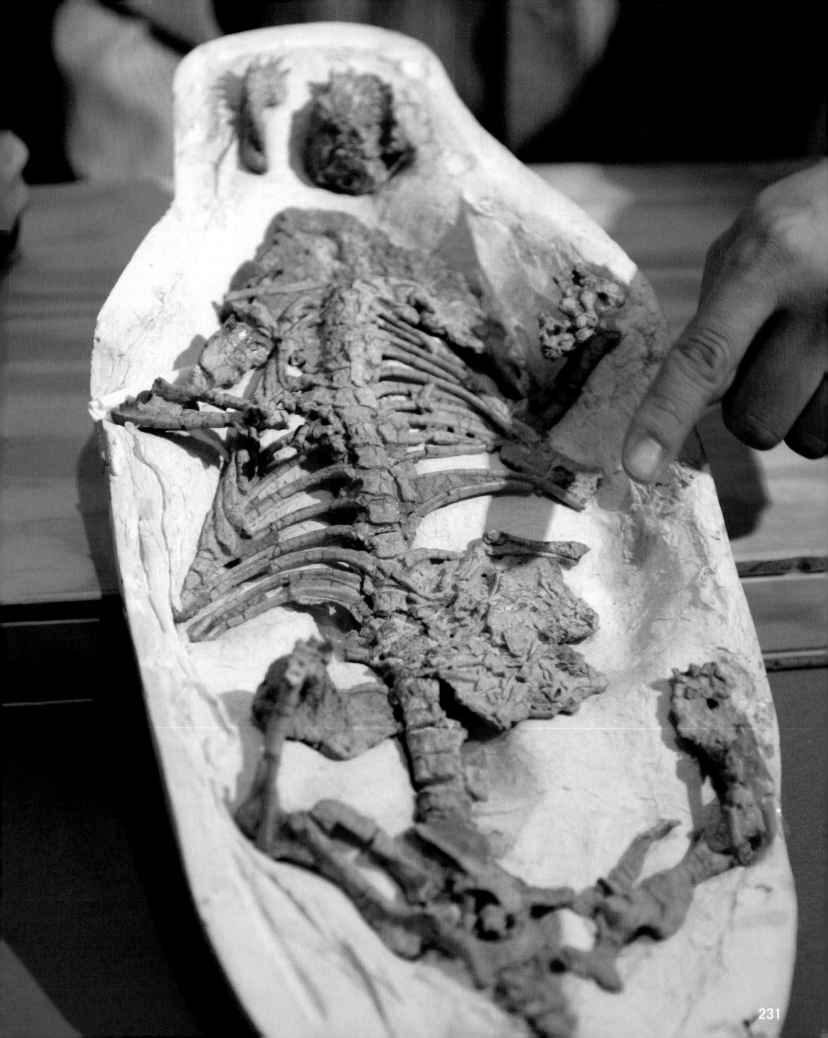

Once we know how to date rocks and other geologic features, can we ever know the age of a long-past event well enough to celebrate its anniversary? Can we pinpoint precisely the moment of the meteorite impact that might have killed off the dinosaurs so that we can commemorate its happening—say, every March 25? The answer is obviously no. Geologists work with such vast lengths of time that it rarely matters whether an event occurred 350 million years ago or 354 million years ago—even though 4 million years is a huge amount of time in human terms—more than the amount of time since the appearance of our first upright-walking, human ancestors.

To tell time geologically, geologists use a variety of ingenious geologic "clocks." Throughout much of this chapter, we will explain how radioactive elements, fossils, trees, and other materials function as geologic clocks. But first we must put geologic time into a human perspective so that we—with our lifespans of 90-plus years—might begin to grasp the magnitude of the Earth's great longevity.

Geologic Time in Perspective

Here are three statements that represent different human views of the age of the Earth:

- "Heaven and Earth, Centre and substance were made in the same instant of time and clouds full of water and man were created by the Trinity on the 23rd of October 4004 B.C. at 9:00 in the Morning."

- "High up in the North in the land called Svithjöd, there stands a rock. It is a hundred miles high and a hundred miles wide. Once every thousand years a little bird comes to this rock to

sharpen its beak. When the rock has been thus worn away, then a single day of eternity will have gone by."

- "The Earth is approximately 4.6 billion years old."

There's something here for everyone. The first quote records the thoughts of Dr. John Lightfoot, Vice Chancellor of Cambridge University. Using the genealogical record of the Old Testament to work back to the biblical account of Creation, Lightfoot, in 1664, dated the Earth as being precisely 5668 years old. (His estimate of the age of the Earth was similar to the calculation by Archbishop James Ussher in 1650.) This view, which appealed to those who value precision and those who seek theological dominion over science, governed scientific thought for several centuries. To the naturalists of the 18th and 19th centuries, however, it proved somewhat "confining." Their observations led them to propose that the Earth would have needed much more time to evolve into its current condition.

The second view comes from a Norwegian folktale recounted by Hendrik Van Loon in his *Story of Mankind.* It should appeal to those among us who fancy ourselves to be deep thinkers. Although imprecise, it gives our finite minds some sense of the great extent of Earth history and perhaps a hint at what a billion years might feel like.

The third view—less precise than the first view and less poetic than the second—is widely accepted by geoscientists today. Yet how can we begin to grasp the magnitude of 4,600,000,000 years? In human terms, that inconceivable length of time might as well be the eternity of the little bird in the Norwegian folktale.

Geology professors will never rest until they find a dramatic way to illustrate to their students the great span of Earth history. One way is to picture a timeline stretching across the United States, 5000 km (3000 mi) long, on which each kilometer represents about 1 million years (Fig. 8-2). No rocks or minerals from

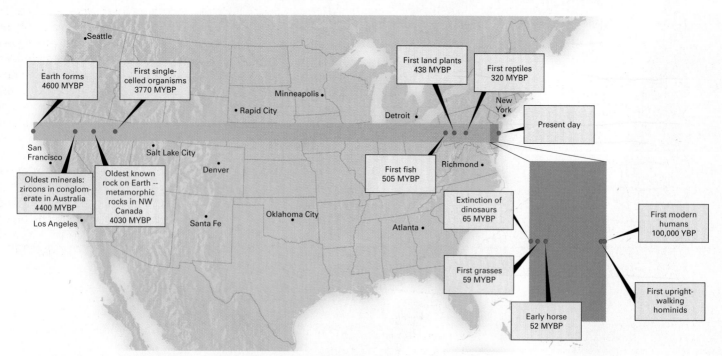

▲ **FIGURE 8-2** Earth history represented by a timeline extending from west to east across the United States, with each kilometer corresponding to about 1 million years. The events that we know most about have generally occurred quite recently in geologic time. "MYBP" stands for "million years before present." "YBP" stands for "years before present."

the Earth's first 200 million years have ever been identified, so no *direct* information is available for this period; the first 200 km (130 mi) of our continent-wide time line is, therefore, barren.

The oldest rocks known, discovered in Canada's Northwest Territories, are dated at 4.03 billion years. That is extremely old, but in western Australia near the city of Perth, a few grains of the mineral zircon ($ZrSiO_4$)—a very tough mineral that resists chemical and mechanical weathering and is, therefore, found as fragments in younger rocks—have been extracted from the region's Jack Hill Conglomerate, a sedimentary rock, and some of those grains have been dated at 4.0–4.4 billion years old. The oldest grain is less than 200 million years younger than the age of the Earth (Fig. 8-3).

The Earth's most ancient rocks do not contain any direct evidence of the planet's earliest life, perhaps because conditions during the planet's first billion years were too hot and toxic to support most forms of life, except perhaps some forms of bacteria that can tolerate such extreme conditions. Alternatively, all evidence for the very earliest life may have been destroyed by weathering, metamorphism, or, more spectacularly, by large meteorite impacts that generated enough heat to vaporize the planet's first bodies of water and kill everything in them. The earliest-known life forms on Earth are microscopic single-celled organisms that lived 3.77 billion years ago. Scientists have identified fossils of these organisms—so primitive that they had not yet developed a cell nucleus—

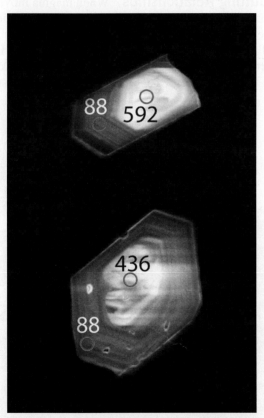

▲ **FIGURE 8-3** Images of two zircons with cores that are hundreds of millions of years older than the rims. The numbers on the crystals are ages in millions of years of different spots in each crystal, as measured by a Super High Resolution Ion Microprobe (SHRIMP). The photographs were taken with an electron microscope to highlight the chemical zoning in the zircons. Each zircon is about 1/10 of a millimeter in length. Both are from a granite in central Turkey. The old cores may have formed in the rock that melted to form the granite, and the younger rims formed during crystallization of the granite. These grains record the geological relationship that characterizes the ancient zircons at Jack Hills, Australia.

in the lithified silt that collected in the bottoms of muddy ponds in what is now northwestern Australia.

Over the next 3 billion years, more complex forms of life began to develop and diversify. What allowed these new forms to survive? Many scientists believe that by about 545 million years ago (4500 km or 2800 mi along our time line), a more hospitable environment for life had finally developed on Earth. The atmosphere had accumulated sufficient oxygen (largely the product of photosynthesis by early forms of algae) to absorb a significant portion of the Sun's ultraviolet radiation. This protective atmosphere allowed an abundance of marine life to thrive. The oxygenated atmosphere also triggered changes in metabolism that allowed for the growth of larger, more complex organisms.

Many momentous events are concentrated in the last 200 million years of Earth history, equal to only a few hundred km of our time line; the rise and fall of the dinosaurs, the onset of the Earth's most recent ice ages, the origin of mammals (including humans), and the arrival of humans on the North American continent. Indeed, *all* of the material taught in a typical American history course at your school represents only 0.0000011% of the Earth's total history!

How did geologists pinpoint the timing of all of the Earth's great events? Why do we know that the planet is 4.6 billion years old? How did we determine the ages of the oldest known rock, the oldest fossil, and the landmark fossils in the evolution of modern species?

Geologists examine time in two ways. First, they use **relative dating,** which compares two or more entities to determine which is older and which is younger. Just as you can usually know the relative ages of people by looking at the faces of infants, toddlers, children, adolescents, adults, and senior citizens, geologists can determine the relative ages of rocks and other geological features by comparing certain characteristics.

Second, geologists analyze time by **numerical dating** (also known as *absolute dating*), which tells you how long ago—*in years*—a particular rock formed or a geologic event occurred. To determine *your* numerical age, someone would have to obtain quantitative information about your date of birth; to prove your numerical age (such as when you're attempting to get admitted to a 21-and-over club), you are asked to provide certain documents, such as a birth certificate or driver's license. But the Earth has no such written record of its age or of its key historical events. Instead, geologists must use the Earth's rocks as the documents for determining such ages, and they've developed a host of accurate geologic "clocks" to study these rocky "documents." These time-telling strategies include such diverse techniques as measuring radioactive decay in a rock's minerals and counting the number of tree rings in an oak tree growing up from an ancient landslide deposit. Let's begin to look at how geologists date things by first looking at the relatively "easy" (and inexpensive) ways to figure out "which is older and which is younger?", the basic questions to be asked when we date things *relatively.*

Determining Relative Age

To compare bodies of rock in the same region and answer the questions "Which is older?" and "Which is younger?", geologists use a few basic rules of observation and logic. All of these principles can be used to date relatively the rocks in a single outcrop, and some of these same methods can even be used to compare rocks that aren't located anywhere near each other. The following are the basic ways we determine what's older and what's younger.

Principles of Relative Dating

Uniformitarianism   Before geologists can even begin to look at rocks to determine their relative ages and origins, they must first ask one key question: Have the geologic processes we see operating today occurred in *exactly* the same way throughout geologic time? Yes, they have, according to 18th-century Scottish geologist James Hutton, widely accepted as the founding parent of modern geology, who first presented his **principle of uniformitarianism** in 1785 at a meeting of the Royal Society of London. Hutton's principle simply states: "*The present is the key to the past.*" Hutton claimed that what we see happening today, happened in the same ways in the past, and thus by looking at modern processes, we hold the key to interpreting the geological past as it is preserved in rocks. This principle is useful for interpreting geological history, and in many cases, it's probably correct: Rivers have *always* flowed downhill. Erupting volcanoes have *always* erupted lava and shot tephra into the air.

We must be careful, however, not to take this simple principle too far. Has the Earth *always* had rivers? *No.* Have volcanoes always erupted the exact same types of lavas as they do today? *No.* The Earth has not always been *exactly* the way we see it today. Early in its history, the planet did not have a well-developed atmosphere, and what little atmosphere it had did not contain much oxygen. There were no oceans, rivers, or continents. There were probably more volcanoes than we see today, and these erupted more frequently and produced extremely hot lava flows, much hotter than typical lavas today. The Earth even spun more rapidly than it does now, so days were shorter and a year contained 440 of them. Thus, the rates and even types of geologic processes have changed since the Earth first formed.

At some point, however, the Earth started to resemble the world as we know it today. It cooled, slowed down its spin, and developed an atmosphere rich in nitrogen and oxygen, liquid water . . . and life. After this time, the principle of uniformitarianism is very useful, and we can use it to explain many ancient geologic events by comparing them with processes we can see today. Thus, armed with the principle of uniformitarianism, we can interpret the origins of the rocks we see and then apply several other principles to date most rocks and events relatively. First, we will look at how rocks are "positioned" relative to one another as a means to determine their relative ages.

Horizontality and Superposition   As we saw in Chapter 6, most sediments settle out from bodies of water and thus are originally deposited as horizontal or nearly horizontal layers. Lava flows also tend to solidify as horizontal layers. The **principle of original horizontality** states that when rock layers are found in non-horizontal positions, they were most likely originally horizontal and were later tilted to achieve their present position.

The **principle of superposition** states that rock materials get deposited on top of earlier, older deposits. Consequently, in any

horizontal sequence of rock layers (or, *strata*), the *youngest stratum will be at the top and the oldest at the bottom.* Here's an analogy. Suppose you're something of a slob around the house, but you're a well-informed slob: You read the newspaper every day. When you finish reading the paper, you throw it on the floor. Over time, the pile grows. Obviously, unless it grows too high and topples over, the recent history of world events is preserved in sequence—with today's paper on top and progressively older editions below. Same is the case with layered rocks—oldest on the bottom, youngest on the top—as long as they haven't been disturbed. The principle of superposition, like the principle of original horizontality, commonly applies to sediments, sedimentary rocks, and lava flows (Fig. 8-4).

When tectonic forces have tilted or even overturned a sequence of rock layers (Fig. 8-5a), we look for identifiable sedimentary structures, such as ripple marks, mudcracks, graded beds, or cross-bedding (all discussed in Chapter 6) to identify the upper surface of any one sedimentary layer (Fig. 8-5b). Similarly, vesicles—the pea-size holes left behind as gas bubbles burst at the tops of some lava flows (see Chapter 3)—may indicate the location of a flow's upper surface. Once we have identified the top of any of the rock layers, we can apply the principle of superposition to date relatively the layers below and above it.

Cross-Cutting Relationships   When magma moves underground, it intrudes other rocks. The rock intruded by the magma had to be there first in order for the magma to intrude it, so the intruded rock must be older than the intrusion. The **principle of cross-cutting relationships** states that any intrusive rock must be younger than the rock it cuts across (Fig. 8-6). Cross-cutting

▲ **FIGURE 8-4** Horizontal beds of sedimentary rock outside of Salt Lake City, Utah. Using the principle of superposition, we can instantly determine that the rocks at the top of this formation are the youngest and those at the bottom are the oldest. Rock layer "A" is younger than all the rocks below it . . . and older than all the rocks above it.

(a)

relationships also provide relative dates for faults—fractures in rocks that have been displaced (discussed in Chapter 9). *Faults must be younger than the rocks they cut and displace* (Fig. 8-7).

Inclusions   During some geologic processes, such as the movement of magma, fragments of one type of rock get incorporated into another. In the case of magmas, chunks of unmelted preexisting rock may become *included* in an intrusion, or in the case of a lava flow, other, older rocks may be picked up and included within the volcanic rock that cools from the lava flow. Inclusions are *always* older than their host rocks because they had to already exist in order to be picked up and incorporated (Fig. 8-8). Such igneous inclusions can be very useful to geologists, especially when they are made up of deep mantle rocks that have been incorporated in a rising magma on its way to eruption at the Earth's surface. These mantle inclusions are one way we get a glimpse of an inaccessible region of the Earth's deep interior.

Fossils and Faunal Succession   Fossils (from the Latin *fossilus,* meaning "something dug up") consist of the remains of ancient organisms, or other evidence of their existence, that became preserved in geologic material. As Figure 8-9 indicates, humankind has apparently been fascinated with fossils for thousands of years. *Paleontologists* study fossils in part to reconstruct the evolution of life on Earth. When we apply the principles of superposition, cross-cutting relationships, and inclusions to fossil-bearing rocks, we can determine the order in which life forms developed and became extinct.

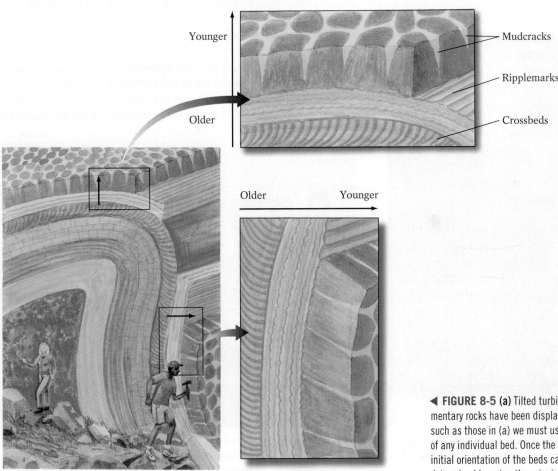

◀ **FIGURE 8-5 (a)** Tilted turbidite beds in Zumaya, Spain. **(b)** When sedimentary rocks have been displaced from their initial horizontal orientation, such as those in (a) we must use sedimentary structures to identify the top of any individual bed. Once the top of a single bed has been determined, the initial orientation of the beds can be reconstructed and their relative ages determined by using the principle of superposition.

235

▲ **FIGURE 8-6** A cross-cutting relationship: A mafic dike cuts across these layered sedimentary rocks in the Grand Canyon, Arizona. The dike is younger than the rocks it cuts across.

Fault

▲ **FIGURE 8-7** The fault that cuts through these sedimentary rocks, found in California's Anza Borrego Desert State Park must be younger than the deposition of any of the rock layers.

▲ **FIGURE 8-8** This granitic intrusion picked up chunks of diorite as it moved through underground rocks. The dark inclusions are older than the surrounding light-colored rock.

Sedimentary rocks, because they form at the Earth's surface, are far more likely to contain fossils than other kinds of rocks. Even in sedimentary rocks, however, fossils are scarce: Only about 1% of all living species that have ever existed are believed to have been preserved as fossils. As Highlight 8-1 explains (see pages 238–240), organisms fossilize only under very special conditions. Many extinct organisms were composed exclusively of soft tissue and thus contained no preservable parts. Others, however, found their way into the fossil record, even without such hard parts.

If you know the position of each rock layer in a sequence of rocks, you can determine the relative ages of the fossils in them using the principles of horizontality and superposition. Once the relative ages of fossils have been worked out in this way, the fossils themselves can then be used to determine the relative ages of other rock layers—even in widely separated locations. The **principle of faunal succession**

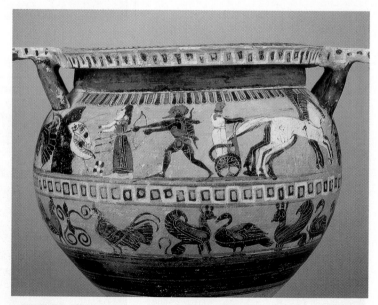

▲ **FIGURE 8-9** This ancient image from Greece indicates that fossils were collected thousands of years ago. Note the skeleton on the upper-left corner of the vase.("Mixing bowl [column krater]." Photograph © 2007 Museum of Fine Arts, Boston)

states that specific groups of fossils follow, or *succeed,* one another in the rock record *in a definite order.* Thus, rocks containing ammonite fossils are younger than rocks containing trilobites (Fig. 8-10).

Certain varieties of organisms have existed (or did exist) for hundreds of millions of years. Sharks, for example, have inhabited the oceans continuously for more than 400 million years. Other organisms lived only during relatively brief, specific periods of Earth history. When paleontologists find fossils of such short-lived organisms, known as **index fossils,** they know that they are looking at rocks of a specific age (Fig. 8-10). The most useful index fossils include those species that were geographically widespread during their short time on Earth. These fossils can be used to date many different—even far-flung—rock formations.

Unconformities

If all the rocks ever formed were still around and we could see them all, we could use the principles of relative dating to determine the sequence of *all* events in Earth history. The Earth is a dynamic place, however, and rocks weather and erode, creating gaps—in some places, huge gaps—in the geologic record of past events. Fortunately for us, the Earth has eroded unevenly, removing rocks of a particular age from one region but leaving them for us to study some place else. We can travel around the world looking at many different rocks to get as complete a picture as possible of past geological events.

Gaps in the geologic record are marked by **unconformities**—boundaries separating rocks of different ages (Fig. 8-11). Unconformities occur where erosion has removed rock layers or where no rocks were produced during a specific period of time. Some unconformities represent only a small amount of "missing" time, such as when a flood strips away soils, sediments, and rocks from a localized region. Others represent millions or even billions of years of time, such as in parts of the Upper Midwest, where 15,000-year-old glacial deposits from the most recent ice age sit directly on top of rocks that are more than 3 billion years old. All record of the rocks and sediments deposited in this region between 3 billion and 15,000 years ago has been removed by erosion. The contact line between the glacial deposits and ancient

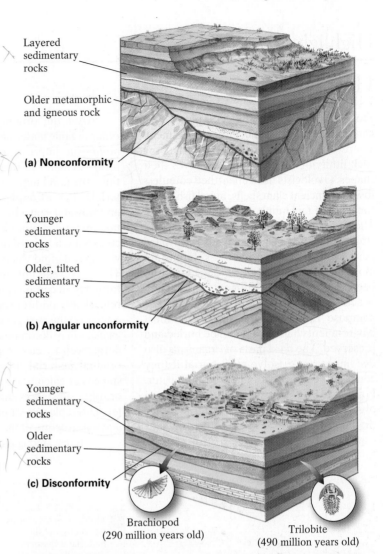

(a) Nonconformity

(b) Angular unconformity

(c) Disconformity

Brachiopod
(290 million years old)

Trilobite
(490 million years old)

▲ **FIGURE 8-11** The three major types of unconformities between rock layers. **(a)** A nonconformity, between metamorphic or igneous rock and overlying sedimentary rock; **(b)** an angular unconformity, between older, deformed sedimentary rock and younger, undeformed sedimentary rock; **(c)** a disconformity, between parallel layers of sedimentary rock.

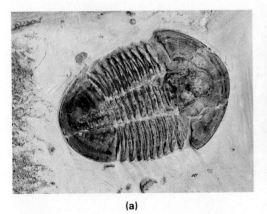
(a)

(b)

(c)

▲ **FIGURE 8-10** Index fossils. Each of these organisms lived during a specific period of Earth history, which enables us to distinguish the rocks in which they are found from rocks formed during other time periods. **(a)** A Cambrian trilobite; **(b)** Pennsylvanian brachiopods; **(c)** a Jurassic ammonite.

Highlight 8-1

How Fossils Form

Most fossils form when the hard parts of organisms become buried in layers of sediment. There they decompose so slowly that the rock lithifying around them preserves their shapes. Geologists have unearthed countless fossils of ancient clam shells, many fish skeletons, and a fair number of dinosaur bones. They have found far fewer remains of worms, slugs, and jellyfish. Organisms with hard parts such as shells and bony skeletons are more likely to remain intact long enough to become preserved in sediment. Those composed solely of soft tissue would likely decompose rapidly at the surface in most environments before they can be buried and preserved. The hard parts of organisms also decompose or dissolve after burial if they come into contact with underground water. For that reason, a fossil is typically composed of some type of replacement material rather than an organism's original biological substance.

After the original bones, shells, and other organic materials have decomposed, a *mold*—an impression of the organism's form—remains in the sediment. At a later time, the mold may be filled with other materials. For example, circulating ion-rich groundwater may enter a mold and precipitate mineral compounds in it. This process creates a solid *cast* of the mold, which is what we typically find when we go fossil-hunting. The cast reveals only the organism's *external* form, but in certain cases the form may be a remarkably faithful replication (see photos a and c on the next page). Some deceased organisms may decompose very slowly and undergo cell-by-cell replacement of their original material by such secondary substances as calcium carbonate, silica, pyrite, or other substances dissolved in groundwater. This replacement process may retain the *internal* structure of the organism, which stirs great interest among paleontologists. The fig-

ure below depicts the processes that lead to the creation of molds and casts and replacement.

Preservation of an organism's original parts occurs only under unusual environmental conditions. For instance, the skeletons of thousands of mammoths, mastodons, saber-toothed cats, and other late Ice Age mammals were preserved in the viscous tar of the La Brea tar pits, L.A.'s ultimate Ice Age "tourist trap," located on Wilshire Boulevard in downtown Los Angeles. La Brea's wonderful array of animals—carnivores, herbivores, and birds alike—became trapped in the tar starting about 44,000 years ago. Intact woolly mammoths have been dug up from the tundras of Alaska and Siberia, where they had become frozen in arctic *permafrost* (permanently frozen ground).

The original bodies of entire insects have been recovered from hardened tree sap, or

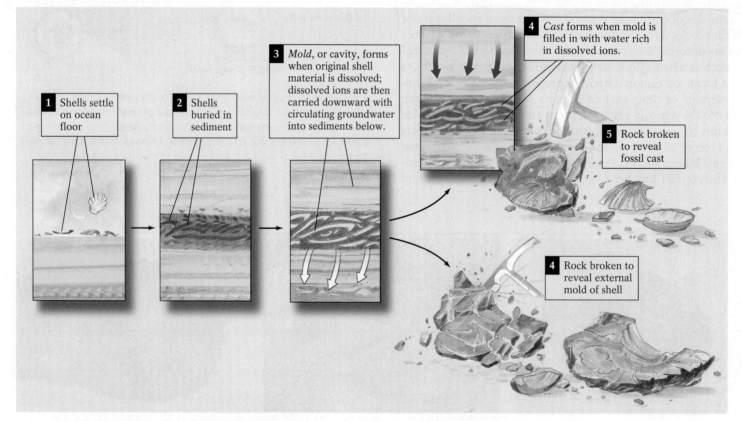

1 Shells settle on ocean floor

2 Shells buried in sediment

3 *Mold*, or cavity, forms when original shell material is dissolved; dissolved ions are then carried downward with circulating groundwater into sediments below.

4 *Cast* forms when mold is filled in with water rich in dissolved ions.

5 Rock broken to reveal fossil cast

4 Rock broken to reveal external mold of shell

▲ An organism that has been buried in sediment may be preserved in the form of a mold or a cast in the resulting rock. A *mold* is an imprint of the organism's external form that remains behind in the surrounding rock long after the organism itself has decayed. After an organism's biological substance has dissolved away, circulating groundwater may precipitate dissolved compounds into the cavity left in the rock, forming a *cast* of the organism.

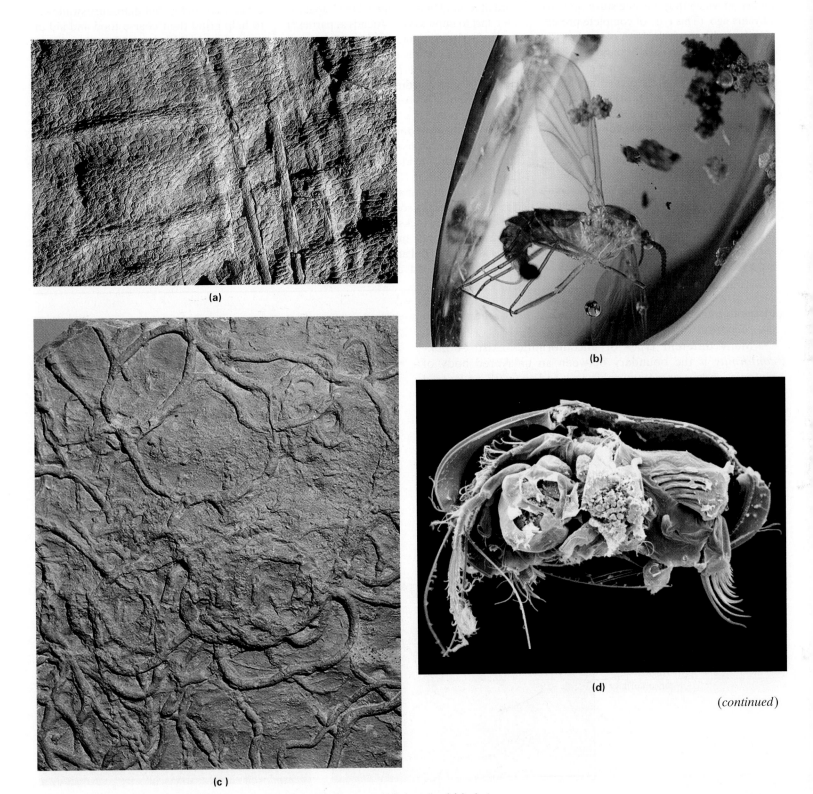

(a)

(b)

(c)

(d)

(continued)

▲ Various types of fossils. **(a)** The skin texture of a dinosaur **(b)** A 3.5-million-year-old fly in amber; **(c)** Ancient worm burrows; **(d)** A 425-million-year-old "male" ostracode.

Highlight 8-1 (continued)

amber, in which they became stuck millions of years ago. (This type of complete preservation of an insect is the basis of the *Jurassic Park* film series.) In each of these cases, the surrounding material did not permit the circulation of water or air, preventing the organism's original material from decomposing. And in the "R"-rated portion of this textbook, paleontologists have recently reported the discovery of the oldest "undeniably male" fossil, a 425-million-year-old sea creature called an *ostracode.* Buried in a fine volcanic ash that allowed for the preservation of all of its soft external parts (such as its gills and eyes), this quarter-inch-long creature (see photo) was named *Colymbosathon eplecticos,* Greek for "amazing swimmer with large penis." (Actually, its penis equals one-third of its total body volume. How's that for a fun fact to impress your friends at parties?) And in 2005, paleontologists have discovered *Tyrannosaurus* bones in Montana that still contain the original *T-Rex* bone marrow. Here is an extraordinary opportunity to study the actual DNA of the cells of a long-extinct dinosaur.

Even when we don't find fossils showing an organism's physical form, we sometimes find other evidence of its presence through *trace fossils.* If an animal burrowed through sediment, it may have left "tunnels"; if it crawled across soft mud, it may have etched delicate tracks into the sediment. Larger animals may have left behind footprints as they walked around. We have even discovered rounded and polished rocks, called *gastroliths,* that dinosaurs swallowed to help grind their coarse food and aid digestion, and lithified digestive "droppings," or *coprolites,* from a variety of creatures.

A final note about fossils: Be careful where you dig them up. In 1996, a commercial fossil dealer, Peter Larson, was convicted of stealing fossils from federal land. He served more than a year in prison and was fined $5000 for illegal excavation of "Sue," a nearly complete *Tyrannosaurus rex* specimen taken from public and private lands in South Dakota (and now on display in Chicago's Field Museum). Apparently, given their considerable value on the museum market, dinosaur fossils are big business. Better ask before you walk off with one.

Precambrian bedrock is an unconformity (that is, the rocks do not "conform" to each other in age).

Many different types of unconformities exist. A *nonconformity* is the boundary between an unlayered body of plutonic-igneous or metamorphic rock and an overlying layered sequence of sedimentary-rock layers (Fig. 8-11a); rock underlying a nonconformity usually shows evidence of erosion that took place before the overlying rock was deposited.

One type of unconformity that most clearly shows that we are missing part of the geologic record is where sedimentary layers have been tilted by a tectonic event, eroded, and then covered by a later, horizontal rock deposit that overlies it (Fig.

8-11b). This is called an *angular unconformity,* and one in particular—that found at Siccar Point in Scotland—has particular historical significance in geology (see Figure 8-12).

A more subtle type of unconformity, a *disconformity,* occurs between *parallel* layers of sedimentary rock (Fig. 8-11c). We know such a boundary is a disconformity if fossil groupings above and

Angular unconformity

▲ **FIGURE 8-12** The angular unconformity at Siccar Point, Scotland. James Hutton, the founder of modern geology (Chapter 1), correctly interpreted this formation as showing layers of rock that had been deposited horizontally, lithified, tilted to a nearly vertical position, eroded smooth at the surface, and then overlain with subsequent horizontally deposited layers that were also later deformed—a graphic example of the dynamics that shape the Earth's features, and the amount of time necessary to achieve them.

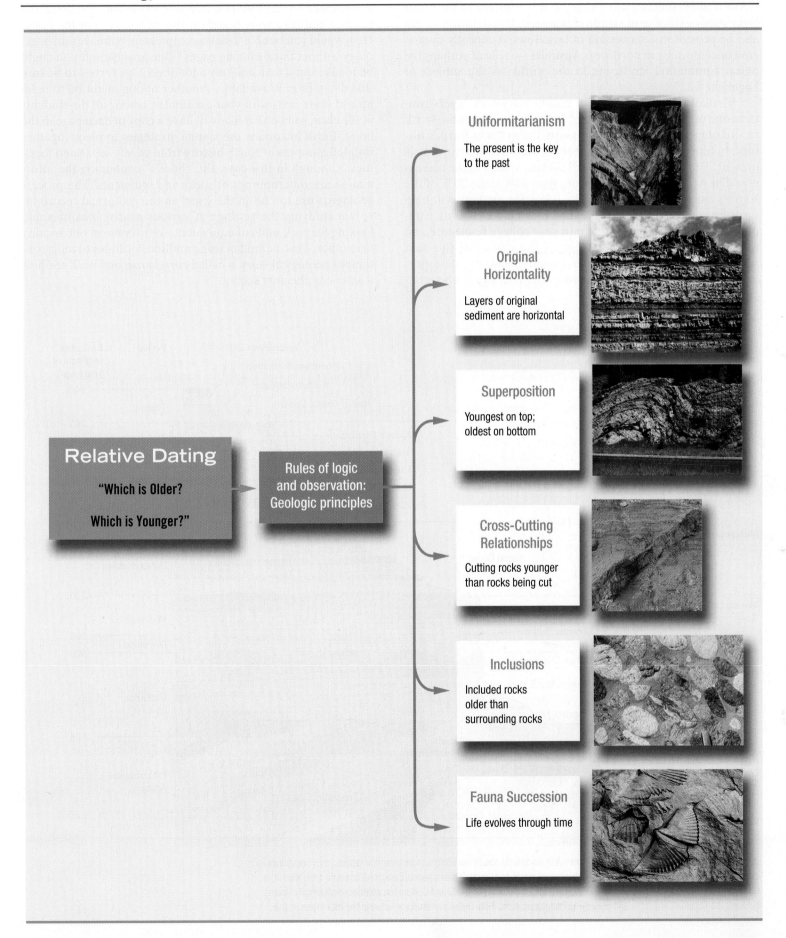

Relative Dating

"Which is Older?

Which is Younger?"

Rules of logic and observation: Geologic principles

Uniformitarianism

The present is the key to the past

Original Horizontality

Layers of original sediment are horizontal

Superposition

Youngest on top; oldest on bottom

Cross-Cutting Relationships

Cutting rocks younger than rocks being cut

Inclusions

Included rocks older than surrounding rocks

Fauna Succession

Life evolves through time

below it differ substantially in age, or if the surface of the lower layer shows evidence of significant erosion. Disconformities can also be present in a succession of lava flows. A recently discovered unconformity in northwest Australia—a relic of perhaps the oldest continental landscape in the world—is the subject of Highlight 8-2.

Virtually every sequence of sedimentary-rock layers contains one or more unconformities. In fact, the sedimentary-rock record of most regions represents only 1% to 5% of Earth's history. In other words, much more of the record is missing than is present. The most complete rock sequences, such as the one exposed in Arizona's Grand Canyon, span only some 20% of the Earth's existence (Fig. 8-13). The following analogy will help show how geologists deal with such incomplete records. Suppose despicable vandals broke into your college bookstore and tore out 80% of the pages from all the copies of this text *randomly* before you bought one. To add to your problem, perhaps you have the only geology professor in the world who is un-

sympathetic to this situation and still insists that you do *all* the reading and take tests based on the information in the book. How would you be able to ace a comprehensive final exam in geology without those missing pages? One possible solution might be to take out a loan and buy a lot of geology texts—to be sure that every page is available. Another option might be to team up and share texts with your classmates; among all the students in the class, someone is likely to have a copy of each page in the book. Earth historians use similar strategies to piece together isolated glimpses of Earth history from widely separated localities, although in this case, the "books" containing the information are outcroppings of rocks and sediments. The process geologists use to "fill in the gaps" in the geological record involves studying the geology at various nearby localities and looking for rock units of equivalent age to connect one locality to another. This method of using multiple localities to tell a continuous geological story is called *correlation,* and we'll see how it's done in the next section.

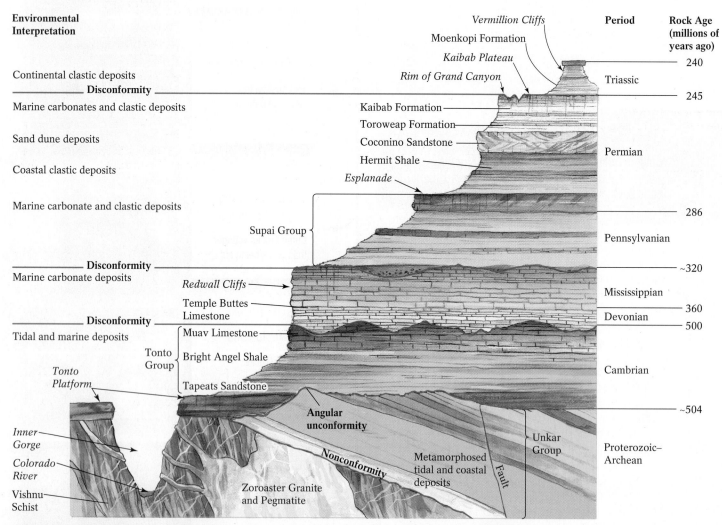

▲ **FIGURE 8-13** The Grand Canyon in northern Arizona affords one of the Earth's most complete exposed rock sequences— almost 2 vertical km (about 1.2 mi) of rock whose age ranges as high as 2 billion years. Even here, however, unconformities represent large gaps in the local rock record. The Great Unconformity in the Grand Canyon, an angular unconformity, occurs between the Tapeats Sandstone and underlying Proterozoic rocks. Note the erosion surface between the rock layers at the unconformity.

Highlight 8-2

The World's Oldest Landscape

By most accounts, the birth of the Earth's continents was a violent affair. Lighter minerals rising from the hot convecting mantle first coalesced as volcanic islands surrounded by lava seas (see figure). These islands, the planet's first parcels of continental real estate, then collided over and over as they bobbed on the churning mantle below. Some probably plunged back to the Earth's interior, only to remelt and erupt again.

With so much planetary remodeling taking place, it is not surprising that very little of the Earth's early continental rock survived. Just recently, however, geologists searching for zinc deposits in northwestern Australia may have found a small piece of the early Earth's continental puzzle. Two bodies of very old continental rock exposed over a distance of some 75 km (46 mi) in Australia's remote, arid Pilbara region are separated by a discernible unconformity. The lower rocks are 3.52-billion-year-old greenstones and granites—a metamorphosed block of old continental crust. These rocks were deformed and eroded before being covered by 3.46-billion-year-old sedimentary rocks, some containing shallow-water ripple marks. The overlying sedimentary rocks contain the Warrawoona cherts, a marine layer that contains fossil cells that provide some of the oldest evidence of life on Earth.

The unconformable surface between these rocks represents a time gap of some 60 million years. It shows tell-tale signs of weathering and erosion, indicating that wind and water must have scoured the rocks *above* sea level at some point. Thus these 3.52-billion-year-old water-worn continental rocks may constitute the Earth's oldest-known *exposed* continental landscape. (Older rocks, such as the 4.03-billion-year-old gneisses of northern Canada, are highly metamorphosed rocks that originated deep in the Earth's interior and probably remained there until more recently exposed.) The rocks of Pilbara (see photo) may contain the first direct evidence of continental lands that stood high and dry above ancient Precambrian seas. What's more, they're a half-billion years older than the next oldest-known continental rocks.

The rocks of the Pilbara unconformity may eventually tell us much about our planet's early atmosphere. During those days,

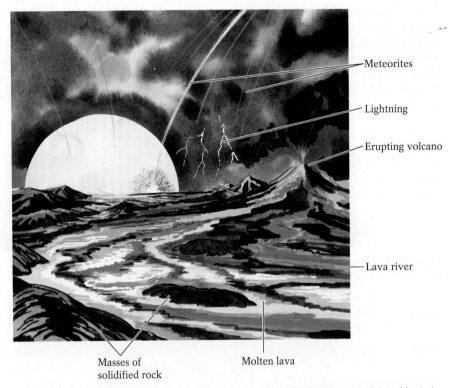

▲ A hypothetical view of the early Earth—about 4 billion years ago. The Earth's greater internal heat at that time would have fueled more extensive volcanism; more frequent meteorite strikes would have periodically ruptured the planet's earliest crust.

Labels: Meteorites; Lightning; Erupting volcano; Lava river; Masses of solidified rock; Molten lava

the young Sun's rays were probably only 70% as strong as they are today. No evidence suggests that the planet was sharply colder back then, so scientists speculate that the Earth's surface may have been warmed by a "super" greenhouse effect, perhaps caused by a very high carbon-dioxide content in the atmosphere. To find traces of the ancient atmosphere, geologists study ancient soils, or paleosols (discussed in Chapter 5). The Pilbara rocks appear to contain such paleosols, and initial analyses of iron-carbonate minerals in them indicate that CO_2 levels at the time may have been 100 to 1000 times higher than those found today. Geologists are also searching these rocks for chemical signs of early life on Earth.

▲ The 3.5-billion-year-old sedimentary rocks of Australia's Pilbara region are some of the oldest surviving remnants of Earth's earliest continental crust.

Correlation

To determine equivalence in age between geographically separate rock units, geologists look for paleontological or other geological similarities between the units, a process known as **correlation.** In Fig. 8-14, we see how the geology of the Grand Canyon in Arizona overlaps with the geology of Zion National Park in western Utah, which in turn overlaps with the geology of Bryce Canyon National Park in central Utah. The Kaibab Limestone ties the geology of the Grand Canyon to Zion and the Navajo Sandstone ties the geology of Zion to Bryce. In this way, the geology of the Grand Canyon can be correlated with the geology of Bryce Canyon.

In general, because specific environmental factors influence the origin of a region's rocks and their subsequent erosion, the farther apart two rock formations are geographically, the less likely they will possess overlapping rock sequences. Yet even formations thousands of km apart may contain individual layers whose similarities suggest that they originated at the same time. The presence of similar fossil groups in different rock units, for example, suggests that the layers are the same age. Geologists also base their correlations on individual index fossils; when we find them—even in widely separated rocks—we conclude that the rocks formed during the same time period.

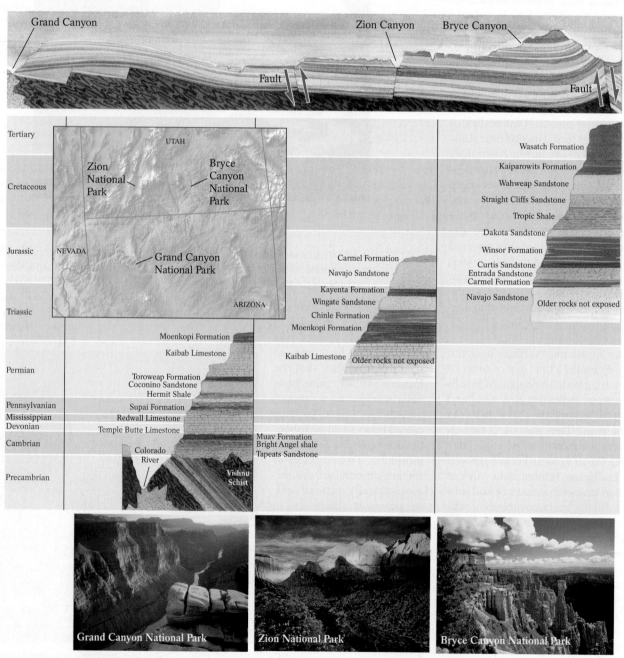

▲ **FIGURE 8-14** Correlation of the rock sequence in the Grand Canyon, Zion, and Bryce Canyon National Parks in Arizona and Utah. The rock sequences at the three sites overlap in ways that enable geologists to see a larger sequence of the region's rocks than appears at a single site.

▲ **FIGURE 8-15** Ash from the eruption of Mount Mazama 7700 years ago, at Crater Lake in Oregon. The lighter-colored ash consists of the frothy, highly felsic material that first erupted from the volcano; it was deposited first. The overlying, darker-colored layer is from a lower position in the magma chamber where slightly denser, less felsic material may have settled during differentiation; it erupted later and thus was deposited on top of the lighter-colored material. Rain and snow then eroded the ash beds to produce the pinnacles. The Mazama ash is a key bed found in same-aged rock units throughout the Pacific Northwest.

Distant rock sequences can also be correlated if they contain the same *key bed,* a distinctive rock layer that appears at several or many locations. A key bed records a geological event *of short duration that affected a wide area.* The eruption of Mount Mazama in the Pacific Northwest about 7700 years ago, for example, produced a whitish-tan, highly felsic ash that was scattered so widely that it can be used to correlate the geologic strata of Oregon, Washington, British Columbia, and Alberta (Fig. 8-15). If you live in any of those areas, you would have a good chance of finding Mazama ash in most any road cut.

Distinctive sedimentary facies of the same age in a wide range of far-flung locations may indicate global environmental conditions during their deposition, such as a widespread climate change or a worldwide rise or fall in sea level. Two-billion-year-old iron-rich redbeds found on many continents mark the time when the Earth's global atmosphere had become sufficiently oxygen-rich to begin to precipitate dissolved iron from the Earth's oceans. The extensive 300-million-year-old coal beds on both the North American and Eurasian continents indicate a distinctively warm episode worldwide that produced substantial vegetation that later lithified as coal. Each of these features enables geologists to correlate rocks from place to place.

Relative Dating by Weathering Features

We saw in Chapter 5 how the processes of mechanical and chemical weathering break down rock to loose sediment. Because rock tends to become increasingly weathered over time, geologists can learn about the relative ages of rocks by comparing their extent of weathering. This method of dating generally is limited to geologic materials that are less than a few million years old.

Crack open a rock that has been sitting at the Earth's surface and you may see *a rind* of weathered material at or near the rock's surface (Fig. 8-16). Such rinds form when water infiltrates rocks, chemically altering their original composition—from the rock's surface inward—into some type of weathering product. In general, the longer a rock is exposed to weathering, the thicker its weathering rind will be. Thus a comparison of weathering rinds—among rocks having the same initial composition—can distinguish, for example, between deposits of different ages. This relative-dating technique is most applicable to basalt, whose high-temperature minerals (olivine, pyroxene, calcium plagioclase) weather relatively quickly at the Earth's surface.

Knowing the degree to which an area's landscape features, from its boulders to its soils, have weathered may enable geologists to date different landscapes relatively (Fig. 8-17). Age is not the only factor that influences weathering; we must also consider such factors as climate and rock composition. After accounting for such variations, it is possible to conclude that, in general, an old, deeply weathered landscape will contain weathered and broken-down boulders, whereas young landscapes will more likely be littered with a high concentration of fresher surface boulders. An experienced geologist can even estimate the relative age of a boulder with a single hearty whack of a geologic hammer: The resounding ring of an unweathered boulder sounds distinctly different from the dull thud of an aged, weathered boulder.

▲ **FIGURE 8-16** A weathering rind developed in a fist-sized cobble of basalt. The reddish outer portion is the weathering rind.

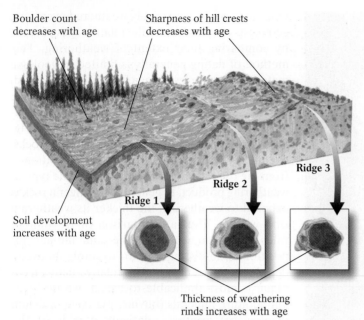

Boulder count decreases with age

Sharpness of hill crests decreases with age

Ridge 3

Ridge 2

Ridge 1

Soil development increases with age

Thickness of weathering rinds increases with age

▲ **FIGURE 8-17** This hypothetical landscape shows the variety of weathering phenomena that can help determine the relative age of an area and its constituent materials. Ridges 1, 2, and 3 are deposits left behind by a retreating glacier. The sharp hill crests and large number of boulders on ridge 3 indicate relatively little exposure to weathering as compared to the crests and boulders on ridge 1; the thickness of the weathering rinds on the boulders in ridge 1 indicates that they are fairly old; the soils in ridges 1 through 3 are progressively less developed. These clues suggest that ridge 1 is the oldest and ridge 3 the most recent of the glacial deposits.

Determining Numerical Age

Relative dating can only tell us that rock A is older or younger than rock B. This knowledge is useful information, but in many cases it is important (and more satisfying) to know how long ago—*in years*—a rock formed or an event occurred. Numerical-dating methods tell us an age in years for a geological material or event. All numerical-dating methods depend on some type of "natural clock"—a process that operates at a constant known rate. For example, as a tree grows, it adds rings to its trunk every year. We can determine a tree's numerical age by counting its rings. Nature has provided a substantial number of such "clocks" that help us determine the ages of a range of geological materials.

Isotope Dating

One method of numerical dating relies on the rate of decay of radioactive isotopes within minerals. We saw in Chapter 2 that the atoms of certain chemical elements exist as *isotopes,* which contain different numbers of neutrons in their nuclei. The nuclei of radioactive isotopes are unstable, and decay from one form—those of the original "parent" isotope—to another, more stable form—those of the "daughter" product. Because this decay occurs at a constant rate, we can determine the age of a rock or mineral by determining the amount of radioactive parent left, and comparing this amount to the amount of daughter produced. The ratio of parent to daughter, along with the decay rate, yields the time at which the parent started decaying (that is, its time of origin).

Radioactive decay produces an entirely different element from the parent isotope—for example, the decay of uranium isotopes produces lead. Alternatively, radioactive decay can transform one isotope of an element into a different isotope of the same element. Loss or gain of *neutrons* converts a parent isotope into a daughter isotope of the same element. Loss or gain of *protons* changes the parent element into an entirely different daughter element having a new set of chemical and physical properties. After a number of intermediate steps, an *unstable* radioactive parent isotope eventually becomes a *stable,* nonradioactive daughter isotope.

Unstable nuclei decay in three primary ways:

- *Alpha Decay* Alpha particles are composed of two protons and two neutrons (essentially they are helium nuclei). With the expulsion of an alpha particle, the atomic mass of an atom decreases by 4 and the atomic number decreases by 2. Of course, this change in atomic number produces an atom of an entirely different element. For example, uranium-238, an isotope of the element uranium, has an atomic number of 92 (that is, there are 92 protons in its nucleus). When an atom of uranium-238 undergoes alpha decay, it loses 2 protons and 2 neutrons and is converted to an atom of an element with 90 protons—that is, thorium. After the loss of the 4 nuclear particles, its atomic mass is 234 (Fig. 8-18a).

- *Beta Decay* Beta particles are essentially electrons. Ordinarily, we picture electrons as tiny negatively charged particles that orbit the nucleus of an atom and are gained, lost, and shared in chemical reactions. The electrons that make up beta particles, however, do not originate from an atom's orbiting cloud, but instead result from the breakdown of a neutron. A neutron is composed of a proton and an electron. Thus, when a neutron decays and its component electron is expelled, an additional proton is left in the nucleus. The addition of this proton *increases* the atomic number, creating an atom of a different element. An example: radioactive potassium-40 (atomic number 19) decays by beta decay to calcium-40 (atomic number 20). Note that while the atomic number changes, the atomic mass remains the same. The loss of the neutron is offset by the addition of the proton, and the departure of the nearly weightless electron leaves the atomic mass essentially unchanged (Fig. 8-18b).

- *Electron (Beta) Capture* In electron capture (also known as beta capture), the nucleus steals an electron from an atom's own orbiting cloud. The newly captured electron joins with a proton—remember, opposites attract—to create an additional neutron. In this case, because a proton is lost by conversion to a neutron, the atomic number *decreases* by 1 but the atomic mass remains the same. For example, when potassium-40 emits beta particles to form calcium-40, some of the unstable potassium atoms capture electrons. The change in the number of protons creates a new element, argon-40 (atomic number 18). Note that the atomic mass remains the same (Fig. 8-18c).

Radioactive isotopes are incorporated into rocks and minerals in various ways. For example, as minerals crystallize in cooling magma, they sometimes trap atoms of radioactive isotopes

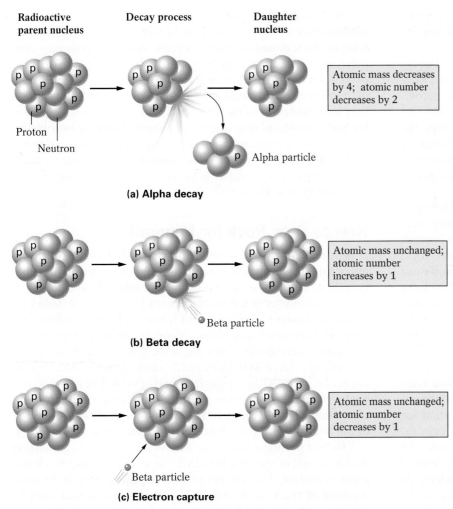

Radioactive parent nucleus — Decay process — Daughter nucleus

Proton
Neutron

Atomic mass decreases by 4; atomic number decreases by 2

Alpha particle

(a) Alpha decay

Atomic mass unchanged; atomic number increases by 1

Beta particle

(b) Beta decay

Atomic mass unchanged; atomic number decreases by 1

Beta particle

(c) Electron capture

▲ **FIGURE 8-18** Various processes by which unstable atomic nuclei decay: **(a)** alpha decay, which involves expulsion of a particle equivalent to a helium nucleus; **(b)** beta decay, which involves expulsion of an electron; **(c)** electron capture, which involves attraction of one of an atom's orbiting electrons to an atomic nucleus where it combines with a proton to form an additional neutron.

ent atoms and 50 daughter atoms after one half-life—and a parent-to-daughter ratio of 1:1. After another half-life, the number of remaining parent atoms (50) would halve again, leaving 25 unstable parent atoms and a total of 75 stable daughters—a parent-to-daughter ratio of 1:3. Note that the total number of parent and daughter atoms remains the same (in this example, 100) throughout the progression. Regardless of which radioactive isotope decays, the proportion of parent-to-daughter atoms exists as a predictable ratio at each half-life (Fig. 8-19).

The half-lives of some isotopes span billions of years. Atoms of uranium-238, for example, take 4.5 billion years to convert to a 1:1 parent-to-daughter ratio with its daughter, lead-206 (that is, for half the U-238 atoms to decay to Pb-206 atoms). Certainly no one sits in a laboratory waiting for that amount of time to elapse to determine the half-life of a given parent–daughter system. Instead, we place samples containing measured quantities of the radioactive isotope in a device that counts the number of nuclear particles emitted per unit of time; we can then extrapolate the decay rate in years. You have probably heard the clicking of a Geiger counter and seen one either in person or in a movie; this is a device that detects radioactivity. Scientists who measure radioactive-decay rates use instruments similar to Geiger counters to detect the energy released when a known amount of a pure sample of a radioactive isotope decays. The counting can take years, and an average rate of decay per year is determined. Long counting times (on a human timescale) are required because most common isotopes used in dating rocks and minerals have very

within their crystal structure, but not any of the daughter element. That way, at the moment the crystal forms, only the parent isotope is present, and any daughter element that accumulates afterwards forms only through radioactive decay. The radioactive isotopes begin decaying immediately in a continuous process. **Isotope dating** uses this continuous decay to measure the amount of time elapsed since the mineral's formation. As time passes, a rock will contain less and less of its initial radioactive parent isotopes and more and more of their daughter products. Uranium-238, for example, commonly found in zircon crystals in granitic rocks, decays to form a stable daughter isotope, lead-206. Over time, a uranium-rich zircon crystal gains lead-206 as it loses uranium-238. Radioactive decay rates are constant, just like the mechanisms of a clock. The decay rates are unaffected by changes in temperature or pressure or by chemical reactions involving the parent isotope. By measuring the ratio of parent-to-daughter isotopes and comparing it with the parent element's known rate of radioactive decay, we can determine the numerical age of a rock.

The time it takes for *half* of the atoms of the parent isotope to decay is known as the isotope's **half-life.** For example, if a mineral has 100 parent atoms when it crystallizes, it will have 50 par-

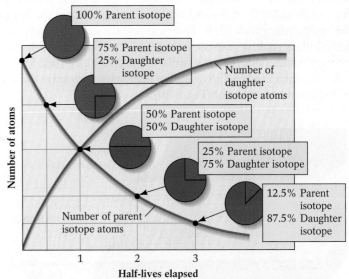

100% Parent isotope

75% Parent isotope
25% Daughter isotope

Number of daughter isotope atoms

50% Parent isotope
50% Daughter isotope

25% Parent isotope
75% Daughter isotope

Number of parent isotope atoms

12.5% Parent isotope
87.5% Daughter isotope

Number of atoms

Half-lives elapsed

▲ **FIGURE 8-19** Radioactive decay converts a radioactive parent isotope to a stable daughter isotope. With the passage of each half-life, the number of atoms of the radioactive isotope is reduced by half, whereas the number of atoms of the daughter isotope increases by the same quantity. If they know the half-life of the isotope, geologists can measure the ratio of parent to daughter isotopes to determine the numerical ages of some rocks.

small decay constants (and very long half-lives). (*Note:* There's a lab in France that has been conducting this type of counting experiment for 15 years!)

Scientists have also experimented to see if the rate of decay can be changed—that is, to see if it *really* is constant. Isotope samples have been heated and cooled to extreme temperatures, compressed to very high pressures, attacked by strong acids, and subjected to powerful magnetic fields and electrical currents. *The rate of decay does not change.* There are no known physical or chemical environments or processes that can alter the decay rate of an isotope; it's a spontaneous unalterable *nuclear* process. The decay constant is truly constant.

Once we know the specific decay rates of individual isotopes, we can then determine the age of a given rock using a *mass spectrometer,* a device that measures the precise quantities of parent and daughter isotopes in a substance such as a mineral, rock, bone, tree, or shell to determine their ratio. From the ratio, we can establish how many half-lives (or fractions of half-lives) have elapsed since the radioactive parent isotope was incorporated into the substance. We then multiply the number of half-lives by the known length of one half-life of that isotope to calculate the substance's numerical age.

Factors Affecting Isotope-Dating Results   If you want to determine the date at which a rock formed, isotope dating is typically more accurate and useful for igneous rocks than for other types of rocks. As an igneous rock crystallizes, it locks radioactive isotopes—but not their daughter elements—into the crystal lattices of its minerals. That action sets the isotopic "clock" within the rock, especially if the rock cools quickly and elements don't diffuse (leak) into or out of the minerals containing the radioactive isotopes. It is more difficult to determine the isotopic age of deposition (or lithification) of a clastic sedimentary rock, simply because these rocks contain mineral fragments of countless preexisting rocks of different ages. Dating these fragments separately yields the ages of their source rocks, but not the age of the sedimentary rock itself. To determine the numerical age of a sedimentary rock, we can use isotope dating of igneous rocks that may cross-cut, underlie, or overlie the sedimentary-rock formation, and then apply relative-dating principles to *bracket* the age of the sedimentary event (Fig. 8-20).

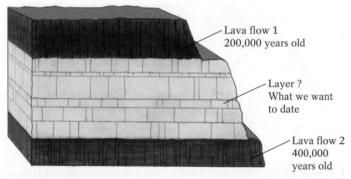

▲ **FIGURE 8-20** Because the numerical age of sedimentary rocks is typically difficult to measure, geologists commonly rely on bracketing ages. Here, the ages of the surrounding layers of datable igneous rocks provide an age range for the intervening limestone layer. Using the principle of superposition, we can see that the sedimentary layer must be older than 200,000 years and younger than 400,000.

Metamorphic rocks can be dated to determine the age of the metamorphic event, but care must be taken when analyzing metamorphosed sedimentary rocks because some minerals may contain isotopes inherited from the parent rock. Also, geologists need to consider the grade of metamorphism because high-temperature metamorphic rocks might not "lock in" their isotopic clock until they have cooled a bit. Heat and circulating liquids and gases during metamorphism may affect the parent–daughter isotopic ratios, essentially resetting the isotopic clock. For example, if atoms of the daughter element leave the heated crystal, the mineral appears younger than it really is because it then has fewer daughter atoms (Fig. 8-21).

How to Date Rock (or Mineral)

If isotopic clocks are actually the ratio of elements found in minute quantities in tiny mineral grains, how do geologists analyze these samples to determine ages of minerals and rocks? A mass spectrometer is the machine used for numerical dating involving radioactive isotopes, and it is capable of detecting isotopes in very small quantities. A major advance in numerical dating has been the development of the *sensitive high-resolution ion microprobe,* or SHRIMP, which allows geologists to analyze particular spots within a single, small crystal. Even if a crystal contains an old-core region inherited from an earlier geological event, the SHRIMP can determine this old age, as well as the age of the later mineral growth at the rim of the mineral (Fig. 8-22).

Before putting a sample in a mass spectrometer, or blasting it with a beam of ions in a SHRIMP, we have to decide which isotopes to analyze. The isotopes present depend in part on the composition of the rock or mineral. For example, if you want to determine the age of a basalt, you would not use an isotopic system involving carbon isotopes because the silicate minerals in basalt don't contain carbon. You could, however, use an isotopic system involving uranium or potassium, as these will be present in extremely small but measurable quantities in the basalt. If you want to date a carbonate rock or organic material, such as a bone or a piece of wood (which contain carbon), you might be able to use a carbon-isotope system.

Choosing an applicable isotopic method also depends on whether the sample is very old (tens to hundreds of *millions* of years old) or very young (tens to hundreds of *thousands* of years). Very young minerals may not have had enough time to produce a sufficient number of daughter atoms to be measured by a mass spectrometer. Thus, if you want to determine the age for a rock that is likely very young, you would not use an isotopic system with a very long half-life, as not enough decay steps would have taken place to produce a measurable quantity of daughter atoms. Instead, you would choose an isotopic system with a short half-life.

What type of isotopic system would you use if you wanted to date a rock that was likely to be very old? Would you use an isotopic system with a short half-life or a long half-life? Keep in mind that there must be a measurable amount of both parent and daughter atoms in the rock or mineral you are analyzing. For an old rock, you would use an isotopic system with a very long half-life, so that not all of the parent atoms would have decayed away. If the isotope had a relatively short half-life, very old rocks and

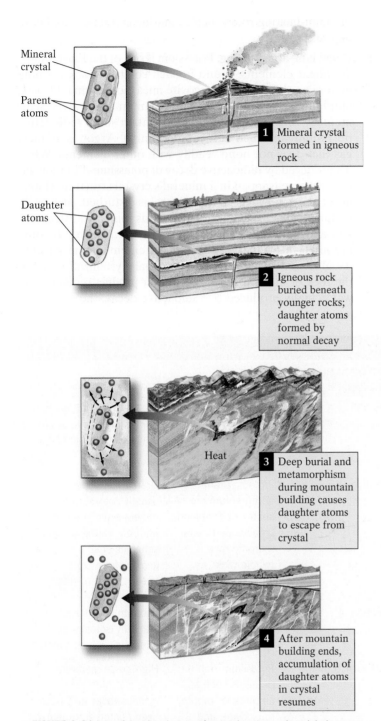

Mineral crystal

Parent atoms

1 Mineral crystal formed in igneous rock

Daughter atoms

2 Igneous rock buried beneath younger rocks; daughter atoms formed by normal decay

Heat

3 Deep burial and metamorphism during mountain building causes daughter atoms to escape from crystal

4 After mountain building ends, accumulation of daughter atoms in crystal resumes

▲ **FIGURE 8-21** Loss of daughter isotopes from rock metamorphosed during mountain building. The heat produced in this process causes mineral crystals to expand, allowing daughter atoms to escape and migrate elsewhere, even as the decay of the parent isotope proceeds normally. The loss of these daughter isotopes during this period skews the resulting parent-to-daughter ratio so that the rock appears younger than it is. After metamorphism ceases and the crystals cool and contract, the daughter isotopes start accumulating again.

minerals may not contain any detectable amount of the original parent isotope.

It might seem like a bit of circular reasoning to say that you have to know something about the age of a rock in order to determine its age, but in practice that is not such a difficult task. If your sample comes from a huge active volcano, it is probably very

▲ **FIGURE 8-22** The oldest rocks yet discovered—4.03-billion-year-old Acasta Gneiss from the Yellowknife area of Canada's Northwest Territories. Uranium and lead isotopes, measured with a sensitive, high-resolution ion microprobe (SHRIMP), were used to date these rocks. Zircon crystals embedded in the Jack Hills Conglomerate of western Australia are the oldest mineral crystals ever found—dated as old as 4.4 billion years.

young (old extinct volcanoes erode away). If your sample comes from a granitic pluton now exposed at the Earth's surface, it may be quite old, since enough time must have passed for the batholith's overlying rocks to erode away.

We'll now look at some of the most common isotope-dating systems. Each of these systems is applied to certain types of rocks and minerals. Some can be applied only to very old rocks; others work best on young rocks and materials. Wherever possible, we use several dating systems to date the same sample. By applying more than one isotopic system or by applying a particular system to many different minerals in one rock, we can learn a lot about the history of the material being dated.

Uranium-, Thorium-, Potassium-, and Rubidium-Dating Systems

Although dozens of radioactive isotopes exist, only a few are useful for dating purposes. Some decay too rapidly, perhaps on the order of thousandths of a second; others decay too slowly to date most rocks, on the order of tens of billions of years. Some migrate into and out of rocks and minerals too readily under ordinary environmental conditions. To satisfy the needs of dating systems, isotopes must be present in common

Earth materials, not be overly susceptible to gains or losses of parent and daughter isotopes, and have half-lives of at least several thousand years. Sometimes rocks contain more than one radioactive-isotope system, allowing us to cross-check the dates yielded by each system.

Table 8.1 shows the most frequently used radioactive isotopes, along with their daughter isotopes, half-lives, and the particular minerals or rocks in which they are commonly found.

1. **Uranium–thorium–lead dating.** Find a rock containing uranium, such as a granite or a limestone, and you have *three* isotope-dating systems with which to work. Virtually all uranium-bearing minerals contain three radioactive isotopes—uranium-238, uranium-235, and thorium-232. Each isotope decays through a series of steps to form a different isotope of another element—lead. Uranium and thorium isotopes have extremely long half-lives, which makes them invaluable in dating some of the Earth's oldest rocks (Fig. 8-22). The uranium-thorium system can also be used to date relatively young

uranium-bearing rocks such as marine carbonates and cave deposits.

2. **Potassium–argon dating.** Potassium is one of the Earth's most abundant elements, found in such common rock-forming minerals as biotite and muscovite mica, potassium feldspars, amphibole, and glauconite (a common mineral in sedimentary rocks). These minerals contain some radioactive potassium-40, which decays to argon-40. Argon is an inert gas that does not bond readily with other elements. When it's produced by radioactive decay of potassium-40, argon gas may become trapped in a mineral's crystal structure. It may also escape, however, if the mineral is disturbed, such as by metamorphic heating (which expands the crystal structure and allows the gas to leak out) or by weathering. If some argon-40 has left the rock, measurement of the parent–daughter ratio will not give the original crystallization age of the rock or mineral. As less daughter product remains than was actually produced by radioactive decay, the age deter-

Table 8.1 The Major Isotopes Used for Isotope Dating

METHOD	PARENT ISOTOPE	DAUGHTER ISOTOPE	HALF-LIFE OF PARENT (YEARS)	EFFECTIVE DATING RANGE (YEARS)	MATERIALS COMMONLY DATED	COMMENTS
Rubidium–strontium	Rb-87	Sr-87	47 billion	10 million–4.6 billion	Calcium-rich minerals such as plagioclase feldspar and calcium-rich garnets.	Useful for dating the Earth's oldest metamorphic and plutonic rocks.
Uranium–lead	U-238	Pb-206	4.5 billion	10 million–4.6 billion	Zircons, uraninite, and uranium ore such as pitchblende; igneous and metamorphic rock (whole-rock analysis)	Uranium isotopes usually coexist in minerals such as zircon. Multiple dating schemes enable geologists to cross-check dating results.
Uranium–lead	U-235	Pb-207	713 million	10 million–4.6 billion		
Thorium–lead	Th-232	Pb-208	14.1 billion	10 million–4.6 billion	Zircons, uraninite	Thorium coexists with uranium isotopes in minerals such as zircon.
Potassium–argon	K-40	Ar-40	1.3 billion	100,000–4.6 billion	Potassium-rich minerals such as amphibole, biotite, muscovite, and potassium feldspar; volcanic rocks (whole-rock analysis)	High-grade metamorphic and plutonic igneous rocks may have been heated sufficiently to allow Ar-40 gas to escape.
Carbon-14	C-14	N-14	5730	100–70,000	Any carbon-bearing material, such as bones, wood, shells, charcoal, cloth, paper, animal droppings; also water, ice, cave deposits	Commonly used to date archaeological sites, recent glacial events, evidence of recent climate change, and environmental effects of human activity.

mined will be younger than that of the initial crystallization of the rock or mineral. These younger ages are nevertheless useful (see the section on Dating Tectonic Events on page 255).

The potassium–argon system is widely used and extremely useful for dating both very old rocks and very young rocks. In East Africa, potassium-bearing minerals in volcanic deposits above and below the sedimentary layers that contain fossils of our earliest human ancestors have yielded potassium–argon dates ranging from 3.6 to 4.0 million years. A geologist at the University of California, Berkeley, has even determined an age of 79 A.D. for volcanic deposits from Italy's Mount Vesuvius, the date of the eruption of Pompeii. This age was, of course, already known from historical sources, but the study was an interesting example of using the K-Ar system to date very young rocks. Knowing the age of the deposits from other means enabled geologists to gauge the accuracy of this method.

3. **Rubidium–strontium dating.** Radioactive rubidium-87, which occurs in such common silicates as plagioclase feldspar and garnet, produces a daughter isotope, strontium-87. This system is used to date igneous and metamorphic rocks. With its 47-*billion*-year half-life, this system is most effective for rocks at least 10 million years old. It is used primarily to date very old intrusive igneous rocks, metamorphic rocks, and Moon rocks.

Carbon-14 Dating   Radioactive carbon-14 has a relatively brief half-life of 5730 years, and therefore **carbon-14 dating** allows us to date materials from 100 to an upper limit of about 70,000 years old. (Note that with C-14's short half-life, 70,000 years represents about 12 half-lives. After that amount of time has passed, only about 1/4096 or 0.00025 of the original parent C-14 remains, too little for even our sophisticated machines to measure accurately.) This time span includes the most recent worldwide glaciation and climate change, the latest rise and fall of worldwide sea levels, and the appearance of our own species, *Homo sapiens*. Unlike the other isotope-dating systems, which help us date rocks, carbon-14 dating is used to date other types of materials—those that contain the carbon-based remnants of past life.

Carbon-14 atoms occur naturally in the atmosphere, where they are produced by cosmic-ray bombardment of nitrogen atoms. These carbon-14 atoms, along with atoms of carbon's stable isotope, carbon-12, then combine with oxygen to form carbon dioxide (CO_2). Some CO_2 that contains carbon-14 dissolves in oceans, lakes, rivers, glaciers, and underground waters, where organisms living in or drinking the water take it up into their cells. Plants incorporate atmospheric CO_2 during photosynthesis, using it to produce sugars and starches. When animals eat the plants, they ingest the carbon-14 as well. Thus, carbon-14 finds its way into the cells of most living organisms, where it continuously decays to its daughter isotope, nitrogen-14. An organism

that is alive and taking in nutrients continuously replenishes its supply of carbon-14. As soon as it dies, however, its supply of carbon-14 is cut off and immediately begins to diminish with time. In this way, its isotopic clock begins to tick. *The greater the time elapsed since the death of the organism, the less carbon-14 (and the more nitrogen-14) it will contain* (Fig. 8-23).

As long as it's not too old, anything that contains organic carbon—such as bones, shells, wood, charcoal, plants, peat, paper, cloth, pollen, and seeds—is suitable for carbon-14 dating. We can date charcoal in ancient campfires, house posts of Native American dwellings, seeds and seafood shells in archaeological refuse dumps, and the bones of our predecessors and their prey. Carbon-14 dating has estimated the age of the linen cloth that was wrapped around the Dead Sea Scrolls (2000 to 2200 years old; Fig. 8-24), the papyrus on which the various books of the Old Testament was written (about 2100 years old). Carbon-14 dating also sets the age of the Shroud of Turin, purported to be the burial cloth of Jesus of Nazareth, at an age of 700 years. These items all have one thing in common: They were once part of a living organism or are composed of materials from once-living organisms.

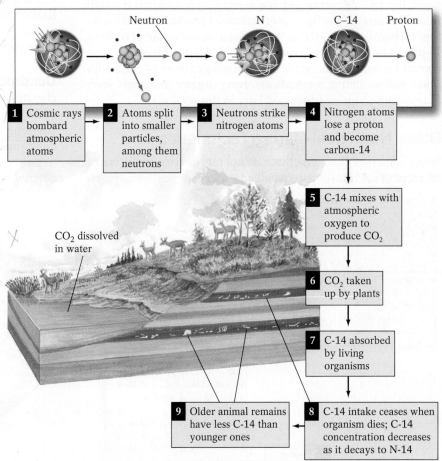

▲ **FIGURE 8-23** The carbon-14 used to date long-dead organisms and ancient artifacts originates in the Earth's atmosphere. Here, carbon-14 combines with O_2 to produce CO_2, much of which dissolves in water or is taken up by plants, and is ultimately ingested by animals. While alive, an organism constantly replenishes the carbon-14 supply in its body. When it dies, its intake of carbon-14 ceases, and the carbon-14 begins to decay to its daughter isotope, nitrogen-14. Thus, the less carbon-14 left in the remains of an organism, the more time has elapsed since it died.

▲ **FIGURE 8-24** The papyrus that comprises the Dead Sea Scrolls, ancient copies of the Old Testament and other Judaic writings, has been dated by C-14 dating. Its age—about 2100 years.

Scientists must be careful when using carbon-14 dating because the rate of cosmic-ray bombardment has varied through time. This rate affects the rate of carbon-14 production and therefore carbon-14 dates as well. Assuming a constant rate of carbon-14 production results in errors in carbon-14 dates. Geologists can correct for some of this error by calibrating carbon-14 dates with other precise dating methods involving organic materials, such as tree-ring counting (discussed later in this section).

Other Numerical-Dating Techniques

Some numerical-dating methods do not rely *directly* on the measurement of the ratio of parent and daughter isotopes. Some measure other physical changes within a mineral crystal brought on by nuclear decay. Other methods rely on countable annual layers (such as tree rings and lake sediments) or a botanical organism that grows at a fairly constant rate (such as lichen, plants that grow directly on rocks).

Fission-Track Dating   *Fission* is the division of a radioactive atom's nucleus into two fragments of approximately equal size. Upon undergoing fission, an atomic nucleus releases several subatomic particles (such as alpha and beta particles). These subatomic particles move at high speed across the orderly rows of atoms within a mineral's crystal lattice, leaving behind a trail of damage within the crystal. These trails are called **fission tracks** (Fig. 8-25). The number of fission tracks in a crystal is a function of the uranium concentration (more uranium, more tracks) and the time elapsed since fission tracks started forming in the crystal (longer time, more tracks). Fission tracks can date minerals that range from 50,000 to billions of years old. This method is particularly useful for dating volcanic glasses and other uranium-trapping mineral grains, such as zircons and apatites $Ca_5(PO_4)_3(F, Cl, OH)$. Fission tracks in zircons are usually erased when their host rocks are heated above 250°C (475°F). Consequently, this method cannot be used to date the crystallization of medium- and high-grade metamorphic rocks, but it can be used to determine the time at which the rock cooled below this temperature.

Dendrochronology (Tree-Ring Dating)   In temperate climates, the annual growth of most trees produces concentric sets of dark and light rings within their trunks and larger branches. By counting its rings, we can determine a tree's age—a numerical-dating method known as **dendrochronology.** This method can help us date recent geological events such as landslides or mudflows, wherever trees have begun to grow on new surfaces. By dating

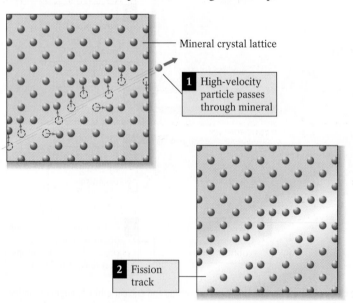

Mineral crystal lattice

1 High-velocity particle passes through mineral

2 Fission track

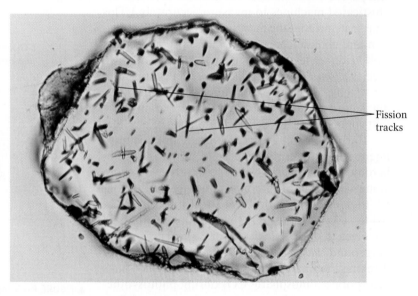

Fission tracks

▲ **FIGURE 8-25** Fission tracks are produced by the decay of radioactive atoms trapped in a mineral's crystal structure. High-velocity decay particles emitted by the radioactive atoms disrupt the orderly arrangement of atoms in the crystal lattice, leaving tracks only a few atoms wide. These tracks are typically flooded with acid and enlarged so that they may be seen (and counted) with an ordinary microscope. When comparing two zircons with the same uranium concentration, the older one will have more tracks.

the tree, we date the *minimum* age of the landslide or mudflow on which the tree is growing (that is, the landform is *at least* as old as the tree, but may be a little older than the tree if a few years had passed before trees had taken root on the new surface). Because tree-ring patterns are preserved even in long-dead trees, dendrochronology can also date geological events older than the oldest living tree on the landscape. Changes in climatic conditions, such as prolonged droughts, create a distinctive pattern of tree-ring variations in all trees living within a given geographical region. By comparing and correlating the ring patterns in trees of different ages, we can see where their life spans (and ring patterns) overlap. We can trace ring patterns from older live trees (some as old as 4770 years) to even older, *dead* trees with overlapping life spans and ring patterns—perhaps those incorporated into archaeological sites (Fig. 8-26). By studying old living trees and dead ones too, tree-ring dating has been extended to as far back as 9000 years in some areas (especially in the American Southwest). Archaeologists use dendrochronology to date wooden artifacts, such as house posts and digging sticks, by comparing their distinctive ring patterns with an identical pattern of known age.

Varve Chronology

Varve Chronology   Deposition in lakes can produce annual layers of sediment that can be counted much like tree rings. These layers, called varves, are best developed in lakes—such as glacial lakes—that freeze partially in the winter and are fed by sediment-rich streams in the summer. **Varves** are paired layers of sediment, typically consisting of a thick, coarse, light-colored layer deposited in summer during periods of high inflow, and a thin, fine, dark-colored layer deposited in winter during periods of low or virtually no inflow (especially if the surface of the lake is frozen) (Fig. 8-27). The thin, dark wintertime layer is composed primarily of tiny clay particles and microscopic bits of organic matter (such as pollen and leaf fragments). These particles may enter the lake in the summer but, because of their sizes, take months to settle out of the water.

To count the varves underlying a lake, geologists drill through the lake mud to extract a *core,* a continuous sequence of lake sediment. The number and nature of the varves in a core reveal how long ago the lake formed and identify events, such as landslides, that affected its existence. For example, a lake produced when a landslide blocks a river valley may immediately begin to produce varves. Counting the varves tells us when the landslide occurred—a good thing to know if we're worried about the history of instability of a region's slopes.

Lichenometry

Lichenometry   **Lichen** are colonies of simple, plantlike organisms that grow directly on exposed rock surfaces. Measurements of the diameters of lichen on rocks is the basis of a dating method known as **lichenometry.** Given similar rocks and climatic conditions, *the larger the lichen colony, the longer the time since the growth surface became exposed.* Larger lichen colonies are found on old tombstones, for example, than on newer ones. Using tombstones and other stone surfaces of known age on which lichen grow (such as buildings, bridge supports, and old mine tailings), we can develop a growth curve, showing the rate at which lichen have grown in a given area (Fig. 8-28). Then we can compare lichen on a surface of unknown age (within the same area) with the growth curve to determine the surface's age. Because lichen colonies eventually grow together and can no longer be measured individually, lichenometry is useful only for surfaces less than about 9000 years old. These organisms provide accurate dates for young glacial deposits, rockfalls, and mudflows—all events that expose new rock surfaces on which lichen can grow.

Surface-Exposure Dating

Surface-Exposure Dating   The dating techniques discussed so far give you an age of a geologic material—a mineral, rock, bone, shell, tree—but other numerical-dating techniques can be used to determine the age of whole landscape features. For example, geologists can date the formation of ridges of glacial debris

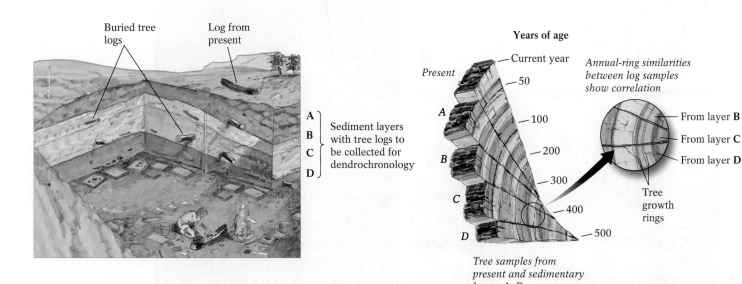

▲ **FIGURE 8-26** Correlation of tree-ring sections in trees or wooden artifacts of overlapping lifespans can establish ages of thousands of years.

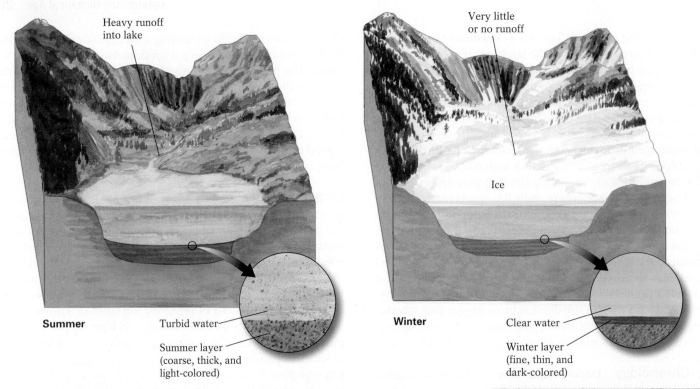

Summer

Heavy runoff into lake

Turbid water

Summer layer (coarse, thick, and light-colored)

Winter

Very little or no runoff

Ice

Clear water

Winter layer (fine, thin, and dark-colored)

▲ **FIGURE 8-27** The origin of lake varves. A typical varve includes a thick, coarse, light-colored summertime layer produced during high runoff (from snowmelt and spring storms) and high sediment influx, and a thin, fine, dark-colored wintertime layer produced during low runoff and low sediment influx (or no influx if the lake is frozen). Note the varves in the photo. Why do you think the varves vary so noticeably in thickness?

▼ **FIGURE 8-28** Estimating the numerical age of a material from the lichen growing on its exposed surface first requires that we determine the rate of lichen growth in the area. After measuring lichen on surfaces of known age, such as the bridge, the church, and the mine entrance (left), geologists plot a growth curve that relates lichen diameters to time. Then they can estimate the age of a surface of unknown age by measuring the lichen on it and finding its position on the growth curve. (right) Lichen colonies on a granite boulder.

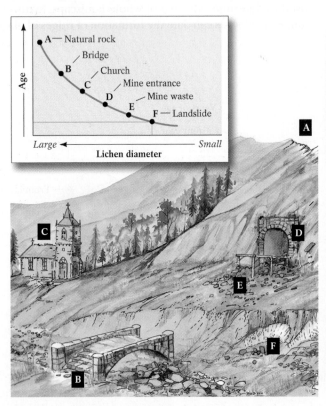

A — Natural rock
B — Bridge
C — Church
D — Mine entrance
E — Mine waste
F — Landslide

Large ← **Lichen diameter** → *Small*

Age

called *moraines* (discussed in Chapter 17), *alluvial fans* (wedges of sediment dropped by rivers when they emerge from mountain valleys), or newly emerging lands rising from the sea, either from plate-tectonic uplift or a relative drop in sea level. In other words, geologists can determine the age of newly exposed land surfaces rather than the age of crystallization of a single rock or mineral. This type of numerical dating uses *cosmogenic* isotopes, which are produced in extremely small quantities at the surfaces of newly exposed rocks and other surface materials from cosmic-ray bombardment.

Intergalactic cosmic radiation, predominantly a stream of atomic nuclei and high-energy electrons, constantly bombards the Earth at phenomenal speeds. (Countless numbers of these particles are passing *through* your body as you read this page.) These particles strike the rocks and soils at the Earth's surface, penetrating to a maximum depth of 2 to 3 m (7 to 10 ft) before being absorbed. As these high-speed particles hit target elements in rocks and soils—such as oxygen, silicon, aluminum, iron, magnesium, calcium, and potassium—they knock other atomic particles from the nuclei of these elements, converting these elements into new "cosmogenic" radioactive isotopes (Fig. 8-29). Oxygen and nitrogen, for example, are typically converted to *beryllium-10;* potassium, calcium, and chlorine become *chlorine-36.*

The atoms of these radioactive substances, like other radioactive atoms, immediately begin to decay. Unlike in the case of non-cosmogenic isotopes, whose number of parent atoms is fixed (such as uranium in the Earth's crust), new atoms of these cosmogenic isotopes are constantly being produced, even as others decay. Thus a simple parent–daughter decay relationship does

▲ **FIGURE 8-30** This massive basaltic boulder was plucked from the Columbia River basalt flows of eastern Washington by a moving glacier about 15,000 years ago. Since the moment it melted from the ice, it has been bombarded by cosmic radiation. Its time of initial exposure can be determined by surface-exposure dating.

yeager Rock

not apply to these isotopes. Instead, we must know and then subtract their production rates and accurately determine their decay rate, or half-life. Chlorine-36 has a half-life of roughly 300,000 years; beryllium-10's half-life is approximately 1.5 million years. Knowing these half-lives enables geologists to date newly exposed surface boulders (Fig. 8-30) whose ages fall between the ranges of carbon-14 dating and other isotope-dating schemes.

Of course, like any new dating method, surface-exposure dating faces its share of technical challenges. The first involves the extremely small amounts of these cosmogenic isotopes present in the rocks. During the past decade, geologists have developed powerful new devices capable of measuring very small numbers of atoms with remarkable accuracy. The accelerated mass spectrometer (AMS), for example, can detect 5 radioactive atoms in a pool of 1,000,000,000,000,000 atoms. This capability would be like finding 5 red garnet grains amidst an entire houseful of white quartz grains. The technology has also been applied to carbon-14 dating, to push back the effective age range of radiocarbon dating. Today geologists can measure very small samples of an organic substance, enabling them to date an object without completely consuming it (as was often the case with past techniques). Thus, a priceless item, such as the Dead Sea scrolls, can be dated by using only a tiny bit of its substance.

Dating Tectonic Events

When we analyze rocks for their numerical ages, we are often trying to answer some tectonic questions, such as: When did a collision of plates occur that produced the mountain range we're studying? How long does it take for such a mountain range to form? How long does it take for a rock to move from deep in the crust to the Earth's surface? We can answer these questions and calculate rates of tectonic and other processes using geochronology.

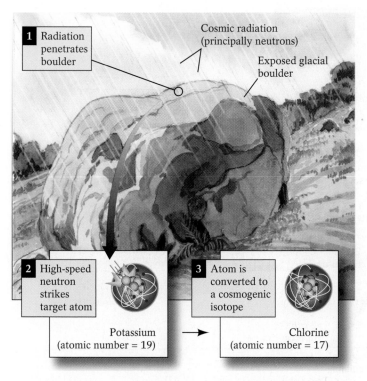

1 Radiation penetrates boulder

Cosmic radiation (principally neutrons)

Exposed glacial boulder

2 High-speed neutron strikes target atom

Potassium (atomic number = 19)

3 Atom is converted to a cosmogenic isotope

Chlorine (atomic number = 17)

▲ **FIGURE 8-29** As soon as it is exposed, any surface rock begins to be bombarded with cosmic radiation. This action converts target atoms into radioactive "cosmogenic" isotopes whose decay rate can then be measured to determine the time of initial exposure.

Consider again the concept of the *closed system* discussed on page 247. When a rock or mineral becomes a closed system, atoms stop migrating in and out of the mineral structure and the isotopic clock (decay of the parent to the daughter, with no loss or gain of atoms of the isotopes) starts ticking. Each isotopic system described in this chapter *closes* at a different temperature: some close to the migration of atoms at high temperatures (example: U-Pb system); others close at lower temperatures (examples: K-Ar system; formation of fission tracks).

If we measured several different isotopic ratios for a volcanic rock, for example, they should give us the same age because the lava cooled quickly and the closure temperatures of all the isotopic systems would have been reached within a short amount of time. Other rocks, such as igneous intrusions and metamorphic rocks, take a longer time to cool from high temperatures at depth in the crust to temperatures low enough for movement of atoms in the minerals to slow down so that the rock or its minerals become closed.

The fact that different isotopic systems close ("lock in") at different temperatures is not a problem; it's actually a useful tool. Imagine that you find an intrusive igneous rock or a metamorphic rock and you want to know how that rock got to the Earth's surface. Figuring out how long it took the rock to get from deep in the Earth to the surface might help you solve this problem. If the rock went from great depth to the surface rapidly, this information would tell you that the mountain range was breaking down quickly. How would you figure this history out? You would first take careful note of the rock's position, observe any evidence of faulting in the field, and identify the rock type and its miner-

als. You would then select the best minerals in the rock for isotopic analysis: perhaps zircon and other U-bearing minerals for U-Pb dating and fission-track analysis, and amphibole or mica for K-Ar dating. If the rock contains garnet, you're in luck—garnet can be dated using most of the isotopic systems mentioned in this chapter. After you analyze each mineral, you have a record of the cooling history of the rock from high temperatures (>700°C) to very low temperatures (<100°C) with dates that record the times when the different temperatures were reached.

In Figure 8-31, we see different stages in the tectonic evolution of a convergent plate boundary. Rocks caught up in the collision will be buried and heated (metamorphosed). Eventually they reach their maximum depth and temperature, and that time is recorded using an isotopic-dating system—such as U-Pb dating on zircons—that "closes" at a high temperature. As time passes, the rocks cool, other isotopes reach their "closing" temperatures, and some combination of erosion and faulting causes them to be exposed at the surface. The rate at which these rocks have moved to the surface and cooled is recorded by other isotopes—such as Ar-dating of biotite, muscovite, and amphibole. If a rock's temperature-time history (the amount of time between the highest-temperature age and the lowest-temperature age for minerals in the rock) shows that it cooled rapidly (the high-temperature and low-temperature ages are close together), the rock may have been brought to the surface by faulting, by unusually intense erosion, or a combination of faulting and erosion. If the rock cooled slowly (high-temperature and low-temperature ages are far apart), the rock probably slowly eroded to the surface without much tectonic activity.

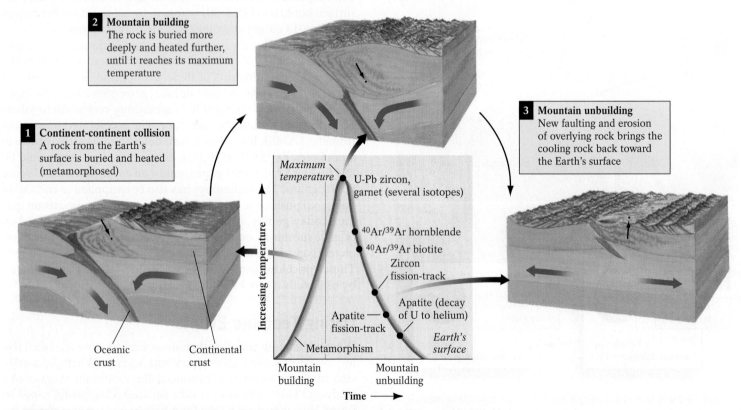

▲ **FIGURE 8-31** The power of the temperature-time dating tool helps geologists to unravel a rock's personal tectonic history. Here we see the "life story" of a rock during a continental collision—from its initial subduction and heating, through its deep burial, and finally to its eventual exposure at the Earth's surface by faulting and erosion.

8.2 Geology at a Glance

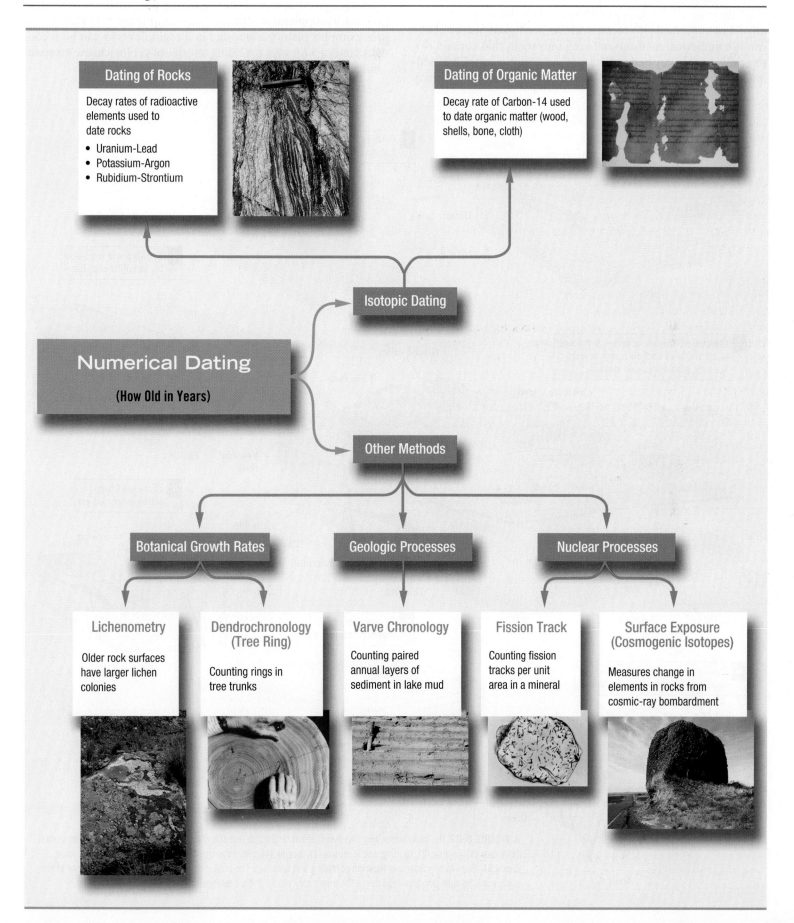

Dating of Rocks

Decay rates of radioactive elements used to date rocks

- Uranium-Lead
- Potassium-Argon
- Rubidium-Strontium

Dating of Organic Matter

Decay rate of Carbon-14 used to date organic matter (wood, shells, bone, cloth)

Isotopic Dating

Numerical Dating

(How Old in Years)

Other Methods

Botanical Growth Rates

Geologic Processes

Nuclear Processes

Lichenometry

Older rock surfaces have larger lichen colonies

Dendrochronology (Tree Ring)

Counting rings in tree trunks

Varve Chronology

Counting paired annual layers of sediment in lake mud

Fission Track

Counting fission tracks per unit area in a mineral

Surface Exposure (Cosmogenic Isotopes)

Measures change in elements in rocks from cosmic-ray bombardment

Combining Relative and Numerical Dating

When geologists and paleontologists study a region for the first time, they use all the dating methods at their disposal—both relative and numerical—to reconstruct its geological history. They can use numerical methods on a region's rocks that contain isotopically datable key beds; other materials, such as fossil-free sedimentary rocks, may only be dated relatively. Geoscientists may write a geological "biography" of some regions, albeit with some chapters missing, by numerically dating every layer that contains radioactive isotopes and then applying the principles of relative dating to supplement this information. Figure 8-32 shows how the complex history concealed in a region's rocks can be reconstructed *step-by-step*, revealing a series of geological events spanning billions of years of Earth history.

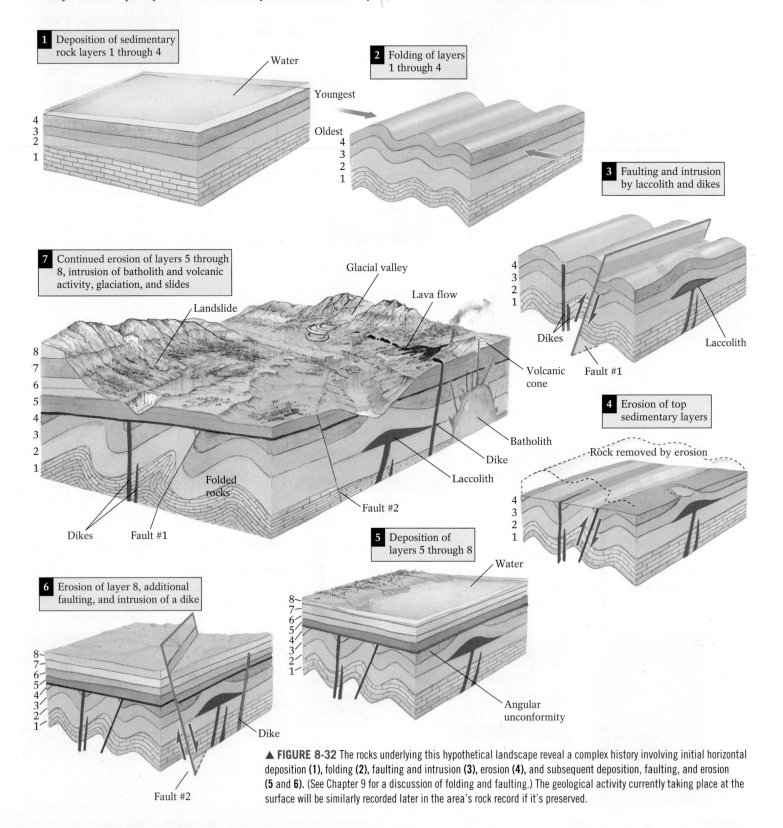

▲ **FIGURE 8-32** The rocks underlying this hypothetical landscape reveal a complex history involving initial horizontal deposition (**1**), folding (**2**), faulting and intrusion (**3**), erosion (**4**), and subsequent deposition, faulting, and erosion (**5** and **6**). (See Chapter 9 for a discussion of folding and faulting.) The geological activity currently taking place at the surface will be similarly recorded later in the area's rock record if it's preserved.

The Geologic Timescale

The **geologic timescale** (Fig. 8-33) organizes all of Earth history into blocks of time during which the planet's major events occurred. The basic divisions of the geologic timescale were established during the 18th, 19th, and early 20th centuries by geologists and paleontologists who identified changing fossil assemblages in rocks and then applied the principles of superposition, faunal succession, and cross-cutting relationships to establish their age sequence. In the last half-century, scientists have used isotope dating to determine numerical-age ranges to these divisions.

The largest time spans shown on the geologic timescale are *eons.* They mark the Earth's truly major developments, such as the first occurrence of life and the worldwide expansion of multi-celled life forms. The earliest is the Hadean ("beneath the Earth") Eon, ranging from the time of the Earth's formation, 4.6 billion years ago, to 3.8 billion years ago. With the exception of the 4.03-billion-year-old rocks of Canada's Northwest Territories, geologists have found virtually no other Hadean rocks or fossils. The earliest-known life forms appear in rock from the Archean ("ancient") Eon, about 3.8 to 2.5 billion years ago (Fig. 8-34).

Eons are subdivided into *eras,* each of which is defined by its dominant life forms. The eras of the timescale typically change when the dominant life forms of one era are swept away, typically by a massive global environmental change. Such periods of extinction and some of their proposed causes are discussed in Highlight 8-3. The Mesozoic Era, for instance, is the Age of Reptiles—the time of the dinosaurs. After the Permo-Triassic extinction—about 250 million years ago—the dinosaurs replaced the invertebrates of the Paleozoic Era. Eras are subdivided into *periods,* which record less dramatic biological changes. Periods, in turn, are subdivided into *epochs,* based on even more subtle changes in the assemblage of living creatures and on some non-biological criteria as well. Each successive subdivision marks a progressively smaller unit of time (that is, eons are larger than eras, which are larger than periods, and so on). The length of time for each division varies, even for subdivisions at the same level, because the key characteristics and geological conditions that define them do not last the same amounts of time. Thus, the Eocene Epoch of the Tertiary Period of the Cenozoic Era lasted for 21 million years, whereas the Oligocene Epoch of the same period and era lasted for only 13 million years.

Following this chapter, on pages 268–287, is a stunning, graphic rendition of the Geologic Timescale. It was adapted for use here from the book *Earth,* published jointly by the Smithsonian Institution and Dorling Kindersley (James F. Luhr, editor). Beginning with the Big Bang and moving forward to the present, the timeline helps lend scale to the immensity of time and order to the key events of Earth history. It is an efficient and absorbing survey of how our planet has come to be and what we see today.

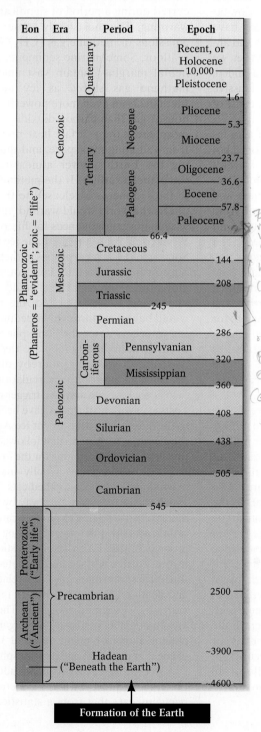

▲ **FIGURE 8-33** The geologic timescale. Numerical ages shown are in millions of years (mya = million years ago). The Hadean Eon represents the time from which we have virtually no direct evidence of the Earth's past.

▼ **FIGURE 8-34** The fossils in this rock, from the Precambrian Period of the Archean Eon, are *stromatolites,* structures created by a very early form of algae.

Highlight 8-3

Death on Earth

Over the past 3.5 billion years, life on Earth has evolved from simple cells to complex animals and plants. Changes in life forms are closely linked to changes in the physical environment at the Earth's surface. Plate-tectonic motions move continents around, create high mountains, drive volcanic activity, and change the composition of the atmosphere. Other events, such as meteorite impacts and global warming and cooling, have also affected life on Earth. Each of these events changes the environment, sometimes in drastic ways, causing the extinction of some organisms and creating new opportunities for others. Let's consider the two largest extinctions in Earth history and ideas about their causes.

The most dramatic catastrophe in the history of life on Earth occurred as the giant supercontinent, Pangaea, broke into separate continents at the end of the Paleozoic Era. This event—known as the Great Dying—occurred at the Permian-Triassic boundary of the geologic timescale. It resulted in the extinction of 90% of marine species and 75% of all land animals as well. Although the ultimate cause of this extinction remains unknown, research has focused on several intriguing possibilities. Two hundred and fifty million years ago, vast amounts of basaltic lava from a hot spot in the Earth's mantle erupted in Siberia, forming layers of lava that were 3 km (2 mi) thick and covered 2.5 million km^2 (965,000 mi^2), an area as large as the entire country of Argentina. Perhaps the sulfurous gases associated with the eruptions reacted with water in the atmosphere to form acid rain so toxic that the global environment became uninhabitable for most life forms. Perhaps the Earth's atmosphere heated up sufficiently to melt vast quantities of frozen methane—a powerful "greenhouse" gas—at the seafloor. Can volcanism itself cause mass extinctions? Recent research has shown that three of the Earth's five largest mass-extinction events coincided with massive outpourings of basaltic lavas.

Recently, a new factor entered the Permo-Triassic extinction debate: a massive meteorite. Geologists studying Permo-Triassic rocks in China, Japan, Hungary, and Australia discovered that they contain extraterrestrial "buckyballs" containing trapped molecules of an alien form of helium. These bits of extraterrestrial carbon—now far flung across the Earth's surface—may have arrived together as a giant meteorite and annihilated most of the life on Earth. As yet, we don't know where this killer may have landed, but some geologists are eyeing the massive Bedout crater off the northwestern coast of Australia as a possible impact site.

The end of the Mesozoic Era also involved massive basaltic eruptions from a hot spot—this time in India. If that weren't enough to create an inhospitable global environment, disaster for the dinosaurs and many other life forms on Earth came from space when a meteorite crashed in the Gulf of Mexico, near the Yucatán Peninsula. The 10-km (6-mi)-wide meteorite punched a 180-km (120-mi)-wide, 30-km (20-mi)-deep hole in the crust. The impact vaporized a vast quantity of rock (including sulfur-bearing evaporites), creating a toxic environment for land- and ocean-dwelling organisms. Life on Earth was devastated by the impact. In less than 1 million years, 75% of all marine species and large land creatures became extinct.

These two mass-extinction events are the largest ever recorded, though others have also occurred. The past 500 million years have been marked by five or six other major extinction events, and about the same number of smaller ones (see figure). Did each one have a different cause, or do the same events lead to mass extinctions? The only mass extinction that we believe *with confidence* was caused in part by a meteorite impact was the Cretaceous–Tertiary event at the end of the Mesozoic Era. Perhaps we just haven't found evidence for other impacts at other times. (The older the event, the more likely the evidence of the crash site has been either eroded to obscurity or buried. Impacts might also have occurred in remote places, such as the middle of an ocean, where evidence for a crater is difficult to find.)

Other extinctions have probably been caused by global climatic change. The extinction event at the Paleocene–Eocene boundary, about 55 million years ago, occurred during a period of significant global warming. The most likely cause: massive emissions of methane gas (CH_4) from the seafloor. Some marine sediments at continental margins contain vast amounts of methane gas stored as icy molecules. Methane is an even more powerful "greenhouse" gas than carbon dioxide. In the atmosphere, it traps surface heat, preventing it from radiating into space, and thus warming the Earth's lower atmosphere near the planet's surface. If the methane stored in the sediments is released to the atmosphere, the world would warm dramatically, with major consequences for life on Earth. What could cause the methane ice to melt? Some geologists have proposed that massive outpourings of basaltic lava could start a global warming trend that could, in turn, melt this ice. Once melted, the methane would bubble up through the sediments and seawater and enter the atmosphere. These events would greatly increase the amount and rate of warming, thereby disrupting oceanic and atmospheric circulation.

Global cooling may also trigger major extinction events. Extinctions have been related to the onset of the most recent Ice Age and to the subsequent warmup that followed the end of glacial times, accounting for the loss of such Ice Age mammals as woolly mammoths, mastodons, and saber-toothed cats. Is life on Earth fragile, easily swept away by environmental change related to geologic catastrophe, or is it actually quite rugged, having survived in some form on Earth despite drastic changes in the Earth's surface and climate through time? These and other questions related to life and death on Earth are topics of intense research by geoscientists and others. The answers are important for predicting how organisms (including humans) might respond to future changes in the Earth and its atmosphere.

Recent NASA-sponsored studies of the potential frequency of massive meteorite impacts indicate that from a statistical standpoint, the Earth can expect a major impact—one that would release about 10 megatons of explosive energy (equivalent to 10 million tons of TNT) and cause widespread death and destruction—at least once every thousand years or so. The 20th century experienced such an event in the atmosphere over Tunguska, Siberia, in 1908. This meteorite exploded in the atmosphere and flattened more than 2000 km^2 (800 mi^2) of Siberian forest. The Tunguska meteorite was only about 55 m across (about 180 ft), a relatively minor event compared to other extinction-level events in Earth history. As just a reminder that such events are always possible, on March 27, 2003, a Volkswagen-size meteorite streaked across the skies above Indiana, Ohio, Wisconsin, and Illinois, sending rocks the size of softballs crashing through the roofs of houses across the southern suburbs of Chicago, Illinois. And in the third week of August 2004, a "close encounter" occurred again as garage-sized meteorite sped by our planet at a distance of 6000 km (4000 mi). The near-miss was detected by a pair of telescopes in New Mexico's White Sands, thanks to LINEAR (Lincoln Near Earth Asteroid Research), a Massachusetts Institute of Technology project of its Lincoln Astrophysics Laboratory that is dedicated to scanning the skies in search of dangerous near-Earth objects.

One final word on mass extinctions: They are a clear example of the interrelatedness of the Earth's various systems. A major tectonic ("lithosphere") event (such as a massive volcanic eruption) or extraterrestrial impact may change the Earth's atmosphere and oceans ("hydrosphere") so profoundly that most species ("the biosphere") die off, leaving very few survivors.

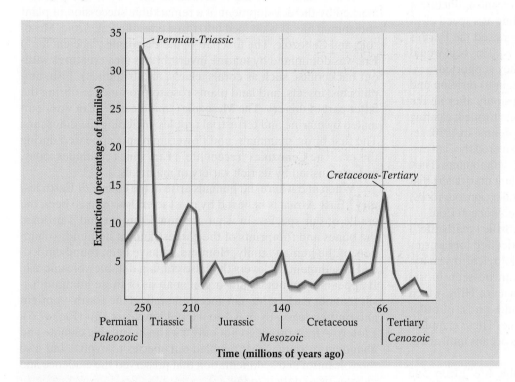

▲ (Top left) The sizes of the peaks on this graph show how widespread the loss of life was on Earth for each extinction. (Top right) An artist's conception of a massive meteorite impact. (Bottom) A softball-size chunk of meteorite that struck Park Forest, Illinois, south of Chicago, on March 27, 2003. About 60 chunks of this size were turned over to authorities by the local residents of the area.

Life on Earth

Many geologists believe that simple life forms may have evolved *here on Earth* during the planet's first billion years. Proponents of this view suggest that all of the ingredients necessary to produce amino acids (carbon, hydrogen, oxygen, and nitrogen)—the precursors to life—existed in the Earth's early atmosphere and oceans. In their minds, a spark of Precambrian volcanic lightning may have energized the reactions that produced organic life on Earth. Alternatively, the nutrients, water, and heat required to create early organic compounds may have existed at oceanic ridges where hot water and minerals react. Others believe that the seeds of life on Earth were *imported*—that is, ferried here by meteorites or comets from around the solar system. Highlight 8-4 (see pages 264–265) discusses some of these recent findings and hypotheses.

Regardless of how the first spark of life was ignited on Earth, it was most likely extinguished shortly thereafter, either by the impact of one of the massive meteors that littered the young developing solar system, or in the wake of one of the early Earth's long and violent periods of widespread explosive volcanism. Picture a meteor 400 km (250 mi) in diameter and traveling at a speed of 15 km (10 mi) per second. Its impact would have raised the Earth's surface temperature to as high as 1870°C (3120°F). This heat would have vaporized the early oceans and any life they supported.

It seems likely that life on Earth could begin to develop and flourish without such catastrophic interruptions only after all the planets and moons of the solar system had finally formed, clearing the surrounding space of most of its remaining debris. In 1980, geologists studying in northwestern Australia discovered the first evidence of what is believed to be Earth's earliest inhabitants. They found simple bacteria-like cells in a layer of sedimentary chert that was dated (by fission-track analysis of the surrounding igneous rocks) at about 3.77 billion years—the beginning of the Archean Eon.

The Archean Eon gave way to the Proterozoic ("early life") Eon, 2.5 billion years ago. The first multicelled organisms appeared during this time of increasing biological diversity. Clear evidence of increasing biodiversity can be seen in the 670-million-year-old rocks in Australia's Ediacara Hills, which contain fossil jellyfish, soft marine worms, and a host of other simple creatures composed entirely of soft tissue. These creatures and their descendants flourished in the world's oceans until about 545 million years ago.

The beginning of the Phanerozoic ("visible life") Eon, 545 million years ago, marks the first point at which we find abundant fossil evidence of marine creatures with hard shells, prominent external spines, and internal skeletons. A number of competing hypotheses have been proposed to explain this development. One hypothesis claims that by this time, algae, which rely on photosynthesis for basic life processes, had released sufficient oxygen (a product of photosynthesis) to form the ozone layer in the atmosphere. This layer protects life forms on Earth from deadly ultraviolet radiation from the Sun. As atmospheric oxygen fueled their metabolism, animals may have grown larger and more complex, requiring a rigid structure to support their larger mass. Another hypothesis suggests that the oceans—and the creatures in them—contained so much calcium at this time that marine creatures were forced to secrete calcium to avoid calcium poisoning. This secreted calcium ultimately became their hard outer shells.

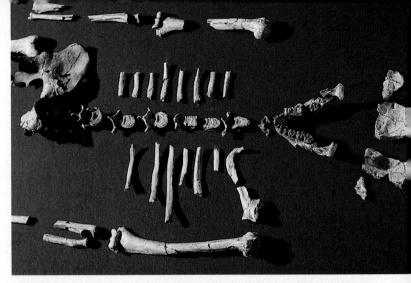

▲ **FIGURE 8-35** The fossil hominid called Lucy, collected in Hadar, Ethiopia, in 1974 by Donald Johanson. Lucy, one of the most complete hominid fossils yet discovered, has been dated by isotope dating of its surrounding tephras. Its age is estimated at 3.18 million years.

The Phanerozoic Eon extends all the way to the present, marked by the development of a remarkable succession of plant and animal life. It is divided into three eras: the Paleozoic, Mesozoic, and Cenozoic. The first part of the **Paleozoic** ("ancient life") **Era** was dominated by marine invertebrates (sea creatures without backbones, such as corals, clams, and trilobites). Fish, amphibians, insects, and land plants rose to prominence during the latter part of this era. The **Mesozoic** ("middle life") **Era** was dominated by marine and terrestrial reptiles, including the dinosaurs. The first birds, mammals, and flowering plants appeared during this era. The **Cenozoic** ("recent life") **Era,** which continues today, is distinguished by its rich variety of mammals.

When and where did humans arise in the course of Earth history? East Africa is believed by most scientists to have been the cradle of our species. In what are now Ethiopia and Tanzania, the bones and footprints of the earliest-known hominids (members of the human family, Hominidae) have been found embedded in sedimentary layers that also contain datable volcanic ash. It appears from potassium–argon analysis of these hominid fossils that our earliest human ancestors walked the Earth—upright and on two legs—from 3.4 to 4.0 million years ago (Fig. 8-35). Eastern Africa provided not only a warm hospitable climate and abundant resources that enabled our species to evolve, but also the volcanism and sedimentation that preserved the evidence and enables us to date it today.

The Age of the Earth

Geologists are fascinated by geologic time and by developing and using new methods for determining the ages of geologic materials. Certainly their enthusiasm for geochronology can be seen from the numerous dating methods mentioned throughout this chapter. Nevertheless, the ability to date the age of the Earth itself—directly and numerically—continues to elude us. This chapter began with the claim that the Earth is 4.6 billion years old. But the oldest Earth rocks discovered thus far still fall short of that age. We may never find a rock that dates from the very origin of our planet. Many of the Earth's oldest rocks are metamorphic—changed from even older, preexisting rocks. The planet's fierce internal heat, recycling of tectonic plates, cata-

▲ FIGURE 8-36 A lunar boulder, as seen from the *Apollo 17* spacecraft in December 1972. (The Earth is visible in the far distant background.) Fragments of such Moon rocks taken back to Earth for analysis have been dated at about 4.5 billion years old.

strophic events such as meteor impacts during the early evolution of our planet, weathering, and a host of other geological processes have worked together over billions of years to change much of the Earth's original material, thereby concealing direct evidence of its true age.

Geologists have had to look elsewhere to determine the age of the Earth. Because all the components of our solar system formed at roughly the same time, the ages of extraterrestrial rocks provide *indirect* evidence of the age of our own planet. Some exciting clues about the Earth's birthdate have come from Moon rocks collected by American astronauts within just a few hundred meters of their landing craft (Fig. 8-36). One of these rocks, dated by both uranium–lead and potassium–argon systems, yielded an age of 4.53 billion years. Others were dated at 4.2 billion years. Remember, the astronauts did not conduct a Moon-wide search for ancient rocks—these rocks just happened to be on the ground where the spacecraft landed. As far as we know, the Moon's rocks were never recycled by plate-tectonic and surface processes, so one of these rocks may record the birth of the Moon. Furthermore, isotope dating of many of the meteorites that have struck the Earth, such as the one shown in Figure 8-37, has yielded a similar age—4.6 billion years. Most meteorites are remnants of the early solar system and thus, they are chunks of rock that have

▼ FIGURE 8-37 The Hoba meteorite, in Namibia. Most of the meteorites that have struck the Earth and been subsequently dated isotopically yield an age of about 4.6 billion years.

never been recycled by planetary tectonics or undergone heating, pressing, weathering, or other planetary forces. In all likelihood, they faithfully record the age of our solar system and, by association, the Earth.

Some clever scientific detective work also enables us to learn something about our planet's date of birth—from the Earth itself. We have seen that uranium and thorium decay to produce three stable lead isotopes: lead-206, lead-207, and lead-208. Lead has a fourth stable isotope, lead-204, that is *not* a product of radioactive decay. As lead-204 is not created by the decay of other elements, its concentration in the Earth has remained constant from the planet's beginning, while the concentrations of the other lead isotopes have been increasing through radioactive decay since the origin of the Earth. The relative proportions of the four lead isotopes are roughly the same in every meteorite that does not also contain a radioactive parent, suggesting that these proportions reflect the mixture of lead originally found in our solar system. By comparing these proportions to those found in Earth rocks, we can determine how much lead of radioactive origin has been added.[1] Scientists have calculated that today's proportions of radioactively produced lead represent about 4.5 billion years of radioactive decay—a good fit with other data collected from Moon rocks and meteorites.

Lunar rocks, meteorites, and the lead isotopes on Earth all provide convincing evidence that the Earth and the rest of our solar system formed roughly 4.6 billion years ago. Since that distant time, geological processes that appear to be excruciatingly slow *from a human perspective* have accomplished incredible feats. It is humbling to realize, standing before Pike's Peak in Colorado or Mount Katahdin in Maine, that the process of erosion, which seems imperceptibly slow by human time standards, could remove those imposing mountains in a geological instant. The next time you sit by a stream and glimpse the current carrying away sand grains, consider that it won't take all that much geologic time for those grains to lithify into a new sedimentary rock and then to crunch it tectonically at the edges of colliding plates to form a new mountain range at the Earth's surface. Now that you've learned how to tell time "geologically," you can see how easily geological change can occur when there's "all the time in the world."

This chapter marks the end of Part 1 in this book, the section devoted to building your foundation of geological basics. In this section we described the basic structure of minerals and rocks, and considered how the major rock groups form—by cooling from a melted state, by lithification of broken-down and dissolved rock fragments into sedimentary rocks, and by metamorphic processes involving the heat and pressure within the Earth. We discussed how plate-tectonic movement affects each of these processes. In this chapter, we examined the vast reaches of geologic time and our methods for measuring it. In Part 2, we will study the dynamic tectonic processes that create the Earth's most massive features, such as its immense interior layers, its continents and ocean basins, and its lofty mountain ranges. We will also focus on earthquakes, one of the most dramatic effects of these dynamic processes.

[1]To develop U-Pb dating techniques, geologists had to create a lead-free lab environment so that they wouldn't analyze any lead that didn't come from the rocks. Back in the 1950s, such a technological task was extremely difficult because of the volume of leaded-gasoline pollution in the air. Clair Patterson, the geologist who developed U-Pb analysis, used his knowledge of the high concentration of toxic lead in the environment to lead the fight against leaded gasoline.

Highlight 8-4

Life on Earth—Are We All Extraterrestrials?

In the 1996 movie *Independence Day,* Earth is almost conquered by swarms of technologically advanced, evil aliens. Exciting stuff. At about the same time as the movie's premiere, a group of planetary geologists and chemists from NASA's Johnson Space Center were publishing their own findings of a different type of proposed alien "invasion." But their aliens were not the stuff of summertime Hollywood blockbusters; they were purported microscopic specks of organic material, 1/100 to 1/1000 the width of a human hair. Despite their tiny size, these specks have ignited a firestorm of debate about life on Mars—and Earth—that is still going on today.

The center of this debate is ALH84001—a 2-kg (4.4-lb), potato-size meteorite retrieved in 1984 during a snowmobile ride in the Far Western Icefield of the Allan Hills, a desolate spot on the vast icy wasteland of Antarctica (see photo at right). (This little rock, by the way, was the subject of the best-selling author Dan Brown's book *Deception Point.*) For eight years this unspectacular chunk of rock gathered dust in a storage cabinet in Houston until analysis of gases emitted during its heating revealed the rock's extraordinary origins.

This rock's life story apparently began about 4.5 billion years ago on the surface of Mars, 78 million km (49 million mi) from Earth. We believe this idea about its origin because the rock's gas content—particularly its mix of oxygen isotopes—is unique to the Martian atmosphere. The very old isotopic age of ALH-84001 has generated great interest, because the rock dates from a time when the Martian surface may have been considerably warmer, wetter, and more hospitable to life than the frozen, barren surface of modern-day Mars.

Over time, Mars lost its atmosphere and, as a consequence, its liquid surface water vaporized. About 16 to 17 million years ago, an asteroid or comet smashed into the Martian surface, hurling great masses of the planet's crust skyward, including ALH84001. This detritus entered into orbit around the Sun, where it remained until about 13,000 years ago, when Earth's gravity tugged at ALH84001, drawing it to our planet's surface, where it landed on the South Pole's Antarctic ice sheet.

What microscopic clues have led some scientists to propose that seeds of life once germinated, if only briefly, within ALH84001?

▲ The meteorite ALH84001.

- The rock contains microscopic globules of carbonates formed from a carbon-rich liquid, such as carbon dioxide *dissolved in water.* The presence of water is generally believed to be a prerequisite for life.
- The black rims surrounding the globules contain iron sulfide and magnetite, which, although nonbiological, are known to be produced by primitive terrestrial bacteria.

Chapter 8 Summary

Geochronology is the study of Earth time and geological dating. To date a rock or sediment, geologists apply either *relative dating* (determining how old it is in relation to its surroundings) or *numerical dating* (determining its actual age in years).

Relative dating relies on several key principles. These include:

- The principle of *uniformitarianism,* which states that modern geological processes resemble those that operated in the past;
- The principle of *original horizontality,* which states that most sedimentary rocks and lava flows are initially deposited in horizontal layers;

- The principle of *superposition,* which states that for tectonically undisturbed sedimentary rocks and lava flows, the uppermost layer in a sequence of rocks is the youngest, with those below it being progressively older;
- The principle of *cross-cutting relationships,* which states that when rock layers are cut across by other layers and features, such as igneous dikes and faults, the layers that are cut must be older than the features that cut them;
- The principle of *inclusions,* which states that rocks that are incorporated within other rocks must be older than the rocks that surround them;
- The principle of *faunal succession,* which states that organisms have succeeded one another through time in a recog-

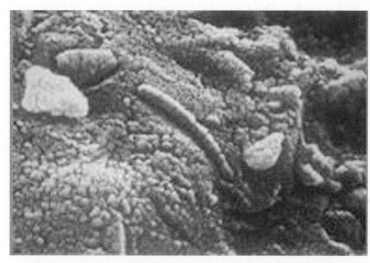
▲ Minute structures believed by some geoscientists to be evidence of primitive bacteria.

- The rock contains polycyclic aromatic hydrocarbons (PAHs). Although these compounds may form from nonbiological chemical reactions, they typically form here on Earth from organic decay.
- Electron microscopy reveals minute worm-shaped and ovoid structures similar to the fossils of primitive bacteria found in some of our planet's oldest rocks (see photo).

Taken individually, none of these findings would amount to absolute proof of past life. Taken collectively, the evidence is suggestive, *but still by no means conclusive.* Shortly after publication of these remarkable findings, other groups of scientists set about refuting them. They argued that nonbiological processes could have created virtually all of the evidence. Carbonates, sulfides, and magnetites have also formed here on Earth by volcanic processes; they do not necessarily require the presence of life-supporting surface water. Similarly, the fossil-like structures may also be inorganic and volcanically produced. Until its batteries petered out in late 1997, NASA's *Pathfinder* probe continued its exploration of the Martian surface, examining rocks and soil samples. In 2004, Martian rovers *Spirit* and *Opportunity* discovered a form of the mineral hematite (Fe_2O_3) believed to have precipitated from water, and layers of salt, most certainly produced by the evaporation of water. Perhaps we will soon have a definite answer to the question of whether organic life has ever existed on Mars, but the saga of ALH84001 has certainly created a great deal of lively scientific debate.

Other meteorites and comets—in addition to those from Mars—might also have delivered carbon-rich materials to Earth at some time in our planet's distant past. Such essential ingredients for life have been detected, for example, in the icy tail of the Hale-Bopp comet, a 40-km (25 mi)-diameter body that came within 120 million miles of Earth on March 22, 1997. Methanol, formaldehyde, carbon monoxide, hydrogen cyanide, hydrogen sulfide, and many other compounds rich in carbon, hydrogen, and oxygen have been detected in the comet's tail. Planetary geologists have even detected non-biogenic amino acids (the molecular units that make up proteins) in certain carbon-rich meteorites. Such ingredients may have been the seeds for our planet's earliest life, and comets or meteorites may have delivered them here.

---

nizable sequence in the geological record that is reproducible anywhere on Earth. Fossils are evidence of past life preserved in rock.

When dating rocks and sediments relatively, we often find that two physically adjacent layers were actually deposited at distinctly separate periods of time. The boundary between such layers, called an *unconformity,* represents a gap in the rock record. To determine the relative histories of geographically distant rock sequences, geologists use a process known as *correlation.* Correlation involves comparing the rocks and fossils in separate rock exposures.

Whenever possible, geologists prefer to date a rock unit numerically. In some cases, the rock or a mineral in the rock may contain a radioactive parent isotope that decays at a constant rate to produce a measurable amount of a nonradioactive daughter isotope. The time it takes for half the atoms of the parent isotope to decay is called the *half-life.* Most igneous rocks contain some measurable proportion of radioactive isotopes, such as uranium-238, which decays into lead-206 (*uranium—lead dating*), or potassium-40, which decays to form argon-40 (*potassium—argon dating*). *Carbon-14 dating* is used to date organic substances as old as 70,000 years.

Fission tracks are produced in a crystal's structure as high-speed particles emitted during radioactive decay pass through the crystal. Some minerals and their host rocks can be dated by measuring the number and length of such fission tracks in their structures. A variety of other numerical-dating techniques principally date rocks and sediments that are less than a few tens of thousands of years old. Some involve counting periodically accumulated layers, such as the annual growth rings in trees (a dating method known as *dendrochronology*) or the seasonally deposited layers of lake-bottom sediment called *varves.*

Scientists have combined these and other techniques to produce the *geologic timescale,* a chronicle of all of Earth history. Use of these dating methods has also enabled us to determine how old the Earth is, how long ago life on Earth originated, and when our early human ancestors lived. The geologic timescale is divided into eons, eras, periods, and epochs. Most of the evidence used to interpret Earth history comes from the Phanerozoic Eon, a segment of time during which evidence of life began to be abundantly preserved as fossils in rocks. The Phanerozoic Eon is divided into:

- The *Paleozoic* ("ancient life") Era; which was dominated by marine invertebrates (such as primitive clams, snails, and corals), and later fish and amphibians;

- The *Mesozoic* ("middle life") Era, which was dominated by reptiles (such as dinosaurs);

- The *Cenozoic* ("recent life") Era, our current era, which is dominated by mammals.

KEY TERMS

carbon-14 dating (p. 251)
Cenozoic Era (p. 262)
correlation (p. 244)
dendrochronology (p. 252)
fission tracks (p. 252)
fossils (p. 235)
geochronology (p. 230)
geologic timescale (p. 259)
half-life (p. 247)

index fossils (p. 237)
isotope dating (p. 247)
lichen (p. 253)
lichenometry (p. 253)
Mesozoic Era (p. 262)
numerical dating (p. 233)
Paleozoic Era (p. 262)
potassium–argon dating (p. 250)

principle of cross-cutting
  relationships (p. 234)
principle of faunal succession
  (p. 236)
principle of original
  horizontality (p. 234)
principle of superposition
  (p. 234)

principle of uniformitarianism
  (p. 234)
relative dating (p. 233)
rubidium–strontium dating
  (p. 251)
unconformities (p. 237)
uranium–thorium–lead dating
  (p. 250)
varves (p. 253)

QUESTIONS FOR REVIEW

1. Briefly explain the difference between relative and numerical dating.

2. Discuss three of the basic principles that serve as the foundation of relative dating.

3. Sketch and label two different types of unconformities.

4. Name three parent–daughter isotopic-dating systems, and give the half-lives of each parent isotope, as well as the rocks or sediments that are most likely to be dated by each.

5. Briefly explain how carbon-14 enters the cells of living organisms.

6. If the oldest rocks found on Earth are 4.03 billion years old, what evidence suggests that the Earth is actually 4.6 billion years old?

7. Look at Figure 8-13 on page 242, the geologic profile of the Grand Canyon. The Cambrian Tapeats Sandstone lies unconformably above several different bodies of rock. Identify two different types of unconformities that separate the Tapeats Sandstone from the underlying rocks.

8. Using the various dating methods discussed in Chapter 8, derive the sequence of events that produced the hypothetical landscape to the left. (Go slowly, and don't jump to premature conclusions. Consider all the principles that we've discussed.) Which of the layers might be dated numerically?

9. Although geologists claim that "the present is the key to the past" (the principle of uniformitarianism), the Earth has certainly changed throughout its 4.6-billion-year history. Think of two geological processes that operate differently today than they did in the past, and discuss how they vary.

10. Suppose you decided not to accept the 4.6-billion-year age of the Earth that geologists have determined (primarily from the ages of Moon rocks and meteorites and the evolution of lead isotopes on Earth). Devise an alternative strategy for determining the age of the Earth, assuming that you have unlimited funds.

LEARNING ACTIVELY

1. Test your knowledge of the various dating methods discussed in this chapter. During the course of a day, think about how you would date the common items you encounter—the trees you walk past on your way to your geology class, the lake near campus, the granite rocks that make up the outside of the campus library, the campus newspaper you glance at before class. Try to identify at least ten datable items you see in your everyday life.

2. If there are exposed rocks, sediments, or soils near your campus, visit them and determine their relative ages using the principles of relative dating.

3. Many college campuses have some type of natural-history museum. If your campus has one, visit its fossil collection. If none exists, stop by the geology department and spend some time with its fossil displays. Try to fit the fossils you see into their rightful place in the geologic timescale.

4. Visit a local stream or lake and look at the evidence of potential future fossils. Has a recent flood partially buried an animal bone or a piece of a tree limb? Can you see footprints of birds or the trails of insects in the fine mud of a lakeshore? Under what circumstances would these traces of life be preserved as fossils?

 ## ONLINE STUDY GUIDE

Practice makes you better. Make the time to take the *Online Study Guide* quiz for this chapter. It should take you about 20 minutes. Automatic grading and instant feedback will help you quickly and accurately identify the concepts you have mastered and the areas in which you need more study.
www.prenhall.com/chernicoff

THE
SCALE
OF
TIME

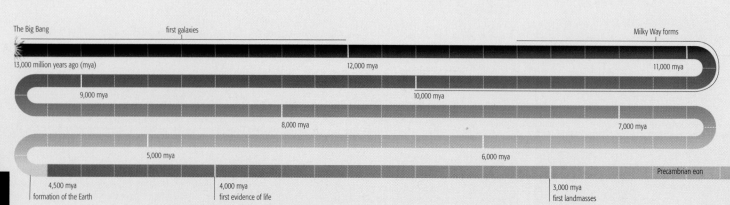

The Big Bang

first galaxies

Milky Way forms

13,000 million years ago (mya)

12,000 mya

11,000 mya

9,000 mya

10,000 mya

8,000 mya

7,000 mya

5,000 mya

6,000 mya

Precambrian eon

4,500 mya
formation of the Earth

4,000 mya
first evidence of life

3,000 mya
first landmasses

EARTH is approximately 4.6 billion years old. This discovery is Geology's great contribution to science. Chemists have given us the theory of the atom; biologists, evolution and DNA; and physicists, the invisible forces of gravity and magnetism and general relativity. What geologists have given us is *deep time*—Earth's almost incomprehensibly long span of time.

As part of your introductory geology course, you will be asked to reflect on this, to somehow comprehend that Earth has evolved—much less *lasted*—over 4.6 billion years. You may find this to be a challenge...for in your reading, millions of years will pass in a phrase. Continents may seem to float effortlessly across the globe's surface in a few pages. The age of dinosaurs may pass in an hour or two of lecture. Treated like this, is it such a mystery that it is not uncommon for educated people to believe that dinosaurs and humans coexisted, even though they are separated by over 60 million years? The true challenge in comprehending time lies in our humanity itself.

Consider this: the average lifespan for a person in a modern, developed country is approximately 75–80 years. If you are an 80-year-old person today, over the course of your life you experienced, among many other things: a World War and countless smaller ones; the rise and fall of the Soviet Union; the discovery of penicillin and the advent of AIDS; the first Winter Olympic games; the eras of Hitler, Ghandi, and Martin Luther King, Jr.; the first rocket flight, *Sputnik*, humanity walking on the Moon, and a remote-control robot sending back pictures from Mars; the first "talkie,"

the first color movie, and the first computer-generated movie; the first computer and the internet revolution; the demise of the telegraph and the invention of the cellphone.

Compare this to some geologic happenings over that same 80 years: Rio de Janeiro and Cairo widened the distance between them from about 9905 km to 9905.0024 km; Mt. Everest rose in elevation from about 8.85 km above sea level to 8.85488 km; and Earth's geographic poles have wobbled about 0.2 degrees—you can picture this, in relative terms, as the distance the second hand on your watch moves in 1/30th of a second.

In comparison to the tumult of the human experience, it's almost as if Earth *has not changed at all*. But the geologic record—evidenced in layers of rock—testifies to the explosion of life on Earth and to the growth of vast supercontinents, all of which challenges us to understand that *Earth does change*. It is a gradual, halting change, made up of minute steps, nearly imperceptible in terms of our human lifespan. It is only when we accept the immensity of geologic time that *it all makes sense*.

On the pages following is a timeline of Earth's history, a graphic depiction of the order of the key events as geologists currently understand them. It is the authors' hope that it will help you grasp both the immensity of time and the immensity of Geology's contribution to science.

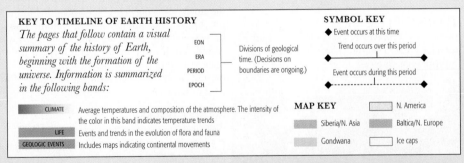

KEY TO TIMELINE OF EARTH HISTORY

The pages that follow contain a visual summary of the history of Earth, beginning with the formation of the universe. Information is summarized in the following bands:

EON
ERA
PERIOD
EPOCH

Divisions of geological time. (Decisions on boundaries are ongoing.)

SYMBOL KEY

◆ Event occurs at this time

Trend occurs over this period

Event occurs during this period

CLIMATE	Average temperatures and composition of the atmosphere. The intensity of the color in this band indicates temperature trends
LIFE	Events and trends in the evolution of flora and fauna
GEOLOGIC EVENTS	Includes maps indicating continental movements

MAP KEY

N. America
Siberia/N. Asia
Baltica/N. Europe
Gondwana
Ice caps

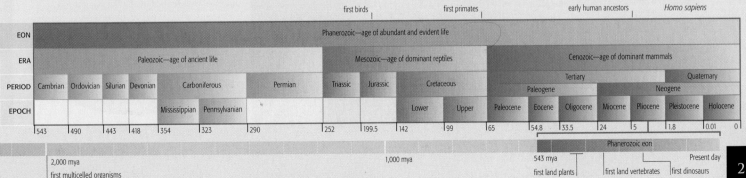

TIME AFTER BIG BANG	1 FEMTOSECOND (10^{-15} SECONDS)	1 PICOSECOND (10^{-12} SECONDS)	1 NANOSECOND (10^{-9} SECONDS)	1 MICROSECOND (10^{-6} SECONDS)	1 MILLISECOND (10^{-3} SECONDS)	1 SECOND	1 MINUTE	1 HOUR	1 DAY	1 YEAR	1 THOUSAND YEARS	1 MILLION YEARS	1 BILLION YEARS
TEMPERATURE OF UNIVERSE (K)	10^{15} 10^{14}	10^{13}	10^{12}	10^{11}	10^{10}	10^{9}	10^{8}	10^{7}	10^{6} 10^{5}	10^{4}	10^{3} 10^{2}	10	
DENSITY OF UNIVERSE (G/CM^3)	10^{25}	10^{20}	10^{15}	10^{10}	10^{5}	10^{0}	10^{-5}	10^{-10}	10^{-15}	10^{-20} 10^{-25}			
DIAMETER OF UNIVERSE (M)	10^{10}	10^{12}	10^{14}	10^{16}	10^{18}	10^{20}	10^{22}	10^{24}	10^{25}				

AFTER THE BIG BANG
In its first billionth of a second, the hot, expanding universe was a "soup" of scores of types of particles that varied enormously in their properties.

QUARKS BIND
Quarks started combining to form protons and neutrons. One "down" and two "up" quarks made a proton; one "up" and two "down" quarks made a neutron.

"up" quark — U
"down" quark — D
PROTON — NEUTRON
electron

BIG BANG NUCLEOSYNTHESIS
Between 1 and 180 seconds after the Big Bang, proton and neutron collisions formed the atomic nuclei of some light elements, mainly helium.

neutron — helium nucleus — energy released by collisions — proton

FIRST ATOMS
After about 300,000 years, protons and helium nuclei (consisting of protons and neutrons) began to capture electrons, forming hydrogen and helium atoms.

electron — proton — HYDROGEN ATOM
electron — nucleus — electron — HELIUM ATOM

BEFORE THE EARTH

To understand the origins of the Earth, it is necessary to go back to the origin of the universe itself, some 13–14 billion years ago in the event described by the Big Bang theory, when matter, time, energy, and space all came into existence. The early universe was small, hot, and dense. Ever since, it has been expanding and cooling. Within a nanosecond (a billionth of a second), it was hundreds of millions of miles in diameter, with a temperature of tens of trillions of degrees Kelvin (K), or degrees above absolute zero (–460°F/–273°C).

BUILDING BLOCKS

At this stage, the universe was a seething "soup" of particles created out of energy, together with vast numbers of photons (little packets of radiant energy). Among the most numerous particles were electrons and quarks, but there were also many other kinds that no longer exist today. Although it contained plenty of electrons, the early universe held neither of the other main building blocks of atoms: protons and neutrons. However, after just one microsecond (a millionth of a second), it had cooled sufficiently for vast quantities of protons and neutrons to form as two different types of quarks combined.

Most of the protons were destined to become the nuclei of hydrogen atoms, but about one second after the Big Bang, collisions between protons and neutrons started to form the nuclei of other light elements—helium and tiny amounts of lithium and beryllium. This process, termed Big Bang nucleosynthesis, was completed in three minutes and formed the nuclei of 98 percent of the helium atoms present in the universe today. It also mopped up all the neutrons.

PARTICLE TRACKS
Today, scientists try to simulate what happened in the Big Bang by using particle accelerators to smash subatomic particles together. The resulting tracks can then be studied.

THE FIRST ATOMS

For the next few hundred thousand years, the universe continued to expand and cool, but it was still too energetic for atoms to form. If electrons and atomic nuclei came together momentarily, they were quickly split apart by photons, which were themselves trapped in a process of continual collision with the particles. Eventually, after 300,000 years, when the temperature had dropped to about 3,000 K, the protons and other atomic nuclei started capturing electrons permanently, forming the first atoms, which were primarily hydrogen and helium. At the same time, the photons were released and streamed freely in all directions. At this stage, the universe became transparent, as the earlier "fog" of particles and energy cleared.

BACKGROUND EVIDENCE

A crucial piece of evidence supporting the Big Bang theory is the existence of Cosmic Microwave Background Radiation (CMBR). This is a faint heat radiation that emanates uniformly from all points in the sky. The only plausible explanation is that this originated in the hot fireball conditions of the early universe, as described by the Big Bang theory.

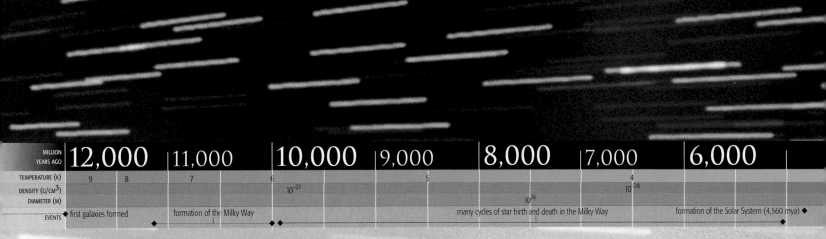

GALAXIES FORM

Gravity now began to cause gas atoms to come together. Over hundreds of millions of years, swirling clouds of hydrogen and helium gas formed and started to extend into long, thin strands. About 13 billion years ago (or 500 million years after the Big Bang), the strands began to clump together to form the first galaxies. Further concentration of matter within galaxies led to the creation of the first stars. The production of energy (including heat and light) in stars began when hydrogen nuclei in their centers started fusing to form helium.

NEW ELEMENTS

When the first galaxies and stars formed, there were still just four chemical elements in the universe: hydrogen, helium, lithium, and beryllium. The formation of the first stars was highly significant, because it is within these that heavier chemical elements are created from lighter ones, through various processes of fusion. Many of the most common chemical elements on Earth, such as oxygen, silicon, and iron, were made in this way. The very heaviest elements, such as lead, cannot be created in ordinary stars but form only in supernovae—the massive explosions of giant stars in their final death

REMNANTS OF A SUPERNOVA
Supernovae are huge star explosions that distribute new elements through galaxies. This is the remnant of Cassiopeia A, which exploded about 10,000 years ago.

throes. These explosions also distributed new elements throughout galaxies, where they were incorporated into new stars and planets.

BIRTH OF THE SOLAR SYSTEM

The exact age of our own galaxy, the Milky Way, is not known, but it was probably formed by 10–11 billion years ago. About 4.56 billion years ago, within our galaxy, a clump of gas and dust, known as a nebula, began to condense into what became the Solar System.

Within this nebula, matter coalesced into a denser central region (the proto-Sun) and more diffuse outer regions. Eventually the nebula shrank into a spinning, disklike object, called the protoplanetary disk. Within the disk, dust and ice collided randomly to form ever larger particles. In the center of the disk, as matter was drawn in by gravity, temperatures rose to the point where hydrogen fused to form helium, and a full-fledged star—our Sun—was born.

In other parts of the disk, the predominant matter was solid particles. Closest to the Sun, rocky materials became the main component. In the colder outer regions of the disk, ice particles were more common. As particles throughout the disk became larger, gravity drew them into collisions. This process, known as accretion, caused larger and larger bodies to merge, eventually forming

NEBULA
The Solar System condensed out of huge clouds of gas and dust, similar to these of the Lagoon Nebula, a star-forming region in the Milky Way.

planetesimals. These were the size of boulders, or larger, and composed of rock or rock and ice.

As well as heat and light, the new Sun also emitted a stream of energetic particles known as the solar wind. This "wind" blew volatile gases from the inner areas into the outer areas of the disk. Most of the remaining planetesimals collided to form the four rocky inner planets. They also formed the cores of the outer planets, around which the volatile gases collected.

Not all of the planetesimals merged to form planets. Some "leftovers" remain in the Solar System as two types of bodies: asteroids and comets. Most of the asteroids orbit in a belt between Mars and Jupiter, but some follow Jupiter's orbit, and some others have paths that cross the Earth's orbit. Comets formed from icy planetesimals in the outer edge of the disk.

FORMATION OF THE SOLAR SYSTEM
The Solar System formed in several stages. First a huge nebula (cloud) of gas and dust condensed into a spinning, disklike object. The center of the disk condensed further to form the Sun, while particles in the outer parts of the disk collided and accreted to form the planets.

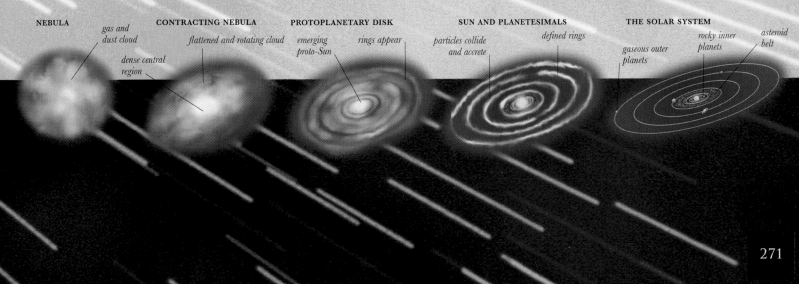

NEBULA
gas and dust cloud
dense central region

CONTRACTING NEBULA
flattened and rotating cloud

PROTOPLANETARY DISK
emerging proto-Sun
rings appear

SUN AND PLANETESIMALS
particles collide and accrete
defined rings

THE SOLAR SYSTEM
gaseous outer planets
rocky inner planets
asteroid belt

CLIMATE		Earth cools; first oceans form ◆		meteorite bombardment vaporizes early oceans			
LIFE		first organic molecules ◆		meteorite bombardment destroys any nascent life		first chemical fossils: carbon in metamorphosed sedimentary rocks, possibly from marine bacteria	
GEOLOGIC EVENTS	◆ formation of the Earth ◆ oldest Moon rock ◆ bombardment by planetesimals ◆ differentiation of the Earth's layers ◆ formation of the moon	oldest minerals (zircons)	oldest rock (Acasta gneiss, Canada) ◆	heavy meteorite bombardment (150 x present level)	zircon (Itsaq Archean gneiss, Greenland) indicates continental crust ◆ first banded iron formation ◆ lavas with pillow structures indicate presence of water		

PRECAMBRIAN EARTH

At one time, the Precambrian Eon was seen as a geological *terra incognita*, of unknown age and without fossil remains. Not until 1956, when the age of the Earth was accurately measured, did the immensity of the Precambrian's 4 billion years become glaringly obvious. Since then, fossils have been found, rocks dated, and the formation of the Solar System unraveled. The Earth's early history is now known to be a time of cataclysmic change. Early accretion and differentiation of the core and mantle, along with bombardment from space, culminated in the Moon's formation. Continued layering was followed by the formation

ZIRCON CRYSTALS
Crystals from rocks in Western Australia have been dated at 4.4–3.9 billion years old.

of a primitive atmosphere, oceans, and possibly life. However, renewed meteorite bombardment, which so scarred the Moon, also devastated the Earth, causing melting of its rocks. Not until about 3.8 billion years ago were an atmosphere and oceans regenerated. The first fossil evidence dates from almost immediately after this time, but these life forms had to tolerate a lack of oxygen and high ultraviolet radiation because there was no ozone layer to protect them. It took time for the atmosphere and oceans to develop to their present condition. Surface temperatures fell slowly and oxygen levels rose slowly as increasing numbers of photosynthesizing microbes produced

FORMATION OF THE MOON
Around 4.5 billion years ago, a planetary body the size of Mars collided with the Earth, tearing away a large volume of rock. The debris was held by the Earth's gravity, and cooled and coalesced to form the satellite we recognize as the Moon.

THE EARTH'S LAYERS FORM
Early accretion of cosmic material formed a larger, growing body, in which melts formed and migrated. Heavy elements concentrated in the core and lighter ones in the overlying mantle.

small bodies and dust accrete to asteroid size

heavy elements
lighter elements form mantle
melts migrate
core forms
molten core

oxygen and a protective ozone layer, aided by the emission of water vapor that split to release oxygen and ozone into the upper atmosphere.

The Precambrian Eon is divided into the Hadean, Archean, and Proterozoic Eras, during which dynamic processes originating from within the Earth generated new ocean-floor rocks and destroyed them elsewhere, while continents grew and were moved through the processes of plate tectonics. Meanwhile, an increasing diversity of marine microorganisms evolved slowly. Major events continued to perturb evolving ecosystems, with large-scale volcanism, impact events, and climate change culminating in runaway glaciations. Although often catastrophic, these upheavals may have stimulated evolution.

BOMBARDMENT OF THE EARTH
Mars-sized impactor
Earth

DEBRIS ORBITS THE EARTH
coalescing rock debris

ACCRETION OF THE MOON
debris melts and forms the Moon

| 3,600 | **3,500** | 3,400 | 3,300 | 3,200 | 3,100 | **3,000** | 2,900 | 2,800 | 2,700 | 2,600 |

high surface temperatures falling gradually

temperatures falling rapidly to well below present average ◆

◆ oldest stromatolites (Warrawoona Group, Australia)
◆ microscopic filaments and spheroids

first prokaryotes and eukaryotes (indicated by chemical fossils)

METEORITE BOMBARDMENT

Dating of the Moon's rocks and the appearance of its heavily cratered surface provide evidence that for some 600 million years after their formation, both the Moon and Earth were subject to intense bombardment from space. The Moon has impact craters, such as the Aitken Basin, up to 1,500 miles (2,500 km) across, and because of its greater size and gravity, the Earth underwent greater bombardment than the Moon. The results were melting of rock and destruction of any nascent life and atmosphere.

A LIFE-SUSTAINING ATMOSPHERE
The Earth's first atmosphere was blasted away by impacts and the solar wind. Volcanism built up a secondary atmosphere by releasing nitrogen, carbon dioxide, and water vapor. The latter was split by ultraviolet light into hydrogen, oxygen, and ozone. The lightest gas, hydrogen, was released into space.

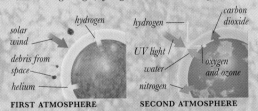

FIRST ATMOSPHERE
solar wind
hydrogen
debris from space
helium

SECOND ATMOSPHERE
hydrogen
UV light
water
nitrogen
carbon dioxide
oxygen and ozone

OCEANS AND CONTINENTS

The Hadean Era, as its name suggests, was hellish: The early Earth's turmoil destroyed any primary atmosphere, oceans, and life. The Archean times that followed saw the water vapor released from volcanoes collecting to form oceans, where salts dissolved to increase salinity gradually. Low-density silicate minerals accumulated to form the Earth's outer crust. But continuing volcanic eruptions resulted in the constant formation of new surface rocks. Since the Earth was no longer expanding through accretion, the cooler and denser rocks, which had been formed previously, sank into the interior to accommodate the new crustal rocks. The Earth's crust was fragmented into a number of plates, with both convergent and divergent margins. Low-density rocks accreted into continents and higher-density rocks formed the ocean floor.

EARLY LIFE

The earliest evidence for life on Earth comes from hydrocarbon residues in metamorphosed sedimentary rocks in Greenland, which date from about 3.8 billion years ago. These residues were derived from living organisms, which were probably aquatic prokaryotes (bacteria) that used

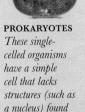

PROKARYOTES
These single-celled organisms have a simple cell that lacks structures (such as a nucleus) found in advanced cells.

light energy from the Sun. Their evolution must have been considerably earlier, perhaps more than 4 billion years ago. It is not known how this early life could have survived the heavy bombardment of the Earth by meteorites that occurred around this time. It is possible that life evolved twice, or was introduced to the Earth by the impacting bodies from space.

By 3.46 billion years ago, microbial photosynthesizers in warm, shallow waters formed mounds known as stromatolites, while other microbes lived off chemicals generated by submarine hot springs. Chemical fossils suggest that the first eukaryotes (organisms with relatively complex cells that contain a nucleus) appeared about 2.7 billion years ago, but the first direct fossil evidence for their existence does not appear until 2.2 billion years ago.

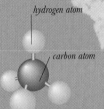

hydrogen atom
carbon atom

METHANE (CH$_4$)
The simplest organic compound is one carbon atom combined with four hydrogen atoms.

CLAIRE PATTERSON

After working on the first atomic bomb and studying radioactivity, US physicist Claire Patterson became interested in calculating the age of the Earth. In 1956, he compared measurements from meteorites and Earth minerals to obtain the age of 4.55 billion years. This was the first accurate estimate.

273

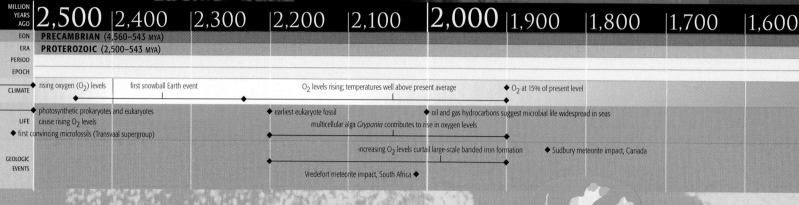

MILLION YEARS AGO	2,500	2,400	2,300	2,200	2,100	2,000	1,900	1,800	1,700	1,600

EON PRECAMBRIAN (4,560–543 MYA)

ERA PROTEROZOIC (2,500–543 MYA)

PERIOD

EPOCH

CLIMATE ◆ rising oxygen (O_2) levels — first snowball Earth event — O_2 levels rising; temperatures well above present average — ◆ O_2 at 15% of present level

LIFE ◆ photosynthetic prokaryotes and eukaryotes cause rising O_2 levels — ◆ first convincing microfossils (Transvaal supergroup) — ◆ earliest eukaryote fossil — multicellular alga *Grypania* contributes to rise in oxygen levels — ◆ oil and gas hydrocarbons suggest microbial life widespread in seas

GEOLOGIC EVENTS increasing O_2 levels curtail large-scale banded iron formation — ◆ Sudbury meteorite impact, Canada — Vredefort meteorite impact, South Africa ◆

OXYGENATING THE ATMOSPHERE

About 2,700 million years ago, primitive photo-synthesizing microorganisms released increasing volumes of oxygen into the early atmosphere. This oxygen was initially used up oxidizing iron in the oceans, and little entered the atmosphere. By 2.2 billion years ago, oxygen levels were still only about 1 percent of present levels (oxygen comprises 21 percent by volume of today's atmosphere), but by 1,900 million years ago oxygen levels were about 15 percent of present levels. Microorganisms preferring the anoxic (low-oxygen) conditions were forced to adapt by retreating within sediments and below ground.

An important development for evolving life was the formation of the protective ozone layer above the atmosphere to filter out incoming ultraviolet rays that are particularly harmful to DNA. Above a 2 percent oxygen level, an ozone layer will begin to form, so it should have been well established by 1.9 billion years ago. An oxygen-rich atmosphere, protected by an ozone shield, permitted new life forms to evolve and thrive. Oxygen-tolerant photosynthesizers inherited the Earth.

CHEMICAL FOSSILS

Even when organisms die and their tissues decay, the organic chemicals (hydrocarbons) of which they are made may survive in sediments, albeit in a degraded form. These are known as chemical fossils. Commonly, these hydrocarbons form mobile oil and gas, but some residues can persist within the rock record. Chemical analysis of these residues can differentiate the organic molecules present, their structural complexity and, broadly, what kind of organisms they are derived from.

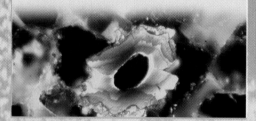

MODERN STROMATOLITES
Known as stromatolites, these mounds of layered sediment with surface films of bacteria are identical to those found in Precambrian sedimentary rocks.

BANDED IRON
These layered iron deposits were formed by oxidation reactions in the oceans before the formation of the atmosphere.

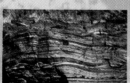

—— early continents formed from low-density rocks

MULTICELLULAR LIFE

Coiled filamentous microfossils from Michigan, called *Grypania*, are 2.2 billion years old and provide the first direct fossil evidence for eukaryotes. These organisms may have had multiple eukaryote cells, but the oldest convincing direct fossil evidence for multicells comes in the form of 1.2-billion-year-old fossils of a red alga called *Bangiomorpha* from arctic Canada. The many cells of these microscopic filaments show some specialization, including structures indicating that they reproduced sexually.

By 1,000 million years ago, toward the end of the Proterozoic Era and following a prolonged evolutionary gestation, the "big bang" of eukaryotic evolution had started. The evolution of multicellular organisms with sexual reproduction (evidence of fossilized embryos have been found in China dating from about 600 million years ago) paved the way for larger and more diverse organisms. It used to be thought that it was the development of sexual reproduction, with its exchange of genetic material, that was the key to this significant diversification. However, it is now known that this innovation alone did not promote eukaryotic diversification of life, as even bacteria exchange genetic material. More important, perhaps, is the ability of multicellular

274

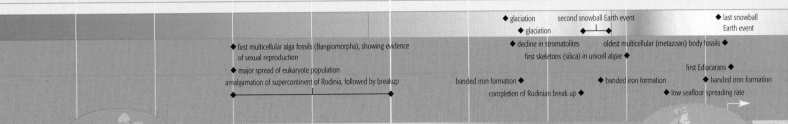
◆ glaciation   second snowball Earth event   ◆ last snowball
◆ glaciation   Earth event

◆ first multicellular alga fossils (Bangiomorpha), showing evidence   ◆ decline in stromatolites   oldest multicellular (metazoan) body fossils ◆
of sexual reproduction   first skeletons (silica) in unicell algae ◆

◆ major spread of eukaryote population   first Ediacarans ◆
amalgamation of supercontinent of Rodinia, followed by breakup   banded iron formation ◆   ◆ banded iron formation   ◆ banded iron formation
                                                                  completion of Rodinian break up ◆   ◆ low seafloor spreading rate

*clustering to
form Rodinian
supercontinent*

organisms to increase in size
beyond the microscopic, with
specialization of certain cells
for certain tasks within the
organism. By 580 million
years ago, the earliest large
animal fossils (known as the
Ediacarans) appear, followed
by the first shelled animals,
Cloudina, in 555-million-year-
old marine sediments
from Namibia.

COMPLEX CELLS EVOLVE
*Unlike the more primitive prokaryotes, eukaryotes have their
nuclear material enclosed in a membrane. Eukaryote cells are
usually part of a larger, multicellular organism, while
prokaryotes are exclusively single-celled organisms and often
incorporate a flagellum for mobility in a watery environment.*

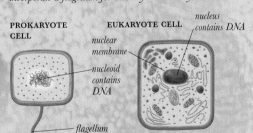

PROKARYOTE
CELL

EUKARYOTE CELL

*nucleus
contains DNA*

*nuclear
membrane*

*nucleoid
contains
DNA*

flagellum

GRYPANIA
*Coiled ribbons, at least
8 in. (20 cm) long, called
Grypania, may be the oldest
multicellular eukaryote algae.*

SNOWBALL EARTH
There is growing evidence for the idea
that the Earth suffered runaway glaciation
in the Precambrian. The theory claims that
ice caps extended from the poles into
the tropics, encompassing most of
the Earth's surface in what are
called snowball events. Proof of
glaciation in low latitudes
requires evidence that glacial
sedimentary rocks were in the
tropics at the appropriate time.
Measurements of magnetically
oriented minerals within
Precambrian glacial sediments
from all over the globe indicate
that 730–580 million years ago
there were several snowball events
(with glacial retreat in between), and
that earlier episodes may have
occurred between 2,450 and 2,220 million
years ago. Important support for the theory
comes from the presence of carbonate sediments,
which are typical of low latitudes, immediately
above glacially related ones. The appearance of
sedimentary iron formations in the oceans
suggests that conditions were anoxic, which is
consistent with glacial events. The cause of such
events is not fully understood but is thought to be
related to the clustering of the continents within
the tropics, raising the amount of light reflected
from the planet and cooling global climates.

SNOWBALL EARTH
*Geological evidence shows
that low-latitude continents
were glaciated; the extent of
sea ice cover is less clear.*

*formation of Gondwanan
supercontinent*

SOFT-BODIED ANOMALIES
The Ediacarans are among the most intriguing
Precambrian fossils. A group of sea-
dwelling organisms that lived 580–543
million years ago, they are the first
known animals. Their bodies were up
to 1.6 ft (.5 m) long, and they looked
like relatives of jellyfish (see below)—
although their relationship to living
animal groups is unclear. They had a
variety of body shapes, including a
flat disk or a frond, while some had
attachment disks and curiously
quilted or serially divided surfaces.
About 100 species are known, all
entirely soft-bodied and preserved as
sediment molds and infills, suggesting
that some lived within the sediment and that
their body tissues were tougher than those of
jellyfish, which are rarely preserved as fossils.
Their first known appearance follows the last
snowball event, and their evolution may have
been stimulated by its meltdown.

EDIACARAN FOSSIL
Spriggina, *an Ediacaran fossil with a
headlike structure, is named after
Australian geologist Reg Spriggs, who
first found the Ediacarans.*

275

PALEOZOIC

MILLION YEARS AGO	550	540	530	520	510	500	490	480	470
EON	PRECAMBRIAN	PHANEROZOIC (543 MYA–PRESENT DAY)							
ERA	PROTEROZOIC	PALEOZOIC (543–252 MYA)							
PERIOD		CAMBRIAN (543–490 MYA)					ORDOVICIAN (490–443 MYA)		
EPOCH									

CLIMATE: O_2 at 18% of present level; temperature high · temperature falling; carbon dioxide (CO_2) to high of over 16 x present level · CO_2 at less than 16 x present level; temperature low, then fluctuating above present average

LIFE: maximum Ediacaran diversity · increasing trace fossils; diverse small shelly fossils · decline in Ediacarans · first metazoan skeletal reefs: archaeocyathan sponges, cyanobacteria · Cambrian explosion: first chordates (animals with a notochord), unarmored vertebrates, arthropods including trilobites · Burgess Shale; conodonts (eel-like marine chordates) and jawless fish · first complete armored jawless fish

GEOLOGIC EVENTS: brief assembly of Pannotia supercontinent · Gondwanan continental assembly stretches from pole to pole · North America separated from Gondwana by the Iapetus ocean · maximum sea-floor spreading rate

THE PALEOZOIC ERA

Paleozoic means "ancient life," and this era saw the appearance of abundant shelly fossils and the development of land plants. It ended with life's biggest-ever extinction event, at the end of the Permian Period, when 90 percent of all life on the Earth was wiped out.

PLATE TECTONICS

The Earth's continents are constantly reconfigured by movements of crustal plates, with new oceans opening and the subduction of the old ocean floor. During the Paleozoic, most continents were joined to form the supercontinent called Gondwana. Over time, this drifted north from the southern hemisphere to form the even larger supercontinent of Pangaea, which by the end of the Paleozoic stretched from pole to pole.

THE CAMBRIAN EXPLOSION

In the mid-19th century, fossil evidence seemed to indicate that life began in the earliest Cambrian with the appearance of a variety of sea-dwelling organisms, such as sponges and trilobites, with

rotation of Gondwana toward south pole

mineralized shells and skeletons. We now know that life is much more ancient and that shelled animals first appear in late Precambrian rocks. However, it still seems that a large variety of fossil shells appear suddenly at the beginning of the Cambrian. Many of these fossils are very small and represent several different kinds of mollusks and armored arthropods (creatures with paired and jointed limbs)—it is as if there was an explosion in the diversity of life.

Nevertheless, some scientists argue that there was no great explosion of life; rather, it looks that way because a large number of organisms suddenly acquired shells and hard parts, which are much better preserved than soft body parts. Fossil faunas of Cambrian age from China, Canada's Burgess Shale, and northern Greenland show that arthropod diversity was already high, with many different habitats filled by specialized animals. The diversity of these organisms could indicate their presence in the late Precambrian, which would imply that the Cambrian explosion was an artifact of preservation. Nevertheless, within Cambrian times there is fossil evidence for

SILURIAN TRILOBITE
Marine arthropods called trilobites, with mineralized exoskeletons, evolved in Cambrian times.

MOLECULAR CLOCKS

The molecular clock estimates the age of groups of organisms, without relying on fossils. By measuring genetic similarity between living groups and assuming a constant rate of evolution, the time of their divergence can be calculated. For instance, according to the molecular clock, sharks evolved 528 million years ago, yet fossils first occur in 385-million-year-old strata. The discrepancy may reflect a lack of fossils, or a misunderstanding about rates of evolution.

CHARLES WALCOTT

American paleontologist Charles Walcott (1850–1927) quarried some 70,000 mid-Cambrian fossils from the Burgess Shale in the Canadian Rocky Mountains, which is now a World Heritage Site. The soft-part preservation of its fossils provides insights into marine life of the time. Walcott became director of the US Geological Survey, then secretary of the Smithsonian Institution in Washington, D.C.

the appearance of most of the major groups of marine invertebrates (those animals that lack a backbone), including a number of groups that subsequently became extinct, such as the trilobites, conodonts, and graptolites. Importantly, from the evolutionary point of view, a recent discovery from China shows that the first fishlike animals with the beginnings of a backbone (a stiffening rod called a notochord) also evolved in the Cambrian. But all this life was confined to the seas; according to the fossil record, fresh waters and land were devoid of preservable life, although it is highly likely that microbial organisms had already invaded these environments.

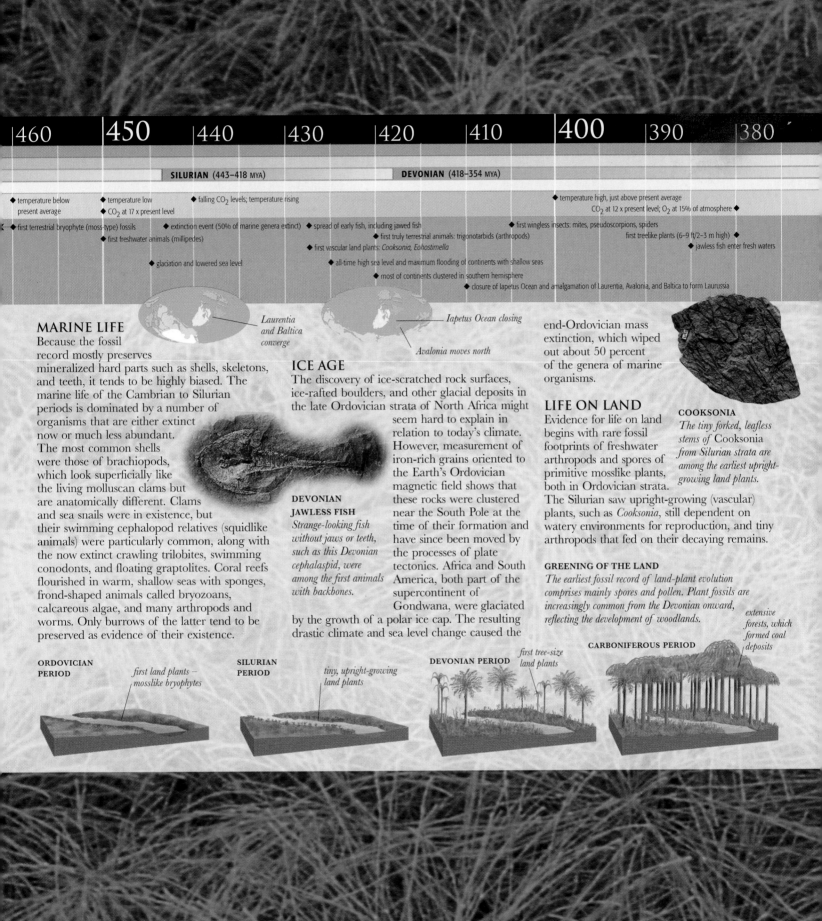

SILURIAN (443–418 MYA)　　　**DEVONIAN (418–354 MYA)**

◆ temperature below present average
◆ temperature low
◆ CO_2 at 17 x present level
◆ falling CO_2 levels; temperature rising
◆ temperature high, just above present average
CO_2 at 12 x present level; O_2 at 15% of atmosphere ◆

◆ first terrestrial bryophyte (moss-type) fossils
◆ extinction event (50% of marine genera extinct)
◆ spread of early fish, including jawed fish
◆ first wingless insects: mites, pseudoscorpions, spiders
first treelike plants (6–9 ft/2–3 m high) ◆

◆ first freshwater animals (millipedes)
◆ first truly terrestrial animals: trigonotarbids (arthropods)
◆ first vascular land plants: *Cooksonia, Eohostimella*
◆ jawless fish enter fresh waters

◆ glaciation and lowered sea level
◆ all-time high sea level and maximum flooding of continents with shallow seas
◆ most of continents clustered in southern hemisphere
◆ closure of Iapetus Ocean and amalgamation of Laurentia, Avalonia, and Baltica to form Laurussia

Laurentia and Baltica converge

Iapetus Ocean closing

Avalonia moves north

MARINE LIFE

Because the fossil record mostly preserves mineralized hard parts such as shells, skeletons, and teeth, it tends to be highly biased. The marine life of the Cambrian to Silurian periods is dominated by a number of organisms that are either extinct now or much less abundant. The most common shells were those of brachiopods, which look superficially like the living molluscan clams but are anatomically different. Clams and sea snails were in existence, but their swimming cephalopod relatives (squidlike animals) were particularly common, along with the now extinct crawling trilobites, swimming conodonts, and floating graptolites. Coral reefs flourished in warm, shallow seas with sponges, frond-shaped animals called bryozoans, calcareous algae, and many arthropods and worms. Only burrows of the latter tend to be preserved as evidence of their existence.

DEVONIAN JAWLESS FISH

Strange-looking fish without jaws or teeth, such as this Devonian cephalaspid, were among the first animals with backbones.

ICE AGE

The discovery of ice-scratched rock surfaces, ice-rafted boulders, and other glacial deposits in the late Ordovician strata of North Africa might seem hard to explain in relation to today's climate. However, measurement of iron-rich grains oriented to the Earth's Ordovician magnetic field shows that these rocks were clustered near the South Pole at the time of their formation and have since been moved by the processes of plate tectonics. Africa and South America, both part of the supercontinent of Gondwana, were glaciated by the growth of a polar ice cap. The resulting drastic climate and sea level change caused the end-Ordovician mass extinction, which wiped out about 50 percent of the genera of marine organisms.

LIFE ON LAND

Evidence for life on land begins with rare fossil footprints of freshwater arthropods and spores of primitive mosslike plants, both in Ordovician strata. The Silurian saw upright-growing (vascular) plants, such as *Cooksonia*, still dependent on watery environments for reproduction, and tiny arthropods that fed on their decaying remains.

COOKSONIA

The tiny forked, leafless stems of Cooksonia *from Silurian strata are among the earliest upright-growing land plants.*

GREENING OF THE LAND

The earliest fossil record of land-plant evolution comprises mainly spores and pollen. Plant fossils are increasingly common from the Devonian onward, reflecting the development of woodlands.

ORDOVICIAN PERIOD

first land plants – mosslike bryophytes

SILURIAN PERIOD

tiny, upright-growing land plants

DEVONIAN PERIOD

first tree-size land plants

CARBONIFEROUS PERIOD

extensive forests, which formed coal deposits

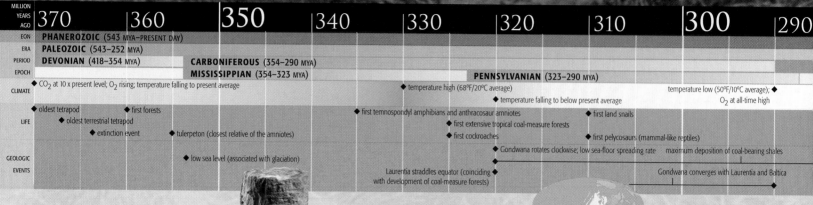

MILLION YEARS AGO	370	360	350	340	330	320	310	300	290

EON	PHANEROZOIC (543 MYA–PRESENT DAY)
ERA	PALEOZOIC (543–252 MYA)
PERIOD	DEVONIAN (418–354 MYA) — CARBONIFEROUS (354–290 MYA)
EPOCH	MISSISSIPPIAN (354–323 MYA) — PENNSYLVANIAN (323–290 MYA)

CLIMATE
- ◆ CO$_2$ at 10 x present level; O$_2$ rising; temperature falling to present average
- ◆ temperature high (68°F/20°C average)
- ◆ temperature falling to below present average
- ◆ temperature low (50°F/10°C average); ◆
- O$_2$ at all-time high

LIFE
- ◆ oldest tetrapod
- ◆ oldest terrestrial tetrapod
- ◆ first forests
- ◆ extinction event
- ◆ tulerpeton (closest relative of the amniotes)
- ◆ first temnospondyl amphibians and anthracosaur amniotes
- ◆ first extensive tropical coal-measure forests
- ◆ first cockroaches
- ◆ first land snails
- ◆ first pelycosaurs (mammal-like reptiles)

GEOLOGIC EVENTS
- ◆ low sea level (associated with glaciation)
- Laurentia straddles equator (coinciding ◆ with development of coal-measure forests)
- ◆ Gondwana rotates clockwise; low sea-floor spreading rate — maximum deposition of coal-bearing shales
- Gondwana converges with Laurentia and Baltica

Gondwana moves north, resulting in South Pole glaciation

The early Devonian brought a rapid diversification of land plants, with accompanying fauna, and by the late Devonian, the first forests covered low-lying wetlands, which were full of fish. Some of these fish evolved four limbs (making them the earliest tetrapods) and ventured onto land to feed, but they had to return to water to breed.

CLIMATE CHANGE
Today the sequence of rock strata preserved within individual continents, such as North America, shows a Paleozoic history of drastic climate change, from the cool marine waters of Ordovician times through Devonian deserts to Carboniferous tropical seas (during the Mississippian epoch), tropical rain forests (during the Pennsylvanian epoch), and back to deserts again in the Permian period. However, this sequence of changes is due

PERMIAN SANDSTONE
Thick piles of sandstone strata were deposited within the vast arid deserts of the Pangaean supercontinent. Structures in the beds show that these were once part of sand dunes.

FOSSILIZED TREE STUMP
Fossil trees from Antarctica indicate times when the climate of the continent was much warmer than it is today.

not, as might be expected, to dramatic changes in global climates, but to the movement of continents through the Earth's climate zones due to plate tectonics. From Ordovician to Permian times, North America (on the Laurentian Plate) moved from the southern hemisphere across the equator into the northern hemisphere.

REPTILE RELATIVES
Early land tetrapods were salamander-like animals about 3 ft (1 m) in length. Little is known of their biology except that they had to return to water to breed. By the mid-Carboniferous, both amphibians and more reptilelike groups had evolved. They were 6–10 ft (2–3 m) long and resembled crocodiles and lizards. However, the critical reptilian feature—the ability to lay a membrane-covered (amniote) egg on dry land—is difficult to recognize in the fossils.

From the late Carboniferous into the Permian Period, however, true reptiles were diversifying. The Pangaean supercontinent was occupied by many short-lived reptile groups, some of which were to eventually evolve into the dinosaurs, others into turtles, and yet others into mammals.

CARNIVOROUS AMPHIBIAN
These early Triassic fossil remains of several meat-eating amphibians were found in South Africa, clustered around those of a plant-eating reptile called Lystrosaurus.

EXTINCTION
At the end of the Permian, an estimated 90 percent of all living organisms (60 percent of genera), both marine and land-living, died out in the biggest mass extinction known. Groups that died out include the Paleozoic corals, trilobites, and water scorpions, while the echinoderm crinoids (sea lilies), brachiopods, clams, and sea snails suffered serious losses, as did most life on land.

Geologists have searched for a single cause, especially a major meteorite impact like that which brought about the end-Cretaceous extinction event (see pp. 280–281), but without success. The extinction did coincide with a vast outpouring of lavas and volcanic gases in Siberia (known as the Siberian Traps), which may have altered global climates. Also, there is evidence for widespread anoxia in the oceans, which may have fatally disrupted food chains, causing many groups to die out. The source of this anoxia may have been a sudden release of huge volumes of carbon dioxide gas from frozen masses of methane, known as gas hydrates, which had lain buried as ice in seabed deposits.

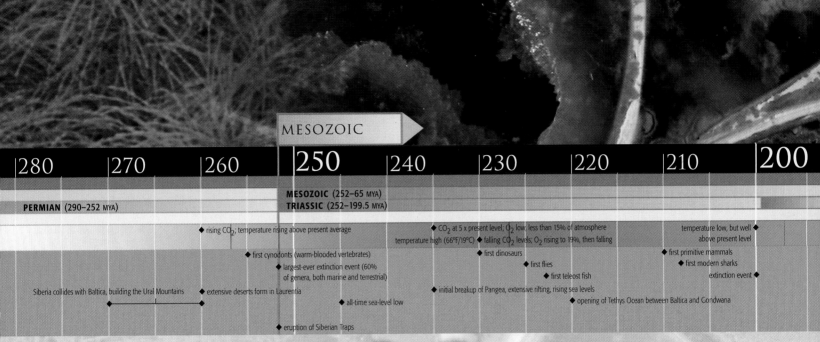

MESOZOIC (252–65 MYA)
TRIASSIC (252–199.5 MYA)

PERMIAN (290–252 MYA)

◆ rising CO_2; temperature rising above present average   ◆ CO_2 at 5 x present level; O_2 low, less than 15% of atmosphere   temperature low, but well ◆
                                                                  temperature high (66°F/19°C) ◆ falling CO_2 levels; O_2 rising to 19%, then falling   above present level

◆ first cynodonts (warm-blooded vertebrates)   ◆ first dinosaurs   ◆ first primitive mammals
◆ largest-ever extinction event (60%   ◆ first flies   ◆ first modern sharks
   of genera, both marine and terrestrial)   ◆ first teleost fish   extinction event ◆

Siberia collides with Baltica, building the Ural Mountains ◆ extensive deserts form in Laurentia   ◆ initial breakup of Pangea, extensive rifting, rising sea levels
                                                         ◆ all-time sea-level low   ◆ opening of Tethys Ocean between Baltica and Gondwana

◆ eruption of Siberian Traps

THE MESOZOIC ERA

Spanning the 187 million years of the Triassic, Jurassic, and Cretaceous periods, the Mesozoic (or "middle life") Era lies between two major extinction events, the second one known as the end-Cretaceous extinction. It is often characterized as the Age of the Reptiles, since many major reptile groups were dominant life forms, including the dinosaurs on land, the ichthyosaurs in the seas, and the pterosaurs in the air.

MARINE LIFE

The end-Permian extinction and collapse of the Paleozoic reefs caused a major turnover in marine life. Eventually, modern reef corals evolved and marine diversity and food chains were reestablished. The more adaptable molluscan clams and snails gradually took over from brachiopods. Both bony and cartilaginous fish diversified, with some sharks becoming top predators. But the waters were also occupied by new marine reptiles, such as turtles and crocodiles, along with the now-extinct plesiosaurs, dolphinlike ichthyosaurs, and mosasaurs.

SEA REPTILES
New predators occupied the seas, such as the fast-swimming ichthyosaurs. Shaped like dolphins with toothed, beaklike jaws, these reptiles lived from the Triassic to the Cretaceous.

PLANT LIFE

During the late Permian and early Triassic, the dominant Paleozoic plants—such as ferns, club mosses, and horsetails—declined. Newly diversifying groups included conifers, which reproduce by means of seeds borne in cones. They became particularly important in Jurassic and Cretaceous times, along with cycads and cycadlike bennettitaleans. Most of the cycads also had seed-bearing cones and were distributed worldwide, even in polar regions when they were free of ice. Flowering plants (angiosperms) evolved through the Cretaceous. They combined a number of features found in other plants of the

Mesozoic, such as flower-type reproductive structures (also found in the bennettitaleans), with the one unique feature that distinguishes flowering plants—unfertilized seeds enclosed in a carpel.

DINOSAURS

For 150 million years, from the late Triassic to the end of the Cretaceous, terrestrial life was dominated by reptiles. One very varied group, known as the dinosaurs, with some 900 genera, diversified from crow-sized bipedal forms to the largest land-living animals ever known—the four-limbed sauropods, at 110 ft (34 m) long.

Dinosaurs were first recognized as a distinct group of reptiles in 1842, their distinguishing characteristic being that they walked with their legs tucked under their bodies, unlike living reptiles, which have splayed legs. Most very large dinosaurs were plant-eaters with long necks to reach the high canopies of Mesozoic trees. However, there were also very large carnivorous predators such as the Jurassic allosaurs, which were up to 40 ft (12 m) long.

MARINE TURTLES
The first turtles with protective shells appeared in the Cretaceous, evolving from the older land-living tortoises of the Triassic.

SURVIVING CYCADS
The few living cycads are survivors of a far larger group of seed-bearing plants that were key members of Jurassic and Cretaceous forests.

PLACERIAS
This late-Triassic 10-ft-(3-m-) long giant from North America was one of the last surviving therapsids (mammal-like reptiles).

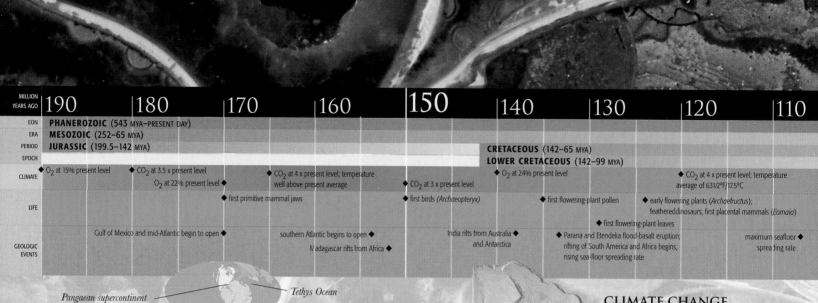

MILLION YEARS AGO	190	180	170	160	150	140	130	120	110
EON	PHANEROZOIC (543 MYA–PRESENT DAY)								
ERA	MESOZOIC (252–65 MYA)								
PERIOD	JURASSIC (199.5–142 MYA)					CRETACEOUS (142–65 MYA)			
EPOCH						LOWER CRETACEOUS (142–99 MYA)			

CLIMATE
- ◆ O₂ at 15% present level
- ◆ CO₂ at 3.5 x present level
- O₂ at 22% present level ◆
- ◆ CO₂ at 4 x present level; temperature well above present average
- ◆ CO₂ at 3 x present level
- ◆ O₂ at 24% present level
- ◆ CO₂ at 4 x present level; temperature average of 63 1/2°F/17.5°C

LIFE
- ◆ first primitive mammal jaws
- ◆ first birds (*Archaeopteryx*)
- ◆ first flowering-plant pollen
- ◆ early flowering plants (*Archaefructus*); feathered dinosaurs; first placental mammals (*Eomaia*)
- ◆ first flowering-plant leaves

GEOLOGIC EVENTS
- ◆ Gulf of Mexico and mid-Atlantic begin to open
- southern Atlantic begins to open ◆
- Madagascar rifts from Africa ◆
- India rifts from Australia ◆ and Antarctica
- ◆ Parana and Etendeka flood-basalt eruption; rifting of South America and Africa begins; rising sea-floor spreading rate
- maximum seafloor ◆ spreading rate

Pangaean supercontinent begins to break up — *Tethys Ocean*

The dinosaurs filled most available habitats of the Mesozoic. There are two distinct types of dinosaur: one with a birdlike pelvis and the other reptilelike. The latter type included a group called the maniraptoran dinosaurs, a number of which were feathered, and one earlier, as yet undiscovered Jurassic, group of these small predators were probably the ancestors of the birds.

MAMMALS AND BIRDS

Mammals (4,250 living species) and birds (9,000 living species) are the two groups of warm-blooded vertebrates that came to dominate Cenozoic landscapes. Although reptiles were

DINOSAUR FOOTPRINTS
These three-toed footprints of dinosaurs, initially mistaken for those of giant birds, are now used to estimate how fast dinosaurs could run.

still abundant (6,000 living species), they were mostly smaller than mammals. Fossils show that the reptile origins of mammals and birds lie deep within the Mesozoic, the former in the late Triassic and the latter in the Jurassic. According to the fossil record, a significant spread of both groups did not occur until after the end-Cretaceous extinction event, which wiped out most of the evolving bird groups and some of the early mammals. However, so many different bird and mammal groups appear immediately after the Mesozoic Era in the Paleocene Period that there were probably more late-Cretaceous ancestral groups than the fossil record preserves.

Primitive Mesozoic mammal groups ranged in size from that of a shrew to a large rat, and most died out. However, the marsupials and egg-laying monotremes have survived, especially in Australasia and the Americas.

MEGAZOSTRODON
Morganucodontids, such as the insect-eating Megazostrodon, *were among the first primitive, shrewlike mammals.*

ARCHAEOPTERYX
The first fossil bird was found in 1862, complete with impressions of flight feathers.

CLIMATE CHANGE

There were no polar ice caps during the Mesozoic, and global temperatures were generally high, averaging 66°F (19°C), but carbon-dioxide levels fluctuated on a downward trend throughout the Mesozoic and the Cenozoic Era that followed. During the end-Cretaceous extinction event 65 million years ago, climates were disturbed on a global scale. The details are still debated, but fossil charcoal indicates there was a short period of global wildfire that devastated life, and perhaps longer-term warming, enhanced by greenhouse gases derived from a massive outpouring of lava known as the Deccan eruptions, which formed the upland region of India known as the Deccan Plateau.

EXTINCTION

Some 75 percent of all species died out at the end of the Cretaceous, in the event known as the K-T extinction event. Victims included the dinosaurs (apart from the birds), the remaining marine and flying reptiles, and ammonites (shelled, squidlike animals). The extinction coincided with the impact of a large meteorite in the Caribbean Sea, off the Yucatan Peninsula at Chicxulub, Mexico, and the Deccan lava eruptions. Both

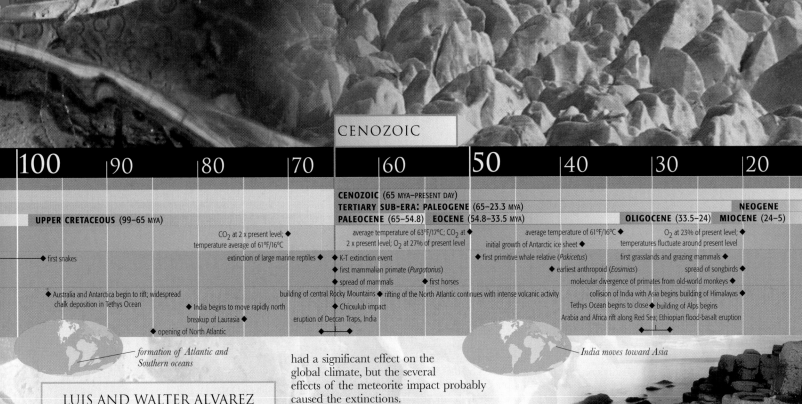

100	90	80	70	60	50	40	30	20

CENOZOIC (65 MYA–PRESENT DAY)
TERTIARY SUB-ERA: PALEOGENE (65–23.3 MYA)   **NEOGENE**
PALEOCENE (65–54.8)   **EOCENE** (54.8–33.5 MYA)   **OLIGOCENE** (33.5–24) **MIOCENE** (24–5)

UPPER CRETACEOUS (99–65 MYA)

CO_2 at 2 x present level; ◆
temperature average of 61°F/16°C

average temperature of 63°F/17°C; CO_2 at ◆
2 x present level; O_2 at 27% of present level

average temperature of 61°F/16°C ◆

O_2 at 23% of present level; ◆
temperatures fluctuate around present level

◆ first snakes

extinction of large marine reptiles ◆

initial growth of Antarctic ice sheet ◆

◆ K-T extinction event

◆ first primitive whale relative (*Pakicetus*)

first grasslands and grazing mammals ◆

◆ first mammalian primate (*Purgatorius*)

◆ earliest anthropoid (*Eosimias*)

spread of songbirds ◆

◆ spread of mammals   ◆ first horses

molecular divergence of primates from old-world monkeys ◆

◆ Australia and Antarctica begin to rift; widespread
chalk deposition in Tethys Ocean

building of central Rocky Mountains ◆ rifting of the North Atlantic continues with intense volcanic activity

collision of India with Asia begins building of Himalayas ◆

◆ India begins to move rapidly north

◆ Chicxulub impact

Tethys Ocean begins to close ◆ building of Alps begins

breakup of Laurasia ◆

eruption of Deccan Traps, India

Arabia and Africa rift along Red Sea; Ethiopian flood-basalt eruption

◆ opening of North Atlantic

*formation of Atlantic and
Southern oceans*

India moves toward Asia

LUIS AND WALTER ALVAREZ

Walter Alvarez (b.1940), an American geologist, found high concentrations of the rare element iridium in a marine clay dated to the Cretaceous-Cenozoic boundary. From this finding, in 1980 he developed the hypothesis with his father Luis (1911–88), a Nobel prize winning physicist, that the iridium concentration is best explained by an extraterrestrial body, such as a very large meteorite, impacting upon the Earth. They further speculated that a global catastrophe accompanying the impact caused the demise of the dinosaurs.

CHICXULUB IMPACT
Sixty-five million years ago, a meteorite crashed into shallow seas that are now part of the Yucatan Peninsula, Mexico. The impact blasted huge volumes of carbonate rock through the atmosphere. On reentry, its radiant heat caused global wildfire.

had a significant effect on the global climate, but the several effects of the meteorite impact probably caused the extinctions.

THE CENOZOIC ERA

The Cenozoic Era extends from 65 million years ago to the present day (Cenozoic means "recent life"). The Era is characterized by a diversification of modern bony fish, flowering plants, pollinating insects, birds, and mammals, including our primate relatives.

RIFTING AND VOLCANISM

Cenozoic movements of the Earth's plates had a considerable influence on the evolution of life. Mountain building and large-scale uplift down the western flank of the Americas transformed regional climates, while the formation of the Himalayas and the Tibetan Plateau generated the Southeast Asian monsoon. Rifting and volcanism opened up the North Atlantic and created the great East African Rift, where many of our primate ancestors flourished.

Cenozoic volcanism, associated with rising plumes of heat from within the Earth, led to

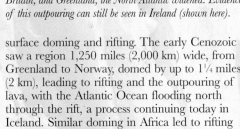

PRESERVED LAVAS AT GIANT'S CAUSEWAY
As early Cenozoic lavas erupted in the rift between Norway, Britain, and Greenland, the North Atlantic widened. Evidence of this outpouring can still be seen in Ireland (shown here).

surface doming and rifting. The early Cenozoic saw a region 1,250 miles (2,000 km) wide, from Greenland to Norway, domed by up to 1¼ miles (2 km), leading to rifting and the outpouring of lava, with the Atlantic Ocean flooding north through the rift, a process continuing today in Iceland. Similar doming in Africa led to rifting and the outpouring of the Ethiopian flood basaltic lava (31–28 million years ago).

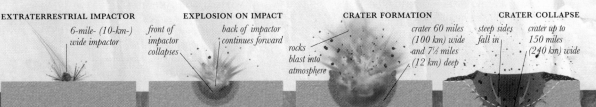

EXTRATERRESTRIAL IMPACTOR
6-mile- (10-km-)
wide impactor

EXPLOSION ON IMPACT
front of
impactor
collapses

back of impactor
continues forward

rocks
blast into
atmosphere

CRATER FORMATION
crater 60 miles
(100 km) wide
and 7½ miles
(12 km) deep

CRATER COLLAPSE
steep sides
fall in

crater up to
150 miles
(240 km) wide

EON	**PHANEROZOIC** (543 MYA–PRESENT DAY)
ERA	**CENOZOIC** (65 MYA–PRESENT DAY)
PERIOD	**TERTIARY SUB-ERA: NEOGENE** (23.3–1.8 MYA)
EPOCH	**MIOCENE** (24–5 MYA)

CLIMATE	temperatures fall then fluctuate just above present levels ◆	temperature high within ◆ general ice-house state

LIFE	◆ spread of primitive apes in Central Africa	molecular divergence of ◆ Orang-Utan from other great apes	African forests ◆ spread of whales diminish and dolphins	first cattle; spread of snakes, ◆ frogs, rats, and mice
	spread of grazing *hipparion* horses in North America and dispersal through Eurasia to Africa		expansion of open grasslands into middle and high latitudes continues throughout Miocene ◆	

GEOLOGIC EVENTS	◆ Australia moves north	uplift of Tibet ◆	◆ Columbia River flood-basalt eruption
	◆ rifting and volcanism in east Africa		

MARINE LIFE

Following the end-Cretaceous extinction, life in the seas and oceans gradually took on a modern appearance, from reef-building animals to the top predators. One of the most important developments was the increased diversity of modern bony fish following the extinction of the large predatory marine reptiles. Turtles, a few crocodiles, some iguanas, and snakes remain as marine reptiles, but most are small. The main innovation among vertebrates was the evolution of marine mammals, especially whales and dolphins (cetaceans). Diversifying Eocene mammals evolved a succession of extinct forms, from which modern whales arose 35 million years ago.

LIFE ON LAND

In early Cenozoic times, mammals took over the habitats of those reptile victims of the end-Cretaceous extinction. However, there were major differences between the emerging mammals, with Australia and South America inheriting the pouched marsupials and egg-laying monotreme mammals, while the rest of the world was soon dominated by placental mammals. The latter give birth to bigger, more advanced young, nurtured for longer within the mother's body via a placenta.

Initially marsupials thrived, evolving both plant-eaters and carnivorous predators, which diversified into forms that were subsequently mimicked by the placental rodents, hippopotamuses, horses, dogs, and big cats. However, as the more advanced placentals evolved and diversified, they tended to diplace the marsupials. Formation of a land bridge between North and South America allowed an interchange of placentals and marsupials between the continents. Today the surviving marsupials of the Americas are the opossums, which are still represented by 63 species. But even more marsupials held out for longer on the isolated continent of Australia. The earliest Cenozoic placental mammals were shrewlike insect-eaters, but soon bigger (sheep-size) browsing and rooting plant-eaters evolved. By the late Paleocene, there were rhinoceros-size browsers and some dog-size carnivores. The number of placental families rose from 21 in the late Cretaceous to 111 in the early Eocene. The Earth's habitats were soon filled by placental mammals, from otters, seals, and whales in the seas, through the vast range of land mammals – from shrews to giant plant-eaters, such as the extinct 8-m (26-ft) long giant rhinoceros *Indricotherium*, and numerous

FLOWERING TREES
By earliest Cenozoic times flowering plants had evolved into a diverse range, including the magnolia (above) and laurels that formed the woodlands.

TARSIER
Today's tiny insect-eating tarsiers are thought to be descended from a primitive Cenozoic primate group.

carnivores – to bats in the air. Environmental and climatic changes also affected mammal evolution, especially decreasing forest cover and increasing grasslands, with the co-evolution of pollinating insects, songbirds, and grazing mammals. These changes are also thought to have affected the evolution of hominoids (higher apes and humans).

MAMMALS OF LAKE MESSEL
Fifty million years ago, the oil-rich muds of a German lake bed preserved an astonishing diversity of its inhabitants, such as this Ailuravus, a rodent slightly larger than a squirrel.

AMBER LIFE
Since Cretaceous times, amber from some resin-producing trees has trapped and preserved small animals, especially insects.

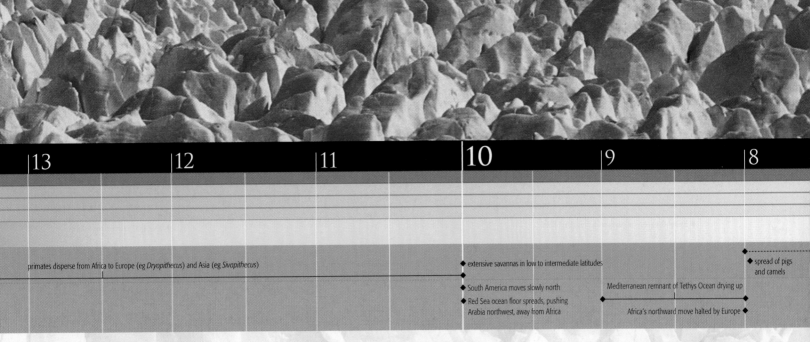

primates disperse from Africa to Europe (eg *Dryopithecus*) and Asia (eg *Sivapithecus*)

◆ extensive savannas in low to intermediate latitudes

◆ spread of pigs and camels

◆ South America moves slowly north

◆ Red Sea ocean floor spreads, pushing Arabia northwest, away from Africa

Mediterranean remnant of Tethys Ocean drying up

Africa's northward move halted by Europe ◆

MOUNTAIN BUILDING

Cenozoic times saw the formation of four major mountain belts around the world: the Andes, Rocky Mountains, European Alps and the Himalayas. The Andes are a classic example of mountain building, resulting from the convergence of an oceanic (Pacific) plate with a continental plate (South America), combined with the subduction of the oceanic plate and frequent earthquakes and volcanoes. By contrast, the Alpine–Himalayan belt resulted from the convergence of the African and Indian plates with Europe and Asia respectively. Africa's northward movement subducted a body of water known as the Tethys Ocean; the intervening rocks were compressed, thickened, and elevated with intense folding and faulting, leading to the formation of the Alps. The breakaway of India from Africa and Antarctica, which began in the Mesozoic, eventually caused the subduction of the eastern Tethys and its convergence with Asia, beginning 20 million years ago. Again, the intervening rocks were compressed, thickened,

SPREADING GRASSLANDS
By the Miocene, cooling climates and greater aridity caused forests to break up and grasslands to expand, along with fleet-footed grazing mammals and their predators, such as big cats.

HIMALAYAS
The spectacular mountain belt of the Himalayas, with the high plateau of Tibet to the north and low sediment-filled floodplains to the south, is the result of the convergence of the Indian and Asian plates.

and elevated with intense folding and faulting, resulting in the formation of the Himalayas. The continuing northward drive of India is thought to have pushed the deeper part of the Indian plate beneath Tibet, thickening and elevating it without folding. The formation of this high and elongated physical barrier caused significant change in the regional climate, with the development of the Southeast Asian monsoon.

THE ORIGIN OF HUMANS

Humans, apes, lemurs, and monkeys are all mammals and were first grouped together as primates in 1758 by the Swedish naturalist Carolus Linnaeus (1707–78). The fossil record shows that the earliest primate-like mammals

originated as small insect-eating animals, such as the shrewlike *Purgatorius* in the early Paleocene. In the early Cenozoic, a group of tree-climbing, squirrellike animals (the plesiadapiforms) had evolved in North America and Europe. They were followed by tarsier- and lemurlike primates, which spread into Africa and Asia along with two groups of higher primates, the New World monkeys and the Old World monkeys and apes. It was from the latter group that hominoids evolved in Africa. These included the 18-million-year-old (early Miocene) tailless and monkeylike *Proconsul*, which could climb trees and walk on all fours. By the late Miocene Epoch, the apes diversified and separated into African, European, and Asian branches, some of whose members increased in size.

BIPEDAL BEGINNINGS
It seems likely that our ape–human ancestor was a knuckle-walker, like today's apes. However, fossil hominids from 6–7 million years ago appear already to have walked on two feet.

283

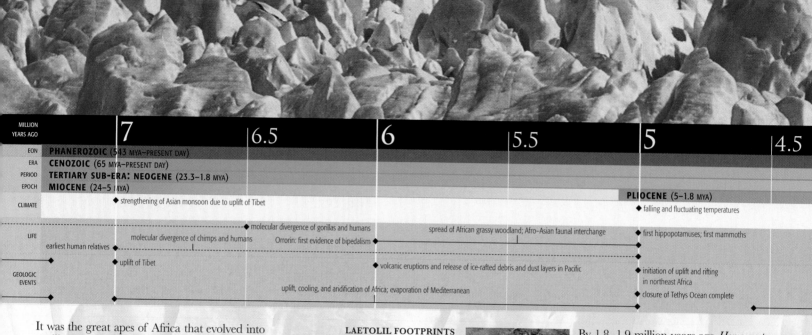

EON	**PHANEROZOIC** (543 MYA–PRESENT DAY)					
ERA	**CENOZOIC** (65 MYA–PRESENT DAY)					
PERIOD	**TERTIARY SUB-ERA: NEOGENE** (23.3–1.8 MYA)					
EPOCH	**MIOCENE** (24–5 MYA)				**PLIOCENE** (5–1.8 MYA)	
CLIMATE	◆ strengthening of Asian monsoon due to uplift of Tibet				◆ falling and fluctuating temperatures	
LIFE	earliest human relatives ◆	molecular divergence of chimps and humans · ◆ molecular divergence of gorillas and humans · Orrorin: first evidence of bipedalism ◆		spread of African grassy woodland; Afro-Asian faunal interchange	◆ first hippopotamuses; first mammoths	
GEOLOGIC EVENTS	◆ · ◆ uplift of Tibet	uplift, cooling, and aridification of Africa; evaporation of Mediterranean	◆ volcanic eruptions and release of ice-rafted debris and dust layers in Pacific		◆ initiation of uplift and rifting in northeast Africa · ◆ closure of Tethys Ocean complete	

It was the great apes of Africa that evolved into gorillas, chimpanzees, and humans. Gorillas diverged first, about 6–8 million years ago, while the chimpanzees and humans share a common ancestor who, according to the molecular clock, lived 5–7 million years ago.

THE SPREAD OF HOMINIDS

Although the fossil record of our human relatives (the hominids) is very sparse, intensive searching over the last 50 years in southern Africa in general and around the Great Rift Valley of east Africa has provided some insight into their evolution. A recent discovery of a fossil skull with some human features (belonging to a hominid called *Sahelanthropus*) is dated at 7 million years old. And the 6-million-year-old *Orrorin* leg bones are claimed to indicate upright walking on two legs (bipedalism), although direct evidence does not appear until the Laetolil footprints (see in next column).

SKULL RECORD

By examining the skulls, it is clear that brain size increases significantly from Australopithecus *to the large-brained* Homo *species:* H. neanderthalensis *and* H. sapiens.

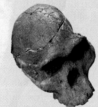

AUSTRALOPITHECUS

HOMO NEANDERTHALENSIS

HOMO SAPIENS

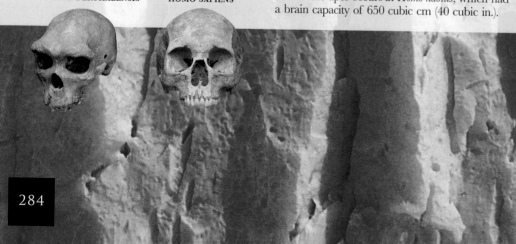

LAETOLIL FOOTPRINTS
Some 3.6 million years ago in Laetolil, east Africa, two upright-walking human-related adults and a juvenile walked on soft mud, leaving the oldest-known footprints.

By 3 million years ago, these early human relatives were definitely small – about 1m (3ft) tall – bipedal apes with small brains (australopithecines). They later split into two branches: large-jawed hominids that fed on plants and small animals, and small-jawed plant-eaters and meat scavengers, which had developing brains and used tools. The latter were the early members of our genus, *Homo*.

HUMAN ATTRIBUTES

The fossil record of human-related bones and stone tools suggests that human attributes were not acquired all at once but at intervals over at least 4 million years. It used to be thought that upright walking was the fundamental human attribute, but by 4 million years ago the australopithecines were bipedal. The first primitive stone tools appear about 2.6 million years ago and are thought to be associated with *Homo habilis*. However, it is equally possible that australopithecines made them, especially since it is now known that chimpanzees use stone tools. The first significant increase in brain size above that of the apes occurs in *Homo habilis*, which had a brain capacity of 650 cubic cm (40 cubic in.).

By 1.8–1.9 million years ago *Homo erectus*, with a larger body, greater mobility, and slightly larger brain size, was the first human relative to migrate from Africa, reaching eastern Asia by about 1.6–1.7 million years ago. The Neanderthals (*Homo neanderthalensis*) and their immediate predecessors were the first human relatives whose brain size reached that of modern humans, at 1,100–1,400 cubic cm (67–86 cubic in.). They occupied Europe and Asia from about 230,000 years ago until 30,000 years ago and were successful game hunters, who could probably speak but had not developed complex language. Some of them buried their dead and made personal ornaments, which are normally regarded as symbolic and cultural attributes associated with the attainment of early modern human consciousness. However,

DEVELOPING SKILLS
The fossil record tends to preserve only stone and bone tools, but some of the most primitive tools were probably digging sticks.

CHARLES LYELL

A Scottish-born barrister, Charles Lyell (1797–1875), took up geology and wrote the *Principles of Geology* (1830–33), an important synthesis of geological ideas that greatly influenced the naturalist Charles Darwin. Lyell believed that the interpretation of the past must be based on an understanding of present processes. He also divided the Tertiary into epochs. Lyell initially disagreed with the theory of evolution proposed by Charles Darwin and Alfred Wallace, but finally, if reluctantly, came to accept it.

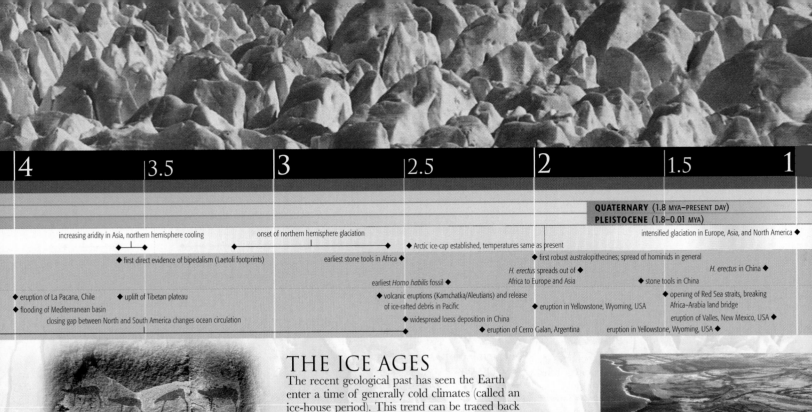

QUATERNARY (1.8 MYA–PRESENT DAY)
PLEISTOCENE (1.8–0.01 MYA)

increasing aridity in Asia, northern hemisphere cooling

onset of northern hemisphere glaciation

intensified glaciation in Europe, Asia, and North America ◆

◆ first direct evidence of bipedalism (Laetoli footprints)

Arctic ice-cap established, temperatures same as present

earliest stone tools in Africa ◆

◆ first robust australopithecines; spread of hominids in general

H. erectus spreads out of
Africa to Europe and Asia

H. erectus in China ◆

earliest *Homo habilis* fossil ◆

◆ stone tools in China

◆ eruption of La Pacana, Chile   ◆ uplift of Tibetan plateau

◆ volcanic eruptions (Kamchatka/Aleutians) and release
of ice-rafted debris in Pacific

◆ opening of Red Sea straits, breaking
Africa–Arabia land bridge

◆ flooding of Mediterranean basin

◆ eruption in Yellowstone, Wyoming, USA

eruption of Valles, New Mexico, USA ◆

closing gap between North and South America changes ocean circulation

◆ widespread loess deposition in China

◆ eruption of Cerro Galan, Argentina

eruption in Yellowstone, Wyoming, USA ◆

THE ICE AGES

The recent geological past has seen the Earth enter a time of generally cold climates (called an ice-house period). This trend can be traced back to the mid-Miocene, about 16 million years ago, when cooling was accompanied by increasing aridity in equatorial regions with the breakup of forests and extension of grassland.

By 10 million years ago, the Antarctic ice sheet had begun to develop. This locked up increasing amounts of water as snow and ice, resulting in falling sea levels. Several independent lines of evidence point to a period of more rapid climate cooling beginning over 3 million years ago. Glaciation of the northern hemisphere intensified approximately 2.7 million years ago.

CAVE ART
Representations of animals, some extinct and others no longer living in the same region, point to the antiquity of much rock art, which is otherwise very difficult to date.

the Neanderthals were replaced by incoming modern humans (*Homo sapiens*), who originated in Africa around 200,000 years ago and first migrated from there about 120,000 years ago. New discoveries in South Africa and from the Congo show that even by 70,000–75,000 years ago *Homo sapiens* were using sophisticated tools of bone and producing symbolic etchings.

HUMAN EVOLUTION
Recent discoveries, mostly in Africa, show that human ancestry extends further back than previously thought. The result is a shrublike evolutionary tree with 20 known species of extinct human relatives, arranged in seven genera.

CORAL TERRACES, PAPUA NEW GUINEA
Layers of coral limestone grown in shallow water and now sitting well above the coastline attest to changes in sea level, seen here on the Huon Peninsula, Papua New Guinea.

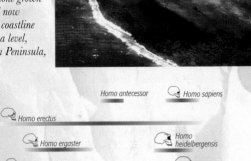

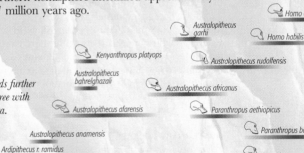
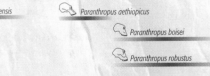

Homo antecessor — Homo sapiens
Homo erectus
Homo ergaster — Homo heidelbergensis
Australopithecus garhi
Homo habilis — Homo neanderthalensis
Kenyanthropus platyops
Australopithecus rudolfensis
Australopithecus bahrelghazali
Australopithecus africanus
Australopithecus afarensis
Paranthropus aethiopicus
Australopithecus anamensis
Paranthropus boisei
Orrorin tugenensis
Ardipithecus ramidus kadabba
Ardipithecus r. ramidus
Paranthropus robustus
Sahelanthropus tchadensis
Chimpanzee

7 million years ago | 6 mya | 5 mya | 4 mya | 3 mya | 2 mya | 1 mya | present day

285

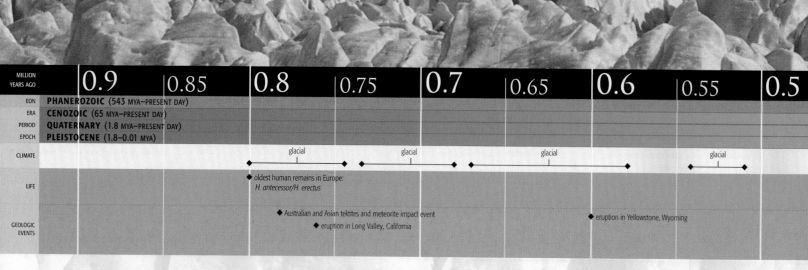

MILLION YEARS AGO	0.9	0.85	0.8	0.75	0.7	0.65	0.6	0.55	0.5

EON — PHANEROZOIC (543 MYA–PRESENT DAY)
ERA — CENOZOIC (65 MYA–PRESENT DAY)
PERIOD — QUATERNARY (1.8 MYA–PRESENT DAY)
EPOCH — PLEISTOCENE (1.8–0.01 MYA)

CLIMATE — glacial · glacial · glacial · glacial

LIFE — ◆ oldest human remains in Europe: *H. antecessor/H. erectus*

GEOLOGIC EVENTS — ◆ Australian and Asian tektites and meteorite impact event · ◆ eruption in Long Valley, California · ◆ eruption in Yellowstone, Wyoming

Exactly what precipitated the global cooling is a matter of intense debate, but many experts think that there was some marked change in the circulation pattern of ocean currents, perhaps set off by plate movements. The temperatures of ocean currents have a significant effect on atmospheric temperatures and humidity. The Quaternary Period, extending from just under 2 million years ago to the present day, was marked by periodic fluctuations between colder glacials and warmer interglacials. The present warm period may just be another interglacial.

ADVANCING GLACIERS

Extensive ice sheets first developed in the northern hemisphere about 2.6 million years ago, with the formation of the Arctic ice cap. This eventually extended as far south as today's New York in North America, and Birmingham, Copenhagen, and St. Petersburg in Europe. Beyond the ice, permanently frozen ground

FORAMINIFERANS

The shell composition of tiny, sea-dwelling single-celled animals called foraminiferans records changes in the chemistry of ocean waters. By recovering such fossil shells from ocean-floor sediments and analyzing their chemistry over successive generations, the changes in ocean chemistry can be reconstructed. Oxygen-isotope ratios in the calcium carbonate of the shells are measured, and from this information past fluctuations in ocean temperature, ice volume, and hence climate change can be recovered.

SEA-LEVEL AND TEMPERATURE

Changes in global average temperatures over the latter part of the Quaternary ice ages, with their fluctuating cold and warm spells, are closely tracked by related sea-level changes.

■ Temperature
■ Sea level

reached the Black and Mediterranean Seas, and mountain glaciers grew even in the tropics. There have been numerous climate fluctuations since, with interglacial temperatures as warm as or even warmer than present. For instance, 125,000 years ago, animals such as elephants and hippopotamuses flourished in England where previously and subsequently there were ice sheets. The periodicity of the climate fluctuations was initially about 40,000 years but changed to a 100,000-year cycle about a million years ago due to changes within the global climate cycle.

QUATERNARY ICE CAPS

Polar ice caps grew and retreated several times during the Quaternary, reaching the maximum extent shown here.

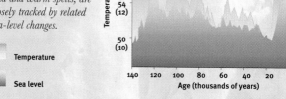

POST-GLACIAL TEMPERATURE

Global temperatures over the last 10,000 years, since the end of the last glacial advance, have been unusually stable, but may not remain so for long.

LIFE ADAPTS TO CHANGE

One of the benefits of sexual reproduction and the associated slow change of genetic material through mutation is that it enables organisms to adapt to new conditions such as changing climates. As the climate cooled with the arrival of the ice ages, animals developed several adaptations for survival. These included insulation (hair, feathers, or clothing), fat reserves for food shortages, and often camouflage for snowy conditions, such as the white coats of Arctic foxes, hares, and polar bears. Typically, cold-

PURPLE SAXIFRAGE
Some flowering plants, such as the Arctic bog cotton and purple saxifrage, are well adapted to tundra conditions.

286

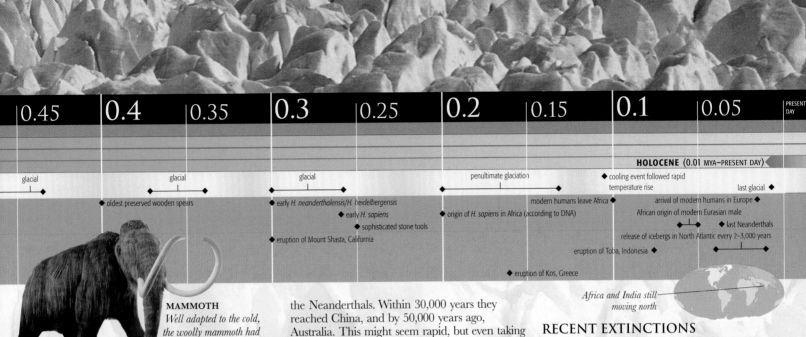

| 0.45 | 0.4 | 0.35 | 0.3 | 0.25 | 0.2 | 0.15 | 0.1 | 0.05 | PRESENT DAY |

HOLOCENE (0.01 MYA–PRESENT DAY)

glacial   glacial   glacial   penultimate glaciation   ◆ cooling event followed rapid temperature rise   last glacial ◆

◆ oldest preserved wooden spears

◆ early *H. neanderthalensis/H. heidelbergensis*

◆ early *H. sapiens*

modern humans leave Africa ◆   arrival of modern humans in Europe ◆

◆ sophisticated stone tools   ◆ origin of *H. sapiens* in Africa (according to DNA)   African origin of modern Eurasian male

◆ eruption of Mount Shasta, California   ◆ last Neanderthals

release of icebergs in North Atlantic every 2–3,000 years

eruption of Toba, Indonesia ◆

◆ eruption of Kos, Greece

Africa and India still moving north

MAMMOTH
Well adapted to the cold, the woolly mammoth had insulating hair, a thick layer of fat, and tusks to clear snow from vegetation.

adapted animals and humans have compact bodies with small limbs and extremities to decrease their surface area and increase volume for more effective heat conservation. Plants have to withstand not only freezing air temperatures but also very short growing seasons, during which reproduction, seed setting, and dispersal must all take place within a matter of a few months. Frozen soil necessitates shallow roots, so even the woody plants, such as dwarf birch, willow and juniper, tend to be ground-hugging to protect themselves from strong winds.

HUMAN MIGRATION
Both archeological and biomolecular evidence shows that the migration of modern humans began in Africa 120,000 years ago. They moved north out of the continent as early as 100,000 years ago, reaching the western Mediterranean by 90,000 years ago, when they first encountered

CARIBOU MIGRATION
Reindeer, also known as caribou, are cold-adapted herd animals that survive in polar regions by migrating over long distances between winter and summer feeding grounds.

the Neanderthals. Within 30,000 years they reached China, and by 50,000 years ago, Australia. This might seem rapid, but even taking the 12,500-mile (20,000-km) coastal route, this works out at less than half a mile (1 km) per year.

Ice-age Europe was more of a physical and climatic barrier and was not reached until 40,000 years ago, and the northeast passage through Siberia into Alaska was even more difficult. The Bering land bridge was exposed between 18,000 and 10,200 years ago by lowered sea levels, allowing migration into the Americas, reaching Chile by 12,500 years ago. The most recent migrations were out into the Pacific, with New Zealand reached only 1,000 years ago.

MIGRATION OF HOMO SAPIENS
Through the collection of fossil records, the pathways by which modern humans (Homo sapiens) *are thought to have achieved global distribution have been reconstructed. The journey is believed to have begun in East Africa about 120,000 years ago.*

RECENT EXTINCTIONS
Since the last glaciation ended 12,000 years ago, there has been a continuing global extinction of the megafauna—animals over 100 lb (45 kg) in weight, such as the mammoth. Another victim was the Neanderthals, whose demise followed the arrival of modern humans about 40,000 years ago. They coexisted for nearly 10,000 years in western Europe, but eventually died out. Analysis of DNA suggests that despite the overlap in time and territory, there was no significant interbreeding between the two human species. The exact cause of the megafauna extinction, whether climatic and environmental change, human hunting, or a combination of the two, is still debated. However, recent accurate dating, especially in Australasia, which suffered little climate change, shows that extinctions closely follow the arrival of modern human hunters. The exception is Africa, where modern humans coexisted with large mammals for much longer, but now rising populations are threatening even Africa's big game.

Of course, human influence on the Earth goes far beyond that of the hunter. The encroachment on almost all natural habitats, from the rain forests to the polar wastes, along with the effects of human-enhanced global warming, pose a far greater threat to the planet's flora and fauna, even threatening humankind itself.

PART 2

Shaping the Earth's Crust

Earth Structures—
Folds, Faults,
and Fabrics

9

Mountains and earthquakes are dramatic evidence for the motion of the Earth's lithosphere. Enormous mountain ranges, such as the Rockies, the Alps, and the Himalaya, show us what happens—on a large scale—when the Earth's plates converge. In these and other places around the world, we can see mountainsides of rock that look twisted, bent, and broken like those in Figure 9-1. Formerly horizontal rock layers are now folded and, in some cases, pushed over on their sides or even turned upside down from their original positions. We might think of rocks as hard and permanent things, but such contorted rocks clearly show that they are not strong enough to withstand the enormous power of plate-tectonic motion.

In Part I of this text, we saw how the movement of the Earth's tectonic plates produces most of the large-scale features on the planet's surface, such as its ocean basins and continents. Mountains also owe their existence to plate tectonics. The same forces that shuffle the Earth's plates for thousands of kilometers tear the edges of those plates apart, rupturing them into huge displaced blocks. These forces also squeeze plate edges together, crumpling them into giant folds of rock.

In this chapter and throughout the rest of Part II, we will focus on how plate movements cause the powerful geological processes that deform the Earth's crust, creating its folds and faults and its other deformational structures. (The branch of geology devoted to crustal deformation is known as **structural geology.**) How and why do rocks fold, bend, and break? What happens when rocks deform? In the case of breaking rocks, the answer to the last question may be quite dramatic and destructive—a powerful earthquake. Further considerations for structural geologists: How do the structures that we see—such as folds and faults—relate to larger-scale tectonic features such as mountains and rifts? What do deformed rocks look like when we examine them closely, both with our eyes and with powerful microscopes?

Geologic structures are important to geologists because they use them to understand active and ancient tectonic events, but they are also important to us in our daily lives. They strongly affect the strength and stability of a rock mass and thus they almost always influence our decisions regarding where to build skyscrapers and high-rise residence halls as we develop our cities, towns, and college campuses. To understand how structures form, we must also consider some concepts related to tectonic forces and the effects these forces have on rocks. Let's begin our study of Earth structures by considering the forces that can

FIGURE 9-1 These crumpled rocks are marine sediments exposed along the Atlantic coast in St. Jean de Luz, France. ▶

▲ **FIGURE 9-2** The person in this photo is standing next to a fold in layers of strong, well-cemented sandstone rock in California's Borrego Desert.

rocks with the marine sedimentary rocks found near Salt Lake City (see Figure 8-4 on page 234) that have lain relatively undisturbed for hundreds of millions of years.

Plates that converge, diverge, or slide past one another subject rocks at their margins to powerful stresses. **Stress** is the force applied to a rock. Consider a point at some depth in the Earth's crust. This point is under pressure—typically lithostatic pressure—the force that acts equally in all directions, pushing inwardly on the rock (see Fig.7-4). The stress at this point is the same as the pressure. Tectonic forces, however, create *differential* pressure, that is, some directions experience higher stress and some lower. These differences in stress—as a function of direction—drive the deformation of rocks.

If a rock is stressed enough, it will deform, changing in shape and typically changing its volume. This change in shape is called **strain,** a rock's response to the stresses applied to it. Strain can be described by a change in length of the material or a change in angles that define the material. Figure 9-3 shows a fossil that is completely undeformed and a fossil that has been deformed by directed stress. The deformed fossil illustrates the effect and the amount of strain because you can see how the shape has changed from the original and how much it's changed.

bend hard rocks into complex intricate folds or break and move them tens of meters or even hundreds of kilometers.

Stress and Strain

For a graphic view of what happens to horizontal marine sediments that are subjected to powerful tectonic forces, take a look at Figure 9-2. The sandstone layers of California's Borrego Desert were crumpled and folded during a plate collision. Compare these

Elastic and Plastic Rocks

When subjected to stress, rocks deform in different ways, depending in part on the amount of stress (Fig. 9-4a). When the stress is minor, a rock may return to its original shape and volume after the stress is removed, much as a stretched rubber band regains its original shape after use. Such a *temporary* change is described as **elastic deformation.** A rock that is deformed *elastically* is not permanently deformed. A greater amount of stress, however, may deform a rock to the point from which it cannot return to its original shape when the stress is removed. The maximum

▲ **FIGURE 9-3** An undeformed trilobite (left) and a "strained" fossil trilobite (right).

stress a rock can withstand before becoming deformed *permanently* is called its **yield point,** or **elastic limit** (Fig. 9-4b and 4c).

The pressure and temperature conditions when stress is applied determine in large part how rocks deform at their yield point. Rocks subjected to great stress *under conditions of low temperature and pressure* (such as rock at or near the Earth's surface) may crack or rupture when they exceed the yield point. This change happens because their mineral bonds break outright all along the zone of maximum stress. Such large-scale bond-breaking results in a type of permanent deformation known as **brittle failure.** Brittle failure also occurs when stress is applied suddenly, leaving little time for atoms to move about and adjust their bonds in response to the applied stress. When struck sharply with a rock hammer, even a rock as hard as granite will crack. In this case, the sudden application of stress causes brittle failure.

Sometimes rocks undergo stress *under conditions of high temperatures and pressures,* such as when they lie buried several kilometers beneath the Earth's surface. Under these conditions, a rock may also exceed its yield point but instead undergo **plastic (or ductile) deformation,** an *irreversible* change in shape that occurs *without the rock breaking.* During plastic deformation, atoms move about and adjust to an applied stress by breaking and reforming their bonds on a microscopic scale. Atoms in the stressed material generally migrate from areas of maximum stress and rebond in areas of lower stress. Think of what happens when you squeeze an uncapped tube of toothpaste. It oozes away from your tightening grasp (the high-stress area) and out the top of the tube (into a lower-stress area—the open air).

Silly Putty is the perfect medium for demonstrating different types of rock deformation. When rolled into a ball and bounced gently on a table top, Silly Putty is subjected to a low level of stress. Because it does not change shape appreciably or permanently, the bouncing putty is exhibiting *elastic behavior.* If you pull the ball slowly in opposite directions until the stress just exceeds Silly Putty's yield point, its shape changes considerably and remains changed even after the stress ceases—that is, it deforms *plastically.* Beyond this point, even small increases in pulling (tensional stress) cause a great deal of plastic deformation, and you can easily pull the putty into long, wispy filaments. Finally, when the original ball of Silly Putty is pulled firmly and abruptly, it fractures into two sharp-edged pieces, exhibiting *brittle failure.* Try this sometime.

Other Factors Affecting Rock Deformation

The magnitude of the stress applied to a rock is not the only factor that determines how rocks will deform. Laboratory experiments that simulate real-world conditions have shown that heat, the amount of time that a stress is applied, and a rock's composition all affect its response to stress.

Heat The bonds between atoms in a mineral stressed at low temperatures—like those found near the Earth's surface—may stretch somewhat without breaking, but will break at fairly low to moderate stress. The result—the rocks undergo brittle failure, break, and generate earthquakes. At high temperatures—generally deeper beneath the Earth's surface—bonds will break in some and reform in others, so the minerals (and therefore the rock that contains them) deform plastically and fold, rather than break. Just as a blacksmith's furnace heats a metal bar so it can be shaped into a horseshoe, so the Earth's internal heat renders rocks plastic and shapeable into folds of rock (see Fig. 9-1).

Time We can simulate in laboratories the heat and pressure applied to rocks deep in the Earth's interior, but we cannot simulate the passage of vast stretches of time. Clues in real-world settings, however, hint at how time affects deformation processes. Consider, for instance, how sturdy wooden bookshelves gradually sag over time under the weight of heavy volumes. The sagging shelves illustrate how a force that is initially insufficient to deform a substance may eventually do so if applied over a long period of time. As we learned from Silly Putty, the relatively slow application of stress allows atoms to move around, typically away from areas of maximum stress and toward areas of minimum stress. Thus, rocks stressed slowly, over long periods of time, may change shape (by deforming plastically) but not break. On the other hand, with rapid application of

▲ **FIGURE 9-4** The relationship between stress and strain in different types of rock deformation. Within the range of elastic deformation **(a)** application of stress produces a proportionate amount of *reversible* strain. Stress that exceeds the *yield point* of a rock causes it to either rupture due to brittle failure **(b)** or deform plastically by flow **(c)**. Within the range of plastic deformation, even small amounts of additional stress create a great deal of irreversible strain. The three cylinders of rock here illustrate undeformed, brittlely fractured, and plastically deformed rock (from left to right).

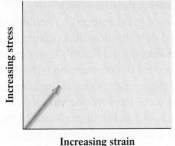

(a) Elastic deformation

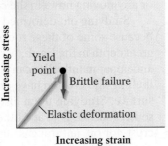

(b) Brittle failure

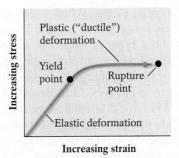

(c) Plastic deformation

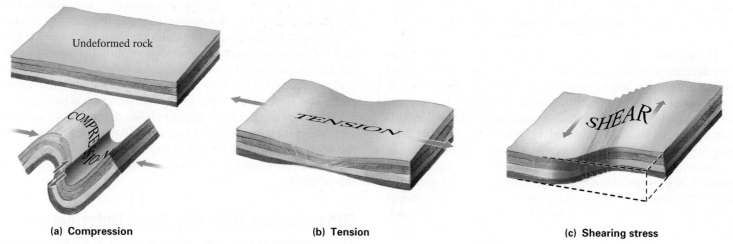

(a) **Compression**   (b) **Tension**   (c) **Shearing stress**

▲ **FIGURE 9-5** The three types of stress applied to rocks occur most often at the edges of the Earth's tectonic plates. The edges of converging plates are generally *compressed* (**a**) thickening vertically and shortening laterally. The edges of diverging plates are subjected to *tension* (**b**) thinning vertically and lengthening laterally. The edges of plates at transform boundaries generally undergo *shear* (**c**) becoming sliced into parallel blocks of rock.

stress (such as the smash of a rock hammer), atoms can't adjust their bonds quickly enough and thus their bonds and the rocks themselves break.

Composition

Some minerals (such as garnet and pyroxene) are strong; others (such as calcite and mica) are weak. Thus, the mineralogy of a rock strongly affects how a rock responds to stress. Another important compositional factor is water. In low-temperature rocks, if water is present along the boundaries of mineral crystals, in the pore spaces between crystals, or even as an inclusion within a mineral crystal, the rock will be relatively weak and is more likely to deform plastically. The influence of composition explains why one rock will flow and another will break, even if temperature, pressure, and the rate of stress application to those rocks are the same. Which rock would you predict would be more likely to fold and which would break—a dry, quartz-rich sandstone or a wet shale consisting largely of clay minerals? Typically, the sandstone would break and the shale would fold; when we find a *folded* quartz sandstone, it may indicate that water was present when it deformed or that deformation was slow and sustained (or both).

Compression, Tension, and Shearing

Rocks may be stressed in one of three principal ways, depending on the orientation of the differential stress. When rocks are pushed together, the largest stress is horizontal and the lowest stress is vertical (Fig. 9-5a). This type of stress is called **compression.** As a result, compressed rocks contract, crumpling and causing them to thicken vertically and shorten laterally. On a small scale, compression creates folds in a rock. On a large scale, compression creates mountains. When rocks are pulled apart (extension), such as at diverging plate margins, the largest stress is vertical and the lowest is horizontal; this type of stress is known as **tension** (Fig. 9-5b). Tensional stress stretches, or extends, rocks so that they thin vertically and lengthen (extend) horizontally.

Tension creates rifts, tears in rocks. When rocks are forced past one another in parallel but opposite directions, such as at transform plate margins, the stress is known as **shear** (Fig. 9-5c). Shear stress typically slices rocks into horizontal blocks, breaking and displacing preexisting rocks and structures.

Interpreting Deformed Rocks

In the field, virtually every rock outcrop contains visible evidence of strain. We can see deformation most readily in sedimentary rocks that have been stressed at plate boundaries, largely because most sedimentary rocks were originally horizontal, and deformation disturbs their horizontality in visually obvious ways. Other layered rocks, such as volcanic deposits and foliated metamorphic rocks, also help us interpret past deformational events. When layered rocks deform plastically, their crumpled appearance makes them easier to distinguish than plastically deformed masses of plutonic igneous rock, such as granite. Likewise, fractures and displacements from brittle failure are also more clearly visible in layered rocks, thanks to the obvious offset of the rock layers.

Geologists look at deformed rocks because they are curious about the type (compression, tension, or shear) and magnitude of stress, the distance that the deformed rocks may have moved from their original positions, and the direction from which the stress was applied. This information tells us how tectonic plates have moved in the past (in the case of rocks no longer being actively deformed) or are moving now (in the case of active deformation).

Studying old deformed rocks is of great interest to geologists because some of these rocks record deformation that occurred at great depth in the Earth, perhaps as deep as the Earth's lower crust and upper mantle. We can see this record of deformation only after faulting and erosion have exposed these deep rocks at the Earth's surface. Studying them also helps us to explore for valuable resources such as gold and other metals and, in the upper levels of the crust, for oil, natural gas, and coal. The manner in which a geologist studies exposed deformed rocks in the field involves making maps that illustrate how rock layers are oriented in three-dimen-

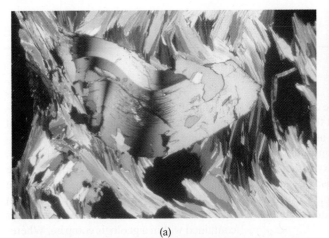

(a)

(b)

(c)

▲ **FIGURE 9-6** Folded rocks of various sizes. **(a)** Folded mineral crystals; **(b)** folded rock strata along Route 23 in Newfoundland, New Jersey; **(c)** folded sedimentary rocks in Dorset, England. All of these folds result from strong compressional forces at a former convergent plate margin.

▼ **FIGURE 9-7** Folding occurs most often at convergent plate boundaries. **(a)** At subduction zones, soft marine sediments are compressed within the oceanic trench and between the trench and its associated volcanic arc, creating extensive folding. **(b)** At continental collision zones, sediments between the plates are intensely folded and metamorphosed.

sional space. The essentials of this type of field study are discussed in Highlight 9-1 on page 296.

Folds

Think about what happens when you push a small rug on a polished hardwood floor. The rug will crumple into a series of folds, with arches separated by troughs. Rocks bend too when they are crumpled, most commonly at convergent plate boundaries. The rocks actually flow as they deform plastically. We can see evidence for this in folded rocks. The most spectacular **folds** appear in layered rocks, where the bending of the layers can be easily seen. Folding of rock can be seen at an extremely wide range of scales—from the microscopic scale of a folded mineral crystal (Fig. 9-6a), to the roadcut scale of a single outcropping of rock (Fig. 9-6b), and finally, to a mountain range–size scale (Fig. 9-6c).

When you see a folded rock, what can you say about its history? Folds are commonly associated with compression, so you would expect to see them at convergent plate boundaries such as continental-collision zones and subduction zones (Fig. 9-7). We find folds in places that at some point experienced high pressures and temperatures (deep in the Earth's crust), in places that deformed very slowly, or in materials that contain lots of flexible

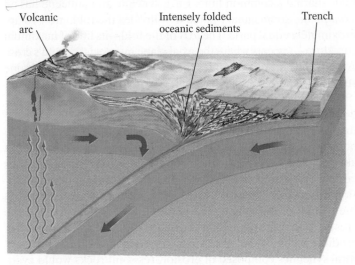

Volcanic arc · Intensely folded oceanic sediments · Trench

(a)

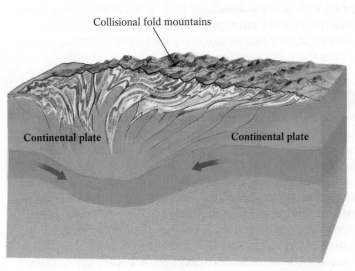

Collisional fold mountains

Continental plate · Continental plate

(b)

Highlight 9-1

Studying Deformed Rocks in the Field

Geologists study the origin of geological structures by first measuring their orientation. The orientation of a geological plane—either a structure or rock layer—in three-dimensional space is determined in relation to the four principal compass directions (north, south, east, and west). One key component of a geological plane's orientation in space is its **strike,** defined as the compass direction of the line that forms at the intersection between the plane (such as a layer of sedimentary rock or a foliation in a metamorphic rock) and an imaginary *horizontal* plane [see figure **(a)**]. We can visualize strike most easily where a geological plane, such as a layer of sedimentary rock, intersects the horizontal plane of a body of water. A compass tells us the direction of the horizontal water line crossing the geological plane—that's the rock's *strike.*

The **dip** of any geological plane is the *angle* at which it is inclined relative to the horizontal. Dip may be measured with an *inclinometer,* a device much like a protractor contained within a geologic compass. Whereas strike is expressed only as a direction, dip requires a measured angle *and* a direction. The directions of strike and dip are always perpendicular to one another. (Dip is measured 90° from strike.) Measuring the strike, dip angle, and dip direction of a geological plane establishes its orientation in three-dimensional space. On geologic maps, these measurements are shown by a *strike-and-dip* symbol [see figure **(b)**]. By making hundreds of strike-and-dip measurements at numerous outcrops and recording the data on a base map or aerial photograph [see figure **(c)**], geologists can determine the three-dimensional *subsurface* form of structures that are barely visible at the Earth's surface.

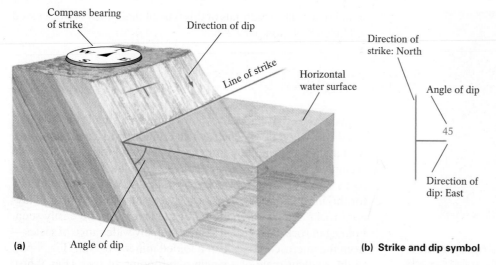

(a)

(b) Strike and dip symbol

▲ Because the surface of a body of water is horizontal, it can help determine the orientation of rock outcrops relative to the Earth's surface. The plane of the water traces a strike line across the exposed geologic surface of a partially submerged outcrop, and provides a horizontal reference against which to measure a plane's angle of dip (using an inclinometer). Note that a plane's strike and its dip are perpendicular to each other. The geologic map here illustrates how strike-and-dip symbols enable geologists to determine underground structures from their surface exposures.

minerals, such as micas and clays. Thus, many folds occur in both metamorphic rocks formed at high pressures and temperatures as well as in wet, clay-rich sedimentary rocks and sediments found at or near the Earth's surface.

Types and Shapes of Folds

The most common types of folded rocks are the tightly folded arches and troughs that occur most often from compression at active convergent plate margins. Concave-upward, troughlike folds are called **synclines** (from the Greek for "inclined together"). Convex-upward, arch-like folds are called **anticlines** (from the Greek for "inclined against") (Fig. 9-8). Folds may have many different geometries: they may have symmetrical sides or uneven sides, may be upright or lying on their sides (recumbent), and may be very tight or more open.

A fold's sides are referred to as its *limbs.* Folding typically produces a series of alternating synclines and anticlines, with adjacent folds sharing a common limb. Each syncline and anticline has an *axial plane,* an imaginary plane that divides the fold into two approximately equal parts. The *axis* of the fold—its line of maximum curvature—appears where the axial plane intersects the fold's crest. An anticline's axis resembles the peaked roofline of an A-frame structure, whereas a syncline's axis is akin to the keel of a sailboat.

Although in diagrams, upright anticlines may look like *topographic* hills and synclines like *topographic* valleys, they are actually structures in the rocks, not surface landforms. Anticlines do not always form hills or ridges in the landscape, nor do synclines always underlie valleys—their topographic expression depends largely on erosion. The layers in the folded rock may resist erosion differently. Thus, if a resistant sandstone were exposed in the trough of a syncline, and readily eroded rocks formed the crests of adjacent anticlines, structural anticlines composed of erodible rocks would tend to form topographic valleys, and structural synclines composed of erosion-resistant rocks would eventually stand out prominently as topographic ridges (Fig. 9-9).

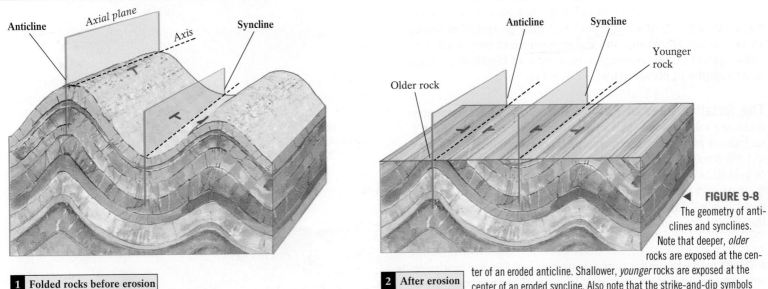

PENNSYLVANIA

(c)

Strike   Dip

Folded rocks before erosion

Anticline

Axial plane

Axis

Syncline

After erosion

Anticline

Syncline

Younger rock

Older rock

◄ **FIGURE 9-8**
The geometry of anti-
clines and synclines.
Note that deeper, *older*
rocks are exposed at the cen-
ter of an eroded anticline. Shallower, *younger* rocks are exposed at the
center of an eroded syncline. Also note that the strike-and-dip symbols
point away from the axis of an anticline and toward the axis of a syncline.

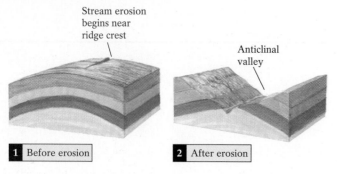

(a) **Anticline with highly erodible surface**

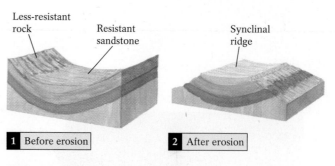

(b) **Syncline with resistant rock at axis**

▲ **FIGURE 9-9** **(a)** Anticlinal valleys and **(b)** synclinal ridges. The composition of exposed bedrock structures strongly controls their susceptibility to weathering and erosion. If weak, erodible rock crops out along the crest of an anticline, its erosion will produce a valley at that site. If the axis of a syncline consists of strong, resistant rock, its surface expression will be a ridge.

Fold Symmetry

Folds vary from simple to complex (Fig. 9-10). Under relatively gentle compression, anticlines and synclines appear as broad *symmetrical*, or *open*, structures with near-vertical axial planes and gently dipping limbs inclined at roughly the same angle. These forms occur most commonly at the edges of mountain belts in areas that are relatively quiet tectonically. Tightly folded anticlines and synclines are more common. They typically form at active or formerly active convergent plate margins where compression may range from moderate to intense. If compression forces one limb to move more than the other, folds may become *asymmetrical*. Prolonged directed pressure may cause the limb of an asymmetrical fold to rotate past vertical, producing an *overturned* fold. Under conditions of extreme compression, a fold may even rotate until its axial plane becomes essentially horizontal, paralleling the Earth's surface. Such *recumbent* folds are found in highly deformed mountain belts, such as the northern Rockies, Appalachians, Himalaya, and Alps.

More complex structures develop when an entire sequence of anticlines and synclines tilts so that the structures' axes actually intersect the Earth's surface. We identify such tilted folds—called *plunging* folds—by the characteristic *zigzag* pattern that forms as they erode. The Valley and Ridge province of the folded Appalachians of Pennsylvania provides some of North America's best examples of plunging folds (Fig. 9-11).

The Relative Ages of Folded Rocks

When sedimentary rocks are folded—and then subsequently eroded at the Earth's surface—a pattern begins to emerge that reveals the relative ages of folded rocks. In all cases, the exposed cores (central portion) of anticlines contain the anticline's oldest rocks, and conversely, the exposed cores of synclines contain the syncline's youngest rocks. Figure 9-8 illustrates this relationship between folds and the ages of their rocks.

▶ **FIGURE 9-10** Folds vary from broad, open structures, with limbs dipping at roughly the same angles, to overturned and recumbent folds. Such evidence of compression generally occurs where continents have collided.

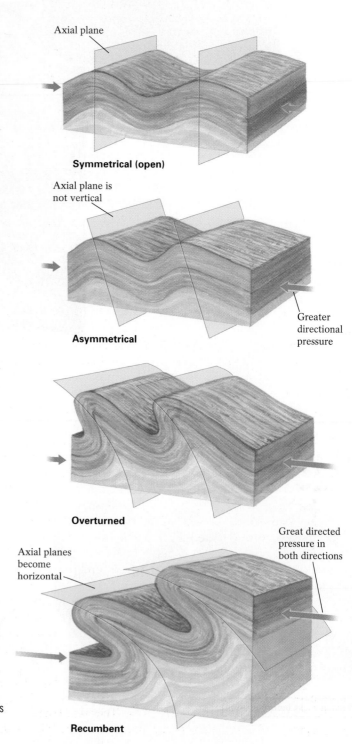

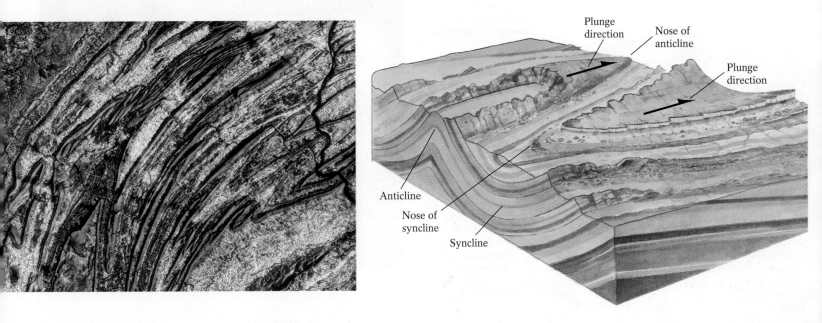

▲ **FIGURE 9-11** (left) Plunging folds in the Valley and Ridge province of central Pennsylvania, near Harrisburg. These structures consist of sedimentary beds that were folded and tilted by the tectonic plate collisions that produced the Appalachian mountains. (right) We can determine the direction in which the folds are plunging by identifying the anticlines and synclines, and then noting the direction in which the *nose* (the area of maximum curvature of the fold, seen on the zigzag pattern of plunging folds) points. The nose of an anticline points toward the plunge direction of a fold, whereas the nose of a syncline points away from its plunge direction.

Domes and Basins

In some cases, folds can plunge outward in all directions—forming a structural **dome**—or inward in all directions, forming a structural **basin** (Fig. 9-12). Domes interest geologists because although they resemble anticlines (a dome viewed in cross-section is an anticline), the processes that form anticlines and domes may be quite different. Domes may form in sedimentary-rock sequences—such as the Nashville dome in Tennessee or some oil-trapping domes in Wyoming—or in metamorphic and intrusive igneous rocks. When formed from metamorphic rocks, they are typically oval in shape and occur in belts or swarms.

Metamorphic domes (gneiss domes) are of particular interest because they are found in every mountain belt and therefore represent a fundamental component of mountain building. For example, gneiss domes are prominent in the Appalachians, the Himalaya, parts of the Alps, and within all regions where Precambrian crust is exposed on the continents. The origin of these domes may be quite different from the origin of anticlines, and interestingly, some clues to their origin may come from domes in a completely different type of rock—salt domes. Salt domes form when layers of salt (originally evaporite layers in sedimentary-rock sequences) are overlain by denser rocks. This creates an unstable situation, similar to that of low-density magma rising because it is less dense than the overlying and surrounding rocks. The low-density salt rises as a solid mass and forms domal structures (Fig. 9-13). Metamorphic rocks may also rise and form domes in a similar way if they are less dense than the rock above them. This will happen if the metamorphic rocks are hot and heat-softened, and thus able to flow.

We see domes more readily than basins. They stand out prominently at the surface, whereas basins usually become filled with sediment and are thus obscured. Noteworthy in North America are the Adirondack dome of northern New York, the Ozark dome of the southern Mississippi River valley, and the Nashville dome of the Tennessee River valley. The scenic Black Hills of South Dakota are actually a large oval dome, whose 2208-m (7242-ft)-high Harney Peak is the highest point in the state (Fig. 9-14). There, the region's sedimentary rocks dip away from an exposed core of Precambrian igneous and metamorphic rocks, which serve as the raw material for the prominent Presidential sculpture atop Mount Rushmore. Notable structural basins include the oil- and coal-bearing Williston basin of North Dakota, the Illinois basin of southern Illinois, and the Michigan basin, which forms

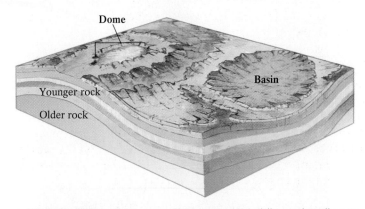

▲ **FIGURE 9-12** Structural domes and basins resemble anticlines and synclines, respectively, especially with regard to the ages of rocks within their cores. Because these structures typically form far from plate boundaries, they may reflect the effects of vertical motion related to crustal density variations rather than plate-edge activity.

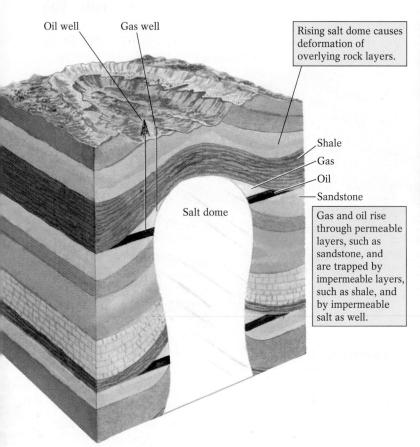

Oil well   Gas well

Rising salt dome causes deformation of overlying rock layers.

Shale
Gas
Oil
Sandstone

Salt dome

Gas and oil rise through permeable layers, such as sandstone, and are trapped by impermeable layers, such as shale, and by impermeable salt as well.

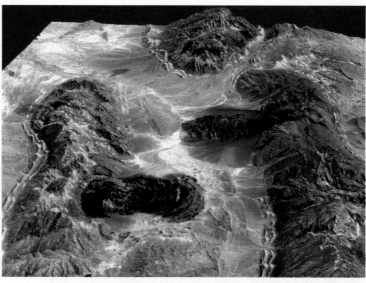

▲ FIGURE 9-13 Salt domes form as low-density flowing salt rises toward the surface, deforming the rocks above them and sometimes breaking through to the surface. The domes here (the dark oval masses) are from the Zagros Mountains of Iran. The domes are about 5 km (3 mi) long.

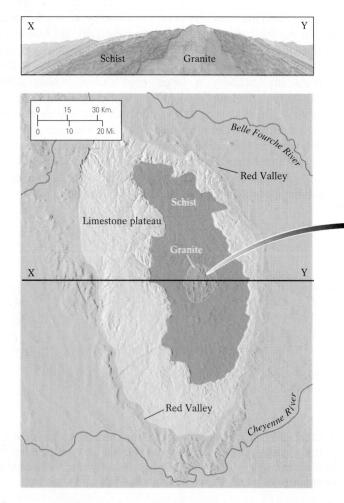

X   Y

Schist   Granite

0   15   30 Km.
0   10   20 Mi.

Belle Fourche River

Red Valley

Schist

Limestone plateau

Granite

X   Y

Red Valley

Cheyenne River

▲ FIGURE 9-14 The Black Hills of South Dakota are a large oval dome whose Harney Peak is the highest point in South Dakota. Photo of the central core of the Black Hills, an area known as "The Needles."

virtually all of lower Michigan (with the campus of Michigan State University sitting at its center). Some structural basins—such as the Gulf of Mexico—are enormous and contain many kilometers of sediment deposited over tens of millions of years.

Faults

Most rocks are brittle at the low temperatures and lithostatic pressures found at or near the Earth's surface. Thus the stresses produced by plate motion, particularly when those stresses are applied rapidly, tend to break these cool, shallow rocks. Rocks may also break at great depths in subduction zones from the stresses created by the moving plates and by metamorphic reactions that dehydrate water-bearing minerals and convert them to dry brittle minerals. Cracks in rocks may also result from *non-tectonic* causes, such as mechanical weathering related to freezing or unloading (discussed in Chapter 5) or the contraction of cooling, solidifying magma (discussed in Chapter 4). Virtually every surface rock contains evidence of brittle failure in the form of cracks, or *fractures*.

The orientation of a rock's fractures tell us much about the stresses applied to the rocks in the past. Some fractures seem to be random and show no evidence of relative movement (Fig. 9-15a). Others are oriented systematically, such as those shown in Figure 9-15b, but still present no evidence of relative movement across the fractures. Rock fractures with no evidence of movement are called **joints.** One common form of jointing in rock is the columnar jointing commonly formed in cooling volcanic rocks (see Fig. 4-12 on page 108).

▼ **FIGURE 9-15** Products of brittle failure in surface rocks. **(a)** These rocks from Acadia National Park, Maine, have been fractured by the application of a variety of stresses, including the weight of the glaciers that once sat upon them, the cracking that occurs during mechanical exfoliation (Chapter 5), the contraction that accompanies the cooling of plutonic igneous rock, and the stress from past plate-edge interactions. **(b)** The patterned appearance of these joints in Arches National Park, Utah, indicates that these rocks experienced some type of regional stress, perhaps related to plate interactions. **(c)** These faulted rocks have clearly moved relative to one another, as can be seen by the displacement of the various colored rock layers.

The most dramatic types of fractures in rocks are **faults.** Along a fault, the rock masses on either side of the fracture (called fault blocks) *have moved relative to each other.* How can we distinguish a joint from a fault? The best clue is evidence of displacement (offset) of layers or other features in rocks on either side of the crack.

Although we may see faults as "lines" on the surface of a rock or on the vertical face of an outcrop (Fig. 9-15c), they are actually planes—called fault planes—that cut through the rocks. Strike-and-dip measurements help determine the orientation of a fault plane and each fault block's relative direction of movement. In Figure 9-16, the two fault blocks marked with arrows appear on either side of a crack in the rock. Just from looking at this picture,

(b)

(a)

(c)

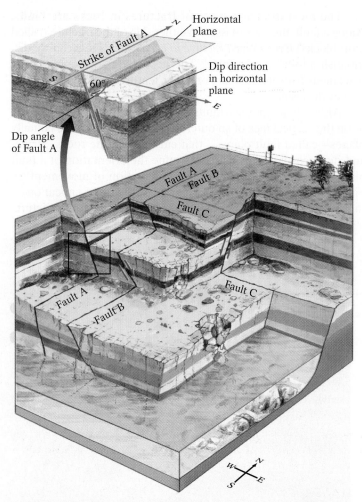

▲ **FIGURE 9-16** Fault planes can be oriented in space by determining their strike and dip. In this illustration of a quarry with exposed faults, fault plane A is oriented with a north–south strike, and dips toward the east at a high angle, perhaps as high as 60°. Can you estimate the strike and dip directions and the angle of dip for fault planes B and C?

we can't tell whether the right side has dropped down, the left side has moved up, or both sides have moved. The fact that motion has occurred, however, means that this structure is a fault.

Another indicator that a rock fracture is in fact a fault is the presence of fine-grained crushed rock in the fault zone. As the fault blocks grind past each other, the rocks along the fault may be crushed, and the surfaces of the fault blocks may become streaked with scratches as the blocks move (Fig. 9-17). These scratches form a shiny surface called "slickensides." In more subtle cases, faults may be identified if a distinctive rock bed disappears abruptly, perhaps reappearing at a greater depth, some distance away. As the two miners in Figure 9-18 would attest, a keen knowledge of faulting might very well determine who strikes it rich.

Types of Faults

All faults form as a result of motion between adjacent blocks of rock where one or both blocks move, or *slip,* relative to their original positions. Different types of stress (compression, tension, shear) cause different types of rock motion and, therefore, different kinds of faults. Each type of stress is associated with a characteristic type of fault. If you find a fault and can figure out what

type it is, you will then know what kind of stress produced the fault and perhaps, what type of plate motion was active when the rock was faulted.

Strike-slip faults are perhaps the easiest to visualize in relation to the stress that creates them because they form by *horizontal* slip of adjacent blocks of rock. That is, each block of rock on either side of the fault moves parallel to the fault, like two trains headed in opposite directions on parallel tracks. The result—surface features that cross the fault are offset sideways. Strike-slip faults generally occur at transform plate boundaries, such as the San Andreas fault in California and the North Anatolian fault in Turkey. The San Andreas underwent one of its greatest recorded displacements on April 18, 1906, the day of the great San Francisco earthquake. If you and a friend had been standing on the shores of Tomales Bay, about 30 km (20 mi) northwest of San Francisco, staring at one another across the San Andreas fault at precisely 5:12 A.M., within seconds you would have moved 7 m (25 ft) to one another's right.

Strike-slip faults do not always move in such dramatic ways. Parts of the San Andreas fault, a strike-slip fault that extends from the Pacific Ocean beyond Cape Mendocino southward into the Gulf of California (Fig. 9-19), creep along daily, millimeter by millimeter. This slow motion causes relatively little damage (except to the creaking houses directly over the fault and to the

▲ **FIGURE 9-17** Slickensides form from the scratching, polishing, and crystallization that occurs along an active fault plane.

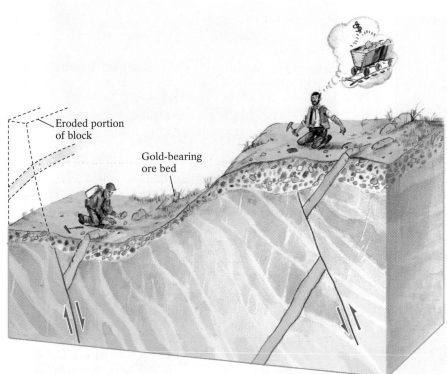

Eroded portion
of block

Gold-bearing
ore bed

◀ **FIGURE 9-18** By offsetting adjacent blocks of rock, faults disrupt the continuity of individual rock layers. One economic implication of faulting is shown in this illustration, in which the relative movement of the blocks is indicated by arrows. At left, the valuable ore bed that was present in the higher fault block has been eroded away, leaving none on that side of the fault. At right, the ore body continues at a slightly greater depth in the lower fault block. If he digs deep enough, the miner at right will find a significant lode of gold, whereas the miner at left will find only a limited amount.

◀ **FIGURE 9-19** (below) A map showing the length of the San Andreas Fault. (left) A section of the fault just south of San Francisco, containing San Andreas Lake and Crystal Spring Reservoir—examples of depressions and erosional basins that commonly occur along strike-slip faults.

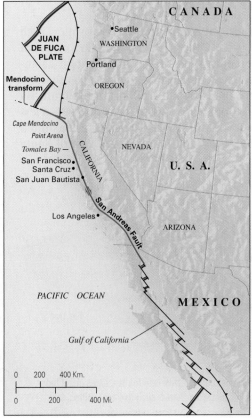

Fault

▲ FIGURE 9-20 Damage to structures from slow fault creep near San Jose, California, within a segment of the San Andreas Fault zone.

sidewalks, curbs, and walls that require frequent renovation, such as you can see in Figure 9-20). Other segments move less often but in large rapid displacements, typically producing great earthquakes. The 1906 quake, for example, offset fences and other linear features along more than 400 km (250 mi) of the fault. By comparison, California's Loma Prieta earthquake of 1989—a less powerful quake—produced far less displacement. Its maximum displacement—2.2 m (7 ft) of motion, measured east of Santa

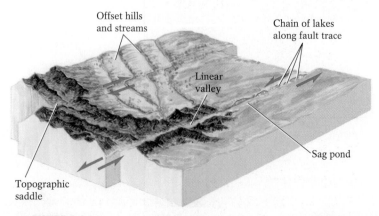

Offset hills
and streams

Chain of lakes
along fault trace

Linear
valley

Sag pond

Topographic
saddle

▲ FIGURE 9-21 Horizontal movement along strike-slip faults creates offset topographic features and distinctive erosional landforms, such as linear valleys, lake chains, and sag ponds.

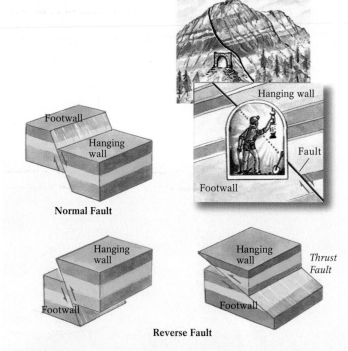

Footwall

Hanging
wall

Normal Fault

Hanging wall

Fault

Footwall

Hanging
wall

Footwall

Reverse Fault

Hanging
wall

*Thrust
Fault*

Footwall

▲ FIGURE 9-22 Dip-slip faults. (above) A dip-slip fault along McGee Creek, south of Mammoth Lakes, California. The rocks in the foreground (in front of the "step" in the landscape) have slipped downward relative to the rocks in the background. Dip-slip faults are classified by the relative movement of their fault blocks. In normal faults, the hanging wall moves downward relative to the footwall; in reverse faults, the hanging wall moves upward relative to the footwall; thrust faults are low-angle reverse faults.

Cruz. This quake's surface rupture stretched for only a few tens of kilometers.

The surface expression of a strike-slip fault may stretch for hundreds of km as low linear features—both ridges and depressions—that disrupt the works of both nature and humanity (see Figure 1-31 on page 28). As the pulverized rock of a strike-slip fault zone erodes, several distinctive landforms may be produced: linear valleys, chains of lakes (such as Loch Ness in Scotland's Great Glen strike-slip fault zone), sag ponds (depressions) caused by uneven ground settling within the fault zone, and topographic saddles (notches in the skyline of faulted hills; Fig. 9-21).

Fault movements are more commonly vertical than horizontal. In **dip-slip faults,** the fault blocks move up and down, parallel to the dip of the fault plane (Fig. 9-22). Such faults are typically marked by steps in the landscape that form where the land surface shifts upward or downward. You will find many old mineshafts along dip-slip faults because that's where valuable metals became concentrated when hot, mineral-rich solutions migrated along the openings within faulted rocks. We identify dip-slip fault blocks by their position relative to the fault plane, using terms derived from mining traditions: The block *above* the fault plane, from which the miners hung their lanterns, is known as the *hanging wall;* the block *below* the fault plane, on which the miners stood, is called the *footwall.*

Various types of dip-slip faults can be distinguished by the relative movement of their fault blocks. In a **normal fault,** the hanging wall has moved downward relative to the footwall (Fig. 9-23). This type of fault is generally associated with tensional stress, which typically occurs where plates are rifting or diverging. Stack some books—each one representing a block of rock, with the surface between two books simulating a fault—and now tip the pile over to one side. The blocks (books) on top of the "fault zones" will slide down relative to the books beneath. They slide down because of gravity, so the faults are "normal."

Tensional stresses initially stretch and thin the Earth's crust, and then eventually fracture rocks into faulted blocks. A fault block that drops during normal faulting may produce a valley called a **graben** (from the German for "grave" or "dig"). The blocks that remain standing above and on either side of a graben are referred to as **horsts** (from the German for "height"). These features characterize the normal faults of East Africa's rift valleys, as well as Earth's 60,000 km (40,000 mi) of mid-ocean divergence zones, including the rifted valleys of central Iceland (Fig. 9-24). Much of the southwestern United States is a normally faulted, tectonically stretched region called the Basin-and-Range province (Fig. 9-25). The basins are the grabens; the ranges are the horsts.

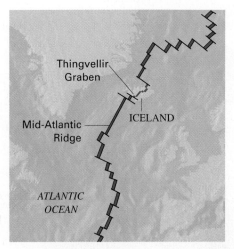

▲ **FIGURE 9-24** Iceland's central valley, the Thingvellir, formed when tensional stress and normal faulting within the diverging mid-Atlantic ridge system created a graben.

▲ **FIGURE 9-23** A normal fault in Utah.

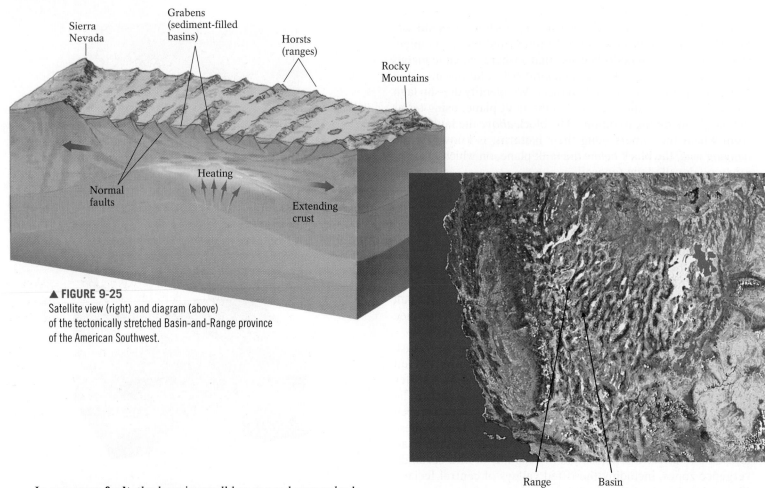

▲ **FIGURE 9-25**
Satellite view (right) and diagram (above)
of the tectonically stretched Basin-and-Range province
of the American Southwest.

In a **reverse fault,** the hanging wall has moved *upward* relative to the footwall (Fig. 9-26). This situation is the "reverse" of what gravity would naturally do with rocks on an inclined plane. To make the upper block of rock (the hanging wall) move up, powerful horizontal compression is required, such as occurs where plates converge. If you experiment with a tipped stack of books again, you will see that you have to push them to make the books on top of the faults move up and over the books beneath. An interesting aspect of reverse faults is that they carry deeper, older rocks and place them atop younger rocks. This violates the principle of superposition (described in Chapter 8), which calls for young rocks to overlie older rocks—as long as the rocks have not been disturbed tectonically.

A **thrust fault** is a type of reverse fault with a relatively low angle (less than 45°). A very-low-angle thrust fault (perhaps as low as 5–10°) produces an *overthrust*, in which enormous slabs of rock move nearly horizontally for tens of kilometers. In the Lewis Overthrust in Glacier National Park, Montana, and Waterton-Lakes National Park in Alberta, a 3-km (2-mi)-thick slab of Precambrian marine sedimentary rock moved more than 50 km (30 mi) eastward, coming to rest atop much younger (Cretaceous) continental sedimentary rock (Fig. 9-27). Billion-year-old rock sitting atop 100-million-year-old rock is a clear violation of the principle of superposition and a clear indicator of reverse faulting.

In many faults, slip is not solely vertical (up or down along the dip of the fault plane) or horizontal (along the fault plane's strike). Such fault motions—called **oblique slip**—combine both strike-slip and dip-slip motion (Fig. 9-28). These faults are very com-

▲ **FIGURE 9-26** A reverse fault in volcanic tuffs.

mon in mountain belts where strike-slip faults have some component of compression or tension as well. The various faults and their relationships to the Earth's plate-tectonic boundaries is summarized for you in the Geology at a Glance on page 308.

Plate Tectonics and Faulting

Each of the three major types of plate margins is associated with a particular type of stress, and thus with a particular type of fault (Fig. 9-29). Normal faults commonly occur where tensional stresses pull apart the Earth's lithosphere, such as along the axes of the world's mid-ocean ridges and at sites where continents are rifting, such as in the north–south Rio Grande valley in New Mexico where North America may be splitting apart.

Reverse and thrust faults, which are produced by compressive stresses, are concentrated along convergent plate boundaries, both where oceanic plates subduct and where continental plates collide. The major faults in Japan, the Philippines, and other western Pacific islands are produced largely by convergence of two ocean plates (with one subducting beneath the other). In contrast, the faults in the Andes of South America and Cascades of Washington, Oregon, and northern California have developed where an ocean plate subducts beneath an adjacent continental plate. Faults in the Alps and Himalaya owe their existence to collisions between continental plates—those in the Alps from past collisions of Africa and Europe, and those in the Himalaya from the ongoing collision of Asia and India.

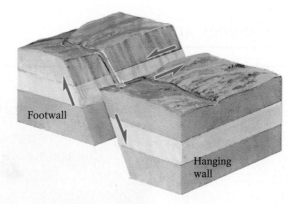

▲ **FIGURE 9-28** Oblique slip is fault motion that involves both dip-slip and strike-slip movement of fault blocks.

Some of the world's longest continuous faults, such as the San Andreas fault in California and the North Anatolian fault in Turkey, are strike-slip faults that coincide with transform boundaries. At these boundaries, plates slide past one another, producing shear stress.

The plate-tectonic forces that produce large dip-slip faults in thick sequences of marine sedimentary rocks, as well as those that yield vast folds in marine rocks, often trap oil and natural gas within rock bodies. Highlight 9-2 on page 311 describes how our knowledge of bedrock structures can aid in the search for these essential fuels.

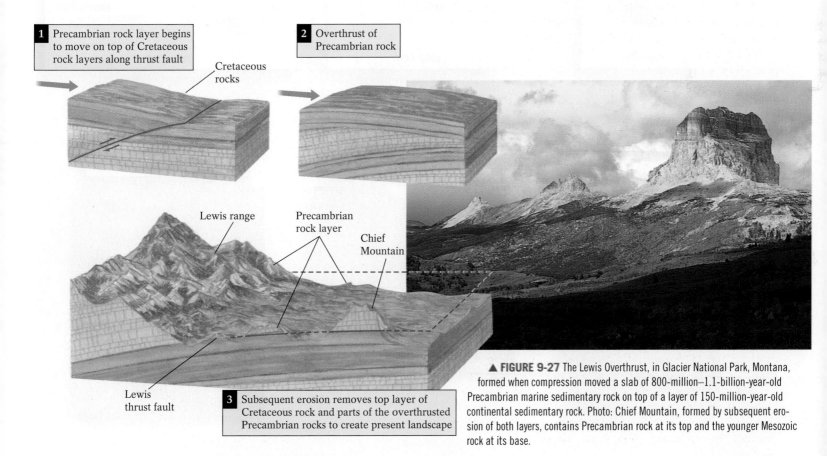

▲ **FIGURE 9-27** The Lewis Overthrust, in Glacier National Park, Montana, formed when compression moved a slab of 800-million–1.1-billion-year-old Precambrian marine sedimentary rock on top of a layer of 150-million-year-old continental sedimentary rock. Photo: Chief Mountain, formed by subsequent erosion of both layers, contains Precambrian rock at its top and the younger Mesozoic rock at its base.

9.1 Geology at a Glance

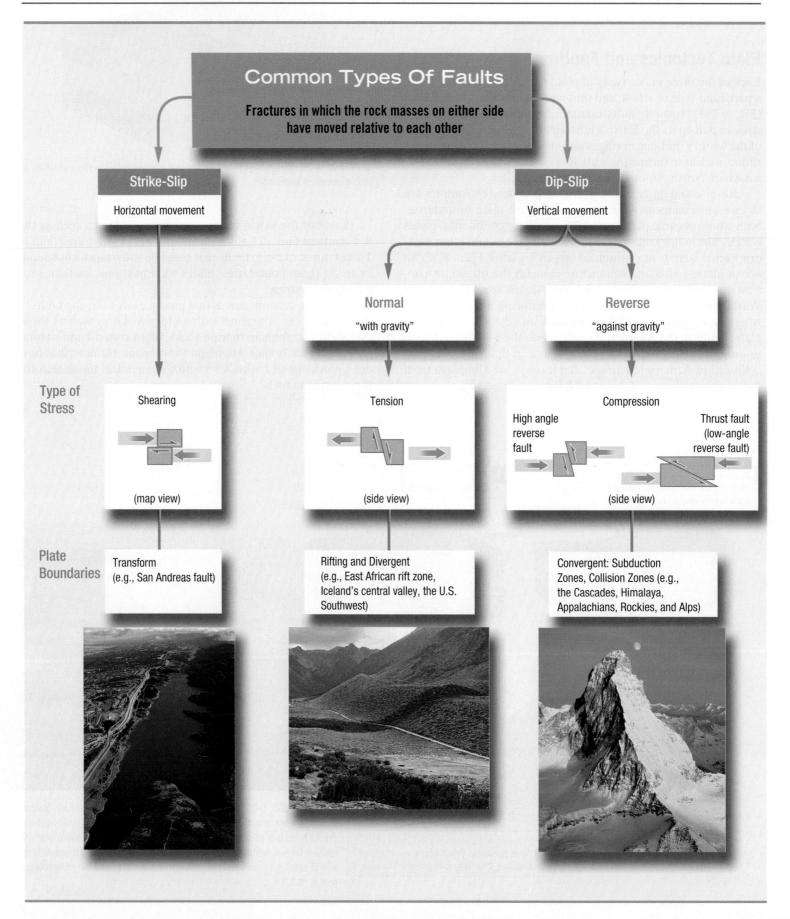

Common Types Of Faults

Fractures in which the rock masses on either side have moved relative to each other

Strike-Slip
Horizontal movement

Dip-Slip
Vertical movement

Normal
"with gravity"

Reverse
"against gravity"

Type of Stress

Shearing

(map view)

Tension

(side view)

Compression

High angle reverse fault

Thrust fault (low-angle reverse fault)

(side view)

Plate Boundaries

Transform (e.g., San Andreas fault)

Rifting and Divergent (e.g., East African rift zone, Iceland's central valley, the U.S. Southwest)

Convergent: Subduction Zones, Collision Zones (e.g., the Cascades, Himalaya, Appalachians, Rockies, and Alps)

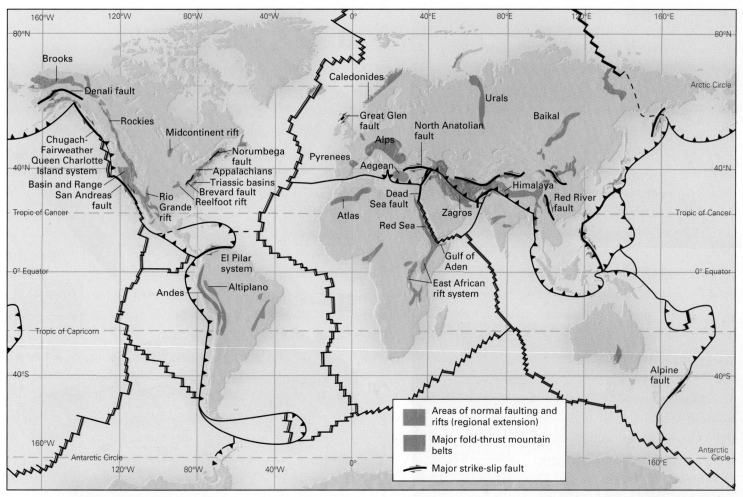

▲ **FIGURE 9-29** The correlation of different fault types with the Earth's plate boundaries. Normal faults typically occur at rifting and divergent zones. Reverse and thrust faults commonly occur at convergent zones, including subduction zones and zones of continental collisions. Strike-slip faults occur principally at transform boundaries.

Seeing Structures

With the exception of the most pristine, undisturbed deposits of young sediments, almost all rocks have been deformed. We determine how a rock was deformed (in what type of tectonic setting?) and at what conditions (shallow or deep in the Earth?) by combining some of the things you learned in earlier chapters about minerals and rocks with the information in this chapter.

First consider what large-scale structures look like and where you can see them. Obviously, mountains contain lots of structures (folds and faults), but what about the cliffs of coastal California, the farmlands of southern Minnesota, and the ridges and valleys of central Pennsylvania?

If you drive down Route 1 along the southern California coast, you will see the dramatic effects of faulting on the landscape. On San Clemente Island, 112 km (70 mi) southwest of and offshore from San Clemente, California, and on the coast north of Santa Cruz, we find a giant stairway of stepped terraces, each of which was cut by waves *at sea level* and then subsequently uplifted (Fig. 9-30). Each terrace remained at sea level during a temporary pause in the tectonic uplift of California's coast, long

enough for the pounding surf to erode the surface before the next pulse of uplift carried it safely above the wave action. The highest terrace, 400 m (1300 ft) above the present sea level, is the oldest. We know the ages of these terraces from cosmogenic-isotope dating of minerals on their newly exposed surfaces (discussed in Chapter 8), and can also date the time *between* each uplift event, giving us a view of the frequency of large faulting events—and their associated earthquakes. Not all cliffs form in this way, but if you see a series of stepped terraces along a coastline, you're probably seeing the action of faulting.

When you drive from east to west across southern Nevada or in a similar direction across parts of Pennsylvania, you will find the road rollercoastering up and down over a series of hills and valleys. Such hills and valleys can result from compression (such as in the case of the folded mountains of Pennsylvania) or extension (such as in the case of the stretched, normally faulted crust of Nevada) during various periods of deformation of the North American continent. You can see evidence for the types of structures in roadcuts that have been blasted through the hills. In Pennsylvania, look for anticlines and synclines (Fig. 9-31), and in the American Southwest, look for prominent normal faults (Fig. 9-32).

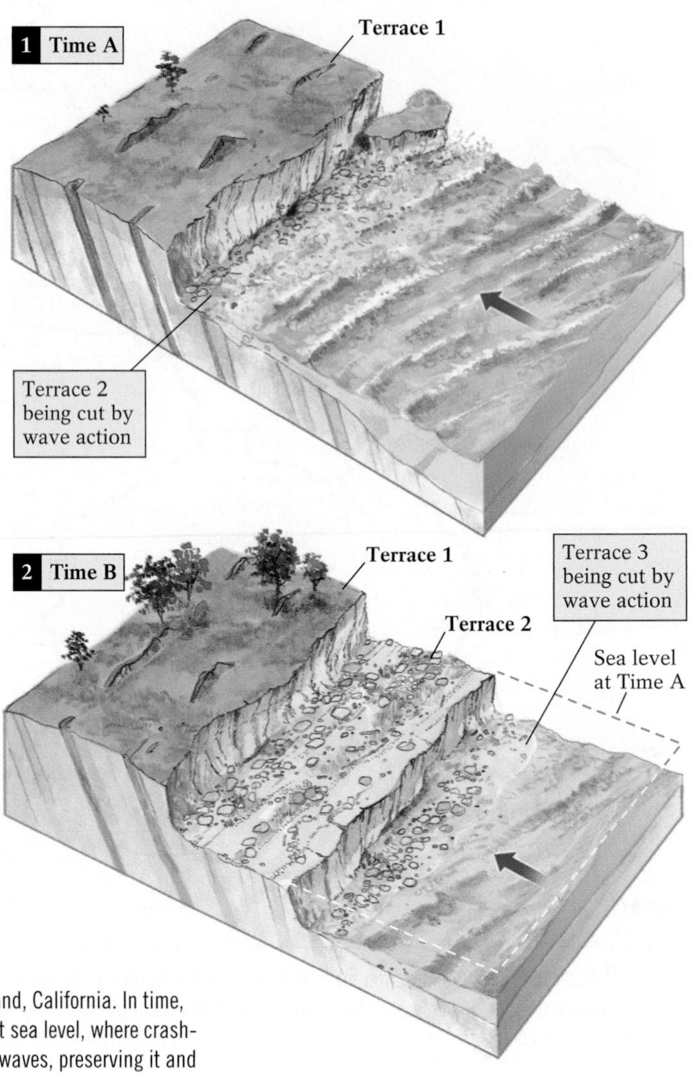

1 Time A

Terrace 1

Terrace 2 being cut by wave action

2 Time B

Terrace 1

Terrace 2

Terrace 3 being cut by wave action

Sea level at Time A

▲ **FIGURE 9-30** Photos above show two views of uplifted marine terraces at San Clemente Island, California. In time, these structures may become mountains. (right) A terrace-producing scenario: Terrace 1 forms at sea level, where crashing surf shapes its surface; subsequent uplift then elevates this terrace beyond the reach of the waves, preserving it and enabling terrace 2 to form at sea level.

▲ **FIGURE 9-31** Folded rocks in the ridge-and-valley province of central Pennsylvania near Harrisburg.

▲ **FIGURE 9-32** Faulted rocks in the basin-and-range province of Nevada. The area in the foreground is a downdropped basin; the area in the background is a high-standing range.

Highlight 9-2

Folds, Faults, and the Search for Fossil Fuels

The search for the Earth's dwindling fossil-fuel reserves focuses on identifying the Earth structures that trap oil and natural gas in recoverable quantities. Heating of the organic material in marine sediments (the so-called "source rocks") produces oil and gas underground. Because they are less dense than water and the surrounding rocks, oil and gas tend to rise toward the surface, migrating through permeable rocks and fractures *until their passage is blocked.* At these locations, the fluids accumulate in the pores and cracks of other rocks (the so-called "reservoir rocks") from which they can then be collected. In folded terrain, the passage of oil and gas may be blocked where impermeable rocks cap the crests of anticlines (see figure at top right). The fluids then accumulate in the pore spaces of permeable sedimentary rock layers. For this reason, searches for oil often concentrate on known oil-producing marine sediments beneath anticlinal crests. Several oil-bearing deposits may even appear at different stratigraphic levels within a single anticline.

Unfortunately for motorists and for residents of cold regions, most of the world's large *exposed* anticlines were drilled by the 1920s, and most of the recoverable oil in *buried* anticlines (probably between 20% and 40% of the total) was collected by the end of the 1950s. New recovery techniques allow some of these anticlines to continue to produce, but today we must seek less-obvious oil traps. Unfortunately, some of these traps are located in isolated, relatively unspoiled arctic regions. Thus we are currently debating whether our need for the oil in these regions—such as the Arctic National Wildlife Refuge—outweighs the potential environmental damage to such pristine lands.

Faulting sometimes places an impermeable bed next to oil-bearing permeable ones, creating a barrier to oil flow (see figure at bottom right). Thus a petroleum geologist also looks at faults to identify sites where expensive test drilling is most likely to be productive. Faults and their associated earthquakes may wreak expensive havoc on property and safety, but they also contribute one important economic benefit: They trap oil. In Chapter 20, we will consider the future of fossil fuels and evaluate the potential for some alternative energy sources.

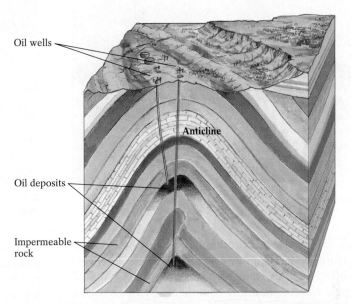

▲ The accumulation of oil and natural gas at the crests of anticlines. The low densities of these fluids allow them to migrate readily through permeable sedimentary bedrock until they encounter an overlying cap of impermeable bedrock. The fluids become concentrated at the crests of anticlines, which are the highest points within the permeable beds.

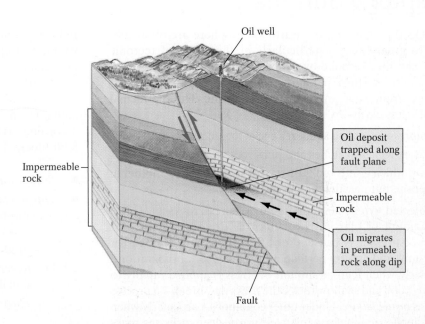

▲ The accumulation of oil along fault planes. Oil migrates upward through permeable sedimentary rock because of its low density, but its path to the surface may become blocked by an impermeable bed moved to that location by faulting.

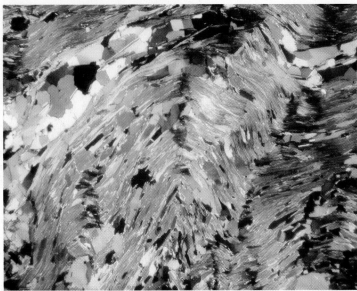

▲ FIGURE 9-33 Folded flakes of muscovite mica.

▲ FIGURE 9-34 The tectonics of these old, eroded mountains can be studied by using the structural fabrics within their rocks and minerals.

Sometimes, the landscape contains no obvious clues that major tectonic forces have been at work, and you then have to look more closely at the rocks themselves. Recall from Chapter 7 that metamorphic rocks commonly contain a *fabric*—that is, a texture formed by minerals such as flaky mica grains that are all aligned within the same plane, or elongated minerals that are all oriented in the same direction. These fabrics form as metamorphic minerals grow during deformation, and are an excellent clue that rocks have been affected by plate-tectonic motion. Consider, for example, Figure 9-33, which shows folded muscovite mica flakes—only several millimeters long. Despite their small size,

these little crystals tell us much about the movement of plates and the growth of mountains. (Their fold axes are typically perpendicular to the major direction of plate convergence.) By looking for clues in the fabrics of rocks, you can imagine the tectonic forces that shaped them even in places where the mountains formed and eroded billions of years ago (Fig. 9-34).

Now that we've discussed how rocks deform—and often break—we can begin to examine the greatest human consequences of breaking rocks—powerful earthquakes—our next topic in the section on "Shaping the Earth's Crust."

Chapter 9 Summary

The rocks that make up the Earth's lithosphere are often subjected to great forces—particularly at the edges of tectonic plates—and they respond by either bending or breaking. When a sufficient stress (force per unit area of rock) is applied to rocks, they may strain, changing in shape or volume. Three principal types of stress exist:

- *compression,* which squeezes rocks together;
- *tension,* which stretches and pulls rocks apart;
- *shear,* which grinds rocks by forcing them past one another in opposite but parallel directions.

Subjected to only a minor amount of stress, rocks undergo a small amount of elastic (reversible) *deformation.* A greater amount of stress may cause a rock to exceed its *yield point* (or elastic limit) and deform permanently. Under certain conditions, particularly when a great amount of stress is applied rapidly to relatively cool, and shallow rocks, the rocks may break—a process known as *brittle failure.* Under other conditions, particularly when stress is applied gradually to relatively warm, deep rocks, the rocks will deform plastically (or ductilely), folding rather than breaking.

When rock layers deform plastically, they form a series of folds, including trough-like synclines and arch-like anticlines.

Folds form most often at convergent plate boundaries, such as within the trenches of subduction zones and at the collision zones between masses of converging continental lithosphere.

Brittle failure may produce randomly distributed fractures, systematically distributed joints, or faults. Faults are fractures along which movement has occurred. Geologists classify faults according to the relative direction in which the affected rocks, or fault blocks, have slipped. Common fault types include:

- *Strike-slip faults,* which are marked by horizontal movement;
- *Dip-slip faults,* which are characterized by vertical movement. The most common types of dip-slip faults are:
  - *Normal faults,* which develop in response to tensional stress;
  - *Reverse faults,* which develop in response to compressional stress;
  - *Thrust faults,* which are low-angle reverse faults.

Strike-slip faults generally occur at transform plate boundaries, normal faults at rift zones and divergent plate boundaries, and reverse faults at convergent plate boundaries.

KEY TERMS

anticlines *(p. 296)*
basin *(p. 299)*
brittle failure *(p. 293)*
compression *(p. 294)*
dip *(p. 296)*
dip-slip faults *(p. 304)*
dome *(p. 299)*

elastic deformation *(p. 292)*
elastic limit *(p. 293)*
faults *(p. 301)*
folds *(p. 295)*
graben *(p. 305)*
horsts *(p. 305)*
joints *(p. 301)*

normal fault *(p. 305)*
oblique slip *(p. 306)*
plastic (ductile) deformation
  (p. 293)
reverse fault *(p. 306)*
shear *(p. 294)*
strain *(p. 292)*

stress *(p. 292)*
strike *(p. 296)*
strike-slip faults *(p. 302)*
structural geology *(p. 290)*
synclines *(p. 296)*
tension *(p. 294)*
thrust fault *(p. 306)*
yield point *(p. 293)*

QUESTIONS FOR REVIEW

1. What are the three principal types of tectonic stress, and at which type of plate boundary does each develop?

2. Determine which way is north and then hold this book so that its direction of strike is 45° west of north, and it dips 45° and toward the northeast; toward the southwest.

3. Draw an anticline and syncline, and label the limbs, axial plane, and axis of each. Why does the core of an eroded anticline contain the formation's oldest rocks, and the core of an eroded syncline hold its youngest rocks?

4. Distinguish between fractures, joints, and faults.

5. What are four ways to identify faults in the field?

6. Draw two simple sketches illustrating the difference between normal faults and reverse faults. Label the hanging wall and footwall; use arrows to show their relative direction of movement.

7. Identify the type of fold and type of fracture that appear in the photo below.

8. Suggest future plate-tectonic scenarios that would result in development of normally faulted grabens in Minnesota and the creation of Appalachian-type folded mountains along the Louisiana coast.

LEARNING ACTIVELY

1. Head over to the local toy store and buy some Silly Putty. Using the putty, simulate compressional and tensional stresses. How does the putty deform? Roll the putty into a ball and bounce it. Here the putty is illustrating elastic behavior. Pull slowly on the putty ball and stretch it. When you stop pulling, the ball has been deformed by tension (it's longer laterally and thinner vertically). You have exceeded the putty's elastic limit, or yield point. If you keep pulling on it, the putty will form thin filaments. Roll the putty into a ball again. Pull on it quickly. It should snap in a brittle way, forming sharp edges. What type of geological feature forms in this way? Note that the rate at which stress is applied affects the deformation of the putty (as well as rocks).

2. Find a thick, flat piece of foam rubber. Crumple it under compression and create folds. Push harder with one hand than the other and produce asymmetrical and overturned folds. Pull on the foam rubber until it tears. Recreate normal, reverse, and strike-slip faults. Note how the foam sticks as you try to move the rough edges past one another. As we will see in Chapter 10, this frictional "sticking" along faults causes stresses to build up before earthquakes occur.

ONLINE STUDY GUIDE

Practice makes you better. Make the time to take the *Online Study Guide* quiz for this chapter. It should take you about 20 minutes. Automatic grading and instant feedback will help you quickly and accurately identify the concepts you have mastered and the areas in which you need more study.
www.prenhall.com/chernicoff

Earthquakes

10

A t 5:03 P.M. Pacific daylight time on October 17, 1989, baseball fans across North America were settling down in front of their television sets to watch game three of the World Series from San Francisco between the Bay Area rivals, the San Francisco Giants and the Oakland Athletics. At that moment, violent movement along a segment of California's San Andreas Fault caused the screens to go blank. When service was restored, instead of baseball, millions of people viewed grim scenes of collapsed buildings and freeways (see Fig. 10-2), broadcast live. Widespread destruction took the lives of dozens of Bay Area residents and injured hundreds more. People in other earthquake-prone regions wondered whether their own towns and cities might be the next to suffer a life-threatening quake.

During the 1989 San Francisco Bay area's Loma Prieta earthquake (named for a mountain peak near the quake's point of origin), tremors sped through the rocks and soils of northern California at 32,000 km/hr (19,000 mph), toppling buildings, freeways, and bridges, rupturing gas mains and igniting fires, and triggering landslides throughout the region. The quake emanated from the 35-km-(20-mi)-long South Santa Cruz mountain segment of the San Andreas Fault, approximately 80 km (50 mi) southeast of San Francisco and 16 km (10 mi) northeast of Santa Cruz. Although public attention was riveted on the fires in the Marina district along San Francisco's waterfront and on the heroic rescue efforts on Oakland's collapsed Nimitz Freeway, the most extensive damage occurred closer to the earthquake's point of origin, in the towns of Santa Cruz, Los Gatos, and Watsonville. In all, the Loma Prieta earthquake caused more than $5 billion in damage and claimed 65 lives.

From 1989 to 2005, Los Angeles, Oakland–San Francisco, Seattle, Alaska, Mexico, Japan, Turkey, Greece, Taiwan, Sumatra, Papua New Guinea, China, India, Iran, Afghanistan, Morocco, Pakistan, and many other regions have experienced cataclysmic earthquakes (Fig. 10-1). In heavily populated and industrialized areas, the damage caused when powerful tremors shake large regions can be devastating. Skyscrapers, highways, hospitals, and homes collapse and thousands of people die in the wreckage. Governments shift abruptly into emergency mode to repair cut-off water supplies, restore snapped electrical lines, and put out the fires produced when gas from severed gas lines ignites. Ironically, the forces that cause such devastation are, in most cases, the very same forces that created our planet's most beautiful mountain scenery, along with the environmental, climatic, and atmospheric conditions that have initiated and sustained life on Earth.

FIGURE 10-1 The tragic destruction of scores of homes on this ridge in Pakistan was caused by the deadly magnitude-7.6 earthquake in October, 2005. Nearly 100,000 people perished in this quake. The tectonics of the region— a continental collision between India and China that created the lofty Himalaya—has triggered many such powerful catastrophic earthquakes. ▶

Deccan Basalts
formed when
India sat atop
the Reunion
hotspot.

*Zagros
Mountains*

Iran

Arabia

Pakistan

Africa

Eurasian
plate

INDIAN OCEAN

India

Himalayas

*Tibetan
Plateau*

Suture
zones

Fold-and-thrust
faulted rocks

Indian lithosphere
(Indian-Australian
plate)

N

Earthquakes are among the greatest catastrophes that the Earth's geological forces can produce. Virtually every culture that has experienced devastating earthquakes has sought to explain their cause. Some cultures tell us of giant creatures that carry the Earth through the sky, occasionally stumbling and jostling the planet. In Japanese tradition, the source of all tremors is a giant catfish; in many Native American traditions, it is a giant tortoise; and for the farmers of Mongolia, the source is an enormous frog. The Wanyamwasi of western Africa believe that the Earth is supported on one side by mountains and on the other by a giant, who causes the Earth to tremble whenever he lets go to hug his wife.

In this chapter, we will discuss the scientific causes of earthquakes and how geologists study and evaluate them. We will also examine how humans have learned to cope with earthquake effects, and how we are trying to predict and prepare for them.

The Causes of Earthquakes

The sudden release of energy accumulated in deformed rocks causes the ground to tremble during an **earthquake.** In Chapter 9 we saw that rocks become deformed—or *strained*—when they're stressed, most often at or near plate boundaries (Fig. 10-3). Deep within the Earth, when rock is stressed beyond

▲ **FIGURE 10-2** The collapsed Cypress Freeway in Oakland, California, in October 1989.

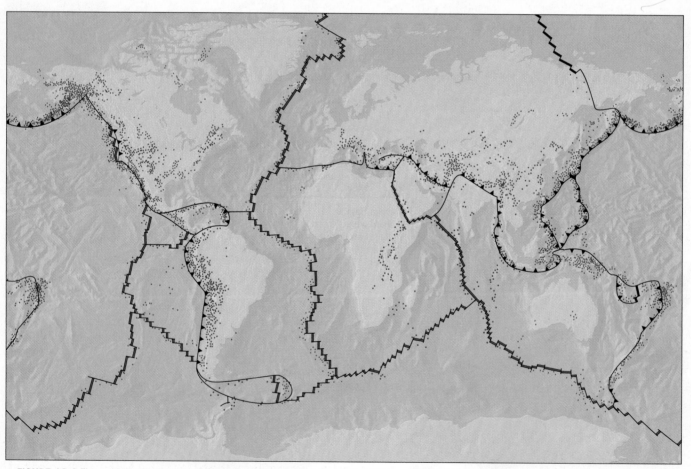

▲ **FIGURE 10-3** The worldwide distribution of earthquakes that occurred during the last 100 years. (The dots represent significant earthquakes—greater than magnitude-4.0) Notice that most earthquakes occur near current (such as western South America), ancient (such as along the Appalachian Mountains front in eastern North America), and newly forming plate boundaries (such as the East African rift).

its yield point (discussed in Chapter 9), it deforms *plastically* to form folds. At shallower depths, closer to the Earth's surface where rock is relatively cool and brittle, it deforms *elastically* until, if the stress persists, it ruptures. At these shallow depths, stressed rocks accumulate *strain energy.* Strain energy builds until the rocks either return to their original shape (upon the removal of the stress) or rupture, much as a stretched rubber band, pulled with increasing force, either snaps back to its original shape when the stress is removed or eventually breaks. Rocks that rupture create new faults; rocks at preexisting faults, although temporarily held in place by friction between the fault blocks, eventually break free and shift under stress. In both cases—new ruptures or slippage along old ones—rocks move suddenly, releasing pent-up strain energy. The release of this energy causes the Earth to quake.

The precise underground location at which rocks begin to rupture or shift is known as an earthquake's **focus** (or *hypocenter*). The quake's **epicenter** is the point *on the Earth's surface* that lies directly above the focus. Figure 10-4 shows the relationship between the earthquake's focus and its epicenter. (An earthquake will typically cause the most damage near the epicenter and less damage farther away from it.) After a major earthquake, the rocks in the vicinity of the focus continue to shake as they adjust to their new positions, producing numerous, generally smaller, earthquakes called **aftershocks.** Aftershocks, which may continue for several years after a powerful quake, can produce significant damage by shaking already weakened structures. In the aftermath of the main shock from southern California's 1994 Northridge earthquake, thousands of aftershocks caused new damage

and added significantly to the quake's $15 billion cleanup and rebuilding costs.

After releasing their stored strain energy during an earthquake, the rocks along a fault stop moving, and friction between the fault blocks locks them in place again. As tectonic stress rebuilds, strain energy accumulates again. The friction holding the fault blocks is eventually overcome, and the rocks then slip suddenly, causing a new earthquake. This alternating slipping and sticking explains why many faults move in periodic jolts, rather than continuously, and why earthquakes occur only now and then. Unfortunately for the citizens of California, seismologists agree that neither the Loma Prieta nor the Northridge quake was the long-dreaded "Big One." California is long overdue for an earthquake of even greater magnitude than either of those events—one powerful enough to relieve the energy that has been building along some segments of the San Andreas and nearby faults for as long as 150 years. Let's now consider how energy is transmitted from an earthquake's focus through the Earth and then to the surface to cause the ground-shaking and other devastating effects of earthquakes.

The Products of Earthquakes— Seismic Waves

When you toss a large stone into a quiet lake, the energy of the falling stone is transferred on impact to the water. Ripples pass through the water in all directions, eventually dying out at some distance from the point of impact. Likewise, when you speak, sound energy produced by the vibration of your vocal cords is transmitted through the air in the form of sound waves. These too die out with greater distance from their source.

When rocks break or a locked fault breaks free, the released energy is similarly transmitted at great speed through the surrounding rocks in all directions. Earthquake energy, like other types of energy, moves from one place to another in the form of waves. At the instant that an earthquake occurs, the released energy is transmitted through the Earth as **seismic waves.** (The term *seismic,* from the Greek for "shaking," refers to anything related to earthquakes.)

Seismology is the study of earthquakes and the Earth's interior. Seismologists use information from two main types of seismic waves: **body waves,** which transmit energy through the Earth's interior in all directions from an earthquake's focus, and **surface waves,** which transmit energy along the Earth's surface, moving outward from a quake's epicenter.

Body Waves

There are two types of body waves, and each one travels through the Earth with a characteristic speed and style of motion. Primary waves, or **P waves,** are the fastest seismic waves; they pass through the Earth's crust at a velocity of 6 to 7 km (about 4 mi) per second. Thus they are the first signal to arrive at an earthquake-

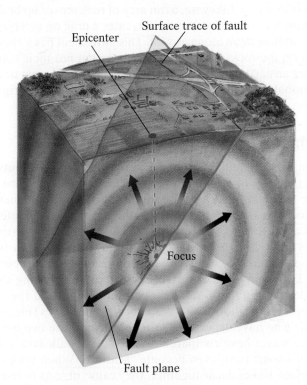

▲ **FIGURE 10-4** At the moment an earthquake occurs, pent-up energy is released at the quake's focus and transmitted through the Earth. The point on the Earth's surface directly above a quake's focus is its epicenter. The surface *trace* of a fault is its surface expression.

Epicenter

Surface trace of fault

Focus

Fault plane

recording station after a quake occurs. The energy released when rocks break or slip near the earthquake's focus compresses neighboring rocks, setting off a P wave. These rocks then expand elastically, or *dilate,* past their original volume after generating the wave, thereby compressing adjacent rocks. The rocks are alternately compressed and dilated again as the next wave of seismic energy passes . . . and so on. In this way, rocks in the path of P waves in turn compress and expand in the same direction as the waves are traveling (Fig. 10-5), much as the coils in a stretched Slinky vibrate back and forth when the toy is struck. As you can see, vibration from P waves is *parallel* to the direction in which the wave travels. During an earthquake, we feel the arrival of P waves at the Earth's surface as a series of sharp jolts that occurs as surface rocks alternately compress and expand with passing waves of energy.

Secondary waves, or **S waves**, move more slowly than P waves, passing through the Earth's crust at a velocity of about 3.5 km (2 mi) per second. Thus, they take longer than P waves to arrive at earthquake-recording stations. The motion of rock in an S wave is similar to that of a rope fixed at one end and snapped sharply up and down at the other end (Fig. 10-6). The rocks move up and down, *perpendicular* to the direction in which the S wave travels. An S wave is generated in much the same way as the once-popular stadium "wave" seen way too often at recent sporting events. As successive individuals stand and then sit, the motion of the components of the wave—the fans—occurs perpendicular (up

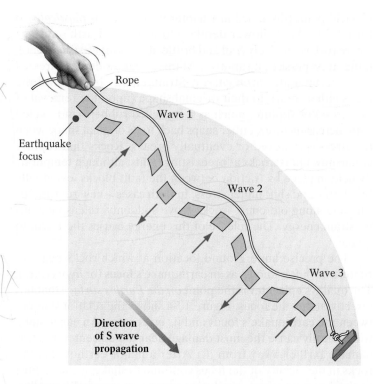

▲ **FIGURE 10-6** S waves, or secondary waves, are shearing waves. They generate a motion that displaces adjacent rocks in a direction perpendicular to the direction of wave propagation. Dark gray rectangles show how rocks deform as the wave passes.

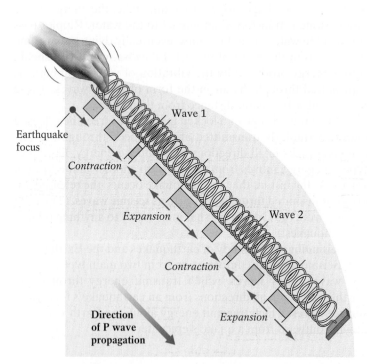

▲ **FIGURE 10-5** Primary waves, or P waves, are compressional. They produce the alternating contraction and expansion of rocks in a direction parallel to the direction of wave propagation. The energy released when you strike one end of the Slinky is transmitted by a compressional wave, like a P wave, that moves from one end of the toy to the other. The back-and-forth motion displaces rocks as P waves pass and can cause power lines and fences to snap and buildings to topple.

and down) to the direction of the wave as it travels around the stadium (sideways). Likewise, as an area of rock moves up or down when an S wave passes, the wave creates a drag on neighboring rocks, pulling them along as well. This movement has a shearing effect on the rocks, changing their shape but not their volume. During an earthquake, P waves strike with a succession of compressional jolts, but S waves impart a continuous wriggling motion.

Surface Waves

When an earthquake occurs, some body waves move outward from the focus and spread up toward the epicenter, where they cause the Earth's surface to vibrate. This vibration generates surface waves, which travel within the upper few kilometers of the Earth's crust. The slowest of the seismic waves, surface waves arrive last at earthquake-recording stations. They travel through the crust at a velocity of about 2.5 km (1.5 mi) per second. Two main types of surface waves exist: one having a side-to-side whipping motion like a writhing snake, and the other having a rolling motion that resembles ocean waves (Fig. 10-7). People who have experienced the second of these two types of surface waves have compared it with a brisk walk across a waterbed—some have even become "seasick." When both types of surface waves occur together, they cause objects to rise and fall while being whipped from side to side. This combination is responsible for most earthquake damage to rigid structures such as buildings, roadways, and bridges.

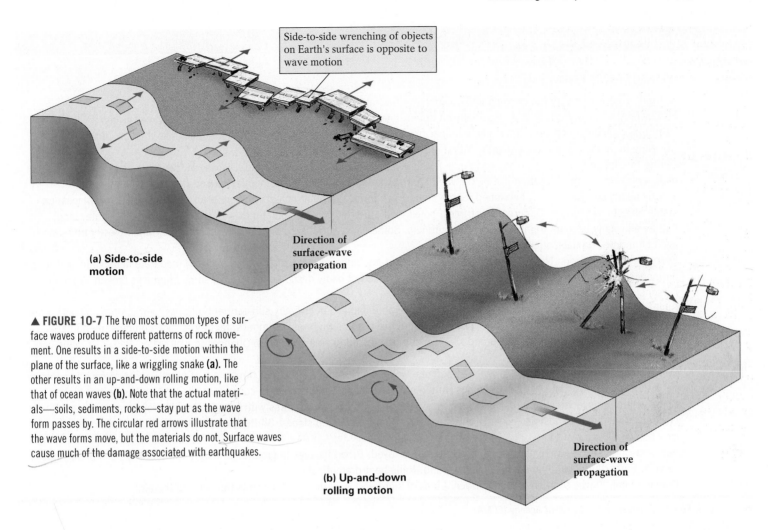

Side-to-side wrenching of objects on Earth's surface is opposite to wave motion

(a) Side-to-side motion

Direction of surface-wave propagation

▲ **FIGURE 10-7** The two most common types of surface waves produce different patterns of rock movement. One results in a side-to-side motion within the plane of the surface, like a wriggling snake **(a)**. The other results in an up-and-down rolling motion, like that of ocean waves **(b)**. Note that the actual materials—soils, sediments, rocks—stay put as the wave form passes by. The circular red arrows illustrate that the wave forms move, but the materials do not. Surface waves cause much of the damage associated with earthquakes.

(b) Up-and-down rolling motion

Direction of surface-wave propagation

Measuring Earthquakes— Which Scale to Use

When an earthquake occurs, citizens and the news media clamor for information about the tremor's strength. City planners and engineers, who need to know how human-made structures fare in quakes of different sizes, are also eager to classify a quake's strength. So too are seismologists, who must estimate how much of a fault's accumulated strain has been released and perhaps help predict the magnitude of future quakes. Calculating an earthquake's strength, however, has proved both challenging for scientists and confusing to the general public. Over the past century or so, geologists have used several different scales to classify the strength of earthquakes.

The Mercalli Intensity Scale

An earthquake's strength depends on how much of the energy stored in the rocks is released. Early efforts to classify earthquakes focused on their destructiveness, or *intensity*, as estimated from observable damage to property and from survivors' descriptions of ground-shaking. In 1902, the Italian seismologist Giuseppe Mercalli developed the **Mercalli intensity scale,** which correlates quake intensity to levels of damage to human-made structures and to eyewitness accounts of the seismic event. His classification scheme appears in Table 10.1.

Although the Mercalli intensity scale is sometimes used in earthquake planning and earthquake-hazard mapping, this type of descriptive scale cannot locate an earthquake's epicenter accurately, nor can it distinguish a strong earthquake that occurs a long distance away from a less severe event that occurs nearby. The Mercalli scale also cannot distinguish damage to structures that results from the actual intensity of ground motion from damage resulting from poor construction quality or related to the nature and stability of the local soil. Moreover, this scale can't be used at all in places where no people live, such as oceans or uninhabited regions of the continents. It is useful, however, in characterizing historical earthquakes that date from the time before sophisticated earthquake-measuring machines. For this reason, whenever possible, seismologists read the vivid written accounts of earthquake survivors from long ago.

Because it puts earthquakes into a human perspective, the Mercalli scale offers a tangible way to characterize the local effects of an earthquake. Several other scales that use *quantitative* measures of the amount of energy released by a quake, however, have proven more useful. Most rely at least in part on the use of a seismograph.

Table 10.1 Modified Mercalli Intensity Scale

INTENSITY VALUE	INTENSITY DESCRIPTION
I	Not felt, except by a very few persons under especially favorable circumstances.
II	Felt by only a few persons at rest, especially on upper floors of buildings. Delicately suspended objects may swing.
III	Felt quite noticeably indoors, especially on upper floors of buildings, but many people do not recognize it as an earthquake. Standing automobiles may rock slightly. Vibration resembles a passing truck.
IV	During the day felt indoors by many people, outdoors by few. At night some awakened. Dishes, windows, doors disturbed; walls make creaking sound. Sensation like heavy truck striking building. Standing automobiles rocked noticeably.
V	Felt by nearly everyone; many awakened. Some dishes, windows, and other objects broken; cracked plaster in a few places; unstable objects overturned. Disturbance of trees, poles, and other tall objects sometimes noticed. Pendulum clocks may stop.
VI	Felt by all; many become frightened and run outdoors. Some heavy furniture moved; a few instances of fallen plaster and damaged chimneys. Damage slight.
VII	Everyone runs outdoors. Negligible damage in buildings of good design and construction; slight to moderate damage in well-built ordinary structures; considerable damage in poorly constructed or badly designed structures; some chimneys broken. Noticed by persons driving cars.
VIII	Slight damage in specially designed structures; considerable damage in ordinary substantial buildings, with partial collapse; great damage in poorly built structures. Panel walls thrown out of frame structures. Chimneys, factory stacks, columns, monuments, walls collapse. Heavy furniture overturned. Sand and mud ejected in small amounts. Changes in well water. Persons driving cars disturbed.
IX	Damage considerable in specially designed structures; well-designed frame structures thrown out of plumb; great damage in substantial buildings, with partial collapse. Buildings shifted off foundations. Ground cracked conspicuously. Underground pipes broken.
X	Some well-built wooden structures destroyed; most masonry and frame structures with foundations destroyed; ground badly cracked. Rails bent. Landslides considerable from river banks and steep slopes. Shifted sand and mud. Water splashed, slopped over banks.
XI	Few (masonry) structures remain standing. Bridges destroyed. Broad fissures in ground. Underground pipelines completely out of service. Earth slumps and land slips in soft ground. Rails bent dramatically.
XII	Damage total. Waves seen on ground surface. Lines of sight and level distorted. Objects thrown into the air.

Source: U.S. Federal Emergency Management Agency (FEMA).

The Modern Seismograph

A **seismograph** is a device that measures the magnitude of an earthquake. It records and graphs the signals it receives from a **seismometer,** the earthquake-wave-detecting instrument that is buried in the Earth. Seismometers amplify seismic-wave motion electronically so that the seismograph can record even weak or distant disturbances. Seismometers are sensitive enough to detect passing traffic, high winds, and nearby crashing surf. They can even detect earthquakes that occur on the opposite side of the globe. Seismometers can also detect underground nuclear explosions and, as we'll see later, are used to monitor compliance with nuclear test-ban treaties.

After the seismometer senses the arrival of earthquake waves, the seismograph produces a **seismogram,** a visual record of the arrival times of the different waves and the magnitude of the shaking associated with them. Figure 10-8 illustrates the components of a seismograph and describes how this device works. Seismographs have been placed at more than 1000 stations around the world. By comparing recordings of a single earthquake at numerous stations, seismologists can locate an earthquake's epicenter and focus and determine how much energy the earthquake released.

The Richter Scale

By 1935 Dr. Charles Richter, head seismologist at the California Institute of Technology, had grown weary of local journalists' constant questions about the relative sizes of California's frequent earthquakes. In response, he devised a way to use seismograms to determine an earthquake's *magnitude,* or the amount of energy released by the tremor. The **Richter scale** defined a quake's magnitude in terms of the amplitude of the largest peak traced on a seismogram for:

- a Wood-Anderson seismograph;
- the location being a standard distance of 100 km (60 mi) from an earthquake epicenter;
- its occurrence being in *California.* (He calibrated his scale to reflect the properties of the rocks of the Earth's crust in California.)

To accommodate the vast range of earthquake sizes, Richter developed a logarithmic scale, in which each successive unit corresponded to a tenfold increase in the amplitude of the seismogram's tracings. By Richter's definition, a tremor of magnitude-1.0 would swing the arm of the Wood-Anderson seismograph one-thousandth of a millimeter. A magnitude-2.0 quake would swing

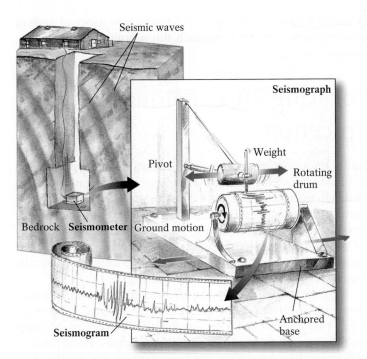

▲ **FIGURE 10-8** The functioning of a traditional seismograph. At its most basic level, a seismograph consists of a mass suspended by a spring or wire from a base that is firmly anchored. (The entire seismograph is usually encased in a protective box and bolted to the Earth's crust meters below the surface.) The base moves with the Earth during an earthquake, while the suspended mass remains motionless. When a seismic wave jostles the seismograph beneath the suspended pen, the pen records the shaking on a roll of paper that turns at a steady rate on a rotating drum anchored to the base, producing a tracing called a seismogram. The amplitude of the seismogram's squiggles is proportional to the amount of energy released by the earthquake. Seismographs can also be oriented vertically to record vertical ground movement.

the arm 10 times as much, or one-hundredth of a millimeter, and so on. Thus a magnitude-4.0 quake would trace a peak with a one-millimeter amplitude, a magnitude-5.0 would trace a 10-mm peak, and a peak of a magnitude-7.0 quake would correspond to a peak of a full meter.

The *actual* amount of energy released during tremors of different Richter magnitudes varies even more than their graphic peaks. Each one-unit increase in the Richter scale corresponds to about a 33-fold increase in energy released. Thus the energy released by a magnitude-6.0 earthquake is approximately 33 times greater than the energy released by a magnitude-5.0 quake and more than 1000 times greater (33 × 33) than the energy released by a magnitude-4.0 event.

Because the Richter scale uses logarithmic measures of energy, it requires some interpretation to compare earthquakes of different magnitudes. To demonstrate the difference in energy released, let's examine earthquakes that have occurred in metropolitan Los Angeles. Some seismologists have proposed that the area is struck by an earthquake of a Richter magnitude greater than 8.0 every 160 years or so. The last quake of this magnitude occurred there in 1857. You might think that an occasional magnitude-6.0 quake would relieve a significant amount of accumulated strain energy and forestall or even prevent the next "Big One." The energy released by a magnitude-8.0 earthquake, however, is 1000 times (33 × 33) that released by a magnitude-6.0

quake. Thus Los Angeles would need a swarm of a thousand earthquakes of magnitude-6.0 to dissipate the accumulated strain energy and avert the next magnitude 8.0 quake. As southern California experiences on average only one quake of magnitude-6.0 every five years, the region can still expect the occurrence of a seismic event 1000 times as violent as any in recent memory.

Although the Richter scale has been used for decades to describe earthquakes, most seismologists now use a different scale for measuring earthquake magnitudes because the Richter scale has several major problems:

- Although theoretically without an upper limit, the scale cannot effectively measure quakes with a magnitude of 7.0 or greater. The Richter scale generally underestimates the energy released by "major" (magnitude-7.0–7.9) and "great" earthquakes (greater than magnitude-8.0).

- More modern devices have replaced Wood-Anderson seismographs.

- The scale is less effective for crustal rocks that differ from those found in California. Because long-traveling seismic waves may pass through numerous regions of different crustal composition, the scale works best for "local" earthquakes within a few hundred kilometers of the seismograph. Seismologists must factor in a number of corrections for earthquakes that occur beyond this distance.

Although the well-known Richter scale continues to be widely cited, it seems to work best for *earthquakes of magnitude-7.0 or less in California.* Consequently, the U.S. Geological Survey and other agencies have recently adopted another scale—the moment-magnitude scale—that consistently measures the magnitude of earthquakes (including truly large ones) and the energy they release, but is unaffected by variations in rock type.

The Moment-Magnitude Scale

The magnitude of an earthquake is a function of the length and depth of fault rupture, the average amount of slip (movement) on the fault, and the strength of the faulted rock. During the past few decades, seismologists have discovered that they can more accurately gauge a large earthquake's total energy by measuring a quantity called the **seismic moment,** defined as:

Moment = (total length of fault rupture)
× (depth of fault rupture)
× (total amount of slip along rupture)
× (strength of rock)

The values produced through this equation generate the **moment-magnitude scale.** In general, the longer the fault, the greater its potential for producing a large quake. Similarly, the stronger the rock, the greater its potential to store strain energy and thus generate large quakes. (Recent studies in southern California indicate that the subsurface presence of soft, warm rocks, such as schists, may actually somewhat reduce the region's potential for extremely high-magnitude earthquakes.)

One principal benefit of the moment-magnitude scale is that seismologists can now measure the size of the earthquake *directly*

Table 10.2 Earthquakes and Their Energy Equivalents

MOMENT MAGNITUDE	APPROXIMATE ENERGY EQUIVALENT
−2	100-watt lightbulb left on for a week
0	1-ton car going 40 km (25 mi) per hour
2	Amount of energy in a lightning bolt
4	1 kiloton (1000 tons) of explosives
6	Hiroshima atomic bomb
8	1980 eruption of Mount St. Helens
10	Annual total U.S. energy consumption (largest recorded earthquake—9.5, Chile, 1960)

from its cause (the rupture in the rocks and the distance that the rocks have moved) rather than solely from its effect (the tracing of seismic waves on a seismograph). To determine length and depth of rupture and total slip requires either field observations (if the rupture reached the surface) or analysis of the precise location of the main shock and its aftershocks (if it failed to break the surface). Because of the time required immediately after an earthquake to complete these analyses, scientists generally first estimate a *preliminary magnitude* based on the seismographic tracings. (They do this by summing the amplitudes of the peaks of these tracings.) Later, after more thorough analysis, an updated moment magnitude will be announced that may differ— although usually not substantially—from the preliminary estimate. Most earthquake magnitudes reported in today's newspapers are moment magnitudes—a more accurate measure of the energy released during a seismic event.

The largest moment-magnitude ever calculated was a 9.5 in the great Chilean quake of 1960 (its Richter magnitude was *only* 8.5). During the 2004 Sumatran quake, the initial moment magnitude was reported as 9.0, but subsequent analysis has led some seismologists to revise the magnitude upward to 9.3, which, given the logarithmic scale, releases three times the energy of a 9.0.

Earthquake Magnitudes on a Human Scale

In their attempts to quantify the strength of earthquakes for the general public, Richter and his seismological colleagues introduced the principle of a logarithmic scale. To simplify matters, Table 10.2 illustrates various earthquake magnitudes and the energy they

Table 10.3 Frequency of Earthquake Occurrence Based on Observations Since 1900

DESCRIPTION	MAGNITUDE	ANNUAL AVERAGE NUMBER OF EVENTS
Great	8.0 or higher	1
Major	7.0–7.9	18
Strong	6.0–6.9	120
Moderate	5.0–5.9	800
Light	4.0–4.9	6200 (estimated)
Minor	3.0–3.9	49,000 (estimated)
Very Minor	2.0–3.0	1000/day
	1.0–2.0	8000/day

release in human terms. Table 10.3 shows how many of each of these events we can expect annually throughout the world.

Locating an Earthquake's Epicenter

During the last 40 years, increased international cooperation in the field of seismology has led seismologists to use similar equipment and procedures and share data freely. Since the early 1960s, stations from many nations have fed their data directly to the Earthquake Information Center of the U.S. Geological Survey in Golden, Colorado. The Center's high-speed computers continuously analyze incoming signals, quickly locating earthquake epicenters and calculating quake magnitudes.

Geologists use the different velocities of the various types of seismic waves to determine the location of a quake's epicenter. Just as two sprinters who set out together but run at different speeds will reach the finish line at different times, P and S waves from an earthquake start at the same moment but reach a seismological station at different times. For example, a seismograph located 600 km (375 mi) from an earthquake's epicenter might begin to record P waves about 100 seconds after the quake begins and S waves about 70 seconds later. Because we know the velocity at which each type of wave travels through various rock types, we can use the difference between their arrival times and our knowledge of the rocks through which they traveled to calculate the distance from the recording stations to the earthquake's epicenter.

By comparing the arrival times of seismic waves from earthquakes of known origin, recorded at stations around the world, geologists have developed *travel-time charts* that convert the lag between P- and S-wave arrival times into the distance to a quake epicenter. For quakes of unknown origin, the gap between the arrival times of the two types of waves is used to calculate the distance from the recording stations to the earthquake's epicenter. Figure 10-9a shows how seismologists work with data on P waves and S waves gathered at different locations to find the quake's epicenter.

The linear distance from an earthquake to a single seismic station will not by itself identify the location of the epicenter; the quake could have occurred at that distance from the station *in any direction*. (This distance may be represented on a map as the radius of a circle that has the station at its center.) When similar circles derived from the lag times in P- and S-wave arrivals are drawn from three different stations (see Figure 10-9b), they intersect at the quake's epicenter. To pinpoint the precise location of an earthquake epicenter, then, at least three seismic stations must record the event. (Note that the three circles will intersect if the focus of the earthquake is very close to the Earth's surface—that is, if the focus lies within a few kilometers. Deeper foci produce a triangle around the quake's epicenter, the size of which is proportional to the earthquake's depth.)

The value of being able to locate an earthquake's epicenter was clearly demonstrated on May 23, 1989, when seismographs around the world detected a "great" earthquake—magnitude-8.3.

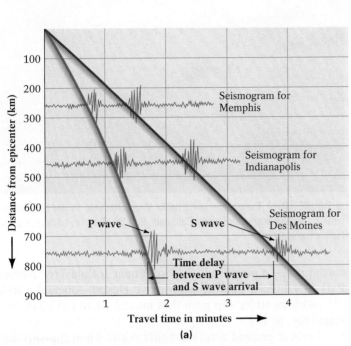

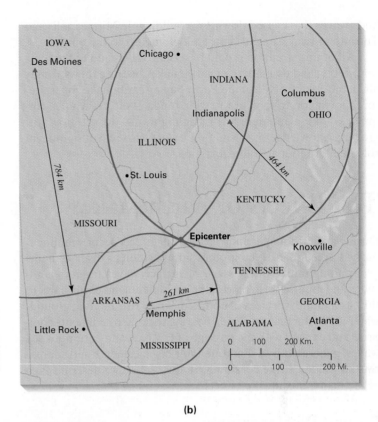

(a)

(b)

▲ **FIGURE 10-9 (a)** Seismograms from various seismographic stations, recording the arrival times of P and S waves at each station from an earthquake in the New Madrid, Missouri, area. The closeness of the two wave types at the Memphis, Tennessee, recording station indicates that the earthquake occurred nearby. Using our knowledge of the velocities of P and S waves through the Earth's crust, and the 120-second and 186-second delays at Indianapolis, Indiana, and Des Moines, Iowa, respectively, we can determine each station's distance from the earthquake epicenter. **(b)** By drawing a circle around each of these stations on a map, with each circle's radius proportional to the station's distance from the quake's epicenter, we can determine the location of the epicenter—it is marked by the single point at which the three circles intersect on the Earth's surface. Computers are used to determine epicenters rapidly and accurately.

Moments after the quake, a worldwide network of more than 100 state-of-the-art seismological facilities transmitted data to high-speed computers at the U.S. Geological Survey facility at Golden, Colorado. Within minutes seismologists determined that the earthquake had occurred along the Macquarie Ridge, a submarine mountain chain 800 km (500 mi) southwest of New Zealand. As we will discuss shortly, a powerful quake that strikes the ocean floor often produces enormous sea waves that can devastate coastal regions. A sea-wave alert was issued for coastal communities around the Pacific Ocean basin, allowing the successful evacuation of all people in those areas.

Earthquake Depth and Magnitude

When an earthquake occurs, news accounts generally identify its location by reporting the nearest town or city. In truth, an earthquake occurs at some depth below the town, at the quake's focus. We describe an earthquake as "shallow" if its focus lies less than 70 km (45 mi) below the surface, "intermediate" if its focus is 70–300 km (45–180 mi) below the surface, and "deep" if its focus

is more than 300 km (180 mi) below the surface. Earthquakes have been known to occur as deep as 700 km (435 mi) below the Earth's surface, though the foci of more than 90% of all earthquakes are less than 100 km (60 mi) deep and the vast majority of catastrophic quakes originate within 60 km (40 mi) of the surface.

Few large earthquakes occur at depths greater than 100 km (60 mi) because heat weakens deeper rocks, causing them to lose some of their ability to store strain energy. The rocks within a few tens of kilometers of the surface are typically brittle and elastic enough to accumulate a vast amount of energy; thus they are far more likely to produce large quakes. The depth of the focus of the 1964 Good Friday earthquake in Alaska (magnitude-9.2), for instance, was 33 km (20 mi). The focus of the Sumatran earthquake of 2004 was only 10 km (6 mi).

General limits apply to the magnitudes of quakes at different depths. Shallow earthquakes have been recorded with magnitudes as high as 9.5, and intermediate ones have magnitudes that can reach 7.5. Deep quakes, however, are rarely recorded with magnitudes greater than 6.9. The magnitude-8.2 earthquake that struck Bolivia in June 1994 did occur at a depth of almost 700 km (400 mi). Seismologists continue to ponder how a quake that deep, occurring in warm, plastic rocks, could store enough energy to be that powerful. On a human note, this quake produced body waves that caused downtown buildings in Minneapolis and Toronto to sway, several thousand kilometers to the north.

All of this attention is paid to seismic waves and to the location, depth, and magnitude of earthquakes because they strongly affect the quality of life in the Earth's many earthquake zones. Before moving on to look at the wide range of effects from earthquakes, take a moment with Highlight 10-1 to see how our growing knowledge of seismic waves can also play a significant role in international political affairs.

The Effects of Earthquakes

During an earthquake, many geological effects occur simultaneously: Large areas of the ground shift position as displacements occur along faults; the passage of surface waves triggers landslides and mudflows; in places, the seemingly solid earth starts to flow as if liquid; surfaces of bodies of water rise and fall; groundwater levels fluctuate as water-bearing rocks and sediments compress and expand with the passage of P waves; and enormous sea waves, generated by submarine fault displacements and submarine landslides, strike coastlines. In this section we describe each of the major earthquake effects.

Ground Displacements

Among the most obvious geological effects of an earthquake is large-scale displacement of the landscape. After fault blocks have unlocked and moved, unleashing pent-up energy, geological features and geographical landmarks—such as those in Figure 10-10—are often noticeably displaced. During the 1906 San Francisco quake, for example, a 432-km (270-mi)-long rupture opened along the strike-slip San Andreas fault. (Strike-slip and dip-slip faults are discussed in Chapter 9.) The west side of the fault then lurched laterally northward, 7 m (25 ft), as determined by measuring displacements of fences and other linear features near the epicenter. More recently, on October 16, 1999, the Hector Mine

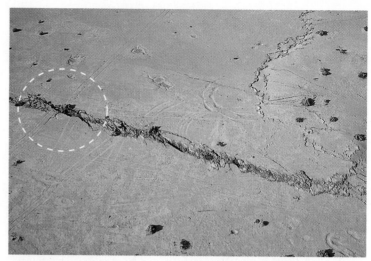

▲ **FIGURE 10-11** The magnitude-7.1 Hector Mine earthquake created this surface break across San Bernardino County, California. Note the offset tire tracks across the fault in the upper left.

earthquake—a magnitude-7.1 event near Joshua Tree National Park in southern California's Mojave Desert—opened a 40-km (25-mi)-long strike-slip gash that moved 4.6 m (15 ft) in an instant (Fig. 10-11).

Vertical ground displacements occur when dip-slip faults move. During the Good Friday earthquake of 1964, reverse faulting thrust 500,000 km^2 (193,000 mi^2) of the Alaskan mainland and the adjacent seafloor upward. The maximum upward land shift totaled 12 m (39 ft). In some places, the seafloor dropped as much as 16 m (52.5 ft). While coastal locations that fell below sea level suffered extensive flooding, uplift at the coast left some harbors, fisheries, and docks high and dry, requiring relocation of those facilities. Similarly, the 2004 Sumatran quake produced an average upward sea-floor shift of 5 m (16.5 ft), displacing the sea surface above the faulting and generating the tragic tsunami.

Landslides and Liquefaction

A large earthquake's violent shaking often dislodges large masses of unstable rocks and soils from hillsides, causing them to rush downslope. Such *landslides* and *rockfalls* (discussed in Chapter 14) may also occur as rock fragments detach from bedrock, thick layers of sedimentary rock slide along weak bedding planes, foliated metamorphic rock shifts along foliation planes, or loose sediment cascades down steep slopes. Such quake-generated landslides can occur on land, as can be seen in Figure 10-12, or on the seafloor. An earthquake that struck the Grand Banks of Newfoundland in 1929 spawned a submarine slide that severed many of the trans-Atlantic telephone cables, halting intercontinental telephone service.

Earthquakes also cause **liquefaction,** the conversion of unconsolidated sediment with some initial cohesiveness and strength into a mass of water-saturated sediment that flows like a liquid *although no water has been added.* When fine-grained, moist sediment is shaken during an earthquake, the shaking increases the pressure on the water between sediment grains, forcing them apart. Without frictional contact between adjacent grains, the

▲ **FIGURE 10-10** Wallace Creek, on the Carrizo Plain in southern California, offset by the San Andreas Fault.

Highlight 10-1

Seismic Eavesdropping—Monitoring the Peace

In an international climate characterized by increasing anxiety regarding "who has them and who doesn't," the world's community of nations adopted the Comprehensive Test Ban Treaty (CTBT) in 1996. It represents a global attempt to monitor all nuclear testing. The product of nearly 40 years of heated discussion and negotiation, this treaty bans *all* nuclear detonations. The difficult task of monitoring compliance with this treaty falls largely to seismologists and their hypersensitive seismometers.

A new 170-station seismic network—the International Monitoring System—forms the backbone of the detection system. At present, the system can detect unmuffled explosions with yields as low as 1 kiloton of energy, roughly equivalent to a magnitude 4.0 earthquake. (A kiloton equals 10^{12} calories of energy; the first U.S. nuclear detonation—the Trinity test in Alamogordo, New Mexico—had a yield of 21 kt.) The system can also pinpoint detonations within an area of 1000 km², equal to a circle with a radius of 18 km (11 mi). The main trick is to distinguish nuclear

blasts from other seismic signals, such as those sent out by actual earthquakes or conventional explosions such as mining blasts.

At higher energies (greater than magnitude-4.0), the seismograms of earthquakes and nuclear detonations differ noticeably. Consider the seismic record of a recent French nuclear test in the South Pacific, equivalent to a magnitude-5.0 quake. The shape of the largest peak reveals the blast's true nature. An explosion, which releases virtually all of its energy in less than a second, generates all of its energy waves instantaneously, causing the peak to be quite sharp. It then tapers off smoothly and uniformly as the energy dissipates. By comparison, the peaks on the seismograms of earthquakes appear "sloppy." The ground may begin to vibrate *before* the main shock (called *foreshocks*) and then continue to vibrate for seconds or minutes after the main shock as the rocks continue to slip and adjust. This pattern creates a complex mix of seismic waves moving away from the quake's focus at different velocities. As a result, the seismogram of a magnitude-

5.0 earthquake starts off slowly, grows to its peak erratically, and then decays unevenly—a pattern easily distinguished from that of a nuclear explosion.

At present, an average of 60 seismic events per day are analyzed by the International Data Center (IDC) in Arlington, Virginia, the prototype for the treaty's detection facility. One recent event in Lop Nor, the Chinese test site in central Asia, was closely scrutinized by the IDC. The relatively small shock, about magnitude-3.5, could have been a low-yield detonation. The seismic network pinpointed its source at a location several hundred kilometers from Lop Nor, and the seismographic pattern indicated that the event was a small earthquake, not a nuclear test. This demonstrated the value of the system that the world community hopes will maintain nuclear peace. This system was again used in May 1998 to determine the number and yield of nuclear detonations by India, a country that did not join the other 149 nations in the test-ban treaty.

▲ **FIGURE 10-12** Aerial view of a landslide spawned by the Denali, Alaska earthquake (magnitude-7.9) in 2002. Slide debris covers the Black Rapids Glacier.

sediment loses its cohesiveness and strength, and once-firm ground becomes a slurry of mud. When buried sand deposits suddenly behave like liquids during violent tremors, they often produce *sandblows,* fountaining geysers of sand and water that become pressurized, burst through overlying dry sediments, and erupt at the surface. These eruptions typically produce a cone of sand shaped like a miniature volcano. Sandblows from great earthquakes may range from 20–60 m (65–195 ft) in diameter and rise to heights of 1 m (3 ft).

During Alaska's 1964 quake, the sediment beneath the Turnagain Heights neighborhood of Anchorage liquefied (Fig. 10-13). The devastated ruptured landscape, which was rendered worthless for future development, was eventually bulldozed into Earthquake Park. During the Loma Prieta quake, liquefaction of the bay muds and landfill soils beneath the Nimitz Freeway in the Oakland area and the Marina district of San Francisco caused many structures to collapse and was responsible for much of that quake's death toll. Liquefaction also contributed significantly to the widespread destruction in the Kobe, Japan, earthquake in 1995 (Fig. 10-14).

The combined effects of liquefaction and landsliding can cause large segments of land surface to move significant distances. On June 7, 1692, a relatively moderate quake sent the *entire* city of Port Royal, on the Caribbean island of Jamaica, sliding into

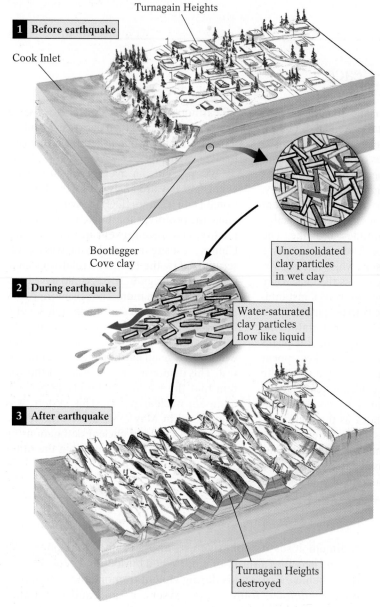

1 **Before earthquake**

Turnagain Heights

Cook Inlet

Bootlegger Cove clay

Unconsolidated clay particles in wet clay

2 **During earthquake**

Water-saturated clay particles flow like liquid

3 **After earthquake**

Turnagain Heights destroyed

▲ **FIGURE 10-13** Liquefaction occurs when the grains in a layer of wet, fine-grained sediment are shaken during an earthquake. In 1964 the clay layer beneath the entire Turnagain Heights area of Anchorage, Alaska, liquefied, producing a chaotic landscape of jumbled homes and asphalt blocks.

the sea. The town finally came to rest beneath 15 m (50 ft) of water. The picturesque city—infamous as a base for numerous pirates and other scoundrels—was built on loose, wet, steeply sloping sediment that was prone to liquefaction. When marine archaeologists explored the underwater remains of the city, it lay protected and virtually intact beneath 3 m (10 ft) of marine silt. In one kitchen, a copper kettle still contained the evening's meal of turtle soup. Eyewitnesses described the scene in the following way:

> "Whole streets (With Inhabitants) were swallowed-up by the opening Earth, which then shutting upon them, squeezed the People to Death. And in that Manner several are left buried with their Heads above Ground; only some Heads the dogs have eaten; others are covered with Dust and Earth, by the People who yet remain in the Place, to avoid the Stench."

Seiches

Seismic waves cause the water in an enclosed or partially enclosed body of water, such as a lake or bay, to move back and forth across its basin, rising and falling as it sloshes about. This phenomenon is called a *seiche* (pronounced "SAYSH"). In Alaska's 1964 Good Friday quake, the water at one end of Kenai Lake, south of Anchorage, rose 9 m (30 ft), overflowed its banks, and flooded inland before reversing its direction. The back-and-forth motion of the water stripped the soils around the lake down to bare bedrock. During Montana's Madison Canyon quake of 1959, the Hebgen Lake dam overflowed because of a seiche that oscillated back and forth across the lake every 17 minutes for 11 hours.

Seiches may develop at great distances from powerful earthquakes—sometimes so far from the epicenter that no other effects of the quake are felt. The great Alaska earthquake, for

(a)

(b)

◀ **FIGURE 10-14** Damage by liquefaction. **(a)** This elevated freeway fell over during the 1995 earthquake in Kobe, Japan, when some of the ground beneath it liquified. **(b)** This apartment house in a Taipei suburb fell over as the ground beneath it liquefied, killing forty people during the tragic earthquake on Taiwan in September 1999. Its architect and builder were arrested for violating Taiwan's strict construction codes.

example, set off seiches in reservoirs as far away as Michigan, Arkansas, Texas, and Louisiana. Water rose and fell 2 m (6.5 ft) in bays along the Texas Gulf Coast, damaging boats by buffeting them with waves. Even some swimming pools in Texas, 6000 km (4000 mi) from the quake's epicenter, developed seiches.

Tsunami

According to a Japanese folktale, a venerable grandfather who owned a rice field at the top of a hill felt the sharp jolt of a tremor one day just before the harvest. From his hilltop vantage point, he saw the sea pull back from the shore, drawing curious villagers to explore the exposed tidal flats and collect shellfish. From experience, the old man knew of the grave danger to his neighbors. With his grandson by his side, he dashed about his fields, setting fire to his crop. The villagers saw the smoke, and hurried up the hill to aid their neighbor. As they beat out the flames, they saw the old man scurrying ahead, setting new fires near the hill's crest. Hoping to prevent him from destroying all of his crops, they rushed up the hill to stop him. Moments later, the villagers saw a tremendous wall of water surging onshore, flooding the flats where they had just been standing, and they then understood that the old man had sacrificed his harvest to save their lives. More than any other country, Japan, with its thousands of kilometers of island coastlines nestled precariously within one of the Earth's principal seismic zones, has suffered from **tsunami** (Japanese for "harbor wave"). In open water, tsunami ("tsunami" is both singular and plural) can move at speeds exceeding 800 km/hr (500 mph), making them far faster than ordinary wind-generated waves.

Out at sea, tsunami are barely perceptible, typically only about 1 m (3 ft) high. As the crests of successive waves can be separated by as much as 160 km (100 mi), these waves appear to be mere bumps on the sea. Only in the shallow waters of enclosed bays and harbors, where drag against the ocean floor causes the fast-moving waves to bunch up, do tsunami become these devastating walls of water. In 1739, at Cape Lopatka on Kamchatka Island in the northeastern Pacific, a tsunami reached a height of 65 m (210 ft), the equivalent of a 20-story building.

Many tsunami originate from a large, rapid displacement of the seafloor during submarine faulting, when the seafloor either drops by normal faulting or rises by reverse faulting. (These waves may also result from submarine landslides.) When a section of the seafloor drops downward, such as typically happens in areas of normal faulting within divergent plate boundaries, a trough in the ocean surface appears, temporarily drawing water from the coast. A bump in the sea surface is created as water rushes in to fill the trough, overcompensates and forms a wave that then travels to land as a potentially catastrophic crest—a tsunami (Fig. 10-15a). When a section of the seafloor is thrust upward, such as

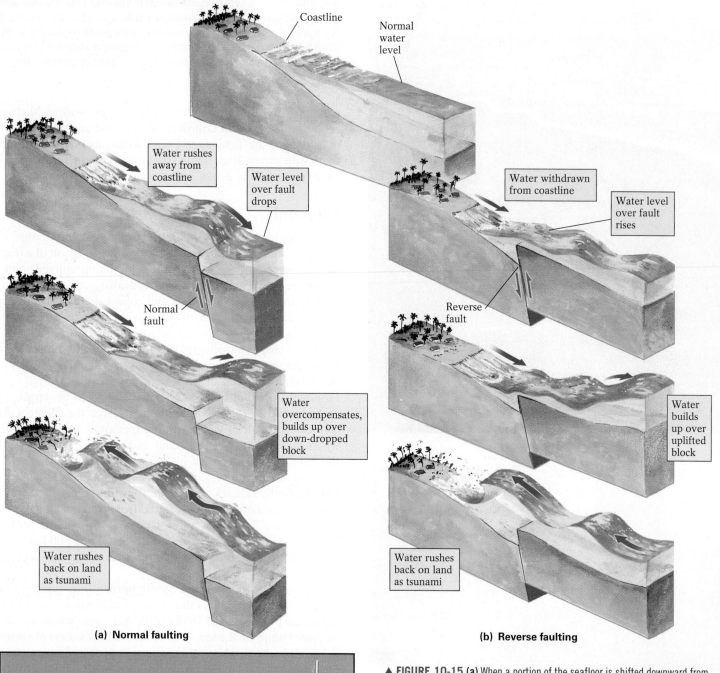

Coastline

Normal water level

Water rushes away from coastline

Water level over fault drops

Water withdrawn from coastline

Water level over fault rises

Normal fault

Reverse fault

Water overcompensates, builds up over down-dropped block

Water builds up over uplifted block

Water rushes back on land as tsunami

Water rushes back on land as tsunami

(a) Normal faulting

(b) Reverse faulting

(c) Incoming wave at Phuket, Thailand

▲ **FIGURE 10-15 (a)** When a portion of the seafloor is shifted downward from normal faulting, a depression forms in the sea surface. A tsunami is generated as water surrounding the depression rushes in to fill it and overcompensates, forming a "bump" in the sea surface that is propagated in all directions as a large wave. **(b)** When seafloor uplift occurs from reverse faulting, the sea surface is displaced upward and water is withdrawn from the shore, forming a tsunami. **(c)** Here is the sight of the incoming tsunami that devastated the coastal resort of Phuket, Thailand, in late December of 2004. Note that the wave appeared as a churning elevated surface, but not as a single giant crest. This tsunami behaved like a rapidly rushing and rapidly rising tide that just kept rising until it reached an elevation of 25 to 30 m (about 80 to 90 ft).

typically happens in areas of reverse faulting within convergent plate boundaries, the movement displaces water upward, pulling along water from the shore; the wave later arrives at the coast as a tsunami (Fig. 10-15b). Because their motion is principally lateral rather than vertical, transform faults on the seafloor don't typically generate tsunami. Such was the case on June 16, 2005, when a magnitude-7.2 quake struck off the coast of Northern California, about 120 km (80 mi) west of Eureka. A tsunami warning was issued because of the size of the quake but it was quickly canceled (within an hour) when seismologists determined that the motion was of the transform-type, thus saving many communities from the need to evacuate low-lying coastal regions.

Recent thoughts about the origins of tsunami have also focused on submarine landslides in addition to submarine faulting as a primary cause. Mapping of the seafloor topography off the coast of New Guinea has revealed a massive carpet of thick landslide deposits accumulated at the time of the devastating 1998 tsunami. This finding suggests that a landslide—even one triggered by a moderate-sized quake—can cause a deadly tsunami.

In most cases, tsunami consist of several waves that arrive at irregular intervals over several hours. Moreover, these giant waves can travel great distances from a quake epicenter and generally go undetected before striking a coast. In 1960, a tsunami generated by a catastrophic quake in southern Chile struck the Hawaiian shore at Hilo seven hours later, taking 61 lives. Twenty-two hours after the quake, the tsunami had traveled 17,000 km (11,000 mi) to reach Honshu and Hokkaido in Japan, where it took 180 lives. Figure 10-16 shows the damage inflicted on the giant sculptures of Easter Island. Highlight 10-2 dramatically illustrates how a far-traveled tsunami that occurred long ago may help prepare today's cities on North America's west coast for its next great earthquake.

Tsunami that strike a coast at high tide often exact a greater human toll than the earthquakes that generate them: Of the 131 deaths attributable to the 1964 Alaskan quake, 109 resulted from tsunami that swamped the southern Alaskan coast. Tsunami from that quake also hit the coasts of Oregon and California, causing 16 deaths there. At Crescent City, California, the first wave arrived at low tide, cresting about 4 m (14 ft) above sea level, destroying several lumber boats and releasing a morass of giant redwood logs. After three additional small waves had passed, residents returned to the coast, believing the worst was over. Unfortunately, a fifth wave struck at high tide, cresting at more than 10 m (33 ft) and washing 12 people out to sea.

Tsunami are certainly a scary prospect for the millions of Americans in the 512 U.S. cities and towns located within range of this quake-related hazard. For their safety, and the safety of millions more who live on the islands and along the coasts of the landmasses that surround the Pacific Ocean basin, the National Oceanographic and Atmospheric Administration (NOAA) has developed a tsunami early-warning system. Consider the value of this project: On November 16, 2003, a magnitude-7.5 earthquake struck the seafloor near Alaska's Little Sitkin island. Within 25 minutes, NOAA issued a tsunami warning using the system's seafloor tsunami-detection sensors (which "feel" the added pressure of a passing tsunami overhead). When the sensor revealed that the tsunami only measured about 2 cm (less than an inch) in height, the warning was canceled and an evacuation of hundreds of communities was averted. NOAA estimated that nearly $70 million would have been lost in Hawai'i alone due to disruptions in the local economies if the evacuation had proceeded. Quite a bargain, given that the entire tsunami-warning program with its six seafloor sensors cost about $18 million.

One final note about tsunami: According to a NOAA official, if you live in a coastal area, you should live by this principle: "If you can *see* the ocean and you *feel* the earthquake, run like hell *away* from the water." Even a two-minute headstart should provide enough time for most people to move inland and upslope far enough to get out of harm's way (Fig. 10-17).

Fires

On September 1, 1923, a powerful earthquake struck the Tokyo–Yokohama area of Japan at midday, overturning thousands of open cooking fires. Fanned by the day's high storm winds, the fires merged and spread until, by evening, the glow

▲ FIGURE 10-16 A tsunami that originated with Chile's powerful 1960 earthquake upended the renowned sculptures of Easter Island, located in the Pacific Ocean west of South America.

▲ FIGURE 10-17 The famous print "Beneath the Waves Off Kanagawa," by Japanese artist Katsushika Hokusai (1760–1849), depicts the helplessness of a group of boatmen in the face of tsunami-size waves.

Highlight 10-2

A Mysterious Tsunami Strikes Japan

Sometime around midnight on January 27, 1700, a 2-m (7-ft)-high tsunami struck Japan's Pacific Coast. The tsunami, cresting at about 3 m (10 ft), swept the wintry sea inland, swamping coastal homes and their surrounding rice paddies. In all, 13 houses were destroyed. Yet Japan itself had not been struck by an earthquake that night. The identity of the unknown seismic culprit—a large earthquake somewhere within the Pacific Basin—had, until recently, never been identified. Research by seismologists at the University of Tokyo and the Geological Survey of Japan now points the guilty finger at the Cascadia subduction zone, a fault system that extends from Vancouver Island, British Columbia, to northern California.

Looking for the source of Japan's mysterious 1700 tsunami, Japanese seismologists searched the historical records of earthquakes in other Pacific Rim regions. None was found, for example, in South America, where Spanish colonists and explorers kept thorough records. They were left with vast areas of Alaska, the Aleutians and Kuril Islands, and western North America as possibilities. Unfortunately, those areas lack written historical records from the time of the quake. The seismologists then consulted the work of 19th-century ethnographers—anthropologists who interview a region's inhabitants and compile their histories through their stories and ancient oral traditions. Although no stories of an earthquake around 1700 come from Alaska or other Northern Pacific lands, dozens of Native American legends describe a quake and tsunami that struck the Pacific Northwest's coast at the appropriate time.

The next piece of evidence came from teams of muck-covered geologists, who combed the marshlands of coastal Washington for physical evidence of a great earthquake that may have struck the Northwest about 300 years ago. In some places, broad areas of the coast have dropped several meters below sea level, swamping and killing stands of red cedar and Sitka spruce. Covering these fallen surfaces is a thin, continuous sheet of sand left behind by what has been estimated as a 10-m (33-ft)-high tsunami. Carbon-14 dating of organic deposits within these sediments and tree-ring dating of the drowned trees places these events within a few decades of 1700, making them a solid suspect in the Japanese earthquake mystery. Compelling evidence also comes from several living trees in the area. Tree rings preserved in the roots of numerous Sitka spruce trees reveal that they were partially submerged sometime between the growing seasons of 1699 and 1700.

Working on the assumption that the Japanese tsunami and the northwestern quake were caused by the same event, Japanese seismologists have estimated that their relatively small tsunami could have been a faint signal of the Northwest's last great quake. Using computer models that derive earthquake magnitudes from the size of resulting tsunami, they estimate that a tsunami spawned by a magnitude-8.0 quake, traveling 8000 km (5000 mi) from the Cascadia subduction zone, would arrive as a 40-cm (16-in.)-high wave 10 hours after the initial tremor. Although such an arrival time is consistent with Native American stories of an early evening quake, the resulting tsunami would be far too small. If a tsunami generated by a magnitude-9.0 quake traveled the same distance, however, it would measure 2 m (7 ft) high, precisely the height of the wave that struck Japan that January night in 1700.

In light of the emerging story of Japan's 1700 tsunami, some North American seismologists now speculate that the 1000-km- (600-mi)-long Cascadia subduction zone may be capable of rupturing simultaneously along its full length, generating an earthquake of magnitude-9.0 or higher. (A similar event occurred off the coast of Chile in 1960 when the Andean subduction zone ruptured, generating history's largest recorded earthquake—magnitude-9.5. The resulting tsunami sped across the Pacific, striking the Japanese coast and taking 140 lives.) In light of this creative seismological detective work, the Pacific Northwest must accept the possibility that an earthquake of magnitude-9.0 or greater that shakes for three to five minutes may send a 10-m-high tsunami onshore in the future.

---

from the blaze was bright enough to read by as far away as 15 km (10 mi) from the city. Before residents managed to extinguish the flames, 500,000 buildings in the mostly wood-frame city were destroyed and 120,000 people perished.

Each year, Japan observes a national day of earthquake preparedness in memory of this tragedy. After the earthquake, Tokyo was rebuilt with heightened earthquake awareness—with broader avenues to aid passage of fire-fighting equipment and concrete (rather than wooden) buildings. Millions of liters of water are now stored underground in quake-resistant structures, and the city's 30,000 taxis are each equipped with fire extinguishers.

Earthquake damage to gas mains, oil tanks, and electrical power lines typically ignites fires. Because water mains are frequently ruptured as well, an area's fire-fighting capability may suffer. During the 1906 earthquake, San Francisco's wooden buildings fed a blaze that caused 80% to 90% of the city's damage (Fig. 10-18). The fire raged uncontrolled for three days, largely because water lines had broken into hundreds of segments that subsequently lost pressure. (Emergency water supplies had apparently been stockpiled, but the San Francisco Fire Chief, the only person with access to their location, was in a hospital, unconscious, from a blow from falling debris.) Finally, workers created a firebreak by dynamiting rows of buildings. Sadly, it is believed that the main blaze was initially set several hours after the quake by a woman on Hayes Street, who, shaken by the events of the day, attempted to return to normal life by preparing a ham-and-eggs breakfast. Unfortunately, the flue of her stove had been damaged during the earthquake, and the resulting fire consumed more than 500 square blocks of the city's central business district. As a side note, San Francisco now stockpiles water underground at its major intersections.

10.1 Geology at a Glance

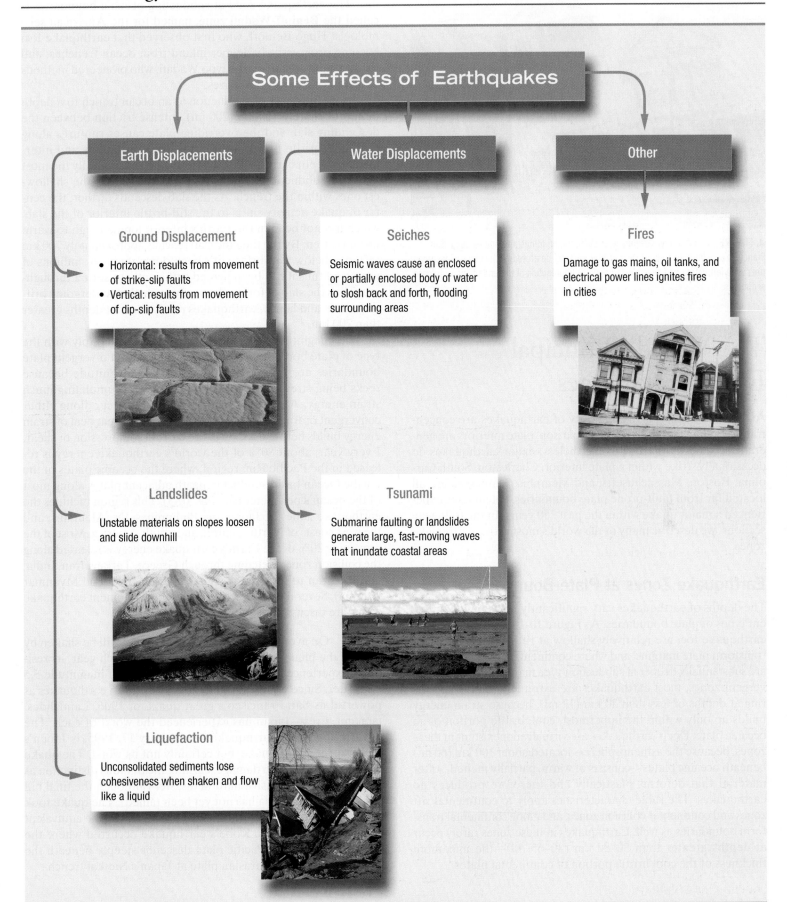

Some Effects of Earthquakes

Earth Displacements

Ground Displacement

- Horizontal: results from movement of strike-slip faults
- Vertical: results from movement of dip-slip faults

Landslides

Unstable materials on slopes loosen and slide downhill

Liquefaction

Unconsolidated sediments lose cohesiveness when shaken and flow like a liquid

Water Displacements

Seiches

Seismic waves cause an enclosed or partially enclosed body of water to slosh back and forth, flooding surrounding areas

Tsunami

Submarine faulting or landslides generate large, fast-moving waves that inundate coastal areas

Other

Fires

Damage to gas mains, oil tanks, and electrical power lines ignites fires in cities

▲ **FIGURE 10-18** In the earthquake of 1906, the buildings and homes along San Francisco's Fisherman's Wharf District that were not destroyed by the city's widespread fires were left listing and severely damaged by liquefaction of the underlying wet, loose sediments.

The World's Principal Earthquake Zones

As we have seen, the vast majority of earthquakes are concentrated at plate boundaries. By comparison, plate interiors are generally seismically inactive. Nevertheless, major earthquakes do occasionally strike within a plate interior. Charleston, South Carolina, Boston, Massachusetts, and Memphis, Tennessee are all located far from fault-prone plate boundaries, yet all have experienced a major quake within the past 250 years. In the following sections, we describe many of the world's most active earthquake zones.

Earthquake Zones at Plate Boundaries

The depths of earthquakes vary significantly between the different types of plate boundaries. As Figure 10-19 shows, in general, earthquake foci are relatively shallow at rifting, divergent, and transform plate margins, and where continental plates collide, but are substantially deeper at subduction-type margins. At oceanic divergent zones, most earthquakes are extremely shallow, occurring at depths of less than 20 km (12 mi), because strain energy builds up only within the uppermost, cool, brittle portion of an oceanic plate. Deep earthquakes are virtually nonexistent in these zones, because the asthenosphere—located about 100 km (60 mi) beneath oceanic plates—consists of warm, partially melted, softer material that deforms plastically and therefore produces no earthquakes. The same characteristics apply to continental rift zones and continental-collision zones, and along continental transform boundaries as well. Earthquakes in these zones rarely occur at depths greater than 50–80 km (30–65 mi)—the maximum thickness of the cool brittle portion of continental plates.

At subduction zones, however, progressively deeper quakes occur within the descending plate to a depth of about 700 km (450 mi). The earthquake zone at subducting oceanic boundaries is called the **Benioff-Wadati zone,** named for the American seismologist Hugo Benioff, who first observed that earthquake foci become progressively deeper inland from ocean trenches, and the Japanese seismologist Kiyoo Wadati, who pioneered methods of comparing earthquake sizes.

From the onset of subduction at an ocean trench to a depth of approximately 300 km (200 mi), intense friction between the descending slab and the overriding plate causes ruptures along the slab's brittle upper surface, resulting in shallow and intermediate-depth earthquakes. These events are generally the most powerful subduction-zone earthquakes, especially the shallowest ones within the trench. As the slab descends farther, the center of quake activity shifts to the still-brittle interior of the slab, which has not been in the Earth's interior long enough to warm up and soften. By the time the slab reaches approximately 700 km (450 mi) below the surface, it's been down there for millions of years and thus it has been heated sufficiently to soften throughout. Thus the slab deforms plastically instead of undergoing brittle failure, and hence, earthquakes rarely occur at depths greater than 700 km.

The magnitude of earthquakes varies considerably with the type of plate boundary involved. Earthquakes at divergent plate boundaries are typically of relatively low magnitude because rocks being stretched tend to break before accumulating much strain energy. Almost all major earthquakes occur along either convergent or transform boundaries because a great deal of strain energy builds before a rock will break from compression or shear. Every year, about 80% of the world's earthquake energy is released in the Pacific Rim region, where the oceanic plates of the Pacific Ocean basin subduct beneath adjacent plates along most of the ocean's perimeter (see Fig. 10-3). This region includes the earthquake zones of Japan, the Philippines, the Aleutians, and the west coasts of North, Central, and South America. Most of the remaining 20% of the Earth's earthquake energy is released along the collision zone stretching through Greece, Turkey, Iran, India, and Pakistan to the Himalayas, southern China, and Myanmar (Burma). Several of the Pacific Rim's most prominent earthquake zones are discussed next.

Japan   On average for any given year, Japan will be shaken by 15% of the planet's released seismic energy. Each year, its residents experience more than 1000 earthquakes of magnitude-3.5 or greater. Since 1900, the country has suffered 25 earthquakes as powerful as San Francisco's great quake of 1906. Landslides, tsunami, fires—Japan has experienced the worst of each. The devastating Kobe earthquake, on January 17, 1995, is Japan's most recent major quake, but certainly not its worst. The quake caused $100 billion in damage, earning the dubious distinction as history's most expensive natural disaster (although the final bill for Hurricane Katrina has not yet been tallied). The quake took 5378 lives, leveled 152,000 buildings, and destroyed the equivalent of 70 U.S. city blocks. Kobe's earthquake occurred where the western edge of the Pacific plate descends steeply beneath the eastern edge of the Eurasian plate at Japan's Nankai trench.

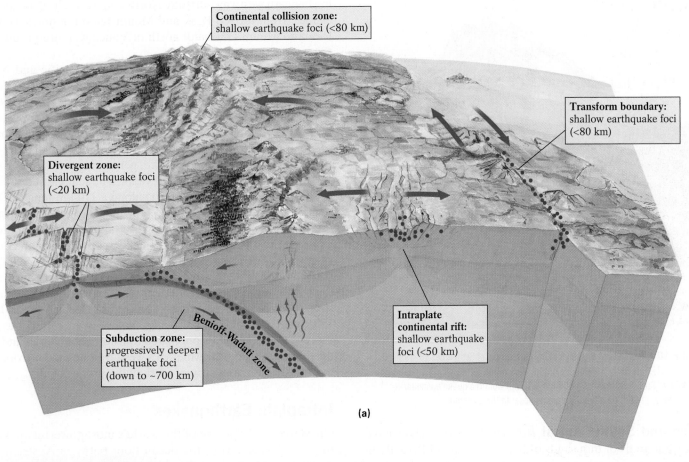

Continental collision zone:
shallow earthquake foci (<80 km)

Transform boundary:
shallow earthquake foci
(<80 km)

Divergent zone:
shallow earthquake foci
(<20 km)

Benioff-Wadati zone

Intraplate
continental rift:
shallow earthquake
foci (<50 km)

Subduction zone:
progressively deeper
earthquake foci
(down to ~700 km)

(a)

(b)

▲ **FIGURE 10-19 (a)** Earthquake depth is directly related to plate-tectonic setting. (The purple dots represent earthquake foci—the actual subsurface locations where rocks broke and earthquakes were generated.) Shallow earthquakes occur where plates are rifting, diverging, undergoing transform motion, and where continental plates collide. Deep-focus earthquakes occur exclusively at subduction zones, within the brittle portion of the descending slab. The slab remains largely intact and brittle (except for possibly some minor melting at its very top) as it passes through the partial-melting zone. It continues to generate earthquakes until it descends to about 700 km (450 mi), at which point it has warmed sufficiently to deform plastically **(b).** An earthquake in June 2000 tore open the ground in Sellfos, Iceland. Iceland sits atop the mid-Atlantic Ridge, a classic divergent zone.

Even with the Kobe earthquake engrained in the national memory, Japan's *next* quake may be its worst. Like other nations located at quake-prone plate boundaries, Japan experiences periodic earthquakes whenever one of its faults accumulates enough energy to overcome the frictional resistance between fault blocks and then breaks free. The more time that passes since the last major earthquake, the more energy that accumulates, and the more likely that the next one will be quite powerful. Historical records from A.D. 818 to the present indicate that Tokyo has averaged one cataclysmic quake every 69 years. Of Japan's nine major quakes since the Tokyo quake in 1923, none has occurred in metropolitan Tokyo, where more than 20 million people now live. The stress on Tokyo's rocks has been building for more than 80 years. Another great quake is probably also overdue in the densely populated Tokai region to the south, where the last major quake released accumulated energy on December 24, 1854.

▲ **FIGURE 10-20** Effects of the Mexico City earthquake of September 19, 1985 (magnitude-8.1). Due to the violent shaking of Mexico City's geological foundation of unconsolidated lake clay, the city—500 km (310 mi) from the quake epicenter—actually suffered greater damage than areas closer to the epicenter.

Mexico and South/Central America
Earthquakes have claimed hundreds of thousands of lives in Chile and Peru along the west coast of South America. There, the Nazca plate (see Figure 1-21 on page 22) flexes downward to form the Peru–Chile trench and the Andes mountains. On May 31, 1970, an earthquake leveled virtually every dwelling along a 70-km (45-mi) strip of the Peruvian coastline, taking 120,000 lives, demolishing 200,000 buildings, and leaving 800,000 people homeless. Similarly, subduction of the Cocos plate produces the tremors that have periodically devastated such cities as Mexico City, Mexico, and Managua, Nicaragua. Figure 10-20 shows some of the tragic aftermath of the Mexico City quake of September 1985.

Western North America
Western North America, from southern California to Alaska, owes much of its extensive earthquake history to the interaction between the Pacific plate and the North American plate. The southern end of the transform boundary between these plates extends for 1000 km (600 mi), from Baja California to Cape Mendocino northwest of San Francisco. Between Cape Mendocino and Vancouver, British Columbia, the Pacific–North American transform boundary is interrupted by two small plates (the Gorda plate off the coast of northern California and Oregon, and the Juan de Fuca plate off the coast of Washington and southern British Columbia). Both are subduct-

ing beneath westward-drifting North America. The Cascade volcanoes, from Lassen Peak and Mount Shasta in northern California to Mount Garibaldi north of Vancouver, are products of this subduction zone.

The transform boundary between the Pacific and North American plates resumes north of Vancouver and extends to Alaska, where the Pacific plate subducts beneath the North American plate to the east and Eurasian plate to the west, producing the volcanic Aleutians. In November of 2002, the Alaskan segment of the transform boundary, called the Denali Fault, was rocked by a magnitude-7.9 quake, generating one of the largest surface ruptures in the past 150 years. The tear caused the land to shift horizontally, 7.8 m (26 ft) along a 210-km-long (130-mi) segment of the fault (Fig. 10-21). The quake triggered thousands of avalanches and rockfalls and tore large crevasses in the region's glaciers. The power of this quake was even felt in the Lower 48 in Seattle's Lake Union houseboat community, 2250 km (1400 mi) to the south, where floating homes were torn away from their walkways and sewerlines ruptured. Fortunately, the quake struck a remote, sparsely populated area near Denali National Park, 150 km (90 mi) south of Fairbanks. No lives were lost, not usually the case along the highly populated transform boundaries and subduction zones of North America's western states and provinces—an almost continuous strip of earthquake-prone country.

Intraplate Earthquakes
On March 25, 1998, one of the world's most powerful quakes in the past decade struck the ocean floor between Australia and Antarctica. Measuring 8.2 on the moment-magnitude scale, the

▲ **FIGURE 10-21** The magnitude-7.9 quake that struck central Alaska on November 3, 2002, ruptured the surface for a distance of about 320 km (200 mi) and displaced the rocks along the strike slip–type Denali Fault about 10 m (33 ft). Photo on the right was taken just south of the Mentasta Junction, Alaska, on the Tok Cutoff Highway on Monday, Nov. 4, 2002.

quake defied the geological conventional wisdom that such large quakes occur only at plate boundaries. In this case, the quake's epicenter was located *within* the Antarctica plate, 350 km (230 mi) from the nearest plate boundary. It is the largest ocean-floor *intraplate* tremor on record. Closer to home, on June 10, 1987, a mid-plate earthquake rattled the citizens of Lawrenceville, Illinois, and 16 surrounding states and Canadian provinces. Registering a 5.0 magnitude, the quake triggered alarms at a nuclear power plant, cut phone service, and shook hospital patients from their beds in Iowa and West Virginia.

Consider the following earthquake facts: In 2002–2004, there were earthquakes in Alabama (M = 4.6), South Dakota (M = 4.0), New York (M = 5.0), Illinois (M = 4.2), and at least three with magnitude ≥4.0 in Wyoming. Though generally shallow—less than about 50 km (30 mi)—mid-plate earthquakes tend to be of lower magnitude than those at active plate margins, apparently because less strain energy builds within the faults at mid-plate locations. On the other hand, the older, colder (and therefore more brittle), less-fractured rocks at mid-plate sites transmit seismic waves more efficiently than the young, warm rocks at active plate margins. Thus mid-plate earthquakes, such as those occurring in eastern North America, are felt over a significantly larger area than are plate-boundary quakes, such as those common in the West.

Locating mid-plate faults *on land* to study their history and predict their future seismic activity can prove difficult, particularly in eastern North America. There, heavy vegetation conceals both the rocks and their faults. Because earthquakes don't happen very often in the East, any surface evidence is usually removed by erosion or buried with the passage of time. A few mid-plate faults, such as the network of faults in the English Hills of southeastern Missouri and northeastern Arkansas, actually break the surface, a real "treat" for eastern geologists.

In the East, subsurface faults are typically detected by seismic-wave studies, which have revealed that the crust of eastern North America is riddled with a deep network of old, near-vertical faults. Those faults along the East Coast may date from the rifting of North America from Europe and Africa during the breakup of the supercontinent Pangaea, 200 million years ago. Faults in the Midwest may date from an older rifting event that occurred some 570 million years ago. At that time, North America apparently rifted, opening the Iapetus Sea, in a manner similar to the rifting that created the present Atlantic Ocean. The so-called Iapetan faults, which are today landlocked because of subsequent plate-tectonic events (principally the formation of the Appalachians) run along the west side of the Appalachians from northeastern Canada, along the St. Lawrence River, through Lake Ontario, and then southward through eastern Tennessee into Alabama. Every so often, these faults remind us that not all feelable earthquakes occur in California. With a deep rumbling like a passing freight train, a magnitude-5.1 quake struck upstate New York on April 20, 2002, near the tiny Adirondack town of Ausable Forks near Plattsburgh within the northern New York–Western Quebec seismic zone. Chimneys fell, water mains broke, mobile homes were shaken from their foundations, and 3000 homes were left without power. The shaking was felt in Boston to the east, Buffalo to the west, and as far south as Baltimore, Maryland.

The exact causes of intraplate earthquakes remain a mystery. Some hypotheses cite processes that are unrelated to plate-tectonic motion. One, for example, proposes that after long-term erosion has removed vast quantities of surface materials, the buoyant unloaded crust rises, generating enough stress to arouse old faults. A related hypothesis for earthquakes may explain their occurrence in or near regions that were covered by ice sheets in the last glaciation, such as in the Great Lakes region of North America. This hypothesis suggests the crust is still rebounding after being bowed down by the immense weight of the ice; as it rebounds, faults move. A third hypothesis proposes that these earthquakes may be triggered by excessive rainfall, which infiltrates and saturates the network of subsurface fractures, reducing friction between adjacent blocks of rocks and reactivating long-dormant faults.

Another hypothesis views intraplate seismic activity as a possible early warning of future rifting. According to this scenario, plate rifting is initiated by rising masses of heat-softened mantle material. Such thermal plumes may have set off quake activity near Boston and Charleston, and seismic studies in those areas have identified extremely young plutons that are considered possible evidence of future rifting.

A simpler model holds that plate motion stresses the weakened and faulted crust at some intraplate sites, triggering infrequent but sometimes powerful quakes. The lateral motion of the westward-drifting North American plate, for example, may create frictional stress within the plate as it rides atop the underlying asthenosphere. Strain energy may accumulate over long periods, until it exceeds the strength of the rocks or the frictional resistance along their faults. Today's northeast-to-southwest trend of the Iapetan faults is oriented so that they absorb the stresses generated by the continual growth of the Atlantic Ocean.

Earthquakes in Eastern North America

Because relatively few earthquakes have released the pent-up seismic energy in the interior of the North American plate, large quakes may be looming in this region. Seismologists working with earthquake records that date back to 1727 predict that there is a better-than-even chance of a magnitude-6.0 earthquake striking east of the Rocky Mountains before the year 2020.

Past eastern earthquakes have included a sharp tremor with an estimated magnitude of 5.0, that struck New York City on August 10, 1884. Felt from Washington, D.C., to Maine, the quake hurled horsecars from their tracks on Fifth Avenue and sent a two-ton safe sliding across the lobby of the Manhattan Beach Hotel in Brooklyn. In 1929, a powerful tremor toppled 250 chimneys in Attica, New York; in 1983, the Adirondack Mountains of New York were rocked by a quake of magnitude-5.2. In Massachusetts, earthquakes occurred near Plymouth in 1638; at Cape Ann 80 km (50 mi) northeast of Boston in 1755; and in Cambridge in 1775. The 1755 quake, which was felt from Nova Scotia to South Carolina, shattered roof gables on colonial homes, destroyed 1500 chimneys, and snapped the gilded-cricket weather vane off the roof of Boston's Faneuil Hall. In Canada, earthquake waves sped outward from fractured rocks 20 km (12 mi) beneath the evergreen forests of the small lumber town of Chicoutimi, 150 km (100 mi) north of Quebec City, on November 25, 1988. Measuring a surprising magnitude-6.0, the quake was felt as far away as Washington, D.C. One of the most powerful eastern earthquakes, estimated at magnitude-7.0, destroyed much of Charleston, South Carolina, in August 1886 and took the lives of 60 residents.

The most powerful recorded seismic events of eastern North America occurred in New Madrid, Missouri, as a series of quakes that began on December 16, 1811, and lasted for 53 days, ending February 7, 1812. The quakes included three main shocks and more than 1500 powerful aftershocks. The most violent aftershock was the last—it shook a 2.5-million-km^2 (1-million-mi^2) area from Quebec to New Orleans and from the Rocky Mountains to the Eastern Seaboard. Church bells rang in Boston, 1600 km (1000 mi) away, windows rattled violently in Washington, D.C., and pendulum clocks stopped in Charleston. Based on observations by fur trappers and an eyewitness account from naturalist John Audubon, geologists believe that each of the three principal tremors may have exceeded a magnitude of 8.5.

Near the estimated epicenter of the New Madrid quake, entire forests were flattened when the violent shaking snapped tree trunks. Large fissures opened, including some too wide to cross on horseback. Sandblows, water, and sulfurous gas erupted as the shaking ground compacted. Thousands of square kilometers of prairie land subsided and flooded, forming St. Francis and

Reelfoot Lakes in northwestern Tennessee (shown in Fig. 10-22) as well as swamps that today stretch 300 km (190 mi) from Cape Girardeau, Missouri, to northern Arkansas. The Mississippi River became dammed in some spots by uplifts and landslides, but flowed wildly elsewhere, overwhelming many riverboats. The flow of the river even reversed direction where uplifts changed lowlands into highlands. Whole river-channel islands disappeared, and offsets along the river produced new cliffs and waterfalls.

Because they occurred at a mid-plate location and produced a great variety of landscape changes, geologists are fascinated by the New Madrid events. It is critical to understand these events because the area could again experience seismicity of the same magnitude—a circumstance for which it is now dangerously unprepared. Recently, geologists and archaeologists have joined forces to determine the frequency with which great earthquakes occur in the mid-continent. Their studies have focused on ancient sandblows that contain pits dug by prehistoric people. These pits contain datable pottery fragments. From the age of these artifacts, we estimate that the last great quake in Missouri, prior to the events of 1812, occurred about 1100 years ago. This finding suggests that the time between great quakes in that region may be on the order of 1000 years. Residents of that area, however, should not rest easy; their region is probably capable of producing more

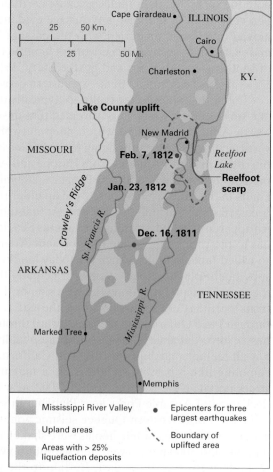

▲ **FIGURE 10-22** The effects of the New Madrid, Missouri, earthquakes of 1811–1812 included liquefaction of sediments, uplift of some areas forming waterfalls and rapids along the region's rivers, and subsidence (sinking of the land surface) of other areas forming several new lakes. The lake shown here, Reelfoot Lake in northern Tennessee, formed when a vast area of prairie land subsided during the New Madrid quakes. A *scarp* is a steep cliff face that forms, in this case, when rocks and sediments are pulled apart at a fault zone.

frequent and strong quakes (magnitude 6.0–6.9) as well as some of the more major (magnitude 7.0–7.9). Dating of sandblows by artifacts and carbon-14 dating of buried wood indicate that tremors of magnitude-6.4 or greater have shaken the region at least four times during the last 2000 years (roughly around the years 500, 900, 1300, and 1600 A.D.). Most definitely, many of today's homes and high-rise office buildings are not built to withstand such a shake. In the nineteenth century, the Mississippi valley was sparsely populated. The same area, from St. Louis to Memphis, is currently home to more than 12 million unprepared citizens.

Coping with the Threat of Earthquakes

The world's population has exploded, doubling in the past 50 years to nearly 6.5 billion earthlings, with nearly half of them living in close quarters in the world's great cities. Roughly 400 of those cities have risen at the seismically active edges of the Earth's tectonic plates. The bad news: An earthquake capable of killing a million people in one tragic moment is statistically likely to occur every century or so. This prediction stems from hundreds of years of earthquake events that have struck catastrophically in areas now populated by tens of millions. Of course, the degree to which our cities prepare for these disasters—by building survivable buildings (which takes lots of money) and planning for the fires, landslides, and other sources of quake-related heartache—will determine the death tolls from place to place.

Fresh memories of the tragic destruction in Kobe, Japan, Taiwan, Izmit, Turkey, and Banda Aceh, Indonesia, have spurred great concern among residents of North America's earthquake zones. Recent studies along the Hayward fault in California's Oakland–Berkeley area and the Seattle fault, which cuts across metropolitan Puget Sound, have raised serious questions about how well we're prepared for such potential urban catastrophes.

Can anything be done to prevent damage from earthquakes and to protect the millions of people in North America and elsewhere who live in vulnerable regions? In this section, we will address this question. Although the U.S. government has prepared a checklist of helpful, logical "do's and don'ts" (Table 10.4), even more can be done.

Limiting Earthquakes Caused by Humans

Any earthquake-defense plan should first eliminate quakes caused by humans, such as those sometimes caused by building dams in earthquake zones. The additional water accumulated in dammed reservoirs adds considerable weight to the underlying rocks and faults over a geologically short time period. Eventually the water may seep into deep, faulted rocks, reducing the friction between adjacent rock masses and activating old, previously stable faults. In this way, dam building may trigger seismic activity in areas that have been inactive for centuries.

The Hoover Dam, which impounds Lake Mead, was built across the Colorado River at the Arizona–Nevada border in an area that had been relatively free of earthquakes for decades. In 1936, shortly after the dam was completed, a series of 600 tremors began that lasted for 10 years. The largest, measuring about magnitude-5.0, was felt throughout the region. It is believed that the 42 km^3 (10 mi^3) of water contained in young Lake Mead depressed fractured crustal blocks and lubricated subsurface fractures in the bedrock, causing the blocks to slip. In a similar way, the massive Three Gorges Dam, currently under construction along China's Yangtze River, could trigger an earthquake of magnitude-6.0 or greater, and endanger the lives of thousands of people living downstream of its vast reservoir if the dam fails.

Another way that humans have caused earthquakes was revealed during the early 1960s in Denver, Colorado. The Mile-High city is built atop fractured billion-year-old granite that had been quake-free for the 80 years prior to 1962. From 1962 until 1965, however, Denver experienced 700 tremors measuring from 3.0 to 4.3 in magnitude. The largest produced significant damage in the area. In 1962, engineers at the U.S. Army's Rocky Mountain arsenal had drilled a 3660-m (12,000-ft)-deep well into faulted bedrock beneath the Denver area. Every month, as much as 16 million liters (4 million gal) of the nasty liquid waste from nerve-gas production were injected into this well under high pressure. The high fluid pressure widened bedrock joints, and the liquid waste lubricated potential slip surfaces along major faults. When the arsenal was told that it had caused the tremors, it discontinued the project. As shown in Figure 10-23, the level of quake activity dropped dramatically once pumping stopped.

Determining Seismic Frequency

Seismologists study the earthquake history of an area to determine the statistical probability that future earthquakes will strike there. If you hear that there is a 40% chance of a magnitude-7.5 earthquake occurring along the Palm Springs section of the San Andreas Fault within the next 30 years, that forecast is based on the frequency and magnitude of past events in that area. In fact, in 2005, for the very first time, the United States Geological Survey is offering Californians 24-hour earthquake forecasts, like weather forecasts, that predict the statistical probability of a quake of a given magnitude. (Check out the website pasadena.wr.usgs.gov/step for today's forecast.)

Seismologists use maps such as the one shown in Figure 10-24 on page 340 to make general predictions by identifying areas that have experienced quakes in the past. Those areas should prepare more seriously for the likelihood of a significant earthquake by adopting strict building codes. The seismic-risk maps do not, however, indicate the frequency of destructive quakes. For example, Charleston, South Carolina, with its one large 1886 event, is represented on such a map in the same way as southern California, which experiences far more frequent tremors. Moreover, because such a map relies on historically recorded and therefore relatively recent events, its information tells us little about the pattern of seismic events over thousands of years. To develop a long-term history for North America, we must attempt to date fault and quake activity older than the 400 to 500 years of recorded North American history.

Table 10.4 Earthquake Do's and Don'ts

What to Do Before an Earthquake

Check for potential fire risks, such as defective wiring and leaky gas connections. Bolt down water heaters and gas appliances.

Know where and how to shut off electricity, gas, and water at main switches and valves.

Place large and heavy objects on lower shelves. Securely fasten shelves to walls. Brace or anchor top-heavy objects.

Do not store bottled goods, glass, china, and other breakables in high places.

Securely anchor all overhead lighting fixtures.

17 Survival Items to Keep on Hand—in a Large, Lockable Plastic Trash Barrel

Portable radio with extra batteries

Portable fire escape ladder for buildings with multiple floors

Flashlight with extra batteries

First-aid kit containing any specific medicines needed by members of the household

First-aid manual

Fire extinguisher

Bottled water—one gallon per person, per day

Adjustable wrench for turning off gas and water

Smoke detector

Matches

Axe (for chopping wood for heating)

Canned and dried foods for one week for each member of household

Nonelectric can opener

Telephone numbers of police, fire, and doctor

Portable stove with propane or charcoal

Sleeping bags for each person/tent or plastic tarps with cord

Cash (banks and automatic teller machines may be closed or inoperable)

Note: The status of these items should be checked every six months.

What to Do During an Earthquake

If you are outdoors, stay outdoors; if you are indoors, stay indoors. During earthquakes, most injuries occur as people enter or leave buildings.

If indoors, take cover under a heavy desk, table, or bench, or in doorways, or halls, or against inside walls. Stay away from glass. Don't use candles, matches, or other open flames either during or after tremors. Extinguish all fires.

If in a high-rise building, don't dash for exits; stairways may be broken or jammed with people. Never use an elevator.

If outdoors, move away from buildings, utility wires, and trees. Once in the open, stay there until the shaking stops.

If in a moving car, drive away from underpasses and overpasses.

Stop as quickly as safety permits, but stay in your vehicle. A car may shake violently on its springs, but it is a good place to stay until tremors stop. When you drive on, watch for fallen objects, downed wires, and broken or undermined roadways.

The Most Common Causes of Earthquake Injuries

Building collapse or damage

Flying glass from broken windows

Falling pieces of furniture, such as bookcases

Fires from broken gas lines, electrical shorts, and other causes, aggravated by a lack of water caused by broken water mains

Fallen power lines

What to Do After an Earthquake

Be prepared for aftershocks.

Check for injuries; do not move seriously injured persons unless they are in danger of sustaining additional injury.

Listen to the radio for the latest emergency bulletins and instructions from local authorities.

Check utilities. If you smell gas, open your windows and shut off the main gas valve. Leave the building and report the gas leakage to the authorities. If electrical wiring is shorting out, shut off the current at the main meter box.

If water pipes are damaged, shut off the supply at the main valve. Emergency water may be obtained from hot-water tanks, toilet tanks, and melted ice cubes.

Check sewage lines before flushing toilets.

Check chimneys for cracks and damage. Undetected damage could lead to fire. Approach chimneys with great caution. The initial check should be from a distance.

Do not touch downed power lines or objects touched by downed lines.

Immediately clean up spilled medicines, drugs, and other potentially harmful materials.

If the power is off, check your freezer and plan meals so as to use foods that will spoil quickly.

Stay out of severely damaged buildings. Aftershocks may shake them down.

If you live along a coast, do not stay in low-lying coastal areas. Do not return to such areas until local authorities tell you that the danger of a tsunami has passed.

Source: U.S. Federal Emergency Management Agency (FEMA)

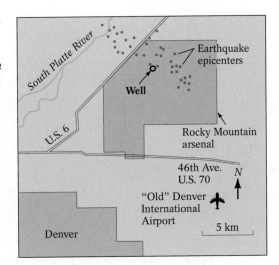

▼ **FIGURE 10-23** Graphs correlating a series of earthquakes that occurred in Denver during the early 1960s with the high-pressure disposal of liquid waste deep in the region's underlying fractured bedrock during those years. Note how the frequency of seismic events coincided with the timing and quantity of waste disposal.

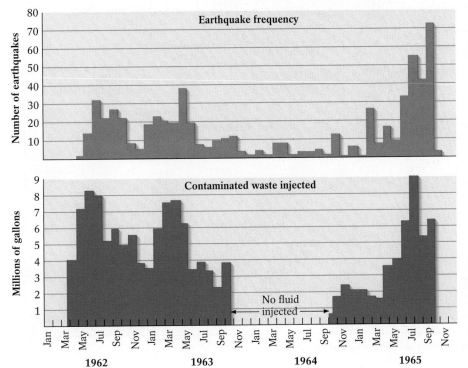

In California, for example, substantial written records exist for only the past 150 years. These data are too brief a set of earthquake observations to determine past quake frequency. To date earlier seismic events in California, we must look for geological evidence of a fault. This evidence may include topographic breaks in the landscape, linear valleys that may follow a fault line, areas of crushed rock, small lakes that form where impermeable crushed rock prevents surface water from seeping underground, or linear features that have been offset by fault movements (such as the orange grove in Figure 1-31 on page 28).

Once geologists find a fault, they then search for evidence of past seismic events related to the fault and try to date them. This work helps them determine the fault's *recurrence interval,* the average amount of time between its quakes. One method requires a backhoe and some knowledge of the principles of relative geologic dating. Trenches excavated across the San Andreas Fault,

such as the one shown in Figure 10-25, reveal a complex stratigraphy that includes vertically offset organic deposits. Geologists apply carbon-14 dating and the principle of cross-cutting relationships to determine when the offsets occurred, enabling them to estimate when past seismic events may have taken place. Northeast of Los Angeles, for example, evidence suggests that nine major seismic events have taken place during the past 1400 years, with an average recurrence interval of 160 years. This fact unnerves many southern Californians given that the last great earthquake in that area occurred in 1857.

Seismologists can estimate the frequency of past earthquakes only for visible faults. Even though we have lasers that can measure fault displacements of hundredths of a millimeter and ultrasensitive seismometers that can detect faint tremors on the opposite side of the Earth, many faults still lay undiscovered beneath downtown Los Angeles and elsewhere. After an earthquake shook Whittier, California, on October 1, 1987, local geologists identified numerous hidden, or "blind," faults in the Los Angeles area (Fig. 10-26). Many of these are thrust faults that have uplifted the mountains surrounding the Los Angeles basin. The faults had little or no surface expression until the quake exposed their locations through compressed pavement, uplift, and fractures at the surface.

One such hidden fault, part of the Oak Ridge fault system, produced the magnitude-6.7 quake that devastated the Northridge area of the northern San Fernando Valley, 30 km (20 mi) northwest of downtown Los Angeles, in January 1994. This tragedy took 63 lives, caused thousands of injuries, produced a massive disruption of daily life, and created more than $15 billion in damage. Although geologists had suspected its existence for at least half a century, the hidden fault's extent and depth remained unknown until surface damage and after-shocks gave clues to its location. Another recently discovered fault, the Elysian Park fault, runs along Wilshire Boulevard through Hollywood toward Santa Monica, threatening such landmarks as Dodger Stadium, Beverly Hills, and the world-famous HOLLYWOOD sign. The discovery of two additional hidden faults in Los Angeles, one through Echo Park and the other through MacArthur Park, has further heightened the city's awareness of its undiscovered danger zones.

And in 1999 yet another hidden fault was discovered—the Puente Hills fault zone—which runs from northern Orange County to downtown L.A. This thrust-fault zone, about 40 km (25 mi) long and about 20 km (12.5 mi) wide, sits at a disturbingly shallow depth of 3 km (2 mi). Because this fault is located directly beneath the city, it may be the *most* dangerous of all. The fault has ruptured at least four times in the last 11,000 years, the last time several thousand year ago. But unlike other, less powerful faults in the region, when the Puente Hills fault ruptures, it tends to do

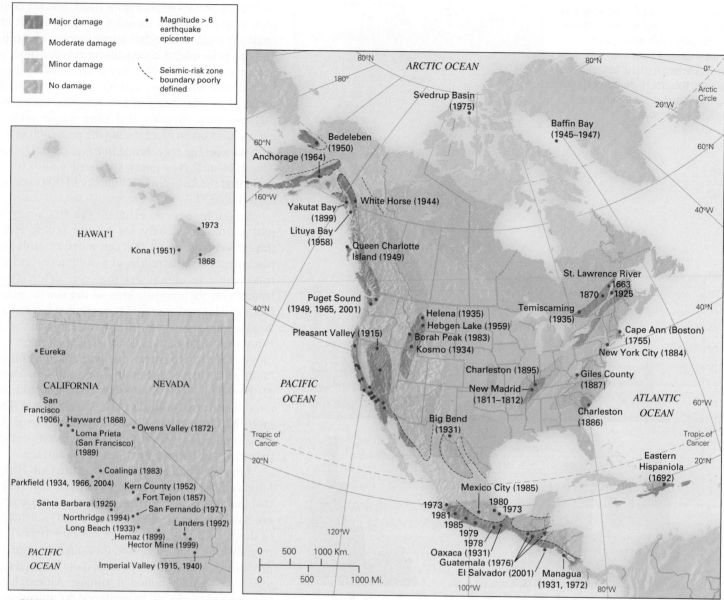

▲ **FIGURE 10-24** A map of North American seismic-risk zones, with locations and dates of some of the continent's major historical earthquakes.

so in a major way, with magnitudes of 7.2–7.5, ten to fifteen times more powerfully than the Northridge quake of 1994, the costliest quake in U.S. history. And sitting atop the fault are some very old, very vulnerable commercial and industrial buildings. The grim prediction: If a 7.5-quake struck during workday hours, as many as 18,000 lives would be lost, with 120,000 injuries and $250 billion in damage. One geoscientist, USC professor James Dolan, described the fault this way: "This fault is in one of the worst places you could think of to put a fault of this size and geometry."

Although the San Andreas Fault is more likely than these recently discovered faults to produce a powerful earthquake of 7.5 to 8.0 in magnitude, it is located 50 km (30 mi) east of downtown Los Angeles. Consequently, its earthquakes may cause less damage than lower-magnitude tremors generated on hidden faults directly beneath the city.

To make matters worse, seismologists studying a 1957 earthquake in Mongolia discovered that that great quake—with a magnitude exceeding 8.0—involved *two* sets of faults in the same region. During that quake, the land slipped suddenly along a 250-km (150-mi) segment of a San Andreas–type strike-slip fault; simultaneously, a nearby 50-km-(30-mi)-long network of thrust faults ruptured. This finding suggests that a major event along the San Andreas Fault could potentially activate one or more of Los Angeles's hidden thrust faults, such as the ones that have uplifted the San Gabriel, Santa Monica, and San Bernardino Mountains.

The ability of faults "to talk to one another" has been further shown by the 1999 Hector Mine quake near Palm Springs, California. Several hours after the quake's main shock, a series of earthquakes struck another fault about 180 km (120 mi) away. Some

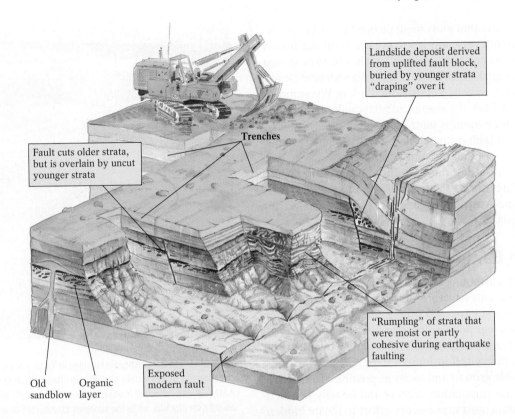

Landslide deposit derived from uplifted fault block, buried by younger strata "draping" over it

Trenches

Fault cuts older strata, but is overlain by uncut younger strata

"Rumpling" of strata that were moist or partly cohesive during earthquake faulting

Old sandblow

Organic layer

Exposed modern fault

▲ **FIGURE 10-25** (above) Trenches excavated across a fault expose the stratigraphy of the fault zone. According to the principle of cross-cutting relationships, any movement along a fault must be younger than the age of the youngest rock or sediment that it displaced, and older than the oldest rock or sediment that it did not displace. Thus, the seismic events shown here can be dated relative to one another, and each event's absolute age can be estimated in relation to the organic deposit. (left) Vertically offset deposits near Palmdale, California, showing evidence of a previous earthquake there. Look at the brown organic layers in the lower part of the exposure.

▶ **FIGURE 10-26** Hidden faults, because they are not evident at the surface, often go undetected until an earthquake occurs. We know now that the Los Angeles area is underlain by numerous hidden faults, such as those recently discovered at Echo, MacArthur, Elysian Parks, and from northern Orange County into downtown L.A.

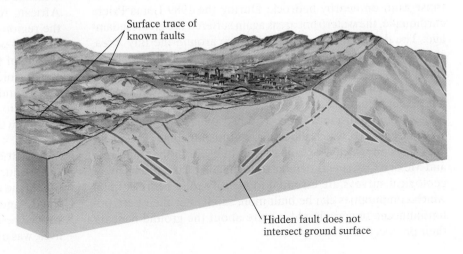

Surface trace of known faults

Hidden fault does not intersect ground surface

area seismologists believe that movement on one fault may set others in motion. We do not yet know if a major quake in one region can cause quakes in faraway places, although it has been proposed that the 1992 Landers quake in southern California's Mojave Desert triggered quakes as far away as Yellowstone Park in Wyoming.

And in 2004 and 2005, the magnitude-9.0 quake in Sumatra was followed only three months later by a magnitude-8.7 quake that struck an adjacent fault on an adjoining plate boundary about 175 km (110 mi) to the southeast. Some seismologists believe that the sudden movement along the plate boundary between the India and Burma plates during the first quake increased the stress on the rocks along the Australian and Sunda plates, perhaps triggering the second quake.

Building in an Earthquake Zone

Safe development in earthquake-prone regions requires that we plan *where* to build various structures and then build them to survive the region's largest expected quake magnitudes. Certain structures, such as nuclear power plants and large dams, are considered *critical facilities,* because their destruction by an earthquake could take a heavy toll in human lives. Critical facilities must be sited on stable ground and as far as possible from active faults as well as major population centers and heavily traveled highways. The builders must make every effort to locate hidden faults prior to construction. They must also safeguard those public facilities that help us cope with a quake disaster, such as fire and police departments, hospitals, and the local communications network. In the following sections, we describe how the stability of the ground and the structures we build determine the likelihood that an earthquake will produce major damage.

Ground Stability When seismic waves pass through adjacent areas of solid bedrock and soft, wet sediment, the ground shakes more violently on loose sediment than on solid bedrock. While the bedrock moves in short sharp jolts, the loose sediment develops a continuous rolling motion that can amplify the quake's intensity. As a result, structures built on unconsolidated materials typically suffer far more damage than comparable ones located on solid bedrock, as illustrated by the collapsed buildings within the old channel deposits shown in Figure 10-27. During the 1906 San Francisco quake, the buildings constructed on wet bay mud and landfill sustained four times as much damage as those built on nearby bedrock. During the 1989 Loma Prieta earthquake, the waterfront areas again suffered the greatest damage. The destruction seemed to bounce around the Bay Area, striking those structures sited on the most unstable materials, including the rubble from the 1906 quake.

Although the epicenter of the 1985 Mexico earthquake was near Acapulco, the damage was greater and the death toll far higher in more distant Mexico City. Acapulco is built on solid volcanic bedrock, which shook far less violently than the ancient lake clays on which Mexico City was built. Both the California and Mexico City events demonstrate that, in quake-prone areas, geological surveys are essential to identify solid bedrock upon which communities can be built more safely. For their part, potential home buyers should inquire about the ground beneath their prospective property.

▲ **FIGURE 10-27** Earthquake damage in Varto, Turkey, in 1953. Variations in the destructiveness of earthquake shaking are often controlled by the underlying geologic materials. The collapsed houses shown here were built on the unconsolidated sands of an old river channel, while the surviving houses rest on a solid bedrock bench.

Another area—one of great importance to a number of the world's great religions—must also deal with the effects of earthquakes on unstable ground. Recent geological and archaeological surveys of the Holy Land of Jerusalem indicate that certain parts of this ancient city are far more susceptible to quake destruction than others. Apparently, the most vulnerable areas rest on a loose foundation of rubble rather than solid bedrock, the remnants of past invasions of ransacking armies. Jerusalem's Old City is most at-risk for destruction by the region's frequent earthquakes. Particularly vulnerable is the elevated plaza that contains two major mosques, including the gold-clad Dome of the Rock—an important place of worship for the region's Muslim community—and the Temple Mount, a holy place to the region's Jewish population. The reason for the region's frequent major earthquakes (about a half-dozen major quakes have struck Jerusalem in the past thousand years) is that it sits at the northern edge of the tectonically active Great Rift Valley between the African, Arabian, and Eurasian plates. The last major quake in the region—a magnitude-6.3 near Jericho in 1927—killed more than 200 people. The Bible tells of many great tremors and the tearing of the hills of this region, some of which may have been responsible for the Old Testament account of the destruction of Sodom and Gomorrah and the New Testament description of the events in Jerusalem shortly after Christ died on the cross (Matthew 27:51–54, "The earth shook and the rocks were split").

Structural Stability The amount of damage that a building suffers during an earthquake reflects the magnitude and duration of the quake, the nature of the ground on which the building stands, and the building's design itself. The earthquake that destroyed several cities in Armenia in 1988, taking more than 50,000 lives, was of the same magnitude as California's 1989 Loma Prieta quake, which took 65 lives. India's catastrophic earthquake of 1993 was of lower magnitude than either of these events, yet took almost 18,000 lives. The greater damage and loss of life in Armenia and India and more recently in Izmit, Turkey (Fig. 10-28) largely resulted from the poor construction of the buildings. The Bay Area's structures were built to meet strict standards with the expectation that they would have to survive high-magnitude quakes. Using superior designs, methods, and materials, they could better withstand the shaking.

The most secure structures are built of strong, flexible, but relatively light materials. Wood, steel, and reinforced concrete (containing steel bars) tend to move *with* the shaking ground, improving their chances for survival. Unreinforced concrete, heavy brickwork and concrete block, stucco, and adobe (mud brick) are generally inflexible. Their components move independently and oppose the shaking, battering one another until their structure collapses. The collapsed houses in Figure 10-29 were made of stone and other inflexible materials. Concrete-frame structures, such as the large parking garage that collapsed during California's 1994 Northridge quake and the I-880 Cypress Expressway that failed disastrously during the 1989 Loma Prieta quake (see Fig. 10-2), are particularly vulnerable to shaking of large earthquakes.

Perhaps unexpectedly, well-designed, well-built skyscrapers may be the safest buildings. A building with structural elements (floors, walls, ceilings) that are joined firmly may sway in a quake like a tree in the wind, but it is unlikely to collapse. Distributing a building's weight with more at the bottom also makes it safer. The momentum that develops when a top-heavy building swings from side to side may topple it. The pyramid-shaped Transamerica Building in San Francisco (Fig. 10-30), which tapers toward the top and therefore has most of its weight concentrated in its lower floors or below-ground foundation, can sway with the shaking Earth but will not develop damaging momentum near its top. The buildings that suffer most dramatically during powerful tremors include those with unsupported roofs that extend across a broad span, such as bowling alleys, supermarkets, and shopping malls.

Another general rule about building in an earthquake zone can be summarized as "the simpler, the better." Projections or decorative elements that are not securely attached may loosen or fall during a tremor. Many nineteenth-century buildings have projecting decorations that are attached only by deteriorating mortar. Realizing the danger, the cities of Los Angeles and San Francisco have recently begun requiring owners to install special braces to secure loose elements.

Earthquakes also play havoc with highways, endangering motorists and severing vital transportation lifelines. Through painful experience, we now know that elevated roadways must be strongly connected to their support columns, that the columns themselves must be of equal height so that they all flex at the same rates, and that, wherever possible, the roadways should be

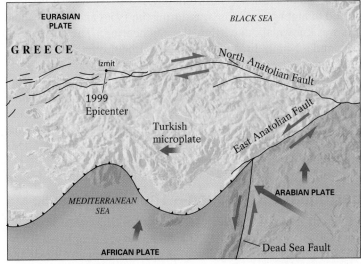

▲ **FIGURE 10-28** (left) The massive destruction and loss of 17,000 lives during the magnitude-7.4 earthquake in Izmit, Turkey (on August 17, 1999), was in part the result of poor building-construction practices and inferior construction materials. (above) Northward movement of the Arabian and African plates squeezes the Turkish microplate, forcing it to migrate westward relative to the Eurasian plate. This motion sets up San Andreas–type strike-slip motion along the North Anatolian fault and causes powerful earthquakes such as the tragic event in 1999.

▲ **FIGURE 10-29** The devastation of villages from the May 27, 1995, earthquake on Sakhalin Island, off the coast of Russia, cost the lives of 2325 of this village's 3200 inhabitants.

▲ **FIGURE 10-30** The Transamerica Building, in San Francisco, California. The pyramidal shape of this structure makes it less likely than more traditional buildings to suffer extensive earthquake damage.

sited on solid bedrock. (Unfortunately, North America's greatest period of highway building was during the 1950s and 1960s, when less was known about earthquake engineering.) The collapse of Oakland's Nimitz Freeway during the 1989 Loma Prieta earthquake clearly demonstrated that multilevel roadways are a poor choice in quake-prone areas, though many such structures remain in California. Costly efforts to increase the strength and stability of California roadways and bridges after the Loma Prieta earthquake did pay off during the 1994 Northridge earthquake; among the dozens of bridges and overpasses that had been upgraded, only one collapsed in that powerful quake.

Earthquake Prediction—The Best Defense

In March 1966, two powerful tremors rocked a broad area 200 km (120 mi) south of Beijing, the capital of the People's Republic of China. The death toll and property damage were so appalling that Premier Chou En Lai declared the "People's War on Earthquakes." For his army, Chou recruited 10,000 seismologists and 100,000 amateur assistants from the ranks of China's college students, telephone operators, and factory workers. Their mission was to set up 250 regional seismic stations and analyze a huge volume of seismic data so that China might better predict and prepare for future earthquakes. The effort paid off handsomely on February 4, 1975.

In late 1974 and early 1975, residents of Liaoning province in southern Manchuria reported that their well water contained high levels of radon gas; among other precursor signals of an impending earthquake, thousands of snakes crept out of their winter hibernation and froze to death on icy roads. Chou's army focused

its attention on Haicheng, the region's center with its 90,000 inhabitants. On February 1, 1975, a series of tremors struck the city. The largest, on the morning of February 4, had a magnitude of 4.8. Early that afternoon Chinese seismologists issued an earthquake advisory, ordering residents to extinguish their cooking fires and evacuate to the city's parks and fields. At 7:36 P.M., a powerful quake (magnitude-7.6) destroyed almost 90% of Haicheng's tile-roofed dwellings. A disaster of enormous proportions was averted, demonstrating the effectiveness of accurate earthquake prediction.

One year later, however, in May 1976, the same network of professional and amateur seismologists failed to predict the greatest natural disaster of the 20th century. A magnitude-8.0 earthquake struck the coal-mining and industrial center of Tangshan, located about 150 km (90 mi) east of Beijing, taking an estimated 250,000 lives and injuring 780,000. Fearing aftershocks, Beijing's residents lived on the streets, in fields, and in parks for three months. These two events—one glorious hit, one tragic miss—illustrate the current uncertain state of earthquake prediction.

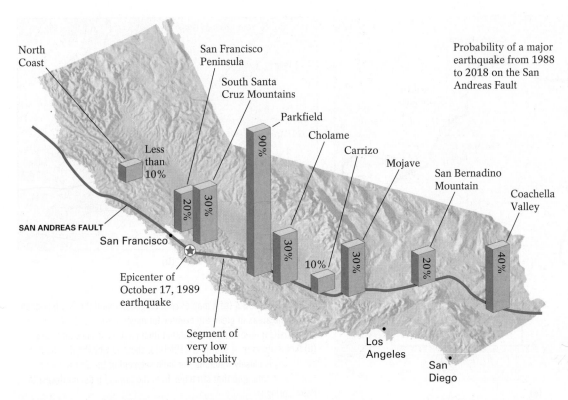

North
Coast

San Francisco
Peninsula

South Santa
Cruz Mountains

Parkfield

Cholame

Carrizo

Mojave

San Bernadino
Mountain

Probability of a major
earthquake from 1988
to 2018 on the San
Andreas Fault

Coachella
Valley

Less
than
10%

90%

20%

30%

30%

10%

30%

20%

40%

SAN ANDREAS FAULT

San Francisco

Epicenter of
October 17, 1989
earthquake

Segment of
very low
probability

Los
Angeles

San
Diego

▲ **FIGURE 10-31** The heights of these bar graphs indicate areas along California's numerous faults that are more likely statistically to suffer major earthquakes within the next few decades.

energy that causes great earthquakes, but it can gradually deform or break any features subjected to it (Fig. 10-32). The creeping Hayward fault in Berkeley, California, runs through the football field at Memorial Stadium on the University of California campus, bending the seats in the bleachers, opening large cracks in the stadium walls, and damaging a large drainage pipe beneath the field.

According to some seismologists, locked fault segments that have not experienced seismic activity for a substantial period of time are prime candidates for major earthquakes. These temporarily locked segments of otherwise active faults are known as **seismic gaps.** Once geologists identify a seismic gap, they estimate the rate of plate movement and determine the amount of strain energy built up in the fault segments. Some seismologists believe that the longer the period of seismic "silence," the more strain energy is being accumulated, and the more powerful the eventual earthquake will likely be. Figure 10-33 shows the areas surrounding the Pacific Ocean basin that have not suffered their expected share of earthquakes in recent times.

To justify an evacuation such as the one in Haicheng, seismologists monitor and analyze certain pre-quake signals, and then attempt to make short-term predictions (on the order of months, weeks, or days). Long-term prediction (on the order of years or decades) is often used to establish building codes and construction standards and to site critical facilities such as nuclear power plants. Recent efforts at long-term prediction have focused on areas where earthquake activity has been unexpectedly absent and is therefore overdue (Fig. 10-31). Recent short-term predictions have involved monitoring the buildup of stress in fault-zone rocks and other potential precursors.

Long-Term Prediction from Seismic Gaps

When two adjacent plates move, strain energy accumulates in the rocks along their shared boundary. Although this energy probably accumulates uniformly throughout the seismically active zone, it may be released irregularly, with earthquakes occurring in different places at different times. Some fault segments release their strain energy continuously as fault blocks move without locking up. This nearly constant motion, called **tectonic creep,** prevents the large buildup of strain

▲ **FIGURE 10-32** A low wall bent by tectonic creep along the Calaveras fault in California. The creeping of the fault continues to confound road-construction engineers, who can't keep the curbs straight.

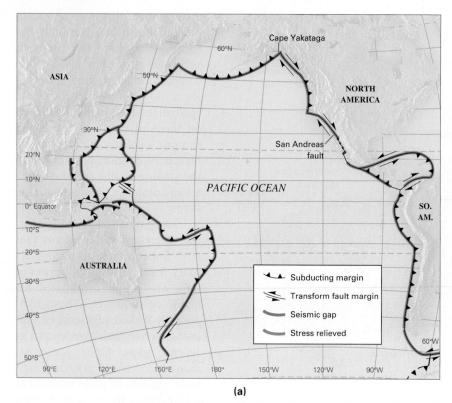

(a)

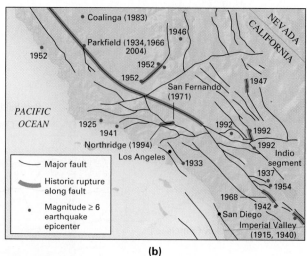

(b)

▲ **FIGURE 10-33 (a)** A map of seismic gaps around the Pacific Ocean basin. The areas in green are overdue for major seismic activity, based on statistical probabilities calculated from past recurrence intervals. **(b)** A closeup map of southern California, showing why the U.S. Geological Survey is closely watching the Indio segment of the San Andreas Fault, a seismic gap that stretches from the town of Indio northwest to Palm Springs.

In southern Alaska, the area around Cape Yakataga has been unnervingly quiet in a seismic sense for more than a century. Based on the amount of accumulated strain in the rocks, geologists predict that a magnitude-8.0 earthquake is due there soon. Fortunately, the area is sparsely populated.

The entire stretch of the San Andreas Fault in densely populated southern California has also been seismically quiet for an unusually long time. The city of Los Angeles itself has suffered only two strong earthquakes (6.0–6.9) during the last two centuries and thus is long overdue for increased earthquake activity. According to the Southern California Earthquake Center's calculations, the city has stored enough strain energy in the last 200 years along its six known fault systems (other faults remain hidden) to generate 17 Northridge-size quakes, leaving enough energy for 15 quakes of magnitude-6.7 sitting beneath the city. These studies also indicate that Los Angeles is overdue for a quake of magnitude-7.2–7.6, one that statistically occurs every 140 years or so. The area's last quake of this size occurred 210 years ago. Thus, contrary to the belief that the 1994 quake has bought them some time, Angelenos are actually enjoying a seismic lull that simply can't last forever.

Recently, however, other seismologists have vigorously challenged the seismic-gap theory. They believe that faults cannot be broken up into discreet segments that build and release strain independently. According to these seismologists, when a fault slips during a large earthquake, it increases the stresses on other nearby faults. These "anti-gap" seismologists suggest that the occurrence of a large quake actually *increases* the likelihood of other large earthquakes. In contrast, proponents of the gap theory argue that once a fault segment has released its accumulated strain energy, it quiets down and becomes less dangerous for a period of

years until significant strain energy rebuilds. Thus they propose that the likelihood of subsequent quakes *decreases*. Given that the seismic-gap theory has been instrumental in estimating future earthquake potential and has greatly influenced building codes and preparedness strategies, this lively debate seems certain to continue for years.

The role played by seismic-gap theory in structural engineering in earthquake zones is dramatically illustrated in the current construction of the 260 m (850-ft) high Tehri Dam in northern India (Fig. 10-34a). The dam is being built within the so-called "central gap," shown in Figure 10-33b. According to seismic-gap theory, this area is overdue for the next great Himalayan earthquake, one whose magnitude may exceed 8.0. Such a massive event would undoubtedly endanger the lives of hundreds of thousands of people in the holy cities of Rishikesh and Hardwar, downstream of the dam in the Bhagirathi River valley.

Short-Term Prediction by Detecting Dilatancy

When rocks become stressed, they develop countless tiny cracks long before they rupture completely. These cracks begin to appear when accumulated strain energy approaches about one-half the amount needed to produce rock failure. As cracking continues, the rocks expand in volume, or *dilate*. **Dilatancy** produces a number of side effects that geologists can monitor to make short-term earthquake forecasts, perhaps on the order of weeks, days, or even hours. Figure 10-35 illustrates how a number of these effects appear before, during, and after an earthquake. The phenomena include the following responses to stress:

1. *Swarms of micro-earthquakes.* As hairline cracks open and rocks dilate, countless micro-earthquakes (magnitude less

(a)

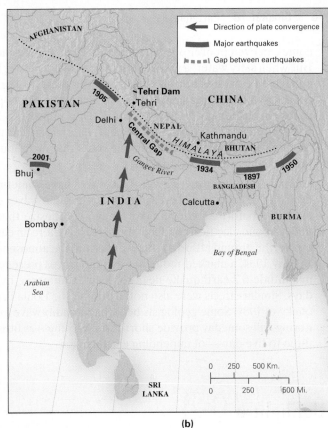
(b)

▲ FIGURE 10-34 (a) Northern India's Tehri Dam project, shown in progress, is highly controversial given its location in the Himalaya's Central Gap region. Because of ongoing compression between India and Asia, this region's seismic gap may be overdue for a major earthquake (b).

than 1) occur that are detectable only with our most sensitive seismometers. A swarm, or cluster, of these tremors sometimes precedes a major quake. Micro-earthquakes that are followed by a more powerful quake are called **foreshocks.**

2. *Tilt or bulge in rocks.* The surface of a dilated rock may bulge upward. Sensitive tiltmeters can measure changes in surface slope, and frequent land surveys can detect changes in ground elevation. In 1964, a major earthquake in Niigata, Japan, was preceded by 10 years of surface bulging near the quake's epicenter.

3. *Changes in seismic-wave velocity.* When rocks dilate and cracks develop, the velocity of seismic waves passing through them declines. Later, if groundwater fills the cracks, the seismic waves speed up again because water transmits seismic waves more efficiently than air does. Seismologists can determine the extent and rate of fracturing within rocks and the presence of groundwater by periodically setting off small explosions nearby and monitoring any changes in seismic-wave velocities.

4. *Variations in electrical conductivity.* Rock is a relatively poor conductor of electricity, and it be-

▶ FIGURE 10-35 Some phenomena caused by the dilatancy of highly stressed rocks and their relationship to seismic activity.

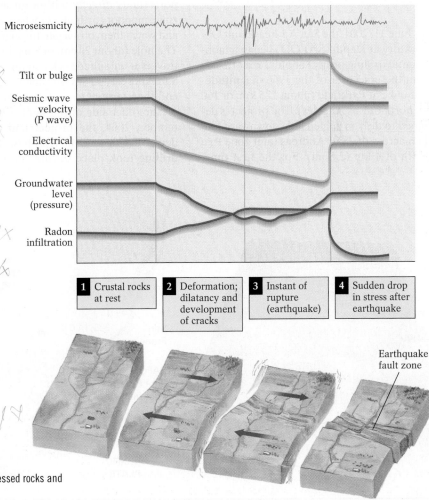

Microseismicity

Tilt or bulge

Seismic wave velocity (P wave)

Electrical conductivity

Groundwater level (pressure)

Radon infiltration

| 1 Crustal rocks at rest | 2 Deformation; dilatancy and development of cracks | 3 Instant of rupture (earthquake) | 4 Sudden drop in stress after earthquake |

Earthquake fault zone

comes an even poorer one when dilated by air-filled cracks. But when salty water, a good conductor, enters a network of connected fractures in a stressed rock, the rock's electrical conductivity increases sharply. This increase sometimes precedes an earthquake by days or hours. Thus improved electrical conductivity may indicate that water is filling new fractures in highly stressed, cracking rocks that are about to fail and generate an earthquake.

5. *Anomalous radio-wave signals.* Geologists have sometimes noticed that the atmosphere contains fewer radio waves several days before an earthquake. They hypothesize that as underground electrical conductivity increases before an earthquake, the ground absorbs radio waves from the atmosphere. In May 1984, a marked reduction in radio waves preceded a magnitude-6.2 earthquake near Hollister, California, by six days. Similar effects were also noted prior to the Loma Prieta quake of 1989. Some geologists believe that radio-wave monitoring will someday provide short-term warning—as little as three to five days—of impending earthquakes.

6. *Changes in groundwater level and chemistry.* As rocks dilate and groundwater seeps into the fresh cracks, the water level in nearby wells may drop perceptibly; some wells may even run dry. Groundwater chemistry is also monitored to detect the increasing presence of elements that normally reside tens of kilometers beneath the surface, but that rise and infiltrate the groundwater in surface wells after stressed rocks dilate and crack. Prior to recent powerful quakes in China and Armenia, geologists observed a sharp increase in radioactive radon gas dissolved in well water near the epicenters. They believe that the radon, produced by the radioactive decay of uranium in deep granitic rocks, migrated through fractures in the dilated rocks and entered the water of near-surface wells. Constant monitoring of the radioactivity of groundwater may therefore provide pre-quake warnings by signaling increased dilatancy in area rocks.

Unusual Animal Behavior   When dilatancy occurs, the cracking rocks emit high-pitched sounds and tiny vibrations that

Highlight 10-3

Drilling the San Andreas

On June 11, 2004, the San Andreas Fault Observatory at Depth (SAFOD) began a monumental drilling project about 2 km (1.2 mi) from the epicenter of the 1966 magnitude-6.6 quake at Parkfield (about 225 km, or 160 mi, north of Los Angeles). The project is designed to drill an angled hole downward and then across the San Andreas fault zone to a depth of 4 km (2.5 mi). For the first time,

geoscientists will be able to sample rocks and pore fluids *directly* within an active fault zone, monitor earthquakes where they occur, and watch them propagate through the rocks. The hole (about 20 cm, or 8 in., in diameter) traverses unfaulted rocks on one side of the San Andreas, passes through the fault zone, and then terminates in the unfaulted rocks on the other side of the fault. By mid-September, 2004, the project had reached a depth of 2760 m (9100 ft). Phase Two of the drilling took place during the summer of

2005. This plan will allow seismologists to measure the accumulating strain in both faulted and unfaulted rocks, monitor microearthquakes, and create 3-D images of the fault zone.

This effort—a component of the National Science Foundation's *Earthscope* initiative—will attempt to address fundamental questions about how earthquakes start, propagate outward, and then stop. What are the physical and chemical conditions that promote fault ruptures? What role does pressurized water play in the faulting process? Are there physical or chemical signals that precede Parkfield's quakes? What are the conditions that cause quakes and fault motion to stop? The Parkfield region was selected because it has ruptured six times in the past 150 years and it is already the most thoroughly instrumented and studied fault anywhere in the world. Our ability to sample the fault immediately before, during, and after a seismic event may help seismologists answer these questions for the first time.

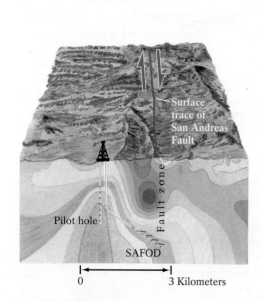

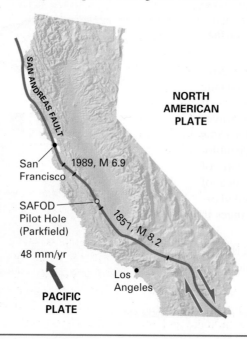

can be heard and felt by animals, but not by humans. Farmers and scientists have often observed unusual animal behavior—perhaps related to the dilatancy events—shortly before earthquakes. Dogs have been known to howl incessantly, as they did in the San Francisco streets the night before the great 1906 earthquake. Cattle, horses, and sheep refuse to enter their corrals, and ordinarily timid rats leave their hideouts and make their way fearlessly through crowded rooms. Shrimp crawl onto dry land, fish jump from water, ants pick up their eggs and migrate, and ground-dwelling birds (such as chickens) fly up into nearby trees. In cold weather, snakes leave the warm ground only to freeze at the icy surface. Such episodes of extraordinary animal behavior may serve as pre-quake warnings.

Predicting Parkfield   Consider the following series of numbers: 1857 . . . 1881 . . . 1901 . . . 1922 . . . 1934 . . . 1966. Those are the years in which Parkfield, California, was struck by a major earthquake on its segment of the San Andreas Fault. The sequence would have looked better if the 1934 was 1944, but the U.S. Geological Survey, overlooking the anomaly, concluded in 1988 that Parkfield was in the midst of a seismic cycle lasting 20 to 22 years. All of the Parkfield earthquakes were remarkably similar in nature and magnitude. Each began with foreshocks that struck within 2 km (1.25 mi) of the epicenter of the main shock, and each had a magnitude in the 5–6 range. For these reasons, in 1988, Parkfield was the subject of the first long-term earthquake prediction made by the U.S. Geological Survey. The Survey stated that there was a 95% statistical probability that a quake of magnitude-5.5 to 6.0 would occur in that area before the end of 1993.

Parkfield, a tiny hamlet of 37 citizens (which greets its visitors with a humorous sign "Now entering the North American Plate"), has been overrun by geophysicists on 24-hour call since the mid-1980s. At the cost of millions of dollars and thousands of research hours, a large concentration of instruments has been installed to "catch an earthquake in the act" by detecting foreshocks and by documenting other precursor signals that may possibly precede a major quake. On November 13, 1993, a magnitude-4.8 earthquake struck Parkfield, prompting the U.S. Geological Survey to issue a high-level alert on November 15 that predicted a magnitude-6.0 earthquake within 72 hours. After three days passed and the quake failed to materialize, the alert was canceled. Finally, on September 29, 2004, the San Andreas Fault rewarded the patience of all those seismologists with a magnitude-6.0 quake that cracked pipes, broke thousands of wine bottles, and knocked pictures off walls. Fortunately, no lives were lost, and a great mass of information was collected, including videos of actual ground motion. At this writing, seismologists are still analyzing their mountain of data.

In our never-ending commitment to learning how we might predict tragic earthquakes, geoscientists have embarked upon a bold new initiative in June of 2004, a project to drill directly into the San Andreas Fault near Parkfield to unlock the secrets of this fascinating segment of California's most famous fault. Highlight 10-3 describes to what lengths (or depths) seismologists will go to study faults and predict their next disasters.

An Early-Warning System: Would an extra two minutes make a difference?   Could an extra few minutes' warning before the deadly shaking of a powerful earthquake spell the difference between life and death? Certainly a minute would be enough time to send small school children ducking and covering beneath desks. Fire engines could roll out of firehouses, ambulances could hit the streets, air-traffic controllers could divert airliners from runways soon to be torn up by seismic waves. Using "Trinet," a network of 155 highly sensitive seismographs in the L.A. area, and newly developed super-fast computers, University of Wisconsin seismologist Richard Allen and Cal Tech geophysicist Hiroo Kanamori (the creator of the moment-magnitude seismic scale) have devised a system to provide those precious extra minutes of earthquake warning. Basically, these scientists have refined our ability to sense the arrival of P waves (typically less powerful and damaging), identify the epicenter of a quake in seconds, and then send out a warning before the arrival of the more-damaging S waves and surface waves. For areas sufficiently distant from an earthquake, the lag in arrival times between relatively non-destructive P waves and devastating S waves could provide enough warning to save hundreds or thousands of lives in a major quake. Here is yet another way that geoscientists are working overtime to prevent earthquake disasters.

And in the wake of the Indian Ocean disaster, the United States, with 150 million of its citizens living in coastal towns and cities, is investing in a much-expanded tsunami-warning system. Thirty-two new tsunami-detection buoys are being emplaced throughout the Pacific Ocean and the Caribbean Sea at a cost of about $40 million. Data from these buoys will be monitored 24 hours per day at two centers—one in Alaska, the other in Hawai'i. (Today there are only six buoys in the Pacific and none in the Caribbean.)

Can Earthquakes Be Controlled?

Although we cannot slow or stop the motion of the Earth's tectonic plates, we may someday be able to control or prevent earthquakes by reducing the energy buildup along segments of individual faults. The U.S. Geological Survey has drawn up a plan to prevent the San Andreas Fault from locking up by converting dangerous seismic gaps into zones of less-threatening tectonic creep. Figure 10-36 illustrates the proposed actions. Hypothetically, a stressed segment could be temporarily locked in place by drilling deep wells at each end of the segment (the two points in panel 1 of Figure 10-36) and withdrawing the pressurized pore water from each end. Then, ever so cautiously, water would be injected into a well in the center of the segment. In theory, this process would lubricate the fault so that its energy would be released gradually by tectonic creep, *and only between the locked ends of the controlled segment.* By tackling small segments along the entire fault, pumping water into locked areas in turn and allowing the fault to inch along, it may be possible to dissipate much of the fault's energy before it accumulates to deadly levels.

Is such an earthquake-prevention strategy really possible? No one knows if injection of water would cause *only* small displacements. What if the proposed procedure unleashed the accumulated energy of more than 130 years and caused the next great quake in the Los Angeles area? The plan's supporters believe it deserves serious consideration *after* the next great San Andreas earthquake, when the injected water would be more likely to unleash only a small accumulation of energy.

10.2 Geology at a Glance

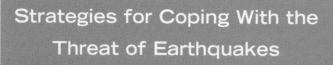

Strategies for Coping With the Threat of Earthquakes

Limit Human-Caused Earthquakes

Avoid constructing dams and reservoirs over old faults: added weight of water can reactivate faults

Refrain from disposing of waste fluids by forcing them into deeply fractured bedrock: fluid lubricates joints

Determine Past Seismic History of an Area

Establish the recurrence interval of known faults

Look for "hidden" faults in the subsurface

Build Safe Structures In an Earthquake Zone

Determine ground stability: more damage occurs on unconsolidated sediments than on bedrock

Design buildings and highways to minimize damage
- Flexible materials and construction
- Weight concentrated at lower levels
- Decorative elements well attached

Earthquake Prediction

Long-term predictions: Seismic gaps are locked segments of a fault that are "overdue" for a rupture

Short-term predictions: effects of dilatancy
- Swarms of microearthquakes
- Tilts and bulges of the land surface
- Changes in seismic-wave velocities
- Changes in electrical conductivity of rocks
- Appearance of anomalous radio waves
- Changes in ground ground water levels and groundwater chemistry (e.g., increase in radon concentrations)
- Anomalous animal behavior

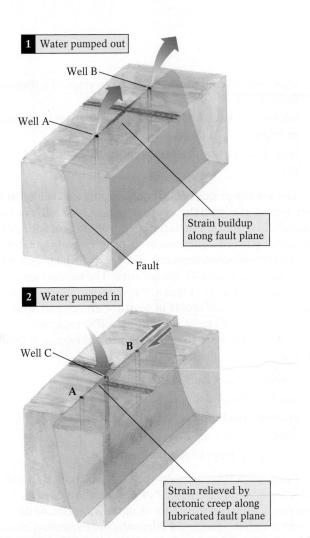

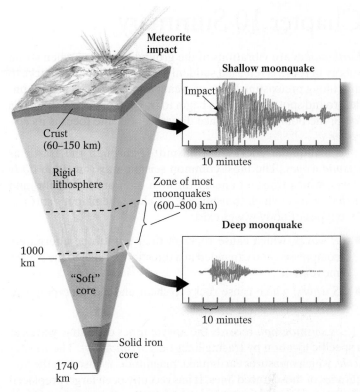

▲ FIGURE 10-37 The Moon has two centers of seismic activity—one at the surface, where meteorite impacts break rocks and generate small tremors, and one in its mantle, where fractured rocks may be stressed in response to rise and fall of the Moon's internal layers (like tides) as it is tugged by the Earth's gravitational pull.

▲ FIGURE 10-36 One U.S. Geological Survey plan to control future earthquake activity along major transform faults, such as the San Andreas, is to pump water into locked fault segments, facilitating tectonic creep and preventing a major buildup of strain energy.

The Moon—A Distant Seismic Outpost

In addition to our growing knowledge of the pattern of Earth-bound earthquakes, geologists are also learning about extraterrestrial quakes. America's *Apollo* astronauts installed highly sensitive seismometers on the Moon that have been sending data back to Earth for more than 30 years. The lunar seismometers are so sensitive that they can detect the impact of a softball-sized meteorite that strikes anywhere on the Moon's surface. Indeed, many of the recorded "moonquakes" are actually meteorite impacts.

Even so, only some 3000 weak moonquakes (magnitude 0.5–1.5) are recorded annually, compared with 1 million or more earthquakes that occur each year on Earth. The low magnitude and relative scarcity of moonquakes indicate that plate-tectonic processes are inactive on the Moon.

Those moonquakes not caused by meteorite impacts appear to be concentrated in the lunar interior at depths of 600 to 800 km (400 to 500 mi) (Fig. 10-37). Most of these deep moonquakes occur when the Moon moves closest to the Earth, suggesting that the Earth's gravitational attraction may cause slight adjustments in deep, fractured lunar rocks, prompting them to release small quantities of strain energy.

This chapter has presented many of the principles underlying the science of seismology. In Chapter 11, we will explore variations in the behavior of seismic body waves, and see what they have taught us about the structure and composition of our planet's otherwise inaccessible interior.

One final note about earthquakes: In early May 2004, one of the major television networks produced an absurd miniseries called "10.5" that attempted to portray a grand what-if scenario surrounding a magnitude-10.5 quake that ruptures the entire West Coast from Seattle to L.A. If you haven't already seen it, give it a look. It is a classic example of how the entertainment business routinely screws up the science on such shows. As a study tool before your geology exam about earthquakes, consider watching this artistic and scientific "disaster" show to see how many things are wrong with the show's earthquake science.

Chapter 10 Summary

Earthquakes are vibrations of the ground that occur when strain energy in rocks is released suddenly. This release causes rocks to shift along preexisting faults or create new faults. The precise underground spot where rocks begin to fracture or shift is a quake's *focus;* the point at the surface directly above the focus is its *epicenter.*

Earthquake energy is transmitted through the Earth as *seismic waves.* The most common seismic waves include *body waves,* which transmit energy through the Earth's interior, and *surface waves,* which transmit energy along the planet's surface. Two types of body waves exist:

- *P waves,* which cause rocks in their path to be alternately compressed and expanded in a direction parallel to the wave's movement;

- *S waves,* which cause rocks to shear and move perpendicularly to the wave's direction.

A *seismograph* records the arrival times of seismic waves at a specific location by tracing lines on a *seismogram.* The *Richter scale,* which measures earthquake magnitude, is based on the amplitudes of those traced lines. It has recently been largely replaced by the *moment-magnitude scale,* which relies on the *seismic moment* (based on the length of fault rupture and other criteria). Because seismic waves travel at different velocities, the lag in arrival times between the faster-moving P waves and slower-moving S waves at at least three seismograph locations can be used to pinpoint the epicenter of an earthquake. Quakes exceeding magnitude-3.5 can be felt by humans, magnitudes of 6.0 denote major quakes, and exceptionally powerful earthquakes that exceed magnitude-8.0 occur once every 5 to 10 years.

Some of the common effects of high-magnitude earthquakes include:

- large ground shifts along faults;
- landslides;

- *liquefaction* (conversion of wet, fine-grained sediment into a fluid mass);
- seiches (the back-and-forth sloshing of water in enclosed bays or lakes);
- large fast-moving sea waves called *tsunami;*
- fires.

The world's principal earthquake zones are located at or near plate boundaries with the deepest quakes generally taking place at subduction zones. Earthquakes at subduction zones exhibit a distinctive pattern—they occur at progressively greater depths landward from ocean trenches. The area in which such a pattern occurs is called a *Benioff-Wadati* zone. The subduction zones surrounding the Pacific Ocean basin (particularly near Japan, Alaska, and the west coasts of North, Central, and South America) are the world's most seismically active regions.

Human attempts to cope with the threat of earthquakes involve:

- limiting quakes caused by humans;
- assessing local seismic history and future risk;
- planning land-use carefully;
- building quake-resistant structures;
- developing ways to predict earthquakes.

Long-term forecasts, on the order of tens to hundreds of years, focus on the frequency of past earthquake events. Short-term predictions, on the order of months, weeks, or days, center around the phenomenon of *dilatancy,* the expansion of rock that occurs as tiny cracks open in highly stressed rocks. Dilatancy may produce tilting or bulging of surface rocks, changes in seismic-wave velocities, electrical conductivity, and groundwater levels and chemistry; anomalies in radiowave signals; and unusual behavior in animals.

KEY TERMS

QUESTIONS FOR REVIEW

1. Draw a simple diagram showing the focus and the epicenter of an earthquake.

2. Describe the key differences between P and S waves.

3. What is the fundamental difference between the Mercalli intensity scale and the moment-magnitude scale? How much more powerful than a magnitude-5.0 earthquake is a magnitude-7.0 earthquake?

4. Draw a simple sketch to illustrate how geologists use P- and S-wave arrival times to locate the epicenter of an earthquake.

5. Describe how earthquakes cause tsunami and liquefaction.

6. Why are the foci of divergent-zone earthquakes limited to depths of less than 100 km, whereas those of subduction zones extend to 700 km? Why don't earthquakes occur below 700 km? If you heard that an earthquake occurred at a depth of 300 km, what could you conclude about its geological setting?

7. How would you determine whether a given location had experienced earthquakes in the past? How could you predict the probability of an earthquake occurring there in the future?

8. Speculate about the nature of the geologic materials (solid bedrock or loose sediment) found in the map area in the figure at right. Where do you think the major faults lie?

9. If you lived in an area that had suffered no earthquakes during its past recorded history, but which then began to experience occasional tremors, how might you explain the apparent sudden onset of seismicity? Propose at least three hypotheses.

10. If you were a city planner in an earthquake-prone region and were responsible for coordinating plans for an urban center that will in-

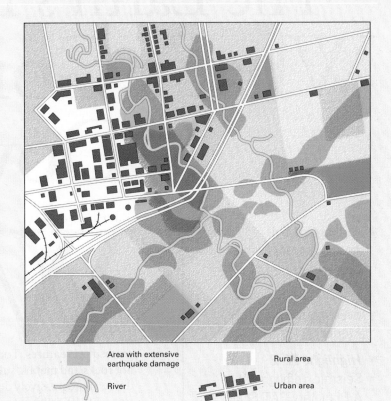

Area with extensive earthquake damage

River

Rural area

Urban area

clude schools, hospitals, fire and police stations, residential areas, parks, commercial areas, a mass-transit system, power plants, dams, and other facilities, how would you proceed in determining the optimal location of these components of the community?

LEARNING ACTIVELY

1. Get a daily newspaper—perhaps *USA Today*—for at least two weeks and check for references to each day's earthquake events. Keep a record of the locations and magnitudes of the earthquakes. Note the locations of the quakes on a world map and then compare them to the locations of plate boundaries (see Figure 1-21 on page 22). What patterns do you see?

2. If you live in an earthquake-prone area, assemble an earthquake-survival kit. If you live in a seismically quiet region, make a list of what you would put in such a kit. (You never know when you might move to California.)

3. Take a tour around your city or town and consider how its crucial buildings (such as hospitals, fire stations, schools, and police headquarters) would fare in a magnitude-7.5 earthquake.

4. Evaluate your living quarters for their degree of earthquake preparedness. Do you have heavy bookshelves or framed posters hanging directly over the head of your bed? In the event of a strong earthquake, would these projectiles come tumbling down on you while you sleep?

ONLINE STUDY GUIDE

Practice makes you better. Make the time to take the *Online Study Guide* quiz for this chapter. It should take you about 20 minutes. Automatic grading and instant feedback will help you quickly and accurately identify the concepts you have mastered and the areas in which you need more study.

www.prenhall.com/chernicoff

The Earth's Interior— What's Going On Down There?

W hat is the Earth's interior like? In 1864, author Jules Verne, in his classic *Journey to the Center of the Earth,* told us of a subterranean world filled with giant sea serpents and other grotesque creatures. Today we know what really fills the Earth's interior—layers of super-hot rocks and metals, subjected to increasing heat and pressure and becoming increasingly dense at progressively deeper levels. For today's scientists, the story of the origin and structure of the Earth's interior is at least as exciting as, if perhaps less colorful than, Verne's fantastic imaginings.

Humans have lived in space for months at a time where they used telescopes to study both our immediate galactic neighborhood and the distant cosmos. They have walked on the Moon, collecting its rocks and conducting numerous experiments on the lunar surface. They have also taken submarines to the floors of the Earth's oceans to study the scientific secrets hidden there beneath many kilometers of seawater. The Earth's deep interior, however, remains a frontier that no one is likely ever to visit *in person.* Even the deepest mines barely scratch the surface of the crust. The deepest hole ever drilled [the Kola Super-Deep Borehole, located in northern Russia and with a mere diameter of 21 cm (8.5 in), or about 8 in. at its base] extends only 0.2% of the distance to the planet's center, piercing about 13 km (8 mi) into the Earth. If we can't visit the Earth's interior *in person* and we can only puncture the very outermost parts of the crust with our drill rigs, can we get inside the Earth any other way? Highlight 11-1 offers an intriguing possibility.

Only in rare circumstances do geologists get to examine rocks that have formed more than a few tens of kilometers below the Earth's surface. Volcanic eruptions that bring up unmelted subterranean rocks within rising deep-origin magmas give us a few direct glimpses into the makeup of the Earth's interior. Rocks from perhaps as deep as 400 km (250 mi) have been found within diamond-bearing kimberlite pipes (Fig. 11-1), which are products of explosive volcanism of magma formed very deep in the Earth. Continental collisions may bury rocks to >100 km (60 mi) in the mantle, and these rocks may rise back to the surface, perhaps in response to their buoyancy relative to the surrounding mantle. Still other rocks from the Earth's upper mantle or lower crust may return to the surface at subduction zones along the major thrust faults that develop between the subducting and over-riding plates. Analysis of these exposures of rocks that have formed in or journeyed to the mantle (and back) tell us about the pressures, temperatures, and compositions of rocks at depth in the Earth, but we are still left with many unanswered questions about our planet's internal structure.

FIGURE 11-1 Because they are carried up from deep within the Earth's mantle, diamonds such as this one provide clues ▶ to the nature of the planet's interior.

Highlight 11-1

Journey to the Center of the Earth, 21st-Century Style

In the 2003 film classic *The Core* (please note sarcasm here), explorers (terranauts) burrow into the center of the Earth to fix a disastrous (and impossible) malfunction of the Earth's core—one that doomed all life on Earth. Apparently, the Earth's core, for some odd reason, had stopped spinning and with this absurd development, the Earth's magnetic field disappeared, causing migrating birds to fly about crazily and dangerously like little missiles and unshielded solar radiation to melt the Golden Gate Bridge. (By this point in the film, thoughtful geology students should be scratching their heads and chuckling.) Although the journey of humans to the core is complete fantasy, and it is technologically impossible to drill to the core (at least in the early 21st century), is there any other way to study the core—the upper surface of which is 2900 km (1800 mi) below the Earth's surface—other than *remotely* from the surface? Might we actually be able to study the core *directly,* perhaps by sending a robot probe down there?

Caltech scientist David Stephenson recently described in the journal *Nature* how a small unstaffed probe (who would volunteer for such a one-way mission?) could—hypothetically—reach the core-mantle boundary. In Stephenson's "thought experiment," scientists would make a large crack in the surface of the Earth, but this wouldn't be any old crack; it would have to be immense, such as the type of fracture that would be produced by a magnitude-7 earthquake or the detonation of a powerful nuclear bomb. The crack would then be filled with 100 million kg (100,000 tons) of molten iron. Ignoring for the moment the difficulty of creating such a crack and producing and pouring 100 million kg of molten iron into it, the rest of the process is relatively simple. Place a small (grapefruit-size) robotic probe inside the molten iron. The probe would have to be made of an alloy that wouldn't melt even at extremely high temperatures. Because the molten iron is much denser than the surrounding crust, it would sink (extending the crack) as it is pulled down by gravity, bringing the probe along with it. The crack would heal above the iron as the iron sinks downward. This would definitely be a one-way trip to the core.

Once below the crust, the iron would continue to sink through the mantle because the iron would still be denser than the ultramafic silicate rocks of the mantle. Once at the boundary between the lowermost mantle and the uppermost portion of the Earth's iron core, the iron encasing the probe would join the core, leaving the probe free to transmit *first-hand* accounts of the nature of the core-mantle boundary, including its temperature and composition. This information might help us learn more about how the core originally formed and perhaps also how the Earth's magnetic field is generated.

How long would the probe's journey take from the Earth's surface to the core-mantle boundary? The answer is, remarkably fast. You can estimate it yourself, assuming the probe travels at about 5 m (about 17 ft) per second to the core-mantle boundary at a depth of 2900 km—the speed that the crack would open under the downward force of 100,000 tons of molten iron. (The answer? roughly a one-week, one-way trip.)

Keep in mind that this was not an entirely serious proposal by Dr. Stephenson, but more of a challenging thought problem, asking, "Could it be done?" It may turn out to be the way scientists one day send a data-transmitting probe to this as yet unreachable frontier. As a taxpayer, you can decide for yourself if it would be worth the time, effort, and vast sum of money, and whether visiting the core is any more or less important than sending probes into space to visit other planets.

This chapter explains how geologists explore the Earth's interior and introduces some remarkable recent discoveries about the deep, physically inaccessible regions of the Earth's interior. You will learn how geoscientists use seismic waves to study the composition and layering of the deep Earth's interior. We then turn to some fundamental properties of the Earth—its internal heat, its gravity, and its magnetic field. We describe these properties, explore how variations in the Earth's internal composition and structure affect them, and see how geologists apply their knowledge of these properties to interpret the Earth's evolution. The magnetism in ancient rocks, for example, enables geologists to trace past movements of the Earth's plates. Because study of the fundamental properties of the Earth's interior depends on the physics of seismic waves, heat flow, gravity, and magnetism, this branch of the geosciences is called **geophysics.**

Investigating the Earth's Interior

Most of what we know about the Earth's interior comes from analysis of variations in the velocity of seismic waves. Like all waves, seismic waves tend to travel in a straight line and at an *unchanging* velocity as long as they continue passing through a homogeneous medium (that is, all one composition) at constant temperature and pressure. Comparisons of data collected at seismograph stations around the world, however, show that seismic waves typically vary in speed as they travel from the focus of an earthquake to the location of the seismograph. The waves slow down as they pass through some parts of the Earth's interior and speed up through others. These changes in speed indicate that the waves are passing through rocks and other materials of varied composition and structure and at different temperatures and pressures. From the differing speeds of seismic waves, we conclude that the Earth's interior is not homogeneous, nor do its temperature and pressure remain the same at all depths.

In the next section, we consider the behavior of seismic waves, and use that information to build a model of the structure and composition of the Earth's internal layers.

The Behavior of Seismic Waves

If the Earth were completely homogeneous in its composition and physical properties (such as the density and temperature of the rocks), seismic waves would travel through the planet at a constant speed in all directions. The waves would also travel along straight paths as they passed through the Earth (Fig. 11-2a). Recordings of seismic waves made at seismic stations around the world reveal that seismic waves must speed up in the deep Earth and that some waves are deflected from straight paths by discontinuities (sharp compositional variations) in the Earth's interior.

In Chapter 10 you learned that P waves compress the materials through which they pass and that they pass through all types of matter—solids, liquids, and gases. Such waves travel most rapidly through rigid, dense materials that easily transmit seismic energy, and more slowly through soft materials and fluids. In contrast, S waves move by shearing, a form of deformation in which material changes shape. Fluids cannot be sheared (you

can't change the "shape" of water, for example), so S waves do not travel through liquids. S waves travel through solids, however. Like P waves, they move more rapidly through rigid, dense rocks and more slowly through less-rigid, less-dense rocks.

When a seismic wave encounters a boundary—such as the contact zone between different layers of the Earth (such as the crust, mantle, and core)—some of the wave's energy will be reflected back toward the Earth's surface and some will continue across the boundary. The seismic waves crossing the boundary will not continue on their original path, however. Instead, they will bend, or *refract* (Fig. 11-2b).

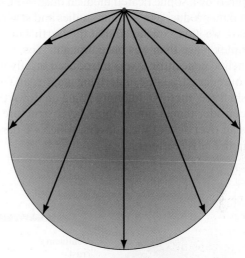

(a) Homogeneous Earth

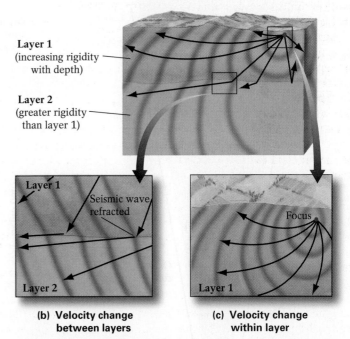

(b) Velocity change between layers   **(c) Velocity change within layer**

▲ **FIGURE 11-2** The paths of seismic waves would be straight if the Earth's interior was homogeneous—that is, made up of just one rock type all at the same temperature and pressure **(a)**. The velocity of seismic waves changes as the waves pass through various media in the Earth's interior that have different compositions and physical properties (such as greater or lesser rigidity). Seismic waves speed up or slow down as they pass *between* two layers, causing the waves to bend, or refract **(b)**. Because rigidity generally increases with increasing depth, seismic waves speed up as they descend *within* individual layers, causing their paths to curve back toward the surface **(c)**.

Using information about the velocities of P and S waves as they travel through the Earth, as well as knowledge of their paths (for example, how and where their paths are refracted), seismologists have calculated the density, thickness, composition, structure, and physical state (solid versus liquid versus *partially* liquid) of each region of the deep Earth. This long-distance approach to studying parts of the Earth that no one has ever seen first-hand (and probably will never see) has produced some dramatic discoveries in recent years as the technology for studying the geophysical properties of the Earth has improved. Highlight 11-2 discusses one such innovative approach to deep-Earth investigation, one inspired by a sophisticated medical diagnostic tool.

Based on data produced by seismic studies and several other new techniques, we have concluded that the Earth's interior is a far more complex place than was once thought. Yes, it contains the three principal concentric layers introduced briefly in Chapter 1—a thin outer crust, a large underlying mantle, and a central core. As we shall see, there's a lot more going on down there for scientists to consider than a simple three-layer structure. Figure 11-3 presents an overview of these layers, which we describe in more detail in the next sections.

The Crust

The crust—the Earth's solid rocky shell—is mostly made up of silicate-rich igneous and metamorphic rocks. It is the most accessible of the Earth's layers. We can sample continental rocks by simply knocking off a piece of any outcropping of rocks, and we can sample the seafloor by drilling holes in it from ships. From these direct methods of investigation, we know that the upper parts of continental crust are very silica-rich, and that the oceanic crust is mafic.

Seismology tells us how the crust changes with depth as well as how thick the crust is. Key information comes from earthquakes that occur at relatively shallow depths in the crust (within 50 km, or 30 mi, of the surface). The seismic waves recorded near the focus of the earthquake have traveled *only* through the crust. Hence their velocities and paths tell us about the density and composition of the crust. If we know the time and location of an earthquake, the distance from the earthquake's focus to the seismograph recording its seismic waves, and the exact time when the waves are first recorded, we can calculate the wave's velocity. By comparing velocities calculated in this way with the velocities

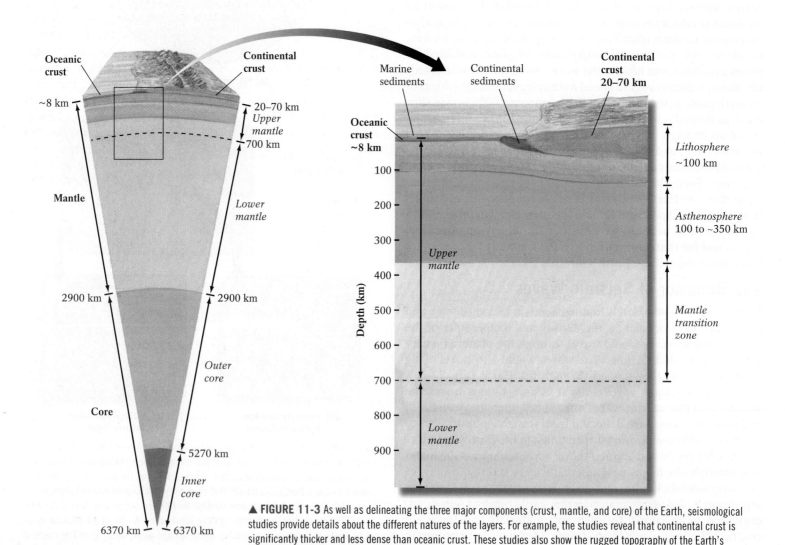

▲ **FIGURE 11-3** As well as delineating the three major components (crust, mantle, and core) of the Earth, seismological studies provide details about the different natures of the layers. For example, the studies reveal that continental crust is significantly thicker and less dense than oceanic crust. These studies also show the rugged topography of the Earth's core–mantle boundary.

of seismic waves determined by laboratory experiments with different rock types, geophysicists have estimated the composition of the deep inaccessible parts of the crust.

When an earthquake occurs in San Francisco, for example, the seismic waves that are recorded close by at Fresno State University in Fresno, California [about 290 km (180 mi) away] have traveled only through continental crust. Because we know the precise timing of some of San Francisco's past earthquakes and the precise arrival times of the P waves recorded in Fresno, we have been able to determine that P waves move through the continental crust between these cities at a velocity of about 6 km (3.7 mi) per second.

Because basaltic oceanic crust is more rigid than continental crust and therefore transmits seismic waves more efficiently, P waves travel faster through oceanic crust. Seismological studies have shown that P waves travel through basalt and its underlying plutonic equivalent, gabbro, at a rate of about 7 km (4.4 mi) per second.

What do these seismic velocities tell us about the composition of continental and oceanic crust? We answer this question next.

Continental Crust

At its thinnest points, where plates are being stretched or rifted, continental crust is generally less than 20 km (12.5 mi) thick. Where continental plates have collided to form mountains, the crust may thicken to 70 km (45 mi). In North America, the thinnest crust—in the stretched basins between mountain ranges in Nevada, Arizona, and Utah—is only 20 to 30 km (12.5–19 mi) thick. The thickest crust—in the Rocky Mountains of Montana and Alberta—is more than 50 km (30 mi) thick.

Slight variations in seismic-wave velocities indicate that the upper portion of continental crust differs in composition from the lower portion. In upper continental crust, P-wave velocity is about 6 km (3.7 mi) per second; this region comprises numerous granitic intrusions, vast subsurface regions of high-grade metamorphic rocks, and a nearly continuous blanket of sediment and sedimentary rock interrupted in places by volcanic deposits. When P waves pass through the deepest parts of continental crust, they accelerate gradually to about 7 km (4.4 mi) per second. The increased velocity is believed to reflect either a change in composition to metamorphosed gabbro or a pressure-induced increase in the lower crust's rigidity. The *average* composition of continental crust, which is characterized by near-surface felsic rocks and deeper mafic rocks, probably matches the composition of andesite or diorite. The density of continental rock varies between 2.7 grams per cubic centimeter (g/cm^3) (nearly three times as dense as water) near the surface and 3.0 g/cm^3 near its mafic base.

Oceanic Crust

Oceanic crust is more difficult to sample than continental crust because it lies below an average depth of about 4 km (2.5 mi) of seawater. This limitation has been largely overcome by new methods of deep-sea drilling that enable geologists to collect ocean-floor rocks, and by seismic studies that help estimate the structure and composition of deeper segments of oceanic crust. Together,

these types of investigations tell us much about the structure and composition of ocean crust. As you will see in Chapters 12 and 13, we can also study oceanic crust *on land* in places where ancient continental collisions have trapped slabs of oceanic crust between collided plates.

Whereas many processes contribute to the formation of continental crust, oceanic crust forms almost exclusively from eruptions of basalt and intrusions of gabbro at oceanic divergent-plate boundaries, as well as deep-sea sedimentation. As a result, the composition of oceanic crust is virtually the same everywhere. From top to bottom, oceanic crust typically consists of an average of 200 m (650 ft) of marine sediment, a 2-km (1.2-mi) layer of submarine pillow basalts, and a 6-km (3.7-mi) layer of gabbro. There are variations—in places the rocks at the seafloor are not sediments but are instead exposed basalts or even ultramafic rocks—and the thickness of the crust varies with the velocity of spreading at the divergent zone. The average density of oceanic crust is about 3.0 g/cm^3 (approximately the density of basalt and gabbro).

The Crust–Mantle Boundary

Consider this scenario: An earthquake occurs somewhere within the Earth's crust, and its seismic waves race away from the focus in all directions. P waves that traveled a fairly direct route—only through crustal rocks—may actually arrive at a seismic station later than P waves that took a longer route through the Earth's mantle (Fig. 11-4). Why? Mantle rocks are much denser and more rigid than crustal rocks, so the P waves that went the mantle route sped up and traveled much faster than the P waves that passed only through the crust. Thus, although they traveled farther, the mantle-traversing waves arrived earlier. This situation is analogous to driving a longer route on an uncrowded freeway at 65 mph rather than taking the shorter but slower route through the stoplights and traffic of city streets at 15 mph. Although you drive farther in terms of mileage, you arrive sooner than you would have if you had traveled the shorter but slower route, stuck in city traffic.

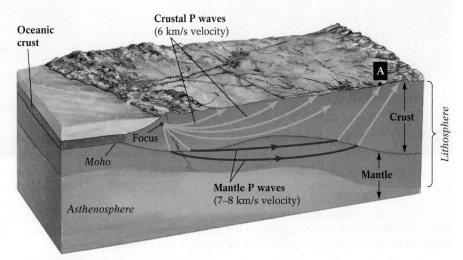

▲ FIGURE 11-4 A seismic wave that travels through the mantle may arrive at a given location faster than one that travels only through the crust—even though the crustal wave travels a shorter distance—because waves move through the mantle more quickly than they do through the crust. Here, the P waves that have traveled through the mantle have already arrived at point A, whereas those that are traveling only through the crust are still en route.

Highlight 11-2

Seismic Tomography—A Planetary CAT Scan

How do we study the Earth's interior? To examine such far-off places as the mantle, the core–mantle boundary, and even the core, geophysicists have developed their own type of CAT (computerized *axial tomography*) scan, based in principle on a diagnostic device originally designed by doctors. In medical scanning, a patient swallows or is injected with a radioactive (but safe) substance. X-ray–like images are then made from numerous angles along successive planes of a portion of the patient's body. A computer calculates the amount of radiation absorbed by the internal organs at every point along each plane, producing a series of three-

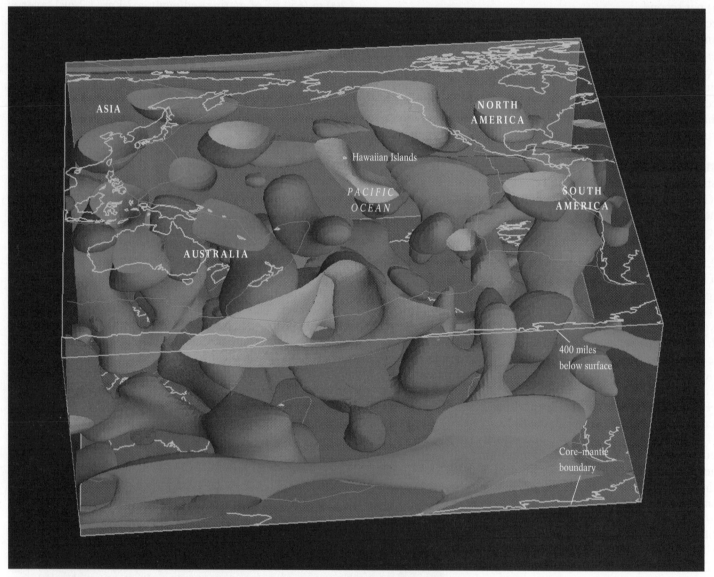

▲ (left) This seismic-tomographic image of the Earth's mantle shows hot regions (shades of reddish orange) and cold regions (blue). The continents are outlined in white. Note the warm regions in the vicinity of the volcanic Hawaiian Islands and along the coast of the Pacific Northwest. Geophysicists are analyzing tomographic data to determine the types of relationships that exist among interior heating, tectonic activity and other processes and surface features. (right) Tomographic images of subduction zones. These images indicate that subducting plates—the cold blue regions—descend to the deep mantle. The top image shows P-wave pattern. The bottom image shows S-wave pattern.

dimensional images. The sequence provides a view of every part of every organ within the scanned part of the body.

Inspired by the CAT-scan technique, geophysicists have developed **seismic tomography,** a technique that uses seismic waves to provide three-dimensional views of the Earth's interior. Here's how the concept works.

When an earthquake (or an underground nuclear blast) occurs, seismic body waves race through the Earth's interior on their way to the thousands of seismographs around the world. When seismologists know the precise timing of a tremor, they fully expect its resulting seismic waves to arrive at each of the world's seismographs at a *specific* moment following the quake, based on each seismograph's distance from the seismic event. During the last few decades, however, seismologists have puzzled over seismic waves that seem to arrive measurably early or late. What could be speeding or slowing the progress of these waves? We now believe that seismic waves slow down when they pass through warmer regions of the interior, where the rocks would be less rigid than surrounding rocks. We also believe that they speed up as they pass through colder regions where the rocks would be more rigid than surrounding rocks.

Seismologists using powerful computers have looked at data from hundreds of thousands of recent earthquakes from around the world to generate a series of cross-sectional views of the Earth. A sequence of such cross sections provides a three-dimensional image of the planet's interior.

Seismic tomography is revolutionizing our view of the Earth's interior. This technique has recently shown, for example, that the outer surface of the Earth's core is not

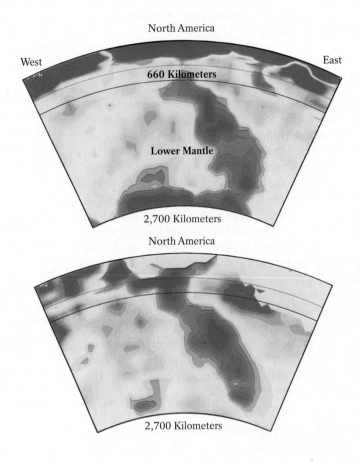

smooth and featureless, as was once believed. Instead, it projects upward into the mantle in some places, and the mantle extends downward into the liquid outer core in others. Seismic tomography has also revealed where heat is concentrated beneath the Earth's surface (such as beneath Hawai'i) and where the interior is anomalously cold (such as where cold subducted slabs of lithosphere reside). It has even demonstrated that subducting plates may descend all the way through the mantle to the core–mantle boundary.

These findings have led geologists to rethink their simple three-layer model of the Earth. As we shall discuss in Chapter 12, they have also challenged us to rethink our models for the driving mechanisms of plate tectonics.

In 1910, Croatian seismologist Andrija Mohorovičić related seismic arrival-time information to the likely paths of the waves through the deep Earth. He proposed that there must be an abrupt change in the composition of rocks beneath the crust. The boundary between the base of the crust and the underlying denser rocks at the top of the mantle is known as the Mohorovičić discontinuity, or simply, the **Moho.**

The Mantle

The mantle, the layer below the crust, is the largest segment of the Earth's interior. Accounting for more than 80% of the planet's volume, it extends to a depth of about 2900 km (1800 mi). At greater depths as pressure increases, the rock compresses and the resulting density of mantle rock increases, from about 3.3 g/cm^3 at the Moho to 5.5 g/cm^3 at the core–mantle boundary. Over the long term, the high temperatures in the mantle cause its rocks to deform and actually flow, though quite slowly. In the short term, such as the time it takes for seismic waves to pass through them, mantle rocks behave like rigid solids, allowing the passage of both compressional P waves and shearing S waves.

The minerals in mantle rocks are dense magnesium-iron silicates, such as olivine and pyroxene, and magnesium-iron oxides as well. The specific minerals and their internal structures vary as pressure and temperature increase with greater depth. Distinct mineralogical changes mark the boundaries between the sublayers of the mantle—the upper mantle, which contains the convecting partially melted mantle of the asthenosphere, the transition zone, and the lower mantle.

The Upper Mantle   Until recently, much of our knowledge about the upper 400 km (250 mi) of the mantle came from the chunks of it incorporated into rising magma or the small amounts brought toward the surface along deep faults where tectonic plates converge. In 1993, researchers successfully collected samples of mantle rock *directly* from one of the few places where the mantle lies exposed at the Earth's surface. On the Pacific Ocean floor, west of South America's Galapagos Islands, lies the Hess Deep—a rifted valley, 5.1 km (3.2 mi) deep. Within this valley, huge blocks of faulted oceanic lithosphere lie broken and tilted. In January 1993, members of a team aboard the scientific research vessel *Resolution* collected cores containing mantle rock, the first time the mantle had been pierced directly by drilling. More recently, a scientific cruise retrieved mantle rocks from the seafloor near the Mid-Atlantic ridge, and visited peridotite (a variety of ultramafic mantle rock) outcrops in a small research submarine.

A combination of techniques, such as analysis of samples from the mantle and experimental work to reproduce mantle conditions in the lab, tell us that the upper mantle is dominated by peridotite, a coarse-grained ultramafic rock containing olivine, pyroxene, garnet, and other minerals that are stable at high temperatures and pressures. Peridotite is solid under most upper-mantle conditions. There are two exceptions: Peridotite flows where it rises up at divergent zones and partially melts as pressure on the rock is reduced, and in subduction zones where the melting temperature is depressed by fluids rising from the subducting plate, enabling the overlying mantle to melt partially.

At the Moho, the velocity of seismic waves increases abruptly as they pass from the low-density rock of the crust to the higher-density rocks of the mantle. The waves' velocity continues to increase until they pass through a depth of approximately 100 km (62 mi) beneath the Earth's surface. At this depth, peridotite weakens and is less rigid than in the overlying mantle. This region, where both P and S waves slow down, is called the **low-velocity zone.** It corresponds to the layer of the Earth called the asthenosphere (Fig. 11-5).

The asthenosphere (where the mantle convects) is located at depths between approximately 100 and 350 km (62 and 217 mi) in the Earth. Although it is mostly solid, the asthenosphere contains a very small amount of melt (1% or so) in pockets and along the boundaries of mineral grains (not enough to stop S waves from passing through but enough to slow them down) (Fig. 11-6). The weakness of this layer is important, because the slow convective movement of the asthenosphere may be what transports the tectonic plates of the overlying, rigid lithosphere.

The Transition Zone   Below the asthenosphere, seismic-wave velocities increase again, indicating that the mantle rocks here are once again completely solid, dense, and rigid. A very few samples from this region of the Earth's interior [some from as deep as 750 km (465 mi)] have made it to the surface and been identified. As expected, they contain minerals that are stable at ultra-high pressures and temperatures, such as microdiamonds and a very-high-pressure variety of garnet. Because samples are so rare (and have been identified only recently), the main way that scientists study the mantle is to simulate the pressures and temperatures of this part of the Earth in laboratory experiments. Both experiments and seismic-wave data indicate that mineral structures collapse into denser, more compact forms in two places in

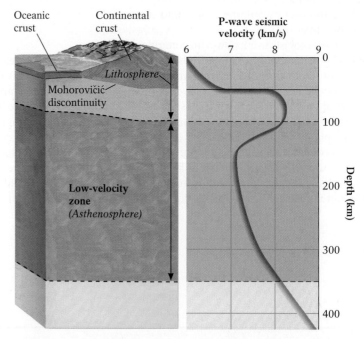

▲ **FIGURE 11-5** The velocities of seismic waves vary across the boundaries of the Earth's interior layers. Seismic discontinuities, such as the Mohorovičić boundary (or Moho) and the low-velocity zone within the upper mantle, result from the changing density, rigidity, and composition of the Earth at various depths.

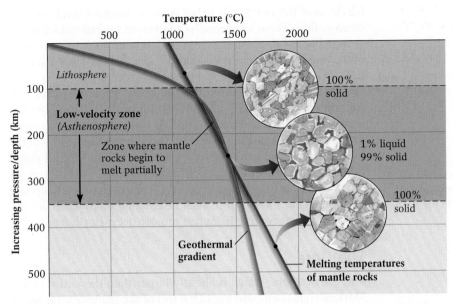

Temperature (°C)

▲ **FIGURE 11-6** The heat-softened, partially melted low-velocity zone (the asthenosphere) of the upper mantle. Above and below this 250-km (155-mi)-thick layer, mantle rocks are completely solid. The red line—the geothermal gradient—shows how the Earth's internal temperature increases rapidly with depth within the lithosphere.

the mantle—at about 400 km (250 mi) and again at 700 km (440 mi). At 400 km, magnesium olivine is compressed to form the structure of the mineral spinel ($MgAl_2O_4$) (Fig. 11-7). At 700 km, spinel and other mantle minerals, such as garnet, are compressed further to form simple magnesium oxides. These two sites of mineral transformation define the mantle **transition zone.**

Until recently, most hypotheses about the structure of the mantle suggested that the transition zone was an impenetrable barrier separating the Earth's upper mantle and crust from the deeper regions of the planet's interior. The discovery of seismic tomography, however, has begun to change this picture of the Earth's deep structure. As we are about to see, subducting lithospheric plates may periodically penetrate the barrier between the

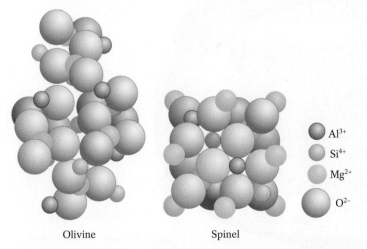

Olivine   Spinel

▲ **FIGURE 11-7** Within the mantle's transition zone, increasing pressure causes ultramafic mantle minerals to collapse into denser, more compact structures. In one such phase change, the structure of olivine gives way to that of spinel, another magnesium-rich mineral with a more compressed structure (and with aluminum ions replacing silicon ions in its crystal structure).

upper and lower mantle and collect deep in the Earth, near the base of the mantle.

The Lower Mantle   The lower mantle extends from 700 to 2900 km (from 440 to 1800 mi) below the Earth's surface. Despite extremely high temperatures, the pressure at such depths—produced by the enormous weight of the overlying crust and upper mantle—is great enough to keep lower-mantle materials solid. It also compresses minerals and crystal structures and, with increasing depth, steadily increases the rigidity of lower-mantle rocks. As a result, the velocity of P waves reaches about 13.6 km (8.5 mi) per second at the base of the mantle. Laboratory studies simulating the pressure-and-temperature conditions of the lower mantle suggest that this region of the Earth probably consists of dense magnesium silicates and oxides, such as *perovskite* [$(Fe, Mg)SiO_3$], perhaps the Earth's single most abundant mineral, and periclase [$(FeMg)O$].

The Core–Mantle Boundary   The boundary between the mantle and the core lies about 2900 km (1800 mi) below the surface, almost halfway to the center of the Earth. The lowermost 200 km (130 mi) of the mantle, in a zone ending at the core-mantle boundary, is called D″ ("D Double Prime"). One of the most fascinating and remote parts of the Earth, D″ contains areas of rough topography and regions of buoyant heat-softened rock. Within the D″ layer, seismic tomography has prompted the discovery of relatively cool slabs of rigid rock that may represent long-subducted plates. Of all the Earth's internal horizons, the core–mantle boundary has the most dramatic effect on seismic waves. In this region, we find the greatest change in the composition and rigidity of Earth materials. At the boundary, the velocity of P waves suddenly decreases from 13.6 km (8.5 mi) per second to 8.1 km (5 mi) per second, and S waves vanish altogether. Because we know that P waves slow substantially when they encounter materials of lower rigidity and that shearing S waves are not transmitted at all by a liquid, we have concluded that the outer portion of the core is a liquid. At the core–mantle boundary, highly compressed ultramafic rock with densities of about 5.5 g/cm³ gives way to a molten iron–nickel mix with densities ranging from 10 to 13 g/cm³.

Until recently, geologists thought that the core–mantle boundary, like the other thresholds between the Earth's principal layers, was a distinct, relatively featureless surface. Seismic tomography, however, has given us glimpses into the topography at this boundary (Fig. 11-8). The boundary itself may have hills and valleys on the scale of a few hundred meters, perhaps as high as 1–2 km, similar in scale to the topography of the gentler portions of the Appalachian Mountains. Using seismic tomography, we have also identified *ultra low-velocity zones* (ULVZs)—vast heat-softened regions of the lower mantle where seismic velocities drop sharply. Some geologists believe that these partially melted zones may be the sources of the great plumes of rising hot mantle material that may be driving the motion of the Earth's plates and fueling the volcanism that occurs at intraplate hot spots, such as the Hawaiian Islands and Yellowstone National Park.

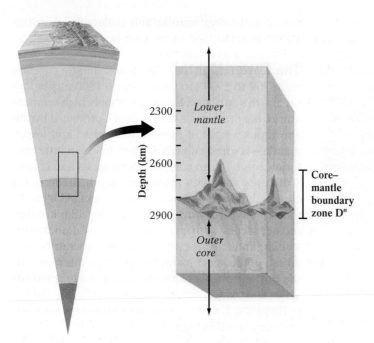

▲ **FIGURE 11-8** The Earth's core-mantle boundary may contain some rugged topography.

With each passing year, improved seismic-tomographic studies (and the emergence of more powerful computers) have enabled geophysicists to sharpen the image of the core–mantle boundary, using both natural earthquake waves and occasionally "artificial" ones. On May 21, 1992, a Chinese nuclear detonation at its testing grounds at Lop Nor generated a magnitude-6.5 shock wave that grazed the core–mantle boundary beneath the Arctic Circle near the northwest tip of Alaska. Because we know the timing of this event so precisely, analysis of seismic velocities associated with this event has provided us with the best resolution of the topography of the core–mantle boundary to date. A slightly early arrival of the seismic waves suggests that they sped up as they passed through a slab of more rigid rock, cooler than surrounding material—128 km (80 mi) thick and 300 km (185 mi) across—that hovers above the liquid outer core. This and other evidence for the presence of other anomalously cool regions in the lowermost mantle has been interpreted as old subducted plates that have piled up at the core-mantle boundary. This part of the Earth's interior may be a graveyard for plates that sink deep into the Earth (Fig. 11-9). More such fascinating discoveries are undoubtedly forthcoming from this remote deep-Earth frontier.

The Core

The Earth's core, which is slightly larger than the entire planet Mars, accounts for 3486 km (2166 mi) of the planet's total radius of 6370 km (3959 mi). Although it constitutes only one-sixth of the Earth's total volume, the core is so dense that it provides more than one-third of the planet's total mass. Because the core is located at such a great depth, pressures within this region are 3 million times higher than atmospheric pressure at the Earth's surface. Core temperatures are believed to exceed 4700°C (8500°F) at the outer core and reach more than 7600°C (13,700°F) in the inner core.

The core most likely consists of a mixture of dense metals, such as iron and nickel (at least 90% iron), along with a few other light elements (perhaps sulfur, silicon, and oxygen). Geologists hypothesize that the Earth has an iron–nickel core for several reasons:

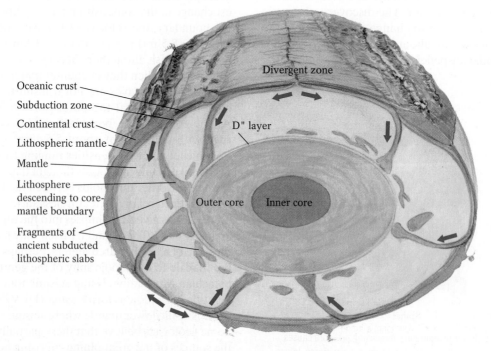

▲ **FIGURE 11-9** D″ layer near the base of the mantle. The areas of relative coolness (in blue) suggest bodies of long-subducted lithosphere.

1. This composition is consistent with seismic-wave data.

2. Some meteorites consist mostly of metallic iron. These iron meteorites are believed to be the cores of asteroids that were broken apart during collisions with other asteroids. The composition of these meteorites is similar to that inferred for the Earth's core.

3. If the entire Earth consisted of crustal material, or even of crust and mantle rocks, its density would be significantly less than the value calculated for the Earth as a whole. If the core consists of an iron–nickel mixture, the predicted density of the Earth matches the calculations. The average density of the core is estimated to be 10.7 g/cm³. Compare this value with the 2.7 g/cm³ density for continental crust.

Although the entire core consists of iron, nickel, and minor amounts of light elements, it has two layers that differ from each other in their physical properties. The pressure-and-temperature conditions in the outer part of the core are such that the metallic iron–nickel is molten. At greater depths [5270 km (3270 mi) below the Earth's surface], the pressure is so great and the temperature is not high enough for melt to exist. Thus the innermost part of the core is solid.

The Liquid Outer Core

Geologists monitoring seismic-wave arrival times after earthquakes have discovered that no direct S waves appear within an arc of 154° lying directly across the globe from an earthquake's epicenter. This region, like the shadow cast by an object that intercepts light waves, is called the **S-wave shadow zone** (Fig. 11-10). This seismic shadow must exist because S waves cannot travel through the outer core. As S waves do not travel through a liquid, geologists have concluded that a liquid region surrounds the Earth's center, and that its outer boundary is indicated by the border of the S-wave shadow zone.

P waves, which do travel through liquids, cast a different kind of shadow. Depending on the angle at which they approach the Earth's liquid outer core, P waves either pass through it and refract (bend) or reflect off its surface. These seismic waves reach the side of the planet opposite the epicenter of an earthquake, but their arrival is somewhat later than we would expect based on their initial speed because they slow down in the liquid outer core. P waves are also missing in the zone between 103° and 143° from the epicenter, for the following reason. When a P wave grazes the core–mantle boundary, it continues on a curved path to the surface, where a seismograph records it at about 103° from the quake's epicenter. When a P wave enters the liquid outer core, it bends as its velocity slows, emerging 143° or more from the epicenter (Fig. 11-11). The region between 103° and 143° of an epicenter, where no P waves arrive, forms the **P-wave shadow zone.**

The Solid Inner Core

Once they confirmed the existence of the liquid outer core, geophysicists turned to another question: Why did the P waves that passed directly through the entire planet arrive on the other side *earlier* than expected? Because P waves travel faster in a solid region, the waves' early arrival indicates that some part of the core is solid.

To test this hypothesis, seismologists analyzed the arrival times of P waves generated by underground nuclear-test blasts for which they knew exactly the time and point of origin. These studies showed conclusively that the velocity of these waves increased abruptly as they entered the inner core, supporting the idea that the inner core is solid. The change in velocity, together with the reflection and refraction patterns of P waves, established

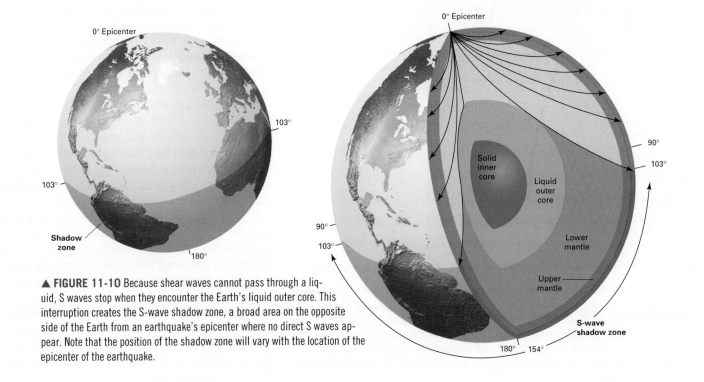

▲ **FIGURE 11-10** Because shear waves cannot pass through a liquid, S waves stop when they encounter the Earth's liquid outer core. This interruption creates the S-wave shadow zone, a broad area on the opposite side of the Earth from an earthquake's epicenter where no direct S waves appear. Note that the position of the shadow zone will vary with the location of the epicenter of the earthquake.

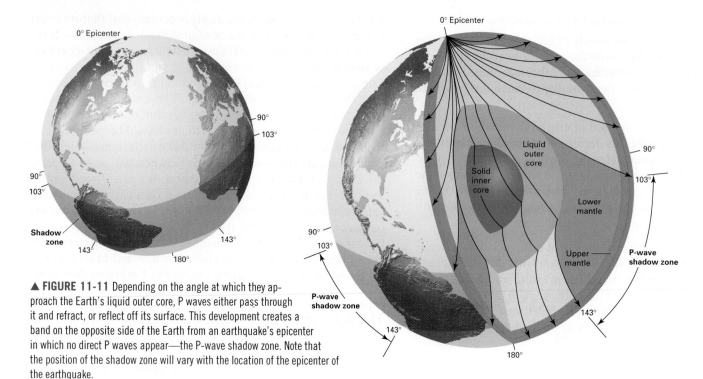

▲ FIGURE 11-11 Depending on the angle at which they approach the Earth's liquid outer core, P waves either pass through it and refract, or reflect off its surface. This development creates a band on the opposite side of the Earth from an earthquake's epicenter in which no direct P waves appear—the P-wave shadow zone. Note that the position of the shadow zone will vary with the location of the epicenter of the earthquake.

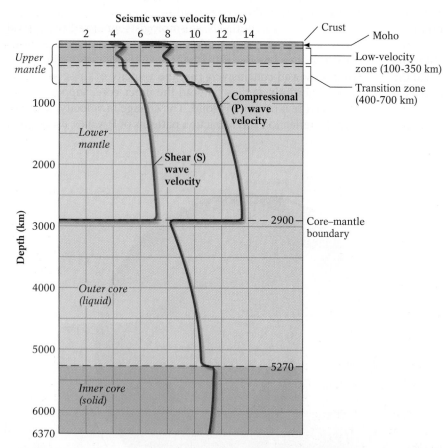

▲ FIGURE 11-12 The behavior of P and S waves as they pass through the Earth's interior layers. P waves accelerate as they cross the crust–mantle boundary (the Moho), slow down as they enter the less rigid, low-velocity zone (or asthenosphere), speed up again as they pass through the lower mantle, slow down sharply as they enter the less-rigid, liquid outer core, and then speed up again as they pass through the solid inner core. Like P waves, S waves accelerate as they cross the crust–mantle boundary and slow down as they enter the asthenosphere; they too speed up again as they enter the lower mantle, but stop altogether when they reach the liquid outer core.

that the outer core–inner core boundary lies 5100 km (3170 mi) beneath the Earth's surface. Figure 11-12 summarizes how P and S waves behave as they pass through the Earth's interior.

After they demonstrated the existence of a solid inner core, geophysicists began to think about how a solid inner core might behave within a liquid outer core. By comparing waves that pass directly through the inner core to those that just barely miss the inner core, geophysicists discovered that the inner core spins faster than the surrounding layers of the Earth. From their calculations, these scientists have concluded that the inner core spins about 1° faster than the rest of the Earth. Figure 11-13 illustrates the extra degree of inner-core rotation that occurred between 1900 and 1996. Using this rate of additional rotation, we estimate that the inner core would experience a full extra rotation every 400 years or so. Much more study is needed to explain fully why the Moon-sized solid inner core spins solo—somewhat like a planet within a planet.

The Earth exhibits some fundamental properties that we can explain with our knowledge of its internal layers. In the sections that follow, we discuss the planet's internal temperatures and the origin of its gravitational and magnetic fields. The study of these physical properties provides us with additional insight into the Earth's internal structure and composition. Before you move on, spend a moment reviewing the properties of the Earth's various internal layers in the Geology at a Glance—The Earth's Interior on page 368.

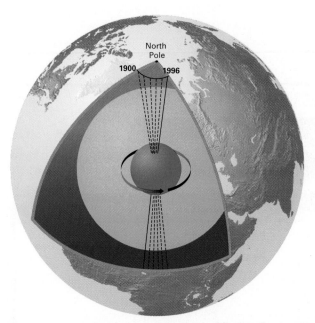

▲ **FIGURE 11-13** The solid inner core (in red) spins independently and faster than the rest of the Earth. Between 1900 and 1996, the inner core rotated an extra one-quarter revolution compared with the other interior layers.

Taking the Earth's Temperature

We can estimate the pressure on any rock in the Earth's interior by knowing its depth; there is a fairly simple relationship between pressure and the thickness and density of the overlying rocks. Yet, although we know that temperature increases with depth in the Earth, no similar simple and predictable relationship between temperature and depth has been identified. Temperature is a function of a host of variables that are more difficult to know for the deep parts of the Earth. To figure out the temperature of the Earth's interior and how it varies from place to place, we must rely on experiments and theoretical models.

We know that the temperature of the Earth's interior increases with depth from the temperatures observed in deep mines and drill holes in the crust. Gold miners in South Africa's Western Deep Levels near Johannesburg report that their mines become hotter as they descend, requiring air-conditioning to offset temperatures of 40°C (104°F) at the mine's deeper levels. In the upper segment of continental crust, the rate at which temperatures increase with depth (*the geothermal gradient,* discussed in Chapter 3) is about 25°C (45°F) per km. At this geothermal gradient, the temperature 1 km beneath your geology classroom would be 25°C (45°F) warmer than the temperature at the base of the building. If the Earth's internal temperature continued to increase at that rate, mantle rock would melt at depths below 100 km (62 mi). From seismological studies, however, we know that the mantle remains solid at that depth. Two principal reasons explain this behavior: First, the pressure at such depths inhibits melting; second, the high concentration of radioactive isotopes in crustal rocks adds heat at and near the surface. The lower con-

centrations of these heat-producing isotopes deeper in the Earth's interior reduce the geothermal gradient so we cannot simply use the near-surface geothermal gradient to estimate temperatures in the mantle.

Determining Interior Temperatures

We know from the travel times of seismic waves whether a region of the Earth's interior is solid or liquid and what the likely composition of the rocks is at depth. Using these data, we can figure out what the *maximum* temperature is for solid rock reasoning that above that temperature, the rock would melt. Similarly, we can estimate the *minimum* temperature for subsurface liquids reasoning that below that temperature, the rock would crystallize as a solid. For example, we know that the mantle is made of peridotite and that it is solid, so we can do experiments on peridotite in our labs to figure out the upper temperature limit for solid peridotite at various mantle depths and pressures. For the parts of the mantle that have small amounts of partial melt, experiments tell us the range of pressure-and-temperature conditions at which peridotite melts partially (approximately 1200°C—or 2200°F—at the pressures of the upper mantle).

We believe that the outer core is a liquid iron-nickel alloy (for the reasons discussed earlier), and we can determine the melting point of this material at extreme pressures of outer-core depths. From our estimates of the outer core's composition and from laboratory experiments, geologists calculate an outer-core temperature of 4800°C (8650°F) or higher. (If it were cooler than this, the outer core would be solid.) In the solid inner core, the effects of pressure and composition cause the material to solidify, keeping the melting points of its materials higher than the prevailing temperatures there. From inferences of the inner core's composition and estimates of its components' melting points at the extremely high prevailing pressures, geologists propose that the temperature at the inner core–outer core boundary is about 7600°C (13,700°F). (If it were hotter than this, the inner core would be liquid.)

In recent years, geologists have developed new experimental methods for estimating the extreme temperature and pressure conditions of both core regions ever more accurately. Highlight 11-3 describes some of these recent technological advances.

Heat at the Earth's Surface

When we measure the temperatures of rocks near the Earth's surface, we see that these temperatures vary from place to place. These temperatures represent the heat flowing out from the Earth's interior to the surface as the planet cools. The temperature of rocks in Minnesota is different from the temperature of rocks on the coast of North Carolina—not because the climate is different, but because the thickness of the crust is different and the types of rocks at these places are different. Rocks such as granite that are rich in radioactive elements (uranium, potassium, thorium) will be hotter than rocks, such as basalt, which contain less of these elements. Because oceanic crust consists of basaltic rocks containing low amounts of radioactive elements, we might then expect it to be relatively cool, but in reality, oceanic crust is relatively thin [8 km (5 mi) or so on average], and thus the

11.1 Geology at a Glance

The Earth's Interior

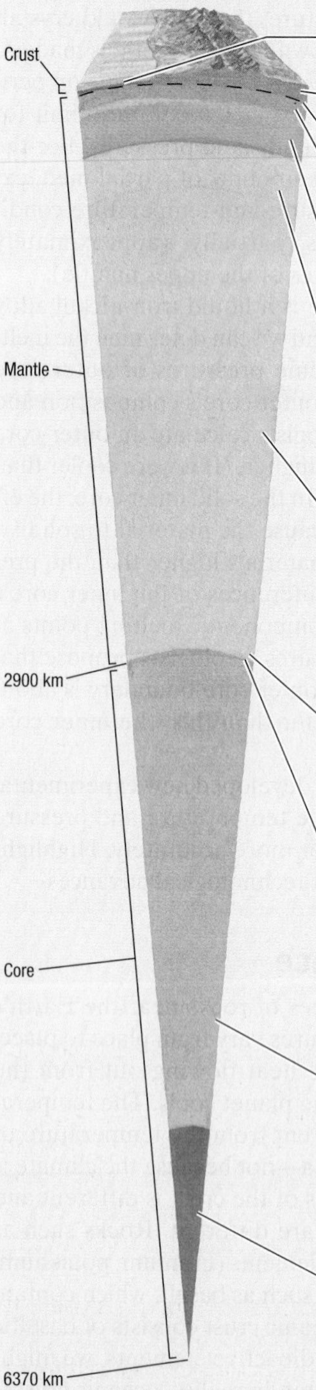

Layer/Depth	Composition	Key Physical Properties	P-wave Velocity
Oceanic crust (ca. 8 km)	Basalt	Density: 3.0 g/cm^3	7 km/sec
Continental crust (ca. 20–70 km)	Diorite	Density: 2.7 g/cm^3	6–7 km/sec
Moho			
Upper mantle (ca. 100 km)	Ultramafic silicate rocks (e.g., peridotite) Key mineral - olivine	Density: 3.3 g/cm^3	8 km/sec
Asthenosphere (aka Low-velocity zone or Partial-melting zone) (ca. 350 km)	Ultramafic silicate rocks (e.g., peridotite)	Partially melted: 1–10% liquid Convective flow	7 km/sec
Transition Zone (ca. 700 km)	Ultramafic silicate rocks Ultramafic silicates (e.g., spinel, garnet, Mg-oxides)	Collapsed mineral structures	8–10 km/sec
Lower Mantle	Ultramafic silicates (Mg- and Fe-silicates, e.g., perovskite) Ultramafic oxides (Mg- and Fe-oxides, e.g., periclase)	Density: 5.5 g/cm^3	13.6 km/sec
D" (ca. 2600-2900 km)	Ultramafic oxides and silicates	Variations in topography and temperature (due to sub-ducted cool lithospheric slabs)	8.1 km/sec
Mantle-Core Boundary (2900 km)		Partial melts (due to contact with hot/liquid outer core ULVZs (Ultra Low Velocity Zones)	
Outer Core (5270 km)	Iron-nickel metal (perhaps with small amounts of sulfur, silicon, oxygen)	Density: 10–13 g/cm^3 Liquid state (S-waves stop) Source of Earth's magnetic field	8.1 km/sec, increasing to 10.4 km/sec
Inner Core (6370 km)	Iron-nickel metal (perhaps with small amounts of sulfur, silicon, oxygen)	Solid (due to high pressure and insufficient temperature to melt)	11.2 km/sec

Crust

Mantle

2900 km

Core

6370 km

Highlight 11-3

Simulating the Earth's Core in the Laboratory

Until the last few decades, laboratory simulations of the Earth's internal heat and pressure could recreate conditions only equal to those at maximum depths of 300 km (200 mi). The jaws of a hydraulic press could produce pressures about 100,000 times atmospheric pressure—but only 1/30 of the estimated pressure at the core. A great advance came in the 1960s with the invention of the diamond-anvil, high-pressure cell. This device squeezes a sample substance between two diamonds and heats it with lasers. It can produce pressures equivalent to 3 million times atmospheric pressure and temperatures as high as 3000°C (5400°F). Although these conditions correspond to those found in the lower mantle, they still fall far short of the estimated conditions at the core.

In 1987, using further technological advances and considerable ingenuity, several laboratories successfully simulated inner-core pressures, which are estimated at 3 to 4 million times atmospheric pressure. By detonating chemical explosives wrapped around rocks that had been fitted with internal electrical sensors, scientists at the University of California at Berkeley, the California Institute of Technology, and the University of Illinois generated shock waves that compressed the rocks to within the range of core pressures. Sensors recorded various properties of the compressed rocks in the few millionths of a second before the shock waves destroyed them.

In another recent experiment, metal and plastic "bullets" were shot into a thin film of iron at a velocity of 27,000 km (17,000 mi) per hour. This study simulated the application of core-level pressures to the materials that most likely make up the core. The results of these pioneering experiments yielded new data about the behavior of Earth materials at the core's extremely high pressures. Based on the new information, scientists even revised their melting-point estimates of these materials. Previous estimates of melting points at the mantle–outer core and outer core–inner core boundaries, and hence of temperatures in those regions, had been significantly lower.

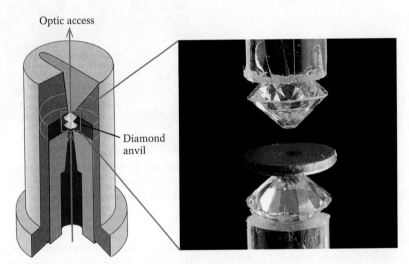

▲ A diamond-anvil high-pressure cell, a device that simulates conditions in the Earth's lower mantle by subjecting samples to extreme pressure and heat. The flat part of the diamond tip is about 0.5 mm across.

underlying hot mantle is closer to the Earth's surface in the ocean basins than on continentals. Therefore, heat at any particular place at the Earth's surface varies with the thickness of the crust (that is, the depth to the hot mantle) *and* the amount of radioactive decay that occurs in the rock (related to its concentration of radioactive isotopes).

Is the Earth Still Cooling?

The Earth has been cooling continually since its origin 4.6 billion years ago. Like all heat, the Earth's internal heat always flows from warmer to cooler areas. In the Earth's interior this transfer is done by convection and conduction, processes that continually transport heat from deep within the Earth to the planet's surface (or at least near it). The heat is then lost to the atmosphere, contributing to the planet's ongoing cooling. Because of the Earth's substantial size and the extremely low heat conductivity of rock, this cooling process has progressed quite slowly throughout Earth history and continues today. Heat from the origin of the Earth is still making its way to the planet's surface, even 4.6 billion years later.

Although the Earth continues to cool, several processes are still *adding* heat. The decay of radioactive isotopes, for example, has been giving off heat since the Earth's formation. Although the amount of radioactive isotopes is continually decreasing and the heat produced diminishing, these materials remain a significant source of heat in the Earth's continental crust and mantle. Ongoing solidification of the core also releases new heat deep within the planet. Friction produced by the motion of the Earth's plates, particularly at subduction zones, also serves as a local heat source. Nevertheless, although additional heat is still being produced locally, the *net* effect of the Earth's cooling and heating processes is that the Earth continues to cool over time.

The Earth's Gravity

The Earth hurtles through space at a speed of approximately 107,000 km (66,500 mi) per hour on its annual journey around the Sun. At the same time, a point on the equator travels around the spinning Earth's axis at a velocity of 1700 km (1060 mi) per hour. So why don't you just fly off your chair? **Gravity**, the force of attraction that all objects exert on one another, pulls you and everything around you toward the Earth's center. This force accounts for every object's *weight* at the Earth's surface. The force of gravity acts in direct proportion to the product of the masses of the attracted objects, and is inversely proportional to the square of the distance from the center of one mass to the center of the other, according to the following expression (the symbol ∝ means "is proportional to"):

$$\text{Force of gravity} \propto \frac{\text{mass of A} \times \text{mass of B}}{\text{distance}^2}$$

Note that two large masses—like the Earth and the Moon—will have a powerful gravitational attraction to one another because of their enormous masses and the relatively small distance between them (thus keeping the Moon in orbit around us). But as the distance between the bodies increases, the gravitational attraction decreases quickly because the force of gravity decreases by the *square* of the distance. Thus, a meteoroid hurtling through space close to the Earth may be drawn to our planet by a greater gravitational force than is exerted by the much larger, but more distant, Sun. The Earth's enormous mass provides the strong gravitational attraction that affects everything on or near it. Let go of that pen you're holding to see this force in action. As we'll see in Chapter 13, the same gravitational pull that caused your pen to fall also drives the Earth's landslides and rockfalls (Fig. 11-14).

Measuring Gravity

A *gravimeter* measures variations in the Earth's gravity. This device contains a mass suspended from a spring so sensitive that it can measure extremely small differences in gravitational attraction to the Earth's center. For example, a gravimeter placed on a thin rug, instead of directly on the floor, would record a reduction in gravitational attraction because of the *increased* distance to the Earth's center.

Thousands of gravimetric measurements have been taken everywhere from mountain tops to ocean floors. They show that the Earth's gravity varies significantly from place to place and from high elevations to low ones (Fig. 11-15). The pull of gravity is stronger at the poles than at the equator because the rotation of

▲ **FIGURE 11-14** Because the mass of the Earth is so much greater than that of any object on its surface, all objects—such as these fallen rocks—are attracted by gravity toward the Earth's center.

the Earth makes the planet bulge at the equator and flatten out at the poles. As a result of this bulge, the equator lies 21 km (13 mi) farther from the center of the Earth than do the poles. The pull of gravity is also somewhat lower at the top of a high mountain than at sea level because of the peak's greater distance from the Earth's center. This difference explains why a person would weigh a bit less when standing on top of a high mountain summit near the equator than when standing at sea level near the North or South Pole. (Note: A person weighing 250 pounds on Earth would weigh 7140 pounds on the Sun and only 13 pounds on Pluto.)

Gravity Anomalies

Even after we account for the effects of altitude and latitude, gravity readings are not the same everywhere on Earth. The differences between actual gravimetric measurements and the *expected* theoretical values are called **gravity anomalies.** Most gravity anomalies stem from variations in the density of the local Earth materials. These variations, in turn, affect the Earth's mass at a given location and consequently, its gravitational pull.

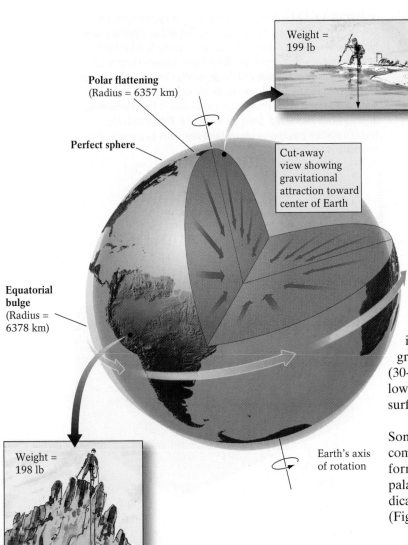

Weight = 199 lb

Polar flattening
(Radius = 6357 km)

Perfect sphere

Cut-away view showing gravitational attraction toward center of Earth

Equatorial bulge
(Radius = 6378 km)

Weight = 198 lb

Earth's axis of rotation

◀ **FIGURE 11-15** The gravitational attraction of the Earth varies from place to place on its surface. Because the planet is not spherical and its surface is irregular, both latitude and elevation determine an object's distance from the Earth's center and consequently affect its weight. A person who weighs 199 pounds at sea level at the North Pole would weigh only 198 pounds in the equatorial mountains of South America, where the gravitational pull is weaker. So if you need to lose a pound or two, head to Cotopaxi volcano, right on the equator in Ecuador.

ample, where high-density mantle rocks lie relatively close to the surface below thin oceanic crust. They also arise where continental crust has thinned substantially, perhaps prior to plate rifting. One such positive anomaly stretches southward along the Minnesota–Wisconsin border from Lake Superior through Iowa and Kansas to Oklahoma (Fig. 11-17 left). Known as the "mid-continental gravity high," this anomaly marks the location of an ancient rift in the North American plate into which basaltic magmas rose 1.1 billion years ago. The rifting stopped before the continent divided completely, leaving mid-continental basaltic flows at the surface. Today, high-gravity readings there record the presence of a shallow, 50-km (30-mi)-wide zone of dense subsurface rock surrounded by the lower-density sedimentary rocks that typify the Midwest's near-surface geology.

Not all positive anomalies are associated with thinning crust. Some may occur where masses of ultramafic mantle rocks become caught at or near the surface between colliding plates and form part of such continental mountains as the Alps and Appalachians. Higher-than-expected gravity readings may also indicate concentrations of dense valuable metallic ores in the crust (Fig. 11-17 right), another practical aspect of gravity studies.

Negative gravity anomalies—attractions *lower* than the expected value—arise where relatively low-density rocks or sediments below the surface lie amid denser surrounding rock. Negative anomalies often occur in continental mountain ranges, which generally have deep roots of low-density felsic rock embedded in the higher-density rocks of the upper mantle. They also occur along oceanic trenches, which are typically filled with water and low-density sediment, and therefore record a lower gravitational attraction than the denser oceanic crust that surrounds them.

Analysis of negative gravity anomalies can have practical applications as well. Along the Gulf Coast of Louisiana and Texas, lower-than-expected gravity helps us locate subsurface low-density salt deposits. These deposits were originally produced by evaporation of ancient oceans and are now surrounded by denser sedimentary rocks (Fig. 11-16). The practical value—such impermeable salt deposits often trap migrating crude oil. As a result, gravity studies are sometimes used to search for oil.

Positive gravity anomalies—attractions *higher* than the expected value—mark the presence of high-density rocks below the surface. These anomalies often occur over ocean basins, for ex-

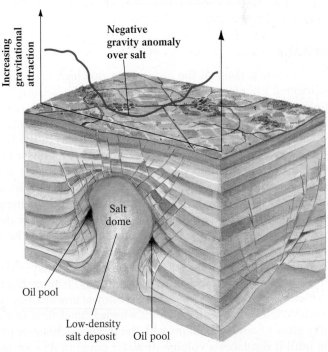

Increasing gravitational attraction

Negative gravity anomaly over salt

Salt dome

Oil pool

Low-density salt deposit

Oil pool

▲ **FIGURE 11-16** A negative gravity anomaly may be caused by a concentration of low-density salt below the surface.

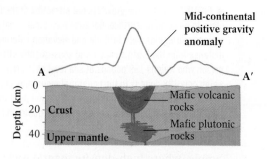

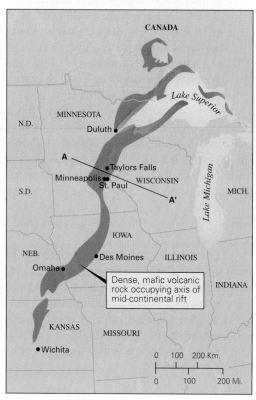

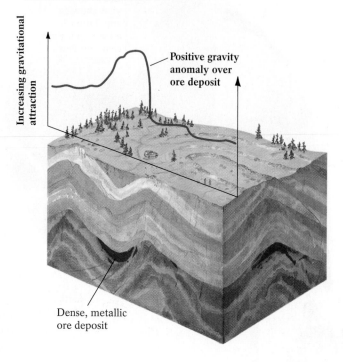

▼ **FIGURE 11-17** Positive gravity anomalies are caused by the presence of high-density rocks near the surface. (left) The mid-continental gravity high, a nearly continent-long positive gravity anomaly found in North America, probably formed when the North American plate began to rift 1.1 billion years ago and a body of dense, mantle-derived, mafic igneous rock intruded into the less-dense continental rocks of the Midwest. The surface expression of this rifting event can be seen in the basaltic flows at Taylors Falls, Minnesota, along the Minnesota–Wisconsin border. (right) A positive gravity anomaly may also be caused by a near-surface concentration of dense metallic ore.

Isostasy

Why do some regions of the lithosphere ride high (such as the continents), whereas others sink low (such as the oceans)? The answer to this question lies in the fact that the Earth's lithosphere "floats" on the underlying denser, heat-softened, partially melted asthenosphere of the upper mantle. Because areas of the lithosphere vary significantly in density and thickness, gravity pulls some portions of the lithosphere (those with more mass) down farther into the asthenosphere. Furthermore, because the weight and dimensions of any given segment of the lithosphere changes over time, its position relative to the mantle changes over time. This equilibrium between lithospheric segments and the asthenosphere beneath them is called **isostasy.**

When an iceberg breaks off a coastal glacier, it may initially plunge below the surface of an adjacent bay. It immediately bobs back up, however. The iceberg is buoyed upward, because ice is less dense than water. (That's why ice cubes float in your water glass.) Isostasy explains why the iceberg adjusts its position until it displaces a volume of water equal to its own total weight. Water is 10% more dense than ice, so a volume of water

equal to the weight of the iceberg will amount to 90% of the iceberg's volume. Thus every iceberg, like every ice cube floating in a water glass, keeps 10% of its volume above the surface of the water and 90% below it (Fig. 11-18 left). If some of the ice melts, the iceberg immediately rises, adjusting its position isostatically to maintain the same 10:90 proportion of ice above and below the water.

In a similar manner, each segment of the Earth's lithosphere, "floating" on the denser underlying asthenosphere, rises or sinks to achieve its own isostatic equilibrium (Fig. 11-18 right). (*Note:* Technically, the lithosphere doesn't actually float like an ice cube on water because the asthenosphere is not a liquid; instead it "rides" as a lower-density layer atop a higher-density, heat-softened, slowly flowing layer beneath it.) Because continental lithosphere is less dense than oceanic lithosphere, it is more buoyant—that is, a larger proportion of it "floats" above the asthenosphere. Because continental lithosphere is much thicker than oceanic lithosphere, however, it extends much deeper into the asthenosphere. If we could look below the surface of the lithosphere, we would see that tall continental mountain ranges have proportionately deep "roots" extending far into the asthenosphere. Oceanic lithosphere, which is much thinner than continental lithosphere, does not extend nearly as far into this layer.

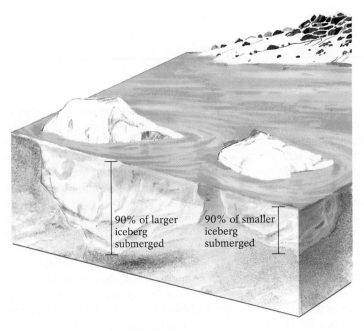

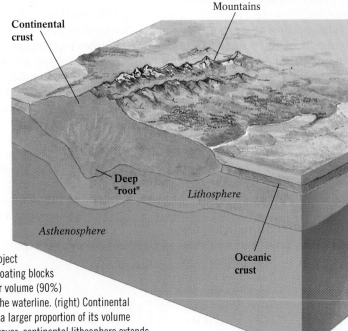

▲ **FIGURE 11-18** The principle of isostasy states that the depth to which a floating object sinks into underlying material depends on the object's density and thickness. (left) All floating blocks of ice have the same density; thus they sink in water so that the same proportion of their volume (90%) becomes submerged. The thicker the block of ice, the greater this volume will be below the waterline. (right) Continental lithosphere, because it is less dense than oceanic lithosphere, is more buoyant and has a larger proportion of its volume above the asthenosphere. Because it is so thick compared with oceanic lithosphere, however, continental lithosphere extends farther into the asthenosphere (that is, it has a deep "root").

Each segment of the lithosphere constantly adjusts isostatically as weight is added to its surface in some places and removed in others. The isostatic response to variations in crustal loads is similar to the response of a ship to the loading or unloading of cargo: The more cargo, the deeper in the water the ship will ride. If we unload the cargo, the ship rises. As lava flows, accumulating sediments, or advancing glaciers add weight to a given segment of the Earth's crust, that segment gradually sinks deeper into the asthenosphere. When materials are removed from a given segment of the crust—through erosion of a hill or melting of a glacier, for example—that segment slowly rises, or *rebounds*. Figure 11-19 shows some of the isostatic adjustments that occur when mass is added to or removed from the lithosphere.

The up-and-down isostatic movements of continents and ocean floors cause the Earth's shoreline to move landward and seaward. Highlight 11-4 considers what seems like a simple, almost silly question: Why is the ocean near the shore? There's actually a bit more geological thought needed to answer this question. Read on.

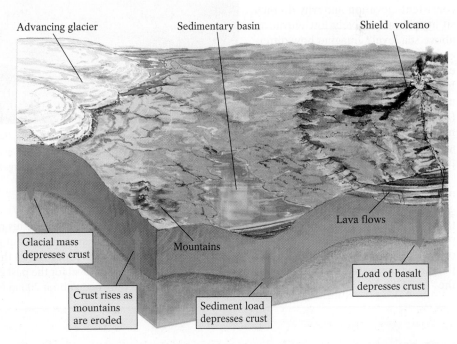

▲ **FIGURE 11-19** Isostatic adjustments in the Earth's crust. As weight is added to a segment of the lithosphere, the segment tends to sink. As weight is removed from a segment of the lithosphere, the segment tends to rise.

Highlight 11-4

Why Is the Shore so Near the Ocean?

Oh I could tell you why
The ocean's near the shore
If I only had a brain
        The Scarecrow in the "Wizard of Oz"

So why is the ocean near the shore? This apparently simple question is related to some surprising facts about the elevation of the continents and the depths of the oceans. The Earth has a tremendous range in elevation, from its highest peak, Mount Everest, more than 8000 m (>29,000 ft) above sea level, to the great depths of the Marianas Trench (a subduction zone), nearly 11,000 m (>35,000 ft) below sea level. Despite this range, the vast majority of the Earth is at one of only two elevations. The seafloor of the world's oceans has a mean depth of about 4000 m (about 13,000 ft), reflecting the fact that oceanic crust has essentially the same thickness and composition worldwide and thus rides isostatically on the Earth's asthenosphere at roughly the same position. What is remarkable, and a bit of a puzzle, is the fact that approximately 60% of the continental crust sits at an elevation of less than 500 m (about 1600 ft).

Why do the continents have such a consistent elevation and why do they sit at such a relatively low elevation above sea level? Imagine how different the world would be if the average elevation of the continents was just a few hundred meters lower than it is now. We would live in a "waterworld" (did you see that Kevin Costner movie?), with only a few islands rising above the sea in places that now have the Earth's highest mountains.

The average height of the continents *is not a random number.* Several geological processes combine to maintain this height, and apparently these processes have operated over most if not all of geologic time. Areas of the Earth's crust that are under water, such as the margins of the continents—areas known as the *continental shelves*—don't erode, but instead are sites of deposition. The elevation of these low-lying regions is built up as sediments are deposited, and may rise further by tectonic events that thicken the crust. In places above sea level, erosion is constantly at work, wearing down high regions until they approach sea level. So, low areas are built up to sea level by sedimentation, and high areas are worn down to sea level by erosion, keeping the continents at a near-constant height, near sea level.

Worldwide sea level changes with time in response to such processes as the advance and retreat of ice sheets. When ice sheets grow (and advance), they do so by taking water from the oceans, thus lowering sea level. When ice sheets melt (and retreat),

▲ Sea level along these coasts and low-lying islands in the Pacific's Solomon Islands has not changed an enormous amount during Earth history. A rise of a hundred meters, however, would submerge the island in the foreground.

they return water to the oceans, thus raising sea level. Yet despite such significant global changes, the geologic record shows that the overall change in sea level for the past 2.5 billion years has only been about 200 m (about 650 ft)—a very small amount compared to the great range in elevations on Earth. This is intriguing because we generally believe that the continents have been growing over time as magma is added to them, for example, at volcanic arcs above subduction zones. The Earth as a whole is *not* expanding, so as the continents grow, they take up more of the Earth's surface and leave less area for the ocean basins. If the ocean basins are losing area, why don't the oceans just overflow their boundaries, drowning the continents?

Consider the following: Since its formation 4.6 billion years ago, the Earth has been cooling. Heat is escaping from the Earth's interior, and through time, radioactive decay provides less and less heat as the Earth's supply of radioactive isotopes is reduced as they decay to non-radioactive daughter isotopes. As the Earth cools, less partial melting takes place at divergent zones, and ocean crust becomes progressively thinner. This thinner oceanic crust is not as buoyant as the thicker oceanic crust of the early Earth, and so the ocean basins deepen over time as less buoyant crust sinks to a deeper position. The deepening of the ocean basins in response to the ongoing cooling of the Earth thus balances the ongoing growth of continents. These deeper oceans can then accommodate the world's ocean waters without swamping the continents. In this way, the continents have remained at or near current sea levels for much of Earth history, even as the planet's oceanic and continental crust have changed substantially in thickness and volume.

This relationship between the continents and oceans is yet another indication of how Earth processes of the planet's various spheres (in this case the lithosphere and hydrosphere) interact to create a balance, even as the Earth changes dramatically through time. And of course, all this affects the biosphere, including us humans, whenever we head off in search of a good beach at our shorelines.

Magnetism

About 2500 years ago, the Greeks noticed that some dark rocks near the Asia Minor city of Magnesia (then part of ancient Greece, but today part of western Turkey) had a strange property: Objects made of iron became attached to them. For centuries magnetic rocks were thought to have wonderful abilities—to comfort the sick, stop hemorrhages, cure toothaches, increase a person's gracefulness, and even repair failed marriages. A truly practical purpose for magnetic rock was discovered 1500 years ago by the Chinese, who observed that a suspended sliver of such rock would turn freely and always come to rest in the same position. This discovery led to the invention of the magnetic compass. In 1600, England's Sir William Gilbert proposed that the Earth itself behaves like a giant magnet, setting the stage for our modern understanding of magnetism.

Magnetism is the force, associated with moving charged particles (such as electrons), that enables certain substances to attract or repel other materials. In our discussion of minerals in Chapter 2, we examined the role of an electrical charge in attracting charged particles. Magnetism in minerals arises not from the charges on the particles, but rather from the *motion* of charged particles. Electrons, for example, are negatively charged particles that are in constant motion; certain patterns of electron motion create magnetic properties in some substances.

Electrons move not only by orbiting the nucleus of an atom, but also by spinning about their own axes. Each orbiting, spinning electron creates and is surrounded by its own region of magnetic influence, known as its **magnetic field.** Every substance contains vast numbers of spinning electrons, so why then are only certain substances magnetic? When electrons spin randomly, their magnetic fields cancel out one another; consequently, the substance containing them is not magnetic. In certain substances, however, electrons align their magnetic fields and tend to spin in one direction more than another. The magnetic fields of these electrons enhance one another, giving the substance itself an overall magnetic field. When magnetic substances are subjected to a strong *external* magnetic field, their internal fields become further aligned. They then become more strongly and, in some cases, permanently magnetized. We measure the intensity of a magnetic field with a sensitive device known as a *magnetometer.*

Although magnetic fields are invisible, we can see the pattern created by their lines of magnetic force by shaking iron filings onto a sheet of paper placed over a bar magnet. The filings form a pattern that shows numerous closed loops of magnetic force emerging from the magnet's north pole and entering its south pole (Fig. 11-20).

Certain compounds of iron, such as the mineral magnetite (Fe_3O_4), are strong natural magnets. The rocks of Magnesia, for example, contain a high concentration of magnetite. Mafic rocks, such as basalt and gabbro, usually contain some magnetite. Granite does as well. Other iron compounds, such as hematite (Fe_2O_3), become temporary magnets after exposure to a strong external magnetic field; they lose their magnetism following the removal of the external field.

▲ **FIGURE 11-20** The magnetic field of a simple dipolar bar magnet includes a north pole, from which the lines of magnetic force emerge, and a south pole, where the lines of magnetic force reenter the magnet. In the same way that electrically charged particles attract opposite charges and repel like charges, magnetic north and south poles attract one another, while like poles repel one another.

A rock can be *demagnetized* by chemical weathering, lightning strikes, and heat, all of which can alter the orientation of the spinning electrons. Heat, for example, causes electrons to vibrate and shift from their aligned positions into random orientations. Magnetite loses its magnetic properties completely when heated to temperatures exceeding 580°C (1075°F), the so-called *Curie point* (named for the French chemist, Pierre Curie).

The Earth's Magnetic Field

The Earth's magnetic field penetrates and surrounds the planet. It extends into space for more than 60,000 km (37,000 mi) beyond the Earth and its atmosphere. Every cubic centimeter of the planet is bathed in this field's invisible lines of magnetic force. Because these lines of force curve, the angle at which they intersect the Earth's surface varies from place to place. As shown in Figure 11-21, they are perpendicular to the surface at the planet's magnetic poles and parallel to the surface at the magnetic equator. At all points in between, the lines of magnetic force intersect the surface at an angle that increases toward the magnetic poles and decreases toward the magnetic equator. As we will see, the Earth's magnetic north and south poles can move and even switch places—a phenomenon that can help geologists tell the history of the Earth's rocks. (*Note:* During a magnetic reversal, the Earth's geographic poles stay put—that is, the planet doesn't physically flip over north to south. Only the invisible magnetic poles and the lines of magnetic force shift polarity.)

Regarding the hard-to-imagine phenomenon of magnetic reversal, there are several important questions to consider: Why does the polarity of the Earth's magnetic field reverse itself from time to time? How long does it take to complete a reversal? How might such a reversal affect life on Earth? To begin to answer these questions, we need to understand why the Earth has a magnetic field in the first place.

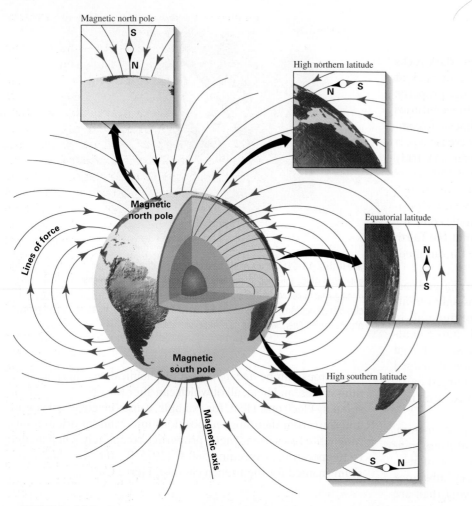

▲ FIGURE 11-21 The Earth's prevailing magnetic field. Unlike those from the simple bar magnet shown in Figure 11-20, lines of magnetic force emerge from the Earth at the planet's magnetic south pole, located near McMurdo Sound in Antarctica, and reenter at the planet's magnetic north pole, near Prince of Wales Island in the Canadian Arctic. Thus, essentially, the Earth's *magnetic south pole actually lies in the Northern Hemisphere* and the *magnetic north pole lies in the Southern Hemisphere*. That's why a magnetic-north-seeking compass needle points toward the *geographic* north—it's actually attracted to the *magnetic* south pole (which, as we just discussed, lies in the northern hemisphere). Note that the planet's magnetic poles do not correspond exactly to its geographic North and South Poles. In aligning itself along the planet's magnetic force lines, a freely suspended magnetic needle at the Earth's surface would orient perpendicular to the surface at the Earth's magnetic poles, parallel to the surface at the magnetic equator, and at various angles to the surface at all points in between.

The Origin of the Earth's Magnetic Field
Although solid iron-bearing metallic minerals are responsible for *local* magnetism at the Earth's surface, the Earth's planetary magnetic field is not caused by a giant solid magnet in the Earth's interior: the core is much too hot. The high temperature of the Earth's deep interior—far higher than the Curie points of all known magnetic minerals—makes a permanent magnet in the Earth's interior impossible. Most naturally occurring magnetic minerals can exist only within the Earth's upper 30 km (20 mi); below that depth, they would be demagnetized by the Earth's internal heat. The rapid and frequent changes in the polarity of the Earth's magnetic field also argue against a permanent internal magnet. Thus, the Earth's global magnetic field cannot be a product of simple rock magnetism. What, then, could generate an Earth-size magnetic field?

Our present understanding of the origin of the Earth's magnetic field comes from our knowledge of the movement of electrons. In the nineteenth century, physicists discovered that a flowing electrical current creates a magnetic field. Such a current can be generated *spontaneously* by moving an electrically conductive substance through an existing magnetic field, in a system known as a *self-exciting dynamo*. Today we have put this knowledge to use to generate electricity in power plants by rotating an electrical conductor in a magnetic field. A self-exciting dynamo induces itself to produce more electricity, which in turn produces a stronger magnetic field, which then produces more electricity, and so on.

The Earth functions as a gigantic self-exciting dynamo, with its magnetic field being generated by an electrical current (Fig. 11-22). The electrical current is produced by movement of electrons in the molten iron in the liquid outer core. The rotation of the Earth sets the liquid in motion, and these moving currents then generate the Earth's magnetic field. The field in turn generates more electrical currents. In this way, the system is sustained and enhanced.

Magnetic Reversals   The Earth's magnetic field changes frequently (in geologic-time terms), perhaps more often than any global feature other than climate and sea level. The location of the magnetic poles changes over time with respect to the geographic poles, although *on average* the magnetic poles do coincide closely with the geographic poles. At present, the magnetic north pole is drifting westward at a rate of about 0.2° per year. At intervals averaging a half-million years, the Earth's magnetic field actually reverses—magnetic north and magnetic south have exchanged places hundreds, perhaps even thousands of times during the Earth's history. The effects of such **magnetic reversals** were first discovered in France at the turn of the century, when scientists noticed that some layers of volcanic rock were magnetized in the opposite direction as other layers.

Magnetic reversals result from variations in outer-core convection. Turbulence arising from this variable flow may strengthen or weaken the magnetic field. During a magnetic reversal, the Earth's field may first weaken substantially for several centuries. The field's polarity then flips, and as a new reversed field gradually develops, its strength may fluctuate erratically. Eventually the new field builds until it becomes well established. Most researchers believe that it takes 1000 to 5000 years to complete a magnetic reversal—from one stable field to another of reversed polarity. A reversal recorded in the layered volcanic rocks of Steens Mountain in southeastern Oregon, however, suggests that reversals may occur even more swiftly, perhaps on the order

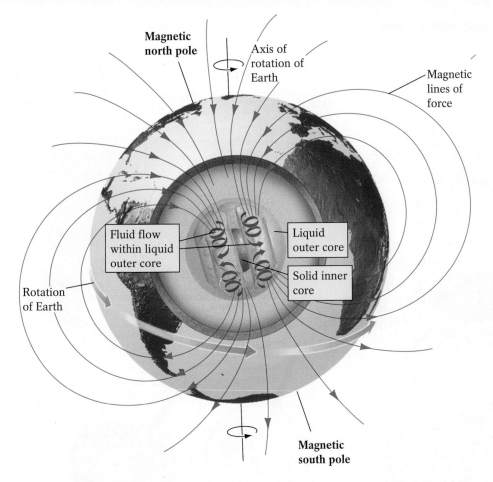

Fluid flow within liquid outer core

Liquid outer core

Solid inner core

Magnetic north pole

Axis of rotation of Earth

Magnetic lines of force

Rotation of Earth

Magnetic south pole

▲ FIGURE 11-22 The Earth's magnetic field is generated by the flow of electrically conductive fluid in its outer core.

emphasis on *may;* we don't know for sure since we've never seen one).

What effects might a reversal in the Earth's magnetic field have on surface-dwellers? Although little is known for certain, such field reversals may profoundly affect life on Earth. The strength of the reversing field may weaken until virtually no magnetic field exists for a brief time. Because the magnetic field shields the Earth's surface from a significant portion of incoming solar radiation, the weakened field and then the brief absence of a field during a reversal may expose life on Earth to a much higher level of ultraviolet radiation for a brief period. Conceivably, people with fair skin might be well advised to use sunblock with an SPF of 1000 during a reversal. After the polarity reverses, the strength of the field rebuilds and the Earth's surface is again protected from ultraviolet radiation.

We watch these magnetic-field changes quite closely because if this trend of a weakening field were to continue, a reversal could occur about 2000 years from now. Such an event could sap the energy from our power grids, expand the ozone holes in the atmosphere, and confuse all sorts of migratory animals—birds and fish, in particular—that navigate by an internal compass that relies on the Earth's magnetic field for guidance. A reversal would probably affect the well-being of all surface-dwelling organisms—but exactly how, we do not yet know.

Paleomagnetism   The geologic record often contains evidence of **paleomagnetism,** a permanent record of past changes in the Earth's magnetic fields. Such changes are most obvious in mafic igneous rocks, such as basalt, and in lake and marine sediments. As mafic lava cools, it forms small crystals of magnetite. The magnetic fields of these crystals, for a while, are free to align themselves with the lines of magnetic force of the Earth's prevailing magnetic field (Fig. 11-23a). By the time the entire lava body has solidified, the magnetite crystals and their magnetic fields have become locked in place, preserving the alignment of the Earth's magnetic field at that time. Similarly, when magnetic grains settle through a relatively still body of water, such as a lake or protected bay, the grains themselves are free to rotate like compass needles and become aligned with the Earth's field as they fall (Fig. 11-23b). Once they are buried by later sedimentation, the grains can no longer rotate freely, and thus their magnetic orientation is locked in place. In this way, magnetite-rich sediments and lava create a long-term record of the Earth's past magnetic fields.

The quality of the magnetic-field record varies from place to place, particularly from land to the seafloor. The terrestrial ("on land") record of the Earth's magnetic-field reversals, for example, is often lost, falling victim to weathering and erosion. Thus, it is, at best, incomplete. The most complete record appears on the

of only hundreds of years. But how do reversals actually occur? Highlight 11-5 illustrates an exciting recent insight into this mysterious process.

Today's polarity, which is characterized by lines of magnetic force that emerge from the magnetic south pole and reenter the magnetic north pole (see Figure 11-21), began about 780,000 years ago; this polarity is referred to as *normal.* (The lines of force for a normal magnetic field—as shown in Figure 11-21—are actually oriented exactly *opposite* to those in a typical bar magnet, as shown in Figure 11-20.) When the Earth's magnetic field reverses, the lines of force emerge from the magnetic north pole and reenter the magnetic south pole, creating what we call *reversed* polarity. At present, the needle in a hand-held compass points toward the north; during a period of reversed polarity, the needle would point toward the south.

Although there's no cause for alarm, recent geomagnetic evidence indicates that about 150 years ago, the Earth's magnetic field began to weaken; its strength has waned as much as 15% and its deterioration appears to be accelerating. Geophysicists debate whether or not this weakening signals an impending magnetic reversal, but to monitor these developments, the European Space Agency plans to launch three satellites in the near future in a project called *Swarm.* Given the timing of the last field reversal—780,000 years ago (when our human ancestors were first learning to fashion stone tools)—and with the half-million-year average for field reversals, it appears that the magnetic field's next reversal is overdue and *may* have already begun (with the

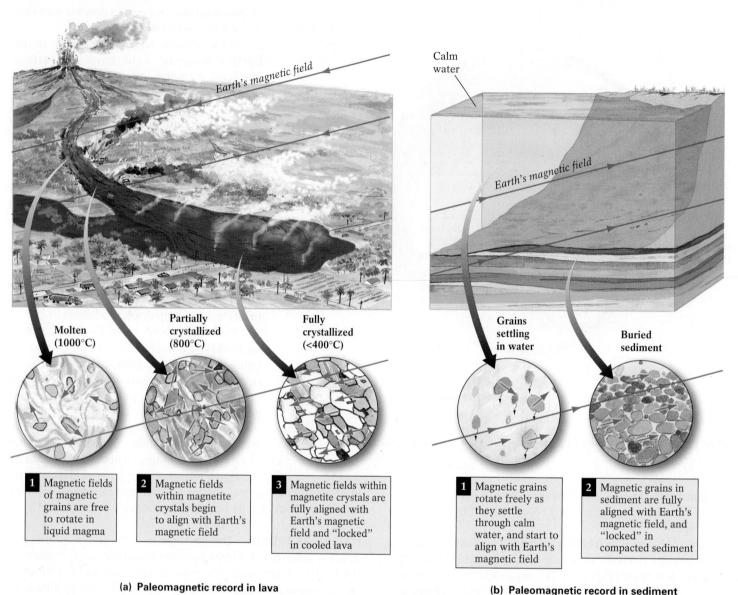

(a) Paleomagnetic record in lava

(b) Paleomagnetic record in sediment

▲ FIGURE 11-23 (a) The magnetic-field polarity within magnetite crystals in molten basaltic flows is free to align with the Earth's prevailing magnetic field. The solidification of the lava fixes the magnetite crystals in place, preserving a record of the Earth's magnetic field at the particular time and place. (b) As they fall, magnetic grains of sediment settling through relatively still bodies of water are free to rotate until they become aligned with the Earth's field at their location. Burial of these grains by subsequent deposits locks them in place, creating a paleomagnetic record of the field.

ocean floor, especially in areas where uninterrupted marine sedimentation has suffered little or no erosion (Fig. 11-24). The growth of oceanic plates by sea-floor spreading provides a unique opportunity to study past magnetic field reversals. These events are recorded in the basaltic lavas that have erupted and cooled continuously at mid-ocean ridges during this spreading.

The ocean-floor record shows that during the past 75 million years alone, the Earth's magnetic field has reversed itself at least 171 times. Periods of stable polarity, either normal or reversed, have lasted from 25,000 to several million years. No definite pattern marks the length of time between magnetic-field reversals, so geophysicists do not know exactly when the next reversal will occur.

Paleomagnetism is a powerful tool that enables geologists to reconstruct past geographic arrangements of the Earth's landmasses. Because the magnetic fields within a rock's magnetite crystals become aligned with the Earth's magnetic field at the time of the rock's formation, the crystals act as a kind of "paleo" compass needle, pointing to their original magnetic poles. We can estimate the rock's original geographic *latitude* by comparing the angle between the magnetic polarity of the crystals and the Earth's surface to the angle at which the lines of force of the Earth's magnetic field intersect the surface. (This angle increases with increasing latitude; see Figure 11-21.) The extent to which any magnetite-bearing rock has been moved from its original position *in the north–south direction* may reflect movement of the

Highlight 11-5

Modeling Reversals

Explorers, sailors, and geoscientists have been measuring the Earth's magnetic field for more than a century. Along the way, they have discovered that the magnetic properties of the Earth are different in different places, and change from year to year. Until recently, we simply didn't know why. In the past, scientists used the theories of physics to explain the origin of the Earth's magnetic field and its complex variations in time and space. More recently, with the development of powerful supercomputers, they have been able to demonstrate how the magnetic field may actually work. Although supercomputer models can show us how the Earth's magnetic field is generated by churning liquid metal in the Earth's outer core, no one had been able to come up with a model that reproduced a major feature of geomagnetism—the periodic reversal of the north and south magnetic poles.

That situation changed in the late 1990s. Geophysicists at the University of California–Santa Cruz, using a supercomputer, were the first to simulate magnetic-polarity reversals successfully. The new computer models predicted this behavior by assuming that the temperature of the mantle rock surrounding the outer core varies over time and varies from place to place at the core–mantle boundary, phenomena recently confirmed by seismic tomography. The model simulates hundreds of thousands of years of molten iron and nickel swirling around in the Earth under enormous pressure and at temperatures nearly as hot as those found on the surface of the Sun. These scientists were also the first to predict that the outer and inner cores rotate out of synch with each other, a finding verified by seismologists (see Figure 11-13). In addition, scientists at Johns Hopkins University have reported that simulations of the geomagnetic field correctly predicted a whirlpool slowly spinning in the Earth's outer core under the North Pole. No such whirlpool has yet been detected at the South Pole, perhaps because there are less data available thus far from the Antarctic region. Geophysicists continue to work on these models in an effort to understand more fully the behavior of the Earth's magnetic field, a fundamental feature of the Earth and one that enables life to continue to exist on the planet's surface by shielding the surface from potentially dangerous levels of solar radiation.

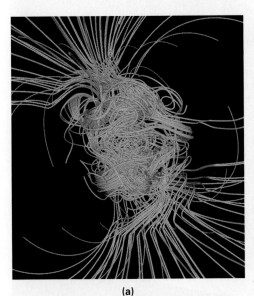
(a)

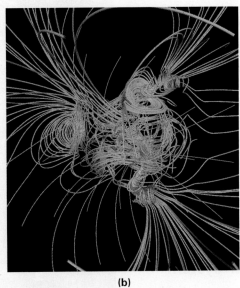
(b)

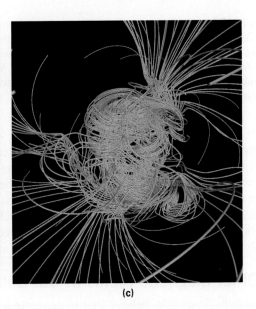
(c)

▲ A supercomputer model of the Earth's magnetic field before, during, and after a polarity reversal. Note that the magnetic field within the Earth is extremely complex (as shown by the jumble of lines in the center) and becomes more orderly outside the Earth's core. The lines that connect to the north and south poles (magnetic lines of force) extend out into space beyond the image and connect with each other. **(a)** Five hundred years before magnetic poles reverse in the model. **(b)** Magnetic lines of force in the middle of a reversal. **(c)** Five hundred years after a reversal. Note that the blue and orange lines have switched places relative to the Earth's poles (compare with **a**).

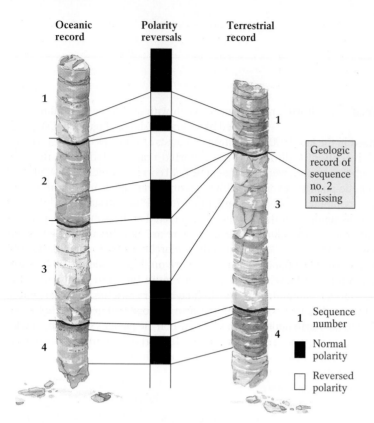

Oceanic record   Polarity reversals   Terrestrial record

1 Sequence number

■ Normal polarity

☐ Reversed polarity

Geologic record of sequence no. 2 missing

▲ FIGURE 11-24 The terrestrial record of magnetic reversals has been interrupted by erosion and the intermittent nature of continental volcanism and sedimentation. The oceanic record, which has been generated by uninterrupted marine sedimentation and continuous volcanism at spreading centers, is more complete. Note that as the rate of sedimentation varies, the thicknesses of the layers corresponding to specific magnetic reversals also vary.

The Earth's Moon, for example, does not appear to contain a liquid interior layer, so we would not expect it to generate a magnetic field. Indeed, it does not. The Moon is partially differentiated, however, and its materials could not have migrated into zones of different densities unless it had been at least partly melted at one time. Magnetometer-toting astronauts and Earth-bound lunar scientists have found aligned magnetite crystals in lunar rocks, suggesting that the Moon's small, iron-rich core was most probably once molten and generated a magnetic field. These rocks, which became magnetized billions of years ago during their formation, have not been reheated and demagnetized since then. This evidence suggests that the Moon has been tectonically inactive for quite some time.

The existence of a magnetic field on Mars has inspired much debate in recent years. Magnetic measurements conducted by the Viking probe in 1976 detected no evidence of a magnetic field on

tectonic plate on which it is located. Thus we can use paleomagnetism to estimate how far a rock (and therefore its plate) has moved with respect to latitude. By studying the magnetite grains in the igneous rocks of the Wrangell Mountains of Alaska, shown in Figure 11-25, geologists have been able to determine that these rocks probably formed near the warm tropical equator and moved to their present frigid polar location by plate motion.

Other Magnetic Fields in Our Solar System

Geophysicists now believe that to generate a magnetic field, a planet must:

■ *Have a liquid layer, such as a metallic core, although it need not be a "core."* (The magnetic fields of Uranus, Neptune, and Jupiter's moons Callisto and Ganymede may originate with a flowing briny solution.)

■ *Rotate rapidly enough around its axis to set the liquid in motion.*

All of the planets and moons of our solar system spin. All probably have differentiated (compositionally layered) interiors as well. But only some appear to contain liquid interior layers. Without this feature, the planet or moon cannot produce a self-exciting magnetic field.

▲ FIGURE 11-25 The Wrangell Mountains of Alaska. Igneous rocks in these mountains contain magnetite grains that indicate that they formed near the equator. This evidence suggests that the mountains were moved a great distance by plate motion.

Mars, suggesting to some that its presumed metallic core might be completely solid. Later measurements made in 1997 during the Mars Global Surveyor mission, however, revealed a weak magnetic field. Although registering only one eight-hundredth the strength of the Earth's magnetic field, Mars' field is still stronger than NASA's scientists had expected. Proof that a more vigorous magnetic field existed on this planet in the past has recently been discovered in the form of magnetic "stripes" on Mars, products of past magnetic reversals. This suggests that the weak current field may be a fading remnant of a much stronger field, perhaps from a time when Mars had a substantial liquid-metal core.

Like Earth, Mercury appears to have undergone extensive planetary differentiation. Because Mercury has an unusually high density, planetary geologists believe that it has an extremely large, iron-rich core, and that all or most of this planet was once molten. Although Mercury rotates so slowly on its axis (one of its days lasts 59 Earth days) that it probably could not generate the necessary currents in its liquid core to form a self-exciting dynamo, the magnetometer on the 1973 Mariner-10 probe did detect a weak magnetic field on Mercury (about 1% as strong as the Earth's field). The pattern of the lines of force on this planet closely resembles that of the Earth's field. The existence of a magnetic field on Mercury supports the hypothesis that the planet's core is at least partly molten. Venus, with its slow rotation, does not possess a detectable magnetic field, even though it is believed to contain a partly liquid iron core.

Neptune and Uranus and several of the moons of Jupiter also have magnetic fields, each more powerful than Earth's. Their fields differ from Earth's in that they are not centered on these planets and moons and not aligned with their rotation axes. These observations have led some planetary scientists to speculate that their fields arise from the flow of a briny fluid in their mantles (not their cores).

Our knowledge of the geophysical properties of planet Earth has grown considerably during recent years, thanks in part to rapid advances in space-satellite technology. In the next chapter, we will see how these and other technological advances are enabling Earth scientists to track the movements of the continents, study the details of the ocean floor, and reconstruct the past positions of the Earth's plates.

Chapter 11 Summary

Geophysics is the study of the fundamental properties of the Earth's interior, such as heat flow, gravity, and magnetism. Much of the data that contribute to this study come from the recordings of the seismic body waves (compressional P and shearing S waves) that emanate from the foci of earthquakes.

By analyzing the arrival times of seismic waves, geophysicists have identified the thickness, density, composition, structure, and physical state of the layers of the Earth's interior. We now believe that the Earth consists of the following layers:

- *continental and oceanic crust,* each of differing composition, thickness, and structure;
- a seismic discontinuity at the base of the crust and the top of the mantle, known as the *Moho;*
- a thick, multilayered *mantle* whose upper portion contains the *low-velocity zone,* a region in which seismic waves slow down temporarily;
- a distinct transition zone separating the upper- and lower-mantle zones where mantle minerals collapse into denser structures under the increasing pressures at greater depths;
- a 200-km-thick layer at the base of the mantle (D'') that may contain cool slabs of subducted lithosphere;
- a seismic discontinuity at the core–mantle boundary;
- a two-part *core* consisting of a liquid outer core and a solid inner core.

Because we have few direct means of measuring the Earth's deep interior temperatures, scientists must estimate them by laboratory experimentation. If the Earth is solid at a particular depth, the temperature there must be lower than the melting point of the material at that pressure. If the Earth is liquid at a particular depth, the temperature there must be higher than the melting point of the material at that pressure.

Gravity is the force of attraction that any object exerts on another object. It is proportional to the product of the objects' masses, and inversely proportional to the square of the distance between their centers. Gravity at the Earth's surface depends on several factors:

- The altitude of the land surface, and thus the distance from the Earth's center of gravity, produces a greater attraction in lowlands (closer to the center) than on mountain tops (farther from the center);
- The Earth's flattened-sphere shape creates a higher gravitational attraction at the poles (closer to the Earth's center) than at the equator;
- The composition of the Earth's underlying interior layers also affects gravity. High-density materials, for example, create a greater gravitational attraction than low-density materials.

The Earth's lithospheric segments are continually responding gravitationally to variations in their density relative to the underlying denser, heat-softened asthenosphere. Segments of equal density (based on the density and thicknesses of their rocks and sediments) will "ride" on the asthenosphere at the same level. This equilibrium between lithospheric segments is called *isostasy.* Each segment of the lithosphere is continually adjusting isostatically to the load placed upon it—subsiding into the asthenosphere with the addition of weight, such as with the arrival of glaciers or new lava flows, or rising, such as occurs when glaciers melt away or rocks erode away.

Magnetism is a property of some materials, associated with the movement of electrons, that causes them to attract or repel other materials having this property. The Earth has a *magnetic field*, similar to that of a simple bar magnet. This field arises from the flow of an electrical current within the Earth's liquid outer core. The Earth's magnetic field periodically weakens and changes the direction of its polarity, a phenomenon known as a *magnetic reversal.*

KEY TERMS

geophysics *(p. 357)*
gravity *(p. 370)*
gravity anomalies *(p. 370)*
isostasy *(p. 372)*

low-velocity zone *(p. 362)*
magnetic field *(p. 375)*
magnetic reversals *(p. 376)*
magnetism *(p. 375)*

Moho *(p. 362)*
P-wave shadow zone *(p. 365)*
paleomagnetism *(p. 377)*
S-wave shadow zone *(p. 365)*

seismic tomography *(p. 361)*
transition zone *(p. 363)*

QUESTIONS FOR REVIEW

1. Describe how the velocity and path of a seismic wave change as it moves from the crust to the upper mantle.

2. What is the Moho, and how was it discovered?

3. What is the major difference between the upper mantle and the lower mantle?

4. What is the S-wave shadow zone, and what causes it?

5. Discuss how and why the Earth's gravity varies with topography and latitude.

6. Describe two geologic settings where you might find positive gravity anomalies and two where you might find negative gravity anomalies.

7. Briefly describe two geological settings where land surfaces rise isostatically and two settings where surfaces subside isostatically.

8. The outer core of the Earth's interior is in a liquid state. Why doesn't the liquid rise to the surface and erupt as lava?

9. In films of the first humans walking on the Moon, the astronauts seemed to be jumping and bounding because of the weak force of gravity. Why is the gravitational attraction of the Moon so much weaker than that of the Earth? (Helpful hint: There are two primary factors.) What type of gravitational attraction would we be likely to find on the surface of a planet that is much larger and composed of much denser material than the Earth?

10. What kind of gravity anomaly would exist on an enormous body of dense platinum ore at the bottom of a canyon in Antarctica? Above a large body of subterranean salt at the top of a lofty mountain at the equator?

11. Has the Earth always had a magnetic field? What evidence would you look for to support your answer?

LEARNING ACTIVELY

1. Pour a large volume of water into a clear glass and place a spoon in it. Looking at the spoon, note its apparent "bend." Speculate about how the velocity of light is changing as it passes through air and water. How are the water and air comparable to the Earth's crust and mantle?

2. Using ice cubes and water, demonstrate the principle of isostasy. Note how much of the cube sits below the water line. Now direct the hot air from a hairdryer onto the cube. As it melts, note how the cube adjusts its position relative to the water line. Speculate about how erosion from the tops of a mountain range would affect the position of the range relative to the Earth's surface.

ONLINE STUDY GUIDE

Plate Tectonics and the Formation of the Earth's Oceans

The entire Earth is in constant motion. The molten iron in the planet's outer core swirls chaotically, producing the Earth's geomagnetic field. The heat-softened rocks in the mantle flow as the Earth's heat escapes upward. The outer layers of the planet are active as well. The solid crust and uppermost mantle are broken into rigid blocks, called plates, that move along with the underlying flowing mantle. Plates crash into each other, rift apart, and slide past each other. This activity creates the Earth's continents and oceans, mountain ranges, volcanoes, and 1000-km long faults. As the plates move, rocks break, causing earthquakes.

We often think of the plate-tectonic system as primarily involving rocks and structures (such as folds and faults). Let's take a different perspective and consider the role of tectonics in life and environments here at the Earth's surface. One prominent idea about the birthplace of life on Earth focuses on oceanic divergent zones—a common plate-tectonic setting (Fig. 12-1). It is there that the first *complex* organic molecules may have formed by reaction of hot seawater and volcanic gases with metallic minerals such as pyrite and other iron sulfides that precipitate in these regions. While plate tectonics may have played an important role in the origin of life, tectonics may have also contributed significantly to the eradication of life; tectonic processes such as rifting, convergence, and their associated volcanic activity may have altered fragile ecosystems in the past, triggering mass extinctions.

The emission of gases from subduction-zone volcanoes can also affect the composition of the Earth's atmosphere and, in turn, modify its climate. Likewise, the tectonically controlled positions of continental landmasses and the creation of high mountains and plateaus on continents affect climate profoundly by blocking and diverting air and ocean currents, thereby altering atmospheric and oceanic circulation. Such effects may have initiated the Earth's ice ages. (See Chapter 17 for more on these effects.)

As mountains are uplifted, weathered, and eroded, the sediments and ions released as rocks break down are subsequently carried to the oceans. These materials may eventually be dragged down to the Earth's interior at a subduction zone. In this way, silica-rich and organic materials are transported to the Earth's mantle. Some of the resulting carbon-bearing materials may eventually become buried deeply enough to be pressed and heated to form diamonds.

FIGURE 12-1 Photo taken out the window of the submersible *Alvin* on dive 2432 (in 1991) at the northern Cleft segment of the Juan de Fuca Ridge, west of Washington State within the Pacific Ocean basin. This is a submarine lava flow that erupted in 1986, the first documented historical eruption on this mid-ocean ridge.

385

Another interesting consequence of subduction relates to the fate of the subducted plates: What happens when subduction "shuts off," such as when two continents collide after the intervening ocean plate is completely subducted? Pieces of old subducted oceanic plates may continue their journey downward, inching their way through the deep mantle as part of the Earth's vast lithosphere-recycling system. Perhaps they end up at the core-mantle boundary, but they can't proceed further—they're impeded by the higher density of the iron-rich core.

These examples show that plate tectonics profoundly influences virtually all aspects of Earth dynamics—from the atmosphere and biosphere at the planet's surface to the composition of the rocks at the core–mantle boundary. Although we know much about the basic mechanics of the plate-tectonic system, geoscientists still actively debate many fundamental and puzzling questions about the Earth's plate-tectonic processes: What drives plate motion? How do divergent zones form in the first place? What happens to subducted plates? What will happen to the Earth as plate-tectonic processes slow down as the planet continues to cool? These and other questions, along with fundamental aspects of the tectonic system, are the focus of this chapter. In the next chapter, we will consider the tectonics of the continents in greater detail, including the origin—and destruction—of mountain belts and the geological biography of North America.

Some Plate-Tectonic Basics

A simple review of some of the key elements of the theory of plate tectonics includes the following key points:

- The Earth's lithosphere consists of rigid plates averaging 100 km (60 mi) in thickness and ranging from about 70 km (43 mi) thick for the oceans to 150 km (90 mi) thick for the continents. (Remember, the lithosphere is the crust + uppermost mantle.)
- Oceanic lithosphere is typically more dense than continental lithosphere.
- The plates are all constantly moving relative to one another by divergence, convergence, or transform motion.
- Oceanic lithosphere forms at divergent plate boundaries and is consumed at subduction zones, one type of convergent plate boundary.
- Convergent plate boundaries can involve subduction or collision. Two types of subduction zones exist—one where the *oceanic* lithosphere of two plates meet (the older, colder, denser one will subduct) and one where the oceanic lithosphere of one plate meets the continental lithosphere of another plate (the oceanic lithosphere will always subduct). Collision zones occur where two or more segments of continental lithosphere collide after an intervening ocean plate has entirely subducted.
- Most earthquake activity, volcanism, faulting, and mountain building take place at plate boundaries.
- The centers of plates tend to be geologically stable (although there are some notable exceptions); they do not deform in their interiors, away from plate boundaries, as they do at plate boundaries.

The theory of plate tectonics addresses the origin of both the continents and the oceans. Its predecessor, Alfred Wegener's continental-drift hypothesis, had generally ignored the ocean basins. Wegener primarily sought to explain past movements of the continents because when he proposed his hypothesis in 1915, the ocean basins were almost entirely unexplored. Like most geologists of his time, Wegener believed that the ocean floors were very old, featureless plains. He thought that the oceans passively surrounded the drifting continents, which moved by plowing through the ocean floors. This lack of knowledge about the geology of the seafloor led to most of the conceptual problems that doomed the continental-drift hypothesis to decades of merciless rejection. Intense deep-sea exploration, which began in the 1950s, brought the discovery of the spreading ocean ridges and the deep ocean trenches and revealed the true nature of the ocean's role in plate motion. Only then could continental drift become a useful concept, underlying today's better understanding of plate tectonics.

Our *modern* understanding of plate motion begins with our current view of the origin of the world's oceans and the plate interactions that occur there.

The Nature and Origin of the Ocean Floor

If all the water were drained from the Earth's ocean basins, we would see valleys as deep as 11,000 m (36,000 ft), volcanic mountain ranges thousands of kilometers long, chains of flat-topped mountains, long linear fractures oozing lava, and faulted cliffs stretching like walls for great distances. Most of these features were discovered and mapped in the early 1940s, a by-product of the U.S. Navy's World War II search for enemy submarines and safe passages. The principal method of mapping at that time involved **echo-sounding sonar.** A ship using this technology emitted a sharp pinging noise, the sound waves travelling at a speed of about 1500 m (5000 ft) per second, the speed of sound in seawater. The time it took for sound waves to bounce off the seafloor and return to a listening device on board was used to calculate the depth of the ocean bottom and map its topographic features. Today, oceanographers map the seafloor with various multibeam echo sounders, such as the SeaBeam system, which employs wide continuous sonar bands to cover broad swaths of the seafloor.

Geoscientists today also map *subsurface* ocean-floor rocks by means of **seismic profiling,** which uses powerful energy waves generated by small targeted explosions. Some of these waves reflect off the seafloor, while others penetrate the surface and reflect off layers of underlying sediment and rock enabling geologists to map the subsurface as well (Fig. 12-2).

Deep-sea drilling and submersible vessels photograph and take samples from the seafloor. Since 1965, a group of universities, research laboratories, and government agencies has jointly funded and staffed seafloor exploration projects, such as the Ocean Drilling Project (ODP), known today as the International

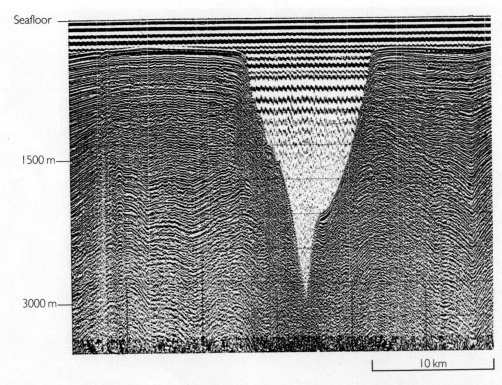

Seafloor

1500 m—

3000 m—

|—————— 10 km ——————|

▲ **FIGURE 12-2** Penetrating sound waves have produced this profile of a submarine canyon off the west coast of Africa. The canyon is about 3000 m (10,000 ft) deep and about 10 km (6.2 mi) wide.

Ocean Drilling Project (IODP). During the ODP-sponsored voyages of the *Glomar Challenger* research vessel, scientists collected more than 96 km (60 mi) of deep-sea cores that have been studied for clues to the origin of the Earth's ocean basins. Other oceanographers, diving to the seafloor in ALVIN, the best-known deep submersible, continue to do "field work" at the seafloor. Along the way, they have discovered new life forms and previously unknown geological processes. More recently, remote-controlled video cameras have permitted even more extensive deep-sea exploration (and studied in detail such deep-sea curiosities as the *HMS Titanic*).

Exciting new findings have come from space satellites that can image the entire ocean floor *indirectly* by bouncing a beam of microwaves between the satellite and the sea surface. NASA's SEASAT, a satellite dedicated to studying the ocean floor, has confirmed that the ocean's surface contains bulges and depressions that correspond to variations in sea-floor topography (Fig. 12-3). The added mass of a chain of submarine mountains, for example, produces a positive gravity anomaly (see Chapter 11 for a discussion of gravity anomalies) that attracts water, creating a sea-surface bulge as high as 30 m (100 ft). Likewise, the lesser mass of a deep-sea trench produces a negative gravity anomaly and a corresponding sea-surface depression as large as 60 m (200 ft) deep. In this way, satellites have charted vast unexplored areas of the world's ocean floors.

And in the latest attempt to work at the frontier of scientific discovery, a plan is in the works to establish a regional seafloor observatory *directly* at the source—an installation built on the floor of the Pacific Ocean (Fig. 12-4). The project involves a 3000-km (2000-mi) network of fiber-optic cables that crisscross

and encircle the Juan de Fuca tectonic plate off the coast of Washington, Oregon, and southwestern British Columbia. A series of 25 stations will dot the cable network—sites where instruments will measure physical, chemical, and biological phenomena, such as seafloor earthquakes, submarine landslides and turbidity flows, seafloor volcanic eruptions, and fish migration. Robotic underwater vehicles, rechargeable at the various stations, will install and monitor seafloor sensor networks. Here, with Project Neptune, scientists will be able to study the seafloor in great detail *on a plate-size scale* for the very first time. Funding from various U.S. and Canadian agencies is moving ahead toward a 2007 start date for operations. Figure 12-5 illustrates the various elements involved in our efforts to learn about the Earth's ocean floors.

Rifting and the Origin of Ocean Basins

Although the seafloor is difficult to study directly, we know much about its topography (the location and elevation of undersea mountains and valleys), its composition, and the age of ocean basins worldwide. We know that the oceans are geologically young compared to the continents—nowhere in the world is there seafloor older than about 200 million years old simply because *all* the older oceanic lithosphere has subducted.

We also know that oceans initially form when a continent breaks apart and two continental fragments drift away from each other. How does rifting begin? Rising mantle over hot spots in the Earth's interior and in the upwelling segments of convection cells may stretch and thin the overlying continental lithosphere, causing its surface to break. When continents break up in this

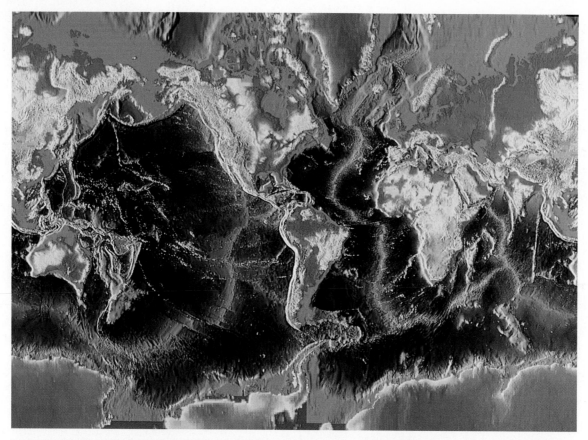

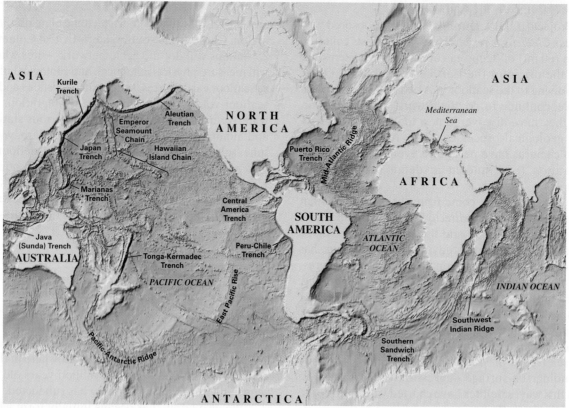

▲ **FIGURE 12-3** This map, which shows the significant variations in the topography of the seafloor, was created using satellite-gathered data on the elevation of the seawater surface. Positive gravity anomalies produced by masses such as submarine mountain ranges attract large quantities of seawater, causing an upward expansion of the ocean surface (light blue areas); conversely, negative gravity anomalies produced by bedrock depressions lower the ocean surface (dark blue areas).

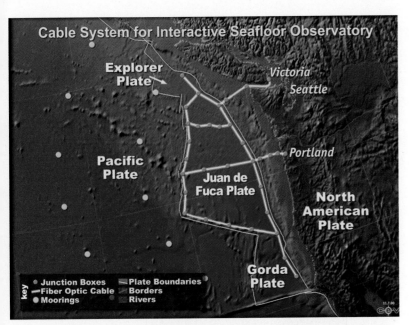

Cable System for Interactive Seafloor Observatory

Explorer Plate

Victoria
Seattle

Pacific Plate

Portland

Juan de Fuca Plate

North American Plate

Gorda Plate

key
- Junction Boxes
- Fiber Optic Cable
- Moorings
- Plate Boundaries
- Borders
- Rivers

▲ **FIGURE 12-4** Project NEPTUNE, a proposed seafloor observatory, involves thousands of kilometers of fiber-optic cable connecting sensors that monitor a host of physical, chemical, and biological properties of the Earth's ocean plates and their boundaries.

way, three fractures form (Fig. 12-6). Geologic evidence shows that two of these three fractures will continue to break and separate, whereas the third one will become inactive, a so-called *failed rift*.

A good modern example of an active rift, seen in Figure 12-7, is the East African rift zone. There, a three-branched fracture has already separated Africa from Arabia. The rift's two *active* arms are the Red Sea and the Gulf of Aden, where divergence continues today. Although GPS measurements indicate that the rift valley, relative to the other branches, is still widening very gradually, the East African branch of the three-part rift, is the so-called failed arm. The Great Rift Valley of East Africa is currently occupied by the large lakes (such as Lakes Kiva, Tanganyika, and Malawi) near which much fossil evidence of early humans has been found. The valley may never evolve into an ocean, but instead may remain a sediment-filled topographic depression that may someday become Africa's next great river system. The Red Sea, on the other hand, is about to become a true ocean. When rifting fully breaches the Sinai Peninsula, the Red Sea will connect the Mediterranean and the Indian Ocean. Its floor has already developed oceanic basaltic crust.

Many of the world's great river valleys, such as South America's Amazon River, Africa's Niger River, and North America's Mississippi River, may occupy ancient failed rifts. Intraplate

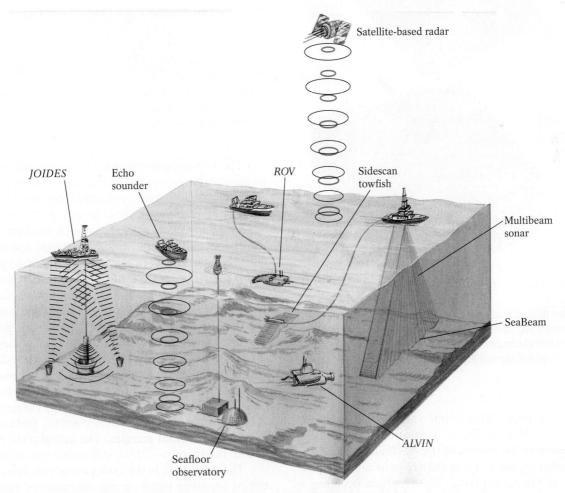

Satellite-based radar

JOIDES

Echo sounder

ROV

Sidescan towfish

Multibeam sonar

SeaBeam

ALVIN

Seafloor observatory

▲ **FIGURE 12-5** Various ways that geoscientists are exploring the Earth's seafloor. JOIDES refers to the deep-sea drilling projects; ROV refers to remotely operated vehicles.

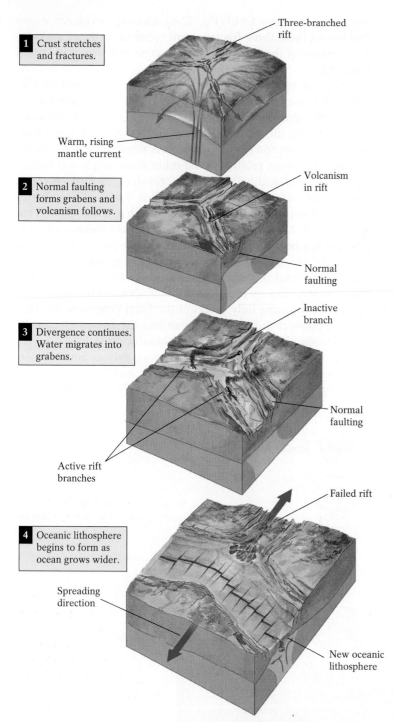

1. Crust stretches and fractures.

Three-branched rift

Warm, rising mantle current

2. Normal faulting forms grabens and volcanism follows.

Volcanism in rift

Normal faulting

3. Divergence continues. Water migrates into grabens.

Inactive branch

Normal faulting

Active rift branches

4. Oceanic lithosphere begins to form as ocean grows wider.

Failed rift

Spreading direction

New oceanic lithosphere

▲ **FIGURE 12-6** Active rifting of a continental plate. Rising warm currents from the Earth's interior flow beneath the continental lithosphere, stretching and then tearing it. The initial branched rift typically consists of three fractures, two of which continue to diverge. The third, the *failed arm,* becomes inactive.

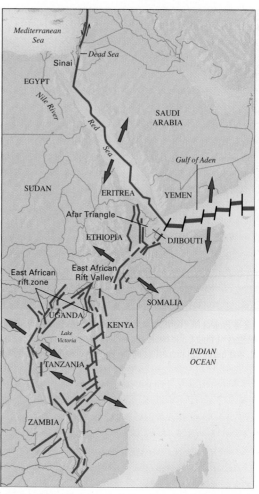

▲ **FIGURE 12-7** The Red Sea and the Gulf of Aden are the actively diverging segments of the three-branched rift in eastern Africa. The valleys and gorges of Kenya and Tanzania are the "failed" branch of the rift.

The two actively rifting fractures of a rift system are marked by high heat flow, normal faulting, frequent shallow earthquakes, and widespread basaltic volcanism. Normal faulting at rift fractures produces steep-walled rift valleys. Sedimentation in these valleys is particularly active because of the sharp topographic contrast between the valley floors (grabens) and the high-standing cliffs (horsts) that mark the edges of the rifted plate. Figure 12-8 shows the stages of the growth of an ocean basin in a rift valley.

As rifting progresses, the rift valley lengthens and widens, possibly expanding until it reaches an adjacent ocean. The waters of the ocean may then flood the low-lying graben. Over time, the high-standing cliffs above a rifted valley may erode, perhaps eventually being reduced to sea level. As the rocks at the rift margins cool, the land surface subsides and the one-time rift edges become the foundations for continental shelves. After rifting ends, the original rift edges are no longer plate margins, nor are they volcanically or seismically active. Instead, they are now ***passive* continental margins.** The actual plate margins are now half an ocean away—at the still-active spreading center.

The final stage in the evolution of rifted continental margins can be observed today on opposite sides of vast oceans, such as the Atlantic Ocean along the east coasts of North and South America and the west coasts of Europe and Africa. Some

earthquakes in the upper Mississippi valley, such as those recorded in the New Madrid area of Missouri and Tennessee, may have resulted from the release of stresses built up along ancient faults. Another example of a modern river located in a possible failed rift is the Connecticut River, which occupies a 180-million-year-old rift valley that dates from the breakup of the Pangaea supercontinent (discussed in Chapter 1).

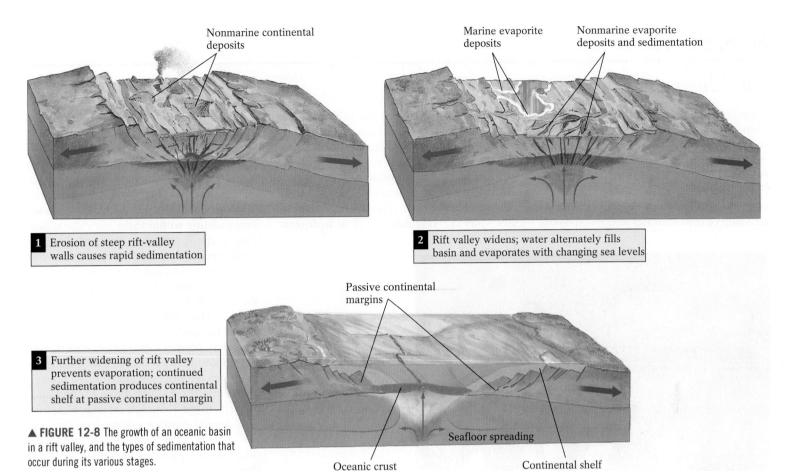

Nonmarine continental deposits

Marine evaporite deposits

Nonmarine evaporite deposits and sedimentation

1 Erosion of steep rift-valley walls causes rapid sedimentation

2 Rift valley widens; water alternately fills basin and evaporates with changing sea levels

Passive continental margins

3 Further widening of rift valley prevents evaporation; continued sedimentation produces continental shelf at passive continental margin

Seafloor spreading

▲ **FIGURE 12-8** The growth of an oceanic basin in a rift valley, and the types of sedimentation that occur during its various stages.

Oceanic crust

Continental shelf

evidence of the Atlantic's initial rifting lies buried beneath the sediment of these continental shelves. On land, much of the rifting record is evident in the fault-bound, 200-million-year-old sediments of the northeastern United States and in the numerous outcrops of basalt found along the Eastern Seaboard, from the Canadian Maritime provinces and New England to North Carolina. Faulted basins and rocks of the same age and composition occur on the opposite side of the Atlantic throughout the British Isles and northwestern Africa. We now know that about 180 million years ago, these British and African rocks were joined to today's North American rocks on the supercontinent of Pangaea (see Figure 1-29). Their journey to their present locations began when a rift separated the Americas from Europe and Africa. It may have taken as many as 20 hot spots to rift Pangaea into today's major continents, most of which had taken shape by about 65 million years ago. Continents are most likely to rift when numerous hot spots are aligned.

Rifting sometimes ceases following an initial period of faulting and volcanism. We do not yet understand why it stops. One such interrupted rift is believed to have begun about 1.1 billion years ago, as an outpouring of mafic lava reached the surface in the middle of the North American continent. Basalts and gabbros of this age are found along the north and south shores of Lake Superior in Minnesota and Wisconsin (see Figure 11-17). Farther south, similar rocks form a linear zone of anomalously high gravity buried beneath Cambrian and younger rocks. This mid-continent gravity high, tens of kilometers wide, stretches through Iowa and Missouri to the Oklahoma panhandle. Had this

rifting continued, people today might be enjoying the ocean beaches of Minnesota and Wisconsin instead of Maine and North Carolina.

Divergent Plate Boundaries

As we discussed back in Chapter 3, new oceanic crust is created at divergent zones, as upwelling ultramafic mantle rock melts partially to produce basaltic magma. But why are divergent zones ridges? That is, why does an undersea mountain range appear at divergent zones? The answer is not simply "because a lot of basalt piles up." Instead, the plates are moving apart and the basalt fills the space in between them. And, because little erosion occurs underwater, just as much basalt is found away from the divergent zone as is located at the active high-standing plate boundary. The active divergent zone is, however, much hotter than the parts of the ocean basin that have long ago moved away from the spreading center. The lithosphere at the divergent zone is hotter and therefore has a lower density than the colder, older crust, so the divergent zone rides higher isostatically, forming an elevated ridge.

Today, continuous oceanic ridges crisscross about 65,000 km (40,000 mi) of the Earth's surface, across all the major ocean basins. Each segment of this ridge system spans an area as wide as 1500 km (900 mi), and in some places ridge peaks rise more than 3 km (2 mi) from the ocean bottom. Ocean ridges are the largest raised topographic features on the Earth's surface, and their length and breadth account for approximately 23% of the Earth's total surface area.

Highlight 12-1

The Unseen World of Divergent Zones

Since the early 1970s, oceanographers using deep-sea submersibles have studied the rift valleys of the mid-Atlantic ridge, the Galápagos ridge off the coast of Ecuador and Peru, and the East Pacific rise south of Baja California. They have photographed the eruption of pillow lavas and the chemical interaction of cold seawater and warm basalt.

During recent dives into the valleys of the East Pacific rise and the Juan de Fuca ridge (off the coast of Washington), oceanographers returned with videos showing black sulfurous "clouds" of mineral-laden water rising from vertical chimney-like structures. These hydrothermal vents emit extremely hot water carrying gases, such as hydrogen sulfide, and clouds of solid metallic sulfides. The emitted water nourishes a complex community of previously unknown life forms, including giant clams and exotic tube worms (see left photo below) that thrive in this high-temperature world, untouched by sunlight.

Vast quantities of valuable minerals are accumulating within these rift valleys. Seawater seeps down into newly formed oceanic crust, becomes heated by the underlying magma reservoir, and then dissolves copper, iron, zinc, cobalt, silver, and cadmium out of the warm mafic rocks before erupting to the surface. These 400°C (750°F) mineral-rich waters rise and gush from the seafloor, creating dark plumes called *black smokers* (see right photo below). Contact with cold seawater precipitates minerals from the plumes, encrusting the basaltic flows around each plume vent and forming a chimney-like structure. Similar massive sulfide deposits occur within many of the Earth's folded mountains; many are likely ancient slabs of oceanic crust that originated in this manner.

▲ Bacteria around volcanic vents at a mid-ocean ridge draw energy from heat-generated chemical reactions involving compounds such as hydrogen sulfide (H_2S). More complex creatures, such as these giant tube worms, live on these bacteria.

▲ Black smokers consist of hot plumes of mineral-rich water vented at volcanically active regions of the seafloor. Accumulations of precipitated minerals form chimney-like structures around the plumes. The "smoke" here consists of hot water and particles of iron, copper, and other sulfide minerals.

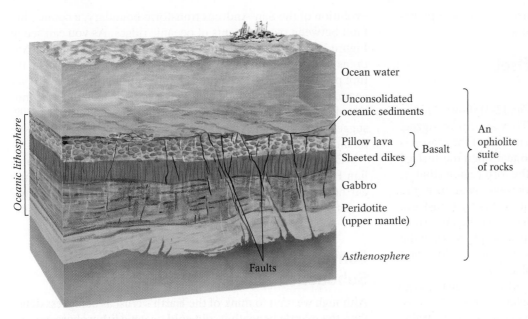

▲ FIGURE 12-9 The layers of an ophiolite suite, which make up oceanic lithosphere.

Some ridges have deep rift valleys at the center of the divergent zone. These valleys are created by normal faulting from the tensional stress that occurs as plates diverge slowly. Some of these valleys are deeper than Arizona's Grand Canyon and three times as wide. Scientists exploring oceanic rift valleys in submersibles have discovered geologic processes never observed on land as well as a host of strange and previously unknown life forms, including creatures that thrive in total darkness and in an extremely hot, mineral-rich environment that would be toxic to most marine life. Highlight 12-1 describes the extreme challenges of life within an oceanic divergent zone and the mineral wealth that accumulates there. Ridges that don't have rift valleys are spreading too rapidly to form a deep rift; any valleys that form are filled in rapidly by basaltic lava flows. The fast-spreading divergent zone that is creating the Pacific Ocean's lithosphere is called the East Pacific *Rise* because it lacks a rift valley.

Let's examine typical oceanic lithosphere—from top to bottom (Fig. 12-9). Its upper 200 m (700 ft) or so commonly consists of unconsolidated sediment of siliceous or carbonate ooze from the remains of microscopic marine organisms, fine reddish-brown clay from weathering of iron-rich marine lavas, or both. Generally, the older the oceanic plate, the thicker its sediment layer becomes as sediment accumulates with time. In some places, though, this sedimentary cover is missing, and basalt is exposed at the seafloor. The basaltic layer of oceanic crust is about 2 km (1.2 mi) thick, and the basalt typically shows the pillow structures characteristic of volcanic eruption under water.

Under the basalt lies 5 to 6 km (3 to 4 mi) of gabbro, formed from slow underground crystallization of mafic magma. Below this, a layer of the ultramafic mantle rock *peridotite* makes up the

rest of the oceanic lithosphere. The mantle rocks immediately below the oceanic lithosphere—within the soft, flowing asthenosphere—are also made largely of peridotite, but they are hotter and weaker than the lithospheric peridotite. There may be some chemical differences between mantle lithosphere and mantle asthenosphere because the lithosphere has had basaltic magma extracted from it during divergent-zone eruptions. The boundary between the two, however, is defined mostly based on the mechanical change from strong rigid lithospheric peridotite to weaker ductile peridotite of the flowing mantle asthenosphere.

This entire sequence of ocean-floor rock may be altered chemically by reaction with the seawater that penetrates its faults and fissures. Water reacts with the pyroxene in the basalt and gabbro to form a hydrous green mineral, chlorite. It also changes magnesium olivine in peridotite to the hydrous magnesium silicates, serpentine and talc. These metamorphic reactions may eventually produce *serpentinite*, a soft, green rock. Therefore, when we see oceanic-lithosphere rocks, their original, igneous minerals may be lost—a consequence of seafloor (hydrothermal) metamorphism.

The geological name for the group of rocks that makes up oceanic lithosphere is the **ophiolite suite** (from the Greek *ophis*, meaning "serpent," and *lithos*, meaning "rock"—a reference to the green snakelike swirls within serpentinites). There are places in the world where some or all of this sequence of rock can be seen *on land* where plate convergence has pushed oceanic lithosphere onto the continents (Fig. 12-10): Oman, Cyprus, Turkey,

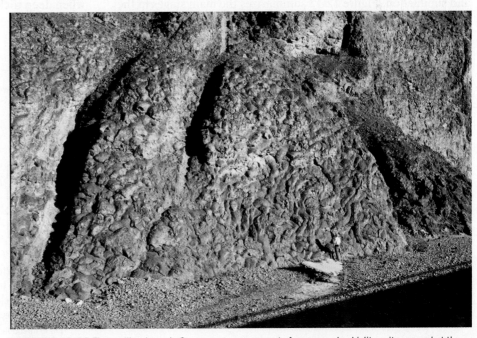

▲ FIGURE 12-10 These pillow lavas in Oman are one component of an exposed ophiolite suite exposed at the surface in the southern portion of the Arabian Peninsula. The rocks were thrust to the surface within a zone of past plate collision.

the Alps, and western North America are some of the places where ophiolite suites can be found on land.

Transform Boundaries and Offset Mid-ocean Ridges

A glance at an image of the seafloor (see Fig. 12-3) shows that the Earth's oceanic ridges are not straight. They appear to zigzag. Why are ocean ridges offset in this manner? Part of the reason is that the Earth is spherical. It is difficult to tear open the surface of a sphere along a straight line; part of the torn region (that is, the divergent zone) will tend to move sideways, away from the tear. At divergent zones, this movement occurs by transform faulting. Some faults may also be inherited from the pattern of fractures during the original breakup of the lithosphere when the divergent zone first formed in part because the rate of rifting and divergence may vary substantially from place to place.

In addition to transform faults, which offset divergent zones, entire plate boundaries themselves may be large transform faults. At transform boundaries, plates move past one another like trains headed in opposite directions on parallel tracks. The plates do not move away from or toward one another—they simply slide by each other. This motion can produce powerful earthquakes as rocks become stressed by shearing and then break along a transform fault. At divergent zones, earthquakes occur *between* the offset segments of the spreading ridge. Beyond the ridge segments, very few earthquakes occur because the segments of the plates there are sliding in the same direction (Fig. 12-11).

California's San Andreas fault zone is a transform boundary—part oceanic, part continental—that extends from the northern end of the East Pacific Rise in the Gulf of California to the southern end of the Gorda Ridge off the coast of southern Oregon and northern California. It was created by a complex sequence of events involving both oceanic transform movement and subduction. Figure 12-12 summarizes the main stages in the evolution of the San Andreas transform boundary, a connecting fault between two segments of oceanic ridges. As you can see in Figure 12-12, prior to about three million years ago, the San Andreas transform fault sat west of the continent, offshore. The future position of the San Andreas Fault may migrate eastward yet again. Recent motion and earthquake activity, including the motion during the powerful Landers earthquake of late June, 1992, suggest that the San Andreas may have already begun its move inland and eastward (Fig. 12-13).

Other important transform faults include the North Anatolian Fault—which periodically slips and produces devastating earthquakes in Turkey—and the Queen Charlotte and related faults offshore from British Columbia and southeastern Alaska.

Convergent Plate Boundaries— Subduction Zones

Although we tend to think of the Earth's crust as being less dense than the mantle beneath it, old cold oceanic lithosphere can become denser than the underlying warm asthenosphere of the upper mantle. When it reaches a point—at some substantial distance from the divergent zone—old, cold, dense oceanic lithosphere will begin to sink and return to mantle as it subducts beneath a continent or beneath younger, less dense oceanic lithosphere. No seafloor is older than about 200 million years because oceanic lithosphere subducts once it is old, cold, and therefore denser than the mantle below it: Huge amounts of oceanic lithosphere have been swallowed up by the mantle during the Earth's history.

Further evidence for ancient subduction is the occurrence of blueschists—the bluish metamorphic rocks that form only in the high pressures and relatively low temperatures of subduction zones. These rocks can become incorporated into mountain belts when deep rocks are brought up toward the surface along faults.

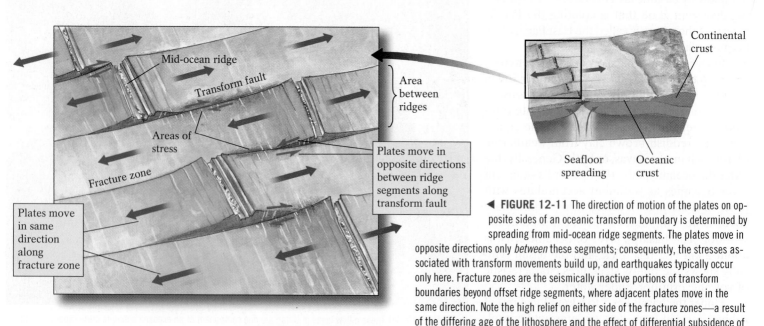

◀ **FIGURE 12-11** The direction of motion of the plates on opposite sides of an oceanic transform boundary is determined by spreading from mid-ocean ridge segments. The plates move in opposite directions only *between* these segments; consequently, the stresses associated with transform movements build up, and earthquakes typically occur only here. Fracture zones are the seismically inactive portions of transform boundaries beyond offset ridge segments, where adjacent plates move in the same direction. Note the high relief on either side of the fracture zones—a result of the differing age of the lithosphere and the effect of differential subsidence of rocks of differing density. (The younger, warmer, buoyant lithosphere at the divergent zone stands prominently above older, colder, denser, subsiding lithosphere across the fracture zone.)

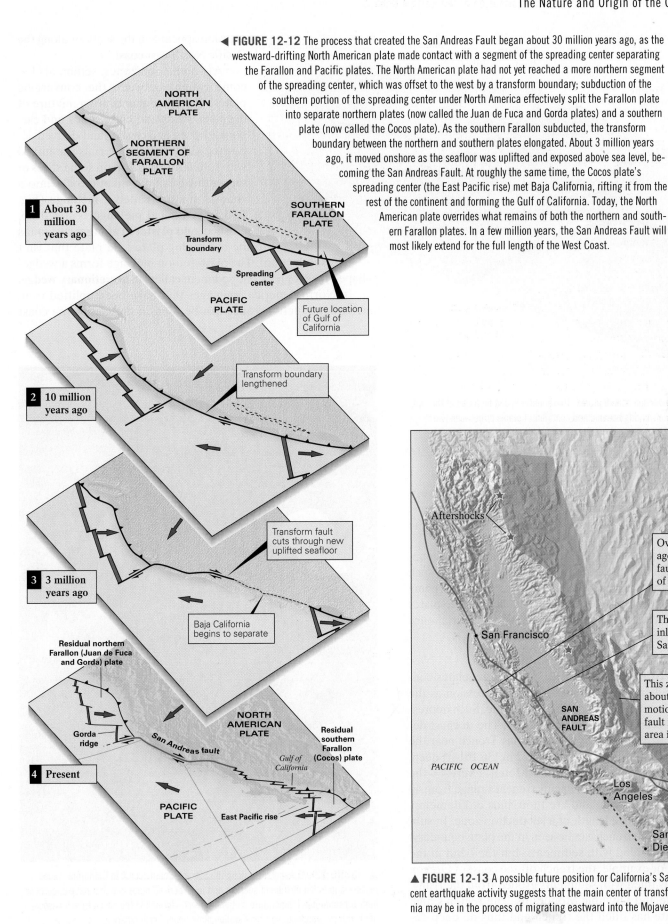

◀ **FIGURE 12-12** The process that created the San Andreas Fault began about 30 million years ago, as the westward-drifting North American plate made contact with a segment of the spreading center separating the Farallon and Pacific plates. The North American plate had not yet reached a more northern segment of the spreading center, which was offset to the west by a transform boundary; subduction of the southern portion of the spreading center under North America effectively split the Farallon plate into separate northern plates (now called the Juan de Fuca and Gorda plates) and a southern plate (now called the Cocos plate). As the southern Farallon subducted, the transform boundary between the northern and southern plates elongated. About 3 million years ago, it moved onshore as the seafloor was uplifted and exposed above sea level, becoming the San Andreas Fault. At roughly the same time, the Cocos plate's spreading center (the East Pacific rise) met Baja California, rifting it from the rest of the continent and forming the Gulf of California. Today, the North American plate overrides what remains of both the northern and southern Farallon plates. In a few million years, the San Andreas Fault will most likely extend for the full length of the West Coast.

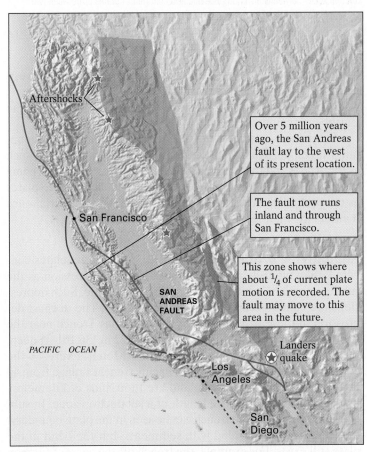

▲ **FIGURE 12-13** A possible future position for California's San Andreas Fault. Recent earthquake activity suggests that the main center of transform motion in California may be in the process of migrating eastward into the Mojave Desert.

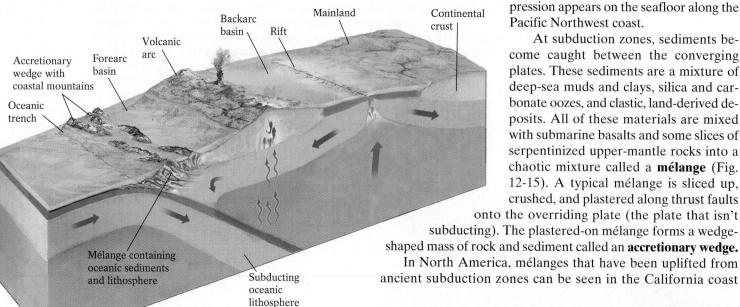

▲ **FIGURE 12-14** Subduction-zone features. These features can be found at the subduction zones between converging oceanic and continental plates or between two plates of oceanic lithosphere.

Worldwide, major regions containing blueschist rocks are western North America (California, Oregon, Mexico), Greece, Turkey, Japan and other islands in the western Pacific, and the Caribbean region.

Subduction-related processes have created many great mountain ranges, particularly those rimming the Pacific Ocean, such as the Andes and the Cascades, because the Pacific is undergoing subduction along most of its vast margin. What happens to plates as they subduct? What happens to the subducted material once it enters the mantle? For some answers, read on.

Anatomy of a Subduction Zone

Where two plates converge, they typically produce a narrow, deep depression as the subducting plate bends downward into the mantle. Such depressions are called **oceanic trenches** (Fig. 12-14). The deepest depression in the Earth's surface is the Marianas Trench near the Pacific island of Guam. This trench is 11,022 m (36,161 ft) deep, making it deeper than the Himalaya's Mount Everest—the world's highest mountain—is tall. Pacific trenches range from 40 to 120 km (25 to 75 mi) wide and are thousands of kilometers long, as they run the entire length of a subduction zone. In subduction zones where no trench can be seen at the plate boundary, the trench has been filled with sediments, but the bend in the plate still exists. For example, the trench off the coast of Oregon and Washington is filled with sediments eroded from the rising, glaciated Cascade and Olympic Mountains. The region's high sedimentation rate has filled the trench and thus no deep de-

pression appears on the seafloor along the Pacific Northwest coast.

At subduction zones, sediments become caught between the converging plates. These sediments are a mixture of deep-sea muds and clays, silica and carbonate oozes, and clastic, land-derived deposits. All of these materials are mixed with submarine basalts and some slices of serpentinized upper-mantle rocks into a chaotic mixture called a **mélange** (Fig. 12-15). A typical mélange is sliced up, crushed, and plastered along thrust faults onto the overriding plate (the plate that isn't subducting). The plastered-on mélange forms a wedge-shaped mass of rock and sediment called an **accretionary wedge.** In North America, mélanges that have been uplifted from ancient subduction zones can be seen in the California coast

▲ **FIGURE 12-15** A mélange along the Sonoma County coast in California. These rocks—a jumbled mixture of seafloor and land-derived materials that have undergone high-pressure/low-temperature metamorphism—formed in the ocean trench associated with the subduction of the Farallon plate under North America.

ranges (Fig. 12-16) and in the Klamath Mountains of Oregon. They consist of a highly deformed sandy and shaly matrix containing chunks of metamorphosed basalts (as blueschists) and other rocks that show evidence of metamorphism within the unique high-pressure, low-temperature environment of subduction zones. On many hillsides in the San Francisco Bay Area, you can see large blueschist boulders standing alone; these blueschists and other high-pressure rocks have resisted weathering more than the rest of the soft, crushed-up melange. Other melanges in North America are found in the Kootenay Mountains of eastern British Columbia, the coastal mountains of south-central Alaska, the Blue and Wallowa Mountains of eastern Oregon, and the Appalachians of New England and the Canadian Maritime provinces. Each melange is strong evidence of past subduction.

A few hundred kilometers away from the plate boundary, a **volcanic arc** typically develops atop and within the overriding plate. At this point, the subducting plate has reached a depth at which it gives off enough fluids through dehydration to melt the overlying mantle *partially*. A chain of volcanoes marks this point. We can use the distance of the volcanic arc from the trench—called the *arc–trench gap*—to figure out how steeply the subducting plate is diving back into the mantle (Fig. 12-17). A steep subduction angle typically indicates rapid descent of old, dense oceanic lithosphere. A gentle subduction angle marks the descent of relatively young, warm, buoyant oceanic lithosphere. (We can also figure out the location of the subducted plate in the mantle from the location of earthquakes, but the arc-trench gap is a quick and easy way to estimate subduction angles from visible features at the Earth's surface.)

Between the accretionary wedge and the volcanic arc is a sediment-trapping depression called the **forearc basin** (see Fig. 12-14). Sediment eroded from both the uplifted accretionary wedge and the eroding volcanic arc accumulates in the forearc basin. The area between Oregon's Cascades and its uplifted Coast Range is a modern example of a forearc basin. Another sediment-trapping depression, the **backarc basin**, forms on the inland side of a volcanic arc. Sediments deposited there are derived from the eroding volcanic arc, and in the case of ocean lithosphere–continental lithosphere convergence, from continental streams flowing toward the sea as well. A good modern example of a backarc basin is the Columbia Basin east of Washington's Cascades (a volcanic arc).

Backarc basins form when, during subduction at an ocean trench, the overriding plate is stretched and thinned on the "back" side of the volcanic arc. This process may occur if the velocity of the overriding plate exceeds the velocity of the sub-

▲ **FIGURE 12-16** Coastal mountains typically begin as an accretionary wedge at a subduction zone. These coastal mountains near Big Sur, California, originated as offshore turbidites of continental origin that became packed in the Farallon trench and were dragged downward tens of kilometers before being uplifted isostatically.

ducting plate with which it is converging. Often, the crust of the overriding plate may stretch so much that it eventually rifts; this, in turn, reduces the pressure on the underlying mantle and lowers the melting point of the mantle's rocks. Currents of basaltic magma may then rise and solidify as new oceanic crust, by a process called **backarc spreading.** Backarc spreading beneath the Sea of Japan, for instance, is causing the volcanic island arc of Japan to move eastward, as the backarc region expands from the creation of new seafloor crust between Japan and China. Such spreading is marked by thinning plates, high heat flow, normal faulting, frequent earthquakes, and basaltic volcanism.

Convergent Plate Boundaries— Collision Zones

As plate convergence and subduction progresses, eventually an ocean plate will be entirely subducted. At this point, any continental mass that was being towed behind the subducting oceanic lithosphere will come face-to-face with the overriding plate. The overriding plate might be another continent or it might be an island arc. In either case, the two masses of buoyant felsic crustal material collide. Because they are relatively light, neither mass subducts entirely. The leading edge of the continent that is attached to the subducting oceanic lithosphere will be dragged down the subduction zone hundreds of kilometers into the mantle, but because it is less dense than the surrounding mantle, the subducted continental lithosphere will then bob back up to the surface (this is discussed further in Chapter 13). While it's down at mantle depths, deep-Earth pressures may convert crustal

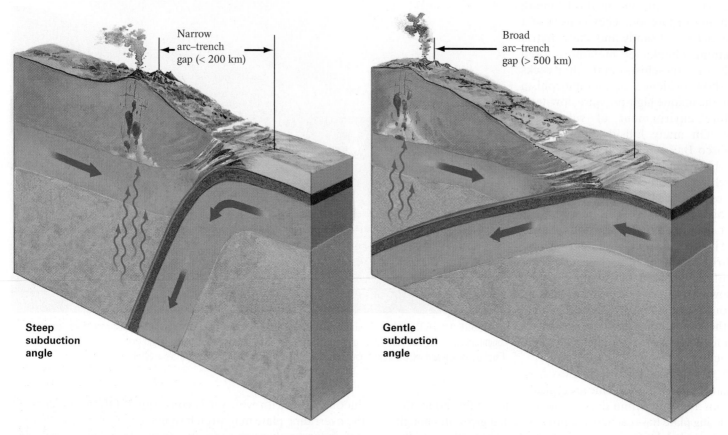

▲ **FIGURE 12-17** The breadth of an arc–trench gap is proportional to the size of the angle of subduction. A steep angle produces a narrow arc–trench gap; a gentle angle produces a broad arc–trench gap.

carbon to that ever popular mineral—the diamond. Highlight 12-2 discusses what happens when continental rocks are dragged down to great depth at subducting plate edges.

The boundary between collided continents—the place where the ocean has been entirely consumed by subduction—is called a **suture zone** (like medical sutures, or stitches, that bind together the edges of a deep cut). Fragments of oceanic crust and upper mantle that become trapped between the colliding continental rocks at a suture zone mark the site of the former ocean. Suture zones are relatively weak structurally. This factor becomes important long after the collision ends, as suture zones represent the most likely sites for future rifting when the cycle of ocean creation by rifting and divergence starts anew.

Now that we have discussed the various tectonic processes associated with both the origin and the consumption of the Earth's oceanic lithosphere, let's consider some current thoughts about the forces and processes that drive plate motion and shuttle oceanic lithosphere from its divergent-zone birthplace to its subduction-zone graveyard. First we will examine ways to observe and measure plate motion, including plate velocity, and then we will conclude this chapter by discussing the forces that likely drive plate motion.

Determining Plate Velocity

A plate's velocity is a measure of both its speed *and* its direction of movement. One key way to estimate a plate's speed in *absolute* terms requires a fixed point outside of the plate to use as a reference (such as a distant star or an orbiting satellite). We can't measure absolute velocity from the ground because every place on Earth is in motion. (This attempt would be like trying to measure the speed of a car from another moving car.) Without a fixed reference point, we can determine only each plate's velocity *relative* to the other plates (Fig. 12-18).

Every year, ground-based laser beams are bounced off the Laser Geodynamics Satellite (LAGEOS) (Fig. 12-19), recording the amount of time required for the beams to make a round-trip journey. With that information, geologists can determine the precise distance between the satellite and the ground station. Because LAGEOS speeds around the Earth at exactly the same velocity as the Earth rotates on its axis, its position over the planet remains fixed. Thus any change in the laser beam's travel time records a change in the geographic position of the landmass where the ground station is located.

Similarly, a network of 24 orbiting satellites that constitutes the Global Positioning System (GPS) can locate precise points

Highlight 12-2

Continental Subduction and Formation of Diamonds

Continental crust is less dense than oceanic crust and much less dense than mantle rocks. Once formed, continental crust may rift, drift, or collide with other continents, but its buoyancy relative to the more mafic rocks makes it difficult for this crust to sink into the mantle. Thus it tends to remain at the Earth's surface. The discovery of ultra-high-pressure minerals—such as diamonds—in continental crust, however, proves that continental crust, at least at its margins, does subduct, at least temporarily. These crustal diamonds and a high-pressure form of quartz called coesite have been found in many orogens in Europe, Asia, Africa, and Greenland, but they have not (yet) been found in North America.

In many cases, the ultra-high-pressure minerals occur in rocks that were originally sediments, indicating that these rocks have had a very long journey—from the surface, down to diamond-forming depths in the mantle (greater than about 150 km/90 mi), and then back to the surface. It is no surprise, however, that these once-deeply-buried rocks return to the surface. Imagine dunking a beach ball below the surface of a pool of water and then letting go of it. The buoyant beach ball will pop back to the surface because it is not dense enough to remain underwater without your help. Similarly, low-density crustal rocks will rise from the mantle once the force dragging them down is no longer capable of overcoming the buoyancy forces.

If these rocks are so buoyant, why do they subduct in the first place? When an ocean plate is entirely consumed by subduction, the leading edge of a continent attached to it—the continental margin, which is thinner than the rest of the continent—is dragged down by the gravitational pull of the descending oceanic slab for some distance—before buoyancy overcomes the downward drag.

An especially interesting aspect of continental subduction is the possibility that not all of the continental crust bobs back up to the surface. If the mantle absorbs some of this material, its addition may affect the mantle's overall chemical composition. The very presence of diamonds in the mantle may be evidence for the addition of continental ingredients to the mantle: The carbon in diamonds may have originally come from organic or other carbonaceous matter that was subducted along with the sedimentary rocks of the continental margin.

Ultra-high-pressure rocks have been found on most continents, but not yet in North America. Who will be the first to find evidence for continental subduction in North America? Or did continental subduction simply not occur in North America (and if not, why not?). Maybe you can use your new knowledge of geology to predict the most likely places to look for evidence for ultra-high-pressure metamorphism in North America. The only tricky part is that the diamonds will probably be "microdiamonds" and therefore would not be visible without the aid of a powerful microscope.

▲ Arrows point to minute diamonds embedded within high-grade gneiss in Kazakhstan. (inset) Photomicrograph of a diamond grain from the rocks of Norway.

on the Earth's moving plates. Much like the triangulation used to locate earthquakes (discussed in Chapter 10), the overlapping beams from a trio of satellites can locate a single geographic point on the Earth's surface (Fig. 12-20). To increase precision in order to measure very small plate motions, signals from more than three satellites are typically used; the more satellites, the more precise the measurement. Using signals from a dozen or more of these satellites, geoscientists can measure the movement of a plate to within a few millimeters. In this way, space-age technology is used to confirm that the Earth's plates do indeed move—and shows that they do not all move at the same rate.

One Earth-bound way to measure our moving plates involves tracking hot-spot volcanoes. **Hot spots** are localized regions where plumes of hot mantle material rise from great depths—perhaps from as deep as the core–mantle boundary—to the base of the lithosphere. Although these plumes may themselves move as the deep

12.1   Geology at a Glance

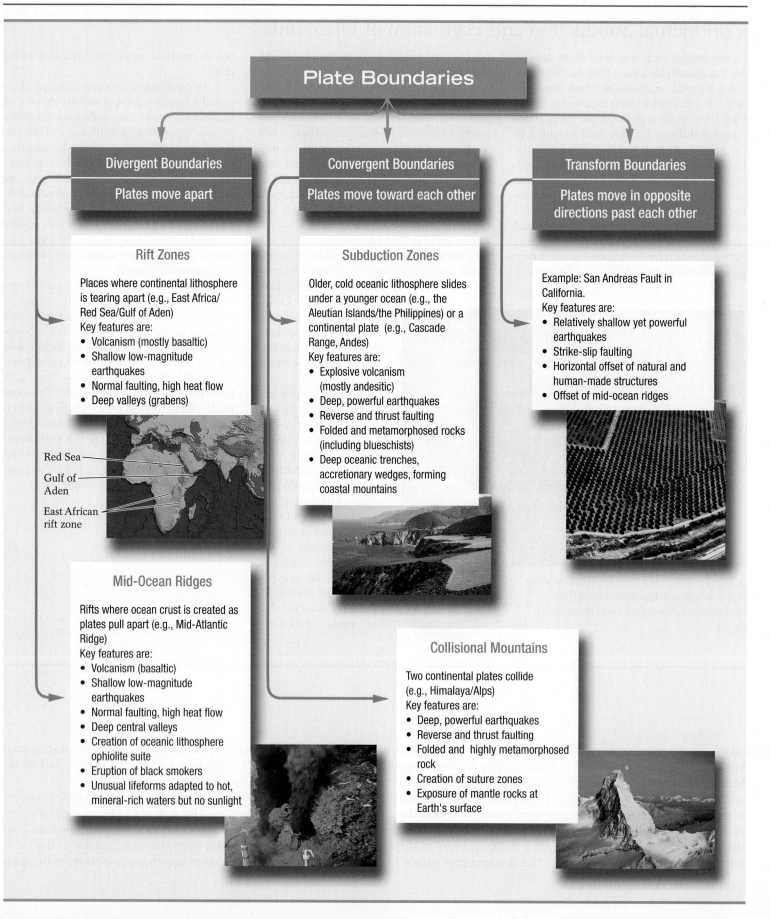

Plate Boundaries

Divergent Boundaries
Plates move apart

Convergent Boundaries
Plates move toward each other

Transform Boundaries
Plates move in opposite directions past each other

Rift Zones

Places where continental lithosphere is tearing apart (e.g., East Africa/Red Sea/Gulf of Aden)
Key features are:
- Volcanism (mostly basaltic)
- Shallow low-magnitude earthquakes
- Normal faulting, high heat flow
- Deep valleys (grabens)

Red Sea
Gulf of Aden
East African rift zone

Subduction Zones

Older, cold oceanic lithosphere slides under a younger ocean (e.g., the Aleutian Islands/the Philippines) or a continental plate (e.g., Cascade Range, Andes)
Key features are:
- Explosive volcanism (mostly andesitic)
- Deep, powerful earthquakes
- Reverse and thrust faulting
- Folded and metamorphosed rocks (including blueschists)
- Deep oceanic trenches, accretionary wedges, forming coastal mountains

Example: San Andreas Fault in California.
Key features are:
- Relatively shallow yet powerful earthquakes
- Strike-slip faulting
- Horizontal offset of natural and human-made structures
- Offset of mid-ocean ridges

Mid-Ocean Ridges

Rifts where ocean crust is created as plates pull apart (e.g., Mid-Atlantic Ridge)
Key features are:
- Volcanism (basaltic)
- Shallow low-magnitude earthquakes
- Normal faulting, high heat flow
- Deep central valleys
- Creation of oceanic lithosphere ophiolite suite
- Eruption of black smokers
- Unusual lifeforms adapted to hot, mineral-rich waters but no sunlight

Collisional Mountains

Two continental plates collide (e.g., Himalaya/Alps)
Key features are:
- Deep, powerful earthquakes
- Reverse and thrust faulting
- Folded and highly metamorphosed rock
- Creation of suture zones
- Exposure of mantle rocks at Earth's surface

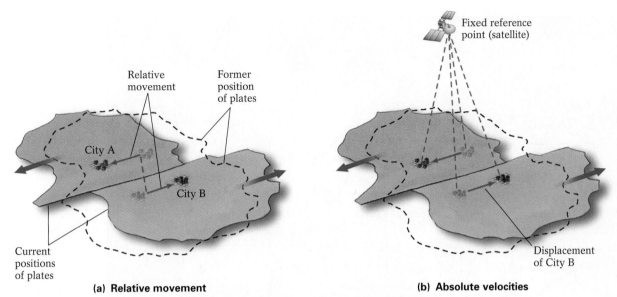

(a) **Relative movement**

(b) **Absolute velocities**

▲ **FIGURE 12-18** Determining plate velocities. **(a)** If two plates are both moving, and we have no fixed reference point by which to measure their displacement, we cannot establish the *absolute* speed of either one. The changing distance between the moving plates reflects only their motion *relative* to one another. Here, we know the two cities are moving apart, but we don't know for sure how much of the movement comes from plate A or plate B. **(b)** By using a fixed reference point outside the Earth's moving plates—here, a satellite—we can determine the plates' absolute velocities.

mantle convects, they move very slowly relative to the overlying plates. The hot spot that fuels Hawaiian volcanism, for example, appears to be moving southeastward, in the opposite direction from the overlying Pacific plate, but at $\frac{1}{10}$ the speed. Thus, after accounting for their own slow motion, hot spots can be used as "fixed" reference points that allow us to estimate both the speed and direction of the plates that move above them. Approximately three dozen hot spots, shown as red dots in Figure 12-21, are active today, and perhaps 100 or so have been active in the past 10 million years. Many appear at or near divergent plate boundaries, but some (such as the one beneath Yellowstone National Park in Wyoming) reside in plate interiors, where they power intraplate volcanism.

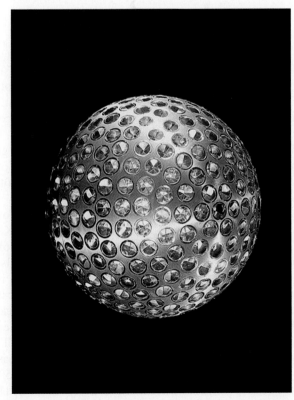

▲ **FIGURE 12-19** The LAGEOS satellite. The craft's numerous reflectors bounce back ground-based laser emissions to determine beam travel times. Changes in these travel times enable geoscientists to measure absolute plate motion.

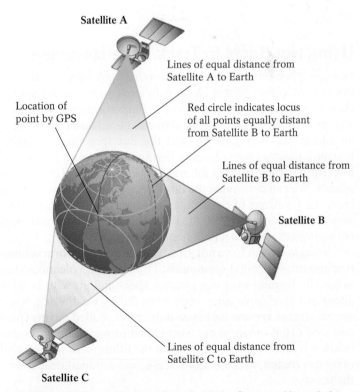

▲ **FIGURE 12-20** The satellite-based Global Positioning System enables geologists to locate changes in precise points on the Earth's surface, confirming that the Earth's plates do indeed move.

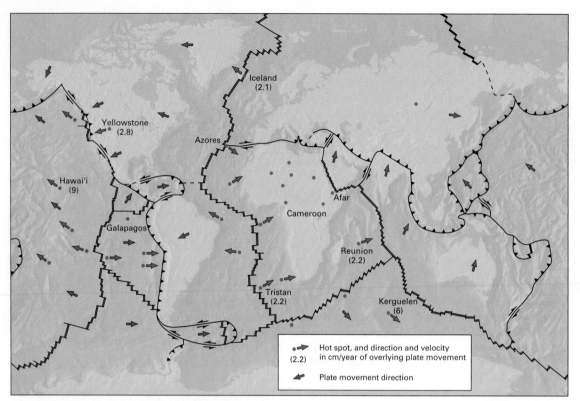

▲ **FIGURE 12-21** Hot spots and absolute plate motion around the world. Hot spots, which may produce volcanoes, usually occur at divergent plate boundaries, but can also be found under plate interiors. Because hot spots are essentially fixed relative to the faster-moving plates above them, geologists can use them as reference points to determine the absolute velocities of the world's tectonic plates. Hot spots originate at great depth in the Earth's interior—perhaps as deep as the core–mantle boundary. Such deep-seated hot spots fuel the volcanism at certain mid-plate locations (such as Hawai'i and Yellowstone). The upwelling hot material also drives divergent motion at continental rift zones (such as the East African rift) and oceanic divergent boundaries (such as the Mid-Atlantic Ridge).

Using Hot Spots to Track Plate Movements

Hot spots beneath oceanic plates produce volcanoes on the sea floor directly above them. As a plate moves across a hot spot, a chain of submarine volcanoes forms. Each volcano may eventually grow to heights that reach above sea level and thus become an island. As they pass beyond the hot spot (and its rising magma), these active volcanoes become extinct, and new volcanoes form in their place at the end of the growing trail of volcanic islands. We can determine the direction of the plate's movement from the locations of the extinct volcanoes, and we can determine the plate's speed from the volcanoes' ages and the distances between each volcano and the hot spot.

Volcanoes that have not grown above sea level form submarine mountains called **seamounts.** Other island volcanoes that originally formed over oceanic hot spots and grew to heights above sea level have since been worn flat by weathering, wave action, stream erosion and mass movement. Called **guyots** (pronounced GHEE-owes), these worn-down forms are found today below sea level, principally because the lithosphere that carried them has cooled, become more dense, and subsided deeper into the mantle.

All oceans contain numerous hot-spot islands, seamounts, and guyots, generally arranged in long chains that form as drift-

ing plates trail away from hot spots. The Hawaiian Island–Emperor Seamount chain is a classic example of such a hot-spot trail (Fig. 12-22). The chain extends first to the northwest, then northward, through more than 6000 km (4000 mi) of the central Pacific, before ending at the Aleutian trench near Alaska. From potassium–argon dating of basalt from the entire chain, geologists have found that these volcanic features developed over a span of more than 75 million years, indicating that this hot spot is long-lived.

The Hawaiian Islands are relatively recent features formed by the central Pacific hot spot. The Hawaiian island of Kaua'i, whose rocks date from 5.6 to 3.8 million years ago, is the oldest, most northwestern island in the chain. Hawai'i, the Big Island, which today experiences nearly continuous eruptions, is the youngest, most southeastern, and the only actively volcanic island in the chain. Between Kaua'i and Hawai'i, the islands' rocks are progressively younger toward the southeast. To the southeast of Hawai'i is the Loihi seamount (Loihi means "the long tall one" in a Polynesian dialect), which will likely become the next island in the chain. Loihi's summit rises roughly 4400 m (14,500 ft) above the seafloor, still several thousand meters below sea level. If Loihi's eruptions continue at their current rate, the volcano will replace the island of Hawai'i as the volcanic centerpiece of

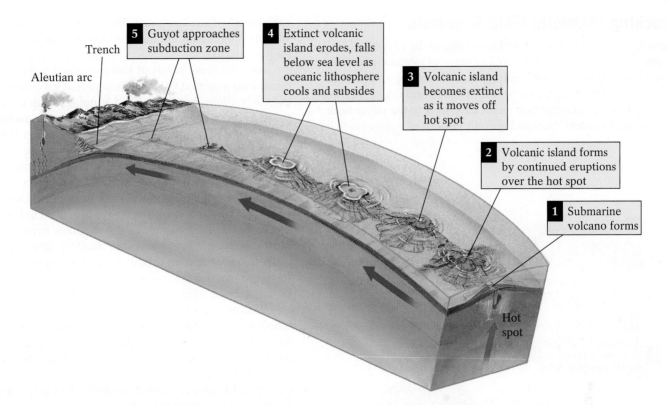

5 Guyot approaches subduction zone

4 Extinct volcanic island erodes, falls below sea level as oceanic lithosphere cools and subsides

3 Volcanic island becomes extinct as it moves off hot spot

2 Volcanic island forms by continued eruptions over the hot spot

1 Submarine volcano forms

Trench

Aleutian arc

Hot spot

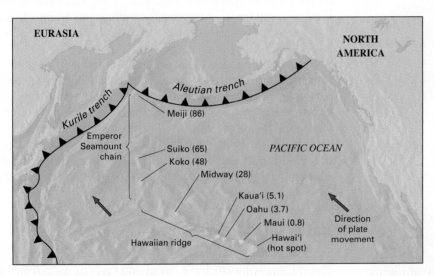

▲ **FIGURE 12-22** The same mid-Pacific hot spot has fueled the eruptions that produced every volcanic island and submarine mountain in the 6000-km (4000-mi)-long Hawaiian Island–Emperor Seamount chain. As the moving plate carries each island away from the hot spot, the volcanoes lose their heat source and become extinct. Weathering and erosion gradually reduce their heights above sea level. Ultimately, they become submerged as the lithosphere cools and subsides isostatically. (Volcano ages, in millions of years, are shown in parentheses.)

the central Pacific sometime during the next few hundred thousand years.

Several other central Pacific seamount chains exist, all of which are marked by the same noticeable bend, formed when the Pacific plate changed its direction from due north to its present northwesterly trend. Basalt from volcanoes at these bends date to about 40 million years ago. Thus we know that the Pacific plate changed direction at that time. We can estimate the rate of Pacific-plate movement by using the Hawaiian hot spot as a fixed reference point. For example, Midway Island is located today about 2700 km (1700 mi) northwest of the hot spot, and its basaltic rocks are 27.2 million years old. Dividing the distance from the hot spot by the age of the rocks, we calculate that the Pacific plate has moved at an average speed of nearly 10 cm (4 in.) per year.

Hot spots beneath continental plates do not always produce such obvious volcanic chains, probably because much of their magma cools and solidifies within the thick continental lithosphere and never reaches the surface. In a few places on continents, however, we can see what appears to be a chain of volcanic rock and associated intrusions that may have formed as the continent moved over a hot spot. For example, we know that a hot spot lies beneath Yellowstone National Park today and that the rocks in the park formed in the past 1 million years. West of Yellowstone, a chain of igneous rocks—1 to 5 million years old—extends from Wyoming through southern Idaho. This chain includes the vast basaltic plains of the Snake River. These lavas might represent the track of the Yellowstone hot spot as the North American plate moved over it.

Tracking Magnetic Field Reversals

Geoscientists can also use their knowledge of the reversals of the Earth's magnetic field to determine rates of oceanic-plate motion. During World War II, military vessels bearing magnetometers traversed the Pacific and Atlantic Oceans seeking metal from the wrecks of submarines that might have sunk to the ocean floor. The magnetometers recorded an odd pattern of magnetic-field variations: Alternating invisible bands of slightly stronger and weaker magnetism appeared along the ocean floor parallel to mid-ocean ridges.

At first, these bands of variable magnetism puzzled technicians, who assumed that their equipment was malfunctioning. In the early 1960s, however, Fred Vine, a young Cambridge University student, named the magnetic stripes **marine magnetic anomalies** and proposed that they indicated sea-floor spreading at divergent plate boundaries. Vine and his associate Drummond Matthews explained that as basaltic lava cools at spreading mid-ocean ridges, it becomes magnetized in the direction of the Earth's magnetic field at the time of the eruption. Basalts that erupt at times of reversed polarity (discussed in Chapter 11) will therefore display evidence of the magnetic reversal (Fig. 12-23). Magnetometers towed above a normally magnetized stripe record a slight strengthening (about 1%) of the magnetic field. When towed above a stripe formed under reversed magnetism, they record a slight weakening of the field (Fig.12-24).

Within a single magnetic stripe on the sea floor, all of the basalt has essentially the same age. Dating a few basalt samples from one stripe, therefore, tells us the age of the basalt along the entire length of the stripe. Because oceanic plates grow continuously at divergent zones, seafloor basalts contain a complete record of the polarity and strength of the Earth's magnetic field for the past two hundred million years—the maximum age of the Earth's ocean floors. After much sampling and dating of seafloor basalts, we now know the age of virtually all marine magnetic anomalies.

Marine magnetic records can also be used to determine rates of plate motion. We can calculate how long it took any particular magnetic stripe to be moved with its drifting plate a known distance from the divergent zone where it formed. For example, if a magnetic stripe in the Atlantic Ocean is 4.5 million years old (with its age being determined by dating the basalt within it) and it is located 90 km (55 mi) from the divergent zone where it originated, the rate of divergence would be 2 cm (0.8 in) per year.

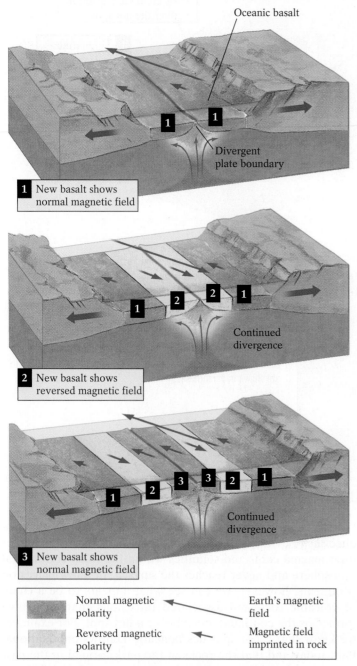

▲ **FIGURE 12-23** Marine magnetic anomalies show the reversals in the Earth's magnetic field. As basaltic lavas cool and solidify at mid-ocean ridges, the magnetic fields of their magnetite crystals become aligned with the prevailing direction of the Earth's magnetic field. Each resulting stripe of basalt has either normal magnetism (like today's field) or reversed magnetism (opposite to today's field).

Rates of Plate Motion

So, how fast do plates move? In examples given earlier in this chapter, geoscientists have calculated rates ranging from 10 cm (about 4 in) per year for the motion of the Pacific plate over the Hawaiian hot spot to 2 cm per year for divergence at the Mid-Atlantic spreading center. Using information from satellites, oceanic hot spots, and marine magnetic anomalies, we can measure the absolute velocity of every plate. Figure 12-25 shows us that the Pacific, Nazca, Cocos, and Australian-Indian plates, among the Earth's fastest-moving, travel more than 10 cm (4 in) per year. All of these plates are composed principally of oceanic lithosphere. The North and South American, Eurasian, and Antarctic plates, among the slowest-moving, shift their positions by 1 to 3 cm (0.4 to 1.2 inches) per year. Each of these plates largely consists of continental lithosphere. Thus it appears that oceanic lithosphere generally moves faster than continental lithosphere.

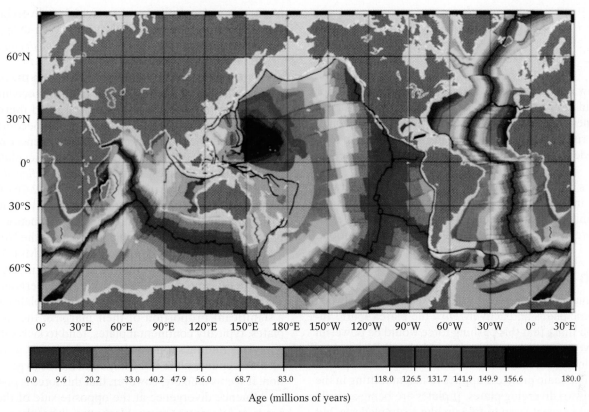

▲ **FIGURE 12-24** The ages of oceanic lithosphere segments around the world, as determined by dating marine magnetic anomalies. The colored stripes represent oceanic lithosphere of the ages indicated in the key. The width of each stripe is proportional to the spreading rate at the mid-ocean divergent plate boundaries. A symmetrical anomaly pattern (upper right) characterizes a mid-ocean spreading center, such as can be seen within the Atlantic Ocean basin. An asymmetrical pattern (center) shows that subduction has consumed part of an oceanic plate along the northwestern coast of North America.

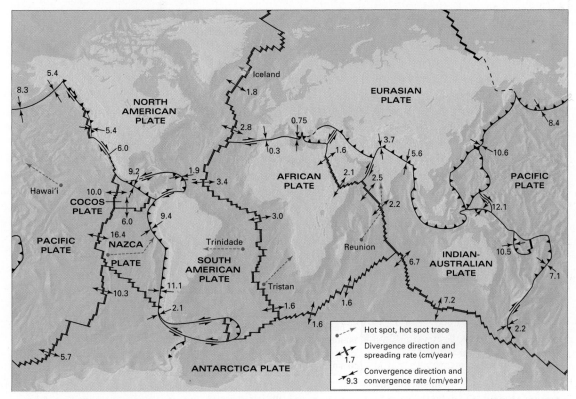

▲ **FIGURE 12-25** The directions and rates at which the Earth's plates move, calculated from marine magnetic anomalies, offset rocks along transform faults, and island distances relative to hot spots.

The Driving Forces of Plate Motion

Geologists know *why* plates move: Heat from the Earth's interior sets the plates in motion as hot, less dense rocks and magma rise at divergent zones, and cold, dense rocks sink at subduction zones. This pattern of rising and sinking in response to variations in temperature and density is called *convection* (first discussed in Chapter 1).

Beyond this basic answer for why plates move, understanding *how* plates move requires discussing whether the plates are passive "hitchhikers" on the underlying flowing mantle, or whether they are actively involved in driving their own motion.

Ridge Push or Slab Pull?

At divergent zones, does the rising magma wedge itself between adjoining oceanic plates and actively *push* the plates apart? Intuitively, do you think that this pushing force would be powerful enough to move a 5000-km (3000-mi)-wide plate? Remember, water that sits in a crack in a rock cannot push the crack open; it just sits there. The same principle applies to magma sitting in the cracks between two diverging plates. If plates are being actively pushed, we might expect them to fold up like a rumpled rug, but we don't see this in the rocks at divergent zones. Instead, we see— with the help of ocean-exploring submersibles—evidence that

plates are in fact pulled and stretched at divergent zones. The thinning lithosphere there is torn into down-dropped blocks by normal (extensional) faulting. Thus, *ridge push cannot* be a major driving force of plate motion.

So if the plates are not being pushed from the ridges, are they being *pulled* at subduction zones? Old, cold oceanic lithosphere is denser than the warm asthenosphere and therefore it sinks. The lower portion of the descending slab becomes even more dense as oceanic crust metamorphoses to denser rocks. This increased density may actively drag oceanic plates into their subduction-zone graves. Thus, **slab pull** *is believed to be the most powerful driving force in plate motion* (Fig. 12-26). Other mechanisms—involving the dynamics of mantle convection—contribute to moving those plates that do not have subduction zones at their boundaries, such as the oceanic segments of the North American and Eurasian plates on opposites sides of the Mid-Atlantic Ridge.

Recent studies show that rates of plate motion are proportional to the amount of a plate's margin undergoing subduction: The fastest-moving plates, typically oceanic plates, subduct along a significant proportion of their margins; the slowest-moving plates, typically continental plates, tend to lack subducting margins. This finding supports the hypothesis that gravitational pull on a descending plate is the key contributor to general plate motion. It seems unlikely, however, that this force is powerful enough to influence divergence at the opposite side of the plate, thousands of kilometers away. Moreover, lithospheric plates would most likely break if such large-scale pulling forces tugged at them for thousands of kilometers. Furthermore, the Atlantic Ocean's

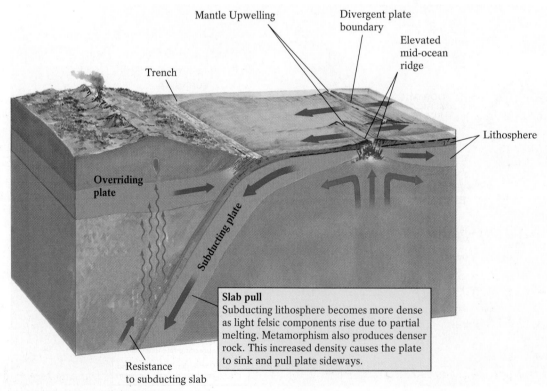

Slab pull
Subducting lithosphere becomes more dense as light felsic components rise due to partial melting. Metamorphism also produces denser rock. This increased density causes the plate to sink and pull plate sideways.

▲ **FIGURE 12-26** Factors that drive plate tectonics: *slab pull* as old, dense plates subduct into the mantle and *upwelling and underflow* at mid-ocean ridges.

floor diverges from the Mid-Atlantic Ridge at a rate of about 2 cm (nearly 1 in.) per year, even though no slabs sink along any of its margins. Some force other than slab pull must therefore produce the slow motion of the Atlantic Ocean segments of the North and South American, Eurasian, and African plates.

The best simple answer to the question of what drives the motion of the Earth's plates is that most plates move in response to mantle flow. The downwelling of the lithosphere at subduction zones is the dominant driving force—pulling plate edges downward into the convecting mantle as cold, dense material sinks into warm, less-dense regions of the mantle. We have also observed that plates without subduction zones at their edges still move, but much more slowly. They move because the whole plate-tectonic system is connected, so as subduction pulls down plates in some places, other plates on the spherical surface of the Earth must also move. Thus, the *downwelling* at subduction zones is balanced by *upwelling* of the mantle at divergent zones, where hot, lower-density mantle material rises up and flows laterally. The rising mantle also melts, producing the basaltic magmas that fuel volcanic eruptions at divergent zones.

The Tectonics of Continents—A Preview

Plate tectonics was an exciting idea several decades ago, but now it is a well-accepted theory that explains many things about how the Earth works. It explains, for example, why volcanoes are found in some places and not others, and why some volcanoes erupt with great explosive force with enormous clouds of felsic tephra while others are marked by non-explosive streams of red-hot mafic lava. Plate theory also explains the locations and magnitudes of earthquakes and the formations of the Earth's largest features—its mountain ranges, ocean basins, and continents.

In this chapter, we have discussed the major features of the plate-tectonics system with an emphasis on the different types of plate boundaries and the driving forces of plate motion. We have not yet discussed the Earth's continents in detail, although continents are a key part of the plate-tectonic system: They are produced by magmatism and metamorphism at convergent plate boundaries, their positions on Earth are rearranged by plate motion, and they change through time as they collide, partially subduct, and are sliced by faults related to plate-tectonic motion.

Tectonic processes in the continents, however, are somewhat different from those predicted by classic plate-tectonic theory. You have seen that earthquakes and volcanoes closely follow plate boundaries, and that much tectonic-related activity (faulting, metamorphism, magmatism) occurs in narrow zones near the plate boundaries. In fact, you could probably draw a pretty good map of the Earth's plate boundaries using only the location of earthquakes and volcanoes as a guide. Such a map of plate boundaries, however, would not predict some important features of the continents, such as the great width of the North American Cordilleran mountains (including the Rocky Mountains, Sierra Nevada, and the Cascade-Coast Range mountains) and the Tibetan Plateau associated with the Himalaya of China, Tibet, and Nepal. In the next chapter, we will consider continents in the context of tectonic processes, and discuss the origin of continents, their evolution, and their relationships to other Earth systems.

Chapter 12 Summary

Our understanding of the Earth's tectonic past comes largely from our knowledge of the changing appearance of the world's ocean floors. Ocean plates form as preexisting continental plates rift and diverge. Rifting begins when a warm current of mantle material rises under a continental plate, stretching and thinning the plate until it tears. Three radiating rift valleys typically form simultaneously; two of these eventually diverge, while the third commonly becomes inactive. After rifting between two new plates ends, the rift edges become inactive tectonically, forming passive continental margins.

As divergence continues, a full-blown seaway develops. New oceanic lithosphere forms at a mid-ocean ridge where mantle-derived basalts erupt to produce a linear chain of volcanic mountains. The typical rock sequence within oceanic lithosphere—called an *ophiolite sequence*—consists of:

- marine sediment;
- underlying layers of pillow basalts and related intrusive gabbros;
- a bottom layer of ultramafic mantle peridotite.

The mid-ocean ridge generally becomes divided into short, offset segments that are cut by oceanic transform boundaries.

A diverging oceanic plate gradually cools and becomes more dense. It may eventually sink back into the Earth's interior by subducting under a less-dense plate far from the mid-ocean ridge. An *ocean trench* develops where a subducting plate flexes downward into the mantle, forming a depression in the Earth's surface. Subduction produces several rock types and geologic structures:

- Chaotic mixtures of oceanic sediments and ophiolite rocks called *mélanges* that pile up within the trench, forming masses of rock called *accretionary wedges;*
- A *volcanic arc,* a chain of volcanoes that forms as the subducting plate sinks into the warm mantle and expels its water, thus promoting partial melting of the overlying mantle. When these magmas erupt, they form the volcanic arc;
- Sediments accumulate on either side of a volcanic arc in topographic depressions called *forearc and backarc basins.*

Once subduction of an oceanic plate is complete, the continental blocks on either side of the plate collide, forming a larger continent. The boundary between them, the *suture zone,* represents a weak spot in the Earth's lithosphere, and thus suture zones are good candidates for future rifting.

Geoscientists have learned much about the origin of oceanic rocks and the tectonic processes that produce them, but they are still exploring the precise mechanisms that drive the plate motion of specific plates. Plate motion is believed to be driven principally by gravity-induced subduction of old, cold, dense plates back into the Earth's mantle.

KEY TERMS

accretionary wedge *(p. 396)*
backarc basin *(p. 397)*
backarc spreading *(p. 397)*
echo-sounding sonar *(p. 386)*
forearc basin *(p. 397)*

guyots *(p. 402)*
hot spots *(p. 399)*
marine magnetic anomalies
  (p. 404)
mélange *(p. 396)*

oceanic trenches *(p. 396)*
ophiolite suite *(p. 393)*
passive continental margins
  (p. 390)
seamounts *(p. 402)*

seismic profiling *(p. 386)*
slab pull *(p. 406)*
suture zone *(p. 398)*
volcanic arc *(p. 397)*

QUESTIONS FOR REVIEW

1. Describe three methods of determining the velocity of plate motion.

2. List four phenomena that accompany continental rifting.

3. Describe what happens at a rift margin, from the onset of rifting to the formation of an ocean.

4. Draw a simple sketch of an ophiolite suite (include all of the rock types and structures). Be sure to include the vertical scale for the layers.

5. Why do the earthquakes associated with oceanic transform boundaries generally occur only between offset oceanic ridge segments?

6. Sketch a convergent boundary between an oceanic plate and a continental plate. Be sure to include all of the important landforms and structures.

7. Of the two patterns of marine magnetic anomalies shown above right, which shows a faster rate of seafloor spreading? Explain.

8. How would the east coast of North America change if the oceanic segments of the plates that make up the Atlantic Ocean basin began to subduct? How would it change if the Atlantic Ocean lithosphere subducted completely?

9. How might the development of a new subduction zone along the East Coast affect plate interactions on the West Coast?

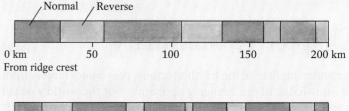

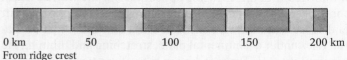

LEARNING ACTIVELY

1. Open an atlas and look at a world map. From the shapes of the continents, speculate about various ways that you might use today's continents to assemble a supercontinent, jigsaw-puzzle style. As you ponder these assemblies, think about what kinds of geological evidence you would need to prove these past configurations.

2. Over the next few weeks, monitor major geological events around the world, focusing on earthquakes and volcanic eruptions. With each event, determine whether it occurred at a plate boundary. If so, using the plate map in Chapter 1, determine the type of plate boundary. Did any of the events occur at a plate interior? Was one associated with a known hot spot?

3. Imagine that you had an *unlimited* budget. How would you go about trying to determine the driving force of plate tectonics?

ONLINE STUDY GUIDE

Practice makes you better. Make the time to take the *Online Study Guide* quiz for this chapter. It should take you about 20 minutes. Automatic grading and instant feedback will help you quickly and accurately identify the concepts you have mastered and the areas in which you need more study.

www.prenhall.com/chernicoff

Continental Tectonics and the Formation of the Earth's Continents

13

The general theory of plate tectonics explains very well how and why plates move and the fascinating processes associated with this motion, such as volcanism, earthquakes, faulting, and mountain building. According to general plate-tectonic wisdom, most geological activity occurs in fairly narrow zones near plate boundaries. We can see all of this plate-edge action at oceanic divergent zones, such as the Mid-Atlantic ridge; transform zones, such as California's San Andreas fault zone; and subduction-related island arcs, such as the Aleutian Islands of Alaska. When we look at the continents, however, we often see *wide* zones of tectonic activity, such as in the mountains of western North America (Fig. 13-1) or the Tibetan Plateau of Asia (Fig. 13-2). Plate-tectonic theory certainly still applies to the continents, but we need to consider some aspects of tectonics that apply specifically to continental lithosphere, which, with its more silica-rich composition, can thicken more easily than oceanic lithosphere and therefore deform in a different way.

The Origin and Shaping of Continents

The continents are mostly made up of very old, thick, felsic crust. Unlike oceanic crust, which subducts completely, only the edges of continents subduct. Thus, most continental crust is not recycled back into the mantle. Consequently, whereas the crust of the ocean basins in the world today does not contain rocks older than about 200 million years old, all continents have some rocks that are billions of years old. If we are curious about Earth history *prior to 200 million years ago* (the time of the initial breakup of the supercontinent Pangaea), we must look to the rocks of the continents.

The Anatomy of Continents

Every continent has the same basic architecture (Fig. 13-3). The oldest parts are the **continental shields,** broad areas of exposed crystalline rock in continental interiors that have not changed much for more than a billion years. Every continent contains at least one large

FIGURE 13-1 A shaded relief map of North America. Note the breadth of North America's western mountains. ▶

▲ **FIGURE 13-2** The Himalaya—the world's highest mountains—lie between China and India, a product of an ongoing collision of continents.

shield area. North America's Canadian Shield extends across much of Canada from Manitoba to the Atlantic coast, dipping into the northern United States from northern Minnesota and Wisconsin to upstate New York's Adirondack Mountains.

Surrounding the continental shield is the **continental platform,** where the shield is covered by younger sedimentary rock. Together, the continental shield and platform make up the continental **craton**—that portion of a continent that has remained tectonically stable for a vast period of time. At the edges of the craton, near the borders of continents, we find mountains, **coastal plains,** and **continental shelves.** The east coast of North America contains all of these elements: the mountains are the Appalachi-

ans, such as the Blue Ridge of Virginia and the White Mountains of New Hampshire; the coastal plain stretches eastward from the Appalachians, forming the Atlantic Coastal Plain, such as the eastern region of North and South Carolina; and the Atlantic continental shelf stretches seaward from the Atlantic coast. The shelf, although underwater, is actually part of the geology of the continent.

The Origin of Continents

In the Earth's first few hundred million years of existence, the planet had no continents, no oceans, no atmosphere, and—

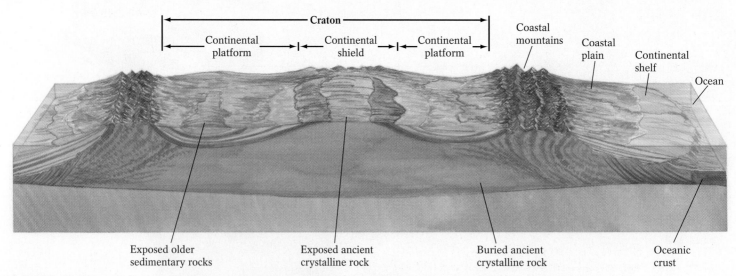

▲ **FIGURE 13-3** The anatomy of a continent. Every continent contains a tectonically stable nucleus, or craton, that consists of one or more ancient crystalline continental shields and a surrounding continental platform of old sedimentary rock. The outer edge of the continent is typically marked by relatively younger coastal mountains and plains and a submarine continental shelf.

obviously—no land or sea life. The planet's surface may have resembled the present-day Moon, pockmarked with craters from the impact of countless fragments of interplanetary debris. Four billion years later (about 400 million years ago), continental landmasses speckled with primitive plants covered 30% of the Earth's surface. Vast oceans, teeming with diverse plant and animal life, covered the remaining 70%. A thick protective atmosphere surrounded the entire planet. How did the Earth change from a lifeless, cratered planet to a place with continents and oceans, both teeming with life?

The continents formed by the same basic processes that differentiated the Earth to form its internal layers: through the melting of rocks and the separation of the different materials by density. When the early Earth melted partially (or, as some geologists think, melted entirely from the heat generated by massive colliding meteorites), the densest material sank to the core (mostly iron and nickel metals) and the lightest material floated to the surface (as silica-rich magma).

This process of differentiation must have occurred more rapidly in the early Earth than it does today. Billions of years ago, the planet was much hotter than it is today, so the mantle probably flowed much more rapidly, bringing greater volumes of magma to the surface at zones of upwelling similar to the hot regions that fuel modern divergent zones. More hot spots might have existed as well, creating even more crust at the surface. The resulting volcanic eruptions—at hot spots, divergent zones, and subduction zones—brought gases up from the Earth's interior. Great clouds of steam, carbon dioxide, and other gases erupted from volcanoes, creating the Earth's early atmosphere. When some of these gases condensed in the atmosphere, they created the planet's first oceans. The origins of the continents, oceans, atmosphere, and eventually life are therefore closely related to one another and to volcanism, which in turn was largely driven by tectonic activity.

Partial melting of mafic and ultramafic rocks to produce felsic and intermediate rocks (discussed in Chapter 3) only partially explains the origin of the continents. The Earth's oldest rocks—found in the continental shields of Australia, Greenland, and North America and dated at 3.8–4.03 billion years old—are metamorphic but were originally sedimentary, suggesting that some type of atmosphere must have existed that could have weathered preexisting rocks to form the sediments. Some rocks even appear to have been transported by and deposited in bodies of water. We can therefore conclude that as early as 4 billion years ago, the Earth had both an atmosphere and bodies of water. This is important because weathered materials may have contributed to the creation of silica-rich continental crust.

The Earth's early mafic crust was probably thinner, hotter, and thus more buoyant than the present-day mafic oceanic lithosphere. Consequently, it was less likely to be subducted. The first felsic *continental* lithosphere probably formed as a result of plate-tectonic processes that differed somewhat from those of today. As you can see in Figure 13-4, before distinct continents existed,

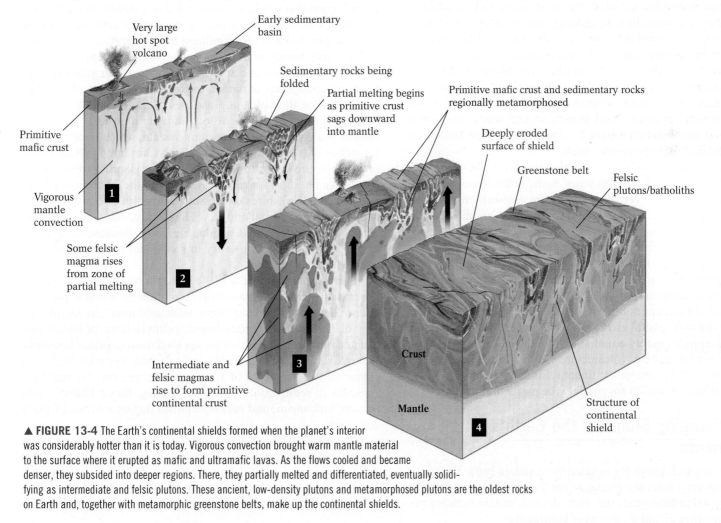

▲ **FIGURE 13-4** The Earth's continental shields formed when the planet's interior was considerably hotter than it is today. Vigorous convection brought warm mantle material to the surface where it erupted as mafic and ultramafic lavas. As the flows cooled and became denser, they subsided into deeper regions. There, they partially melted and differentiated, eventually solidifying as intermediate and felsic plutons. These ancient, low-density plutons and metamorphosed plutons are the oldest rocks on Earth and, together with metamorphic greenstone belts, make up the continental shields.

early landmasses probably consisted of huge volcanic islands that became enlarged as great volumes of mafic (basaltic) and ultramafic lava erupted from the mantle. As the eruptions continued, the islands grew and thickened, and early sedimentary basins developed around them. Older sections of the early lithosphere cooled and became more dense until some parts sagged downward into the hotter interior. There the sagging lithosphere folded up and dragged the sedimentary basins downward along with them. The basalts, ultramafic rocks, and sediments melted *partially* and then differentiated, producing the first intermediate and felsic magmas—the makings of the first continental lithosphere. Continued partial melting and differentiation of these rock materials eventually created new sets of volcanic islands that consisted largely of basalts and felsic plutons of varying compositions. These basalts later metamorphosed, forming greenstones, nonfoliated metamorphic rock with a greenish color derived largely from the minerals chlorite and epidote (discussed in Chapter 7). *Greenstone belts,* which include metamorphic basalts and various other metamorphosed sedimentary rocks, are thought to have accumulated next to very old island arcs. This combination of rock types characterizes most ancient continental shields.

After these small, lower-density felsic landmasses developed, subduction began to occur where the dense oceanic lithosphere and the light continental lithosphere converged. Recycling of primitive oceanic lithosphere increased the volume of andesitic rocks. In time, the continental nuclei became enlarged as drifting buoyant continental plates collided, the crust thickened by folding, thrusting, and magma addition, mountains were uplifted, and forearc mélanges (discussed in Chapter 12) accumulated. These masses of ancient continental lithosphere have since been weathered and eroded over the ages, processes that have removed the overlying sediments and exposed their deep plutonic and metamorphic cores. These felsic landmasses subsequently rifted, creating numerous centers of continental growth around which additional subduction and accretion could take place. Today, these centers of growth form the Earth's stable shields.

Much of the Earth's early continental lithosphere formed within the planet's first billion years. Some remnants of very old continental crust still remain, such as the 4.03-billion-year-old Acasta Gneiss of the Yellowknife Lake in the Canadian Arctic (Fig. 13-5). In addition to a slowing of continental-crust-building processes, probably because the Earth's internal-heat supply has diminished after the first couple of billion years of Earth history, very ancient rocks are scarce because of recycling by processes such as weathering and subduction.

While some continental lithosphere returns to the mantle through the continental erosion–marine deposition–subduction cycle, an approximately equal amount is produced by the subduction–partial melting–volcanic-arc-eruption cycle. Hence, the Earth's overall volume of continental lithosphere has probably remained fairly constant for roughly the past 3 billion years.

The Changing Shape of the Earth's Continents

Most continental shields are rimmed by mountain belts. The rocks in the mountain belts are younger than the shield rocks, typically less than 600 million years old. How does mountain building relate to the growth and evolution of continents?

▲ **FIGURE 13-5** The 4 billion-year-old Canadian shield, and the Acasta Gneiss near Yellowknife Lake in the Northwest Territories of Canada.

We all know a mountain when we see one, but geologists have a specific way of defining them. A *mountain* is a part of the Earth's crust that stands more than 300 m (1000 ft) above the surrounding landscape, has a discernible top, or *summit,* and possesses sloping sides. Every continent has mountains, as does every ocean basin. Some mountains, such as Georgia's Stone Mountain, stand alone, isolated and towering above their surroundings. Some group together in *ranges,* a succession of high-peaked structures, such as the volcanic Cascades of the Pacific Northwest. Some ranges form continent-long mountain groups, or *systems,* such as the Appalachian mountain system, which comprises the Great Smokies of Tennessee and North Carolina, the Blue Ridge of Virginia, the Catoctin Mountains of Maryland, the Poconos of Pennsylvania, the Catskills of New York, the Taconics of Connecticut, the Berkshires of Massachusetts, the Green and White Mountains of Vermont and New Hampshire, and other ranges in Maine and northeastern Canada (Fig. 13-6). Because of their different origins, mountains vary in shape and rock composition, even within a system.

Some mountains, such as the Catskills of upstate New York, are actually just the uplands that remain after streams have cut deep valleys into a plateau. Some, such as Mauna Loa and Mauna Kea of Hawai'i, are undeformed accumulations of basalt that erupted initially and built above sea level from sea-floor hot spots. Complex systems, such as the Appalachians of eastern North America and the Alps of southern Europe, involve multiple episodes of sedimentation, intense folding, thrust faulting, plutonism, volcanism, and metamorphism during a series of plate collisions.

Mountains formed during most of the Earth's past, and they continue to form today. The Himalaya of India, China, and Tibet are very young and still rising, and the collision between the Eurasian plate and the Indian plate that created them continues as the Indian plate converges with Eurasia. The Appalachians

Mt. Katahdin, ME

White Mountains, NH

Green Mountains, VT

Berkshire Mountains, MA

Great Smoky Mountains, SC

Appalachian Plateau

Valley and Ridge

Piedmont

Coastal Plain

Blue Ridge

ATLANTIC OCEAN

Gulf of Mexico

0   250   500 Km.

0   250   500 Mi.

▲ **FIGURE 13-6** Map of the Appalachian system and photos keyed to the ranges.

and the Urals of Russia and central Asia are so old that the primary tectonic forces that created them ceased to operate hundreds of millions of years ago. The principal processes affecting these mountains today are weathering and erosion. Some mountains are even older, so old in fact that they have completely worn away and are now just flat regions in continental interiors. The continental shields are the eroded remnants of once great mountains.

Mountain Building and Mountain "Unbuilding"

Many different processes can form mountains, including volcanism and even erosion. For the rest of this chapter, however, we will focus on the range of processes that build mountains where plates converge, especially those associated with continental collisions, the most important tectonic setting for forming the Earth's major mountain belts. Geologists refer to the processes of mountain building as **orogenesis** (from the Greek *oro,* meaning "mountain," and *genesis,* meaning "birth"), and they use the term even when the *topographic* expression of the mountain building is long gone. So, for example, we can talk about *orogens* (the belts of deformed rocks—typically metamorphic and plutonic—associated with mountain building) within the continental shields, even though they are now flat.

When continents collide, the Earth's crust is compressed and thickened into mountain belts. Different processes affect rocks at different depths. Deep rocks metamorphose and fold and probably melt; shallow rocks and sediments break (or bend) and may be moved long distances along giant thrust faults; rocks at the surface are weathered and eroded. All of these processes are linked. The creation of areas of high elevation affects erosion rates and mechanisms, and the movement of rocks and sediments by erosion from mountain tops to valley floors or ocean basins changes the thickness of the crust in the mountains. The interrelationship of all these processes means that even as mountains are building up by the squeezing of the crust caught between colliding plates, they are also "falling down" as a result of erosion and faulting (Fig. 13-7).

The early stages of orogeny related to tectonism are characterized by mountain building and the thickening of continental crust. This means that the processes that thicken the crust (folding and faulting) are proceeding at a faster rate than the processes that break down and transport rocks (weathering and erosion). Once the crust has thickened, some mountain systems maintain a balance, at least for a time, between building and unbuilding, and therefore maintain their shape and size as orogeny proceeds. In other mountain systems, the crust may erode or be thinned by faulting, creating deep valleys and lower elevations overall so that the mountains "collapse" or are never built up very high in the first place.

In recent years, geologists have come to appreciate the link between tectonics and climate by studying mountains. As we discussed in Highlight 5-2 on page 158, mountains influence climate in various ways, but climate also influences mountains. Mountains in rainy climates (or those located on the sides of mountain belts exposed to high rainfall) may have a different geologic history than drier mountains because erosion rates will be higher

for the former. Such higher rates bring once-deep rocks to the surface more quickly, and the distribution of high-grade metamorphic rocks exposed at the Earth's surface may relate to such climate patterns or to the locations of rivers and glaciers.

There is only one place in the world today where major continents are actively colliding—the mountain belt that extends from the eastern Mediterranean (where the Arabian peninsula is colliding with southeastern Turkey) to the Himalaya (where India is crashing into Asia) causing such tragedies as the Pakistani earthquake of October, 2005. Highlight 13-1 discusses this most active and often deadly stretch of the Earth's surface.

Collisions have produced the Earth's largest mountains, including the Alps of southern Europe and the Himalaya of Asia. Older mountains that have eroded down to less-spectacular elevations than the young Himalayan range, but were probably once far loftier, include the Appalachians of eastern North America and the Urals of Russia.

Supercontinents

The geologic record becomes progressively more difficult to read as we go back in time. Much of what we know about the Earth's plate-tectonic history occurred within the last 200 million years, after the breakup of the supercontinent Pangaea. Paleomagnetic and other geologic studies, however, strongly suggest that there have been other supercontinents at various times in Earth history, and some scientists have even suggested (somewhat controversially) that supercontinents occur in cycles. What did these supercontinents look like? How big were they? Why did they form? Keep in mind that before the breakup of Pangaea, the continents did not have the same shapes as those we recognize today.

Exciting recent research suggests that about 700 million years ago, a proposed supercontinent named *Rodinia* (from the Russian for "motherland") occupied much of the Southern Hemisphere. And growing evidence—particularly paleomagnetic records—indicates that 700 million years ago, North America may have sat near the South Pole *wedged between Antarctica and Australia on the west and South America on the east* (Fig. 13-8). Thus, the Precambrian rocks of today's hot, dry American Southwest may have once been linked to rocks that today dot the icy wastelands of Antarctica's Transantarctic Mountains. This idea is known as the SWEAT (Southwest-East Antarctic) hypothesis. The breakup of Rodinia, which apparently occurred about 650 million years ago, may have left a 1200 km² (500 mi²) chunk of what is now Alabama and Arkansas attached to present-day western Peru. (The geology in those far-removed places is quite similar.) At the time, Peru may have been North America's neighbor to the east. The fossils found in the western foothills of the Peruvian Andes are identical to those found in Alabama, but appear completely unrelated to those found throughout the rest of South America.

Five hundred million years ago (well before Pangaea formed), Gondwana, another proposed supercontinent, perhaps a remnant of Rodinia, may have occupied the Southern Hemisphere somewhere near the South Pole (Fig. 13-9). Containing all of the Southern Hemisphere landmasses, Gondwana would eventually become the southern part of Pangaea. At the same time, three northern landmasses also existed, each probably separated from the others

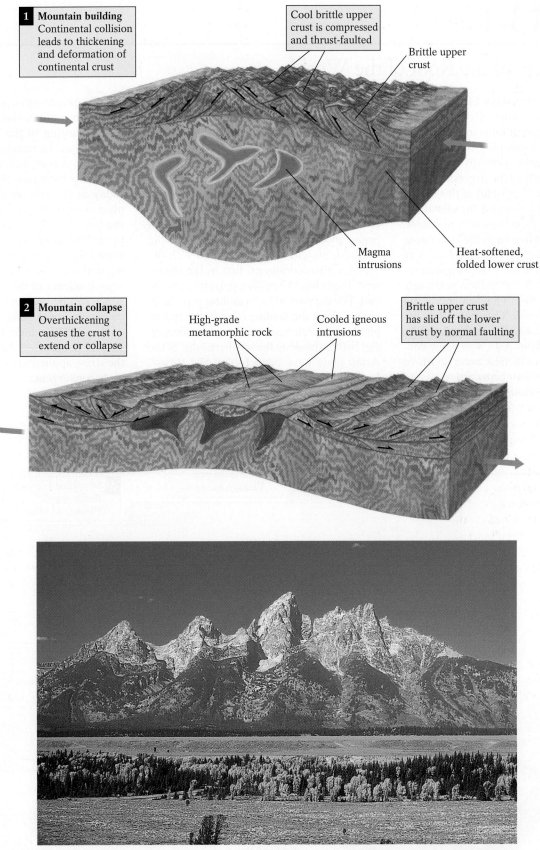

1 **Mountain building** Continental collision leads to thickening and deformation of continental crust

Cool brittle upper crust is compressed and thrust-faulted

Brittle upper crust

Magma intrusions

Heat-softened, folded lower crust

2 **Mountain collapse** Overthickening causes the crust to extend or collapse

High-grade metamorphic rock

Cooled igneous intrusions

Brittle upper crust has slid off the lower crust by normal faulting

▲ **FIGURE 13-7** The evolution of a collisional mountain range—from building up by reverse faulting, folding, and crustal thickening—to tearing down by normal faulting, crustal thinning, and erosion. The Grand Tetons in Wyoming are an example of a mountain range that is undergoing "unbuilding." The low flat region in the foreground is Jackson Hole, the downdropped block of rock associated with normal faulting.

Highlight 13-1

The Geology of the Roof of the World

The origin of the Himalaya Mountains, illustrated below, is a story of convergence, collision, and suturing of India to Asia. These events have had far-reaching consequences for the entire Earth, perhaps even contributing significantly to the triggering of our most recent ice age. The uplift of this mountain chain certainly changed the climate of Asia by isolating it from the southern oceans.

The events leading up to the Himalayan orogeny began during the Mesozoic Era when India separated from the Pangaean supercontinent. About 180 million years ago, India broke away from Pangaea and began its solo drift northward. The ocean plate beneath the Tethys Ocean, which separated the Indian continent from Asia, was being consumed along a subduction zone south of Asia. As India drifted northward, it passed over the stationary Reunion hot spot 65 mil-

lion years ago, triggering a period of intense volcanism, marked by the eruption of the extensive Deccan basaltic flows. Today, the Reunion hot spot lies about 5000 km (3000 mi) southwest of India in the Indian Ocean.

For the next 15 million years, India continued to move rapidly northward as subduction along the southern edge of Asia continued. The collision of India with Asia began about 50 million years ago when the northern margin of India collided with the Tibetan microcontinent, first in the northwest, then about 10 million years later, in the east. For the past 40 to 50 million years, sediments along the leading edge of the Indian plate and accretionary wedges, volcanic arcs, and batholiths along the southern edge of the Asian plate have been folded and thrust-faulted up onto both continents to form the modern Himalayan mountain belt.

Continued convergence of India against the Asian continent has resulted in repeated thrust faulting of the leading edge of the Indian plate, increasing the thickness of the continental crust beneath the rising Himalaya. This increased thickness is expressed today as the Tibetan Plateau—the highest plateau on Earth. Several suture zones mark the boundary between the colliding continents and continental fragments trapped within the collisional zone. India is still moving into Asia today at nearly 5 cm (2 in.) per year. Evidence of this ongoing plate convergence can be seen in this region's frequent devastating earthquakes. The Pakistani earthquake of 2005, which cost the lives of nearly 100,000 people, was a direct consequence of the stress applied to the rocks in this zone of plate convergence.

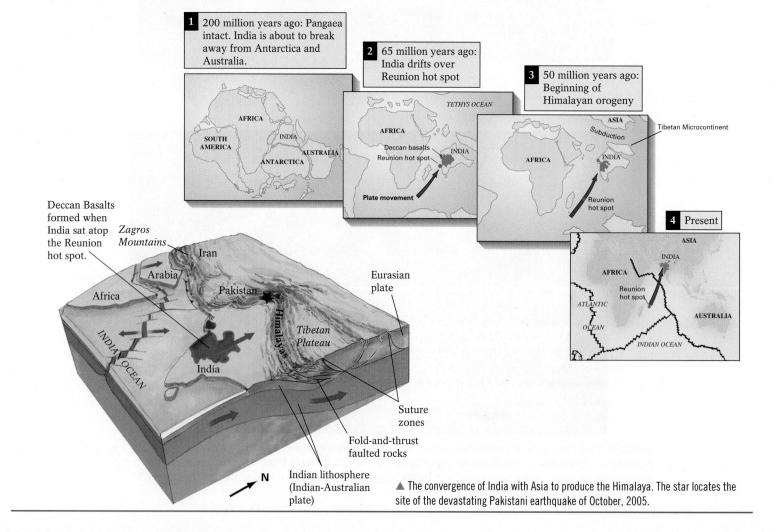

1 200 million years ago: Pangaea intact. India is about to break away from Antarctica and Australia.

2 65 million years ago: India drifts over Reunion hot spot

3 50 million years ago: Beginning of Himalayan orogeny

4 Present

▲ The convergence of India with Asia to produce the Himalaya. The star locates the site of the devastating Pakistani earthquake of October, 2005.

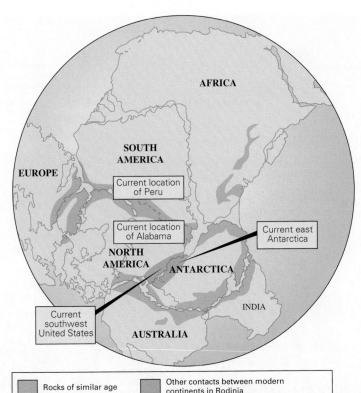

Rocks of similar age	Other contacts between modern continents in Rodinia

▲ **FIGURE 13-8** A supercontinent—Rodinia—as it may have looked 700 million years ago. Note how the American Southwest sits nestled against east Antarctica. Note too how the east coast of North America rested against the west coast of South America.

The Tectonic History of the North American Continent

North America is a classic example of a continent. It has all the components of a typical continent arranged more or less concentrically about the ancient core of the continent. Keep in mind that the North American *plate* extends from California to the Mid-Atlantic Ridge (Fig. 13-10), and the North American *continent* is just one part of this plate.

The North American continent varies in crustal thickness from about 25 km (15 mi) near the coasts to about 55 km (34 mi) in the thickened crust of the Rocky Mountains. Most of the tectonic processes that created the continent took place a long time ago, but there are still many active tectonic processes occurring that affect the landscape. In this section, we briefly review some of the "great *geological* moments" in the evolution of North America.

by a sizable ocean. These independent continents—which would become Laurasia, the northern half of Pangaea—included most of what is now North America, northern Europe, and a combination of southern Europe and parts of Africa and Siberia.

The pre-Laurasian landmasses and Gondwana began converging about 420 million years ago. Their initial collision produced the northern Appalachians and corresponding ranges in Greenland, the British Isles, and Norway. Meanwhile, Siberia collided with northern Europe, completing the formation of Laurasia and producing the Ural Mountains of central Russia.

During the next 100 million years, Laurasia and Gondwana collided and formed Pangaea. The final collision in the assembly of the supercontinent involved Africa, a separate continent that is now southeastern North America, and the rest of North America. This last stage in the formation of Pangaea created the southern Appalachians. By about 225 million years ago, at the close of the Paleozoic Era, Pangaea was a single vast landmass that stretched from the North pole to South pole. Virtually all of the Earth's continental lithosphere remained joined together for the next 25 to 50 million years, until Pangaea began to rift and water flowed in to form the modern Atlantic, Indian, and Antarctic oceans. Well, that's a brief look at the world tectonic view. Let's turn our attention now to our own corner of the planet—North America.

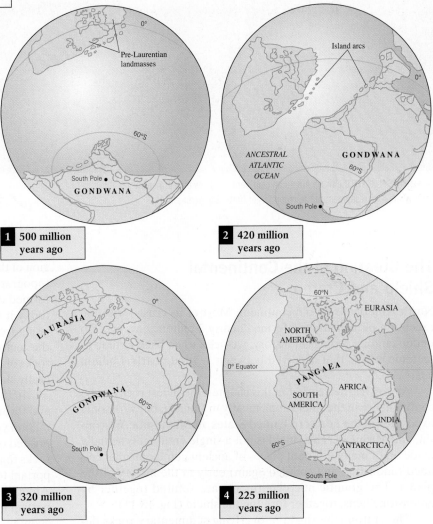

1 500 million years ago	**2** 420 million years ago
3 320 million years ago	**4** 225 million years ago

▲ **FIGURE 13-9** The origin of the supercontinent Pangaea.

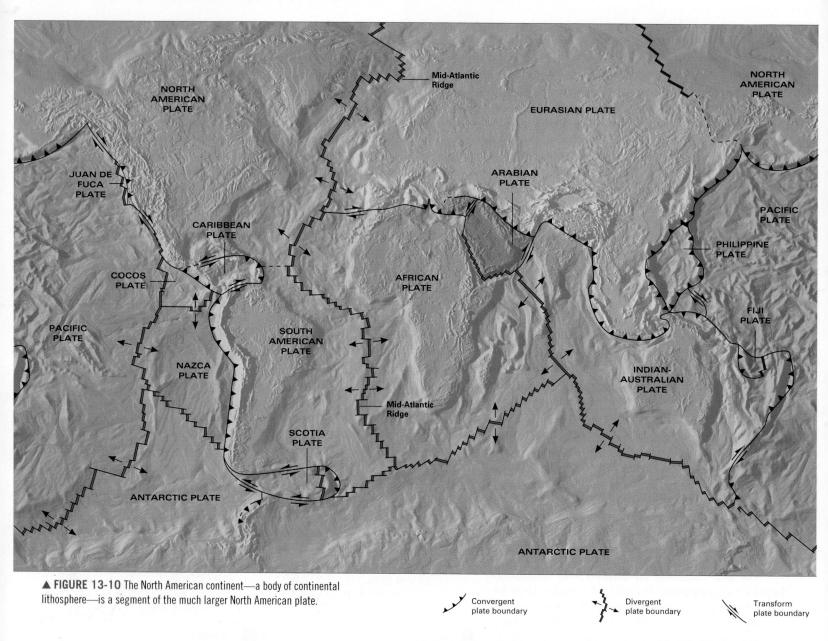

▲ FIGURE 13-10 The North American continent—a body of continental lithosphere—is a segment of the much larger North American plate.

Convergent plate boundary

Divergent plate boundary

Transform plate boundary

The Creation of the Continental Shield and Craton

North America is an old continent. Most of its crust—with the exception of its margins—formed nearly 2 billion years ago. To understand its geological history, we need to go back in time to tectonic events that occurred between 1.7 and nearly 4 billion years ago.

As you learned earlier in this chapter, the oldest rocks in the continent are in the shield area, mostly in Canada, but with some exposures in the northern United States, in Minnesota, Wisconsin, and Michigan. This shield is not a single entity; it is actually made up of many different pieces of ancient continents that collided (and rifted and collided again) early in the history of the planet. The granitic microcontinents are sutured together by greenstone belts, together forming the shield (Fig. 13-11).

The craton (the shield + overlying sedimentary rocks deposited hundreds of millions of years ago) forms the central re-

gion of the continent and is characterized by low and relatively flat topography (think Kansas and Nebraska). The mountains that formed during collisions billions of years ago are long gone, and today it takes a healthy imagination to picture towering peaks rising from the plains of North Dakota and Manitoba. Because the tectonic events that created the metamorphic and igneous rocks of the craton are long over, this part of the continent is generally stable tectonically.

In the late Precambrian (~1 billion to 545 million years ago) and throughout much of the Paleozoic era (between 545 and 225 million years ago), marine and other sedimentary rocks were deposited on top of the shield rocks, forming the thick sequence of rocks that makes up the craton. These sedimentary rocks are important to geologists for the information they contain about North America's ancient environments and processes, and they are important in general, because some of them contain deposits of great economic value, including energy resources such as oil, coal, and uranium.

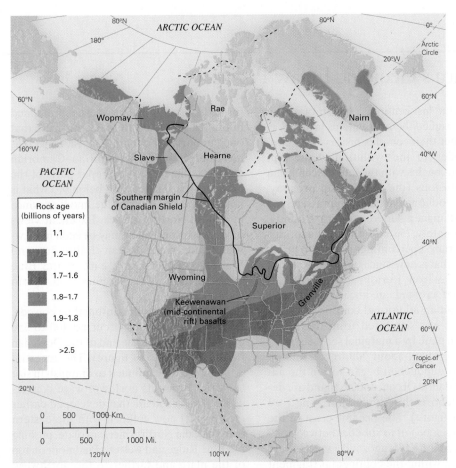

▲ **FIGURE 13-11** The geological provinces of North America. Rocks of the Canadian Shield, North America's ancient crystalline nucleus, are exposed throughout central and southern Ontario and Quebec as well as in northern Michigan, Wisconsin, and Minnesota. In most other places, however, younger rock covers the provinces.

The late-Precambrian sedimentary rocks in western Montana, including those in Glacier National Park, are particularly fascinating because they were deposited in a very quiet (low-energy) environment that preserved many structures in the fine-grained rocks. In Glacier Park you can see fine ripples, mudcracks, and even raindrop impressions that formed when scattered drops fell on moist fine sediments and were then gently covered by subsequent deposits and preserved. Thus, quite remarkably, these rocks preserve a billion-year-old rainstorm. These rocks of Glacier Park also contain fossil evidence of ancient life called *stromatolites*—mats of blue-green algae—some of the earliest forms of life on Earth.

Sedimentary rocks of this age are also found in the Wasatch Range of Utah, Death Valley and other parts of southern California, Pikes Peak, Colorado, and Arizona, including the lower strata of the Grand Canyon. The Grand Canyon is a particularly excellent place to view the Paleozoic history of western North America. The canyon is a mile-deep, river-cut slice through a part of the continent that has been fairly stable tectonically for hundreds of millions of years (Fig. 13-12).

As the craton formed, North America did not have its present shape or size, and it was not even in the same position on the planet. Continents drift, and North America is no exception. Geologists have used paleomagnetism and inferences from the environments in which ancient sedimentary rocks formed to determine that at times early in its history, the continent was likely located at or near the warm wet equator.

Mountain Building in the East: The Appalachian Events

The Paleozoic in North America was not simply a time of deposition of sedimentary rocks—exciting tectonic events were occurring in the eastern part of the continent as the ocean that separated North America from continents and other landmasses to the east began to subduct. This now-vanished ocean (called the Iapetus Sea) was not the modern Atlantic, of course, because the Atlantic is no older than about 200 million years old. The subduction of this ancient oceanic plate and the resulting collisions with westward-drifting landmasses is the story of the formation of the various ranges of the Appalachian Mountains. Subduction of ancient ocean plates also brought to North America many of the foreign terranes that make up the land that stretches from the Canadian Maritime provinces of Labrador and Nova Scotia to the hills of Georgia and Alabama. This complex tale is summarized in the following highlighted discussion about the East Coast's rugged highlands.

▲ **FIGURE 13-12** Inner Gorge of the Grand Canyon of the Colorado River. These rocks—high-grade metamorphic rocks—are 2 billion years old.

Highlight 13-2

The Appalachians—North America's Geologic Jigsaw Puzzle

The Appalachians of eastern North America are a mosaic of folded, thrusted, and metamorphosed provinces that evolved over nearly a billion years of Earth history. The North American segment of the system extends 3000 km (2000 mi), with a northeast–southwest orientation, from Lewis Hill [810 m (2672 ft) high] in eastern Newfoundland to Cheaha Mountain [730 m (2407 ft) high]—the highest point in Alabama—75 km (40 mi) east of Birmingham. When geologists reconstruct the Northern Hemisphere's plate boundaries, however, we find that the Appalachians actually extend a good deal farther to the northeast, all the way to the Caledonides of Western Europe and on to Norway. To the west of North America's Appalachian Mountains lies the Appalachian plateau, a region of relatively unfolded, unmetamorphosed, coal-bearing rocks. During Paleozoic time, this plateau consisted of lushly vegetated wetlands located in the shadow of the rising Appalachians. To the east lie the more recent deposits of the Atlantic coastal plain.

The Appalachians' considerable breadth, about 600 km (400 mi), covers three principal provinces, separated by major thrust faults. Each province developed at different times, in different places, and each has distinct rock formations, but all were affected by the same series of mountain-building episodes. West to east, the Valley and Ridge province consists of a thick sequence of relatively unmetamorphosed Paleozoic sediments that were folded and thrust to the northwest by compression from the southeast. The Blue Ridge province includes highly metamorphosed Precambrian and Cambrian crystalline rocks. The Piedmont province consists of metamorphosed Precambrian and Paleozoic sediments and volcanic rocks that were intruded by granitic plutons (see map in Fig. 13-15, p. 425).

For the last century, a legion of geologists has studied these mountains in an effort to decipher their complex history. One popular model for the evolution of the Appalachians begins with the late Precambrian plate collisions (the Grenvillian orogeny, about 1.1 billion years ago) that assembled the pre-Pangaea supercontinent Rodinia. Approximately 750 million years ago, this Precambrian supercontinent began to break up, with what is now much of North America rifting away from Eurasia and Africa. This movement created an ancestral Atlantic Ocean and an additional continental fragment that was separated from the North American plate by a marginal sea (see part 1 of figure). Then, about 700 to 600 million years ago, subduction began within the ancestral Atlantic Ocean, and an island arc developed above this subduction zone (see part 2 of figure). The oceanic crust in the marginal sea began to subduct between 600 and 500 million years ago, and a volcanic arc developed within the continental fragment (see part 3 of figure). Meanwhile, subduction continued beneath the island arc, and the ancestral Atlantic Ocean began to shrink.

By 500 million years ago, the marginal sea had entirely subducted, allowing the continental fragment to collide with the eastern edge of the North American plate (see part 4 of figure). In this collision, the "foreign" rocks of the continental fragment were thrust-faulted northwesterly over younger continental and marine sediments along the continental shelf of the North American plate.

This first mountain-building episode in the long geologic evolution of the Appalachian system is called the *Taconic orogeny*. Today's Blue Ridge Mountains eventually developed on top of the structures formed during the Taconic orogeny. Preserved in these mountains are highly deformed masses of gabbro, serpentinite, and pillow basalts from the oceanic crust that once underlay the marginal sea, as well as metamorphosed Precambrian–Cambrian marine sediments from deeper parts of the marginal sea. These rocks belong to what is now the western or inner Piedmont province in eastern North America. Excellent views of these exposed rocks may be seen in Newfoundland, metropolitan Baltimore and Washington, D.C., and east of the Blue Ridge Parkway in Virginia and North Carolina.

The second mountain-building episode, the *Acadian orogeny,* occurred 400 to 350 million years ago, at the end of the Devonian Period (see part 5 of figure). The ancestral Atlantic Ocean continued to diminish in size as Africa began its northwestward convergence. The island arc that developed earlier (see part 2) now collided with the eastern margin of the North American plate, pushing the Blue Ridge and western Piedmont on top of the plate and farther to the northwest. These rocks form the eastern Piedmont province. Much of New England's widespread metamorphism (discussed in Chapter 7) and many of the East Coast's granite plutons (such as those in Acadia National Park in Maine) resulted from this collision and the subduction of the ancestral Atlantic Ocean crust beneath the eastern edge of the North American plate.

The third and final mountain-building collision, the *Allegheny orogeny,* occurred between 350 and 270 million years ago (see part 6 of figure). As the ancestral Atlantic Ocean continued to subduct, the continental edge of Africa approached North America—an episode recorded in the plutonic intrusive rocks found today along the eastern edge of Piedmont province. The orogeny culminated when the continental crust of Africa collided with North America, finally closing the

Not long after the last mountain-building event of the Appalachians, during the final assembly of Pangaea, the supercontinent began to break apart and the Atlantic Ocean was born. We don't know why this happened, but the rifting may have involved heating of the thickened continental lithosphere. This heating, in turn, may have weakened the thickened lithosphere to the point that it could no longer support the differential topographic stresses of its mountains, causing the lithosphere to collapse, extend laterally, and thus rift apart. This rifting event created some of the large north-south-trending basins that mark the East Coast such as the Newark Basin of New Jersey, the Hartford Basin of Connecticut, and the Durham Basin of North Carolina. These basins contain thick sequences of mafic volcanic rock (such as the Palisades sill along the New York–New Jersey

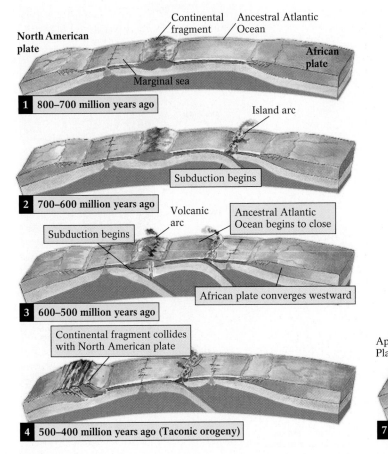

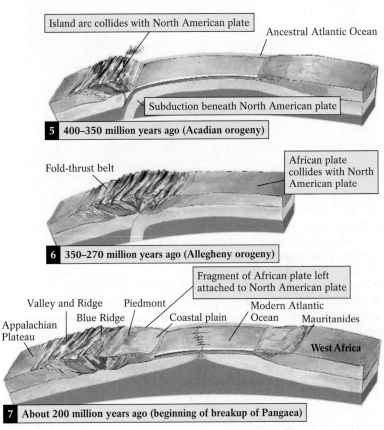

ancestral Atlantic Ocean and forming the supercontinent Pangaea. This collision produced the fold-and-thrust structures of the Valley and Ridge province, and thrust the Blue Ridge and Piedmont provinces farther inland. In Africa, a corresponding orogenic belt is marked by the Mauritanide Mountains.

The last great East Coast tectonic event involved the opening of the modern Atlantic, some 200 million years ago (see part 7 of fig-

ure). At this time, the African plate pulled away from the North American plate, but a healthy chunk of the African plate remained attached to eastern North America from New York City to Florida. Thus, the three plate collisions that produced the Appalachians added to the East Coast an ancient (Precambrian) fragment of North America that had rifted away earlier, an island arc that originated on the African–European side of the

ancestral Atlantic basin, and a massive slice of the African plate.

The Appalachians' complex geology and long history of multiple orogenic events are characteristic of the fold-and-thrust mountain systems created by continental collisions. Despite geologists' extensive studies of the Appalachians, many unresolved questions and much exciting work remain.

Hudson River border and the Holyoke basalt of western Massachusetts) and sedimentary rock (such as the bright red arkosic sandstone that can be seen along Interstate 91 in Connecticut). Some of these arkoses contain long trails of dinosaur footprints (Fig. 13-13). After rifting, the eastern margin of North America became tectonically inactive—the passive continental margin that persists today.

Mountain Building in the West: Westward Growth of the North American Continent

Clearly, the North American continent has grown—and continues to grow—since the craton stabilized. How did the rocks at the western margin, from Montana to the West Coast, form? The answer is a bit surprising because many of the rocks in the

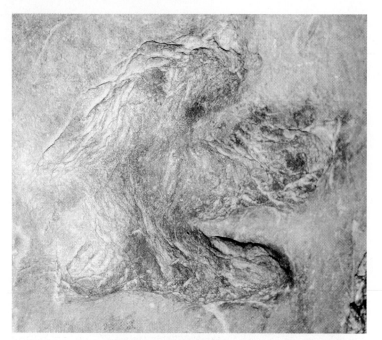

▲ **FIGURE 13-13** Dinosaur footprints from Mount Tom, near Holyoke, Massachusetts. These creatures were roaming around 200 million years ago in the great rift basins created by the breakup of Pangaea. This footprint is about 25 cm (10 in) across.

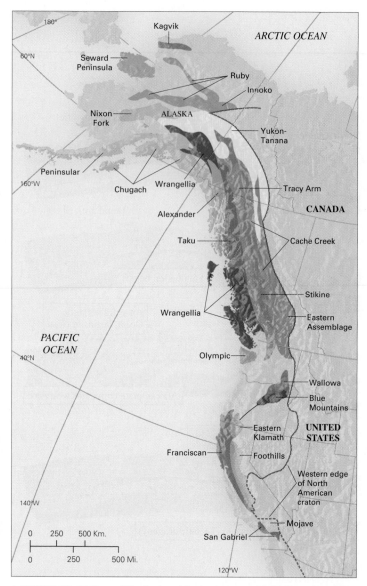

▲ **FIGURE 13-14** Some of the displaced foreign terranes attached to western North America. These structural blocks are believed to be the remnants of assorted microcontinents, islands, and other unsubductable landmasses that became attached to the continent during subduction of earlier Pacific plates. All the lands west of the dashed line have been added during the last 200 million years or so.

mountainous regions along both coasts are not simply crumpled-up cratonic rocks or young magmatic rocks added during plate subduction and convergence. In many cases, the rocks are *exotic* to North America—that is, the rocks originated somewhere else—on another continent or as an island. These exotic rocks have been stuck on, or *accreted,* to the edges of the continent during plate convergence (after subduction and collision). Such fault-bounded blocks of related rocks that formed elsewhere but have been subsequently added to a different continent are called **terranes.** (Note that the spelling of terrane is different from the more familiar word *terrain,* which means a landscape or region.)

We can distinguish accreted terranes from their surrounding rocks by their differing ages, geological structures, stratigraphies, fossil assemblages, and magnetic properties. Some may have been island arcs that formed in an ocean basin and were then towed to a continental margin by subduction of the intervening ocean plate. Others are **microcontinents,** or pieces of continental lithosphere broken from distant continents by rifting or transform faulting. For example, Baja, California, and the portion of California west of the San Andreas Fault may eventually separate from the rest of western North America along the transform boundary of the San Andreas Fault and become an independently moving microcontinent. Geologists locate and map accreted terranes by searching for geological evidence that marks former plate margins. As we have seen, former convergent boundaries may contain evidence of subduction and suturing in the form of mélanges, blueschists, and slices of ophiolite rocks (oceanic lithosphere).

In the West, more than 100 fault-bounded regions of various sizes, accreted to western North America from Alaska to California, have been identified as such exotic terranes (Fig. 13-14). Most were swept against the western boundary of the westward-drifting North American plate starting about 200 million years ago when

Pangaea rifted apart and the Mid-Atlantic Ridge began to develop. Subduction of an ancient plate of the Pacific basin beneath the continent's western edge swept dozens of buoyant fragments against North America's western shore. More than 200 million years ago, before these foreign terranes were added, the continent's west coast sat farther east, extending southward from present-day central British Columbia through Idaho to Utah or Arizona. Virtually all lands farther west (except for some younger volcanic rocks and sediments) are largely accreted terranes. Two hundred million years ago, Montana may have had some very nice beaches because that was the continent's west coast back then.

One of the most extensive accreted terranes of western North America is called *Wrangellia,* a structural block of foreign geology that stretches more than 3000 km (about 2000 mi) from the Wrangell Mountains east of Anchorage, Alaska, through Vancouver Island, British Columbia, to Hells Canyon, Idaho. Wrangellia's rocks and fossils are distinctive: They boast thick sequences of basalts and shallow marine limestones with the fossil clam *Daonella,* a species found in rocks native to Asia but nowhere else in North America. The magnetic record of the Wrangellian basalts suggests that they formed near the equator. About 100 million years ago, this micro-continent may have drifted northeast from equatorial Asia for 4000 km (2500 mi) or more before colliding with North America.

But where did Wrangellia first collide along the western margin of North America? Geoscientists who study the foreign terranes of western North America are engaged in a lively debate over this question. One view, based largely on the paleomagnet-ism of 90-million-year-old rocks in southwestern British Columbia, proposes that Wrangellia collided with North America at about 25°N latitude, near what is now Baja, California. According to this hypothesis, between about 83 and 45 million years ago, this large block of land—nicknamed Baja British Columbia—journeyed north roughly 3000 km (2000 mi) along a continent-long San Andreas–type transform fault. Similar long-traveled rocks in southeastern Alaska tell the same story.

The accreted terranes of eastern North America (Fig. 13-15) have an older, longer, and more complex tectonic history. As we just discussed in Highlight 13-2, the Appalachian moun-

tain system is believed to contain foreign island arcs and pieces of Africa that were added to North America when an ancient (pre-Atlantic) ocean closed in several stages between 500 and 250 million years ago. In the eastern Canadian maritime provinces of New Brunswick, Newfoundland, Prince Edward Island, and Nova Scotia, in adjacent parts of Maine, Massachusetts, and Rhode Island, and farther south, these and other accreted terranes record closing oceans, accreted island arcs, continental collisions, transform faulting, and rifting, all dating from 600 to 150 million years ago.

In summary, both the eastern and western margins of the continent have grown by the addition of foreign accreted terranes—first in the East (Appalachian orogenies date from 480 to 250 million years ago) and more recently in the West (during the past 200 million years and related to the westward drift of the North American plate that dates from the breakup of Pangaea).

More Mountain Building in the West

The accretion of terranes to western North America wasn't the only process that built and shaped the continent during the past few hundred million years. Subducting plates of oceanic litho-sphere in the Pacific region, some of them now entirely consumed, ferried the terranes to western North America. As they subducted, they also created volcanic arcs and vast regions of granitic and other intrusive rocks from Alaska to Mexico. The Sierra Nevada and the North Cascades—Coast Range mountains formed during this time (mostly between 120 and about 50 million years ago), intruding the accreted terranes and adding new continental crust to North America. Magmatism was accompa-nied by metamorphism and deformation as plates converged, causing folding, thrusting, and melting of the crust.

At the same time that mountains were building along the western edge of North America, mountains were also being cre-ated further inland—the Rocky Mountains. The Rockies, a bit of an enigma, are marked by sharply varying styles of mountain building represented within the mountain system and, even more perplexing, the Rockies surprisingly have risen far from the west-ern plate boundary. In terms of tectonic styles, the Rockies are characterized by giant thrusts of sedimentary rocks (such as those seen in Glacier National Park in Montana), indicators of plate convergence and structural compression, as well as normal-fault-bound ranges of igneous and metamorphic rocks (such as the Black Hills of South Dakota and the Tetons and Bighorn Range of Wyoming), indicators of extensional (pull-apart) tec-tonics (Fig. 13-16).

The origin of the Rockies is much debated by geologists. One hypothesis for their origin proposes that the plate subducting under western North America in the late Cretaceous and early Tertiary did not dive down steeply and deeply into the mantle but instead bumped along under the continent at a shallow angle of subduction, causing uplift and faulting. The fact that the con-tinent to the west had been assembled from accreted terranes may also play an important role in creating a mountain chain so far inland. Perhaps when the mountains began to form, the sub-duction zone wasn't quite as far west as it is today.

During the Rockies' mountain-building events, the crust of western North America became extremely thick, perhaps as thick

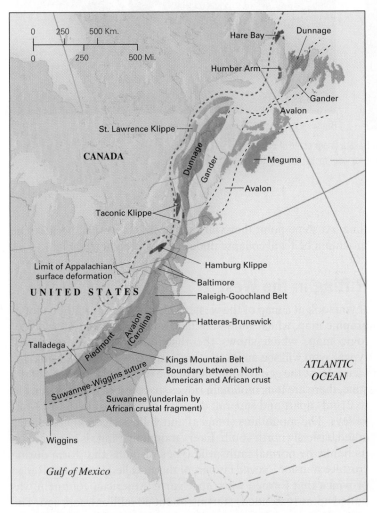

▲ **FIGURE 13-15** The displaced foreign terranes of eastern North America, many dat-ing from the assembly of the supercontinent of Pangaea. They remained attached to North America after Pangaea rifted to form the modern Atlantic Ocean.

Black Hills, S.D.

Bighorn Range, WY

▲ **FIGURE 13-16** Regional map of the Rockies, including (top right) the Black Hills of South Dakota (inset photo of metamorphic rocks within the Needles). (bottom right) The Bighorn Range of Wyoming (inset photo of regional rocks). Both of these ranges contain igneous and metamorphic cores bounded by normal faults.

as the modern Himalaya (~70 km/45 mi), and may have been associated with a vast high plateau similar to Tibet today. Eventually this hot, thick crust couldn't bear its own weight and thus collapsed along normal faults and was subsequently eroded, so that ancient igneous and metamorphic rocks that used to reside deep in the crust beneath these mountains are now exposed at the surface.

The collapse of mountains can occur quite rapidly from a geologic-time perspective. Mountains in collision zones might become so high that their extreme elevations cannot be supported by the underlying crust. At a certain point, the crust becomes too thick and collapses. Lowering of the mountains occurs by normal faulting near the surface where the rocks are brittle, and by flowing and folding of rocks at depth where the rocks deform plastically. Very thick crust that contains some volume of magma might be very weak and will flow away from the region of thickened crust, bringing the mountains down with it. To illustrate this process, consider that even if India continues to collide with Eurasia for tens of millions of years, Mount Everest and the other Himalayan peaks will not continue to gain elevation for that entire

time. At some point they will start to lose elevation, because the mountain belt will collapse under its own weight (see Fig. 13-7).

Rifting in the West

If you look at a map of the western United States—either a topographic map, which shows the land's hills and valleys, or a geologic map, which shows the distribution of different rock types—you will see an interesting pattern (Fig. 13-17). Look, for example, at the geology and landscape of Nevada: In much of the state, there are long mountain ridges spaced about 25 to 35 km (15 to 21 mi) apart and separated by 10 to 20-km-wide (6 to 13-mi) valleys. The mountains (ranges) and valleys (basins) are all oriented basically north-south. Every mountain range is bordered on its flanks by normal faults, the types of faults that form during crustal extension. Nevada is part of the so-called Basin-and-Range province that formed when the North American continent was pulled apart (extended) starting about 30 million years ago. The entire Basin-and-Range province is about 1000 km wide (about 600 mi); it is located between the Colorado Plateau of Colorado

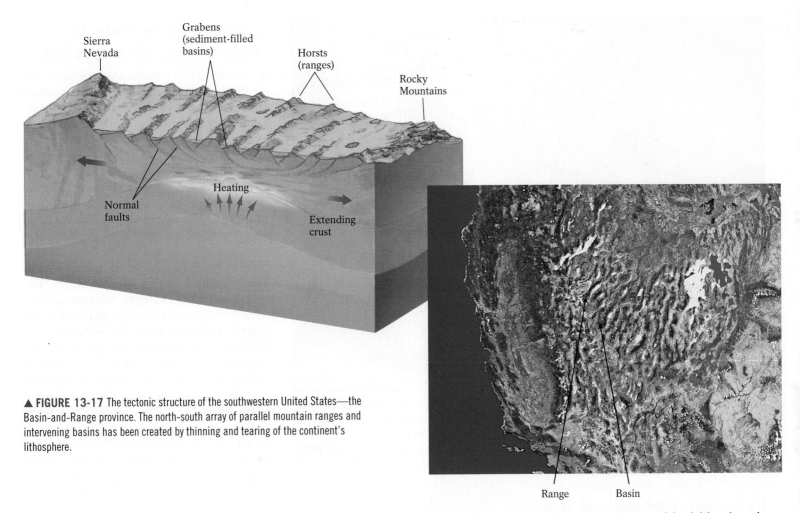

▲ FIGURE 13-17 The tectonic structure of the southwestern United States—the Basin-and-Range province. The north-south array of parallel mountain ranges and intervening basins has been created by thinning and tearing of the continent's lithosphere.

to the east and the Sierra Nevada of California to the west, and from Idaho to the north to Mexico to the south.

The extending part of the continent is characterized by frequent earthquakes, occasional volcanism (such as the 1000-year-old Sunset Crater near Flagstaff, Arizona), and high surface elevations despite unusually thin crust. These features suggest that the continent is being stretched and thinned while it is being heated by rising mantle from below. For several practical reasons, it is very important to understand why this region is so geologically active. We need to be able to forecast what might happen geologically in this region for the foreseeable future because, as we'll see in Chapter 20, the federal government has proposed burying an enormous quantity of highly radioactive nuclear waste in the region at Yucca Mountain, Nevada.

Unfortunately for those who would like to predict geological events with some certainty, and fortunately for scientists and students who like an intriguing puzzle, the causes of Basin-and-Range extension are not yet well understood. One hypothesis for which there is growing evidence suggests that part of the old divergent zone for the Pacific Plate (that is, the East Pacific Rise) was partially subducted about 30 million years ago. This event occurred when western California's tectonics changed from the subduction and convergence that produced the plutonic rocks of the Sierra Nevada to the transform-style tectonics of the modern San Andreas fault system. The evidence focuses on the pattern of magnetic stripes in the eastern Pacific, where part of a divergent

ridge and its magnetic-stripe pattern has vanished (that is, subducted) as the ridge was overridden by the westward-drifting North American plate.

What happens when a divergent zone subducts? It does not keep diverging by the same mechanisms as a divergent zone at the Earth's surface, but the subducted ridge might be a zone of weakness where the subducting plate will break. If a subducting plate breaks, hot mantle will rise up to fill the gap. This rising hot mantle could heat the overlying continental crust, causing it to stretch and thin, but still remain high in elevation because of the rock's expanded volume associated with its heat (Fig. 13-18).

Most Recent Tectonic Events in North America

Western North America remains very active tectonically. Strike-slip faults, including the San Andreas, slice through California and parts of British Columbia and Alaska, as the Pacific plate moves northward relative to the North American plate. The active volcanoes of the Cascade Range from northern California to southern British Columbia continue to erupt as the small Gorda plate (off the coast of northern California and Oregon) and equally small Juan de Fuca plate (off the coast of Washington and southern British Columbia) subduct beneath the western edge of the continent. (Mount St. Helens in Washington is actually erupting at the precise moment of this writing.) These plates

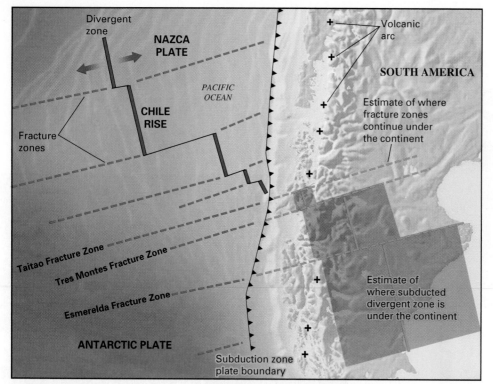

▲ **FIGURE 13-18** The Basin and Range of North America may have evolved in part in a way similar to the modern-day subduction of a ridge at the west coast of South America. This illustration shows a map view of subduction of an oceanic ridge beneath a continent, and an estimate of where the subducted ridge is likely to be (under the continent). Oceanic fracture zones (dashed lines) are also subducted. The scale of this map is hundreds of kilometers in length and width, and the crosses represent volcanoes formed by subduction-related magmatism.

Looking into the Plate-Tectonic Future: A Global Perspective

By extrapolating current plate motions into the future and speculating about the location of future rifts and subduction zones, geologists have developed a possible view of the plate-tectonic world 50 to 100 million years from now (see Figure 1-41 on page 35). Australia will likely collide with Southeast Asia. Transform motion along the San Andreas Fault will cleave western California completely from the North American continent, allowing it to move northward as a separate microcontinent toward an eventual collision with Alaska. Other "non-North American" terranes will join us along Pacific Coast subduction zones. The glamorous beaches of the Mediterranean's Riviera may evolve into rugged landlocked deserts, wedged between the colliding edges of the African and Eurasian plates. Rifting may continue in East Africa and begin in earnest in the American Southwest. And subduction may resume along the east coast of North America, bringing explosive volcanism and powerful earthquakes to such cities as New York City, Boston, and Philadelphia. For a glimpse at this fascinating possibility, spend a moment with the following highlight.

are remnants of the once much larger plate. Because they are being consumed faster than they are being created at the divergent zone they share with the Pacific plate to the west (Fig. 13-19), they will be completely subducted within the next few million years. But for the time being, subduction along North America's Pacific coast continues to trigger the region's powerful earthquakes and spawn occasional but potentially catastrophic tsunami.

So is North America still growing? Yes. New material is being added as plate motion continues. For example, the Olympic Peninsula of Washington State and the Coast Ranges along the Oregon coast are fairly recent additions to North America as subduction of the oceanic lithosphere to the West continues and seafloor volcanoes and their surrounding sediments slice off the subducting plate and are pasted to the edge of the continent. The next addition will probably be a group of undersea mountains—the Cobb Seamounts—that currently lie due west of Seattle and Vancouver, British Columbia.

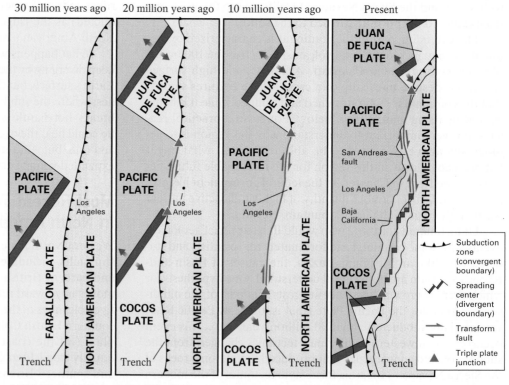

▲ **FIGURE 13-19** The evolution of the Farallon plate and the resulting origin of the modern Juan de Fuca and Cocos plates.

Highlight 13-3

Volcanoes in New Jersey and Georgia? When Will the Atlantic Subduct Again?

Unlike the Pacific Ocean, which is ringed by subduction zones, the Atlantic Ocean is bounded by continental margins that are passively drifting along with their plates away from the Mid-Atlantic Ridge. With the exception of the Atlantic-Caribbean boundary, *the margins of the Atlantic Ocean are not plate boundaries.* The major plate boundary in the Atlantic—a divergent zone—is in the middle of the ocean, and the Atlantic has been growing wider for the past 200 million years. Will the Atlantic keep diverging, or will it someday subduct again?

The oceanic lithosphere at the edges of the Atlantic Ocean is among the oldest ocean floors in the world. We know that oceanic lithosphere doesn't remain at or near the Earth's surface for much more than 200 million years because by that point, it has moved far enough from the divergent zone where it formed and become very cold and dense—denser than the convecting warm mantle below it. Once the oceanic lithosphere becomes denser than the rock under it, it will sink—that is, subduct.

Future subduction of the Atlantic will be driven in large part by the densification of oceanic lithosphere as it cools, but there may be other factors that trigger subduction as well. Current ideas involve the effect of water seeping into the oceanic crust. The oceanic crust is sliced by fractures that develop from transform faulting and other deformation of the rigid crust, so seawater can seep into the basalt and other oceanic rocks. This water will react with the oceanic crust, creating hydrous minerals that make the overall rock weaker and more able to bend and slide. Water-rock reactions occur at the divergent zone, but they also occur in the oceanic crust once it leaves the divergent zone, with the effects of loading with a thick pile of sediment and water from those sediments becoming more important close to the continents.

We believe that the Atlantic will subduct in the not-too-distant geological future and will most likely form deep trenches near the margins of North and South America, Europe, and Africa. If you live near the Atlantic coastal margin, in places such as Atlantic City, New Jersey and Savannah, Georgia, you probably don't need to worry about a stratovolcano forming in your backyard just yet; this process simply doesn't happen quickly on a human timescale. Geophysicists in Minnesota and Switzerland calculate that subduction probably won't even start for another 3 to 10 million years, and it will take another 3 million years or so to really get going.

When the Atlantic finally does subduct, it will likely do so along one of the most spectacular subduction zones on the planet. If subduction follows the coastlines of the modern continents, trenches and volcanic arcs will be thousands of kilometers long. Great earthquakes and tsunami will become common events in the Atlantic basin. High steep stratovolcanoes (much like those of the Andes Mountains of South America) will form on the continents, and volcanism will undoubtedly be explosive. In North America, the volcanoes will likely form about 200 km (120 mi) from the trench, which will be offshore at the boundary between the oceanic and continental lithosphere (see figure). Predictions based on this typical arc-trench gap could place the volcanoes in downtown Albany, New York, Washington, D.C., and Raleigh, North Carolina. Interesting prospect, but one that's still perhaps at least 6 million years off.

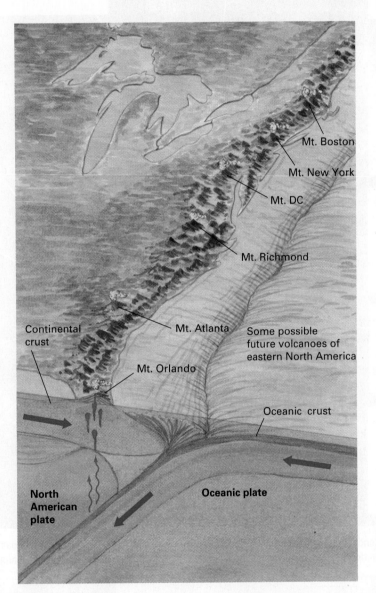

Continental crust

Mt. Boston
Mt. New York
Mt. DC
Mt. Richmond
Mt. Atlanta
Mt. Orlando

Some possible future volcanoes of eastern North America

Oceanic crust

North American plate

Oceanic plate

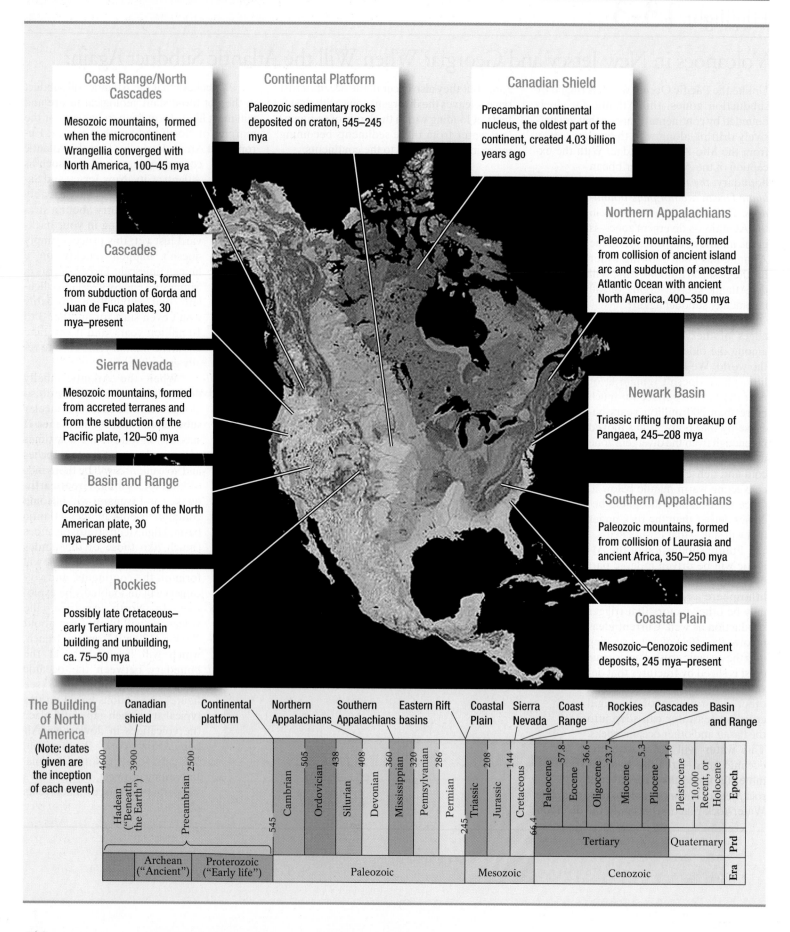

Coast Range/North Cascades

Mesozoic mountains, formed when the microcontinent Wrangellia converged with North America, 100–45 mya

Continental Platform

Paleozoic sedimentary rocks deposited on craton, 545–245 mya

Canadian Shield

Precambrian continental nucleus, the oldest part of the continent, created 4.03 billion years ago

Northern Appalachians

Paleozoic mountains, formed from collision of ancient island arc and subduction of ancestral Atlantic Ocean with ancient North America, 400–350 mya

Cascades

Cenozoic mountains, formed from subduction of Gorda and Juan de Fuca plates, 30 mya–present

Sierra Nevada

Mesozoic mountains, formed from accreted terranes and from the subduction of the Pacific plate, 120–50 mya

Newark Basin

Triassic rifting from breakup of Pangaea, 245–208 mya

Basin and Range

Cenozoic extension of the North American plate, 30 mya–present

Southern Appalachians

Paleozoic mountains, formed from collision of Laurasia and ancient Africa, 350–250 mya

Rockies

Possibly late Cretaceous–early Tertiary mountain building and unbuilding, ca. 75–50 mya

Coastal Plain

Mesozoic–Cenozoic sediment deposits, 245 mya–present

The Building of North America

(Note: dates given are the inception of each event)

Canadian shield · Continental platform · Northern Appalachians · Southern Appalachians · Eastern Rift basins · Coastal Plain · Sierra Nevada · Coast Range · Rockies · Cascades · Basin and Range

Timeline dates: –4600, –3900, 2500, 545, 505, 438, 408, 360, 320, 286, 245, 208, 144, 66.4, 57.8, 36.6, 23.7, 5.3, 1.6, 10,000

Hadean ("Beneath the Earth") · Precambrian · Cambrian · Ordovician · Silurian · Devonian · Mississippian · Pennsylvanian · Permian · Triassic · Jurassic · Cretaceous · Paleocene · Eocene · Oligocene · Miocene · Pliocene · Pleistocene · Recent, or Holocene — Epoch

Archean ("Ancient") · Proterozoic ("Early life") · Tertiary · Quaternary — Prd

Archean · Proterozoic · Paleozoic · Mesozoic · Cenozoic — Era

Scientists have constructed computer models that predict that within the next few hundred million years, all of the Earth's landmasses could very well reunite as another Pangaea-like supercontinent. As you can see on a modern-day map, India has already attached itself to the nucleus of a potential next supercontinent. Australia is moving northward, about to collide with Indonesia and join Southeast Asia. And Africa is moving northward, ultimately to collide with southern Europe and the Near East. As you read this, the next supercontinent is slowly assembling. Yet another indication that we shouldn't grow too attached to maps of the Earth—they are indeed changing, thanks to plate-tectonic motion.

Chapter 13 Summary

Once subduction of an oceanic plate is complete, the continental blocks on either side of the plate collide, forming a larger continent. The boundary between them—a suture zone—is typically marked by a collisional mountain range such as the Himalaya between China and India. Because they represent weak spots in the Earth's lithosphere, suture zones are good candidates for subsequent rifting.

Because the Earth's oldest *oceanic* rocks date from only 200 million years ago, we must rely on the older, more complex geology of the continents to study the Earth's earlier tectonic history. The Earth's continents typically consist of:

- A *continental shield,* the nucleus of a continent, which consists of the most ancient bodies of rock;
- A *continental platform,* a surrounding region of younger sedimentary rocks.

Together the shield and platform make up the *continental craton,* the tectonically stable portion of a continent.

In the Earth's first few hundred million years of existence, the planet had no continents, no oceans, no atmosphere, and—obviously—no land or sea life. The rocks of the current continents formed by the same basic processes that differentiated the Earth to form its internal layers—the melting of rocks and the separation of the different materials by density and composition.

The Earth's early mafic crust was probably thinner, hotter, and thus more buoyant than the present-day *mafic* oceanic lithosphere. Consequently, it was less likely to be subducted. The first felsic *continental* lithosphere probably formed as a result of plate-tectonic processes that differed somewhat from those of today. Before distinct continents existed, early landmasses probably consisted of huge volcanic islands that became enlarged as great volumes of mafic and ultramafic lava erupted from the mantle. As the eruptions continued, the islands grew and thickened, and early sedimentary basins developed around them. Older sections of the early lithosphere cooled and became more dense, until some parts sagged downward into the hotter interior, folding and dragging down the sedimentary basins along with them. The basalts, ultramafic rocks, and sediments partially melted and then differentiated, producing the first intermediate and felsic magmas. As the underling mantle convected, these early lower-density felsic masses drifted tectonically at the Earth's surface—eventually to collide with other low-density lithospheric masses, forming the Earth's shields.

The geologic record of plate collisions and continent formation is more difficult to read as we go back in time. Much of what we know about the Earth's plate-tectonic history occurred only within the last 200 million years after the breakup of the supercontinent Pangaea. About 750 million years ago, a supercontinent named *Rodinia* (from the Russian for "motherland") occupied much of the Southern Hemisphere.

Five hundred million years ago (well before Pangaea formed), *Gondwana,* another proposed supercontinent that may have been a remnant of Rodinia, occupied the Southern Hemisphere somewhere near the South Pole. Containing all of the Southern Hemisphere landmasses, Gondwana would eventually become the southern part of Pangaea. At that time, three northern landmasses also existed, each probably separated from the others by a sizable ocean. These independent continents—which would become Laurasia, the northern half of Pangaea—included most of what is now North America, northern Europe, and a combination of southern Europe and parts of Africa and Siberia. During the next 100 million years, Laurasia and Gondwana collided and formed Pangaea. By about 225 million years ago, at the close of the Paleozoic Era, Pangaea was a single vast landmass that stretched from pole to pole. Virtually all of the Earth's continental lithosphere remained joined together for the next 25 to 50 million years until Pangaea began to rift and water flowed in to form the modern Atlantic, Indian, and Antarctic oceans.

The North American continent has grown substantially since the breakup of Pangaea. Many of the rocks of western North America are relative newcomers to the continent. In many cases, the rocks originated somewhere else—on another continent or as an island—and these exotic rocks have been stuck on, or *accreted,* to the edges of the continent during plate convergence (after subduction and collision). Fault-bounded rocks or blocks of related rocks that formed elsewhere but have been subsequently added to a different continent are called *terranes.* More than 100 fault-bounded regions of various sizes have been accreted to western North America from Alaska to California during the last 200 million years. Most were swept against the western boundary of the westward-drifting North American plate after about 200 million years when Pangaea rifted apart and the mid-Atlantic ridge began to develop.

The West is also undergoing processes of thinning and rifting. These processes produce the Basin-and-Range province, which formed when the continent began to be pulled apart (extended), starting about 30 million years ago. This part of the continent is characterized by frequent seismic activity, rare but occasional volcanism, and high elevations despite unusually thin crust.

KEY TERMS

coastal plains *(p. 412)*   continental shelves *(p. 412)*   craton *(p. 412)*   orogenesis *(p. 416)*

continental platform *(p. 412)*   continental shields *(p.410)*   microcontinents *(p. 424)*   terranes *(p. 424)*

QUESTIONS FOR REVIEW

1. Draw a simple sketch of the structure of a typical continent. Include the shields, platforms, cratons, and continental shelves.

2. Briefly explain the process by which the Earth's early mafic and ultramafic lithosphere evolved into the Earth's first felsic and intermediate lithosphere that became the rocks of the continental shields.

3. How was the North American continent positioned differently within the supercontinent Rodinia from its position within Pangaea?

4. Briefly discuss the three principal orogenic events that produced the geological structure of the Appalachians in eastern North America.

5. What are geological terranes? How do they differ from microcontinents?

6. Where in the world would you find Wrangellia? Where did Wrangellia form, and how did it find its way to its present location?

7. Briefly discuss how some geologists explain why the Rocky Mountains sit so far east of the western margin of North America.

8. How do geologists explain the origin of the Basin-and-Range province of the southwestern United States?

9. How would the east coast of North America change if the oceanic segments of the plates that make up the Atlantic Ocean basin began to subduct? How would it change if the Atlantic Ocean lithosphere subducted completely?

10. How might the development of a new subduction zone along the East Coast affect plate interactions on the West Coast?

LEARNING ACTIVELY

1. Open an atlas and look at a world map. From the shapes of the continents, speculate about various ways that you might use today's continents to assemble a supercontinent, jigsaw-puzzle style. As you ponder these assemblies, think about what kinds of geological evidence you would need to prove these past configurations.

2. Over the next few weeks, monitor major geological events around the world, focusing on earthquakes and volcanic eruptions. With each event, determine whether it occurred at a plate boundary. If

so, using the plate map in Chapter 1, determine the type of plate boundary. Did any of the events occur at a plate interior? Was one associated with a known hot spot?

3. You've been given an unlimited travel budget, but you're required to search for the oldest rocks on the face of the Earth. Look at a map of the world and, using your knowledge of continental tectonics, plan a trip—to all seven major continents—that would take you to see the Earth's oldest rocks. Do you live near any of these places? Have you ever visited any of these places?

ONLINE STUDY GUIDE

Practice makes you better. Make the time to take the *Online Study Guide* quiz for this chapter. It should take you about 20 minutes. Automatic grading and instant feedback will help you quickly and accurately identify the concepts you have mastered and the areas in which you need more study.
www.prenhall.com/chernicoff.

3 Sculpting the Earth's Surface

Mass Movement

14

A t 11:37 P.M. on August 17, 1959, a magnitude-7.1 earthquake shook thousands of square kilometers near West Yellowstone, Montana. The soil surface rolled like sea waves, and lakes and rivers sloshed back and forth. A wall of water rushed across Hebgen Lake and over its dam, sweeping away campgrounds on its way downstream. The quake also triggered a landslide of more than 80 million tons of rock and weathered regolith—the loose rock blanket of fragmented material that covers much of the Earth's land surface—that hurtled downslope at speeds reaching 160 km (100 mi) per hour. As the slide tumbled into Madison Canyon, it compressed and expelled the air in its path, producing hurricane-force winds that battered the valley. Two-ton cars were blown into the air. One flew more than 10 m (33 ft) before smashing against a tree. By the time the slide mass finally came to rest, it had covered the valley floor with 45 m (150 ft) of bouldery rubble, buried U.S. Highway 287, and taken the lives of 28 campers.

The Madison Canyon disaster represents an extreme example of **mass movement**, the relentless process that transports Earth material (such as bedrock, loose sediment, and soil) down slopes—from high places to low places—by the pull of gravity (Fig. 14-1). *Every* slope is susceptible to mass movement. Sometimes the movement is extremely fast and involves a great mass of material that may travel more than 80 km (50 mi) from its source. More often the movement is so slow that you may not even know it's happening. It may involve only the upper few centimeters of loose soil on a gentle hillside.

Of the 20,000 lives lost as a result of all natural disasters in the United States from 1925 to 1975, fewer than 1000 deaths were a direct result of mass movement. Mass movement, however, accounted for a staggering $75 billion in damage, compared with only $20 billion in damage for all other natural catastrophes. Thus, while few of us are likely to die in a mass-movement event, we are much more likely to lose something valuable because of it, perhaps from a cracked home foundation or a living room filled with flowing mud. In this chapter, we will look at both the underlying and immediate causes of mass movement, various types of mass movements, and some ways to prevent these disasters.

FIGURE 14-1 Mass-movement destruction, January, 2005, Laguna Hills, California. Hundreds of residents were driven from their homes and eleven homes were crushed by this event. ▶

What Causes Mass Movement?

The underlying cause of all mass movement is the same: *Gravity* pulls materials on an incline downslope. Some factors loosen materials and promote downslope movement; others resist such movement. In a general sense, *mass movement occurs when the factors that drive materials downslope overcome the factors that resist downslope movement.*

Two principal factors provide resistance to mass movement: the friction between a slope and the loose material at its surface, and the strength and cohesiveness of the material composing the slope, which prevents the slope from breaking apart and slipping at its surface. Counteracting these forces are steep slopes, the material's water content, a lack of vegetative cover, and a slope's history of human and other animal disturbances.

Gravity, Friction, and Slope Angle

Look at the boulder in Figure 14-2. Why does it just sit there without toppling over? Two main factors determine whether the boulder will stay put or hurtle down the slope: gravity and friction. Gravity promotes the boulder's downslope movement; friction resists it. Gravity pulls on the weight (W) of the boulder, drawing it downward toward the Earth's center. On a slope, the force of gravity on the boulder actually has two components: one parallel to the slope (G_d, for gravity$_{downslope}$), which works to pull surface materials downslope, and one perpendicular, or *normal*, to the slope (G_p, for gravity$_{perpendicular}$). The normal force, G_p, contributes significantly to the effects of friction (F), the force that opposes motion between two bodies in contact. The amount of friction between a boulder and a slope depends on the magnitude of G_p, the area of contact between the boulder and the underlying slope, and the roughnesses of the surfaces in contact.

The boulder in Figure 14-2 is likely to stay put if G_d is relatively low and/or G_p and its related factor, F, are relatively high. This scenario typically occurs when a slope is gentle and/or its surface is very rough. The boulder is likely to roll down if G_d is relatively high and/or G_p (and F) are relatively low. This situation typically arises when the slope is steep and/or its surface is very smooth.

Several natural and artificial processes can create steeper slopes. Faulting, folding, and tilting of strata, river-cutting, glacial erosion, and coastal-wave cutting are all natural slope-steepening processes; quarrying, road cutting, and waste dumping are some human activities that can steepen and destabilize slopes (Fig. 14-3).

Slope Composition

The materials that make up the slope also affect the likelihood of mass movement. A slope may be composed of any combination of solid bedrock, unconsolidated ("loose") material, including weathered bedrock, soil, and vegetation, and a variable amount of water. Under certain conditions, some of these materials promote slope stability; in other situations, they cause instability and spur mass movement.

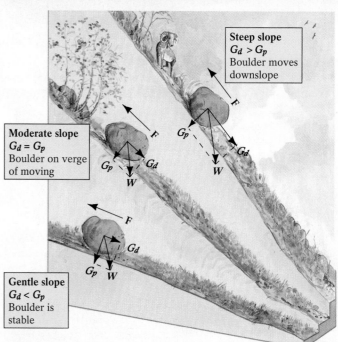

▲ **FIGURE 14-2** (top) This boulder sits on a steep hillside in Acadia National Park, Maine, without sliding off. Friction holds it in place by exceeding the gravitational force that would drive it downslope. (bottom) The principal forces tending to hold materials in place on a slope are friction (**F**) and **Gp**, the component of the force of gravity that acts perpendicular (or *normal*) to the slope's surface. The principal force tending to drive materials downslope is the parallel component of gravity (**G_d**). These forces are affected by several factors, including the steepness and slipperiness of the slope as well as the weight (**W**) and water content of the material on the slope.

Solid Bedrock Solid rock tends to be highly stable, even when it forms a vertical cliff. Its degree of cementation (if it's a clastic sedimentary rock) or the interlocking pattern of its crystals (if it's an igneous, metamorphic, or chemical sedimentary rock) strengthens the rock so that it resists the downslope force trying to pull it apart. Solid rock becomes less stable under any of the following geological conditions:

▲ FIGURE 14-3 Some common processes that steepen slopes.

- Tectonic deformation breaks rock into a network of joints, fractures, or faults.
- The mechanical-weathering processes of freeze–thaw, thermal expansion and contraction, exfoliation, or root penetration open significant cracks in the rock.
- The rock is sedimentary (such as a sandstone) that had developed bedding planes during deposition.

- The rock is soluble (such as limestone) and formed large cavities through dissolution.
- The rock is igneous and developed a joint pattern during cooling (such as columnar basalt).
- The rock is a foliated metamorphic rock with either pronounced rock cleavage (such as slate) or schistosity (such as schist).

When a plane of weakness—whether a fault, crack, joint, or bedding plane—lies roughly parallel to the surface of a steep slope, the rock is even more likely to break along that plane and slide downslope (Fig. 14-4).

Unconsolidated Materials A slope composed of loose, dry material remains stable only if the friction between its components and underlying materials exceeds the downslope component of gravity (G_d). Different materials form stable slopes at different slope angles. If you pile up dry sand, for example, it forms a small conical hill. When the hill reaches a certain slope, the slope angle remains constant; you may add more sand to the hill, but the extra material simply slides down the slope, gradually piling up around its base. The maximum angle at which unconsolidated material is stable is known as that material's **angle of repose**. For dry sand, like that found in sand dunes, the angle of repose generally ranges between 30° and 35° (Fig. 14-5a).

The angle of repose of any loose material depends partly on the size and shape of its particles. Large, flat, angular grains with rough, textured surfaces generally have steeper angles of repose than smaller, rounded, smooth grains. Imagine stacking a pile of highly polished marbles, and another pile of chocolate-chip cookies (with big chips). The flatter shape and bumpiness of each cookie's surface enable you to stack the cookies more steeply than the marbles. In nature, the steepest slopes of loose materials, with angles of repose greater than 40°, consist of large, highly angular boulders. These large boulder piles, called **talus slopes**, typically form at the foot of cliffs that have been weathered by frost wedging and other mechanical-weathering mechanisms (Fig. 14-5b).

The tendency of loose material to move also depends on the arrangement of its particles. Rapidly deposited particles are generally arranged somewhat chaotically, resembling a house of cards (Fig. 14-6a). Although this structure appears unstable, the friction between the particles at their contact points may add enough strength to resist mass movement for centuries. Loose particles that are deposited slowly as they settle calmly through still water generally arrange themselves in a more organized fashion, in a structure that resembles a stack of pancakes (Fig. 14-6b). Here, the particles tend to slide past one another, promoting mass movement.

Vegetation Vegetation, especially the deep root networks of large shrubs and trees, binds and stabilizes loose, unconsolidated material. Removal of trees by forest fire or clear-cutting for timber or farming compels loose material to move downslope, especially shortly after a rainstorm. Several decades ago in the town of Menton, France, farmers decided to remove deeply

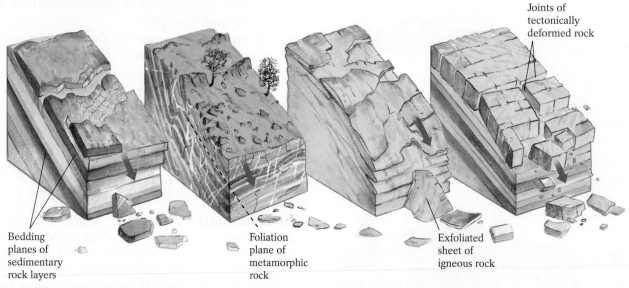

Joints of tectonically deformed rock

Bedding planes of sedimentary rock layers

Foliation plane of metamorphic rock

Exfoliated sheet of igneous rock

▲ **FIGURE 14-4** Slopes such as these, with planes of weakness parallel to their surface, are especially susceptible to mass movement.

rooted olive trees from the area's steep slopes to plant more profitable, shallow-rooted carnations. This miscalculation contributed to landslides that took 11 lives. Similarly, widespread tree-cutting in Philippine forests was a primary cause of the tragic landslides that claimed 3400 lives in November 1991. More recently, vegetation losses from the 2002 wildfires in the hills near Durango, Colorado, and more fires near San Bernardino, California, in 2003, triggered several tragic mudflows during the subsequent rainy seasons. Without the roots of growing trees to keep soils open and absorbent, the heavy rains simply stripped both regions' soils, boulders, and downed trees into torrents of mud (Fig. 14-7).

Water Water, *more than any other factor,* is most likely to cause previously stable slopes to fail and slide. Initially, a small amount of water actually increases the cohesiveness of loose material: It binds adjacent particles by surface tension, a weak electrical force on the surface of a drop of water that attracts and holds other

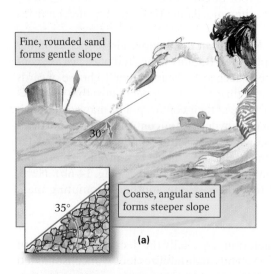

Fine, rounded sand forms gentle slope

30°

35°

Coarse, angular sand forms steeper slope

(a)

Angular boulders form steep talus slope

(b)

▲ **FIGURE 14-5** The angle of repose of unconsolidated materials depends largely on particle size and shape. **(a)** Coarse, angular sand forms a steeper slope than does fine, rounded sand. **(b)** A talus slope, composed of large and irregularly shaped boulders, can form slopes greater than 40°. (photo) Talus slope at Wheeler Peak, Great Basin National Park, Nevada.

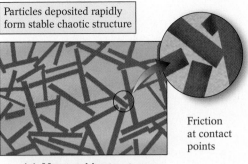

Particles deposited rapidly form stable chaotic structure

Friction at contact points

(a) More stable structure

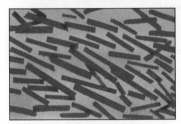

Particles deposited slowly form unstable layered structure that allows particles to slide past one another more easily

(b) Less stable structure

▲ **FIGURE 14-6** The effect of particle arrangement on stability. **(a)** When loose material is deposited rapidly, its particles become oriented at angles to one another, producing friction between the grains and strengthening the material. **(b)** When particles are deposited slowly and in parallel planes, both the friction and the strength of the material are reduced.

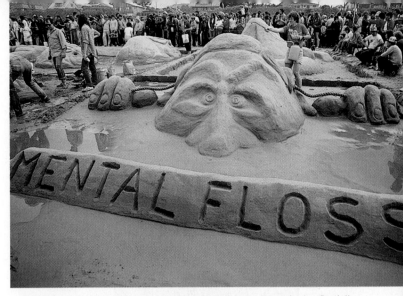

▲ **FIGURE 14-8** A little water can initially make sediments more cohesive. Partially saturated sand, for example, can be molded into intricate sand sculptures, such as this one at Cannon Beach, Oregon.

particles, such as sand grains (Fig. 14-8). Water also fosters the growth of stabilizing vegetation. On the other hand, excessive water can promote slope failure when:

- It reduces the friction between surface materials and underlying rocks, or between adjacent grains of unconsolidated sediment.
- It adds weight to the slope, enhancing G_d.
- It counteracts some or all of G_p—the perpendicular component of gravity—by buoying upward the weight of slope materials. (*Pressurized* water counteracts the normal component of gravity even more effectively.)

If you have ever slid down a giant water-park slide, you have experienced firsthand the role water plays in reducing friction between an object and a slope. A thin film of water on the slide reduces friction dramatically, allowing you to slide down at a high speed. Even solid bedrock can be moved with the aid of water: Infiltrating water reduces the friction between adjacent rock masses that are separated by some type of plane of weakness, such as a bedding plane, fault, or metamorphic foliation (Fig. 14-9).

Saturation with water also reduces the cohesiveness *between individual sediment grains*. In general, an unconsolidated sediment's cohesiveness stems from the internal friction between grains in contact and the weak attractive forces (surface tension) between grains. When water completely surrounds sediment grains, it isolates them from adjacent grains and effectively eliminates the friction and the surface tension between them. Thus, although damp soil may be more cohesive than dry soil, *saturated* soil may become a formless mass of flowing mud as water forces apart individual grains (Fig. 14-10).

Setting Off a Mass-Movement Event

Before a stable slope becomes unstable and fails, it may first develop a fragile balance, or equilibrium, between the forces that tend to drive movement downslope and the forces that tend to resist such movement. Some event may then tip the delicate balance in favor of the driving forces, setting off downslope

▲ **FIGURE 14-7** Mudflow in Devore, California, on Christmas Day, 2003. The root cause of the mudflow was the widespread wildfires that devegetated the region during several preceding summers.

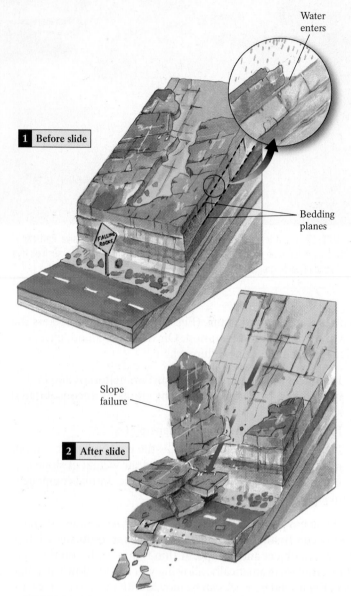

▲ **FIGURE 14-10** Saturation with water promotes mass movement in unconsolidated slope materials by decreasing the internal friction and surface tension between particles.

▲ **FIGURE 14-9** Water can cause failure in slopes of solid bedrock by reducing friction along planes of weakness in the rock.

movement. The trigger mechanisms, or immediate causes, of mass movement can be either natural or human-induced.

Natural Triggers

A number of climatic and geological events can set off mass movements. Torrential rains, earthquakes, and volcanic eruptions can all send loose materials downslope. Heavy rainfalls, for example, can quickly saturate a thick body of clay-rich regolith that may have taken thousands of years to accumulate. At some point, the slope's sodden material can no longer withstand the downslope force of gravity, and it slides downhill. In 1967, a three-hour downpour in central Brazil triggered hundreds of slope failures, taking more than 1700 lives. More recently, nearly 10,000 Venezuelans lost their lives in late December 1999 during rain-driven mudflows along Venezuela's north coast. And in the aftermath of Hurricane Ivan and other

powerful storms that struck the East Coast in mid-September 2004, thousands of landslides and mudflows caused enormous losses of life and property on the Caribbean islands of Grenada, Trinidad and Tobago, the Caymans, and in the Appalachian regions of western North Carolina (Fig. 14-11) and the hill country of West Virginia.

An earthquake can dislodge massive blocks of bedrock and enormous volumes of unconsolidated material. The series of powerful earthquakes that struck New Madrid, Missouri, between December 1811 and February 1812 initiated hundreds of mass-movement events within a 13,000-km^2 (5000 mi^2) area of the Mississippi Valley. In the most tragic event of its kind within the past forty years, a 1970 Peruvian earthquake of magnitude-7.7 shook the Andes so violently that one landslide sent more than 150 million m^3 (5 billion ft^3) of ice, rock, and soil tumbling, burying several mountainside towns and their 30,000 inhabitants (see Figure 14-25). Earthquakes may also cause wet, unconsolidated materials to liquefy, as we saw in the 1964 quake in Anchorage's Turnagain Heights neighborhood (see Chapter 10).

When volcanic eruptions propel large quantities of hot ash onto snow- or glacier-covered slopes, they produce enormous volumes of meltwater. This water may then mix with fresh ash and other surface debris to form a muddy slurry, or *lahar* (see Chapter 4) that flows rapidly down steep volcanic slopes, burying the landscapes and communities below.

Human-Induced Triggers

Human activity can also set off mass movements. Mismanagement of water and vegetation, cutting, steepening, and overloading of slopes, mining miscalculations, and even loud sounds can all trigger slope failure.

When we overirrigate slopes for farming, install septic fields that leak sewage, divert surface water onto sensitive slopes at construction sites, or overwater our sloping lawns, we introduce liquids that destabilize slopes by reducing friction. One Los Angeles family went on vacation and forgot to shut off their lawn sprinklers, which malfunctioned and saturated their lawn. When they returned, they found their hillside lawn *and their home* sitting on the highway in the valley below.

When we clear-cut forested slopes or accidentally set forest fires, we eliminate the deep, extensive root networks that bind loose materials together (Fig. 14-12). When we cut into the bases of sensitive slopes to clear land for homes or roads, we steepen the slopes and disturb their equilibrium. Building housing developments on hillsides or cliffsides—a common practice in California—can trigger mass movement, especially when slopes bearing the extra weight of roads and buildings later become weakened by heavy rainfall.

The rumble of a passing train, the crack of an aircraft's sonic boom, and the blasting that accompanies mining, quarrying, and road construction can trigger mass movement as well. Vibrations from these activities can shake loose grains of sediment, forcing them apart and reducing or eliminating the friction between them.

Sometimes human activities, such as mining, may combine with natural factors to increase the chances of mass movement. Such a situation led to the Turtle Mountain landslide of 1903 near the Canadian Rockies town of Frank, Alberta, which claimed 70 lives (Fig. 14-13). Removal of a large volume of coal near the foot of the mountain apparently undermined its steep slope, weakening its already unstable fractured structure. In addition, the month preceding the slide had been a particularly wet one in the southern Alberta Rockies, and water from extensive snowmelt probably entered the fractures in Turtle Mountain. This water added weight to the slope and reduced friction between fracture surfaces, further increasing the risk of a catastrophic mass movement.

Before we move on to discuss the various types of mass movements, take a moment and think back to the causes of mass move-

▲ FIGURE 14-12 Forest fires, such as this one in Montana in 2000, remove vegetation and its binding roots and trigger mass movement during subsequent rainy seasons.

▲ FIGURE 14-11 Mass-movement events in western North Carolina associated with the torrential rains of Hurricane Ivan, September 17, 2004.

ment and note how the Earth's systems interact to promote downslope movement. Perhaps as much as any other group of geologic processes, the processes of mass movement involve:

- The state of the *lithosphere:* Are the rocks fractured, bedded, or weakened by weathering? Have they been uplifted, tilted, or steepened by tectonic forces?
- The roles of the *atmosphere* and the *hydrosphere:* Are the rocks and soils saturated with rainwater, snowmelt, or underground water? Have streams and waves cut overly steep slopes?
- The role of the *biosphere:* Do the roots of trees, plants, and grasses bind up the soils and hold them in place? Have animals—even we humans—disturbed the slopes and set them in motion?

Virtually all of the Earth's surficial geologic processes are clear examples of the interplay among several or all of the Earth's various systems.

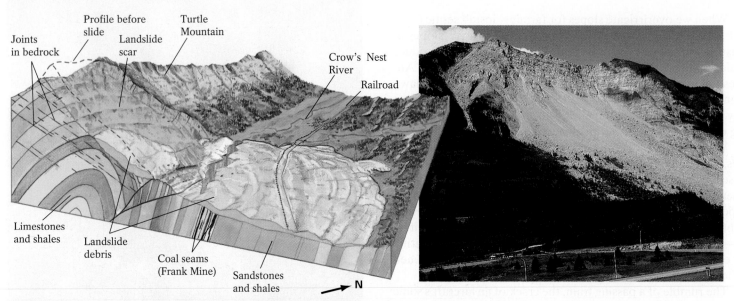

▲ **FIGURE 14-13** The 1903 Turtle Mountain landslide in Frank, Alberta. (left) The disaster began when an enormous mass of limestone broke free and moved rapidly downslope to a protruding rock ledge, where it was launched airborne toward the valley below. After a 900-m (3000-ft) drop, the rock struck the weak shales and coal seams in the valley, shattering into a great avalanche of crushed rock that spread at speeds as high as 100 km/hr (60 mph) to a distance of 3 km (2 mi) across the valley. The mass of crushed rock had such momentum that it actually rose 120 m (400 ft) up the opposite side of the valley. Ironically, 16 men working in the mines survived by digging their way out through a soft coal seam. (right) Turtle Mountain as it looks today, showing evidence of more recent rockfalls within the scar from the 1903 slide.

Types of Mass Movement

Ask a soil scientist, a geologist, and a civil engineer to classify a mudflow (a rapidly flowing slurry of mud and water), and you'll likely receive three different answers. The reason? No universally accepted classification scheme exists for mass movements. Geologists, however, generally classify mass movements based on their speed and the manner in which materials move downslope.

Slow Mass Movement

Creep, which is measured in millimeters or centimeters *per year,* is the slowest form of mass movement. It occurs virtually everywhere, even on the gentlest slopes. Creep generally moves unconsolidated materials, such as soil or regolith. It rarely extends downward to more than a few meters in depth.

Because friction increases with depth as a result of increased compaction, a mass of creeping material tends to move faster at its surface and more slowly at greater depth. The result? Most originally upright objects set in the creep zone (such as lamp posts and telephone poles) become tilted with the passage of time. Telephone poles inserted vertically in a creeping slope near Yellowstone National Park are now found inclined 8° after 10 years. In cemeteries, older gravestones are generally more tilted than younger ones because they have been creeping for a longer period. Trees on creeping slopes continually adjust their shape to remain vertical so they can capture maximum sunlight; as a result their trunks are typically bent (Fig. 14-14).

The grains within loose creeping material are continuously rearranged as each individual grain is pulled downslope by gravity. A particle may be dislodged and set in motion in any number of ways: the burrowing of an insect or larger animal, surface trampling of passing animals, the splash of raindrops. Even the swaying of plants and trees in the breeze can cause their roots to move enough to shake loose soil and send it creeping downslope.

In cold environments where soil water freezes periodically and ice develops in the ground, the volume of the near-surface materials expands upon freezing and displaces soil particles at the slope surface. When thawing causes the near-surface materials to contract, gravity pulls the displaced surface particles slightly downslope before they settle back into a stable position (Fig. 14-15). In clay-rich soils, periodic wetting and drying swell and shrink soil clays, moving particles in a similar zigzagging manner. Each of the countless particles in such soils gradually but constantly creeps downhill, even on a gentle slope.

A special variety of creep, seen principally in very cold regions such as Alaska's tundra, is **solifluction** ("soil flow"). This relatively faster form of creep develops where the warm sun of the Arctic's brief summer season thaws the upper meter or two of soil or regolith. Because the underlying *permafrost* (permanently frozen ground) can't absorb much water, the thawed soil becomes waterlogged and flows downslope at rates of 5 to 15 cm (2 to 6 in.) per year (Fig. 14-16).

Rapid Mass Movement

Rapid mass movements can send materials rushing downslope at speeds measured in kilometers per hour or even meters per

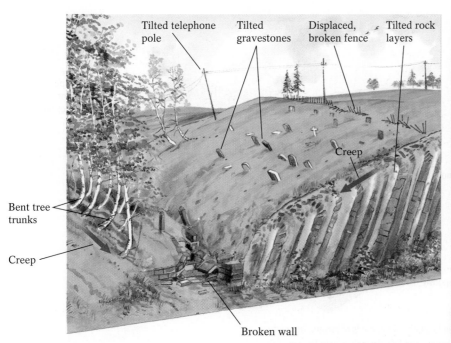

▲ **FIGURE 14-14** Creep affects every near-surface feature, including soil, rock, vegetation (lower photo), and a variety of shallow-rooted, human-made items such as fences, gravestones, and utility poles. In the upper photo, note the bend in the rock layers near the top of the exposure. The layers are bending as their upper portion moves downslope by creep.

second. They are classified according to their type of motion—falls, slides, slumps, and flows (Fig. 14-17). Each type of movement can occur at different speeds and include a variety of materials. Often one type of rapid mass movement spawns another.

Falls

A **fall** is the fastest type of rapid mass movement. It occurs when loose rock or sediment breaks free, typically at a plane of weakness, and drops through the air from very steep or vertical slopes. The falling material plunges to the ground or water below, traveling at the acceleration of gravity— 9.8 m/s^2 (32 ft/s^2) (Fig. 14-18). Even a relatively small rockfall can prove deadly; large falls are often disastrous. On July 10, 1996, in the Happy Isles area of Yosemite National Park, a 162,000-ton mass of diorite fell from the Yosemite Valley's steep canyon wall and accelerated to a speed of 260 km/hr (160 mph), killing one hiker and injuring 20 others. The powerful blast of air that preceded the falling mass knocked down more than 2500 trees in its path. Fortunately, the fall occurred at 6:52 P.M., after the Happy Isles nature center and snack shop had closed for the day. Most day-hikers had already come off the Mist Trail, the site of the rockfall.

▶ **FIGURE 14-15** Frost-induced creep. Ice that forms beneath soil particles displaces the particles, pushing them toward the surface. Gravity carries the displaced particles a short distance downslope as the frost thaws, and the particles sink back down.

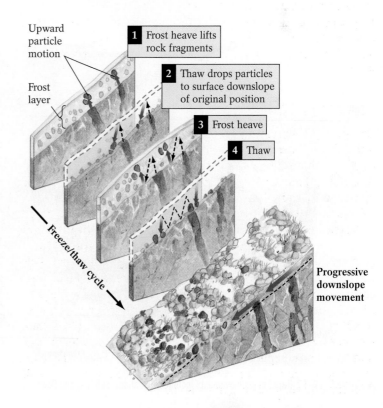

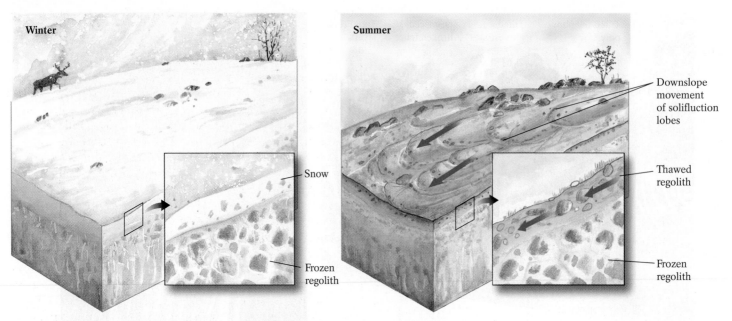

▲ **FIGURE 14-16** Solifluction occurs where the soil and regolith are frozen to depths of hundreds of meters. During the brief summer season, the surface materials thaw out to a depth of several meters and flow downslope with the aid of the water released by melting. Solifluction typically creates lobe-shaped masses of slowly moving sediment.

A massive fall that occurs at the coast may create an enormous sea wave. In 1958, a spectacular coastal rockfall, set off by an earthquake, sent 31 million m³ (1 billion ft³) of bedrock plummeting 900 m (3000 ft) into Lituya Bay near Anchorage, Alaska. An enormous wave, 528 m (1722 ft) high, raced across the bay. The wave lifted boats anchored immediately inside the bay mouth and transported them over an island and then out to sea. Fortunately, the region affected by this monumental wave was sparsely populated. A far greater disaster looms along both coastlines of the Atlantic Ocean, the highly populated coastal regions of North Africa, the Caribbean Islands, and the population centers of the Eastern Seaboards of North and South America. Highlight 14-1 on page 450 offers some insight into this potential mass-movement disaster.

Slides and Slumps   A **slide** occurs when rock, sediment, or unconsolidated material breaks loose and moves in contact with the underlying slope along a preexisting plane of weakness, such as a fault, fracture, or bedding plane. A slide moves *as a single, intact mass.* A slide may involve a small displacement of soil over solid bedrock, amounting to just a few meters of movement, or a huge displacement of an entire rocky mountainside. Large slides may be preceded by days, months, or even years of accelerated creep. The word *landslide* is the general term for any gravity-driven downslope movement where rock and soil have detached from the underlying materials along a plane of weakness.

In the largest events, a great mass of material detaches along an identifiable plane of weakness, or **slip plane.** The nature of a slip plane varies with the local geology. Unconsolidated sediments often develop slip planes where sedimentary material changes—for example, along the boundary between a layer of sand and a layer of clay. Slip planes also appear where unconsolidated sediments make contact with the underlying bedrock. Highlight 14-2 on page 452 describes a damaging slide in southern California that occurred along the contact between loose and solid slope materials, a typical geologic situation along the California coast.

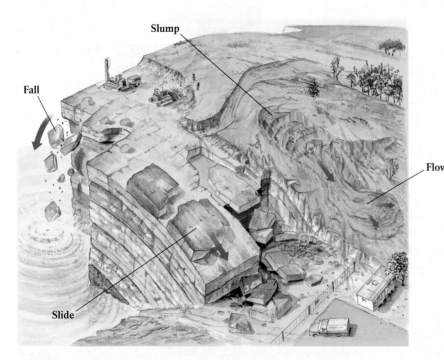

▲ **FIGURE 14-17** Rapid mass movements include falls, slides, slumps, and flows.

▲ **FIGURE 14-18** This massive rockfall occurred on January 1997 along Highway 140, several kilometers east of the Arch Rock entrance to Yosemite National Park, California.

In solid bedrock, slip planes develop where planes of weakness exist within the rock itself (see Figure 14-4). In sedimentary rocks, the slip plane is typically a bedding plane within a weak rock layer—for example, a thinly bedded shale that inclines toward an open space, such as a river valley (Fig. 14-19). In plutonic igneous rocks, the slip plane may be a large joint produced by the mechanical-weathering process of exfoliation. In metamorphic rocks, slip is likely to occur along foliation surfaces.

A slide's slip plane is generally flat, or planar. A **slump** is a type of slide that separates along a *concave* slip surface, moving in a downward and outward motion. Whereas slides typically move along a preexisting slip plane, slumps generally create their own slip plane within a body of unconsolidated material. They are distinguished by the crescent-shaped scar that forms in the landscape where the slump detaches. The steep, exposed cliff face that forms where the slumping mass pulls away from the rest of the slope is known as a **scarp.** Slumps do not usually travel far from their point of origin, nor do they typically move at high velocities. That explains why the slump block itself usually remains intact (Fig. 14-20).

Flows   A **flow** occurs when a mixture of rock fragments, soil, and sediment moves downslope as a highly viscous fluid. Flows typically move more swiftly and are more dangerous when they have a high water content. In February 1969, drenching rainstorms moved into southern California from the Pacific, dumping 25 cm (10 in.) of rain in nine days. In the next six weeks, the Los Angeles area received an additional 60 cm (25 in.) of rain, more than twice the normal amount for its rainy season. The previous summer in Los Angeles had seen a prolonged drought, sparking widespread brush fires that removed the vegetation from hundreds of thousands of acres in the Santa Monica Mountains. The torrential rains in 1969 triggered numerous flows as saturated clayey soils, stripped of binding vegetation by fire and drought, flowed over the area's clay-rich shaly bedrock. One hundred lives were lost, 15,000 people were forced out of their mud-covered homes, and $1 billion in property damage occurred.

Like other mass movements, flows are classified by their composition and velocity. A flow may contain a wide variety of materials, such as loose rocks, soil, trees and other vegetation, water, snow, and ice. Its velocity is determined by its water content, the nature of its component materials, the nature of the underlying materials, and the angle of the underlying slope.

Earthflows are relatively dry masses of clayey or silty regolith. Because of their high viscosity, they typically move as slowly as a meter or two per hour (though they may move as rapidly as several meters per minute when mixed with water). Their slow motion ensures that they rarely threaten life, but they may easily destroy all structures in their paths.

A typical earthflow is faster than creep but slower than a *mudflow,* a swifter-flowing slurry of regolith mixed with water, typically composed of sand-size and finer materials. Heavy rainstorms that soak into a thick blanket of relatively fine-grained regolith generally trigger mudflows. In October 2005, the torrential rains of Hurricane Stan (kind of embarrassing for one of your authors) triggered hundreds of deadly mudflows in Guatemala and Mexico. The consistency of mudflows can vary from that of wet concrete to muddy water (Fig. 14-21). Because they are typically saturated with water, these flows tend to travel through topographic low spots, such as valleys and canyons, just as streams do, and they can move considerable distances on slopes as gentle as 1° to 2°.

Mudflows are most likely to develop after heavy rainfalls on sparsely vegetated slopes with abundant loose regolith. Such conditions are common in the canyons and gullies of semi-arid mountains, where loose sediment accumulates between infrequent storms and few roots bind the loose materials. The occasional cloudburst washes large quantities of sediment into arroyos (dry stream channels) and canyons. The saturated sediment then rushes rapidly—and dangerously—down the channel.

Occasionally, a seemingly solid body of waterlogged clay-rich sediment almost instantaneously transforms into a highly fluid mudflow, or **quick clay** (Fig. 14-22). Quick clays occur when ground vibrations, perhaps caused by an earthquake, explosion, thunder, or even the passing of heavy vehicles, increase the water pressure between sediment particles. The particles separate from one another, reducing the friction between them. Quick clays may even occur spontaneously without triggering vibrations when groundwater dissolves away the salts that

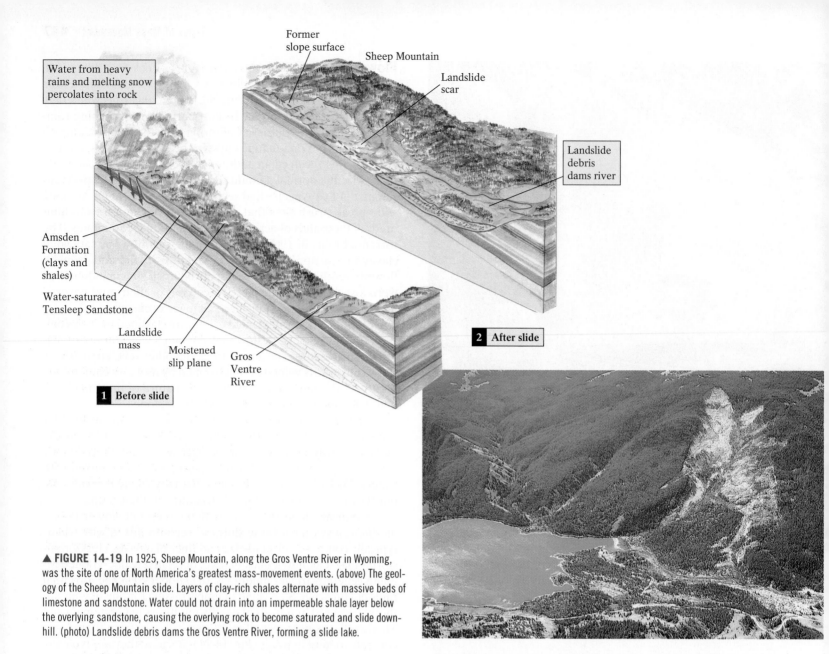

Water from heavy rains and melting snow percolates into rock

Former slope surface

Sheep Mountain

Landslide scar

Landslide debris dams river

Amsden Formation (clays and shales)

Water-saturated Tensleep Sandstone

Landslide mass

Moistened slip plane

Gros Ventre River

1 Before slide

2 After slide

▲ FIGURE 14-19 In 1925, Sheep Mountain, along the Gros Ventre River in Wyoming, was the site of one of North America's greatest mass-movement events. (above) The geology of the Sheep Mountain slide. Layers of clay-rich shales alternate with massive beds of limestone and sandstone. Water could not drain into an impermeable shale layer below the overlying sandstone, causing the overlying rock to become saturated and slide downhill. (photo) Landslide debris dams the Gros Ventre River, forming a slide lake.

typically enhance the electrostatic charges that bind *marine*-clay deposits together.

Certain regions of North America are particularly vulnerable to quick-clay flows. Fifteen thousand years ago, much of eastern Canada and New England was covered by an ice sheet several kilometers thick whose weight depressed the land beneath it. When the ice finally melted, the land had been depressed by hundreds of meters. The salty seawater that flowed into the depressed areas left behind marine-clay deposits—a loose, disorganized frame-work of platy clay particles held together in part by salt. On November 12, 1955, near the town of Nicolet, Quebec, along the Gulf of St. Lawrence, 165,000 m^3 (5.8 million ft^3) of this marine clay liquefied spontaneously into quick clay that flowed toward the Nicolet River valley. The principal cause? Its binding salt had been dissolved and carried away by thousands of years of groundwater flow through the marine clays. Miraculously, the flow stopped just meters before engulfing the crowded local cathedral,

▶ FIGURE 14-20 Slumping on cliffs in Dorset, England.

448

▲ **FIGURE 14-21** Tragic mudflow in Sarno, Italy, on May 6, 1998. This mudflow, which took more than 100 lives, formed when water from heavy spring rainfall mixed with rock and soil on nearby slopes, producing a muddy slurry that overwhelmed the low-lying sections of the village.

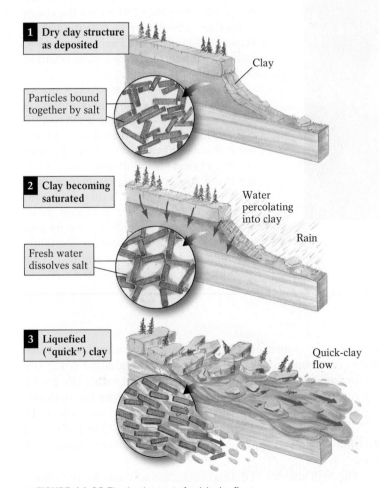

1 Dry clay structure as deposited

Clay

Particles bound together by salt

2 Clay becoming saturated

Water percolating into clay

Rain

Fresh water dissolves salt

3 Liquefied ("quick") clay

Quick-clay flow

▲ **FIGURE 14-22** The development of quick-clay flows.

and only three people died. A similar but much larger and more catastrophic quick-clay flow occurred on May 4, 1971, in the Quebec village of Saint-Jean-Vianney, where a layer of marine clay liquefied instantaneously; 6.9 million m³ (240 million ft³) of clay buried 40 homes and took 31 lives. Figure 14-23 shows the aftermath of this tragedy.

Volcanic eruptions can also produce catastrophic mudflows, known as *lahars*. Lahars develop when hot tephra melts snow or glacial ice or when the rainfall that commonly accompanies an eruption saturates fresh tephra and the accumulated soil and tephra layers from previous eruptions. This type of flow may rush downslope at very high speeds. The 1980 eruption of Mount St. Helens in Washington state set off a lahar that traveled between 29 and 55 km (18 to 35 mi) per hour, sweeping away homes, bridges, and everything else in its path.

Lahars can also occur during a volcano's noneruptive periods. Washington's Mount Rainier has generated more than 60 lahars during the last 10,000 years with many of them occurring during volcanically quiet periods. A major one took place about 5000 years ago when the summit of the mountain became unstable after thousands of years of steam emissions had converted solid volcanic andesite to loose clay. Today Mount Rainier continues to vent steam and convert its andesitic summit to clay, raising concerns about the possibility of another noneruptive lahar in the near future. For this reason, the U.S. Geological Survey monitors this volcano closely and has drafted an evacuation plan for nearby communities, where more than 50,000 people live today.

Debris flows, like mudflows, are common in sparsely vegetated mountains in semi-arid climates. They are typically triggered by the sudden introduction of large amounts of water, and tend to follow topographic low spots. These flows, however, consist of particles that are generally *coarser than sand-size.* Often

▲ **FIGURE 14-23** Tragic results of the devastating quick-clay flow in Saint-Jean Vianney, Quebec, May 4, 1971.

Highlight 14-1

Keeping an Eye on a Canary

Miners in years gone by often carried along canaries down their mineshafts to check for the presence of poisonous gases in their mines. Geologists today are casting a very wary eye toward another kind of Canary—La Palma, the east-central Atlantic's Canary Island, home to the extremely dangerous Cumbre Vieja volcano (see photo). Cumbre Vieja is the most active volcano in the Canary Islands, last erupting in 1971. But it's the volcano's 1949 eruption that is causing such great alarm because, during that event,

the western flank of the Cumbre Vieja "detached" from the volcano and slipped roughly 4 m seaward (14 ft) and then stopped. Geologists now believe that a future eruption, either in a few years or a few hundred years, may dislodge a block of rock 20 km wide (13 mi) and 20 km long and send it sliding into the adjacent Atlantic Ocean at a speed of 350 km/hr (about 200 mph). The energy release by this avalanche of rock would equal the entire electricity consumption of the U.S. for six months. The result? A

giant sea wave, 900 m high (about 3000 ft), traveling at about 800 km/hr (about 500 mph) across the Atlantic Ocean.

The greatest effect of this monumental wave would be felt along the coast of the west Sahara of Africa (see map), which would be struck by waves 100 m (330 ft) high within the hour. The coasts of Spain and the United Kingdom would be swamped by 10-m-high waves (33 ft) between two to five hours after the collapse of the volcano's flank. And after nine hours, a train of waves—perhaps a dozen—would sweep across the Caribbean islands and then strike the east coast of the U.S., each wave estimated conservatively to be about 20 to 25 m high (66 to 83 ft) and some perhaps 50 m high (165 ft). These waves would then wash inland several to tens of kilometers. Skyscrapers would topple, bridges would snap, and many coastal cities would be swept away in this onslaught. Evidence of past events of this magnitude litter the seafloor off the coasts of the Canary Islands in the Atlantic and the Hawaiian Islands in the Pacific. Perhaps as many as a dozen such events have occurred in the past 200,000 years.

Geologists who are studying Cumbre Vieja believe that the volcano erupts every 20 to 200 years or so, so while it's very difficult to predict exactly when this event will occur, it seems likely that it's more a matter of "when" than "if." Will nine hours warning afford enough time to evacuate millions in the East Coast's cities? Is there a plan for such an event? This is a potential mass-movement event for which there is no prevention or human intervention. Certainly something to think about if you live along the populous Eastern Seaboard.

▲ Cumbre Vieja volcano

they contain boulders 1 m or more in diameter (Fig. 14-24). Hence, debris flows require a steeper slope than other types of flow to trigger their movement. A mass movement that begins as a fall or slide can eventually break into a mass of rubble that continues downslope as a debris flow. Debris-flow velocities typically range from 2 to 40 km/hr (1 to 25 mph).

The swiftest and most dangerous type of flow is the *debris avalanche*, common in the Appalachians (such as New England's Green and White Mountains) and the Pacific Northwest's Cascades and Olympic Mountains. These mass movements occur

on very steep slopes, are triggered during and after heavy rains, and become enhanced by the removal of vegetative cover (by fire or logging, for example). During an avalanche, the entire thickness of soil and regolith may pull away from the underlying bedrock and rush downslope through narrow valleys. In 1973, when Hurricane Camille dropped 60 cm (25 in.) of rain in Virginia's Blue Ridge Mountains, debris avalanches claimed 150 lives.

The high velocity of debris avalanches accounts for their great destructiveness. Sometimes they move so fast that they can hurl

▲ FIGURE 14-24 A debris flow caused by a one-night flood in a small stream, in Britannia Beach, British Columbia.

debris through the air. In Yungay, Peru, a powerful earthquake on May 31, 1970, dislodged an avalanche of debris that rushed downward at speeds exceeding 200 km/hr (120 mph) (Fig. 14-25). On patches of the slope in the avalanche's path, grass and flowers—including shrubs more than 1 m (3.3 ft) high—were left undisturbed. This suggested that the debris flew through the air above them. Some geologists believe that compressed air, trapped below rapidly moving debris, can form a supporting air cushion that reduces the friction between the debris mass and the underlying surface. Eventually, as gravity pulls the flying debris back down to the slope's surface, a powerful wind gust is created as the air cushion is expelled. Tangles of downed trees that appear to have been uprooted by a great blast of air commonly mark the edges of large debris avalanches.

▲ FIGURE 14-25 Triggered by a major earthquake, the Yungay debris avalanche of May 1970 buried several Peruvian towns, taking 30,000 lives. (left) Before the avalanche. (right) After the avalanche.

Highlight 14-2

The La Conchita Mudslide

On September 29, 1994, residents of La Conchita, California, an unincorporated community located along the Pacific Coast Highway near Santa Barbara, met with a team of local geologists and engineers. The scientists informed the community that cracks had developed in the unconsolidated materials that form the steep slopes surrounding the town. They warned that the slopes could fail at any time, as they had several times during the past decade. La Conchita sits atop a massive 10,000-year-old landslide that overlies westward-dipping sedimentary rocks. Extensive irrigation of the avocado and citrus groves at the head of the slope had enhanced the region's slide potential.

In January and February of 1995, a succession of drenching Pacific storms (see figure) saturated the unconsolidated material and created pools of water at the heads of the slope's numerous cracks and fissures. On February 9, 1995, geologists returned to warn the residents of La Conchita that creep on the surrounding slopes had accelerated significantly and the fissures had become enlarged. According to scientists, there was a 50–50 chance that the slopes would fail soon.

At 2:10 P.M. on Saturday, March 4, 1995, a 600-ton wall of mud—more than 80 m (260 ft) wide and 10 m (33 ft) high—slid a distance of 300 m (1000 ft), roaring down upon La Conchita. Nine homes were destroyed; some were engulfed in mud; others were shoved from their foundations. Two hundred residents in 60 other homes were evacuated.

Local authorities continued to monitor La Conchita's slopes, and the area remained on a slide alert, particularly in those areas that lost their binding vegetation during the region's tragic wildfires of 2001–2004. The combination of earthquakes, fires, drenching rains, steep slopes, dipping rocks, and loose soils in this part of California makes it exceedingly vulnerable to mass movement. A small retaining wall was constructed to hold back future slides. This structure may have created a false sense of security in the town, a security that was shattered on January 10, 2005, in the wake of southern California's greatest rainstorms on record.

At 2:05 P.M., the retaining wall gave way, overwhelmed by the collapse of the town's 180-m (600-ft)-high rain-weakened bluff (see photo). Ten La Conchita residents died, 15 homes of 150 in the community were destroyed, 16 more were severely damaged. Defying the odds against future such disasters, the surviving residents confirmed their intention to remain in La Conchita and state officials called for efforts to rebuild at the same site—a *very questionable decision* given the town's topography, geology, climate, and recent history.

Note here again how the Earth's various systems (atmospheric rains, biospheric vegetation, hydrospheric groundwater, and lithospheric earthquakes and soils) combined to produce La Conchita's mudslides.

▲ La Conchita mudslide of January 2005.

▼ Storm tracks during the winter of 1994–1995 were diverted to southern California from their normal path across the Pacific Northwest by that year's El Niño effect. At the same time, the polar jet stream in eastern North America was displaced northward; as a result, cold arctic air failed to penetrate southward, allowing the east coast to experience a significantly warmer-than-average winter.

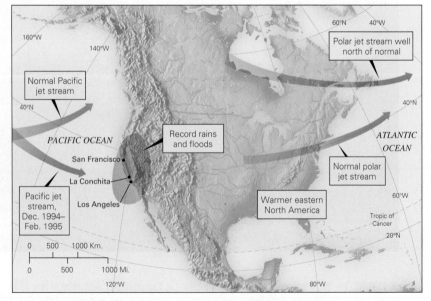

14.1 Geology at a Glance

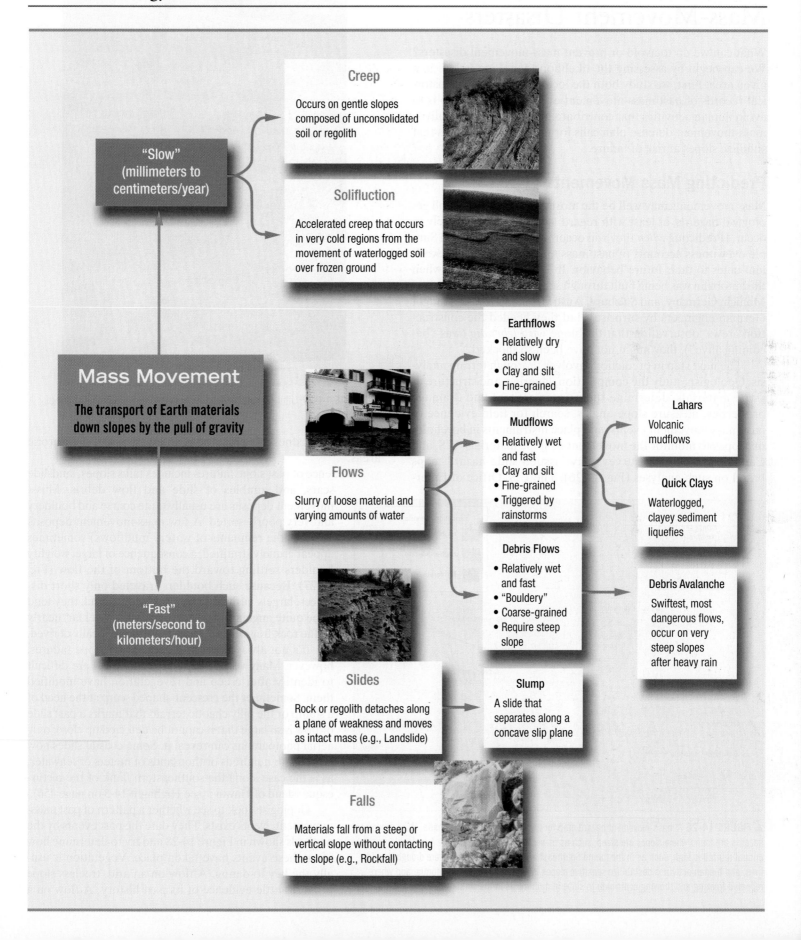

Creep

Occurs on gentle slopes composed of unconsolidated soil or regolith

Solifluction

Accelerated creep that occurs in very cold regions from the movement of waterlogged soil over frozen ground

"Slow" (millimeters to centimeters/year)

Mass Movement

The transport of Earth materials down slopes by the pull of gravity

"Fast" (meters/second to kilometers/hour)

Flows

Slurry of loose material and varying amounts of water

Earthflows
- Relatively dry and slow
- Clay and silt
- Fine-grained

Mudflows
- Relatively wet and fast
- Clay and silt
- Fine-grained
- Triggered by rainstorms

Debris Flows
- Relatively wet and fast
- "Bouldery"
- Coarse-grained
- Require steep slope

Lahars

Volcanic mudflows

Quick Clays

Waterlogged, clayey sediment liquefies

Debris Avalanche

Swiftest, most dangerous flows, occur on very steep slopes after heavy rain

Slides

Rock or regolith detaches along a plane of weakness and moves as intact mass (e.g., Landslide)

Slump

A slide that separates along a concave slip plane

Falls

Materials fall from a steep or vertical slope without contacting the slope (e.g., Rockfall)

Avoiding and Preventing Mass-Movement Disasters

What can we do to avoid or prevent mass-movement disasters? We can begin by assessing the likelihood for slope failure in a given area. First, we study both the local geology and the historical records of past mass-movement events. The next step is to avoid human activities that contribute to slope failure. Finally, a mass-movement defense plan calls for preventive measures that stabilize slopes at risk of failure.

Predicting Mass Movements

Mass movements may well be the most easily predicted of all geological hazards, at least with regard to *where* they are likely to occur. (Predicting *when* they will occur is far more difficult.) Simple eyewitness accounts of past mass movements provide excellent clues to their future behavior. In the spring of 1935, when the autobahn was being built through some clay deposits between Munich, Germany, and Salzburg, Austria, a series of slides caught German engineers by surprise. Had they heeded the construction crews' observation that the slope *wird lebendig* (was "becoming alive"), they might not have been so shocked.

The next step in prediction involves thorough terrain analysis. Geologists study the composition, layering, and structure of slope materials, determine their water content and drainage properties, measure slope angles, search for field evidence of past mass-movement events, and place instruments in boreholes on slopes to monitor the movement of slope materials. U.S. and Canadian geological surveys have issued slide-hazard maps based on such analyses (Fig. 14-26). The U.S. Office of Emer-

▲ **FIGURE 14-27** Lithified mudflow deposits from the Pliocene Epoch, in Borrego, California.

gency Preparedness also maintains an inventory of slide-prone areas.

Evidence of past slope failures includes talus slopes, landslide scars, and jumbles of slide and flow debris. Mass-movement deposits are usually quite coarse and bouldery and very poorly sorted. A few mass-movement deposits (such as the remnants of watery mudflows) sometimes appear crudely stratified, a consequence of large, weighty boulders settling toward the bottom of the flow (Fig. 14-27). Because such boulders traveled only short distances, largely cushioned by surrounding mud, they tend to be quite angular. Because they don't travel far, nearly all the rock fragments in such deposits are locally derived.

It's not always easy to detect earlier slope failures, however. Many very large slumps and slides are difficult to identify after creep and revegetation have modified them. Sometimes the crescent-shaped scarp at the head of a slump or the hilly chaotic terrain that marks a past slide or flow is so large that it cannot be detected up close; only aerial photographs can reveal it. Some coastal slides now lie beneath hundreds or thousands of meters of seawater, as is the case along the southeastern flank of the picturesque island of Hawai'i (see Highlight 14-3 on page 456).

Geologists look to see whether a pattern of past mass-movement events exists. They date the past events by the methods shown in Figure 14-28 and try to determine how often these events have taken place. Vegetation is usually the key to dating. A flow on an arid, treeless slope provides little evidence of its past history. A flow on a

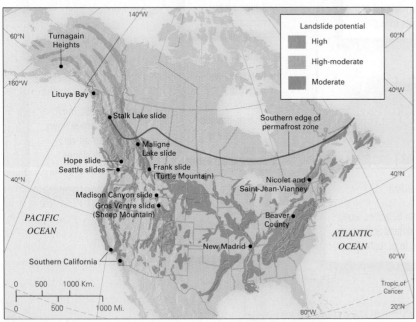

▲ **FIGURE 14-26** A mass-movement hazard map for the United States and Canada. Mass-movement hazards are found where slopes are steep, such as in western mountains and the Appalachians; where annual rainfall is high, such as in the humid Southeast and misty Northwest; where droughts, wildfires, and human activities destabilize sensitive slopes, such as in southern California; and where repeated freezing and thawing contribute to slope instability, as in the permafrost regions of Canada.

▲ FIGURE 14-28 Methods of dating mass-movement events. Carbon-14 dating is helpful if the moving mass incorporated trees or other types of vegetation. Lichenometry can be used if lichens have colonized bedrock exposed or boulders shattered by the event. Dendrochronology is useful if the moving mass overran living trees that survived the event: The rings of an affected tree grow asymmetrically from the moment the mudflow bends it, so we can determine how long ago the mudflow occurred by counting the asymmetrical rings. Newly exposed boulders begin to accumulate cosmogenic isotopes at the surfaces immediately after the mass-movement event.

vegetated slope, however, often picks up organic material that can then be dated by carbon-14 dating. Dendrochronology, lichenometry, and surface-exposure studies (all discussed in Chapter 8) can also help date a mass-movement event. A tree engulfed by a flow or slide may become bent by the flowing mass and reveal a datable disruption in its tree-ring pattern (we can count the rings that have formed since the tree was bent). The extent of lichen growth on a shattered rock can reveal how long ago the rock was broken. The buildup of cosmogenic isotopes on the surface of a freshly exposed boulder in a landslide or rockfall can also help geologists date mass-movement events.

After we determine the dates of past mass-movement events, we can compare them to the dates of other events that might have triggered the mass movement. If heavy rain is a suspected trigger, mass-movement dates are compared with climatic records. If the records agree, we can predict how an area's slopes will respond to various amounts of rainfall in the future. We know, for example, that most slides in the San Francisco Bay area occur after a storm drops more than 15 cm (6 in.) of rain on steep, already saturated slopes in a closely spaced series of storms.

Although we can never predict *precisely* when a mass-movement event will occur, we can stay alert to the potential dangers of specific slopes under specific conditions. Where property is valuable, or where mass movement poses a particular threat to human safety, it makes sense to install expensive landslide-warning devices that signal a slide or flow in progress. In the slide-prone southern Rockies, the Denver and Rio Grande Railroad has installed electric slide detectors upslope from certain bridges. These detectors are wired to respond to the high pressure asso-ciated with an encroaching slide mass. Contact with a moving mass breaks an electrical circuit in the detectors, sending a stop signal to approaching trains.

A less expensive early-warning system calls for observing changes in animal behavior. Shortly before the entire Swiss village of Goldau was destroyed in 1806, the town's livestock acted nervously and the beekeeper's bees abandoned their hives. Within hours, a block of rock 2 km (1.2 mi) long and 300 m (1000 ft) wide broke loose from a steep valley wall, burying the town and its 457 inhabitants in a massive rockfall. Apparently the area's animals and insects perceived the impending failure of the rocks and the onset of the tragic rockfall before it actually occurred.

Avoiding Mass Movements

It may seem obvious that we shouldn't build on sensitive slopes, especially because few insurance companies cover homes and property for mass-movement losses. But even people who live in definite slide-prone areas still won't leave their homelands, commercially or agriculturally valuable properties, and scenic hillsides. Thus, they refuse to avoid certain mass movement, as the overdevelopment of southern California's steep canyon slopes and unstable oceanfront cliffs clearly shows. Highlight 14-4 on page 460 examines the importance of mass-movement avoidance to prospective home buyers.

People in developing nations often overcultivate unstable slopes—by necessity—to feed the hungry. Their cattle may overgraze already sparse vegetation. They may also clear-cut forested slopes to provide wood to build homes and heat them.

Highlight 14-3

Crumbling K'ilauea

The seafloor along the southeastern shore of the Big Island of Hawai'i is littered with the lost remnants of the K'ilauea volcano. Even as geologists map K'ilauea's new surface lava flows, the mountain itself is tearing away from the rest of the island (see photo) and sliding toward the ocean at a pace of roughly 10 to 20 cm (4 to 8 in.) per year. Although seemingly slow by human standards, this rate is a brisk gallop by geologic-time standards; the southeastern flank of K'ilauea is probably the fastest-moving piece of real estate on Earth.

A check of the seafloor off the coast confirms that the crumbling of this volcano has been ongoing for thousands of years. A sonar-produced map reveals a dump site for slides of monumental proportions (see figure). Slices of past volcanoes, some larger than Manhattan Island, have come to rest after massive slides—some as far as 200 km (120 mi) from their starting point. Slides of such magnitudes probably triggered tsunami that may have reached heights equal to that of the Empire State Building.

The process that appears intent on dismantling the volcano comes from the very magma that creates it. As magma fills the volcano's subterranean magma chamber, K'ilauea inflates like a balloon and pushes outward against its faulted southeastern flank; the flank then detaches along the fault and slides.

At K'ilauea, geologists from the Hawaiian Volcano Observatory are using recent earthquake data to map the hidden fault that may trigger the next great landslide. They continue to monitor the flank's nonstop march toward the sea by using a Global Positioning System (GPS) to measure the precise locations of selected benchmarks on the volcano's slopes. As the flank slides southeastward, its bottom portion is apparently bulldozing the seafloor, piling up rubble like a mound of snow in front of a snowplow.

When the current flank eventually gives way, Hawai'i will undoubtedly export some of the disaster to other Pacific Rim locales. A massive Pacific-wide tsunami is a strong likelihood. To understand the potential magnitude of such a wave, consider that scientists have retrieved fragments of coral, which grows below sea level, from as high as 365 m (1200 ft) on the slopes of a mountain on the Hawaiian island of Lanai. These bits of coral were most likely thrown to those great heights by a giant sea wave triggered by a massive landslide on the Big Island.

When will this great catastrophe take place? The best estimates suggest that such events occur every 100,000 to 200,000 years. Researchers are now attempting to date past slides to estimate how much time has elapsed since the last great slide. Yet even if Hawai'i's next slide is a long way off, movement along K'ilauea's massive southeastern fault has apparently triggered the islands' largest historic earthquakes, some with magnitudes as great as 8.0. Such events also generate life-threatening tsunami, reason enough to pursue the research that will let us better understand how the Hawaiian islands have grown.

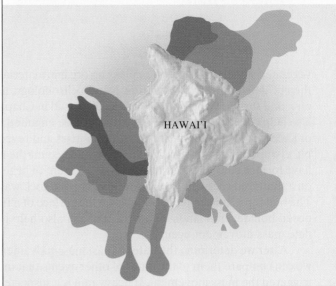

▲ Large fractures on the slopes of Hawai'i's K'ilauea volcano indicate that its southeastern slopes are being pulled apart by large-scale mass movements. Map shows the distribution of past massive slides into the adjacent Pacific Ocean.

They act out of necessity, but unfortunately, all these actions promote mass movement. As long as we continue to use unstable slopes in these ways, *prevention* will remain the most practical way to deal with the threat of mass movement.

Preventing Mass Movements

It is far less expensive and much safer to stabilize a slope *before* it fails than to clean up *after* it fails. Unfortunately, the dollars lost to mass movement continue to exceed by more than 50 times the dollars spent to prevent them. Civil engineers estimate that we could avoid more than 95% of all slope failures by undertaking regional, local, and site surveys and pursuing aggressive preventive measures.

The first step in developing a prevention plan for any site is to understand clearly its subsurface geology and identify its potential trigger mechanisms. Unfortunately, detailed studies require costly drilling and sophisticated analytical techniques. To compli-

cate matters further (and increase costs), some settings are highly variable geologically, and thus require drilling and testing at numerous locations. In other settings, the local stratigraphy is similar virtually everywhere within the region. For example, the entire city of Seattle, one of North America's urban landslide capitals, is highly susceptible to slope failure because the geology of much of the Puget Sound region of Washington state (including Seattle) contains a layer of permeable sand overlying a layer of impermeable clay. This layering promotes instability of the city's steep slopes as excess water builds up near the surface when infiltrating rainwater is trapped between the sand and clay, causing the region's damaging landslides and mudflows (Fig. 14-29).

The second step in preventing slope failure is to determine how best to enhance the forces that *resist* mass movement or reduce the forces that promote it. Geologists use several methods, often simultaneously, to increase slope stability. Nonstructural methods, which do not require costly engineering efforts, include management of vegetation and introduction of

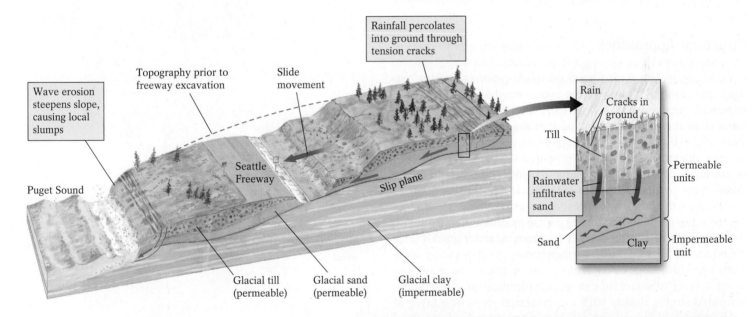

▲ **FIGURE 14-29** The slide-prone geology of Seattle, Washington. (above) Virtually all of this city is built on 15,000-year-old glacial deposits consisting, in part, of a permeable layer of sand overlying an impermeable layer of clay. Water from Seattle's well-known rainfall seeps through the sand and filters down until it reaches the clay. There the water becomes trapped, high water pressures build up, and the overlying sand layer is buoyed upward. The sand layer is then likely to slip, especially during the rainy months of January, February, and March, taking some of the underlying clay along with it. In addition to the effect of water pressure, Seattle's steep hills, occasional earthquakes, and high level of human activity on slopes increase the city's potential for slope failure. (photo) Tragic slide on Bainbridge Island, Washington, January 1997, that took the lives of a family of four whose home was pushed without warning into Puget Sound.

soil-strengthening agents. Structural methods, which are usually very expensive, involve building structures such as retaining walls and drains or actively modifying potentially unstable slopes.

Nonstructural Approaches   One nonstructural method of preventing mass movement involves planting fast-growing trees and other vegetation. The plan here is to develop extensive, deep-root networks that bind near-surface soils. Broad-leaved trees, such as elms, maples, and oaks, protect slopes further with their large leafy canopies; some falling rain lands on the leaves and evaporates, reducing the likelihood of soil saturation and mass movement.

Another nonstructural approach adds chemical solutions to soil or regolith to bind clay-mineral particles and increase the strength of loose slope materials. This step can even be taken after a flow has begun: In Norway, a quick-clay flow was halted by bulldozing sodium-chloride and magnesium-chloride salts into the moving mass. Cement injected into a slide or flow mass also lends strength and cohesiveness to unconsolidated sediment.

Structural Approaches   Geologists also attempt to prevent mass movement by modifying a slope, reducing its water content, or building supports for it. One slope-modification tactic is to "unload" the slope by removing the excess weight of buildings and loose soil. An overly steep slope can be "graded" by moving material from the top and spreading it at the toe. This reduces the slope angle (Fig. 14-30a). More aggressive approaches involve removing all material overlying a potential slip plane (Fig. 14-30b) or cutting flat terraces into a slope to reduce weight and create areas where small slide masses can come to rest (Fig. 14-30c).

Perhaps the best way to prevent mass movement is by reducing the water content of a slope. This can be done either by not allowing water to enter the slope or by removing water quickly after it has entered. Erecting small earthen barriers to divert water away from a sensitive slope or covering a slope with thin sheets of plastic or a layer of concrete can prevent infiltration. In already-saturated slopes, drains, such as perforated pipes or a layer of coarse permeable gravel, can collect water underground and channel it away. It is also possible to pump water from a slope to reduce the weight overlying potential slip planes. Drying out the slope also increases friction along the planes and decreases swelling and weakening of soil clay. In Pacific Palisades, California, engineers excavated tunnels through an unstable slope and circulated hot air through them to dry the subterranean pore water.

Slopes can also be frozen to stabilize them temporarily. During construction of the Grand Coulee Dam in Washington, the circulation of liquid nitrogen froze 377 points in the dam site's slope. A section of slope, 12 m (40 ft) high, 30 m (100 ft) long, and 6 m (20 ft) thick was frozen every day.

Bolts or pins can hold a potential rockslide in place (Fig. 14-31a), but their installation is costly and thus this approach is used only in particularly dangerous situations or in cases where valuable property is at risk. Retaining walls, like those used to shore up excavation walls at construction sites, can provide support to the base of a slope. Their installation requires driving steel

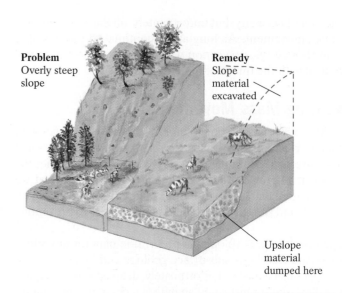

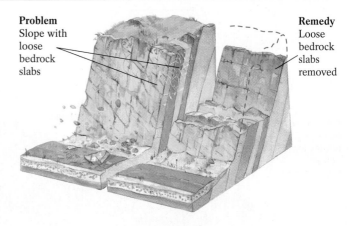

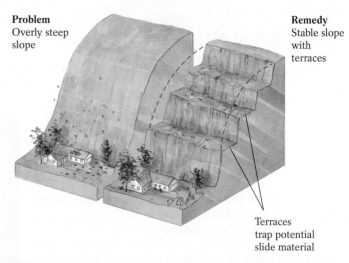

▲ **FIGURE 14-30** Structural methods of slide prevention attempt to reduce the forces driving downslope movement or enhance the forces resisting movement. (top) Grading a slope requires moving material from its top to its base to decrease the slope angle. (middle) Removing loose slope material helps prevent unexpected falls. (bottom) Terraces cut into a slope reduce a steep slope angle and catch any falling debris.

▼ **FIGURE 14-31** Structural supports to prevent slope failure include rock bolts and retaining walls. **(a)** Rock bolts stabilize loose-jointed granite on Storm King Mountain, overlooking Storm King Highway in Westchester County, New York. **(b)** A retaining wall supports the ground underlying Route 659, near Blacksburg, Virginia.

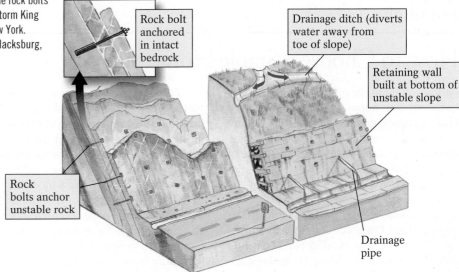

Rock bolt anchored in intact bedrock

Drainage ditch (diverts water away from toe of slope)

Retaining wall built at bottom of unstable slope

Rock bolts anchor unstable rock

Drainage pipe

(a)

(b)

or wooden piles through loose surface materials into the underlying stable materials (Fig. 14-31b).

Extraterrestrial Mass Movement

Images from NASA's space probes reveal that mass movement occurs on two of Earth's nearest neighbors—the Moon and Mars. Even though the Moon's gravity is only one-sixth that of Earth, mass movement is one of the few ways in which its surface has been modified. With the exception of recently discovered ice patches in some of the Moon's sun-shaded craters near the poles, the Moon's surface is a very dry place, and thus mass movement remains largely limited to dry processes. Meteorites are a prime trigger; numerous daily impacts set off avalanches of rock and regolith, even in the absence of an atmosphere. After an impact crater forms, slumps occur on its inner walls, which are generally steeper than the angle of repose of loose lunar regolith. Marks left by rolling and sliding boulders may remain unchanged for millions of years on the Moon's surface, which is virtually free of most forms of mechanical and chemical weathering (Fig. 14-32).

The surface of Mars exhibits a wider range of mass-movement features. Trigger mechanisms for Martian slides and flows include meteorite impacts, "marsquakes," and related faulting. In the past, periodic melting of the planet's subsurface permafrost

Highlight 14-4

How to Choose a Stable Home Site

At some point in the future, you may wish to own your own home (if you don't already). Having studied physical geology, you will want to ensure that your dream house doesn't slide down a steep slope and tumble into a pile of rubble or fall from a cliff and land in the sea.

Suppose you're exploring southern California's scenic beachfront locales. You happen upon the mosaic sign for the town of Portuguese Bend, but notice that the beautiful ceramic signpost is cracked in two. That find should raise your newly grown geologic antenna. You check the local real estate listings and note a house priced at $50,000 that should be worth $500,000. Your intuition warns you that something must be wrong. Perhaps the owner is trying to unload a disintegrating house on a worthless, sliding property. What should you look for? How can you tell—in southern California or anywhere else in the world—if you're in mass-movement country?

First, examine the property itself. Do you see any signs of an old mud or debris flow? Look for exposed rocks or fresh breaks in the landscape that reveal the underlying geologic material. Next, investigate the entire neighborhood. Is there evidence of a slump scar upslope where a block may have broken away? Is the site of your dream house in the middle of a large potential slide mass? Evidence of current or past activity may be hundreds of meters or even several kilometers away from your dream house. As you drive through the community, look for fences that are out of alignment, and for power and telephone lines that seem too slack in some places and too taut in others (see figure).

Look carefully at the house itself, and if possible at neighboring ones. Do you see large cracks in the foundation (small cracks may form during the initial drying and settling of the concrete)? Doors and windows that stick may indicate that once-straight structural features are now out-of-alignment, although poor craftsmanship or high moisture content may also be the culprits that cause a stuck window. A cracked pool lining might explain why a swimming pool can't hold its water. Finding only one such problem may not indicate danger, but the presence of several problems should send you a strong warning.

If the geology, topography, and hydrology of a home site all raise questions about slope stability, the location may well be undergoing active slope failure. You might confirm your suspicions by checking newspaper accounts and the records of the local housing authority, by contacting the state geological survey, and by interviewing the property's neighbors. Of course, anyone buying property on a steep slope would be well-advised to seek a professional geological inspection before finalizing the deal.

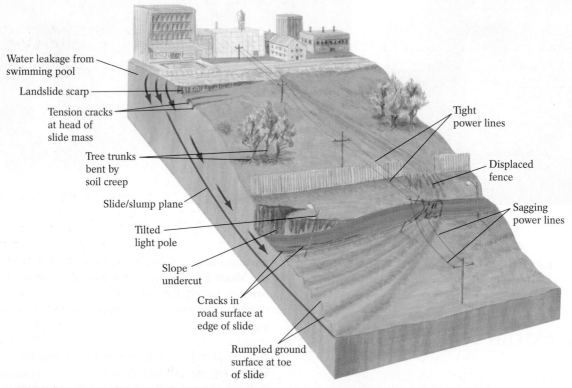

Water leakage from swimming pool

Landslide scarp

Tension cracks at head of slide mass

Tree trunks bent by soil creep

Slide/slump plane

Tilted light pole

Slope undercut

Cracks in road surface at edge of slide

Rumpled ground surface at toe of slide

Tight power lines

Displaced fence

Sagging power lines

▲ Various signs of past, current, and potential future mass movement in an urban area. If the power lines in a neighborhood are very loose, the poles that hold them may have moved closer together, as one might expect at the toe of a slide where the slope is bunching up like a rumpled carpet. At the head of the slide, tight lines may indicate that the poles on the slide mass are moving away from those upslope. The poles in the middle of a slide mass may keep their original spacing because the mass may not be deforming much internally.

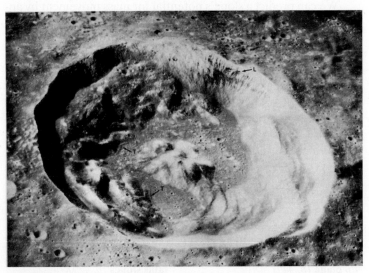

▲ **FIGURE 14-32** Massive slumps from the steep walls within the Euler impact crater on the Moon's surface.

▲ **FIGURE 14-33** Mass movement on the Martian surface is evident in the Valles Marineris, a 2-km (1.2-mi)-deep canyon. Satellite images show that some of the flows are similar in form to those on Earth and may have traveled as far as 30 km (20 mi) across the canyon floor. Slumps that closely resemble their counterparts on Earth are visible along the margin of the canyon.

may have released large amounts of water that oversteepened slopes and promoted development of massive slides, slumps, and debris flows. Mass movement, perhaps in the form of debris flows or debris avalanches, may have enlarged all or parts of the Valles Marineris, one of Mars' largest canyons (Fig. 14-33).

In this chapter we have explored the many ways that gravity drives erosion and modifies the Earth's surface. In Chapters 15 through 17, we will look at how water—as stream water, unseen groundwater, and frozen glacial water—transports weathered materials and modifies the Earth's landscapes.

Chapter 14 Summary

Mass movement is the process in which the pull of gravity transports masses of Earth materials downslope. It occurs when the factors that drive material downslope overcome the factors that tend to resist this movement. The principal factors that control mass movement include:

- gravity, the principal driving force behind mass movement, largely controlled by the steepness of the slope;
- the strength and cohesiveness of slope materials;
- friction between slope materials and the underlying slope;
- the amount of water within slope materials;
- the nature and extent of a slope's vegetative cover;
- the slope's history of human activities and other disturbances.

Of these factors, slope steepness (and its effect on the downslope component of gravity) and water content (and its effect on friction) largely control a slope's mass-movement potential. The effect of slope steepness is particularly pronounced for loose, unconsolidated material, which is stable only until it reaches its

angle of repose, the maximum angle that it can maintain without downslope particle motion. In general, large, irregularly shaped particles have a higher angle of repose. Water contributes to mass movement by increasing the weight of slope materials, reducing friction between planes of weakness, and reducing internal friction between loose grains in unconsolidated sediment.

A variety of natural and artificial causes trigger mass movement. Common *natural* triggers include:

- earthquakes;
- volcanic eruptions;
- heavy rains and snowmelt.

Common human-caused triggers include:

- ill-advised excavation of unstable slopes;
- overloading of unstable slopes by dumping or overbuilding.

Mass-movement processes are classified by their velocity and composition. Slow mass movement is called *creep.* It occurs on virtually all slopes composed of unconsolidated soil or regolith.

A special type of creep occurring principally in very cold regions is *solifluction,* the movement of waterlogged soil over frozen ground.

Several types of rapid mass movement have been identified. These include:

- *Falls,* where materials become dislodged and fall from a steep or vertical slope—through the air—without contacting the slope;

- *Slides,* where rock or regolith detaches along a plane of weakness, or slip plane, and moves as an intact mass downslope. A *slump* is a type of slide that separates along a *concave* slip plane, leaving behind a crescent-shaped scarp—the steep, exposed cliff face that remains where the slide mass pulls away;

- *Flows,* which are slurries of loose material and varying amounts of water. Flows may contain a wide variety of materials, such as loose rocks, soil, trees and other vegetation, water, and ice. *Earthflows* are relatively dry and quite slow-moving; mudflows are faster-moving slurries composed of regolith mixed with water. *Debris flows* consist of particles coarser than sand-size, and may contain boulders one meter or more in diameter.

We can predict a slope's potential for mass movement by analyzing its stability, examining evidence of past events and historical records, and using landslide-warning devices. Avoiding mass-movement events requires people to bypass building on sensitive slopes. Prevention of mass movement may follow a nonstructural strategy, such as increased planting of vegetation with extensive root systems and adding chemicals to a soil to bind its particles. Structural approaches include unloading slopes, grading slopes, building retaining walls, sealing slopes from water infiltration, and draining water from slopes.

KEY TERMS

angle of repose *(p. 439)*	**flow** *(p. 447)*	**slide** *(p. 446)*	**solifluction** *(p. 444)*
creep *(p. 444)*	**mass movement** *(p. 436)*	**slip plane** *(p. 446)*	**talus slopes** *(p. 439)*
debris flows *(p. 449)*	**quick clay** *(p. 447)*	**slump** *(p. 447)*	
fall *(p. 445)*	**scarp** *(p. 447)*		

QUESTIONS FOR REVIEW

1. Draw a simple sketch that illustrates the components of the force of gravity that act on an object located on a slope. Include the effect of friction.

2. List at least one type of weakness plane that can be found within each of the three common types of rock (igneous, metamorphic, and sedimentary).

3. Describe three ways that water contributes to mass movement.

4. List three different mass-movement trigger mechanisms.

5. What is the fundamental difference between a flow and a slide?

6. Describe two ways that you might date a landslide.

7. If you were thinking of moving into a slide-prone region, how would you determine whether your prospective home was located on a stable slope?

8. How would the prospects for mass movement change if a completely clear-cut slope were replanted and became densely reforested? How might the recent forest fires in southern California, Montana, and Colorado have affected the stability of the regions' slopes?

9. In which North American states or provinces would you expect to find the most mass movement? Explain your reasoning.

10. What type of mass movement is likely to happen in the photo at right? What steps would you take to stabilize this material?

LEARNING ACTIVELY

1. It's time to get on your bike or in your car and look around your community at its mass-movement situation. Check out the hill-sides in a neighborhood park: Are they creeping? Are the tree trunks bent? Look for evidence of past slides and mudflows. Visit again after several days of rain have occurred. How have the conditions changed? Look around your community at the utility wires (first reread Highlight 14-4). Do some neighborhoods appear to be "on the move"?

2. Spend some time looking at the measures taken by your community to deal with mass movement along roadsides. Have the hills been graded? Are there drainage ditches at the toes of the slopes? Are the slopes covered with plastic or planted with protective vegetation? Has your community invested in expensive retaining walls?

3. Open up the local Yellow Pages and check for Geotechnical/Soils Engineering/Geological Engineering companies. Most of these companies deal with slope stability. The number of companies in the Yellow Pages directory will be proportional to the size of your community's mass-movement concerns. What does the number of such companies tell you about the area's mass-movement prospects? If you don't already live there, visit an online Yellow Pages for a community in coastal southern California (such as Pacific Palisades or Laguna Hills) and compare the listings there for geo-engineers to those in your community. What does this tell you about southern California's mass-movement prospects?

ONLINE STUDY GUIDE

Practice makes you better. Make the time to take the *Online Study Guide* quiz for this chapter. It should take you about 20 minutes. Automatic grading and instant feedback will help you quickly and accurately identify the concepts you have mastered and the areas in which you need more study.
www.prenhall.com/chernicoff

Streams and Floods

15

- **The Earth's Water**
- **Streams**
- **The Geological Work of Streams**
- ***Highlight 15-1*** The Mississippi River and Its Delta
- **Controlling Floods**
- ***Highlight 15-2*** The Great Midwestern Floods of 1993
- ***Highlight 15-3*** The Tragedy of Katrina
- **Stream Evolution and Plate Tectonics**
- **Extraterrestrial Stream Activity—Evidence from Mars**

Look down from an airplane on a clear day and you will almost certainly see a river. Rivers are perhaps the most common geological feature on the Earth's surface. They flow in virtually every geological and geographical setting, with the exception of the entire frozen Antarctic. They rush down mountainsides, meander across farmlands, and cut through or border almost all major cities and towns. Most major cities are located on the banks of rivers, having been founded at convenient river crossings or sites where waterfalls or rapids prevented boats from moving upstream. Historically, most cities owe their prosperity to the size of their rivers.

At best, however, life along the banks of a river is a "good news–bad news" proposition. *The good news?* Rivers provide a steady supply of water for home and industrial use and, through irrigation, enable us to make a desert bloom. Crops grown in the fertile soils near rivers feed more than one-third of the Earth's human population. Rivers enable cities and towns to transport goods inexpensively and, in modern times, provide a source of clean hydroelectric power. Of course, we also fish, swim, sail, windsurf, and water-ski in rivers, and even the laziest among us enjoy just sitting on a river's banks, watching it flow by. *The bad news?* Rivers bring floods, the most universal of natural disasters (Fig. 15-1). Many streams flood *every* year, and some flood every two or three years, causing injuries, drowning deaths, and starvation due to loss of crops and livestock. Floods damage property, disrupt transportation networks, and spread diseases caused by insects and microorganisms in contaminated floodwaters.

The health and welfare of much of humankind are intimately tied to the rivers, streams, and creeks that flow by and through our farms, towns, and cities. Thus we should learn about how they form, how they work, and how they affect our daily lives. We will begin by considering the origin of the water that keeps these Earth-surface arteries flowing.

The Earth's Water

The amount of water *in, on,* and *above* the Earth totals an estimated 1.36 billion km^3 (326 million mi^3). This volume has apparently remained fairly constant for more than a billion years, although some scientists have recently suggested that incoming icy comets continue to contribute to the Earth's water supply over time. Approximately 97.2% of the

FIGURE 15-1 Floodwaters from Hurricane Katrina cover a portion of New Orleans, Tuesday, August 30, 2005. ▶

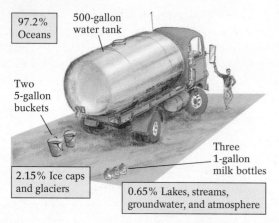

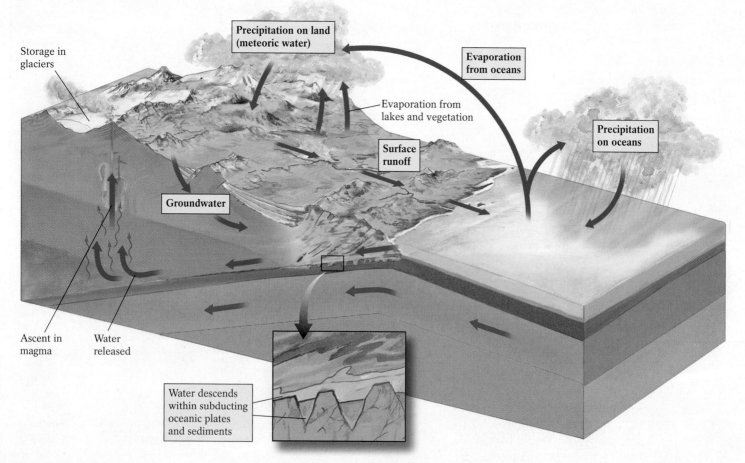

▲ FIGURE 15-2 Distribution of the Earth's water, by relative volume.

Earth's water resides in its oceans; 2.15% is frozen in ice caps and glaciers; and the remaining 0.65% occupies our lakes, streams, groundwater, and the atmosphere (Fig. 15-2). Most of the Earth's water originated in its mantle and came to the surface as the hot magma-derived water vapor expelled during volcanic eruptions. Some water may eventually return to the interior as water-rich oceanic plates subduct and a great deal of water seeps from the surface into the subsurface groundwater system.

Water heated by the Sun's rays evaporates from the Earth's surface and rises into the atmosphere. Because the atmosphere cools at higher altitudes, this water vapor cools as it rises, and be-

cause cold air can hold less moisture than warm air, the water vapor condenses around microscopic ice particles, creating the water droplets that form clouds. When the droplets become too heavy to remain suspended, they precipitate as rain, snow, sleet, or hail, depending on the air temperatures they encounter as they descend. Some of the water falls back into bodies of water—primarily the Earth's oceans—and the rest onto land.

The water that falls on land may then head in one of several directions. Snow on a high peak may be stored in the snowpack or within a glacier for tens, hundreds, or even thousands of years. Likewise, rain that falls into a lake may remain stored there for many years. Water is also "locked up" temporarily as the moisture that adheres to the clay particles within soils. Rainfall may also evaporate back into the atmosphere, run across the landscape as surface flow, or **runoff,** either as a sheet of water or in distinct channels, or as underground flow after the water has been absorbed into the ground.

Once water has been absorbed into the ground, it may be picked up by plant roots, absorbed by soil particles, or pass into and flow through the underground water system. Ultimately, much of the water that strikes the Earth's surface, even if it is stored there for many years, will eventually flow back to the ocean—either in a surface stream or as groundwater. Thus the Earth's water is constantly moving between its oceans, atmosphere, and lands, and, via plate subduction and volcanism, between its mantle and its surface. This relentless movement of water is referred to as the **hydrologic cycle** (Fig. 15-3).

▲ FIGURE 15-3 The hydrologic cycle. All the water that falls from the atmosphere onto the Earth's surface eventually enters the vast oceanic reservoir through one or more of the pathways of the cycle.

In this chapter, we will explore the surface-water part of the hydrologic cycle and learn how rivers flow and create distinctive landforms by erosion and deposition. We will also examine why they flood and discuss our efforts to deal with flooding, the most universal and devastating of natural hazards.

Streams

Any surface water whose flow is confined to a channel (a linear topographic depression) is called a **stream,** although the locals may know it as a river, bayou, creek, brook, or run. The United States has an estimated 2 million streams, ranging in size from the smallest creeks to the mighty Mississippi River.

A stream's flow is maintained by two sources of water—the precipitation that falls on the land and then runs off into a stream, and water that flows into the stream from the underground water system. This latter source of water, known as **base flow,** enables a stream to continue to flow between storms and even during periods of prolonged drought when no surface runoff is maintaining a stream's water supply.

Every stream is fed by a water-collecting area known as its **drainage basin** or **watershed**. A drainage basin consists of the total area from which overland flow of precipitation (rain and snowmelt) reaches a stream. Some streams receive water from a network of smaller streams, called **tributaries,** within their drainage basin. In North America, watersheds range from as small as a square kilometer (the area drained by a small tributary) to as large as 3.2 million km² (1.25 million mi²)—the area drained by the Mississippi River system. Every drainage basin is bounded by an area of higher topography, called a **drainage divide,** that separates it from an adjacent drainage basin. Divides may be low ridges between two small tributaries or continent-spanning mountain ranges (Fig. 15-4). The Rocky Mountain portion of North America's continental divide (The Great Divide), for example, stretches from the Canadian Yukon to New Mexico and south to Panama. Rain falling anywhere along this divide flows either westward to the Pacific Ocean, eastward to the Atlantic Ocean (via the Gulf of Mexico), or northward toward the Arctic Ocean.

At a drainage divide, rainfall initially flows downslope as *overland flow*—the unchannelized water that flows across the surface wherever more precipitation falls than can seep into the ground. These flows near the divide are a stream's *headwaters.* Wherever the headwaters encounter slight surface depressions or changes in surface composition, they erode narrow depressions called *rills.* (Once the water begins to flow in even the smallest channels, it's called runoff.) Throughout the stream's *transport area,* flowing water shapes and travels through progressively larger channels. Downslope, where the rills carry more water, they merge into larger, branching channels. Like the branches of a tree, these tributaries eventually join together to form the main, or **trunk stream.** The trunk stream crosses the greatest part of the transport area and carries the most water and sediment. At the *mouth* of the trunk stream, at sea level, its water and sediment load begin to disperse. Numerous small channels, or **distributaries**, may branch off, carrying water and sediment across lowlands to an ocean (Fig. 15-5).

Stream Flow and Discharge

The flow of a stream is driven by its **gradient**, or slope—its vertical drop in elevation over a given horizontal distance. (Gradient is expressed in terms of meters of vertical drop per kilometer of horizontal distance.) The steeper its gradient, the more rapidly a stream flows. A stream's gradient depends on the topography over which it flows, but generally decreases from its headwaters to its mouth. In North America, gradients range from 66 m/km (350 ft/mi) for the upper part of the Uncompahgre River in the Colorado Rockies to 0.1 m/km (0.5 ft/mi) for the lower Mississippi River downstream of Cairo, Illinois. All streams—even those that appear to be calm—flow *turbulently* (from the Latin for "turmoil"), with their water moving in swirls and eddies. Although the overall direction of flow is downstream, some water may swirl upward, like dry leaves caught in autumn gusts, or descend violently, like the whirlpool produced as the last of your bathwater disappears down the drain.

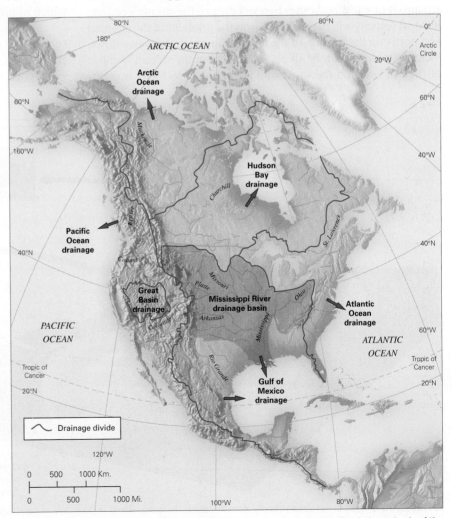

▲ **FIGURE 15-4** North America's drainage divides and major drainage basins. The drainage basin of the Mississippi River includes roughly 40% of the lower 48 states in the United States.

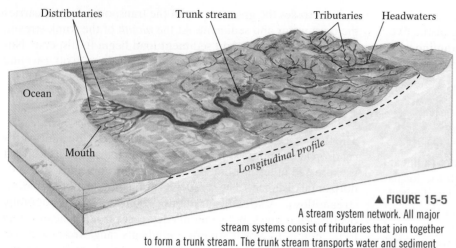

▲ **FIGURE 15-5**
A stream system network. All major stream systems consist of tributaries that join together to form a trunk stream. The trunk stream transports water and sediment downstream, until it splits into a network of smaller distributaries that deliver the water and sediment to the sea.

A stream's velocity is the distance that a stream's water travels in a given amount of time. It is usually expressed in meters (or feet) per second. Slow-moving streams have velocities of less than 0.27 m/s (1.0 ft/s); swift-flowing streams can have velocities exceeding 10 m/s (33 ft/s).

The velocity of the water varies from place to place *within* a stream. If no rocks and other barriers disturb the stream's course, its *local* velocity will depend partly on the gradient of the channel and partly on the portion of the channel in which the water flows. Velocity is slowest at the sides and bottom of the stream because of friction between the water and the channel. Velocity is great-

est in the center of a stream, in a straight segment of the channel, about halfway down from the water's surface. There the water encounters the least amount of friction from either the bed or the atmosphere. Where a stream curves, velocity is greatest at the outside of the curve and slowest at the inside (Fig. 15-6).

A stream's velocity increases wherever its channel narrows. This happens because a narrower, deeper channel would reduce the amount of contact between the stream's water and its bed, and thus there would be less frictional resistance to flow. With less resistance, the stream flows faster.

The texture of a stream's bed also determines its velocity. Upstream in the headwaters, boulders in the channel bed create significant frictional drag, sending turbulent waters swirling upward, downward, and sideways. Even though it's flowing swiftly and turbulently, the stream expends most of its energy against its bed, and thus its actual *downstream* velocity is relatively low. Farther down the stream, actual *downstream* velocity generally increases *despite a decrease in gradient*, for two reasons: The volume of water grows as tributaries join the trunk stream, and the stream flows over a smoother bed of sand, silt, and clay, which minimizes frictional resistance to flow.

Suppose you wanted to compare two very different streams—such as the mighty Mississippi and a neighborhood creek. What could these two streams possibly have in common? When geologists compare streams, they often do so *quantitatively* by

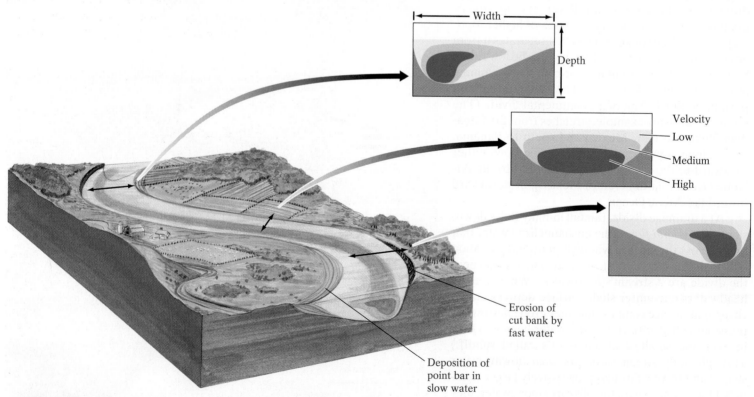

▲ **FIGURE 15-6** Variations in streams velocity along its course. The fastest flow is in the middle of the stream about halfway down from the surface—the area of minimal friction with the stream's banks and streambed. Within a bend in a stream's course, faster flow is along the outer portion of the curve; slower flow occurs along the inner portion of the bend.

measuring the **stream discharge,** the volume of water passing a given point on the stream bank per unit of time. All flowing streams have a measurable discharge usually expressed in cubic meters (or feet) of water per second. Discharge can be calculated when the stream's width, depth, and velocity are known. Multiplying the width and depth offers an estimate of the stream's cross-sectional area. The formula for a hypothetical *square* stream channel is:

$$Discharge = width \times maximum\ depth \times velocity.$$

Because stream channels are rarely square, a corrected estimate for a more realistic stream cross section is:

$$Discharge = (width \times depth \times velocity)/2.$$

A stream's discharge is controlled principally by the size of its drainage basin and the amount of precipitation that the basin receives. The discharge of minor tributaries can be as small as 5 to 10 m^3 (180 to 360 ft^3) per second. In contrast, the discharge of the largest river in North America, the Mississippi, is about 18,000 m^3 (600,000 ft^3) per second. The discharge of the largest river on Earth, the Amazon of South America, which drains an area equal to nearly 75% of the continental United States, is roughly 200,000 m^3 (7 million ft^3) per second. This discharge equals one-fifth of the Earth's entire freshwater stream flow. To appreciate the size of this massive amount of water, consider that *one day's* discharge from the Amazon would satisfy New York City's freshwater needs for *more than five years.*

A stream's daily discharge is also influenced by such climatic conditions as the amount and timing of precipitation and the volume of snowmelt. The ability of local soils to absorb water affects discharge levels as well.

The Geological Work of Streams

Streams carry only about one-millionth of the Earth's water. Nevertheless, because they can erode, carry, and deposit sediment, they are the planet's most important geological agent of surface change. A stream cuts down through uplifted land toward its **base level,** the lowest level to which it can erode its channel. For most streams, the *ultimate* base level is sea level. Streams that flow across land areas *below* sea level—such as the Jordan River in the Middle East—may actually have base levels *below* sea level. Streams may also reach *temporary* base levels, such as when they cut down to an extremely hard layer of bedrock that halts further downcutting for a considerable length of time.

When global sea levels fall, coastal streams respond quickly by cutting downward to their newly lowered base levels. This situation typically arises during periods of glaciation, when a great volume of seawater converts to continental ice. During periods of warming and glacial melting, sea level rises and streams deposit their sediments farther inland, where they meet the rising sea.

Graded Streams

Streams are among the most dynamic of geological agents; they respond *immediately* to changes in their environment. When a drenching storm dumps 25 cm (10 in.) of rain on a drainage basin, the streams in that region rise quickly, flow more rapidly, erode greater volumes of sediment, and deposit that sediment downstream wherever stream flow becomes blocked or slowed. In a similar fashion, streams respond immediately to local tectonic events. If, for example, normal faulting occurs across a stream's course and the *downstream* fault block drops, a waterfall immediately forms. The stream then uses the increased energy of the falling water to erode away the newly uplifted land (Fig. 15-7).

In a hypothetical unchanging environment, a stream would maintain its gradient at precisely the angle that it needs to flow just swiftly enough to transport all the sediment supplied to it from its drainage basin. Thus, there would be little net erosion or deposition. In reality, however, a stream's environment is always changing. Thus the stream's gradient is constantly adjusting to achieve a balance between erosion and deposition. A stream in such a state of *dynamic* equilibrium is called a **graded stream.** The equilibrium of any graded stream is merely temporary, lasting only until the next change in its environment. For example, a sudden increase in sediment load, such as from an inflowing volcanic mudflow or the collapse of a stream bank, may disturb a stream's equilibrium. Similarly, a rapid increase in discharge, such as from a heavy rainstorm, will force the stream to adjust. A graded stream responds *instantaneously* to any such change in sediment load, discharge, or base level so as to reduce the change's effects.

Let's say that several loads of coarse gravel are dumped into a graded stream that previously transported only fine sands and silt. The stream initially cannot transport this new material; it's flowing too slowly. The coarse sediment settles onto the stream

Fault

▲ **FIGURE 15-7** Streams flowing over and eroding a newly formed fault scarp that cuts across an alluvial fan in Hanpan Canyon in the Panamint Mountains of Death Valley, California. The fault scarp is the sharp linear break on the surface in the foreground of this photo.

bed, however, immediately steepening the channel gradient, in a process known as **aggradation** (Fig. 15-8a). As the new, steeper gradient increases the stream's velocity, the added sediment is eroded until the graded equilibrium is restored. The streams of central California aggraded in this fashion during the Gold Rush of the 19th century, when gold mining operations dumped millions of tons of mining waste into them.

What happens when the sediment load of a graded stream suddenly decreases? Look at Figure 15-8b and note how sediment accumulates in the still water behind the dam in part 2. When the water is released on the downstream side of the dam, it has no sediment load and thus the stream now flows on a steeper gradient and with a resulting swifter flow than is needed to carry the new reduced load. As a result, it will scour its bed, eroding more sediment and thus reducing its gradient. This process is called **degradation.** Degradation often occurs downstream from a dam. Construction of the Hoover Dam on the Lower Colorado River caused the river to degrade its slope downstream as far south as Yuma, Arizona, a distance of 560 km (350 mi) from the dam. To prevent scoured slopes from undermining dams, engineers must design and build structures that protect the dam from degradation.

Stream Erosion

The principal result of the erosive power of streams is the creation of valleys, the Earth's most common continental landform. The composition of a valley's rocks and sediments strongly af-

fects the rate at which it erodes. It could take thousands or even millions of years for a stream to cut a valley through a granitic batholith, whereas a single powerful flood can erode a sizeable valley in loose, uncemented sands.

Streams cut vertically downward through uplifted terrain like a saw cutting through a board. Ideally, they should form a chasm with near-vertical walls, so why do most river valleys have a distinctive V-shape (Fig. 15-9)? As the river carves out the valley, overland flow and mass movement remove loosened material from the valley's steep slopes. For example, although downcutting by the Colorado River largely produced the 1810 m (5940 ft) of *vertical* relief in the Grand Canyon (*relief* is the difference in elevation between an area's highest and lowest points), overland flow and mass movement have eroded most of the 21-km (13-mi) width *across* the canyon.

In temperate and humid regions, chemical weathering contributes to the development of V-shaped valleys by breaking down exposed bedrock. In such climates, stream valleys typically have gradually sloping walls. In arid regions, only a small amount of chemical weathering takes place, and thus, with little loose sediment to remove, valleys in those areas are typically steep-walled.

A stream's erosive power increases as it flows faster. The increase in the erosion rate is approximately equal to the *square* of the increase in velocity. Thus, if a stream's velocity doubles, its erosive power increases by a factor of four ($2^2 = 4\times$). For this reason, the erosive power of a flooding stream—with its sharp rise in velocity—increases dramatically.

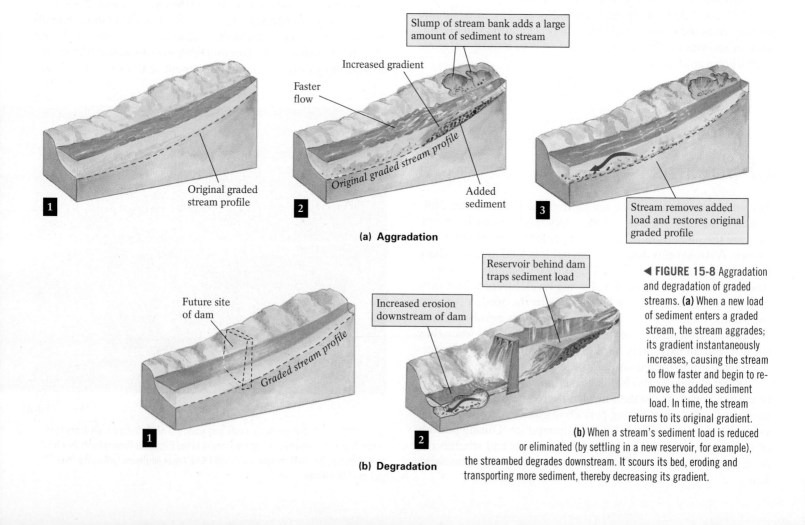

Slump of stream bank adds a large amount of sediment to stream

Increased gradient

Faster flow

Original graded stream profile

Original graded stream profile

Added sediment

Stream removes added load and restores original graded profile

(a) Aggradation

Future site of dam

Graded stream profile

Increased erosion downstream of dam

Reservoir behind dam traps sediment load

◀ **FIGURE 15-8** Aggradation and degradation of graded streams. **(a)** When a new load of sediment enters a graded stream, the stream aggrades; its gradient instantaneously increases, causing the stream to flow faster and begin to remove the added sediment load. In time, the stream returns to its original gradient. **(b)** When a stream's sediment load is reduced or eliminated (by settling in a new reservoir, for example), the streambed degrades downstream. It scours its bed, eroding and transporting more sediment, thereby decreasing its gradient.

(b) Degradation

▲ FIGURE 15-9 The Yellowstone River in Wyoming cuts down rapidly and vertically through the uplifted rocks of its canyon. Mass movement in the upper portion of the canyon helps produce the valley's characteristic V shape.

Processes of Stream Erosion

Streams erode their channels by abrasion, hydraulic lifting, and dissolution. **Abrasion** is the scouring of a stream's bed by transported particles. Fine particles suspended in the water constantly scrape at the bed's surface. Some large pebbles rotate in swirling eddies, carving circular depressions called *potholes* into the bedrock (Fig. 15-10); others bounce against the underlying bedrock. Larger rocks may roll along the bed, abrading the sediment and bedrock there. Erosion by abrasion is most efficient in swift-flowing floodwaters carrying a load of coarse sediment and rocks.

Hydraulic lifting—erosion by water pressure—occurs when turbulent streamflow dislodges sediment grains and loosens large chunks of rock from the stream bed. Like abrasion, hydraulic lifting is most active during high-velocity floods. To illustrate the power of hydraulic lifting, consider that during the 1923 flood along the Wasatch Mountain front of central Utah, the turbulent floodwaters lifted 90-ton boulders and transported them more than 8 km (5 mi) downstream.

When a stream flows across soluble bedrock (such as limestones, dolostones, and evaporites), some of the bedrock may dissolve, yet another form of stream erosion. Every year, *dissolution* removes an estimated 3.5 million tons of rock from the continents and carries it out to sea.

Drainage Patterns

Streams erode their networks of tributary valleys in distinctive **drainage patterns.** The patterns of stream channels are determined largely by topography, composition, and geological structure of the terrain. A geologist can learn much about a landscape simply by looking at its stream drainage pattern from an airplane or on aerial photographs, or by examining the web of fine blue stream lines that form the pattern on a topographic map.

When the rocks or sediments that underlie a drainage basin are uniform in composition and undeformed, streams tend to erode them in a *dendritic* (from the Greek *dendron,* or "tree") drainage pattern, resembling the branches of a mature tree. In general, dendritic patterns develop on relatively flat sedimentary rocks, newly exposed coastal muds, and newly exposed unjointed, massive plutonic igneous or metamorphic rocks (Fig. 15-11a).

In a landscape with topographic peaks, such as young volcanic cones, lava domes, and structural domes (discussed in Chapter 9), a network of streams typically drains from a central high area in a pattern resembling the spokes of a wheel. This pattern is called *radial* drainage (Fig. 15-11b).

A *rectangular* drainage pattern, which looks like a grid of city streets (Fig. 15-11c), forms when stream erosion enlarges systematic joints, fractures, and faults in bedrock. Because these features usually develop in perpendicular sets, the stream pattern shows many right-angle bends.

A *trellis* drainage pattern develops where narrow valleys, underlain by soft, easily eroded rocks, are separated by parallel ridges that consist of resistant durable rocks. Short, steep tributaries, resembling the parallel slats of a garden trellis, drain down the ridge slopes perpendicular to the main stream valleys (Fig. 15-11d).

▲ FIGURE 15-10 A stream and potholes in Jasper National Park, Canada. The potholes were abraded in the solid bedrock by stones swirling in the stream. Potholes can be as deep as 5 m (17 ft) and as large as 2 m (7 ft) in diameter. The stones that ground the potholes typically lie at their bottoms.

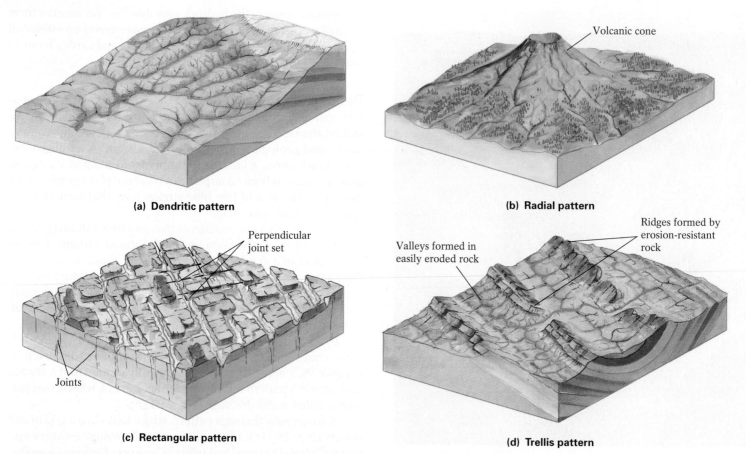

(a) Dendritic pattern

(b) Radial pattern

Volcanic cone

Perpendicular joint set

Joints

(c) Rectangular pattern

Ridges formed by erosion-resistant rock

Valleys formed in easily eroded rock

(d) Trellis pattern

▲ **FIGURE 15-11** Four types of drainage patterns, produced by stream erosion of various bedrock types and under various structural conditions. **(a)** Dendritic drainage patterns often develop on undeformed sedimentary rock of uniform composition. **(b)** Radial drainage patterns usually form on volcanic cones of homogeneous composition. **(c)** Rectangular drainage patterns commonly appear where bedrock is cut by perpendicular joints. **(d)** The best place to look for trellis drainage is in an area with long parallel folds of sedimentary rock, such as the Ridge and Valley province of the Appalachians.

Stream Piracy The slopes of the two sides of a ridge—their steepness, their geology, and even their climate—usually differ. A stream on the steeper slope of a divide tends to flow more swiftly *and erosively* than one on the opposing gentler slope. One side of a divide may also be eroded more vigorously if it contains less-resistant surface rocks, sediments, or soils, perhaps localized by structural weaknesses in the bedrock. Similarly, if one side of a divide receives more rain than the other, its streams will have greater, more erosive discharges.

When two streams on opposite slopes erode headward toward a common drainage divide, the stream that erodes more vigorously will reach the crest of the ridge first and may then erode a channel *through the divide itself.* It may then capture the headwaters of the stream on the other side. This process, colorfully named *stream piracy,* (or, stream capture) is illustrated in Figure 15-12. A stream that has lost its headwaters in this fashion is described as *beheaded.* A pirating stream can create a *water gap,* a steep-walled rocky gorge cut through a ridge or mountain range. (If the pirating stream later becomes diverted from its course and abandons this gorge, the water gap becomes a dry notch in the ridge, a landform known as a *wind gap.*)

Superposed and Antecedent Streams A *superposed stream* cuts downward through a number of rock layers of dif-

ferent compositions and structures *yet maintains the drainage pattern established at the original surface.* Picture a dendritic pattern of streams that becomes established on top of flat, undeformed sedimentary rock overlying folded subsurface rocks or masses of plutonic igneous rock. As the sedimentary rock erodes away, its dendritic drainage pattern may be superposed on the layers underneath, which are quite different. A good example of a superposed stream drainage can be found in Wind River Canyon across the Owl Creek Mountains of Wyoming (Fig. 15-13).

Antecedent streams are older streams that cut through recently uplifted landscapes. If uplift occurs faster than a stream can cut down through the rising bedrock, a topographic barrier forms that simply diverts the stream to another course around the barrier. If, however, uplift is slow and steady, and if the rocks underlying the stream are weak (perhaps fractured and jointed) and highly erodible, stream-cutting may match its pace, allowing the stream to maintain its original course and cut a water gap through the rising ridge or mountain range (Fig. 15-14).

Channel Patterns Stream channels can be *straight, braided,* or *meandering.* A single stream may contain segments of all three shapes between its headwaters and its mouth. Straight channels are rare, but can occur where linear fractures, joints, and faults in resistant bedrock control a stream's path. Straight segments also

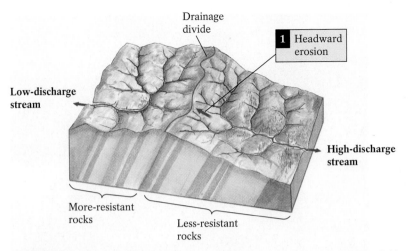

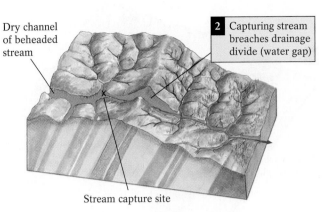

FIGURE 15-12 Stream piracy. When two streams erode headward toward a drainage divide, the stream that erodes more vigorously may breach the divide and capture, or "pirate," the headwaters of the other. The stream that loses its headwaters is "beheaded." Photo: A beheaded stream in the Bighorn Mountains of Wyoming.

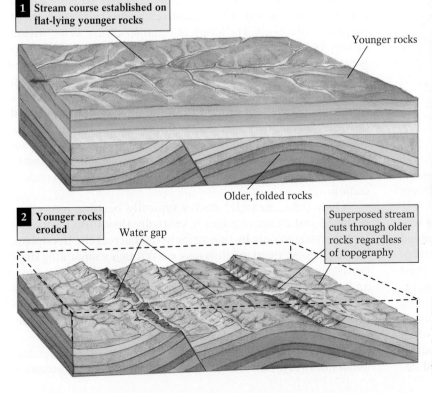

FIGURE 15-13 Superposed streams. A drainage pattern that seems inconsistent with the surrounding topography, such as a dendritic pattern cutting across a folded landscape, may indicate that the stream was established on a different, overlying layer of rock that has since eroded away. Photo: A superposed stream drainage, Wind River Canyon across the Owl Creek Mountains of Wyoming.

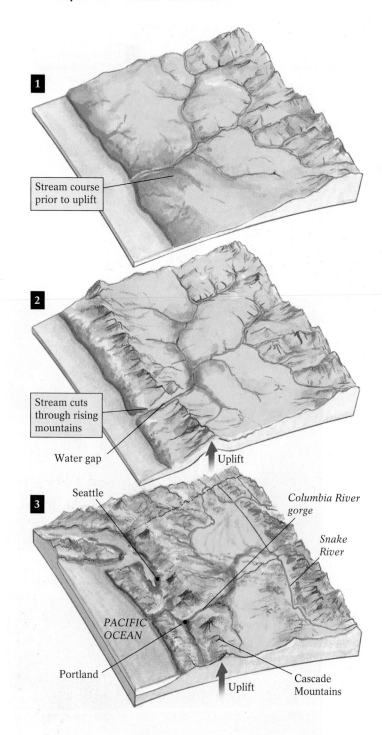

1 Stream course prior to uplift

2 Stream cuts through rising mountains

Water gap

Uplift

3 Seattle

Columbia River gorge

Snake River

PACIFIC OCEAN

Portland

Uplift

Cascade Mountains

▲ FIGURE 15-14 Antecedent streams. The Columbia River gorge, which cuts across the Cascades mountain range and forms the state line between Washington and Oregon, is a typical antecedent stream valley.

creases abruptly (such as at the foot of a mountain), reducing the stream's ability to carry its large sediment load. Braided streams also form where a stream's discharge drops sharply, such as in deserts, where much of a stream's water evaporates or seeps into the sand.

The flat land immediately adjacent to a stream channel is submerged whenever the stream overflows its banks and is called its **floodplain.** A **meandering stream** (from the Latin *maendere,* meaning "to wander"), the most common type of stream-channel pattern, winds its way across these relatively flat floodplains in evenly spaced loops. The loops, or *meanders,* form as the stream erodes its banks; erosion and deposition then combine to exaggerate these curves (Fig. 15-16). Because flow velocity increases toward the outside bank of a curve, the water is most turbulent there and erodes the most sediment in that portion of the stream's course. The sediment eroded from the outside bank is carried downstream and deposited on the *inside* bank of the next curve, where velocity and turbulence are minimal. In this way, a meandering stream transfers sediment from its erosional outer banks to its depositional inner banks, with very little net erosion or deposition.

Over time, meander curves typically become more pronounced and closer together. Eventually they form a series of loops separated only by thin strips of land across the floodplain. During a flood, overland flow may cut through a separating strip, bypassing and thus abandoning an entire loop. The isolated meander loop thus becomes a crescent-shaped body of standing water called an **oxbow lake,** so named because its shape resembles the curved collar worn by a farmer's ox. Subsequent flood deposits may fill the oxbow lakes, forming *meander scars.*

From 1765 to 1932, the Mississippi River cut off nineteen meanders between Cairo, Illinois, and Baton Rouge, Louisiana—

form in areas of active uplift, where the headwaters of a stream flow rapidly and directly down steep slopes.

Networks of converging and diverging stream channels separated by narrow sand and gravel bars are known as **braided streams** (Fig. 15-15). They develop where streams with exceptionally high sediment loads deposit excess sediment in their channels. These deposits build up during periods of high water flow until they reach the water surface. Lowered water levels then expose these masses of sediment, leaving them standing above the stream as islands that divert the flow. Braided streams often occur where a stream's banks consist of loose, highly erodible sediment. They also form where the slope of a streambed de-

▲ **FIGURE 15-15** A braided channel of the Toklat River, in Denali National Park, Alaska.

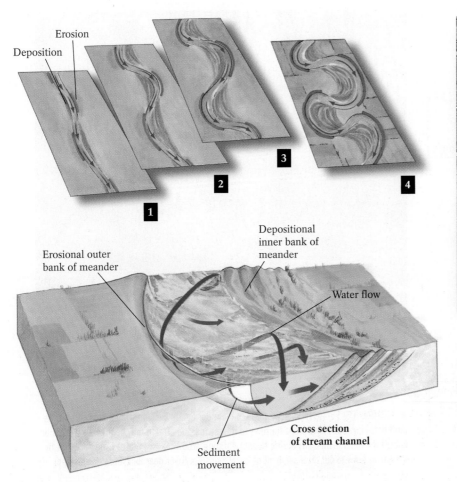

▲ **FIGURE 15-16** The evolution of meandering streams. As a somewhat curving stream flows, the high-velocity water at the outside of each bend produces a cross-channel, corkscrewlike flow that erodes the outer bank. The eroded material is deposited downstream at the inner bank of the next bend. Meander bends typically grow more pronounced as progressively more sediment is removed from their outside banks and added to their inside banks. Photo: The meandering Little Bear River, in Utah.

equal to about 400 km (250 mi) of the river's course. During the Civil War, General Ulysses S. Grant, commander of the Union Army, tried unsuccessfully to *create* a cutoff artificially so that Union forces traveling by river could avoid the Confederate guns at Vicksburg, Mississippi. Nature finally achieved that task in 1876, the year of the United States' 100th birthday, when the so-called "Centennial cutoff" separated Vicksburg from the river's active channel (Fig. 15-17). So, with all of these cutoffs, is the course of the Mississippi River shorter today? Before the answer, check out what the 19th-century humorist Mark Twain had to say about the Mississippi and this cutoff process in his work, *Life on the Mississippi,* published in 1883.

In the space of one hundred and seventy-six years the Lower Mississippi has shortened itself two hundred and forty-two miles. That is an average of a trifle over one mile and a third per year. Therefore, any calm person, who is not blind or idiotic, can see that in the Old Oolitic Silurian Period, just a million years ago next November, the Lower Mississippi River was upwards of one million three hundred thousand miles long, and stuck out over the Gulf of Mexico like a fishing-rod. And by the same token any person can see that seven hundred and forty-two years from now the Lower Mississippi will be only a mile and three-quarters long, and Cairo and New Orleans will have joined their streets together, and be plodding comfortably along under a single mayor and a mutual board of alderman. There is something fascinating about science. One gets such wholesale returns of conjecture out of such a trifling investment of fact.

Notwithstanding what our friend Twain imagined, in truth, despite the presence of so many cutoffs, the Mississippi course has *not* been shortened appreciably. Existing meander curves have expanded, and once-straight cutoff sections have themselves begun to meander and lengthen. So, in the long run, the cutoff process rarely affects the length of a river very much.

In tectonically active areas where the Earth's surface is rising, a stream's meandering pattern may become deeply etched into the landscape. As the surface rises, the meandering channel, with all its twists and turns, cuts vertically into the bedrock. This down-cutting produces **incised meanders**—a meandering steep-sided valley pattern with virtually no surrounding floodplain (Fig. 15-18).

Waterfalls and Rapids
Waterfalls and rapids form at sharp step-like breaks in the land surface along a stream's course. Typically they occur where erosion has removed softer sections of bedrock, leaving behind a ledge of more resistant rock as a step in the stream's profile. Waterfalls and rapids also form, sometimes instantaneously, where faulting lowers or raises a portion of the stream's profile. Whitewater rapids commonly represent the remains of waterfalls whose steps were eroded into irregular rocky beds over which water rushes turbulently.

A waterfall lasts only as long as the conditions that created it last. The energetic plunging water of a waterfall erodes a deep pool at its base, called a *plunge pool,* that undermines the step from which the water falls. As the step is eroded, the waterfall

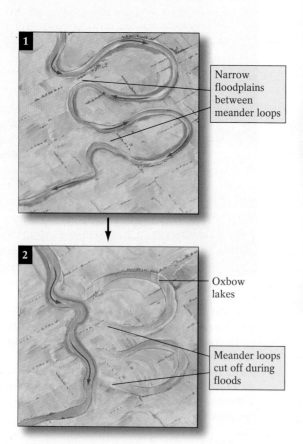

Narrow floodplains between meander loops

Oxbow lakes

Meander loops cut off during floods

▲ **FIGURE 15-17** (left) When a stream's meander bends become so pronounced that the land between successive loops narrows to thin strips, the stream may completely bypass or cut off the bend (especially during floods), forming oxbow lakes. (above) Failed attempt by the Union Army to cut off a segment of the Mississippi River near Vicksburg, Mississippi, during the American Civil War.

◀ **FIGURE 15-18** A meandering-stream pattern will be incised into a tectonically uplifted landscape. Here the Colorado River above Salt Creek Canyon, Canyonlands National Park, Utah, is deeply incised into the rising Colorado Plateau. The deep incision occurs where the intersection of bedrock fractures makes the rocks more erodible.

ans into the soft sediments of the Atlantic coastal plain. The hard rocks form the step.

Best known of the eastern falls is, of course, Niagara. This favorite of honeymooners and tourists alike is roughly 55 m (176 ft) high and 670 m (2200 ft) wide (Fig. 15-20). It began to form about 12,000 years ago, when the last great North American ice sheet retreated north of the Great Lakes region. Its withdrawal uncovered a ridge of resistant dolostone. Since then, the falls has retreated southward 11 km (7 mi) from its point of origin at Lake Ontario. Billions of liters of water annually thunder through the 50-km (30-mi)-long gorge of the Niagara River between Lake Erie and Lake Ontario, plunge over the falls, and erode the shale bed below the resistant cap of dolostone, undercutting the falls.

The future of Niagara Falls is uncertain. At its past rate of retreat, averaging about 1 m (3 ft) per year, Niagara Falls would have eroded the remaining 30 km (20 mi) of dolostone all the way to Lake Erie in 30,000 years. The United States and Canada, however, divert approximately 75% of the river's discharge into four large tunnels to turn giant turbines and generate thousands of megawatts of hydroelectric power, enough to supply more than 2 million homes. The falls has also been "closed for repairs," its flow shut off so that engineers could install reinforcement materials to stabilize the falls and slow its retreat. These measures have slowed the retreat of the falls to less than 0.5 m (less than 2 ft) per year, possibly doubling Niagara's remaining life.

Stream Terraces   When a stream's discharge increases over a long period of time (perhaps because of climate change) or when its base level drops (perhaps because of a global sea-level change or local tectonic movement), it typically erodes its channel to a lower depth. By doing so, its floodplain is left high and dry above the new channel, forming a gently sloping topographic bench known as a **stream terrace.** A terrace usually stands high enough above a stream that even the most extreme flood cannot touch it. You can usually see several terraces, like a small flight of stairs within a single stream valley. Each terrace is a remnant of a once continuous floodplain surface that spanned the entire valley before being cut down to progressively lower levels (Fig. 15-21 on page 480). Geologists typically date the origin of a sequence of terraces by cosmogenic-isotope dating (discussed in Chapter 8).

migrates upstream. Ultimately, as shown in Figure 15-19, when the step is finally removed, the stream reestablishes a graded profile. One Montana waterfall, created by fault motion during Yellowstone's Madison Canyon earthquake of August 1959, lasted barely a year. The faulting forced a tributary to the Madison River—Cabin Creek—to flow over a fresh 3-m (10-ft)-high fault scarp. By June 1960, the scarp had almost completely eroded, and a small set of rapids replaced the waterfall. Five years later no trace of the rapids or the waterfall could be found.

Thousands of waterfalls and rapids are scattered across North America, from picture-postcard falls such as Yosemite Falls in California's Sierra Nevada, to countless small falls and rapids that dot the flood-swollen springtime streams of the eastern United States. Many of the eastern falls occur where streams descend from the hard igneous and metamorphic rocks of the Appalachi-

Stream Transport

The world's streams deliver about 45 *trillion* cubic meters of water to the sea every year, along with 9 to 10 billion tons of sediment. The Mississippi River alone carries more than 1 million tons of sediment *daily* to the Gulf of Mexico. The amount and type of sediment that a stream transports are largely determined by the

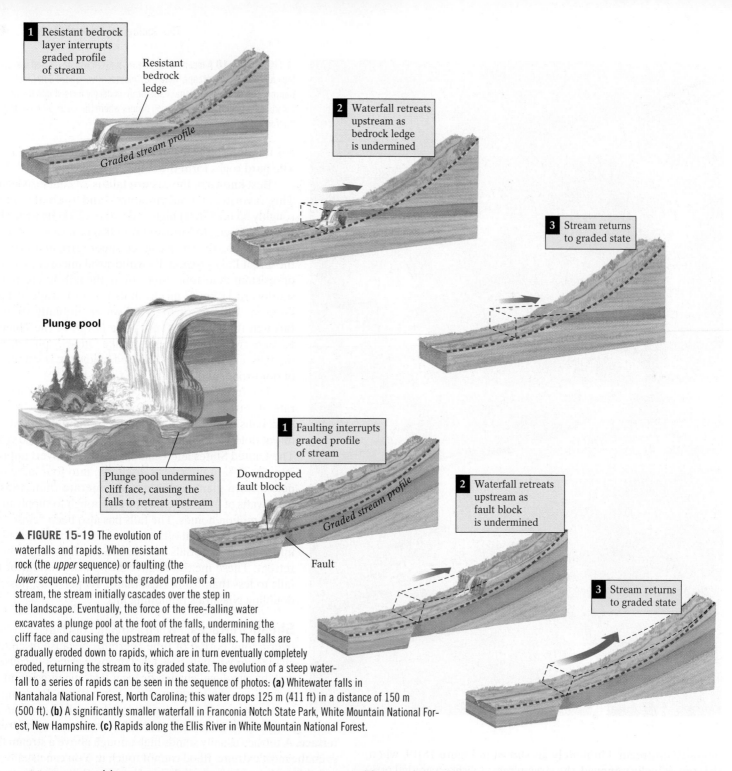

1 Resistant bedrock layer interrupts graded profile of stream

Resistant bedrock ledge

Graded stream profile

2 Waterfall retreats upstream as bedrock ledge is undermined

3 Stream returns to graded state

Plunge pool

Plunge pool undermines cliff face, causing the falls to retreat upstream

1 Faulting interrupts graded profile of stream

Downdropped fault block

Graded stream profile

Fault

2 Waterfall retreats upstream as fault block is undermined

3 Stream returns to graded state

▲ **FIGURE 15-19** The evolution of waterfalls and rapids. When resistant rock (the *upper* sequence) or faulting (the *lower* sequence) interrupts the graded profile of a stream, the stream initially cascades over the step in the landscape. Eventually, the force of the free-falling water excavates a plunge pool at the foot of the falls, undermining the cliff face and causing the upstream retreat of the falls. The falls are gradually eroded down to rapids, which are in turn eventually completely eroded, returning the stream to its graded state. The evolution of a steep water- fall to a series of rapids can be seen in the sequence of photos: **(a)** Whitewater falls in Nantahala National Forest, North Carolina; this water drops 125 m (411 ft) in a distance of 150 m (500 ft). **(b)** A significantly smaller waterfall in Franconia Notch State Park, White Mountain National For- est, New Hampshire. **(c)** Rapids along the Ellis River in White Mountain National Forest.

(a)   (b)   (c)

▲ **FIGURE 15-20** Niagara Falls at the New York–Ontario border.

flow's velocity, the composition and texture of its sediment, and the characteristics of the bedrock that it crosses. High-velocity streams can erode and transport large boulders. High velocity is also required to pick up small, flat clay particles and mica flakes, because electrical charges on the surfaces of these sediment particles cause them to cling to the stream bed. Once in the streamflow, however, these particles can travel long distances, often remaining suspended in the water all the way to the sea.

The maximum amount of sediment that a stream can transport *at a given discharge* is defined as its **capacity.** Capacity is expressed as the *volume* of sediment passing a certain point on the channel bank in a given amount of time. Capacity tends to increase as discharge increases; the more water flowing in the channel per second, the greater the volume of sediment that can be transported in that time.

The diameter of the largest particle that a stream can transport describes its **competence.** Competence is proportional to the *square of a stream's velocity:* The greater a stream's velocity, the larger the particles that it can transport. Thus, when velocity doubles—for example, in a flood—competence increases by a factor of four. This relationship explains why a stream that ordinarily carries only fine gravel can sweep large boulders downstream during a flood. During a 1933 flood in California, the Tehachapi River carried steam locomotives from the Santa Fe Railroad railyards hundreds of meters downstream and then buried them under tons of bouldery stream gravel.

Depending on a stream's discharge, its velocity, and local geological conditions (such as the composition of the stream banks), four combinations of capacity and competence are possible:

- *High capacity, high competence.* A rapidly flowing, sediment-laden stream transporting large particles, such as in a flood.
- *High capacity, low competence.* A slow-moving, sediment-laden stream, such as one flowing in a broad, downstream valley through a channel composed of easily eroded sediment.
- *Low capacity, high competence.* A swift-flowing, relatively sediment-free stream transporting large boulders along its bed, such as in a stream's mountainous headwaters, particularly in a region of hard, erosion-resistant bedrock.
- *Low capacity, low competence.* A slow-moving stream on a relatively gentle slope flowing through resistant, relatively unweathered bedrock.

Sediment Load   Streams transport sediment in different ways, depending on the particle size involved (Fig. 15-22). Very fine solid particles are carried up in the water column as **suspended load.** Coarse particles that move along the stream bottom form the **bed load.** Sediment carried invisibly as dissolved ions in the water makes up the **dissolved load.**

Most of the world's stream-carried sediment travels as suspended load. These fine particles generally drift along in a turbulent, fast-moving stream, like confetti caught in swirling air currents at a parade. The Colorado River derives its characteristic red color from suspended particles of silt and sand eroded from red formations in the Grand Canyon and other red rocks in Arizona.

More than 10% of the world's stream sediment is transported by the remarkable Huang He (Yellow River) of central China. In fact, half of the Huang He's flow volume consists of sediment. During floods, sediment may represent as much as 90% of the river, reducing its flow rate so much that it looks like flowing molasses. Much of this sediment comes from the 200-m (650-ft)-thick blanket of yellow, windblown silt through which the river flows. Little vegetation anchors the loose silt in this extremely arid region, allowing periodic storms to wash huge volumes of sediment from the adjacent slopes. The Huang He's annual sediment load could build a wall 25 m (80 ft) high and 25 m wide that would encircle the planet.

Although a stream's velocity and discharge largely determine the volume of its suspended load, particle shape and density also play a role. Both flat and low-density particles are likely to remain suspended, somewhat like a Frisbee floating on a cushion of air. A flat particle of low-density muscovite mica, for example, can travel thousands of kilometers without dropping onto the stream bed. Denser rounder particles settle to the stream bed more quickly. A gold particle, for example, whose density is 19 times that of water (density = 19.3 g/cm^3), falls to the stream bed much more rapidly than a grain of quartz sand (density = 2.65 g/cm^3).

Unlike its suspended load, which moves constantly, a stream's *bed load* moves only when the water's velocity is rapid enough to bounce, lift, or roll particles. The coarse material that makes up a stream's bed load moves in several ways: The sandy portion skips along the bed with a bouncing motion known as *saltation* (from the Latin *saltare,* meaning "to jump," from the same root as somer*sault*). Saltating sand grains rise from the bed

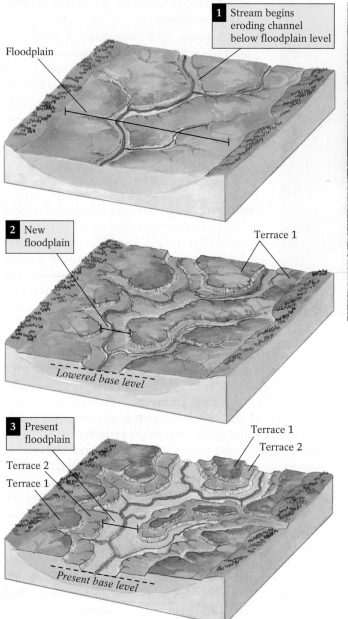

1 Stream begins eroding channel below floodplain level

Floodplain

2 New floodplain

Terrace 1

Lowered base level

3 Present floodplain

Terrace 1
Terrace 2

Terrace 2
Terrace 1

Terrace 1
Terrace 2

Present base level

▲ **FIGURE 15-21** The creation of stream terraces. **(1)** A terrace forms when a stream erodes its channel below the level of its floodplain, either because its discharge has increased or because its base level has fallen. **(2, 3)** When a stream undergoes a series of such changes, it may create a flight of terraces in a single valley. Photo: Terraces along the Strathmore River, in Sutherland, Scotland.

by hydraulic lifting or by collisions with other saltating grains. Once bounced into the stream, these grains are carried upward and forward by the stream's turbulence and velocity until the pull of gravity returns them to the bed. Larger, heavier cobbles and boulders usually move by a rolling motion called *traction*. The largest boulders only move during major floods. These rocks often feel slippery because they settle in and become covered with algae; few algae would grow if the boulders were in constant motion.

Natural waters are never pure H_2O; instead, they contain ions dissolved from surrounding bedrock. Streams carry these ions as their *dissolved load*. The most common dissolved ions are bicarbonate, calcium, sodium, and magnesium, although smaller amounts of chlorides, sulfates, and silica may be present as well. The amounts and kinds of dissolved ions in stream water depend on the factors that control chemical weathering—such as climate,

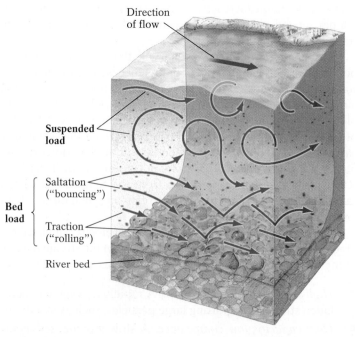

Direction of flow

Suspended load

Saltation ("bouncing")

Bed load

Traction ("rolling")

River bed

▲ **FIGURE 15-22** Sediment distribution and movement within a stream.

vegetation, and bedrock composition—and the acidity of the water. Velocity has no effect on a stream's ability to carry dissolved load, although it might influence the rate at which material enters solution. In general, higher dissolved loads are found in streams that flow through warm, moist regions of low relief, lush vegetation, and outcrops of soluble bedrock, such as limestones and evaporites.

The volume of a stream's dissolved load also depends on the drainage basin's topography. The steepness of the landscape determines whether precipitation runs off rapidly as sheet flow and stream flow (on steep slopes) or instead seeps into the groundwater system (on gentle slopes or plains). Dissolved ions are most abundant in local groundwater whose slow movement through soluble bedrock promotes dissolution more effectively than contact of a fast-flowing stream with its bed.

Stream Deposition

When the downward pull of gravity on stream particles overcomes the upward motion of stream velocity and turbulence, the particles become deposited on the stream bed, with the heaviest particles settling out first. Deposited stream sediments are described collectively as **alluvium** (a word of French and Latin derivation, meaning "washed over"). Some of a stream's alluvium drops on land in its channel and on its floodplain, but most of it settles out where the stream enters the sea. In this section we describe the various ways that streams deposit their sediment loads.

In-Channel Deposition

During normal (non-flooding) flow, a stream with a large amount of coarse bed-load sediment typically deposits some of its load *in its channel* as midchannel bars. As sand and gravel accumulate, the stream is diverted around the channel bars and becomes braided (see Fig. 15-15). The bars become increasingly resistant to erosion as vegetation begins to grow on them and plant roots bind up the bars' loose sediments.

In-channel deposits may also take the form of **point bars,** accumulations of the sediment scoured from the outside banks of meanders or any bend along the stream's course. These sediments are carried downstream and then dropped along the stream's less-turbulent inner banks (see Fig. 15-16). As a point bar grows, the water flowing over it becomes shallower, thus increasing the friction between the water and the surface of the bar. This increased friction in turn decreases stream velocity even further, promoting even more deposition. Figure 15-23 shows an example of point bars in Arizona. Because meandering streams drop their heaviest particles there, point bars are excellent places to search for such heavy minerals as gold, platinum, and silver. The gold deposits in the rivers of the Yukon and Alaska, which drew thousands during the gold rushes of the nineteenth century, were formed either in this manner or by the settling out of the heavy flakes within the channel-bar sediments of braided streams.

Floodplain Deposition

When a rising stream overflows its banks, it spills onto the surrounding floodplain. As noted earlier, because of its increased velocity, floodwater can carry a large volume of sediment. As this water escapes from the channel and spreads over the relatively flat floodplain, its velocity decreases and the floodwaters drop their sediment load. Floodplain depo-

▲ **FIGURE 15-23** Point bars on the Colorado River, Arizona.

sition also occurs as the flood wanes and its stagnant floodwater drops its fine suspended sediment. Some typical features associated with floodplains are illustrated in Figure 15-24.

Levees (from the French *lever,* meaning "to raise") are ridges of sediment deposited on both banks of a flooding stream. As floodwaters decrease in velocity, turbulence, and competence, they begin to drop their sediment. The coarser sediment accumulates adjacent to the channel and forms the levees; the finer sediment settles farther out on the floodplain, enriching agriculture for miles around with fresh nutrient-rich silt. The levees, whose heights increase with each flood, tend to be the highest points on a floodplain. If they reach sufficient heights, they can sometimes prevent lower-volume flows from overflowing the channel banks. Between floods, trees that prefer well-drained growth sites may take root on and stabilize the coarse-grained levee deposits (Fig. 15-25). Towns and farms tend to prosper along the gentle, relatively dry, landward slopes of levees—that is, until they're swamped by the next levee-topping flood.

Growing levees can divert tributaries away from the main stream channel, either temporarily or over the long term. Diverted tributaries may be forced to flow parallel to the main channel for tens or hundreds of kilometers before they can cross a low spot in the levee and join the trunk stream. Such parallel streams, which remain isolated at the margin of a floodplain, are called *yazoo streams*—named for the Yazoo River, which shadows the Mississippi River for 320 km (200 mi) before finally connecting with it near Vicksburg, Mississippi.

Adjacent to the levee slopes lies the **backswamp,** that portion of the floodplain where deposits of fine silts and clays settle from standing waters following a flood. After flooding subsides, the water left behind on the floodplain typically either evaporates or slowly seeps into the groundwater system. Wherever drainage to the groundwater system is poor or where the floodwaters become ponded in surface depressions, water may remain at the surface for an extended period of time as environmentally valuable

▼ **FIGURE 15-24** Floodplain features, produced when streams overflow their banks and deposit their sediment on the surrounding land.

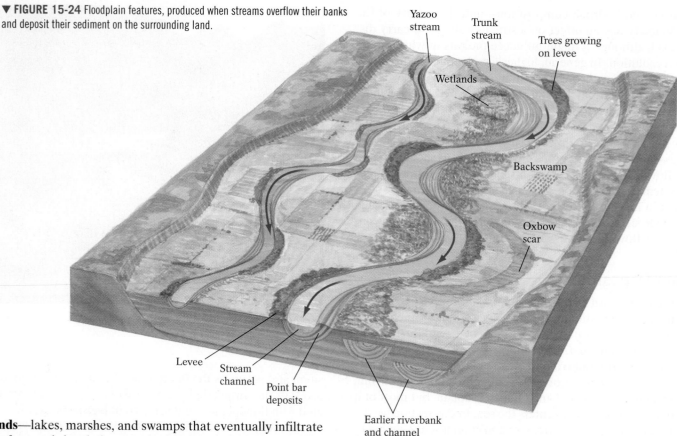

wetlands—lakes, marshes, and swamps that eventually infiltrate the surface and thus help to replenish the groundwater system. Wetlands may also serve as habitats for migrating birds and other wildlife. In the past, these areas were routinely drained to allow farming or other land uses, such as industry or housing. Today, conversion of environmentally valuable wetlands to such uses often sparks hot debate and may meet with sharp opposition.

▲ **FIGURE 15-25** A vegetated levee on the bank of the Mississippi River—the line of trees between the river and the white-topped road. Natural levees, composed of coarse sediment, are generally the best-drained parts of the floodplain. Consequently, they are most capable of supporting trees that cannot tolerate saturated soil conditions.

Flooding is not the only force that reshapes floodplain landscapes. A stream's point bars deposited within a stream's channel, for example, will eventually become major parts of a floodplain's geological record. As a meandering stream erodes the outside curves and deposits sediment on the inside curves of its meanders, its channel migrates snakelike through its own previous floodplain deposits. A mature floodplain may consist of a succession of such migrated, partially eroded point bars (see Fig. 15-21).

As their positions shift within the floodplain, meanders may eventually threaten the prosperity of their communities. Consider the town of New Harmony, in southwestern Indiana, whose citizens waited nervously for years as the meandering Wabash River approached it. The river would have reached the town site and eliminated it in 1994 if civil engineers had not halted the river's migration. The engineers built a concrete-block barrier less than a kilometer to the north of New Harmony that stabilized the Wabash River's eroding banks and prevented further erosion, saving the town.

Deposition Where Stream Valleys Widen When a stream leaves a narrow mountain valley and flows out onto a wide valley floor, it loses velocity quickly. In turn, its transport energy is also sharply reduced. Free from the confinement of the narrow valley, the stream spreads out across the lowland plain and deposits much of its coarse sediment load, forming a braided, fan-shaped deposit known as an **alluvial fan** (Fig. 15-26 on page 484).

15.1 Geology at a Glance

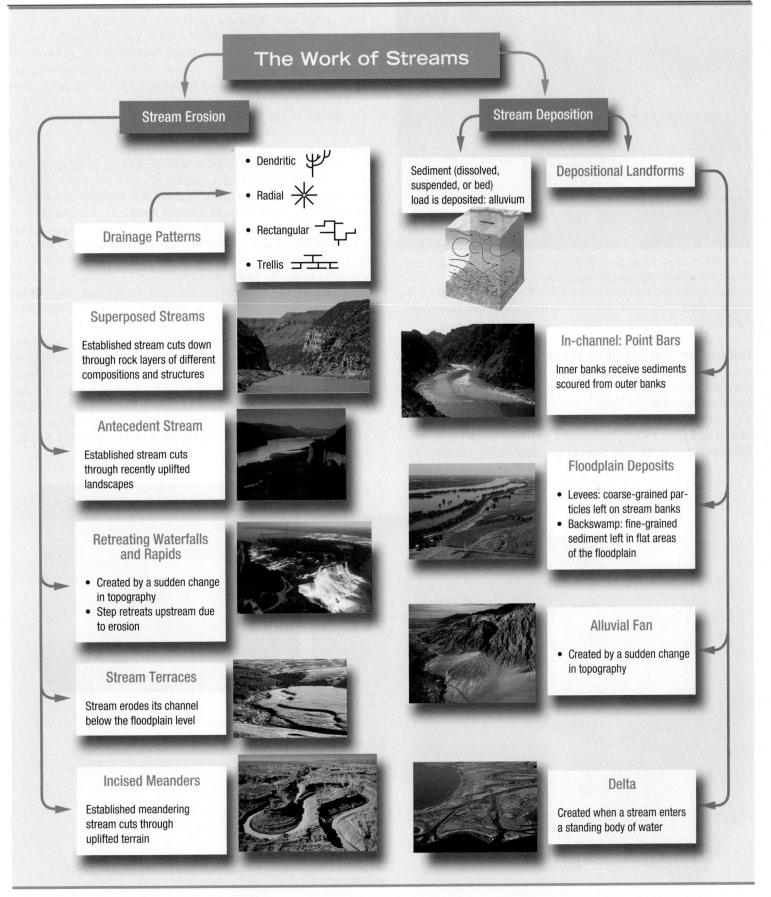

The Work of Streams

Stream Erosion

Drainage Patterns
- Dendritic
- Radial
- Rectangular
- Trellis

Superposed Streams

Established stream cuts down through rock layers of different compositions and structures

Antecedent Stream

Established stream cuts through recently uplifted landscapes

Retreating Waterfalls and Rapids
- Created by a sudden change in topography
- Step retreats upstream due to erosion

Stream Terraces

Stream erodes its channel below the floodplain level

Incised Meanders

Established meandering stream cuts through uplifted terrain

Stream Deposition

Sediment (dissolved, suspended, or bed) load is deposited: alluvium

Depositional Landforms

In-channel: Point Bars

Inner banks receive sediments scoured from outer banks

Floodplain Deposits
- Levees: coarse-grained particles left on stream banks
- Backswamp: fine-grained sediment left in flat areas of the floodplain

Alluvial Fan
- Created by a sudden change in topography

Delta

Created when a stream enters a standing body of water

▲ **FIGURE 15-26** An alluvial fan in Death Valley, California. Alluvial fans form when streams flowing from steep mountain slopes encounter a sharp reduction in slope at the foot of the mountain and a substantial widening of their valleys. They flow over this plain more slowly and less energetically, dropping their sediment loads as fan-shaped deposits.

Deposition into Standing Water

Where a stream enters a standing body of water, such as a lake or ocean, its flow decreases abruptly. As a result, it typically deposits its suspended load of fine sand, silt, and clay in the standing water as a **delta,** a roughly triangle-shaped alluvial deposit that fans outward from the mouth of a stream (Fig. 15-27). [The Greek historian Herodotus named this formation in the fifth century B.C., after noting that the sed-

iment deposit at the mouth of the Nile River looked like the uppercase Greek letter *delta* (Δ).]

As a delta forms, the coarser stream sediment is deposited first, close to the mouth of the stream, producing the delta's angled *foreset beds.* Finer sediments settle farther from the mouth, in more gently dipping, sometimes horizontal layers known as *bottomset beds.* The greater volume of the nearshore sediment gradually expands the foreset beds outward, extending the delta and burying previously deposited bottomset beds. Flooding at the surface of the expanding delta spreads thin, largely horizontally layered sediments on top of the inclined foresets, forming *topset beds.*

A delta continues to grow seaward as long as its stream deposits more sediment than is eroded from the delta by waves and shoreline currents. The Earth's great deltas typically occur where major rivers—such as the Mississippi, the Ganges in Bangladesh, the Indus in Pakistan, and the Nile in northern Africa—deliver large sediment loads to relatively quiet coastal waters. Deltas do not form at the mouths of some other large rivers, such as the Amazon in South America and the Niger in Africa, because vigorous waves and currents immediately sweep away most of the sediments. The largest river in North America's Pacific Northwest, the Columbia River, has no delta because the Pacific's powerful winter storm waves carry away virtually all of its sediment. The St. Lawrence River, which rises in Lake Ontario and flows past Montreal and Quebec, has no delta for two reasons: (1) Much of the region's sediment settles to the bottom of the Great Lakes and never enters the river; and (2) the river does not accumulate a delta-building load on its relatively short, 600-km (380-mi) journey from the Great Lakes to the Gulf of St. Lawrence and the northern Atlantic.

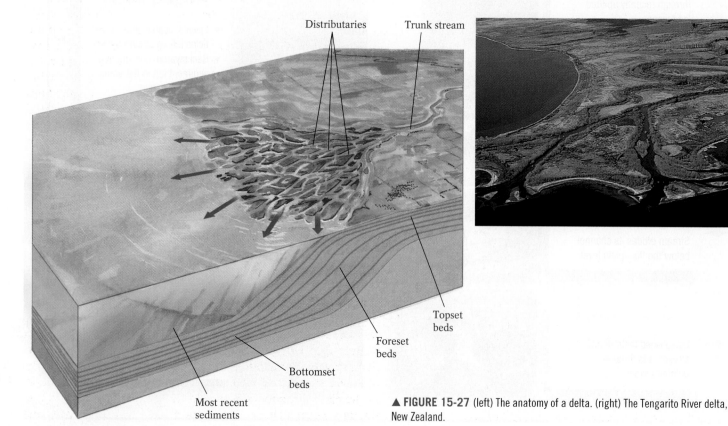

Distributaries

Trunk stream

Topset beds

Foreset beds

Bottomset beds

Most recent sediments

▲ **FIGURE 15-27** (left) The anatomy of a delta. (right) The Tengarito River delta, New Zealand.

As a stream's delta grows outward, its surface gradient decreases. The stream then flows more slowly and gradually loses the capacity to carry its sediment load. As the stream drops more sediment in its channel, it becomes clogged and the stream's flow then splits into a network of branching *distributaries*. The channel also becomes shallower, increasing the likelihood that the stream will flood and break through its levees. Released from its original channel, the stream may discover a new, more direct route to the sea, usually one with a steeper gradient. The stream's new main distributary then begins to build its own active *lobe* of the larger delta at its new outlet, forming a young, wedge-shaped body of fresh sediment (Fig. 15-28). As the younger distributaries capture more of the sediment load, a new lobe of the delta grows at the expense of the older, abandoned lobes. The abandoned lobes, which now lack a replenishing sediment supply, then begin to erode away. Old abandoned lobes of a large delta may also subside as their underlying sediments compact. Thus, the shapes and locations of a stream's active and abandoned delta lobes change continuously.

Offering fertile soils of fresh floodplain silt and intricate networks of navigable waterways, deltas have served for thousands of years as coastal centers of agriculture and commerce. Ancient deltas represent valuable sources of oil and natural gas (from the decomposition of marine fauna) and coal (from accumulation of plant remains in stagnant deltaic swamps). Highlight 15-1 focuses on the history and economic significance of North America's greatest delta—that of the Mississippi River.

Before moving on to discuss humankind's most devastating form of natural hazard—*flooding*—you might want to spend a few minutes reviewing the various landforms and stream features associated with both stream erosion and stream deposition in the Geology at a Glance on page 483.

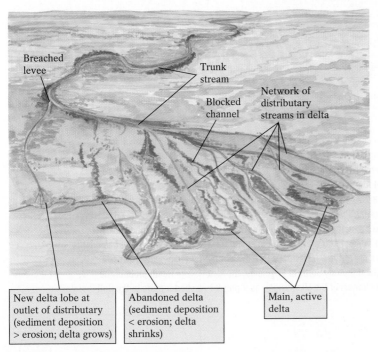

Breached levee

Trunk stream

Blocked channel

Network of distributary streams in delta

New delta lobe at outlet of distributary (sediment deposition > erosion; delta grows)

Abandoned delta (sediment deposition < erosion; delta shrinks)

Main, active delta

▲ **FIGURE 15-28** The large sediment loads and gentle gradients of deltaic streams promote in-channel deposition. The streams then either diverge around these deposits or overflow their channel walls, forming new distributaries and delta lobes.

Controlling Floods

Some streams flood *every* year. Many other streams overflow their banks every two or three years. Floods usually result from local weather conditions—simply too much rain falls or snow melts, producing more water than the channel can carry. A succession of storms typically increases the severity of the ensuing floods. If rain falls on dry unsaturated ground, the soil may soak up the water and delay its entry into streams. After the ground becomes saturated, however, it cannot absorb the next downpour. The excess water then runs off the surface without delay, *directly* into local streams, causing them to rise rapidly and flood.

The geology and topography of the landscape largely determine whether water runs rapidly into streams or seeps into the groundwater system, where it may remain long enough to avert flooding. Are the soils and the underlying bedrock near the surface *permeable* (that is, they soak up water easily, allowing it to pass through), like dry sandy soils or coarse sandstone? Or are they relatively impermeable like wet clayey soils or crystalline granitic bedrock? Is the rock highly fractured and faulted? (Fractured rocks can more readily absorb water.) Are the surface soils clay-rich types that swell when they become wet and seal the surface against infiltration? Is the local topography steep or gently sloping? (Steep slopes promote runoff and enhance the prospects for flooding; gentle slopes enhance infiltration and thus reduce the likelihood of flooding.) The steepness of the Rocky Mountain slopes were largely responsible for the catastrophic Big Thompson Creek flood of 1976. Big Thompson Creek, in the Colorado Rockies northwest of Denver, received more than 30 cm (12 in.) of rain on the evening of July 31, 1976. This volume of precipitation, which nearly equaled the average annual precipitation for the area, could not be absorbed by the steep bedrock of the canyon. As a result, discharge in the creek rose rapidly to more than four times its previous record. The surface of the creek rose from two-thirds of a meter (about 2 ft) to more than 6 m (20 ft). By the time the floodwaters receded, the flood had taken 144 lives. (Many victims would have survived had they climbed upslope instead of trying to outrun the wall of water that thundered down the narrow gorge.)

Flooding becomes more likely when heavy rains fall on an area where extensive forest fires, droughts, or widespread clearcutting has removed a region's vegetation. Trees and their root systems tend to keep soils open, enhancing water infiltration and thus reducing the chance of runoff and flooding. As communities expand into unpaved, undeveloped areas, new roads, buildings, and parking lots replace vegetated areas with paved-over lands. This change prevents surface water from infiltrating the groundwater system, increasing the potential for flooding. Flooding is also strongly and often tragically enhanced because more communities are built today on river floodplains than ever before. August 2002 saw the largest floods in central Europe since the Middle Ages, causing $30 billion in damage and taking 94 lives. Germany's Elbe River rose to a level 9.4 m (31 ft) above flood stage—the highest in 500 years. The rising Elbe in the city of Prague in the Czech Republic forced the emergency evacuation of the city's 400 zoo animals. Did this extraordinary period of flooding signal a change in Europe's climate or an unfortunate

Highlight 15-1

The Mississippi River and Its Delta

The Mississippi River begins its 3750-km (2350-mi) journey as a creek a few meters wide and 10 cm (4 in.) deep that trickles out of Lake Itasca in the pine forests of northern Minnesota. In its first 2100 km (1300 mi), the *upper* Mississippi cascades over bedrock falls and rapids wherever it is not dammed. South of Cape Girardeau, Missouri, the Ohio joins the Mississippi. Together, the two rivers form a majestic alluvial valley—the much tamer *lower* Mississippi valley—that stretches for more than 1600 km (1000 mi) and reaches widths of 48 to 200 km (30 to 125 mi). By the time the Mississippi empties into the Gulf of Mexico 200 km (120 mi) south of New Orleans, it carries water and sediment from the Missouri, the Ohio, the Arkansas, and more than 100,000 other streams. Along its way, it has deposited fertile silt over more than 770,000 km² (300,000 mi²) of floodplain farmland (and, occasionally, over water-logged cities).

Much of the lower Mississippi valley may have once been a shallow marine bay located in a failed rift arm that formed approximately 200 million years ago when North America began to split from Europe and Africa (or perhaps from South America as some geologists suggest). The Mississippi River's first delta grew southward into this bay, filling the ancient rift with a succession of southward-growing deltas. As a series of ice ages caused the Earth's water to accumulate in glaciers during the last 3 million years, global sea levels fell and the Mississippi River's marine bay receded. During this period, the Mississippi River transported an enormous supply of sediment eroded from the deposits left behind by the vast ice sheets that blanketed Canada and the Great Lakes region of the United States. With such a large sediment source, the delta grew steadily. Beginning 12,000 years ago, after the last period of worldwide glaciation ended, the Earth's climate warmed and sea levels rose. Approximately 9000 years ago, the Mississippi River began to construct its present-day delta.

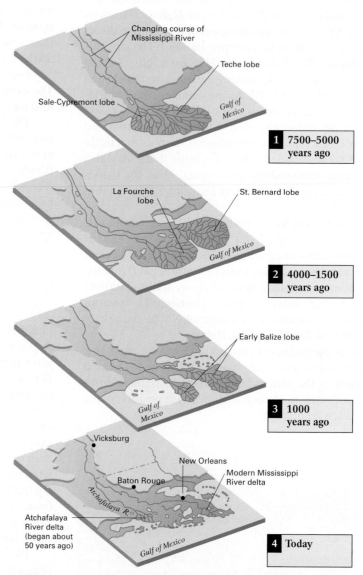

The evolution of the Mississippi River delta plain. **(1)** The Sale-Cypremont delta (active 7500 through 5000 years ago), one of the earliest-known deltas of the Mississippi, was abandoned when the river diverted its sediment to the Teche lobe (active 5500 to 3800 years ago). **(2)** Delta growth shifted farther east when the St. Bernard lobe (active 4000 to 2000 years ago) was deposited. Delta building moved west again, forming the La Fourche lobe (active 2500 to 1500 years ago). **(3)** By some 1000 years ago, the modern bird's-foot (Balize) lobe had begun to be constructed. **(4)** In the last 100 years or so, the Atchafalaya River, a distributary, has begun to carry more of the Mississippi's flow. Were it not for flood-control structures built by the U.S. Army Corps of Engineers, the Atchafalaya would probably become the main channel of the Mississippi, as it presents a shorter route to the Gulf of Mexico.

Today the Mississippi delta consists of at least seven distinct lobes, formed over the past 4000 to 5000 years, each of which was once the river's active delta. Together, these lobes amount to about 40,000 km² (15,000 mi²) of real estate added to southern Louisiana. The active portion of today's delta, known as the Balize lobe, dates from 1000 years ago. Its bird's-foot shape has been built by three major distributaries whose gradients are now too gentle to transport sediment effectively. The inactive segments of the Balize lobe were each abandoned as the river sought shorter, steeper routes to the Gulf. Wave action has eroded away many more abandoned lobes, subsidence caused by the weight of overlying sediments has depressed others below sea level. Most of the very large Balize lobe now lies under water. The Mississippi River *bayous,* the colorful network of lakes and minor streams in southern Louisiana, consist of abandoned channels that crisscross inactive lobes.

For several decades, the Mississippi River has been trying to cut a steeper, shorter course along the Atchafalaya River, located 100 km (60 mi) to the west of its present channel. If and when it succeeds, the lower 500 km (300 mi) of the present river will be cut off, endangering the livelihood of Baton Rouge and New Orleans and jeopardizing the great investments made in navigational improvements along its course.

Thus far, human intervention has managed to preserve the river's course. In 1973, a winter of unusually high precipitation produced raging floods in the lower Mississippi valley. These floods nearly opened the Atchafalaya channel and threatened to bypass New Orleans, leaving it a sleepy bayou town instead of a center of river commerce. Floodgates diverted water from the main channel to Lake Pontchartrain, however, preventing the river from breaching its levee and taking the Atchafalaya course. Continued dredging to deepen the lower Mississippi channel enough to accommodate the river's flow has saved Baton Rouge and New Orleans *for now,* but the river may yet have its way.

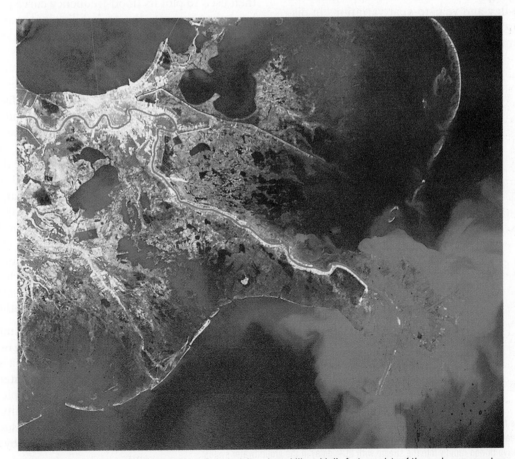

▲ The modern delta of the Mississippi River. Today's delta, shaped like a bird's foot, consists of the modern segment centered at Head of Passes, Louisiana, and a number of inactive, abandoned lobes that no longer receive sediment. During the past century, the river has attempted to shift away from this lobe and establish a new course along the Atchafalaya River. Unless human intervention succeeds in controlling rivers, it will eventually do so.

consequence of questionable land-use planning? Climate scientists believe that extreme summer flooding in Europe actually occurred more frequently in the past, so such floods should have been expected and land-use planning should have minimized damage and loss of life. Extensive building on and overpopulation of Europe's floodplains was more likely the culprit in these disasters than climate change.

Flood Prediction

Floods usually occur during heavy spring rains and snowmelt in temperate climates and during rainy seasons elsewhere. Because most streams flood annually, the residents of a flood-prone region may choose to be thoroughly prepared—by keeping rowboats in basements, building structures on stilts or cinderblocks, and maintaining supplies of sandbags. *Unusually large* floods, however, cause the greatest damage. Can such events be predicted?

Because forecasting the weather is still a less-than-perfect science, flood prediction is imprecise as well. Our best hope lies in analyzing the frequency of past floods and using those data to determine the *statistical* probability of a major flood occurring within a given period of time. Stream discharge, for example, is monitored over the long term by creating *hydrographs,* graphical plots that show day-to-day and year-to-year variations in discharge. Hydrographs also illustrate how human activities such as urbanization and forest clear-cutting affect discharge and thus

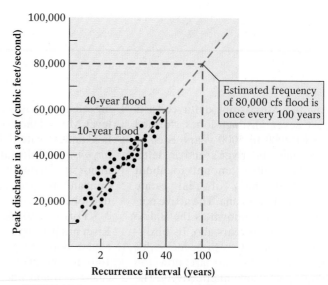

▲ **FIGURE 15-30** A flood-frequency curve for a hypothetical stream showing how often large discharges and floods of various magnitudes have occurred throughout the stream's recorded history. Hydrologists may extrapolate the estimated discharge of the 100-year event from the stream's 40-year monitoring record.

flood potential (Fig. 15-29). Hydrograph data for a stream are then used to plot its flood-frequency curve (Fig. 15-30). This allows us to estimate the probability that a specific flooding discharge will be equaled or exceeded in that stream in any given year.

Consider the following example: After recording a stream's discharge over a 100-year period, we determine that a peak discharge as great as 2000 cfs (cubic feet per second) occurs once every 10 years. Thus the chance of such a discharge arising in any particular year is 1 in 10 (a 10% probability). When that discharge does take place, we call the event a 10-year flood. Similarly, recorded data may show that the discharge of the same river reaches 3500 cfs once every 100 years. When there is a 1-in-100 chance (a 1% probability) of such a discharge occurring in any particular year, the event is termed a 100-year flood. These probabilities are updated as the discharge record grows. Several cautions are necessary regarding flood predictions based on this method, however:

- Few places in North America have peak-discharge records spanning more than 100 years. Thus our calculation of a stream's 100-year flood must be extrapolated from a shorter record. Such estimates typically *underestimate* a flood's size. A community whose structures are engineered based on such flood estimates may be rudely awakened by much larger flood events.

- Substantial changes in a watershed throughout the period of stream monitoring, *primarily from human development,* typically increase stream discharge substantially. (Paving over soils and sediments reduces surface infiltration of water

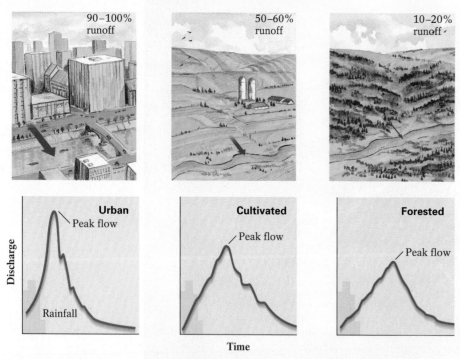

▲ **FIGURE 15-29** Hydrographs record how a stream's discharge varies with time. A sharply peaked hydrograph curve indicates that a stream's discharge has risen dramatically in a brief time, suggesting that precipitation and other surface water have flowed rapidly into the stream with little delay in the groundwater system. Because urban areas are largely paved over with asphalt and cement, little surface water can infiltrate the groundwater system, and much of it runs off into streams; thus, the hydrographs of urban streams—and their streams' flood elevations—tend to be steeper. By comparison, the peak on the hydrographs of streams in undeveloped areas are generally less steep (changes in discharge occur more gradually). The amount of rainfall (and the volume of water transported) and the total discharge for these areas of land use is the same (the area under their curves is equal).

to the underground water system, in turn increasing runoff to streams and decreasing the lag time between precipitation and stream rise.) Thus, the discharge of a stream today during its so-called 100-year flood *may be significantly higher* than the peak discharge encountered decades ago, before extensive development of the watershed.

The discharge associated with the 100-year flood of one stream is not necessarily the same as that of any other stream. Every stream possesses a unique flood-frequency curve, with specific discharges associated with floods of particular magnitudes.

Moreover, nature seldom cooperates with our statistical predictions. An area that endures one 100-year flood will probably not enjoy a 99-year break before the next one. Residents living along the banks of the Mississippi River, for example, had to cope with 100-year floods in 1943, 1944, 1947, and 1951. Moreover, because of inadequate early recordkeeping, flood frequency is sometimes hard to define. Some agencies have classified the 1993 Mississippi flood (see Highlight 15-2) as a 200-year flood; others have called it a 500-year event.

Flood Prevention   Communities have tried a number of defensive strategies—structural, nonstructural, and combinations of both—to try to curb the catastrophic effects of flooding. The structural approach involves building some form of artificial construction that attempts to contain a stream within its banks, divert some of its water to temporary storage facilities, or accelerate flow through the stream system.

The most common stream-containment structures are *artificial levees*—earthen mounds built on the banks of a stream's channel to increase the volume of water that the channel can hold. Such levees have been built along the Mississippi River to protect croplands since the eighteenth century. Spillways cut through the levees allow rising water to escape into human-made diversion channels, reducing the likelihood of a flood downstream. Concrete *flood walls,* which are more expensive, have also been constructed by the U.S. Army Corps of Engineers to prevent overflow from the channel at strategic locations, such as along the commercial centers of riverside cities.

To be effective, artificial levees and flood walls must protect both sides of a stream and must extend for its full length (Fig. 15-31). If these structures are built *only upstream,* flooding downstream will be far more catastrophic because the flooding stream has no access to its floodplain—the natural place for floodwaters to go—along the protected upstream section. If artificial levees and flood walls are built *only downstream,* the upstream floodwater will wash over the floodplain and then head downstream across the very area the levees were built to protect. If they are constructed on *only one side of a stream,* the floodplain on the unprotected side receives a catastrophic double share of floodwater.

Flood-control dams are often built upstream from densely populated or economically important areas. These earthen or concrete structures transfer excess water to temporary storage facilities called impoundment basins to prevent, or at least slow, channel overflow. The excess water in the basins may either evaporate or seep into the groundwater system. Such structures also have uses unrelated to flood control. Dams along the Tennessee, Mississippi, Missouri, Colorado, and Columbia Rivers, for example, generate hydroelectricity, provide irrigation water for agriculture, supply municipal water to numerous communities, and create regional recreational facilities.

Diversion channels draw off some of a river's water and divert it to another channel to reduce a river's discharge before it has a chance to flood. After the catastrophic flood of 1950, the city of Winnipeg in the Canadian province of Manitoba (population, 620,000) decided to undertake a major structural change in the Red River's local drainage. The city excavated a massive trench—45 km (28 mi) long, 155 m (500 ft) wide, and 9 m (30 ft) deep. The plan was to divert as much as 50% of the river's floodwaters around and away from the city, allowing the water to rejoin the river north of the city. Even with this structural protection, during the region's 1997 floods (Fig. 15-32 on page 492), the city built

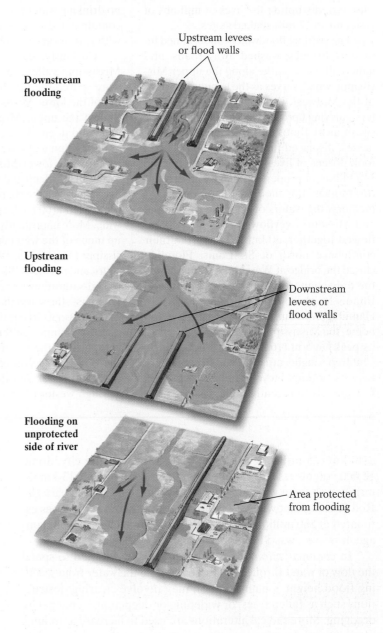

▲ FIGURE 15-31 Artificial levees and flood walls must protect the full length of a river and be built on both of its banks; otherwise, the structures will merely divert the flooding to unprotected areas.

Highlight 15-2

The Great Midwestern Floods of 1993

On June 11, 1993, a foot of rain fell in southern Minnesota. Four days later, more than 11 more inches fell in roughly the same area. Thus began the wettest June and North America's worst flood since compilation of weather records began in 1878. Before the wet weather ended two months later, the upper Mississippi River had surged over its banks from St. Paul, Minnesota, to St. Louis, Missouri, disrupting the lives of millions of residents in 12 rain-soaked states.

The swirling floodwaters undermined interstate bridges, washed away roads, and halted all barge traffic along the upper Mississippi, which serves as the economic artery of the Midwest. Rushing, muddy floodwaters, carrying uprooted trees and junked cars, swept away thousands of homes and businesses, displaced more than 50,000 people, took dozens of lives, and caused more than $10 billion in property damage and agricultural losses. For the first time in anyone's memory, the waters of both the Mississippi and Missouri overflowed their levees and flowed together, 30 km (20 mi) from their confluence north of St. Louis. Flooding closed the bridge at West Quincy, Missouri—the only connection between Missouri and Illinois for more than 300 km (200 mi). In Hannibal, Missouri, Mark Twain's boyhood home, the Mississippi River *crested* (reached its peak) at 5 m (16.5 ft) above flood stage. Children caught catfish on city streets where they had ridden bicycles only days before. Fortunately, a recently built, 11-m (33-ft)-high flood wall saved the historic downtown area and the Twain homestead. In Davenport, Iowa, where residents had chosen in the past to preserve their river views rather than build levees and flood walls, ducks swam lazily in the 4 m (14 ft) of water covering the outfield of Davenport Stadium, the home of the minor-league baseball Quad City River Bandits. In Des Moines, 250,000 Iowans had no drinking water for days after floodwaters contaminated their municipal water supply with raw sewage and chemical fertilizers.

Curiously, the residents of the lower Mississippi valley, south of Cairo, Illinois, were spared the flooding. Why? The channel of the lower Mississippi, compared with that of the upper Mississippi, is relatively wide and deep; thus it can accommodate a greater volume of water. At Vicksburg, Mississippi, the lower Mississippi's channel is 37 m (123 ft) deep and 970 m (3200 ft) wide; in comparison, the channel north of St. Louis is only 18 m (60 ft) deep and 575 m (1900 ft) wide. More importantly, the states that make up much of the watershed of the upper Mississippi (Minnesota, Wisconsin, Iowa, Missouri, and northern Illinois) received months of soil-saturating storms; Nebraska and the Dakotas—the states that form the watershed of the Missouri River, the upper Mississippi's major tributary—suffered a similar fate. By comparison, the lower Mississippi's watershed, which receives 65% of its water from the Ohio River and its tributaries flowing through western Pennsylvania, Ohio, West Virginia, Indiana, Kentucky, and southern Illinois, experienced below-normal precipitation. As a result, the lower Mississippi valley states of Arkansas, Tennessee, Mississippi, and Louisiana were spared this disaster.

Some people maintain that human activity—65 years of "managing" the Mississippi River to prevent floods—actually contributed to the magnitude and tragedy of the flood of 1993. Prior to the last 100 years or so, the Mississippi and its tributaries determined their own boundaries. During periods of high water, the rivers broke through or overflowed natural levees, flooding tens of thousands of cubic meters of the surrounding, largely uninhabited lands. In the twentieth century, however, millions of people migrated to the region, building cities, towns, and large farms along the rivers' banks. When a flood in 1927 took 214 lives, Congress enacted the first Mississippi River Flood Control Act, assigning the task of confining the river to its channel to the U.S. Army Corps of Engineers.

The Corps of Engineers built nearly 300 dams and reservoirs and thousands of kilometers of artificial levees and concrete flood walls, all designed to prevent the river from spilling onto its natural floodplain. The system also includes numerous pumping stations, spillways, and diversion channels that shift water into temporary holding basins for storage. The events of 1993 dramatically illustrated that the Corps' efforts have largely

a 40-km (25-mi)-long artificial levee to protect the city. Some 12,000 volunteers laid 4 million sandbags, along with a 13-km (8-mi)-long chain of wrecked school buses and cars to hold back the Red River's floodwaters. The total cost of these projects was estimated at $60 million. The estimated total savings in flood damage, however, was $1 billion.

In *channelization,* a stream's channel is modified to speed the flow of water through it, thus preventing the water from reaching flood height. Channelization may involve clearing obstructions such as fallen trees, or widening or deepening a channel by dredging. More radical alterations are used to increase a stream's gradient and therefore its velocity, such as cutting off meanders to straighten a stream. The shorter, straight channel will have a steeper gradient than the previous channel, and its increased velocity will enable the transport of more water, perhaps enough to prevent flooding.

Channelization can also produce a host of negative effects. The Blackwater River, a meandering tributary of the Missouri River located southeast of Kansas City, has flooded regularly and destructively for years, largely because it flowed too slowly to accommodate large storm discharges. In 1910, engineers excavated a straight channel that increased the river's gradient and velocity but also caused the river to scour its bed more energetically. The channel subsequently expanded from about 4 m (14 ft) to more than 12 m (39 ft) deep, and from 9 m (30 ft) to more than 71 m (233 ft) wide. As the river widened, it undermined its bridges, which had to be replaced. Only constant maintenance now keeps the Blackwater River straight.

▲ Extent of flooding in Davenport, Iowa, June 1993. Left: Standing water near the town's River Drive casino. Right: Three weeks later at the same location after floodwaters had receded.

failed, perhaps because such flood-control structures simply cannot contain such an *extraordinary* flood.

By confining the massive discharges to a channel, the retention structures actually forced the swollen river to flow more rapidly and violently, damaging the very structures designed to restrain it. By denying the river access to its natural floodplain, they caused the river to rise higher than it would have naturally, ensuring that the floods produced by levee breaches would be more damaging. Furthermore, the existence of artificial levees and flood walls created a false sense of security that encouraged the growth of cities, towns, and farms closer to the river banks than was really safe.

What does the future hold for the residents of the upper Mississippi valley? Certainly more flooding, but perhaps less human interference with the river's natural behavior. Some communities have proposed to remove all flood-retention systems and zoning limits. They also plan to reduce future development close to the region's rivers, at least within their 100-year floodways. Others would like to build more structures like St. Louis's 16-m (52-ft)-high concrete flood wall, which saved that city's downtown business district when the Mississippi reached its record crest at 14.2 m (47 ft). Still others have simply moved off the floodplain. The debate continues between those who believe we can tame the mighty Mississippi and those who know we cannot.

Structural solutions to flood problems are almost always expensive. Such defenses can also give local residents a false sense of security; the more protected they feel, the closer to a stream they tend to build their homes.

And as the hundreds of thousands of displaced citizens of New Orleans will attest in the wake of Hurricane Katrina in August of 2005, flood-control structures may only work effectively when they are well-maintained and augmented by the constant attention of highly skilled engineers. The following Highlighted discussion recounts some of Hurricane Katrina's human-caused misery.

Because of the drawbacks to structural solutions, a *nonstructural* approach is almost always preferable. In developing a nonstructural defense against flooding, we identify high-risk areas, regulate zoning to minimize development in them, and manage natural resources thoughtfully to restrict the amount of water entering a stream channel at any one time. This approach first requires a thorough geological and botanical survey to identify flood risk. Mapping the natural distribution of water-loving plants (phreatophytes) helps identify areas that have flooded regularly in the past. Geologists can map and date ancient flood deposits and analyze their thicknesses and textures to estimate the timing and magnitude of prehistoric floods.

Geologists then use these data to create a flood-hazard map showing *floodplain zones,* or *floodways*—the corridors of land adjacent to the stream that would be covered by floods of different magnitudes (Fig. 15-33). Communities can then use this information to draft zoning ordinances. For example, a 12-km

▶ FIGURE 15-32 Damage from floodwaters and fires along the Red River in Grand Forks, North Dakota, in April 1997.

Flood-hazard map

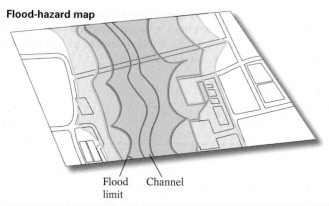

Flood   Channel
limit

Actual landscape

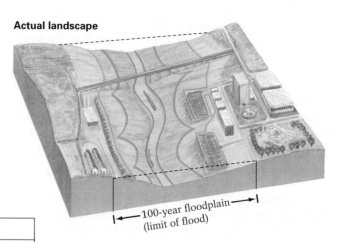

◀————100-year floodplain————▶
(limit of flood)

Floodplain zones

◻ "Floodway district": regulated; kept open at all times; no flood-control measures, no building

◻ "Floodway fringe district": flood-prone, protected by flood-control measures; building allowed if protected by "flood-proofing"

◻ "Floodway limit": area subject to flooding only by very large discharge (100-year flood)

▲ FIGURE 15-33 Geologists construct a flood-hazard map from historical data to predict the frequency and magnitude of future floods and identify areas that will be affected. These data are used to establish floodways, the areas of a river valley that floods of various magnitudes would cover.

(7-mi)-wide floodway surrounding the Rapid River was designated a high-risk area after the disastrous Rapid City, South Dakota, flood of June 10, 1972. Many buildings were then moved from this region. Today the floodway is used mostly for ballfields and golf courses. Artificial levees and flood walls protect structures in the floodway that could not be moved or abandoned.

Nonstructural flood prevention also includes protection and wise management of forests and a prompt response to forest fires. As we have seen, vegetation cover and their root systems promote absorption of water into the groundwater system and reduces stream discharge, so maintaining our forests is an important part of any flood-prevention plan. The community of Los Alamos, New Mexico has had to deal with a higher chance of flooding over the past few years. The raging fires that burned during May 2000 destroyed much of the vegetation on the forested slopes of the Jemez Mountains that surround Los Alamos (Fig. 15-34). The loss of the forest's underbrush, which typically slows waterflow off the mountain slopes, has definitely enhanced the region's flood potential. Ironically, the fire was set *intentionally* by the National Park Service to burn away the underbrush as a fire-prevention measure.

How have we fared in our efforts to prevent floods? During June and July of 1993, a succession of storms dumped an extraordinary amount of rain on the drainage basin of the Mississippi River in Minnesota, Iowa, Illinois, and Missouri, triggering the river's worst flood on record and causing billions of dollars in damage to croplands and other property. Highlight 15-2 on pages 490 and 491 focuses on this tragic event and illustrates how nature can confound our best-laid flood-prevention plans.

Artificial Floods

Sometimes a flood is beneficial—even desirable. Such was the case at dawn on March 26, 1996, when Secretary of the Interior Bruce Babbitt turned a lever at the pumphouse of Arizona's Glen Canyon dam, releasing a white torrent of water—45,000 cfs—*intentionally* flooding the Colorado River valley in the Grand Canyon region. If floods cause widespread damage and misery, why would a high governmental official unleash one on this popular tourist destination?

Before the completion of Arizona's Glen Canyon Dam in 1963, the Colorado River "managed" its own ecosystem through annual flooding. During the spring months, when snows melted in the southern Rockies, the extremely high discharge (roughly 125,000 cfs) flooded the lower Colorado River valley. At these times, the turbulent waters ran red and muddy as they transported roughly 65 million tons of sediment eroded from the surrounding bare-rock plateaus. Early settlers described the river in flood as "too thick to drink and too thin to plow." When the flooding ended, the deposition of these sediments replenished beaches and sandbars along the river, providing sheltering habitats for the region's plants and animals. The raging floods also cleared debris from the Colorado's tributaries, keeping these backwaters open for fish to spawn and creating ideal "nurseries" in which young fish could mature without facing the rigors of life in the main channel.

The construction of the Glen Canyon Dam drastically altered the ecology of the Colorado River. The river was tamed to suit the needs of hydropower cooperatives that supplied energy to 20 million residents of nearby states. Dam managers controlled the river's flow, allowing increased flow through the dam only during heavy-use periods, such as during the evening dinner rush (when ovens were in full use) and on hot afternoons (when air-conditioner use peaked). Because seasonal flooding was eliminated, far less sediment was available to restore the river's natural habitats, which quickly began to erode away. With more than 90% of the watershed's sediment remaining on the bottom of Lake Powell behind the dam, the once red, muddy Colorado flowed green and clear and cold (too cold for many native fish species to survive because colder bottom water was being released during controlled flow).

After more than 10 years of lively debate and dozens of scientific studies, the U.S. government and the managers of the Glen Canyon Dam decided to try to reverse the damage produced by the dam's operation. In an attempt to renew the health of the Colorado River's ecosystem, 117 billion gallons of water were released over a weeklong period. This *artificial* spring flood temporarily restored the Colorado River to a semblance of its former self with some significant results.

As four great arcs of water shot from the dam's outlet (Fig. 15-35 on page 496), the Colorado crept higher up the salmon-colored sandstone walls surrounding the channel below the dam. The river rose 5 m (17 ft) in the Grand Canyon, which is located 25 km (15 mi) downstream of the dam. The rushing torrent was designed to scour out sediment and vegetation that had accumulated over 30 years in the river bed and in tributaries and side canyons, converting those clogged backwaters into future spawning grounds for native fish. As the river overflowed its banks for the first time in decades, eroded sediment was transferred to the river's banks, creating 55 new beaches and sandbars where vegetation could take root to feed the region's birds. Backwater swamps formed behind these bars, providing ideal breeding grounds for the insects that nourish the river's fish. Even the

▲ **FIGURE 15-34** The Cerro Grande fire in Bandelier National Forest in Los Alamos, New Mexico, on July 5, 2000. Loss of vegetation will enhance future flood potential.

Highlight 15-3

The Tragedy of Hurricane Katrina

As this highlight essay is being written, the grim story of North America's greatest natural disaster is just beginning to unfold. Grim death tolls, astronomical rebuilding costs, and myriad accounts of heroism and tragedy are still emerging, but one thing seems clear, yet again: Our human-made structures are simply no match for the relentless power of flowing water. The following describes the current state of affairs along America's Gulf Coast. By the time you read this, many more thoughts will have been shared about why this tragedy took place, what could have been done to prevent it, and what should be done to ensure that an event of its magnitude will be far less likely in the future.

Katrina's History

The hurricane started out as a tropical storm over the Bahamas in late August 2005. It crossed Florida as a relatively weak hurricane, but intensified greatly over the warm waters of the Gulf of Mexico. It was briefly a Category 5 hurricane (on the scale of 1 to 5, with 5 having the highest wind speeds), with one of the lowest barometric pressures ever recorded in the Western Hemisphere, wind speeds of 280 kph (175 mph), and gusts over 320 kph (200 mph). On August 29, Katrina made landfall in Louisiana and Mississippi as a Category 4 hurricane with winds of 225 kph (140 mph), soon weakening to a Category 3 hurricane with winds of 200 kph (125 mph). It was the third strongest hurricane on record to make landfall in the United States. The 9-m-high (30-ft) storm surge along the Mississippi coast, centered on Biloxi, is the highest ever recorded in the U.S.

Damage

As Katrina slammed into Mississippi and Louisiana, its high winds and roiling waters left a trail of massive destruction across the region. Although strong winds caused severe damage in coastal Mississippi, the destruction of New Orleans, Louisiana, had a different cause: city-wide flooding that followed the rupture of several of the city's artificial levees. Fearing that the hurricane would pass directly over New Orleans, government officials ordered people to evacuate the city (in practical terms, although this means that the citizens of New Orleans were told to leave, those who could not afford to do so remained in the city). New Orleans was spared the brunt of the hurricane's direct winds, but was not spared the devastation resulting from high water levels created by the storm.

Much of the city of New Orleans lies below sea level, including some regions that reside as much as 3 to 4 m (10 to 15 ft) below sea level. New Orleans has sunk roughly 1 m (3 ft) in just the last century. The city has been described as a bowl sitting between the Mississippi River, which runs through the city, and Lake Pontchartrain, a large inlet from the Gulf of Mexico along the northern edge of the city (see figure). In 1989, John McPhee wrote in his book *The Control of Nature:* "The [Mississippi] river goes through New Orleans like an elevated highway. Jackson Square, in the French Quarter, is on high ground with respect to the rest of New Orleans, but even from the benches of Jackson Square one looks *up* across the levee at the hulls of passing ships. Their keels are higher than the Astro Turf in the Superdome, and if somehow the ships could turn and move at river level into the city and into the stadium they would hover above the playing fields like blimps."

The city has existed for centuries owing to an extensive system of high walls (levees) and pumps. When the city was first settled, people built homes on natural levees—ridges of sand and silt built up by periodic flooding of the river. As the city grew, owing to its important location near the outlet of the Mississippi River, these natural levees had to be

raised and eventually were replaced entirely by human-constructed levees. Cypress swamps bordering Lake Pontchartrain were filled in and settled, and these new (sinking) regions of the city required their own levees as protection from the lake. Today the levee system has grown to comprise hundreds of kilometers of barriers between water and land. Even with the high levees, pumps are needed to keep the city from flooding. Because of the bowl-shape of the city, water has no natural way to flow out, and must be pumped up and over the edge of the bowl into the lake or river.

Although the levees have long protected the city, a consequence of the levee system is the blocking of river silt from replenishing the floodplain and marshes and from creating barrier islands on the coast. In a natural system, these regions act as a buffer from powerful storm surges. Once the levees were built, however, they needed constant maintenance and additional construction to protect the growing but sinking city and outlying industrial areas that are critical to the nation's economy. More than 15 years ago, John McPhee wrote: "Among the five hundred miles of levee deficiencies now calling for attention along the Mississippi River, the most serious happen to be in New Orleans." Maintenance of the levee system was further harmed by recent major reductions in levee repair funds to the U.S. Army Corps of Engineers, the organization responsible for levee construction and repair. Adding to the danger, the levees were built to withstand a Category-3 hurricane, but not a Category 4 or 5.

Although Hurricane Katrina slid by the city to the East with only minor damage from the strong winds, on August 30, some of the levees between the city and Lake Pontchartrain broke apart, the pumps were overwhelmed and failed, and 80-90% of the city filled with water, submerging houses and flooding multi-story buildings to their rooftops. Some of the levee breaches were the length of several city blocks, and could not be easily or rapidly repaired. At this writing (fall 2005), the city is still largely uninhabited owing to the massive water damage and health hazards caused by contaminated water.

The Future

It is too soon to know or predict what the future of New Orleans and surrounding communities will be. Will the levees be repaired and the city pumped dry? Geologists, engineers, and government officials have long known that this region's very existence depends on controlling nature, that is, controlling the course of the Mississippi River. Will the city be relocated to a less vulnerable site?

Geologically, Louisiana would not exist if the Mississippi River had not changed its course through time, depositing sand and silt as it meandered back and forth over a region about 320 km (200 mi) wide. Once New Orleans became an important port and its population expanded, the course of the river had to be contained to prevent destructive floods.

Nevertheless, the city has experienced major flooding in 1735, 1785, 1884, 1890, 1891, 1897, 1898, 1903, 1912, 1913, 1922, 1927 and so on, but with decreasing frequency (1965, 1973, 1983) as the Atlantic storm cycle entered a relatively quiet period and as the levees grew higher and longer. Despite the apparent surprise of some governmental officials, the floods of 2005 were widely expected, although perhaps they might not have been held back by levees under any circumstances, given the elevation difference between the level of Lake Pontchartrain and the low levels of the ever-subsiding city of New Orleans. The big question today: Should the city—or any community—be rebuilt in such a geologically challenging place and in the face of perpetual threat from hurricanes and floods?

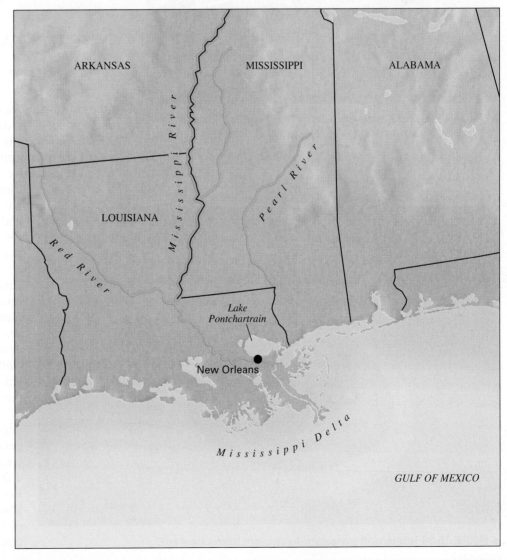

▲ **FIGURE 15-35** In 1996, water released at the Glen Canyon Dam flooded through the Colorado River valley, restoring some of the river's pre-dam ecosystem.

river's human visitors benefited, as river rafters could enjoy new camping areas along the river.

The results of this first attempt to repair the Colorado River's ecosystem, however, yielded at best, mixed results. Rather than stir up sediment from the bottom of the river, the 1996's artificial flood apparently just shifted sand from existing channel-side beaches to create others. Undaunted, the Department of Interior tried again in late November 2004. A greater torrent was released in an attempt to stir up 800,000 metric tons of sediment during a more intense, 90-hour flood. At this writing in early October 2005, ecologists are studying the effects of this second attempt at restoring some of the river's pre-dam conditions.

The Colorado River's artificial floods of 1996 and 2004 may have ushered in a new era of dam management in North America—one that has preservation of a river's ecosystem as a major goal. Scientists hope to continue to repeat the Colorado's artificial flooding at five-to-ten-year intervals. Managers of other hydroelectric-power systems are discussing plans for similar restoration floods in the Columbia River Basin of the Pacific Northwest, in Montana's Missouri River Basin, and along several major rivers in Western Canada. The governments of Japan, Turkey, and Pakistan are also considering such dam-management policies.

Stream Evolution and Plate Tectonics

Our growing knowledge of plate tectonics has added to our understanding of the development of stream systems. Plate-margin stresses produce much of the local uplift that downcutting streams erode, creating deeply incised river channels, spectacular waterfalls, and distinctive terraces (Fig. 15-36). Tectonic uplift also produces the rising mountains through which antecedent streams flow, such as the Columbia and Snake rivers of the Pacific Northwest's Cascade Mountains at the convergent boundary between the North American and Pacific plates.

Tectonic activity commonly determines the character of the regional stream-drainage pattern. For example, tectonic stresses cause the faults and fractures that produce a rectangular drainage pattern, whereas folded mountain belts at convergent plate boundaries tend to display trellis drainage. Radial drainage patterns develop as streams dissect the young volcanoes that grow at active plate boundaries.

Many large stream systems begin their journeys in the folded mountains found at convergent margins. They cross stable mid-plate interiors and then build distributary systems and deltas at passive plate margins. The Amazon River, for example, rises in the Andes of western South America, where the South Pacific's Nazca plate subducts beneath the South American plate. The river flows eastward across the conti-

▲ **FIGURE 15-36** Tectonic uplift is responsible for the stream terraces shown here.

nent for thousands of kilometers before emptying into the Atlantic Ocean. Similar tectonically produced drainage divides appear in the Alps of southern Europe, which rose from the convergence of the African and Eurasian plates; the Himalaya, produced by convergence of the Indian and Eurasian plates; and the North American Rockies, originating from convergence of the North American and Pacific plates.

Extraterrestrial Stream Activity—Evidence from Mars

Photographs taken by the 1971 *Mariner 9* spacecraft and the 1976 *Viking* probe clearly reveal braided channels, stream-modified islands, dendritic drainage patterns, and catastrophic-flood topography on the surface of Mars. These findings were confirmed by the Mars *Pathfinder* mission in 1997 and again in 1999 by the remarkably sharp images sent back to Earth by the Mars *Global Surveyor* mission. Figure 15-37 shows us a segment of the walls of a long Martian canyon with rills and gullies reminiscent of many such features we find here on Earth that we know are created by flowing surface water. Mars' surface, however, contains virtually no free water; all of the *liquid* water at its surface and in its atmosphere would barely fill a small swimming hole. The low atmospheric pressure on Mars allows the atoms on a liquid's surface to escape easily, so any surface water vaporizes immediately. Moreover, the planet's thin atmosphere does not trap enough heat reradiated from the surface to keep the surface warm. As a result, the Martian surface is too cold for water to exist in liquid form. How then can we explain the spectacular stream-carved landforms photographed on this planet?

Some geologists propose that water is trapped as ice below ground in the pores of the Martian regolith. Catastrophic events may release this water from time to time, for example, when the ice is melted by intrusions of magma or by heat generated by meteorite impacts. Many of what appear to be major flood channels occur in Mars' southern volcanic highlands. There, magma may have melted a vast volume of underground ice, causing the surface to collapse into a jumble of large, angular blocks. The ensuing flood of meltwater may have carved the region's deep canyons. When the floodwaters evaporated soon afterward, they left braided channels separated by teardrop-shaped islands of sediment. Although we cannot determine precisely how long ago such catastrophic floods took place, the numerous meteorite-impact craters superimposed on the channels suggest that they may be at least hundreds of millions of years old.

Although no rain falls on Mars today, in the early life of the solar system, water may have accumulated on its surface from volcanic outgassing of the planet's interior just as water did on Earth. The effects of volcanism may have been enhanced by extensive meteorite impacts that would have generated magma and also cracked the Martian crust, allowing subsurface magmas to rise to the surface. This activity likely created a more substantial Martian atmosphere at that time, as well as greater atmospheric pressure and some form of warming greenhouse effect. Thus, water could have survived at the surface during Mars' early years. The planet's streamlike features include braided and meandering channels and dendritic drainage patterns such as those shown in Figure 15-38. Their presence suggests that 3 to 4 billion years ago, lakes and streams may have dotted the Martian surface. These bodies of water may have been fed by frequent rainstorms. During the last few billion years, since the space in our solar system has been swept relatively clear of meteorites, Mars has experienced

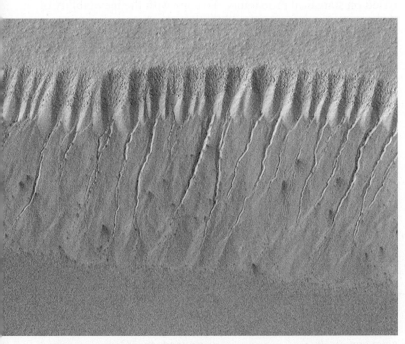

▲ **FIGURE 15-37** This 1999 photograph taken by the Mars *Global Surveyor* shows rills and gullies on the Martian surface that suggest surface flow of liquid water.

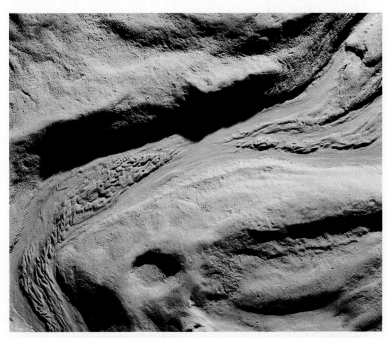

▲ **FIGURE 15-38** A 3-mi-wide channel north of the Martian equator, photographed in 1972 by *Mariner 9*. This and other apparently stream-formed features on Mars may have formed billions of years ago, when the planet's climate and atmosphere allowed free water to exist and flow at its surface.

fewer impacts and less volcanism. Thus its present atmosphere is cool, thin, and dry, and its surface is frigid, arid, and stream-free.

For most of us, streams are the most visible component of the Earth's hydrologic cycle and have the greatest impact on our daily lives. In the future, as surface waters suffer increasing contamination from growing human populations, the cycle's largely unseen groundwater component may become our best source of clean water. In the following chapter, we will examine how groundwater accumulates and flows, and how we can tap the reservoir beneath our feet while still trying to preserve its quality.

Chapter 15 Summary

Virtually all of the Earth's water initially comes from the planet's interior, principally as the steam that accompanies volcanic eruptions. All water on the planet eventually follows a path through its oceans, land, and atmosphere and between its mantle and its surface, a path called the *hydrologic cycle.*

Any surface water whose flow is confined to a narrow topographic depression, or channel, is called a *stream.* All streams are surrounded by *drainage basins,* or *watersheds,* which are the areas of land that supply their water. The topographic highland that separates two adjacent drainage basins is called a *drainage divide.* As streams flow, they erode a network of small *tributaries* that feed water into a main, or trunk, stream. Downstream, the *trunk stream* may split into a network of small channels, or *distributaries,* that usually empty into an ocean.

Stream velocity is controlled principally by two factors:

■ the slope, *or gradient,* of the stream bed,
■ the bed's roughness, which may produce friction and impede the water's flow.

The stream's velocity is directly related to *stream discharge,* the volume of water passing a point on the stream bank during a given unit of time (such as cubic feet of water per second).

Streams tend to cut downward into sediment or bedrock until they reach their *base level,* the lowest point to which a stream can erode (usually sea level). When a stream flows just swiftly enough to transport its sediment load with very little net erosion or deposition, it enters a temporary state of equilibrium and is called a *graded stream.*

Erosion, combined with mass movement, produces a stream's characteristic V-shaped valley. The rate of erosion is directly proportional to the stream's velocity. A stream erodes its bed by *abrasion,* hydraulic lifting of loose particles, and dissolution of soluble rock. Streams develop characteristic drainage patterns, formed as networks of interconnecting tributaries, trunk streams, and distributaries. These patterns reflect the drainage basin's underlying geology.

Responding to the local topography, climate, and geology, streams may be straight, braided, or meandering. When a stream erodes downward to a new base level, it often leaves behind a fairly flat-lying surface, or stream terrace, that marks the former position of the stream's *floodplain.*

A stream's ability to transport sediment varies according to several characteristics. These include:

■ The volume of sediment transported by a stream—controlled principally by its discharge—called its *capacity.*
■ The maximum size of its transported particles—largely a function of stream velocity—defined as a stream's *competence.*

The sediment in a stream can travel as a *suspended load* in the water column, bounce or roll along the stream bed as a part of its *bed load,* or move in solution as a *dissolved load.*

Floods occur when waterflow into a stream exceeds the channel capacity. They are especially likely where water can't enter the ground, such as in areas of exposed impermeable bedrock, in extensively paved urban areas, and where the ground is saturated from previous rainfalls. To predict floods, geologists document the historical pattern of floods in a region and develop a model based on statistical probability. To cope with the inevitability of floods, communities may build flood-control structures (such as artificial *levees* and flood walls), alter channels to allow water to pass through more swiftly (a strategy called channelization), and designate floodway zones in which structural development is restricted.

Plate tectonics plays an important role in the development of streams. Plate motions may control, among other things, a stream's base level, disruptions to its profile (by faulting and uplift), and its supply of sediment (particularly in volcanic areas).

KEY TERMS

abrasion (p. 471)
aggradation (p. 470)
alluvial fan (p. 482)
alluvium (p. 481)
backswamp (p. 481)
base flow (p. 467)
base level (p. 469)
bed load (p. 479)
braided streams (p. 474)
capacity (p. 479)

competence (p. 479)
degradation (p. 470)
delta (p. 484)
dissolved load (p. 479)
distributaries (p. 467)
drainage basin (p. 467)
drainage divide (p. 467)
drainage patterns (p. 471)
floodplain (p. 474)
graded stream (p. 469)

gradient (p. 467)
hydraulic lifting (p. 471)
hydrologic cycle (p. 466)
incised meanders (p. 476)
levees (p. 481)
meandering stream (p. 474)
oxbow lake (p. 474)
point bars (p. 481)
runoff (p. 466)

stream (p. 467)
stream discharge (p. 469)
stream terrace (p. 477)
suspended load (p. 479)
tributaries (p. 467)
trunk stream (p. 467)
watershed (p. 467)
wetlands (p. 482)

QUESTIONS FOR REVIEW

1. Briefly describe the main elements of the hydrologic cycle, including the processes responsible for water's changes in state. List three ways water can be stored temporarily on land before eventually reaching an ocean.

2. How do we measure stream discharge? How do changes in stream velocity, width, and depth affect stream discharge? Calculate the discharge of a river that is 70 m wide, 20 m deep, and flows at an average velocity of 10 m/s.

3. Briefly explain how one stream might "pirate" water from another stream.

4. Draw a sketch illustrating the origin of an oxbow lake. How does an oxbow lake become a meander scar?

5. How do the sediment particles that are most likely to be transported in suspension differ from those transported by traction? What is the difference between a stream's competence and its capacity?

6. What are the principal causes of flooding?

7. Explain how structural and nonstructural approaches to flood control differ.

8. List three ways in which plate tectonics may affect the evolution of a stream profile.

9. How would the Earth's fluvial landscapes change if plate-tectonic activity ceased suddenly?

10. How would the world's streams respond if climatic warming melted the Antarctic ice sheet?

11. How would the potential for flooding change over a long period of time if we removed the dams along the Mississippi River?

12. What kind of deposits do you think underlie the farmland shown in the photo below?

LEARNING ACTIVELY

1. Regardless of where you live, a stream is probably nearby. Pay a visit to the stream and study it as a geologist might. You can estimate your stream's discharge by estimating its width (by sight) and depth (with a stick, pole, or a quick swim, unless it's the Mississippi, Missouri, Colorado, Ohio, or Hudson River) and by making a simple calculation of its velocity. To estimate the velocity, toss a leaf into the stream and time the leaf for 5 seconds, noting how far it traveled. Measure the distance by pacing it off. (Each pace equals the two-step distance covered when you walk from right foot to right foot again—that is, right, left, right. A person of average height has a pace between 5.5 and 6.5 ft.) Thus, if your leaf traveled 60 ft in the 5 seconds (equal to 10 paces or so), its velocity is 12 ft/s. Multiply the width times the depth times the velocity to get the stream's discharge in cubic feet per second. Come back after a major rainstorm and try this activity again, focusing on the change in velocity.

2. Look again at your stream and try to determine whether its channel is straight, braided, or meandering. Look for levees, point bars, and floodplain deposits. Visit your stream after it has flooded (but definitely not *during* a flood). Look for the effects of the flood's high currents (such as the sizes of the particles dropped on the floodplain).

3. Interview people who have lived in the area near your stream for a long time and create an oral history of the stream's flooding. Estimate its flood frequency from these interviews. Look at the local newspaper archives and assess the stream's flood-frequency interval from the data. Note all references to 100-year floods and other statistical measures of flooding.

4. Look at a newspaper every day for a month and note the frequency of articles about flood disasters around the world. One or more major floods should be mentioned virtually every week.

5. The next time you fly in an airplane on a clear day, ask for a window seat, and take some time out from your reading or the movie to catalogue the various stream features you can see from 30,000 ft. Having studied geology and read this chapter, you'll be amazed how much you know about streams and their landforms. Pay particular attention to the drainage patterns. You can expect to see a lot of dendritic patterns.

ONLINE STUDY GUIDE

Practice makes you better. Make the time to take the *Online Study Guide* quiz for this chapter. It should take you about 20 minutes. Automatic grading and instant feedback will help you quickly and accurately identify the concepts you have mastered and the areas in which you need more study.
www.prenhall.com/chernicoff

Groundwater

16

- **Groundwater Infiltration and Flow**
- **Tapping the Groundwater Reservoir**
- **Finding and Managing Groundwater**
- ***Highlight 16-1*** Should You Hire a Water Witch?
- ***Highlight 16-2*** Who Owns Groundwater?
- **Maintaining Groundwater Quality**
- ***Highlight 16-3*** Yucca Mountain—A Safe Place for N-Wastes?
- **Caves and Karst—The Products of Groundwater Erosion**
- **Caves**
- **Karst Topography**

For thousands of years, wells and natural springs have supplied clean, abundant groundwater to human communities throughout the world (Fig. 16-1). Located in the centers of many villages, wells have been the key meeting place for commerce and for sharing the news of the day. As in the past, today an adequate supply of uncontaminated groundwater often determines whether a town or city will grow and prosper. Pure water has even become a big money-maker: Check the shelves of your local supermarket and you'll find bottles of rather ordinary groundwater with fancy names selling for unusually high prices.

Humans can survive for weeks—even months—without food. But because virtually all biological processes take place in water, we can survive without water for only a few days. Our bodies require about 4 liters (1 gal) of water per day. Thus, we must draw 1.6 billion liters (400 million gals) of water from the continent's freshwater reserves *each day* just to meet the biological needs of North America's 400 million people. Furthermore, we use a thousand times that amount—more than 1.8 *trillion* liters (450 billion gallons) of water *per day*—for domestic purposes (cooking, bathing and other sanitary needs, and lawn watering), agriculture (livestock and irrigation), and industry (processing natural resources and manufacturing the goods that maintain our way of life).

The Earth's *hydrosphere* extends from the top of the atmosphere to approximately 10 km (6 mi) below the Earth's surface and includes the planet's oceans, glaciers, rivers and lakes, atmospheric water vapor, and groundwater. This water constantly moves through the various sites within the hydrologic cycle (see Figure 15-3). The largely unseen groundwater part represents only 0.6% of the world's water. It, however, accounts for 97% of the Earth's supply of unfrozen freshwater, provides more than 50% of our drinking water, 40% of our irrigation water, and 25% of the water used in industry. This water fills the countless tiny pore spaces of rocks and sediments, like water that fills a wet sponge.

Groundwater use doubled between 1955 and 1985 as the U.S. population grew and the nation's economy became more industrialized. Throughout North America, we are withdrawing groundwater from reserves that have accumulated over thousands of years, and groundwater supplies (particularly in the Southwest) are most definitely running low. On the bright side, in many areas climate and local geology are still capable of providing an ample supply of clean groundwater for present and future generations—*if* we learn to manage and preserve this critical resource. In other areas, however, human, animal, and

FIGURE 16-1 This well in a village in Yemen is the center of town life. It is the only source of drinkable water in the ▶ town.

501

industrial wastes have contaminated the groundwater. Some urgent questions facing the world's citizens relate to groundwater: Where do we find it? How do we keep it clean? Who owns it?

To answer these questions, we need to understand how groundwater seeps through the Earth's surface and how it flows from one place to another—topics covered in this chapter's first section. After examining this issue, we discuss the *water table*—the underground surface beneath which water fills all cracks, crevices, and pore spaces. Then, we turn to the problems of finding and managing groundwater and maintaining its quality. Finally, we discuss the origins and structural aspects of the geological products of flowing groundwater in limestone regions—our fascinating caves and karstlands.

Groundwater Infiltration and Flow

Wherever the ground is receptive, surface water, mostly from precipitation, may seep into, or *infiltrate*, the Earth's surface and become part of the groundwater system. After infiltration, groundwater flows principally downward under the influence of gravity, through soils, sediments, and rocks. Some of the water eventually reaches the water table, flows as groundwater, and ar-

rives back at the surface again as groundwater *discharge,* as streams, springs, or the waters of wetlands. Some of the water may remain stored underground perhaps for thousands of years.

As shown in Figure 16-2, the amount of water that infiltrates and eventually arrives at the water table and the groundwater system depends on a host of factors, including: the condition and type of local surface rocks and soils; the nature and abundance of vegetation in the area; the area's topography; and, the region's amount of precipitation. We describe each of these factors next.

Condition and Type of Surface Materials   The abundant and relatively large pore spaces in loose soils and unconsolidated sands and gravels enable water to seep into the Earth's surface. Water can also enter if a region's exposed bedrock is fractured or jointed, or inherently porous, such as a coarse sandstone. On the other hand, the tiny pore spaces within unconsolidated clay limit infiltration. Infiltration is also poor in exposed *unfractured* crystalline bedrock, such as plutonic igneous rocks (for example, granite and diorite), high-grade metamorphic rocks (such as gneiss), fine-grained sedimentary rocks such as shales, and the concrete and asphalt surface humans produce when they build their cities and towns.

Vegetation   Much of the precipitation that falls in heavily vegetated areas lands on the leafy canopies of trees and evaporates back into the atmosphere. Some of the precipitation that reaches the ground is taken up by plant roots and then rises through the

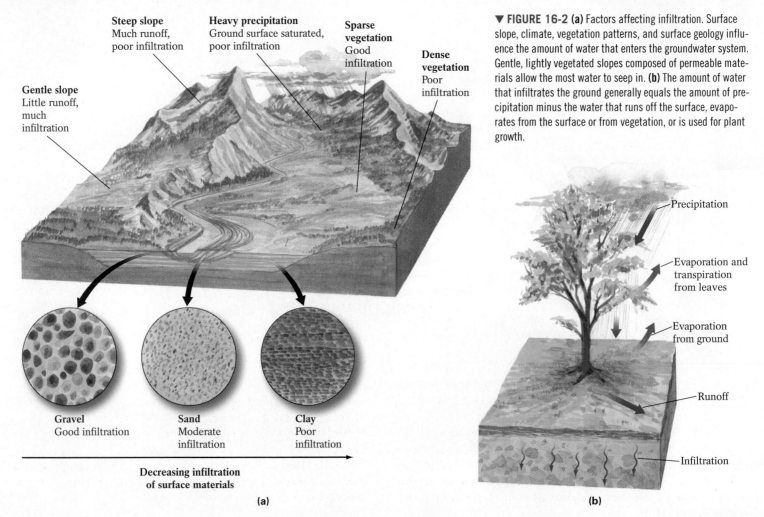

▼ **FIGURE 16-2 (a)** Factors affecting infiltration. Surface slope, climate, vegetation patterns, and surface geology influence the amount of water that enters the groundwater system. Gentle, lightly vegetated slopes composed of permeable materials allow the most water to seep in. **(b)** The amount of water that infiltrates the ground generally equals the amount of precipitation minus the water that runs off the surface, evaporates from the surface or from vegetation, or is used for plant growth.

Steep slope Much runoff, poor infiltration

Heavy precipitation Ground surface saturated, poor infiltration

Sparse vegetation Good infiltration

Dense vegetation Poor infiltration

Gentle slope Little runoff, much infiltration

Gravel Good infiltration

Sand Moderate infiltration

Clay Poor infiltration

Decreasing infiltration of surface materials

(a)

Precipitation

Evaporation and transpiration from leaves

Evaporation from ground

Runoff

Infiltration

(b)

plant to its leaves. There the water then evaporates from the small openings on the underside of the plant's leaves (called *stomata*) and returns to the atmosphere, a process called *transpiration* (Fig. 16-2b). Approximately 25% of falling precipitation actually reaches the ground—unaffected by vegetation—but it then either evaporates from the ground surface, runs off to streams, or is temporarily stored in lakes or glaciers. The remaining precipitation may then seep into the ground, *if surface conditions permit.* Plant and tree roots enhance infiltration by opening pathways in soil into which water can seep.

Topography   Surface water runs slowly down gently sloping terrains, so it has plenty of time to seep into the ground. On steep slopes and cliff faces, the water runs rapidly downslope and enters nearby streams before much of it can infiltrate the soil.

Precipitation   The amount of precipitation that falls over time affects the amount of groundwater infiltration *in any kind of terrain.* Extended droughts may slow groundwater infiltration significantly for several years or more. Variations in the amounts of spring rains and snowmelt may have a short-term, seasonal effect on the groundwater supply.

The type of precipitation also affects infiltration. A driving, hard-pounding rainstorm packs surface soils and washes fine clayey particles into soil pores, clogging them and thus limiting infiltration. Likewise, water from the last in a series of drenching storms tends to favor runoff; the first storm saturates the surface, and water from subsequent storms cannot find pathways into the soils.

Movement and Distribution of Groundwater

Once water has entered the Earth's surface, gravity pulls it downward until it reaches the water table. The types of soils, sediments, and rocks that the water encounters determine the depth to which it descends. Figure 16-3 illustrates the subsurface levels through which water travels and collects.

As water passes through the weathered regolith, some water molecules become attracted and bound to clay minerals in the soil or are taken up by plants. The attraction between water and the charged surfaces of clay minerals reduces both evaporation and infiltration from this near-surface region as water is held tightly by the clay particles. The water that is not bound by clay minerals, used by plants, or lost to evaporation moves downward through a **zone of aeration,** or unsaturated zone (also known as the *vadose zone*), where the pore spaces in rocks and soils contain both water and air. Some water descends farther, into the **zone of saturation** (also known as the *phreatic zone*), where water fills *every* available pore space. The **water table**—the upper surface of the zone of saturation—separates the zones of aeration and saturation. The lower part of the aeration zone, which can range from a few centimeters to about 1 m above the water table, is called the **capillary fringe.** In this region, the gravity-driven downward flow of water is partly countered by the attraction of the water molecules to mineral surfaces and to other water molecules. This attraction causes a small volume of water to rise *upward* under saturated conditions from the water table *against* gravity and be held in the unsaturated zone. This phenomenon,

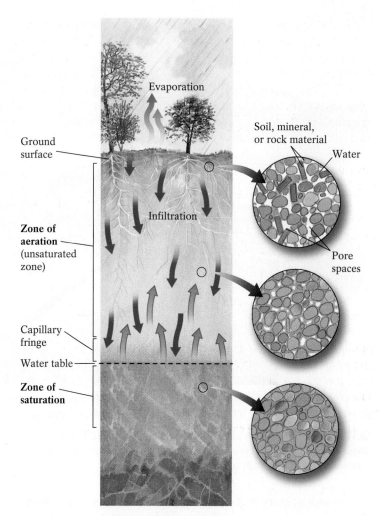

▲ **FIGURE 16-3** The subsurface distribution of water.

known as *capillary action,* can be seen when you dip the corner of a paper towel into a glass of water and the water rises in the towel against gravity.

The saturation zones in most places do not extend to great depths because the pressure of overlying rock closes deep fractures and recrystallizes some rock into crystalline metamorphic rocks with very low permeabilities. For these reasons most groundwater is confined to the upper 1000 m (0.6 mi) of continental crust. Given adequate precipitation, the availability of groundwater is governed by two characteristics of the soil or rock unit into which the water might infiltrate—its porosity and its permeability.

Porosity   **Porosity** is the volume of pore space compared to the total volume of a soil, rock, or sediment. Porosity (which is expressed as a percentage) determines *how much water the material can hold.*

Primary porosity is the porosity that develops as a rock forms. For example, the primary porosity of basalt results from the presence of *vesicles*—the small gas-bubble holes that remain near the top of a lava flow after a lava's gases have escaped and the lava has cooled and solidified. *Secondary* porosity develops *after* a rock has formed, usually as a result of faulting, fracturing, or dissolution.

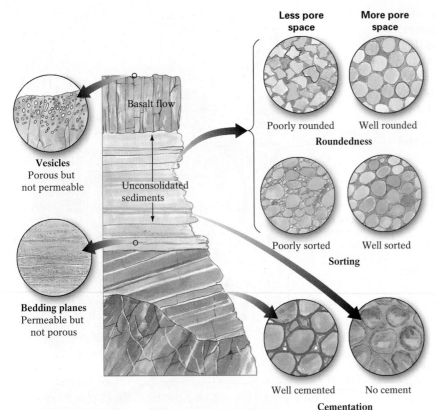

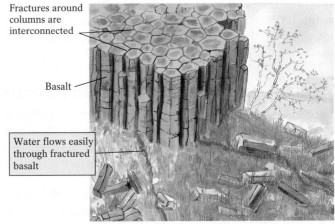

▲ **FIGURE 16-4** The primary porosity of sedimentary rocks is affected by the grain shape, sorting, and particle cementation. The presence of vesicles and bedding planes also increases a rock's primary porosity.

In sedimentary rocks, primary porosity is determined by such factors as grain rounding, sorting, and cementation. As you can see in Figure 16-4, well-rounded grains tend to have larger pore spaces and therefore hold more water. When a sediment contains grains of various sizes (that is, when it is poorly sorted), the finer particles tend to fill the voids between the coarser particles, clogging the pores and thus reducing porosity. When cementation converts loose sediments to sedimentary rock, the cement fills the pore spaces and further diminishes porosity. Fine clayey muds (though they slow groundwater infiltration) hold much more water when saturated than coarse sediments because the clay contains a higher percentage of pores than the coarse sands. Fresh mud from the Mississippi River's delta, south of New Orleans, may consist of 80% water. Yet, because the extremely small size of the pores slows flow, this water is very hard to extract from the ground.

Permeability The crucial factor that determines the availability of groundwater is not how much water the ground holds, but whether the water can flow *through* soil and rock pores. **Permeability** is the capability of a rock, sediment, or soil to allow the passage of a fluid. The permeability of rock or sediment is controlled by the size of its pore spaces and the extent to which these spaces are connected. If pores are very small, as in clay-rich sediment, water flows through them extremely slowly. Some water molecules may stick as a fine film to adjacent particles, narrowing the pore spaces and slowing the flow even further. Only when pores are relatively large can water flow easily. The pores between grains of clean sand are more than 1000 times larger than

the pores in clay, explaining why sand is so much more permeable than clay.

The passage of groundwater may be limited even in a porous material if the pores are sealed off from one another. For example, the pores in a very porous limestone may not be connected, and thus water does not flow readily through this rock. In other cases, a material may be essentially nonporous, such as the basalt shown in Figure 16-5, but still be permeable, because it contains a network of interconnected fractures that accommodate a modest flow of water. A rock such as basalt, which often develops a highly interconnected fracture system as it cools, can therefore serve as a reliable source of groundwater.

How Groundwater Flows Gravity causes surface water in a stream to move down its channel's slope from higher to lower elevations. Gravity causes groundwater to flow from high to low areas as well, but its movement is also driven by differences in pressure on the water caused by the weight of overlying water and, at greater depth, from the weight of the overlying rocks. The **potential** of groundwater flow at any depth below the water table combines the influence of gravity and the pressure on the

▲ **FIGURE 16-5** Pore connection and permeability. Although basalt is a generally nonporous igneous rock, its connected network of columnar joints makes this basaltic lava flow highly permeable. In contrast, the fractures in the granitic bedrock may be poorly interconnected; thus this granitic outcrop is relatively impermeable.

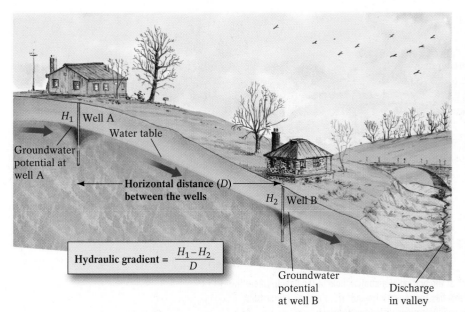

$$\text{Hydraulic gradient} = \frac{H_1 - H_2}{D}$$

▲ **FIGURE 16-6** The hydraulic gradient is the difference in potential between two areas $(H_1 - H_2)$ divided by the distance between them (D). Groundwater flows from areas of high potential (beneath hills) to areas of low potential (beneath valleys); the rate of flow is generally greater where a shorter distance separates the two areas.

water at that depth. Groundwater generally moves from areas of high potential, such as the high spots in water tables under hills, to areas of low potential, such as the low spots in water tables under valleys.

Groundwater flows between any two points that differ in elevation and pressure. As shown in Figure 16-6, the flow rate is partially controlled by the **hydraulic gradient,** the difference in the potentials of two points divided by the lateral distance between those points. Groundwater tends to travel most swiftly between two points when one site has a much greater potential than the other, and the distance between the points is small. (These conditions resemble surface-water flow on a very steep slope.)

The flow rate of groundwater is also affected by the material through which it flows. Every rock, sediment, and soil possesses a characteristic **hydraulic conductivity** (a way to *quantify* permeability) that reflects, among other things, the sizes, shapes, and degree of sorting of its grains. Coarse, well-rounded, well-sorted gravel, for example, has high hydraulic conductivity (that is, it lets water pass through easily), whereas fine-grained, angular, poorly sorted sediment has low hydraulic conductivity (that is, it impedes groundwater flow).

Henry Darcy, a 19th-century French hydraulic engineer, was the first person to calculate the flow rate of groundwater. Through his experiments, Darcy demonstrated that the rate at which groundwater flows between two wells is directly proportional to the difference in potential between the wells and to the hydraulic conductivity of the materials through which the water must travel. Where the hydraulic gradient and the hydraulic conductivity are high, groundwater flows more swiftly than where the hydraulic gradient and hydraulic conductivity are low.

Geologists use many methods to measure groundwater flow, including two methods that can measure the rate of groundwater flow *directly* at any location. When the water's path is known, geologists inject a harmless dye into a well. They then compare the

time of injection with the time of the dye's arrival at a nearby well downslope. Dividing the distance between the wells by the time elapsed yields the flow rate.

To determine groundwater flow rates over much longer distances, geologists sometimes use radiocarbon dating. Rainwater incorporates carbon-14 as it passes through the atmosphere and soils. As the groundwater moves slowly for a long distance underground, its carbon-14 undergoes radioactive decay. The distance between the groundwater-*recharge* site (the place where the water enters the ground) and the well where the water is sampled, divided by the carbon-14 age of the water, yields an estimate of the flow rate, assuming the water hasn't become enriched in carbon isotopes along its route (for example, by encountering underground layers of coal).

Groundwater flow is typically quite slow, averaging between 0.5 and 1.5 cm (0.2 and 0.6 in) per day through moderately permeable material, such as poorly sorted sand. Flow through unfractured crystalline rocks such as granites and gneisses may be only a few tens of centimeters *per year* or less. The swiftest flow—through well-sorted, coarse, uncemented gravels or highly fractured basalts—can reach 100 m (330 ft) per day.

The slow movement of groundwater accounts for its availability for human use. If this water traveled as rapidly as rivers, it would not be stored for long in the ground. The groundwater we draw today from a deep well may have actually fallen as rain thousands of years ago. Deep groundwater drawn from wells in the hot, dry lands of southern Arizona, for example, is more than 10,000 years old. This water apparently precipitated during the last worldwide glaciation when Arizona's climate was cool and cloudy enough for rainwater to infiltrate the groundwater system instead of evaporating, as much of it does today.

Tapping the Groundwater Reservoir

We rely heavily on groundwater to meet our wide range of water needs—and how better to tap a region's groundwater supply than to locate the local water table. The first clue—lakes, swamps, and year-round streams appear where the water table intersects the Earth's surface. To determine the depth of an area's water table where no such surface water arises, we can dig a hole or drill a well. We know we have reached the local water table when water begins to seep into the hole or well, indicating that the surrounding material is saturated with water. If you've ever dug a hole at the beach, you probably noticed that water begins to accumulate at the hole's bottom almost immediately—considerably less than a meter below the surface. Inland, in moist climates, the water table typically occurs within about 1 to 2 m (3.3 to 6.6 ft) of the surface. To confirm this fact, you could dig a hole in a backyard garden and watch the seeping groundwater pond up at the bottom of the hole. Don't bother digging in arid locales because

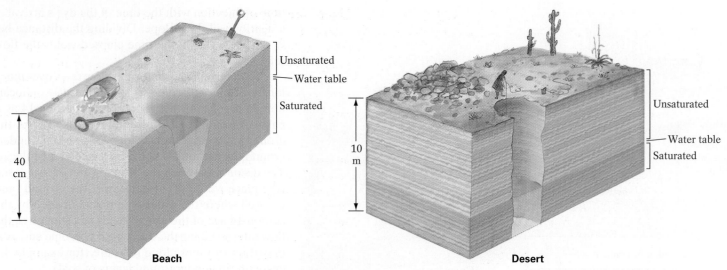

▲ **FIGURE 16-7** Variations in water-table depth. Groundwater pools at the bottom of a shallow hole at the beach, where the water table lies close to the surface. In the desert, where the water table lies far beneath the surface, the hole intersecting the water table must be much deeper. Note that the scales for these two illustrations differ substantially.

water may not appear until the hole reaches tens of meters below the surface. Figure 16-7 compares the typical depth of the water table at two different locations: a beach and a desert.

The depth of the water table varies according to the local topography and climate. The water table often mimics the surface topography, although its highs and lows are less pronounced than those of the overlying landscape (Fig. 16-8). (In moist, temperate regions, the water table is essentially a subdued replica of the surface topography.) Its depth generally reflects a long-term balance between recharge and discharge. The water table fluctuates seasonally—it tends to be higher during the rainy winter months in

temperate mid-latitude places and lower during the warm, sunny summer months. Major climatic events, such as storms and droughts, can raise or lower the water table *temporarily,* as can variations in our rate of extraction of water from the ground.

A prolonged dry spell, such as the one that struck the northeastern United States during the early 1960s, can lower the water table dramatically. During that drought, the region experienced a 2-m (6.6-ft) drop of its water table. Some unethical developers took advantage of these temporary conditions and constructed many homes in areas that normally have high groundwater levels. When the drought ended in the mid-1970s and recharge increased, the water table rose to the region's normal pre-drought levels. Many unhappy homeowners in Long Island and New Jersey soon discovered that their flooded basements sat squarely within the zone of saturation.

Sometimes a *local* impermeable layer (such as a thin layer of fine clay) may lie within and obstruct downward flow of water through a region's zone of aeration. Water may then accumulate above this barrier, saturating part of the overlying zone. In this case, what may appear to be the regional water table is actually a **perched water table,** a zone of saturation that lies within the zone of aeration (Fig. 16-9).

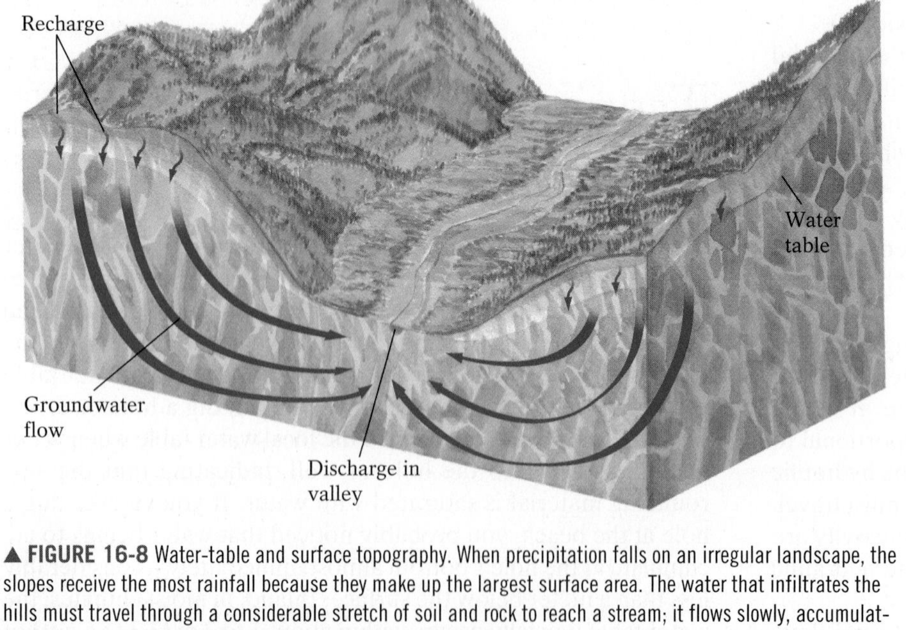

Aquifers—Water-bearing Rock Units

Aquifers (from Latin, meaning "to bear water") are permeable bodies of rock, sediment, or soil that transmit groundwater readily. They are the underground bodies from which we withdraw the groundwater we use to meet our domestic, agricultural, and industrial needs. The most productive aquifers consist of unconsolidated sand and gravel, well-sorted, poorly cemented sandstones, and highly jointed limestones and basalts.

▲ **FIGURE 16-8** Water-table and surface topography. When precipitation falls on an irregular landscape, the slopes receive the most rainfall because they make up the largest surface area. The water that infiltrates the hills must travel through a considerable stretch of soil and rock to reach a stream; it flows slowly, accumulating below the surface of the hills. If groundwater recharge were to cease for a long time—for instance, during a drought—the water-table highs would flatten out under the hilly areas as groundwater flow continued. In most humid areas, however, rainfall is able to replenish the supply beneath the hills and maintain the ups and downs of the water table.

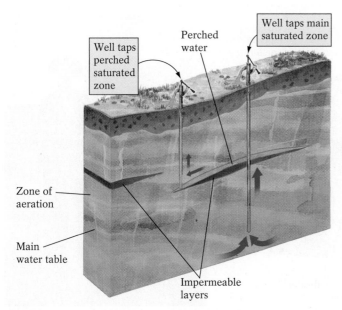

▲ **FIGURE 16-9** Perched water forms in the zone of aeration when a local impermeable layer blocks the downward flow of water. To ensure a steady, year-round flow of groundwater, wells must be drilled to a depth below the main—not the perched—water table.

Unconfined and Confined Aquifers

Unconfined aquifers extend nearly to the Earth's surface. Because the regional water table lies *within* them, these shallow resources are relatively easy to tap. The water in unconfined aquifers is generally "local" in origin, flowing primarily through the loose slope material, sands, gravels, and floodplain deposits left behind by streams and rivers, sands and gravels transported by recent glaciers, and young, jointed lava flows such as those in Hawai'i and the Pacific Northwest. In the unconfined aquifer shown in Figure 16-10, water flows through the glacial debris and river alluvium. The Ogallala Formation—a major unconfined aquifer—consists of a poorly sorted, generally uncemented mixture of clay, silt, sand, and gravel. A component of the High Plains regional aquifer system, the Ogallala aquifer underlies much of Nebraska and parts of South Dakota, Wyoming, Colorado, New Mexico, Kansas, Oklahoma, and Texas. It supplies 1.2 trillion liters (317 billion gallons) of water per year to irrigate 14 million naturally arid acres. Lands irrigated by water from this aquifer yield 25% of all U.S. feed-grain exports and 40% of the country's flour and cotton exports.

Sometimes groundwater resides at greater depth, flowing through deeper layers of rock or sediment, typically far below the water table and the near-surface unconfined aquifer, and separated from them. Such *confined aquifers* comprise water-transmitting layers of rock or sediment sandwiched between (that is, confined by) other rock layers that are either effectively impermeable or have very low permeability. Such low-permeability confining layers are called **aquitards.** Unlike the locally derived water in most unconfined aquifers, the water in confined aquifers may have traveled a great distance from its recharge site. One such confined aquifer, the Dakota Sandstone, gets its water supply from the Black Hills of South Dakota but then flows at depth beneath the Great Plains of the eastern Dakotas and Nebraska. Aquifers can even be found at great depths under deserts, such as in Nevada, eastern Utah, and southern Arizona. The water in the deep confined aquifers comes from precipitation that entered the groundwater system through exposed permeable rocks in mountain ranges, hundreds of kilometers away.

Artesian Aquifers

Under certain geological conditions, such as those illustrated in Figure 16-11, the water in a confined aquifer may rise *against* the downward pull of gravity and even gush unaided from the ground. This situation occurs where the confined aquifer is tilted at an

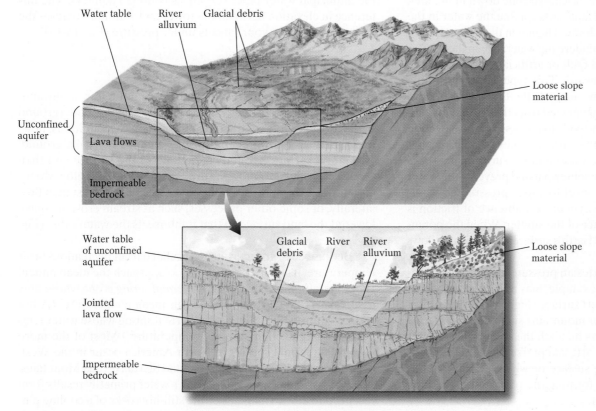

▲ **FIGURE 16-10** This composite landscape shows an unconfined aquifer composed of four different types of materials: glacial debris, loose slope material, jointed lava flows, and river alluvium.

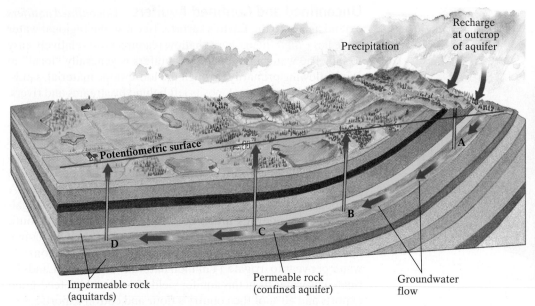

▲ **FIGURE 16-11** Artesian aquifers. (above) High pressure causes groundwater to rise in the wells at points B, C, and D, where the potentiometric surface is significantly higher than the well sites. Water does not flow unaided from well A, the top of which lies above the potentiometric surface. (photo) Water gushing from a natural artesian well in the Dakota Sandstone aquifer (circa 1910). When this aquifer was first discovered, water gushed freely from wells under unusually high artesian pressures. During the last 75 years, more than 10,000 wells have been drilled to tap the Dakota Sandstone aquifer. The extensive withdrawal of water has reduced the pressure within the aquifer to the point that the potentiometric surface is no longer higher than the land surface of the eastern Dakotas. Today, water must be actively pumped in areas where it once flowed out forcefully and unassisted.

angle to the Earth's surface so that it becomes exposed at the surface high in the mountains. Thus the aquifer's recharge zone is at a much higher elevation than its discharge zone down in the lowland. Because the surrounding aquitards confine the water in the aquifer, the weight of the water located higher in the inclined water column presses strongly on the underlying water. When the aquifer is tapped (either naturally by a fault or artificially by a well), the pressure drives the water upward. This type of self-pumping aquifer is referred to as **artesian**—named after the town of Artois, France, where water has flowed unaided from the ground for centuries. Although artesian water tends to rise *toward* the elevation of the point of infiltration at the surface, it never quite reaches this point; friction between the water and the surrounding rocks and sediments inhibits the flow somewhat and prevents the water from reaching this level. The level to which pressurized water would rise in a tall pipe in the *hypothetical* absence of friction is called the **potentiometric surface** of the aquifer. When the potentiometric surface lies above the Earth's surface, groundwater flows freely from the ground.

Sometimes water under artesian pressure may surface in unexpected places. An oasis, for example, may arise where pools of artesian water appear at a desert surface (Fig. 16-12). An aquifer that receives its recharge from mountains surrounding a desert may extend hundreds of meters beneath the arid surface. Water may flow to the surface under artesian pressure where the rocks of the aquifer crop out at the surface or where a fault extends down to the aquifer, thereby forming the oasis and nourishing the palm trees that surround it.

Most modern municipal water-supply systems are designed to simulate artesian conditions. Have you ever wondered why water

flows at high pressure when you turn on a faucet? The tall water towers in your city or town act like elevated recharge areas, with the municipal water pipes serving as confined aquifers. The difference in elevation between the tower and your home causes the water to flow from your faucets under pressure (Fig. 16-13).

Natural Springs and Geysers

With the exception of artesian aquifers and the flow of groundwater-fed streams, groundwater typically remains underground until we drill wells to extract it. Some geological conditions, however, allow groundwater to "spring" unaided from the ground. At **natural springs,** groundwater flows to the surface and then freely from the ground. Natural springs may develop where groundwater encounters an impermeable bed and must then flow laterally, or some other processes, such as stream erosion—have lowered the land surface so that it intersects the water table (Fig. 16-14).

Because deep groundwater is insulated from fluctuations in air temperature, its temperature tends to approach the mean annual air temperature for the region. A *thermal spring* is one whose flow is at least 6°C (11°F) higher than this mean temperature. (A hot spring is informally defined as a thermal spring whose water temperature exceeds human body temperature.) Most of the more than 1000 thermal springs in North America occur in the West, near the Rocky, Cascade, Olympic, and Sierra Nevada Mountains.

The warmth of a thermal spring's water primarily results from recent contact with magmas or the still-hot rocks of a cooling pluton. After moving downward for hundreds of meters, the groundwater is heated by warm igneous rocks, in some places to its

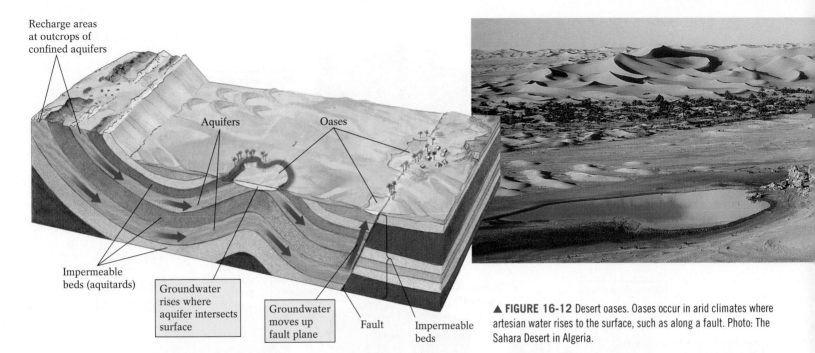

Recharge areas
at outcrops of
confined aquifers

Aquifers

Oases

Impermeable
beds (aquitards)

Groundwater
rises where
aquifer intersects
surface

Groundwater
moves up
fault plane

Fault

Impermeable
beds

▲ **FIGURE 16-12** Desert oases. Oases occur in arid climates where artesian water rises to the surface, such as along a fault. Photo: The Sahara Desert in Algeria.

boiling point. As the hot water rises, it usually dissolves large quantities of soluble rock. Upon reaching the surface, the water evaporates, depositing the minerals as *sinter* (if it's silica) or *travertine* (if it's calcite) (Fig. 16-15). Sometimes the hot water dissolves sulfur on its way to the surface; in that case, the spring gives off the distinctive rotten-egg odor of hydrogen sulfide, and its deposits have a yellowish tinge.

Most eastern hot springs, such as those found in the Appalachians and Ouachitas, are not heated by hot igneous rocks. Instead, deeply circulating groundwater passes through rocks warmed by the Earth's geothermal heat flow, hundreds of meters below the Earth's surface. For example, the water that flows from Warm Springs, about 90 km (60 mi) south of Atlanta, Georgia, enters the Hollis Formation at Pine Mountain at an average temperature of 16.5°C (62°F). After an underground journey of 3 km (2 mi) along a curving path that descends to a depth of 1100 m (3300 ft), the warmed-up water flows out at Warm Springs under artesian conditions with a temperature of 36.5°C (98°F).

An *intermittent* surface emission of hot water and steam is called a **geyser** (from the Icelandic *geysir*, meaning "to rush forth"). Geysers occur where surface water seeps into the ground, descends through fractured permeable rock, and is warmed by an underlying heat source, such as a shallow magma chamber or a body of young, warm igneous rock. It is then pushed back to the surface by rising steam under great pressure (Fig. 16-16). The pressure exerted by the overlying water column raises the boiling point of water to more than 100°C (212°F). The expansion of the hot water near the heat source pushes some water out of the geyser opening, which in turn reduces the pressure on the deep water and thus lowers its boiling point. The deep water—already heated to temperatures exceeding its boiling point—then flashes to steam, forcefully expelling all the water from the column. A period of recharging and reheating then occurs, followed by another eruption. The time lapse between eruptions varies with the groundwater supply, heat supply, permeability of the rock, and complexity of the network of fractures that delivers the water to the heat source.

The best-known geyser fields lie in areas of current or recent volcanism. Volcanism at the divergent zone that bisects Iceland and at the subduction zone under New Zealand produces geysers. In North America, geysers in northern California derive their heat from the subduction that fuels the volcanism of Mount Shasta and Lassen Peak.

In Yellowstone National Park, hot-spot volcanism powers the park's numerous geysers. Yellowstone's Old Faithful geyser, so named because of its once remarkable punctuality, used to shoot a jet of steam and boiling water to a height of 50 m (170 ft) at roughly one-hour intervals. Recent earthquake activity in the region, however, has altered the geyser's plumbing, affecting its recharge time. Nowadays Old Faithful erupts *on average* every 90 minutes, but intervals between eruptions may vary from 35 to 120 minutes, with eruption

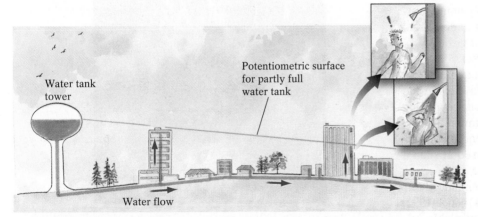

Water tank
tower

Potentiometric surface
for partly full
water tank

Water flow

▲ **FIGURE 16-13** A municipal water tower functions in much the same way as the recharge area of an artesian aquifer. Water on the lower floor of a building rises under high pressure, because its outlet is located below the potentiometric surface. Water pressure is low on the top floor because its outlet is located at a higher elevation than the potentiometric surface.

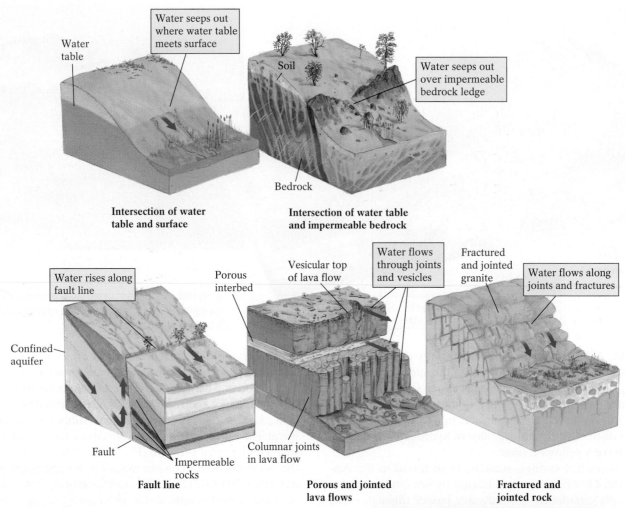

Intersection of water
table and surface

Intersection of water table
and impermeable bedrock

Fault line

Porous and jointed
lava flows

Fractured and
jointed rock

▲ **FIGURE 16-14** Natural springs. (above) Various geological settings in which natural springs occur. (photo) Water gushing from limestone rock in Switzerland.

durations of 1.5 to 5.0 minutes and eruptive heights from 27 to 55 m (90 to 184 ft). Today, the geyser is far less "faithful."

In November 2002, an earthquake again altered the daily schedule of Yellowstone's geysers, but this time the quake in question occurred in the Denali area of Alaska, 3100 km (2000 mi) to the northwest. As the surface waves from this magnitude-7.9 quake passed through Montana and Wyoming, normally quiet hot springs suddenly began to boil over and spout to heights exceeding 1 m. Normally clear-water springs began to flow muddy. Eight of 22 of the park's largest geysers began to erupt more frequently, while a few erupted less frequently. Researchers suggest that the quake's vibrations may have disturbed some of the mineral deposits that line the geyser vents, changing the rates of water flow within their subterranean plumbing. These events mark North America's first recorded occurrence of quake-related changes in geysers and hot springs triggered at such a great distance from an earthquake epicenter.

Recent technological advances in optics have enhanced our understanding of how Yellowstone's geysers work. Since 1992, when miniature video cameras first became widely available, scientists at Yellowstone have been examining firsthand Old Faithful's "arteries." They have produced a 15-m (50-ft)-long profile of Old Faithful's vent by using a 5-cm (2-in) insulated video

▲ **FIGURE 16-15** Travertine deposited at Mammoth Hot Springs in Yellowstone National Park, Wyoming. (left) Hot groundwater has infiltrated and dissolved carbonate deposits and then precipitated them at the surface to create white cliffs of travertine. (right) Steam vents in the travertine.

▼ **FIGURE 16-16** Geysers develop when groundwater encounters a shallow heat source and erupts at the surface as boiling water and steam. Water pressure at the bottom of the water column inhibits boiling initially. As water near the top of the column is pushed out by expansion of the heated bottom water, the pressure on the column diminishes and the "overheated" water turns to steam. Photo: Strokkur geyser in Iceland.

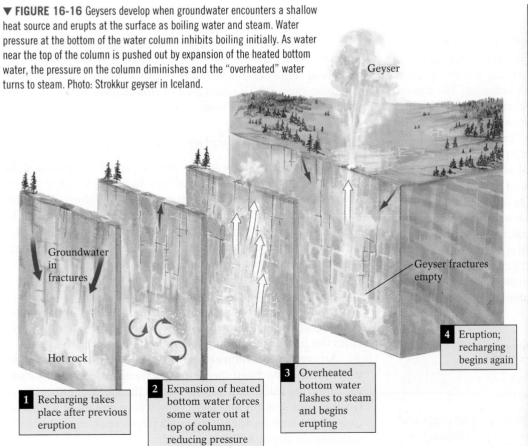

Geyser

Groundwater in fractures

Geyser fractures empty

Hot rock

1 Recharging takes place after previous eruption

2 Expansion of heated bottom water forces some water out at top of column, reducing pressure

3 Overheated bottom water flashes to steam and begins erupting

4 Eruption; recharging begins again

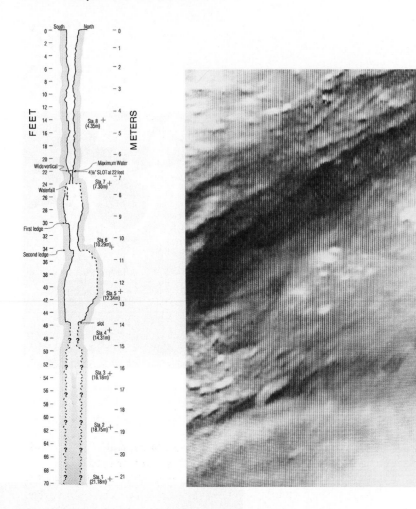

Constriction

◀ **FIGURE 16-17** A heat-tolerant fiber-optic camera has penetrated the Old Faithful geyser in Yellowstone National Park, revealing its complicated vent shape. The constriction in the vent at the 7-m (22-ft) depth, produced by the ongoing buildup of precipitated silica, may soon choke off Old Faithful's geyser stream.

camera to monitor the vent between eruptions (Fig. 16-17). This investigation has uncovered a far more complicated vent shape than was originally expected. The vent widens and constricts in ways that may critically affect the geyser's behavior. The camera has also revealed vent walls filled with the cracks that allow water to enter the column continuously at various depths. On a sobering note, the watchful eye of Yellowstone's camera has also shown that the vent continues to narrow as silica precipitates from the mineral-charged water. In the geological equivalent of hardening of the arteries, the vent of Old Faithful appears destined to clog and the geyser fated to die, as have many of its neighbors in the Upper Geyser Basin. That day will be a sad one for tourism at the Old Faithful Lodge, unless earthquakes—local or distant— somehow manage to clean out Old Faithful's arteries.

Many countries with hot springs and geysers are developing ways to harness this geothermal energy to help meet their energy needs; we will discuss some of these efforts in Chapter 20.

Finding and Managing Groundwater

The first part of this chapter focused on the mechanics of groundwater flow—how it enters the ground, how it travels through subsurface rocks and sediments, and how, in certain places, it erupts boiling hot from the ground. Given your current understanding of how the groundwater system works, we can now address some key questions: How do we find groundwater? How can we safeguard our supply of this valuable resource and maintain its quality? Who owns it? The next part of this chapter will be devoted to answering these questions.

Locating Groundwater

To find and use groundwater in most areas, we usually dig wells that intersect the water table. But how do we know where to dig? The best strategy for locating groundwater and digging productive wells begins with detailed knowledge of the local water table. Figure 16-18 depicts the various ways geologists attempt to locate the water table in an unconfined aquifer. They begin by mapping the distribution of surface water because most bodies of surface water represent an intersection with the water table. Thus, geologists begin their water search by plotting all of an area's lakes, year-round streams, natural springs, and wetland swamps. A topographic map showing the locations of these features provides the first key to finding the underlying water table.

An area's vegetation also provides clues to the location of its water table. Some plant species have extensive deep-root systems for capturing water that lies far below the surface. Others have shallow root systems and grow only where abundant water lies close to the surface. Identifying such plants enables geologists to estimate the depth of the water table.

Even with these clues, it may still be necessary to drill a series of small test wells to identify the best place for the final well.

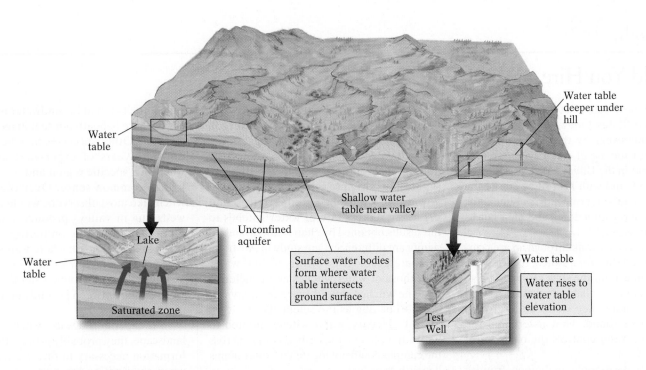

▲ **FIGURE 16-18** The search for water. Geologists estimate the configuration of the local water table by noting the distribution of surface water and water-loving plants and by monitoring water levels in test wells.

Because the water in an uncased well (that is, one that is not lined with a casing) dug in an unconfined aquifer rises to the level of the water table, test wells show the depth of the water table at particular places. With the well data and the surface-water distribution—and with the knowledge that the water table lies deeper under hills than under valleys—we can reasonably estimate the water table's configuration.

Some people forgo the measured scientific approach in favor of hiring water witches, or *dowsers,* to search for water. Water witches claim to locate water by using forked willow sticks or metal rods. Can a water witch really find water with such devices? Highlight 16-1 provides some insights.

Threats to the Groundwater Supply

An area's groundwater supply is threatened with depletion when its population increases its demand for water and lowers the water table. Rural villages have modest groundwater needs and usually have been able to meet them by withdrawing water from shallow wells in unconfined aquifers or by tapping perched water. Many cities, however, have been forced to find new sources of uncontaminated groundwater. Residents in fast-growing suburbs have also had to drill deeper wells to reach higher-yield confined aquifers. As urban sprawl continues, water needs rise sharply. Population growth, the arrival of new industries, the expansion of agriculture, and other changes in the local economy

have increased the need for water so dramatically that groundwater supplies in many areas are being severely taxed.

Consider the example of the Everglades of central and southern Florida. Although the Everglades appear to be broad areas of standing surface water, they are actually fed by groundwater that is flowing, at most times very slowly, toward Florida's southern coast (Fig. 16-19). More of the Everglades' groundwater must be pumped as Florida's population grows and as the Everglades' waters are transferred to central Florida's farmlands for irrigation. This increased pumping of groundwater has drastically

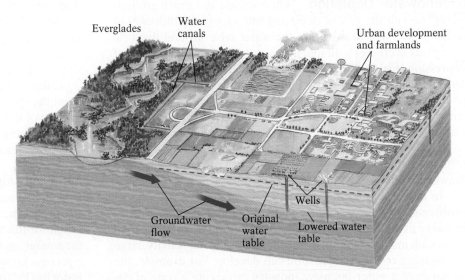

▲ **FIGURE 16-19** Overuse of groundwater. The water table beneath southern Florida once intersected the land surface across a broad area, creating the Everglades. Recently, withdrawal of water from Florida's groundwater reservoir has lowered the water table and dried out much of the Everglades. The lesson is clear: When the water table sits at the surface, it takes only a slight alteration in its height to produce far-reaching ecological disruption.

Highlight 16-1

Should You Hire a Water Witch?

There are more than 25,000 water witches in the United States today—and a crowd of people who swear by them. An early description of the art of water witching may even appear in the Bible: "And Moses lifted up his hand, and with his rod he smote the rock twice; and water came forth abundantly, and the congregation drank and their beasts also" (Numbers 20:11).

Water witches walk through an area carrying a forked stick (or a switch, rod, or wire) in their hands until the stick seems to twist, dip, or jerk uncontrollably downward toward the place where they claim underground water will be found. They assert that the presence of water controls the behavior of the stick.

Are these people malicious frauds? Probably most are not: They truly believe that they possess a supernatural ability to "feel" the presence of water. Many dowsers don't even charge for their services. But can they really find water? Skeptics suggest that the water witch manipulates the rod by holding it in delicate balance so that the slightest small-muscle movement turns it downward. Controlled studies have compared the success rate of water witches to the results obtained by chance and found no significant difference. One study in Australia compared geologists' efforts to find water with those of water witches and concluded that the witches caused twice as many dry holes to be dug as the scientists. At Iowa State University, water witches invited to find water along a prescribed course across the campus could not locate the water mains beneath their feet.

Yet water witches do often locate water. Geologists believe their success may reflect the fact that groundwater is virtually everywhere beneath our feet in most humid climates. Other successes may be due to a dowser's years of experience at finding water in a specific region and to some geological common sense. Over the years, a witch, like most observers, would learn that wells dug in valleys produce water more often than those dug on hilltops, and that certain plants flourish where water lies near the surface. The dowser may even acquire an intuition about groundwater flow and the relationship between groundwater and permeable rocks.

As experienced water witches walk the landscape, they probably process the *real* information necessary to find water. Perhaps when they notice the best conditions for groundwater, their muscles flex subconsciously and the rod responds.

---

lowered the regional water table. Today, much of the Everglades' formerly wet, marshy surface remains dry much of the year, with fires in the dried-out vegetation and organic muck smoldering for months.

When groundwater withdrawal exceeds recharge, the water table may become lowered over a broad geographic range. As a result, the land located above the depleted water source may sink, or salty marine water may invade the freshwater supply.

Groundwater Depletion When we use too much groundwater, we run the risk of depleting our supply. Withdrawal of water from a well draws down the water table around the well. This process creates a dimple in the water table called a **cone of depression** (Fig. 16-20). You can simulate such a depression by inserting a drinking straw a few millimeters into a milkshake. When you begin to drink, the surface of the milkshake dips around the straw; when you stop drinking, the milkshake flows back immediately, eliminating the depression. Because groundwater flow is so slow, however, a cone of depression in a water table does not refill immediately. Instead, it lasts as long as groundwater withdrawal continues.

In moist, temperate regions, cones of depression pose no problems as long as domestic use is moderate and natural recharge exceeds the modest rate of groundwater withdrawal. In areas with major irrigation or substantial industrial activity, large overlapping cones of depression can develop. For example, the demands of a water-intensive industry can create a cone of depression that lowers the water table over a sizable area, leaving neighboring wells dry (Fig. 16-21). As a result, neighboring water users must drill deeper (and more expensive) wells to reach the water below the cone of depression.

In many areas of North America, the rate of groundwater recharge can no longer keep pace with our increased demand for groundwater. When groundwater is withdrawn at a rate that exceeds natural recharge, it is essentially being "mined." Eventually the water table will fall below the bases of local unconfined aquifers. At that point, the aquifers are essentially depleted. Groundwater mining is a particularly serious issue in arid regions such as Phoenix, Arizona, where extensive irrigation, domestic needs of growing populations, and enormous losses to evaporation place an impossible demand on the region's groundwater reservoir.

The effect of a single overdrawn well on neighboring wells raises an important legal issue: Who owns groundwater? Unlike other natural resources that stay in one place and whose ownership is clear (such as gold, or coal, or diamonds), groundwater flows from beneath one property to another. Our ability to determine the direction of groundwater flow enables us to estimate the size of this water supply. But even when we know how much water is there, the question remains: Who owns it? Highlight 16-2 provides some answers.

Land Subsidence When more groundwater is withdrawn from an aquifer than is replenished by recharge, less water is available to support the overlying load of rock, sediment, and soil. As the weight of this unsupported material pushes down, it compresses the aquifer, and as a result, the elevation of the land surface drops—a process known as **subsidence.** Excessive withdrawal of groundwater has led to subsidence of more than 3 m (10 ft) in New Orleans (with dire consequences during Hurricane Katrina), more than 1 m (3 ft) in Las Vegas, more than 7 m (23 ft) in Mexico City, and more than 8 m (26 ft) in central California (Fig. 16-22). Subsidence damages structures such as roads, buildings, and water and sewer lines.

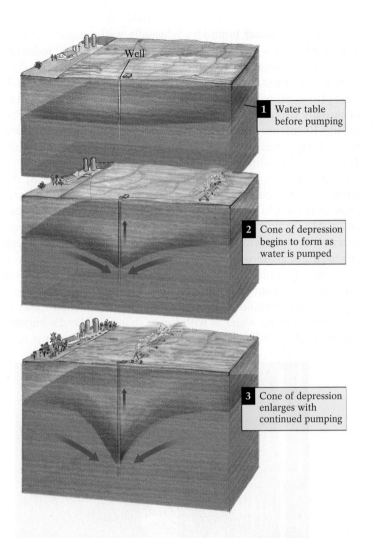

◀ **FIGURE 16-20** The water table around a well is drawn down when water is pumped out, creating a cone of depression.

1 Water table before pumping

2 Cone of depression begins to form as water is pumped

3 Cone of depression enlarges with continued pumping

Well

In coastal areas, subsidence may cause flooding during high tides and storms. For example, Houston, Texas, and Venice, Italy, are both coastal cities built on unconsolidated deltaic deposits. After years of withdrawing groundwater for domestic and industrial uses, these deposits and their aquifers have become compressed, causing the land surface to subside (Fig. 16-23). Venice, which once sat *at* sea level, has now subsided *below* sea level. The city has subsided more than 3 m (10 ft) during the last 1500 years. Invading saltwater has severely damaged many of the city's historic and architectural treasures. To combat this ever-worsening problem (aggravated by the worldwide rise in sea level that has accompanied global warming) and save the city's eroding civic treasures, Venice has embarked upon a 4.5-billion-dollar scheme to construct 78 movable, underwater gates that will rise and block off the advancing Adriatic Sea during periods of unusually high tides and storm surges. The Houston–Galveston area of the Gulf Coast of Texas, one of North America's fastest-growing metropolises, is also subsiding, by as much as 1 to 2 m (3.3 to 6.6 ft) since 1906. This process has caused severe flooding problems in this area, which is frequently battered by hurricanes.

Saltwater Intrusion Aquifers in coastal communities, particularly those on islands and peninsulas, face a constant threat from the intrusion of saltwater. At the same time that a coastal aquifer is infiltrated and replenished by precipitation falling on nearby lands, it is also being invaded by salty marine water from the seaward side of the landmass. Because freshwater is less dense than saltwater and therefore floats above the marine water, the

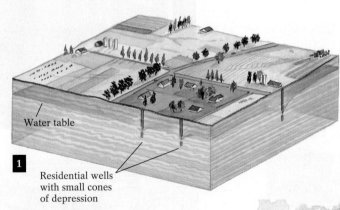

▼ **FIGURE 16-21** The effect of development on a water table. Stage 1: The community consists of farms and some suburban homes having small wells. Stage 2: Industrial development begins, replacing some of the farms, and groundwater withdrawal increases. Stage 3: A large industrial complex covers much of the farmland, lowering the water table to a level below the depth of residential wells.

Water table

1

Residential wells with small cones of depression

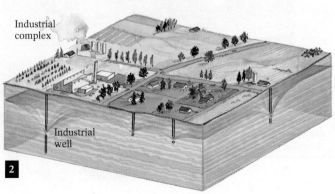

Industrial complex

Industrial well

2

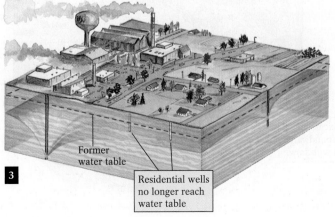

Former water table

Residential wells no longer reach water table

3

▲ **FIGURE 16-22** Subsidence of the surface from excessive groundwater withdrawal causes these fissures in the surface in Antelope Valley, CA. The subsided area is about 625 m (2000 ft) long.

freshwater–saltwater boundary is usually well-defined. As long as groundwater withdrawal remains moderate, natural recharge provides a constant pool of freshwater above the saltwater. Any increase in withdrawal that disturbs the recharge–discharge balance allows more saltwater to enter the aquifers, where it may eventually completely replace the freshwater (Fig. 16-24). The result? Residences and businesses that were formerly supplied by freshwater wells turn on their taps and out comes saltwater. They must then import their water at considerable expense.

This saltwater problem has recently emerged in one of the unlikeliest of locales—Hawai'i's island paradise, Maui. There, some of the world's most persistent rains nourish the lush greenery of Maui's volcanic peaks. Yet as commercial development, tourism, and agriculture grow, the islanders' thirst for fresh groundwater grows as well. Maui's population—150,000 today— may exceed a million within the next 50 years. And as tourism expands, hotels and golf courses (real water guzzlers) will place even greater demands on Maui's main aquifer, which today is shrinking as more water is withdrawn than the island's impressive natural rainfall can replenish. The result? Saltwater is now

▼ **FIGURE 16-23** Subsidence in coastal areas. Excessive groundwater withdrawal in a coastal city such as Venice, Italy, compresses its aquifer and causes the land to sink below sea level. Photo: In Venice today, canals serve as streets. Recent efforts to repressurize Venice's aquifer through artificial recharge (pumping water into the aquifer) and to regulate the withdrawal of groundwater have slowed its subsidence rate somewhat, but at a considerable cost.

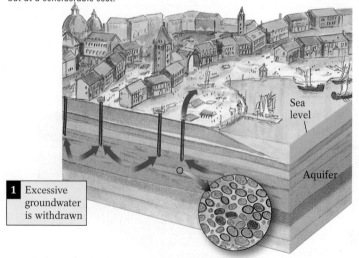

1 Excessive groundwater is withdrawn

Sea level

Aquifer

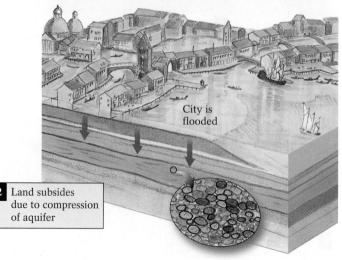

2 Land subsides due to compression of aquifer

City is flooded

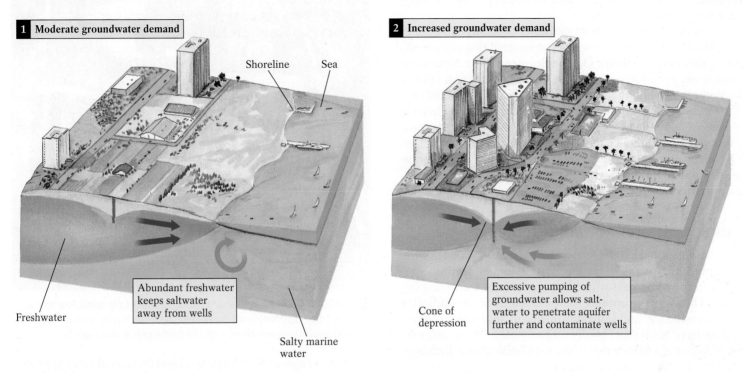

1 | Moderate groundwater demand

Shoreline   Sea

Abundant freshwater keeps saltwater away from wells

Freshwater

Salty marine water

2 | Increased groundwater demand

Cone of depression

Excessive pumping of groundwater allows saltwater to penetrate aquifer further and contaminate wells

▲ **FIGURE 16-24** Saltwater intrusion of an aquifer in a coastal setting. Excessive withdrawal of groundwater allows marine water to penetrate aquifers at such sites, and saltwater flows from the faucets in coastal residences as a result.

Highlight 16-2

Who Owns Groundwater?

The United States has no established *uniform* code outlining groundwater rights. Instead, legal ownership of water is determined by state and local statutes and by the application of a few broad underlying principles. In most of the arid western states, a doctrine known as *prior appropriation* gives priority to the region's earliest users of groundwater, with latecomers having subordinate rights. With an accurate chronology of first use, few disputes can arise over rights to groundwater under this doctrine. During a shortage, the last user of the groundwater becomes the first to lose it. In most of the eastern United States, the *riparian doctrine* gives all users equal (and virtually unrestricted) rights, regardless of the point at which their ownership began.

Can property owners legally exploit their groundwater to the detriment of their neighbors? In colonial America, water disputes were decided by the traditional English Rule of Capture, or Absolute Ownership, which stated that the water under the land is part of the land. This rule allowed any property owner to use as much water as she or he saw fit, even to the detriment of other people. Only if damage to others could be shown to result from *malicious* intent would the excessive user be held liable.

In most states today, landowners can use water *only on the land from which it was taken,* and they are liable for any sale or transfer of water from their land that harms a neighbor's water supply. Also fundamental to the U.S. rules of water use is the principle of "reasonable and beneficial use." Domestic use (within a household) has always been considered reasonable and beneficial, as has watering livestock in cattle-raising states and irrigation in the arid West (but not necessarily where no compelling need to irrigate exists). As industrialization of the nation progressed, manufacturing joined the list of reasonable uses. In California, the use of water to mine gold is still classified as reasonable (a relic of the Gold Rush), as is the use of water for oil exploration in Texas and Louisiana.

In recent years, a few states, including California, have established a doctrine of correlative rights. Under this doctrine, each user's share of groundwater is proportional to her or his share of the overlying land. Again, the principle of "reasonable and beneficial" use is applied. Thus the answer to the question "Who owns groundwater?" depends in some cases on the state in which the question is asked.

encroaching from below as the island's dwindling 150-m (400-ft)-thick lens of freshwater thins.

During the 1930s, the high groundwater demands of the 3 million residents of Brooklyn, New York, converted that city's freshwater aquifer into an Atlantic Ocean saltwater aquifer. Since then, Brooklyn has been forced to import its freshwater from upstate New York. In neighboring Nassau County, New York, rapid suburbanization in the 1950s and 1960s allowed saltwater to intrude into many of Long Island's wells. Residents of the county must now pay for their freshwater and for the pipelines that transport it from the mainland. The fresh groundwater supplies of coastal communities along the Gulf of Mexico, the Carolinas, Georgia, Florida, and California are similarly threatened by invading seawater.

Dealing with Groundwater Problems

Hydrogeology, the scientific study of groundwater, is one of the most rapidly growing fields in the Earth sciences. Hydrogeologists try to solve the subsurface water problems we've just described. They have proposed some remedies to groundwater mining, subsidence, and saltwater intrusion, and a few of their solutions have been successfully implemented. Most solutions seek to enhance local recharge, reduce withdrawal and discharge, or develop alternative water sources.

Enhanced Recharge We now have the means to supplement inadequate natural recharge in a community and thereby increase the local groundwater supply, although such remedies tend to be quite costly. We can, for example, grade a steep slope (that is, make it less steep) so that water infiltrates the ground and replenishes storage basins or depleted aquifers, rather than running off into streams. In areas of moderate to heavy rainfall, communities can construct open recharge basins that store rainwater and gradually allows it to seep into the groundwater system (Fig. 16-25).

In drier regions, or in areas where withdrawal consistently exceeds precipitation, freshwater can be imported and pumped into the ground to replenish dwindling groundwater supplies. This technique is used in and around Chicago, where the region's considerable natural recharge is nevertheless inadequate to meet the needs of the city's large population. To reduce the number of shortages, engineers pump freshwater from Lake Michigan into the ground. Similarly, surface water from California's High Sierras is transported by aqueduct to the arid San Fernando Valley of Los Angeles, where it is injected into the groundwater system to maintain the position of the water table. In Long Island, New York, even the wastewater from industrial air-conditioners is recaptured and returned to the ground.

Some communities with overtaxed groundwater supplies choose to borrow or buy groundwater from areas of abundance—an arrangement known as an *interbasin transfer.* But how can water be transported long distances without substantial losses through seepage and evaporation? One solution is to build pipelines to carry the water. For example, the rapidly growing city of Denver, Colorado, receives much of its water through pipelines and tunnels that stretch from the western slopes of the Rockies. Similarly, New York City draws most of its municipal water supply from rural upstate regions through an extensive pipeline and reservoir system.

▲ **FIGURE 16-25** A recharge basin/Little League field in Long Island, New York. To counteract high groundwater-withdrawal rates, basins such as this one are excavated to collect precipitation and release it into the groundwater system gradually.

In the cases of New York and Denver, groundwater appropriations are primarily *intrastate* transfers. When such water transfers cross state lines and international boundaries, however, the costs typically skyrocket and politicians get involved. How can the donor community be convinced to give up some of its water? When southern California wants water from the Columbia River of Oregon and Washington, the outcry from Portland's residents can be heard in San Diego. The Great Lakes states have already declared their lakes to be "off limits" to thirsty southwestern states. And it will take remarkable international cooperation—and bushels of money—to convince the underpopulated, water-rich areas of Canada to share their water with high-demand areas in the arid regions of their neighbors to the south, the United States and Mexico.

Conservation During dry spells, local authorities often declare water emergencies. They may prohibit car washing, limit lawn watering to early morning or evening hours (to reduce evaporation), discontinue serving water in restaurants except upon request, and encourage us to turn off faucets while brushing teeth and shaving. Table 16.1 shows how much water is consumed by some common household activities. New bathroom-fixture technologies—low-flow toilets and showerheads—use much less water; some cities even offer rebates on water bills if customers install these devices.

Table 16.1 Water Use in Household Activities

ACTIVITY	VOLUME OF WATER USE
Dishwashing	40 liters (10 gal)
Toilet flushing	12 liters (3 gal)
Showering	80–120 liters (20–30 gal)
Bathing	120–160 liters (30–40 gal)
Washing-machine load	80–120 liters (20–30 gal)
Drinking and food preparation	4–8 liters (1–2 gal)

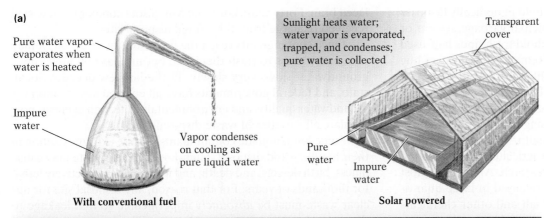

(a)

Pure water vapor evaporates when water is heated

Impure water

Vapor condenses on cooling as pure liquid water

With conventional fuel

Sunlight heats water; water vapor is evaporated, trapped, and condenses; pure water is collected

Transparent cover

Pure water

Impure water

Solar powered

(b)

▲ **FIGURE 16-26 (a)** The process of desalination. **(b)** A desalination plant on the Netherlands Antilles island of Curaçao, off the northwest coast of Venezuela.

▲ **FIGURE 16-27** Massive icebergs, like this one, broken and drifted from an Antarctic ice shelf, may someday provide a source of fresh water to parched communities located at lower latitudes.

The authorities may also mandate industrial conservation, with stiff penalties for excessive use.

Other Water Sources Another solution to water shortages may be *desalination*, the removal of salt from seawater. In this process, seawater is either evaporated or boiled, whenever possible by using inexpensive, nonpolluting solar radiation (Fig. 16-26a). The resulting salt-free vapor condenses and can be collected. Alternatively, the seawater may be passed through a filter that removes the salt. Desalination is costly, but it can be relatively cost-effective in arid lands with extensive seacoasts, especially given the value of freshwater in such places. Desalination plants, such as the one shown in Figure 16-26b, have operated for some time in Israel and Egypt. In the future, the process may become economically feasible in many other nations.

Another means of providing a new water source was the subject of an international symposium at Iowa State University back in the late '70s. Could it be possible to tow massive icebergs from Antarctica (Fig. 16-27) to water-desperate countries in the arid Middle East? Because 90% of the ice would melt or evaporate in transit, researchers proposed that only icebergs at least 10 km (6 mi) across should be considered worthwhile. The cost and possible environmental repercussions of such a venture make it unlikely that the iceberg solution will be a practical option in the foreseeable future although the possibilities are still being studied today.

Maintaining Groundwater Quality

Groundwater for drinking and most other human uses must be relatively pure. As we discussed in Chapter 5, because most groundwater is slightly acidic, it dissolves the soluble components of the rock and sediments through which it passes. Thus groundwater naturally contains a variety of dissolved ions—some safe (such as those resulting from the dissolution of various carbonates and chlorides) and some potentially quite harmful (such as arsenic, mercury, and selenium).

Groundwater may also contain manufactured contaminants that have reached the groundwater system in various ways. On a small local scale, virtually every household periodically disposes of toxic substances through the municipal garbage system or storm drains. A check of typical household trash finds half-used cans of paint, cleansers, and solvents. Remember that can of bug spray you threw out, the burned-out light bulbs, and all your medicine cabinet's expired prescriptions? Poisons and heavy metals from these materials may leach (dissolve) from city dumps and pass into the groundwater system, posing a significant threat to groundwater quality and community health. The volume of these toxins, however, is small compared with agricultural and industrial wastes or the by-products of medical research. Among the most threatening—because they are often released in large quantities—are insecticides and fertilizers, salt and other chemicals washed from roadways, highly toxic industrial by-products from factories (such as PCBs, or polychlorinated biphenyls), biological wastes from cattle feedlots and slaughterhouses, and sewage from overworked septic fields and broken sewer lines (Fig. 16-28).

Pollutants enter the groundwater system when we dispose of wastes improperly, such as by placing them in receptacles that rust and leak, by tossing them haphazardly into poorly designed landfills, or by dumping them illegally on unmonitored land and allowing them to seep directly to the water table or into an aquifer. Groundwater contaminants are often detected only after they reach a well; by that time, however, they may have spread throughout the entire aquifer. In western Minnesota in 1972, a number of employees of a company became seriously ill from arsenic poisoning, even though the firm did not use arsenic in its operations. Environmental detective work traced the arsenic to the company's new well. During the Dust Bowl years of the 1930s, the western plains had been struck by an infestation of locusts. To protect their crops, farmers at that time mixed bran with arsenic and scattered it over their fields to kill the insects. In 1934, after the locusts disappeared, the farmers buried the unused arsenic-

laced bran. Forty years later, the Minnesota company's new well received recharge from this buried disposal site.

Once contaminants reach the groundwater reservoir, it may take many years to flush them out, because, as you now know, groundwater flows very slowly. In the last few decades, local, state, and federal governments have all enacted laws to monitor groundwater quality, and environmental-protection agencies now regulate all varieties of waste disposal.

Some human-made contaminants require extreme caution in their handling and disposal. High-level nuclear waste may cause illness, birth defects, and death, and it remains radioactively lethal for thousands of years. For that reason, any disposal site for nuclear waste must be *absolutely* impermeable so that leakage to the groundwater system cannot occur. The recent selection of Yucca Mountain, Nevada, as the prime candidate for the United States' nuclear-waste dump has focused considerable attention on that region's groundwater conditions. Highlight 16-3 discusses some of the key issues confronting nuclear-waste disposal.

Natural Groundwater Purification

Some groundwater systems are self-cleansing. Indeed, most deep groundwater passes so slowly through rocks of low permeability, that it emerges in a relatively purified state. Three natural purification processes combine to eliminate such toxins as sewage bacteria, larger viruses, and other suspended solids:

- *Filtration.* Some contaminants adhere to clay particles as the water passes through the soil.
- *Decomposition.* Some contaminants decompose completely by oxidizing in the soil.
- *Bacterial action.* Many organic solids are consumed by various microorganisms.

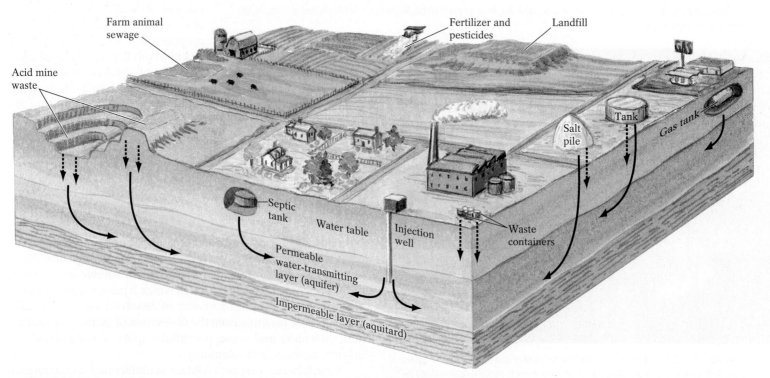

▲ **FIGURE 16-28** Some of the various sources of potential contamination of an unconfined aquifer.

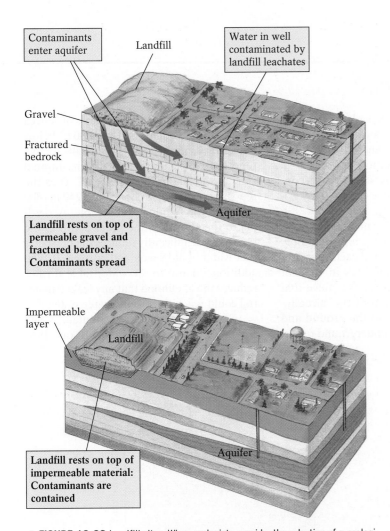

Raw sewage that passes through soil, even if only for 30 m or so (about 100 ft) of moderately permeable materials, is still partially purified naturally. (This process actually accounts for the only treatment of sewage discharge from a septic tank or drain field.)

The cleansing ability of a soil, sediment, or rock depends primarily on whether its permeability allows impurities in the groundwater to adhere to mineral surfaces. If an aquifer is too permeable, water passes through it too quickly to remove the contaminants. Geologists estimate the permeability of a section of an unconfined aquifer by conducting a *percolation test.* In this test, they dig a hole, fill it with water, and observe how long it takes for the water to seep from the hole. If the water vanishes quickly—for example, within minutes or even hours—the soil may be too permeable to cleanse contaminated water. In rural areas, the municipal water-quality agency may not be willing to issue building permits unless percolation tests show that the soils are suitable for household septic systems.

Some rock types remove contaminants less effectively than others. Certain bedrock aquifers, such as those containing large joints, cavities, and faults, allow water to pass through too swiftly to be cleansed. Consequently, major pollution sources such as sewage plants, landfills, and cattle feedlots should be located only where fine-grained impermeable soils overlie unfractured, low-permeability bedrock. Figure 16-29 compares two different landfill sites and predicts the likelihood of groundwater contamination for each.

The Prairie Pothole region of southern Canada and the northern United States (Fig. 16-30) is a 775,000-km² (300,000-mi²) area of shallow, glacially produced ponds and lakes that functions as a *natural* groundwater-treatment facility. Many of the chemical impurities introduced through the fertilization of the surrounding agricultural lands are filtered out or broken down chemically and biologically as water seeps from these pools and infiltrates the underlying soils and unconfined aquifers. In a recent study, this system was shown to remove 80% to 90% of the harmful nitrates introduced by fertilizers. Recently, however, nearly half of these beneficial wetlands have been drained of their waters, principally to develop more agriculture land. Minnesota alone has lost 95% of its original prairie potholes. Elimination of this natural cleansing system will undoubtedly lead to a sharp decline in groundwater quality. How, then, can we improve the quality of our groundwater?

▲ FIGURE 16-29 Landfill sites. When geologists consider the selection of a geologically sensitive site for a municipal landfill, especially in a humid climate with a shallow water table, they try to minimize the potential for groundwater contamination by avoiding areas where coarse, highly permeable gravels overlie highly permeable bedrock. Basaltic flows with unseen lava tubes, limestone terrains with extensive cavern systems, and granites with complex and connected fracture systems would all promote the swift passage of polluted groundwater directly to a well. The ideal solution is to build the landfill on a thick, relatively impermeable layer of soil and sediment overlying an area of impermeable bedrock.

▼ FIGURE 16-30 The water in these bodies of standing water—the Prairie Potholes of the Coteau des Prairies, near McClusky, North Dakota—are naturally purified as they pass through the glacially deposited soils. The contaminants introduced during agriculture, such as fertilizers and pesticides, are largely filtered out or broken down chemically.

Highlight 16-3

Yucca Mountain—A Safe Place for N-Wastes?

In December 1997, hydrogeologists at Richland, Washington, discovered high-level radioactive waste—principally from the spent fuel rods from nuclear power plants and the by-products of nuclear-weapons production—leaking from temporary underground storage tanks. This material has already entered the local groundwater system and threatens to discharge into the mighty Columbia River, the Northwest's largest waterway and one of its principal sources of irrigation water, commercial transportation, salmon spawning, and recreation. This alarming discovery dramatizes the nation's urgent need for a permanent safe nuclear-waste repository, one that will isolate *without fail* this hazardous waste from contact with all biological organisms *for at least 10,000 years.*

In 1976, nearly 20 years after commercial nuclear power first came on-line, a long-overdue federal program was launched to study the feasibility of underground nuclear-waste storage. (The spent fuel rods from the 40 years of civilian nuclear power have been temporarily stored in cooling pools located adjacent to the nation's reactors.) In 1983, the U.S. Department of Energy selected nine locations in six states as potential sites for the nation's waste repository. After three years of exhaustive study, only three candidates remained: Deaf Smith County, Texas; Hanford, Washington; and Yucca Mountain, Nevada. In December 1987, the government selected Yucca Mountain. Since then, more than 1400 Department of Energy employees, including 700 scientists and engineers, have migrated to southern Nevada to study this site. By the time the study phase was completed in 2001, more than $6 billion had been spent on the site, a small fraction of the total amount necessary to build the repository.

Why Yucca Mountain? We can answer this question by examining the site's groundwater conditions: Specifically, how does water move through the mountain, and how might groundwater affect the repository?

The climate of southern Nevada is extremely dry and, as a result, the local groundwater table is unusually deep. These factors, coupled with the composition of the local bedrock, have led many scientists to conclude that the nuclear waste can be safely isolated at Yucca Mountain.

Yucca Mountain is located in the southern portion of the Great Basin, where only about 15 cm (6 in) of rain fall each year, and most of it quickly evaporates in the hot desert sun. The site's scientists believe that only a very small fraction of this already-meager rainfall soaks into the ground and could actually reach the underground repository. Moreover, the water table at Yucca Mountain lies roughly 540 m (1800 ft) below the surface, and the site's plan calls for the repository to be built within the unsaturated zone at the 300-m (1000-ft) level. Thus the repository will be perched about 240 m (800 ft) above the regional water table. According to the scientists, the great depth to the repository significantly reduces the prospects that water could reach and corrode it. The additional depth to the water table further reduces the likelihood that any leaked material could reach the groundwater system.

The geology at Yucca Mountain, which is composed largely of welded tuffs (ash-flow

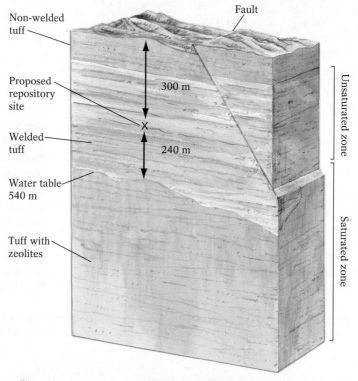

▲ The stratigraphy at Yucca Mountain, Nevada. Note that the proposed nuclear-waste repository is located 300 m (1000 ft) from the potential surface-water infiltration and is isolated from the region's zone of saturation by an additional 240 m (800 ft) of welded volcanic tuff.

deposits), may also bolster the site's ability to prevent long-distance migration of a waste leak. The tuffs are rich in zeolites, water-rich minerals that bond with and thus capture migrating radioactive-waste particles.

Despite the positive geologic factors that have drawn this army of scientists to Yucca Mountain, some skeptics argue that the site will never be *completely* safe and should be abandoned. Critics cite the following concerns about the proposed repository location:

■ The area is riddled with faults, many of which have been active recently. In June 1992, for example, a magnitude-5.6 earthquake occurred at Little Skull Mountain, only 20 km (12 mi) from Yucca Mountain. The quake shattered glass, cracked plaster, and dislodged concrete blocks at the site's headquarters. How would the underground repository fare during an even larger quake? Would such a tremor rupture the containment tanks, releasing the radioactive wastes?

■ The economic potential of the site—believed by some to be a source of precious metals, oil, and natural gas—may someday entice commercial interests to explore it. Such exploration could release radioactive wastes into the environment.

■ Climatic change to wetter conditions during the next 10,000 years could dramatically alter groundwater flow. If more water passes through Yucca Mountain, the increased flow in the unsaturated zone could accelerate corrosion of the storage tanks, increasing the likelihood of leakage. Increased flow would also cause the water table to rise, reducing the travel time between the repository and the site's saturation zone.

Ongoing concerns about the site's past groundwater history were heightened substantially with the excavation of Trench 14, a 100-m (330-ft)-long rectangular furrow plowed along the mountain's eastern slope. The trench walls are cut by thick, deep-running veins of precipitated calcite, a residue of greater groundwater flow in the past. To some site hydrogeologists, the presence of these veins suggests that the deep position of the water table is far less constant than it appears, and that the water table has risen dramatically in the past—perhaps all the way to the surface—and thus *through* the proposed repository site. A recent report by a team of scientists from the National Academy of Sciences disagrees. These scientists argue that the veins were deposited by infiltrating rainwater and not a rapidly rising water table. Such dissenting opinions clearly show that controversy continues to swirl around the state of the hydrological conditions at Yucca Mountain.

What should be done about the Yucca Mountain repository? Can it ever be judged truly safe? Is there a geologically safer place to store these deadly materials? At present, it is our nation's only developing plan to remove decades of accumulated life-threatening nuclear waste at 70 temporary storage facilities in 35 states, such as the failing one at Richland, Washington. The fate of this site poses a decidedly tough dilemma for scientists and legislators, but one that must be resolved *much* sooner than later.

▲ These deep veins of precipitated calcite on the eastern slopes of Yucca Mountain indicate more extensive groundwater infiltration in the past, a cause for concern in the minds of some geologists who are assessing the site's development as a safe repository for nuclear waste.

Cleansing Contaminated Aquifers

When pollution is found in an aquifer, the first step is to identify and, if possible, eliminate the pollution source. We can pinpoint the site at which contaminants enter the groundwater system by taking samples from local streams, digging monitoring wells and sampling the well water, and investigating the waste-disposal policies of local industries. If the source of the chemical pollutants was buried years earlier, or if the contaminants have entered the complex regional-groundwater system by illegal dumping some distance away, detection may prove more difficult and time-consuming.

The second step is to discontinue groundwater use while cleansing of the aquifer takes place, assuming that another water source is available. Cleansing can take decades, given the average rate of groundwater flow. Several methods for cleansing groundwater systems exist. The tainted water can be pumped out, or clean water can be pumped in to dilute the contaminants. An experimental groundwater-cleansing project has been under way for several years at the Massachusetts Military Reservation. There, polluted groundwater passes through a barrier of iron filings designed to react with and decontaminate the water. Introducing chemicals into the groundwater system that either remove or neutralize the contaminants is also being attempted; such treatment is seldom feasible, however, because groundwater usually flows too slowly for the agents to spread throughout the system. Instead, contaminated groundwater is often pumped to a surface holding pond for treatment and then returned to the groundwater system. Whatever method is employed, restoration of a polluted aquifer to good health is both costly and technically demanding. And in most cases, while our efforts may *improve* groundwater quality, the waters of a contaminated aquifer can rarely be restored to purity. It is far better to prevent contamination in the first place.

Groundwater not only is essential to the well-being of human communities but also accomplishes a considerable amount of geological work as it passes slowly through certain types of subsurface rock. In the next sections of this chapter, we explore the principal geological product of groundwater flow—the caves that develop when naturally acidic groundwater passes slowly through soluble carbonate bedrock.

Caves and Karst—The Products of Groundwater Erosion

Caves are unique geological structures that form when groundwater slowly dissolves and gradually enlarges tiny, hairline openings in soluble bedrock. The result is sometimes spectacular enough to attract the attention of tourists, artists, photographers, and scientists alike (Fig. 16-31).

The features created when groundwater dissolves soluble rocks underground or when surface water dissolves exposed soluble rocks are known collectively as **karst** (from the Slavic *kars*, meaning "a bleak, waterless place"). Beneath the surface, karst may be a single cavern or a complex network of caves, some with spectacular deposits of precipitated limestone. At the surface, karst often appears as a spongelike landscape, one pocked with

so many circular depressions that a stream can barely cross it without plunging into a depression and disappearing from the surface. Before we look at the fascinating, exotic products of erosion by dissolution, let's first turn our attention to how slow-moving water can do so much work.

How Limestone Dissolves

Although almost all rocks can dissolve under certain conditions, very few rocks dissolve easily and rapidly. Limestone—by far the most abundant soluble rock—is the material from which nearly all caves and karst landscapes form. Calcite, the major component of most limestones, is a very soluble mineral.

Although limestone is relatively insoluble in *pure* water, it dissolves readily in natural rainwater, which acquires carbon

▲ **FIGURE 16-31** Spectacular cave formations in Carlsbad Caverns National Park, New Mexico.

▲ **FIGURE 16-32** Calcite deposits in pipes, the product of carbonate-rich "hard" water.

dioxide (CO_2) as it passes through the atmosphere and soil. (The concentration of CO_2 in soils, produced largely by the respiration of soil organisms and decomposition of the remains of plants, may reach 100 times the levels found in the atmosphere.) Water and CO_2 combine to form carbonic acid (H_2CO_3), which attacks and dissolves limestone by chemical weathering, through the process of carbonation. In the carbonation reaction, carbonic acid reacts with calcite ($CaCO_3$) in limestone to produce calcium ions (Ca^{2+}) and bicarbonate ions (HCO_3^-). (The equation for this reaction appears on page 149 in Chapter 5.) When these ions are carried away in solution, voids develop in the bedrock. Most of the Ca^{2+} and HCO_3^- ions produced by limestone dissolution are then carried out to sea. There, marine organisms use these ions to manufacture their shells and skeletons. Some of the Ca^{2+} and HCO_3^- may be also precipitated as inorganic limestone.

On a practical note, water containing dissolved calcium bicarbonate ("hard water") makes it difficult to lather soap and leaves an unpleasant residue on anything it washes. It also produces the hard, white, scaly calcium carbonate deposits found in pipes, water heaters, and tea kettles (Fig. 16-32). Water softeners that exchange sodium ions for calcium ions can, however, solve hard-water problems in household water supplies.

Factors Determining Dissolution Rates
Most limestone dissolution occurs within a few hundred meters of the surface. Along with the concentration of carbonic acid in groundwater and the local climatic conditions of warmth and moisture, the extent of a rock's fracturing, jointing, and bedding also determines the rate of limestone dissolution. Bedrock generally dissolves more rapidly when it includes extensive fractures or joints that enable more water to infiltrate and increase the surface area open to dissolution. In time, as water continues to seep into joints, fractures, and bedding planes, a barely perceptible web of tiny cracks can evolve into an intricately branched network of large caves and connecting passageways.

Climatically, karst is rare in polar regions because the water there is frozen and the thick permafrost layer is a barrier to water infiltration. Furthermore, the relatively thin soils and regolith of polar regions generally contain little CO_2 because what little vegetation exists there decays very slowly. Likewise, karst is rare in hot desert regions where the sparse vegetation and lack of water tends to *preserve* rather than dissolve limestone. Caves found in these environments are usually relics of some past period of a rainier, more temperate climate.

Estimating Dissolution Rates   The factors that control calcium-carbonate dissolution, and thus the local rate of cave formation, vary significantly from place to place. For example, in a churchyard cemetery in cool, damp England, gravestones cut from fossil-rich limestone dissolve at a rate of about 5 mm (0.2 in) per 100 years (Fig. 16-33a). Because the fossil sea-lily stems in

▼ **FIGURE 16-33** Differential dissolution. **(a)** A less-soluble fossil of an ancient sea-lily projects above a partially dissolved limestone surface. **(b)** The height of the limestone pedestals beneath their protective sandstone boulders provides a measure of the local dissolution rates in Yorkshire, England.

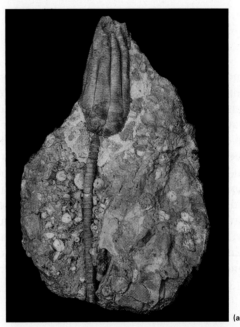

(a)

(b)

these rocks are more resistant to dissolution than the surrounding fine-grained limestone, they project above the gravestone's surface (assuming that dissolution of the fossils themselves has been minimal). If we know the year that a particular gravestone was cut, and we then measure the difference between the relief of the fossils and the stone's surface, we can estimate the rate of the limestone's dissolution.

We can use a similar approach—on a somewhat larger scale—to determine some *regional* rates of dissolution. Yorkshire, England, for example, was covered by ice sheets until about 12,000 years ago. As the glaciers melted, they left behind large boulders of relatively insoluble sandstone that protected the limestone surface immediately beneath them. Eventually, the unprotected surface dissolved, and the protected areas beneath the boulders became 50-cm (20-in)-high pedestals on which the boulders now rest (Fig. 16-33b). The dissolution rate of this limestone is approximately 4.2 mm (0.17 in) per 100 years (50 cm divided by 12,000 years), similar to the rate determined for the gravestone fossils about 100 km (60 mi) away.

Caves

Caves are natural underground cavities and the most common geological products of limestone dissolution. Almost every state and province in North America contains some type of cave, and geologists believe that more than half the caves in North America have not yet been discovered. Major systems can be found in the Black Hills of South Dakota; in northern Florida; in the mountains of Montana and Wyoming; in the Guadalupe Mountains of New Mexico and Texas; in the Appalachians of Pennsylvania, New York, Maryland, Virginia, West Virginia, Tennessee, and Alabama; in the Ozarks of Arkansas, Missouri, and Oklahoma; in western Kentucky; and in the Great Lakes region of Indiana, Illinois, Iowa, and Minnesota. In central and eastern Canada and much of New England, caves tend to be relatively smaller and

less extensive than elsewhere in North America, largely because much of the bedrock in these regions is crystalline and contains little limestone.

Cave Formation

The early stages of cave development progress slowly. Thousands of years may pass before small amounts of circulating water dissolve enough bedrock to enlarge tiny bedrock cracks to a significant extent. As the cracks grow larger, water seeps into the cracks in larger amounts and flows more steadily, vastly accelerating the rate of dissolution. As the enlarging underground openings capture water from other narrower cracks, a few primary passages begin to dominate the underground drainage. Eventually, these hollows become the main caverns and connecting passageways of a growing cave system.

When caves first begin to form, dissolution occurs primarily in the zone of saturation, probably at or just below the water table (Fig. 16-34a). In that region, groundwater combines with inflowing CO_2-rich water that can readily dissolve limestone. This period of intense dissolution along joints, fractures, and bedding planes below the water table forms the intricate honeycomb pattern of many major cave systems.

The second stage of cave formation occurs when the water table drops below its original level, perhaps in response to tectonic uplift, by downcutting of regional streams, or by climatic change to drier conditions. This leaves the cave high and relatively dry above the water table and the cave environment becomes exposed to the open air (Fig. 16-34b). With the water table lowered, erosion by fast-flowing subterranean streams may begin to carve potholes and canyonlike slots into a cave's bedrock floor (Fig. 16-35).

Cave Deposits

Once a network of cave chambers and connecting passageways has formed and the water table has dropped below the base of the system, water moving downward from the surface will enter open-

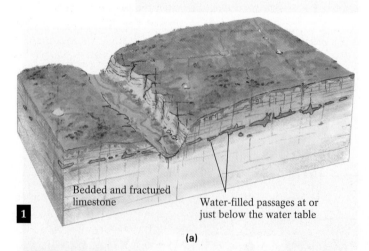

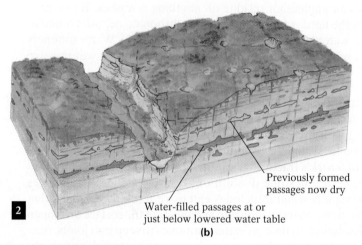

1 Bedded and fractured limestone | Water-filled passages at or just below the water table

(a)

2 Water-filled passages at or just below lowered water table | Previously formed passages now dry

(b)

▲ **FIGURE 16-34** The two-stage process of cave formation. **(a)** Stage 1: Acidified groundwater fills all joints, fractures, and bedding planes below the water table, dissolves limestone, and forms large caverns and connecting passageways. **(b)** Stage 2: The water table drops below the caverns and passageways, leaving them in an open-air environment where speleothems can then develop.

air spaces. Because this water is generally rich in calcium bicarbonate, two things can happen:

1. In the open-air environment of the cave, some water may evaporate, increasing the concentration of Ca^{2+} and HCO_3^- ions in the remaining water until it becomes saturated;

2. Much of the CO_2 dissolved in the water may escape into the cave's air, reversing the carbonation process and converting carbonic acid into CO_2 and water (see Chapter 5).

The high calcium bicarbonate $[Ca(HCO_3)_2]$ concentrations and the CO_2 loss cause calcium carbonate $(CaCO_3)$ to precipitate from solution. The precipitate is deposited on cave surfaces and builds up a variety of formations known as **speleothems.** Speleothems consist largely of the rock **travertine,** the name applied to calcium carbonate when it forms cave deposits.

Stalactites and Stalagmites

Stalactites (from the Greek *stalaktos,* meaning "oozing out in drops") are stony travertine structures, resembling icicles, that hang from cave ceilings. Stalactites form as water, one drop at a time, oozes from a crack or joint and enters a cave. As the water loses carbon dioxide or evaporates, it leaves behind a trace of calcium carbonate. Initially, the center of each growing stalactite is a hollow tube, like a soda straw, through which the next drop emerges. Eventually precipitated travertine clogs the tube, forcing the drops of water to find an alternative route into the cave, usually at the top or side of the stalactite. As the water drips down the stalactite's outer surface, it deposits more travertine and creates the irregular ornate shape of a typical stalactite (Fig. 16-36).

Stalagmites (from the Greek *stalagmos,* meaning "dripping") are the travertine deposits that accumulate on the floors of caves from water that drips from their ceilings. Some of the water that drips from joints and cracks falls onto the cave floor, where it evaporates or releases its CO_2, precipitating travertine. Eventually, these stalagmites may grow as high as 30 m (100 ft). Sometimes they merge with stalactites to form single floor-to-ceiling speleothems called *columns.* (Some people mix up stalactites and stalagmites. You can distinguish between them by thinking of the "t" in stala*c*tites as representing a structure growing from the *top* of the cave, or the "c" representing the *c*eiling. Similarly, the "g" in stalagmites represents a structure growing on the cave's *ground.*)

Other Cave Deposits

Speleothems come in a variety of other delicate and beautiful shapes (Fig. 16-37). Speleothems that form below a joint or crack in the ceiling of a cave may produce *banded draperies,* or *drip curtains. Helictites* are twiglike structures whose central tubes are so narrow that

▲ **FIGURE 16-35** An underground river carved this deep channel by dissolution and abrasion.

Water entering cave

Travertine mound
Stalactite
Stalagmite

Column

▲ **FIGURE 16-36** The creation of stalactites, stalagmites, and columns from precipitated travertine.

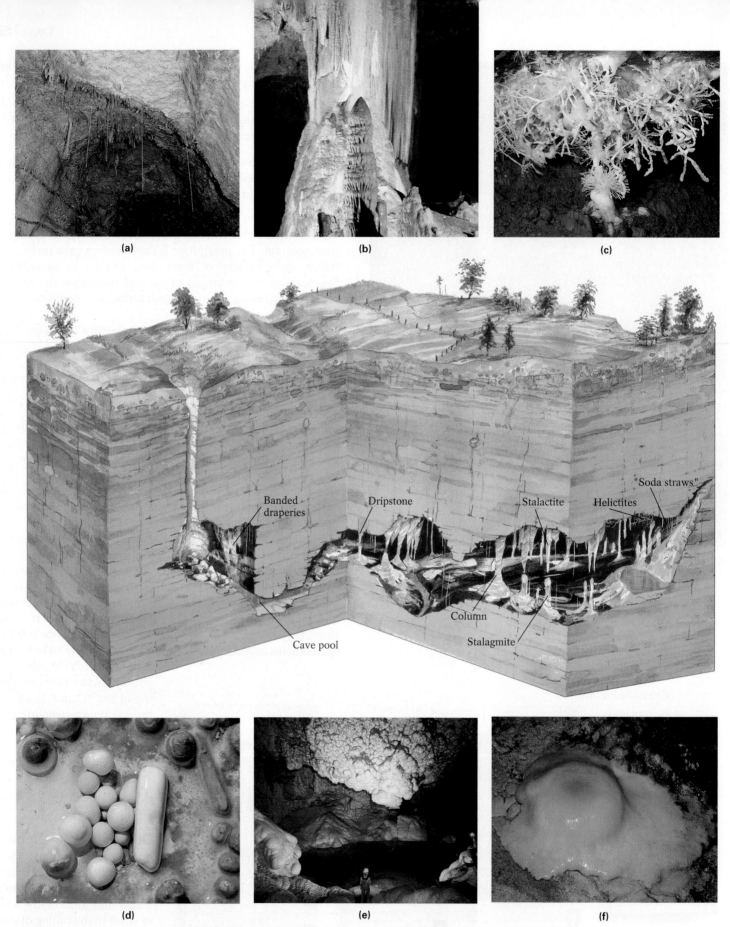

▲ **FIGURE 16-37** Various types of speleothems. **(a)** Stalactites in their early stages of formation resemble hanging soda straws. **(b)** Stalactites hanging from the cave ceiling and stalagmites growing upward from the floor merge to form a column. **(c)** Helictites form by capillary action and can grow in any direction. **(d)** Cave pearls form when layers of travertine precipitate around a sand grain or similar particle. **(e)** Giant cave popcorn forms when travertine precipitates at the surface openings of large speleothems. **(f)** A speleothem resembling a fried egg.

water passes through them only by capillary action. Because capillary action does not depend on gravity, helictites can grow from any point on a cave ceiling, wall, or floor, and *in any direction.*

Cave pearls, another type of unusual speleothem, form as calcium carbonate precipitates in concentric layers around a sand grain or other sand-sized particle. These "pearls" form when water drips from the cave ceiling into a bicarbonate-rich pool of water. The constant drip agitates the pool's water, which releases carbon dioxide and precipitates travertine onto small rock fragments at the pool's bottom. The constant agitation keeps the growing pearl in motion, preventing it from adhering to any surface. The travertine can therefore be added in uniform layers.

Cave popcorn grows on the sides of a larger speleothem as water within the speleothem moves to its surface and evaporates, leaving small clumps of precipitated travertine around surface openings.

Speleothem Growth   The same factors that enhance dissolution—increased acidity and infiltration of groundwater—also promote speleothem growth: solubility and porosity of surface rocks, climate, topography, and vegetation. Conditions in caves in warm, wet, lushly vegetated regions (such as Virginia, Kentucky, and Florida) favor the formation of speleothems, which may grow as rapidly as 1 cm (0.4 in) per year. Speleothem growth is much slower, occurring at a rate of less than 1 cm per *100 years,* in regions with relatively impermeable, less soluble rocks, cold or arid climates, and limited vegetation (such as northern Alberta and British Columbia).

Cut through a speleothem and you can observe its growth record. A horizontal slice of a stalactite, for example, displays concentric layers of travertine. Although not annual records like tree rings, these layers do provide information about the stalactite's growth history. In caves under landscapes that were glaciated during the most recent ice age, speleothems display cracked travertine layers. These cracked layers indicate interruptions in speleothem growth that occurred when the ground froze or became covered with ice, thus halting water infiltration. As a result, the caves dried out, and the speleothems became coated with dust blown about by cave winds. When the climate later warmed, the ice departed, the ground thawed, groundwater again infiltrated, and speleothem growth resumed. By applying isotopic dating techniques to the travertine on either side of a cracked, dusty layer within a speleothem, we can estimate the time and duration of a past glacial period (Fig. 16-38). Studies in France and England, for example, indicate that speleothem growth in these areas ceased about 90,000 years ago and resumed some 15,000 years ago, spanning the period of the last worldwide glacial expansion.

Even speleothems in tropical coastal regions that were never covered by glaciers can be used to date past ice ages. In these areas, travertine deposits record the rise and fall of the sea level rather than the comings and goings of ice. During a period of worldwide glacial expansion, marine water is converted to snow and ice on land, which causes sea levels to fall worldwide. A drop in the water level in coastal caves produces the air-filled cave

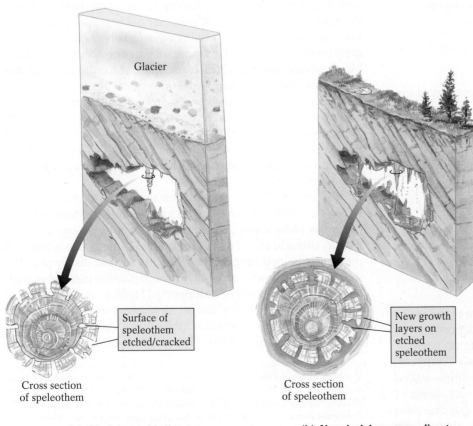

(a) **Glacial or cold climate**

Glacier

Surface of
speleothem
etched/cracked

Cross section
of speleothem

(b) **Nonglacial or warm climate**

New growth
layers on
etched
speleothem

Cross section
of speleothem

▼ **FIGURE 16-38** The effect of climate on speleothem growth. **(a)** During cold periods, when the ground is frozen or when ice covers the land above a cave, very little water can infiltrate the limestone bedrock, and speleothem growth essentially ceases. Without water, speleothems dry out, their surfaces crack, and cold winds from the surface spread a layer of dust on them. **(b)** During warm periods, water is free to infiltrate the ground and speleothem growth resumes. The time and duration of glacial and nonglacial periods can be estimated by dating the layers within speleothems. Photo: A sliced speleothem from Kokaweaf Cavern, California. The specimen is 50 cm (about 18 in) across.

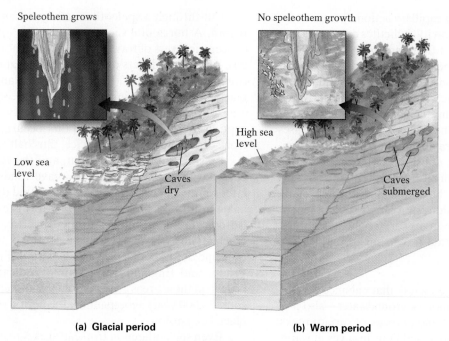

Speleothem grows

No speleothem growth

High sea
level

Low sea
level

Caves
dry

Caves
submerged

(a) Glacial period   **(b) Warm period**

▲ **FIGURE 16-39** The rise and fall of sea level affects speleothem growth in low-latitude and equatorial coastal caves. **(a)** During worldwide glacial periods, global sea level is lower, coastal caves dry out, and speleothems tend to grow. **(b)** During warm periods, glaciers melt, sea level rises, and speleothem growth ceases because caves become filled with water.

environment required for speleothem growth (Fig. 16-39). When glaciers melt away during warmer periods, sea levels rise, and caves located at or below the new higher sea level fill once again with seawater. No speleothem growth takes place during the period of relative global warmth because the water has no air-filled cavity into which it can drip and thus precipitate travertine.

Karst Topography

Karst topography is the *surface* expression of the geology of dissolved limestone and the work of near-surface groundwater. Any of the millions of square kilometers of soluble bedrock that lie at the Earth's surface are likely to become or may already be karst. Extensive karst appears in China, Southeast Asia, Australia, Puerto Rico, Cuba and elsewhere in the Caribbean, the Yucatán peninsula of Mexico, and much of southern Europe. Nearly 25% of the area of the continental United States is karst land, including substantial parts of Virginia, Alabama, Florida, Kentucky, Indiana, Iowa, and Missouri. Figure 16-40 shows the worldwide distribution of karst landforms.

Karst Landforms

Karst landforms, shown in Figure 16-42, range from broad plains studded with small, almost imperceptible, surface depressions, to such unusual features as stream valleys containing no streams, streams that flow at the surface and then suddenly disappear, spectacular natural rock bridges, and huge monoliths—called karst towers (Fig. 16-41)—composed of insoluble rock. Some of these karst landforms occur virtually everywhere that extensive surface carbonate bedrock exists. Others, particularly those

caused by rapid dissolution acting over long periods, occur only in tropical environments.[1]

Sinkholes   The circular surface depressions that appear in most limestone terrains are called **sinkholes.** Streams flowing across a well-developed karst plain generally drain into sinkholes and then through passageways that descend far underground, commonly reaching the caverns below. Sinkholes usually occur together in great numbers over broad expanses of limestone bedrock containing numerous joints and fractures. Central Kentucky alone has 60,000 sinkholes; more than 300,000 are present in southern Indiana.

Solution sinkholes form when groundwater rich in carbonic acid dissolves limestone at or just below the surface (Fig. 16-43). Solution sinkholes generally develop on flat or gently sloping landscapes. Rather than promoting runoff, such surfaces keep acidified water in contact with the soluble rock, fostering extensive dissolution.

Collapse sinkholes occur when the roofs of caves collapse under the weight of overlying rocks (Fig. 16-44). Many collapse sinkholes form when the regional water table drops during a lengthy drought (Fig. 16-45a). In such conditions, the water that previously helped support the overlying surface disappears. If a soaking rain follows a drought, the precipitation can add considerable weight to a newly unsupported surface, causing it to fall into the caverns below. Collapse sinkholes can also form when communities lower their water tables by excessive withdrawal of groundwater (Fig. 16-45b). This problem has plagued certain

[1]For great views of tower-karst landscapes, you can check out several James Bond movies, such as *GoldenEye* (1995), where 007 saves the Earth by destroying the Arecibo telescope in the tower-karst region of Puerto Rico; and more recently, *Tomorrow Never Dies* (1997), set in the karstlands near what is today the tsunami-ravaged coastal town of Phuket, Thailand.

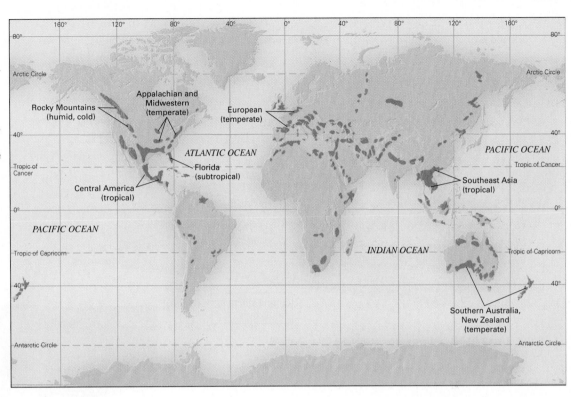

► **FIGURE 16-40** The worldwide distribution of karst landforms. Little or no karst is found in arctic regions where water is often frozen and hence unavailable for dissolution. Whatever surface water is available (during the brief summertime thaw) is prevented from infiltrating the ground by the region's thick, impermeable permafrost. Furthermore, polar soils contain little organic matter and thus less carbonic acid to dissolve limestone rock. In temperate regions and in cool, humid, mid-latitude regions, water falls as rain or snow, and the ground remains unfrozen most of the year, readily absorbing water. In these moist areas, dissolution reactions proceed at a relatively rapid pace. In low-latitude deserts and other arid regions, low precipitation levels and high rates of evaporation combine to protect limestone from dissolution. These regions contain virtually no karst. The heavy rainfall and lush vegetation in the humid tropics provide the optimal environment for limestone dissolution, and karst landscapes are highly developed here.

▲ **FIGURE 16-41** Tower Karst in Guilin, China.

communities in the southeastern United States, where the population has grown rapidly in recent years. Many of these communities have extracted a great volume of groundwater to meet the rising demand for freshwater. Lowering of the regional water table has left the surfaces in many areas unsupported. The result: thousands of new collapse sinkholes.

Sinkholes can play an important role in groundwater management. They can also be used to control the water-table level and flood potential of local streams. In Springfield, Missouri, an area prone to periodic flooding, excess surface water is channeled toward a few selected sinkholes for subsurface disposal. At Kentucky's Bowling Green airport, engineers devised a system of drainage ditches that terminate at specific sinkholes, safely transferring excess water underground. In this way, the groundwater system is replenished and damaging floods are averted.

Disappearing Streams Water soaks rapidly into an absorbent karst plain. As a result, many streams drain completely into sinkholes after flowing only a short distance across the surface. Such flows are called **disappearing streams** (Fig. 16-46). Because disappearing streams appear only briefly at the surface, they rarely meander and seldom contribute to flooding. The climate of central and western Pennsylvania, for example, would normally promote frequent flooding; few floods occur in this region, however, because most of the surface water disappears into the area's many sinkholes.

Some disappearing streams eventually return to the surface as natural springs. How can we tell if the water gushing from a natural spring is the same water that disappeared down a sinkhole some kilometers away? Geological researchers sometimes trace the water of a disappearing stream by placing a harmless chemical or biological dye in it. Volunteers stationed at springs throughout the

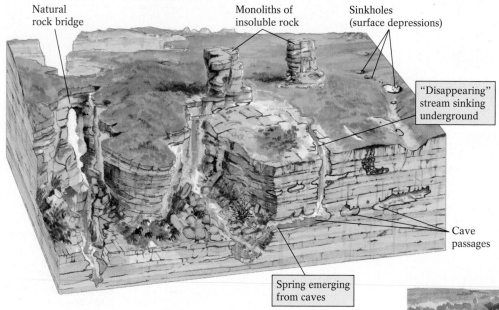

▲ FIGURE 16-42 Typical landforms associated with karst topography.

▲ FIGURE 16-43 A solution sinkhole in Minnesota.

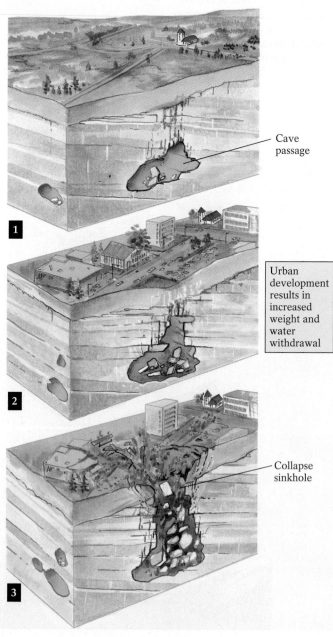

▲ FIGURE 16-44 The origin of one type of collapse sinkhole.

(a)

(b)

▲ FIGURE 16-45 A drop in the level of the water table is a common cause of collapse sinkholes. **(a)** A drought-induced sinkhole in Winter Park, Florida. This sinkhole formed in 1981, when the roof of a large subterranean cavity caved in without warning, swallowing a three-bedroom bungalow, a portion of a Porsche dealership, and half of a municipal swimming pool. A drop in the local water table brought on by a lengthy drought had removed the support supplied by groundwater to the roof of the cavity. The resulting crater was approximately 122 m (400 ft) wide and 38 m (125 ft) deep. **(b)** The December Giant, a collapse sinkhole in an isolated section of Shelby County, Alabama. The crater, 140 m (450 ft) across and 50 m (165 ft) deep, formed with an earth-shaking rumble in December 1962, after heavy November rains soaked a surface left unsupported by heavy groundwater pumping nearby.

▲ FIGURE 16-46 Streams that flow across karst landscapes often disappear suddenly down sinkholes that open into caverns and passageways below the surface. In southern Indiana, streams with suggestive names, such as Sinking Creek and the Lost River, flow underground for more than 10 km (6 mi).

area then await the reemergence of the dyed water (Fig. 16-47). If a subterranean stream system's passageways are well connected, the water may travel distances of several kilometers in only a few hours, far more rapidly than the typical groundwater flow.

Natural Bridges As a series of neighboring sinkholes expands and joins together, the surface overlying broad sections of underground stream channels may collapse. The segments of surface material that do *not* collapse form a **natural bridge** over the exposed channel. Perhaps the most famous natural bridge in North America is located in the Blue Ridge Mountains of Virginia, where a remnant of massive dolomitic limestone forms a bridge approximately 30 m (100 ft) long and 40 m (130 ft) high (Fig. 16-48). Thomas Jefferson purchased the bridge in 1774 (from King George III, for the equivalent of $2.50!) to preserve it as a natural wonder and resource. U.S. Highway 11 now runs across it, saving millions in bridge-construction dollars. Quite a smart and visionary move by our third President.

Protecting Karst Environments

Well-developed cave and karst environments typically contain numerous sinkholes that connect with a vast underground network

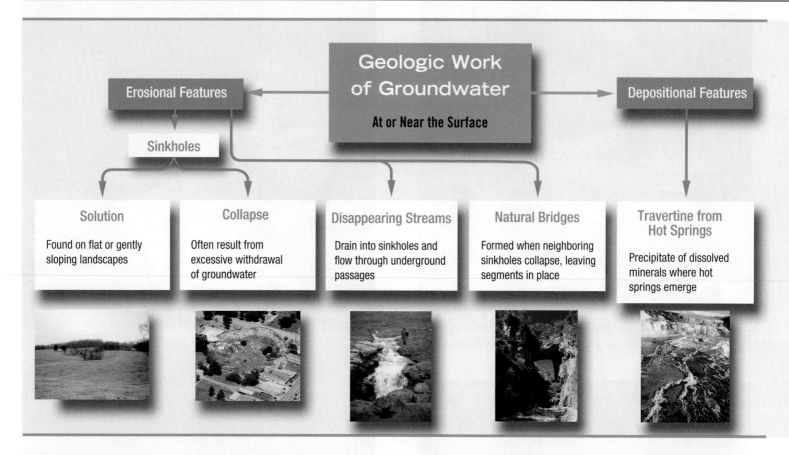

Geologic Work of Groundwater
At or Near the Surface

Erosional Features ← → **Depositional Features**

Sinkholes

Solution	Collapse	Disappearing Streams	Natural Bridges	Travertine from Hot Springs
Found on flat or gently sloping landscapes	Often result from excessive withdrawal of groundwater	Drain into sinkholes and flow through underground passages	Formed when neighboring sinkholes collapse, leaving segments in place	Precipitate of dissolved minerals where hot springs emerge

of large caverns and passageways. In such places, water enters the ground readily and moves through it so rapidly that soils and impermeable bedrock have few opportunities to filter out contaminants. Thus cave and karst landscapes are *extremely* sensitive to careless handling of waste.

Garbage dumped into a sinkhole will quickly contaminate both the local and regional water supply (Fig. 16-49). During the 1960s, the community of Alton, Missouri, in beautiful Ozark country, discarded much of its garbage in a 15-m (50-ft)-deep dry sinkhole near the town. In 1969, after an unusually rainy winter

▼ **FIGURE 16-47** Tracing disappearing stream water. (left) Dye is poured into a sinkhole. (right) Colored water emerges several hours later from a natural spring several kilometers away.

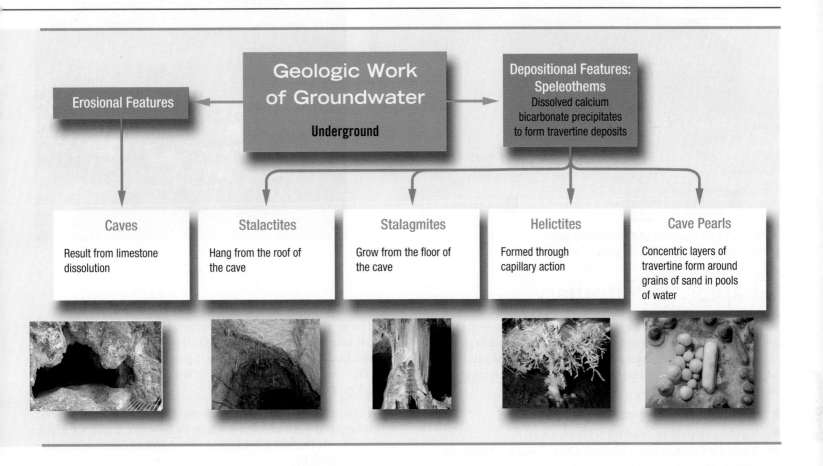

Geologic Work of Groundwater

Underground

Erosional Features

Depositional Features: Speleothems
Dissolved calcium bicarbonate precipitates to form travertine deposits

Caves
Result from limestone dissolution

Stalactites
Hang from the roof of the cave

Stalagmites
Grow from the floor of the cave

Helictites
Formed through capillary action

Cave Pearls
Concentric layers of travertine form around grains of sand in pools of water

▼ **FIGURE 16-48** Natural Bridge, in the Blue Ridge Mountains near Lexington, Virginia, spans Cedar Gorge Creek.

and spring, 9 m (30 ft) of water accumulated in the sinkhole. Concerned about the effects of Alton's sinkhole disposal, the U.S. Forest Service poured thousands of gallons of dyed water into the sinkhole's standing water and surveyed groundwater sources for miles around. Three months later, dyed water containing a high concentration of undesirable material leached from Alton's dump was detected 25 km (16 mi) away at the neighboring town of Morgan Springs. Alton's careless dumping had adversely affected the entire region's groundwater quality.

In karst regions, polluted surface waters reach the groundwater through *numerous* entry points. A large cave system may underlie hundreds or even thousands of sinkholes—each one being a potential point source of pollution. And because complex cave systems typically hold a huge volume of water, chemical treatment of contaminated karst aquifers is not feasible; there's simply too much water to clean. Thus, because they are so open to contamination and impossible to treat effectively, aquifers in karst regions should not be used for waste disposal *of any kind*. Likewise, they should not be considered reliable sources of drinkable water.

In the last two chapters, we have examined the surface and subsurface geological work of flowing water that passes through various segments of the hydrologic cycle. In the next chapter, we will look at how flowing *frozen* water—in the form of glaciers—has substantially altered the landscape in regions cold enough to freeze the Earth's surface water.

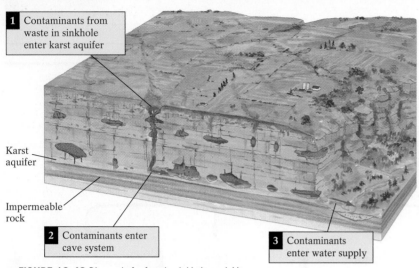

1 Contaminants from waste in sinkhole enter karst aquifer

Karst aquifer

Impermeable rock

2 Contaminants enter cave system

3 Contaminants enter water supply

▲ **FIGURE 16-49** Disposal of refuse in sinkholes quickly contaminates karst aquifers. Photo: A garbage-filled sinkhole.

Chapter 16 Summary

Surface water—flowing in rivers, standing in lakes, or falling as precipitation—enters the ground wherever topography, geological composition, and vegetation cover permit infiltration. The *best* conditions for development of a large reservoir of groundwater include:

- a well-vegetated landscape;
- a gently sloping or near-level landscape;
- a surface characterized geologically by fractured bedrock or coarse, well-sorted sediment.

Water enters the ground under the influence of gravity, which carries it downward through soils, sediments, and rocks until every available connected pore space is filled. Surface evaporation, plant use, and insufficient groundwater infiltration promote formation of a *zone of aeration* near the Earth's surface. This zone contains many air-filled pores. Below it lies the *zone of saturation* in which all available pore spaces are water-filled. The upper surface of the saturated zone is called the *water table*.

Water flows underground where the geological materials are sufficiently permeable. *Porosity,* the percentage of pore space relative to the total volume of soil, rock, or sediment, measures how much water can be held by Earth materials. *Permeability* is a measure of the ability of rock or sediment to transmit a fluid. The presence of connected fractures in solid bedrock and the coarse, well-sorted texture of unconsolidated sediment and soil both increase permeability.

Groundwater flows between two points because of differences in *potential,* the energy that arises from variations in the points' elevation and water pressure. The rate at which the potential changes over a lateral distance is called the *hydraulic gradient.* Because of variations in permeability, every material has a unique ability to transmit groundwater, a property known as its *hydraulic conductivity.*

Aquifers are permeable, water-bearing bodies of geologic materials. Aquifers not overlain by an impermeable cap rock are called unconfined aquifers. Confined aquifers, found at greater depth, are sandwiched between low-permeability rock layers called *aquitards.* When an aquifer's water is under high pressure

from large elevation differences between recharge and discharge sites, it may rise above the level of the aquifer and gush from the ground. Geologists describe such self-pumping aquifers as *artesian.*

The regional water table can often be located from the following methods:

- Locating and mapping surface-water features such as rivers, lakes, and natural springs, places where the Earth's surface intersects the water table and groundwater flows at the surface naturally;
- Drilling test wells to determine the local depth to the water table by applying the knowledge that the water level in wells approximates the position of the local groundwater table.

Human activity can disturb the groundwater system through excessive withdrawal and contamination. In areas of rapid population growth and intensified industrial development, groundwater demand rises sharply, leading the water table to drop significantly. Large *cones of depression*—local depressions in the water table—may develop around wells in such areas. Lowered water tables make it necessary to dig ever-deeper wells. They also promote subsidence, as depleted aquifers become compressed and the overlying land surface falls. In coastal regions, excessive groundwater withdrawal may cause saltwater to infiltrate subsurface aquifers.

Karst terrains develop when soluble bedrock dissolves both above or below the Earth's surface as naturally acidic water passes through it. Nearly all karst results from the dissolution of the calcite in limestone by carbonic acid.

Extensive cave systems typically occur below the surface of a karst landscape. *Caves* are natural underground cavities that generally form as carbonate bedrock dissolves along preexisting joints, fractures, faults, and bedding planes. Cave deposits, called *speleothems,* include stalactites, which grow downward from the cave ceiling, and *stalagmites,* which grow upward from the cave floor.

Karst topography—the surface expression of *karst*—is dominated by:

- circular depressions called *sinkholes* that typically occur by the thousands in areas characterized by exposed flat-lying limestone bedrock. Collapse sinkholes form when the roofs of caves give way under the weight of overlying materials, often following the withdrawal of water that helps support the overlying surface;

■ *disappearing streams* that form because sinkholes usually occur in large numbers and close together, and thus surface water rarely traverses a karst plain completely. Instead, streams disappear from the surface when they encounter sinkholes.

■ Towers of limestone that stand high above gorges that have been carved into highly soluble bedrock, commonly along regional joints in the bedrock.

Because surface water moves so rapidly into and through the groundwater system in karst regions, contaminants cannot be filtered out or consumed by microorganisms. Consequently, karst regions are particularly sensitive to groundwater contamination.

KEY TERMS

aquifer *(p. 506)*
aquitards *(p. 507)*
artesian *(p. 508)*
capillary fringe *(p. 503)*
caves *(p. 526)*
cone of depression *(p. 514)*
disappearing streams *(p. 531)*

geyser *(p. 509)*
hydraulic conductivity *(p. 505)*
hydraulic gradient *(p. 505)*
karst *(p. 524)*
natural bridge *(p. 533)*
natural springs *(p. 508)*
perched water table *(p. 506)*

permeability *(p. 504)*
porosity *(p. 503)*
potential *(p. 504)*
potentiometric surface *(p. 508)*
sinkholes *(p. 530)*
speleothems *(p. 527)*
stalactites *(p. 527)*

stalagmites *(p. 527)*
subsidence *(p. 514)*
travertine *(p. 527)*
water table *(p. 503)*
zone of aeration *(p. 503)*
zone of saturation *(p. 503)*

QUESTIONS FOR REVIEW

1. Describe three factors that affect groundwater infiltration.

2. What is the difference between primary and secondary porosity? List three factors that control a rock's primary porosity. Describe a rock that would have high permeability but low porosity. Low permeability but high porosity?

3. What is an aquifer? Where would you look for an unconfined aquifer? A confined aquifer?

4. Draw a simple sketch showing the conditions necessary to create an artesian groundwater system.

5. Describe two negative consequences of unwise groundwater management in coastal settings.

6. How would you solve the groundwater problems facing the Florida Everglades?

7. Suppose you are charged with satisfying the groundwater needs of Los Angeles, California, in the 21st century. Name five problems you would face and the solutions you would propose.

8. List four factors that affect the rate of limestone dissolution.

9. Briefly describe how caves form. Where is the water table located during the two stages of cave formation?

10. Why is a karst landscape particularly susceptible to groundwater pollution?

11. Discuss three reasons why you would not expect to find caves in polar regions.

LEARNING ACTIVELY

1. Grab a shovel and dig some holes—some on tops of hills, others at the bottoms of hills. How deep do you have to dig before water begins to pond up in the bottom of the hole? If you live in the arid Southwest, you won't hit the water table, so don't bother digging for too long (although you might try digging next to a stream or lake). If you live in the damp Northeast or Southeast, you should reach the water table within a meter of the surface. Estimate the thickness of the zone of aeration and the depth to the water table.

2. Take a walk or drive around your community and look for municipal water towers. You probably have seen them hundreds of times but never gave them much thought. Note the locations of the water tanks. Are they found on hills or in valleys?

3. Fill several tall kitchen glasses with sands and soils of different textures. Pour some dyed water into the glasses and observe how the water seeps into the sediment. Compare the time it takes for the dyed water to reach the bottom. Here you have observed variations in hydraulic conductivity. Which of the sediments have the highest hydraulic conductivity? The lowest?

4. Grab that shovel again and dig a hole, but dig one that's not as deep as the water-table depth. Pour in a bucket of water and do a percolation test. Note how long it takes for all of the water to seep out of the hole. Speculate about whether the soil you just tested would be a good candidate for a landfill or a septic field.

 ## ONLINE STUDY GUIDE

Practice makes you better. Make the time to take the *Online Study Guide* quiz for this chapter. It should take you about 20 minutes. Automatic grading and instant feedback will help you quickly and accurately identify the concepts you have mastered and the areas in which you need more study.
www.prenhall.com/chernicoff

Glaciers and Ice Ages

17

- **Glacier Formation and Growth**
- **The Work of Glaciers**
- ***Highlight 17-1*** The World's Greatest Flood—The Channeled Scablands of Eastern Washington
- **Reconstructing Ice Age Climates**
- **The Causes of Glaciation**
- **The Earth's Glacial Past**
- ***Highlight 17-2*** The Snowball Earth—When the Whole Planet Froze
- ***Highlight 17-3*** Opening up the Northwest Passage–A By-product of Global Warming?
- **Glaciation on Other Planets and Moons**

Glaciers create some of the most spectacular scenery on Earth (Fig. 17-1), drawing millions of visitors each year to Alaska and other high-latitude regions. Aside from stream erosion and mass movement, the Earth's surface has probably been changed more by glaciation than by any other process. Without glaciers, the fjords of Norway would not exist. North America would have no Great Lakes, Niagara Falls, Hudson Bay, Puget Sound, or the 11,842 lakes of Minnesota. There would be no Cape Cod in Massachusetts and no fertile rolling hills in the Midwest and southern Canada. The peaks of the Rockies and Cascades would be rounder and less impressively jagged, and California's Yosemite Valley would lack its sheer-faced cliffs. Rivers such as the Missouri and Ohio would drain north to the Arctic and Atlantic Oceans rather than south to the Mississippi River and the Gulf of Mexico. If the Earth had no glaciers today, the shapes of the continents themselves would differ substantially because sea level would be about 70 m (230 ft) higher than it is today. Landlocked cities such as Memphis, Tennessee, and Sacramento, California, would be seaports, and San Francisco, New York, and many other coastal cities would be mostly under water.

Where do we find the glaciers that have changed so much of our continents? How can such a small fraction of the hydrologic cycle (see Figure 15-2) accomplish so much change? The answers lie in the occurrence of **ice ages,** the dozen or so periods of Earth history—each lasting millions of years—when the planet's climate was significantly cooler than for most of the rest of Earth history. During these cold times, glaciers covered a much larger portion of the Earth's land surface than they do today. During an ice age, climatic fluctuations cause glaciers to grow and advance during *glacial periods,* and then thaw and retreat during *interglacial periods.*

The Earth is currently in the midst of an ice age—commonly referred to as the **Quaternary ice age,** named for the Quaternary Period of the Cenozoic Era (although this ice age actually began to take shape during the latter portion of the Pliocene Epoch about 2.5 million years ago). The Quaternary Period has spanned the last 1.6 million years of the planet's history, and its ice age has been marked by many episodes of glacial advance and retreat, primarily during the **Pleistocene Epoch** of the Quaternary Period, which ended

FIGURE 17-1 The Sawyer Glacier entering Tracy Arm, a glacial fjord in Tongass National Forest, Alaska. ▶

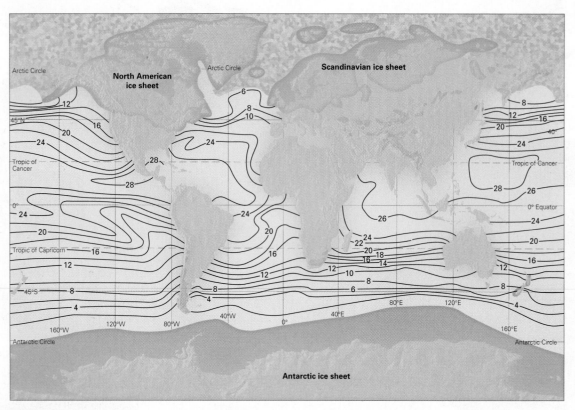

▲ **FIGURE 17-2** A "climap" showing glacier distribution during the last major period of worldwide glacial expansion. During this expansion, which reached its peak about 20,000 years ago, ice covered approximately 30% of the Earth's land surface. The contour lines indicate ocean temperatures in degrees centigrade. Note how the temperatures drop near glacial margins. (The "Climap" project reconstructs and maps world climate at various times during the last 50,000 years using various geological, geophysical, chemical, and biological climatic indicators.)

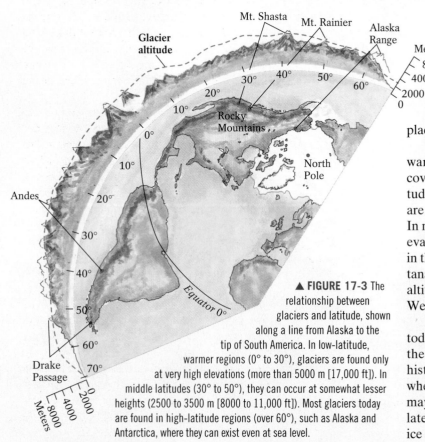

▲ **FIGURE 17-3** The relationship between glaciers and latitude, shown along a line from Alaska to the tip of South America. In low-latitude, warmer regions (0° to 30°), glaciers are found only at very high elevations (more than 5000 m [17,000 ft]). In middle latitudes (30° to 50°), they can occur at somewhat lesser heights (2500 to 3500 m [8000 to 11,000 ft]). Most glaciers today are found in high-latitude regions (over 60°), such as Alaska and Antarctica, where they can exist even at sea level.

some 10,000 years ago. At the height of glacial expansion in the late Pleistocene about 20,000 years ago, ice covered approximately 30% of the planet's land surface, including most of northern Europe, northwestern Asia, Canada, and the northern United States (Fig. 17-2). In some places, the ice was as much as 4 km (2.5 mi) thick.

We now live in a warm *interglacial* period of an ice age. This warmer period began about 11,000 years ago. Today glaciers cover only 10% of the world's land surface. At very high latitudes, such as in the Arctic, Antarctica, and Greenland, glaciers are fairly common and can exist at any elevation, even at sea level. In mid-latitude regions, however, they are found only at high elevations (at roughly 2500 to 3000 m [8000 to 10,000 ft]), such as in the Northern Rockies of Alberta, British Columbia, and Montana. Figure 17-3 illustrates the relationship between latitude and altitude and shows how both affect where ice masses exist in the Western Hemisphere.

In this chapter, we discuss what glaciers are, where they exist today and where they were found in the past, and how they shape the landscape. We also consider why certain periods of Earth history have been marked by worldwide glacial expansions whereas others have not, and we speculate about when glaciers may again cover vast areas of North America. We'll also speculate a bit about what might happen if some or all of the Earth's ice melts.

Glacier Formation and Growth

A **glacier** is a moving body of ice that forms *on land* from the accumulation and compaction of snow. Glaciers up in the mountains flow downslope under the influence of gravity and the pressure of their own weight. Glaciers that cover lowland areas flow outward from where their surfaces are high (typically in the centers of the ice mass) to where their surfaces are low (typically the margins of the ice).

Glacier formation begins with a snowfall and the accumulation of snowflakes, delicate hexagonal (six-pointed) crystals of frozen water (Fig. 17-4). The initial snowpack tends to be fluffy because the pointy shapes of the snowflakes initially keep them separated; about 90% of a pile of newly fallen snow consists of space between the crystals. The density of fresh snow is only 0.1 g/cm³. In contrast, the density of liquid water is 1.0 g/cm³.

Soon air circulating through the spaces in the snowpack causes the fragile points of some snowflakes to *sublimate,* a process by which a solid changes directly into a gas. In this way, some snow is directly converted to water vapor, which then recrystallizes in the open spaces between flakes. As more snow falls, the weight of the overlying snow melts some of the contact points between crystals by *pressure melting,* the melting of ice that occurs from the pressure-induced lowering of its melting point. The water produced by pressure melting migrates to open, low-pressure spaces in the snowpack and refreezes, binding the snow crystals together. (Pressure melting also occurs under the blades of an ice skater. The skater's weight, pressing downward on the narrow blades at each moment, melts a small amount of ice that

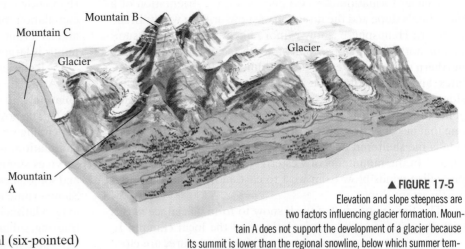

▲ FIGURE 17-5
Elevation and slope steepness are two factors influencing glacier formation. Mountain A does not support the development of a glacier because its summit is lower than the regional snowline, below which summer temperatures are too warm to support glaciers. Mountain B, whose summit is well above the snowline, also does not support glacier development because it is too steep and snow slides downhill rather than accumulating. Mountain C has a gently sloped shape and much of its summit is above the snowline, making it an optimal site for glacier development.

refreezes immediately when the weight is removed. As a result, the skater always glides on a thin film of water.)

Eventually, the paired processes of sublimation and recrystallization, and pressure melting and refreezing, convert large pointed snowflakes to small, tightly packed, and nearly spherical crystals, called *corn snow.* This is the snow that remains on the slopes—late in the skiing–snowboarding season. Corn snow gives way to **firn** (from the German for "last year"), the well-packed snow that survives a summer melting season; its density is about 0.4 g/cm³. Dense firn consists of interlocking crystals of glacial ice that form after nearly all the air has been squeezed out by repeated pressure melting and refreezing. As firn recrystallizes and nearly all of its trapped air is expelled, the density of glacial ice increases to 0.85 to 0.90 g/cm³, about the same as the density of an ice cube but still less dense than water.

Glaciers form and grow when snow accumulation exceeds losses caused by summer melting. Despite frigid temperatures and lots of snow, glaciers do not occur in locations such as modern-day Minnesota, where summer warmth completely removes even heavy winter accumulations of snow. Cool, cloudy, brief summers that minimize melting probably contribute more to glacier growth than very low winter temperatures or thick accumulations of snow.

In mountainous regions, glaciers form and grow best where snow falls at high elevations. Mountainside glaciers typically form at or above the **snowline,** *the lowest topographic limit of year-round snow cover.* Snow can persist above the snowline, but any snow that falls *below* the snowline, where temperatures are warmer, typically melts in summer. Furthermore, snow tends to accumulate more on gentle slopes, where it is less likely to slide downslope or blow away (Fig. 17-5).

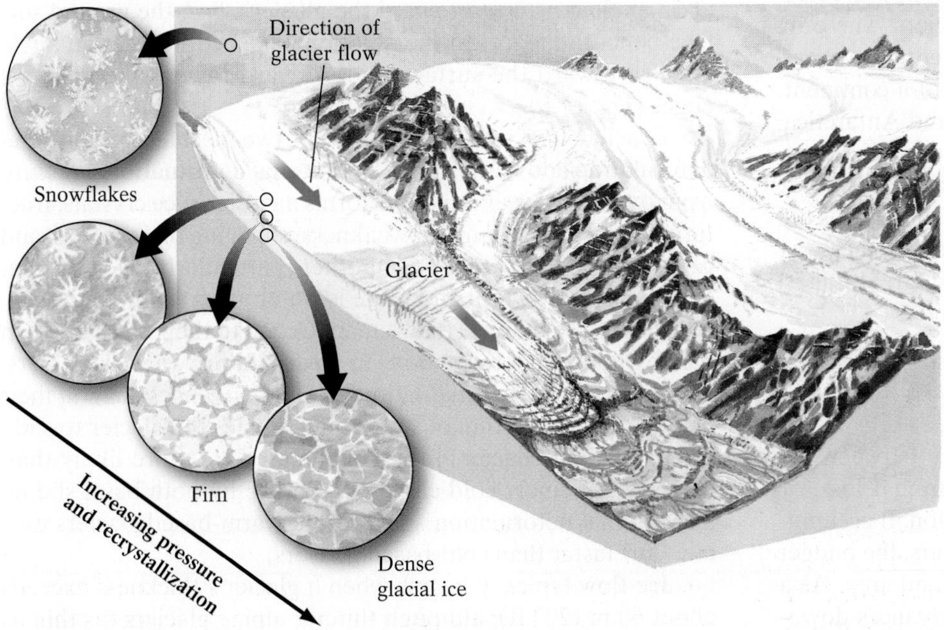

▲ FIGURE 17-4 The formation of glacial ice from snow. Hexagonal snowflakes that fall on a glacier are gradually changed into rounded crystals. As overlying snow buries the crystals and exerts pressure on them, they become packed into increasingly dense firn and finally glacial ice.

Glacier formation can also depend on the orientation of a mountain's slope and the direction of snow-drifting winds. In the Northern Hemisphere, snow is more likely to survive the summer on north-facing slopes than on south-facing ones because the northern slopes receive less direct sunlight. More snow accumulates, and glacier formation is also more likely, on the leeward slopes of mountains, where snow drifts from the windward side generally settle. (The *windward* side of a mountain range is the one that lies directly in the path of oncoming winds. The *leeward* side is the sheltered backside—that is, the opposite side of the range.) In the mountains of western North America, where the prevailing winds blow from west to east, most glaciers form on northeastern slopes.

The time required for fresh snow to form glacial ice varies with the rate of snow accumulation and the local climate. In snowy regions where average annual air temperatures are close to the melting point of ice (0°C [32°F]), snow may be converted to ice in only a few decades by melting and refreezing. In extremely cold dry polar settings, where snowfall is typically sparse and little melting takes place, glacial-ice formation may take hundreds or even thousands of years.

Classifying Glaciers

Glaciers are classified broadly by whether the local topography confines them or allows them to move freely (Fig. 17-6). **Alpine glaciers** are confined by surrounding bedrock highlands. As a consequence, they are relatively small. There are three basic types of alpine glaciers: **cirque glaciers,** which create and occupy semicircular basins on mountainsides, usually near the heads of valleys; **valley glaciers,** which flow through and modify preexisting stream valleys; and **ice caps,** which form at the tops of mountains.

A *piedmont* ("foot of the mountain") glacier begins as an alpine glacier but then flows onto an adjacent lowland where, unconfined, it spreads laterally. Piedmont glaciers that flow to coastlines and into seawater are called *tidewater* glaciers. The only *completely* unconfined form of glacier is a **continental ice sheet,** an ice mass so large that it blankets much or all of a continent. Modern continental ice sheets cover Greenland and Antarctica. The Antarctic ice sheet actually consists of two ice sheets separated by the Transantarctic Mountains. It is 4.3 km (3 mi) thick in some spots and occupies an area about 1.5 times as large as the continental United States. Twenty thousand years ago, such vast ice sheets covered North America to south of the Great Lakes, and western Europe south to Germany and Poland.

The Budget of a Glacier

The *budget* of a glacier consists of the difference between the glacier's annual gain of snow and ice and its annual loss, or *ablation,* of snow and ice by melting and sublimation. If accumulation exceeds ablation for several consecutive years, the budget is positive, and the glacier increases in thickness and area. As a glacier expands, its outer margin, or **terminus,** advances downslope in confined glaciers and outward in unconfined glaciers. If ablation exceeds accumulation for several years, the budget is negative. The glacier decreases in size and its terminus retreats

(by melting, not by running uphill in defiance of gravity). If accumulation roughly equals ablation, a glacier's budget is balanced and its terminus remains stationary. Thus the position of the terminus is largely determined by the glacier's long-term budget.

All glaciers—from the smallest cirque glacier in Montana to the Antarctic ice sheet—have a **zone of accumulation,** a **zone of ablation,** and an **equilibrium line** separating the two zones (Fig. 17-7). The position of the equilibrium line can change every year, depending on the gain or loss of glacial volume. The zone of accumulation, which can be identified by a blanket of snow that survives summer melting, is nourished principally by snowfall, and sometimes adds snow from avalanches off surrounding slopes. Summertime melting causes ablation in middle and low latitudes; in high latitudes, where summer temperatures may remain below freezing, sublimation causes ablation. The zone of ablation is recognizable in summer by its expanse of bare ice.

Glaciers that terminate in bodies of water, such as tidewater glaciers, can also lose ice—in the form of icebergs—by *calving,* a process in which chunks of ice break off by simply falling from a steep ice cliff or by flexing from wave or tidal action. Calving also occurs as sea level rises and buoys up an ice mass; as it begins to float, icebergs tend to snap off the main ice mass.

Glacial Flow

Newcomers to the study of the geology of glaciers must learn an important fact: Regardless of whether a glacier's *terminus* is advancing, retreating, or remaining stationary, the ice within the glacier still flows forward from the accumulation zone toward the ablation zone. Thus, the position of the terminus reflects the glacier's long-term budget (positive, negative, or at equilibrium), even as the glacier's flow may reflect such short-term factors as accumulation of new snow, conversion of this snow to glacial ice, and flow of ice under the influence of gravity. In the accumulation zone, as the snow becomes compacted into ice, it begins to flow at an angle *downward* toward the glacier's *bed,* the ground surface beneath the ice. In the ablation zone, ice flows at an angle *upward* toward the surface and outward toward the glacier's edges (Fig. 17-8).

Glaciers flow by a combination of two mechanisms: internal deformation and basal sliding. In **internal deformation,** pressure from overlying ice and snow deforms a glacier's ice crystals, fracturing them along planes of weakness, changing their shapes, and causing them to slip past one another (particularly when the glacier is relatively warm and there is water at crystal boundaries). All glaciers, even those that remain frozen to their beds, move in part by internal deformation. In **basal sliding,** warmer glaciers, such as those found in mid-latitude mountain ranges, thaw at their bases, producing a film of water that enables the glacier to slide across its bed. Glaciers in warm climates are more likely than those in extremely cold climates to move by both basal sliding and internal deformation. As a result, warm-based glaciers usually flow faster than cold-based glaciers.

Ice flow typically begins when a glacier's thickness exceeds about 60 m (200 ft), although thinner alpine glaciers (as thin as 10 m) may begin to flow when water at their bases enhances basal sliding. At 60 m, the weight of the overlying ice begins to deform ice crystals internally. This process largely explains why glacial

▲ **FIGURE 17-6** Types of glaciers. **(a)** Angel Glacier, a cirque glacier on Mount Edith Cavell, Jasper National Park, Canada.
(b) A valley glacier in Tongass National Forest, Alaska. **(c)** An ice cap in the Sentinel Range, part of the Antarctic continental
glacier. **(d)** A tidewater glacier at Kenai Fjords National Park, Alaska.

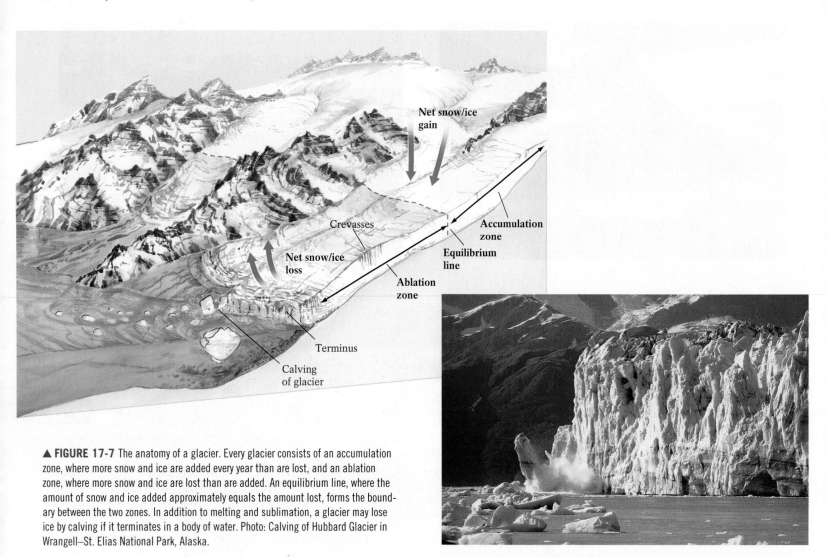

▲ **FIGURE 17-7** The anatomy of a glacier. Every glacier consists of an accumulation zone, where more snow and ice are added every year than are lost, and an ablation zone, where more snow and ice are lost than are added. An equilibrium line, where the amount of snow and ice added approximately equals the amount lost, forms the boundary between the two zones. In addition to melting and sublimation, a glacier may lose ice by calving if it terminates in a body of water. Photo: Calving of Hubbard Glacier in Wrangell–St. Elias National Park, Alaska.

▶ **FIGURE 17-8** The mechanics of glacial flow. In general, particles within a glacier move downward in the accumulation zone, parallel to the glacier's bed at the equilibrium line, and upward in the ablation zone. Movement of the glacier as a whole results from a combination of internal deformation of its ice crystals and basal sliding.

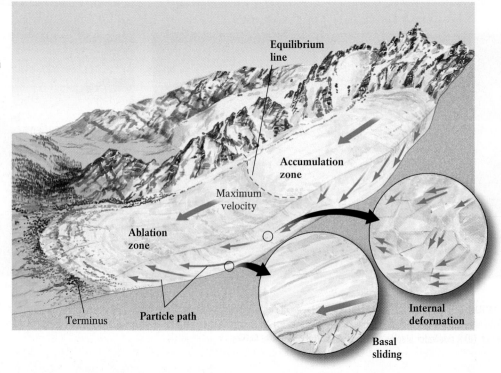

crystal deformation. The steepness of the bedrock slope (under alpine glaciers) or the ice-surface slope (for continental ice sheets) also affects glacial flow velocity by increasing gravity's effect on basal sliding. Typically, cold-based, nonsliding ice sheets flow slowly, usually advancing only a few meters per year. Warm-based alpine glaciers on steep slopes can flow 300 m (1000 ft) or more annually.

Some glaciers periodically **surge,** or accelerate, usually in brief episodes lasting several months to a few years. Periods of surging are typically separated by longer periods (10 to 100 years) of normal flow. Surging glaciers can move as much as 100 times faster than their normal velocity. Although there is much debate and several hypotheses to explain surging, many appear to occur when a large volume of water accumulates at the base of a glacier. According to this hypothesis, water at the base reduces friction between the glacier and its bed, and increases basal sliding. The ice is lifted from its bed by high water pressure, much like a car that hydroplanes on wet highway pavement. Although we cannot observe the action of the water directly, the large quantity of water that flows from a glacier's terminus as a surge comes to an end supports this hypothesis.

The most rapid surge ever recorded occurred in 1953 on the Kutiah Glacier in northern Pakistan. The surge reached a peak velocity of 110 m (350 ft) *per day* and lasted for several months. Another newsworthy glacier surge took place during the summer of 1986 on the Hubbard Glacier in the St. Elias Mountains near Glacier Bay National Park in southeastern Alaska (Fig. 17-10). Its velocity increased from its usual 30 to 100 m (100 to 300 ft) per year to about 5 km (3 mi) per year. Advancing across the mouth of Russell Fjord, the surging glacier blocked the outlet of a large marine bay, turning it into a rapidly filling freshwater lake and separating thousands of marine mammals from their saltwater habitat. As the lake's level rose to about 20 m (65 ft), the water swamped trees along the shore and threatened to overflow the Situk River. Fortunately, the surging ceased and the ice in the fjord broke up, restoring the bay to its pre-surge condition.

▲ FIGURE 17-9 Crevasses such as these appear in the brittle upper portion of a glacier wherever the glacier's surface is forced to bend (typically when the glacier crosses a substantial bump at its bed or traverses a steep cliff). The depth of crevasses is limited to about 50-60 m; below this depth, the flowing ice heals the crevasse.

▼ FIGURE 17-10 The Hubbard Glacier surging. Basal melting caused the glacier to surge in the summer of 1986, clogging the mouth of Russell Fjord with ice and turning a large marine bay into a lake.

crevasses—the deep, wide tears in the ice surface that pose great hazards to glacier visitors—are rarely more than 60 m (200 ft) deep (Fig. 17-9). Crevasses form because the near-surface ice is cold and brittle, and thus capable of fracturing. At depths greater than 60 m, flowing ice closes the crevasses.

Velocity of Glacial Flow
A glacier's velocity depends on both the physical properties of the ice and the glacier's location. The ice temperature controls the availability of water for basal sliding and ice-crystal slippage. In general, the closer the temperature of the ice is to its melting point, the more rapidly it tends to flow. A glacier's thickness contributes to its flow by providing the pressure that causes ice-

The Work of Glaciers

Glacial erosion and deposition, which we describe in this section, have been more effective than almost any other geological process in shaping the features of North America, northern and central Europe, and Asia, especially the alpine highlands on these continents. Let's look at some of the many ways that glaciers shape our landscape into some of its most fascinating forms.

Glacial Erosion

Glaciers remove soils, sediments, and rocks from the landscape by erosion. The greatest amount of erosion occurs when glaciers flow largely by basal sliding, particularly within the accumulation zone immediately upglacier of the equilibrium line. There the ice flows most rapidly and at a downward angle, impinging on the glacier's bed. Glaciers erode by abrasion and by quarrying. **Glacial abrasion** occurs when rock fragments embedded in the base of a glacier scrape the surface of underlying bedrock like sandpaper on wood. A glacier whose basal ice contains fine, gritty rock fragments may eventually polish the rock surface below it to a high shine. Ice carrying coarser rock fragments may cut long **striations,** or scratches, into the bedrock surface, like those seen in Figure 17-11. These striations may range in size from barely visible bedrock scratches to sizable grooves, meters wide and deep. *Striations typically line up parallel to the direction that the ice flowed.*

Glacial abrasion is enhanced when: (1) the glacier's base has a steady supply of fragments; (2) the glacier's base contains frag-ments that are harder than the underlying bedrock; (3) the sliding velocity of the ice is rapid (a circumstance typically associated with warm-based glaciers in relatively temperate environments); and (4) the underlying bed consists of readily eroded materials. Deep striations occur when a rapidly sliding glacier carries a large supply of hard pebbles or cobbles (such as granite or gneiss) over soft sedimentary bedrock (such as shale or limestone). The abrading stones themselves become abraded and may eventually be pulverized into a silty rock powder called *glacial flour.* This fine sediment, like talcum powder, adds a distinctive gray, milky tinge to rivers that flow through glaciated regions. **Glacial quarrying** occurs when a glacier lifts masses of bedrock from its bed. It is most likely to move bedrock that is jointed or fractured and where subglacial meltwater refreezes in fractures in the subglacial bedrock.

Working together, abrasion and quarrying can shape a rock mass into a distinctive asymmetrical form—a *roche moutonnée* (from the French, meaning "sheeplike rock") (Fig. 17-12). A glacier typically abrades the upglacier side of the rock mass (the side from which the glacier advances) and deeply quarries the downglacier side (the side toward which the glacier flows). A roche moutonnée is aligned lengthwise in the direction of glacial flow. Its upglacier side consists of a gentle humpbacked slope, usually well-striated, whereas the downglacier side is typically steep and jagged. A roche moutonnée's asymmetry provides another indicator of the direction of past glacier flow.

Erosional Features of Alpine Glaciation Alpine glacial erosion can sculpt smooth mountain slopes into spectacular peaks and deep gorges. The process typically begins with the creation

(a)   (b)

▲ **FIGURE 17-11** Bedrock striations produced by glacial abrasion. **(a)** Glacially striated bedrock in Blue Mounds State Park in southwestern Minnesota. **(b)** Large glacial grooves in bedrock on Slate Island in Lake Superior, Ontario, Canada.

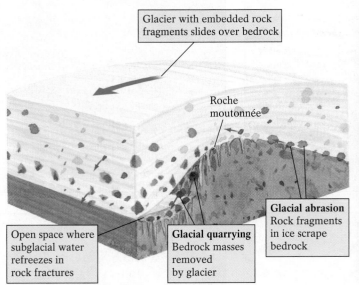

▲ **FIGURE 17-12** Glacial abrasion and quarrying combine to shape a roche moutonnée, like this one in Yosemite National Park, California. This distinctive erosional feature is produced when a glacier overrides a bedrock hill, abrading its upglacier side and quarrying its downglacier side. In this photo, the glacier flowed toward the left.

of a **cirque,** a horseshoe-shaped basin that develops in a shady depression at or above the local snowline. In these areas, lack of sunlight and accumulation of wind-drifted snow support the growth of persistent snowfields. During periods of partial thaw, meltwater from these snowfields seeps into the surrounding bedrock. When the weather grows colder, the water refreezes, loosening rock fragments by the mechanical-weathering process of frost wedging. Mass movement and surface runoff subsequently remove some of the loosened debris, enlarging the basin. As snow accumulates in the basin, it may be eventually converted into glacial ice, forming a *cirque glacier.* Quarrying and abrasion by the glacier's flow continue to erode the cirque, making it progressively deeper and longer.

Several cirque glaciers can form on the same mountain at the same time, producing various unique alpine landforms (Fig. 17-13). When two cirque glaciers on opposite sides of a mountain erode headward and converge, they form a sharp ridge of rock between them called an **arête** (from the French, meaning "fishbone," so-named because a series of arêtes resembles a fish skeleton when viewed from the air). Continued headward erosion of these cirques can remove part of an arête, producing a mountain pass, or **col** (from the Latin *collum,* meaning "neck"). Cols sometimes provide natural routes through imposing mountain ranges. Colorado's State Route 40, for example, traverses Berthoud Pass, a col through the Rockies near Winter Park. When three or more cirque glaciers erode a mountain, a steep peak, or **horn,** develops. Although the Matterhorn in the Swiss Alps is perhaps the best known of these forms, horns also dot the skyline of Banff and Jasper National Parks in the Canadian Rockies and Glacier National Park in northwestern Montana.

If the climate warms over time, a cirque glacier may melt away completely, leaving a basin that may later fill with water to form a cirque lake, or **tarn.** If the climate cools and the glacier's budget remains consistently positive, a cirque glacier may spread beyond its basin and flow downslope along a preexisting stream valley. As it moves, the glacier will erode the valley floor and walls, ultimately transforming the narrow V-shaped stream valley into a broad-floored U-shaped glaciated valley (Fig. 17-13b).

If cooling continues, small glaciers may develop in tributary stream valleys and eventually join together, forming a main valley glacier. The main glacier will be thicker and flow faster than its tributaries, eroding its underlying bedrock more rapidly. Eventually, it may undercut the tributary valleys, leaving them perched above the main glacier as hanging valleys. If a warm period then causes melting, waterfalls will cascade from the hanging valleys. Yosemite Falls in Yosemite National Park, with a vertical drop of 735 m (2425 ft), is one example of this phenomenon (Fig. 17-13c).

When seawater submerges a U-shaped, glaciated *coastal* valley, it forms a deep, saltwater-filled **fjord** (Fig. 17-13d). Some fjords were once thought to be stream valleys drowned by rising sea levels. The depth of most fjords, however, far exceeds the postglacial rise in sea level. Thus we believe that now-vanished tidewater glaciers eroded their valleys *below* sea level.

Erosional Features of Continental Ice Sheets   Erosion by continental ice sheets produces striations and roches moutonnées as well as several types of much larger features. An ice sheet can even abrade entire mountains, producing large-scale landforms called *whalebacks,* which are sculpted on both their upglacier and downglacier sides into a shape reminiscent—as their name suggests—of a whale's back. In addition, as a continental ice sheet flows through preglacial stream valleys, it carves monumental U-shaped troughs. The Finger Lakes of western New York—11 long, deep troughs—were excavated by ice-sheet erosion of preglacial valleys (Fig. 17-14a), as were North America's five Great Lakes (Fig. 17-14b), the northern section of Puget Sound in Washington, and Scotland's Loch Ness.

Continental ice sheets also erode on a continental scale. Much of a glacier's erosion occurs in the downglacier segment of the accumulation zone near the equilibrium line, where the glacier's

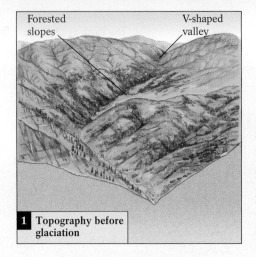

Forested slopes · V-shaped valley

1 Topography before glaciation

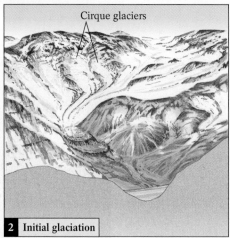

Cirque glaciers

2 Initial glaciation

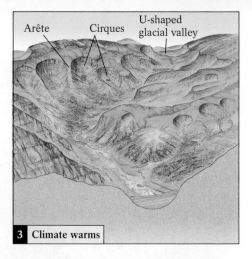

Arête · Cirques · U-shaped glacial valley

3 Climate warms

▶ **FIGURE 17-13** Effects of alpine glacial erosion. **(a)** Growth of cirques on mountainsides may eventually produce horns, arêtes, cols, hanging valleys, and tarns. Another common indicator of glacial erosion is the conversion of V-shaped stream valleys into U-shaped glaciated valleys. Although a single period of glaciation can produce all of these alpine glacial features, multiple glaciations (as shown here) accentuate and sharpen these landforms. **(b)** A U-shaped valley in Tracy Wilderness, southeastern Alaska. **(c)** Yosemite Falls. **(d,** on facing page**)** U-shaped valleys that become flooded with seawater produce fjords; this one is Bela Bela Fjord in British Columbia, Canada.

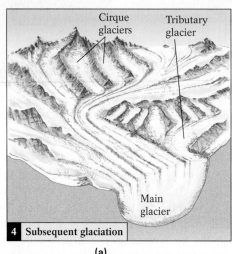

Cirque glaciers · Tributary glacier · Main glacier

4 Subsequent glaciation

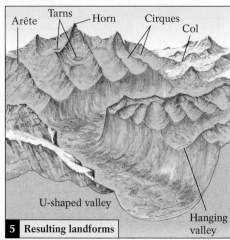

Arête · Tarns · Horn · Cirques · Col · U-shaped valley · Hanging valley

5 Resulting landforms

(a)

(b)

(c)

(d)

velocity is highest. Thus the ice sheet that covered much of northern North America about 20,000 years ago scoured the bedrock beneath this part of the ice sheet's accumulation zone, stripping it bare of loose rock and soil. This area today corresponds to the southern portion of the Canadian Shield and the region of soft sedimentary rocks to the south. This area includes southern Ontario and Quebec, Minnesota, Wisconsin, Michigan, Illinois, and New York.

Glacial Transport and Deposition

A glacier eventually deposits the load of debris it has eroded, usually when the ice begins to melt. Glacial deposits collectively are called **glacial drift.** (The name originated in the 19th century, when people thought that such deposits had been carried to their resting places by icebergs *drifting* on Noah's biblical floodwaters.)

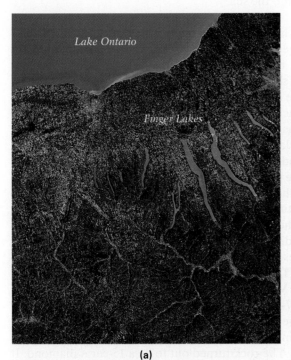

(a)

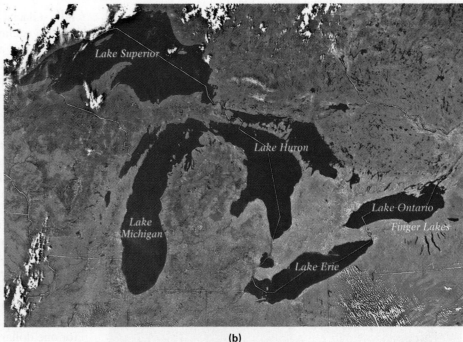

(b)

▲ **FIGURE 17-14** Continental ice sheets erode preglacial lowlands to leave deep basins and valleys. **(a)** During the Pleistocene Epoch, ice accumulating in the Lake Ontario basin (in what is now upstate New York) overflowed and moved south over the landscape, enlarging preexisting stream and river valleys to form the deep linear troughs known collectively today as the Finger Lakes. Lake Ontario, itself a glacially carved basin, lies to the north of the Finger Lakes in this satellite photo. **(b)** A satellite photo showing the five Great Lakes of North America (from left to right, Lakes Superior, Michigan, Huron, Erie, and Ontario). The much smaller Finger Lakes are shown at lower right, south-southeast of Lake Ontario. State and international boundaries are added for reference. The Great Lakes are more than 400 m (1300 ft) deep in places.

In contrast to flowing water and wind, where sediment transport is highly selective, transport by glaciers is essentially nonselective. The size of the particles that a given flow of water or air can carry depends on its flow velocity and turbulence. Deposits left by water or air are typically *well sorted*—that is, their range of sizes is limited to a narrow range. (Under normal conditions, water can carry up to gravel sizes; wind only up to silt and fine-sand sizes.) Although glaciers flow very slowly and certainly not turbulently, because of the high viscosity of ice, even extremely large masses of rock cannot settle through the ice, and thus are transported by glaciers once they're picked up. As a result, glacial deposits are typically *poorly sorted,* often ranging in size from fine clays to house-size boulders.

Most glacial drift is deposited after being carried only a few kilometers, but some materials can be transported for remarkably long distances. We can sometimes trace a specific rock in a glacial deposit to its source outcrop and thus estimate its transport distance. For example, chunks of copper have been found in glacial deposits in southern Illinois, but copper is not native to Illinois. Rather, the copper in Illinois' drift comes from Michigan's Upper Peninsula, *960 km (580 mi) to the north.* An exceptionally large, glacially transported boulder that has been eroded from one type of bedrock and then deposited atop another is known as a **glacial erratic** (Fig. 17-15).

Some glacial debris is deposited directly by the ice that carried it. Other debris is picked up and deposited later by meltwater streams that flow on top of, within, beneath, and in front of the

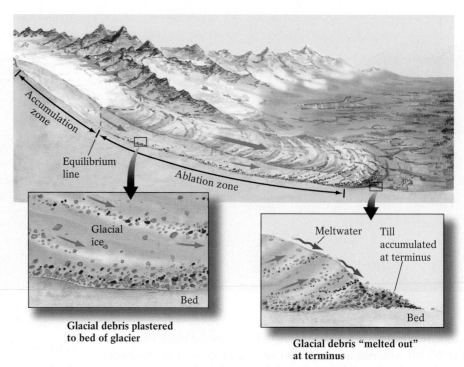

▲ **FIGURE 17-16** Deposition of glacial till. Tills are deposited downglacier from the equilibrium line, either by being plastered to the glacial bed by the weight of the glacier or by "melting out" at the glacier's terminus.

melting ice. Meltwater streams deposit glacial sediment wherever their paths take them—on land, into the lakes that form in front of the ice, or into the sea.

Till Glacial drift deposited directly from glacial ice forms a distinctive sediment called **glacial till** (from the Scottish word for "obstinate," a reference to Scotland's plow-resisting rock-and-clay soils). As shown in Figure 17-16, till is deposited almost exclusively in the ablation zone, either by being plastered onto the underlying glacial bed by flowing ice or by sloughing off the glacier's surface as the ice melts. Tills are generally unstratified (that is, they're not layered) and commonly consist of large rock fragments surrounded by a finer-grained matrix of sand, silt, and clay (that is, they're poorly sorted). Because the particles in a till were eroded from the bedrock over which the glacier passed, geologists can identify the rock types that are located along the glacier's route. They can then use this information to determine the glacier's flow direction.

Sometimes the rocks and minerals in till are quite valuable economically. Diamonds, for example, have been discovered in the tills of the Midwest and Great Lakes region at numerous localities (Fig. 17-17). In 1876, in a southeastern Wisconsin town, a well digger unearthed a pale yellow rock and gave it to his employer. Thinking it was common quartz or topaz, he later sold it for one dollar. The rock turned out to be a 15-carat diamond. It was eventually sold to the American Museum of Natural History in New York City (from which it was stolen in 1964 in a notorious heist by a character named Jack "Murf the Surf" Murphy and never recovered. *Google* him for more on this famous jewel thief and his other illegal "mineral-collecting" efforts.).

Glacial till often forms a **moraine,** a landform that typically accumulates at the margin of a glacier. (The name comes from

▲ **FIGURE 17-15** South Bubble Rock glacial erratic, Acadia National Park, Maine. This large granitic boulder was left behind by the melting Laurentide ice sheet about 15,000 years ago.

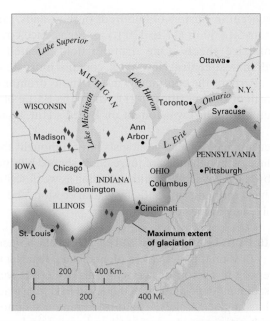

▲ **FIGURE 17-17** Known locations of diamond-bearing tills around the Great Lakes.

morena, a word used by farmers to describe ridges of rocky debris in the French Alps.) Moraines occur as bands of hills marking the various advances and retreats of a glacier (Fig. 17-18). The front of the glacier may bulldoze sediment as it advances, even when the terminus of a glacier remains stationary because ice continues to flow from the accumulation zone to the ablation zone, it erodes and transports sediment to the terminus where the load piles up. In general, the longer the terminus remains stationary, the more till the glacier will deposit in one place and the larger the resulting moraine. As the climate warms and ablation begins to exceed accumulation, the glacier recedes, leaving behind a *terminal,* or *end, moraine* that indicates the farthest advance of the ice.

If the climate again deteriorates and returns to glacial conditions, glacial recession halts and the terminus of the receding glacier may stop re-

treating and become stationary again, perhaps for a few hundred years or more. During this period, the glacier may deposit a *recessional moraine* upglacier from and usually parallel to its terminal moraine. A glacier that recedes intermittently may deposit several recessional moraines. As you can see in Figure 17-19, continental ice sheets, like smaller alpine glaciers, produce terminal and recessional moraines, such as the ones on Long Island and Cape Cod.

Two other types of moraines—lateral and medial—are formed only by alpine glaciers. Because they are topographically confined by their valleys, these glaciers have distinct sides, located adjacent to the valley walls. Till is deposited in these sites, forming *lateral moraines.* Lateral moraines also contain material that has fallen from the valley walls into the crevices between the ice and the walls. When two valley glaciers merge, their lateral moraines, which are carried downvalley along the sides of the two distinct ice streams, merge as well. The merger of these lateral moraines forms *medial moraines*—moraines located *within* a large valley glacier formed by two ice streams. You can see several medial moraines in Figure 17-20, where they appear as the dark-colored ridges interspersed between the streams of lighter-colored ice.

If the climate cools significantly after glaciers have remained stationary for a long period of moraine building, the glaciers may advance over previously constructed moraines, incorporating their sediments and redepositing them at new terminal moraines. When ice sheets—thicker and more massive than alpine glaciers—override such moraines, they may apply enough pressure to reshape the moraines into gently rounded, elongated hills called **drumlins** (from the Gaelic *druim,* meaning "ridge" or "hill"). A drumlin has an asymmetrical streamlined profile similar

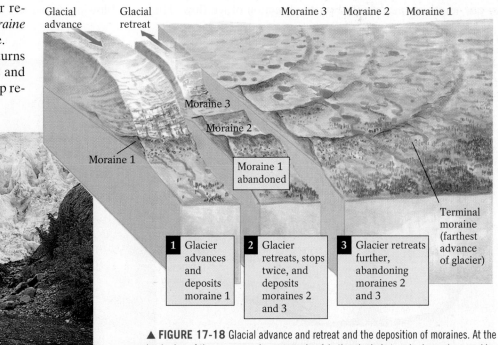

▲ **FIGURE 17-18** Glacial advance and retreat and the deposition of moraines. At the beginning of the sequence shown, moraine 1 is the glacier's terminal moraine, marking the maximum extent of the glacier. Moraines 2 and 3 are recessional moraines, marking the positions of the glacier's terminus during periods when the glacier's retreat was halted by temporary cooling trends. Photo: A moraine deposited by Exit Glacier at Kenai Fjord in Alaska.

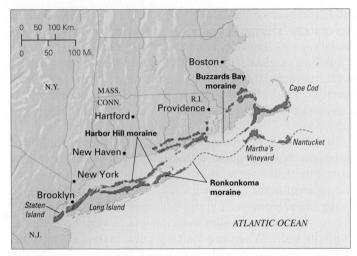

▲ **FIGURE 17-19** Terminal and recessional moraines were deposited over a large area of the northeastern United States by the North American ice sheet. The recessional Harbor Hill moraine, deposited about 14,000 years ago, passes through Prospect Park in Brooklyn, New York, continues across the north shore of Long Island, and then appears farther north, where it forms the mid-island hills of Cape Cod. The terminal Ronkonkoma moraine, deposited about 20,000 years ago, forms central Long Island and continues east and then north to Martha's Vineyard and Nantucket in Massachusetts Bay. The map indicates local names for the moraines, such as Buzzards Bay moraine.

to that of the inverted bowl of a kitchen spoon (Fig. 17-21). All drumlins are characteristically blunt on their upglacier ends and gently sloping on their downglacier ends—a shape that some geologists suggest is a consequence of their formation beneath actively flowing ice.

Drumlins are typically aligned with their long axes parallel to one another and oriented in the direction of ice flow. They

generally occur in clusters of hundreds or thousands, ranging in height from 5 to 50 m (17 to 170 ft), in areas once covered by continental ice sheets. One drumlin field, located east of Rochester in west-central New York, contains more than 10,000 drumlins. Perhaps the continent's most famous drumlins can be found north of Boston (Fig. 17-21), where the Battle of Bunker Hill (itself a drumlin) was fought on Breed's Hill, an adjacent drumlin.

In recent years, glacial geologists have rethought the origin of drumlins. Although many of these oddly shaped hills may have formed by glacial overriding and reshaping as we described, others may have formed principally by glacial erosion, and still others by subglacial floods of meltwater.

Deposits from Glacial Meltwater Streams
Glacial meltwater may flow on top of, in front of, and beneath a glacier. Sediment deposited on top of a glacier fills in depressions in its surface; sediment deposited immediately in front of a glacier may form deltas in glacial lakes. The most common sediment from glacial meltwater, a mixture of sand and gravel, is deposited by braided streams downstream of a glacier's terminus as **outwash.** These deposits, like all water-borne deposits, are typically well sorted and stratified (layered). In these ways, they differ sharply from tills, which are unsorted and unstratified.

Powerful glacial-age winds erode drying outwash, producing a large volume of fine silt. The fertile plains along the Mississippi River near Vicksburg, Mississippi, though far removed from any *direct* effects of glaciation, accumulated about 20,000 years ago as a thick layer of silt eroded from the outwash of the North American ice sheet. Although the ice sheet extended only as far south as northern Iowa, its outwash flowed down the Mississippi River to the Gulf of Mexico and was deposited along its banks. Winds blowing across the drying outwash eroded and transported the

▲ **FIGURE 17-20** Medial moraines form when adjacent glaciers converge, causing the lateral moraines at their edges to run together. Here several medial moraines are developing along the branches of the Kennicott Glacier in the Wrangell–St. Elias National Park, southeastern Alaska.

17.1 Geology at a Glance

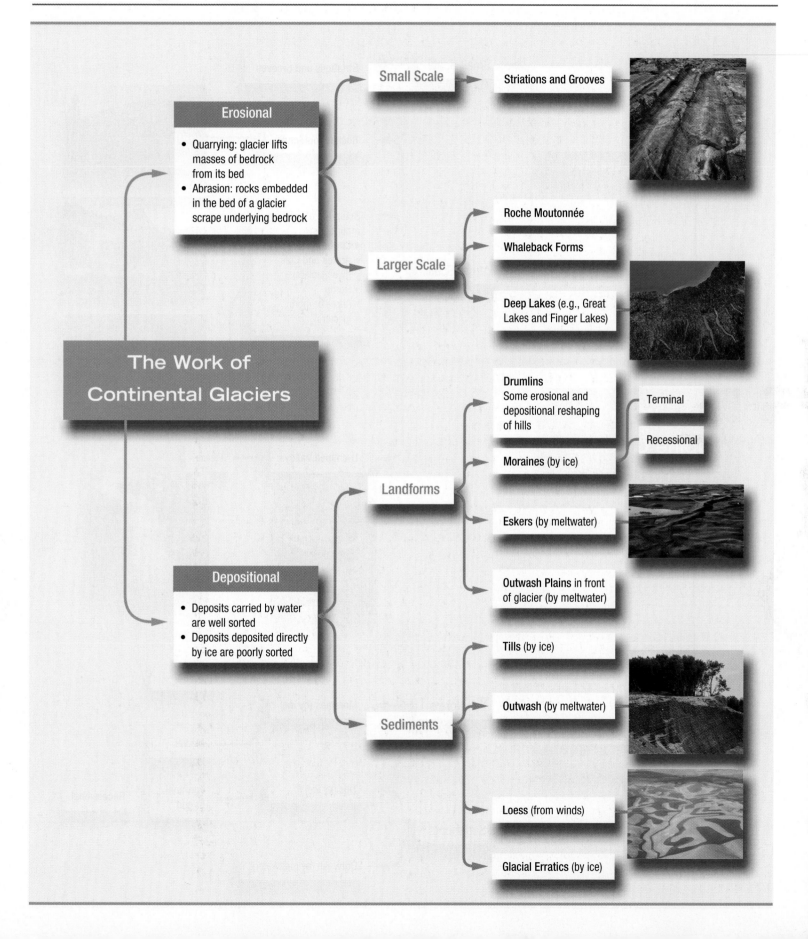

The Work of Continental Glaciers

Erosional
- Quarrying: glacier lifts masses of bedrock from its bed
- Abrasion: rocks embedded in the bed of a glacier scrape underlying bedrock

Small Scale → Striations and Grooves

Larger Scale
- Roche Moutonnée
- Whaleback Forms
- Deep Lakes (e.g., Great Lakes and Finger Lakes)

Depositional
- Deposits carried by water are well sorted
- Deposits deposited directly by ice are poorly sorted

Landforms
- Drumlins Some erosional and depositional reshaping of hills
- Moraines (by ice) → Terminal / Recessional
- Eskers (by meltwater)
- Outwash Plains in front of glacier (by meltwater)

Sediments
- Tills (by ice)
- Outwash (by meltwater)
- Loess (from winds)
- Glacial Erratics (by ice)

17.2 Geology at a Glance

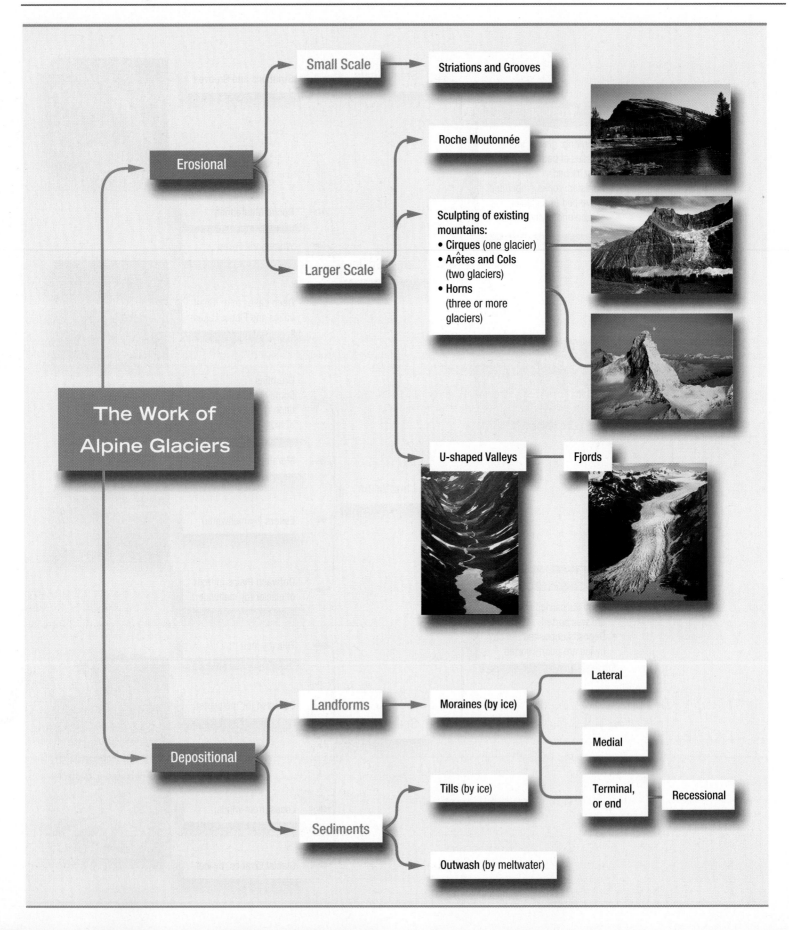

The Work of Alpine Glaciers

Erosional

Small Scale → Striations and Grooves

Larger Scale → Roche Moutonnée

Sculpting of existing mountains:
- Cirques (one glacier)
- Arêtes and Cols (two glaciers)
- Horns (three or more glaciers)

U-shaped Valleys → Fjords

Depositional

Landforms → Moraines (by ice) → Lateral, Medial, Terminal, or end → Recessional

Sediments → Tills (by ice)

Outwash (by meltwater)

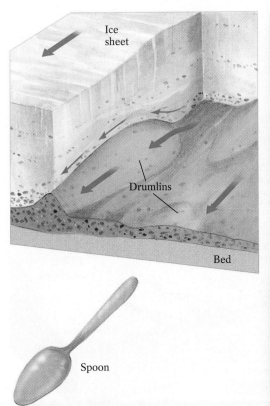

▲ **FIGURE 17-21** (above) Ice sheets passing over preexisting moraines can exert enough pressure to reshape them, forming low oval hills called drumlins. Drumlins are usually 5 to 50 m (17 to 150 ft) high, 0.5 to 1 km (0.3 to 0.6 mi) long, and 400 to 600 m (1300 to 2000 ft) wide. Note the similarity between the shape of a spoon and the shape of drumlins. Photo: A drumlin field east of Rochester, New York.

fine particles to the surrounding land. Such windblown silt deposits, known as **loess** (from the German for "loose"), commonly appear downwind from exposed, drying outwash. Throughout the Mississippi River valley, we see buff-colored, near-vertical bluffs of loess. (We will discuss loess in greater detail in Chapter 18.) The loess eroded from outwash plains makes up some of North America's finest farmland.

Meltwater deposits of sand and gravel can also accumulate underneath glacial ice, forming sinuous (S-shaped) ridges known as **eskers.** Eskers form beneath the ablation zone, as frictional heat from turbulently flowing subglacial streams melts tunnels in the overlying ice. Later, when the glacier wastes away and meltwater flow diminishes, sand and gravel are deposited in these tunnels (Fig. 17-22). Flowing meltwater may also carve bedrock channels into soft subglacial sediment. Eskers are found throughout southern Canada and in much of the northern United States, including Maine, Michigan, Wisconsin, Minnesota, North and South Dakota, and eastern Washington. If

you're interested in seeing an esker, you'd better act quickly; eskers are being mined extensively for their commercially valuable sand and gravel and thus are disappearing rapidly from the landscape.

Other Effects of Glaciation

In addition to the *direct* effects of glacial erosion, transport, and deposition, glaciers produce a variety of other direct and indirect changes on land, at sea, and in the atmosphere. They may even contribute to the evolution, migration, and extinction of plants and animals. In this section, we describe some signs of past worldwide glacial expansion.

Glacial Effects on the Landscape
Glaciers and the conditions associated with glacial periods can alter the landscape in ways other than erosion and deposition. The movement of the

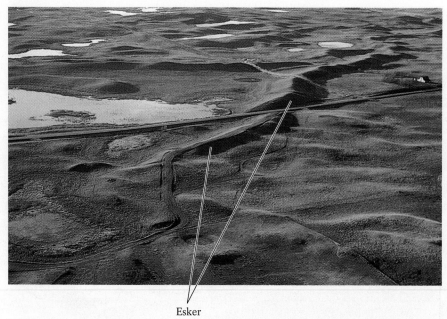

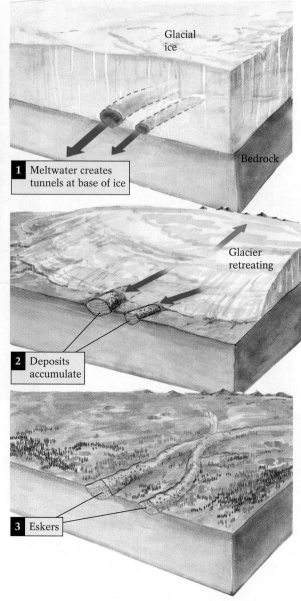

▲ **FIGURE 17-22** The origin of eskers. Eskers typically form under the ablation zone of a glacier, where meltwater erodes curved channels in the ice and deposits its load en route. When the glacier retreats, the meltwater flows away, leaving behind meandering ridges of stratified, cross-bedded sand and gravel. Eskers may be less than a kilometer long or more than 150 km (100 mi) long, and can reach 30 m (100 ft) high.
Photo: Eskers in Coteau des Prairies hills, South Dakota.

ice can disrupt preexisting stream systems, and its weight can depress the lithosphere on which it sits. Changes in climate can create vast areas of frozen ground and large landlocked bodies of water.

Expanding ice sheets can act like continent-wide dams, altering the courses of preexisting streams. The Missouri River and its tributaries, for example, once flowed north to Hudson Bay until the North American ice sheet blocked and diverted them. The Missouri River's present-day route, established during the last period of worldwide glacial expansion, follows the southern terminus of the departed ice sheet and then joins the Mississippi River on its way south to the Gulf of Mexico. Figure 17-23 shows the sequence of events that led to the Missouri River's diversion. Highlight 17-1 chronicles another drainage disruption in the Pacific Northwest that produced some of the most monumental floods in Earth history.

A continental ice sheet weighs so heavily on the Earth's lithosphere that it pushes the asthenosphere (the heat-softened, flowable layer of the upper mantle discussed in Chapter 11) away from the center of the ice sheet and toward its margins. This weight creates a crustal depression under the glacier and topographic bulges along its borders. When the climate warms and the ice sheet melts, the displaced asthenosphere *gradually* returns to its original position, the depressed crust slowly rebounds, and the marginal bulges disappear. (The changes in the Earth's crust are analogous to what happens to the surface of a water bed when

you lie on it. The weight of your body pushes the water from beneath you, forming a depression in the bed's surface and a bulge toward the edges of the bed. When you get up, the water at the edges flows back to its original position, eliminating the depression and bulge.) We observe the effects of crustal rebound throughout Scandinavia and around the Hudson Bay region of east-central Canada in the form of uplifted terraces (Fig. 17-24). The process of isostatic rebound (discussed in Chapter 11) continues today in those areas, thousands of years after the departure of the ice. The reason the rebound occurs so slowly is that the asthenosphere is extremely viscous, and thus quite slow-flowing. The rebound process and its activation of old faults may actually be responsible for some of New England's earthquakes, such as the one that rattled Boston on April 6, 2005.

When a crustal depression causes ice-sheet margins to bulge and then later subside, the subsiding region may then experience serious coastal floods. The southwestern Netherlands, located near the margin of the former European ice sheet, once stood

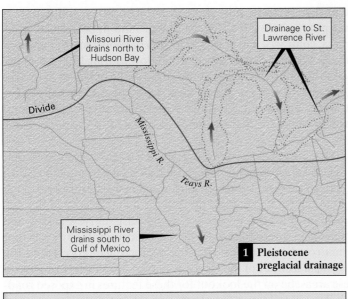

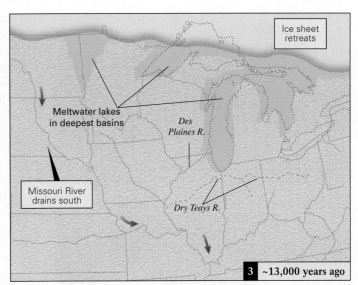

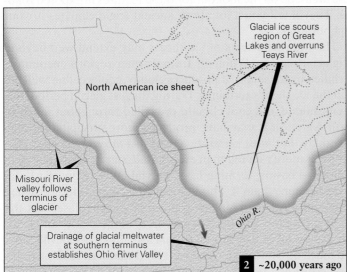

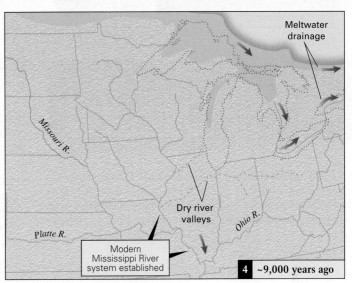

▲ **FIGURE 17-23** The effect of the North American ice sheet on the Great Lakes region. **(1)** Before the period of glacial expansion in the Pleistocene, streams flowed through the broad valleys that today are the Great Lakes. The major east–west tributary of the Mississippi River was the Teays River, which traversed what are now Illinois, Indiana, and Ohio. North of the Teays was an elevation that served as a divide; north of this divide rivers such as the Missouri flowed northward. **(2)** At the maximum extent of Pleistocene glaciation about 20,000 years ago, the North American ice sheet completely covered the Great Lakes region, deepening the lakes' basins by erosion. It also blocked the flow of the Missouri River, diverting it southward. **(3)** With warming, the ice sheet melted and retreated northward, leaving accumulations of meltwater in the deepest basins. By about 13,000 years ago, the Great Lakes and the Mississippi and its major tributaries were filled. Water in shallower basins, such as the Teays River, flowed away or evaporated, leaving dry valleys. **(4)** By 9000 years ago, the ice sheet had retreated from what is now the United States, and the current drainage system of this region was well established.

well above sea level atop one such crustal bulge. Since the departure of the ice sheet about 12,000 years ago, the depressed lithosphere has rebounded and its companion bulge has subsided, causing the southwestern Netherlands to drop, in some places to a depth below sea level. Since the 13th century, only Holland's famous dikes—and a lot of pumping of water—have kept the North Atlantic from flooding its valuable farmland.

The nonglaciated areas adjacent to glaciers and ice sheets are extremely cold. Icy winds and widespread frost action mark these *periglacial* (near-glacial) environments, producing a distinctive set of landscape features. (These features also appear in some

frigid high-latitude regions, such as northern Alaska, Arctic Canada, and Siberia, even where they're far from glaciers.) Perennially frozen ground, or **permafrost,** is found throughout high-latitude periglacial environments. Today permafrost underlies about 25% of the Earth's land surface, mostly in remote polar regions. In Siberia, permafrost extends to a depth of 1500 m (5000 ft); sections in the Canadian Arctic and Alaska reach depths of 1000 m (3300 ft) and 600 m (2000 ft), respectively. Geologists believe that most of the world's permafrost developed during the last worldwide glacial period, between 90,000 and 12,000 years ago, although some permafrost may still remain from earlier

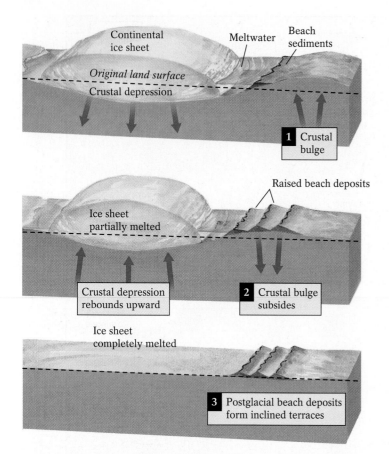

▲ **FIGURE 17-24** Formation of terraces from crustal rebound. Meltwater accumulates in the space between a retreating ice sheet and the crustal bulge at its margin, leaving behind sediments. As the ice sheet retreats farther and the crust slowly returns to its original position, the bulge subsides, leaving the sediment deposits as slightly inclined terraces.

glacial periods of the Pleistocene Ice Age, as far back as 800,000 years ago. The discovery of frozen woolly mammoths and other carbon sources in the near-surface layers of the permafrost of Siberia and Alaska has enabled carbon-14 dating of those deposits (within the range of C-14 dating).

A permafrost layer consists of soil, sediment, and bedrock that is continually at or below 0°C (32°F). During the summer—even brief, high-latitude summers—the thin upper zone of permafrost, or *active layer,* thaws and becomes wet and extremely unstable. As we discussed in Chapter 14, this waterlogged soil tends to flow. Buildings and other structures in permafrost areas must be designed to minimize the flow of heat from the structures to the ground. Oil pipelines (carrying warm crude oil) and heated buildings are commonly elevated above the ground surface, allowing cold air to circulate beneath them and keep the surface frozen.

The ground surface in periglacial areas is characterized by a variety of unique features, such as those seen in Figure 17-25. For example, as periglacial regolith alternately freezes and thaws, the frozen subsurface materials expand upward, slowly pushing stones up until they surface. The newly exposed stones often form surprising arrangements: *Sorted circles,* for instance, form as stones

are pushed laterally into circular patterns. When soil temperatures plunge below 0°C (32°F), the ground in periglacial areas freezes, contracts, and cracks to form large polygons, or *patterned ground.* When the surface water that flows during brief warm periods drains into the cracks at the edges of the polygons, it freezes and forms *ice wedges.* When these ice wedges eventually thaw, the water drains out, leaving cavities that may be filled by in-washing sediment that becomes compacted to form *ice-wedge casts.*

The presence of *relict* periglacial features indicates that periglacial environments once existed along the front of the North American ice sheet from New Jersey and Pennsylvania west to the Dakotas and at higher elevations south along the crest of the Appalachians. Pleistocene-age ice-wedge casts in New Jersey and Connecticut, more than 1500 km (1000 mi) south of the current permafrost limit, suggest that this region once represented the periglacial frontier of the North American ice sheet. If you live in these areas, be on the lookout for these features in exposed road-cuts. They'll tell you a lot about ancient frigid climates in your neighborhood.

The indirect effects of ice sheets may extend well beyond periglacial regions. When air from warmer regions encounters air chilled by ice sheets, it produces cloudy, cool weather well beyond the ice sheet's terminus. During the last glacial period, this type of humid and cool climate prevailed throughout much of North America. It may have produced more precipitation than falls today, *but most certainly less of this moisture evaporated.* Wherever more rain fell than evaporated, runoff accumulated in landlocked basins, forming large **pluvial lakes** (from the Latin *pluvia,* meaning "rain"). This indirect effect of glaciation reached as far south as Death Valley, California and the deserts of the American Southwest, covering vast areas of those now-arid lands with shallow lakes. The largest pluvial lake, Lake Bonneville, covered much of Utah, eastern Nevada, and southern Idaho. Utah's Great Salt Lake represents a small remnant of this water body (see Figure 6-20).

Glacial Effects on Sea Levels

Because the water in glacial ice ultimately comes from the oceans, global sea levels drop sharply when glaciers expand worldwide. As a result, during glaciations, much currently submerged land—specifically, large areas of the continental shelves—become exposed and the coastlines of the continents change dramatically. Twenty thousand years ago during the last glacial maximum, global sea level was about 130 m (430 ft) lower than it is today. As a result, North America's Atlantic coastline extended more than 150 km (100 mi) east of present-day New York City (Fig. 17-26, p. 562). Fossilized tree stumps and mastodon bones recovered from North America's continental shelves reveal that the exposed continental shelves were covered by spruce and pine forests and were home to herds of migrating ice-age mammals.

Lowered sea levels also expose "landbridges," connections between landmasses that are separated by shallow seaways in nonglacial times. Between about 100,000 and 40,000 years ago, during a period of worldwide glacial expansion, Great Britain and France were connected by a landbridge that today is covered by the shallow English Channel. Humans migrated from the

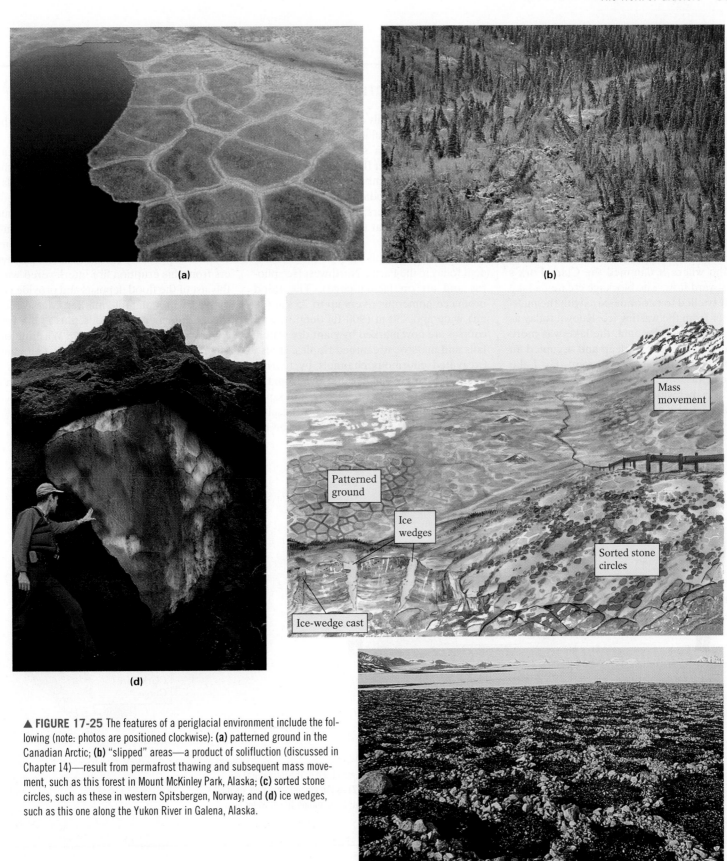

(a)

(b)

(d)

Mass movement

Patterned ground

Ice wedges

Sorted stone circles

Ice-wedge cast

(c)

▲ **FIGURE 17-25** The features of a periglacial environment include the following (note: photos are positioned clockwise): **(a)** patterned ground in the Canadian Arctic; **(b)** "slipped" areas—a product of solifluction (discussed in Chapter 14)—result from permafrost thawing and subsequent mass movement, such as this forest in Mount McKinley Park, Alaska; **(c)** sorted stone circles, such as these in western Spitsbergen, Norway; and **(d)** ice wedges, such as this one along the Yukon River in Galena, Alaska.

Highlight 17-1

The World's Greatest Flood—The Channeled Scablands of Eastern Washington

The Clark Fork River of northwestern Montana today is a freely flowing mountain stream. Some 13,000 years ago, however, it became blocked by a broad lobe of ice protruding from a retreating Cordilleran ice sheet, which covered western Canada to the Canadian Rockies (see Figure 1). The lobe of ice, known as the Purcell Lobe, flowed through a wide U-shaped valley extending from southern British Columbia to northern Idaho where it dammed the Clark Fork's westward flow. The blockage created a lake that swelled to enormous size with the meltwater from the wasting ice. Known today as Glacial Lake Missoula, the lake was more than 300 m (1000 ft) deep and occupied an area about the size of present-day Lake Michigan.

As meltwater continued to flow into it, Glacial Lake Missoula rose until it overflowed the ice dam, carving deep channels into the ice. Eventually the ice dam ruptured, releasing the accumulated waters of Glacial Lake Missoula in a sudden torrent (see Figure 2). More water flooded eastern Washington than the combined flow of all the rivers in the world today. Based on the sizes of deposited boulders, the flood's velocity has been estimated at from 50 to 75 km (30 to 50 mi) per hour.

The effects of this episode are still evident today in the Pacific Northwest (see photos and art on facing page). The flood produced numerous rivers up to 75 km (48 mi) wide and 250 m (900 ft) deep, whose courses are now marked by giant dry waterfalls and massive streambed ripples. It also excavated enormous channels by tearing hundreds of meters of basaltic lava from the Columbia River plateau, leaving thousands of remnant lava *scabs*—masses of basaltic rock that survived the flood intact. The region is aptly named the Channeled Scablands.

Geologists speculate that such great floods occurred repeatedly in eastern Washington during the Pleistocene ice age, as glaciers periodically advanced and retreated. We know that the last flood took place about 13,000 years ago because—in an amazing geological coincidence—Mount St. Helens erupted at precisely the same time. Ash layers from this eruption are interlayered with the last of the flood deposits and provide geologists with a datable marker. Massive ice sheets, unimaginable floods, plus a fiery volcanic blast, would have made Washington State of 13,000 years ago a geologist's dream: The evidence left behind gives geologists a storehouse of information about the region's dynamic recent past.

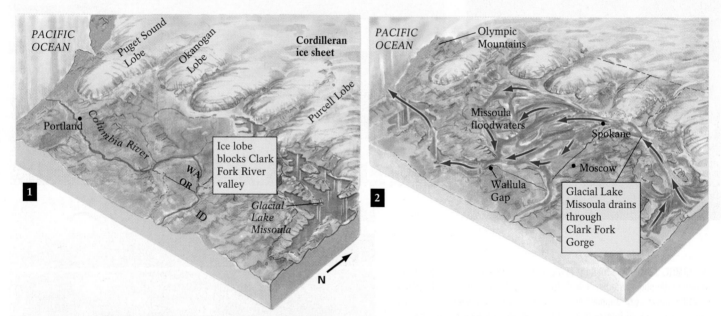

▲ The creation and draining of Glacial Lake Missoula. **(1)** About 13,000 years ago, the Purcell Lobe of the retreating Cordilleran ice sheet blocked the Clark Fork River valley, creating Glacial Lake Missoula. **(2)** As the glacier retreated, its meltwater swelled the lake until it finally overflowed its dam, releasing more than 20 million m³ (700 million ft³) of water per second.

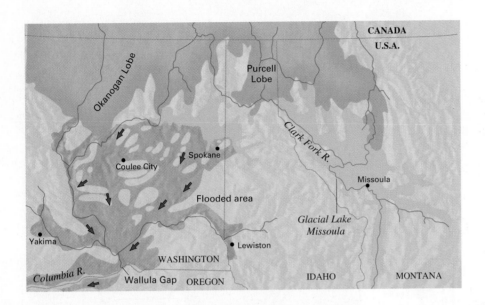

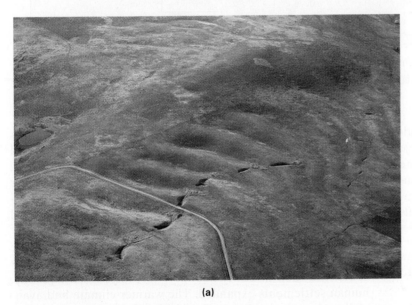

(a)

(b)

▲ Effects of the Missoula flooding. **(a)** Massive rivers created by the flooding produced giant ripples (about 10 m or 33 ft high), such as these in northwestern Montana. **(b)** This basalt bedrock near Coulee City, Washington, was carved into dry channels and massive dry falls where floodwaters once rushed at 30 to 50 mph. The cliffs in the background are 200 m (700 ft) high.

▲ **FIGURE 17-26** Twenty thousand years ago, falling sea levels exposed vast areas of the continental shelves. Some of these strips of land served as bridges between landmasses. One exposed land strip in particular altered the course of human history in the Western Hemisphere—the landbridge now covered by the shallow Bering Strait, separating Alaska and Siberia. Until about 12,000 years ago, this landbridge permitted human and animal traffic to cross in both directions. Many American archaeologists believe that it provided the principal route by which the first human inhabitants of North America arrived from Asia.

European mainland to the British Isles by walking over this landbridge. Similarly, as recently as 12,000 years ago, a landbridge linked Siberia in Asia to Alaska in North America. Today the waters of the Bering Strait cover this area. Over this route, humans and giant mammals such as mammoths and mastodons migrated into North America, and other animals, such as camels and horses, migrated from North America into Asia and Europe.

Glacial Effects on Flora and Fauna When climate changes and glaciers alter natural habitats, animals may not be able to find sufficient food and nesting sites. Likewise, the life cycles of plants may be disrupted. During periods of glacial expansion, ice sheets overrun vast stretches of land, temperatures become colder, skies become cloudier, and sunlight decreases. Plants and animals must either migrate to more suitable environments, evolve in ways that enhance their ability to survive the changed conditions, or become extinct.

As ice advanced to its last major maximum about 20,000 years ago, many of the plants and animals of North America and Europe migrated as much as 3000 km (2000 mi) southward, where they then had to compete with the local flora and fauna for space and food. As the climate warmed and the ice sheets finally melted, environments changed again, leading to another series of migrations and adaptations. Rising seas drowned the formerly habitable continental shelves, grasslands gave way to returning forests, and plants and animals began to populate newly deglaciated but still-

barren lands. Organisms that could readily adapt to these environmental changes survived both the Pleistocene's cold glacial and warm interglacial periods. Others, such as the ground sloth, mammoth, and saber-toothed cat could not adapt to the warmer postglacial climate and became extinct. (They also fell prey to migrating human hunters who had crossed ice-age landbridges and arrived on new continents to find an abundant supply of meat.)

Humankind was affected as well. With the warmer climate that followed the end of the last glacial period, migrating humans could find food in newly habitable territories; consequently, human settlements expanded. The warmer climate and availability of new animal and plant species that could be exploited for food may have spurred the development of new tool technologies and brought about the discovery of agriculture. As surpluses accumulated and less time was spent hunting and gathering food, trade developed and complex societies emerged in South and North America, Africa, Europe, and the Near East.

Reconstructing Ice Age Climates

Sixty-five million years ago, average global temperatures were as much as 10°C (18°F) warmer than today, and few, if any, ice masses existed on the Earth, even in the planet's polar regions.

Over the last 65 million years, however, the Earth's climate gradually cooled, as landmasses moved toward the poles and mountain ranges, such as the Himalaya, rose to great heights. About 2.5 million years ago, the planet entered an ice age from which it has yet to emerge.

During the last worldwide glacial expansion, some 20,000 years ago, temperatures in New England averaged about 5° to 7°C (9° to 12°F) colder than today, making what is now Boston, Massachusetts, as cold as Anchorage, Alaska. Temperatures in continental interiors, such as in the American Midwest, may have been as much as 10° to 15°C (18° to 27°F) colder, because these regions do not benefit from being close to oceans and their warming currents.

Given the absence of "fossil thermometers," where do geologists go to measure such past temperatures? The terrestrial geological record of climate change is unfortunately incomplete, largely because of the glacial processes of erosion and deposition. One geologist has compared the terrestrial record to a blackboard that has been partially erased and written over repeatedly. The geological record in various places has either been removed by glacial erosion or lies buried beneath thick layers of younger glacial and other deposits. Yet because the terrestrial record is the most accessible, geologists still study it in an effort to reconstruct past climates.

In Chapter 16, we saw how geologists use the growth of speleothems in caves to distinguish between alternating glacial and nonglacial conditions, both in glaciated regions and in tropical seacoast caves. Geologists may also map the distribution of "fossil" periglacial features, such as sorted circles and ice-wedge casts. Knowing the mean annual air temperatures in which these features form today enables geologists to estimate past temperatures in areas where we find relict periglacial evidence.

Some geologists and biologists have studied the distribution of climate-dependent biological organisms, such as fossil pollen or insect remains (beetles, for example), to reconstruct past climates. Their reasoning is quite simple: By identifying the preserved fossil remains extracted from sediment cores and comparing them with organisms that live in modern environmental conditions, we can reconstruct temperatures and precipitation characteristics of past environments. Furthermore, we can date these carbon-based materials as long as their ages fall within the range of carbon-14 dating.

Palynologists (geologists who study fossil pollen to reconstruct past climates) have used pollen extracted from lake cores to prove that plants that prosper today in cold, dry, south-central Canada lived during ice-age time in what is now warm, wet Louisiana. From the discovery of such pollen, they conclude that the environmental conditions in Louisiana at the time of deposition were comparable to those in Canada today.

Knowing modern temperatures and precipitation levels enables palynologists to estimate past conditions. Their main tool is a *pollen diagram,* such as the one in Figure 17-27 drawn from the percentages of fossil pollen collected from datable horizons in a sediment core. A pollen diagram shows how a region's pollen, and thus its vegetation (and climate), has varied through time. Pollen studies often go hand-in-hand with varve studies (discussed in Chapter 8). Geologists use varves to date lake deposits and then study the varve's pollen to learn about the climate during the time they were deposited.

Until recently, however, we could not reconstruct past climates sufficiently to develop a *continuous* record of ice-age climate change. We have instead been forced to search for a geological record that was neither erased nor buried by subsequent events. The search has produced two candidates, but both pose unique scientific and human challenges: One lies at the bottom of the Earth's oceans, and the other is found in the bitter cold of Greenland and Antarctica at the centers of those continents' long-lived ice sheets.

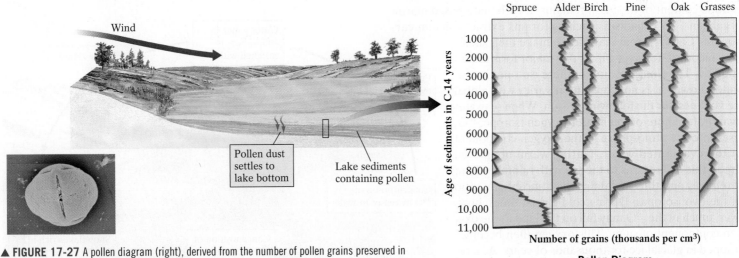

▲ **FIGURE 17-27** A pollen diagram (right), derived from the number of pollen grains preserved in datable sediment layers, can determine which plant species occupied a particular location and when. It allows us to estimate past climatic conditions. For example, pollen from numerous lake beds throughout the temperate, treeless prairies of North America's Central Plains indicates that during the last major glaciation, this part of the continent was covered with lush forests of spruce, tamarack, alder, and birch. Because these trees grow today farther north in southern Canada, we can conclude that the mid-continent region was significantly colder and moister 15,000 to 20,000 years ago than it is today. Photo, (left): Micrograph of pollen grain.

Deep-Sea Studies

Although the completeness of the past climatic record in the glaciated regions of the world has proved to be less than perfect, one part of the Earth's surface has remained immune to the effects of glacier-driven erosion and deposition—the seafloor. If the terrestrial record of glaciation resembles a partially erased blackboard, the marine record can be thought of as a thick paint chip from an apartment wall whose many colors record all the times the wall was painted. Because the oceans were not glaciated, the continuous sedimentation at the seafloor provides the most complete record of worldwide glacial advances and retreats—*if you know what to look for.* Unlike the terrestrial record, which at best retains evidence of four or five periods of worldwide glacial expansion, seafloor sediments reveal the occurrence of as many as 30 such events during the Pleistocene Epoch alone (1.6 million to 10,000 years ago).

What evidence supports geologists' claim of fickle glaciers that expand and overrun the landscape, only to contract again, frequently and somewhat regularly? One way to identify past climates is by comparing the oceanic distribution of fossil and living foraminifera (*forams* for short), microscopic marine organisms that produce distinctive coiled shells out of calcium carbonate extracted from seawater. Because forams are sensitive to water temperatures, knowing their current oceanic distribution and their preferred water temperatures enables geologists to reconstruct past water temperatures by charting forams extracted from cores of deep-sea sediments. For some foram species, the direction of coiling of their shells is largely controlled by water temperature. Thus, as shown in Figure 17-28, geologists can use the direction of coiling of *fossil* foram shells in marine sediment to estimate water temperatures at the time of their deposition. These factors, in turn, can then be correlated with periods of global cooling and warming.

Another powerful tool that geologists use to reconstruct the global movements of glaciers involves analysis of the oxygen-isotope content of the fossil remains of carbonate-based marine organisms. As organisms such as forams extract calcium carbonate ($CaCO_3$) from seawater to make their shells, they incorporate oxygen in its two major forms—^{16}O and ^{18}O. The concentration of these isotopes in seawater, however, is not constant; it varies with global ice volumes, for the following reason. When seawater evaporates, more of the lighter oxygen isotope, ^{16}O, rises into the atmosphere as the oxygen component of the evaporated water. During warm nonglacial times, this ^{16}O returns to the sea with rainfall, where it remixes and thus maintains the atmospheric ratio of the two isotopes. During cold glacial times, however, much of the ^{16}O may fall on the land as the oxygen component of snow. This oxygen then becomes trapped in glacial ice for thousands of years. As a result, as ^{16}O is trapped on land, the concentration of ^{18}O rises in seawater during glacial times. Thus, at the same time, ^{18}O concentrations rise in the carbonate shells of forams as well (Fig. 17-29). By analyzing the oxygen-isotope ratios in forams throughout a core of marine sediments that spans the last 2 million years or

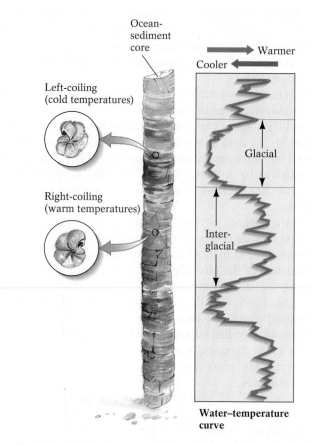

▲ **FIGURE 17-28** The coiling patterns of fossil forams can be used to identify ancient climate changes.

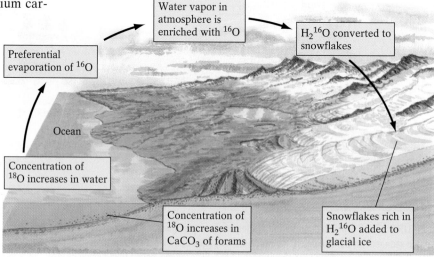

▲ **FIGURE 17-29** Oxygen-isotope concentrations in the shells of deep-sea organisms vary with global ice volumes. During cold periods, such as times of worldwide glacial expansion, ^{16}O is preferentially removed from seawater by evaporation. This isotope is then incorporated into snow that falls on land and is trapped in glacial ice, perhaps for thousands of years. Consequently, ^{18}O concentrations in seawater increase during glaciations and in foram shells as well.

so, geologists have been able to count glaciations and inter-glaciations. Figure 17-30 shows the 30 or so glacial–interglacial cycles that characterize the climate over this time period.

To date these global climate events, geologists use a combination of methods: carbon-14 dating of organic remains found in the more-recent upper sections of cores of deep-sea muds; isotope dating of trace quantities of uranium incorporated from seawater into the calcium-carbonate shells of microscopic marine organisms (uranium then decays to one of its daughter products thorium); potassium-argon dating of ash layers within seafloor stratigraphy; and geomagnetic-stratigraphy dating of the core (see Chapter 11 for a discussion of this dating technique).

Although these deep-sea studies have revolutionized our understanding of ice-age events, a recent discovery has provided geologists with another means of reconstructing past climates—one that enables us to decipher the chemistry of the ice-age atmosphere, gauge the windiness of the ice-age climate, and even detect large ancient volcanic eruptions. What is the source of this wealth of information? The Earth's vast polar ice sheets.

Ice-Core Studies

For many years, a refrigerated van sat in a parking lot on the University of New Hampshire campus that contained a potential key to unlocking the mysteries of the Earth's ice-age climate. The van housed row upon row of 1-m (3.3-ft)-long tubes of ice drilled from the center of Greenland's ice sheet—a richly detailed frozen archive of the planet's last 110,000 years of climatic history. (The cores have since moved a bit farther north with their collector, Dr. Paul Mayewski, to the University of Maine in Orono.) After decades of trial and error, glaciologists (scientists who study the snow and ice in glaciers) have accomplished a remarkable technological achievement: They can now extract cores of ice, several kilometers long, by boring through the Earth's surviving ice sheets down to the underlying bedrock. Examination of the physical and chemical properties of these ice cores reveals that they contain a *continuous* record of climate change for much of the Earth's current ice age.

Much of the upper part of the United States' GISP-2 (Greenland Ice Sheet Project) core is marked by layers consisting of now-icy remnants of each year's snowfall. Almost 80,000 annual layers have been identified. Scientists have already made the following discoveries:

■ Measurements of the dust content in the ice layers show that as much as 12 times as much airborne dust accumulated during the cold times associated with glacial expansions as during warm interglacial periods. This finding suggests that the atmosphere was far more turbulent during glacial times than it is today. The stronger-than-normal winds during glacial periods may have resulted from frequent clashes of warm and cold air masses; such winds might have swept across the continents as well as exposed areas of the continental shelves, eroding dried-out mud and depositing it on glaciers with their accumulating snows.

■ Samples of the trapped air in each layer's bubbles indicate that the CO_2 content of the atmosphere at the time of snowfall was relatively low during cold times and significantly higher during warm times. Carbon dioxide is, of course, the gaseous component of the atmosphere that contributes most to the greenhouse effect.

■ Analysis of a layer's oxygen-isotope concentrations has revealed the presence of a geochemical "thermometer." Unlike oxygen-isotope studies of deep-sea sediments, which reveal global ice volumes, analyses of ice allow geologists to estimate the mean annual air temperature, for the following reason. During extremely cold periods, water vapor containing the lighter ^{16}O isotope condenses from the atmosphere more readily than vapor containing the heavier ^{18}O. Thus falling snow contains proportionately more ^{16}O during these cold times. Such studies enable geologists to identify frigid periods and estimate the air temperatures at these times.

As an interesting by-product of ice-core studies, geologists have been able to identify specific volcanic ash layers from the

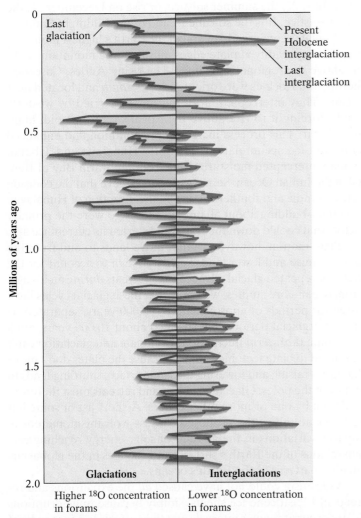

▲ **FIGURE 17-30** Changes in climate during the Quaternary ice age as revealed by oxygen-isotope studies of deep-sea sediments. This period of Earth history was marked by alternating periods of glacial expansion and warm interglacial periods. During the last million years or so, the periods of glaciation have lasted about 90,000 to 100,000 years and the intervening interglacials have been relatively brief, lasting roughly 10,000 years. (The Holocene interglacial period has lasted from about 10,000 years ago to the present.)

Figure labels:
Last glaciation
Present Holocene interglaciation
Last interglaciation
Millions of years ago
0 — 0.5 — 1.0 — 1.5 — 2.0
Glaciations | Interglaciations
Higher ^{18}O concentration in forams | Lower ^{18}O concentration in forams

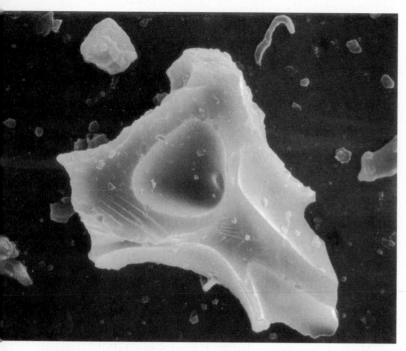

▲ **FIGURE 17-31** A tephra particle embedded in glacial ice.

datable annual ice layers (Fig. 17-31). For example, the GISP-2 core may have provided long-awaited proof of the timing of the catastrophic eruption that struck the island of Thera in the Aegean Sea—the blast that apparently wiped out the Minoan civilization and formed the basis for the legend of Atlantis recounted in Plato's dialogues. To the delight of archaeologists who have long debated the question, the studies revealed that Thera apparently erupted in the year 1645 B.C. (with an error margin of seven years). We now await confirmation of this date from the tephra itself, as geologists analyze the chemistry of the volcanic glass and compare it with known tephras from the eruption of Santorini, Thera's volcano.

Ice-core studies are just beginning to reveal remarkable new "truths" about the Earth's climate, including some with dramatic implications for the well-being of humankind. We now know that sharp climatic shifts can occur very quickly; the relative warmth we are enjoying during our current interglacial period could be replaced by a prolonged, sharply colder climate in the space of a mere decade or so—a condition that undoubtedly would affect growing seasons and food supplies throughout the world.

The Causes of Glaciation

What could cause a climatic shift so dramatic that it could bury New York City beneath 1000 m (3000 ft) of ice? What geological conditions enable ice to expand around the world, even to now-temperate mid-latitude regions such as Boston, Philadelphia, Cleveland, Chicago, Detroit, Minneapolis, and Seattle? What factors spur continent-size sheets of ice to advance and retreat numerous times *during* an ice age? The answers to these questions are complex, and no true consensus exists among those who study them. One frustrated student concluded that there must be at

least as many hypotheses to explain the causes of glaciation as there are glacial geologists.

Many geologists have concluded that most ice ages share three requirements: sizable landmasses at or near the poles; land surfaces with relatively high average elevations; and nearby oceans to provide the moisture that falls as snow. These conditions increase the likelihood of an ice age by putting landmasses at latitudes and elevations where the climate is colder and glaciers tend to grow. The phenomenon that can move landmasses to polar regions, raise them to higher elevations, and manipulate the positions of the Earth's ocean basins is, of course, plate tectonics.

The slow global cooling that started some 50 million years ago and led to the onset of the most recent ice age began once Antarctica had arrived at the South Pole. Other major landmasses, such as the northern parts of Greenland, Russia, Scandinavia, and Canada, had simultaneously traveled to positions north of the Arctic Circle. (The Arctic Circle, located at 66°33′ north of the equator, is the southernmost latitude where the sun doesn't set on June 21, the summer solstice, or rise on December 21, the winter solstice. It is the so-called Land of the Midnight Sun.)

As plates converged, several of the Earth's highest mountain ranges rose steadily at plate edges, raising those mountain peaks far above the regional snowlines. The rising Andes, Cascades, Sierras, and Rockies were oriented *north–south* and located near oceans. They intercepted moist marine air carried by westerly winds, forcing it to rise above their peaks into the cold, high-altitude air. This process increased precipitation and enhanced snow and ice accumulation. The lofty Himalaya and Tibetan plateau intercepted moisture rising from the warm Bay of Bengal of the Indian Ocean. Some geologists believe that the collision between India and southern Asia and the inception of Himalayan mountain-building about 50 million years ago were the primary factors that cooled down the Earth and started its current ice age.

Plate motion, however, cannot move lands to and from the poles or raise and lower lands swiftly enough to account for the cycles of repeated glacial advances and retreats *during* an ice age. From ocean-core studies, we know that these glacial cycles have lasted for periods of approximately 100,000 years, separated by brief interglacial periods that spanned about 10,000 years. Such systematic *short-term* fluctuations in climate and glaciation can't be related to plate tectonics simply because the plates don't move that fast—raising and lowering the land surface, shuttling lands to and from the poles. Glacial advances and retreats must therefore be driven by one or more other causes. A short list of some but certainly not all of the proposed causes—volcanism, meteorite impacts, variations in the amount of solar energy reaching certain regions of the Earth's surface, and changes in the global circulation patterns of the Earth's oceans.

Volcanism could not have caused *all* the expansions and retreats of Pleistocene ice sheets, simply because huge eruptions do not occur at regular intervals and their effects last for too brief a time. Large volcanic eruptions can lower world temperatures by as much as 1° to 2°C (2° to 4°F) for several years as volcanic ash and gas released into the atmosphere reflect and absorb solar radiation, causing less sunlight to reach the Earth. No consistent relationship between major eruptions and ensuing ice expansions has been proven, however, perhaps because the lowered temperatures are so short-lived. Worldwide glacial expansion, for example, did not follow such explosive eruptions as those associated with Mount Mazama (about 7700 years ago in southern Oregon) or Krakatoa (in 1883 in Indonesia).

Large meteorite impacts hurl veils of dust into the atmosphere and screen out significant solar radiation, sharply lowering global temperatures. Like volcanic eruptions, however, they occur sporadically and their effects are also short-lived. Thus they are also unlikely candidates to be the principal cause of the cycles of glacial expansion and contraction that mark the last 2 million years of Earth history. Nevertheless, a major meteorite strike could enhance climatic cooling that has already developed in response to other factors and push Earth closer to the deep freeze, but just not cyclically, as the glacial record indicates.

Variations in the Earth's Orbital Mechanics

Most glacial geologists believe that the primary cause of glacial fluctuations during an ice age involves the variations in the Earth's position and orientation relative to the Sun. This hypothesis was first proposed by British physicist John Croll in the late nineteenth century and then further refined by Yugoslavian astronomer Milutin Milankovitch around 1930. These mechanisms explain why even though the total amount of solar radiation the Earth receives may vary only a little, periodic changes in *where* and *when* that radiation strikes have profoundly affected global climate.

The so-called Milankovitch hypothesis proposes that the critical region for determining global glacial advances and retreats during ice ages is the vicinity of 65°N latitude. There, ice sheets have developed repeatedly during the past several million years, although today the region is largely ice free. A glance at a world map reveals that, when compared with other latitudes, 65°N—with North America, northern Europe, Scandinavia, and Asia—has the Earth's greatest concentration of land above sea level. Thus, this region provides large land surfaces upon which to grow ice. When this area receives less solar radiation and becomes sufficiently cold, enough winter snow accumulates to survive the summer melting season. The snowpack thickens from year to year until glaciers form. By comparison, the region in the vicinity of 65°S latitude is less likely to support glacier growth simply because there's hardly any land there (check out a world map). Falling snow there would just enter the southern oceans and melt. Once a global glacial expansion begins and the world's climate cools, glaciers do grow in Southern Hemisphere highlands such as the Andes.

Milankovitch calculated the effects of three periodic astronomical factors to show how they might combine to lower the radiation levels of 65°N, and why they happen at roughly 100,000-year intervals—the recurrence interval of worldwide glacial expansions during the last million years. These three factors include variations in the shape of the Earth's orbit around the Sun from circular to elliptical (oval-shaped), variations in the tilt of the Earth's axis toward or away from the Sun, and the shift, or *precession,* of the spring and autumn equinoxes caused by the wobble of the Earth's axis as the Earth spins like a top (Fig.

17-32). Each factor varies in a unique cycle, but according to Milankovitch the three cycles sometimes coincide, producing centuries or even millennia of low summertime radiation at 65°N. When this situation arises, summers are cool and brief, snow stays around from year to year, and alpine glaciers and continental ice sheets expand.

For years, glacial geologists have debated whether these orbital factors alone are sufficient to produce a full-blown glacial expansion. Climate experts estimate that, taken together, these factors can reduce solar radiation enough to lower global temperatures about 4°C (7°F). Although such a temperature reduction is certainly significant, it falls far short of the ice-age temperatures that geologists have detected in the geological and biological record (5° to 8°C [9° to 14°F] at mid-latitude coasts; 15° to 20°C [27° to 36°F] in continental interiors). Thus, additional factors must contribute to the ice-age climatic deterioration that leads to the worldwide expansion of glaciers.

Terrestrial Factors That Amplify Ice Age Climate Change

Once the positions and altitudes of continental lands are arranged tectonically in ways that promote global cooling, and the Milankovitch orbital factors conspire to produce a solar-radiation minimum at high northern latitudes during the summer melting season, numerous other factors may contribute to the trend toward global cooling or warming, perhaps even switching these climatic events on and off. These factors include (but are not limited to) the reflectivity of the Earth's surface and changes in oceanic salinity that alter large-scale oceanic circulation.

After tectonic and astronomical factors trigger a period of glaciation and the Earth's snow and ice cover expand, white snow and ice (which reflect radiation) cover and thus replace dark-colored rocks and soils (which absorb radiation rather than reflect it) and vegetation (which is nonreflective) at the Earth's surface. (Remember, you should wear light-colored clothing in the summer, not black clothing. Same effect.) As the snow and ice cover rises, so rises the Earth's *albedo*—the percentage of incoming solar radiation reflected from surfaces back into space. As more incoming radiation is reflected off the white snow and ice, less remains available to warm the Earth's surface. Consequently, the Earth's climate cools further.

Another factor that substantially affects the course of glacial expansions and contractions involves the salinity of ocean currents in the North Atlantic. During warm periods, the water flowing northward by ocean currents from the equator (such as the Gulf Stream) evaporates significantly as it passes through warm climates. Through this process, the water becomes saltier. As this salty water passes through colder northerly environments, it cools sharply. By the time these currents reach the North Atlantic, their waters—*cold and salty and thus more dense*—begin to sink, creating North Atlantic Deep Water. As this water sinks (and begins its return flow toward the equator), the resulting void pulls additional equatorial currents northward, thereby maintaining the steady flow of warm tropical water to the North. This watery "conveyor belt," as shown in Figure 17-33, helps to moderate

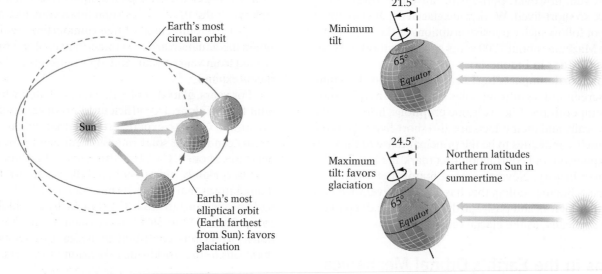

(a) Orbital effects

(b) Tilt of Earth's axis of rotation

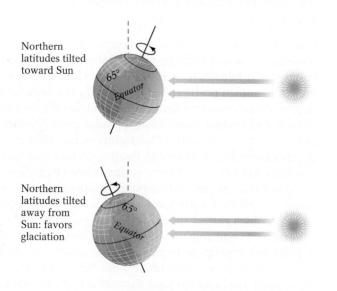

(c) Wobble of Earth's axis of rotation (precession)

▲ **FIGURE 17-32** Milutin Milankovitch proposed that three astronomical factors interact to affect the amount of solar radiation striking the Earth's high northern latitudes, possibly triggering the periodic advance and retreat of glaciers around 65°N latitude. The climate at 65°N latitude cools: **(a)** when the Earth's orbit around the Sun is at its most elliptical (especially when the Earth is located farthest from the Sun during northern summers); **(b)** when the tilt of the Earth's axis positions the high northern latitudes at a lower angle to incoming solar radiation; and **(c)** when the "wobble" of the Earth's spin axis carries this crucial latitude out of the path of direct solar radiation. Each of these three factors has a different period of recurrence. When they have coincided in the past, glaciations have occurred.

global climatic conditions by bringing warm water to the cold Arctic. This warm water warms the air above it.

But what happens when this conveyor belt switches off? This happens when the global climate cools and seawater evaporation diminishes or when a great influx of fresh glacial meltwater flows into the adjacent North Atlantic, further reducing its salinity and thus its tendency to sink. Without the transfer of warm tropical waters to the Arctic, the climate there deteriorates sharply, producing much colder conditions. We can observe the effect of such declines in oceanic salinity in the geological record—terrestrial, ice-core, and deep-sea—during the last episode of worldwide glacial expansion 11,700 years ago. At that time, even though most of the world's glaciers retreated in response to higher solar radiation brought on by the Milankovitch factors, the planet plunged into the deep freeze one last time. Here's why.

When the North American ice sheet retreated, it created an enormous meltwater lake in south-central Canada—larger than

Lake Superior—that flooded through the newly opened St. Lawrence valley (exposed by glacial retreat) and into the North Atlantic. The massive influx of freshwater from the draining lake reduced the ocean's salinity levels and temporarily cut off the North Atlantic Deep Water current. This brief but dramatic return to glacial conditions lasted for a few centuries. As the world climate warmed further, evaporation increased, and oceanic salinity restored the North Atlantic Deep Water conveyor.

In a similar way, periods of ice-sheet surging into the Atlantic may have decreased salinity and shut down the warming currents. The North Atlantic's oceanic-sedimentary record contains numerous pebbly layers that document melting of vast numbers of floating icebergs that release their sediment loads when they melt. Melting of these *freshwater* icebergs—and their contribution to reduced salinity—apparently brought about colder conditions. These episodes marked by vast influxes of climate-changing icebergs, which may have occurred every 10,000 years or so, are

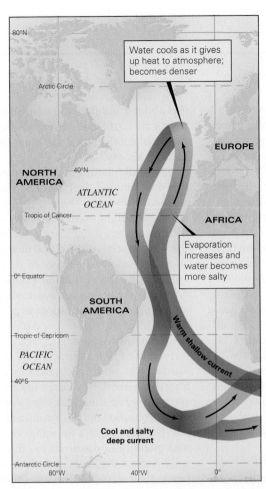

▲ **FIGURE 17-33** The origin of North Atlantic Deep Water. As warm equatorial water heads northward, some evaporates in the warm subtropical air, increasing the ocean's salinity. As the water reaches more northerly climates, it cools and grows more dense. As this cold, salty water sinks, it draws more water up from the equator, maintaining the warm moderating effect of the current. When this water conveyor shuts off, cold times ensue.

called *Heinrich events,* named for the German oceanographer who discovered them.

There is clearly more to the causes of glaciation and climate change than initially meets the eye. Geologists expect even more discoveries as we probe the mysteries of the remarkable Greenland ice cores.

The Earth's Glacial Past

The Earth has experienced numerous ice ages throughout its 4.6-billion-year history. Unfortunately, erosion of the ancient geologic record and its burial beneath younger rocks and sediments limit geologists' ability to study the earliest ice ages. Nevertheless, the geologic record yields clear evidence of at least three ice ages in Precambrian time. The oldest recorded Precambrian ice age, which occurred about 2.2 billion years ago, is identifiable from layers of *tillite,* a sedimentary rock produced by lithification of till. Geologists studying this Precambrian event have discovered signs that indicate that during this glaciation, ice may have ex-

tended from the Earth's poles to the tropics. During this monumental ice age, virtually the entire planet may have been locked in a glacial deep freeze. Another proposed worldwide deep freeze, perhaps the most amazing one, is discussed in Highlight 17-2.

Geologic evidence of global glaciation after the Precambrian is more widespread and better preserved. Layers of tillites and "fossilized" moraines and eskers show that during the Paleozoic Era, about 500 million years ago, an ice sheet occupied what is now one of the world's hottest and driest lands, the Sahara of northern Africa. Before plate tectonics moved the continents to their present locations, the Sahara was located at the South Pole. Ironically, one may now swelter in the merciless African sun while standing atop an ancient esker.

An extensive glaciation that occurred during the Permian Period, about 245 to 286 million years ago, left the most complete record of any pre-Pleistocene glaciation. This ice sheet striated the bedrock and deposited extensive tillites in regions that now form Australia, South America, India, and South Africa (Fig. 17-34). At the time, these lands constituted the supercontinent of Gondwana, located in the vicinity of the South Pole.

▲ **FIGURE 17-34** This outcrop of an ancient glacial till—now lithified as a tillite—is exposed in the walls of the Orange River canyon where it runs along the border of South Africa and Namibia, near the Namibia border town of Noordoewer. The till was deposited by an ice sheet during a time when Africa sat near the South Pole.

Highlight 17–2

The Snowball Earth—When the Whole Planet Froze

On the eastern slopes of Virginia's Blue Ridge Mountains, near the town of Batesville, geologists have been studying a layer of 700-million-year-old rock that has become one important piece of an amazing global puzzle. This layer, which is a mixture of fine mud and grapefruit-size cobbles, is a classic tillite, a body of lithified glacial till and a clear indicator of long-ago glaciation. What's astounding about this rock layer is its relationship to other rock layers of the same-age: Nearly identical rock is found in northern Namibia in Africa, in northwestern Canada, in Australia, and in the Arctic islands of Svalbard in northern Europe.

Ancient tills on far-flung continents and with the same 700-million-year age. Clearly, the late Proterozoic Earth was held in the icy grip of a worldwide ice age at that time. What is even more surprising and perplexing to geologists is finding that, based on geomagnetic evidence, many of these tills were deposited somewhere between 15° and 30° of the Earth's equator. To picture glaciers forming in such ancient environments, think of massive glaciers advancing across Miami Beach (at latitude 26°).

All of these pieces of the late Proterozoic Earth were parts of Rodinia, an ancient supercontinent that is believed to have sat mostly in a low-latitude, equator-straddling position. How could the Earth have grown so cold that such equatorial landmasses were completely covered with ice? The answer, ac-cording to the fascinating "Snowball Earth" hypothesis, may involve an Earth completely encased in ice, with frozen oceans as well. The key to this remarkable development is carbon dioxide.

As we discussed in Chapter 5, when the atmosphere is rich in carbon dioxide, a green-house effect follows, and the Earth warms up. But when the atmosphere has been stripped of nearly all of its carbon dioxide, the reverse effect develops—the "icehouse effect." Apparently, prior to the 700-million-year-old global glaciation, microscopic plant life—plankton—flourished in great abundance in the Earth's warm oceans. Like all plants, these blooming plankton used CO_2 for photosynthesis. This rampant photosynthesis may have cleansed the atmosphere of nearly all of its carbon dioxide, which then became buried in sea-floor sediments within the cells of dead plankton, and thus was locked away from the atmosphere for millions of years.

Consequently, atmospheric CO_2 concentration fell so sharply that the Earth could no longer retain its surface heat. Much of this heat escaped into space, and as a result, snow and ice could accumulate virtually every-where. Most of the Earth's oceans froze solid. Once ice began to develop in mid-latitude oceans, its vast whiteness created a "super-albedo" effect, reflecting even more solar radiation away from the Earth. Add to this mix the fact that the late Proterozoic Sun was about 7% dimmer than it is today, ren-dering the Earth more susceptible to freezing over completely than at any other time in the planet's history.

How did the Earth finally emerge from this monumental deep freeze? The answer again involves CO_2. The breakup of Rodinia, about 650 million years ago, was accompa-nied by a massive increase in volcanism as the supercontinent rifted and new divergent zones formed. Over millions of years, a great volume of carbon dioxide poured out of divergent-zone volcanoes, accumulating in the atmosphere at concentrations as much as 350 times higher than those found today. At that point, a resulting supercharged greenhouse ef-fect sent the Earth's temperatures sharply up-ward and the world's ice rapidly melted away. Shortly thereafter, an extreme increase in chemical weathering associated with the run-away production of CO_2-produced carbonic acid filled the Earth's warming oceans with a vast quantity of dissolved calcium. The re-sult? A great thickness of limestone was de-posited in the world's oceans following the "Snowball Earth." In the photograph from Namibia, note the student standing next to the obvious tillite and check out the large layer of light-colored rock above: That's the limestone deposited in the Earth's warm, postglacial oceans.

Thus, the geological record of the late Proterozoic Era tells us of the days when the Earth's climate swung wildly between a frigid global icebox and a stifling global green-house. The effect on life caused by these unimaginable swings would certainly have been profoundly stressful, and only the hardi-est life forms survived. Some geologists and biologists believe, however, that from these extreme environmental swings came new evolutionary developments that ultimately led to the complex multicellular life forms that now inhabit our planet, including all of us humans. Here again is yet another dra-matic example of the interaction of all four of the Earth's systems: The biosphere (in the form of plankton) changes the chemistry of the atmosphere (its CO_2 content). As a di-rect result, the hydrosphere (in the form of global ice) spreads and adds a fascinating rock record (the tillites and limestones) to the lithosphere.

▲ The boulder-filled layer next to the geologist is a 700-million-year-old tillite from Namibia, Africa that marks the global ice cover of the "Snowball Earth."

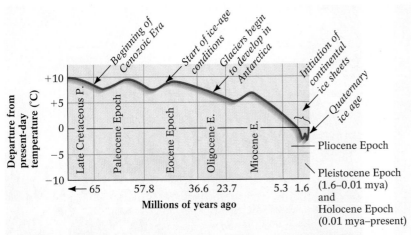

▲ **FIGURE 17-35** Changes in global climate during the Cenozoic Era. Although the most recent (Quaternary) ice age is associated with the start of the Pleistocene 1.6 million years ago, the Earth had actually been cooling for about 50 million years prior to that time.

Geologists have found *no* evidence of glaciation from the Mesozoic Era, which lasted from 225 to 65 million years ago. Apparently the Mesozoic was a time of great global warming and increased plate motion associated with the rifting and breakup of Pangaea. Many of the Earth's landmasses moved from the poles to subtropical and tropical latitudes, and episodes of volcanism at the vast network of rift zones and mid-ocean ridges spewed great volumes of warming greenhouse gases into the atmosphere. Mesozoic rocks in northern Alaska, for example, contain remnants of coral reefs, tropical vegetation, and dinosaur fossils.

Although the Earth's recent ice age is associated with the end of the Pliocene Epoch (about 2.5 million years ago) and into the Pleistocene Epoch, 1.6 million years ago, recent evidence from forams and other marine species suggests that Earth's surface waters actually began to cool at least 50 million years ago (Fig. 17-35) during the Eocene Epoch. Between 50 million and 20 million years ago, world temperatures dropped approximately 5°C (9°F). At about this time, large domes of ice may have begun to grow on Antarctica; striated stones, presumably carried northward by floating icebergs, are found today in the south Pacific in the layers of muddy oceanic sediments that date from that period. Gradual global cooling continued, and by 12 to 10 million years ago ice had begun to form in the mountains of Alaska. By 2.5 million years ago, an ice sheet had buried the island of Greenland, and 500,000 years later, ice was accumulating on the high plateaus and mountains of North America and Europe. The Quaternary ice age, with its alternating mid-latitude ice-sheet advances and retreats, was poised to begin.

Glaciation in the Pleistocene

As noted earlier, the Earth has experienced a succession of alternating glacial and interglacial periods during the past 1.6 million years. During the glacial periods, the ice cover expanded from the polar regions to the middle latitudes. In North America, it reached as far south as the Great Plains. Tills and outwash deposits exposed in Kansas and Nebraska have been strongly modified by weathering and erosion over their long existence since the early Pleistocene; the slopes of these old moraines have been gently rounded and subdued by mass movement, and their tills have been extremely oxidized. Deposits from more recent Pleistocene glacial events, especially those occurring in the last 100,000 years, remain relatively fresh and relatively unweathered and unmodified by mass movement.

The location of North America's *young* glacial deposits indicates that the continent's last ice sheet extended from the eastern Canadian Arctic south to New York City and west to the foothills of the Rocky Mountains in Montana, an area of more than 15 million km² (6 million mi²). The source of these deposits, the *Laurentide* ice sheet, probably began to grow in Canada west of Hudson Bay and in Labrador and Newfoundland, where the relatively high topography and cold climates fostered snow and ice accumulation. Nourished by moisture from the Atlantic Coast and the Gulf of Mexico, the Laurentide ice sheet advanced east and south. Another glacial center, the *Cordilleran* ice sheet, formed farther west between the Canadian Rockies and the Coast Range of British Columbia. It probably began as a network of alpine glaciers that expanded and merged to form an ice cap that eventually covered the adjacent valleys and plains.

At the time of the last maximum glacial expansion, a continuous wall of ice stretched 6400 km (4000 mi) across the full breadth of southern Canada and the northern United States (Fig. 17-36). This ice sheet stripped an average of 15 to 25 m (50 to 80 ft) of

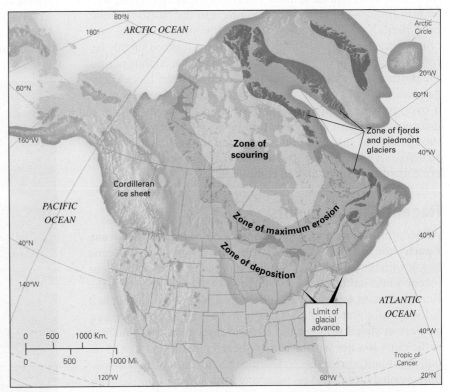

▲ **FIGURE 17-36** The effect of the North American ice sheet, which reached its greatest extent about 20,000 years ago. Within the zone of scouring, the bedrock has been stripped of its soils and the rocks are highly striated.

▲ FIGURE 17-37 A log in glacial till in Two Creeks, Wisconsin. The log was buried about 11,700 years ago when a temporary cooling period caused the retreating North American ice sheet to readvance over newly established forests.

▲ FIGURE 17-38 Frozen canals in Holland during the Little Ice Age, about 400 years ago (c. 1600). (Hendrick Averkamp *Winter Landscape with Iceskaters*, 1608. Rijksmuseum, Amsterdam)

regolith and bedrock from central and southern Canada and the northern Great Lakes region and deposited it from Cape Cod to northern Washington as glacial drift averaging about 15 m (50 ft) deep.

Between 13,000 and 12,000 years ago, the world's climate warmed, ice sheets began to melt, and the North American ice sheet retreated from the United States and southern Canada. Perhaps in response to meltwater flooding into the North Atlantic, brief periods of sharp cooling temporarily interrupted the general warming trend, and some glaciers even advanced slightly before resuming their final retreat. During one such cooling period, the portion of the ice sheet that flowed through the Lake Michigan basin moved south of Green Bay, Wisconsin. In the process, it overran trees that had begun to grow in the warmer climate, incorporating them into a new thin layer of till (Fig. 17-37). Although the ice sheet had then retreated into Canada by about 10,000 years ago, another 3000 to 4000 years elapsed before northern Canada's ice caps melted completely.

Recent Glacial Events

Alternating warm and cool periods also mark the **Holocene Epoch**—the last 10,000 years of the Earth's history. The Holocene began with the postglacial warming that melted most of the Pleistocene's mid-latitude ice. From about 8000 to 6000 years ago, the Earth's climate was about 2°C (3.6°F) warmer on average than it is today. Moderate cooling since then has produced many new glaciers in alpine settings, but temperatures have not remained sufficiently low nor has the climate been cold long enough to rebuild mid-latitude ice sheets. During the most recent cold snap—the so-called Little Ice Age, occurring from about 700 to 150 years ago—winters were colder and snowier than average and summers relatively cool and wet. Indeed, some winters were so cold that 18th-century New Yorkers could walk across the frozen Hudson

River to New Jersey, and the rivers and canals throughout Europe froze solid as well (Fig. 17-38). Snowlines in alpine areas descended sufficiently to cause significant glacial expansion in the world's major mid-latitude mountain ranges. In some mountains, such as the Alps, ice overran entire villages.

Sustained warming began by 1850 and continues today, although remnant icebergs from the Little Ice Age lingered long enough in the North Atlantic to doom the maiden voyage of the *Titanic* on April 15, 1912. By 1920, far less floating ice was observed south of the Arctic Circle. Today most glaciers in North America and Europe are still retreating (Fig. 17-39). To establish the amount and rate of glacial retreat, we can compare modern

▲ FIGURE 17-39 The retreating Athabasca Glacier in the Columbia Icefield of western Canada. The signposts indicate past positions of the glacier's terminus.

glacial positions to those shown in paintings and photos of the early- to mid-1800s. A recent field and photographic survey of 200 glaciers in Alaska found 7% advancing, 63% retreating, and 30% virtually standing still. One of these retreating glaciers, the Columbia Glacier, flows from a deep coastal fjord into Prince William Sound where it sheds icebergs into the Sound's shipping lanes, posing a direct threat to Alaska's economy.

The Effects of Human Activity on Glaciers
As you've probably read in the news, human activities may be contributing substantially to global warming and speeding up the melting of the Earth's remaining ice masses. In recent years, our use of refrigeration, air-conditioning, fire extinguishers, and aerosol cans has introduced a vast quantity of gases called chlorofluorocarbons (CFCs) into the atmosphere. These gases, together with increased volumes of atmospheric carbon dioxide generated from our burning of coal, oil, and natural gas, are believed to be shrinking the atmosphere's ozone layer (which shields the Earth from most solar ultraviolet radiation) and increasing the *greenhouse effect.*

By allowing the entry of solar radiation and preventing the escape of heat, the greenhouse effect promotes global warming in roughly the same way that a glass-walled botanical greenhouse creates a warm environment for plants to grow. (Look back at the figure in Highlight 5-2). Carbon dioxide in the Earth's atmosphere allows solar radiation to pass through it (as short-wave ultraviolet radiation) and warm rocks, soils, water, and vegetation on the Earth's surface. It then absorbs some of the heat (as long-wave infrared radiation) that radiates from the surface (Fig. 17-40). This trapped heat warms the Earth below. Any mechanism that increases the concentration of atmospheric CO_2, methane, and CFCs promotes climatic warming and accelerates the melting of glaciers.

Human activity has increased atmospheric CO_2 levels in two principal ways: by introducing vast amounts of CO_2 from the burning of carbon-rich fossil fuels and by destroying forests and reducing agriculture, especially the vast tropical rainforests (which release CO_2 from soils and decaying vegetation). Since 1850, as the use of fossil fuels has surged worldwide, the atmospheric CO_2 concentration has risen by 30%. If this rate of CO_2 production continues, people in the next century may have to live with CO_2 levels double those in the pre-industrial era. Such levels could cause global temperatures to rise by 2° to 4°C (3.5° to 7°F).

Such a global temperature increase would be sufficient to melt a significant amount of the Earth's polar ice and permafrost and raise global sea levels by as much as 50 to 100 cm (1 to 3 ft) by the year 2030. Sea level has already risen 20 to 25 cm (8 to 10 in) since 1900 through the melting of glacial ice and the thermal expansion of seawater. The predicted additional rise in sea level of 25 to 75 cm (10 to 30 in) would surely flood large areas of such low-lying cities as New Orleans, Louisiana, and Miami, Florida, during major storms. (Note: This was initially written before Hurricane Katrina. It makes us sound like we can predict the future.)

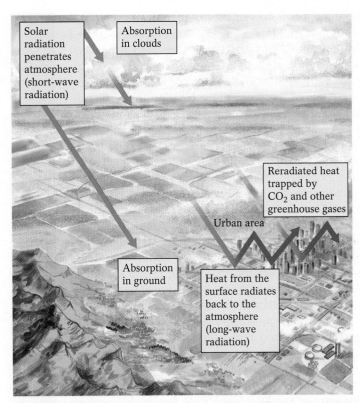

▲ **FIGURE 17-40** Human activity can contribute to global warming—and thereby glacial melting—in various ways. The greenhouse effect develops when CO_2 and other lower-atmosphere gases absorb heat radiating from the Earth's surface. If too much heat is trapped in the atmosphere, the lower atmosphere becomes increasingly warm, causing temperatures on Earth to rise.

Recent developments indicate that the effects of global warming on the Earth's climates are becoming more obvious with each passing day. A clear indication of the problem: In 1938, renowned author Ernest Hemingway wrote "The Snows of Kilimanjaro," a short story about a mountain in east Africa. In March 2001, global warming was rapidly melting the snows of Kilimanjaro at the summit of the 5895-m-high (19,340-ft) peak. Scientists predict that all of Kilimanjaro's glaciers will be gone within 15 years (Fig. 17-41). At the same time, massive icebergs broken

▲ **FIGURE 17-41** This small remnant of Mount Kilimanjaro's glaciers in northeast Tanzania has now completely melted away, clear evidence of global warming.

from the ice shelves of West Antarctica have drifted north and are threatening the shipping lanes south of South America's Cape Horn. (Ice shelves are vast bodies of floating ice that extend from and remain attached to a continent.)

Thousands of kilometers to the north, the Greenland ice sheet is also feeling the effects of global warming. A National Aeronautics and Space Administration (NASA) aerial survey published in July 2000 indicates that more than 46 km^3 (11 mi^3) of glacial meltwater have been streaming off this ice sheet into the surrounding North Atlantic each year. (The 50 billion tons of water produced by this melting could supply the water needs of all 120 million American households for roughly 140 days.) This substantial amount of meltwater is responsible for about 7% of the world's annual sea-level rise. And recent studies in Alaska indicate that the rapid melting of its glaciers is responsible for about 10% of worldwide sea level rise.

Perhaps the most telling indicator of global warming was discovered during the second week of August 2000 at the Earth's geographic North Pole. A Russian icebreaker, returning from a tourist cruise to the area after a six-year absence, made a startling discovery: *open ocean at the North Pole.* During the previous voyage, the icebreaker had to work its way slowly through continuous sea ice, 2 to 3 m (6 to 9 ft) thick. This time, the ship's passengers were astounded to find a mile-wide, ice-free patch of ocean. They may have been the first humans to ever see open water at the North Pole (which has probably been completely frozen for at least the past 3 million years and perhaps for tens of millions before then). Eventually, the ship had to steam 10 km (6 mi) away to find ice thick enough for its tourists to walk out onto the ice and be able to claim that they stood at the North Pole. These changes in polar sea ice may profoundly affect worldwide commerce in the next decade or so. Highlight 17-3 offers a glimpse of one economically critical effect of an ice-free Arctic.

If human activities caused all of the world's ice to melt within a few thousand years, global sea level would rise approximately 70 m (230 ft) (Fig. 17-42). The water in New York Harbor would reach to the armpits of the Statue of Liberty, and most of the Atlantic Coast from New Jersey to the Carolinas would be submerged. The entire state of Florida would be covered by the Atlantic Ocean; similarly, much of Alabama, Mississippi, and Louisiana by the Gulf of Mexico and much of interior California would be submerged by the Pacific Ocean as well. The rising sea would completely drown such low-lying places as the Netherlands, Bangladesh, Bermuda, and numerous Pacific island nations, effectively removing them from the world map.

Global warming may also expand the world's deserts, triggering major shifts in agriculture and forestry. Of course, this trend would create national winners and losers. Canada would do well: Cold regions now unsuitable for agriculture would become productive as the climate that now supports the Midwestern corn belt of Iowa and Minnesota migrated north to Ontario and Manitoba. In contrast, the Great Plains wheat belt of the

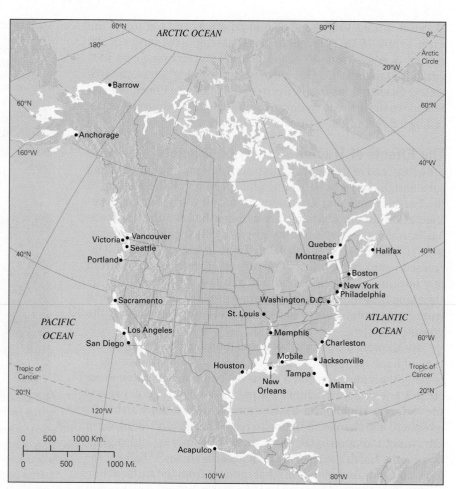

▲ **FIGURE 17-42** The effect on North America if all the ice on Earth today were to melt completely. The global sea level would rise about 70 m (230 ft), submerging most coastal areas and even some inland areas. The light-colored areas of the continents, such as Florida, represent places that would be submerged if all the Earth's ice melted.

United States—America's Breadbasket—might become too warm and dry for productive agriculture. Recent Midwestern droughts and the resulting food shortages and price hikes may offer an early glimpse into the Earth's balmy future.

The Future of Our Current Ice Age

How can we predict when the next period of worldwide glacial expansion or melting will occur when we can't even predict with absolute certainty whether tomorrow's ball game will be rained out? The Pleistocene record suggests that interglacial periods have an average length of 8000 to 12,000 years, followed by 10,000 to 20,000 years of slow, intermittent cooling before glacial conditions return. Full glacial conditions may then last—with intermittent brief periods of warming—for another 60,000 years or so. We are now in an interglacial period that began about 10,000 years ago. This period may be drawing to a close, and a new period of slow cooling may be on its way. Conversely, some climate scientists believe that a major glacial expansion may still be 10,000 years away. Either way, the ice will most certainly return.

If human activities are indeed producing a significant greenhouse effect, this condition is likely to prove temporary (especially when we run out of fossil fuels in the next few hundred years or if we succeed in reversing CO_2 and CFC buildup in the

Highlight 17–3

Opening up the Northwest Passage–A By-product of Global Warming?

For centuries, merchants and sailors journeyed along the long, tedious, and expensive southerly shipping routes around Cape Horn (the southern tip of South America) or, more recently, through the Panama Canal and the Cape of Good Hope (southern tip of Africa), carrying goods from the sources to their markets. All the while, they dreamed of a northern shortcut between Europe, Asia, and North America across the Northwest Passage—a route across the Arctic. Alas, the Earth's Ice Age climate and the Arctic's frozen seas thwarted their dreams, until now. It seems that with the protracted global warming of the early 21st century, the sailors' dream will come true: Commercial shipping across a northern Arctic route may be less than a decade away.

Climate scientists have shown in the last few years that the Arctic may be warming faster—and its ice melting more rapidly—than other regions of the planet. Arctic temperatures today are believed to be the warmest in at least 700 years. The area covered by Arctic sea ice has diminished in the past 20 years by about 1,036,000 km^2 (400,000 mi^2), roughly an area the size of Texas. According to various climate models, the Earth's polar ice caps and the Arctic's span of sea ice may be all but gone by the end of this century, and navigable ice-free corridors may be available for shipping within five to 10 years. Several different routes are developing—one on the Russian side of the Arctic (the Northern Sea Route, or Northeast Passage), a shortcut between Scandinavia and Japan, and another on the Canadian side of the Arctic (the Northwest Passage). Both routes, which may be open for as long as three months during the Arctic summers, dramatically shorten shipping distances. The difference between a Europe-to-Asia sail through the Northwest Passage as compared to one through the Panama Canal is 11,000 km (6800 mi). The difference between a northern route and a southern route around Cape Horn (supertankers, for example, must travel the long way around the horn because they can't pass through the canal) is 19,000 km (11,800 mi)!

Of course, other factors will come into play before shipping begins along these northerly routes. Oil tankers must be designed (probably with double hulls) to withstand rough seas and stray ice. Northern governments remain concerned about potential pollution, disturbances to animal habitats (seals, walruses, whales, polar bears), and disruption to the cultures of the indigenous peoples along the routes. International legal issues must also be resolved: The United States considers Arctic waters to be *international* and thus free for all nations to use; Canada claims dominion over the Northwest Passage; Russia asserts its sovereignty over its northern seas.

If these problems can be overcome, northerly ship travel between Europe and Asia will be cut by at least five days, resulting in a substantial cost savings. Of course, Panama and the folks who own and operate the canal will be less than thrilled with their financial losses, but this development may constitute one of the few *potentially positive* by-products of global warming.

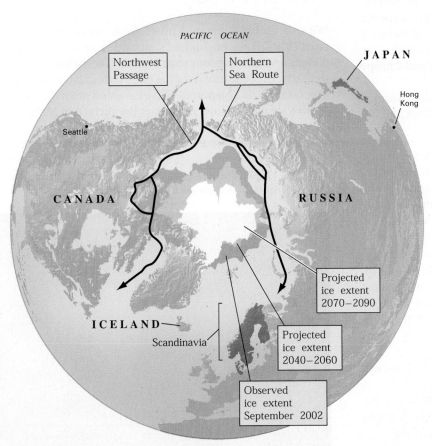

▲ A map of the Arctic and its potential shipping routes if the northern passages become ice-free in the next few decades in response to global warming.

▲ FIGURE 17-43 The South Cascade Glacier in the Washington Cascades has shrunk substantially between 1928 and 2000.

atmosphere through international agreements like the Kyoto Accords). If we let CO_2 production go unchecked for the next 2000 years, however, the very high level of CO_2 buildup will coincide with a warming trend predicted by Milankovitch's Earth–Sun orbital factors. The result could be an unusually intense warming and a period of *superinterglaciation*. If human-induced warming can be curtailed, then the Milankovitch factors predict that ice sheets will again cover vast areas of North America and Europe about 23,000 years in the future.

In the short run, however, if the Earth continues to warm at its current rate, a few possibilities can be predicted for the immediate climatic future:

- Modern plants and animals will migrate into new territories, primarily northward in the Northern Hemisphere, following their preferred temperatures. Plants will flower and eggs will hatch earlier in the spring. The range of some butterfly species in Europe has already shifted 50 to 100 km (30 to 60 mi) northward.

- Glaciers in Alaska will continue to shrink, and those in lower-latitude mountains in Washington and Montana will disappear altogether (Fig. 17-43). Montana's Glacier National Park—former home to 150 glaciers (and thus the name)—has but 26 remaining, shrinking glaciers. All of these ice masses may be nothing more than a memory within two or three decades.

- With the loss of mountain glaciers, many western states will find themselves dealing with great summertime droughts as their surface water supplies drop by as much as 30% by 2050. Irrigation, hydroelectric power, salmon spawning–all will be affected by reduced groundwater flow and streamflow.

- Antarctic ice sheets will accelerate their flow into adjacent oceans. Atmospheric warming will produce a greater volume of subglacial meltwater that will in turn enhance basal sliding of the ice sheets. The risk of a catastrophic rise in sea level will increase.

These fascinating, and in some cases alarming, developments are being studied and monitored by a legion of climate scientists, biogeographers, glacial geologists, and the like, and much controversy surrounds their findings and predictions.

Glaciation on Other Planets and Moons

We can find glaciers and other forms of ice (such as the icy crust/lithosphere on Jupiter's moon Europa) on several other planets and moons in our solar system. Recent satellite surveys of the surfaces of the moons of the large outer planets—Jupiter, Saturn, Uranus, and Neptune—have revealed polar ice caps and extensive ice cover. Closer to home, the surface of Mars is marked by white patches near the poles that appear to grow and shrink seasonally, much like the Earth's polar ice (Fig. 17-44).

▼ FIGURE 17-44 An ice cap near the south pole of Mars, photographed by the *Viking Orbiter 2* during Mars' southern summer in 1972.

Mars' polar ice is probably only a few meters thick, and it may largely consist of frozen CO_2 (dry ice), the main component of its atmosphere. In the Martian summer, the northern hemisphere ice cap recedes rapidly at first and then more slowly. The rapid shrinkage may reflect sublimation of frozen CO_2 at the edges of the ice cap, where it is thinnest. Any ice that lasts through the summer is probably frozen water, which evaporates very slowly in Mars' atmosphere.

Satellite photographs of Mars reveal U-shaped, striated troughs and ridges in the polar regions that resemble the glaciated valleys and large moraines on Earth. The terrain around the ice caps consists of alternating horizontal layers of light and dark materials, which may be stratified outwash or loess. As many as 50 layers, each of which is 15 to 35 m (50 to 115 ft) thick, have been counted at a single site. Confirmation of a complex history of multiple glaciations will probably have to wait for a visit to Mars by a glacial geologist.

Chapter 17 Summary

Glaciers are bodies of ice, composed of recrystallized snow, that flow downslope or spread radially under the influence of gravity. *Ice ages* are the dozen or so known periods of Earth history when the planet's climate was substantially cooler than today and glaciers covered a significant portion of the land surface. Climate fluctuations during an ice age cause the Earth's glaciers to grow and advance during glacial periods, and then thaw and retreat during interglacial periods. The Earth's current ice age began during the latter portion of the Pliocene Period about 2.5 million years ago. Much of the geological record of this period of Earth history is commonly referred to as the *Quaternary ice age,* which spans the last 1.6 million years of the planet's history. During the Pleistocene Epoch of the Quaternary Period (1.6 million–10,000 years ago), ice sheets expanded numerous times to cover southern Canada and the northern United States, as well as northern and central Europe.

Glaciers form when snow that has survived a summer melting season becomes well-packed *firn* and then recrystallizes to ice. They tend to form when the climate is marked by moist, snowy winters and, especially, cool cloudy summers that minimize melting. Relatively small *alpine glaciers* include *cirque glaciers, valley glaciers,* and *ice caps. Continental ice sheets* cover vast topographic lowlands; they exist today only at very high latitudes, such as in Antarctica and Greenland.

The *terminus* of a glacier is its outer margin. In a glacier's *zone of accumulation,* the addition of snow exceeds the volume lost; in the *zone of ablation,* more snow melts than is added. An *equilibrium line* between the two zones marks the point at which accumulation equals ablation. Overall glacier flow occurs either by *internal deformation* in which individual ice crystals fracture or slip past one another, by *basal sliding* across a thin film of meltwater that accumulates at the glacier's base, or some combination of the two.

Small-scale glacial erosion, or *glacial abrasion,* is carried out by small particles embedded in a glacier's bed; these materials make linear scratches, or *striations,* in the underlying bedrock surface. On a larger scale, *glacial quarrying* breaks off and removes large masses of bedrock. Glacial erosion generally occurs in the down-glacier segment of the accumulation zone near the equilibrium line, as downward-flowing ice impinges on the glacier's bed. Alpine glaciers carve smooth, unglaciated slopes into sharp, jagged peaks that may display the following landforms:

- horseshoe-shaped depressions called *cirques;*
- sharp pointy peaks called *horns;*
- long serrated ridges called *arêtes;*
- distinct breaks in the ridgeline called *cols.*

Together, glacial abrasion and quarrying remove a large quantity of bedrock and sediment, producing the asymmetry of glacially eroded hills, and converting V-shaped stream valleys into U-shaped glaciated valleys.

Glaciers tend to deposit their loads in their ablation zones, where melting releases the debris held in the ice. Deposition can occur directly from glacial ice or after transportation by meltwater. All glacial deposits are collectively known as *glacial drift.* Deposits of glacial drift include:

- *glacial till,* a type of drift that consists of unsorted, unstratified sediments deposited directly from the ice;
- *outwash,* a type of drift that consists of sorted, stratified sediments deposited from meltwater streams.

Meltwater sediments deposited in channels at the glacier's base form sinuous ridges called *eskers.* Tills and outwash deposited at a glacier's margin combine to form a series of hills called *moraines.*

Continental glaciers depress the crust beneath them and disrupt preglacial drainage systems. Other effects associated with glacial periods, which can appear beyond the actual glaciated areas, include:

- freezing of soils and regolith to form permafrost and other periglacial features;
- creation of *pluvial lakes* in now-arid regions;
- fluctuations in global sea levels;
- the evolution, migration, and extinction of various flora and fauna, including our human ancestors.

Plate tectonics has moved the Earth's continents toward the poles and raised mountains above *snowlines,* preparing the Earth for the development of ice ages. Milankovitch hypothesized that three astronomical factors—variations in the shape of the Earth's orbit around the Sun, the tilt of its axis, and the wobble of its equatorial plane as it rotates on its axis—affect the amount of solar radiation that reaches the Earth. When these factors coincide, they cause worldwide cooling and allow glaciers to grow and advance. The high albedo of ice-covered glacial regions causes them to reflect heat away from the Earth, enhancing cooling; the low albedo of dark, ice-free regions causes them to absorb incoming solar radiation, enhancing warming. Thus, the more ice-covered the land, the cooler the climate.

Ice ages have occurred in the Proterozoic and Paleozoic Eras and during the *Pleistocene Epoch* of the Cenozoic Era. For the last 10,000 years—the *Holocene Epoch*—Earth has been in an interglacial period of generally moderate warming, with a few cool periods lasting only a few hundred years each.

KEY TERMS

alpine glaciers *(p. 542)*
arête *(p. 547)*
basal sliding *(p. 542)*
cirque *(p. 547)*
cirque glaciers *(p. 542)*
col *(p. 547)*
continental ice sheet *(p. 542)*
drumlins *(p. 551)*
equilibrium line *(p. 542)*
eskers *(p. 555)*

firn *(p. 541)*
fjord *(p. 547)*
glacial abrasion *(p. 546)*
glacial drift *(p. 549)*
glacial erratic *(p. 550)*
glacial quarrying *(p. 546)*
glacial till *(p. 550)*
glacier *(p. 541)*
Holocene Epoch *(p. 572)*
horn *(p. 547)*

ice ages *(p. 538)*
ice caps *(p. 542)*
internal deformation *(p. 542)*
loess *(p. 555)*
moraine *(p. 550)*
outwash *(p. 552)*
permafrost *(p. 557)*
Pleistocene Epoch *(p. 538)*
pluvial lakes *(p. 558)*
Quaternary ice age *(p. 538)*

snowline *(p. 541)*
striations *(p. 546)*
surge *(p. 545)*
tarn *(p. 547)*
terminus *(p. 542)*
valley glaciers *(p. 542)*
zone of ablation *(p. 542)*
zone of accumulation *(p. 542)*

QUESTIONS FOR REVIEW

1. Draw a simple diagram showing the basic components of all glaciers: the accumulation zone, ablation zone, and equilibrium line. Indicate where glacial erosion and deposition are most likely to occur.

2. Explain how the two glacial flow mechanisms—basal sliding and internal deformation—work. How would increased warming of the atmosphere affect these two flow mechanisms?

3. List five prominent erosional features produced by alpine glaciers.

4. Describe the origin and location of terminal, recessional, lateral, and medial moraines.

5. Describe the difference in texture and appearance of till and outwash. Why are they so different?

6. Describe three major effects of glaciers that occur during ice ages in areas far removed from the glaciers.

7. What factors are believed to cause (a) ice ages; (b) fluctuations in ice volume *during* an ice age?

8. Explain three methods that geologists use to reconstruct past climates.

9. In the Northern Hemisphere, the lateral moraines of alpine glaciers generally are larger on the south-facing sides of east–west oriented valleys. Why?

10. Speculate about what might happen to the world's climate and glaciers if the Antarctic ice sheet surged into the surrounding oceans and floating ice covered vast areas of the ocean.

11. Identify the "mystery glacial landform" in the photograph below.

LEARNING ACTIVELY

1. Look at a map of North America and check the coastlines for areas with land elevations that measure less than about 70 m (230 ft). All of those areas would be under water if the Earth's current ice masses melted. What would happen to the state of Florida? To coastal Texas? To New York City, Boston, and Washington, D.C.? The result might not be quite as dramatic as the giant-wave scene in the movies *Deep Impact* or *The Day After Tomorrow,* but the long-term effect would be similar.

2. Watch the movie *The Day After Tomorrow.* In light of what you know about climate change, assess the geological logic of this film. Could such events happen as quickly as the movie suggests?

3. Look at a map of North America and imagine that the Earth has returned to a full glacial expansion like the one 20,000 years ago. Ice covers the United States as far south as a line that stretches from Boston and New York City, through Columbus, Ohio, to Des Moines, Iowa. How would your community be affected by this development? How about if you lived in Lincoln, Nebraska; St. Louis, Missouri, or even Salt Lake City, Utah?

4. If you live near one of the continent's coastlines, visit the beach and imagine that glacial times have returned. Because of the lower sea level, the coastline would lie as many as 160 km (100 mi) seaward. Think about the glacial-age environment on the continental shelf seaward from where you are now standing. What types of animals and plants would inhabit what is now the sea floor?

5. If you have access to a waterbed (such as in the bedding section of a department store), lie down on it and note how the water beneath you migrates from beneath your weight. Note the bulges on the edges of the bed. Imagine what happens when an ice sheet, 2 mi thick, sits on the Earth's lithosphere. What would happen to the soft, flowable asthenosphere beneath the ice? Note what happens when you arise from the waterbed. Where have the bulges gone? Where has the depression gone that was just beneath your weight? Why did the surface of the waterbed return to its original position so quickly? How would the water in the waterbed compare with the flowability of the asthenosphere? How would the difference in flowability between water and the asthenosphere affect how quickly the Earth's surface returned to "normal" after the ice sheets melted away?

 ## ONLINE STUDY GUIDE

Practice makes you better. Make the time to take the *Online Study Guide* quiz for this chapter. It should take you about 20 minutes. Automatic grading and instant feedback will help you quickly and accurately identify the concepts you have mastered and the areas in which you need more study.
www.prenhall.com/chernicoff.

Deserts and the Work of Winds

18

The landscapes of Death Valley National Park, located in southeastern California (Fig. 18-1), owe their beautiful, unique appearance to their extreme lack of water. With little water, little chemical weathering takes place—soils are thin, dry, and crumbly—and winds readily sweep loose particles into dunes or use them to sandblast exposed rock surfaces. These characteristics are typical of a **desert,** a region that receives very little annual rainfall and is generally sparsely vegetated.

Every major continent, although surrounded by vast oceans, contains at least one extensive desert region. Deserts account for as much as one-third of the Earth's land surface, more area than any other type of geographical environment. Relatively few of these regions, however, resemble the popular Hollywood image of endless tracts of drifting sand and caravans of camels. In North Africa's Sahara, the world's largest desert (*sahara* means "desert" in Arabic so calling it the Sahara Desert would be redundant), only 10% of the surface is sand-covered. Even the Arabian Desert, the Earth's sandiest, is only 30% sand-covered. Desert climates, which are actually characterized by their dryness and not by their sandiness, can be found in frigid polar regions as well as in sizzling low-latitude regions. Such diverse landscapes as ice-bound Antarctica, the fog-shrouded coasts of Peru and Chile, and the near-continuous 8000-km (5000-mi) stretch of land across northern Africa and the Arabian peninsula to southern Iran are all deserts, as defined by their extreme lack of surface water.

The term *desert* is misleading if we believe that "deserted" means devoid of life. Hot deserts are home to some of the Earth's hardiest plants and animals, a couple of which are shown in Figure 18-2. These animals and plants have evolved over thousands of years to adapt to the extremely dry conditions. Some hot-desert plants produce seeds that can endure 50 years of drought. Some have small, waxy leaves that minimize water loss to evaporation. Most have thick, spongy stems that store water from the occasional cloudbursts, and they produce deep root systems to tap groundwater supplies. Some desert plants may resemble dead twigs for months or even years on end, only bursting into a brief but memorable bloom when an occasional downpour arouses them.

Animals that live in hot deserts include insects, reptiles, birds, and mammals. Some birds found in these environments may fly hundreds of kilometers to find water, which they then carry back to their young in their absorbent abdominal feathers. Some desert rodents live their entire lives without a single sip of water, absorbing needed moisture instead

FIGURE 18-1 Wind-swept sand dunes at Mesquite Flats, Death Valley National Park, California. ▶

(a)   (b)

▲ **FIGURE 18-2** Some desert life forms. **(a)** A roadrunner in the Sonoran Desert, Arizona. **(b)** A thorny devil lizard (*Moloch horridus*) from Rainbow Valley, Alice Springs, Australia.

from the plants they eat. Most desert animals are nocturnal, avoiding activity during the hottest and driest times of the day, and venturing out only after temperatures have cooled down considerably.

In this chapter, we will examine the processes that shape a desert's unique landscape. We will see how water's brief, intermittent appearances actually contribute significantly to the evolution of desert landforms. Our discussion will also emphasize the role of wind, another major agent of surface change in deserts. We also will consider how human activity contributes to the expansion of deserts and describe efforts to use desert lands for agriculture and human habitation. Finally, we will journey to Mars to view one of our planetary neighbor's wind-swept deserts.

Identifying Deserts

How do we decide if an area is dry enough to be called a desert? One common guideline defines a desert as any region that receives less than 25 cm (10 in) of precipitation annually—the minimum amount needed to sustain crops *without irrigation.* The world's driest deserts, however, fall well short of the defining 25-cm mark. A very arid region may receive all of its annual rainfall in one sudden cloudburst; the water from this downpour may evaporate or be absorbed by the desert regolith within minutes without a trace. After the cloudburst, no precipitation may fall for months or even years.

The world's driest region is the Sahara, which on average receives only 0.04 cm (0.016 in) of precipitation annually. The Sahara may actually receive no measurable precipitation for years; instead, a few storms over the course of a decade may determine its annual average. The northern Sahara was virtually rainless between 1970 and 1990.

A more useful measure used to identify deserts is the **aridity index,** a ratio of a region's potential annual evaporation (a function of its yearly receipt of solar radiation) to its average annual

precipitation. An area with an aridity index of 1.0 has an annual amount of precipitation equal to the area's potential for evaporation. Such an area would have a humid climate. An area with an aridity index greater than 4.0, where the potential for evaporation is at least four times greater than the annual precipitation, is classified as a desert. Two of the driest spots on Earth, the eastern Sahara of Africa and the Atacama Desert of Chile, have an aridity index of 200. These areas are described as *hyper-arid.* An area with an aridity index between 1.5 and 4.0 is classified as *semi-arid.* Semi-arid regions, such as the Great Plains to the east of the Colorado Rockies, can support a greater diversity of life than deserts can. As we will see later in this chapter, these regions are climatically fragile; even a relatively slight increase in temperature and decrease in precipitation may convert them to true deserts.

Types of Deserts

Deserts may be cold, temperate, or hot. In a cold desert, such as in northern Scandinavia, the 25-cm (10-in) annual precipitation may even support a dense forest because the region's low temperatures minimize evaporation. Thus some moisture is available for tree growth. In contrast, no forests grow in warm southern Nevada, which has roughly the same amount of annual precipitation. There, the warm climate causes substantial evaporation. The best-known deserts are far hotter and drier than most of their other water-starved counterparts. Highlight 18-1 describes the deserts of northern Africa, which are among the hottest, driest places on Earth.

Subtropical Deserts   Although deserts can be found at all latitudes, most occur in the subtropical belts between latitudes 20° and 30° on either side of the equator. Why do we find extremely dry, subtropical deserts next to equatorial areas that are commonly drenched in tropical rain?

Direct solar radiation strikes the equatorial zone virtually year-round, and the warm air there evaporates a large volume of

Highlight 18-1

How Hot Is Hot?

Although deserts may occur in any climatic zone, most—and the best known—appear in extremely hot places. Where are the hottest places on Earth? At El Azizia, Egypt, a desolate outpost in the Libyan Desert of northeastern Africa, the air temperature rose to a blistering 58°C (136°F) *in the shade* on September 13, 1922. The North American record, 56°C (133°F), was recorded in Death Valley, California, in July 1913.

In such oppressive temperatures, the human body can lose a gallon of perspiration between sunrise and sunset. If water intake is not maintained, severe dehydration occurs as the body rapidly draws on the water reserves stored in a body's fat and blood. Sweat glands become overworked, the body cannot maintain its normal temperature, and a high fever results. In a few

hours, blood circulation slows. Ultimately, the person dies.

Where there is no surface moisture to evaporate, humidity remains low and few clouds form to filter daytime sunlight or retain the heat that radiates at night from rock surfaces. Consequently, the daytimes in such places become even hotter, and nighttime temperatures may plummet by as much as 50°C (90°F) from daytime highs. At night, heat is lost rapidly to the atmosphere and temperatures often drop below freezing.

In 1974, the young British journalist Richard Trench vividly described life in the thermally fickle Sahara of Algeria:

The writer will not soon forget arising at 3:00 A.M.—in the Algerian Sahara in early September. The bucket of water

for washing presented a thin film of ice, and a heavy wool sweater and leather jacket were comfortable while riding in an open jeep. By 9:30 A.M., the air temperature was nearing 85°F; by 11:30 A.M., a pocket thermometer registered over 105°F and at 2:00 P.M. the same thermometer registered 127°F. What passed for a local pub in a nearby oasis cooled its beer by wrapping the bottles in wet sacking and laying them in the sun. Evaporation occurs at an almost unbelievable rate under such circumstances and effective ground moisture levels are fantastically low. The foregoing account amounts to a record of a daily temperature variation of at least 95°F. The beer was delicious.

---

moisture from equatorial oceans. As the air warms, it expands, becoming less dense. As this air begins to rise, it carries water vapor upward, like steam rising from a pot of boiling water. The moist air immediately begins to cool as it ascends. By the time it reaches an altitude of about 10 km (6 mi), the air has cooled enough to reduce sharply its capacity to hold water vapor. Consequently, the moist air releases its water as torrential rains that fall on equatorial lands.

As the now-dry tropical air continues to rise and cool, it becomes denser. Once it is no longer capable of rising, it begins to spread laterally. By the time it reaches a latitude of 20° to 30° north or south of the equator, it is dense enough to sink back to the Earth's surface. There the air warms and its capacity to hold water vapor increases, enabling it to evaporate a vast amount of surface water, and thereby dry out the landscape. Under the pressure of overlying descending air, some of the warm air approaching the surface spreads across the surface toward the equator, where it again becomes heated, evaporates a great deal of water, and reenters the cycle of rising warm moist air and sinking cool dry air. As shown in Figure 18-3, this cycle of heating, rising, cooling, sinking, and reheating air creates an atmospheric *convection cell,* quite similar to the mantle convection cells that help drive plate motion.

These air-circulation patterns produce the two subtropical belts of desert lands shown in Figure 18-4. Just north of the Tropic of Cancer in the Northern Hemisphere lie the Sahara, the Arabian Desert of the Middle East, and the parched landscapes of Mexico and the American Southwest. Straddling the Tropic of Capricorn in the Southern Hemisphere are the Kalahari and Namib Deserts of southwestern Africa,

the Atacama Desert of Chile, and the interior deserts of the Australian outback.

Rain-Shadow ("Orographic") Deserts Many smaller deserts form on the leeward sides of mountain ranges. (The *windward* side of a mountain range is the one that lies directly in the path of oncoming winds. The *leeward* side is the sheltered

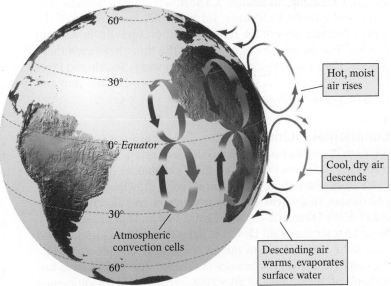

▲ **FIGURE 18-3** Atmospheric convection cells form as hot, moist equatorial air rises and cools when it reaches about 10 km (6 mi) above the surface. There it releases its moisture as tropical rain, spreads northward and southward from the equator, and then descends in the subtropical latitudes 20° to 30° north and south of the equator. The now-dry air warms as it descends and evaporates ground moisture rapidly, producing desert conditions on land.

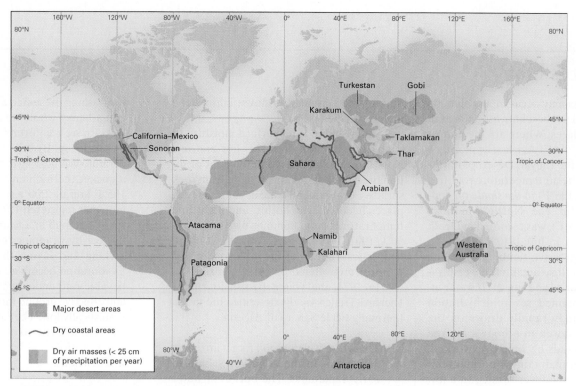

▲ **FIGURE 18-4** The worldwide distribution of the Earth's deserts. Most deserts lie in the two subtropical belts between latitudes 20° and 30° north and south of the equator. Polar deserts are omitted here.

backside—that is, the opposite side of the range.) As moist air on the windward side of a mountain range rises and cools, it loses much of its capacity to hold water. This moisture thus precipitates on the windward slope. The leeward slopes and the regions beyond them are left dry. This process (shown in Figure 18-5) causes moist air to be wrung dry by mountains and creates the **rain-shadow effect,** also known as the *orographic effect* (*oro,* from the Latin meaning "mountain"). The desert lands of Nevada, Arizona, and eastern California, for example, are *leeward* of the Sierra Nevada mountain range and therefore lie in its rain shadow. Pacific air masses are dry by the time they reach these deserts. The same relationship keeps rain-swept Seattle and Portland lushly vegetated, while parched eastern Washington and Oregon remain only sparsely vegetated.

Continental Interior Deserts
Deserts also form deep in the interiors of continents, far from oceanic moisture sources. In central Asia, the Gobi Desert and the Taklamakan (meaning "the place from which there is no return") exist principally because of two factors: their great distance from water and the rain-shadow effect of the Himalaya. Look at Figure 18-4 and find these deserts. Note how far the Gobi Desert is from the Indian Ocean. It also sits on the leeward side of the great moisture barrier created by the lofty Himalaya. Air masses that reach these areas have lost virtually all their moisture after crossing thousands of kilometers of land. The location of these Asian deserts is a direct consequence of plate tectonics: They developed about 40 million years ago after the Indian subcontinent collided with China, which enlarged the continent and left central Asia several thousand kilometers away from the nearest ocean.

Deserts Near Cold Ocean Currents
Some deserts in warm subtropical regions lie close to oceans and next to cold ocean currents. Thus, they create the strange unexpected sight of a bone-dry desert astride a vast cold ocean. The cold air above these cold ocean currents holds little water. As shown in Figure 18-6, when this dry air moves over land, it becomes warmer, enabling it to hold more water. Thus it evaporates almost all of the surface water in these coastal desert regions. The Atacama Desert of Chile owes its extreme aridity in large part to the Pacific Ocean's cold Humboldt current, which flows northward toward the equator from the icy Antarctic, hugging the coast of western South America. Coastal Chile, which seldom experiences more than a drizzle, has an annual precipitation level of less than 1 mm (0.04 in).

Polar Deserts
Some places are both extremely arid and frigidly cold. Because they receive little solar radiation, such high-latitude locations as northern Greenland, Arctic Canada, northern Alaska, and the vast continent of Antarctica have temperatures that remain below freezing year-round, even during their brief summers when the sun never sets. Global atmospheric circulation, however, brings only cold dry air to these lands; thus they qualify as deserts.

Weathering in Deserts
Very little water produces relatively little chemical weathering in deserts. (For a quick refresher on the role of water in weathering, look back at Chapter 5, Weathering—The Breakdown of Rocks.) This simple principle explains why we find steep angular cliffs, sharp-edged stones, and relatively thin soils in most deserts.

▼ **FIGURE 18-5** The rain-shadow effect arises when moist air masses lose their moisture as they pass over a large topographic barrier, such as a mountain range. The region in the lee of such a barrier dries out. In western North America, moist Pacific Ocean air masses rise west of the Olympic Mountains, cool, and precipitate over the Olympic rainforest of coastal Washington, where annual precipitation exceeds 250 cm (100 in). After picking up additional moisture over Puget Sound, the air masses cool as they rise, causing rain to fall on the western side of the Cascade Mountains. The now-dehydrated air descends on the mountains' lee side, where it warms and becomes able to hold more moisture, producing rapid evaporation, dryness, and the treeless landscape of the arid Columbia Plateau, where annual precipitation is less than 25 cm (10 in).

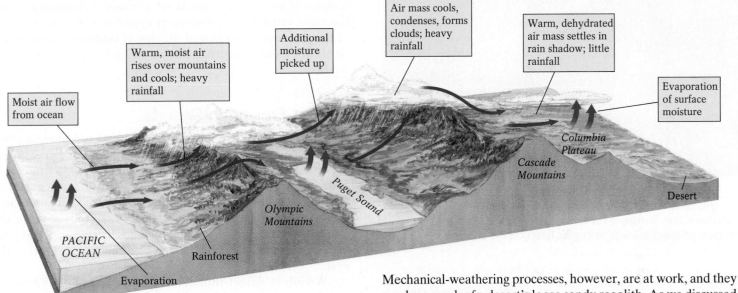

Mechanical-weathering processes, however, are at work, and they produce much of a desert's loose sandy regolith. As we discussed in Chapter 5, extreme daily temperature fluctuations cause rock and mineral surfaces to expand and contract repeatedly. Eventually individual grains flake off the surface and join a desert's supply of shifting sands. At the same time, ongoing evaporation of any available water precipitates salt crystals within rock fractures. As these salt crystals grow, they exert enormous pressure on adjacent mineral grains, acting like tiny crowbars to force the grains apart.

Because water is necessary for chemical reactions, geologists long believed that the lack of water in deserts meant that little chemical weathering could occur there. Cold desert nights, however, promote condensation of dew on rocks, and some deserts do have a brief rainy season. The little moisture present in these areas, along with the organic acids produced by desert vegetation, support a small amount of chemical weathering. Even hyper-arid regions may contain enough moisture to oxidize iron-rich silicates, producing colorful desert rocks (Fig. 18-7).

Many desert rock surfaces, such as the petroglyphs in Figure 18-8, are covered with **desert varnish,** a thin, shiny, red-brown or black layer of manganese oxides and iron oxides. How can this coating cover pure-white quartzite rocks that contain no manganese or iron? According to one hypothesis, high winds transport manganese- and iron-rich clays from nearby sources and forcefully plaster them onto dew-dampened rocks. The clays stick to their surfaces and become oxidized by oxygenated dew. Another hypothesis proposes that microbes in the desert soil concentrate manganese on rock surfaces. Whatever the process involved, as you can see in Figure 18-8, it is clearly a slow one; the images etched in the varnish hundreds or thousands of years ago by ancient North Americans have not yet been revarnished.

▲ **FIGURE 18-6** Cool, dry air from above the cold Humboldt current blows onshore along the Chilean coast. The air warms over the land surface, and, because its capacity to hold water increases, it evaporates virtually all surface water from coastal Chile. The result is one of the Earth's driest places—the Atacama Desert, just west of the Andes.

▲ **FIGURE 18-7** Although oxidation, like other forms of chemical weathering, proceeds slowly in deserts, it has created colorful rocks such as these of the Chinle formation in Utah. Oxidation of iron-bearing silicates in the area's sandstones produces red-orange iron oxide, which gives these rocks their distinctive colors.

▲ **FIGURE 18-8** Petroglyphs etched into desert-varnished rocks by native North Americans 2000 years ago. Because petroglyphs such as these have not been revarnished since they were etched, we deduce that desert varnish forms quite slowly.

The Work of Water in Deserts

Although desert landscapes are vastly different from those in other environments, they are nevertheless shaped by many of the same processes that act on virtually all land surfaces. Despite conditions of extreme aridity, water still serves as the primary sculptor of desert landforms. Brief, occasional cloudbursts cause the rapid erosion and subsequent deposition that create many of the prominent features of the desert landscape. Figure 18-9 illustrates these landforms, and we will now describe how they develop.

Stream Erosion in Deserts

In arid regions, only a minimal network of soil-binding plant roots anchors loose sediment. Thus, flowing water can easily remove those grains. Moreover, infrequent but intense desert storms produce immediate and powerful surface runoff (so-called *flash floods*) that is not slowed by vegetation as it would be in grasslands or forests. Over thousands of years, the rapidly moving water of short-lived desert streams erode numerous **arroyos**—stream channels that remain dry most of the year.

A violent desert thunderstorm can drop 5 cm (2 in.) of rain in minutes and trigger a flash flood. Because the water cannot be readily absorbed by the compacted, sun-baked ground, it rapidly overflows any existing stream channels. A single storm can send a 3-m (10-ft)-high wall of water and sediment sweeping through an arroyo with virtually no warning. These rushing streams excavate their channels vertically, forming long, narrow, steep-sided canyons. Anyone wandering in an arroyo when a flash flood hits would have great difficulty escaping up the near-vertical canyon walls (Fig. 18-10). Such was the case in August 1997, when a flash flood in Antelope Canyon near Page, Arizona, swept away 11 unsuspecting tourists.

As a storm subsides and the velocity of its floodwaters decreases, the water infiltrates loose desert sands or evaporates, leaving behind a high volume of sandy sediment that covers the canyon floor. Little happens between floods to change these deposits.

When desert streams remove sediment from a mountain, an erosional surface forms called a **pediment** (from the Latin, meaning "foot"). Pediments lie at an angle of 5° or less and may extend for many kilometers from the base of an eroding mountain. Pediments enlarge slowly as the repeated action of running water causes mountain slopes to recede. This process continues until nearly all of a mountain's mass has been eroded. Very large pediments may have begun to form under the moist conditions of the Pleistocene ice age, a period characterized by more rainfall, runoff, and weathering in deserts than occur today. It is unlikely that the arid environment in which pediments currently appear could have provided sufficient water to erode that much sediment, even over several thousand years. The reduced desert waterflow today may only be moving materials already loosened by weathering in a past wetter climate.

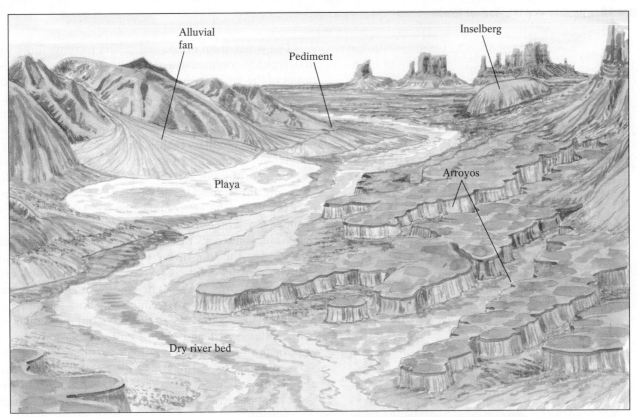

Alluvial fan

Pediment

Inselberg

Playa

Arroyos

Dry river bed

▲ FIGURE 18-9 Desert landforms produced by water. Intermittent surface-water flow in deserts produces arroyos, pediments, inselbergs, and playas. Although the arroyo pictured here is dry, it fills rapidly with rushing water during a desert cloudburst. Deposition from desert streams creates alluvial fans and playas.

▼ FIGURE 18-10 Water-carved Antelope Canyon, Page, Arizona. The steep walls of this canyon prevented tourists from escaping the tragic flash flood of 1997.

After advanced pediment development, part of the mountain may remain as a resistant, steep-sided bedrock knob. Such a residual erosional feature is called an **inselberg** (from the German, meaning "island mountain") and is typically composed of a hard durable rock such as granite, gneiss, or well-cemented sandstone. Another variety of inselberg, shown in Figure 18-11, may develop after a long period of surface erosion uncovers a body of resistant rock. Because inselbergs stand prominently above the surrounding plain, their smooth, steep slopes shed water quickly. This action minimizes prospects for further weathering and accelerates the weathering and erosion rate of the inselbergs' surrounding rocks and sediments.

Stream Deposition in Deserts

Whenever erosion takes place, deposition is sure to follow. In moist, temperate regions, most products of erosion are carried to and deposited in the nearest ocean. In arid regions, short-lived desert streams can deposit thousands of meters of sediment in subsiding basins and build great alluvial fans (introduced in Chapter 6) that slope gently away from the remnant mountains to the desert floor. Because few streams can flow for long distances through the desert environment, their sediments are typically deposited close to their erosion site rather than at the bottom of some distant ocean. Wherever surface water evaporates or infiltrates, it drops the sediment load washed from the surrounding steep mountain slopes, forming alluvial fans and dry lake beds.

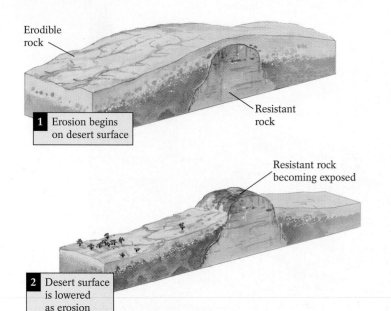

1 Erosion begins on desert surface

Erodible rock

Resistant rock

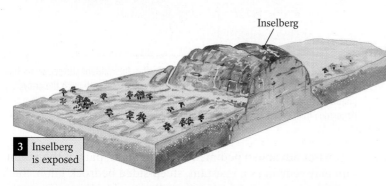

2 Desert surface is lowered as erosion continues

Resistant rock becoming exposed

3 Inselberg is exposed

Inselberg

▲ **FIGURE 18-11** Inselbergs, such as Ayers Rock in Australia shown here, rise abruptly from desert plains. They form when soft rock is eroded and leaves a mass of more resistant rock standing prominently above the surface.

The water from occasional desert torrents flows rapidly at first, typically confined to a narrow arroyo with steep walls. At the outlet of the arroyo, where the canyon walls end, the flow slows appreciably as the then-unconfined water spreads out. As the water's velocity drops, the stream loses its ability to transport sediment and deposits its load in the form of an alluvial fan. Each mountain stream that reaches the desert plain produces its own alluvial fan, proportional in size to the stream's drainage area. The slope of the fan surface—about 10° at the top—decreases gradually until it merges with the desert floor. The steeper slope near the top of the fan contains the coarsest sediment, which is dropped first as the stream slows.

Water that seeps into the coarse upper-fan sediments may reemerge at the foot of the fan as a spring. In the American Southwest, especially in Arizona, such springs are often marked by groves of mesquite trees (something to remember if you ever find yourself roaming around and thirsty in that region). Alluvial-fan sediments may contain enough water to irrigate the desert and permit cultivation, as Native Americans in the Southwest have done for more than 1500 years. Such alluvial-fan springs may also contribute to the water needs of a nearby city. San Bernardino, California, for example, draws some of its water from local alluvial-fan deposits.

During brief periods of higher-than-average precipitation in a desert's interior, water drains toward topographically low, closed basins, where it may collect as temporary lakes. As shown in Figure 18-12, such a lake may evaporate in a few days or weeks, leaving behind a **playa** (from the Spanish, meaning "beach") or a dry lake bed on the desert floor.

A playa's dry bed typically consists of precipitated salts plus fine-grained clastic sediment. The salts, commonly a mixture of sodium and potassium carbonates, borates, chlorides, and sulfates, may form a blinding-white residue or a bizarre landscape filled with jagged pinnacles. Industrial chemicals have been mined from the playas of the Southwest for more than 100 years. Death Valley has long been a source of sodium borate, or *borax,* a salt used in pottery glazes and household cleansers, and as a hardening agent for alloys and a shielding material against nuclear radiation.

If its inflowing water does not contain much dissolved rock, a playa may contain only fine-grained, clay-rich sediment. The desert sun may bake such a surface until it becomes so hard that it can serve as a natural landing strip for aircraft. Rogers Dry Lake, on Edwards Air Force Base in California's Mojave Desert north of Los Angeles, is a frequent landing site for NASA's space shuttles.

When precipitation and runoff from surrounding mountains exceed evaporation and infiltration on the desert floor over an extended time, a large lake may form that can last for years or centuries, even in the ultradry desert air. Because they lack outlets from which water can drain, most desert lakes become extremely saline (salt-rich) as evaporation continually concentrates the dissolved salts. Large saline water bodies of this type include the Great Salt Lake in Utah, Mono Lake in southern California, and the Dead Sea, a basin that sits 392 m (1286 ft) below sea level (Fig. 18-13).

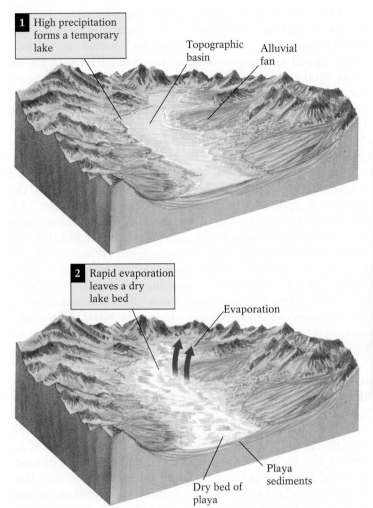

1 High precipitation forms a temporary lake

Topographic basin

Alluvial fan

2 Rapid evaporation leaves a dry lake bed

Evaporation

Dry bed of playa

Playa sediments

▲ FIGURE 18-12 A playa is a dry basin in the desert floor from which a temporary lake has evaporated. Photo: Evaporite deposits in Devil's Golf Course, Death Valley, California. 90 m (295 ft) below sea level, Devil's Golf Course is a playa that posed a significant obstacle to westward migration in the late 19th century.

▲ FIGURE 18-13 The Middle East's Dead Sea, which is essentially a desert lake, is so salty that swimmers float noticeably higher in it than they would in normal saltwater.

The Work of Winds in Deserts

After flowing water, winds are the second most powerful force in shaping deserts. Winds are air currents set in motion by heat-induced changes in air pressure. High-pressure air is cool and generally flows *as wind* into regions of lower-pressure, warmer air. Near the equator, a zone of low pressure is created as Sun-warmed air becomes less dense and rises. As we saw earlier in our discussion of the origin of subtropical deserts, this air cools and becomes more dense as it ascends. Eventually it loses its buoyancy and, while still at high altitudes, spreads north and south from the equator. At latitudes 20° to 30° in both directions, the cool, dense, dry air sinks back toward the surface. The result is two permanent subtropical high-pressure zones known as the *horse latitudes*, which are noted for their calm, clear, warm weather. (The name originated from floating horse carcasses that littered the subtropical seas during the 18th and 19th centuries. When sailing ships stalled without a driving breeze to propel them, and food and drinking water became exhausted, most of the ships' cargoes of horses were apparently thrown overboard to conserve water supplies. The rest became the sailors' next dinner.)

The high-pressure air of the horse latitudes ultimately spreads outward, close to the Earth's surface, and moves back toward the equator, filling the void created by the rising low-pressure equatorial air. Some of the high-pressure air also heads the other way, toward the poles, drawn by the relatively low air pressure near 60° of latitude.

Global wind patterns do not follow a strictly north–south or south–north path, however. The Earth's rotation about its axis also sets in motion an east–west component to the flow of air above its surface for the following reason. As the Earth rotates counterclockwise, it turns with a greater velocity at the equator than at the poles. (Because each of the Earth's rotations takes 24 hours, the planet's surface at the equator—with its larger circumference—must speed along more swiftly than the surface closer to the poles, simply to cover the greater distance within the same 24-hour period.) Moving air masses *above* the spinning planet thus appear either to "move ahead" or "lag behind" the underlying Earth, causing air masses to curve *relative to the surface.* Figure 18-14 illustrates these global wind patterns. The air masses that have more velocity than the underlying Earth (nearer to the slower-spinning poles) move ahead of the Earth's counterclockwise rotation, producing *westerlies.* The air masses that have less velocity than the underlying Earth (nearer to the faster-spinning equator) lag behind the Earth's rotation, producing the trade winds, or *easterlies.*

This apparent deflection of moving air masses (or any freely moving body, such as an ocean current or a speeding missile) in response to the Earth's rotation is called the *Coriolis effect.* Spend a moment studying the wind patterns in Figure 18-14. Notice that in the Northern Hemisphere, descending air currents (such as those spreading north and south of latitude 30°N) are deflected *toward the right* of the direction in which they are moving; in the Southern Hemisphere, they are deflected *toward their left*—a simple way to remember the windy implications of the Coriolis effect.

In the Northern Hemisphere, the descending air of the subtropical high (at 30° latitude) that returns south toward the equator is deflected to its right, or from east to west. This northeasterly deflection produces the *trade winds.* These winds blow along the warm, low-latitude routes taken in the 18th and 19th centuries by merchant sailing vessels. (Winds are designated by the direction from which they are blowing; hence a "northeasterly wind" blows *from* the northeast, a southerly wind blows *from* the south, and so on.) The descending air of the subtropical high that turns toward the North Pole is also deflected to the right, or from west to east, by the Earth's rotation. This deflection produces the mid-latitude winds, or *westerlies.* The westerlies prevail between about 30° and 60° of latitude in both hemispheres; in the Northern Hemisphere, they dominate the wind pattern for most of North America, Europe, and Asia.

Global wind patterns have probably remained relatively constant throughout the Earth's history because the orientation of the Earth's axis, its counterclockwise spin, and the planet's distance from the Sun have not changed dramatically. Consistent differences in the relative amount of heat received by the equator and poles have probably always produced the Earth's pattern of high- and low-pressure zones, which in turn generate the global winds. The direction of the Earth's rotation has always been the same, so the Coriolis effect has always deflected winds in the same way as it does today.

Low-altitude winds, such as the pleasant onshore and offshore breezes that moderate coastal climates, result from *local* differential heating of the atmosphere. Figure 18-15a shows how differences in land and water temperatures may produce such breezes. When this heating differential occurs on a continental scale, it sometimes has catastrophic results. As you can see in Figure 18-15b, differential heating of land and ocean draws moist air from the cooler, higher-pressure zone of the Gulf of Mexico into the warmer, lower-pressure zone of the Mississippi Valley during the warm spring and summer months. This process produces warm, moist southerly winds that, when they encounter cold, dry air descending from the Arctic, can trigger powerful tornadoes. Differential heating similarly causes the gusty winds of deserts. During the day, heated surface air rises to create a low-pressure zone; in the evening, relatively cooler and higher-pressure surface air rushes in, often at a velocity of 100 km (60 mi) per hour or more, to replace it. Such winds produce blinding desert sandstorms.

Both local and global wind patterns affect desert areas in several significant ways. Winds can erode desert landforms, transport sediments, and deposit sediments. We describe these wind-driven processes next.

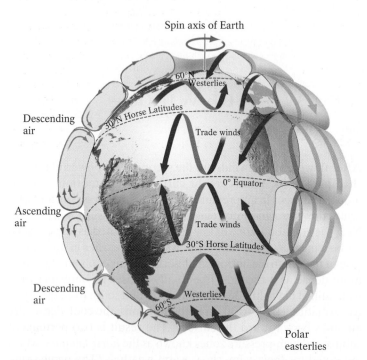

▲ **FIGURE 18-14** The Earth's global wind patterns. If the Earth didn't spin at all, all airflow would simply move north-south due to atmospheric convection. The Earth's rotation, however, deflects moving currents of air (and water as well). In the Northern Hemisphere, laterally spreading air masses are deflected to their right; in the Southern Hemisphere, they are deflected to their left. This deflection is known as the Coriolis effect.

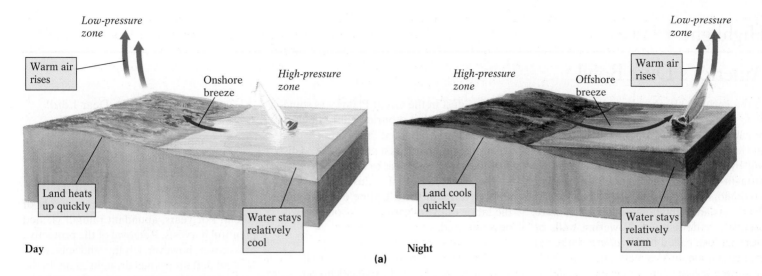

Day

Night

(a)

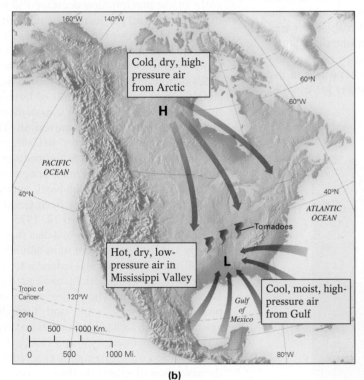

(b)

▲ **FIGURE 18-15** On both local and continental scales, onshore and offshore winds develop because solid ground warms and cools quickly, whereas water warms and cools more gradually. **(a)** On a warm afternoon, land heats up more rapidly than the ocean, and a zone of low pressure forms over the land; cooler, high-pressure air then blows from the sea to the land as an onshore breeze. At night, the sea cools slowly, retaining more daytime heat than the land does. As the land cools, higher air pressure develops over it, sending an offshore breeze out to the low-pressure zone over the ocean. **(b)** During the summer months, the entire North American continent heats up, producing a large temporary zone of low pressure overhead; higher pressure develops over the relatively cooler surrounding ocean. (This scenario differs from the winter pattern of cold land/high pressure and relatively warm ocean/lower pressure.) The pressure differential produces southerly winds that funnel moist air from the Gulf of Mexico up the Mississippi Valley, where it encounters cold air descending from the Arctic. The resulting clash of cold, dry Arctic air and warmer, moist Gulf air produces the springtime tornadoes in the middle of the continent.

Erosion by Wind

When you walk along a beach on a blustery day and feel the sting of blowing sand against your legs, you are experiencing the power of wind erosion. The erosive power of wind was readily evident when a fierce windstorm swept through California's San Joaquin Valley on December 12, 1977. Winds blowing at 300 km (190 mi) per hour removed nearly 100 million tons of topsoil from a 2000-km^2 (775-mi^2) area. The blinding dust-and-sand storm severely damaged crops and buildings. During a period of zero visibility, five motorists were killed in accidents and many more were injured.

Wind by itself cannot erode solid rock in the same way as flowing water or ice, by dislodging cemented grains directly from sedimentary rock or by quarrying crystalline bedrock from out-

crops of fractured igneous and metamorphic rock. Desert landscapes, however, are more vulnerable to wind erosion than most other geographical settings. Because deserts contain little vegetation to anchor the soil and hold its particles down, they are generally covered with *loose* materials. Loose sand grains lifted from the surface by turbulent winds can then be used as tools with which to etch exposed rocks.

The lifting of these loose grains—a process called *deflation*—and their use as tools to etch exposed rocks—a process called *abrasion*—are discussed in the following sections.

Deflation   Wind gusts blowing across a dry, treeless desert lift sand- and silt-size particles from the surface, but leave behind larger, heavier pebbles and cobbles. Clay-sized particles may also remain

Highlight 18-2

America's Dust Bowl

In the early 1930s, the plowed wheat fields from the Texas Panhandle to the prairie provinces of Canada (Alberta, Saskatchewan, and Manitoba) were stripped of their rich topsoil by strong winds that swept across the drought-parched landscape. Southeastern Colorado, western Kansas and Nebraska, and the Texas and Oklahoma panhandles were particularly devastated. Towering walls of swirling dust, called *black rollers,* darkened the sky at noon. In May 1934, powerful winds lasting for 36 hours lifted great masses of dust from the Great Plains into a dense black cloud that cast a 2000-km (1200-mi)-long shadow across the eastern half of the continent. In autumn, prairie dust fell in upstate New York as "black rain"; in winter, it covered the mountains of Vermont as "black snow."

The dust in the Great Plains was sometimes so dense that it buried entire fields of crops. Many people and farm animals died from suffocation or "dust pneumonia," a disease like miner's silicosis (a condition caused by the inhalation of mine dust). The strong winds even sent the ultrafine dust through the fabric of every garment. As one eyewitness reported:

These storms were like rolling black smoke. We had to keep the lights on all day. We went to school with the headlights on and with dust masks on. I saw a woman who thought the world was coming to an end. She dropped down to her knees in the middle of Main Street in Amarillo and prayed out loud: Dear Lord! Please give them a second chance.

A combination of natural and human factors produced the Dust Bowl. Much of the Great Plains had long been overcultivated. Settlers had plowed vast areas of the thick tough prairie sod to plant great fields of wheat. For years, abundant rainfall yielded plentiful harvests. Removal of the protective grass cover, however, left the land vulnerable to wind deflation when drought came. In the early 1930s, precipitation decreased to less than 50 cm (20 in) per year, and winds gusted regularly at speeds exceeding 15 km (9 mi) per hour. Crops failed everywhere. With no roots from trees or crops to anchor it, hundreds of millions of tons of rich topsoil were simply blown away. Thousands of farmers, their crops parched and their soils depleted, abandoned their lands and moved on. The westward migration of "Okies" to California was immortalized in John Steinbeck's epic novel, *The Grapes of Wrath.* Songwriter Woody Guthrie captured the bleak outlook of the times in *So Long, It's Been Good to Know You,* written on April 14, 1935, the date of one of the decade's worst dust storms.

By 1939, a few years of more abundant rainfall, along with state and federal attempts to improve farming practices, had ended the Dust Bowl tragedy. The prairie states and provinces became livable and farmable again. Although damaging droughts occurred several times between 1950 and 1980, extensive irrigation and crop rotation ensured that there was less of a threat to the region's soils and agricultural industry. Residents of the Plains have learned to adjust to naturally recurring dry periods.

▲ Ominous clouds of dust obscured the midday sun during the Dust Bowl years of the 1930s. This photo was taken in April 1935, in Mills, New Mexico.

behind, stuck to the surface by the electrical charges on their surfaces. **Deflation** involves the removal of loose material by wind. It may excavate distinct depressions in a desert floor, forming depressions that vary in size and shape, largely depending on the force of the wind and the physical properties of the deflating sediment.

Deflation usually lowers the landscape slowly, by only a few tens of centimeters per thousand years. In extraordinary cases, however, such as occurred during North America's Dust Bowl of the 1930s (see Highlight 18-2), deflation can strip away as much as 1 m (3.3 ft) of fine-grained topsoil in just a few years. Land surfaces may deflate over broad expanses or in local areas only.

Small lowered regions, known as *blowouts,* form where animal trampling, overgrazing, range fires, drought, or human activity has disturbed surface vegetation. Blowouts are typically little more than 3 m (10 ft) across and 1 m (3.3 ft) deep. Because water helps bind loose deflatable material into a more cohesive mass, blowouts seldom extend below the local groundwater table. Thousands of blowouts, such as the one shown in Figure 18-16, dimple the semi-arid Great Plains of North America from Texas to Saskatchewan.

Deflation basins are large blowouts that occur where local bedrock is particularly soft or where faulting has produced a

▲ **FIGURE 18-16** A blowout caused by deflation in Sand Hills State Park, Texas.

broad area of crushed bedrock. Sizable deflation basins are shaped by the winds that blow across the western Great Plains, for example. Near Laramie, in southeastern Wyoming, soft fine-grained bedrock has been deflated to produce Big Hollow, a depression 15 km (9 mi) long, 5 km (3 mi) wide, and 30 to 50 m (100 to 160 ft) deep.

Abrasion   **Wind abrasion** occurs when a wind-borne supply of eroded particles is hurled against a surface and sandblasts it. In a Saharan sandstorm, for example, several hours of wind abrasion can strip the paint from a jeep and pit the windshield until it is no longer transparent. Wind-eroded sand consists largely of quartz (with a hardness of 7 on the Mohs mineral-hardness scale). The collision of hard quartz grains with other grains or bedrock surfaces can weaken the cement between sedimentary grains, fracture grains in igneous and metamorphic rocks, and dislodge particles previously loosened by weathering.

Wind abrasion produces several distinctive desert features, such as the rock pedestal shown in Figure 18-17. It also creates wind-shaped stones with flat sharp-edged faces, called **ventifacts** (from the Latin, meaning "wind-made"). Figure 18-18 illustrates how ventifacts form. Their polished facets show a pitted "orange-peel" texture produced when prolonged wind abrasion attacks the side of a ventifact that faces the prevailing wind direction. The orientation of the sandblasted facets on a group of ventifacts indicates a desert's dominant wind direction. Ventifacts with more than one faceted side may have been shaped by winds that changed direction or blew from multiple directions. The position of such rocks may have also changed, exposing a new surface to wind abrasion. A stone's position can readily be disturbed by the hooves of passing animals, powerful storm winds, floodwaters, a cycle of frost wedging, or undercutting at the base of the stone by wind abrasion.

Wind abrasion also produces large-scale topographic features such as the streamlined desert ridges called *yardangs* (from the Turkish *yar,* meaning "steep bank"). Yardangs most commonly form

▲ **FIGURE 18-17** In desert areas, such as this one in Goblin Valley, Utah, it is common to see balanced rocks perched precariously on narrow pedestals that have been sculpted by wind abrasion.

where strong, one-directional winds abrade soft sedimentary bedrock layers, leaving behind more resistant layers. As you can see in Figure 18-19, the resulting landform, which resembles the inverted hull of a sailboat, is surrounded by wind-abraded depressions. Yardangs are usually wider on the windward side and taper leeward. They may be as much as tens of km long and tens of meters high.

Transport by Wind

In deserts, wind-transported particles bounce, or *saltate* (from the same Latin root as somer*sault*), along the surface much like the sand grains near the bottom of a stream. They may also travel several to tens of meters above the surface, temporarily suspended by wind turbulence. Two interacting and opposing factors influence the movement of even the smallest particle of dust: the driving force of the wind, which is controlled by the air's velocity and turbulence, and the particle's *inertia,* or its tendency to stay in place, which is controlled by gravity.

Other modes of transport, such as rivers and glaciers, can lift and move a particle whenever their driving force exceeds the particle's resistance to movement. Wind, however, must achieve a much greater relative velocity because it cannot lift a loose particle *directly* from the surface. The viscosity of air is quite low relative to the density of sediment particles. Thus moving air is far

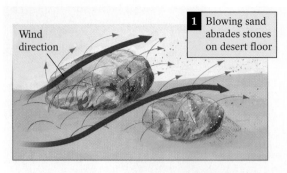

1 Blowing sand abrades stones on desert floor

Wind direction

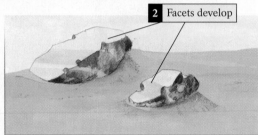

2 Facets develop

3 More facets develop as new surfaces are exposed

Original shape of stone

▲ **FIGURE 18-18** Ventifacts form when wind blowing predominantly from one direction abrades desert-floor stones, creating flat surfaces and sharp edges. As the wind changes direction or the stones shift position, other surfaces are exposed to wind abrasion, and more facets form on the newly exposed surfaces. Photo: A wind-etched ventifact.

▲ **FIGURE 18-19** Yardangs from the White Desert, Egypt.

less likely to lift and carry particles than are more viscous fluids, such as water, glacial ice, or lava. Vegetation and coarse sediment on the surface also interrupt air flow just above the surface, creating a thin layer of virtually "dead air."

To move particles in this still layer, the velocity of the wind must be capable of overcoming the inertia of those grains that are large enough to extend above this dead-air layer. When winds reach this velocity, these particles begin to move by rolling. They then collide with other particles, propelling them upward (Fig. 18-20a). Once airborne, particles move forward as the wind carries them in curved paths until gravity pulls them back down to the surface. Large saltating grains are overcome by gravity fairly quickly, fall back to the surface, and strike other grains, setting them in motion. When the wind velocity drops below the level required to move large grains, transport of mid-size saltating grains continues until the wind dies down sufficiently to prevent even the smallest grains from saltating. The wind-driven chain reaction that set the particles in motion then comes to an end until the next gust.

Bed Load A desert's *bed load*, like the water-borne sediment that moves along the bottom of a stream, is the wind-transported sediment that moves on or close to the ground. Air is not dense or viscous enough to lift and carry a windstream's bed load, and consequently, it must move particles by saltation. The nature of the surface, the shape and density of the particle, and the force of the collision that sets it in motion determine the height to which a grain will saltate. A grain propelled from a soft bed of sand can attain a height of only 50 cm (20 in) or less, whereas a grain from a surface covered by a continuous stone pavement may saltate to a height of 2 m (6.6 ft). (To see what accounts for this difference, toss some unpopped popcorn kernels on a thick-pile carpet and then again on a polished hardwood floor. Which ones bounced higher?)

A violent wind can churn up a rock-strewn desert. Even when the sediment supply lacks silt- and clay-size particles, high winds

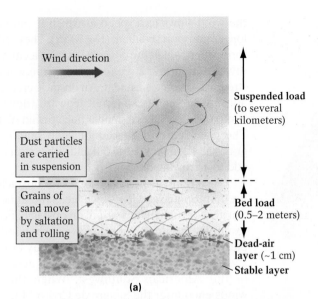

▲ **FIGURE 18-20** Transport of wind-borne sediment. **(a)** Like stream-borne particles, wind-borne particles may travel by bouncing or saltating along the surface or by being carried in suspension above the surface. Greater velocity is needed to transport particles in wind, however, because of the low viscosity of air and because of the presence of obstructions at ground level that slow air movement, creating a dead-air layer. **(b)** On a relatively windy day, little dust rises from a dry country road because the particles do not extend above the dead-air layer, and thus they remain relatively unaffected, even by strong wind gusts. When a passing truck or a tractor disturbs the dead-air layer, it throws dust up above the zone of motionless air and into the windstream, and a thick cloud of dust rises behind its tires. The dust remains suspended in the wind long after the truck and its turbulence are gone.

can produce a bed load that extends upward to a height of 1 to 2 m (3.3 to 6.6 ft). The air above this dense, surface-hugging cloud of particles typically remains quite clear. A person caught in such a sandstorm appears to be submerged in a swirling pool of sand, like the one described in this eyewitness account by British journalist Richard Trench, who survived a Saharan sandstorm in 1974:

> Suddenly the wind began to rise, blowing in short and powerful gusts. The surface of the desert, normally so still, was growing restless. As the wind rose, so the desert rose too. It was dancing about my feet. It hurled itself against my calves, whirled around my body, and beat at my bare arms. Nor did it stop there. It grabbed my neck, stung my face, encrusted itself in my throat, blocked my nostrils, and blinded my eyes. I felt alone and feared death by drowning.

Suspended Load Like the water-borne sediment carried higher up in the water column above a stream bed, a desert's *suspended load* moves long distances without settling back to the surface. Most of the suspended load in an airstream consists of *dust,* particles about 0.15 mm (0.006 in) or less in diameter. Dust may include silts and clays deflated from soils, volcanic ash, pollen grains, airborne bacteria, tiny plant fragments, charcoal from forest fires, fly ash from the burning of coal, small salt crystals from evaporation of sea spray, or finely crushed glacial flour deflated from exposed dry outwash.

Many dust particles are relatively flat, with a large surface area relative to their volume. This shape aids uplift, much as the wings on an airplane carry it aloft. Most dust particles therefore counter the pull of gravity effectively and can be lifted above the dead-air layer by turbulent winds and swept into the upper atmosphere, thousands of meters high. The dust may remain there for several years, their flat shapes (like so many tiny frisbees) re-

sisting gravity's pull back to Earth. Suspended dust may travel thousands of kilometers. Distinctive red dust from the Sahara, for example, has been found 4500 km (2800 mi) away on the Caribbean island of Barbados. Recent studies point to these Africa dust particles, and their "hitchhiking" microbes, as the primary culprits in the death of coral reefs off the Florida and Caribbean coasts (Fig. 18-21). And in March, 1935, dust from eastern Colorado was carried by winds with sustained speeds of

▲ **FIGURE 18-21** Dust particles and related microbes may be responsible for the demise of this coral within the Florida Keys National Marine Sanctuary in southern Florida.

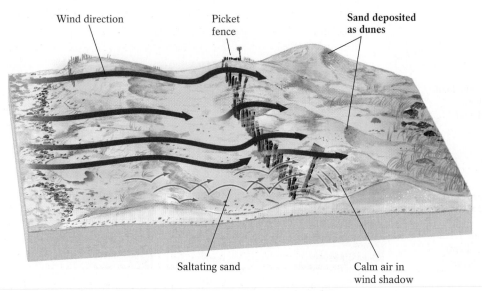

Wind direction   Picket fence   **Sand deposited as dunes**

Saltating sand   Calm air in wind shadow

▲ **FIGURE 18-22** Sand dunes typically form when a surface obstacle, such as a boulder, hill, or picket fence, interrupts the flow of migrating, saltating sand. A wind shadow is created leeward of such an object, and sand settles and accumulates on the ground.

80 km (50 mi) per hour to upstate New York, 3000 km (1800 mi) away. There it darkened the midday skies for days. Highlight 18-2 further describes America's tragic Dust Bowl years.

Deposition by Wind

If winds can excavate blowouts and deflation basins and darken the sky with a blinding veil of dust, what happens when the winds stop blowing? Airborne particles begin to fall, with the largest, heaviest, saltating grains dropping closest to their source, and the flattest, smallest, least-dense particles of dust dropping farthest downwind. In this way, wind *sorts* its deposits. The nature and location of the resulting depositional landforms are determined by the size and amount of sediment (whether particles are carried as bed or suspended load), the constancy and direction of the wind, and the presence or absence of stabilizing vegetation. We now turn to two of the most common products of wind deposition—sand dunes (the products of bed-load deposition) and loess sheets (the primary products of suspended-load deposition).

Dunes—Bed-Load Deposition
Dunes are wind-built mounds or ridges of sand that form in both arid and humid climates. They develop wherever there is a sufficient supply of sand that is initially not stabilized by vegetation and strong winds that blow constantly. Vegetation often gains a foothold later, stabilizing the dunes. Dunes appear along the sandy shores of oceans, seas, and large lakes (such as Lake Michigan) and near large, dry, sandy floodplains. North America boasts dunes in such scenic places as Florence, Oregon; Alamosa, Colorado; Padre Island, Texas; Burns Harbor, Indiana; Long Island, New York; Cape Cod, Massachusetts; Ogunquit, Maine; and Cape Breton, Nova Scotia.

Dunes typically form where an obstacle, such as a hill or picket fence, interrupts the flow of saltating sand, or when

the wind slows to a speed at which it can no longer transport sediment (Fig. 18-22). They also form where a narrow obstacle, such as a clump of vegetation or a large rock, forces the windstream to diverge around it. An obstacle creates a *wind shadow* of calmer air leeward (downwind) of it, as well as a smaller one upwind of it. The calm leeward air cannot keep saltating grains aloft, so they settle out and either start or add to a growing dune.

As a dune grows, it becomes an obstacle itself, trapping sand in its own lee. Sand dunes can also develop windward of an obstacle. For example, in North America, westerly winds pick up large quantities of sand as they blow across the barren sandstone cliffs and deserts of the American Southwest. When these winds encounter the Sangre de Cristo Mountains of southern Colorado, they slow and drop their load onto the windward side of the mountains, forming the massive dunes of the Great Sand Dunes National Park (Fig. 18-23).

Dunes accumulate more rapidly on sand-covered surfaces than on a stone-strewn desert floor. They continue to grow as long as sand-carrying winds blow from the same general direction, or until they reach the height limit imposed by the local wind pattern. This limit is reached when the wind speed required to carry a load of sand over the dune and drop it on the lee side becomes too great. Under those high-wind conditions rather than depositing additional sand, the wind's load is simply carried past the dune. Most dunes measure from 10 to 25 m (35 to 80 ft) high. Some in Saudi Arabia and China exceed 200 m (660 ft), but they are actually complexes of many dunes piled one atop another (Fig. 18-24).

Most of the sand-size particles that make up dunes are weathering-resistant quartz grains. The countless collisions that

▲ **FIGURE 18-23** The dune field of the Great Sand Dunes National Park continues to grow as loose material eroded from the arid lands to the west and carried by westerly winds is dropped against the Sangre de Cristo Mountains of southern Colorado.

▲ **FIGURE 18-24** Complexes of dunes in the Gobi Desert, in western China.

these grains endure typically pit their surfaces, giving them a frosted appearance when viewed under a microscope. Gypsum grains, eroded from evaporite deposits on the slopes of the San Andres Mountains to the west, form the dunes at White Sands, New Mexico; the dunes on the island of Bermuda are made up of the broken shells of marine organisms. Thus, dunes can be made of whatever sand-size materials are available.

Most dunes are asymmetrical, sloping gently (about 10°) on the windward side and more steeply (about 34°) on the leeward

side at the angle of repose of dry sand (discussed in Chapter 14). As sand grains saltate up the gentle windward slope and reach the dune crest, they spill over into the leeward wind shadow. In this area, the decreased wind velocity quickly deposits the particles on the steep leeward slope, or **slip face.**

As sand saltates up the exposed windward side and becomes deposited on the sheltered leeward side, the dune migrates downwind. Figure 18-25 illustrates the phenomenon of dune migration. Small dunes tend to migrate faster and farther because less sand needs to be moved to shift them. The rate of migration ranges from a few meters per year for large dunes in vast sandy deserts with gentle variable winds, to hundreds of meters per year for small dunes on bare, rocky desert floors with strong one-directional winds. Coastal dunes may even migrate seaward and extend a coastline offshore by several kilometers. You can see a dramatic example of this type of seaward migration of coastal dunes in Figure 18-26.

Migrating dunes can bury most any structures or objects downwind of them, whether buildings, forests, roads, or railroads. Dune migration across major interstate highways requires

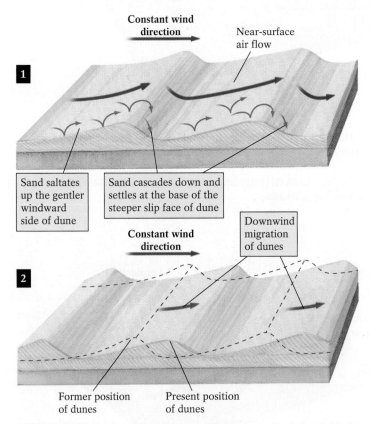

▲ **FIGURE 18-25** Dune migration. **(1)** Sand saltates up a dune's windward side, then cascades down its slip face. **(2)** Transport of sand from the windward to the leeward side of dunes causes downwind migration of the dunes.

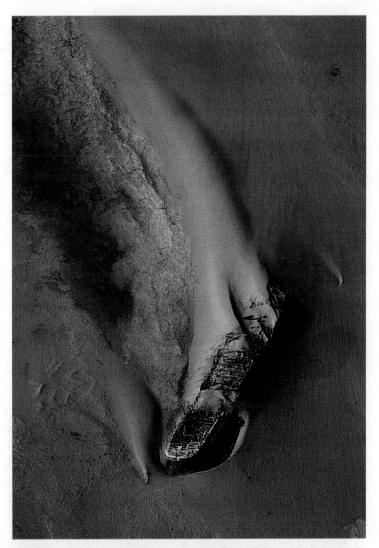

▲ **FIGURE 18-26** The stranded German freighter *Eduard Bohlen,* which ran aground off Africa's southwestern coast in 1912, now lies sand-locked more than a kilometer inland. The migrating and drifting sands of the Namib Desert have extended the coast seaward, engulfing the freighter.

▲ **FIGURE 18-27** Sand encroaching on Interstate 84, in the Columbia River Gorge, Oregon. (© Gary Braach)

frequent and costly sand removal to keep the roads open (Fig. 18-27). To halt sand movement, a continuous grass cover can be planted on a dune where the climate permits. If vegetation is removed—perhaps by motorcycles, dune buggies, dirt bikes, or all-terrain vehicles—wind gusts may erode new blowouts and dunes may resume their downwind march.

Dune Shapes
Dune shapes are determined by local conditions, such as the type of sand available, the degree of aridity, the nature of the prevailing winds, and the type and amount of vegetation present. Dune shapes are classified as transverse, longitudinal, barchan, parabolic, and star.

Transverse dunes, shown in Figure 18-28a, consist of a series of parallel ridges that typically occur in arid and semi-arid regions where sand is plentiful, wind direction constant, and vegetation scarce. These dunes form *perpendicular* to the prevailing wind direction and possess gentle windward slopes and steep leeward slip faces. In the Sahara, these dunes can be as large as 100 km (60 mi) long, 100 to 200 m (330 to 660 ft) high, and 1 to 3 km (0.6 to

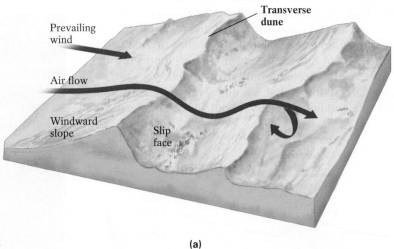

(a)

▲ **FIGURE 18-28** Common dune types. **(a)** Transverse dunes on Mesquite Flat, Death Valley, California.

▼ **FIGURE 18-28 (b)** Longitudinal dunes at Stovepipe Wells, Death Valley National Park, California.

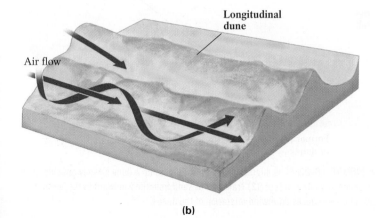

(b)

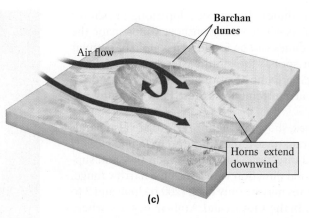

(c)

◀ **FIGURE 18-28 (c)** Barchan dunes in the Baja Desert of Baja California.

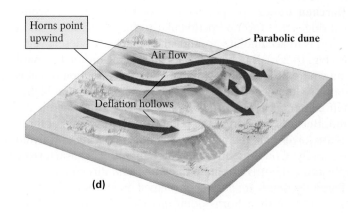

(d)

▲ **FIGURE 18-28 (d)** Parabolic dunes.

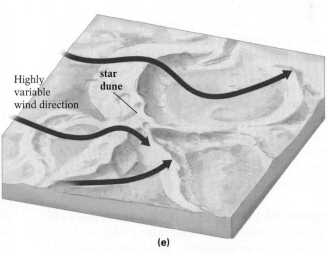

(e)

◀ **FIGURE 18-28 (e)** Star dunes (in foreground), Death Valley National Park, California.

2 mi) wide. Transverse dunes also can develop along the shores of oceans and large lakes where strong onshore winds shape the abundant sand. Such dunes can be found at a couple of popular Midwestern vacation spots—the southeastern shore of Lake Michigan at the Indiana Dunes National Lakeshore in northwestern Indiana, and at Warren Dunes State Park in southwestern Michigan.

Longitudinal dunes, shown in Figure 18-28b, are also parallel ridges, but unlike transverse dunes, these are oriented *parallel* to the prevailing wind direction. They form when the sand supply is moderate and wind direction varies within a narrow range. Small longitudinal dunes may be only 60 m (200 ft) long and 3 to 5 m (10 to 16 ft) high. In the Libyan and Arabian deserts, where strong winds blowing from several directions converge over the dune crest, longitudinal dunes can reach 100 km (60 mi) long and 100 m (330 ft) high.

Barchan dunes (pronounced "BAR-kane") are crescent-shaped ridges that form perpendicular to the prevailing wind as sand begins to accumulate around small patches of desert vegetation (Fig. 18-28c). Barchans develop in arid regions on flat, hard ground where little sand is available and wind direction remains constant. As they grow, these dunes become thicker and higher in their centers. There, air flow is slowed and more sand is deposited. Because the horns (the points of the crescents) of these dunes are lower than their centers, the horns migrate downwind more rapidly. Consequently, the barchan and its characteristic sharply pointed horns extend in the *downwind* direction.

Parabolic dunes are horseshoe-shaped dunes that differ from barchans in that their horns point *upwind* (Fig. 18-28d). They commonly form along sandy ocean and lake shores, the only major dune areas outside of deserts. Parabolic dunes develop when transverse dunes become exposed to accelerated wind deflation, especially after the removal of some vegetation. A small deflation hollow forms on the transverse dune's windward side, allowing the wind-excavated sand to pile up downwind. As the hollow grows, the wind becomes concentrated at its center, speeding the migration of that portion of the dune. The horns—a remnant of the original transverse dune—are usually still covered by vegetation and remain anchored in place; the rest of the crescent-shaped dune continues to migrate downwind, forming a horseshoe shape that can become quite elongated.

Star dunes, shown in Figure 18-28e, are the most complex dune types. They form when winds blow from three or more principal directions, or when wind direction constantly shifts. Star dunes tend to grow vertically to a high central point and may develop three or four arms radiating from the center.

Loess—Suspended-Load Deposition

Loess is a windborne silt deposit that resembles fine-grained lake mud (Fig. 18-29). Unlike water-borne sediments, which are typically deposited at topographic low spots, loess deposits blanket hills, slopes, and valleys evenly. Most loess grains consist of quartz, feldspar, mica, or calcite. Slight oxidation of accessory iron minerals gives the deposit its characteristic yellow-brown color. The *terrestrial* origin of loess is confirmed by the presence of shells from air-breathing snails, as well as by the scattered mammal bones, worm burrows, and plant-root tubes that loess commonly contains.

▲ **FIGURE 18-29** Loess (windblown silt) from the Columbia River basin, Washington.

Loess almost always appears *downwind* of a plentiful supply of loose dry silt. Much of the loess in America's Midwest, for example, is a yellow-brown deposit that originated when coarse-to-medium silt particles (0.01 to 0.06 mm in diameter) were eroded from the drying glacial outwash deposited during the last period of ice expansion about 15,000 years ago and then deposited downwind. The streams draining from the melting ice carried large volumes of silt-sized glacial flour, the product of subglacial abrasion. As the outwash plains dried, the prevailing winds from the west swept up the silt and deposited it downwind to the east as loess. More than 500,000 km^2 (200,000 mi^2) of the land east of the Missouri and Mississippi rivers is covered by loess, including much of the rich farmland in Iowa, Illinois, Indiana, and Missouri (Fig. 18-30).

Some loess—for example, that found *west* of the Mississippi River in eastern Kansas and Nebraska—originated in *nonglacial* environments. Wind apparently eroded these deposits from the ancient dust-storm deposits that formed the sand hills of northwestern Nebraska. The world's largest loess deposits occur in northern China. Their phenomenal 300-m (1000-ft) thickness comes from the nearly endless supply of silt from the vast Gobi Desert to the west. This airborne dust settles and washes into the Huang He (Yellow River) and Yellow Sea, giving them their distinctive color. Many of the local residents of this region literally live within this deposit in human-excavated Earth-sheltered housing (Fig. 18-31).

Loess particles are typically remarkably fresh because most loess originates in glacial environments and arid deserts, both places where little chemical weathering occurs. Thus, these unweathered loess particles retain most of their original ions, and for this reason, loess-covered lands, such as those in Iowa, are extremely fertile. Furthermore, because they are airborne, loess particles are typically quite angular. Unlike stream-borne particles, they have not experienced the numerous collisions that tend to round particle edges. This angularity makes loess deposits quite porous (because their angles prevent them from packing together too tightly), which accounts for their high moisture-holding capacity. Moist, mineral-rich loess is commonly a Midwestern farmer's favorite soil. In fact, nearly all of the major cities of the

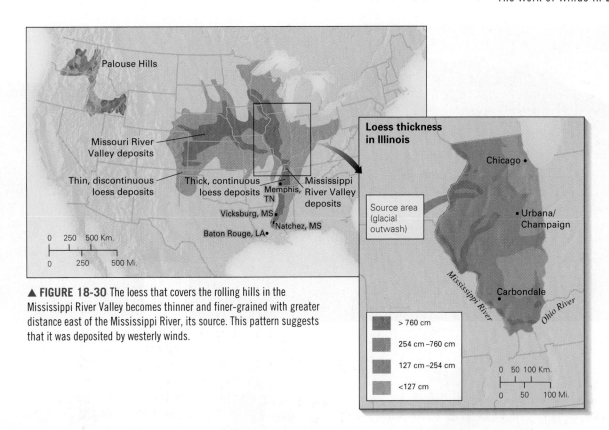

▲ FIGURE 18-30 The loess that covers the rolling hills in the Mississippi River Valley becomes thinner and finer-grained with greater distance east of the Mississippi River, its source. This pattern suggests that it was deposited by westerly winds.

Lower Mississippi Valley are on the eastern, loess-covered bank of the river. Memphis, Vicksburg, Natchez, and Baton Rouge are all built atop the eastern bank's fertile loess bluffs.

Another product related to the deposition of fine airborne dust is a feature called a **desert pavement** (Fig. 18-32a), a surface layer of closely packed stones that sometimes blankets a desert's surface. Geologists once viewed a desert pavement as a *lag deposit,* a coarse, stony remnant left behind by wind deflation as finer particles were carried away, leaving behind a concentrated layer of large stones too large to be carried aloft. We now believe, however, that such an armor of stones is produced largely by deposition and accumulation of wind-blown dust between the larger stones on a desert floor. Over hundreds or thousands of years, the wetting and drying of clay minerals in this dust by occasional desert rains swells and shrinks the moistened dust particles, in turn pushing the larger stones to the surface. This process

has a similar effect as the freezing and thawing process in colder regions that raises buried stones to the surface in temperate regions such as New England (where stony fields make life difficult on farmers) and in polar regions, where rock-strewn patterned ground develops (see page 559). Figure 18-32b illustrates how a desert pavement forms.

It is important to remain aware of how fragile a desert pavement is, often just a stone or two thick. The passage of hikers, dirt bikers, or grazing animals can disturb a surface that has been thousands of years in the making. Once the pavement is breached, winds can then deflate the exposed fine dusty sediment beneath, sometimes with some adverse human consequences. Airborne dust causes or aggravates a host of respiratory ailments, especially asthma. A dramatic example of desert-pavement disturbance took place in 1991 when the tanks of the invading Iraqi army destroyed the pavement of the desertlands of Kuwait,

◄ FIGURE 18-31 Earth-sheltered dwellings excavated from thick loess deposits in central China. These homes are insulated from the sharp variations in temperatures and thus are warm in the region's frigid winters and cool during the stifling summers. Their only drawback—such deposits tend to liquefy in powerful earthquakes. Hundreds of thousands of people perished in this region during a massive earthquake in February of 1552 when their dwellings liquefied.

Clay-rich dust is deposited around these surface stones.

Occasional desert rains wet the clay, causing it to swell.

Swelling clays around the stones push stones to the surface where they become concentrated as a desert pavement.

▲ **FIGURE 18-32** Desert pavements form as the shrinking and swelling of wetted, dust-sized clay particles raise and concentrate stones at a desert's surface. **Photo and inset:** Baja California, Mexico.

churning up huge clouds of choking dust. The ruts left behind in the Kuwaiti desert pavement have yet to heal, and this region's inhabitants suffer from ongoing health problems related to large amounts of airborne dust.

Reconstructing Paleowind Directions

Winds of the past, or *paleowinds,* have left traces of their intensity and prevailing direction in ancient dunes and loess deposits, some that have long since turned to stone. These traces reveal much about the geologic past. When geologists study a sandstone formation, they search for clues to determine whether it was deposited by wind or water. Excellent sorting, the existence of cross-bedded sands displaying the angle of repose of dry sand, the occasional discovery of sharply faceted ventifacts, the presence of sand grains polished and pitted by mid-air collisions, and the ab-

sence of ultralight mica flakes (which are easily swept away in suspension) all suggest a wind-deposited sandstone.

After geologists confirm a wind-related origin, they use their knowledge of wind-generated landforms and deposits to reconstruct ancient wind directions. They base their hypotheses on the shape of yardangs, the position of ventifact facets, the asymmetry of dunes, the thickness and texture of loess deposits, and most importantly, on the orientation of cross-beds.

Although whole sand dunes rarely survive intact in sedimentary rocks, parts of them, particularly their internal structures, are sometimes buried by later dune migration. Although the surface expression of a gentle windward slope and the steep leeward slip face of a dune may vanish, the sands of advancing dunes may nevertheless bury the inclined bedding of the dune's internal structure, preserving dune-type cross-bedding. The leeward slip-face beds, which are inclined at the angle of repose of dry

18.1 Geology at a Glance

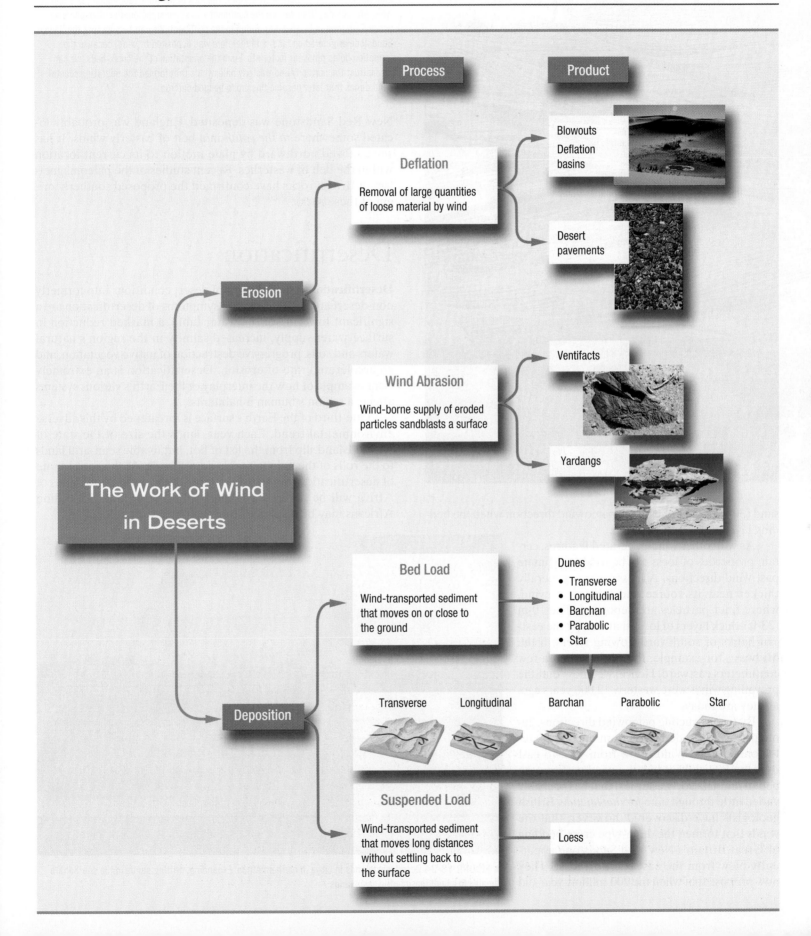

Process

Product

The Work of Wind in Deserts

Erosion

Deflation
Removal of large quantities of loose material by wind

Blowouts
Deflation basins

Desert pavements

Wind Abrasion
Wind-borne supply of eroded particles sandblasts a surface

Ventifacts

Yardangs

Deposition

Bed Load
Wind-transported sediment that moves on or close to the ground

Dunes
• Transverse
• Longitudinal
• Barchan
• Parabolic
• Star

Transverse   Longitudinal   Barchan   Parabolic   Star

Suspended Load
Wind-transported sediment that moves long distances without settling back to the surface

Loess

Numerous ancient lithified sand dunes appear in the national parks of the arid Southwest. The Navajo Sandstone, for example, crops out throughout Zion National Park in southwestern Utah. The rock dates from 180 million years ago, when present-day Utah was on the continent's west coast, before the lithosphere on which California is located was added to the continent. We can be sure that the Navajo Sandstone originated on land, and therefore was deposited by wind, because the formation holds dinosaur footprints. From the orientation of the cross-beds, we can reconstruct the northerly and westerly paleowinds that formed the migrating coastal sand dunes that later became this cross-bedded outcrop.

New Red Sandstone was deposited, England was probably located somewhere *to the south* in a belt of easterly winds. It has since moved northward by plate motion to its current location within the belt of westerlies. Recent studies of the paleomagnetism of British rocks have confirmed the proposed southerly origin of these rocks.

Desertification

Desertification is the invasion of desert conditions into formerly non-desert areas. The common symptoms of desertification are a significant lowering of the water table, a marked reduction in surface-water supply, increased salinity in the region's natural waters and soils, progressive destruction of native vegetation, and an accelerated rate of erosion. Desertification is an extremely clear example of how the interplay of the Earth's various systems affects a region's human inhabitants.

One-third of the Earth's surface is threatened by this adverse environmental trend. Each year, lands the size of the state of Rhode Island slip from the list of hot, but livable semi-arid lands to the rolls of the uninhabitable desert lands. At the current rate of desertification, by 2025, two thirds of the agricultural lands of Africa will be desert wastelands (Fig. 18-34) and 135 million Africans may be displaced from their ancestral homelands.

sand (30° to 34°), indicate the paleowind direction when the bedding layer formed (Fig. 18-33).

As with other wind-generated features, certain properties of loess can be used to estimate past wind directions. A loess layer is generally thicker near its source and thins downwind, where finer particles are deposited. The 10-m (33-ft)-thick layers of loess that occupy the eastern banks of southward-flowing rivers in the Midwest, for example, thin to less than a few centimeters eastward. Hence, we know that the prevailing winds were westerly 12,000 years ago, as they are today.

By reconstructing paleowind directions, geologists can also deduce certain ancient plate-tectonic events. Winds blow from west-to-east over the British Isles today, a wind pattern (the belt of mid-latitude westerlies) that has probably varied little through time *for that latitude*. British geologists have discovered, however, that the winds that formed the dune-type cross-bedding of Great Britain's New Red Sandstone apparently blew from the east and northeast. They now propose that when the 200-million-year-old

▲ **FIGURE 18-34** Desert wastelands in Libya in northern Africa. Expanding, drifting sands in the sub-Sahara threaten to engulf these fertile oasis lands.

▲ **FIGURE 18-35** This cave painting depicts falling rain. It is found today in the parched Namib Desert of southwestern Africa, but was painted thousands of years ago when the climate here was moist and temperate.

Northern and western Africa have been experiencing rapid desertification for 2000 years. Approximately 8000 years ago, several thousand years after the end of the last major worldwide glaciation, the Namib Desert of southwestern Africa was a lush savannah that supported advanced Stone Age societies (Fig. 18-35). It remained a fertile grassland, teeming with wildlife, for the next 6000 years.

Geologists reconstruct the land features of that ancient world by using radar imaging. Radar waves cannot penetrate moist soil, so in humid areas, radar is either absorbed by soil moisture or reflected, creating an image of the surface. The bone-dry regolith of hyper-arid deserts, however, is virtually transparent to radar; the reflected image shows the contours of the underlying bedrock. Space-shuttle flights cruising 200 km (120 mi) above the eastern Sahara have used radar to map the configuration of the bedrock *beneath* the extremely dry Selima sand sheet. The radar waves penetrated this vast, flat, featureless sea of sand, revealing an underlying ancient surface marked by rocky hills and deep, wide river valleys, as shown in Figure 18-36.

Two forces have caused the lush northern African savannah and grassland to turn into the hyper-arid modern Sahara. The first, drought, is natural and unavoidable; the other, overpopulation and land mismanagement, is human-caused. Drought, overpopulation, or both can start the process of desertification. When a land's inhabitants cannot produce enough food and other raw materials to provide for their needs, they tend to overgraze their cattle, plant crops without replenishing the soil, cut down trees for shelter and fuel, and draw more water from springs and other sources than is naturally replenished. Gradually, soils become depleted of their nutrients, and the removed trees and their roots fail

▲ **FIGURE 18-36** A satellite image of the Selima sand sheet, enhanced by space-shuttle radar imagery (the "blue" area). The topography of the bedrock under the sand shows several stream valleys, currently hidden from view by the modern sand sheet.

to grow back. Without vegetation, the soils cannot hold water or prevent wind erosion. Water from occasional cloudbursts washes away the unbound topsoil and, in hot climates, evaporates before it can enter and replenish the groundwater supply. Eventually the parched land can produce nothing, forcing the land's inhabitants to migrate in search of food, water, and shelter.

As much as 35% of the world's land sits within the semi-arid margins of deserts. In the next few decades, overpopulation and drought may convert these lands from dry, *but habitable* grasslands to desertlands unable to sustain human populations. As many as 70 nations are affected by this trend. About half of these countries are in Africa, where the Sahara is advancing southward by as much as 50 km (30 mi) per year. Choking dust storms and newly formed dune fields already threaten some northern and western African cities.

The Sahel—the region of northern Africa immediately south of the Sahara—has been particularly hard-hit by desertification in recent decades. When the worst drought of the century struck this area in the early 1970s, it displaced 20 million people and their herds of cattle, sheep, goats, and camels. Animals that were not led away, died. Although the drought was the immediate cause of the disaster, land mismanagement had set the stage for it decades earlier. From 1935 to 1970, the Sahel's population doubled and progressively larger herds grazed on the marginally productive land. A few abnormally wet years in the early 1960s increased the region's productivity temporarily, encouraging farmers to plant more crops and expand their herds and grazing lands. Nomadic herdsmen from the south rushed to the Sahel, which had become an area of scattered groves of trees (for fuel and shelter), abundant seasonal grasses (for grazing), and agricultural fertility (yielding the region's primary crops: cotton, beans, millet, sorghum, and maize).

But in 1970, no rain fell in the Sahel. The large herds quickly consumed the existing grasses. The rich topsoils, without binding

Highlight 18-3

The Great Plains—North America's Once and Future Sahara?

Images of the tragic Dust Bowl years of the 1930s—especially if you live in Oklahoma, Nebraska, and the Dakotas—are etched in the collective memories of friends and relatives of recent generations. Was that crop-killing drought a climatic anomaly or merely the region's norm? Will global warming soon desertify America's Breadbasket, converting the fields of grain that feed millions into a vast expanse of shifting, windblown sand?

Recent research by Dr. Kathleen Laird, an ecologist at Queen's University in Ontario, suggests that the Great Plains region has endured numerous droughts over the past 10,000 years, including some far more extreme than the Dust Bowl years. Furthermore, Laird believes that the climate in that region has actually been uncommonly *moist* during the past 700 years, an ominous finding for sure.

To arrive at the conclusion that much drier times may lie ahead, Laird studied the region's water-dwelling algae called *diatoms*. These microscopic plants produce highly distinctive, decorative cell walls composed of the silica extracted from their watery surroundings. Apparently, different species of diatoms prefer water with different salinities ("saltiness"). By studying the diatoms of 53 lakes, each with a differing salinity, Laird correlated *past* lake salinity from cores of lake-bottom mud with the presence of the various diatoms. She reasoned that past drier climates would have increased evaporation, which in turn would have increased salinity (as concentrations of dissolved salt increased in the diminishing water), thus altering past diatom populations in the region's lakes.

Once a connection had been established between specific diatoms and differing salinities, Laird's group sought a small isolated lake with no incoming or outgoing streams—one whose salinity-level history recorded variations in rainfall and evaporation alone rather than inflow or outflow of stream water. Laird selected Moon Lake in North Dakota and proceeded to extract cores of the lake-bottom sediment, which had accumulated since the departure of the Laurentide ice sheet had created the lake, roughly 12,000 years ago. The appearance of the various salt-dependent diatoms within the carbon-dated lake muds indicated past droughts and wet times. Laird discovered that periods of extreme dryness persisted for centuries and occurred quite frequently. Her research indicates that the most extreme droughts occurred during the years 200 to 370 A.D., 700 to 850 A.D., and 1000 to 1200 A.D.

The likelihood of desertlike conditions spreading throughout the now semi-arid Great Plains is further confirmed by scientists at the United States Geological Survey in Denver. Dr. Richard Madole reports that at some point between 1000 years ago and the arrival of European settlers, wind-driven sand dunes buried much of eastern Colorado and adjacent Nebraska and Kansas, as they still do today in isolated spots on the Plains. This period of desertification, far more severe than the worst droughts of recorded history, was marked by sand-dune expansion throughout the region. It also correlates well with Laird's proposed drought between 1000 and 1200 A.D. According to Madole, relatively small changes in temperature and precipitation converted the semi-arid prairies into vast expanses of arid desertlike sand seas. The loss of the landscape's vegetation to the drought conditions accelerated the movement of the rootless sand. Any agriculture would have been impossible at this time.

What does the future hold for America's invaluable wheat-producing plains? Apparently, the climatic threshold to the region's future desertification looms close at hand. The causes of the cycles of drought in the past remain unknown, but we are increasingly aware that our actions today and their effects on climate may inadvertently trigger a drought in the future. The possible outcome—lots of empty shelves at the local market—provides yet another reason to monitor closely and reduce the ongoing human-caused negative impacts on global climate.

▲ Remnant dunes on the southeast margin of the Nebraska Sand Hills in the semi-arid Great Plains, Eldred Ranch, Garden County, western Nebraska.

roots, were easily eroded by wind. By the time the next year's rain fell, the ground was baked so hard that the subsequent surface runoff only accelerated soil erosion. Without crops, starving people were forced to eat their grain seeds, eliminating all hope of new planting. Foraging animals and people consumed the few remaining trees, eroding the soil further and depriving the region of fuel and building materials. Animals died by the millions. The starving population migrated to the region's larger cities, which tripled in size as refugee camps sprang up around them. Worldwide relief efforts were "too little, too late" to prevent the deaths of hundreds of thousands of people from starvation, malnutrition, and associated diseases. The drought continued into the 1980s, ending only in the mid-1990s. Figure 18-37 shows the area threatened by desertification and the effects of the process.

In the United States, overpopulation and urban growth, overgrazing, and excessive groundwater withdrawal—particularly in the arid and semi-arid Sunbelt states—have already increased soil salinity and accelerated erosion. Today approximately 27% of U.S. non-desert lands face encroaching desertification. Of great concern is the prospect of desertlike conditions replacing the "amber waves of grain" that mark America's Great Plains. Highlight 18-3 looks back at the region's prehistoric past to attempt to predict its climatic future.

Reversing Desertification

To reverse desertification, available water must be delivered where it is needed. To supplement the scant rainfall typical of arid regions, large volumes of water can be pumped from deep aquifers, conveyed from distant lakes or rivers, or drawn from the ocean and desalinated (converted from saltwater to freshwater). Of course, as we discussed in Chapter 16, the water supply within an aquifer will eventually be exhausted if more water is pumped out than seeps in. Thus, desert communities must be aware that they may be rapidly depleting their aquifers when they rely too heavily on groundwater to reverse desertification.

Some desert communities, such as Palm Springs, California, are naturally supplied by deep groundwater that originates as rain or snow on moist highlands hundreds of kilometers away. This water can be pumped for use in irrigation. In Africa, rain falling in the eastern highlands near the equator enters the Nubian Sandstone, an aquifer that extends more than 3000 km (1800 mi) north, where it lies deep beneath the Sahara. Faulting and folding have brought this aquifer's waters to the surface in the form of artesian springs where oases flourish. The rate of flow through the aquifer determines the quantity and quality of artesian water at an oasis. Swift flow brings water that is relatively pure and supports the growth of date palms and other vegetation. Slow flow through rocks may dissolve a large volume of salt and render the water undrinkably salty—even toxic—by the time it surfaces at an oasis.

In some extremely dry places, communities have developed technologies to reclaim land recently desertified and by doing so make portions of ancient deserts more habitable for humans. Systems of wells, channels, and collecting pools are used in the Middle East to collect and store the infrequent storm runoff *underground,* thereby protecting it from evaporation. Throughout the region, farmers apply a specially designed plastic mulch to their farmlands or cover their fields with plastic sheeting to retain irrigation moisture. Computers monitor and regulate water flow through pipes and canals, delivering water where it is needed at times when evaporation is minimal. In Israel, "drip" agriculture, using perforated garden hoses that snake amid plantings, delivers water constantly to individual plants, literally drop by drop. Water is even drawn from the Mediterranean Sea and desalinated for agricultural use.

In the American Southwest, billions of liters of water from the Colorado River have been diverted to the Sonoran Desert of southern Arizona, the aqueducts of southern California, and the Canal Central of Baja California, to irrigate the naturally arid lands that produce food for millions of people. Figure 18-38 shows the results of this irrigation program. Such diversions of water, however, do not lack negative consequences. By the time the Colorado River reaches Mexico, its flow has been reduced to a mere trickle, stopping altogether about 30 km (20 mi) short of the Gulf of California and depriving Mexico of its fair share of the river's water. Moreover, when the diminished river finally reaches the Mexican border, excess agricultural runoff in Arizona has made it so salty that the United States must desalinate the water before it enters Mexico.

Efforts to reverse desertification can also focus on controlling wind erosion. Trees serve as an effective windbreak to halt shifting sands and help to retain both soil cover and groundwater. Migrating dunes can be stabilized by planting fast-growing trees, such as poplars, which have deep, sand-binding roots. In northeastern China, an entire forest, 500 by 800 km (300 by 500 mi), was planted upwind of 90,000 acres of prime farmland. Dunes in China have

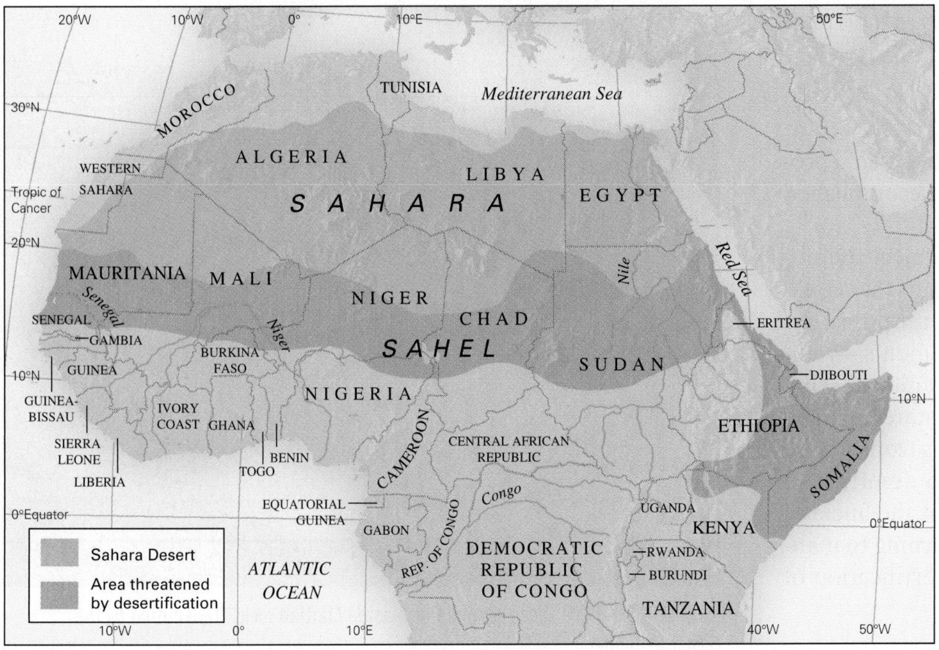

▲ **FIGURE 18-37** Map of the Sahel, northern Africa

▲ FIGURE 18-38 A circle of irrigated land in the Yuma Valley of Arizona's Sonoran Desert. Irrigation is largely responsible for the agricultural productivity and rapid population growth of this region.

even been leveled *by hand* and covered with topsoil to create productive new farmland, testimony to what can be accomplished with a cooperative population of 1.3 billion people.

Many ecology-minded individuals, however, reject the notion that deserts are lifeless environmental wastelands that could and should be modified by humans for their own purposes. Instead, they view deserts as vibrant natural environments—valuable ecosystems teeming with distinctive plants and animals—that should be protected and preserved in their natural state, in much the same way we have attempted to save the Earth's tropical rain forests. Perhaps the best approach we can take to deserts is to appreciate the beauty and ecological uniqueness of the ancient, naturally occurring water-deprived lands, while learning to manage their marginal lands more wisely to avoid desertification of now-habitable semi-arid regions.

Wind Action on Mars

The Earth is not the only windswept planet in the solar system. All planets are heated differentially by the Sun's rays, so those with atmospheres also experience wind action. As the *Pathfinder* and *Global Surveyor* missions have revealed, Mars is an especially windy planet.

Martian windstorms have been studied from close up for several decades. The *Mariner 9* space probe began its orbit around Mars on November 13, 1971, after a five-month journey from Earth. Its first photographs depicted, disappointingly, a planet-wide dust storm. For three months, dust that rose as high as 6 km (4 mi) obscured the entire Martian surface. On Earth, where the air is more than 100 times as dense as on Mars, it would have taken winds blowing in excess of 160 km/hr (100 mph) to raise dust to that height. On Mars, however, the weak gravitational force enables dust particles to ascend four times as high as those carried by comparable wind gusts on Earth. In fact, the fine oxidized particles perpetually suspended in the Martian atmosphere contribute substantially to the planet's pink-red appearance.

Mars' global dust storms apparently start when summer comes to its southern hemisphere. As the edges of the Martian frozen-carbon-dioxide polar cap rapidly sublimate (pass from ice directly to vapor without melting to a liquid phase), large temperature differences arise between the still-solid polar cap remnants and the warming surrounding landscape. These temperature differences produce differential air pressures. The resulting high winds trigger dust storms that spread northward until they engulf the planet's entire surface for months.

More recently, in 1997, the robotic "geologist" Sojourner, of the *Pathfinder* mission, provided crisp, clear images that con-

▲ FIGURE 18-39 Close-up of sand-blasted Martian rocks transmitted by the *Sojourner* lander.

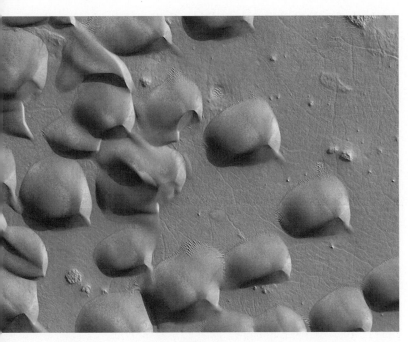

▲ **FIGURE 18-40** Sand dunes on Mars photographed by *Mars Global Surveyor*, 2003.

firmed Mars' windy past. Its close-up photos of wind-scored rocks suggest that powerful gusts—exceeding 200 km (120 mi) per hour—have sand-blasted exposed rocks, carving deep gouges from their surfaces (Fig. 18-39). And in 2003, ultraclear images of Martian sand dunes were sent back to NASA scientists by way of the *Global Surveyor* cameras (Fig 18-40).

Now that we've explored the hottest and driest corners of the globe, we now turn to the world's wettest environments—the Earth's vast oceans—in the next chapter.

Chapter 18 Summary

Desert is a term used to describe dry regions where vegetation is typically sparse. It is commonly defined as an area that receives less than 25 cm (10 in) of precipitation annually. Five principal types of deserts exist:

- the subtropical deserts that lie between about 20° to 30° latitude both north and south of the equator;
- deserts on the dry leeward side of the major mountain ranges, created by the rain-shadow effect;
- interior deserts, found in the centers of continents far from moisture sources;
- coastal deserts that develop where prevailing onshore winds have been cooled by cold oceanic currents;
- polar deserts, which are simultaneously extremely cold and dry.

In the dry desert environment, such processes as weathering and sediment transport operate in different ways and at different rates than in moist temperate climates. Weathering in deserts takes a mostly mechanical form. What little chemical weathering exists consists of manganese and iron oxidation.

Although one seldom sees surface water in deserts, many of their characteristic landforms have been carved by rapidly flowing, short-lived streams produced by occasional storms. Streams move through and deepen the numerous channels that cross a desert. The mountains that border many deserts recede as surface-water erosion produces large-scale, gently inclined surfaces called *pediments*. An *inselberg* is a steep-sided knob of durable rock—the remains of a mountain following advanced pediment development.

In deserts, the transporting ability of surface runoff diminishes rapidly because of evaporation or infiltration. While surface runoff lasts, it carries sediments that it later deposits to form distinctive features. Alluvial fans accumulate where a slope ends and the desert floor begins. A *playa* is a dry lake bed that develops when a desert-floor lake evaporates.

Differential heating of the Earth's atmosphere creates the pressure differences that generate winds. The resulting atmospheric convection cells set north–south winds into motion. These winds are deflected by the Earth's rotation. Because soil-binding moisture and vegetation are scarce in deserts, their land surfaces are particularly susceptible to modification by wind action. Wind erodes by *deflation*, removing fine particles from the surface. Wind-carried particles are hurled against rock surfaces, effectively sandblasting the rock in a process called *wind abrasion*.

Wind generally slows down on the leeward side of a surface obstacle, forming *dunes*—hills of loose, wind-borne sand. The size, shape, and orientation of a sand dune are determined by:

- the amount of sand available;
- the extent of local vegetational cover;
- the intensity, direction, and constancy of winds.

Most dunes are asymmetrical, with a gradual windward slope and steep leeward slip face. Five principal types of dunes exist:

- *transverse dunes* are parallel ridges that develop perpendicular to the prevailing wind direction;
- *longitudinal dunes* are parallel ridges that develop parallel to the prevailing wind direction;

- *barchan dunes* are crescent-shaped dunes that lie perpendicular to prevailing winds with their horns pointing downwind;
- *parabolic dunes* are horseshoe-shaped dunes that also lie perpendicular to prevailing winds but whose horns point upwind;
- *star dunes* grow vertically to produce a high central point with three or four radiating arms.

Wind-borne silt is deposited as loess. Because their grain size and bed thickness diminish downwind, loess sheets provide clues to paleowind directions, as do the cross-beds in lithified dunes.

In the late 20th century, conditions that promote the growth of deserts are overtaking many formerly semi-arid regions. Drought, overpopulation, and land mismanagement can cause *desertification*—the invasion of desert conditions into formerly non-desert areas. Desertification is occurring on every continent except Antarctica. Its symptoms include significant lowering of the water table, marked reduction in surface-water supply, increased salinity in natural waters and soils, progressive destruction of native vegetation, and accelerated rates of erosion.

KEY TERMS

aridity index (p. 582)
arroyos (p. 586)
barchan dunes (p. 600)
deflation (p. 592)
desert (p. 580)

desertification (p. 604)
desert pavement (p. 601)
desert varnish (p. 585)
dunes (p. 596)
inselberg (p. 587)

longitudinal dunes (p. 600)
parabolic dunes (p. 600)
pediment (p. 586)
playa (p. 588)
rain-shadow effect (p. 584)

slip face (p. 597)
star dunes (p. 600)
transverse dunes (p. 598)
ventifacts (p. 593)
wind abrasion (p. 593)

QUESTIONS FOR REVIEW

1. Briefly describe three different types of deserts and explain how they form, and why they're so arid.

2. Give examples that show that both chemical and mechanical weathering occur in arid regions.

3. Describe three landforms that are formed by the work of water in arid regions.

4. Briefly describe how a desert pavement forms.

5. Sketch the basic shape of a transverse dune, viewed from the side. What is the essential difference between a barchan dune and a parabolic dune?

6. Describe three ways in which geologists can reconstruct paleowind directions.

7. What are two causes of desertification? How do they produce their effects?

8. Speculate about how changes in the configuration of the Earth's continents through plate-tectonic activity might increase or decrease the total area occupied by arid regions.

9. What would happen to the distribution of deserts if significant global warming takes place?

10. Determine the prevailing wind direction of the landscape in the photograph to the right.

11. What actions would you favor in combating the global trend toward increased desertification? List both the advantages and disadvantages of your proposed actions.

LEARNING ACTIVELY

1. If one is nearby, visit a beach—preferably at the ocean and preferably on a windy day. If you can't get to an ocean, then a large lake (such as one of the Great Lakes) will do (or even, for that matter, a playground sandbox). There, you should be able to observe the effects of wind on loose sand. Take a handful of sand and let the sand fall slowly to the beach surface, building up a small mound. Notice that once the mound has built to a certain height and slope steepness, no additional sand will accumulate at the top of the pile (it'll just slide down the slope). This point is the sand's angle of repose. If it's a windy day, watch what happens to the sand on the mound's surface. It will begin to move up the windward slope relative to the direction in which the wind is blowing, and cascade down the steep leeward slope, or slip face. Here you are seeing how wind erodes, transports, and deposits sand, forming dunes. If you're patient, create a little obstruction on the sand surface (try using a soda can) and watch sand accumulate on the leeward side of the obstruction.

2. If you can't visit a beach, acquire a bagful of sand (perhaps along a riverbank) or from a plant nursery. Let it dry, and then follow the same process of building a sand pile to see the sand's angle of repose. If one is available, use a hair dryer to blow the sand on your pile. Start with the dryer set on "low" and see whether the sand will move—by saltating—up the windward side of your pile. Now try it on "high" and see how the pile changes. Has some of the sand moved up and over the crest of the pile to the leeward side? Here you are seeing how and why sand dunes migrate downwind.

 ## ONLINE STUDY GUIDE

Practice makes you better. Make the time to take the *Online Study Guide* quiz for this chapter. It should take you about 20 minutes. Automatic grading and instant feedback will help you quickly and accurately identify the concepts you have mastered and the areas in which you need more study.
www.prenhall.com/chernicoff.

Shores and Coastal Processes

19

- **Waves, Currents, and Tides**
- **Processes That Shape Coasts**
- ***Highlight 19-1*** Lake Michigan's Vanishing Shoreline
- ***Highlight 19-2*** Can (and Should) We Control Nature's Coasts?
- **Types of Coasts**
- ***Highlight 19-3*** The Future of North Carolina's Coast—Some Good News, Some Bad
- **Plate-Tectonic Settings of Coasts**
- **Sea-Level Fluctuations and Coastal Evolution**
- ***Highlight 19-4*** The Drowning of Tuvalu—A Tiny Country Not Long for This World

North Americans are passionate about their shores and coasts. Indeed, more than half of us live within 80 km (50 mi) of the Atlantic or Pacific Ocean, the Gulf Coast, or near one of the Great Lakes. Our shores and coasts, both scenic and educational, are wonderful places to observe natural processes, particularly the action of waves, tides, and near-shore currents. They are also places that can pose great danger to neighboring communities, as we saw in August 2005 when Hurricane Katrina drowned the Gulf Coast of Mississippi and Lousiana.

Natural processes change all shores *constantly.* Sometimes those processes act rapidly and dramatically: Powerful crashing waves, such as those shown in Figure 19-1, produce immediate changes to a shoreline. On January 2, 1987, for example, waves driven by a powerful winter storm gouged more than 20 m (65 ft) of dunes and beaches from Nauset Beach, at the bend in the "elbow" of Cape Cod along the Atlantic coast of Massachusetts. The beach, a 20-km (12-mi)-long pile of sand that waves and tides had scraped from the sea floor over the past 4000 years, had sheltered the oceanside town of Chatham and its fishing fleets for centuries. On that night, Nauset was breached by 6-m (20-ft)-high waves that also swept away nearly 1 km (0.6 mi) of the cape's offshore islands, eliminating the barrier that had protected Chatham from the ferocity of the Atlantic's waves and storms.

These natural modifications to Nauset Beach paid unexpected dividends to the local economy and environment. Chatham's fishing fleet acquired a shortcut to its Atlantic fishing grounds. Even more importantly, newly increased wave and tidal activity flushed nitrates and phosphates out of septic fields adjacent to the ocean. Those chemicals had contaminated or driven off much of the edible sea life in the area for decades. The influx of seawater through the breach left cleaner water and restored the local aquatic life. Today bay scallops flourish in an environment with lower nitrate levels, and striped bass and bluefish have returned after years of pollution-caused infertility.

Nauset Beach illustrates that coasts are some of the Earth's most dynamic geologic settings (Fig. 19-2). A **coast**—the area of contact between land and sea—is the entire region bordering a body of water. Coasts extend inland until they encounter a different geographical setting, such as a mountain range or a high plateau. A **shoreline** is the precise boundary where a body of water meets the adjacent dry land. A *shore,* the strip of coast closest to a sea or lake, often includes a sandy strip of land, or *beach.*

FIGURE 19-1 Crashing surf along the central Oregon coast, at Cape Kiwanda near Pacific City. ▶

613

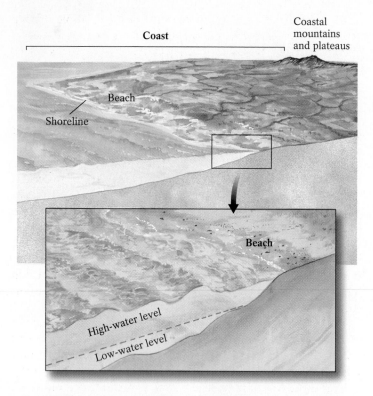

▲ **FIGURE 19-2** A typical coast consists of a shoreline, where the ocean meets the land, and a beach, where water from breaking waves washes a narrow strip of the coast. The area between high and low tides, discussed later in this chapter, is called the foreshore.

In this chapter we consider the processes that shape and change our coasts—the waves, currents, and tides that erode and deposit coastal materials, and the plate movements that raise or lower coasts relative to sea level. We also describe the types of coasts that result from these processes. Finally, we consider how human activity affects the development of coasts, and how people have attempted—most often ineffectively—to control coastal environments.

Waves, Currents, and Tides

Moving water is the major agent of geologic change at the Earth's coasts. Water is set in motion by wind, which produces most waves and surface currents. Water is also moved by the gravitational pull of the Moon and Sun and the rotation of the Earth, which alternately raise and lower water surfaces, producing the Earth's tides.

Waves and Currents

All waves—whether seismic, sound, or water—transport energy. The *ultimate* source of the energy transported by water waves is solar radiation because it produces wind. As we saw in Chapter 18, solar radiation heats the atmosphere more in some regions (such as near the equator) than in other regions (such as near the poles). Variable heating creates zones of low atmospheric pressure (more heated) and high atmospheric pressure (less heated). Air flows from high-pressure to low-pressure zones *in the form of*

wind, dragging across the surface of any water body in its path, reshaping the water surface into waves.

Wind drag raises the surface water of an ocean or lake, forming *crests* and then falling to form *troughs* (Fig. 19-3). A wave's *height* is the vertical distance between its crest and its adjacent trough. In the open ocean, waves commonly reach heights of 2 to 5 m (6.5 to 17 ft); during hurricanes, wave heights may exceed 30 m (100 ft). Wave*length* is the distance between two adjacent waves, measured from crest to crest or trough to trough (Fig. 19-3). Ocean waves are typically 10 to 200 m (33 to 720 ft) apart. The time required for one wavelength to pass a stationary point is called the wave's *period.* Ocean waves commonly have a period of a few seconds. Wave *velocity,* the speed at which an individual wave travels, typically ranges from 30 to 90 km (20 to 60 mi) per hour in mid-ocean.

To understand how waves affect a coastline, we must first understand how waves move. We will then describe the impact that waves have on the coastline and shore.

How Waves Move Blow forcefully across the surface of a bowl of water, and you will create waves on the water's surface. As you gradually reduce the force of your breath, the height of the waves diminishes. In nature, a breeze of 3 km (2 mi) per hour is needed to set a perfectly calm water surface rippling into small waves. Wave heights, lengths, periods, and velocities are determined by the speed and duration of the wind, the constancy of wind direction, and the distance that the wind travels across the water surface.

The area of water over which the wind travels is called its *fetch.* A brisk wind that blows for a long time from the same direction *across a large body of water* produces a series of closely spaced, large, fast-moving waves. Fetch is largest and wave heights greatest when waves can travel long distances without encountering landmasses. Pacific Ocean waves, for example, tend to be very large, in part because this ocean is so huge and its landmasses so widely spaced. Thus, the fetch of its winds is quite large. The band of the global ocean system that lies to the south of the southern tip of South America is virtually uninterrupted by any landmasses. This globe-circling fetch generates some of the world's largest waves, commonly exceeding 15 m (50 ft) in height. The largest wind-generated wave ever observed, sighted in the

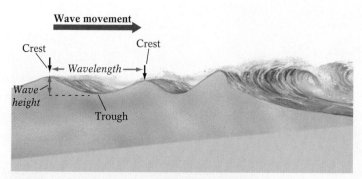

▲ **FIGURE 19-3** The components of a wave. A wave's height is the vertical distance between its crest (the peak of a wave) and the adjacent trough (the low spot between waves). Wavelength is the distance between two adjacent waves, measured from crest to crest or from trough to trough.

northern Pacific in 1933 by the crew of the U.S.S. *Ramapo,* was 34 m (115 ft) high. It had been generated by persistent gale-force winds.

When waves from several storms blowing from different directions and distances converge in one area of an ocean, complex sets of waves of various heights and periods arrive at the shore at irregular intervals. Regardless of their point of origin, waves with similar velocities and periods can become synchronized, reinforcing one another to produce large waves. This process is called *constructive interference.* Waves with different velocities and periods may also become unsynchronized, thereby partly canceling one another and producing small waves. This action is *destructive interference.* Some of the big waves favored by surfers at Malibu Beach, California, may actually originate in arctic storms thousands of kilometers away and then grow even larger thanks to constructive interference by waves from Hawaiian storms. When you watch waves from the ocean shore, look for irregular arrival intervals and sharply variable wave heights; they reveal the cumulative effect of storms blowing throughout the ocean basin, perhaps some raging thousands of kilometers from your beachside viewpoint.

Although mid-ocean waves move outward from a wind source, only the wave *form* moves outward; the actual water *within* the wave essentially remains in place. The water follows a rolling circular path, rising and falling as the wave passes but moving only a short distance from its original position. If you have

ever gone surfing, or watched one of those incredible surfing movies, you've probably noticed how surfers, waiting patiently on their boards, bobbing up and down over small passing waves without changing position, scan the horizon in search of that one perfect wave to ride. This type of wave motion, where the wave form passes by but the water that makes up the wave stays put is known as **oscillatory motion** (Fig. 19-4). Oscillatory motion dies out with depth beneath the surface. Indeed, it is virtually absent below the *wave base,* a depth equal to one-half the wavelength (defined earlier as the distance between two successive crests or troughs). For example, if the wavelength for a series of waves is 150 m (500 ft), the depth to which the waves' circular path would extend would be about 75 m (250 ft). Water below the wave base remains undisturbed by the waves passing above. For this reason, scuba divers in deep water can swim along calmly, even when the water surface churns with the force of powerful storm waves. In deep water, surface waves have no effect on the seafloor; in shallow water, however, a wave base may intersect the seafloor, and the waves do disturb the loose seafloor sediments.

After a journey of perhaps thousands of kilometers across the deep open ocean, waves may begin to cross a continental shelf as they approach a coast. As the shelf becomes shallower landward, the seafloor depth decreases to the wave base, and the seafloor eventually interrupts the water's oscillatory motion.

As waves touch the seafloor, their velocity decreases. Consequently, as they slow, their lengths decrease and their heights

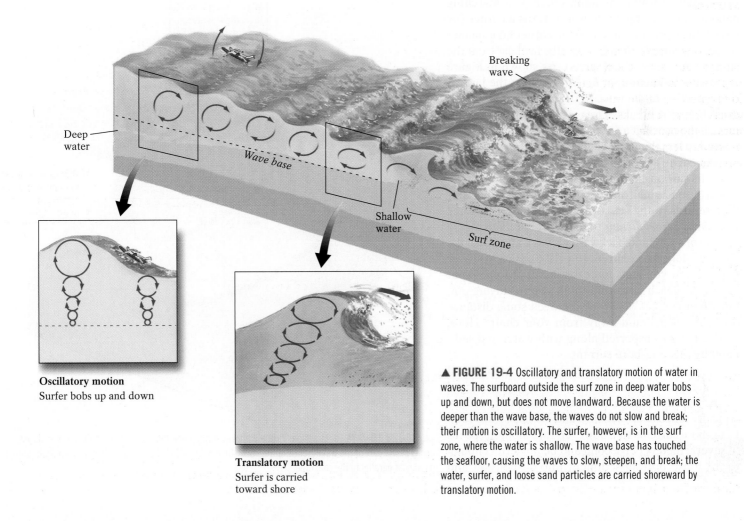

Oscillatory motion
Surfer bobs up and down

Translatory motion
Surfer is carried
toward shore

▲ **FIGURE 19-4** Oscillatory and translatory motion of water in waves. The surfboard outside the surf zone in deep water bobs up and down, but does not move landward. Because the water is deeper than the wave base, the waves do not slow and break; their motion is oscillatory. The surfer, however, is in the surf zone, where the water is shallow. The wave base has touched the seafloor, causing the waves to slow, steepen, and break; the water, surfer, and loose sand particles are carried shoreward by translatory motion.

increase. In other words, the waves bunch up and steepen. As a wave continues into the near-shore shallows, its crest—the top of the wave—begins to move faster than its bottom. Eventually, the crest overruns the rest of the wave, topples over, and *breaks.* The water in a breaking wave then moves landward as *surf,* turbulent foamy water. The surf moves by **translatory motion,** meaning the water itself actually surges forward, carrying along body surfers and churned-up seafloor sediments alike. Figure 19-4 shows these changes in wave motion from oscillatory to translatory motion.

The water that hurtles up the beach after a wave breaks is known as *swash.* The water that returns to the sea is called *backwash.* Both swash and backwash roll and push sand back and forth along the shore, eroding some material from coastal rocks and beaches and bringing some sand inland from points farther out at sea.

Wave Refraction and Coastal Currents

Except for the rare times when winds blow *precisely* perpendicular to the shore, waves generally strike a coast at some angle, depending on the direction from which the wind is blowing. Because a wave almost always approaches the shore at an angle, part of it enters shallower water sooner than the rest. When the first-arriving part touches bottom, it slows down and steepens, and its crest breaks. The rest of the wave, still in deeper water, continues to move at a higher speed, causing the wave to pivot around the slow, shallow-water segment, much as a marching band maintains straight lines as it turns a corner (by having the marchers closest to the corner take smaller slower steps while their bandmates farthest from the corner take larger faster steps). As shown in Figure 19-5, **wave refraction,** or *bending,* causes the last arriving portion of the wave to be *nearly* parallel to the coast before it breaks. Thus, waves that initially approach the beach at a 50° to 60° angle are typically refracted to less than a 5° angle.

As each refracted wave breaks and strikes the coast, its surf pushes the swash ahead of it and up the beach at some small angle to the shoreline. The water then returns to the sea as backwash under the force of gravity, *perpendicular* to the shoreline. This zigzag motion of swash and backwash moves water *down* the coast as a **longshore current** (Fig. 19-6). Have you ever noticed how you may enter the ocean at some point near your beach chair and think you're staying put, but when you get out, you've moved some distance down the beach and away from your chair? Here, you've been transported along with water and sediment by the longshore current.

Geologists and oceanographers have monitored longshore currents using fluorescent dye or colored grains of sand, and have learned that their velocity typically ranges between 0.25 and 1.0 m/s (1 to 3 ft/s). Longshore currents can, however, race along at speeds of several meters per second and be powerful enough to lift loose bottom sediment into suspension and

transport it down the coast for great distances. Mineral studies have shown that sand that originated along the rocky coast of Maine can be found along the Outer Banks of North Carolina, some 1500 km (1000 mi) to the south.

Rip currents, unlike longshore currents that flow parallel to the coast, flow straight out to sea, moving water and sediment *perpendicular* to the shoreline. Figure 19-7 shows the path of a rip current. These currents occur when water that builds up in the surf zone by translatory motion *moves seaward.* These currents commonly interfere with incoming waves, causing them to break before they reach the beach or even preventing them from breaking altogether. Swimmers caught in powerful rip currents can bypass the dangerous, seaward-moving current by swimming *parallel* to the shore for a few tens of meters. Surfers, however, may choose to ride rip currents seaward for a swift, effortless return to the breaker zone.

Tides

Tides are the daily rises and falls of the surfaces of oceans and, to a lesser degree, large lakes. Tides move shorelines alternately onshore (during high tides) and offshore (during low tides). In a general sense, tides are bulges in the surface of a water body that result from the combined effects of the gravitational pull of the Moon and the Sun and from the effect the spinning Earth has on water.

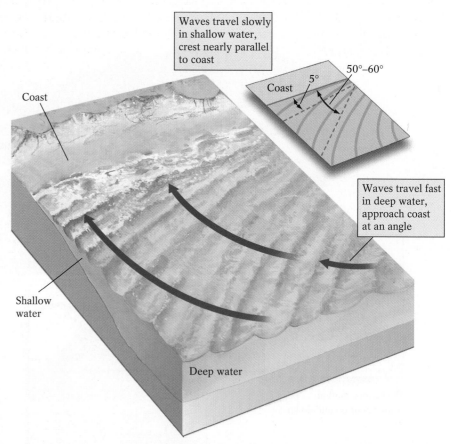

Waves travel slowly in shallow water, crest nearly parallel to coast

Coast

50°–60°

5°

Coast

Waves travel fast in deep water, approach coast at an angle

Shallow water

Deep water

▲ **FIGURE 19-5** Wave refraction. Virtually all waves approach the coast at an angle, but the slower velocity of the part of the wave front that reaches shallow water first bends the incoming wave front until it is almost parallel to the coastline.

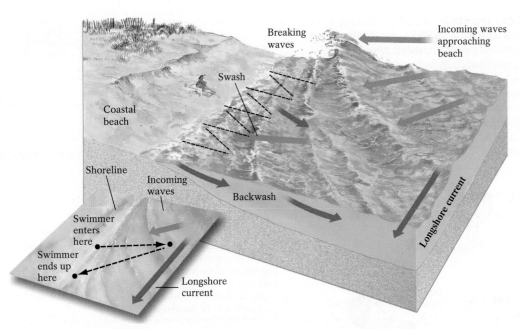

▲ **FIGURE 19-6** The swash and backwash of breaking waves combine to produce a longshore current, which travels parallel to the shoreline and transports sediment along the coast. If you have ever gone swimming in the ocean and been unable to locate the place where you entered the water, you were probably carried down the beach by the longshore current.

All objects on or in the Earth are attracted by gravity to the center of mass of the Earth–Moon system. Everything on or in the Earth is also attracted gravitationally to the Sun, although the Sun is so far away that the attraction is less—about 46% of the Earth–Moon attraction. The Moon is closer to the Earth than the Sun, so objects on or in the Earth feel this attraction more strongly in ways that we can see every day. Tides in the Earth's oceans are the most obvious example.

Tides occur because the spin of the Earth brings some parts of the Earth closer to the Moon than other parts. The regions of the Earth closest to the Moon will feel the gravitational tug from the Moon, and because this tug is not even over the entire Earth, the Earth (both the solid and liquid parts) is slightly distorted. The solid Earth distorts a very small amount—an amount we cannot see with our eyes but can measure with sensitive instruments. The water in the Earth's vast oceans, however, is more easily distorted and thus moves enough for us to observe: this motion of the water is the tide.

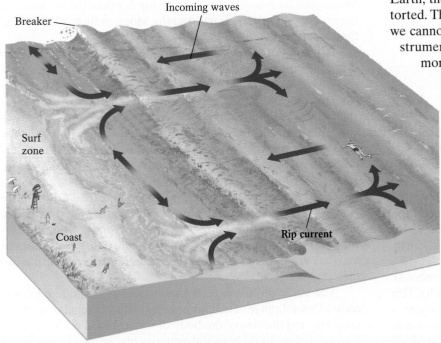

▲ **FIGURE 19-7** A rip current occurs when backwash builds up in the surf zone and then flows seaward through the incoming waves. An ocean swimmer who cannot return to shore because of a strong seaward rip current should swim parallel to the coast for a short distance to escape it.

If the Earth were covered completely with one large ocean of the same depth everywhere, the gravitational attraction of the Moon would create two regions with elevated water surfaces (tidal bulges)—on opposite sides of the Earth (Fig. 19-8a). One bulge would reside on the side of the Earth closest to the Moon at any particular time. The other bulge, on the side of the Earth opposite the Moon, is related to the fact that like the Earth's water, the solid Earth is also being pulled toward the Moon. As the solid Earth is pulled Moonward, it is pulled away from the water on the opposite side of the Earth, which is so far from the Moon's gravitational tug that it does not move as much toward the Moon. Thus a bulge in the water surface forms there too. As the Earth rotates under these bulges, each place on the surface will experience two high tides *daily* (when a given location is experiencing either of the two tidal bulges) and two low tides daily (when that location has rotated 90° from the bulge).

▼ **FIGURE 19-8** Tidal bulges. **(a)** The Earth's high tides occur where the gravitational attraction of the Moon is strongest (the side where the Moon is closest) and where it is weakest (where the Moon is farthest away). As the solid Earth is drawn toward the Moon and away from the water surface, the water surface bulges. **(b)** The Sun also affects the Earth tides. The Earth has its highest tides (called spring tides) twice per month at the New and Full Moon phases when the Moon and the Sun are aligned. The Earth has its lowest tides when the Moon and the Sun are 90° out of phase (at one-quarter and three-quarter Moon phases).

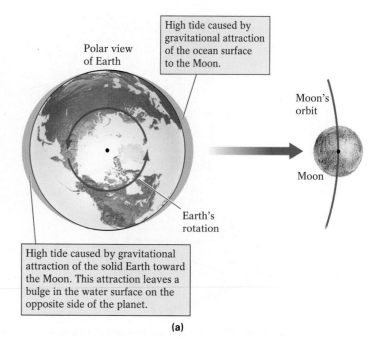

High tide caused by gravitational attraction of the ocean surface to the Moon.

Polar view of Earth

Moon's orbit

Moon

Earth's rotation

High tide caused by gravitational attraction of the solid Earth toward the Moon. This attraction leaves a bulge in the water surface on the opposite side of the planet.

(a)

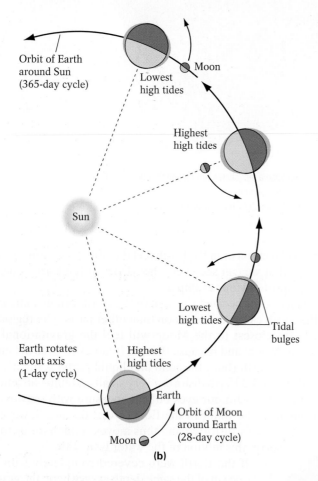

Orbit of Earth around Sun (365-day cycle)

Moon

Lowest high tides

Highest high tides

Sun

Lowest high tides

Tidal bulges

Earth rotates about axis (1-day cycle)

Highest high tides

Earth

Orbit of Moon around Earth (28-day cycle)

Moon

(b)

Suppose you plan a walk along a beach during low tide to see starfish and sea anemones and other assorted marine life, so you consult a tide table—the schedule that predicts the times of high and low tides in a particular place. Tide tables also tell you the phase of the Moon on a given day. The reason? The highest tides of the month, called *spring tides* (from the German *springen*, meaning to "rise," or "bounce up"), typically correspond with a new Moon and a full Moon because at these times the Earth, Moon, and Sun are aligned (Fig. 19-8b). At these times, the gravitational attraction of both the Sun and the Moon are combined, raising the water to its highest tidal level (twice per month). This lining up of the Earth, Moon, and Sun has an impressive name—*syzygy* (a great word for Scrabble players, but try pronouncing it. It's *SIZ-i-jee*). During a full Moon, the Earth is between the Moon and the Sun, and during a new Moon, the Moon is between the Sun and the Earth. Conversely, when the Sun and Moon are 90° out of phase (at the quarter- and three-quarter Moon), their grav-

itational attractions partially counteract one another, resulting in *neap tides,* the lowest tides (lowest high tides and lowest low tides) of the month. (*Neap* is a Middle English word that apparently signifies "without power.")

Of course, the Earth does not have a uniform ocean over its entire surface. You may know from your own experiences that some beaches experience enormous tides, whereas others have only very small changes in the position of the shoreline with time. The timing and magnitude of the tidal bulges as well as the shape of the tidal bulges are affected by the locations and shapes of continents and their coastlines, the varying depth of the oceans, and the motion of both the Earth and the Moon as they orbit around the Sun.

A rising tide, or **flood tide,** elevates the water surface of an ocean, advancing the shoreline landward (Fig. 19-9). On a gently sloping coast, a high tide advances farther inland than on a steeply sloping coast. A rising tide can also migrate up a river. For instance, after each low tide, a tidal bulge moves into New York City's harbor and travels more than 200 km (120 mi) up the Hudson River to the cities of Troy and Albany. Flood-tide currents can reach speeds of 25 km (15 mi) per hour and can even reverse the downstream surface flow of rivers. A flood tide with a rapid and turbulent upstream movement is called a **tidal bore** (Fig. 19-9). A falling tide, or **ebb tide,** lowers the water surface of an ocean, causing the shoreline to move seaward. Ebb-tide currents transport sediment and organic nutrients from coastal marshes and lagoons into the open sea, nourishing marine life offshore.

The vertical difference between local high and low tides is an area's *tidal range.* Tidal ranges vary with the regularity of the coastline and the size of the body of water. Flood tides funneled into narrow enclosed bays and estuaries pile up a large volume of water in a small area. In this situation the tidal range is generally quite large. (As we shall see in Chapter 20, a large tidal range can be put to work, spinning a turbine and generating electrical

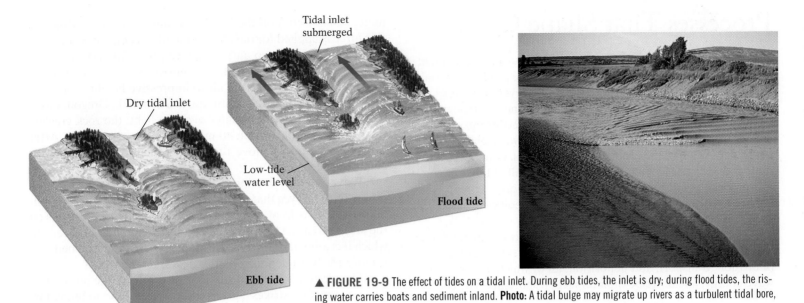

Dry tidal inlet

Tidal inlet submerged

Low-tide water level

Flood tide

Ebb tide

▲ **FIGURE 19-9** The effect of tides on a tidal inlet. During ebb tides, the inlet is dry; during flood tides, the rising water carries boats and sediment inland. **Photo:** A tidal bulge may migrate up rivers as a turbulent tidal bore, such as this one in the River Hebert in Nova Scotia.

power.) The tidal range at Seattle, in the relatively narrow constriction of Puget Sound, is about 3 to 4 m (10 to 13 ft). In the Bay of Fundy, it is even larger, reaching 20 m (75 ft) (Fig. 19-10). In areas where the coastline is relatively straight, the additional water carried by a flood tide is spread over a large area. The resulting tidal range is usually small, as little as 0.5 m (1.5 ft), as is observed in the open Pacific along Hawaiian beaches. Lakes and other inland water bodies, especially large ones such as the Great Lakes, also experience flood and ebb tides although they generally have a very modest tidal range of only 5 cm (2 in) or so.

Tidal currents rarely produce *major* geological effects. They can, however, scour the bottoms of shallow tidal inlets—the channels between coastal islands—and transport fine-grained sediment landward and seaward with the tides.

(a)

(b)

▲ **FIGURE 19-10** The tidal range at the Bay of Fundy, between the Canadian Maritime provinces of New Brunswick and Nova Scotia. **(a)** Eroded columns of rock at high tide. **(b)** The same columns at low tide. The people provide a sense of scale.

Processes That Shape Coasts

The constant battering of waves can erode loose unconsolidated sediment and solid bedrock alike, transport the eroded material, and then deposit it in quieter water. If you've ever been knocked off your feet by pounding surf, you've experienced the power of sea waves. Wherever wave energy is concentrated, erosion of rock and sediment typically takes place. Wherever wave energy dissipates, deposition of the eroded material occurs. Other processes that modify coasts include the daily rise and fall of tides, biological processes that form coastal swamps and organic reefs, the long-term rise and fall of global sea level, and geological processes such as tectonic uplift and subsidence.

The energy transported in waves by sea is the major force that sculpts the Earth's coastlines. Let's turn first to the processes of erosion, transport, and deposition, all of which are driven by wave energy, and describe their impact on shorelines and coasts.

Coastal Erosion

When winds set the sea into motion, the moving water applies stress against its container—that is, the shoreline. Breaking waves can hurl thousands of tons of water against coastal rock and sediment with a force powerful enough to be recorded on nearby seismographs. On average, 14,000 waves strike the exposed rocks and beaches at a given coast every day. As shown in Figure 19-11, waves erode coasts principally by forcing water and air under high pressure into rock crevices. A small, 2 m (6.5 ft)-high wave, for example, exerts nearly 15 metric tons of pressure per square meter (2000 lb/ft^2) on the rock or sediment surface. Although this force is applied for only a fraction of a second, it is repeated every six seconds or so and can prove powerful enough to dislodge large masses of bedrock or sediment. Storm waves can also hurl large rocks through the air to impressive heights. During a monumental storm at the lighthouse at Tillamook, Oregon, waves tossed a 61 kg (130 lb) boulder *over* the light; the rock crashed through the keeper's roof, 40 m (130 ft) above sea level. After waves loosen and remove rocks, they keep hurling the rocks against the coastline.

Wave-induced erosion is driven not only by the sizes and power of the waves that pound our coasts but also by the erodibility of local rocks or sediments. Erodibility, in turn, depends on the strength of the exposed rock or sediment and the extent to which it is jointed or fractured. Softer or more fractured bedrock is more easily eroded than harder, less fractured rocks.

The slope of the local seafloor also affects the amount of wave energy that strikes a given coast. Waves tend to break farther offshore when the seafloor slopes gradually, conditions typically associated with a broad, gently sloping continental shelf. In such locations, wave bases intersect the seafloor well before a wave reaches the shore, and thus waves then break, dissipating much of the wave energy offshore. When the seafloor slope is steep, however, incoming waves may never touch bottom. Instead, they strike the coast head-on with their full force. This situation is particularly common along irregular coastlines that are marked by prominent **headlands**—cliffs that jut out into deep water.

The orientation of a coastline relative to prevailing storm winds influences coastal erosion as well. A coast that is oriented

▲ **FIGURE 19-11** Crashing surf erodes solid bedrock by forcing water and air into crevices in the rock. **Photo:** Waves crashing against bedrock at Schoodic Point, Acadia National Park, Maine.

Breaker in surf zone

Water and air forced into fractured rock

the past decade, largely by storms such as July 1996's Hurricane Bertha and 1999's Hurricanes Floyd and Dennis. A lighthouse on Morris Island near Charleston, South Carolina, which once stood on dry land, is now located about 500 m (1700 ft) offshore and surrounded by water (Fig. 19-13), a product of recent coastal erosion.

Large inland lakes also are susceptible to coastal erosion. In North America, the eastern shores of the Great Lakes, which are often battered by powerful late-autumn storm waves, have suffered from rapid coastal erosion in recent years. Highlight 19-1 focuses on recent erosion problems along Lake Michigan's eastern shore.

Landforms Produced by Coastal Erosion   The prominent headlands that bear the full brunt of the great energy of waves wear down rapidly into a series of distinctive erosional features. Figure 19-14a illustrates this process. Initially, a *wave-cut notch* forms at the base of the cliff. Continued wave action enlarges the notch until it undercuts the cliff, removing the foundation of the overlying rock masses and causing them to fall into the surf. After enough rock has fallen, the remaining cliff base becomes a

▲ FIGURE 19-12 El Niño–induced erosion and subsequent repairs along California's scenic Highway 1, north of Big Sur, winter 1998.

perpendicular to the most frequent wind direction will generally receive a greater share of the force of incoming waves. For example, any segment of the west coast of North America that is oriented north to south lies directly in the path of waves driven by Pacific westerlies. Any segment oriented east to west is more likely to receive Pacific storm waves at less than full force.

Erosion along California's 1100 km (700 mi) of Pacific coastline claims an average of 15 to 75 cm (0.5 to 2.5 ft) of the coast annually. Such erosion often produces substantial damage, like the eroded oceanfront highway shown in Figure 19-12. Erosion is also claiming vast tracts of land along the Gulf Coast of Texas and Louisiana. On the East Coast, the beaches of North and South Carolina have been worn back by as much as 20 m (70 ft) during

▲ FIGURE 19-13 Lighthouse at Morris Island, South Carolina. Coastal beach erosion has left this formerly land-based structure submerged and stranded off the coast.

Highlight 19-1

Lake Michigan's Vanishing Shoreline

Thousands of Michiganders live in towns from Benton Harbor to Ludington on the eastern shore of Lake Michigan, and thousands more from nearby Chicago vacation there. For the past two decades, residents and visitors have watched helplessly as erosion has removed lakefront beaches and caused the lake's eastern bluffs to recede. This beach and cliff erosion is occurring at an unusually rapid pace for two reasons: The area is composed largely of soft, readily eroded glacial deposits, and powerful winter-storm winds push large waves against the lake's eastern shore. Adding to the problem, recent balmy winters have reduced the volume of lake ice that typically protects the bluffs from winter storm waves. Furthermore, several years of unusually high precipitation have produced record-high lake levels, so that water now covers part of the shore. In 1964, when the water level of the Great Lakes was 1.5 to 2.0 m (5 to 7.5 ft) lower than today, wide beaches dissipated much of the wave energy. Today, the wave energy strikes the bluffs directly. The few relatively dry years that have occurred did not lower lake levels enough to reduce the rate of erosion. Consequently, the bluffs continue to recede about 0.4 m (1.5 ft) annually. It would take approximately five years of drought to restore the lakes to their pre-1964 levels.

Human activity has also sped up erosion of the lake's beaches and cliffs. Overdevelopment—building too many homes along the lakeshore—has removed much of the pro-tective vegetation from bluff tops and thinned the vegetation that stabilizes the sands of energy-absorbing dunes. Meanwhile, disposal of wastewater and sewage in septic fields and on bluff-top farmlands has saturated cliff slopes, loosening their sediments and promoting mass movement.

What does the future hold for Lake Michigan's scenic bluffs? The warmer, wetter climate of the past two decades may continue. In fact, the drier period of the past century may have been an anomaly, in which case the lake appears to be returning to its normal, long-term level. If so, accelerated bluff erosion and shore loss may pose an ongoing challenge to those who choose to live along Lake Michigan's eastern shore.

▲ Waves generated by westerly winds easily erode the loose glacial sediments of this coastal bluff at St. Joseph, Michigan, on the eastern shore of Lake Michigan.

---

platform called a **wave-cut bench.** Meanwhile, the debris-laden surf further abrades the cliff face and enlarges the bench. Eventually, the cliff retreats so far inland that it no longer protrudes out into the breaker zone. The wide bench may initially protect the cliff from further erosion, except during large storms.

Over time, the bench may actually increase headland erosion by refracting incoming waves against the headland's flanks. Here's how: As waves approach a wave-cut bench, the wave front encounters the shallow bottom and breaks over the bench. The segments of the wave front in the adjacent deeper-water bays pass the headland without touching bottom. Because they can travel more rapidly, they refract toward the flanks of the exposed headland and concentrate their energy there.

At first, battering waves erode the cliff rock to form **sea caves** (Fig. 19-14b), typically attacking weaker rocks, such as those composed of softer materials or those containing more fractures, faults, and joints. Further wave action may excavate caves on both sides of the headland until the caves join, forming a **sea arch.** Continued erosion by waves can wear away the supporting foundation of the arch until it finally collapses. Only an isolated remnant of the original headland, a **sea stack,** may then remain.

Protecting Against Coastal Erosion Human efforts to minimize the effects of coastal erosion usually involve building structures that deflect waves so they do not strike the coast with full force. In some spots along the southern shore of New York's Long Island, an unsightly mix of driftwood, old tires, rusted cars, discarded refrigerators, and even Christmas trees has been piled up to fortify the area's natural dunes.

Sand dunes can provide a more attractive first line of defense against the encroaching sea by absorbing the energy of waves, but they tend to migrate with the wind unless they are stabilized

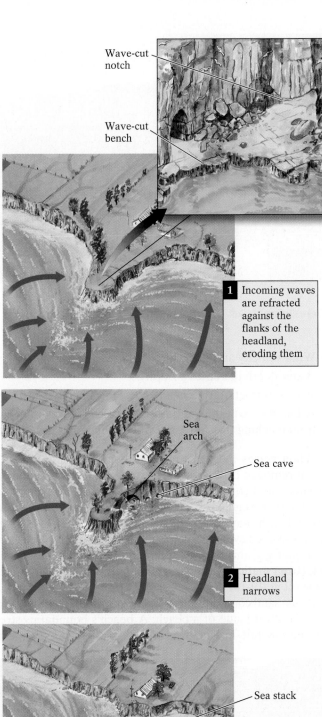

Wave-cut notch

Wave-cut bench

1 Incoming waves are refracted against the flanks of the headland, eroding them

Sea arch

Sea cave

2 Headland narrows

Sea stack

3 Much of headland has been removed

(a)

(b)

▲ **FIGURE 19-14** The evolution of erosional coastal landforms. **(a)** Wave fronts are refracted against the flanks of headlands, forming wave-cut notches and benches. As the headland narrows, the notches erode to form sea caves. Two sea caves, eroding at both sides of a headland, form a sea arch. Finally, the arch collapses, isolating a sea stack. **(b)** Sea caves on Cape Kildare, Prince Edward Island, Canada.

in some way. On the Outer Banks of North Carolina, residents have planted hardy vegetation whose roots can hold soil against wind erosion, and thus sand has been added to create a higher and continuous dune line.

Larger protective structures reduce coastal erosion by absorbing the brunt of wave energy. **Riprap** is a heap of large, angular to subangular boulders piled along the shoreline (Fig. 19-15a). Although breaking waves remove the riprap eventually, they take several years or even decades to erode it, and riprap is relatively easy and inexpensive to rebuild.

Seawalls are sturdy, longer-lasting structures, generally built parallel to and attached to the shore (Fig. 19-15b). They are designed to withstand the pounding of the highest recorded waves in an area. Large, solid seawalls repel waves seaward, thereby diverting some of their energy.

Seawalls are expensive to build and maintain, however, and they often cause other erosional problems. A $12 million, 6.4-m (20-ft)-high seawall was completed in 1905 at Galveston, Texas, after the hurricane of 1900, the worst natural disaster in North American history (the storm killed 6000 people and destroyed two-thirds of Galveston's buildings). The wave energy diverted by the seawall quickly eroded the sandy beaches directly in front of the smooth-faced wall. Without the beach, the waves struck the wall directly, breaching it in several places. Although the wall remains in service, its damaged portions must be rebuilt periodically to maintain its effectiveness.

Perhaps the greatest problem with seawalls is their deflection of wave energy to other locations. Although they may protect a specific site from wave erosion, these walls accelerate

(a)

(b)

▲ **FIGURE 19-15** **(a)** Riprap at Carlsbad, California, about 30 km (20 mi) north of San Diego. **(b)** Seawall along the Gulf Coast of Louisiana.

erosion where the wave energy has been refocused. They also interrupt the longshore current, thereby disrupting the natural movement of sediment that nourishes coastal beaches. This action thus accelerates coastal erosion in areas down the coast from the seawall. In light of these problems, some states (such as North and South Carolina) have prohibited seawall construction and other "hard" stabilization solutions to coastal erosion. We discuss "soft" stabilization solutions, such as beach nourishment, later in this chapter.

Coastal Transport and Deposition

Deposition occurs in coastal settings for much the same reasons as it does where wind, rivers, and glaciers are at work: The supply of sediment exceeds the ability of a current to transport it. When the energy needed to transport sediment is greater than the energy of the breaking waves, waves can no longer carry the sediment and deposition takes place. Wave energy may be lost or interrupted for several reasons:

- Wind velocity and wave force vary seasonally.
- Water depth increases abruptly.
- Waves, refracted from headlands, are channeled into bays.
- A barrier—whether natural or artificial—prevents waves from reaching the shore, interrupting the longshore current.

When waves deliver more sediment to a shore than the amount removed by near-shore currents, the excess sediment is deposited, most commonly as a beach. A **beach** is a dynamic, relatively narrow segment of a coast that is washed by waves and tides. A beach may consist of only sand, or it may also contain coarse gravel or even cobbles. Sand beaches are typically white (from quartz grains or shell fragments), but they may even be black (from mafic volcanic fragments) or green (from grains of olivine) (Fig. 19-16).

Beach sand commonly accumulates as windblown sand *dunes* on the landward side of a beach. A beach's boundaries stretch from the low-tide line landward, ending where the topography

(a)

(b)

▲ **FIGURE 19-16** **(a)** Black Sand Beach, Kalapana, Hawai'i. This beach is made of sand-size grains of basalt. **(b)** Green Sand Beach, Ka Lae, Hawai'i with inset of green olivine sand.

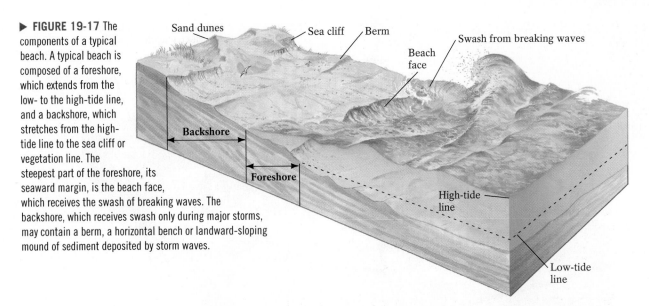

► **FIGURE 19-17** The components of a typical beach. A typical beach is composed of a foreshore, which extends from the low- to the high-tide line, and a backshore, which stretches from the high-tide line to the sea cliff or vegetation line. The steepest part of the foreshore, its seaward margin, is the beach face, which receives the swash of breaking waves. The backshore, which receives swash only during major storms, may contain a berm, a horizontal bench or landward-sloping mound of sediment deposited by storm waves.

changes—for example, at a sea cliff or sand-dune field, or at the point where permanent vegetation begins. A typical beach (Fig. 19-17) is composed of a **foreshore,** which extends from the low- to the high-tide line, and a **backshore,** which stretches from the high-tide line to the sea cliff or vegetation line. A beach's profile is made up of several components with variable slopes. The steepest part of the foreshore is its seaward margin, called the **beach face,** which receives the swash of breaking waves. The backshore, which receives swash only during major storms, may contain a **berm,** a horizontal bench or landward-sloping mound of sediment deposited by storm waves. Some beaches contain several parallel berms from different storms. Others show no berms, depending on the extent of recent storm activity.

More than 90% of beach sediment originates from inland and upland sources and is delivered to the coast by coastal streams. Other sediment comes from erosion of headlands and beach cliffs. A smaller amount is pushed onland from the offshore surf zone. After sediment arrives at the shoreline, it begins its longshore journey as soon as it encounters its first breaking wave. If human activity doesn't interfere with the process, a beach is maintained naturally when the amount of sediment received by the shore balances the amount exported by longshore currents. Typically, the volume of sediment transported offshore by rip currents and storm waves or to backshore dunes by sea breezes is relatively small compared with the volume of sediment swept along by the longshore current.

We now explain the role of sediment transport and deposition in changing shorelines and coasts. We also examine how human actions affect the destruction and restoration of coastlines.

Longshore Transport   The energy for transport of coastal sediment comes primarily from waves and longshore currents. As we have seen, when waves break at some oblique angle to the shore, they create a longshore current that flows parallel to the coast, transporting sediment in the same general direction as the wind and waves. The current moves sediment either on the beach face, as **beach drift,** or immediately offshore within the surf zone, as **longshore drift** (Fig. 19-18). In either case, waves lift the

grains and move them by swash at an oblique angle to the shoreline. Backwash then returns the grains seaward, perpendicular to the slope of the beach face. In a single storm, a grain on the beach face can travel down-current as far as 1000 m (3300 ft). Longshore drift can transport enormous quantities of sand. For example, 750,000 tons of sand, some eroded from granitic cliffs on the Maine coast, pass by Sandy Hook, New Jersey, each year.

Landforms Produced by Coastal Deposition   When a longshore current suddenly encounters deeper water, such as where the entrance to a bay interrupts a shoreline, the current is interrupted as well. The incoming waves no longer touch bottom in the bay (whose depth may exceed the wave base). Consequently, the incoming waves do not break, and thus they fail to produce the swash and backwash that drive the current. The entire sediment load is deposited there as a **spit,** a fingerlike ridge of sand that projects from the coast into the open water of the bay entrance (Fig. 19-19). With the deposition of more sediment, the spit can grow by tens of meters per year, with the actual growth rate depending on the supply of sand and the intensity of wave energy. Where particularly strong waves or currents sweep into a bay, a growing spit may become curved, forming a **hook.** A spit may ultimately become a **baymouth bar** if it extends completely across a bay entrance.

Coastal deposition also occurs when the waves that drive the longshore current find their path to the shore interrupted, perhaps by the shallow continental shelf or another natural offshore landform, or perhaps by a human-made structure. Longshore currents slow down, for example, where the continental shelf has a gentle slope and waves lose much of their energy by breaking farther offshore. In such a case, the waves do not remove the sediment loads delivered by streams to the beaches, and the beaches tend to widen.

One natural structure that intercepts incoming waves, thereby interrupting longshore currents, is a sea stack. Waves typically break on the seaward face of the stack, leaving the water calm on its landward side. Deposition of wave-borne sand takes place in the wave-free zone on the landward side of the stack because no wave energy carries it farther down the coast. Instead, the

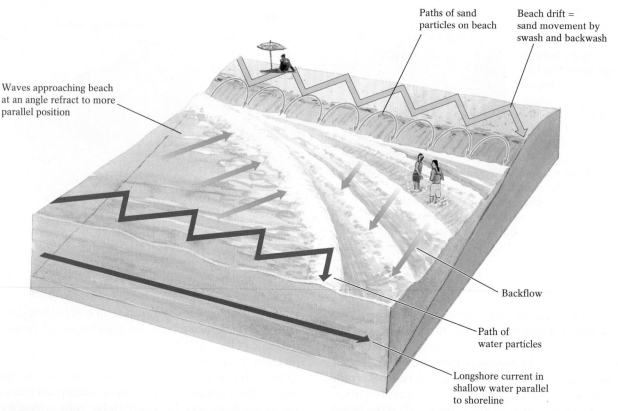

Paths of sand particles on beach

Beach drift = sand movement by swash and backwash

Waves approaching beach at an angle refract to more parallel position

Backflow

Path of water particles

Longshore current in shallow water parallel to shoreline

▲ **FIGURE 19-18** Beach drift and longshore drift transport sediment parallel to the coastline. Beach drift moves sand along the beach; longshift drift moves sand within the surf zone. The zigzagging motion is produced by swash (which moves water and sand landward at an oblique angle to the coast) and backwash (which moves water and sand seaward perpendicular to the coast).

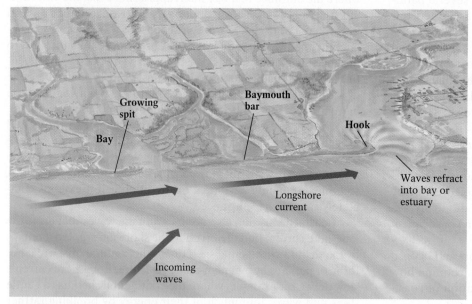

Growing spit

Baymouth bar

Hook

Bay

Longshore current

Waves refract into bay or estuary

Incoming waves

(a)

(b)

▲ **FIGURE 19-19** Deposition of spits, hooks, and baymouth bars. These features typically develop when the current deposits its sediment load where water velocity is reduced, such as at the mouth of a bay. **(a)** A spit at Cape Henlopen, Delaware. **(b)** Baymouth bars on the south shore of Martha's Vineyard, Massachusetts.

▲ **FIGURE 19-20** Deposition of a tombolo landward of a sea stack. The sea stack absorbs the wave energy, interrupting the longshore current and causing it to drop its load of sand. **Photo:** Deposition of tombolo in the lee of a large sea stack, Big Sur, California.

sediment accumulates as a *tombolo,* a sandy landform that grows from and connects the mainland to an offshore sea stack (Fig. 19-20).

Human-Induced Coastal Deposition
Some human-made structures disrupt the natural balance between the amount of sediment delivered to the shore and the amount removed by the longshore current. These structures may cause beaches to grow in some places and to shrink—sometimes drastically—in others.

Communities may build **breakwaters,** walls designed to intercept incoming waves, to create quiet, wave-free zones that minimize coastal erosion and protect boats in harbors. These structures are typically oriented parallel to the coasts. A breakwater was constructed offshore at Santa Barbara, California, about 75 years ago to protect private boats from being battered by Pacific waves. Predictably, because the structure interrupted the longshore current, sediment was deposited seaward of the beach (Fig. 19-21). After 30 years, the beach had widened by

(a) (b)

▲ **FIGURE 19-21 (a)** Santa Barbara Harbor in 1931. **(b)** Santa Barbara Harbor in 1977. The breakwater at Santa Barbara was built in the 1920s to provide safe harborage for pleasure boats. It interrupted the longshore current, causing sand to be deposited behind it and on the seaward edge of the beach. Note the bulge in the beach shape. This deposition extended the beach but closed off the passage from the marina to the sea. In addition, wave action immediately down the coast from the breakwater caused beach erosion there. To solve the dual problems of excessive deposition behind the breakwater and beach erosion down the coast, sand is regularly dredged from behind the breakwater and pumped down-current to the depleted beach area, a costly substitution of human energy and money for the interrupted wave energy.

hundreds of meters, clogging the harborage, requiring dredging operations to keep the harbor open.

Groins are shore-protection structures that jut out perpendicular to the shoreline (Fig. 19-22a). They are designed to interrupt longshore drift and trap sand, thus restoring an eroding beach. Most groins are built where a wide sandy beach, such as Miami Beach, is vital to a community's economic life. They are, however, a mixed blessing. Groins rob sand from one place in favor of another; once one is built, others typically begin to appear all along a coastline to intercept migrating sand.

Jetties are structures, typically built in pairs, that extend the banks of a stream channel or tidal outlet beyond the coastline (Fig. 19-22b). They direct and confine channel flow to keep the channel open and sediment moving, preventing it from filling the channel. Like groins, jetties broaden *up-current beaches* by trapping sand there. Also like groins, they remove sand from the longshore transport system, eroding and narrowing *down-current beaches.*

Beach Nourishment—Attempting to Restore Shrinking Beaches

Humans may cause beaches to shrink either when our breakwaters, groins, and jetties accelerate erosion or when our inland dams on rivers intercept beach-bound sediment. Flood-control dams built in the last 35 to 40 years in the Missouri–

Mississippi River system, for example, have trapped half of the sediment bound for the Gulf of Mexico. In the process, they have limited the natural replenishment of the sediment removed by longshore currents from the Louisiana coast and delivered to Gulf Coast islands off of Texas. Similar beach losses occur along the California coast wherever dams intercept and retain the sediments of ocean-bound rivers.

To compensate for the sediment loss at a shrinking beach, we can import sand, but usually at very high costs. The sand for such *beach nourishment*—a "soft" stabilization approach (avoiding the construction of "hard" structures, such as breakwaters)—is typically dredged from nearby lagoons or transferred from inland sand dunes and offshore sandbars. In one successful beach-nourishment project along Mississippi's Gulf Coast at Biloxi and Gulfport, a mixture of sand and mud was pumped from offshore sites to create the world's longest artificial beach, stretching 30 km (20 mi). The gulf's gentle waves remove the fine mud particles, leaving behind a sandy beach that had transformed the area into a popular vacation spot until the devastation wrought by Hurricane Katrina in 2005 eliminated much of the project's sandy beach.

Many beachfront resort communities find, however, that the costs of beach nourishment far outweigh the benefits. For example, a beach-nourishment program in Miami Beach, Florida,

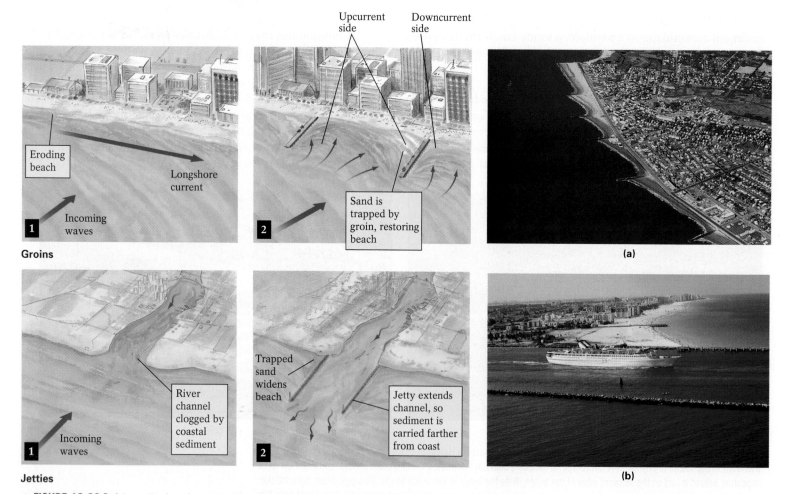

▲ **FIGURE 19-22** By interrupting longshore currents, groins and jetties cause longshore drift to accumulate on their up-current side, broadening beaches there. **(a)** Groins off Cape May, New Jersey. By looking at the sediment accumulating adjacent to the groins, in which direction is the longshore current flowing? **(b)** A cruise ship passes through a jetty as it leaves Miami Beach, Florida.

The Work of Waves and Tides

Erosion of Cliffed Headlands

- Wave-cut benches
- Sea caves
- Sea arches
- Sea stacks

Deposition

- Beaches
- Berms
- Spits and hooks
- Baymouth bars
- Tombolos

halted decades of beach erosion and widened the beach to 100 m (330 ft). The sand, which was imported from the Everglades at a cost of $64 million over a 10-year period, consisted of quartz and fine clay particles rather than the coarse shell fragments of Miami's natural beach. When wave erosion carried off the clay, the offshore water became quite cloudy, damaging coral reefs (which flourish only in clear, sunlit water).

In the hopes of economic gain, communities continue to pursue beach nourishment, even in the face of past failures. Consider the current beach-nourishment project under way at Sea Bright, New Jersey. During the late 19th and early 20th centuries, historic Sea Bright was a thriving coastal resort, drawing thousands of bathers to its beautiful beaches from the nearby sweltering streets of New York and Philadelphia. In the early 1900s, however, the state constructed a sturdy seawall that stood 5 m high (18 ft) and stretched for 10 km (6 mi) to protect the local railroad that shuttled bathers back and forth between Sea Bright and Sandy Hook. Deflected storm-wave energy off this gray, fortresslike barrier quickly eliminated Sea Bright's wonderful beach, starving the resort economy of the town. Decades later, the largest beach-nourishment project in history is under way to reverse this damage. Over the next 40 years, the government will spend more than $1.8 billion to dredge sand from offshore and rebuild Sea Bright's beach. The first section—a 100-m (300-ft)-wide swath of sand—has cost $226 million, with 70% coming from federal tax dollars. A spokesperson for the Corps of Engineers boasted, "This year we had traffic jams on the Garden State Parkway leading to New Jersey beaches. In 1994, traffic wasn't a problem; we didn't have a beach here then."

This optimism notwithstanding, beach nourishment remains a difficult and costly strategy for maintaining beaches. Although it may be less damaging than the variety of "hard" solutions, this tactic still rarely succeeds in the long run. It sometimes appears that virtually every time we try to control natural coastal processes, we create a host of new problems. Highlight 19-2 illustrates this paradox.

Highlight 19-2

Can (and Should) We Control Nature's Coasts?

A powerful hurricane hits—as when Hurricanes Frances, Charley, and Jeanne struck in rapid succession in September 2004—and up and down the East Coast, heavily damaged coastal communities beg state and federal governments for protection and relief. Each bout of storm-driven destruction adds more fuel to a raging debate: Can humans control the winds, waves, and shifting sands that alternately sustain and sap the financial life of coastal communities? *Can* we engineer permanent structures that will forever protect our beaches from the ravages of nature? If we can, *should* we? And finally, who should foot the enormous bills for such projects?

No place more clearly dramatizes this passionate debate than North Carolina's Bodie Island, a segment of the Outer Banks. There, the sands at Oregon Inlet, one of the hundreds of breaks in the chain of coastal islands that line the East Coast from Maine to Florida, are shifting southward. The movement threatens to close the outlet between Albemarle and Pamlico Sounds and the Atlantic Ocean.

During the past six years, the mouth of the inlet has marched progressively southward, carried by the waves of the region's longshore current. To keep the 150-year-old inlet open and halt the natural down-current movement of sand, the Army Corps of Engineers has proposed construction of two jetties that would project out to sea, block the sand's southward migration, and thus maintain the inlet at its present position.

For some in the region, the decision to proceed with the project and protect their investment in tourism along the Outer Banks is a classic no-brainer. "This is our economy," said one local official. "In the post-industrial world, where will the jobs be? It's a leisure world now." For others, the Oregon Inlet project is just another expensive and illogical attempt to control "the uncontrollable." Duke University geoscience professor Orrin Pilkey, the region's expert on coastal-sediment transport, is one vocal critic of the project. He believes, as do most geologists, that structures that interrupt the region's natural movement of sand, simply rob one beach to save another *temporarily*. Pilkey points out that the project will undoubtedly halt sand movement, causing it to accumulate north of the jetties. At the same time, the project will definitely accelerate erosion of sand-deprived beaches located farther south. How will we preserve those beaches if we cut off their replenishing supply of sand?

Instead of spending huge sums of federal money on projects that are destined to create a domino effect of problems, Pilkey and other scientists have proposed an inexpensive but, for some, a highly radical and unacceptable solution: "strategic retreat" from the coastline by human development. Pilkey believes that implementation of these expensive yet questionable shoreline-protection projects encourages and promotes construction of even more homes, hotels, and other businesses that stand directly in harm's way. If global sea level continues to rise, as it is expected to do, future storms will take more lives and generate enormous economic hardship. Pilkey contends that hurricanes and other large storms are the natural—even beneficial—way that nature maintains its dynamic coasts. These storms wash over the islands, nourishing vegetation, building dunes, and periodically cleaning accumulated sediment (and human pollution) from inland sounds and waterways. Such storms are "only a natural disaster when there's a bunch of houses," says Pilkey, a view that apparently has not fully registered with developers and governmental agencies.

Of course, strategic retreat is unlikely in densely populated, highly developed areas such as Atlantic City, New Jersey, and Miami Beach, Florida. Such a strategy would undoubtedly prove painful and disruptive to communities that might be forced to move. Yet, as global warming raises the levels of the Earth's seas, coastal communities will face increasing threats from storm-driven erosion. In the long run, no matter what the costs, the sea will force us to move out of its way.

Before we move on to look at the various types of coasts, take a moment to review the different landforms produced by coastal erosion and deposition illustrated in the Geology at a Glance: "The Work of Water and Tides" on page 629.

Types of Coasts

Every coast is shaped by a combination of erosional and depositional processes that mark its geological setting. To see how coasts vary, look at Figure 19-23. Notice how dramatically the coastline at Crescent Beach, Oregon differs from the coastline of Cape Hatteras, North Carolina. So far, we have considered such processes as wave-driven longshore and rip currents, rising and falling tides, and sediment delivery by coastal streams. A host of other erosional and depositional processes also produce and modify coasts.

A **primary coast** is shaped principally by nonmarine processes, such as those illustrated in Figure 19-24. Glacial erosion, for example, produced such *primary* coasts as the fjords of Alaska and British Columbia. Glacial deposition of terminal and recessional moraines shaped the northeast Atlantic from Cape Cod, Massachusetts, to the southwestern edge of Long Island at Brooklyn, New York. Stream deposition has created primary coasts such as the Gulf Coast of Louisiana, where stream deltas extend into the marine environment. Long-term rising sea levels produce bays and estuaries by flooding river systems; the postglacial sea-level rise that began about 13,000 years ago created the primary coasts of Chesapeake Bay and Delaware Bay along the mid-Atlantic coast of North America. Furthermore, the interaction of biological processes with geological processes may shape primary coasts, such as carbonate-reef coasts. Volcanic activity can also produce a primary coast wherever new lava flows enter the ocean and create a new coastline. The lavas of the Big Island of Hawai'i and the island of Iceland are good examples of this type of primary coast.

A **secondary coast** is shaped predominantly by ongoing marine erosion or deposition, such as the processes described in the previous section. Secondary *erosional* coasts contain cliffed headlands, wave-cut terraces, and an assortment of sea caves, stacks, and arches. Secondary *depositional* coasts include beaches, spits, hooks, and tombolos. A secondary coast can be both erosional *and* depositional, as sometimes occurs with the formation of offshore barrier islands. As we shall see in the following discussion, this type of coast is not easily classified as either primary or secondary.

Barrier Islands

Barrier islands are long, narrow continuous ridges of sand, located parallel to a main coast but separated from it by a bay or lagoon. Barrier islands can rise as much as 6 m (20 ft) above sea level and are 10 to 100 km (6 to 60 mi) long and 1 to 5 km (0.6 to 3 mi) wide. Most such islands are made up of a relatively narrow beach, perhaps 50 m (165 yds) wide, and a broader zone of inland dunes that accounts for most of the island.

Barrier islands are the most common North American coastal feature (Fig. 19-25). They line the East Coast for 1300 km (850 mi), stretching from eastern Long Island, New York, south to Florida. They then continue for another 1300 km along the Gulf Coast to eastern Texas. Only occasional tidal inlets or the flow of major streams (such as the Hudson River at New York Harbor or the Delaware River at Delaware Bay) interrupt this string of 295 islands. Padre Island, Texas, North America's longest beach, is a classic barrier island.

How did these widespread islands form? Are they secondary coasts consisting of mostly depositional landforms? Are they secondary coasts consisting of combined erosional–depositional landforms? Or are they primary coasts that formed during the last period of worldwide glaciation?

Supporters of the depositional model think of barrier islands as elongated spits, projecting from bends in the coastline. This

(a)

(b)

▲ **FIGURE 19-23** Crescent Beach, Ecola State Park, Oregon **(a)** is an erosional coast marked by scores of sea stacks and arches and a rugged, bouldery shoreline. The processes that shaped it differ from those active at Cape Hatteras, North Carolina **(b)** which is a depositional coast characterized by sandy beaches, dunes, and barrier islands. The endangered Cape Hatteras lighthouse has been moved inland since this photo was taken.

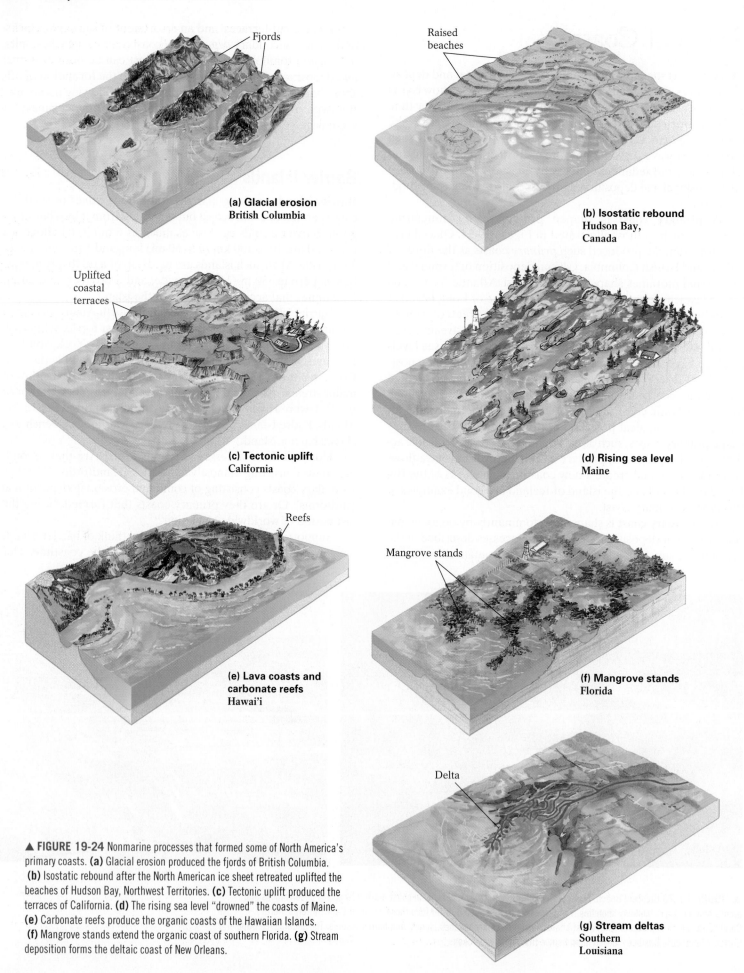

Fjords

(a) Glacial erosion
British Columbia

Raised beaches

(b) Isostatic rebound
Hudson Bay,
Canada

Uplifted coastal terraces

(c) Tectonic uplift
California

(d) Rising sea level
Maine

Reefs

(e) Lava coasts and carbonate reefs
Hawai'i

Mangrove stands

(f) Mangrove stands
Florida

Delta

(g) Stream deltas
Southern
Louisiana

▲ **FIGURE 19-24** Nonmarine processes that formed some of North America's primary coasts. **(a)** Glacial erosion produced the fjords of British Columbia. **(b)** Isostatic rebound after the North American ice sheet retreated uplifted the beaches of Hudson Bay, Northwest Territories. **(c)** Tectonic uplift produced the terraces of California. **(d)** The rising sea level "drowned" the coasts of Maine. **(e)** Carbonate reefs produce the organic coasts of the Hawaiian Islands. **(f)** Mangrove stands extend the organic coast of southern Florida. **(g)** Stream deposition forms the deltaic coast of New Orleans.

▲ **FIGURE 19-25** Barrier Island with inland lagoon north of Cape Hatteras (Pea Island, North Carolina).

ing extraordinary storms. The moving island does not usually overrun the mainland coast because the shoreline behind the island also retreats as sea level rises. As the two landmasses retreat together *in step*, the narrow barrier island remains an island, separated from the mainland by a lagoon. And to the relief of bathers, vacationers, and sun worshippers, the process also preserves these islands' recreational beaches. As the islands retreat under the onslaught of the rising sea, their beaches retreat as well, maintaining the same width while the island itself narrows.

Human Impact on Barrier Islands North America's coast-hugging chain of barrier islands has enormous value as the first line of defense against storm waves in the hurricane-prone Southeast and Gulf Coast. These islands are also popular with city dwellers as prime destinations for vacations and weekend getaways. Many barrier islands have been set aside for recreation and as wildlife preserves.

model proposes that these spits have been breached by tidal currents or powerful storm waves to form the now-separate islands. In contrast, the erosional–depositional model proposes that waves breaking offshore above the broad, gently sloping continental shelf move sand from the bottom and deposit it as long sandbars that continue to grow until they rise above sea level. Storm waves then breach these elongated bars to form the islands. According to this model, the islands trend parallel to the coast because the refracted waves that lift and deposit sand to form them follow the same direction.

Some geologists firmly believe that most barrier islands originated as primary coasts, and then evolved later as secondary coasts. This model views the islands as remnants of a sand-dune system that bordered the continent during the last period of worldwide glacial expansion. Because sea level at that time was as much as 140 m (460 ft) lower than the level today, coastal dunes accumulated some distance seaward of the present coastline. As the glaciers melted, the rising sea surrounded and isolated the dunes as islands.

Whatever the origin of barrier islands, once they began to form, erosional and depositional processes combined to shape them. Since the 1800s, most of these strips of sand have been narrowing, reflecting the global rise in sea level. (Because their seaward slope is so gentle, even a small sea-level rise strongly affects the width of barrier islands.) This thinning, however, may actually assist the islands' long-term prospects for survival.

According to the prevailing geological wisdom, as a barrier island thins, it reaches a critical width that allows storm waves to wash completely over the island (Fig. 19-26). At this point, the island's narrowing ceases—the transfer of sand from the seaward side to the landward side maintains the island's width (Fig. 19-27). By this process, the island simply migrates landward roughly on pace with the rising sea. The landward migration rate for barrier islands along the Atlantic Coast ranges from 0.5 to 2 m (1.5 to 6.5 ft) per year, although it increases considerably dur-

▲ **FIGURE 19-26** A new inlet was created by hurricane waves that struck the North Carolina coast in 1999 at Portsmouth Island.

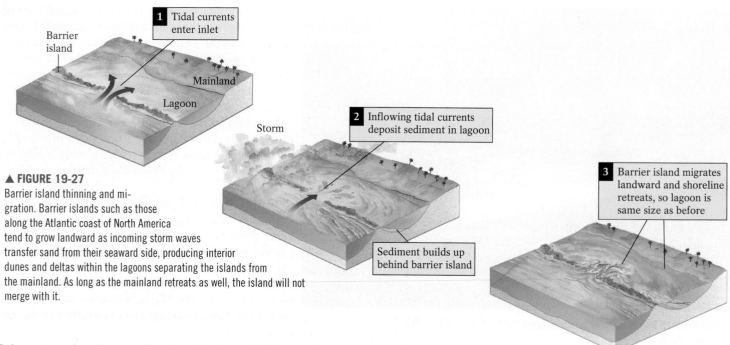

1 Tidal currents enter inlet

Barrier island

Mainland

Lagoon

Storm

2 Inflowing tidal currents deposit sediment in lagoon

Sediment builds up behind barrier island

3 Barrier island migrates landward and shoreline retreats, so lagoon is same size as before

▲ **FIGURE 19-27**
Barrier island thinning and migration. Barrier islands such as those along the Atlantic coast of North America tend to grow landward as incoming storm waves transfer sand from their seaward side, producing interior dunes and deltas within the lagoons separating the islands from the mainland. As long as the mainland retreats as well, the island will not merge with it.

Others are privately owned and undeveloped. More than 25% now serve as vacation resorts, overbuilt with condominia, hotels, and casinos. Rehoboth Beach, Delaware; Ocean City, Maryland; Virginia Beach, Virginia; Hilton Head, South Carolina; Miami Beach, Florida; and Galveston, Texas, are but a few of the well-known barrier-island communities developed as commercial properties in this century. Miami Beach, which had a population of 644 permanent residents in 1920, had a population of 87,933 in 2004, along with 187 hotels, all crammed into an 18.4 km^2 (7.1 mi^2) sliver of sand.

Developers at Ocean City, Maryland, have constructed so many luxury hotels and condominium apartment buildings that this recently wild, natural seashore is now valued at more than $500 million per mile. Unfortunately, their ill-conceived alterations to the beach are also jeopardizing their investment. To give every room an ocean view, developers leveled many of the island's protective dunes, causing massive erosion. A costly beach-nourishment program, which may never restore the beaches to their predisturbance dimensions, is now under way.

Only 3 km (2 mi) south of Ocean City, Assateague Island—home to herds of rare wild ponies—has been set aside as a national seashore and wildlife habitat. Because the law protects the island's frontal-dune line, Assateague Island's beaches have suffered relatively little erosion. Recent construction of jetties to the north (to widen the beaches at Ocean City), however, is now beginning to intercept the sand destined for Assateague's beaches. Consequently, the northern portion of the island is thinning.

The protective role played by barrier islands, the processes that shape them, and the costs of human interference with natural coastal processes can also be seen on North Carolina's Outer Banks, discussed in Highlight 19-3.

Organic Coasts

Some coasts, particularly those in mangrove swamps and along reefs, form when erosion and deposition act together with biological processes. Mangrove trees live in standing tidal water in tropical climates. They grow from an extensive web of long roots that rise above the water. The presence of this root system absorbs much of the energy of waves that enter the swampy mangrove forest, creating a quiet-water environment. The roots also trap fine sediment, expanding an island's area. Hardy mangrove seedlings take root in the growing tidal mud flats, enabling the forest to grow seaward (Fig. 19-28). In North America, mangrove swamps are expanding along the southern tip of the Florida mainland. These types of coasts also occur on some Caribbean islands and in the tropics of Southeast Asia.

▲ **FIGURE 19-28** A typical mangrove coast in the Florida Everglades. Mangrove coasts form as the exposed roots of mangrove trees trap sediment.

Highlight 19-3

The Future of North Carolina's Coast—Some Good News, Some Bad

The Outer Banks of North Carolina provide a wonderful outdoor laboratory for judging the effects of human interference with an efficient natural system. This string of barrier islands includes both a developed coast, the Cape Hatteras National Seashore, and an undeveloped coast, the pristine Cape Lookout National Seashore.

The Cape Hatteras coast has been stabilized for more than 50 years by a 10 m (33 ft)-high, human-made line of dunes that protects State Highway 12 from flooding by winter storms and periodic hurricanes. Built beyond the reach of damaging salt spray and wind- and water-borne sand, the artificial dunes are lushly vegetated. During storms, however, they deflect wave energy much as

a seawall does, redirecting it toward the seaward beaches and accelerating erosion there. Since the dunes were built, erosion has narrowed the island's 200 m (650 ft)-wide beaches to only 30 m (100 ft). The inland side of the Outer Banks along Pamlico Sound is also being eroded, largely because the high artificial dunes intercept the sand that would otherwise wash over lower natural dunes. This limitation restricts the natural tendency of a barrier island to migrate landward. Disruption of the natural barrier-island system along the Hatteras section has forced North Carolina to compensate with an expensive beach-nourishment program that has thus far failed to halt erosion. If erosion of that segment of the Outer Banks continues at its

present rate, little of Cape Hatteras may be left in the future.

At Cape Lookout, the banks have naturally adjusted to the area's periodic storms. Broad beaches and low dunes absorb erosive energy from storm waves. In this natural system, the vegetation growing along salt marshes has evolved a resistance to salt spray and flooding. It traps landward-migrating sand, replenishing the natural interior dunes. When a powerful storm hits, the island migrates landward, instead of eroding. Thus, here we see two very different approaches to coastal management, one beset by human-made problems, the other thriving in its natural system.

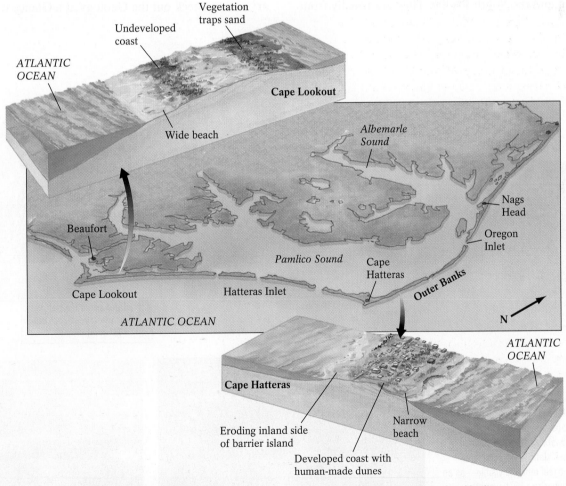

▲ Various sections of the Outer Banks of North Carolina have evolved differently, partly reflecting the role played by the National Park Service. Human-made frontal dunes protect the narrow beaches of Cape Hatteras; the undisturbed Cape Lookout area, on the other hand, has wide beaches.

Reefs grow near continental or island shores from the carbonate remains of corals, algae, and sponges. The warm, near-surface water in which reefs flourish is limited to a geographic band that stretches from about latitude 30°N to 25°S. Clear water is necessary for a reef's growth because most reef organisms are filter-feeders that suck in a large quantity of water and filter out microscopic plankton to eat. They simply could not survive if they ingested the suspended particles in muddy water. Reef organisms also require seawater of normal salinity (about 35 parts salt per thousand parts water). Thus, reefs are unlikely to form where a large influx of freshwater dilutes the local seawater, or where evaporation concentrates the salt in seawater, increasing its salinity.

The active portion of a reef is generally located near the sea surface, where sunlight and the algae on which reef organisms feed are plentiful. The active zone can extend downward only as far as sunlight penetrates, to a depth of approximately 100 m (330 ft). When sea level rises, as is happening today, reefs grow vertically to keep pace with the rising sea and to satisfy the sunlight needs of their organisms.

Three basic types of carbonate reefs exist: fringing, barrier, and atoll (Fig. 19-29). **Fringing reefs** are built directly against the coast of a landmass, such as along the margins of volcanic islands in the Caribbean and the South Pacific. They are usually from 0.5 to 1.0 km (0.3 to 0.6 mi) wide and grow seaward toward their organisms' food supply.

Barrier reefs occur on the local continental shelf and are separated from the mainland, typically by a wide lagoon. Incoming waves break against the reef, which protects the mainland coast from wave erosion. Some barrier reefs begin as a fringing reef around a subsiding landmass, such as a volcanic island. As the island subsides and much of its area descends below sea level, the reef grows vertically, keeping the surface of its active zone within a few meters of sea level.

Atolls, the most common type of reefs, are circular structures that extend from a very great depth to the sea surface and enclose relatively shallow lagoons. These barrier reefs once surrounded oceanic islands that have since subsided completely below sea level. This explanation of the origin of atolls was first proposed by Charles Darwin in 1859. It was confirmed in the 1940s by deep drilling of the Bikini and Eniwetok atolls in the South Pacific prior to their selection as atomic-bomb test sites. Bikini was found to consist of more than 700 m (2300 ft) of coralline limestone resting atop a submerged volcanic island. Surf crashing against an atoll continues to pile the reef debris above sea level.

For a summary of the various types of primary and secondary coasts, check out the Geology at a Glance on page 629.

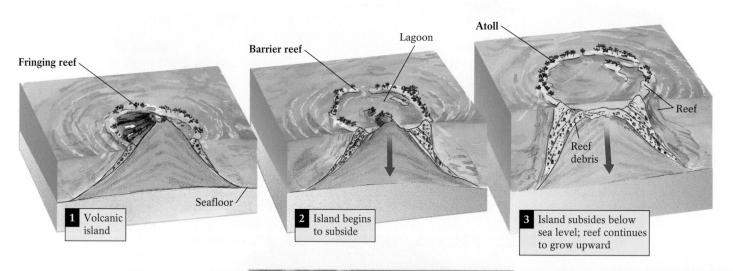

1 Volcanic island

2 Island begins to subside

3 Island subsides below sea level; reef continues to grow upward

▲ **FIGURE 19-29** The evolution of carbonate reefs. A fringing reef forms on the side of a subsiding volcanic island. As the island subsides, the fringing reef becomes a barrier reef, separated from the volcano by a circular lagoon. After the island sinks completely below sea level, only the reef remains visible—as an atoll—because it continues to grow toward the water's surface. **Photos: (left)** Great Barrier Reef, Australia. **(right)** Wake Island, a coral atoll in the Pacific Ocean.

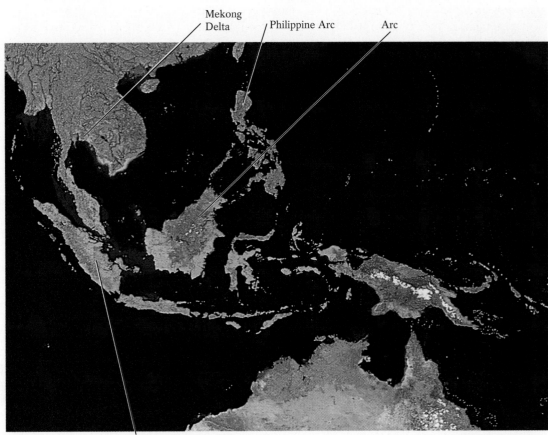

Mekong Delta

Philippine Arc

Arc

Malaysian Arc

▶ **FIGURE 19-30** Protection of a coast by an island arc. Formation of the Mekong Delta in the lee of the Philippine and Malaysian arcs.

Plate-Tectonic Settings of Coasts

The different characteristics of North America's East, West, and Gulf coasts result, in large part, from their distinctive plate-tectonic settings. Plate setting strongly determines the steepness of the local continental shelf. This factor in turn governs the behavior of waves and the prospects for coastal erosion and deposition. Every major type of plate boundary is associated with a distinct coastal style.

Divergent plate boundaries that have recently rifted, such as those found in the Gulf of California between Baja California and the Mexican mainland, are marked by the high scarps produced by normal faulting. Their coasts have steep continental shelves, are vulnerable to the head-on attack of wave action, and feature recently formed erosional landforms such as wave-cut benches and prominent sea stacks. As rifted plate margins move away from a warm, spreading center, they subside and eventually become tectonically quiet and relatively flat. These *passive continental margins* have broad, gently sloped continental shelves. Incoming waves break offshore on the shallow shelf, typically creating depositional features such as spits, beaches, and barrier islands. The East Coast of North America, with its depositional

landforms, clearly shows the results of coastal development at a passive continental margin.

Coasts that remain shielded from the direct onslaught of oceanic waves by an offshore subduction-produced island arc develop principally by marine deposition. This is especially true where the local climate enhances weathering and where rivers transport and deliver a large volume of sediment to the coast. In such an environment, large rivers can extend coastlines by depositing enough sediment to produce deltas. For example, the Mekong River delta in tropical southeast Asia survives in the South China Sea, where it is relatively protected in the lee of the island arcs of the Philippines and Malaysia (Fig. 19-30).

Along convergent plate boundaries, coasts generally have narrow, steep continental shelves, and tectonic uplift continuously exposes new sea cliffs beyond the shoreline. Incoming waves strike the coast directly, forming such erosional features as sea stacks, sea arches, and wave-cut benches. The northwest coast of North America, with its assortment of erosional landforms, highlights coastal evolution at a convergent plate margin (see Figure 19-23a).

The East Coast of North America
North America's East Coast is bounded by a broad, tectonically quiet continental shelf that is slowly subsiding under the weight of continent-derived, river-transported sediment. Moreover, the entire coast has been

◀ **FIGURE 19-31** East Coast shorelines, such as this one at Currituck Sound, North Carolina, are characterized by wide sandy beaches and offshore barrier islands.

slowly submerging for the past century as global sea level has risen. Rising sea level largely accounts for the numerous "drowned" valleys, such as New York's Hudson Valley, that appear along the length of the Eastern Seaboard.

Variations in the East's coastal topography result primarily from its local glacial history and its diverse bedrock composition. Glacial erosion carved deep fjords and impressive U-shaped valleys into the soft sedimentary rocks along the northeastern coast of Canada and on the eastern coasts of Baffin Island and the Labrador Peninsula. To the south, glacial erosion of exposed granitic plutons produced the low, rocky coasts of the Canadian Maritime provinces and of Maine and New Hampshire. As the local ice sheets melted, isostatic rebound of the land surface lifted (and continues to lift) these coasts and continental shelves, allowing present-day oncoming waves to shape these coasts by erosion.

To the South, the East Coast is marked by a zone of glacial deposition and terminal-moraine formation that stretches from Boston to Long Island, New York. These coasts consist primarily of loose glacial sediments. Vulnerable to attack by crashing surf, these deposits have been eroded and then carried by longshore transport until they are redeposited farther down the coast as spits, baymouth bars, barrier islands, and beaches. Figure 19-31 is an example of the sandy beaches typical of this coastal region. From the glacial terminus in New York City south to the Florida Keys, the East Coast features large bays such as Delaware Bay, Chesapeake Bay, and Pamlico Sound in the mid-Atlantic states. Barrier islands and inland lagoons and marshes parallel the coastline for more than 1000 km (620 mi). Reefs form much of the coast in the tropical Florida Keys.

The West Coast of North America The West Coast from Alaska through California differs markedly from the East Coast. Along much of its length, converging plates and tectonic uplift have produced a steep, narrow continental shelf and rising coastal mountains (Fig. 19-32). The stretch of coast from Alaska to northern Washington, which was extensively glaciated during recent ice expansions, is marked by primary coasts formed by major westward-draining fjords and U-shaped valleys. South of the glacial terminus, from southern Washington through Oregon and then to southern California, wave erosion has cut into uplifted terraces, shaping headlands into a near-continuous chain of cliffs, sea stacks, and rugged offshore islands. The eroded sediment from the battered headlands is deposited as narrow beaches in protected bays.

Southern California's beaches remain narrow, in part because flood-control and irrigation dams on coastward-draining rivers in the dry Central Valley trap stream sediment and prevent it from reaching and replenishing the coastal beaches. The narrow beaches, in turn, expose sea cliffs to greater storm-wave erosion than wide beaches, which would absorb some of the wave energy. In recent years, the West Coast has also sustained accelerated erosion attributed to El Niño, the warm ocean current that shifts periodically to the eastern Pacific. In the early 1980s, El Niño caused a 10- to 15-cm (4–6-in.) rise in the Pacific sea level, unusually high tides, severe winter storm waves, and accelerated erosion. These very same conditions revisited the California coast upon El Niño's return in 1997 and again quite tragically in the winter of 2005 (see Chapter 14, The La Conchita Mudslide).

▲ **FIGURE 19-32** This terrace at the Point Vincente Lighthouse, Palos Verde, California, is typical of a western U.S. coast uplifted by plate tectonics. The interaction of the Pacific and North American plates is tectonically lifting California's coast, but this tectonic activity occurs only sporadically, separated by periods of relative tectonic stability during which the wave-cut benches have time to develop. The next episode of tectonic uplift raises the bench above the oncoming waves, forming a marine terrace.

Sea-Level Fluctuations and Coastal Evolution

Sea level is one of the Earth's most dynamic features—always changing, although slowly by human standards. This change may be local, affecting just a single stretch of coastline, or it may be global, affecting all the Earth's oceans.

Local changes may result from the following factors:

■ *Tectonic movement at plate edges.* Changes such as uplift of the land at a convergent boundary produce a relative drop in sea level.

■ *Isostatic movements caused by growth or shrinkage of ice masses.* The weight of a growing glacier depresses the Earth's crust, prompting a relative rise in sea level. Melting of an ice mass and the removal of some or all of its weight allows the crust to rebound, producing a relative drop in sea level.

■ *Isostatic movements caused by accumulation or erosion of sediments.* Accumulating sediments weigh down the crust, producing a relative rise in sea level. Erosion of the crust removes weight, enabling the land to rise and the sea to fall relative to it.

■ *Withdrawal of large volumes of groundwater or oil from aquifers, causing their compression.* When aquifers become compressed, the land surface subsides, producing a relative rise in sea level (Fig. 19-33).

Global, or *"eustatic,"* changes (the term *eustatic* refers to worldwide changes in sea level) may occur in the following situations:

■ *The shape of the oceanic "container" changes.* The shape of an ocean basin changes when sediments are transferred from land to sea, forming deltas and partially filling the basin. The oceanic container may also be altered when rates of mid-ocean-ridge spreading change. For example, when the rate of ridge spreading and growth increases, a broader, more elevated ridge occupies more space on the seafloor, displacing water landward in the form of a sea-level rise. (Think of what happens in a hot tub when a crowd jumps in; their body volumes cause the water level to rise, sometimes overflowing the tub.)

■ *The amount of water in the ocean basin changes.* Water levels in ocean basins change sharply during glacial and interglacial cycles. The global sea level has risen dramatically during the last 10,000 years and continues to rise today, partly because a substantial volume of the world's glacial ice has melted during the current interglacial period.

■ *The physical properties of seawater change.* These properties may change during periods of global warming or cooling. Global warming, which many scientists have linked to our widespread use of fossil fuels and the resulting greenhouse effect, has contributed significantly to the recent rise in global sea level. The warming atmosphere causes the upper few meters of the oceans to warm and expand, thereby raising sea levels.

A dramatic example of how eustatic sea-level change can affect coastal development occurred during the last period of maximum glacial expansion. At that time, about 20,000 years ago, sea level was approximately 130 m (430 ft) lower than the current level. Water that evaporated from the ocean basins became stored in the continent-size Pleistocene ice sheets and, as a result, sea levels fell and vast areas of the Earth's continental shelves were exposed. At the end of the last glacial maximum about 13,000 years ago, melting of those continental ice sheets returned the stored water to the world's oceans. The influx of meltwater caused global sea levels to rise, rapidly at first and then more gradually thereafter, until they stabilized near modern levels between 4000 and 5000 years ago. Today we find the remains of terrestrial animals, entire forests, and archaeological sites on the continental shelves, now submerged beneath the rising post-glacial seas of the last 10,000 years.

The factors that combine to drive local sea-level fluctuations can be observed along the coasts of Alaska, Iceland, and Acadia National Park in southern Maine (Fig. 19-34). Coastal Alaska is currently rising, tectonically and isostatically, a result of ongoing convergence of the Pacific and North American plates and minor crustal rebound produced by the shrinking of Alaska's glaciers. Worldwide sea level, however, is rising faster than Alaska's landscape, and thus *relative* sea level there is rising as well. The resulting drowning of the Alaskan coast has produced its breathtaking fjord scenery.

Iceland is also rising rapidly, for several reasons: Post-glacial crustal rebound following the melting of the island's ice, active volcanism, and the island's position atop the growing mid-Atlantic divergent ridge all contribute to a rising Iceland. Because the rate at which Iceland is rising exceeds the pace at which worldwide sea level is rising, Iceland is emerging from the North Atlantic. Thus, the local relative sea level is falling. Acadia National Park in Maine is located at a passive continental margin and thus is not rising much tectonically. It is, however, rising isostatically, because it was heavily glaciated during the Pleistocene ice expansion. Nevertheless, like Alaska, Acadia's uplift rate

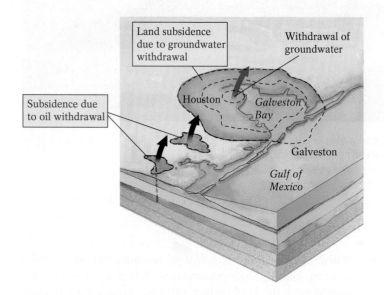

◄ **FIGURE 19-33** Subsidence of the Galveston, Texas, area from excessive groundwater and oil withdrawal has caused the local sea level to rise relative to the coast. A massive seawall had to be built to protect the city from the rising Gulf of Mexico. (Courtesy of the Rosenberg Library, Galveston, Texas)

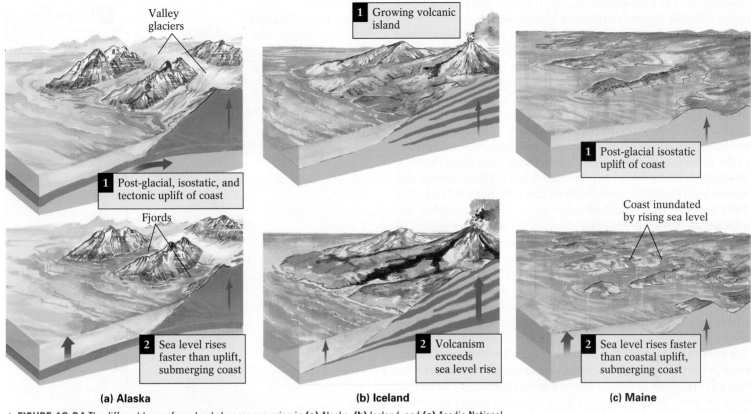

▲ **FIGURE 19-34** The different types of sea-level changes occurring in **(a)** Alaska, **(b)** Iceland, and **(c)** Acadia National Park along the Maine coast. These changes illustrate how tectonics, recent glacial history, and global sea-level change combine to determine local sea levels.

Highlight 19-4

The Drowning of Tuvalu–A Tiny Country Not Long for This World?

Sitting about 2000 km (1250 mi) north of New Zealand and 4000 km (2500 mi) northeast of Australia is a tiny group of islands—really nine bits of coral reef just barely above sea level—known collectively as Tuvalu. (Tuvalu's total area is about one-tenth the size of the Washington, D.C.) This former British colony is home to 10,000 native Tuvaluans, and it and some of its island neighbors are in real danger of disappearing completely from the world map as global sea level rises. The plight of these inhabitants has thus far failed to draw sufficient attention to the *human* consequences of global warming and its resulting rise in worldwide sea levels.

That the Pacific is rising and swallowing up these islands is beyond debate: In 2003,

with the arrival of its annual highest tides, the central areas of the islands became flooded to knee-high depths, a level of flooding that had never occurred until the last decade. If the global warming trend continues and the seas continue to rise, the islands will be gone within the next 50 years or so. They will be completely covered by water, and the people of Tuvalu and their culture will also disappear, perhaps dispersed to other islands.

The islands of Tuvalu are environmentally pristine, its people subsisting on farming, fishing, and the sale of textiles, soap, and collectors' interest in their stamps and coins. The accelerating flooding of their lands threatens to eliminate their native crops, such

as tava, a sweet-potato-like root that grows beneath the surface and is being contaminated by infiltrating saltwater. If global warming proceeds and sea level continues to rise, the culture and lifestyle of these people, thoroughly linked to their homeland islands, will vanish beneath the waves. Here again is yet another opportunity to observe the interplay of Earth's various systems—the chemical composition of the *atmosphere* affects world climate, which in turn affects the nature of the oceanic segment of the *hydrosphere* (causing sea level to rise). The end result will be some profound and potentially tragic effects to the *biosphere*—the people of Tuvalu.

falls short of the post-glacial rise in worldwide sea level, and the rising Atlantic is slowly drowning Maine's coast.

Several studies have shown that eustatic sea level has risen by about 10 cm (4 in) during the past century. With global temperatures forecast to increase by 0.6° to 1.0°C (1° to 1.8°F) over the next 40 years, the rate at which sea level is rising will accelerate. Some climate experts predict an elevation rise of 30 to 70 cm (12 to 28 in) over the next century. In some places, such as the gently sloping, low-lying Atlantic and Gulf Coasts, this change will accelerate coastal erosion and cause a significant landward advance of the shoreline. Higher sea levels increase erosion along the coast by enabling storm waves to penetrate farther inland, threatening homes and other structures. In some cases, higher sea levels may force an actual change in the Earth's maps, with the complete disappearance of several low-lying islands in the southern Pacific. The Highlight on page 641 focuses on the coming crisis that may wipe these islands and their cultures from the face of the Earth.

Because the Atlantic Coastal Plain slopes very gently, a sea-level rise of as little as 30 cm (12 in) along the coast of Pamlico Sound in North Carolina would shift the shoreline nearly 3 km (2 mi) inland. The higher sea levels would invade the mouths of rivers, extending coastal estuaries farther inland and upstream. For every 10-cm (4-in) rise in sea level, the freshwater–saltwater interface in Atlantic estuaries would migrate approximately 1 km (0.6 mi) upstream. If world climate warmed sufficiently to melt *all* of the Earth's glacial ice, worldwide sea level would rise about 70 m (230 ft), enough to cause widespread geographical chaos. In North America, the Florida peninsula would be almost completely submerged, as would New York, Boston, and Philadelphia. Huge bays would penetrate the continent on both coasts, and new coastal cities would arise in such now-inland locations as St. Louis, Missouri; Memphis, Tennessee; San Antonio, Texas; and Fresno, California.

While coastal erosion removes prominent headlands, deposition of beaches in adjacent bays and construction of baymouth bars that trap continental sediments fill in coastal bays. Overall, these processes, as shown in Figure 19-35, should straighten out the world's coasts. Why, then, do many of the world's coasts remain irregular? As with other geological systems, the processes that shape coasts rarely proceed to completion. In the long term, they are almost always interrupted.

Protecting Coasts and Coastal Environments

Individuals have little power to control the long-term geological conditions that shape our coasts. Our governments, however, can act to protect us from the negative consequences of coastal change. A few states offer low-interest loans to property owners willing to relocate oceanfront homes to less-vulnerable sites. Twenty-nine of the 30 coastal and Great Lake states have instituted coastal-zone management programs. Many

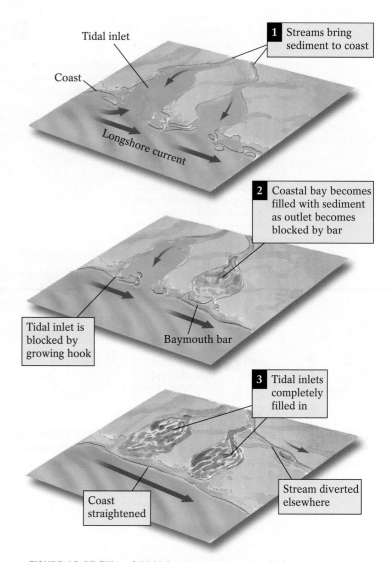

▲ **FIGURE 19-35** Filling of tidal inlets is one process that, if left uninterrupted by tectonic and isostatic movements, would tend to straighten coasts.

states prohibit construction of permanent structures that might be vulnerable to swamping by rising sea levels. They also prohibit construction of barriers that interfere with natural longshore currents.

Environmental organizations can also influence coastal evolution, sometimes by purchasing environmentally sensitive coastal property to preserve and protect it. The Nature Conservancy of Washington, D.C., has bought long stretches of the barrier islands off Virginia to prevent future overdevelopment of these fragile coastal environments. This approach may prove to be the best means of protecting coastal communities from the onslaught of rising sea levels and from themselves.

Chapter 19 Summary

Shorelines are the places where bodies of water meet dry land. Landward of ocean shorelines are *coasts,* the strips of land bordering an ocean that extend inland to a different geographical setting. The shapes of coasts depend on the sediments delivered to them, primarily by streams, and the waves that erode and redeposit those sediments.

Waves form when wind drags across the sea surface. A wave moving across deep, open ocean exhibits *oscillatory motion,* in which water moves in circular orbits. As a wave moves toward the coast, it usually first encounters the continental shelf, where the ocean floor rises. In the shallow zone above the continental shelf, the ocean floor restricts a wave's circular orbit, causing it to slow and steepen. Eventually, the steeper upper part of the wave overruns its base and the wave breaks. Its water then no longer moves in a circular path but instead is sent hurtling forward by *translatory motion.*

When a wave approaches the coast at an angle, the first part of the wave front to encounter shallow water slows, causing the rest of the fast-moving wave front to swing around until it becomes nearly parallel to the coastline. This process is called *wave refraction.* Virtually all waves strike the coast at some angle. Their water moves up the beach at an angle, but takes a perpendicular path back to the ocean. This zigzag motion generates a *longshore current,* which moves parallel to the coast.

Tides, the twice-daily rise and fall of the ocean surface, are primarily caused by the gravitational pull on the ocean surface by the Moon and from the force created as the Earth spins on its axis. Flood tides elevate the sea surface and cause the shoreline to move inland, sometimes migrating upriver. Falling or *ebb tides* lower the sea surface and cause the shoreline to move seaward, carrying land-derived sediment out to sea.

Most coastal erosion occurs when waves strike a coast with enough energy to remove loose materials. When the water immediately offshore is deep, the waves crash against coastal *headlands* with their full force. Erosion of a headland eventually undercuts the headland's cliff, removing the support from overlying rock masses, which fall into the surf. This erosional process produces a series of geomorphic features, including:

- *wave-cut benches,* the terraces washed by breaking waves that form from the remains of a cliff base;
- *sea caves* that form as wave refraction at headlands bends incoming waves against the sides of the headlands, eroding large notches in the flanks of the headlands;
- *sea arches* that forms when two sea caves erode completely through the headland;

- *sea stacks* that form when sea arches collapse, leaving isolated offshore bedrock knobs.

Coastal deposition produces a variety of landforms, the most common being a *beach*—a dynamic, relatively narrow segment of a coast washed by waves or tides. Beaches consist of:

- a *foreshore* area, which extends from the low-tide line to the high-tide line;
- a *backshore* area, which stretches from the high-tide line to the sea cliff or inland vegetation line.

Sediment transported along the beach face by swash and backwash moves by beach drift down the coast. Sediment transported offshore within the surf zone moves by *longshore drift.*

When a longshore current suddenly encounters deeper water, such as at the entrance to a bay, it deposits its sediment load as a *spit,* a fingerlike ridge of sand that extends from the land into open water. A spit that grows completely across a bay entrance is called a *baymouth bar.* Coastal deposition occurs when wave energy falls below a critical level. Wave energy can be dissipated by wind variation, sudden increases in water depth, or the interception of the waves by natural or artificial structures offshore. Human-built structures that intercept waves or currents, causing deposition, include *breakwaters* (built parallel to the coast), and *groins* and *jetties* (built perpendicular to the coast).

Primary coasts form from nonmarine processes such as glaciation, stream deposition, tectonic activity, flooded river systems, and biological activities. Secondary coasts are products of coastal erosion and deposition. *Barrier islands* are nearly continuous ridges of sand, parallel to the main coast but separated from it by a bay or lagoon.

Organic coasts include:

- those that develop around mangrove tree roots in swamps;
- those that are formed by carbonate reefs.

Carbonate reefs commonly appear along the margins of volcanic islands in the Caribbean and the South Pacific. Carbonate reefs include:

- *Fringing reefs,* that initially surround land and grow seaward toward their organisms' food supply;
- *Barrier reefs,* separated from the coast by a wide lagoon, that form and grow vertically as the island subsides;
- *Atolls,* circular structures that extend from great depth to the sea surface and enclose a relatively shallow lagoon. Atolls form after a volcanic island completely subsides and its barrier reef becomes isolated as a circular ridge of limestone.

KEY TERMS

atolls (p. 636)
backshore (p. 625)
barrier islands (p. 631)
barrier reefs (p. 636)
baymouth bar (p. 625)
beach (p. 624)
beach drift (p. 625)
beach face (p. 625)
berm (p. 625)
breakwaters (p. 627)

coast (p. 612)
ebb tide (p. 618)
flood tide (p. 618)
foreshore (p. 625)
fringing reefs (p. 636)
groins (p. 628)
headlands (p. 620)
hook (p. 625)
jetties (p. 628)
longshore current (p. 616)

longshore drift (p. 625)
oscillatory motion (p. 615)
primary coast (p. 631)
rip currents (p. 616)
riprap (p. 623)
sea arch (p. 622)
sea caves (p. 622)
sea stack (p. 622)
seawalls (p. 623)
secondary coast (p. 631)

shoreline (p. 612)
spit (p. 625)
tidal bore (p. 618)
tides (p. 616)
translatory motion (p. 616)
wave refraction (p. 616)
wave-cut bench (p. 622)

QUESTIONS FOR REVIEW

1. Draw a simple diagram showing wave crests, wave troughs, wave height and amplitude, and wavelength. Define wave period and fetch.

2. Why are most wave fronts nearly parallel to the shoreline on arrival?

3. How does a longshore current develop?

4. Explain why the Earth's coasts generally experience two high tides and two low tides each day.

5. Describe how the following depositional coastal landforms develop: spit, hook, baymouth bar, tombolo.

6. Discuss how breakwaters, groins, and jetties interrupt longshore transport and affect coastal erosion and deposition. With respect to these structures, where is sand deposited? Where is sand eroded?

7. Why do you suppose surfing is more popular on the West Coast than on the East Coast of North America? Speculate about the prospects for surfing on the western coast of Europe or northern Africa.

8. Suppose you have acquired some valuable beachfront property in Oregon. How would you protect this land from coastal erosion? (Note: You have unlimited financial resources.) In your plan, try to minimize the negative secondary effects that follow most cases of human interference with natural coastal systems.

9. What would be the effect on California's beaches if all of the state's dams were removed?

10. How would the eastern coast of North America change if widespread subduction resumed there?

LEARNING ACTIVELY

1. If you live close enough, visit a beach on a calm summer day and again on a blustery winter day. Compare the sizes and velocities of the waves. Note the condition of the beach. Is the nearshore region rocky and pebbly or sandy? The constant, powerful beat of large winter waves removes sand, leaving behind coarse pebbles. Does the beach show evidence of this sand-removing effect?

2. Look at a map of North America and focus your attention on the coasts. Compare the coasts of Washington, Oregon, Delaware, Georgia, Alabama, and Louisiana. Just by looking at the map,

speculate about whether erosion or deposition dominates the coastal processes at these localities.

3. The next time August or September rolls around, follow the progress and outcomes of Atlantic and Gulf Coast hurricanes. Note how news coverage in newspapers and on television portrays the human tragedy of these storms. Note the scenes of destruction from coastal erosion—beach houses lying shattered in the surf. In the news reports, who gets the blame—the fury of Mother Nature or the unfortunate homeowners who built houses too close to the shoreline?

ONLINE STUDY GUIDE

Practice makes you better. Make the time to take the *Online Study Guide* quiz for this chapter. It should take you about 20 minutes. Automatic grading and instant feedback will help you quickly and

accurately identify the concepts you have mastered and the areas in which you need more study.
www.prenhall.com/chernicoff

Human Use of the Earth's Resources

20

- **Reserves and Resources**
- **Fossil Fuels**
- **Alternative Energy Resources**
- ***Highlight 20-1***
  Gas Hydrates—A Major 21st-Century Fuel or a Potential Global Environmental Disaster?
- ***Highlight 20-2***
  Hydrogen—The Solution to All Our 21st-Century Energy Needs?
- **Mineral Resources**
- **Future Use of Natural Resources**

In earlier chapters, we have often described the many natural resources that are useful or essential in our modern lives. Stop for a moment and think about the vast array of natural resources needed to produce and deliver a pizza to your room. They include the energy and materials to power the farms and to harvest the crops that make a pizza's ingredients. Then throw in the metals and other materials to make the pizza oven and other necessary appliances, along with the metals and other materials to manufacture the delivery vehicle. Now add in the gasoline used to deliver your pizza and the materials that make up the pen with which you sign the credit-card receipt, and you can see that the list of resources necessary to feed you that pizza is incredibly long.

Until recently, most people believed that the Earth's resources, such as those being mined in Figure 20-1, were unlimited. Today we face serious shortages of many essential materials. For example, scientists predict that the world's *recoverable* supply of crude oil, from which we get our gasoline, may last only another 50 to 100 years at the current rate of use. How have we exhausted our stores so quickly? How can we compensate for these losses?

The answers to these questions depend largely on the growth rate of the world's population, the quantity of natural resources each individual uses, and our success or failure in our search for alternative resources. In the United States alone, each person directly or indirectly uses about 10,000 kg (22,000 lb) of raw materials each year, most of which consists of stone and cement for the construction of roads and buildings, but which also includes about 500 kg (1100 lb) of steel, 25 kg (55 lb) of aluminum, and 200 kg (440 lb) of industrial salt (mostly for cold-weather road maintenance). Each American also uses nearly 3800 liters (1000 gal) of oil per year. Collectively, Americans consume about 30% of the world's oil. The United States, with a little less than 5% of the world's population, uses nearly 30% of its minerals, metals, and energy. A single American may use 30 times as much material and energy as a person in a developing nation.

Natural resources are distributed unevenly among nations and continents. Valuable materials are abundant in some geological settings, but elsewhere they are in short supply. Some nations, such as Canada, the United States, and Russia, possess a wealth of varied natural resources. Some of these resources are readily accessible, others require a great deal

FIGURE 20-1 Pursuit of the Earth's dwindling resources forces us to mine in remote areas. This gold mine is being excavated out of a jungle in northeastern Brazil using a large amount of cheap manpower. ▶

647

▲ FIGURE 20-2 Giant dredges are used to search for gold in Siberian mines.

of effort and technical skill to exploit (Fig. 20-2). Other nations, such as Japan, have few resources. Some *former* resource producers have virtually exhausted their domestic supplies and must now import from other nations. Great Britain, once a great mining nation that exported tin, copper, lead, and iron, must now import those metals. Meanwhile, some smaller, developing nations have vast supplies of a few important materials. Guyana and Suriname, on the northeast coast of South America, and Jamaica, in the Caribbean, possess some of the world's richest supplies of aluminum (a product of extreme chemical weathering in warm, wet, tropical conditions). Nevertheless, *no nation is self-sufficient with regard to all essential resources.*

Resource consumption worldwide is accelerating as the world population increases (now approaching 7 billion—more than three times the population in 1920—and expected to double by 2040). People everywhere are striving to benefit from technological development. Unless we identify new supplies of depleted resources, find substitutes for them, and manage industrial development to limit resource depletion, impending shortages will force people *everywhere* to make drastic changes in their ways of life in the not-too-distant future.

Reserves and Resources

Reserves are natural resources that have already been discovered and *can be exploited for profit with existing technology and under prevailing economic conditions.* We know where the reserves are buried, and we can extract them. Most importantly, their economic value in the marketplace *exceeds* their cost of extraction. **Resources** are deposits that we know or believe to exist, *but that are not exploitable today,* whether for technological, economic, or political reasons. We can estimate the location of resources hidden beneath the surface by exploratory drilling,

geophysical modeling, and extrapolation from known reserves. To illustrate the difference between reserves and resources, consider that world oil *reserves* are estimated at 700 billion barrels (a *barrel* is a volume equaling 159 liters, or 42 gal), whereas world oil *resources* are thought to total about 2 trillion barrels.

Resources may become reserves if extracting them becomes more profitable. In the 1970s, when the price of gold surged to $800 per ounce, deposits that had not been considered worth developing at lower prices immediately became highly profitable ores. (An **ore** is a mineral deposit that can be mined for a profit; this word is basically an economic, not a geologic, term.) And in 2005, as the price of oil rose sharply (you know this from the nearly $3.00 you paid at the pump that summer), a host of oil resources have been converted to sought-after reserves. Billions are now being invested to mine oil from the tar sands of Alberta, Canada; new wells are being drilled north of Russia's Arctic Circle. In fact, in response to the higher prices, Russia has increased its oil production so that it is now the world's leading producer—greater than the top OPEC producer, Saudi Arabia.

On the other hand, a price reduction for a material on world markets can lead some countries to import it rather than mine and develop it themselves. Thus, a price change can transform a profitable reserve into an unprofitable resource. Such a shift can be seen in the recent decline of the U.S. steel industry, which has struggled since it became cheaper to import steel from South Korea than to produce steel domestically.

Access to reserves is often controlled by politics: The Persian Gulf War of 1991, for example, produced a spike in oil prices as political conditions converted some Middle Eastern oil reserves

▲ FIGURE 20-3 During the 1991 Persian Gulf War, the Iraqi military ignited oil in thousands of Kuwaiti wells. Political conditions in certain oil-producing nations have prompted oil-poor industrial nations to seek alternative energy sources.

to unavailable resources (Fig. 20-3). Other deposits that might be extracted profitably today are located in national parks or wilderness areas. They would be reserves if they could be exploited, but they remain resources because environmental legislation protects them from development. Political events have been affecting resource development for millennia, as has changing technology. Five thousand years ago, humans made tools of stone, especially chert and obsidian; when copper, iron, and bronze came into use, and especially after smelting developed, stone for toolmaking lost value.

Some natural resources are *renewable*—that is, they are naturally replenished over relatively short time spans (such as trees) or available continuously (such as sunlight). *Nonrenewable* natural resources form so slowly that they are typically consumed much more quickly than nature can replenish them. They include fossil fuels such as coal, oil, and natural gas, and metals such as copper, bronze, iron, aluminum, gold, and silver. Nonrenewable materials are like crops that can be harvested only once. We cannot "grow" another crop of nonrenewable resources, although many, such as copper and aluminum, can be recycled for reuse. Some resources may be either renewable or nonrenewable, depending on how we use them. Soil, for example, is a renewable resource if we follow sound agricultural practices. Soil becomes nonrenewable when we deplete its nutrients by overplanting or when it is allowed to blow away after being overgrazed, overplanted, or deforested.

To maintain our standard of living and to protect the global environment in the 21st century, we must all become more knowledgeable about the Earth's energy and mineral resources. To use dwindling resources wisely, we need to understand how they form, where they are abundant, what environmental consequences are associated with their use, and how long the known supplies are likely to last. This chapter addresses these and other crucial issues in regard to fossil fuels, alternative energy sources, and both metallic and nonmetallic minerals.

Fossil Fuels

The Earth's 6.5 billion inhabitants get relatively little of their energy from renewable sources such as solar energy, the energy produced by the rise and fall of tides, wind power, and power from flowing streams. Instead, we obtain most of our energy from nonrenewable **fossil fuels** derived from the combustible organic remains of past life. The principal fossil fuels are oil, natural gas, and coal. Oil and natural gas have now passed their peak production period in the United States. Extraction of the world's reserves increases each year as demand rises, and at the current rate of use *and price*, these reserves will be virtually exhausted within the next century or so. Nevertheless, for economic and technological reasons, the nations of the world continue to draw nearly 95% of their total energy from a dwindling supply of fossil fuels that are nonrenewable, at least on a human time scale.

In this section we describe how various hydrocarbons form including petroleum, natural gas, oil shale and oil sand, coal, and peat and how they're used for fuel. We also discuss a host of environmental problems caused by the human use of fossil fuels.

Hydrocarbons

Hydrocarbons—molecules consisting entirely of hydrogen and carbon—comprise the most common and versatile group of fuels that we currently use to heat our homes and power our factories and modes of transportation. Hydrocarbons include coal, natural gas, and **petroleum,** a group of gaseous, liquid, and semisolid hydrocarbon compounds. Typically pumped from the ground in the form of dark, viscous crude oil, petroleum is refined to produce propane for camp stoves, motor oil and gasoline for cars, tar and asphalt for roads, and the natural gas and heating oil that warm our homes and workplaces. It is also the major ingredient in plastics, synthetic fibers, dyes, cosmetics, explosives, certain medicines, certain fertilizers, and videos, tapes, and compact discs.

Humans have used petroleum for thousands of years. Approximately 4500 years ago, Babylonians collected crude oil bubbling from natural pools to make glue for attaching metal projectile points to spears. In ancient Iraq, oil seeping from rocks in the valleys of the Tigris and Euphrates Rivers was used in mortar for setting bricks, in grout to set tiles in ancient mosaics, and in waterproofing materials for boats. People in China have been using natural gas, discovered while drilling for salt, to fuel lamps and stoves for nearly a thousand years,

In the United States, modern use of petroleum began in 1816, when gas extracted from coal was first used in Baltimore's gaslights. Combustible gas was discovered in 1821, when a water well in Fredonia, New York, was accidentally ignited, producing a spectacular flame. Wooden pipes were installed to carry the gas to 66 gaslights in downtown Fredonia. Commercial use of oil began in 1847, when a Pittsburgh merchant bottled and sold natural "rock oil" as a lubricant for machines in the home and workplace. In 1852, Canadian chemists, using oil from what is now Oil Springs, Ontario, first refined kerosene from rock oil for use in home lamps. This discovery soon eliminated much of the candle and whale-oil industries. The oil-well industry was born on Sunday, August 27, 1859, when Edwin Drake of Titusville, Pennsylvania, pumped oil from the first true oil well (Fig. 20-4).

The Origin of Petroleum Petroleum generally forms in marine basins in tropical environments where a rich abundance of microscopic plants and animals live (Fig. 20-5). After the organisms die, they start to decay by oxidation; eventually the oxygen in the bottom waters becomes depleted, however, and decay stops. Layers of sediment and additional organic material may bury the organic remains, thus preventing their subsequent decay. As sediments accumulate, pressure and heat convert the organic molecules to a substance called **kerogen,** a solid, waxy organic material. Kerogen in turn is converted to various liquid and gaseous hydrocarbons at temperatures between 50° and 100°C (122° to 212°F) and at a depth of 2 to 10 km (1.2 to 6 mi).

At the start of this process, kerogen's large, complex organic molecules form highly viscous hydrocarbons such as tar. With increasing heat, these molecules break down to form smaller, simpler, less viscous molecules, such as those found in diesel oil, kerosene, and gasoline. At temperatures exceeding 100°C (212°F), liquid petroleum converts to a variety of natural gases, ranging from those with relatively complex molecules, such as

▲ FIGURE 20-4 The first commercial oil well was developed in 1859 in Titusville, Pennsylvania, by Edwin Drake (at right). Drake's well, which was 21.2 m (70 ft) deep, yielded 35 barrels of oil per day.

butane, to the simplest, lightest natural gases—propane, ethane, and methane. At temperatures of about 200°C (400°F) and at depths of 7 km (4 mi) or more, methane, the lightest gas, breaks down completely to carbon dioxide and water, and the rocks no longer contain hydrocarbons. Thus, with respect to geothermal temperatures and pressures, a limited "window of opportunity" exists for the conversion of organic remains to hydrocarbon fuels. Otherwise, we would probably have a great deal more oil and the gasoline at the corner pump would be cheaper.

The rocks in which hydrocarbons form are called **source rocks.** They are typically shales and siltstones lithified under reducing (oxygen-poor) conditions from fine-grained, organic-rich muds. Oil and gas are rarely found in their source rocks, however, because most liquid and gaseous hydrocarbons are readily expelled from these compacting muds. These fluids tend to migrate upward into adjacent *permeable* **reservoir rocks,** such as well-sorted, coarse-grained sandstones and highly fractured or porous limestones. (The reservoir rocks into which these fluids migrate can be the very same rocks that function as the groundwater aquifers discussed in Chapter 16.) The hydrocarbons continue to migrate upward until they are trapped by an *impermeable* cap rock or reach the surface and ooze out as an oil seep.

Geological activity can both create and destroy **oil traps,** the subsurface sites where migrating oil accumulates as its flow is blocked. (See Highlight 9-2 to refresh your memory of these structural traps.) Uplift and erosion can remove the trapping cap rock, allowing oil or gas to escape at the surface. Such an escape continues to occur at the La Brea Tar Pits along Los Angeles' Wilshire Boulevard. A new fault, or the extension of an old one, can breach an oil trap and allow its oil to seep out. As a result of these processes, much of the oil and gas formed before about 65 million years ago has long since escaped from its traps to be oxidized at the surface. More than 60% of today's oil-producing wells can be found in rocks that are less than 65 million years old.

Geologists estimate that less than 0.1% of all marine organic matter buried at the seafloor eventually is trapped as usable petroleum. Some settings may lack sufficient heat to convert

▶ FIGURE 20-5 Petroleum begins as a large accumulation of partially decayed microorganisms in marine mud. The remains of the organisms are eventually converted to kerogen as the marine mud lithifies to become source rocks. Geothermal heat cooks the kerogen, which converts to petroleum as its organic molecules break down into a variety of hydrocarbons. Petroleum is expelled from its source rocks and migrates into permeable reservoir rocks, accumulating there when further movement is blocked by a structural or stratigraphic feature, such as an impermeable cap rock. Erosion and faulting may breach oil traps, enabling oil to escape to the surface.

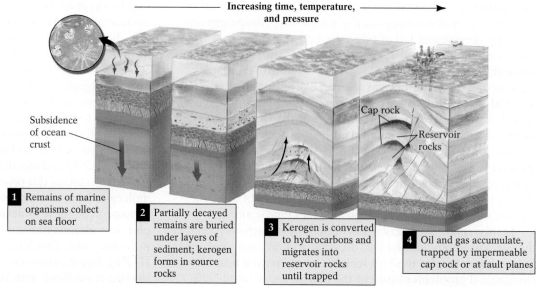

Increasing time, temperature, and pressure ⟶

Subsidence of ocean crust

1 Remains of marine organisms collect on sea floor

2 Partially decayed remains are buried under layers of sediment; kerogen forms in source rocks

3 Kerogen is converted to hydrocarbons and migrates into reservoir rocks until trapped

Cap rock

Reservoir rocks

4 Oil and gas accumulate, trapped by impermeable cap rock or at fault planes

organic matter to kerogen and petroleum. In others, deposits may have been heated enough but not at great enough depth, enabling shallow-forming hydrocarbons to escape without ever becoming trapped. The conditions required to produce, trap, and retain hydrocarbons in rocks rarely occur together, which explains why most marine rocks are petroleum-free. We do not know how long it takes for oil and gas to form. No known petroleum sources are less than 1 to 2 million years old, so the process must take at least that much time.

Oil Shale and Oil Sand

Oil shale is a black-to-brown clastic sedimentary rock consisting of a mixture of waxy kerogen and fine mineral grains. Because it was never buried deeply enough to raise its temperature high enough to convert the kerogen to oil, oil shale has retained its hydrocarbons, locked in the rock. Based on current and anticipated increased rates of usage, the estimated 2 to 5 *trillion* barrels of shale oil (the oil extracted from oil shale) underlying the United States could supply America's petroleum needs for more than 500 years (Fig. 20-6). (The U.S. possesses about two-thirds of the world's oil shale.) According to conservative estimates, this country's largely untapped oil shales contain more than 10 times the known oil reserves of the Middle East. Economics, technology, and our desire to protect the environment currently keep us from using this oil to heat our homes and fuel our cars. Current world prices for crude oil are not yet high enough to justify the cost of developing this resource.

To develop oil shale, the rock must be mined, its trapped kerogen removed and then processed, and the waste rock disposed of in an environmentally sound fashion. After mining, the shale must be crushed (which increases the volume of the rock by about 30%) and then heated to more than 500°C (930°F) to vaporize its kerogen, which then condenses as crude oil. Like popped popcorn, which occupies much more space than its kernels, fragments of oil shale expand in volume by as much as 20% when heated. Thus oil shale's waste rock, whose volume is approximately 50% greater than that of the original rock after crushing and heating, simply won't fit in the hole from which it was mined. If the United States were to obtain its entire oil supply from oil shale, it would need to dispose of 13 billion tons of waste rock each year. This is enough to fill 130 *million* freight-train cars. In addition, enormous amounts of water would be needed to extract and refine this resource. And because much of the oil shale is located in the arid West, the water needed to mine it would be awfully hard to find.

Someday, technology may solve this problem. Microwaves or radiowaves could perhaps heat kerogen in the shale without removing and crushing it. Given the cost, oil-shale deposits will be exploited only after we exhaust more accessible, less expensive sources or oil prices rise so steeply that the potential profits outweigh the costs of development.

Oil sand (or tar sand) is a mixture of *unconsolidated* sand and clay with a semisolid, tarlike hydrocarbon called **bitumen.** Bitumen can be refined to produce gasoline, fuel oil, or other hydrocarbon products. Bitumen sticks so strongly to the mineral grains in oil sand that it can't flow and thus it can't be pumped from the ground. Oil sand must therefore be mined and then heated with hot water and steam to release the bitumen in a more fluid state. Oil sand, or any other fossil fuel, is developed only when the extracted fuel generates more energy than is consumed in mining and processing it.

The world's largest oil sand deposit, located about 400 km (250 mi) north of Edmonton in Alberta, Canada (Fig. 20-7), has been producing crude oil since 1967. Today it yields nearly 200,000 barrels per day, approximately 16% of Canada's oil

▼ **FIGURE 20-6** North American oil, gas, oil shale, and oil sand deposits, most of which are located in the United States. The world's greatest body of exploitable oil shale, the Green River Shale of western Wyoming and northern Colorado and Utah, originated as a deposit from a freshwater lake that occupied about 40,000 km² (15,000 mi²) 50 million years ago. In places, this rock formation is 650 m (2100 ft) thick. **Photo:** A hand specimen of oil shale from the Green River Formation.

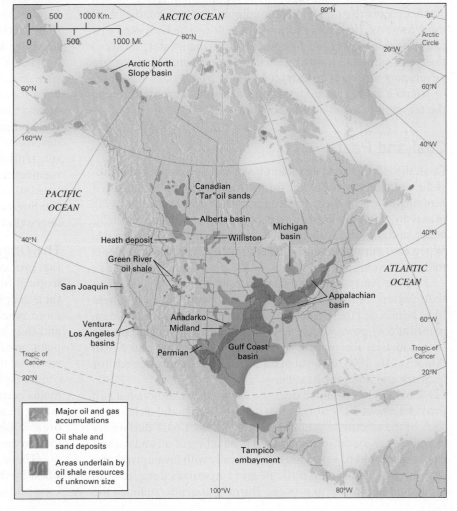

▶ **FIGURE 20-7** Cliffs of tar sand along the Athabasca River, Alberta, Canada.

needs. This deposit is commercially exploitable because it lies close enough to the surface to be extracted by surface mining, which involves the removal of overlying sediment layers—a less costly process than underground mining. Surface mining of oil sand, however, produces enormous piles of residual oily sand that can contaminate local surface water and groundwater. To avoid this environmental hazard, Canadian petroleum engineers are developing a process to extract bitumen by injecting hot water or steam *directly* into underground deposits, thereby softening the bitumen so it can be pumped out.

Coal and Peat

Coal, the Earth's most abundant fossil fuel, is a combustible biogenic sedimentary rock that forms from the highly compressed remains of land plants. It contains the energy stored in living plants by photosynthesis, the process through which sunlight, water, and carbon dioxide produce the materials for plant growth. When coal burns, it releases the energy that was stored in plants millions of years ago.

Native Americans used coal thousands of years ago to fire their pottery. A thousand years ago, Europeans mined it to heat their homes and fuel the fires of their smelting industries. Cheap, plentiful coal powered the newly invented steam engine during the Industrial Revolution of 19th-century Europe and North America. By 1900, coal supplied 90% of U.S. energy needs for industry and domestic heating.

Coal use has declined over the last 40 years, although it still provides about 23% of U.S. total energy needs and as much as 55% of the electricity generated in the nation. Coal's decline occurred largely because of increased production of oil and natural gas, which could be extracted economically with fewer problems. In contrast, coal can be difficult and sometimes dangerous to mine, and it is costly to process. Burning coal also pollutes the air more than burning oil or gas. As its petroleum replacements become depleted, however, coal will likely rebound as our principal fossil fuel, especially in electricity-producing power plants. New technology enables us to convert coal to liquid and gaseous fuels. In these forms, existing world resources could meet energy needs for hundreds of years.

The United States has more than 30% of the world's accessible coal. What may be the world's largest coal field has recently been discovered in Antarctica, stretching for hundreds of kilometers along the eastern side of the Transantarctic Mountains. If ever mined (the region has been declared off-limits for resource development through international treaty), this coal would add substantially to the world's coal inventories. (As a side note, remember that coal represents the remains of an enormous mass of vegetation, evidence of past swampy, often tropical, conditions. Finding vast coal deposits in frigid Antarctica today is solid evidence that our continents move around great distances.)

The Origin of Peat and Coal The abundant plant growth that produced the world's coal deposits likely occurred mainly in tropical or semitropical swamps. These organic deposits were then deeply buried, either by the growth and accumulation of more vegetation or by overlying clastic sediments. Once buried and thus largely separated from oxygen-rich air and water, the plant remains could not completely decompose by oxidation. The weight of overlying deposits squeezed water from the porous mass of incompletely decomposed vegetation. And as the plant remains became more deeply buried, increasing pressure, geothermal heat, and bacterial action removed additional water and the organic gases, such as CO_2 and CH_4 (methane), produced by oxidation and bacterial activity.

These processes created a series of fossil fuels that are classified by their increasing carbon and decreasing water content. Low-pressure conditions produce **peat,** the first fuel to form from buried vegetation. A soft brown mass of compressed, still recognizable plant structures, peat contains substantial amounts of water, organic acids, and incorporated hydrogen, nitrogen, and oxygen. It is only about 50% carbon. For hundreds of years and still today, peat is dried and then burned as fuel for home heating and other domestic uses in rural Ireland, England, and elsewhere in Europe.

As bacterial action continues in peat, it eventually breaks down to form a kerogen-rich muck that becomes compressed as **lignite.** This soft, brown coal consists of about 70% carbon plus 20% water and 10% oxygen. Lignite's higher carbon content makes it a more concentrated heat source than peat. Deep burial of lignite subjects it to increased pressure and geothermal heat, converting it to lustrous black **bituminous coal** (Fig. 20-8), which has a carbon content of 80% to 93%. This type of coal produces more heat, with much less smoke, than peat or lignite.

As increasing heat and pressure reach metamorphic conditions, lignite and bituminous coal are gradually transformed into a metamorphic coal—**anthracite**—a hard jet-black coal that contains 93% to 98% carbon. Anthracite burns with an extremely hot flame and very little smoke. Unlike lignite and bituminous coal, which occur widely, anthracite only occurs in low-grade metamorphic zones in mountainous regions, notably the Appalachians of northeastern Pennsylvania. Anthracite seams commonly occur within the steeply dipping limbs of folded sedimentary rocks where most coal cannot be strip-mined. (*Strip mining* is the extraction process in which a relatively flat-lying layer of resource-bearing rock or sediment is exposed by stripping away the overlying soils and regolith.) Thus, anthracite requires more difficult, dangerous, and costly deep underground coal mining to extract it.

Peat, lignite, and bituminous coal typically occur as distinct seams or beds in sequences of clastic sedimentary rocks, particularly those that accumulated in warm, moist, coastal environments. A series of alternating clastic sediments and coal beds indicates cycles of rising and falling sea level. As the seas rise, the coasts are submerged and clastic sediments are deposited atop heavily vegetated coastal swamplands. As the seas fall, the land emerges and vegetation in these lush coal-producing swamps resume their growth. Much of North America's vast coal deposits formed during ancient episodes of high worldwide sea level, when shallow seas invaded the continent's interior. The coal of Utah, Montana, and the Dakotas formed at the swampy margins of such inland seas. Coal deposits have also formed on vegetated river floodplains. The coal found in Wyoming's Powder River basin occurs as layers between floodplain sediments.

Peat and coal have probably been accumulating continuously since land plants first appeared on Earth, about 380 million years ago. Warmer-than-average global climatic conditions and a high proportion of tropical or semitropical landmasses promoted

▼ **FIGURE 20-8** Major coal deposits of North America. Photo: A coal seam in Alaska.

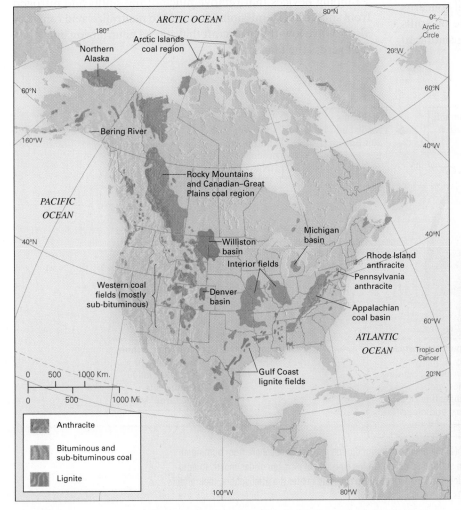

extensive coal formation. During the Mississippian and Pennsylvanian periods of the Paleozoic Era (360 to 286 million years ago), giant ferns and scale trees in great swamps and forests covered the tropical equatorial lowlands of the supercontinent Pangaea. During this so-called "Carboniferous" Period, the great bituminous coal beds of Europe (Great Britain, Germany, and Poland) and the anthracites and bituminous deposits of eastern North America's Appalachians (such as those in West Virginia) were laid down. A second great period of coal formation occurred between about 135 and 30 million years ago, producing lignites and bituminous coals from *flowering* plants similar to those that exist today. Some of these deposits contain well-preserved fossils that identify the types of plants from which they formed. These coals extend from New Mexico through Colorado to the Great Plains of North Dakota and Saskatchewan.

Coal is also forming today. In the Great Dismal Swamp of coastal Virginia and North Carolina, a 2-m (7-ft)-thick layer of vegetation has accumulated over a 5000 km² (2000 mi²) area during the last 5000 years. The next major rise in sea level will bury these organic deposits with clastic sediment and halt their further decomposition. Eventually pressure and heat may convert them to peat and coal.

Fossil Fuels and the Environment

Widespread use of coal, oil shale, and other fossil fuels can produce a host of environmentally damaging effects. Real concerns about acid rain, global warming, and massive marine oil spills have led to increased public awareness and recent legislation to reduce or prevent such effects.

Acid Rain The burning of fossil fuels releases harmful compounds into the atmosphere that reduce the quality of the air we breathe. When coal is burned, coal ash—fragments of noncombustible silicates and toxic metals—enters the atmosphere. Burning sulfur- and nitrogen-rich hydrocarbons releases sulfur and nitrogen oxides, which then combine with water in the atmosphere to form sulfuric acid (H_2SO_4) and nitric acid (HNO_3), the principal components of **acid rain** (Fig. 20-9). Most scientists hold acid rain, along with acidic mine water, responsible for damaging forests and crops, killing aquatic life in lakes, and accelerating the destructive weathering of human-made structures. The most pronounced damage from acid rain occurs downwind of major coal-burning industrial regions (Fig. 20-10).

In some settings, environmental damage from acid rain is lessened by the area's geology. In areas with calcium-rich soils or exposed carbonate bedrock (limestone, dolostone), the carbonates react with the acids and neutralize them before they can do much harm. Without this neutralizing effect, such as where thin, calcium-poor soils overlie granitic bedrock, acid-rain damage to the ecosystem may be more severe. Such is the case in the lake and forest country of the Canadian Shield of the northern Great Lakes and adjacent southeastern Canada. Eastern Canada and the Adirondack Mountains of northern New York receive the acid rain produced by the coal-burning smokestacks of the steel and automotive industries of the American Midwest and from the Midwest's coal-burning power plants. In addition to installing scrubbing devices that remove much of the sulfur and nitrogen from smokestack emissions, attempts to neutralize acid rain may also involve low-flying aircraft that spread neutralizing limestone dust across the landscape.

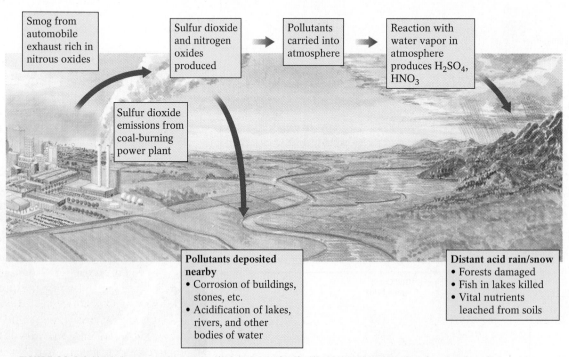

Smog from automobile exhaust rich in nitrous oxides

Sulfur dioxide and nitrogen oxides produced

Pollutants carried into atmosphere

Reaction with water vapor in atmosphere produces H_2SO_4, HNO_3

Sulfur dioxide emissions from coal-burning power plant

Pollutants deposited nearby
• Corrosion of buildings, stones, etc.
• Acidification of lakes, rivers, and other bodies of water

Distant acid rain/snow
• Forests damaged
• Fish in lakes killed
• Vital nutrients leached from soils

▲ **FIGURE 20-9** Acid rain has two major causes: (1) the release of sulfur dioxide into the atmosphere from burning of sulfur-rich coal, which combines with water to form sulfuric acid; and (2) the emission of nitrogen oxides mostly in automobile exhaust, which combine with water to produce nitric acid. Once these pollutants enter the atmosphere, their effects can spread to nearby and even distant environments.

▲ **FIGURE 20-12** One problem associated with oil exploration and development is the potential for oil spills, which can cause large-scale environmental damage to oceans and coastal habitats. The negative impact on marine life and habitats can be profound, as dramatized by the oil-covered wildlife found following the 1989 breakup of the supertanker *Exxon Valdez* in Alaska's Prince William Sound; the effects of this spill are still being felt today.

▲ **FIGURE 20-10** Acid-rain damage to statuary

Global Warming Burning all types of fossil fuels increases the volume of carbon dioxide (CO_2) in the atmosphere. Measurements made from 1958 to 1984 at the top of Hawai'i's Mauna Loa volcano, *far from industrial pollution sources and large population centers,* showed a 9% increase in global atmospheric CO_2. As discussed in Chapters 5 and 17, atmospheric CO_2 absorbs and traps heat from the Earth's surface, creating a greenhouse effect that is proposed by many scientists to have caused substantial global warming. Some climatologists predict that if atmospheric CO_2 levels continue to rise at their present rate, atmospheric temperatures could rise 1.5° to 4.5°C (2.5° to 8°F) by the middle of the 21st century. The increased temperatures could cause droughts in some regions and increase desertification in others. It could also lead to melting of a substantial portion of the Earth's ice masses (Fig. 20-11). Along with thermal expansion of warming seawater, this melting would raise sea levels worldwide, perhaps by 1 to 2 m (3.3 to 6.6 ft) per 100 years for the next 200 to 500 years. A rise of 2 to 10 m (6.6 to 33 ft) would accelerate coastal erosion and submerge a substantial amount of coastal land, including major port cities on every continent.

Marine Oil Spills Occasionally, an oil tanker runs aground and leaks its cargo of crude oil into the sea. Extraordinary environmental damage was caused by one such marine oil spill in March 1989, when the supertanker *Exxon Valdez* broke apart in Alaska's Prince William Sound. More than 10.2 million gallons of crude oil flowed from its cracked hull into the sea, much of which washed up onshore in layers tens of centimeters thick. The oil killed hundreds of thousands of birds and thousands of marine mammals (Fig. 20-12), halted herring and

▲ **FIGURE 20-11** Recent visits to arctic regions have revealed that areas that were formerly covered with continuous sea ice are now areas of open water. The Canadian icebreaker *Des Groseilliers* is shown here in 1998 visiting an ice station at 75° N in the Arctic Ocean. Note the open water and thinness of the ice—an indication of progressive global warming.

salmon fishing during peak season, and coated hundreds of kilometers of Alaska's coastline with a thick smear of oil.

Oil spills in calm seas can usually be contained successfully within floating barriers placed around the oil, allowing workers to skim the oil from the surface. In Alaska, unfortunately, stormy weather caused rough seas that breached the floating barriers. The Exxon Company recovered only about 500,000 gallons of its oil. Efforts to ignite and burn the spilled oil failed as well, although some of the lightest hydrocarbons evaporated and some were consumed by oil-eating bacteria. Millions of gallons washed onshore, triggering an enormous cleanup operation. Standing pools of oil were soaked up with peat moss, wood shavings, and even chicken feathers. High-powered steam hoses removed some of the oil coating on the rocks. Workers even scrubbed rocks *individually by hand.* (Imagine how much time and money that took.) After several months and some $4 billion in cleanup costs, more than 85% of the spilled oil was gone. The remainder consisted of thick asphalt clumps that could not be scrubbed, did not oxidize in sunlight, and would not decompose by bacterial action. These clumps fouled breeding grounds and other wildlife habitats on land, as well as major fishing grounds.

And at the time of this writing—sixteen years *after* the spill—these problems linger. Hidden pools of oil, tucked away beneath boulders and below mussel beds, have continued to escape oxidation by sunlight and removal by wave action. These pools continue to degrade the biological environment, and as a result, pink-salmon eggs continue to die and the growth of sea otters and harlequin ducks, which consume toxic clams contaminated by the oil, has remained retarded.

The *Exxon Valdez* disaster dramatized the need for improved methods to respond to such events, sparked renewed congressional demands for (very expensive) double-hulled, spill-resistant tankers, and prompted research that may someday yield a new strain of microbes that hungrily consumes spilled petroleum. In 2005, Congress actively debated a proposed plan to drill for oil in another environmentally sensitive area, the Arctic National Wildlife Refuge in the wilds of Alaska.

A recently discovered source of a new type of fossil fuel holds both the promise of a great store of energy and the threat of profound environmental disruption. Highlight 20-1 introduces this potential source of energy—and its problems.

Alternative Energy Resources

As fossil-fuel reserves dwindle and environmental damage related to their use increases, responsible governments and industries continue to seek alternative ways to meet our growing energy needs. Some alternative energy sources already in use, such as solar and wind power, are renewable; others rely on nonrenew-

▲ **FIGURE 20-13** Icelandic geothermal energy.

able natural resources, such as uranium, the element that powers nuclear energy.

Renewable Alternative Energy Sources

Renewable alternative energy resources are those that can be used virtually without depletion or that are replenished over a relatively short time span. These include geothermal, hydroelectric, tidal, solar, and wind energy and the energy produced by burning such renewable organic materials as trees and agricultural waste.

Geothermal Energy   Reykjavik, the capital of Iceland, is relatively pollution-free because it has a clean, inexpensive source of energy: Heat from shallow hot rock and Iceland's ever-present magma beneath the surface of the island's divergent zone is used to convert groundwater to hot water and steam (Fig. 20-13). The hot water is then circulated through pipes and radiators to heat homes and municipal buildings. The steam also drives electrical generators (Fig. 20-14a).

Since such *geothermal heat* was first tapped as an energy source in Larderello, Italy, in 1904, approximately 20 countries have taken advantage of it, including the United States (Fig. 20-14b). More nations would use this relatively inexpensive, nonpolluting energy if they could. Unlike oil, coal, and natural gas, however, geothermal energy cannot be exported and must be used close to its sources. Every nation using geothermal energy is located on a currently or recently active plate margin or near an intraplate hot spot, where magmatic heat has not yet dissipated.

Some areas, such as the American Southwest, have significant subterranean heat but little water to transfer it to the surface. Some of these localities are experimenting with "dry-rock" geothermal projects. Explosives fracture warm, shallow rocks to

Highlight 20-1

Gas Hydrates—A Major 21st-Century Fuel or a Potential Global Environmental Disaster?

In the years to come, the name *methane hydrate* will be heard increasingly in the offices of oil companies and in the chambers of national legislatures. In the 21st century, this substance may hold a key to the world's future energy needs as other fossil fuels become exhausted. Methane hydrate is a crystalline combination of natural gas and water, locked together in a substance that looks like ice but burns when ignited. Immense amounts of methane gas (CH_4) are trapped in tiny cages of ice (the "hydrates") within marine sediments along the world's continental margins at depths greater than 300 m (1000 ft). This gassy sediment layer may be as thick as 1 km (0.62 mi).

The origin of this hydrocarbon gas begins with seafloor bacteria. As these microscopic organisms consume bits of plant and animal matter, they excrete methane in much the same way that cows and sheep do on a farm. The principal difference—when conditions are cold and pressure is high (such as at the bottom of deep water), the bacterial gas becomes locked in hydrate crystals and preserved in enormous quantities.

Until recently, methane hydrate was seen as just a nuisance—something that plugged up our natural-gas pipelines. Some geologists now propose that the world supply of this gassy ice could provide more than twice as much energy as all the world's oil, coal, and gas deposits combined. The U.S.

Geological Survey estimates that the methane hydrates trapped within U.S. continental shelves alone contain roughly 200 trillion ft^3 of natural gas, enough to supply the nation's energy needs for more than 2000 years at our current rate of use. But one question plagues both scientists and eager lawmakers: From an environmental standpoint, can we tap this resource safely without causing widespread environmental disruption?

During this era of global warming, scientists worry that shallow methane deposits could melt, releasing millions of tons of this gas to the atmosphere. Our attempts to warm these icy gases so that we can collect them could have the same effect. Methane is a greenhouse gas, so escaping methane could profoundly affect global climate. In fact, catastrophic release of methane from melting gas hydrates has been implicated in extreme global warming events in the past.

Gas hydrates also cement the sediment grains of continental-shelf sediments, so their melting could strongly affect the stability of the continental margins. Release of the binding gas hydrates from these sediments could cause massive submarine landslides along continental slopes, and in the process, trigger huge, life-threatening tsunami.

Today, gas hydrates are being studied in at least two ways—by measuring the velocity of seismic waves traveling through

continental-margin sediments, and by drilling through sediments to bring samples (see photo) to the surface. Sediment layers cemented with gas hydrates have a higher seismic velocity than those without this hydrate cement. Thus, gas hydrate–rich layers can be imaged by their seismic properties. Direct sampling by drilling is useful for detailed chemical analysis of gas hydrates and their host sediments. The drilling remains extremely challenging, however, because the methane ices are stable only at low temperatures and high pressures, very different conditions from those found at the Earth's surface. If these conditions are not maintained, gas hydrates tend to melt and could explode when brought to the surface. Special pressurized drilling devices are required to bring these gases up to the surface in a stable form.

Because gas hydrates are stable at cold temperatures and high pressures, they may be destabilized and therefore released from the sediments in two very different ways—by raising their temperature (such as by warming of the oceans) or by decreasing their pressure (such as by falling sea level or by landsliding of the overlying sediments). Sea level falls during ice ages when glaciers store a large amount of the Earth's water. Imagine that during an ice age, sea level falls sharply, the pressure on gas hydrates drops sharply, and the gas hydrates begin to break down, releasing methane to the atmosphere. The resulting greenhouse effect would, in turn, warm the Earth's climate, melt the glaciers, and restore sea level to pre-ice age levels. This scenario sounds like a plausible way for the Earth to regulate its temperature. If too much methane is released, however, runaway greenhouse conditions might develop—as they apparently have in the Earth's past. This intense global-warming mechanism has been proposed as the cause of a major warming event—and its resulting *marine* mass extinction—55 million years ago. (Land animals were far less affected.) Clearly more research needs to be done to understand gas hydrates and their role in climate change and mass extinctions before we attempt to extract them as a way to solve our pressing energy needs.

▲ Samples of gas hydrates collected in the Canadian Arctic.

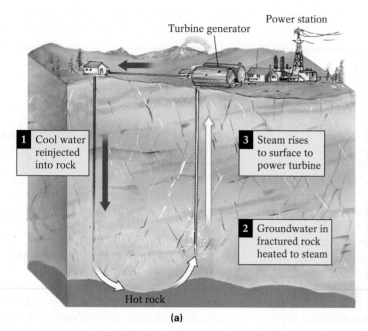

(a)

(b)

▲ FIGURE 20-14 Geothermal heat provides nearly nonpolluting energy (although some SO_2 and H_2S can be released from vented steam). **(a)** Groundwater heated by shallow magma or hot rock is converted to steam, which drives turbines, generating electricity. Cooled water is generally reinjected into the system to keep the cycle going. **(b)** The Geysers geothermal plant, 140 km (90 mi) north of San Francisco in Sonoma County, is the world's largest geothermal operation. It uses heat from subterranean rocks warmed by recent volcanic activity to supply the energy needs of 500,000 homes in the Bay Area.

increase their permeability. Imported water is then injected into the fractures, creating hot water or steam (Fig. 20-15). In the Jemez Mountains of north-central New Mexico, water is injected to a depth of 10 km (6 mi) into young volcanic rocks whose temperature is about 200°C (400°F). The resulting steam drives turbines and produces electricity.

Keep in mind that geothermal energy remains a renewable resource only as long as the fluids are returned to the hot rocks of the system. If steam is vented to the atmosphere or the hot water dispersed to area streams and lakes (a source of "thermal" pollution), the resource may be depleted, especially in areas where an arid local climate limits groundwater supplies.

Hydroelectric Energy
Falling water has been used for centuries as an energy source to mill flour and saw logs. Today, hydroelectric facilities use falling water to produce electricity. To generate hydroelectric power, a high-discharge stream is impounded by a dam to create a vertical drop sufficient to rotate the blades of large turbines.

Hydroelectric power is widely available. Since 1983, nearly one-third of all *new* electricity-generating plants in the United States have been hydroelectric installations. At present, however, the United States generates only about 15% of its current electricity output in this way. The Federal Power Commission estimates that if every sizable river in the United States were used, hydroelectric power could supply 50% of our total electricity needs. Global hydroelectric development lags even further; only about 6% of the world's hydroelectric potential is being used. In South America and Africa, where this source has the greatest potential, only 1% has been developed. Canada, at the other extreme, gets 75% of its electricity from this clean resource.

Although hydroelectric power is nonpolluting, dams can disrupt the local ecological balance by altering or destroying wildlife habitats. Dams also affect natural erosion processes; their reservoirs eventually fill with sediment that would otherwise replenish coastal beaches. Decisions to build dams must balance

Geothermal energy

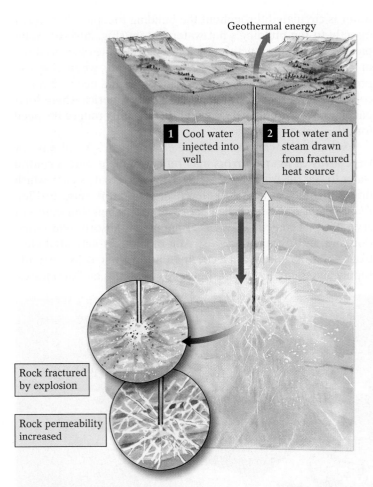

1 Cool water injected into well

2 Hot water and steam drawn from fractured heat source

Rock fractured by explosion

Rock permeability increased

▲ **FIGURE 20-15** Geothermal energy can be harnessed even in arid regions by injecting imported water into warm rocks fractured by explosives.

their environmental costs against their energy yield. Several dams in the United States have recently been removed to restore natural river conditions.

Tidal Power In coastal areas with a high *tidal range*—defined as the difference in the water-surface level between high and low tide—we can harness energy from rising and falling water levels by building a dam across a narrow bay or inlet. Here's how: During rising tides, the dam's gates remain open, allowing water to enter the bay. The flow of this water can be channeled through the system's turbines to generate electricity. When the water in the bay reaches its maximum height, the dam's gates are closed, trapping the water. The elevated water is then channeled seaward through the same turbines, producing more renewable, pollution-free energy (Fig. 20-16). The world's largest tide-powered plant, the Rance River project in France, provides virtually all the electricity used by the French province of Brittany.

Tidal power, however, is not without its own problems. It requires a minimum tidal range of 8 m (26 ft), and it also disturbs the ecology of estuarine habitats. As yet, no tidal-power facilities have been built in North America, although Passamaquoddy Bay in northeastern Maine, with a tidal range of 15 m (50 ft), is a strong candidate for future development. Likewise, the Bay of Fundy in New Brunswick, Canada, with a tidal range of about 20 m (66 ft), is a potential tidal-power site. Maximum development of the United States's potential tidal power would provide only 1% of our total electricity needs, although it could become a significant supplement to other local energy sources. The worldwide potential, which is only slightly better, is estimated at 2%. Still, the power is clean, reliable (tides are always rising and falling), and relatively inexpensive to maintain. Certainly worth the effort.

Solar Energy Solar-powered pocket calculators and wristwatches use an energy source that requires no expensive drilling or destructive strip mining, cannot be monopolized by unfriendly political regimes, and produces no hazardous wastes or air pollution. The Sun, expected to shine for another 5 billion years or

Power plant

Turbine housing

Incoming tide moves through open gates

1 Flood tide

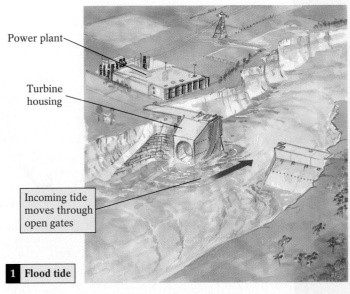

Electricity generated by turbines

Gates closed

Water forced through turbines

2 Ebb tide

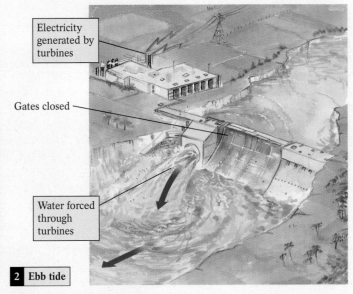

▲ **FIGURE 20-16** Producing electricity through tidal power. Here, tidal power taps the energy of falling water by trapping water at high, or flood, tide and then releasing it seaward through electricity-producing turbines at low, or ebb, tide.

so, is a totally renewable energy source. Solar power can heat buildings and living spaces and generate electricity—energy needs that together account for two-thirds of North America's total energy consumption.

Solar heating can be either passive or active. *Passive* solar heating distributes the Sun's heat naturally by radiation, conduction, and convection. At mid-northern latitudes, the simplest way to heat spaces passively is to construct buildings with windows facing south. As Figure 20-17a shows, sunlight passes through the window glass and heats objects within the room. Their heat then radiates to warm the air. Such an architectural design, coupled with efficient insulation, sharply reduces both air pollution and the cost of heating with fossil fuels.

Active solar heating uses water-filled, roof-mounted panels with black linings to absorb maximum sunlight. The solar-heated

water is circulated throughout the building for space heating or directly to the building's hot-water system (Fig. 20-17b). Solar panels are most productive in mild, sunny climates such as those in Florida, Texas, the Southwest, and California, where they can provide as much as 90% of a building's heating needs. Even in colder regions, such as northeastern North America where solar panels are less productive, they can significantly reduce the need for other energy sources.

Solar energy can also generate electricity in several ways. A large array of many mirrors may reflect sunlight onto a central water tower; the water there is then heated to create steam, which drives turbines attached to electrical generators (Fig. 20-17c). *Photovoltaic cells* use sunlight more directly. Their thin wafers of crystalline silicon coated with certain metals absorb solar radiation and, in turn, produce a stream of sunlight-activated electrons—that is, a flow of electricity. Photovoltaic cells currently supply electricity to tens of thousands of homes and businesses.

▼ **FIGURE 20-17** Solar power can be used in several ways. **(a)** Solar energy passively heats a single dwelling. **(b)** Water-filled panels on rooftops provide hot water and space heating, a type of active solar heating. **(c)** This experimental facility near sunny Daggett, California, uses solar energy to generate electricity. Reflecting mirrors focus on a water tower, concentrating the Sun's energy and converting the water to turbine-driving steam.

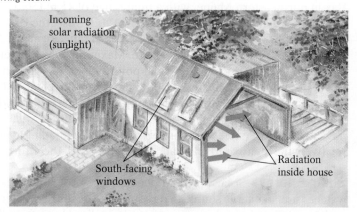

(a)

(b)

(c)

They are becoming increasingly popular at remote sites, largely because they do not require expensive transmission lines to deliver the electricity to the consumer. Satellites are also powered by photovoltaic cells. Roadside emergency phones are powered by these cells as well (Fig. 20-18).

Solar energy can generate electricity more efficiently in some regions than in others, but the technology to maximize its effectiveness at low cost is not yet widely available. To meet current U.S. electricity needs with solar power would require a system of collecting mirrors that would occupy about 25,000 km^2—about one-tenth the size of the state of Nevada—which is clearly not feasible. Substantial use of solar energy to generate electricity probably remains decades away.

Wind Power
Like falling water, tidal motion, and sunlight, wind power is a clean, renewable, nonpolluting energy source with a long history of use. The picturesque windmills of the Netherlands and those of rural midwestern North America have pumped groundwater and powered sawmills and flour mills for centuries. Wind power, however, is rarely cost-effective on a large scale: Only in a few sparsely populated mountain passes do winds blow constantly, forcefully, and from a consistent direction—all requirements for a practical application.

During the energy crisis of the 1970s, when fuel costs rose because of limited supplies of imported oil, engineers in the Department of Energy studied wind power as a possible way to decrease U.S. energy dependence on foreign oil. In the 1980s, a pilot project at Altamont Pass, east of San Francisco, connected 2000 wind turbines to electrical generators (Fig. 20-19). Altamont has

▲ **FIGURE 20-18** Power for roadside phones, far from electricity sources, is often provided by photovoltaic cells.

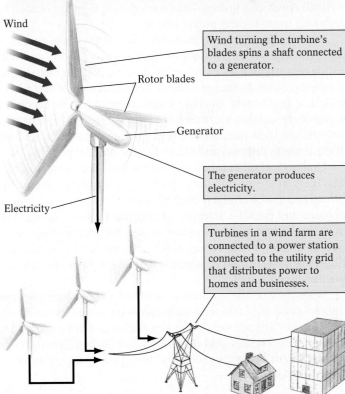

Wind

Wind turning the turbine's blades spins a shaft connected to a generator.

Rotor blades

Generator

Electricity

The generator produces electricity.

Turbines in a wind farm are connected to a power station connected to the utility grid that distributes power to homes and businesses.

▲ **FIGURE 20-19** The productivity of this wind farm at Altamont Pass, California, is made possible by sustained westerly winds that blow through California's Coast Range.

contributed substantially to the Bay Area's electricity needs, and California's government harvests as much as 8% of the state's electricity from wind farms.

Drawbacks to wind power include its limited geographic application and the large area needed to develop it. In North America, the wind force in the East and industrial Midwest does not meet the minimum requirements, nor do those densely populated areas have sufficient available land on which to locate wind-power facilities. The gusty Great Plains of Nebraska, Kansas, Oklahoma, and Colorado are possible candidates, although California, with its many large, undeveloped, windy passes remains the best possibility. Northern Oregon and southern Washington have recently begun development of wind power along the windswept Columbia River Gorge. Wind farms have also sprung up along Interstate 80 in eastern Wyoming.

Biomass

In developing countries, as much as 35% of the energy used for cooking and heating comes from burning wood and animal dung. These fuels, derived from plants and animals, are known collectively as **biomass fuels.** Biomass fuels also include grain alcohol (used as an additive to gasoline), methane gas that rises from decaying garbage in landfills, urban trash, and plant waste from crops such as sugar cane, peanuts, and corn. The most widely used biomass fuel is wood, which today heats about 10% of North America's homes—more residences than are heated by electricity from nuclear power plants.

Biomass fuels are a renewable resource—when managed well. Although trees grow slowly, continuous planting and harvesting can produce a steady supply. Many countries have begun to develop biomass-energy sources in anticipation of the end of the era of fossil fuels. Burning trash for electricity is becoming increasingly common around the world. In Rotterdam, the Netherlands, trash fuels an electrical generator that provides both efficient waste disposal and electricity for 250,000 people.

Unlike most other renewable energy resources, however, biomass fuels create air-pollution and global-warming problems. As with oil and coal, burning these fuels introduces potentially harmful gases and particles into the air, further reducing air quality. Moreover, in arid regions, overuse of trees for energy removes the root systems that help retain water and soil. This removal contributes to desertification, which in turn kills off the animals that provide dung for fuel.

What may be the ultimate renewable, non-polluting energy source—good old-fashioned hydrogen—is thought by many scientists to be "the answer"—the key source of future energy for our homes, businesses, and cars. The following highlight provides a status report on hydrogen fuel cells, a potential revolution in energy production.

Nuclear Energy—A Nonrenewable Alternative

In the mid-20th century, physicists harnessed the energy released when the nuclei of radioactive isotopes split apart spontaneously. Now they are attempting to tap the energy produced when atomic nuclei are fused. These energy sources may provide yet another 21st-century solution to the inevitable loss of fossil fuels.

Energy from Fission When the nuclei of the atoms of certain isotopes of heavy elements are bombarded with neutrons, they split into several lighter elements. In this reaction, they release part of the binding energy of an atom's nucleus, additional neutrons, and an enormous quantity of heat. This process is called **nuclear fission.** The heat generated by the reaction can convert water to steam, driving turbines and producing electricity. The released neutrons can in turn bombard other heavy nuclei, setting off a chain reaction.

In commercial nuclear reactors, the naturally decaying nuclei of uranium-235 atoms release neutrons that in turn bombard other U-235 nuclei, which then undergo fission (splitting). This fission produces heat and, most importantly, more neutrons, which then bombard and split neighboring U-235 nuclei (Fig. 20-20). Nuclear reactors are designed to control the production of neutrons and their flow to the fissionable material, thereby preventing uncontrolled chain reactions (as in a nuclear bomb). Neutron-absorbing control rods are moved in and out of the uranium fuel to regulate the rate of reactions. The enormous amount of heat generated by nuclear reactions is transferred to a coolant, usually water, which is converted to steam to drive turbines.

Uranium, a relatively uncommon element in the Earth's crust, in places becomes concentrated in granitic magmas in the late stages of cooling. Thus uranium sometimes occurs in pegmatites as a crystalline uranium oxide, such as the mineral uraninite (UO_2). This element is highly soluble, so it readily dissolves when granitic igneous rocks are weathered. Uranium is then transported by groundwater into permeable sediments and sedimentary rocks, where it bonds to the surfaces of organic-matter particles. These concentrations serve as the source of most uranium ore. Uranium ore is commonly mined from ancient stream sands and gravels, either in granular form (as the black, shiny, noncrystalline uranium oxide *pitchblende*) or as a canary-yellow encrustation on sand grains (as the potassium–uranium–vanadium mineral *carnotite*).

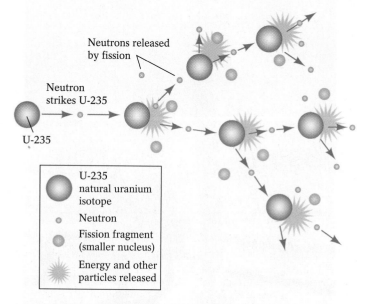

▲ **FIGURE 20-20** A neutron released by the natural decay of a U-235 nucleus initiates a chain reaction.

Highlight 20-2

Hydrogen—The Solution to All Our 21st-Century Energy Needs?

Hydrogen is everywhere throughout our environment—in the water we drink, in our global oceans, in the groundwater beneath our feet. And in the next decade or two, this vastly abundant element may power our cars, heat our homes, fuel our farm tractors, and power our factories. And, amazingly, the only waste product of all this energy production is simple H_2O. The secret to this potential revolution in energy production is the hydrogen fuel cell, and here's how it works.

First things first: Electricity is simply a flow of electrons—those negatively charged, subatomic particles. So where can we acquire an unlimited supply of electrons? From the hydrogen atom, which, as we discussed in Chapter 2, consists of a single proton in its nucleus and a single orbiting electron. If we can strip the electrons from a stream of hydrogen atoms, we can produce an unlimited supply of electricity using just unspectacular, garden-variety hydrogen. The device to do this is the hydrogen fuel cell (see figure).

In its basic form, the fuel cell has two streams passing through channels within its charged electrodes—one hydrogen gas, the other just air. Hydrogen gas is passed through the cell's *positively* charged electrode, which attracts and collects the electrons from the hydrogen atoms in the gas (remember, "opposites attract"). These flowing electrons create an electrical current that can then be used to do work. Air—21% of which is oxygen—is passed through the cell's *negatively* charged electrode and then attracts the positively charged hydrogen (H^+) ions, which are essentially the atom's single-proton nucleus now missing its stripped-away electron. The oxygen in the air stream flowing through the negatively charged electrode reacts with these discarded hydrogen nuclei and with the electrons that are guided back to the negatively charged electrode (after doing their work) to produce H_2O, the waste product of these reactions. Thus, the Earth's virtually unlimited supply of hydrogen gas

and a similarly unlimited supply of air combine to produce electricity and water—an energy prospect as exciting as the unfulfilled promise of cold fusion or the fanciful dream of a perpetual-motion machine.

So why aren't all our cars powered and all our homes heated by such fuel cells today? Because at present, the process is way too expensive and the fuel cells themselves way too large for our modern cars. (NASA has used this technology on satellites, space shuttles, and space stations; they are fine in the weightlessness of space, but they are not yet ready for widespread commercial use.) Another key stumbling block is that the cell itself requires a large supply of platinum to coat the electrodes that act as catalysts for the cell's electron-stripping and the water-producing reactions. And the cost of platinum at the present time renders the cost of producing energy this way prohibitively expensive. Many research labs around the world are tackling these problems and searching for alternatives to the platinum catalyst and finding ways to produce more compact, practical cells. For now, the hydrogen fuel cell remains a tantalizing prospect, one that may eliminate the need for fossil fuels and one that holds the promise of pollution-free energy. As one fuel-cell expert stated: "The exhaust pipe of your car would be just like a tea kettle." That image is certainly appealing to those among us who live and breathe in smog-shrouded urban centers.

Fuel cell

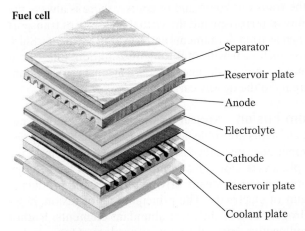

- Separator
- Reservoir plate
- Anode
- Electrolyte
- Cathode
- Reservoir plate
- Coolant plate

An electrochemical reaction
In the cells, the electrons are stripped of hydrogen atoms in the platinum catalyst.
The electrons leave the cell, creating current.

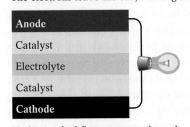

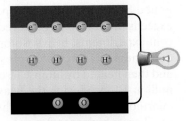

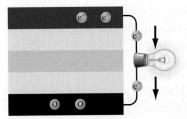

Hydrogen fuel flows as a gas through channels in the reservoir plate to the anode on one side of the fuel cell. Oxygen from air flows through channels to the cathode on the other side of the cell.

A platinum catalyst in the anode causes hydrogen to split into protons (positively charged H ions) and electrons (negatively charged particles).

Electrons cannot enter the electrolyte; they travel through an external circuit, creating an electrical current.

Electrons travel back to the cathode, where they combine with hydrogen ions and oxygen to make water. The water then flows out of the cell.

In North America, uranium deposits are found in the Mesozoic stream gravels of Colorado, New Mexico, Wyoming, and Texas, often associated with fossilized plant and wood remains. (The uranium in circulating groundwater bonds with and becomes concentrated in dead plant cells.) During the uranium boom of the 1950s, western prospectors toting Geiger counters tested tens of thousands of petrified logs.

Uranium-235, the only naturally occurring radioactive isotope that can maintain a nuclear chain reaction, accounts for only 0.7% of natural uranium; nonfissionable U-238 makes up the remaining 99.3%. Because U-235 is consumed during fission, at its current rate of use, recoverable reserves of U-235 may be sufficient to power the world's 575 nuclear power plants for only another 30 years or so.

To address the problem of dwindling U-235 supplies, nuclear engineers are experimenting with techniques that are *safe and economical* to produce fissionable plutonium-239 from the widely abundant but *nonfissionable* U-238. In one such process, uranium-238 is placed in a reactor with a small amount of U-235. The U-235 and its neutrons bombard the U-238 nuclei, converting them to a form of plutonium, Pu-239. This process takes place in a so-called **breeder reactor,** a chamber in which fission of the manufactured plutonium atoms produces a surplus of neutrons that can then be used to create, or "breed," more fissionable material than they consume.

Currently, few breeder reactors are on-line, primarily because Pu-239 is potentially a weapons-grade nuclear material. Funding for Tennessee's Clinch River breeder reactor was canceled in the mid-1980s, in large part because of public concern about a global increase in nuclear weapons. The Super Phoenix breeder reactor in France is currently the world's largest breeder facility.

In 1974, the U.S. Geological Survey predicted that the United States would get 60% of its electricity from nuclear plants by 2000. In 1993, this percentage was about 16% and dropping. Meanwhile, western Europe and Japan, both of which have little homegrown fossil fuel, were expanding their nuclear-energy facilities. France, for example, derives 65% of its electricity from its many reactors. A number of reasons explain the United States's reluctance to develop nuclear energy. In addition to concern about weapons-grade fuels falling into the wrong hands, there are technological problems of reactor safety and radioactive-waste disposal, and numerous other economic, political, and psychological problems associated with the growth of nuclear energy.

At Three Mile Island, Pennsylvania, in 1979, an instrument malfunction led plant operators to conclude that too much water was flushing through the reactor's cooling system. They responded by reducing water flow, leaving the reactor core uncovered for several hours and allowing it to overheat. Quite fortunately for Pennsylvanians, although the reactor core suffered damage, little measurable radiation escaped.

In 1986 in the former Soviet Union, technicians working at a poorly designed nuclear facility at Chernobyl accidentally allowed a runaway chain reaction to develop. (Apparently, many of the plant's safety systems had been disabled to allow Soviet engineers to experiment with the reactor core.) Two small explosions blew the roof off the building, showering the immediate area with radioactive material. Eastern Europe and Scandinavia were covered with a cloud of radioactive vapor. Numerous cases of radiation sickness and cancer developed in the aftermath of this accident. The event disrupted life in Chernobyl's immediate vicinity as well as agriculture and commerce over a wide area.

Radioactive nuclear waste is so toxic that it must be isolated from all life for at least 10,000 years. Thus, safe disposal is the industry's greatest technological (and political) problem. Spent fuel rods, internal machinery from reactor cores, and the waste products of nuclear-fuel processing must be stored in a way that prevents *any* leakage to the atmosphere or groundwater system. A burial site must be seismically stable—an earthquake could damage the facility or a new fault could breach the repository—and be completely isolated from groundwater.

In 1991, Congress selected Yucca Mountain, about 160 km (100 mi) northwest of Las Vegas in southwestern Nevada as the first U.S. nuclear-waste repository (Fig. 20-21). Located within the Nevada Test Site (where the first atomic bomb was tested), this area is largely uninhabited, has been relatively free of earthquakes during recorded time, and is so arid that its water table is very deep. In 2002, despite ongoing strong political opposition by the governor of Nevada, construction of the repository at Yucca Mountain continued apace with a planned opening date sometime in 2010. For the time being, this deadly material is stored in *temporary* facilities in 39 states—certainly a precarious situation in light of the potential for a terrorist attack. Once the Yucca Mountain facility is available, roughly 80,000 tons of high-level radioactive waste will be transported by rail and truck to Nevada, another cause for great concern. Millions of Americans live along the transport route and of course, there is always the real possibility of terrorism and the virtual certainty of transport accidents, such as train derailments and trucking accidents. Shipping distance each year will total roughly 2 million km—about 1.2 million miles—quite a distance to guard against assorted dangers and threats to the deadly cargo.

Energy from Fusion
Nuclear fusion occurs when atomic nuclei of light elements are subjected to such extremely high pressure and temperature that they fuse and form heavier atoms. Like fission, this process releases an enormous amount of heat energy that can convert water to steam, driving turbines and producing a vast amount of electricity. The principal fuel of fusion, however, is hydrogen, one of the most abundant elements. Rather than toxic radioactive waste, the primary product of fusion is helium, a harmless inert gas.

Fusion's potential is enormous: The energy that could be generated by the hydrogen isotopes in 1 km^3 (0.24 mi^3) of seawater exceeds the total energy stored in the world's oil reserves. In recent experiments, heavy isotopes of hydrogen (tritium, ^3H, and deuterium, ^2H) were compressed in a powerful magnetic field and then heated by intense laser bursts to about 50 million°C (90 million°F), producing the heavier element helium (He) and an enormous amount of heat. Physicists have been able to sustain small-scale hydrogen fusion to generate energy, although it is not yet economically feasible; the amount of energy needed to achieve fusion still exceeds the amount produced by it. Further study and expensive experimentation will be needed to make this process commercially viable.

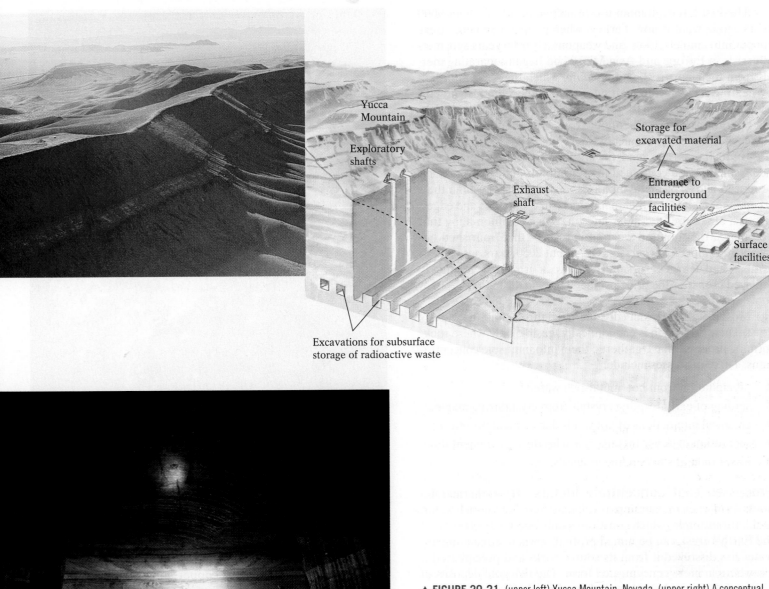

▲ **FIGURE 20-21** (upper left) Yucca Mountain, Nevada. (upper right) A conceptual sketch of the United States' first permanent repository for high-level nuclear waste at Yucca Mountain, Nevada. When preparation of the site finishes, according to the construction schedule in 2010, radioactive waste from around the country will arrive in 3-m (10-ft)-long containers and be transferred to a cavern excavated into a thick layer of relatively impermeable ash-flow tuff, 300 m (1000 ft) below the surface (lower left).

Mineral Resources

Aside from the energy resources that heat our homes, power our cars, and sustain our industries and farms, modern life requires *vast* supplies of minerals. As we discussed way back in Chapter 2, minerals are naturally occurring inorganic solids with their chemical elements in specific proportions and their atoms arranged in a systematic internal pattern. Of the 3000 or so known minerals, only a few dozen have economic value. These include various metallic minerals, such as the oxides and sulfides of iron, copper, and aluminum; certain metals, such as gold, silver, and platinum, which commonly occur uncombined with other elements; and various nonmetallic rock and mineral materials, such as sand and gravel (used as building materials), limestone (for cement), halite (common table salt), and a few highly prized gemstones (such as diamonds, rubies, and emeralds).

In this section we describe how these valuable metals and nonmetals form and how they are used. We also discuss some environmental problems that develop when we mine them.

Metals

Most metals combine readily either with oxygen to form oxides or with sulfur to form sulfides. Common oxides include those of tin, iron, aluminum, and manganese; common sulfides include those of zinc, lead, iron, copper, and molybdenum. *Native metals,* such as gold, platinum, and silver, do not combine with other elements.

The first known human use of metals occurred about 9000 years ago in what is now Turkey, when people hammered pure copper into amulets, tools, and weapons. By 6000 years ago, metallurgists in Europe and Asia Minor had begun separating metals from their host rocks by *smelting,* which they accomplished by heating rocks to the melting points of their metals and collecting the molten metal for further processing. The earliest smelting operations extracted copper from copper sulfides, such as chalcocite (Cu_2S) and chalcopyrite ($CuFeS_2$).

By 5000 years ago, lead, tin, and zinc, as well as copper, were being smelted and combined in their molten state. This process produced **alloys,** or metal mixtures, that were harder than their component metals and that maintained sharper points and edges. Bronze, a mixture of copper and tin, was a common alloy that dominated the tool-and-weapon industries that flourished between 5000 and 2650 years ago. By about 2650 years ago, metallurgists had learned to smelt and work iron, which quickly replaced bronze as the principal metal used in the manufacture of tools, weapons, and armor.

Most metals are widely disseminated throughout the Earth's crust. Their extraction is economically feasible only where rock-forming processes have gathered them into mineable concentrations. These processes include:

- precipitation from hot, metal-rich water;
- settling of early-forming crystals from crystallizing magma;
- chemical interactions of hot fluids during metamorphism;
- separation of dense metallic particles during sedimentation;
- dissolution of surrounding minerals.

Processes That Concentrate Metals

Hydrothermal deposits form when metallic ions precipitate from hot ion-rich water. Gold, for example, which constitutes only about 0.0000002% of the Earth's crust, can be mined profitably where circulating hot water has dissolved it from its source rocks and precipitated it elsewhere in more concentrated form. The rich gold deposits of California's Sierra Nevada and along Cripple Creek in the Colorado Rockies precipitated from hydrothermal solutions. Similar processes produced the silver deposits of Coeur D'Alene, Idaho; the copper of Butte, Montana and the Keweenaw Peninsula of northern Michigan; and the lead–zinc deposits of the tri-state area of Missouri, Oklahoma, and Arkansas (Fig. 20-22).

A common source of metal-rich hot water is a cooling magma that contains abundant metallic ions after most silicate minerals have crystallized. Such hot solutions may contain copper, lead, zinc, gold, silver, platinum, and uranium (none of which bonds readily with silicate tetrahedra), as well as some uncrystallized silica (SiO_2). The hot water typically seeps into faults, fractures, and bedding planes of surrounding rocks, where it cools further and deposits its dissolved minerals as silica veins that may contain visible masses of metals (Fig. 20-23a). Some of the world's most productive gold and silver mines are hydrothermal vein deposits that cut across rocks adjacent to granitic batholiths.

Porphyry copper deposits may also occur in plutonic rocks. They form when the water from a cooling body of water-rich felsic or intermediate magma is converted to pressurized steam. As it expands outward, the steam ruptures the newly crystallized

▲ **FIGURE 20-22** A chunk of silvery galena (a lead sulfide) on dolomite (calcium-magnesium carbonate) with chalcopyrite (a copper-iron sulfide) from the Sweetwater Mine, in Milliken, Reynolds County, Missouri. The specimen is $7 \times 4 \times 5$ cm (about $3 \times 1.5 \times 2$ in).

rock at the magma's edge, creating numerous hairline fractures at the margin of a batholith. Hydrothermal solutions then fill these fractures, precipitating copper and other metals as they cool.

Hydrothermal deposits also originate from deep-circulating groundwater that has been heated through close contact with shallow magmas or young, warm, plutonic rocks. As the heated water rises, it dissolves metals from the pluton and surrounding rocks. It then flows into fractures, joints, bedding planes, and faults, where it precipitates its dissolved metals as concentrated vein deposits. Other hydrothermal solutions pass through porous rocks and cool slowly, precipitating their metals in microscopic pore spaces over large areas. These deposits are much more dispersed and thus they are far less cost-effective to mine.

Massive sulfide deposits form where cold seawater enters the fractures at divergent zones and then becomes heated by shallow basaltic magma. As the hot water rises through the fractures, it dissolves some of the trace metals in the basaltic rocks of the mid-ocean ridge. It also picks up sulfurous gases typical of basaltic magma. The resulting hydrothermal solution contains sulfides of copper, zinc, manganese, lead, and iron (Fig. 20-23b). When the solution erupts at the seafloor and rapidly cools, it precipitates its metals in a black jetlike plume called a *black smoker.* The "smoke" is actually fine black particles composed largely of sul-

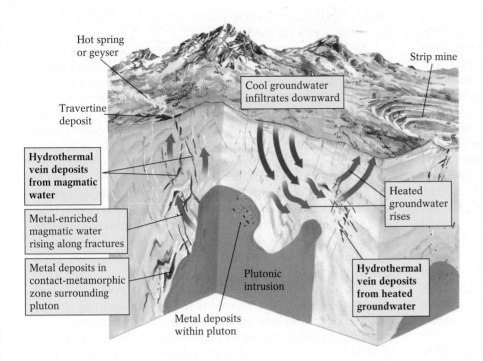

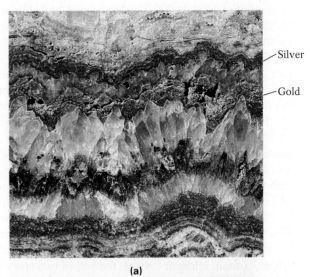

Silver

Gold

(a)

(b)

▲ **FIGURE 20-23** Hydrothermal processes that concentrate valuable metals. Metals may precipitate from magmatic water or from circulating groundwater or seawater heated by the magma. **(a)** These veins of gold and silver ore formed when hot magmatic solutions of the metals entered cracks around the cooling magma. **(b)** These massive sulfide deposits in Quebec formed at an ancient mid-ocean ridge where seawater heated by basaltic magma emerged at hot springs and precipitated large quantities of metals on the ocean floor.

fide minerals such as pyrite (FeS_2), chalcopyrite ($CuFeS_2$), sphalerite (ZnS), and galena (PbS).

The processes that crystallize minerals from magma also produce metal deposits. In *gravity settling,* dense, *early-crystallizing* minerals sink through the magma to the bottom of the magma chamber, where they accumulate in layers. In *filter pressing,* large, early-forming crystals stick to the walls of the magma chamber when tectonic forces compress the chamber and force the still-liquid portion of the magma into fractures in adjacent rocks. Gravity settling and filter pressing of mafic and ultramafic magmas have produced valuable ore bodies of iron (in magnetite, Fe_3O_4), chromium (in chromite, $FeCr_2O_4$), titanium (in ilmenite, $FeTiO_3$), and nickel [in pentlandite, $(Fe, Ni)_9S_8$)]. The rich chromite deposits of the Stillwater intrusive complex of northwestern Montana and the Bushveld complex of southern Africa formed from gravity settling of chromite crystals from mafic and ultramafic magmas (Fig. 20-24). The iron ore of the Kiruna district of northern Sweden (60% iron) resulted from filter pressing

▲ **FIGURE 20-24** Deposits of chromite (black layers) in the Bushveld complex of southern Africa. In addition to its vast deposits of chromite, the Bushveld complex contains 70% of the world's platinum, one of the most precious metals. This huge sill, 240 km (150 mi) by 480 km (300 mi) in area and 8 km (5 mi) thick, was the site of gravity settling of heavy early-forming crystals.

of magnetite crystals. The world's largest nickel deposit is found at Sudbury, Ontario, near the shores of Lake Superior. It is thought to have formed as droplets of nickel sulfide rose from the Earth's mantle and then settled through a ring-shaped system of fractures that resulted from the impact of an enormous meteorite about 1.9 billion years ago. Some of Sudbury's nickel deposits may contain the remains of the impacting meteorite.

Felsic magmas also are a potential source of valuable ore bodies. Late in the cooling of granitic magma, a substantial amount of water remains along with uncrystallized silica (SiO_2) and some rare elements that have not yet crystallized. These elements may include lithium, beryllium, boron, uranium, fluorine, and cesium. In this watery, highly fluid, late-stage magma, unbonded ions migrate freely and bond readily to growing crystal structures. The resulting pegmatites typically contain very large crystals. Pegmatites at Kings Mountain, North Carolina, for example, contain feldspar crystals the size of two-story houses, and those in the Black Hills of South Dakota contain crystals of the lithium-rich mineral spodumene as large as telephone poles. Some pegmatites also contain precious crystals of beryllium-rich emerald and aquamarine and of boron-rich tourmaline.

When hot ion-rich fluids move through rock bodies, their heat can produce contact-metamorphic mineral alterations in the host rock, and metallic ores typically result. For example, when hot ion-rich fluids pass through impure limestones and dolostones that contain aluminum-rich clay, chemical reactions in the contact zone release CO_2. The CO_2 then migrates outward and produces an extensive metamorphic aureole that contains several valuable aluminum oxides (such as corundum, Al_2O_3) and other metal-rich minerals. Metallic ions from the hot fluids may replace a large amount of the calcium in limestone. In the vicinity of a basaltic intrusion, the replacement ions are typically iron (in magnetite). Other metallic deposits produced by contact metamorphism in-

clude zinc (in the mineral sphalerite), lead (in galena), and copper (in chalcopyrite and bornite).

Sedimentary processes also produce valuable metal deposits, ranging from the gold nuggets and gem crystals that settle out from the slow-moving currents of rivers to the vast iron deposits precipitated on the floors of ancient oceans. During transport, water-borne heavy materials, such as gold, platinum, and tin, are sorted out by their density and durability, becoming concentrated as **placer deposits** wherever flowing water slows. Placer deposits commonly occur in potholes in the stream bed, along the inside banks of meander bends, downstream from narrow sections in the channel, at places where two or more streams merge, and at coasts, where wave action cannot move the heavy materials (Fig. 20-25). Extremely durable minerals, such as diamonds, sapphires, rubies, and emeralds, also form placer deposits because they resist abrasion and can survive long journeys that wear away softer minerals.

The most valuable placer deposits consist of gold. Gold's high density (19 g/cm^3) causes it to sink from its transporting stream. Nuggets of gold weathered and eroded from hydrothermal vein deposits in California's Sierra Nevada batholiths have been carried down the range's numerous streams for hundreds of thousands of years. (The largest gold nugget ever found—discovered in 1869 in the southeastern Australia state of Victoria—weighed 70 kg, or 155 lbs. A smaller one—the Hand of Faith nugget found in 1980—weighed 27 kg, or 60 lbs. This nugget sold for more than $1 million dollars and sits on display in the Golden Nugget Casino in Las Vegas.) Placer deposits are generally discovered first; lucky prospectors then attempt to retrace the gold's journey upstream to the so-called mother lode from which the placers eroded, assuming that the source vein has not completely eroded away.

Most sedimentary iron deposits are found in a series of layered Precambrian sediments that were deposited in shallow

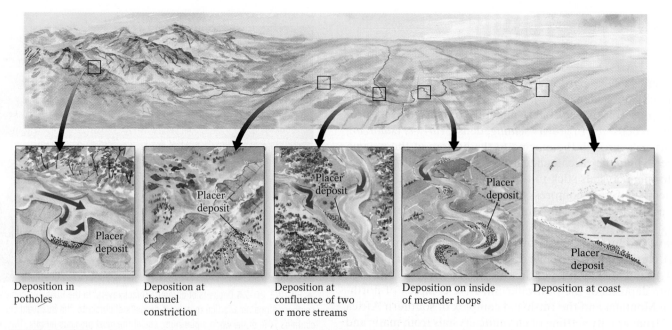

Deposition in potholes

Deposition at channel constriction

Deposition at confluence of two or more streams

Deposition on inside of meander loops

Deposition at coast

▲ **FIGURE 20-25** Placer deposits develop wherever the velocity of flowing water significantly declines, such as at potholes in a stream bed, downstream from a narrow spot in a channel, at places where two or more streams merge, at the inside of a meander bend, and at coasts, where dense minerals settle from the water and are reworked by wave action.

▲ **FIGURE 20-26** This banded iron formation in Australia began to form about 2 billion years ago when enough oxygen from photosynthesizing plants accumulated in the Earth's atmosphere to trigger oxidation and precipitation of iron.

content of the Earth's atmosphere and surface waters, which had previously been quite low, began to increase. Iron in solution could then precipitate in the form of such oxides as magnetite and hematite. Iron is very soluble in an oxygen-poor environment, and thus any iron that weathered from continental rocks prior to this time would have remained in solution during its transport to the oceans. Expansion of plant life on Earth added oxygen to the chemical composition of the planet's air and water, triggering precipitation of iron and producing the widespread iron-oxide deposits that account for more than 90% of the iron ore mined every year.

Metals also become concentrated either when other minerals weather and dissolve, leaving a metallic residue at or just below the surface, or when metals dissolved by groundwater precipitate elsewhere in concentrated form. These **secondary enrichment** processes develop best in a warm climate with abundant atmospheric water, permeable bedrock (that promotes groundwater flow), and a soluble host rock that surrounds economically valuable materials.

The most prominent weathering-produced ore is *bauxite* ($Al_2O_3 \cdot H_2O$), an aluminum oxide that is the most abundant source of aluminum. Sizable deposits of this valuable ore form only where extreme chemical weathering in warm, humid climates breaks down the aluminum-rich feldspars in granitic rocks and the aluminum-rich clays in impure limestones. The aluminum, which is highly insoluble, is left behind as an oxide after much of the rest of the rock has dissolved away. Bauxite is found today in tropical locations such as Jamaica. Deposits of this ore also occur in nontropical regions, such as France, Australia, and Arkansas (Fig. 20-27a), indicators of tropical environments in the past.

marine basins on nearly every continent about 2 billion years ago. These **banded iron formations** consist of alternating layers of light-colored chert and dark-colored highly concentrated iron oxides (Fig. 20-26). They are found in Labrador, eastern Canada, Brazil, western Australia, Russia, India, the western African nations of Gabon, Mauritania, and Liberia, and around Lake Superior in Minnesota, Wisconsin, and Michigan. Two billion years ago, when these sediments were deposited, vegetation (such as microscopic algae) had become plentiful enough to produce a significant amount of oxygen through photosynthesis. The oxygen

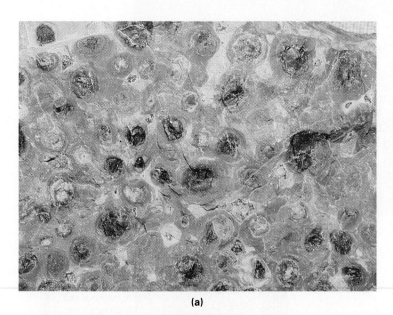

(a)

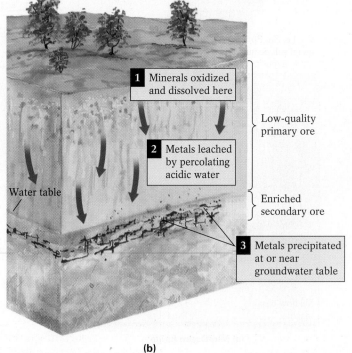

(b)

▲ **FIGURE 20-27** Secondary enrichment occurs when groundwater either **(a)** dissolves the host rock around valuable metals, leaving them as a residual deposit, such as this Arkansas bauxite, a main source of aluminum, or **(b)** dissolves the metals themselves and transports them in solution down to the groundwater table, where they precipitate. Copper is particularly susceptible to such secondary enrichment.

Soluble metals dissolved by acidic groundwater may be transported downward to the water table where they commonly precipitate in concentrated form. When sulfur is released from weathered sulfide minerals such as pyrite (FeS_2), the sulfides oxidize to produce sulfuric acid. The acid dissolves some metals, which then move downward in solution to the less acidic zone below the water table. There, the metals are precipitated. In this way, metals originally scattered through-

out a body of rock can become concentrated in the subsurface (Fig. 20-27b).

Metals and Plate Boundaries Most of the world's major metal deposits occur at past or present plate boundaries. At such locations, magmas are generated, hot hydrothermal solutions circulate through rocks, and plate collisions thrust metal-rich, seafloor rocks upward above sea level.

▼ **FIGURE 20-28** The processes that occur at each type of plate boundary often determine the type of metal found at specific locations. **(a)** Divergence produces massive sulfides and oxides rich in copper, lead, zinc, manganese, iron, and nickel. **(b)** Subduction may separate metals into distinct bands: (1) Typically, the first to appear are iron-rich. (2) Farther inland, gold and porphyry copper deposits, associated with felsic-to-intermediate intrusions, are found. (3) The next band of deposits generally contains hydrothermal veins of lead, zinc, silver, and copper. (4) Tin and molybdenum typically appear farthest inland. **(c)** Deeply seated deposits formed by plate divergence or subduction may be thrust to the surface by a continental collision.

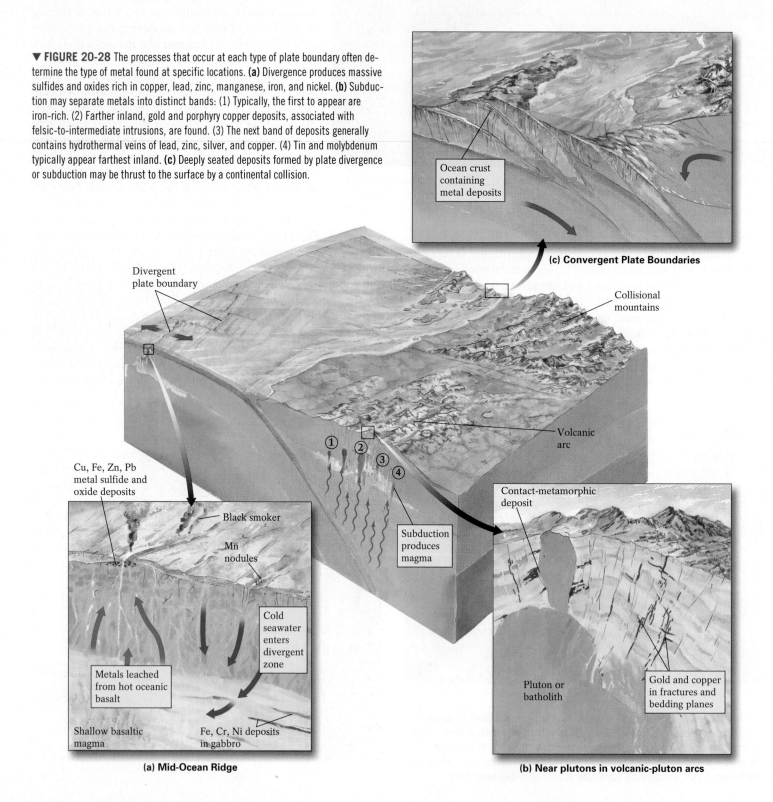

(c) Convergent Plate Boundaries

(a) Mid-Ocean Ridge

(b) Near plutons in volcanic-pluton arcs

Massive sulfide and oxide deposits—rich in copper, zinc, lead, and manganese—typically precipitate at or near mid-ocean ridges. There, plate divergence opens an extensive network of fractures through which cold seawater circulates into warm oceanic lithosphere (Fig. 20-28a). In the mid-1960s, geologists collected a hot, salty solution mixed with a black powdery substance at the floor of northwestern Africa's Red Sea, which occupies the recent rift between the African and Arabian plates. The black powder was rich in various sulfides and oxides of iron, zinc, and copper.

The sulfide deposits that form at divergent zones then become incorporated into the oceanic lithosphere that spreads away from the mid-ocean ridge. As spreading continues, these deposits eventually reach a subduction zone and descend to depths where higher subterranean temperatures reach the melting points of these metals. As subduction-produced magmas rise and cool, they incorporate these metals in subduction-zone batholiths, such as the Sierra Nevada of California. This process has produced the rich metal ores found along the western edges of the Americas from Alaska to Chile. Different metals appear at different positions within subduction-zone batholiths, sometimes as distinct bands, such as those depicted in Fig. 20-28b. The segregation of these metals largely occurs because of the concentrations of specific metals in their source magmas and because of local hydrothermal conditions.

After subduction consumes an oceanic plate, the extreme compression produced when plates collide may thrust slices of the subducted seafloor back to the Earth's surface. For this reason, continental-collision zones often contain masses of metal-rich rock that originated at a mid-ocean ridge (Fig. 20-28c). The island of Cyprus, for example, has risen from the floor of the eastern Mediterranean as the African and Eurasian plates have collided. Cyprus' vast copper deposits (from which the island derives its name) originally formed at the Mediterranean's mid-ocean ridge.

Environmental Problems Caused by Mining
Most major metal deposits are mined from open pits by operations that begin by removing millions of tons of *overburden,* the rock and regolith that cover ore deposits. These mine tailings, along with the great volume of waste from ore mills, are typically piled into huge hills and often left without covering vegetation. In this form, they are highly susceptible to rapid erosion and mass movement. Surface-water runoff from waste piles may pollute regional streams by clogging them with silt and poisoning them with toxic levels of dissolved metals and acid (Fig. 20-29). In the mine pits, ore rocks may come in contact with surface water, which may then become contaminated with sulfides. Sulfides in turn oxidize to form sulfuric acid, producing acidic ground and surface water.

These and other problems have prompted the U.S. Congress to pass legislation requiring monitoring and control of mine-water discharge. State and federal regulations also require land reclamation after a mining operation has been completed. The mine operator must isolate the sulfides from the groundwater system, restore the topography, and replant vegetation to prevent erosion (Fig. 20-30).

▲ **FIGURE 20-29** Acid mine water, Dola, West Virginia. Acid water is produced by the sulfur leached from the mine's coal (forming sulfuric acid).

Nonmetals

In the past, the search for gold, silver, and platinum has typically unleashed a stampede of prospectors and altered the histories of entire states and nations. Although no "gravel rush" or "phosphate rush" has occurred to match the gold rushes of 19th-century America, such nonmetals are at least as useful and valuable as those shiny precious metals. Nonmetal resources range from the building materials derived from common rocks to the natural mineral fertilizers that help to produce our food supply.

Nonmetal Building Materials
Only petroleum exceeds the combined economic value of natural building materials. Limestone, the most widely used building material, provides crushed stone for road beds as well as cut stone for monuments and other stately buildings. It is also a key ingredient in Portland cement, a staple of the construction industry. To make Portland cement, finely ground limestone and shale are mixed and then heated to 1480°C (2700°F) to drive off limestone's carbon dioxide. The resulting quicklime (CaO) reacts with clays in the shale to produce a calcium slag that is then mixed with a little gypsum ($CaSO_4 \cdot 2\,H_2O$). This cement hardens slowly when mixed with water.

Gypsum, which is soft and soluble, is rarely used as a building stone. Heating gypsum to 177°C (351°F), however, drives off about 75% of its water, leaving behind a powdery substance known as plaster of Paris (named for the gypsum quarries near Paris that produce high-quality plaster). After water is added to plaster of Paris, the wet mixture hardens. Tiny crystals of gypsum form within the plaster, creating a firm solid that is used to fashion smooth interior walls. This plaster has numerous other applications in such fields as art, dentistry, and orthopedics.

Mixing sand and gravel with cement forms concrete, used principally in building foundations and roads. A single kilometer

▲ **FIGURE 20-30** Before-and-after images of a strip-mine reclamation project.

of four-lane highway, for example, requires about 40 tons of gravel for its concrete. Common sources of sand and gravel include river channel and sandbar deposits, coastal offshore bars, beach deposits, sand dunes, and glacial eskers and outwash (Fig. 20-31). Increased urban construction worldwide over the last 25 years has doubled the demand for sand and gravel, depleting many sources. In Scandinavia, entire eskers have been removed from the landscape to satisfy the growing need for these materials.

Clay minerals are the end products of the chemical weathering of feldspars and therefore remain relatively stable in the Earth's surface-weathering environment. Wet clays can be shaped into a variety of building materials, such as decorative terracotta bricks, tiles, and pipes. These materials harden when they are fired in a kiln.

Various types of stone are quarried for specific building purposes. Slate is used for roofing, flooring, fireplaces, and patios. Slabs of polished granite, diorite, gneiss, and other attractive coarse-grained rocks decorate the walls of office and government buildings. Retaining walls that support unstable slopes are made of blocks of basalt, and tons of crushed limestone, granite, marble, and schist lie beneath highways as road-bed fill.

Nonmetals for Agriculture and Industry
As we said earlier, the number of people on Earth is approaching 7 billion. To feed so many people using soil that has, in many areas, already been depleted of nutrients by overcultivation will require a vast supply of chemical fertilizers to enhance plant growth and increase sugar and starch production.

Phosphorus and potassium are two of the main elements in agricultural fertilizers. Phosphorus is derived primarily from the mineral apatite $[Ca_5(PO_4)_3(OH, F, Cl)]$, which is found in certain marine sedimentary rocks, in guano deposits (bird and bat droppings) in tropical caves and on islands (like the Galapagos Islands of southern South America), and in some calcium-rich igneous rocks. In North America, most phosphorus comes from

ancient sedimentary deposits in Montana and Wyoming and from more recent sediments along coastal North Carolina and Florida. Potassium comes primarily from sylvite (KCl), an evaporite mineral deposited in shallow marine basins during periods of climatic warming. Large sylvite deposits are found in New Mexico, Utah, Colorado, Montana, and Saskatchewan.

Sulfur has applications in both agriculture and industry. In the form of sulfuric acid, it is added to alkaline soils to maintain optimal pH conditions for plant growth. Sulfuric acid is also used to treat rubber, explosives, and wood pulp (for the manufacture of paper). Although some sulfur is mined in native form from the deposits that crystallize from escaping sulfurous gases around the vents of volcanoes, most is refined from the sulfide mineral pyrite (FeS_2), the sulfate mineral gypsum, or solid sulfur removed and purified from sulfur-bearing petroleum.

A very common nonmetal resource is quartz sand, which is melted and then cooled rapidly ("quenched") to produce glass. Ancient Egyptians were the first to manufacture glass from melted sand, 4000 years ago. Today, 30 million tons of sand are quarried each year in North America from the purest sandstone deposits, such as the St. Peter sandstone of Illinois and Missouri (which consists of 95% quartz). Slight impurities in the sand impart color to the resulting glass.

Another non-metallic resource with multiple industrial uses is *asbestos* (the generic term for a variety of magnesium-rich, white, gray, and green fibrous silicates, including serpentine and some amphibole minerals). (See Highlight 7-2 for a review of asbestos and its environmental concerns.) This nonflammable material is used primarily in fireproof clothing, in electrical insulation, and in the linings of automobile brakes. Its use in the construction industry has been curtailed, however, because asbestos dust has been linked to several lung and digestive-tract diseases. The asbestiform silicates are produced primarily by hydrothermal metamorphism of olivine and pyroxene in ultramafic mantle rocks of continental-collision zones. Large asbestos reserves occur in the metamorphic rocks of the northern Appalachians of Vermont and Quebec and in hills surrounding Libby, Montana.

Future Use of Natural Resources

The United States has about 50 billion barrels of known oil reserves that can be recovered by current methods. It may also have an as-yet-undiscovered 35 billion more barrels. The current rate of use is 7.5 billion barrels per year, about 50% of which is imported. Thus, the domestic supply, supplemented with imports, will last only about 30 years. The impending petroleum shortage will surely force many changes in our consumption of this energy source. Even if technological advances substantially improve oil recovery, you may very well be around when the last drop of economically accessible *American* oil trickles from the pumps.

Large new petroleum finds on American soil are unlikely, because virtually all potential oil-producing rock formations have been explored. Indeed, most of the Earth's potential oil-bearing regions have been thoroughly investigated. Even if large new

▲ **FIGURE 20-31** A gravel operation in glacial outwash in Ontario, Canada.

20.1 Geology at a Glance

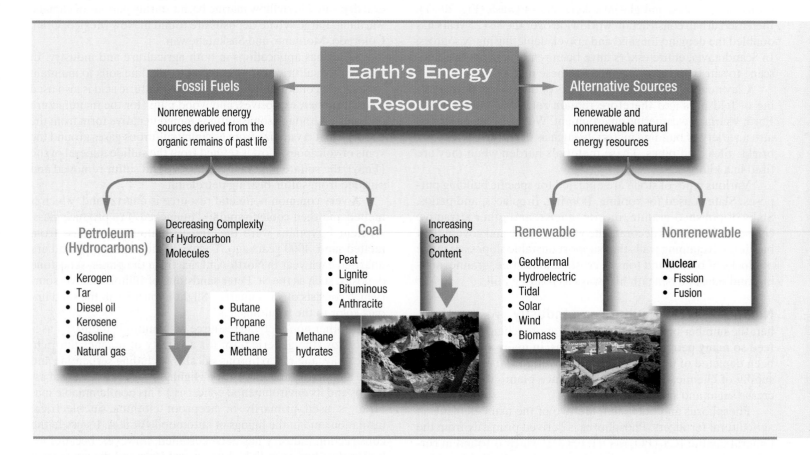

reserves were discovered, at the current rate of use they would be exhausted rapidly. The most recent significant domestic discovery, about 10 billion barrels at Alaska's North Slope, would satisfy less than two years of current total U.S. oil needs.

At the worldwide petroleum-use rate of about 21 billion barrels per year, the world's 700-billion-barrel reserves would last only 35 years or so. Unless oil prices skyrocket and reduce demand (a likely scenario because it's already happening), sometime during the 21st century, petroleum-burning cars, buses, and trucks everywhere will likely stop in place and rust along the side of the road. This scenario appears unavoidable unless more effort, money, and scientific ingenuity are devoted to finding new oil reserves, improving the percentage we can economically recover, or, more likely, finding ways to extend the *finite* crude oil supply or develop alternatives (such as the hydrogen fuel cell) or the wider distribution of gas-electric hybrids.

Some countries have already exhausted their supplies of crucial resources and turned from producers to importers. Others are just beginning to feel the effects of critical shortages. What does the future hold for humankind as the Earth's nonrenewable resources dwindle?

Meeting Future Energy Needs

Substantial oil shortages and price increases in developed countries (what are you paying for "unleaded" these days?) may reduce consumption, thereby extending the life of current supplies. Widespread use of gasoline additives, such as ethanol (a form of

alcohol), could reduce gasoline consumption by as much as 25%. (Of course, we might all consider buying one of those gas-and-electric hybrids, like the Prius, which get 55 mpg of gas, but the waiting list today is about a mile long.)

One way to maximize our oil reserves is to extract as much as possible from known reservoir rocks. In newly drilled wells, oil usually gushes out due to the confining pressure exerted on the reservoir rocks by overlying rocks. Collection of this first flow of oil is called *primary recovery*. As reservoir rocks become partially depleted and no longer oil-saturated, pressure on the remaining oil diminishes and the flow gradually slows. In the past, wells were abandoned when they no longer gushed, or they were subjected to *secondary recovery* methods, such as injection of water into the reservoir rocks to buoy oil upward. Even after secondary recovery, however, one-half to two-thirds of the oil remains in the ground.

In the United States alone, more than 300 billion barrels of unrecovered oil may still lie in the reservoir rocks of known oil fields, perhaps enough to provide for an additional 50 years of use. Some abandoned wells are now being reopened and subjected to *enhanced recovery* methods. In one such method, compressed carbon-dioxide gas and superheated steam are injected into the rocks to make the petroleum less viscous, enabling it to flow more readily. In another technique, the petroleum is burned *in place* to produce electricity at the source. Although enhanced recovery is more expensive than primary and secondary recovery, it becomes economically feasible as oil supplies dwindle and prices rise.

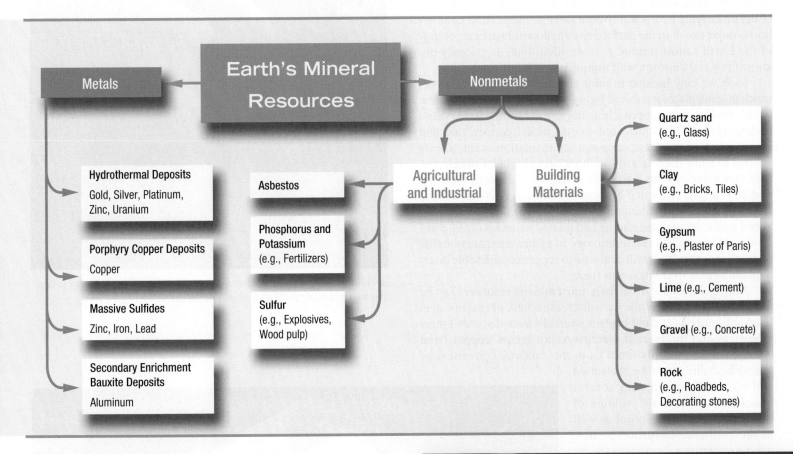

Earth's Mineral Resources

Metals

- Hydrothermal Deposits
  Gold, Silver, Platinum, Zinc, Uranium
- Porphyry Copper Deposits
  Copper
- Massive Sulfides
  Zinc, Iron, Lead
- Secondary Enrichment Bauxite Deposits
  Aluminum

Nonmetals

Agricultural and Industrial

- Asbestos
- Phosphorus and Potassium (e.g., Fertilizers)
- Sulfur (e.g., Explosives, Wood pulp)

Building Materials

- Quartz sand (e.g., Glass)
- Clay (e.g., Bricks, Tiles)
- Gypsum (e.g., Plaster of Paris)
- Lime (e.g., Cement)
- Gravel (e.g., Concrete)
- Rock (e.g., Roadbeds, Decorating stones)

Other technological initiatives seek alternatives to petroleum as the primary energy provider for transportation. Liquefied and gasified coal, for example, show promise as transportation fuel. Most major car makers have begun to develop electric cars with long-distance cruising capacity, capable of superhighway speeds. Other attempts to prepare for the inevitable disappearance of petroleum include the development of solar-powered buses and trains propelled by powerful electromagnets (Fig. 20-32).

Meeting Future Mineral Needs

For now, our supplies of mineral resources remain adequate, although some deposits are located in some pretty remote places. Unlike our search for fossil fuels, geologists continue to discover new and valuable deposits of important metals and nonmetals. Major ore bodies have recently been unearthed in Chile, Australia, and Siberia. Extensive deposits rich in zinc, lead, copper, nickel, cobalt, and uranium have also been found in Antarctica, along with a 120-km (70-mi)-long, 100-m (330-ft)-thick deposit of iron, an ore body large enough to meet the world's iron demand for 200 years.

Mineral prospecting has ceased to be a pick-and-shovel operation. To find metals buried beneath more than 150 m (500 ft) of sediment on the seafloor, gravimeters and magnetometers are employed to identify gravity and magnetic anomalies. Gravimeters towed by airplanes and helicopters have even detected bodies of chromite, magnetite, and nickel beneath the ice of

▲ **FIGURE 20-32** Magnetic-levitation trains are one alternative to petroleum-based transportation systems.

Antarctica (as mentioned earlier, an international treaty has placed this area off-limits for resource development). Mass spectrometers can analyze the chemistry of soils, and even that of the gases trapped between soil particles to determine the composition of the underlying bedrock. Infrared satellite images and satellite-borne radar can map the surface and shallow-subsurface geology of the Earth's most remote regions, identifying previously unknown mineral deposits with mining potential.

Soon we may be able to mine the deposits on the seafloor, which in some places is covered by billions of nodules of manganese oxide (MnO_2), some approaching the size of bowling balls. These nodules (Fig. 20-33), which contain iron, nickel, copper, zinc, and cobalt, accumulate where deep-sea sedimentation occurs slowly enough to allow them to grow without being buried. Collectively, they may represent the Earth's largest mineral deposit. Over the past 30 years, three United Nations conventions (UNCLOS—the United Nations Convention on the Law of the Sea) have been convened to discuss seafloor mining and related issues involving ownership of the seas. As the technology to gather manganese-oxide nodules improves, they will continue to trigger considerable international debate over who owns them.

And again, unlike fossil fuels, most mineral resources can be recycled and reused. While we can't reuse a tank of gas, the steel from automobiles and old bridges, mercury from discarded thermometers and fluorescent mercury-vapor lamps, copper from electrical wire, and platinum from the catalytic converters of abandoned cars can all be reclaimed and reused. Recycling has several benefits: It reduces the volume of waste that requires disposal as well as the land area disturbed by new mining operations. In many cases, recycling also reduces the amount of energy that would be needed to mine and refine new ores. Scrap-aluminum recycling, for example, uses only one-twentieth the energy needed to mine and process an equivalent amount of new aluminum from bauxite.

Of course, some metals are difficult to recycle. Consider all the aluminum and steel that is intermingled with other materials in complex manufactured objects as refrigerators and lawn mowers. Precious metals, such as gold and silver, are the easiest to reuse. The gold fashioned today into a new pair of earrings may have been mined thousands of years ago by ancient Romans or Peru's Incas and used in their ceremonial objects.

The discarded, seemingly worthless waste-rock piles from old mining operations (called *mine tailings*) have even emerged as a source of valuable minerals. Acids and other solutions can be flushed through mine tailings

▲ **FIGURE 20-33** Nodules of manganese oxide, found abundantly on the ocean floor, may become a source of valuable minerals in the near future.

▲ **FIGURE 20-34** This garbage dump near Mount Everest illustrates that virtually no place on Earth escapes the environmental impact of careless human behavior. This trash will remain here for decades or centuries because decomposition occurs only slowly in the cold, dry climate of the region.

to leach metals from them. This process concentrates substantial quantities of valuable minerals from the mine waste, which in turn is itself reduced in volume.

Everything we humans have ever had or have ever made comes from the Earth. Now that you understand the processes that have built, moved, and shaped the continents and ocean floors, and the vast amount of time over which the Earth and its resources developed, you can appreciate why we should try to manage wisely the resources that make our modern lives possible. The rate at which the Earth makes more of these things is painfully slow from a

human time perspective. If we waste these resources today, most assuredly little will remain for future generations. Industry, local governments, and the world community must cooperate to ensure that the search for, development of, and use of the planet's resources do not squander those resources or irreversibly damage our shared environment (Fig. 20-34). With the knowledge you have acquired from your geology training this semester, you are now prepared to contribute to the ongoing debate about the need to balance resource development and environmental protection. By all means, use your training and be heard.

Chapter 20 Summary

Natural resources include all types of fuels and a variety of useful metals and nonmetals. The availability of natural resources depends on their abundance, the technology needed to extract and develop them, and the economic forces of supply and demand. *Reserves* are quantities of natural resources that have been discovered and can be exploited *profitably* with current technology. *Resources* are deposits believed to exist but that are not considered exploitable today for various technological, economic, or political reasons. *Renewable* resources are either replenished naturally over a short time span, such as trees, or can be used continuously without being depleted, such as sunlight. *Nonrenewable* resources form so slowly that they are consumed much more quickly than nature can replenish them. Examples of nonrenewable resources include all fossil fuels, which are derived from the organic remains of long-past life and metals such as gold, that form extremely slowly.

Most of our current energy needs are satisfied by gaseous, liquid, and semisolid organic compounds known collectively as *hydrocarbons*. These *fossil fuels* are complex organic molecules composed entirely of hydrogen and carbon. *Petroleum* typically forms in a marine basin where abundant nutrient input and warm shallow water nourish microscopic organisms that are eventually buried by younger sediments. Pressure and geothermal heat transform the organic remains first into *kerogen,* a solid waxy material, and then into oil, gas, and other hydrocarbons. Petroleum forms in source rocks (typically marine muds), then migrates into permeable reservoir rocks.

Coal, a sedimentary rock, consists of the combustible, highly compressed remains of land plants. It begins as *peat,* a soft brown mass of compressed, largely undecomposed plant materials. Bacterial action within the sediment gradually breaks down the vegetation to an organic muck. Pressure from the weight of overlying deposits squeezes water from this muck, increasing the proportion of carbon, and forming a series of coals of different grades. These include:

- *lignite,* which is 70% carbon;
- *bituminous coal,* which forms when deep burial of lignite is subjected to increased pressure and geothermal heat and is thus converted to a lustrous black coal that is 80% to 93% carbon;
- *anthracite,* which forms when metamorphic conditions transform lignite and bituminous coal into a very hard coal that is 93% to 98% carbon.

The dwindling of fossil-fuel supplies makes it necessary to discover and develop alternative energy sources. Communities and industries today tap a variety of alternative sources of energy. These include:

- the Earth's *geothermal energy,* which flows from the Earth's hot interior;
- *hydroelectric power,* the power of flowing streams;
- *tidal power* from the harnessing of the power of rising tides;
- *solar energy,* a variety of direct and indirect ways to tap the Sun's energy;
- *wind power,* a clean, renewable, nonpolluting energy;
- burning of *biomass fuels,* derived from plants and animals;
- *nuclear energy,* derived either by capturing the energy released from the splitting of nuclei of certain atoms, a process known as *nuclear fission,* or by fusing atomic nuclei together, a process called *nuclear fusion.*

The other principal group of resources—minerals—includes metals such as iron, copper, aluminum, gold, and silver. Nonmetals include:

- sand and gravel for building;
- limestone for cement;
- potassium and phosphorus for fertilizers;
- gemstones, such as diamonds and emeralds.

For metals to be economically useful, some geological process must have concentrated them somewhere near the Earth's surface. Metals may precipitate from magma-derived water or from circulating groundwater or seawater to produce concentrated hydrothermal deposits. Massive sulfide deposits form principally at oceanic divergent zones, where seawater enters cracks in hot, young basaltic rock and becomes heated, dissolves metals, then rises to the seafloor, depositing the metals around surface vents.

Sedimentary and weathering processes may concentrate valuable metal deposits as well. Minerals accumulate as placer deposits where the velocity of surface waters decreases, forcing dense particles of the transported sediment (including precious metals, such as gold and silver) to settle out of the current. Dissolution of metals by chemical weathering, followed by precipitation from groundwater, also concentrates metallic ore bodies by processes of secondary enrichment. Most major metal deposits occur near past or present plate boundaries, likely sites for the generation of magmas and hot hydrothermal solutions.

Nonmetals are widely used as building materials (such as sand and gravel for concrete, limestone for cement, gypsum for plaster of Paris, and clay for bricks) and agricultural fertilizers (such as phosphorus and potassium). Nonmetals commonly used in industry include pure quartz sand (used for glass production) and asbestos (used for fireproofing clothing and electrical insulation).

As our use of energy and other resources increases, future resource needs will be met only through the development of alternative energy sources, conservation, and enhanced recovery of known reserves. The search for new mineral supplies will take us to such remote places as the seafloor (for manganese nodules) and possibly Antarctica.

KEY TERMS

acid rain *(p. 654)*
alloys *(p. 666)*
anthracite *(p. 653)*
banded iron formations *(p. 669)*
biomass fuels *(p. 662)*
bitumen *(p. 651)*
bituminous coal *(p. 653)*
breeder reactor *(p. 664)*

fossil fuels *(p. 649)*
hydrocarbons *(p. 649)*
hydrothermal deposits *(p. 666)*
kerogen *(p. 649)*
lignite *(p. 653)*
massive sulfide deposits *(p. 666)*
nuclear fission *(p. 662)*
nuclear fusion *(p. 664)*

oil sand *(p. 651)*
oil shale *(p. 651)*
oil traps *(p. 650)*
ore *(p. 648)*
peat *(p. 652)*
petroleum *(p. 649)*
placer deposits *(p. 668)*

porphyry copper deposits *(p. 666)*
reserves *(p. 648)*
reservoir rocks *(p. 650)*
resources *(p. 648)*
secondary enrichment *(p. 669)*
source rocks *(p. 650)*

QUESTIONS FOR REVIEW

1. How do resources differ from reserves? Give an example of each.

2. List three fossil fuels and three alternative energy sources. List three metallic and three nonmetallic resources.

3. What are the key physical properties of a petroleum reservoir rock? Is oil shale a source rock or a reservoir rock?

4. Briefly explain how to harness tidal power.

5. Discuss two ways in which geological processes may concentrate dispersed metals in rocks.

6. Name three types of locations where placer deposits may be found.

7. How does plate tectonics affect the distribution of the Earth's metal deposits?

8. Using your general knowledge of the geology of North America, explain why aquatic life in the lakes of Yosemite National Park,

California (see Chapter 3), would be more susceptible to the effects of acid rain than aquatic life in the lakes of Florida and Indiana (see Chapter 15).

9. Speculate on why copper mining in the United States has been so drastically reduced in recent years. Under what circumstances might U.S. mining companies resume extensive copper mining?

10. Propose a U.S. energy plan for the year 2050. Which energy sources would you develop? Why?

11. Suppose the Earth's natural resources become completely exhausted and it is decided to prospect on neighboring planets. Which planet might hold extensive iron, chromium, and platinum deposits? Which might hold sand and gravel deposits? Could any of our planetary neighbors provide us with limestone or gypsum?

LEARNING ACTIVELY

1. Look around your room or home. Pick up any object and evaluate the resources that went into its manufacture. Check the labels on any food product. Note the various minerals and other natural substances that make up the food item.

2. Check your local newspaper and monitor the price of gasoline. Why does it rise during the spring and summer months? Note the changes in oil and gas prices during periods of political unrest in the Middle East. Check the financial pages for the prices of such precious metals as gold, silver, and platinum over the course of a month or so. What external factors might affect these prices?

3. Visit a construction site in your town or city periodically and follow its progress. Note the various materials used to construct the building. You should see piles of steel reinforcement bar (from iron ore), sand and gravel (to make cement), concrete, petroleum products for making tar and asphalt, and decorative stone for building facings. Speculate about the amounts of such materials needed to build the roads, homes, schools, and other structures in your community.

ONLINE STUDY GUIDE

Practice makes you better. Make the time to take the *Online Study Guide* quiz for this chapter. It should take you about 20 minutes. Automatic grading and instant feedback will help you quickly and accurately identify the concepts you have mastered and the areas in which you need more study.
www.prenhall.com/chernicoff

APPENDIX A

Conversion Factors for English and Metric Units

Length	1 centimeter	=	0.3937 inches
	1 inch	=	2.54 centimeters
	1 meter	=	3.2808 feet
	1 foot	=	0.3048 meters
	1 yard	=	0.9144 meters
	1 kilometer	=	0.6214 miles (statute)
	1 kilometer	=	3281 feet
	1 mile (statute)	=	1.6093 kilometers
Velocity	1 kilometer/hour	=	0.2778 meters/second
	1 mile/hour	=	0.4471 meters/second
Area	1 square centimeter	=	0.16 square inches
	1 square inch	=	6.45 square centimeters
	1 square meter	=	10.76 square feet
	1 square meter	=	1.20 square yards
	1 square foot	=	0.093 square meters
	1 square kilometer	=	0.386 square miles
	1 square mile	=	2.59 square kilometers
	1 acre (U.S.)	=	4840 square yards
Volume	1 cubic centimeter	=	0.06 cubic inches
	1 cubic inch	=	16.39 cubic centimeters
	1 cubic meter	=	35.31 cubic feet
	1 cubic foot	=	0.028 cubic meters
	1 cubic meter	=	1.31 cubic yards
	1 cubic yard	=	0.76 cubic meters
	1 liter	=	1000 cubic centimeters
	1 liter	=	1.06 quarts (U.S. liquid)
	1 gallon (U.S. liquid)	=	3.79 liters
Mass	1 gram	=	0.035 ounces
	1 ounce	=	28.35 grams
	1 kilogram	=	2.205 pounds
	1 pound	=	0.45 kilograms
Pressure	1 kilogram/square centimeter	=	0.97 atmospheres
	1 kilogram/square centimeter	=	14.22 pounds/square inch
	1 kilogram/square centimeter	=	0.98 bars
	1 bar	=	0.99 atmospheres
Temperature	°F (degrees Fahrenheit)	=	°C(9/5) + 32
	°C (degrees Celsius)	=	(°F − 32)(5/9)

SURFACE AREAS	
Landmasses	150,142,300 kilometers^2 (57,970,000 miles^2)
Oceans and Seas	362,032,000 kilometers^2 (138,781,000 miles^2)
Entire Earth	512,175,090 kilometers^2 (197,751,500 miles^2)

DISTRIBUTION OF WATER, BY VOLUME	
Oceans and Seas	1.37×10^9 kilometers^3 (3.3×10^8 miles^3)
Glaciers	2.5×10^7 kilometers^3 (7×10^6 miles^3)
Groundwater	8.4×10^6 kilometers^3 (2×10^6 miles^3)
Lakes	1.25×10^5 kilometers^3 (3×10^4 miles^3)
Rivers	1.25×10^3 kilometers^3 (3×10^2 miles^3)

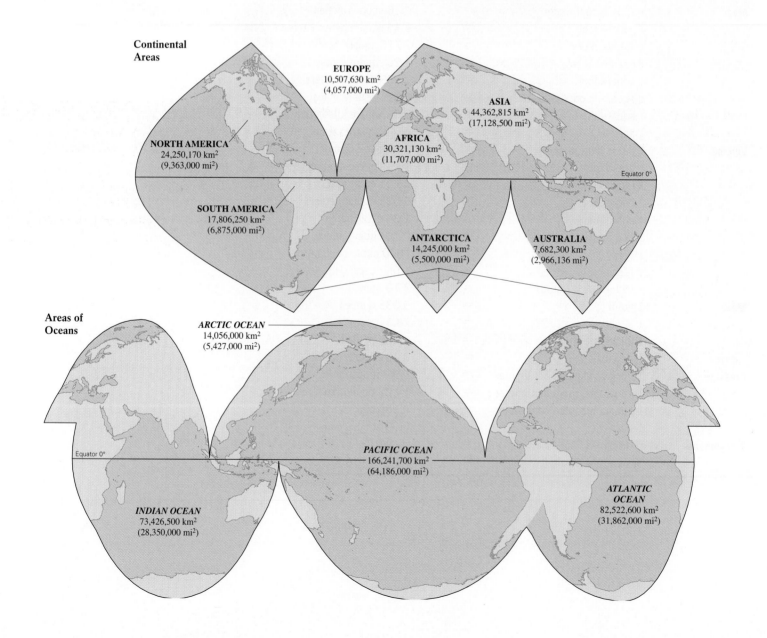

Continental Areas

EUROPE
10,507,630 km^2
(4,057,000 mi^2)

ASIA
44,362,815 km^2
(17,128,500 mi^2)

NORTH AMERICA
24,250,170 km^2
(9,363,000 mi^2)

AFRICA
30,321,130 km^2
(11,707,000 mi^2)

Equator 0°

SOUTH AMERICA
17,806,250 km^2
(6,875,000 mi^2)

ANTARCTICA
14,245,000 km^2
(5,500,000 mi^2)

AUSTRALIA
7,682,300 km^2
(2,966,136 mi^2)

Areas of Oceans

ARCTIC OCEAN
14,056,000 km^2
(5,427,000 mi^2)

Equator 0°

PACIFIC OCEAN
166,241,700 km^2
(64,186,000 mi^2)

ATLANTIC OCEAN
82,522,600 km^2
(31,862,000 mi^2)

INDIAN OCEAN
73,426,500 km^2
(28,350,000 mi^2)

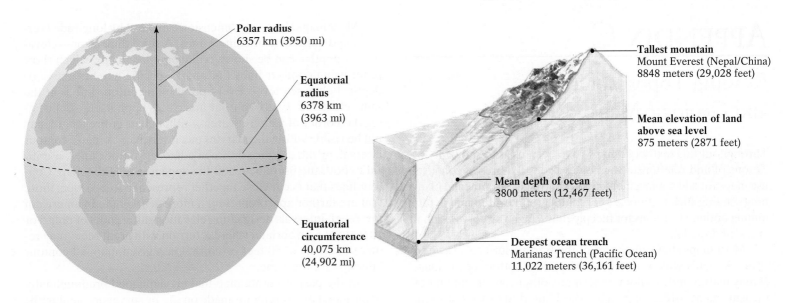

Polar radius
6357 km (3950 mi)

Equatorial
radius
6378 km
(3963 mi)

Equatorial
circumference
40,075 km
(24,902 mi)

Tallest mountain
Mount Everest (Nepal/China)
8848 meters (29,028 feet)

Mean elevation of land
above sea level
875 meters (2871 feet)

Mean depth of ocean
3800 meters (12,467 feet)

Deepest ocean trench
Marianas Trench (Pacific Ocean)
11,022 meters (36,161 feet)

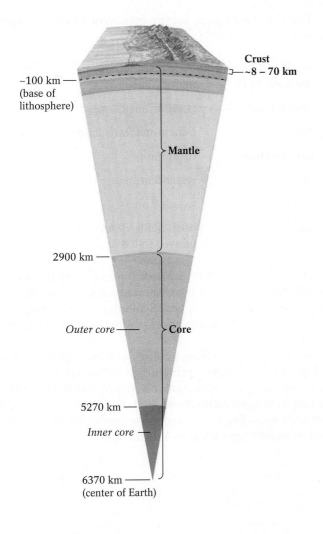

Crust
~8 – 70 km

~100 km
(base of
lithosphere)

Mantle

2900 km

Outer core ——— Core

5270 km

Inner core

6370 km
(center of Earth)

Density of the Earth by Layer	
	DENSITY*
Continental crust	2.7 grams/centimeter^3
Oceanic crust	3.0 grams/centimeter^3
Mantle	4.5 grams/centimeter^3
Core	10.7 grams/centimeter^3
Whole Earth	5.5 grams/centimeter^3 (mean density)

APPENDIX C

Reading Topographic and Geologic Maps

Throughout this textbook, maps have been used to show the locations of and relationships among various features. Geologists use maps for soil management, flood control, environmental planning, finding such resources as ores and groundwater, and determining optimal locations for fuel pipelines, highways, recreational areas, and the like.

Most maps show all or part of the Earth's surface, drawn to scale. A map's scale is usually shown at the bottom of the map. It may feature miles, kilometers, or any other convenient unit of measurement. Most maps from the United States Geological Survey are drawn to a scale of 1:24,000, meaning that 1 centimeter = 240 meters or 1 inch = 2000 feet; the latter would be read as "one inch on this map is equal to two thousand feet in real-world distance." The U.S. Geological Survey has been phasing in metric topographic maps using a scale of 1:100,000.

Most maps show two dimensions, marked by longitude (vertical) and latitude (horizontal) lines. A third dimension—elevation or depth—can be shown on *topographic maps,* which show relief using contour lines that mark specific heights above or depths below sea level. Every point on a contour line is at the same elevation. Every contour line closes upon itself—delineating the border of a discrete area—although its entire length may not be visible within the margins of a given map. A map's contour interval, or the distance between adjacent contour lines, is critical to both the usefulness and the appearance of the map: Contour lines that occur close together represent a steep slope; those that are farther apart represent a more gentle one. In an area of low relief, intervals designating elevation differences of only 5 or 10 feet are appropriate, whereas terrain with great relief may require intervals of 50 feet or more. On a given map, all contour intervals are the same.

In the past, accurate map-making depended predominantly upon actual measurements made on site by surveyors, geologists, and others. The advent of aerial photography, and then of space-based satellite photography, has made accurate standardized revisions possible.

The U.S. Geological Survey uses color to indicate specific features:

black and/or red solid lines	: major roads
brown lines	: contour lines
black	: human-made structures, names
light red lines	: town limits
blue	: water features
green	: wooded areas
white	: open fields, deserts, and other nonvegetated areas

Dotted or dashed lines are used for temporary features (that is, solid blue represents a lake, whereas a dashed blue line marks a seasonal stream). Any special symbols used on a map are explained in its legend, which appears with the scale in the bottom margin.

A region's underlying geology can be represented on a *geologic map.* Symbols representing various types of rock have been standardized; such commonly accepted rock symbols have been used throughout this book. Standardized colors are used to show rock ages. The key to a geologic map's labeling, colors, and symbols usually appears at its side margins.

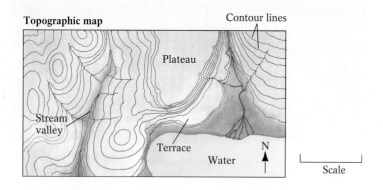

Topographic map

Contour lines
Plateau
Stream valley
Terrace
Water
N
Scale

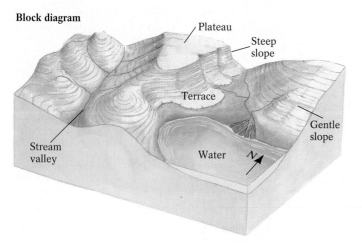

Block diagram

Plateau
Steep slope
Terrace
Gentle slope
Stream valley
Water
N

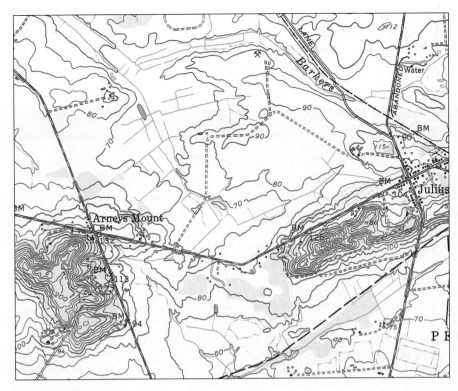

Topographic map

Tertiary Period Deposits

Tch	Cohansey sand *Sand, quartz, light-gray to yellow-brown, medium- to coarse-grained, pebbly, ilmenitic, micaceous, stratified*
Tkw	Kirkwood formation *Sand, quartz, light-gray to tan, fine- to very fine-grained, clayey, micaceous, ilmenitic, kaolinitic, sparingly lignitic, massive-bedded*
Tmq	Manasquan formation *Sand, quartz, dark green-gray, medium- to coarse-grained, glauconitic, clayey*
Tvt	Vincentown formation *Upper member—calcarenite, quartz and glauconite, dusky-yellow to pale-olive, clayey; lower member—sand, quartz, dark-gray, poorly sorted, fine to coarse, clayey, glauconitic, entire formation very fossiliferous*
Tht	Hornerstown sand *Sand, glauconite, dusky-green, medium- to coarse-grained, clayey, massive-bedded*

Cretaceous Period Deposits

Krb	Red Bank sand *Lower member—sand, glauconite, dark grayish-black, coarse-grained, very clayey, micaceous, lignitic, massive-bedded (upper member not present)*
Kns	Navesink formation *Sand, glauconite, varying amounts of quartz, greenish-black to brown, medium- to coarse-grained, clayey*
Kml	Mount Laurel sand *Sand, quartz, reddish-brown to green-gray, poorly sorted, fine- to coarse-grained, glauconitic, massive-bedded*

Geologic map

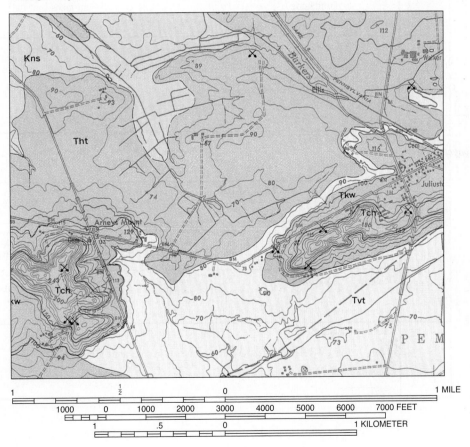

APPENDIX D

Mineral-Identification Charts

Most Common Rock-Forming Minerals

		MINERAL OR GROUP NAME	COMPOSITION/ VARIETIES	CRYSTAL FORM/OTHER DIAGNOSTIC FEATURES	CLEAVAGE/ FRACTURE	USUAL COLOR/ LUSTER	STREAK	HARDNESS
Light-colored; abundant in all rock types	SILICATES / Framework	Feldspar	Potassium (orthoclase) feldspar ($KAlSi_3O_8$)	Coarse crystals or fine grains	Good cleavage in two directions at 90°	Commonly salmon-pink; white to gray, with pearly luster	White	6
			Sodium, calcium (plagioclase) feldspar Albite: $NaAlSi_3O_8$ Anorthite: $CaAl_2Si_2O_8$		Good cleavage in two directions at 90°; cleavage surfaces striated	White to gray	White	
		Quartz	SiO_2	Six-sided crystals; individual or in masses; also in irregular masses	No cleavage; conchoidal fracture	Colorless or slightly smoky gray, pink, sometimes milky yellow, purple	White	7
	SILICATES / Sheet	Mica	Muscovite $KAl_3Si_3O_{10}(OH)_2$	Platy crystals	One perfect cleavage plane; splits into thin sheets	Colorless or slightly gray, green, or brown, with vitreous luster	White	2–2½
			Biotite $K(Mg, Fe)_3AlSi_3O_{10}(OH)_2$	Platy crystals	One perfect cleavage plane; splits into thin sheets	Black, or brown, with vitreous luster	White or gray	2½–3
		Chlorite	$(Mg, Fe)_5(Al, Fe)_2Si_3O_{10}(OH)_8$			Black to dark green	White or colorless	2–2½
Dark-colored; abundant in metamorphic and igneous rocks	SILICATES / Double Chain	Amphibole	Actinolite $Ca_2(Mg, Fe)_5Si_8O_{22}(OH)_2$	Long, tabular, maybe fibrous (needle-like)	Two cleavage planes at 56° and 124°; slightly "stepped" surfaces (not flat)	Green with vitreous luster Dark green to black	Pale green or white	5–6
			Hornblende $(Ca, Na)_{2-3}(Mg, Fe, Al)_5Si_6(Si, Al)_2O_{22}(OH)_2$					
	SILICATES / Single Chain	Pyroxene	Augite $(Ca, Na)(Mg, Fe, Al)(Al, Si)_2O_6$	Short, blocky crystals	Two cleavage planes at about 90°; slightly "stepped" surfaces (not flat)	Black, light to dark green, with vitreous luster	Pale green	5–6
			Diopside $CaMg(Si_2O_6)$			White to light green, with vitreous luster	White or pale green	
			Orthopyroxene $MgSiO_3$			Pale gray, green, brown, with vitreous luster	White	
	SILICATES / Single Tetrahedra	Olivine	$(Mg, Fe)_2SiO_4$	Small grains and granular masses	No cleavage; conchoidal fracture	Green; maybe green-brown or yellowish with vitreous, glassy luster	White	6½–7
		Garnet	$(Fe, Mn, Mg, Ca)_3Al_2Si_3O_{12}$ $Ca_3Al_2Si_3O_{12}$	12- or 24-sided crystals	No cleavage; conchoidal fracture	Commonly deep red, with vitreous to resinous luster	White	6–8
Light-colored; abundant in sedimentary rocks	CARBONATES	Calcite	$CaCO_3$	Fine to coarsely crystalline. Effervesces rapidly in HCl.	Three oblique cleavage planes, forming rhombohedral cleavage pieces	White, gray or yellow, with pearly luster	White	3
		Dolomite	$CaMg(CO_3)_2$	Fine to coarsely crystalline. Effervesces slowly in HCl when powdered.		Colorless, white, or pink, and may be tinted by impurities, with pearly luster	White to pale gray	3½–4
	CLAY MINERALS (Hydrous alumino-silicates)	Kaolinite	$Al_2Si_2O_5(OH)_4$	Very fine grains; found as bedded masses in soils and sedimentary rocks; earthy odor	Earthy fracture	White to buff, or tinted gray by impurities	White, off-white, or colorless	1½–2½
		Illite	$K_{0.8}Al_2(Si_{3.2}Al_{0.8})O_{10}(OH)_2$					
		Smectite	$Na_{0.3}Al_2(Si_{3.7}Al_{0.3})O_{10}(OH)_2$					

Accessory or Less-Abundant Rock-Forming Minerals

		MINERAL OR GROUP NAME	COMPOSITION/ VARIETIES	CRYSTAL FORM/OTHER DIAGNOSTIC FEATURES	CLEAVAGE/ FRACTURE	USUAL COLOR/ LUSTER	STREAK	HARDNESS
Light-colored; common in sedimentary rocks	SULFATES	Gypsum	$CaSO_4 \cdot 2\,H_2O$	Tabular crystals in fine-to-granular masses	One perfect cleavage plane, forming thin sheets; also two other good cleavage planes	Colorless to white, with vitreous luster	White	1–2½
		Anhydrite	$CaSO_4$	Granular masses	Three good to perfect cleavage planes at 90°	White, gray, or blue-gray, with pearly to vitreous luster	White	3–3½
		Halite	NaCl	Perfect cubic crystals, soluble in water. Tastes salty.	Three excellent cleavage planes at 90°	White or gray, with pearly luster	White	2½
Light-colored; mainly in metamorphic	ALUMINO-SILICATES	Kyanite	Al_2SiO_5	Long, bladed, or tabular crystals	One perfect cleavage plane, parallel to length of crystals	Commonly blue; may be pale blue or white with vitreous luster	White	5 along cleavage plane, 6½–7 across cleavage plane
		Sillimanite		Long, slender crystals or fibrous masses		White to gray, with vitreous luster	White	6–7
		Andalusite		Tabular crystals with nearly square ends	Weak cleavage	Red, reddish brown, or green, with vitreous luster	White	5–6
		Serpentine	$Mg_3Si_2O_5(OH)_4$	Fibrous or platy masses	Splintery fracture	Light to dark green, brownish yellow and black with pearly luster	White	4–6
		Talc	$Mg_3Si_4O_{10}(OH)_2$	Foliated masses. Feels soapy.	Perfect in one direction, forming thin flakes	White to pale green, with pearly or greasy luster	White	1–1½
		Corundum	Al_2O_3	Short, six-sided crystals	Irregular fracture	Usually red/pink or blue, with adamantine luster	None	9
		Fluorite	CaF_2	Octahedral or cubic crystals	Cleaves easily in four directions	White, yellow, green, or purple, with vitreous luster	White	4
Dark-colored; common in metamorphic rocks	SILICATES	Epidote (paired tetrahedra)	$Ca_2(Al,Fe)Al_2 Si_3O_{12}(OH)$	Usually granular masses; also slender prisms	One good cleavage direction, one poor	Yellow to dark green, with vitreous luster	White or gray	6–7
		Staurolite (single tetrahedra)	$Fe_2Al_9Si_4O_{22} (O,OH)_2$	Tabular crystals, some cross-shaped	One poor cleavage direction	Brown or reddish brown to black, with vitreous luster	Off-white to white	7
		Graphite	C	Scaley, foliated masses. Feels greasy.	One direction of cleavage	Steel gray to black, with vitreous or pearly luster	Gray or black	1–2
Dark-colored; common in all rock types		Apatite	$Ca_5(PO_4)_3 (OH,F,Cl)$	Small single crystals	Poor cleavage	Green, brown, yellow, or red, with adamantine or greasy luster	White	5
		Magnetite	Fe_3O_4	Granular masses and octohedral crystals	Uneven fracture	Black, with metallic luster	Black	5½
		Hematite	Fe_2O_3	Granular masses	Uneven fracture	Brown-red to black, with earthy, dull, to metallic silvery luster	Brick red	5½–6½
		Limonite	$2\,Fe_2O_3 \cdot 3\,H_2O$	Earthy masses	Uneven fracture	Yellowish brown to black	Brownish yellow	5–5½
Metallic luster; common in all rock types	SULFIDES	Pyrite	FeS_2	Cubic crystals or granular masses	Uneven fracture	Pale brass yellow, with metallic luster	Greenish black	6–6½
		Galena	PbS	Cubic crystals or granular masses	Three perfect cleavage planes at 90°	Silver gray with metallic luster	Gray	2½
		Sphalerite	ZnS	Granular masses or pyramid-shaped crystals (4 to 6 sided)	Six perfect cleavage planes at 60°	Typically brown, golden brown, or black, with resinous to submetallic luster	Reddish brown to yellow brown	3½–4
		Chalcopyrite	$CuFeS_2$	Granular masses	Irregular fracture	Brass yellow	Greenish black	3½–4

Minerals and Elements of Industrial or Economic Importance

MINERAL OR GROUP NAME	COMPOSITION	USUAL COLOR/ LUSTER	STREAK	HARDNESS	OTHER PROPERTIES/ COMMENTS	ORIGIN
Asbestos (Chrysotile)	$Mg_3Si_2O_5(OH)_4$	White to pale green, with pearly luster	White	$1-2\frac{1}{2}$	Flexible, nonflammable fibers	Typically occurs in hydrothermally metamorphosed ultramafic rock
Bauxite	$Al(OH)_3$	Reddish to brown, with dull luster	Pale reddish brown	$1\frac{1}{2}-3\frac{1}{2}$	Found in earthy, clay-like masses. Principal source of commercial aluminum.	Weathering of many rock types
Chalcopyrite	$CuFeS_2$	Yellow, with metallic luster	Greenish black	$3\frac{1}{2}-4\frac{1}{2}$	Uneven fracture; softer than pyrite. Irridescent tarnish. Most common ore of copper.	Hydrothermal veins and porphyry copper deposits
Chromite	$FeCr_2O_4$	Black, with metallic or submetallic luster	Dark brown	$5\frac{1}{2}$	Massive; granular; compact. Most common ore of chromium. Used in making steel.	Ultramafic igneous rocks
Copper (native)	Cu	Red, with metallic luster	Red	$2\frac{1}{2}-3$	Malleable and ductile. Tarnishes easily.	Mafic igneous rock (basaltic lavas); also in oxidized ore deposits
Galena	PbS	Silver-gray, with metallic luster	Gray	$2\frac{1}{2}$	Perfect cubic cleavage. Most important ore of lead; also commonly contains silver.	Hydrothermal veins
Gold (native)	Au	Yellow, with metallic luster	Yellow	$2\frac{1}{2}-3$	Malleable and ductile	Sedimentary placer deposits; hydrothermal veins
Hematite	Fe_2O_3	Brown-red to black, with earthy, dull, or metallic luster	Dark red	$5\frac{1}{2}-6\frac{1}{2}$	Granular, massive. Most important source of iron.	Found in all types of rocks. Commonly in contact metamorphic aureoles around mafic igneous sills.
Magnetite	Fe_3O_4	Black, with metallic luster	Black	$5\frac{1}{2}$	Uneven fracture. An important source of iron. Strongly magnetic.	Found in all types of rocks
Platinum (native)	Pt	Steel-gray or silver-white, with metallic luster	Off-white, metallic	$4-4\frac{1}{2}$	Malleable and ductile. Occasionally magnetic.	Mafic igneous rocks and sedimentary placer deposits
Silver (native)	Ag	Silver-white, with metallic luster	Silver-white	$2\frac{1}{2}-3$	Malleable and ductile. Tarnishes to dull gray or black.	Mostly in hydrothermal veins; also in oxidized ore deposits
Sphalerite	ZnS	Shades of brown and red, with resinous to adamantine luster	Reddish brown	$3\frac{1}{2}-4$	Perfect cleavage in six directions at 120°. Most important ore of zinc.	Hydrothermal veins

Precious and Semi-Precious Gems*

MINERAL OR GROUP NAME	COMPOSITION	USUAL COLOR/ LUSTER	STREAK	HARDNESS	OTHER PROPERTIES	ORIGIN
Beryl (aquamarine, emerald)	$Be_3Al_2(SiO_3)_6$	Blue, green, yellow, or pink, with vitreous luster	White	$7\frac{1}{2}$–8	Uneven fracture; hexagonal crystals	Cavities in granites and pegmatites, and schist
Corundum (ruby, sapphire)	Al_2O_3	Gray, red (ruby), blue (sapphire), with adamantine luster	None	9	Short, six-sided crystals; irregular, occasionally cleavage-like fracture ("parting")	Metamorphic rocks; some igneous rocks
Diamond	C	Colorless or with pale tints, with adamantine luster	None	10	Octahedral crystals	Peridotite; kimberlite; sedimentary placer deposits
(S) Garnet	$(Ca, Mg, Mn, Fe)_3 \cdot (Al)(SiO_3)_2$	Nearly every color except blue with vitreous to resinous luster	White	$6\frac{1}{2}$–$7\frac{1}{2}$	No cleavage; 12- or 24-sided crystals	Contact-metamorphic and regionally metamorphosed rocks, and sedimentary placer deposits
(S) Jadeite (jade)	$NaAl(Si_2O_6)$	Green, with vitreous luster; can be white	White or pale green	$6\frac{1}{2}$–7	Compact fibrous aggregates	High-pressure metamorphic rocks
Olivine (peridot)	$(MgFe)_2SiO_4$	Light to dark green, with vitreous, glassy luster	White	$6\frac{1}{2}$–7	Uneven fracture, often in granular masses	Basalt, peridotite
(S) Opal	$SiO_2 \cdot nH_2O$	White with various other colors, with vitreous, pearly luster	White	$5\frac{1}{2}$–$6\frac{1}{2}$	Conchoidal fracture; amorphous; tinged with various colors in bands	Low-temperature hot springs; weathered near-surface deposits
(S) Quartz (includes amethyst, citrine, agate, onyx, bloodstone, jasper, etc.)	SiO_2	Colorless, white, or tinted by impurities, with luster depending on variety	White	7	No cleavage; six-sided crystals	Origin specific to gem variety. (Quartz found in all rock types except ultramafic igneous rocks.)
Topaz	$Al_2SiO_4(F, OH)_2$	Colorless, white, or pale pink or blue, with vitreous luster	White	8	Cleavage in one direction; conchoidal fracture	Pegmatite, granite, rhyolite
Tourmaline	$(Na, Ca)(Li, Mg, Al)$- $(Al, Fe, Mn)_6$ $(BO_3)_3$-(Si_6O_{18}) $(OH)_4$	Black, brown, green, or pink, with vitreous luster	White to gray	7–$7\frac{1}{2}$	Poor cleavage; uneven fracture; striated crystals	Metamorphic rocks; pegmatite; granite
(S) Turquoise	$CuAl_6(PO_4)_4 \cdot (OH)_8 \cdot 5H_2O$	Blue-green, with waxy luster	Blue-green or white	5–6	Massive	Hydrothermal veins

*Gems are classified as semi-precious (S) if they are more accessible than precious gems and/or their properties are somewhat less valued.

Glossary

abrasion A form of *mechanical weathering* that occurs when loose fragments or particles of rocks and minerals that are being transported, as by water or air, collide with each other or scrape the surfaces of stationary rocks.

accretionary wedge A mass of *sediment* and oceanic *lithosphere* that is transferred from a *subducting* plate to the less dense, overriding plate with which it *converges.*

acid rain Rain that contains such acidic compounds as sulfuric acid and nitric acid, which are produced by the combination of atmospheric water with oxides released when *hydrocarbons* are burned. Acid rain is widely considered responsible for damaging forests, crops, and human-made structures, and for killing aquatic life.

aftershock A ground tremor caused by the repositioning of rocks after an *earthquake.* Aftershocks may continue to occur for as long as two years after the initial earthquake. The intensity of an earthquake's aftershocks decreases over time.

aggradation The process by which a stream's *gradient* steepens due to increased deposition of *sediment.*

alloy A metal that is manufactured by combining two or more molten metals. An alloy is always harder than its component metals. Bronze is an alloy of copper and tin.

alluvial fan A triangular deposit of *sediment* left by a stream that has lost velocity upon entering a broad, relatively flat valley.

alluvium A deposit of *sediment* left by a stream on the stream's channel or *floodplain.*

alpine glacier A mountain *glacier* that is confined by highlands.

andesite The dark, *aphanitic, extrusive rock* that has a silica content of about 60% and is the second most abundant volcanic rock. Andesites are found in large quantities in the Andes Mountains.

angle of repose The maximum angle at which a pile of unconsolidated material can remain stable.

anthracite A hard, jet-black coal that develops from *lignite* and *bituminous coal* through *metamorphism,* has a carbon content of 92% to 98%, and contains little or no gas. Anthracite burns with an extremely hot, blue flame and very little smoke, but it is difficult to ignite and both difficult and dangerous to mine.

anticline A convex *fold* in rock, the central part of which contains the oldest section of rock. See also *syncline.*

aquifer A *permeable* body of rock or regolith that both stores and transports groundwater.

aquitard A layer of rock having low permeability that stores groundwater but delays its flow.

arête A sharp ridge of erosion-resistant rock formed between adjacent *cirque glaciers.*

aridity index The ratio of a region's potential annual evaporation, as determined by its receipt of solar radiation, to its average annual precipitation.

arroyo A small, deep, usually dry channel eroded by a short-lived or intermittent desert stream.

artesian Of, being, or concerning an *aquifer* in which water rises to the surface due to pressure from overlying water.

asthenosphere A layer of soft but solid, mobile rock comprising the lower part of the upper *mantle* from about 100 to 350 kilometers beneath the Earth's surface. See also *lithosphere.*

atoll A circular reef that encloses a relatively shallow lagoon and extends from a very great depth to the sea surface. An atoll forms when an oceanic island ringed by a *barrier reef* sinks below sea level.

atom The smallest particle that retains all the chemical *properties* of a given *element.*

atomic mass 1. The sum of *protons* and *neutrons* in an atom's nucleus. 2. The combined mass of all the particles in a given atom.

atomic number The number of *protons* in the nucleus of a given atom. *Elements* are distinguished from each other by their atomic numbers.

backarc basin A depression landward of a *volcanic arc* in a *subduction* zone, which is lined with trapped *sediment* from the volcanic arc and the plate interior. See also *forearc basin.*

backarc spreading The process by which the overriding plate in a *subduction* zone becomes stretched to the point of *rifting,* so that *magma* can then rise into the gap created by the rift. Backarc spreading typically occurs when the subducting plate sinks more rapidly than the overriding plate moves forward.

backshore The portion of a *beach* that extends from the high-tide line inland to the sea cliff or vegetation line. Swash, the movement of water from a breaking wave up the beach face, reaches the backshore only during major storms.

backswamp The section of a *floodplain* where deposits of fine silts and clays settle after a flood. Backswamps usually lie behind a stream's natural levees.

banded iron formation A rock that is made up of alternating light silica-rich layers and dark-colored layers of iron-rich minerals, which were deposited in marine basins on every continent about 2 billion years ago.

barchan dune A crescent-shaped *dune* that forms around a small patch of vegetation; lies perpendicular to the prevailing wind direction; and has a gentle, convex windward slope and a steep, concave leeward slope. Barchan dunes typically form in arid, inland deserts with stable wind direction and relatively little sand.

barrier island A ridge of sand that runs parallel to the main coast but is separated from it by a bay or lagoon. Barrier islands range from 10 to 100 kilometers in length and from 2 to 5 kilometers in width. A barrier island may be as high as 6 meters above sea level.

barrier reef A long, narrow reef that runs parallel to the main coast but is separated from it by a wide lagoon.

basal sliding The process by which a *glacier* undergoes thawing at its base, producing a film of water along which the glacier then flows. Basal sliding primarily affects glaciers in warm climates or mid-latitude mountain ranges.

basalt The dark, dense, *aphanitic, extrusive rock* that has a silica content of 40% to 50% and makes up most of the ocean floor. Basalt is the most abundant volcanic rock in the Earth's crust.

base level The lowest level to which a *stream* can erode the channel through which it flows, generally equal to the prevailing global sea level.

basin A round or oval depression in the Earth's surface, containing the youngest section of rock in its lowest, central part.

batholith A massive discordant *pluton* with a surface area greater than 100 square kilometers, typically having a depth of about 30 kilometers. Batholiths are generally found in elongated mountain ranges after the country rock above them has eroded.

baymouth bar A narrow ridge of sand that stretches completely across the mouth of a bay. (Also called "bay bar" and "bay barrier.")

beach The part of a *coast* that is washed by waves or tides, which cover it with *sediments* of various sizes and composition, such as sand or pebbles.

beach drift 1. The process by which swash and backwash move *sediment* along a *beach face.* 2. The sediment so moved. Beach drift typically consists of sand, gravel, shell fragments, and pebbles. See also *longshore drift.*

beach face The portion of a *foreshore* that lies nearest to the sea and regularly receives the swash of breaking waves. The beach face is the steepest part of the foreshore.

bed load A body of coarse particles that move along the bottom of a *stream.*

bedding The division of *sediment* or *sedimentary rock* into parallel layers (beds) that can be distinguished from each other by such features as chemical composition and grain size.

Benioff-Wadati zone A region where the *subduction* of oceanic plates causes earthquakes, the *foci* of which are deeper the farther inland they are.

berm A low, narrow layer or mound of *sediment* deposited on a *backshore* by storm waves.

biogenic chemical sediment Sedimentary rocks derived from living organisms. Common examples of this include fossiliferous limestones and coal.

biomass fuel A renewable fuel derived from a living organism or the by-product of a living organism. Biomass fuels include wood, dung, methane gas, and grain alcohol.

bitumen Any of a group of solid and semi-solid *hydrocarbons* that can be converted into liquid form by heating. Bitumens can be refined to produce such commercial products as gasoline, fuel oil, and asphalt.

bituminous coal A shiny black coal that develops from deeply buried *lignite* through heat and pressure. Bituminous coal has a carbon content of 80% to 93%, which makes it a more efficient heating fuel than lignite.

body wave A type of *seismic wave* that transmits energy from an earthquake's *focus* through the Earth's interior in all directions. See also *surface wave.*

bond To combine one atom, by means of chemical reaction, with another atom to form a compound. When an atom bonds with another, it either loses, gains, or shares electrons with the other atom.

Bowen's reaction series The sequence of *igneous rocks* formed from a mafic *magma,* assuming mineral crystals that have already formed continue to react with the liquid magma and so evolve into new minerals, thereby creating the next rock in the sequence.

braided stream A network of converging and diverging *streams* separated from each other by narrow strips of sand and gravel.

breakwater A wall built seaward of a coast to intercept incoming waves and so protect a harbor or shore. Breakwaters are typically built parallel to the coast.

breccia A *clastic rock* composed of particles more than 2 millimeters in diameter and marked by the angularity of its component grains and rock fragments.

breeder reactor A nuclear reactor that manufactures more fissionable isotopes than it consumes. Breeder reactors use the widely available, nonfissionable uranium isotope U-238, together with small amounts of fissionable U-235, to produce a fissionable isotope of plutonium, Pu-239.

brittle failure The rupture of cool, near-surface rocks, caused by relatively low stress.

burial metamorphism A form of *regional metamorphism* that acts on rocks covered by 5 to 10 kilometers of rock or sediment, caused by heat from the Earth's interior and *lithostatic pressure*.

caldera A vast depression at the top of a *volcanic cone*, formed when an eruption substantially empties the reservoir of *magma* beneath the cone's summit. Eventually the summit collapses inward, creating a caldera. A caldera may be more than 15 kilometers in diameter and more than 1,000 meters deep.

caliche A white *soil horizon* consisting of calcium carbonate, typical of arid and semi-arid areas. Brief, heavy rains dissolve calcium carbonate in the upper layers of soil and transport it downward; the rainwater then evaporates rapidly, leaving the calcium carbonate to form a new, solid layer of soil.

capacity The ability of a given *stream* to carry *sediment*, measured as the maximum quantity it can transport past a given point on the channel bank in a given amount of time. See also *competence*.

capillary fringe The lowest part of the *zone of aeration*, marked by the rising of water from the *water table* due to the attraction of the water molecules to mineral surfaces and other molecules, and to pressure from the *zone of saturation* below.

carbon-14 dating A form of *isotope radiometric dating* that relies on the 5,730-year half-life of radioactive carbon-14, which decays into nitrogen-14, to determine the age of rocks in which carbon-14 is present. Carbon-14 dating is used for rocks from 100 to 100,000 years old.

carbonate One of several minerals containing one central carbon atom with strong *covalent bonds* to three oxygen atoms, and typically having *ionic bonds* to one or more positive ions.

catastrophism The hypothesis that a series of immense, brief, worldwide upheavals changed the Earth's crust greatly and can account for the development of mountains, valleys, and other features of the Earth. See also *uniformitarianism*.

cave A naturally formed opening beneath the surface of the Earth, generally formed by *dissolution* of carbonate bedrock. Caves may also form by erosion of coastal bedrock, partial melting of glaciers, or solidification of lava into hollow tubes.

cementation The *diagenetic* process by which sediment grains are bound together by *precipitated* minerals originally dissolved during the chemical weathering of preexisting rocks.

Cenozoic Era The latest era of the Phanerozoic Eon, following the *Mesozoic Era* and continuing to the present time, and marked by the presence of a wide variety of mammals, including the first hominids.

chemical sediment *Sediment* that is composed of previously dissolved minerals that have either precipitated from evaporated water or been extracted from water by living organisms and deposited when the organisms died or discarded their shells.

chemical weathering The process by which chemical reactions alter the chemical composition of rocks and minerals that are unstable at the Earth's surface and convert them into more stable substances; *weathering* that changes the chemical makeup of a rock or mineral. See also *mechanical weathering*.

chert A member of a group of *sedimentary rocks* that consist primarily of microscopic silica crystals. Chert may be either organic or inorganic, but the most common forms are inorganic.

chondrules Small nuggets of rocky material that exist in certain meteorites. These droplets of matter are believed to have condensed from our solar system's original nebula about 5 billion years ago. Their primary element is iron.

cinder cone A *pyroclastic cone* composed primarily of *cinders*.

cirque A deep, semicircular basin eroded out of a mountain by an *alpine glacier*.

cirque glacier A small *alpine glacier* that forms inside a *cirque*, typically near the head of a valley.

clastic Being or pertaining to a *sedimentary rock* composed primarily from fragments of preexisting rocks or fossils.

cleavage The tendency of certain minerals to break along distinct planes in their *crystal structures* where the bonds are weakest. Cleavage is tested by striking or hammering a mineral, and is classified by the number of surfaces it produces and the angles between adjacent surfaces.

coal A member of a group of easily combustible, organic *sedimentary rocks* composed mostly of plant remains and containing a high proportion of carbon.

coast The area of dry land that borders on a body of water.

coastal plain The gently sloping marginal region of a continent, seaward of the continental craton and the coastal mountain range.

col A high mountain pass that forms when part of an *arête* erodes.

color One of the physical properties of a mineral, although typically one of the least diagnostic.

compaction The *diagenetic* process by which the volume or thickness of sediment is reduced due to pressure from overlying layers of sediment.

competence The ability of a given *stream* to carry *sediment*, measured as the diameter of the largest particle that the stream can transport. See also *capacity*.

composite cone See *stratovolcano*.

compound An electrically neutral substance that consists of two or more *elements* combined in specific, constant proportions. A compound typically has physical characteristics different from those of its constituent elements.

compression *Stress* that reduces the volume or length of a rock, as that produced by the *convergence* of plate margins.

cone of depression An area in a *water table* along which water has descended into a well to replace water drawn out, leaving a gap shaped like an inverted cone.

conglomerate A *clastic* rock composed of particles more than 2 millimeters in diameter and marked by the roundness of its component grains and rock fragments.

contact metamorphism *Metamorphism* that is caused by heat from a magmatic intrusion.

continental collision The *convergence* of two continental plates, resulting in the formation of mountain ranges.

continental drift The hypothesis, proposed by Alfred Wegener, that today's continents broke off from a single supercontinent and then plowed through the ocean floors into their present positions. This explanation of the shapes and locations of Earth's current continents evolved into the theory of *plate tectonics*.

continental ice sheet An unconfined *glacier* that covers much or all of a continent.

continental platform Continental platforms are the regions adjacent to and surrounding continental shields. They are typically a relatively thin veneer of sedimentary rock that buries the edges of the shields.

continental shelf The gently sloping seaward edge of a continent located beyond the edge of a continental coastal plain. The shelf—although it's underwater—is actually part of the geology of the continent.

continental shield Broad areas of exposed ancient crystalline rocks in the cores of the Earth's continents. These rocks are typically the oldest on the continents, often more than 2.5 billion years old.

convection cell The cycle of movement in the *asthenosphere* that causes the plates of the *lithosphere* to move. Heated material in the asthenosphere becomes less dense and rises toward the solid lithosphere, through which it cannot rise further. It therefore begins to move horizontally, dragging the lithosphere along with it and pushing forward the cooler, denser material in its path. The cooler material eventually sinks down lower into the mantle, becoming heated there and rising up again, continuing the cycle. See also *plate tectonics*.

convergence The coming together of two lithospheric plates. Convergence causes *subduction* when one or both plates is oceanic, and mountain formation when both plates are continental. See also *divergence*.

core The innermost layer of the Earth, consisting primarily of pure metals such as iron and nickel. The core is the densest layer of the Earth, and is divided into the outer core, which is believed to be liquid, and the inner core, which is believed to be solid. See also *crust* and *mantle*.

correlation The process of determining that two or more geographically distant rocks or rock strata originated in the same time period.

covalent bond The combination of two or more atoms by sharing electrons so as to achieve chemical stability under the *octet rule*. Atoms that form covalent bonds generally have outer energy levels containing three, four, or five electrons. Covalent bonds are generally stronger than other bonds.

craton The segment of the Earth's continents that have remained tectonically stable and relatively earthquake-free for a vast period of time. The craton is composed of the continental shield and the surrounding continental platform.

creep The slowest form of mass movement, measured in millimeters or centimeters per year and occurring on virtually all slopes.

cross-bed A bed made up of particles dropped from a moving current, as of wind or water, and marked by a downward slope that indicates the direction of the current that deposited them.

crust The outermost layer of the Earth, consisting of relatively low-density rocks. See also *core* and *mantle*.

crystal A mineral in which the systematic internal arrangement of atoms is outwardly reflected as a latticework of repeated three-dimensional units that form a geometric solid with a surface consisting of symmetrical planes.

crystal structure 1. The geometric pattern created by the systematic internal arrangement of atoms in a mineral. 2. The systematic internal arrangement of atoms in a mineral. See also *crystal*.

daughter isotope An *isotope* that forms from the radioactive decay of a *parent isotope*. A daughter isotope may or may not be of the same element as its parent. If the daughter isotope is radioactive, it will eventually become the parent isotope of a new daughter isotope. The last daughter isotope to form from this process will be stable and nonradioactive.

debris flow 1. The rapid, downward mass movement of particles coarser than sand, often including boulders 1 meter or more in diameter, at a rate ranging from 2 to 40 kilometers per hour. Debris flows occur along fairly steep slopes. 2. The material that descends in such a flow.

deflation The process by which wind erodes bedrock by picking up and transporting loose rock particles.

degradation The process by which a stream's *gradient* becomes less steep, due to the *erosion* of *sediment* from the streambed. Such erosion generally follows a sharp reduction in the amount of sediment entering the stream.

delta An *alluvial fan* having its apex at the mouth of a *stream*.

dendrochronology A method of *numerical dating* that uses the number of tree rings found in a cross section of a tree trunk or branch to determine the age of the tree.

density The measure of a mineral's mass for a given volume of the mineral, typically measured in grams per cubic centimeter.

desert A region with an average annual rainfall of 10 inches or less and sparse vegetation, typically having thin, dry, and crumbly soil. A desert has an *aridity index* greater than 4.0.

desertification The process through which a desert takes over a formerly non-desert area. When a region begins to undergo desertification, the new conditions typically include a significantly lowered *water table*, a reduced supply of surface water, increased salinity in natural waters and soils, progressive destruction of native vegetation, and an accelerated rate of erosion.

desert pavement A closely packed layer of rock fragments concentrated in a layer along the Earth's surface by the *deflation* of finer particles.

desert varnish A thin, shiny, red-brown or black layer, principally composed of iron manganese oxides, that coats the surfaces of many exposed desert rocks.

detrital sediment *Sediment* that is composed of transported solid fragments of preexisting igneous, sedimentary, or metamorphic rocks.

diagenesis The set of processes that cause physical and chemical changes in sediment after it has been deposited and buried under another layer of sediment. Diagenesis may culminate in *lithification*.

diatreme A variety of volcanic neck that forms from deep-seated, extremely gaseous, highly explosive eruptions. Diatremes contain highly fragmented chunks of lava mixed with rock ripped from the vent wall, and masses of rock from the Earth's deep crust and mantle.

dike A discordant *pluton* that is substantially wider than it is thick. Dikes are often steeply inclined or nearly vertical. See also *sill*.

dilatancy The expansion of a rock's volume caused by *stress* and deformation.

diorite Any of a group of dark, phaneritic, *intrusive rocks* that are the *plutonic* equivalents of *andesite*.

dip The angle formed by the inclined plane of a geological structure and the horizontal plane of the Earth's surface.

dip-slip fault A *fault* in which two sections of rock have moved apart vertically, parallel to the *dip* of the fault plane.

directed pressure Force exerted on a rock along one plane, flattening the rock in that plane and lengthening it in the perpendicular plane.

disappearing stream A surface *stream* that drains rapidly and completely into a *sinkhole*.

displaced terrane Displaced terrane is a fault-bounded body of rock—sometimes thousands of km² in area—that originated elsewhere geographically and has then moved, perhaps long distances, by plate motion.

dissolution A form of *chemical weathering* in which water molecules, sometimes in combination with acid or another compound in the environment, attract and remove oppositely charged ions or ion groups from a mineral or rock.

dissolved load A body of sediment carried by a *stream* in the form of *ions* that have dissolved in the water.

distributary One of a network of small *streams* carrying water and sediment from a *trunk stream* into an ocean.

divergence The process by which two lithospheric plates separated by *rifting* move farther apart, with soft mantle rock rising between them and forming new oceanic *lithosphere*. See also *convergence*.

dolostone A *sedimentary rock* composed primarily of dolomite, a mineral made up of calcium, magnesium, carbon, and oxygen. Dolostone is thought to form when magnesium ions replace some of the calcium ions in *limestone*, to which dolostone is similar in both appearance and chemical structure.

dome A round or oval bulge on the Earth's surface, containing the oldest section of rock in its raised, central part. See also *basin*.

drainage basin The area from which water flows into a *stream*. Also called a *watershed*.

drainage divide An area of raised, dry land separating two adjacent *drainage basins*.

drainage pattern The arrangement in which a *stream* erodes the channels of its network of *tributaries*.

drumlin A long, spoon-shaped hill that develops when pressure from an overriding *glacier* reshapes a *moraine*. Drumlins range in height from 5 to 50 meters and in length from 400 to 2,000 meters. They slope down in the direction of the ice flow.

dune A usually asymmetrical mound or ridge of sand that has been transported and deposited by wind. Dunes form in both arid and humid climates.

dynamothermal metamorphism A form of *regional metamorphism* that acts on rocks caught between two *converging* plates and is initially caused by *directed pressure* from the plates, which causes some of the rocks to rise and others to sink, sometimes by tens of kilometers. The rocks that fall then experience further dynamothermal metamorphism, this time caused by heat from the Earth's interior and *lithostatic pressure* from overlying rocks.

Earth systems The Earth is made up of four basic systems: the lithosphere (the rocks of the Earth); the hydrosphere (the waters of the Earth); the atmosphere (the gases that surround the Earth); and the biosphere (the life on Earth). These systems interact to produce most of the geological processes that occur on Earth. An event involving one of these systems may affect some or all of the others.

earthquake A movement within the Earth's *crust* or *mantle*, caused by the sudden rupture or repositioning of underground rocks as they release *stress*.

ebb tide A *tide* that lowers the water surface of an ocean and moves the shoreline farther seaward.

echo-sounding sonar The mapping of ocean topography based on the time required for sound waves to reach the sea floor and return to the research ship that emits them.

elastic deformation A temporary *stress*-induced change in the shape or volume of a rock, after which the rock returns to its original shape and volume.

elastic limit See *yield point*.

electron A negatively charged particle that orbits rapidly around the *nucleus* of an *atom*. See also *proton*.

element A form of matter that cannot be broken down into a chemically simpler form by heating, cooling, or chemical reactions. There are 115 known elements: 92 of them natural and 23 manmade. Elements are represented by one- or two-letter abbreviations. See also *atom* and *atomic number*.

energy level The path of a given electron's orbit around a nucleus, marked by a constant distance from the nucleus.

epicenter The point on the Earth's surface that is located directly above the *focus* of an *earthquake*.

equilibrium line The point in a *glacier* where overall gain in volume equals overall loss, so that the net volume remains stable. The equilibrium line marks the border between the *zone of accumulation* and the *zone of ablation*.

erosion The process by which particles of rock and soil are loosened, as by *weathering*, and then transported elsewhere, as by wind, water, ice, or gravity.

esker A ridge of *sediment* that forms under a glacier's *zone of ablation*, made up of sand and gravel deposited by meltwater. An esker may be less than 100 meters or more than 500 kilometers long, and may be anywhere from 3 to over 300 meters high.

estuary The mouth of a river where fresh river water mixes with salty marine water as a river flows into an ocean.

evaporite An inorganic *chemical sediment* that precipitates when the salty water in which it had dissolved evaporates.

extrusive rock An *igneous rock* formed from *lava* that has flowed out onto the Earth's surface, characterized by rapid solidification and grains that are so small as to be barely visible to the naked eye.

fall The fastest form of mass movement, occurring when rock or sediment breaks off from a steep or vertical slope and descends at a rate of 9.8 meters per second. A fall can be extremely dangerous.

fault A *fracture* dividing a rock into two sections that have visibly moved relative to each other.

fault block A section of rock separated from other rock by one or more *faults.*

fault-block mountain A mountain containing tall *horsts* interspersed with much lower *grabens* and bounded on at least one side by a high-angle *normal fault.*

fault-zone metamorphism The *metamorphism* that acts on rocks grinding past one another along a fault and is caused by *directed pressure* and frictional heat.

firn Firmly packed snow that has survived a summer melting season. Firn has a density of about 0.4 gram per cubic centimeter. Ultimately, firn turns into glacial ice.

fission tracks Marks left in the latticework of a mineral crystal by subatomic particles released during the fission of a radioactive atom trapped inside the crystal.

fjord A deep, steep-walled, *U*-shaped valley formed by erosion by a *glacier* and submerged with seawater.

flood tide A *tide* that raises the water surface of an ocean and moves the *shoreline* farther inland.

floodplain The flat land that surrounds a *stream* and becomes submerged when the stream overflows its banks.

flow A relatively rapid mass-movement process that involves a mixture of rock, soil, vegetation, and water moving downslope as a viscous fluid. Within a flow (such as a mudflow), each particle, regardless of its size, moves independently.

fluorescence Emission of visible light by a substance, such as a mineral, that is currently exposed to ultraviolet light and absorbs radiation from it. The light appears in the form of glowing, distinctive colors. The emission ends when the exposure to ultraviolet light ends.

focus (plural **foci**) The precise point within the Earth's *crust* or *mantle* where rocks begin to rupture or move in an *earthquake.*

fold A bend that develops in an initially horizontal layer of rock, usually caused by *plastic deformation.* Folds occur most frequently in *sedimentary rocks.*

foliation The arrangement of a set of minerals in parallel, sheet-like layers that lie perpendicular to the flattened plane of a rock. Foliation occurs in *metamorphic rocks* on which *directed pressure* has been exerted.

forearc basin A depression in the sea floor located between an *accretionary wedge* and a *volcanic arc* in a *subduction* zone, and lined with trapped *sediment.* See also *backarc basin.*

foreshock A minor, barely detectable *earthquake,* generally preceding a full-scale earthquake with approximately the same *focus.* Major quakes may follow a cluster of foreshocks by as little as a few seconds or as much as several weeks.

foreshore The portion of a *beach* that lies nearest to the sea, extending from the low-tide line to the high-tide line.

fossil The remains of ancient organisms, or other evidence of their existence, that has been preserved in geologic material.

fossil fuel A nonrenewable energy source, such as oil, gas, or coal, that derives from the organic remains of past life. Fossil fuels consist primarily of *hydrocarbons.*

fractional crystallization The process by which a *magma* produces crystals that then separate from the original magma, so that the chemical composition of the magma changes with each generation of crystals, producing *igneous rocks* of different compositions. The silica content of the magma becomes proportionately higher after each crystallization.

fracture 1. A crack or break in a rock. 2. To break in random places instead of *cleaving.* Said of minerals.

fringing reef A reef that forms against or near an island or continental *coast* and grows seaward, sloping sharply toward the sea floor. Fringing reefs usually range from 0.5 to 1.0 or more kilometers in width.

frost wedging A form of *mechanical weathering* caused by the freezing of water that has entered a pore or crack in a rock. The water expands as it freezes, widening the cracks or pores and often loosening or dislodging rock fragments.

gabbro Any of a group of dark, dense, phaneritic, *intrusive rocks* that are the *plutonic* equivalent to *basalt.*

geochronology The study of the relationship between the history of the Earth and time.

geologic time scale The division of all of Earth history into blocks of time distinguished by geologic and evolutionary events, ordered sequentially and arranged into eons made up of eras, which are in turn made up of periods, which are in turn made up of epochs.

geology The scientific study of the Earth, its origins and evolution, the materials it is made up of, and the processes that act on it.

geophysics The branch of *geology* that studies the physics of the Earth, using the physical principles underlying such phenomena as *seismic waves,* heat flow, *gravity,* and *magnetism* to investigate planetary properties.

geyser A *natural spring* marked by the intermittent escape of hot water and steam.

glacial abrasion The process by which a glacier erodes the underlying bedrock through contact between the bedrock and rock fragments embedded in the base of the glacier. See also *glacial quarrying.*

glacial drift A load of rock material transported and deposited by a glacier. Glacial drift is usually deposited when the glacier begins to melt.

glacial erratic A rock or rock fragment transported by a glacier and deposited on bedrock of different composition. Glacial erratics range from a few millimeters to several yards in diameter.

glacial quarrying The process by which a glacier erodes the underlying bedrock by loosening and ultimately detaching blocks of rock from the bedrock and attaching them instead to the glacier, which then bears the rock fragments away. See also *glacial abrasion.*

glacial till *Drift* that is deposited directly from glacial ice and, therefore, not *sorted.* Also called "till." See also *glacial drift.*

glacier A moving body of ice that forms on land from the accumulation and compaction of snow, and that flows downslope or outward due to *gravity* and the pressure of its own weight.

gneiss A coarse-grained, *foliated metamorphic rock* marked by bands of light-colored minerals such as quartz and feldspar that alternate with bands of dark-colored minerals. This alternation develops through *metamorphic differentiation.*

graben A block of rock that lies between two *faults* and has moved downward to form a depression between the two adjacent fault blocks. See also *horst.*

graded bed A bed formed by the deposition of sediment in relatively still water, marked by the presence of particles that vary in size, density, and shape. The particles settle in a gradual slope with the coarsest particles at the bottom and the finest at the top.

graded stream A stream maintaining an equilibrium between the processes of erosion and deposition, and therefore between *aggradation* and *degradation.*

gradient The vertical drop in a stream's elevation over a given horizontal distance, expressed as an angle.

granite A pink-colored, felsic, *plutonic rock* that contains potassium and usually sodium feldspars, and has a quartz content of about 10%. Granite is commonly found on continents but is virtually absent from the ocean basins.

gravity 1. The force of attraction exerted by one body in the universe on another. Gravity is directly proportional to the product of the masses of the two attracted bodies. 2. The force of attraction exerted by the Earth on bodies on or near its surface, tending to pull them toward the Earth's center.

gravity anomaly The difference between an actual measurement of *gravity* at a given location and the measurement predicted by theoretical calculation.

groin A structure that juts out into a body of water perpendicular to the *shoreline* and is built to restore an eroding beach by intercepting *longshore drift* and trapping sand.

guyot A *seamount,* the top of which has been flattened by *weathering,* wave action, or stream *erosion.*

half-life The time necessary for half of the atoms of a *parent isotope* to decay into the *daughter isotope.*

hardness The degree of resistance of a given mineral to scratching, indicating the strength of the bonds that hold the mineral's atoms together. The hardness of a mineral is measured by rubbing it with substances of known hardness.

headland A cliff that projects out from a *coast* into deep water.

Holocene Epoch The second epoch of the Quaternary Period, beginning approximately 10,000 years ago and continuing to the present time. See also *Pleistocene Epoch.*

hook A spit that curves sharply at its coastal end.

horn A high mountain peak that forms when the walls of three or more *cirques* intersect.

hornfels A hard, dark-colored, dense *metamorphic rock* that forms from the intrusion of magma into *shale* or *basalt.*

horst A block of rock that lies between two *faults* and has moved upward relative to the two adjacent fault blocks. See also *graben.*

hot spot An area in the upper *mantle,* ranging from 100 to 200 kilometers in width, from which magma rises in a plume to form *volcanoes.* A hot spot may endure for 10 million years or more.

hydraulic conductivity The extent to which a given substance allows water to flow through it, determined by such factors as sorting and grain size and shape.

hydraulic gradient The difference in *potential* between two points, divided by the lateral distance between the points.

hydraulic lifting The *erosion* of a streambed by water pressure.

hydrocarbon A molecule that is entirely made up of hydrogen and carbon.

hydrogen bond An intermolecular bond formed with hydrogen.

hydrologic cycle The perpetual movement of water among the mantle, oceans, land, and atmosphere of the Earth.

hydrolysis A form of *chemical weathering* in which ions from water replace equivalently charged ions from a mineral, especially a silicate.

hydrothermal deposit A mineral deposit formed by the precipitation of metallic ions from water ranging in temperature from 50° to 700°C.

hydrothermal metamorphism The chemical alteration of preexisting rocks by hot water that typically occurs in the faults and fractures at divergent plate boundaries.

hypothesis A tentative explanation of a given set of data that is expected to remain valid after future observation and experimentation. See also *theory*.

ice age A period during which the Earth is substantially cooler than usual and a significant portion of its land surface is covered by *glaciers*. Ice ages generally last tens of millions of years.

ice cap An *alpine glacier* that covers the peak of a mountain.

igneous rock A rock made from molten (melted) or partly molten material that has cooled and solidified.

index fossil The fossil of an organism known to have existed for a relatively short period of time. Index fossils are used to date the rocks in which they are found.

inorganic chemical sediment Sedimentary rocks—such as limestone, chert, and evaporates—that form when the dissolved products of chemical weathering precipitate from solution. The ornate columns of travertine in caves are examples of inorganic chemical sedimentary rock.

inselberg A steep ridge or hill left when a mountain has eroded. Inselbergs are found in otherwise flat, typically desert, plains.

intermolecular bonding The act or process by which two or more groups of atoms or molecules combine due to weak positive or negative charges that develop at various points within each group of atoms due to uneven distribution of their electrons. The side of the molecule where electrons are more likely to be found will have a slight negative charge, and the side where they are less likely to be found will have a slight positive charge. Such charged regions attract oppositely charged regions of nearby molecules, forming relatively weak bonds.

internal deformation The rearrangement of the planes within ice crystals, due to pressure from overlying ice and snow, that causes the downward or outward flow of a *glacier*.

intrusive rock An *igneous rock* formed by the entrance of *magma* into preexisting rock.

ion An atom that has lost or gained one or more electrons, thereby becoming electrically charged.

ionic bond The combination of an atom that has a strong tendency to lose electrons with an atom that has a strong tendency to gain electrons, such that the former transfers one or more electrons to the latter and each achieves chemical stability under the *octet rule*. The atom that loses electrons acquires a positive electric charge and the atom that gains electrons acquires a negative electric charge, so that the resulting compound is electrically neutral.

ionic bonding The act or process of forming an ionic bond.

ionic substitution The replacement of one type of ion in a mineral by another that is similar to the first in size and charge.

isostasy The equilibrium maintained between the gravity tending to depress and the buoyancy tending to raise a given segment of the *lithosphere* as it floats above the *asthenosphere*.

isotope One of two or more forms of a single element; the atoms of each isotope have the same number of protons but different numbers of neutrons in their nuclei. Thus isotopes have the same *atomic number* but differ in *atomic mass*.

isotope dating The process of using relative proportions of *parent* to *daughter isotopes* in radioactive decay to determine the age of a given rock or rock stratum.

jetty A structure built along the bank of a stream channel or tidal outlet to direct the flow of a stream or tide and keep the sediment moving so that it cannot build up and fill the channel. Jetties are typically built in parallel pairs along both banks of the channel. Jetties that are built perpendicular to a *coast* tend to interrupt *longshore drift* and thus widen *beaches*.

joint A *fracture* dividing a rock into two sections that have not visibly moved relative to each other. See also *fault*.

juvenile water The steam that accompanies *volcanic* eruptions.

karst A topography characterized by *caves, sinkholes, disappearing streams,* and underground drainage. Karst forms when groundwater dissolves pockets of limestone, dolomite, or gypsum in bedrock.

kerogen A solid, waxy, organic substance that forms when pressure and heat from the Earth act on the remains of plants and animals. Kerogen converts to various liquid and gaseous *hydrocarbons* at a depth of 7 or more kilometers and a temperature between 50° and 100°C.

laccolith A large concordant *pluton* that is shaped like a dome or a mushroom. Laccoliths tend to form at relatively shallow depths and are typically composed of granite. The *country rock* above them often erodes away completely.

lahar A *flow* of *pyroclastic* material mixed with water. A lahar is often produced when a snowcapped volcano erupts and hot pyroclastics melt a large amount of snow or ice.

lava *Magma* that comes to the Earth's surface through a *volcano* or fissure.

leach To dissolve from a rock. For example, when acidic water passes through fractured rocks, soluble minerals leach, or dissolve, from the rocks.

levee A protective barrier built along the banks of a stream to prevent flooding.

lichen Plantlike colonies of fungi and algae that grow on the exposed surface of a rock. Lichen grow at a constant rate within a single geographic area.

lichenometry A method of *numerical dating* that uses the size of *lichen* colonies on a rock surface to determine the surface's age. Lichenometry is used for rock surfaces less than about 9,000 years old.

lignite A soft, brownish *coal* that develops from *peat* through bacterial action, is rich in *kerogen*, and has a carbon content of 70%, which makes it a more efficient heating fuel than peat.

limestone A *sedimentary rock* composed primarily of calcium carbonate. Some 10% to 15% of all sedimentary rocks are limestones. Limestone is usually organic, but it may also be inorganic.

liquefaction The conversion of moderately cohesive, unconsolidated *sediment* into a fluid, water-saturated mass.

lithification The conversion of loose *sediment* into solid *sedimentary rock*.

lithosphere A layer of solid, brittle rock making up the outer 100 kilometers of the Earth, encompassing both the crust and the outermost part of the upper *mantle*. See also *asthenosphere*.

lithostatic pressure The force exerted on a rock buried deep within the Earth by overlying rocks. Because lithostatic pressure is exerted equally from all sides of a rock, it compresses the rock into a smaller, denser form without altering the rock's shape.

loess A load of *silt* that is produced by the erosion of *outwash* and transported by wind. Much loess found in the Mississippi Valley, China, and Europe is believed to have been deposited during the *Pleistocene Epoch*.

longitudinal dune One of a series of long, narrow *dunes* lying parallel both to each other and to the prevailing wind direction. Longitudinal dunes range from 60 meters to 100 kilometers in length, and from 3 to 50 meters in height.

longshore current An ocean current that flows close and almost parallel to the *shoreline* and is caused by the rush of waves toward the shore.

longshore drift 1. The process by which a current moves sediment along a surf zone. 2. The sediment so moved. Longshore drift typically consists of sand, gravel, shell fragments, and pebbles. See also *beach drift*.

lopolith A saucer-shaped intrusive body of igneous rock. Lopoliths are typically mafic in composition.

low-velocity zone An area within the Earth's upper *mantle* in which both *P waves* and *S waves* travel at markedly slower velocities than in the outermost part of the upper mantle. The low-velocity zone occurs in the range between 100 and 350 kilometers of depth.

luster 1. The reflection of light on a given mineral's surface, classified by intensity and quality. 2. The appearance of a given mineral as characterized by the intensity and quality with which it reflects light.

magma Molten (melted) rock that forms naturally within the Earth. Magma may be either a liquid or a fluid mixture of liquid, crystals, and dissolved gases.

magnetic field The region within which the *magnetism* of a given substance or particle affects other substances.

magnetic reversal The process by which the Earth's magnetic north pole and its magnetic south pole reverse their positions over time.

magnetism The property, possessed by certain materials, to attract or repel similar materials. Magnetism is associated with moving electricity.

mantle The middle layer of the Earth, lying just below the *crust* and consisting of relatively dense rocks. The mantle is divided into two sections: the upper mantle and the lower mantle. The lower mantle has greater density than the upper mantle. See also *core* and *crust*.

marble A coarse-grained, nonfoliated *metamorphic rock* derived from *limestone* or *dolostone*.

marine magnetic anomaly An irregularity in magnetic strength along the ocean floor that reflects *sea-floor spreading* during periods of *magnetic reversal*.

mass movement The relentless process that transports Earth material (such as bedrock, loose sediment, and soil) down slopes—from high places to low places—by the pull of gravity.

massive sulfide deposit An unusually large deposit of sulfide minerals.

meandering stream A *stream* that traverses relatively flat land in fairly evenly spaced loops and separated from each other by narrow strips of *floodplain.*

mechanical exfoliation A form of *mechanical weathering* in which successive layers of a large *plutonic rock* break loose and fall when the erosion of overlying material permits the rock to expand upward. The thin slabs of rock that break off fall parallel to the exposed surface of the rock, creating the long, broad steps that can be found on many mountains.

mechanical weathering The process by which a rock or mineral is broken down into smaller fragments without altering its chemical makeup. Mechanical weathering is *weathering* that affects only physical characteristics. See also *chemical weathering.*

mélange A body of rock that forms along the inner wall of an *ocean trench* and is made up of fragments of *lithosphere* and oceanic *sediment* that have undergone *metamorphism.*

Mercalli intensity scale A scale designed to measure the degree of intensity of *earthquakes,* ranging from I for the lowest intensity to XII for the highest. The classifications are based on human perceptions.

Mesozoic Era The intermediate era of the Phanerozoic Eon, following the *Paleozoic Era* and preceding the *Cenozoic Era,* and marked by the dominance of marine and terrestrial reptiles, as well as the appearance of birds, mammals, and flowering plants.

metallic bonding The act or process by which two or more atoms of electron-donating elements pack so closely together that some of their electrons begin to wander among the nuclei rather than orbiting the nucleus of a single atom. Metallic bonding is responsible for the distinctive properties of metals.

metamorphic differentiation The process by which minerals from a chemically uniform rock separate from each other during *metamorphism* and form individual layers within a new *metamorphic rock.*

metamorphic grade A measure used to identify the degree to which a *metamorphic rock* has changed from its parent rock. A metamorphic grade provides some indication of the circumstances under which the metamorphism took place.

metamorphic index mineral One of a set of minerals found in *metamorphic rocks* and used as indicators of the temperature and pressure conditions at which the metamorphism occurred. A metamorphic index mineral is stable only within a narrow range of temperatures and pressures, and the metamorphism that produces it must take place within that range.

metamorphic rock A rock that has undergone chemical or structural changes. Heat, pressure, or a chemical reaction may cause such changes.

metamorphism The process by which conditions within the Earth, below the zone of *diagenesis,* alter the mineral content, chemical composition, and structure of solid rock without melting it. *Igneous, sedimentary,* and *metamorphic rocks* may all undergo metamorphism.

meteoric water The precipitation of condensed water from clouds as rain, snow, sleet, or hail.

microcontinent A section of continental *lithosphere* that has broken off from a larger, distant continent, as by *rifting.*

migmatite A high-grade metamorphic rocks that contains granitic layers or regions that in most cases formed by partial melting of the rock.

mineral A naturally occurring, usually inorganic, solid consisting of either a single element or a *compound,* and having a definite chemical composition and a systematic internal arrangement of atoms.

mineraloid A naturally occurring, usually inorganic, solid consisting of either a single element or a *compound,* and having a definite chemical composition but lacking a systemic internal arrangement of atoms. See also *mineral.*

Moho (abbreviation for Mohorovičić) The seismic discontinuity between the base of the Earth's *crust* and the top of the *mantle. P waves* passing through the Moho change their velocity by approximately 1 kilometer per second, with the higher velocity occurring in the mantle and the lower in the crust.

moment-magnitude scale A recently developed alternative to the Richter scale used to measure more accurately the amount of energy released by large earthquakes. This scale involves measurement of an earthquake's seismic moment.

moraine A single, large mass of *glacial till* that accumulates, typically at the edge of a glacier.

mudcrack A fracture that develops at the top of a layer of fine-grained, muddy sediment when it is exposed to the air, dries out, and then shrinks.

mudstone A detrital sedimentary rock composed of clay-sized particles (<0.002 mm in diameter).

natural bridge An arch-shaped stretch of bedrock remaining in a *karst* region when the surrounding bedrock has dissolved.

natural spring A place where groundwater flows to the surface and issues freely from the ground.

neutron A particle that is found in the *nucleus* of an *atom,* has a mass approximately equal to that of a *proton,* and has no electric charge.

normal fault A *dip-slip fault* marked by a generally steep *dip* along which the hanging wall has moved downward relative to the footwall.

nuclear fission The division of the *nuclei* of *isotopes* of certain heavy *elements,* such as uranium and plutonium, effected by bombardment with *neutrons.* Nuclear fission causes the release of energy, additional neutrons, and an enormous quantity of heat. Nuclear fission is used in nuclear power plants and nuclear weapons. A by-product of nuclear fission is toxic radioactive waste. See also *nuclear fusion.*

nuclear fusion The combination of the *nuclei* of certain extremely light *elements,* especially hydrogen, effected by the application of high temperature and pressure. Nuclear fusion causes the release of an enormous amount of heat energy, comparable to that released by *nuclear fission.* The principal by-product of nuclear fusion is helium.

nucleus (plural nuclei) The central part of an *atom,* containing most of the atom's mass and having a positive charge due to the presence of *protons.*

nuée ardente A sometimes glowing cloud of gas and *pyroclastics* erupted from a *volcano* and moving swiftly down its slopes. Also called a *pyroclastic flow.*

numerical dating The fixing of a geological structure or event in time, as by counting tree rings.

oblique slip Fault motion that involves both dip-slip and strike-slip movement of fault blocks.

ocean trench A deep, linear, relatively narrow depression in the sea floor, formed by the *subduction* of oceanic plates.

octet rule A scientific law stating that all atoms, except those of hydrogen and helium, require eight electrons in the outermost *energy level* to maintain chemical stability.

oil sand A mixture of unconsolidated sand and clay that contains a semi-solid *bitumen.*

oil shale A brown or black *clastic source rock* containing *kerogen.*

oil trap Subsurface structures—such as folds and faults—that impede the flow of migrating oil, allowing it to accumulate.

ophiolite suite The group of *sediments, sedimentary rocks,* and mafic and ultramafic *igneous rocks* that makes up the oceanic *lithosphere.*

ore A mineral deposit that can be mined for a profit.

orogenesis Mountain formation, as caused by *volcanism, subduction,* plate divergence, *folding,* or the movement of *fault blocks.* Also called "orogeny."

oscillatory motion The circular movement of water up and down, with little or no change in position, as a wave passes.

outwash A load of *sediment,* consisting of sand and gravel, that is deposited by meltwater in front of a *glacier.*

oxbow lake A crescent-shaped body of standing water formed from a single loop that was cut off from a *meandering stream,* typically by a flood that allowed the stream to flow through its *floodplain* and bypass the loop.

oxidation The process of combining with oxygen ions. A mineral that is exposed to air may undergo oxidation as a form of *chemical weathering.*

oxide One of several minerals containing negative oxygen ions bonded to one or more positive metallic ions.

paleomagnetism 1. The fixed orientation of a rock's crystals, based on the Earth's *magnetic field* at the time of the rock's formation, that remains constant even when the magnetic field changes. 2. The study of such phenomena as indicators of the Earth's magnetic history.

paleosol An ancient, buried soil whose composition may reflect a climate significantly different from the climate now prevalent in the area where the soil is found.

Paleozoic Era The earliest era of the Phanerozoic Eon, marked by the presence of marine invertebrates, fish, amphibians, insects, and land plants.

parabolic dune A horseshoe-shaped *dune* having a concave windward slope and a convex leeward slope. Parabolic dunes tend to form along sandy ocean shores and lakeshores. They may also develop from *transverse dunes* through *deflation.*

parent isotope A radioactive *isotope* that changes into a different isotope when its nucleus decays. See also *daughter isotope.*

parent material The source from which a given soil is chiefly derived, generally consisting of bedrock or *sediment.*

partial melting The incomplete melting of a rock composed of minerals with differing melting points. When partial melting occurs, the minerals with higher melting points remain solid while the minerals whose melting points have been reached turn to *magma.*

passive continental margin A border that lies between continental and oceanic *lithosphere,* but is not a plate margin. It is marked by lack of seismic and volcanic activity.

peat A soft, brown mass of compressed, partially decomposed vegetation that forms in a water-saturated environment and has a carbon content of 50%. Dried peat can be burned as fuel.

pediment A broad surface at the base of a receding mountain. The pediment develops when running water erodes most of the mass of the mountain.

pegmatite A coarse-grained *igneous rock* with exceptionally large *crystals,* formed from a *magma* that contains a high proportion of water.

perched water table A saturated area that lies within a *zone of aeration.*

peridotite An *igneous rock* composed primarily of the iron–magnesium *silicate* olivine and having a silica content of less than 40%.

permafrost Permanently frozen regolith, ranging in thickness from 30 centimeters to over 1,000 meters.

permeability The capability of a given substance to allow the passage of a fluid. Permeability depends on the size and degree of the connections among a substance's pores.

petroleum Any of a group of naturally occurring substances made up of *hydrocarbons.* These substances may be gaseous, liquid, or semi-solid.

phosphorescence Emission of visible light by a substance, such as a mineral, that is exposed to ultraviolet light and absorbs radiation from it. The light appears in the form of glowing, distinctive colors. The emission continues after the exposure to ultraviolet light ends.

phyllite A *foliated metamorphic rock* that develops from *slate* and is marked by a silky sheen and medium grain size.

placer deposit A deposit of heavy or durable minerals, such as gold or diamonds, typically found where the flow of water abruptly slows.

plastic deformation A permanent *strain* that entails no rupture.

plate tectonics The theory that the Earth's *lithosphere* consists of large, rigid plates that move horizontally in response to the flow of the *asthenosphere* beneath them, and that interactions among the plates at their borders cause most major geologic activity, including the creation of oceans, continents, mountains, volcanoes, and earthquakes.

playa A dry lake basin found in a desert.

Pleistocene Epoch The first epoch of the Quaternary Period, beginning 2 to 3 million years ago and ending approximately 10,000 years ago. See also *Holocene Epoch.*

pluton An *intrusive rock,* as distinguished from the preexisting *country rock* that surrounds it.

plutonic rock An *intrusive rock* formed inside the Earth.

pluvial lake A lake that formed from rainwater falling into a landlocked basin during a glacial period marked by greater precipitation than is found in the region in prior or subsequent periods.

point bar A low ridge of *sediment* that forms along the inner bank of a *meandering stream.*

polymorph A mineral that is identical to another mineral in chemical composition but differs from it in *crystal structure.*

porosity The percentage of a soil, rock, or sediment's volume that is made up of pores.

porphyry copper deposit A crystallized rock, typically porphyritic, having hairline fractures that contain copper and other metals.

potassium–argon dating A form of *isotope dating* that relies on the extremely long *half-life* of radioactive isotopes of potassium, which decay into isotopes of argon, to determine the age of rocks in which argon is present. Potassium–argon dating is used for rocks between 100,000 and 4 billion years old.

potential The combined influence of *gravity* and water pressure on groundwater flow at a given depth.

potentiometric surface The level to which the water in an *artesian aquifer* would rise if unaffected by friction with the surrounding rocks and sediments.

primary coast A *coast* shaped primarily by nonmarine processes, such as *glacial erosion* or biological processes.

principle of cross-cutting relationships The *scientific law* stating that a *pluton* is always younger than the rock that surrounds it.

principle of faunal succession The *scientific law* stating that specific groups of animals have followed, or succeeded, one another in a definite sequence through Earth's history.

principle of original horizontality The *scientific law* stating that *sediments* settling out from bodies of water are deposited horizontally or nearly horizontally in layers that lie parallel or nearly parallel to the Earth's surface.

principle of superposition The *scientific law* stating that, in any unaltered sequence of rock strata, each stratum is younger than the one beneath it and older than the one above it, so that the youngest stratum will be at the top of the sequence and the oldest at the bottom.

principle of uniformitarianism The *scientific law* stating that the geological processes taking place in the present operated similarly in the past and can therefore be used to explain past geologic events.

property A characteristic that distinguishes one substance from another.

proton A positively charged particle that is found in the *nucleus* of an *atom* and has a mass approximately 1,836 times that of an *electron.*

P wave (abbreviation for **primary wave**) A *body wave* that causes the *compression* of rocks when its energy acts upon them. When the P wave moves past a rock, the rock expands beyond its original volume, only to be compressed again by the next P wave. P waves are the fastest of all *seismic waves.* See also *S wave.*

P-wave shadow zone The region that extends from 103° to 143° from the *epicenter* of an *earthquake* and is marked by the absence of *P waves.* The P-wave shadow zone is due to the refraction of *seismic waves* in the liquid outer *core.* See also *S-wave shadow zone.*

pyroclastic Being or pertaining to rock fragments formed in a volcanic eruption.

pyroclastic cone A usually steep, conic *volcano* composed almost entirely of an accumulation of loose *pyroclastic* material. Pyroclastic cones are usually less than 450 meters high. Because no *lava* binds the *pyroclastics,* pyroclastic cones erode easily.

pyroclastic eruption A volcanic eruption of *viscous,* gas-rich magma. Pyroclastic eruptions tend to produce a great deal of solid volcanic fragments rather than fluid *lava.*

pyroclastic flow A rapid, extremely hot, downward stream of *pyroclastics,* air, gases, and ash ejected from an erupting *volcano.* A pyroclastic flow may be as hot as 800°C or more, and may move at speeds exceeding 150 kilometers per hour.

pyroclastics (used only in the plural) Particles and chunks of *igneous rock* ejected from a *volcanic vent* during an eruption.

quartzite An extremely durable, *nonfoliated metamorphic rock* derived from pure *sandstone* and consisting primarily of quartz.

Quaternary ice age An *ice age* that began approximately 1.6 million years ago and continues to the present time.

quick clay *Sediment* that sets off a sudden mudflow by changing rapidly from solid to liquid form, as after an earthquake, an explosion, or thunder.

rain-shadow effect The result of the process by which moist air on the windward side of a mountain rises and cools, causing precipitation and leaving the leeward side of the mountain dry.

recrystallization The *diagenetic* process by which unstable minerals in buried sediment are transformed into stable ones.

regional metamorphism *Metamorphism* that affects rocks over vast geographic areas stretching for thousands of square kilometers.

regolith The weathering-produced layer of fragmented material that covers much of the Earth's land surface.

relative dating The fixing of a geologic structure or event in a chronological sequence relative to other geologic structures or events. See also *numerical dating.*

reserve A known *resource* that can be exploited for profit with available technology under existing political and economic conditions.

reservoir rock A permeable rock containing oil or gas.

resource A mineral or fuel deposit, known or not yet discovered, that may be or become available for human exploitation.

reverse fault A *dip-slip fault* marked by a hanging wall that has moved upward relative to the footwall. Reverse faults are often caused by the *convergence* of lithospheric plates.

rhyolite Any of a group of felsic *igneous rocks* that are the *extrusive* equivalents of *granite.*

Richter scale A logarithmic scale that measures the amount of energy released during an *earthquake* on the basis of the amplitude of the highest peak recorded on a *seismogram.* Each unit increase in the Richter scale represents a 10-fold increase in the amplitude recorded on the seismogram and a 30-fold increase in energy released by the earthquake. Theoretically, the Richter scale has no upper limit, but the *yield point* of the Earth's rocks imposes an effective limit between 9.0 and 9.5.

rifting The tearing apart of a plate to form a depression in the Earth's *crust* and often eventually separating the plate into two or more smaller plates.

rip current A strong, rapid, and brief current that flows out to sea, moving perpendicular to the *shoreline.*

ripple marks A pattern of wavy lines formed along the top of a *bed* by wind, water currents, or waves.

riprap A pile of large, angular boulders built seaward of the *shoreline* to prevent erosion by waves or currents. See also *seawall.*

rock A naturally formed aggregate of usually inorganic materials from within the Earth.

rock cycle A series of events through which a *rock* changes, over time, between *igneous, sedimentary,* and *metamorphic* forms.

rock-forming mineral One of the 20 or so minerals contained in the rock that composes the Earth's crust and mantle.

rubidium–strontium dating A form of *isotope dating* that relies on the 47-billion-year half-life of radioactive isotopes of rubidium, which decay into isotopes of strontium, to determine the age of rocks in which strontium is present. Rubidium–strontium dating is used for rocks that are at least 10 million years old, deep-Earth plutonic rocks, and Moon rocks.

runoff Surface water flow, either as a sheet of water or in distinct stream channels.

sandstone A *clastic rock* composed of particles that range in diameter from 1/16 millimeter to 2 millimeters in diameter. Sandstones make up about 25% of all sedimentary rocks.

scarp The steep cliff face that is formed by a *slump.*

schist A coarse-grained, strongly *foliated metamorphic rock* that develops from *phyllite* and splits easily into flat, parallel slabs.

scientific law 1. A natural phenomenon that has been proven to occur invariably whenever certain conditions are met. 2. A formal statement describing such a phenomenon and the conditions under which it occurs. Also called "law."

scientific methods Techniques that involve gathering all available data on a subject, forming a *hypothesis* to explain the data, conducting experiments to test the hypothesis, and modifying or confirming the hypothesis as necessary to account for the experimental results.

sea arch A landform produced by coastal erosion of a prominent headland. Sea arches form when sea caves are excavated so deeply by crashing waves that two caves eroding on opposite sides of the headland become joined. The overlying rocky roof is left as an arch.

sea cave The notches in the sides of a prominent coastal rocky headland eroded by crashing waves.

sea-floor spreading The formation and growth of oceans that occurs following *rifting* and is characterized by eruptions along *mid-ocean ridges,* forming new oceanic *lithosphere,* and expanding ocean basins. See also *divergence.*

seamount A conical underwater mountain formed by a *volcano* and rising 1,000 meters or more from the sea floor.

sea stack A steep, isolated island of rock separated from a *headland* by the action of waves, as when the overhanging section of a sea arch is eroded.

seawall A wall of stone, concrete, or other sturdy material built along the *shoreline* to prevent erosion even by the strongest and highest of waves. See also *riprap.*

secondary coast A *coast* shaped primarily by erosion or deposition by sea currents and waves.

secondary enrichment The process by which a metal deposit becomes concentrated when other minerals are eliminated from the deposit, as through *dissolution,* precipitation, or *weathering.*

sediment A collection of transported fragments or precipitated materials that accumulate, typically in loose layers, as of sand or mud.

sedimentary environment The continental, oceanic, or coastal surroundings in which sediment accumulates.

sedimentary facies 1. A set of characteristics that distinguish a given section of sedimentary rock from nearby sections. Such characteristics include mineral content, grain size, shape, and density. 2. A section of sedimentary rock so characterized.

sedimentary rock A *rock* made from the consolidation of solid fragments, as of other rocks or organic remains, or by precipitation of minerals from solution.

sedimentary structure A physical characteristic of a *detrital sediment* that reflects the conditions under which the sediment was deposited.

seismic gap A locked fault segment that has not experienced seismic activity for a long time. Because *stress* tends to accumulate in seismic gaps, they often become the sites of major *earthquakes.*

seismic moment Seismic moment is a numerical means of measuring an earthquake's total energy release. It is calculated by measuring the total length of fault rupture and then factoring in the depth of rupture, total slip along the rupture, and the strength of the faulted rock.

seismic profiling The mapping of rocks lying along and beneath the ocean floor by recording the reflections and refractions of *seismic waves.*

seismic tomography The process whereby a computer first synthesizes data on the velocities of *seismic waves* from thousands of recent earthquakes to make a series of images depicting successive planes within the Earth, and then uses these images to construct a three-dimensional representation of the Earth's interior.

seismic wave One of a series of progressive disturbances that reverberate through the Earth to transmit the energy released from an *earthquake.*

seismogram A visual record produced by a *seismograph* and showing the arrival times and magnitudes of various *seismic waves.*

seismograph A machine for measuring the intensity of *earthquakes* by recording the *seismic waves* that they generate.

seismology The study of *earthquakes* and the structure of the Earth, based on data from *seismic waves.*

shale A *sedimentary rock* composed of *detrital sediment* particles less than 0.004 millimeter in diameter. Shale tends to be red, brown, black, or gray, and usually originates in relatively still waters.

shearing stress *Stress* that slices rocks into parallel blocks that slide in opposite directions along their adjacent sides. Shearing stress may be caused by *transform motion.*

shield volcano A low, broad, gently sloping, dome-shaped structure that forms over time as repeated eruptions eject *basaltic lava* through one or more *vents* and the lava solidifies in approximately the same volume all around.

shock metamorphism The *metamorphism* that results when a meteorite strikes rocks at the Earth's surface. The meteoric impact generates tremendous pressure and extremely high temperatures that cause minerals to shatter and recrystallize, producing new minerals that cannot arise under any other circumstances.

shoreline The boundary between a body of water and dry land.

silicate One of several rock-forming minerals that contain silicon, oxygen, and usually one or more other common elements.

silicon–oxygen tetrahedron A four-sided geometric form created by the tight bonding of four oxygen atoms to each other, and also to a single silicon atom that lies in the middle of the form.

sill A concordant *pluton* that is substantially wider than it is thick. Sills form within a few kilometers of the Earth's surface. See also *dike.*

sinkhole A circular, often funnel-shaped depression in the ground that forms when soluble rocks dissolve.

slab pull The process by which old, dense oceanic lithosphere, denser than the warm asthenosphere beneath, sinks back toward the Earth's interior at a subduction zone. Believed to be the principal driving force of plate motion.

slate A fine-grained, *foliated metamorphic rock* that develops from *shale* and tends to break into thin, flat sheets.

slide The mass movement of a single, intact mass of rock, soil, or unconsolidated material along a weak plane, such as a *fault, fracture,* or *bedding* plane. A slide may involve as little as a minor displacement of soil or as much as the displacement of an entire mountainside.

slip face The steep leeward slope of a *dune.*

slip plane A weak plane in a rock mass from which material is likely to break off in a *slide.*

slump 1. A downward and outward *slide* occurring along a concave *slip plane.* 2. The material that breaks off in such a slide.

snowline The lowest point at which snow remains year-round.

soil The top few meters of regolith, generally including some organic matter derived from plants.

soil horizon A layer of soil that can be distinguished from the surrounding soil by such features as chemical composition, color, and texture.

soil profile A vertical strip of soil stretching from the surface down to the bedrock and including all of the successive *soil horizons.*

solifluction A form of *creep* in which soil flows downslope at a rate of 0.5 to 15 centimeters per year. Solifluction occurs in relatively cold regions when the brief warmth of summer thaws only the upper meter or two of regolith, which becomes waterlogged because the underlying ground remains frozen and, therefore, the water cannot drain down into it.

sorting The process by which a given transport medium separates out certain particles, as on the basis of size, shape, or density.

source rock A rock in which *hydrocarbons* originate.

speleothem A mineral deposit of calcium carbonate that precipitates from solution in a *cave.*

spheroidal weathering The process by which *chemical weathering,* especially by water, decomposes the angles and edges of a rock or boulder, leaving a rounded form from which concentric layers are then stripped away as the weathering continues.

spit A narrow, fingerlike ridge of sand that extends from land into open water.

stalactite An icicle-like mineral formation that hangs from the ceiling of a *cave* and is usually made up of travertine, which precipitates as water rich in dissolved limestone drips down from the cave's ceiling. See also *stalagmite.*

stalagmite A cone-shaped mineral deposit that forms on the floor of a *cave* and is usually made up of travertine, which precipitates as water rich in dissolved limestone drips down from the cave's ceiling. See also *stalactite.*

star dune A *dune* with three or four arms radiating from its usually higher center so that it resembles a star in shape. Star dunes form when winds blow from three or four directions, or when the wind direction shifts frequently.

strain The change in the shape or volume of a rock that results from *stress.*

stratovolcano A cone-shaped *volcano* built from alternating layers of *pyroclastics* and viscous *andesitic lava.* Stratovolcanos tend to be very large and steep.

streak The color of a mineral in its powdered form. This color is usually determined by rubbing the mineral against an unglazed porcelain slab and observing the mark made by it on the slab.

stream A body of water found on the Earth's surface and confined to a narrow topographic depression, down which it flows and transports rock particles, sediment, and dissolved particles. Rivers, creeks, brooks, and runs are all streams.

stream discharge The volume of water to pass a given point on a stream bank per unit of time, usually expressed in cubic meters of water per second.

stream terrace A level plain lying above and running parallel to a streambed. A stream terrace is formed when a stream's bed erodes to a substantially lower level, leaving its floodplain high above it.

stress The force acting on a rock or another solid to deform it, measured in kilograms per square centimeter or pounds per square inch.

striation One of a group of usually parallel scratches engraved in bedrock by a *glacier* or other geological agent.

strike 1. The horizontal line marking the intersection between the inclined plane of a solid geological structure and the Earth's surface. 2. The compass direction of this line, measured in degrees from true north.

strike-slip fault A *fault* in which two sections of rock have moved horizontally in opposite directions, parallel to the line of the *fracture* that divided them. Strike-slip faults are caused by *shearing stress*.

structural geology The scientific study of the geological processes that deform the Earth's crust and create mountains.

subduction The sinking of an oceanic plate edge as a result of *convergence* with a plate of lesser density. Subduction often causes *earthquakes* and creates *volcano* chains.

subsidence The lowering of the Earth's surface, caused by such factors as compaction, a decrease in groundwater, or the pumping of oil.

sulfate One of several minerals containing positive sulfur ions bonded to negative oxygen ions.

sulfide One of several minerals containing negative sulfur ions bonded to one or more positive metallic ions.

surface wave One of a series of *seismic waves* that transmits energy from an earthquake's *epicenter* along the Earth's surface. See also *body wave*.

surge To flow more rapidly than usually. Said of a *glacier*.

suspended load A body of fine, solid particles, typically of sand, clay, and silt, that travels with stream water without coming in contact with the streambed.

suture zone The area where two continental plates have joined together through *continental collision*. Suture zones are marked by extremely high mountain ranges, such as the Himalayas and the Alps.

S wave (abbreviation for secondary wave) A *body wave* that causes the rocks along which it passes to move up and down perpendicular to the direction of its own movement. See also *P wave*.

S-wave shadow zone The region within an arc of 154° directly opposite an earthquake's *epicenter* that is marked by the absence of *S* waves. The S-wave shadow zone is due to the fact that S waves cannot penetrate the liquid outer core. See also *P-wave shadow zone*.

syncline A concave *fold*, the central part of which contains the youngest section of rock. See also *anticline*.

talus slope The large pile of rocky boulders that accumulates at the foot of a cliff, typically by the mechanical-weathering process of frost-wedging.

tarn A deep, typically circular lake that forms when a *cirque glacier* melts.

tectonic creep The almost constant movement of certain *fault blocks* that allows strain energy to be released without major *earthquakes*.

tension *Stress* that stretches or extends rocks, so that they become thinner vertically and longer laterally. Tension may be caused by *divergence* or *rifting*.

tephra (plural noun) *Pyroclastic* materials that fly from an erupting volcano through the air before cooling, and range in size from fine dust to massive blocks.

tephra cone See *pyroclastic cone*.

terminus The outer margin of a *glacier*.

terrane Fault-bounded blocks of related rocks that formed elsewhere but have been subsequently added to a different continent. These exotic rocks have been stuck on, or *accreted,* to the edges of the continent during plate convergence.

theory A comprehensive explanation of a given set of data that has been repeatedly confirmed by observation and experimentation and has gained general acceptance within the scientific community but has not yet been decisively proven. See also *hypothesis* and *scientific law*.

thermal contraction A form of *mechanical weathering* in which cold causes a mineral's crystal structure to contract.

thermal expansion A form of *mechanical weathering* in which heat causes a mineral's crystal structure to enlarge.

thermal plume A vertical column of upwelling *mantle* material, 100 to 250 kilometers in diameter, that rises from beneath a continent or ocean and can be perceived at the Earth's surface as a *hot spot*. Thermal plumes carry enough energy to move a plate, and they may be found both at plate boundaries and plate interiors.

thrust fault A *reverse fault* marked by a *dip* of 45° or less.

tidal bore A turbulent, abrupt, wall-like wave that is caused by a *flood tide*.

tide 1. The cycle of alternate rising and falling of the surface of an ocean or large lake, caused by the gravitational pull of the Sun and especially Moon in interaction with the Earth's rotation. Tides occur on a regular basis, twice every day on most of the Earth. 2. A single rise or fall within this cycle.

topography The set of physical features, such as mountains, valleys, and the shapes of landforms, that characterizes a given landscape.

transform motion The movement of two adjacent lithospheric plates in opposite directions along a parallel line at their common edge. Transform motion often causes *earthquakes*.

transition zone The seismic discontinuity located in the upper *mantle* just beneath the *asthenosphere* and characterized by a marked increase in the velocity of *seismic waves*.

translatory motion The movement of water over a significant distance in the direction of a wave.

transverse dune One of a series of *dunes* having an especially steep *slip face* and a gentle windward slope and standing perpendicular to the prevailing wind direction and parallel to each other. Transverse dunes typically form in arid and semi-arid regions with plentiful sand, stable wind direction, and scarce vegetation. A transverse dune may be as much as 100 kilometers long, 200 meters high, and 3 kilometers wide.

travertine The name applied to calcium carbonate when it forms cave deposits.

tributary A *stream* that supplies water to a larger stream.

trunk stream A large stream into which *tributaries* carry water and sediment.

tsunami (plural tsunami) A vast sea wave caused by the sudden dropping or rising of a section of the sea floor following an *earthquake*. Tsunami may be as much as 30 meters high and 200 kilometers long, may move as fast as 250 kilometers per hour, and may continue to occur for as long as a few days.

unconformity A boundary separating two or more rocks of markedly different ages, marking a gap in the geologic record.

uniformitarianism The hypothesis that current geologic processes, such as the slow erosion of a coast under the impact of waves, have been occurring in a similar manner throughout the Earth's history and that these processes can account for past geologic events. See also *catastrophism*.

upwarped mountain A mountain consisting of a broad area of the Earth's *crust* that has moved gently upward without much apparent deformation, and usually containing *sedimentary, igneous,* and *metamorphic rocks.*

uranium–thorium–lead dating A form of *isotope dating* that relies on the extremely long *half-life* of radioactive isotopes of uranium, which decay into isotopes of lead, to determine the age of rocks in which uranium and lead are present.

valley glacier An *alpine glacier* that flows through a preexisting stream valley.

van der Waals bond A relatively weak kind of *intermolecular bond* that forms when one side of a molecule develops a slight negative charge because a number of *electrons* have temporarily moved to that side of the molecule. This negative charge attracts the *nuclei* of the *atoms* of a neighboring molecule. The side of the molecule with fewer electrons develops a slight positive charge that attracts the electrons of the atoms of neighboring molecules.

varve A pair of sediment beds deposited by a lake on its floor, typically consisting of a thick, coarse, light-colored bed deposited in the summer and a thin, fine-grained, dark-colored bed deposited in the winter. Varves are most often found in lakes that freeze in the winter. The number and nature of varves on the bottom of a lake provide information about the lake's age and geologic events that affected the lake's development.

vent An opening in the Earth's surface through which *lava,* gases, and hot particles are expelled. Also called *volcanic vent* and *volcano.*

ventifact A stone that has been flattened and sharpened by *wind abrasion*. Ventifacts are commonly found strewn across a *desert* floor.

viscosity A fluid's resistance to flow. Viscosity increases as temperatures decrease.

volcanic arc A chain of *volcanoes* fueled by magma that rises from an underlying *subducting* plate.

volcanic cone A cone-shaped mountain that forms around a *vent* from the debris of *pyroclastics* and *lava* ejected by numerous eruptions over time.

volcanic crater A steep, bowl-shaped depression surrounding a *vent*. A volcanic crater forms when the walls of a vent collapse inward following an eruption.

volcanic dome A bulb-shaped solid that forms over a *vent* when *lava* so *viscous* that it cannot flow out of the *volcanic crater* cools and hardens. When a volcanic dome forms, it traps the volcano's gases beneath it. They either escape along a side vent of the volcano or build pressure that causes another eruption and shatters the volcanic dome.

volcanic rock See *extrusive rock.*

volcanism The set of geological processes that result in the expulsion of *lava, pyroclastics,* and gases at the Earth's surface.

volcano The solid structure created when lava, gases, and hot particles escape to the Earth's surface through *vents.* Volcanoes are usually conical. A volcano is "active" when it is erupting or has erupted recently. Volcanoes that have not erupted recently but are considered likely to erupt in the future are said to be "dormant." A volcano that has not erupted for a long time and is not expected to erupt in the future is "*extinct.*"

watershed See *drainage basin.*

water table The surface that lies between the *zone of aeration* and the underlying *zone of saturation.*

wave-cut bench A relatively level surface formed when waves erode the base of a cliff, causing the overlying rock to fall into the surf. A wave-cut bench stands above the water and extends seaward from what remains of the cliff.

wave refraction The process by which a wave approaching the shore changes direction due to slowing of those parts of the wave that enter shallow water first, causing a sharp decrease in the angle at which the wave approaches until the wave is almost parallel to the coast.

weathering The process by which exposure to atmospheric agents, such as air or moisture, causes *rocks* and *minerals* to break down. This process takes place at or near the Earth's surface. Weathering entails little or no movement of the material that it loosens from the rocks and minerals. See also *erosion.*

welded tuff A volcanic igneous rock that forms when still-warm tephra accumulates at the Earth's surface. Because the particles are still warm and soft, they can weld together under the weight of overlying deposits, forming a hard rock.

wetland A lake, marsh, or swamp that supports wildlife and replenishes the groundwater system.

wind abrasion The process by which wind erodes bedrock through contact between the bedrock and rock particles carried by the wind.

xenolith A preexisting rock embedded in a newer *igneous rock.* Xenoliths are formed when a rising *magma* incorporates the preexisting rock. If the preexisting rock does not melt, it will not be assimilated into the magma and will therefore remain distinct from the new igneous rock that surrounds it.

X-ray diffraction The scattering of X-rays passed through a mineral sample so as to form a pattern peculiar to the given mineral.

yield point The maximum *stress* that a given rock can withstand without becoming permanently deformed.

zone of ablation The part of a *glacier* in which there is greater overall loss than gain in volume. A zone of ablation can be identified in the summer by an expanse of bare ice. See also *zone of accumulation.*

zone of accumulation The part of a *glacier* in which there is greater overall gain than loss in volume. A zone of accumulation can be identified by a blanket of snow that survives summer melting. See also *zone of ablation.*

zone of aeration A region below the Earth's surface that is marked by the presence of both water and air in the pores of rocks and soil. Also called *aeration zone.*

zone of saturation A region that lies below the *zone of aeration* and is marked by the presence of water and the absence of air in the pores of rocks and soil.

Photo Credits

(b) Paul H. Moser/Geological Survey of Alabama; **Figure 16-46** Arthur N. Palmer; **Geology at a Glance (1)** Samuel S. Frushour/Indiana Geological Survey; **Geology at a Glance (2)** Timothy O'Keefe/Bruce Coleman Inc.; **Geology at a Glance (3)** Arthur N. Palmer; **Geology at a Glance (4)** Steve Solum/Bruce Coleman Inc.; **Geology at a Glance (5)** Farrell Grehan/Science/Source/Photo Researchers, Inc.; **Figure 16-47 (left)** Dylan Brian Boyle; **Figure 16-47 (right)** E. Calvin Alexander, Jr.; **Geology at a Glance (6)** Ian Aitken © Rough Guides; **Geology at a Glance (7)** Robert & Linda Mitchell Photography; **Geology at a Glance (8)** Albert Copley/Visuals Unlimited; **Geology at a Glance (9)** Jeff Lepore Photo Researchers, Inc.; **Geology at a Glance (10)** Alvarez Photography; **Figure 16-48** Steve Solum/Bruce Coleman Inc.; **Figure 16-49** Dr. Pete Coxon, April 2005.

Chapter 17   Figure 17-1 Tom Bean/Corbis/Stock Market; **Figure 17-6 (a)** Comstock Images; **Figure 17-6 (b)** Tom & Susan Bean, Inc.; **Figure 17-6 (c)** Betty Crowell/Faraway Places; **Figure 17-6 (d)** Jim Wark/Peter Arnold, Inc.; **Figure 17-7** Tom & Susan Bean, Inc.; **Figure 17-9** Galen Rowell/Mountain Light Photography, Inc.; **Figure 17-10** Steve McCutcheon/Visuals Unlimited; **Figure 17-11 (a)** Tom & Susan Bean, Inc.; **Figure 17-11 (b)** Dan Guravich/Photo Researchers, Inc.; **Figure 17-12 (right)** William E. Ferguson; **Figure 17-13 (b)** Bill Kamin/Visuals Unlimited; **Figure 17-13 (c)** John Kieffer/Peter Arnold, Inc.; **Figure 17-13 (d)** Betty Crowell/Faraway Places; **Figure 17-14 (a)** ©Advanced Satellite Productions, Inc. 1993; **Figure 17-14 (b)** U.S. Dept. of Interior/EROS Data Center, U.S. Geological Survey; **Figure 17-15** Dick Keen/Visuals Unlimited; **Figure 17-18 (left)** Art Gingert/Comstock Images; **Figure 17-20 (right)** Tom Bean/DRK Photo; **Geology at a Glance 17.1 (1)** Dan Guravich/Photo Researchers, Inc.; **Geology at a Glance 17.1 (2)** Advanced Satellite Productions, Inc.; **Geology at a Glance** 17.1 **(3)** John S. Shelton; **Geology at a Glance 17.1 (4)** William E. Ferguson; **Geology at a Glance 17.1 (5)** John S. Shelton **Geology at a Glance 17.2 (1)** William E. Ferguson; **Geology at a Glance 17.2 (2)** Comstock Images; **Geology at a Glance 17.2 (3)** James Balog/Tony Stone Images; **Geology at a Glance 17.2 (4)** Bill Kamin/Visuals Unlimited; **Geology at a Glance 17.2 (5)** Breck P. Kent; **Figure 17-21 (top)** John S. Shelton; **Figure 17-21 (bottom)** John S. Shelton; **Figure 17-22** John S. Shelton; **Figure 17-25 (a)** Peter Dunwiddie/Visuals Unlimited; **Figure 17-25 (b)** Glenn Oliver/Visuals Unlimited; **Figure 17-25 (c)** Professor Berhard Hallet/Prof. Berhard Hallet; **Figure 17-25 (d)** Troy L. Pewe; **Highlight 17-1 (left)** John S. Shelton; **Highlight 17-1 (right)** John S. Shelton; **Figure 17-27 (inset)** Stanley Chernicoff; **Figure 17-31** Courtesy of Mark Germani, PhD./MicroMaterials Research, Inc.; **Figure 17-34** David L. Reid; **Highlight 17-2** Paul F. Hoffman; **Figure 17-37** John W. Attig/Wisconsin Geological Survey; **Figure 17-38** Hendrick Averkapm "Winter Landscape with Iceskater's", 1608, oil on panel, 87.5 x 132 cm. Rijksmuseum, Amsterdam; **Figure 17-39**

Marli Miller; **Figure 17-41** AP Wide World Photos; **Figure 17-43** Austin Post/U.S. Geological Survey, Denver; **Figure 17-44** NASA Headquarters; **Questions for Review** Robert Fulton Nature Picture Library;

Chapter 18   Figure 18-1 David Muench/Corbis/Bettmann; **Figure 18-2 (a)** Jeff Foott/Bruce Coleman Inc.; **Figure 18-2 (b)** Michael Fogden/DRK Photo; **Figure 18-7** Linda Waldhofer/Getty Images, Inc – Liaison; **Figure 18-8** Tom Bean/Getty Images Inc. - Stone Allstock; **Figure 18-10** Breck P. Kent Animals Animals/Earth Scenes; **Figure 18-11** David Ball/Getty Images Inc. - Stone Allstock; **Figure 18-12** Betty Crowell/Faraway Places; **Figure 18-13** Norman Weiser; **Highlight 18-2** U.S. Farm Security Administration, Collection Prints and Photographs Division, Library of Congress; **Figure 18-16** Scott Berner/Visuals Unlimited; **Figure 18-17** John Gerlach/Visuals Unlimited; **Figure 18-18** John D. Cunningham/Visuals Unlimited; **Figure 18-19** Peter M. Wilson/Corbis/Bettmann; **Figure 18-21** Tom Stack & Associates, Inc.; **Figure 18-23** M. Dillon/Corbis/Bettmann; **Figure 18-24** Carl Purcell/Photo Researchers, Inc.; **Figure 18-26** Georg Gerster/Comstock Images; **Figure 18-27** ©Gary Braasch; **Figure 18-28 (a)** Peter L. Kresan Photography; **Figure 18-28 (b)** Art Wolfe/Photo Researchers, Inc.; **Figure 18-28 (c)** Michael E. Long/National Geographic Image Collection; **Figure 18-28** Marli Miller; **Figure 18-29** William E. Ferguson; **Figure 18-31** Julia Waterlow/Eye Ubiquitous/CORBIS- NY; **Figure 18-32 (left)** Marli Miller; **Figure 18-32 (inset)** Stephenie Ferguson/William E. Ferguson; **Geology at a Glance (1)** Scott Berner/Visuals Unlimited; **Geology at a Glance (2)** Stephenie Ferguson/William E. Ferguson; **Geology at a Glance (3)** John D. Cunningham/Visuals Unlimited; **Geology at a Glance (4)** Peter M. Wilson/Corbis/Bettmann; **Figure 18-33** Stephenie S. Ferguson/William E. Ferguson; **Figure 18-34** Voltchev/UNEP/Peter Arnold, Inc.; **Figure 18-35** M&R Borland/Bruce Coleman Inc.; **Figure 18-36** NASA Headquarters; **Highlight 18-3** James Swinehart/Conservation & Survey Division/School of Natural Resources/University of Nebraska – Lincoln; **Figure 18-38** Georg Gerster/Comstock Images; **Figure 18-39** NASA Headquarters; **Figure 18-40** NASA/Jet Propulsion Laboratory; **Questions for Review** John S. Shelton;

Chapter 19   Figure 19-01 Craig Tuttle/Corbis/Stock Market; **Figure 19-09** Clyde H. Smith/Peter Arnold, Inc.; **Figure 19-10a** William E. Ferguson; **Figure 19-10b** William E. Ferguson; **Figure 19-11** Dick Poe/Visuals Unlimited; **Figure 19-12** Paul Sequeira/Photo Researchers, Inc.; **Figure 19-13** Donald Carter; **Highlight 19-1** Michael J. Chrzastowski/Illinois Geological Survey; **Figure 19-14** John Elk/Bruce Coleman Inc.; **Figure 19-15a** Jack Dermid/Photo Researchers, Inc.; **Figure 19-15b** Marli Miller; **Figure 19-16a** Breck P. Kent; **Figure 19-16b** Chlaus Lotscher/Peter Arnold, Inc.; **Figure 19-16 (inset)** Dr. Joseph J. Gerencher, Jr.; **Figure 19-19a** Stock State College/Prof. Stewart Farrell; **Figure 19-19b** John S.

Shelton; **Figure 19-20** Cliff Wassmann Photography; **Figure 19-21a** The Benjamin and Gladys Thomas Air Photo Archives/Department of Geography,UCLA; **Figure 19-21b** John S. Shelton; **Figure 19-22a** John S. Shelton; **Figure 19-22b** Townsend P. Dickinson; **Geology at a Glance (top left)** Jim Brandenburg/Minden Pictures; **Geology at a Glance (bottom left)** Breck P. Kent; **Geology at a Glance (right 1)** Alan Keohane Rough Guides/Dorling Kindersley; **Geology at a Glance (right 2)** Stockton State College/Professor Stewart Farrell; **Geology at a Glance (right 3)** John S. Shelton; **Geology at a Glance (right 4)** Cliff Wasserman Photography; **Figure 19-23a** Breck P. Kent; **Figure 19-23b** Peter L. Kresan Photography; **Figure 19-25** Breck P. Kent; **Figure 19-26** Chris Seward/Raleigh News & Observer/AP Wide World Photos; **Figure 19-28** S.J. Krasemann/Peter Arnold, Inc.; **Figure 19-29 (left)** David Ball/Corbis/Stock Market; **Figure 19-29 (right)** William E. Ferguson; **Figure 19-30** Tom Van Sant/The GeoSphere Project; **Figure 19-31** Peter L. Kresan Photography; **Figure 19-32** Lloyd Cluff/Corbis/Bettmann; **Figure 19-33** Courtesy of The Rosenberg Library, Galveston, Texas.

Chapter 20   Figure 20-1 H. Collart-Odinetz/Getty Images, Inc – Liaison; **Figure 20-2** Novosti/Getty Images, Inc – Liaison; **Figure 20-3** Laurent van der Stockt/Getty Images, Inc – Liaison; **Figure 20-4** From Pennsylvania Historical & Museum Commission, Drake Well Museum, Titusville, PA; **Figure 20-6** Silvia Dinale ©1994; **Figure 20-7** Visuals Unlimited; **Figure 20-8** Steve McCutcheon/Visuals Unlimited; **Figure 20-10** Adam Hart-Davis/Photo Researchers, Inc.; **Figure 20-11** SHEBA Project Office/University of Washington School of Oceanography; **Figure 20-12** Ben Osborne/Getty Images Inc. - Stone Allstock; **Figure 20-13** Stanley Chernicoff; **Highlight 20-1** USGS/Tom Pantages; **Figure 20-14 (b)** Nicholas deVore III/Bruce Coleman Inc.; **Figure 20-17 (b)** Martin Bond/Science Photo Library/Photo Researchers, Inc.; **Figure 20-17 (c)** William E. Ferguson; **Figure 20-18** George D. Lepp/Corbis/Bettmann; **Figure 20-19** Kevin Schafer/Getty Images Inc. - Stone Allstock; **Figure 20-21 (inset)** U.S. Department of Energy; **Figure 20-21 (left)** Tom Reese/Seattle Times; **Figure 20-22** Mark A. Schneider/Photo Researchers, Inc.; **Figure 20-23 (a)** Peter L. Kresan Photography; **Figure 20-23 (b)** Rona Bruce Coleman Inc.; **Figure 20-24** Spencer R. Titley; **Figure 20-26** Spencer R. Titley; **Figure 20-27** William E. Ferguson; **Figure 20-29** William Campbell/Peter Arnold, Inc.; **Figure 20-30 (top)** Mineral Information Institute and Department of Environmental Protection/Bureau of Mining & Reclamation/Commonwealth of PA; **Figure 20-30 (bottom)** Mineral Information Institute and Department of Environmental Protection/Bureau of Mining & Reclamation/Commonwealth of PA; **Figure 20-31** William E. Ferguson; **Geology at a Glance (1)** Steve McCutcheon/Visuals Unlimited; **Geology at a Glance (2)** Martin Bond/Science Photo Library/Photo Researchers, Inc.; **Figure 20-32** Thyssen Henschel/Siemens AG; **Figure 20-33** Bruce Dale/National Geographic Image Collection; **Figure 20-34** Seny Norasingh.

Index